THE DICTIONARY OF
Physical Geography

THE DICTIONARY OF

Physical Geography

Fourth Edition

Edited by

David S.G. Thomas

International Advisory Panel

David Dunkerley (Australia)
Giles Foody (UK)
Andrew Goudie (UK)
Michael Meadows (South Africa)
David Montgomery (USA)
Sharon Nicholson (USA)
Tom Spencer (UK)

WILEY Blackwell

Library of Congress Cataloging-in-Publication Data

Names: Thomas, David S. G.
Title: The dictionary of physical geography / edited by David S.G. Thomas.
Description: Fourth Edition. | Hoboken, NJ : John Wiley & Sons Inc., [2016] |
    Includes index.
Identifiers: LCCN 2015031859 | ISBN 9781118782347 (Cloth) | ISBN 9781118782330
    (Paper)
Subjects: LCSH: Physical geography–Dictionaries.
Classification: LCC GB10 .D53 2016 | DDC 910/.0203–dc23 LC record available
at http://lccn.loc.gov/2015031859

A catalogue record for this book is available from the British Library.

Wiley also publishes its books in a variety of electronic formats. Some content that appears in print may not be available in electronic books.

Cover image: Erg Chebbi dunes in Morocco, Sahara desert © Getty Images

Set in 9/10pt PlantinStd by Thomson Digital, Noida, India

1   2016

Contents

List of Contributors   vi

Preface to the Fourth Edition   x

Preface to the First Edition   xi

Preface to the Second Edition   xii

Preface to the Third Edition   xiii

Acknowledgments   xiv

About the Companion Website   xvi

Introduction   xvii

THE DICTIONARY OF PHYSICAL GEOGRAPHY   1

Index   581

Contributors

NOTE: This list of contributors, and their affiliations, was updated during production of the fourth edition. Some contributions were made for the first and subsequent editions, and contributor affiliations may have since changed. These have been updated where possible, but in some cases the affiliation at the time of contribution may remain.

AB	Alison Banwell University of Cambridge	BAK	Barbara A Kennedy Formerly of University of Oxford. Deceased.
AD	Angus Duncan University of Bedfordshire	BGT	Bruce G Thom University of Sydney
AHP	Allen H Perry Formerly of Swansea University	BJS	Bernie J Smith Formerly of Queen's University of Belfast. Deceased
AH-S	Ann Henderson-Sellers Emeritus, MacQuarie University		
ALH	A Louise Heathwaite University of Lancaster	BWA	Bruce W Atkinson Emeritus, Queen Mary University of London
AMG	Angela M Gurnell Queen Mary University of London	CAMK	Cuchlain A M King Formerly University of Nottingham
AP	Adrian Parker Oxford Brookes University	CDC	Chris D Clark University of Sheffield
ARH	Alan R Hill Formerly of York University, Ontario	CTA	Clive T Agnew University of Manchester
		DB	Denys Brunsden Emeritus, King's College London
ASG	Andrew S Goudie Emeritus, University of Oxford		
AW	Andrew Watson DAI, Washington, DC	DES	David Sugden Emeritus, University of Edinburgh
AWE	Andrew W Ellis Virginia Tech	DEW	Des E Walling Emeritus, University of Exeter

DW David Watts
Formerly of University of Hull.
Deceased.

DGT David G Tout
Formerly of University of
Manchester

DH David Higgitt
University of Singapore

DJN David J Nash
University of Brighton

DJS Douglas Sherman
Texas A&M University

DLD David L Dunkerley
Monash University

DRM David R Montgomery
University of Washington

DSGT David S G Thomas
University of Oxford

DTP David T Pugh
IOS Deacon Laboratory,
Godalming

ECB Eric C Barrett
Formerly of University of Bristol

GCN Gerald C Nanson
University of Wollongong

GFSW Giles F S Wiggs
University of Oxford

GMF Giles M Foody
University of Nottingham

HAV Heather Viles
University of Oxford

HMF Hugh M French
Emeritus, University of Ottawa

HVL Harry van Loon
Formerly of National Center for
Atmospheric Research, Boulder,
CO

IAC Ian A Campbell
Formerly of University of
Alberta. Deceased.

IB Ian Burton
Emeritus, Meteorological
Service of Canada

IE Ian Evans
University of Durham

IGS Ian G Simmons
University of Durham

DSB Doreen S Boyd
University of Nottingham

JAD John A Dearing
University of Southampton

JAM John A Matthews
Formerly of Swansea University

JAS John Shaffer
Arizona State University

JEA John E Allen
Formerly Queen Mary
University of London

JEL Julie E Laity
California State University

JET John E Thornes
Emeritus, University of
Birmingham

JGL John G Lockwood
Formerly of University of Leeds

JL	John Lewin Emeritus, Aberystwyth University	MDB	Mark D Bateman University of Sheffield
JLB	Jim Best University of Leeds	MEM	Michael E Meadows University of Cape Town
JM	Judith Maizels Formerly University of Aberdeen	MFT	Michael F Thomas Emeritus, University of Stirling
JO	Julian Orford Queen's University of Belfast	MHU	Michael H Unsworth Emeritus, Oregon State University
JSAG	John S A Green Formerly University of East Anglia	MJK	Mike J Kirkby University of Leeds
KB	Katherine Brown University of Tasmania	MJS	Michael J Selby Emeritus, University of Waikato
KEB	Keith Barber University of Southampton	MLH	Mark L Hidebrandt Arizona State University
KEI	Keith Idso Arizona State University	NSA	Neil S Arnold University of Cambridge
KJG	Kenneth J Gregory University of Southampton	NJM	Nick J Middleton University of Oxford
KJW	Keith J Weston University of Edinburgh	NJS	Nancy J Selover Arizona State University
KS	Keith Smith Emeritus, University of Stirling	PAB	Peter A Bull University of Oxford
KSR	Keith S Richards Emeritus, University of Cambridge	PAF	Peter A Furley Emeritus, University of Edinburgh
LN	Lynn Newman Glendale Community College, Arizona	PAS	Philip A Stott Formerly School of Oriental and African Studies
MAS	Mike A Summerfield University of Edinburgh	PHA	Patrick H Armstrong University of Western Australia
		PJC	Paul J Curran City University, London

PS	Peter Smithson Emeritus, University of Sheffield	SEN	Sharon E Nicholson Florida State University
PSH	Paul A Shaw Formerly of University of the West Indies	SLO	Sarah L O'Hara University of Nottingham
PWW	Paul W Williams University of Auckland	SMP	Susan M Parker London
RCB	Robert C Balling Jr Arizona State University	SMW	Stephen M Wise University of Sheffield
RGB	Roger G Barry University of Colorado, Boulder	SNL	Stuart N Lane University of Lausanne
RHS	Rodney H Squires University of Minnesota, Minneapolis	SS	Stephen Stokes Formerly of University of Oxford. Deceased
RH-Y	Roy Haines-Young University of Nottingham	TRO	Timothy R Oke University of British Columbia, Vancouver
RID	Ronald I Dorn Arizona State University	TS	Tom Spencer University of Cambridge
RLJ	Robert L Jones Formerly Coventry University	VM	Vashu Mistra Florida State University
RR	Ross Reynolds University of Reading	WBW	W Brian Whalley University of Sheffield
RSC	Randell S Cerveny Arizona State University	WDS	William D Sellers Formerly of University of Arizona. Deceased.
RSW	Richard S Washington University of Oxford	WLG	William L Graf Emeritus. Arizona State University
SAC	Stanley A Changnon Formerly Illinois Department of Energy, Champaign, IL. Deceased		

Preface to the Fourth Edition

It is 15 years since the last edition of this dictionary was published. This edition represents a substantial revision. The entry list has been fully evaluated by the new International Advisory Panel. This has resulted in 152 entries being removed, 347 being updated or fully rewritten and 191 new entries being added. Most others have had their references updated, but with an eye kept on the fact that older references, especially formative ones, can provide critical and essential insights into a topic or a definition.

The decisions to remove or add were based on the group's views of what terms are now less or little used (or too regional in usage to warrant inclusion) and what is new in physical geography. As the volume is finite in length, terms cannot simply keep being added: some have to be lost or reduced in length to make way for the new additions. The entries that have been removed have been retained, however, in a section in the online version of the dictionary.

There is an element of personal choice in what is included, but it is based on expert opinion: someone will inevitably write and say that 'such and such' should have been left in, or included, but as the saying goes, 'the editor's decision is final!' The vast majority of rewriting and writing of new entries has been carried out by the advisory group; but thank you to the small number of additional new authors: as always, authors are identified by their initials and the new names are added to the list of contributors.

As was the case with the third edition, this has been a challenging and time consuming task. Thank you to the advisors, your help has been considerable and support most welcome. Thank you to Andrew Goudie, who started this Dictionary off in 1985. And thank you to Ian Francis, Kelvin Matthews and Delia Sandford at the publishers. Finally, why do we need this dictionary in the age of the internet and online lists? Well it is authoritative, and can't be altered by random contributors!

DSGT

Preface to the First Edition

The preparation of a dictionary of this complexity has involved many people, and all deserve thanks for the efficiency with which they have prepared their material on time and in the format required. We have been fortunate in having as a model our companion volume, *The Dictionary of Human Geography*, which was so expertly edited by R.J. Johnston and his team. I would like to express particular thanks to Janet Godden for having taken over so much of the organizational burden, and to Andrew Watson for being willing to prepare many of the short entries.

ASG

Preface to the Second Edition

In this second edition we have taken the opportunity to update many of the entries and their illustrations, and have added a substantial number of new entries. These new entries include some that should doubtless have been in the first edition, but most are entries that relate to new developments that have taken place in the discipline, especially with respect to increasing concerns over major environmental issues. We have also made substantial additions to the list of acronyms and abbreviations, and have updated many of the references and guides to further reading.

ASG

Preface to the Third Edition

This edition of this dictionary represents substantial evolution from the second edition. Following consultation with the international advisory panel, whose composition reflects many key areas of physical geography, including biogeography, climatology, environmental change and key areas of geomorphology, 200 entries from the second edition have been removed and replaced with 450 new entries. These were chosen from an original list of possible new entries over twice this length, with the final selection representing changes within the discipline, an increased international flavour, and the need to maintain the final volume at a certain length. The total list of contributors is increased by 34, with new experts drawn in to add their knowledge to the volume.

Managing a volume of this size is a complex task, the size of which I did not quite realize when approached by John Davey, formerly of Blackwell Publishers, and Andrew Goudie. Completion of the task has been made much simpler due to the help of the advisory panel, the goodwill of contributors, and especially to the assistance of Jill Landeryou and Sarah Falkus at the publishers, and, in the final stages, the considerable help given in the preparation of the final manuscript by Lucy Heath. All are thanked enormously.

DSGT

Acknowledgements

The editors and the publishers wish to thank the following for permission to use copyright material or photographs in this fourth edition.

Alice Thomas for the photograph in **arches, natural.**

American Physical Society for the figure in **core.**

American Meteorological Society for the figure in **El Nino.**

Andrew Goudie for the figures in **albedo, beach, eustasy, floristic realms, tropical cyclones and volcano.**

Cambridge University Press for the figures in **general atmospheric circulation, sea/land breeze.**

Cory Matthews for the figure in **antecedent drainage.**

David Brigland for the figure in **river terrace,** from Brigland *et al.* (eds) The Quaternary of the lower reaches of the Thames, a field guide. QRA 1995.

David Evans for the figure in **crag and tails forms,** from Benn and Evans, *Glaciers and glaciation, 2010.*

David Thomas for the photographs in **alp, current ripples, entrenched meander, ephemeral stream, eutrophication, fairy circles, hoodoo, inselberg, insolation weathering, kopje, lava, mallee, palaeosol, pan, pediment, sand trap, slip face, strandline, talus, tephra, wadi, wetland.**

Elsevier (and associated imprints) for the figures in **Darcy's law, North Atlantic deep water, pollen analysis, supercontinent, Walker circulation.**

European Geophysical Union for the figure in **Greenland ice sheet.**

Geological Society of London for the figure in **abyssal plain.**

Intergovernmental Panel on Climate Change (IPCC) for the figure in **Antarctic ice sheet.**

Joel Cracraft and Niles Eldredge for the figure in **cladistics.**

John Wiley and Sons (and associated imprints) for the figures in **anabranching, Antarctic bottom water, biofilm, carrying capacity, connectivity, drylands, El Nino, plate tectonics, rift valley, stress, sand seas.**

Mair Thomas for the photograph in **geyser.**

M.A. Wilson for the photograph in **bioerosion.**

Oxford University Press for the figures in **advection, drainage network, landslide, mass movement types, permafrost.**

Peter Furley and the late Walter Newey for the figures in **biogeochemical cycles, estuary, ocean.**

Roy Haynes-Young for the figure in **millennium ecosystem assessment.**

Sharon Nicholson for the photograph in **mesoscale cellular convection.**

Society for Sedimentary Geology (SEPM) for the figure in **delta.**

Springer for the figure in **carbonate budget, of coral reefs.**

Tom Spencer for the photographs in **algal ridge, flume, makatea, ridge and runnel topography.**

About the Companion Website

This book is accompanied by a companion website:

www.wiley.com/go/thomas/physicalgeography

The website includes:
- Pdfs of entries from the third edition, not included in this book, for downloading
- Powerpoints of all figures from the book for downloading
- Pdfs of all tables from the book for downloading

Introduction

This dictionary provides definitions of terms and explanations of key ideas, concepts and issues in physical geography. It draws upon the wealth of knowledge of over 100 contributors and is aimed for the use of professionals, students, teachers and researchers in geography and allied environmental and life sciences.

Entries are organized alphabetically, but to aid further understanding, they are, where appropriate, cross-referenced to other relevant entries, which are shown in small capitals in the text. An index allows the identification of other entries in which a term is referred to, allowing a wider sense of its usage to be gained.

Many entries are referenced and/or accompanied by suggestions for further reading. Together, this allows source material, examples of usage and extended explanations to be explored.

A

abiotic The abiotic components of an ECO-SYSTEM are those which are not living. These include mineral soil particles, water, atmospheric gases and inorganic salts; sometimes, simple organic substances that have resulted from excretion or decomposition may be included. The term abiotic is also used for physical and chemical influences upon organisms; for example, humidity, temperature, pH and salinity. An abiotic environment is one that is devoid of life.   PHA

ablation The process by which snow or ice is lost from a GLACIER, floating ice or snow. Examples are melting and run-off, calving of icebergs, evaporation, sublimation and removal of snow by wind. Melting followed by refreezing at another part of a glacier is not regarded as ablation because the glacier does not lose mass. Melting is the most important process in temperate and subpolar regions and accounts for seasonal and diurnal meltwater floods. Most such ablation occurs at the glacier surface, and at the snouts of glaciers in many mid-latitude areas it lowers the ice surface by the order of 10 m each year. A small amount of melting occurs within and beneath glaciers whose ice is at the pressure melting point. In the Antarctic the most important ablation process is the calving of ice shelves, though considerable losses may also occur through bottom melting of ice shelves and the removal of snow by offshore katabatic winds. DES

Reading
Benn, D.I. and Evans, D.J.A. (2010) *Glaciers & glaciation*. London: Hodder Education.

abrasion The process of wearing down or wearing away by friction, as by windborne sand, material frozen into glacial ice, or sediment entrained in flowing water.   DRM

abrupt climate change Climate is the long-term mean of weather, but considerable variation exists around this mean state. Even extreme conditions such as intense droughts or severe floods exist within the climatic 'mean'. However, nonlinear systems such as the atmosphere tend to fluctuate between two or more internal 'mean' states. Certain systems, termed 'almost intransitive' by the theoretician Ed Lorenz, can abruptly jump from one state to another and back again with no external forcing of the change. According to Lorenz, the atmosphere is such a system.

The possibility of abrupt change is very relevant to questions of global climate change. While the concept is largely theoretical, climate models do simulate such changes, at least when boundary conditions are changed. One possible example is the case of the emergence of the Sahara. Model simulations and some field evidence suggest that the savanna that prevailed in North Africa during the Holocene may have changed to extreme desert within a few hundred years. The idea is controversial, with other field evidence indicating a more gradual change. However, the concept receives support from ecosystem models that similarly predict that positive feedbacks within the system can lead to abrupt and even irreversible change of the surface vegetation cover.   SEN

Reading
Kröpelin, S., Verschuren, D., Lézine, A.M., *et al.* (2008) Climate-driven ecosystem succession in the Sahara: the past 6000 years. *Science*, **320**, 765–768. · Claussen, M., Kubatzki, C., Brovkin, V., *et al.* (1999) Simulation of an abrupt change in Saharan vegetation in the mid-Holocene. *Geophysical Research Letters*, **26**, 2037–2040. · Scheffer, M.M. and Carpenter, S.R. (2003) Catastrophic regime shifts in ecosystems: linking theory to observation. *Trends in Ecology and Evolution*, **18**, 648–656.

absolute age The age of an event or rock, mineral or fossil, measured in years. In physical geography, and Quaternary studies in particular, this often relates to an age determined by a radiometric dating technique such as RADIOCARBON DATING, URANIUM SERIES DATING or LUMINESCENCE DATING. The term *absolute* (as opposed to RELATIVE DATING) is, in reality, a misnomer since such ages are rarely absolute, being subject to various factors that contribute to uncertainty and errors.   DSGT

---

abstraction, of water   The removal or diversion of water from the water environment, on either a temporary or permanent basis. It can be carried out by a variety of means, including a pump, pipes, an engineering structure in a watercourse, a borehole or a well. Depending on the environmental legislation in the relevant country, controls may be placed on abstraction to limit the amount of water that can be removed. Overabstraction can lead to low river levels (or even complete loss of water conveyance) and/or the level of groundwater aquifers reducing unacceptably for the maintenance of human livelihoods, biodiversity and ECOSYSTEM GOODS AND SERVICES.   TS

abyss   *a.* A deep part of the ocean, especially one more than about 3000 m below sea level. *b.* A ravine or deep gorge.

abyssal plain   Areas of the deep ocean floor that are virtually flat and have very low gradients (~0.05°). The subdued relief and low slope of these plains, of which about 75 are known, arise from the blanketing of the sea floor by sediment delivered by turbidity currents draining from the continental slope or rise flanking a neighbouring landmass. Oceanic trenches that intervene between the continental margin and the deep ocean (as around the Pacific) consequently limit the development of abyssal plains. On the other hand, abyssal plains are very large around Antarctica, where, in addition

to the absence of trenches, ice erosion delivers abundant sediment supply.   DLD

Reference
Weaver, P.P.E. and Thomson J. (eds) (1987) *Geology and geochemistry of abyssal plains.* Geological Society Special Publication 31. Blackwell Scientific.

accommodation space   The space available, in both a vertical and a lateral sense, within which sediments can accumulate. Over long-term timescales, geological and climatic processes can determine accommodation space. Thus, for oceanic coral reefs, the long-term subsidence of oceanic plates creates an accommodation space for coral growth over millions of years. During glacial periods, emergent reefs are subject to subaerial solution, thus increasing the vertical accommodation space available for reef regrowth on renewed inundation during interglacial periods. On short-term timescales, accommodation space is mediated by hydrodynamic processes at both upper (e.g. limit of wave overwash on beaches, limit of tidal sedimentation on saltmarshes) and lower boundaries (e.g. wave action to limit the seaward margin of marshes, coral reef growth control by waves). It can be periodically, and temporarily redefined by storm and cyclone impacts and progressively changed by sea level rise, which determines the upper margin of accommodation space in coastal and estuarine environments. It has been argued that rates of sea level rise of ~0.5 m by AD 2100 might create

The distribution of abyssal plains.
Source: Weaver and Thomson (1987). Reproduced with permission of the Geological Society of London.

new accommodation space and switch reef vertical accretion back on, with carbonate production for the entire Great Barrier Reef rising from the current estimated $50\,\mathrm{Mt\,a^{-1}}$ to $70\,\mathrm{Mt\,a^{-1}}$ (Kinsey and Hopley, 1991). Human activities, such as the construction and maintenance of defence structures that prevent the full occupation of floodplains by floodwaters or the landward migration of coastal systems under sea-level rise, represent an artificial constriction of natural accommodation space. The setting back of coastal defences to more landward positions, or 'managed realignment', recreates intertidal accommodation space and reduces flood hazard. TS

Reading and References
French, J.R. (2006) Tidal marsh sediment trapping and resilience to environmental change: exploratory modelling of tidal, sea-level and sediment supply forcing in predominantly allochthonous systems. *Marine Geology*, **235**, 119–136. · Kinsey, D.W. and Hopley, D. (1991) The significance of coral reefs as global carbon sinks – response to greenhouse. *Palaeogeography, Palaeoclimatology, Palaeoecology*, **89**, 363–377. · Townend, I. and Pethick, J. (2002) Estuarine flooding and managed retreat. *Philosophical Transactions of the Royal Society of London, Series A*, **360**, 1477–1495.

accordant junctions, law of John Playfair's 1802 statement of the observation, known as Playfair's law, that tributary rivers generally join main rivers at the same elevation, with no sudden drop or waterfall at their confluence. Offered as direct evidence that rivers cut the valleys through which they flowed.

Every river appears to consist of a main trunk, fed from a variety of branches, each running in a valley proportioned to its size, and all of them together forming a system of vallies, communicating with one another, and having such a nice adjustment of their declivities, that none of them joins the principal valley, either on too high or too low a level; a circumstance which would be infinitely improbable, if each of these vallies were not the work of the stream that flows in it (Playfair, 1802: 102). DRM

Reference
Playfair, J. (1802) *Illustrations of the Huttonian theory of the Earth*. London/Edinburgh: Cadell and Davies/William Creech.

accordant summits Parts of the landscape that are of low relief and which appear from their elevations and directions of slope to have formerly been connected as a part of a prior landsurface. When cutting across contrasting rock types, the interpretation that accordant summits reflect the dissection of a former low-relief land surface is

quite secure. However, within a geologically uniform region, it is possible that accordant summits develop in the ordinary course of landscape evolution, as a consequence of mutual adjustment of slope processes and stream incision rates within the context of uniform valley spacing. DLD

Reading
Römer, W. (2008) Accordant summit heights, summit levels and the origin of the 'upper denudation level' in the Serra do Mar (SE-Brazil, Sao Paulo): a study of hillslope forms and processes. *Geomorphology*, **100**, 312–327.

accretion *a.* The gradual increase in the area of land as a result of sedimentation.
b. The process by which inorganic objects increase in size through the attachment of additional material to their surface as with the growth of hailstones.

accumulated temperature Normally the total number of days (or hours) since a given date during which the mean temperature has been above or below a given threshold. The threshold value for agriculture is usually $6\,°\mathrm{C}$, and accumulated mean temperatures above this value can be correlated with the growth of vegetation. For heating purposes the threshold is usually $15.5\,°\mathrm{C}$, and accumulated mean temperatures below this value can be correlated with energy use. Generally, accumulated temperature is used in agriculture, and DEGREE DAY is used in energy management. JET

accuracy An indication of the quality of a data set. Accuracy indicates the degree of correctness or freedom from error in the data set measured relative to the truth.

Accuracy can be assessed from a range of perspectives and be estimated in a variety of ways to suit the particular needs of a study and the nature of the data being used. For example, accuracy may be estimated for continuous data (e.g. model-based predictions of river flow rate) from the difference between the observed and true values, and be summarized in terms of a measure such as the root-mean-square error. Alternatively, with nominal level data (e.g. land cover classes), accuracy may be estimated by comparing the observed class labels against truth and summarized by measures derived from their cross-tabulation in a confusion or error matrix. The latter is common in the evaluation of a THEMATIC MAP produced via a SUPERVISED CLASSIFICATION analysis of images acquired by REMOTE SENSING. A rigorous and credible accuracy assessment typically requires careful sampling and a gold standard reference to represent the truth. Deviation from the assumed conditions for an accuracy assessment can yield erroneous and misleading estimates of accuracy.   GF

acid deposition   Rainfall is naturally acid (pH 5.65) owing to the dissolution of carbon dioxide (CO_2) in water (H_2O) to give carbonic acid (H_2CO_3):

$$CO_2 + H_2O \rightleftharpoons H_2CO_3$$

Therefore, the functional definition of acid precipitation is a pH <5.65. Acidity is generated by the presence of hydrogen ions (H^+) and is measured in units of pH on a logarithmic scale. A one-unit difference in solution pH is equivalent to a 10 times difference in the concentration of H^+.

Acidifying substances may be deposited from the atmosphere by two main pathways: (1) wet deposition of material entrained in rain, snow and fog-water – often referred to as ACID PRECIPITATION; (2) dry deposition of particulate aerosols (any solid particulate matter transported through the atmosphere), and including uptake of certain gases by vegetation, soil and water surfaces. Atmospheric inputs occur mainly from point sources and are derived from human activities, such as fossil fuel combustion and intensive livestock holdings, and from natural sources, such as volcanic emissions. The balance between wet and dry deposition pathways varies geographically and according to the dominant wind direction. In general, wet deposition is more important in upland areas with higher rainfall amounts. In the north and west of the UK, for example, inputs of H^+ in precipitation may exceed $1\,kg\,ha^{-1}\,a^{-1}$. Dry deposition of sulphur dioxide (SO_2) may exceed $2.4\,kg\,ha^{-1}\,a^{-1}$ in industrial areas of the midlands and northern England and the major urban areas of the Scottish central lowlands (Fowler *et al.*, 1985). In Europe, the relative contributions from wet and dry deposited sulphur change with distance from the coast, with wet deposition becoming relatively less important with distance south from the North Sea.

Wet deposition may occur via two processes, depending on where atmosphere scavenging occurs. Rainout describes acid inputs, which originate within the cloud system; washout describes the removal of solutes by falling precipitation. Mist and fog are generally more acidic than rain; thus, at high altitudes, acid deposition is enhanced by the contribution of mist, fog and cloud water.

Most of the excess acidity in precipitation is generated from sulphuric and nitric acids. The presence of excess sulphur and nitrogen in the atmosphere is largely derived from anthropogenic sources from the oxidation of SO_2 and NO_x (nitrous oxide [NO]+nitrogen dioxide [NO_2]). Key sources of SO_2 include fossil fuel burning and metal smelting. Nitrogen oxides are generated during combustion by oxidation of atmospheric nitrogen; the main anthropogenic source is vehicle emissions. Whilst sources of SO_2 have been controlled and atmospheric concentrations are decreasing, atmospheric NO_x continues to rise.

Dry deposition occurs in the interval between precipitation events. The physics of the mechanisms involved in dry deposition are complex and include gravitational settling and filtering of particulate aerosols together with the direct uptake of gases such as SO_2 and NO_x onto vegetation, water and/or soil surfaces. Factors such as surface wetness, and vegetation size, growth patterns and surface roughness are important controls on the magnitude of dry deposition. Acidity derived from dry deposition is often generated by secondary chemical reactions. Thus, dry-deposited SO_2 is oxidized to the anion sulphite (SO_3^{2-}) and rapidly oxidized to sulphate (SO_4^{2-}); electrochemical balancing releases an equivalent number of H^+, which in turn generates acidity. Similarly, dry-deposited NO_x may be oxidized to nitrate (NO_3^-), which again generates H^+. The atmospheric sources are not necessarily acidic in themselves; thus, acidity may be generated at the ground surface where gaseous ammonia or the cation ammonium (NH_4^+) is deposited and oxidized to NO_3^-, which releases H^+. Quantitative estimation of dry fluxes is difficult; thus, this pathway of acidification is less understood than that of wet deposition. Problems in measurement include wide spatial and temporal variations in atmospheric gases and particulates, and difficulties in estimating rates of deposition to natural ecosystems with inherently complex structures. Dry deposition is more affected by distance from emission source and tends to dominate close to source.

Acid deposition is not the only source of acidification of terrestrial and aquatic environments. Other sources include afforestation and forest clearance, livestock grazing, cultivation techniques on agricultural land and the use of fertilizers.   ALH

References

Fowler, D., Cape, J.N. and Leith, I.D. (1985) Acid inputs from the atmosphere in the United Kingdom. *Soil Use and Management*, **1**, 70–72. · Howells, G. (1990) *Acid rain and acid waters*. Chichester: Ellis Horwood.

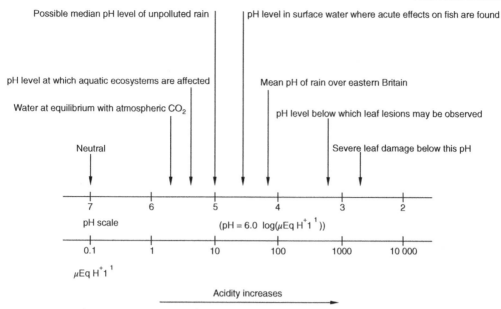

Possible median pH level of unpolluted rain

pH level in surface water where acute effects on fish are found

pH level at which aquatic ecosystems are affected

Mean pH of rain over eastern Britain

Water at equilibrium with atmospheric CO_2

pH level below which leaf lesions may be observed

Neutral

Severe leaf damage below this pH

| 7 | 6 | 5 | 4 | 3 | 2 |

pH scale

$(pH = 6.0 \ log(\mu Eq \ H^+1^1))$

| 0.1 | 1 | 10 | 100 | 1000 | 10 000 |

$\mu Eq \ H^+1^1$

Acidity increases

Scale of pH and H^+ concentration; acidity (H^+) increases, while pH decreases on a 'log scale' (i.e. $10:1$).
Source: Howells (1990).

acid precipitation   Precipitation is naturally slightly acid (pH 5.65) owing to the dissolution of carbon dioxide in water Any form of precipitation (usually rain or snow) with a pH of less than 5.65 is described as acid. The slight natural acidity of precipitation is largely due to weak carbonic acid, formed by dissolved atmospheric carbon dioxide, and to sulphur compounds from volcanic eruptions which are converted to sulphuric acid in the atmosphere. The chemical analysis and dating of fossil ice has revealed that, some two centuries ago, precipitation possessed a pH that was generally in excess of 5. Since that time, industrial–urban development, particularly in northern-hemisphere mid-latitudes, has resulted in the release of increasing quantities of sulphur and nitrogen oxides into the atmosphere. These emissions are caused by fossil fuel burning and sulphide ore smelting, the oxides being transformed into sulphuric and nitric acids in the atmosphere. These relatively strong acids undergo ionic separation in weakly acidic natural precipitation, with the dissociated hydrogen ions causing its pH to fall below 5.6. (Likens *et al.*, 1979). *Sensu stricto*, acid precipitation is thus wet deposition. However, a related process, dry deposition, whereby oxides of sulphur and nitrogen fall out from the atmosphere either as dry gases or adsorbed on other AEROSOLS such as soot, is also operational. These particles become acidic when they join with moisture: fog or surface water.

Atmospheric circulation patterns mean that pollutants (POLLUTION) can travel substantial distances before being deposited as acid precipitation. As the loci of acidic pollution are within the westerly wind-belt, their discharges are usually routed eastward. The rate and distance of movement are associated with the height of pollutant emission (Kemp, 1990). Tall stacks enhance long-distance transfer with upper westerly winds or jet streams more effective than boundary-layer circulation, both in this respect and in increasing the residence time of pollutants in the atmosphere and increasing trans-boundary impacts. Quantitative estimation of dry fluxes is difficult; thus, this pathway of acidification is less understood than that of wet deposition. Problems in measurement include wide spatial and temporal variations in atmospheric gases and particulates, and difficulties in estimating rates of deposition to natural ecosystems with inherently complex structures. Dry deposition is more affected by distance from emission source and tends to dominate close to source.

The phenomenon of acid precipitation was first recognized in England during the mid-nineteenth century, but has been studied in detail only in the last 50 years (Park, 1987). Its role is still imperfectly understood and is complicated by interactions between environmental factors (geology, hydrology and land use, for example) and precipitation. Aquatic ecosystems appear to

5

respond more rapidly to acidification than terrestrial ones do. In 1991 it was estimated that 5% of all lakes in New England, USA, were excessively acidified, and this was widely attributed to acid precipitation effects. While often viewed as a problem of 'old' industrial nations, acid precipitation is a growing issue in emerging economies, such as China (Larssen *et al.*, 2006) and India (Rodhe *et al.*, 2002). Other impacts include reductions in biological productivity in water bodies, tree die-off, building stone damage, and human health issues associated with drinking water quality and inhalation.   RLJ/DSGT

Reading and References
Kemp, D.D. (1990) *Global environmental issues: a climatological approach*. London: Routledge. · Lane, C.N. (2003) *Acid rain: overview and abstracts*. New York: Nova Science Publishers (especially chapter 1). · Larssen, T., Lydersen, E., Tang, D. *et al.* (2006) Acid rain in China. *Environmental Science and Technology*, **40**, 418–425. · Likens, G.E., Wright, R.F., Galloway, J.N. and Butler, T.J. (1979) Acid rain. *Scientific American*, **241**, 39–47. · Park, C.C. (1987) *Acid rain: rhetoric and reality*. London: Methuen. · Rodhe, H., Dentener, F. and Schilz, M. (2002) The global distribution of acidifying wet deposition. *Environmental Science and Technology*, **36**, 4382–4388.

acid rain   See ACID PRECIPITATION.

acid rocks   Commonly used term for igneous rocks that contain more than 66% silica, free or combined, or any igneous rock composed predominantly of highly siliceous minerals (see BASIC ROCKS).

acid   susceptibility/acid   neutralizing capacity   Acid neutralizing capacity is the ability of a soil or water body to exhibit relatively stable pH in the face of additions of acidic compounds. Acid neutralizing capacity may be derived from carbonate compounds that can be dissolved and so raise the pH, with the release of carbonate and bicarbonate ions. Ion-exchange processes involving the clay minerals may also provide acid neutralizing capacity in soils. Soil materials that lack carbonates and clays, such as those derived from the weathering of granitic rocks, may have low acid neutralizing capacity. Acid susceptibility, the lack of acid neutralizing capacity, may also be very high in soils and sediments containing pyrite or other forms of iron sulphide. In the absence of oxygen, pyrite minerals remain stable, but when exposed to oxygen they form sulphuric acid. Drainage works, lowering of the groundwater table from the effects of mining or quarrying, or the use of sulphide soils in earthworks at the surface may all contribute to the oxidation of the iron sulphides. The pH reduction then produced as a result of insufficient acid neutralizing capacity subsequently increases the solubility of other materials, including aluminium and heavy metals. These can have toxic effects when leached into waterways.   DLD

Reading
Mason, B.J. (1990) *The surface waters acidification programme*. Cambridge: Cambridge University Press. · Cresser, M. and Edwards, A. (1987) *Acidification of freshwaters*. *Cambridge*: Cambridge University Press; p. 136. · Sammut, J., Melville, M.D., Callinan, R.B. and Fraser, G.C. (1995) Estuarine acidification: impacts on aquatic biota of draining acid sulphate soils. *Geographical Research*, **33**, 89–100. · Unland, N.P., Taylor, H.L., Bolton, B.R. and Cartwright, I. (2012) Assessing the hydrogeochemical impact and distribution of acid sulphate soils, Heart Morass, West Gippsland, Victoria. *Applied Geochemistry*, **27**, 2001–2009.

acidity profile   The acid concentration in ice core layers as a function of depth as determined from electrical measurements. The magnitudes of some volcanic eruptions in the northern hemisphere have been estimated from the acidity of annual layers in ice cores taken in Greenland. This methodology is sometimes referred to as 'acidity signal' or 'acidity record'.   ASG

active layer   The top layer of ground above the permafrost table that thaws each summer and refreezes each autumn. In temperature terms, it is the layer that fluctuates above and below 0 °C during the year. In permafrost areas, seasonally frozen and thawed ground can be equated with the active layer. Other synonyms include 'depth of thaw', 'depth to permafrost' and 'annually thawed layer'. These terms are acceptable in areas where the active layer extends downwards to the permafrost table, but they are misleading where the active layer is separated from the permafrost by a layer of ground that remains in an unfrozen state throughout the year. The thickness of the active layer varies from as little as 15–30 cm in high latitudes to over 1.5 m in subarctic continental regions. Thickness depends on many factors, including the degree and orientation of the slope, vegetation, drainage, snow cover, soil and rock type, and ground moisture conditions.

Processes operating in the active layer include FROST CREEP and FROST HEAVE or cryoturbation, the lateral and vertical displacement of soil that accompanies seasonal and/or diurnal freezing and thawing. During thaw, water movement through the active layer assists various mass-wasting processes, especially GELIFLUCTION. Most patterned ground phenomena form in the active layer.   HMF

Reading
Brown, R.J.E. and Kupsch, W.O. (1974) *Permafrost terminology*. Publication 14274. Ottawa: National Research Council of Canada. · French, H.M. (2007) *The periglacial environment*. Chichester: John Wiley & Sons, Ltd.

active margin   A continental margin that coincides approximately with a plate boundary. It has much more tectonic and igneous activity than a PASSIVE MARGIN does, which lies within a plate and on the edge of a spreading ocean basin such as the Atlantic, Indian, Arctic or Antarctic Oceans. Active margins rim most of the Pacific, where the volcanic activity gave rise to the term 'Ring of Fire': they show EARTHQUAKE activity at various depths in BENIOFF ZONES or at shallow depths along transform faults. Plate motions are either convergent, with subduction at an orogen (the Andes and Central America) or at an island-arc (the Kurile Islands, Japan and the Philippines), or oblique, with a transform fault (the San Andreas in California or the Alpine in New Zealand).   IE

Reading
Summerfield, M.A. (2013) *Global geomorphology*. London: Routledge; chapter 3.

actualism   See UNIFORMITARIANISM.

adaptation   The ability of organisms to adapt to conditions in a particular environment as a result of hereditary features that have resulted from natural selection. Some organisms, called 'extremophiles', display the ability to withstand various relatively extreme environments – arid, hot, wet, acidic, salty, etc. The adaptive traits of organisms may be structural, behavioural or physiological. Structural adaptations are physical features of an organism (shape, body covering, etc.) Behavioural adaptations include miscellaneous instincts. Physiological adaptations permit the organism to perform special functions (for instance, making venom), but also more general functions such as growth and development and temperature regulation. The concept has been described as 'slippery' and 'controversial' and is still the subject of debate. One issue of current concern is the speed with which organisms may adapt to future climate changes.   ASG

Reading
Hoffmann, A.A. and Sgrò, C.M. (2011) Climate change and evolutionary adaptation. *Nature*, **470**, 479–485. · Reeve, H. K. and Sherman, P.W. (1993) Adaptation and the goals of evolutionary research. *Quarterly Review of Biology*, **68**, 1–32. · Rose, M.R. and Lauder, G.V. (eds) (1996) *Adaptation*. San Diego, CA: Academic Press.

adaptive capacity   The capacity of a system to adapt if the environment where the system exists is changing. Adaptive capacity confers resilience to perturbation, giving ecological and human social systems the ability to reconfigure themselves with minimum loss of function. In ecological systems, this resilience shows as net primary productivity and maintenance of biomass and biodiversity, and the stability of hydrological cycles. The development of adaptive capacity is particularly important in the context of societal responses to global environmental change. See also PANARCHY.   TS

adhesion ripple   An irregular sand ridge transverse to wind direction, formed when dry sand is blown across a smooth, moist surface. It may be 30–40 cm long and a few centimetres high. The crest is symmetrical and migrates upwind. The stoss (windward) side is steeper than the lee side.   ASG

adhesion warts   Small-scale aeolian BEDFORMS resulting from SALTATION over a wet sand surface. The moisture captures the falling grains, and capillary action wets that deposit. With unidirectional winds, adhesion ripples may form and migrate upwind. With polymodal winds, irregularly shaped and oriented adhesion warts result.   DJS

Reading
Goodall, T.M., North, C.P. and Glennie, K.W. (2000) Surface and subsurface sedimentary structures produced by salt crusts. *Sedimentology*, **47**, 99–118.

adiabatic   An adiabatic process is a thermodynamic change of state of a system in which there is no transfer of heat or mass across the boundaries of the system. In the atmosphere the most commonly related variables in the adiabatic process are temperature and pressure. If a mass of air experiences lower pressure than in its initial condition it will expand and do mechanical work on the surrounding air. The energy required to do this work is taken from the heat energy of the air mass and, consequently, the temperature of the air falls. Conversely, when pressure increases, work is done on the mass of air and the temperature rises. A diabatic process is a thermodynamic change of state of a system in which there is transfer of heat across the boundaries of the system.   BWA

adsorption   The process in which substances, often ions, leave a liquid and accumulate upon the surface of a solid. In the soil, adsorption occurs when ions and other charged entities move from the soil water and attach themselves close to the surfaces of CLAY particles. This takes place because the clay surface normally carries an excess negative charge arising from ionic substitution within the lattice structure, and because of unsatisfied valence at crystal edges. Adsorbed materials are most commonly cations like Ca^{2+} and Mg^{2+}, but may also include oxyanions like NO_3^- as well as organic compounds. Adsorbed materials may be difficult to differentiate from precipitates, so that in some cases materials held on a surface may be referred to in a more all-encompassing way as *sorbed*.   DLD

Reading
Bradl, H.B. (2004) Adsorption of heavy metal ions on soils and soils constituents. *Journal of Colloid Science*, **277**, 1–18. · Calvet, R. (1989) Adsorption of organic chemicals in soils. *Environmental Health Perspectives*, **83**, 145–177. · Sposito, G. (1984) *The surface chemistry of soils*. Oxford: Oxford University Press.

advection   The lateral transport of matter or energy; used to indicate the contrast with CONVEC-TION, which is the vertical transport of matter or energy. For instance, advection of air, sensible heat and moisture commonly occur in local atmospheric circulations such as land and sea breezes. Advection of momentum, heat and moisture contribute to the melting of patchy snow cover where darker vegetated surfaces separate snow patches. Likewise, the advection of heat and moisture from ocean to land on a larger scale is involved in the formation of monsoonal circulation systems. The advection of warm air from urban areas may affect evaporation rates in surrounding vegetated areas. At larger scales, the advection of moisture onto the continents, for instance by persistent trade winds

Advection: the effect of precipitation recycling and moisture advection on rainfall over the tropical forests. Annual rainfall declines steeply inland in the dry transects, but remains constant or even increases in some of the wet continental locations.
Source: Sheil and Murdiyaso (2009). Reproduced with permission from Oxford University Press.

or cyclones, is critical for the occurrence of precipitation over land. Within extensive vegetated areas like the tropical forests of South America, advection of moisture is pivotal to the process of precipitation recycling. DLD

Reading
Drost, F. and England, M.H. (2008) Twentieth century trends in moisture advection over Australia. *Meteorology and Atmospheric Physics*, **100**, 243–256. · Sheil, D. and Murdiyaso, D. (2009) How forests attract rain: an examination of anew hypothesis. *BioScience*, **59**, 341–347. · Trenberth, K.E. (1999) Atmospheric moisture recycling: role of advection and local evaporation. *Journal of Climate*, **12**, 1368–1381.

advection fog   See FOG.

advective processes
Processes for which the upslope drainage area influences rates of sediment transport through entrainment of material that moves along with (i.e. is advected by) the flow. Sediment transport Q_s by advective processes is often cast in landscape evolution models as $Q_s = KA^m S^n$, where K is a (fluvial) rate parameter, A is drainage area (which may be replaced by discharge Q in some models), S is the slope, and m and n vary with the process under consideration. Advective processes tend to create relief through the incision of valleys. DRM

aeolation
It has been suggested that in the late nineteenth and early twentieth centuries there was a tendency to identify AEOLIAN processes as the dominant geomorphic mechanism that shaped the world's DESERTs, at the expense of considering the role of water and runoff. While for some desert environments, such as the American Southwest, the power of aeolian erosion may have been overemphasized, the past importance attached to the aeolation paradigm may itself have been overstressed in recent years. Today, aeolation and run-off erosion are not viewed as competing paradigms in the explanation of the formation of dryland landscapes. The expansion of process-based studies at the expense of simple description, the increase in data sources and the availability of high-resolution aerial imagery all allow the relative importance of different processes in space and time to be considered, rather than a unitary explanation of dryland evolution to be sought. DSGT

Reading
Goudie, A.S. (2012) Charles Rolin Keyes and extravagant aeolation. *Aeolian Research*, **4**, 51–53.

aeolian (eolian in the USA)   Of the wind; hence aeolian processes and aeolian sediments and landforms. Derived from Aeolus, Greek god of the winds. DSGT

aeolianite   Cemented dune sand, calcium carbonate being the most frequent cement. The degree of cementation is very variable, the end product being a hardened dune rock with total occlusion of pore space. Aeolianite of Quaternary age is generally found in coastal areas within 40° of the equator, especially those that experience at least one dry season. The balance between leaching and lime production is the prime control of this overall distribution. Most examples contain between 30 and 60% calcium carbonate, although not all of this may occur as cement. According to Yaalon (1967) a minimum of 8% calcium carbonate is required for cementation of dune sands under semi-arid conditions. Sources of calcium carbonate include biogenic skeletal fragments, dust, spray and groundwater. ASG

Reading and Reference
McLaren, S.J. (2007) Aeolianite. In D.J. Nash and S. McLaren (eds), *Geochemical sediments and landscapes*. Chichester: John Wiley & Sons, Ltd; pp. 149–172. · Yaalon, D. (1967) Factors affecting the lithification of aeolianite and interpretation of its environmental significance in the coastline plain of Israel. *Journal of Sedimentary Petrology*, **37**, 1189–1199.

Aeolianite in old coastal dunes, Agulhas Peninsula, South Africa. Photograph by David Thomas.

aerial photography The action of taking photographs, of the Earth's surface, from an airborne platform furnished with a camera placed at some distance above that surface. It is a tried and tested method of REMOTE SENSING. Common platforms deployed include aircraft, helicopters, kites, blimps or unmanned aerial vehicles (i.e. drones). It is the most widely used type of remote sensing, developed extensively during the Second World War as a means of intelligence acquisition, and now most familiar to the public via a virtual globe, such as Google Earth. The photography is taken from three main angles: vertical, low oblique and high oblique. Vertical aerial photography results when the camera axis is pointing vertically downwards, and oblique aerial photography is the result when the camera axis is pointing obliquely downwards. Low oblique aerial photography incorporates the horizon into the photograph, while high oblique aerial photography does not. The latter provide a realistic view of the objects on the Earth's surface and coverage that is many times more than that in vertical aerial photographs at the same acquisition conditions (i.e. height of capture and focal length). Oblique photography is useful for applications such as wayfinding, but the variance in scale across the resultant photograph make measurements difficult. It is the vertical aerial photographs that are most widely used for mapping and measurement, owing to their constant scale, assuming a flat surface across the photograph. Vertical photographs that are captured with an overlap between adjacent photographs afford a three-dimensional (3D) perspective of the Earth's surface: a stereoscopic view can be created and measurements taken both horizontally and vertically. Recently, a new computing vision technique, structure from motion (SfM), was introduced that allows the extraction of 3D structure of an object by analysing motion signals over time. The SfM technique can be applied to large collections of overlapping photographs to obtain sparse point clouds for a wide range of objects, such as buildings and sculptures.

Aerial photographs are acquired at a wide range of spatial resolutions (scales), with a small-scale photograph at 1 : 50,000 affording a synoptic, low spatial resolution overview of a large area, and a large-scale aerial photograph at 1 : 2000 affording a detailed and high spatial resolution view of a small area. Traditionally, black-and-white film cameras were the sensor of choice; this has been superseded by colour film cameras, of which there are two main types: conventional colour film, where the near-infrared response appears red, red response appears green and green response appears blue. Once obtained, aerial photographs are interpreted for the identification of objects and assessment of their significance. During this process the interpreters usually undertake several tasks of detection, recognition and identification, analysis, deduction, classification, idealization and accuracy determination. Recognition and identification of objects or areas comprise the most important link in this chain of events. An interpreter uses seven characteristics of the aerial photography to help with this stage: tone, texture, pattern, place, shape, shadow and size. Tone is the single most important characteristic of the aerial photograph as it represents a record of the electromagnetic radiation that has been reflected from the Earth's surface onto the aerial film. Texture is the frequency of tonal changes that arise within an aerial photograph when several features are viewed together. Pattern is the spatial arrangement of objects on the aerial photograph. Place is a statement of an object's position on the aerial photograph in relation to others in its vicinity. Shape is a qualitative statement of the general form, configuration or outline of an object on an aerial photograph. Shadows of objects on an aerial photograph are used to help in identifying them; for example, by enhancing geological boundaries. Size of an object is a function of the scale of the aerial photograph. The sizes of objects can be estimated by comparing them with objects for which the size is known. The common film cameras are now being replaced by digital aerial cameras adopting two main semiconductor technologies – charge-coupled device and complementary metal oxide semiconductor – that are used in airborne imaging. These airborne digital cameras are classified as either (i) small-format (up to 16 megapixels), (ii) medium-format (from 16 up to 50 megapixels) or (iii) large-format (greater than 50 megapixels) cameras and are rapidly changing the processing chain and applications using aerial photographic technology. DSB

Reading
Arnold, R.H. (2004) *Interpretation of airphotos and remotely sensed imagery.* New York: Prentice Hall. · Falkner, E. and Morgan, D. (2001) *Aerial mapping: method and applications,* 2nd edition. Boca Raton, FL: CRC Press. · Paine, D.P. and Kiser, J.D. (2003) *Aerial photography and image interpretation,* 2nd edition. New Jersey: John Wiley & Sons, Inc. · Petrie, G. and Walker, A. (2007) Airborne digital imaging technology: a new overview. *The Photogrammetric Record,* **22,** 203–225.

aerobic Describes either conditions of the environment or of a metabolic process in which oxygen is freely available. Most frequently it is the term applied to a form of RESPIRATION in which the gaseous or dissolved form of oxygen is the principal agent of oxidation, as opposed to the ANAEROBIC form in which other substances participate in the energy release.   MEM

aerobiology

The study of the conveyance and behaviour of passive airborne organic particles, including pollen, spores, bacteria and very small insects. These particles are AEROSOLS (a term that also includes nonbiological particles) and may be viable or nonviable. They are dispersed in air and move, dependent upon atmospheric properties and conditions. Both living and nonliving organisms are considered in an aerobiological context. (See also POLLEN ANALYSIS.)   RLJ/DSGT

Reading
Frenguelli, G. (2013) New trends in aerobiology. *Review of Allergy and Immunology*, **23**, 3–9.

aerodynamic ripples

These are a type of aeolian bedform that are thought to differ in formation from other ripple types. Normal or CURRENT RIPPLES form at right angles to the direction of fluid flow, whether the fluid is water or air, and their formation is likely to relate to the impact of saltating particles – hence the term impact ripple also being prevalent in the aeolian literature. By way of contrast, Wilson (1972) believed that aerodynamic ripples (wavelength ~0.015–0.25 m, as opposed to a range of 0.05–2.0 m for normal ripples) could form either transverse or parallel to the wind direction, as a result of wave-like instability in the turbulent elements of secondary flow close to the ground surface. As Cooke *et al.* (1993) note, there is little if any empirical support for this theory.   DSGT

References
Cooke, R.U., Warren, A. and Goudie, A.S. (1993) *Desert geomorphology*. London: UCL Press. · Wilson, I.G. (1972) Aeolian bedforms: their development and origins. *Sedimentology*, **19**, 173–210.

aerosol

An intimate mixture of two substances, one of which is in the liquid or solid state dispersed uniformly within a gas. The term is normally used to describe smoke, condensation nuclei, freezing nuclei or fog contained within the atmosphere, or other pollutants such as droplets containing sulphur dioxide or nitrogen dioxide. In recent years, DUST aerosols have been a particular focus of interest in geography and related disciplines. Dust emitted from the land surface, especially in dry regions, can be entrained high into the Earth's atmosphere and has a major impact on processes that affect climate and weather.

Aerosols tend to obscure visibility by scattering light. They tend to vary in size between about a nanometre (10^{-9} m) and a micrometre (10^{-6} m). Clouds are not normally considered to be aerosols because the droplets are too large and tend to fall due to gravity. Aerosols can remain in the atmosphere for long periods, collisions with air molecules keeping them aloft.   JET/DSGT

Reading
Mahowald, N., Albani, S., Kok, J.F., *et al.* (2014) The size distribution of desert dust aerosols and its impact on the Earth system. *Aeolian Research*, **15**, 53–71.

aestivation

The dormancy of certain animals during the summer season, the dry season or prolonged droughts. It is an important means of adaptation for desert animals, and can be contrasted with hibernation – a state of dormancy in the winter months.

afforestation

Planting of new forest on lands that, historically, have not been forested. Not to be confused with reafforestation (the replanting and/or restocking of previously forested areas). Afforestation can relieve pressure on indigenous forest reserves, and it has been argued that large-scale afforestation could successfully absorb the CO_2 emissions generated by fossil fuel burning. The vast areas of afforestation required to achieve this aim would, however, most probably result in many negative environmental impacts. These include loss of, and alien invasions into, valuable wetlands and grasslands by new forests, impacts on the margins of existing forests, including changes in fire regime, and regional reductions in run-off and water availability.   TS

affluent

A stream or river flowing into another; a tributary.

aftershocks

A series of small earthquakes following a major tremor and originating at or near its focus. Aftershocks generally decrease in frequency over time but may occur over a period of several days or months. (See also EARTHQUAKE.)

ageostrophic

Atmospheric motion in which the horizontal pressure gradient is not balanced with the force (CORIOLIS FORCE) due to the Earth's rotation. In the boundary layer, where friction plays a role, flow is ageostrophic. This permits air to converge into low-pressure regions, creating vertical motion, and generally clouds and precipitation. Other flows, such as those in the vicinity of jet-stream cores, also have ageostrophic components that similarly alter the patterns of vertical motion.   SN

agglomerate A rock composed of angular fragments of lava, generally more than 20 mm in diameter, which have been fused by heat.

aggradation The accumulation of sediment at the Earth's surface. Aggradation refers to sediment accumulation, over significant periods of time, which increases the elevation of the surface on which deposition occurs. Short-term deposition flowed by scour, as might occur during the course of a single flood event, does not constitute aggradation. Aggradation can be contrasted with progradation, which is lateral growth arising from sediment accumulation. Thus, a floodplain that receives increments of sediment from overbank flow may experience aggradation, while a shoreline or delta that is being built outwards because of the delivery of sediment from river systems is said to be prograding. Aggradation along floodplain streams can increase the elevation of the alluvial ridge above the distal parts of the floodplain, and so increase the likelihood of avulsive channel relocation. Aggradation can result from climate change, wildfire or the delivery of large amounts of sediment by mass movement, perhaps triggered by seismic activity, or fluvial or glacial erosion. DLD

Reading
Sadler, P.M. and Jerolmack, D.J. (2015) Scaling laws for aggradation, denudation, and progradation rates: the case for time-scale invariance at sediment sources and sinks. In D. G. Smith, R. J. Bailey, P. M. Burgess and A. J. Fraser (eds), *Strata and time: probing the gaps in our understanding*. Geological Society, London, Special Publications, vol. 404. London: Geological Society of London; pp. 69–88.

aggregation ratio The aggregation ratio of a soil is the ratio of the percentage weight of clay minerals, determined by mineralogical analysis, to the percentage weight of clay particles, determined by sedimentation methods. The ratio is meant to account for problems with the activity of a soil in which clay mineral particles aggregate and act as clays, but have a size corresponding to a value in the silt size range.   WBW

agonic line A shifting, irregular imaginary line running through the Earth's north and south magnetic poles along which the compass needle points to true north; hence the line of no magnetic variation.

aggressivity In the context of limestone solution, the propensity of water to dissolve calcium carbonate. When water comes in contact with air it dissolves an amount of carbon dioxide (CO_2) into the water. The resultant carbonic acid can dissolve calcium carbonate (via the theoretical compound $CaHCO_3$) until the aggressiveness of the water diminishes (when the CO_2 is used up). It is then said to be saturated with respect to calcium carbonate. When two saturated solutions of calcium carbonate are mixed it is possible for the resulting water to be aggressive. This phenomenon, first identified by Bögli (1971, English version) is called mixing CORROSION.   PAB

Reference
Bögli, A. (1971) Corrosion by mixing of karst water. *Transactions of the Cave Research Group of Great Britain*, **13**, 109–114.

agricultural drought A particular consideration or calculation of DROUGHT conditions that assesses the effect of moisture deficits upon agricultural productivity. Various definitions exist. For example, it may simply be considered as a period of dry weather of sufficient duration to cause at least partial crop failure. Unlike meteorological drought, which specifically reflects precipitation deficits, agricultural drought focuses upon soil moisture deficits during the growing season (Wigley and Atkinson, 1977), so that precipitation and potential evaporation are assessed in conjunction with soil moisture levels. To develop an index of agricultural drought requires data on seasonal crop water requirements, as well as on the water holding capacity of the soil, which is influenced by texture, depth, surface cover and other factors. The timing of crop planting and harvesting and the occurrence of rain during the year are also important considerations, perhaps more so than absolute rainfall deficits.   DSGT

Reading and Reference
Jayanthi, H. and Husak, G. (2013) *A probabilistic approach to assess agricultural drought*. Geneva: United Nations Office for Disaster Risk Reduction. · Wigley, T.M.L. and Atkinson, T.C. (1977) Dry years in south-east England since 1698. *Nature*, **265**, 431–434.

agroclimatology The study of the interaction between climatological and hydrological factors and agriculture, including animal husbandry and forestry. Its aim is to apply climatological information for the purpose of improving farming practices and increasing agricultural productivity in quantity and in quality. Agroclimatology and AGROMETEOROLOGY share nearly the same aims, scope and methodology. However, in their application the latter tends to emphasize weather forecasting in dealing with daily problems, whereas the former is more concerned with the use of mean data as a guide to long-range planning. With contemporary and future climate change, the challenges of linking improved agriculture to climate are challenging. ASG

Reading
Dinar, A. and Mendelsohn, R. (eds) (2013) *Handbook on climate change and agriculture*. Cheltenham: Edward Elgar.

agroforestry Any system where trees are deliberately left, planted or encouraged on land where crops are grown or animals grazed. It includes practices as diverse as slash-and-burn agriculture, the growth of shade trees and the use of living fences either to contain or to exclude animals. Deep-rooted trees tap nutrient sources that are out of reach of most crops; these nutrients become readily available when the leaves fall. Leguminous trees improve soil fertility directly through nitrogen fixation. Tree roots help to bind the soil and increase aeration. Mixtures of trees and crops provide more complete ground cover, which helps to prevent soil erosion and weed invasion while making full and productive use of available solar radiation. Leaf litter from the trees adds organic matter to the soil and acts as a mulch. Tree cover helps to regulate temperatures, reducing extremes. Farming communities benefit from a regular supply of wood and other tree products. Multipurpose trees can provide fodder for livestock, edible fruits and nuts, fuel, timber, supports for climbing vegetables and medicinal products.

During the last century, social and economic changes – including the amalgamation of holdings and large-scale mechanization – have led to the loss of many agroforestry systems in Europe. New forms of agroforestry, including alley cropping with rows of different crop plants grown between rows of trees, are seen as a means of mitigating the risk of diminished productivity arising from climate change. Increasingly, ecosystem services are seen as additional important benefits of agroforestry systems. DLD

Reading
Jose, S. (2009) Agroforestry for ecosystem services and environmental benefits: an overview. *Agroforestry Systems*, 76, 1–10. · Quinkenstein, A., Wöllecke, J., Böhm, C., *et al.* (2009) Ecological benefits of the alley cropping agroforestry system in sensitive regions of Europe. *Environmental Science & Policy*, 12, 1112–1121.

agrometeorology The science concerned with the application of meteorology to the measurement and analysis of the physical environment in agricultural systems. The influence of the weather on agriculture can be on a wide range of scales in space and time, and this is reflected in the scope of agrometeorology. At the smallest scale the subject involves the study of microscale processes taking place within the layers of air adjacent to leaves of crops, soil surfaces or animals' coats (see MICROCLIMATE and MICROMETEOROLOGY). These processes determine rates of exchange of energy, and mass, between the surface and the surrounding air. Such exchange rates are the essential link between the biological response and the physical environment. For example, the capture of radiant energy and its use to convert carbon dioxide and water into carbohydrates are essential elements in crop growth. Agrometeorologists have studied how the structure of leaf canopies affects the capture of light and how measurements of the atmospheric carbon dioxide concentration may be used to determine rates of crop growth.

On a broader scale, agrometeorologists attempt to use standard weather records to analyse and predict responses of plants and animals. An area of particular interest concerns the estimation of water use by crops as a basis for planning irrigation requirements. Methods based on empirical correlations with wind speed, sunshine, temperature and humidity have been superseded by methods with a sounder physical basis, often using new measuring techniques (see EVAPOTRANSPIRATION). Other examples of this scale of agrometeorology include procedures for forecasting the occurrence of damaging frosts or of plant or animal diseases. The agrometeorologist is also commonly interested in the soil environment because of the large influence that the weather can have on soil temperature and on the availability of water and nutrients to plant roots. MHU

Reading
Griffiths, J.F. (1994) *Handbook of agricultural meteorology*. Oxford: Oxford University Press.

aiguille A sharply pointed rock outcrop or mountain peak. Often applied to pinnacles that are the products of frost action.

air mass A body of air that is quasi-homogeneous in terms of TEMPERATURE and HUMIDITY characteristics in the horizontal plane and has similar LAPSE RATE features. An ideal air mass is a barotropic fluid in which isobaric and isosteric (constant specific volume) surfaces do not intersect. Air mass classification (see diagram) is based on the nature of the source area (e.g. polar,

Table 1 Air mass classification

	Tropical	Polar	Arctic/Antarctic
Maritime	Maritime tropical (mT) Warm and very moist Near Azores in N. Atlantic	Maritime polar (mP) Cool and fairly moist Atlantic south of Greenland	Arctic (A) or Antarctic (AA) Very cold and dry Frozen Arctic Ocean
Continental	Continental tropical (cT) Hot and dry Sahara desert	Continental polar (Cp) Cold and dry Siberia in winter	Central Antarctica

tropical) and the characteristics of the surface during the outward trajectory (e.g. maritime, continental) (Table 1). Thermodynamic and dynamic factors will modify the properties of air masses in their transit from source areas. AHP

Reading
Ahrens, C.D. (2012) *Essentials of meteorology*, 7th edn. Stanford: Cengage Learning; chapter 8.

Air mass classification and properties with examples of source regions.

air parcel An imaginary body of air to which may be assigned any or all of the dynamic and thermodynamic properties of the atmosphere. Investigation of the STABILITY of the atmosphere is made most simply by the 'parcel method' in which it is hypothesized that a test parcel of air moves vertically with respect to its environment as represented by an ascent curve on a TEPHIGRAM. AHP

Reading
Barry, R.G. and Chorley, R.J. (2010) *Atmosphere, weather and climate*, 9th edn. London: Routledge.

air pollution Substances in the air that can be harmful to humans, animals or the ecosystem. Pollutants can be solid particles, liquids or gases and can be natural or anthropogenic. Heavy concentrations of pollutants are found in urban regions, but also in many indoor locations. In many cases, the natural level in the atmosphere may be benign, but that amount added by humans may be high enough to produce adverse effects. Carbon monoxide, for example, occurs naturally in the atmosphere. However, its concentration in many urban areas is high enough to produce respiratory and heart problems. Some of the other most common pollutants include oxides of sulphur and nitrogen, methane, various organic compounds, chlorofluorocarbons, toxic metals such as lead and mercury, ozone and particulates. Typical sources include fossil fuel burning (e.g. in power plants, factories, by vehicles), biomass burning, fumes from aerosol sprays and various solvents, and waste deposition in landfills. Wind erosion also produces tremendous amounts of particulates. Adverse effects on humans are numerous: asthma and other respiratory problems, heart disease, stroke and cancer, among other concerns. SEN

Reading
Vallero, D. (2007) *Fundamentals of air pollution*. San Diego, CA: Elsevier/Academic Press.

Airy wave theory Linear wave theory, named after the British mathematician and astronomer George Biddell Airy (1801–1892), that treats waves as a sinusoidal form. The theory is powerful because it allows wave characteristics to be calculated from simple measurements of wave height and wave period and predicts subsurface water particle motion to be derived from surface characteristics. The dispersion equation in Airy wave theory expresses the relation between wavelength and wave period, which in turn allows wave phase velocity to be derived; both deep-water (water depth greater than twice the wavelength) and shallow-water approximations (water depth less than 1/20th of the wavelength) are available. As Airy waves propagate across the sea surface, the water particles complete a closed orbit equal in diameter to the wave height. The diameters of the orbits decrease rapidly with increases in water depth; thus, at depths greater than half the wavelength the orbital diameters, and corresponding orbital velocities, are less than 5% those at the surface. Deep water waves, therefore, are very much a surface phenomenon. Linear wave theory also shows that potential energy (the deviation of the wave form from the still water level) and kinetic energy (resulting from the orbital motion of the water particles) are equal and that the total wave energy present depends upon wave height. As wave energy depends upon the square of the wave height, a doubling of wave height results in a fourfold increase in wave energy. Taken together, these relationships explain wave shoaling: how wavelengths shorten, wave velocity declines, wave period remains constant and wave height increases in shallow water prior to wave breaking. However, in very shallow water, wave asymmetry develops with increased differences between wave crests and wave troughs. In such settings, higher order wave theories are required to explain wave dynamics, such as Stokes and Cnoidal wave theory. TS

Reading
Komar, P.D. (1997) *Beach processes and sedimentation*, 2nd edition. Upper Saddle River, NJ: Prentice-Hall.

aklé A network of sand dunes, comprising overlapping TRANSVERSE DUNE ridges that totally enclose the interdune areas. Sometimes described as producing a fishscale pattern. See DUNE NETWORK. DSGT

alas A steep-sided, flat-floored depression, sometimes containing a lake, found in areas where local melting of permafrost has taken place. It is one manifestation of a THERMOKARST.

albedo A measure of the reflectivity of a body or surface; derived from the Latin *albus*, meaning white. The albedo is defined as the total RADIATION reflected by the body divided by the total incident radiation. Numerical values are expressed between the ranges of either 0–1 or 0–100%. It is therefore wavelength integrated across the full solar spectrum, while the term reflectivity is generally associated with a single wavelength or narrow waveband; that is, a spectral reflectivity. The term albedo was already in common use by astronomers at the beginning of the twentieth century (e.g. Russell, 1916), when it referred to the whole-planet value. The Earth's albedo is close to 0.3, contrasting strongly with its highly reflective neighbour Venus, which has an albedo of 0.7. The advent of orbiting satellites permitted measurement of reflected radiation at the top of the atmosphere above specific geographical locations. Maps of albedo are now produced illustrating the variation of reflectivity over the globe. This top-of-the-atmosphere albedo is termed the *planetary* or *system albedo*. The satellite-sensed radiation is composed of the reflected beams from the surface, the atmosphere and clouds.

CLOUDS generally have high albedos, though they too exhibit a considerable range, depending on depth and composition. While *cumuliform* clouds can have an albedo of 50% or more, some *cirriform* clouds barely reflect solar radiation, making their detection from space very difficult.

Strictly, surface albedos can be measured only with albedo-meters mounted close to the surface, but they may also be calculated from clear-sky albedos measured by satellites. Surface albedos range in value: oceans, 0.07; dense forests, 0.10; grass and farmlands, 0.16–0.20; bright deserts, 0.25–0.50 or higher; and highly reflective ice, 0.40–0.60, and snow surfaces, 0.50–1.0. The albedo is determined by composition and structure of the surface. Water has a low albedo because it is transparent. Moisture content also greatly influences the albedo of bare soils, which increases with decreasing water content. Tropical forests have a low albedo because the radiation is reflected several times off leaf surfaces, getting trapped in the canopy. AH-S

Note also that the angle at which the sun's rays strike a water surface greatly affects the albedo value. When the sun is low on the horizon, albedo can approach 100%. The effect is particularly pronounced for water surfaces, which can act like mirrors at very low sun angles. Albedo is at a minimum when the sun is nearly overhead. SEN

Reading and References
Goudie, A.S. (2001) *The nature of the environment*, 4th edition. Oxford: Blackwell. · Henderson-Sellers, A. (2010) Surface albedo data for climatic modeling. *Reviews of Geophysics*, **21**, 174–1778. · Houldcroft, C.J., Barnsley, M., Taylor, C.M., *et al.* (2009) New vegetation albedo parameters and global fields of soil background albedo derived from MODIS for use in a climate model. *Journal of Hydrometeorology*, **10**, 183–198. · Russell, H.N. (1916) On the albedo of the planets and their satellites. *Astrophysical Journal*, **43** (3), 173–196.

alcove A steep-sided, arcuate cavity on the flank of a rock outcrop, typically produced by water erosion processes such as solution (see SOLUTION, LIMESTONE) and spring SAPPING or seepage. DJN

The albedo of the earth's surface. The fraction of the total radiation from the sun that is reflected by a surface is called its albedo. The albedo for the earth as a whole, called the planetary albedo, is about 53%. The albedo varies for different surface types. Note also that the angle at which the sun's rays strike a water surface greatly affects the albedo value. *Source:* Goudie (2001). Reproduced with permission of Andrew Goudie.

alcrete   Aluminium-rich duricrusts, often in the form of indurated bauxites. Generally, the products of the accumulation of aluminium sesquioxides within the zone of weathering.

alfisol   Relatively young, acid soils characterized by a clay-enriched B horizon, commonly occurring beneath deciduous forest in humid, subhumid, temperate and subtropical climates. A soil order of the US Seventh Approximation.

algae   A large group of photosynthesizing organisms, many of which are unicellular and microscopic, although some are multicellular and conspicuous, such as the seaweeds. They are most frequently aquatic, common in either marine or freshwater habitats; some are adapted to lower levels of moisture and may be found, for example, as the green coating of trees or in the soil. They are subdivided into several distinctive phyla on the basis of structure, pigmentation and characteristics of the cell wall. The major groupings are the red algae (e.g. the coralline species associated with coral reefs), brown algae (mainly marine; for example, all the common seaweeds, including kelp, which may exceed 60 m in length), diatoms and green algae (frequently occurring as the PLANKTON of freshwater habitats).   MEM

Reading
Sze, P. (1997) *The biology of the algae.* New York: McGraw-Hill.

algal bloom   A spontaneous proliferation of microscopic algae in water bodies as a result of changes in water temperature or chemistry. Algal blooms may be characteristic of lakes where eutrophication has been caused by the addition of pollutants.

algal mat   A layer of usually filamentous algae on marine or freshwater soft sediments. It may be considered one of many types of microbial mats. (See also BIOFILM/EXTRACELLULAR POLYMERIC SUBSTANCES (EPSS) and BIOSTABILIZATION.)   TS

algal ridge, algal rim   Prominent feature of the reef margin on many Indo-Pacific reefs, composed of coralline algae (genera include *Hydrolithon* and *Porolithon*) with little or no coral cover. Algal ridges are typically 5–15 m in width and at least 0.3 m above the landward reef flat surface; under higher energy regimes they can be cut by surge channels and show blowholes. They are best developed on windward coasts under a trade-wind regime. Such features are not as well developed on Caribbean reef margins, although analogous features termed 'boilers' have been described from Bermuda. Similar coralline algal and vermetid accretions form features known as 'cornice' and 'trottoir', and staircases of algal-rimmed pools known as 'plate-formes à vasques', on rocky Mediterranean coasts; they have also been described from tropical rocky coasts.

As these structures are restricted to the surf and wave splash zone – with a vertical range of ~1 m in relation to mean sea level – they have been used as indicators of palaeo sea levels.   TS

Reading
Adey, W.H. and Burke, R.B. (1976) Holocene bioherms (algal ridges and bank barrier reefs) of the eastern Caribbean. *Bulletin of the Geological Society of America,* **87,** 95–109. · Focke, J.W. (1978) Limestone cliff morphology on Curaçao (Netherlands Antilles) with special attention to the origin of notches and vermetid/coralline algal surf benches ('cornices', 'trottoirs'). *Zeitschrift für Geomorphologie, NF,* **22,** 329–349. · Laborel, J. and Laborel-Deguen, F. (1994) Biological indicators of relative sea-level variation and of co-seismic displacements in the Mediterranean

area. *Journal of Coastal Research*, **10**, 395–415. · Tracey, J.I., Ladd, J.S. and Hoffmeister, J.E. (1948) Reefs of Bikini, Marshall Islands. *Bulletin of the Geological Society of America*, **59**, 861–878.

Algal ridge, windward coast of Guam, western Pacific Ocean. Reproduced with permission of Tom Spencer.

allelopathy   The production of chemicals by plants in order to inhibit or suppress the growth, survival and reproduction of competing plants. The phenomenon is widespread in plant communities as diverse as desert shrubs, tropical and temperate forests and heathlands. Examples include the checking of spruce growth on moorland by the production of a chemical by the heather roots that inhibits the growth of the mycorrhizal fungi essential for good growth of the trees; the inhibition of herbaceous species by the shrubs of the Californian chaparral, and the suppression of their own seedlings (autotoxicity) by a number of forest trees, such as black walnut (*Juglans nigra*) and the silky oak (*Grevillea robusta*). Several plants also use chemical defences against herbivores; for example, oak leaves with a high tannin content are less palatable to certain defoliating caterpillars.   KEB/MEM

Reading
Cheema Z.A., Farooq, M. and Wahid, A. (eds) (2012) *Allelopathy: current trends and future applications.* Berlin: Springer-Verlag.

Allen's rule   The rule states that the relative size of the limbs and other appendages of warm-blooded animals tends to decrease away from the equator. This correlates with the increased need to conserve body heat.

aliens   Organisms either deliberately or accidentally introduced by people into regions beyond their natural distribution range. The term is commonly used to describe those exotic species able to propagate themselves in their new habitats without human intervention, so that they are invasive and may become serious pests. Alien species during the twentieth century, in parallel with the increased scale of human impact, have become major factors in land transformation and degradation, have disrupted ecosystem functioning and threaten biological diversity. Organisms transported to remote regions by people often arrive without the significant natural enemies that control population numbers within their natural ranges. In certain cases, these organisms can become successful invaders of natural and semi-natural habitats in their new localities and, as so-called 'environmental weeds', become hazardous to the invaded communities. Mediterranean-type ecosystems appear to have been especially susceptible to such invasions. For example, the species-rich fynbos heathlands of the southwestern Cape of South Africa has been severely impacted by invasive alien trees, among the most significant of which are several Australian varieties of the genus *Acacia*. Islands are also strongly impacted by aliens; the introduction of the European rabbit to Australia has had serious ecological consequences, and the unique flora and fauna of the Galápagos are threatened by the introduction of mammals such as domestic cats and dogs. More recently, attempts to control the populations of alien species have been intensified, and biological control techniques are now frequently utilized.
MEM

Reading
Cox, D.W. (2004) *Alien species and evolution: the evolutionary ecology of exotic plants, animals, microbes and interacting native species.* London: Island Press. · Rotherham, I.D. and Lambert, R.A. (2011) *Invasive and introduced plants and animals: human perceptions, attitudes and approaches to management.* London: Earthscan.

Allerød   The name given to an INTERSTADIAL of the LATE GLACIAL of the last glaciation of the Pleistocene in Europe. The classic threefold division into two cold zones (I and III) separated by a milder interstadial (zone II) emanates from a type section at Allerød, north of Copenhagen, where an organic lake mud was exposed between an upper and lower clay, both of which contained pollen of *Dryas octopetula*, a plant tolerant of severely cold climates. The lake muds contained a cool-temperature flora, and the milder stage which they represented was called the Allerød interstadial. The interstadial itself and the following Younger Dryas temperature reversal are sometimes called the Allerød oscillation. The classic date for the interstadial is 11,350–12,000 BP, but its exact date and status are in dispute.   ASG

Reading
Lowe, J.J. and Gray, M.J. (1980) The stratigraphic subdivision of the Lateglacial of NW Europe: a discussion. In J. J. Lowe, M. J. Gray and J. E. Robinson (eds), *Studies in the Lateglacial of north-west Europe*. Oxford: Pergamon; pp. 157–175. · Mercer, J.H. (1969) The Allerød oscillation: a European climatic anomaly. *Arctic and Alpine Research*, **1**, 227–234.

allochthonous Refers to the material forming rocks that have been transported to the site of deposition, whereas an AUTOCHTHONOUS sediment is one in which the main constituents have been formed in situ (e.g. evaporites, coal).

allogenic stream A stream that derives its discharge from outside the local area. The term is particularly used where local conditions do not generate much streamflow; for example, in arid areas or ones with permeable rocks. Here, streamflow may be derived from distant parts of the topographic catchment where precipitation and run-off are effective. Appearances can sometimes be misleading because streamflow may be augmented by contributions from local groundwater that are not readily appreciated.   JL

allogenic succession This process is caused by an external environmental factor rather than by the organisms themselves. Instances are the change in vegetation induced by the inflow and accumulation of sediment in a pond (a geomorphological process), or a change in regional climate. (See also AUTOGENIC SUCCESSION and CLISERE.)   JAM

allometric growth A condition in which a change in size of the whole is accompanied by scale-related changes in the proportions of aspects of the object under study. Initially a biological concept, which derives from 'the study of proportional changes correlated with variation in size of either the total organism or the part under consideration' (Gould, 1966: 629), it has been utilized in physical geography, especially in the study of landforms. Woldenberg (1966) explicitly introduced the concept to geomorphic analysis, Church and Mark (1980) provided a still-relevant overview and Evans (2006) is a relatively recent application. A focus of geomorphological applications has been on scale-related distortions in geometric relationships. If no such distortions occur, *isometric* growth has taken place. For example, on the one hand, it is widely observed that the gradient of the principal stream channel in a drainage basin is reduced at an ever-decreasing rate as the drainage area enlarges; on the other hand, the relationship between channel width and the wavelength of meanders appears to be roughly constant, regardless of actual channel dimensions.

Allometry refers to a proportional relationship of the form $A_1/A_2 = b$. If the resulting ratio is constant for all values of A_2, isometry exists. If A_1 increases at a faster rate than A_2, there is positive allometric growth; if the reverse, it is negative. It is often the case that b represents the exponent in the general form of the power equation $y = ax^b$ (where a is a constant dependent upon the units of measurement): if y and x have the same scale dimensions, b will be 1.0. The relationship is positively allometric for $b > 1.0$ and negatively allometric if $b < 1.0$. If y and x have different scale dimensions, then the value of b indicative of isometry will vary accordingly. (For example, if y is a length L^1 and x is an area L^2, a value of b of 0.5 would indicate isometry.) A clear departure from isometry is also indicated if the relationship between x and y, plotted as a power function, is curved rather than linear.

Church and Mark (1980) indicate that *dynamic* and *static* allometry should be distinguished. The former deals with the changing proportions of an individual landform over time; in the latter, data are taken from a number of individuals of different sizes at one moment. The interpretation of the results of studies that aim to investigate static allometry can be very readily complicated by extraneous sources of inter-individual variation, particularly those due to differences in materials and detailed history.   BAK/DSGT

References
Church, M.A. and Mark, D.M. (1980) On size and scale in geomorphology. *Progress in Physical Geography*, **4**, 342–390. · Evans, I.S. (2006) Allometric development of glacial cirque form: geological, relief and regional effects on the cirques of Wales. *Geomorphology*, **80**, 245–266. · Gould, S.J. (1966) Allometry and size in ontogeny and phylogeny. *Biological Review*, **41**, 587–640. · Woldenberg, M.J. (1966) Horton's laws justified in terms of allometric growth and steady state in open systems. *Geological Society of America Bulletin*, **77**, 431–434.

allopatric Descriptive of the condition whereby two closely related species (a 'species pair') or subspecies occur in separate geographical areas; for example, *Plantago ovata* (Canary Islands to India across North Africa) and *Plantago insularis* (southwestern USA). It is also used to describe the process of speciation when populations of the same species become isolated from each other through the establishment of a geographical 'barrier'; for example, as a result of climatic or geological change. The resulting genetic isolation allows the separate populations to evolve independently so that, even upon removal of the geographical barrier, the populations are sufficiently distinctive as to be classified as separate species.   MEM

allophane   An amorphous hydrated alumino-silicate gel. The chemical composition is highly variable. The name is applied to any amorphous substance in clays.

alluvial channel   A river channel that is cut in ALLUVIUM. This applies to most larger natural rivers; ones that are entirely developed in bedrock may be found in high-relief areas, but even there incising streams transporting the material they have eroded may have a discontinuous veneer of alluvial material. More generally, even those channels that have bedrock on the floors of their deepest scour pools have banks in alluvial materials that have been deposited during floods or during the lateral movement of such rivers.

Non-alluvial channels may be different in form and development from alluvial rivers. Bedrock channels may be constrained by rock outcrop, while meltwater streams flowing on glacier ice create channels by removing ice rather than interacting with the bed materials. This is an essential characteristic of alluvial channels: they are self-formed in their own transportable sediments and can adjust their morphology according to discharges and the sediment sizes and loads present.

Alluvial channels can be characterized in terms of their cross-section, planform and long profile properties. These are interrelated. For example, a BRAIDED RIVER planform pattern is usually associated with a shallow cross-section and with relatively steep gradients. But these form elements are also dependent on river discharge and stream power and on sediment properties. Alluvial channel systems are therefore complex ones to study. In the short term, stable or equilibrium forms may be developed and these may alter when controlling conditions vary. Unfortunately, equilibrium states are not easy to define, nor are changes precisely predictable. The study of alluvial channels is a large area of scientific enquiry. (See also CHANNEL CLASSIFICATION.)   JL

Reading
Charlton, R. (2008) *Fundamentals of fluvial geomorphology*. Abingdon: Routledge; chapter 8. · Schumm, S.A. (1977) *The fluvial system*. New York: John Wiley & Sons, Inc.

alluvial fans   See FANS.

alluvial fill   Sedimentary material deposited by water flowing in stream channels. During the seventeenth century the term included all water-laid deposits (including marine sediments), but in 1830 Lyell restricted its use to materials deposited by rivers (Stamp, 1961). Particle sizes range from fine clays deposited by overbank waters to boulders deposited in the channel bed by large floods. Materials may be massively bedded if deposited by a single event, or they may occur in a variety of bed forms related to flow variation or location related to the active channel.

Reading and Reference
Happ, S.C. (1971) Genetic classification of valley sediment deposits. *American Society of Civil Engineers: Journal of the Hydraulics Division*, **97**, 43–53. · Schumm, S.A. (1977) *The fluvial system*. New York: John Wiley& Sons, Inc. · Stamp, L.D. (1961) *A glossary of geographical terms*. London: Longman.

alluvial terrace   See RIVER TERRACE.

alluvium   Material deposited by running water. The term is not usually applied to lake or marine sediments and may be restricted to unlithified, size-sorted fine sediments (silt and clay). Fine material of marine and fluvial origin may not in practice be easy to distinguish; on geological maps this is often not attempted and both are classed together. Coarser sediment may not, by convention, be included, but there is no good reason for this if no particular grain size is intended. The term can be modfied by 'fine-grained' or 'coarse-grained'. Studies of alluvial channels do not imply any particular grain sizes.

Distinguishing characteristics of alluvium are its stratification, sorting and structure. Coarser sediment deposited on the channel bed or in bars is overlain by finer materials deposited from suspension, either in channel slackwater areas or as an OVERBANK DEPOSIT following floods. In detail, sedimentary structures may be complex and dependent on the type of river activity. Large-scale contrasts are often drawn (e.g. between the deposits of braided and meandering rivers), while different scales and types of CURRENT BEDDING may be present. The size of sediments involved may depend on that supplied to streams from slopes or BANK EROSION and on the distance from such sources that a particular reach may be, because rivers sort and modify alluvial materials as they are transported. Thus, fine-grained alluvium can be dominant in the lower courses of present rivers; coarser alluvium can be found close to the supply points for such material (e.g. in alluvial cones in high-relief semi-arid areas or at glacier margins) and in earlier Pleistocene deposits in mid-latitudes. Here, the slope- and glacier-derived coarser materials produced under former cold-climate conditions contrast with the finer alluvium coming from more recent slope inputs.   JL

Reading
Allen, J.R.L. (1970) *Physical processes of sedimentation.* London: Allen & Unwin. · Reading, H.G. (ed.) (1996) *Sedimentary environments and facies*, 3rd edition. Oxford: Blackwell Scientific.

alp   *a.* A shoulder high on the side of a glacial trough.
b. (In Switzerland) a summer pasture below the snow level.

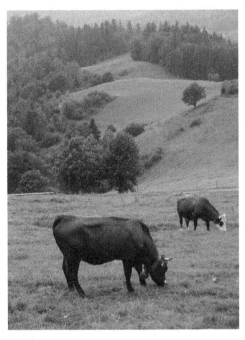

Swiss summer pastures, or alps. Photograph by David Thomas.

alpha diversity   The richness of species *within* habitats; that is, the number of species located within relatively small, environmentally coherent, geographical areas. This contrasts with beta diversity, which measures species turnover *between* habitats, and gamma diversity, which measures turnover between different geographical areas. Alpha diversity is effectively a measure of the numbers of different types of organisms living in a particular locality under similar circumstances. For example, the total number of plant species found in a small (say, $100\,m^2$) sample plot of Mediterranean shrubland in southern France represents its alpha diversity. If the sample plot size is very small ($1\,m^2$), then the measure of richness may be referred to as point diversity. Because of the positive relationship between species richness and area, greater sample plot sizes are likely to contain higher species numbers. Consequently, scale is an integral component of this kind of diversity measure. There is, however, a maximum richness value related to the total number of species available in the geographical area concerned (the species pool), which varies principally with latitude (the tropics have generally greater species complements), taxonomic group under consideration (some groups are richer at the poles) and habitat circumstances (some habitats are especially harsh). Alpha diversity is spatially highly variable, but is especially elevated in the neotropical rain forests, approaching up to 400 different species of tree alone in a sample plot of $100\,m \times 100\,m$.   MEM

Reading
Groombridge, B. (1992) *Global biodiversity: status of the Earth's living resources.* London: Chapman and Hall.

alpine   The zone of a mountain above the tree line and below the level of permanent snow.

alpine orogeny   A phase of OROGENY (mountain building) in the Late Mesozoic and Cenozoic that formed mountain ranges from the Atlas in the west through to the Himalayas in the east. With the UK being over 1000 km from the collision zone, here only minor structures record the orogeny, such as localized folds and faults in southern England. It also led to the formation of the Weald–Artois anticline in southern England and northern France. It occurred when the continental plates of Africa and India collided (from the south) with the Eurasian plate in the north. It saw he closure of the Tethys Ocean as oceanic lithosphere was subducted northwards beneath the Eurasian plate, leaving today what we now know as the Mediterranean Sea. The major phases of mountain building began in the Palaeocene to Eocene. Currently, the process still continues in some of the mountain ranges – a phenomenon called NEOTECTONICS. The Alpine orogeny is one of the three major phases of orogeny in Europe. The other two were the Caledonian orogeny in the early Palaeozoic, and the Hercynian or Variscan orogeny that occurred when Gondwana and the Old Red Sandstone Continent collided in the middle to late Palaeozoic.   ASG

Reading
Ollier, C. and Pain, C. (2000) *The origin of mountains.* London: Routledge; pp. 61–95. · Owen, L.A. (2004) Cenozoic evolution of global mountain systems. In

P. N. Owens and O. Slaymaker (eds), *Mountain geomorphology*. London: Arnold; pp. 33–58.

altiplanation

A form of solifluction, i.e. earth movement in cold regions, that produces terraces and flat summits that consist of accumulations of loose rock. An alternative term is cryoplanation.

altithermal

During the Holocene there was a phase, of varying date, when conditions were warmer (perhaps by 1–3 °C) than at present. In the Camp Century Ice Core (Greenland) a warm phase lasted from 4100 to 8000 BP, whereas in the Dome Ice Core (Antarctica) the warmest phase was between 11,000 and 8000 BP (Dansgaard *et al.*, 1970; Lorius *et al.*, 1979). Rainfall conditions may also have changed, aridity having triggered off both renewed sand dune activity and a decline in human population levels in areas such as the High Plains from Texas to Nebraska, USA. ASG

References

Dansgaard, W., Johnsen, S.J. and Clausen, H.B. (1970) Ice cores and paleoclimatology. In I. U. Ollson (ed.), *Radiocarbon variation and absolute chronology*. New York: John Wiley & Sons, Inc; pp. 337–351. · Lorius, C., Merlivat, L., Jouzel, J. and Pourclet, M. (1979) A 30,000-year isotope climatic record from Antarctic ice. *Nature*, **280**, 644–648.

altocumulus

See CLOUDS.

altostratus

See CLOUDS.

alveolar

Alveolar or honeycomb weathering features take the form of small hollows in rock surfaces that may occur as individual features but more commonly are found in clusters. They may be related to TAFONI; indeed, the two forms can often be seen on the same rock surfaces, suggesting that alveoles (~1–50 cm in diameter) may evolve into larger cavernous features in some circumstances (Mellor *et al.*, 1997). There is no clear morphometric boundary between the two, and the terms have sometimes been used in an interrelated manner. There may be a tendency for alveoles to form on vertical or near-vertical surfaces (distinguishing them from GNAMMAS, which occur in more horizontal surfaces). Although alveoles have often been cited as a particular feature of sandstones and of drier and coastal environments, they can be found in many different lithologies and in many environments. There is a tendency towards explaining their development in terms of SALT WEATHERING and granular disintegration (Mustoe, 1982), which may favour their occurrence in drylands and coastal situations. DSGT

Reading and References

Mellor, A., Short, J. and Kirkby, S.J. (1997) Tafoni in the El Chorro area, Andalucia, southern Spain. *Earth Surface Processes and Landforms*, **22**, 817–833. · Mustoe, G.E. (1982) The origin of honeycomb weathering. *Geological Society of America*, Bulletin, **93**, 108–115. · Viles, H.E. (2011) Weathering systems. In D. S. G. Thomas (ed.), *Arid zone geomorphology: process, form and change in drylands*, 3rd edition. Chichester: John Wiley & Sons, Ltd; 87–100.

ambient

Preceding or surrounding a phenomenon; for example, ambient temperature refers to the temperature of the surrounding atmosphere, water or soil.

amensalism

A kind of interspecific interaction in which there is a negative influence of one species (the inhibitor) but no reciprocal negative impact upon the other (the amensal). This kind of interaction is illustrated by the African proverb: 'When elephants fight, the grass suffers'. In essence, one organism harms the other as a by-product of its activities. Such interactions are common. Indeed, in the case of humans, many of our interactions with other organisms could be described as amensal; for example, the negative impact of acid rain on coniferous forests. MEM

amino acid racemization

A method in GEO-CHRONOLOGY. This dating technique is based on the fact that the amino acid building blocks that make up proteins in skeletal remains of animals undergo time-dependent chemical reactions. Amino acid racemization dating is a relative-age method involving measurement of the extent to which certain types of amino acids within protein residues have transformed from one of two chemically identical forms (stereoisomers) to the other. Materials that may preserve such protein residues within sediment bodies include bones and other body components (e.g. mummies), mollusc shells and eggshells. At formation, only L-form amino acids are present. Over time, and in part controlled by temperature and other factors (hydrolysis, pH), some of the L-form acids are converted to a D-form until an equilibrium is reached. *Racemization* and *epimerization* reactions differ, in that racemization involves only amino acids with a single chiral carbon atom, whereas epimerization involves amino acids with two chiral carbon atoms (Aitken, 1990). The use of a range of amino acids that undergo racemization or epimerization over a range of timescales makes it possible to apply the method over timescales ranging from a few years to hundreds of thousands. The degree of change in amino acid composition, however, depends on factors other than time (e.g. temperature), which

21

may result in substantial errors in amino acid dates.   SS

Reference
Aitken, M.J. (1990) *Science-based dating in archaeology.* London: Longman.

amphidromic point The node around which KELVIN WAVES rotate. The rotation is caused by the CORIOLIS FORCE acting on the flow induced by a standing tidal wave in large basins. The amplitude of the wave is nil at the amphidromic point and at a maximum at the boundaries of the tidal basin. Rotation of the TIDES is counter-clockwise in the northern hemisphere and reversed in the southern hemisphere. *Cotidal* lines connect points experiencing simultaneous high tides, and these radiate from the amphidromic point. *Corange* lines connect points of equal tidal range, and these are concentric about the amphidromic point. *Amphidromic* points are approximately constant in location, but may show seasonal or other periodic migrations.   DJS

Reading
Cartwright, D.E. (1999) *Tides: a scientific history.* Cambridge: Cambridge University Press. · Thurman, H.V. (1991) *Introductory oceanography*, 6th edition. New York: Macmillan Publishing Company.

AMS radiocarbon dating   See RADIOCARBON DATING.

anabatic flows Upslope winds usually produced by local heating of the ground during the day. The most common type is the VALLEY WIND. Anabatic flows develop best on east- or west-facing slopes on days with clear skies. The air along the slope is heated by contact with the warm surface much more rapidly than air at the same elevation away from the slope. The resulting temperature difference sets up a thermal circulation with the air ascending along the slope and descending over the adjoining plain or valley. Under ideal conditions anabatic winds can reach speeds of $10-15\,\mathrm{m\,s^{-1}}$ and can be a factor in the spreading of forest fires in dry weather. If the air is moist the anabatic flow may produce anabatic clouds above the crest of the slope.   WDS

Reading
Atkinson, B.W. (1981) *Meso-scale atmospheric circulations.* New York: Academic Press. · Geiger, R. (1965) *The climate near the ground.* Cambridge, MA: Harvard University Press. · Oke, T.R. (1987) *Boundary layer climates*, 2nd edition. London: Routledge.

anabranching A channel planform that resembles braiding but where the individual channels (anabranches) are separated by islands that divide flow at bankfull rather than by bars in a braided channel that are overtopped below bankfull (Nanson and Knighton, 1996). Brice (1984) describes anabranching as having islands of more than three times the width of the river at average

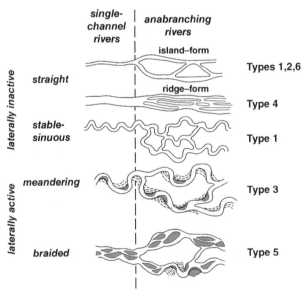

A classification of channels including anabranching rivers.
Source: Nanson and Knighton (1996). Reproduced with permission from John Wiley and Sons.

discharge (see CHANNEL CLASSIFICATION). Individual anabranches can be meandering, braided or straight. *Anastomosing* is a term commonly used by sedimentologists to define a group of fine grained, low-energy anabranching rivers with a distinctive alluvial architecture sometimes associated with coal and hydrocarbon preservation. Study of very large rivers suggests that the anabranching form is characteristic of 'mega rivers', those whose mean discharge exceeds $17,000 \, m^3 \, s^{-1}$ and whose gradient is $<0.000\,07$ (Latrubesse, 2008).   DLD

References

Brice, J.C. (1984) Planform properties of meandering rivers. In: C. M. Elliot (ed.), *River meandering: proceedings of the conference Rivers '83*. New Orleans, LA: American Society of Civil Engineers; pp. 1–15. · Latrubesse, E.M. (2008) Patterns of anabranching channels: the ultimate end-member adjustment of mega rivers. *Geomorphology*, **101**, 130–145. · Nanson, G.C. and Knighton, A.D. (1996) Anabranching rivers: their cause, character and classification. *Earth Surface Processes and Landforms*, **21**, 217–239.

anaclinal   Refers to a feature, especially a river or valley, which is transverse to strike and against the dip of strata.

anaerobic   Term used to describe conditions in which oxygen is absent or, in the case of metabolic processes, respiration in the absence of oxygen (see AEROBIC). Most organisms obtain their energy through aerobic respiration involving the breakdown of sugars in the presence of oxygen. However, some organisms (e.g. certain types of bacteria or mould) do so in the absence of oxygen. Such organisms are referred to as obligate anaerobes if they live permanently in oxygen-deficient conditions and facultative anaerobes if they respire aerobically when oxygen is present and resort to anaerobic respiration if oxygen is scarce or absent. Also known as fermentation, the process generates carbon dioxide and either organic acids (e.g. lactic acid in the case of anaerobic respiration in vertebrate skeletal muscles) or alcohols (e.g. ethanol production from yeast). Fermentation of yeast is perhaps the most familiar form of anaerobic respiration, since it is responsible for the production of alcohol for human consumption in the form of beer or wine. Biochemically, the forms of respiration may be compared as follows; note the variation in energy production (identified here in terms of kilojoules), which indicates the relative efficiency of aerobic respiration:

Anaerobic respiration with ethanol formation (alcoholic fermentation)

$$C_6H_{12}O_6 \rightarrow 2\,CH_3CH_2OH + 210 \, kJ$$
$$\text{ethanol}$$

Anaerobic respiration with lactic acid formation

$$C_6H_{12}O_6 \rightarrow 2\,CH_3CH(OH)COOH + 150 \, kJ$$
$$\text{lactic acid}$$

Aerobic respiration

$$C_6H_{12}O_6 + O_2 \rightarrow 6H_2O + 6CO_2 + 2880 \, kJ$$

In still other organisms, such as bacteria of the genus *Sulfobolus*, energy is obtained anaerobically through the oxidation of inorganic substances, for example hydrogen sulphide.   MEM

ana-front   A front that has ascending air at one side, particularly one experiencing the unusual phenomenon of rising cold air.

analemna   A scale drawn on a globe to show the daily declination of the sun, enabling the determination of those parallels where the sun is directly overhead at any specific time of year.

analogue approach   When QUATERNARY palaeoenvironmental reconstructions are made using PROXY data, it is necessary to interpret the data in terms of its environmental meaning (e.g. in terms of rainfall, temperature, seasonality). To do this, scientists usually assume that the controls on a variable operating now were the same as in the past (e.g. on the distribution of a particular plant or animal, or on the formation of a landform). This is based on James Hutton's maxim that 'the present is the key to the past'. Critics would argue that this is an untestable hypothesis. Additionally, for many plant species, and even for the development of some landforms, the *precise* controls on distribution or formation are unknown or at best poorly understood. In essence, and this is sometimes forgotten, the best Quaternary science needs to work hand in hand with analyses of contemporary phenomena, otherwise additional uncertainties are added to interpretations.   DSGT

anamolistic cycle   The tidal cycle, normally taken as lasting 27.5 days, which is related to the varying distance between the Earth and the Moon.

anaseism   The vertical component of the waves moving upwards from the focus of an earthquake.

anastomosing river   See ANABRANCHING and BRAIDED RIVER.

andosols   Dark soils developed on volcanic rock and ash.

andromy   The migration of some fish species from salt water to fresh water for breeding.

anemograph   A self-recording instrument for measuring the speed and sometimes the direction of the wind.

anemometer   A mechanical or electronic device for measuring wind speed. The simplest of these devices is the cup-anemometer which consists of a rotating set of cups (or vanes) measuring the 'run of wind' on a continuous counter. This is the distance the wind travels per unit time and is commonly expressed as kilometres per day. Cup-anemometers suffer from inertia and friction, both of which limit their accuracy in low-velocity winds or TURBULENT FLOW, particularly when measurements are required over short timespans on the order of seconds. Wholly accurate measurements are inherently problematic to obtain with all mechanical anemometers as it is always necessary to place the sensor in the wind, hence disrupting it. This problem is solved through the use of non-mechanical types, such as hot-wire, laser, sonic and pressure anemometers. Hot-wire anemometers, for example, determine high-frequency velocity fluctuations from the resistance of a thin wire placed in the flow, the electroconductivity of which is controlled by its temperature. In practice, the voltage required to keep the wire at a constant temperature is measured and this is calibrated against the velocity of flow. Laser Doppler anemometry measures the velocity of airflow by sensing the change in frequency of light scattered back from tiny particles introduced to the flow. This technique is expensive and generally used only in wind tunnels. Sonic anemometers measure the time between pulses of ultrasonic waves and can measure both horizontal and vertical velocity. Pressure anemometers calculate wind from the force (i.e. pressure changes) it produces.   GFSW/SEN

angiosperms   Flowering plants, a subdivision of seed-producing plants (Spermatophyta), whose main characteristic is the presence of the flower. Angiosperms are distinguished from the 'naked-seed' plants (gymnosperms) by the fact that, following fertilization, the seeds are developed in a protected ovary. Flowers are essentially modified shoots comprising four series arranged as whorls. From the outside of the flower inwards, these series are (1) sepals, (2) petals, (3) stamens (the pollen-producing male component) and (4) carpels (female structures from which the seeds develop). Flowering plants represent the dominant group of plants today, and there are approximately 250,000 described species, including all the commonly occurring herbs, shrubs and trees. Of the two groups of seed plants, angiosperns appear much later in the fossil record, although they have diversified to occupy virtually every ecological NICHE and, moreover, are of major economic importance in that they provide much of human food supply as wheat, rice, maize, etc. Gymnosperms, of which there are approximately 700 species, have a more ancient evolutionary history and have been outcompeted in most contemporary environments by their more successful flowering relatives, although they remain prominent in certain environments (e.g. the boreal forests of North America and Eurasia). Angiosperms are divided into two classes based on the number of leaves in the embryo, one in the case of the Monocotyledoneae (e.g. lily) and two in the Dicotyledoneae (e.g. oak).   MEM

Reading
Rudall, P. (1993) *Anatomy of flowering plants: an introduction to structure and development*. Cambridge: Cambridge University Press.

angle of dilation (θ)   The angle by which the grains of a granular material are displaced and reorientated on a shearing surface (SHEAR STRENGTH) cutting through the mass of particles. The reorientation movement is a response to the interlocking of particles which provide the frictional resistance or shear strength. The angle θ is related to the ANGLE OF INTERNAL SHEARING RESISTANCE (ϕ) and the static ANGLE OF PLANE SLIDING FRICTION (ϕ_{us}) by $\theta = \phi - \phi_{us}$.   WBW

Reading
Statham, I. (1977) *Earth surface sediment transport*. Oxford: Clarendon Press.

angle of initial yield (φ_i)   The angle of a slope of granular material at which movement, often as 'avalanching', is seen to start. The angle depends upon the type of material, particularly its packing and bulk density, which together create interlocking particles. It has a higher value than the ANGLE OF RESIDUAL SHEAR (φ_r).   WBW

angle of internal shearing resistance (ϕ)   The angle, usually measured in a TRIAXIAL APPARATUS or SHEAR BOX apparatus to give the friction angle φ, for granular materials, of the MOHR–COULOMB EQUATION. It is not a constant for any material but depends upon the VOID RATIO or POROSITY, as well as other frictional properties that relate to the interlocking of particles.   WBW

Reading
Statham, I. (1977) *Earth surface sediment transport*. Oxford: Clarendon Press.

angle of plane sliding friction (ϕ_{us}) The angle at which noncohesive (granular) particles just begin to slide down a surface. Strictly, this is the static angle; if it is the angle at which particles just stop moving it is the dynamic angle. It can apply to individual granular particles, a mass of such particles or a slab of rock. In the latter case it is related to joint friction. For any of these, the static angle ϕ_{us} and the dynamic angle ϕ_{ud} are approximately constant, but ϕ_{us} is greater than ϕ_{ud}.   WBW

Reading
Statham, I. (1977) *Earth surface sediment transport*. Oxford: Clarendon Press.

angle of repose The angle at which granular material comes to rest (also called the angle of rest); it approximates to the angle of scree slopes. Strictly, natural slopes for one material may have a variable angle of repose according to whether the material has just come to rest or is about to move; hence, it is also related to the ANGLE OF INITIAL YIELD (φ_i) or the ANGLE OF RESIDUAL SHEAR (φ_r).   WBW

angle of residual shear (φ_r) The angle at which granular material comes to rest after movement. The angle is less than the ANGLE OF INITIAL YIELD (φ_i) and is comparable to the ANGLE OF INTERNAL SHEARING RESISTANCE (ϕ) for cohesionless material in its most loosely packed state; thus, the difference between φ_r and φ_i represents a loss of strength of the material to define a residual shear strength.   WBW

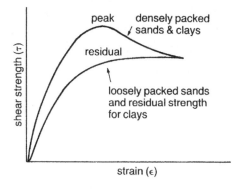

Reading
Statham, I. (1977) *Earth surface sediment transport*. Oxford: Clarendon Press.

angular momentum The product of the mass of a particle, the distance to an axis, and tangential components of velocity. It can be shown that forces directed towards (or away from) the axis cannot change the angular momentum. Many natural systems, including hurricanes and tornadoes, are dominated by forces acting towards a centre of low pressure. The (large) tangential components of velocity V at distance r are constrained by conservation of angular momentum to have V nearly proportional to $1/r$. Angular momentum on the global scale is similarly constrained. (See also MOMENTUM BUDGET.)   JSAG/SN

angular unconformity A stratigraphic unconformity represented by younger strata overlying older strata that dip at a different angle, usually a steeper one.

annual series A term used in flood frequency analysis for the series of discharges obtained by selecting the maximum instantaneous discharge from each year of the period of record. The annual series is therefore equal to the number of years of hydrological record analysed. Analyses of large global sets of annual flow data have explored other measures of annual streamflow, including the 90th percentile and 10th percentile annual flows. Results suggest that there are regional differences in the behaviour of annual flows. (See also FLOOD FREQUENCY.)   KJG/DLD

Reading
McMahon, T.A., Vogel, R.M., Peel, M.C. and Pegram, G.G.S. (2007) Global streamflows – part 1: characteristics of annual streamflows. *Journal of Hydrology*, 347, 243–259.

annular drainage A circular or ring-like drainage pattern produced when streams and rivers drain a dissected dome or basin.

Antarctic Bottom Water (AABW) An oceanic water mass formed in the seas around Antarctica. AABW is very dense, with a temperature range from −0.8 to +2 °C, salinity from 34.6 to 34.7 and a high oxygen content. It has several main source locations around the coast of Antarctica; source waters include Weddell Sea Bottom Water (WSBW), Ross Sea Bottom Water and Adelie Land Bottom Water, with WSBW making up the densest component of AABW. The cold, dense water is formed primarily by deep convection driven by brine rejection during sea ice formation, and strong evaporation and cooling driven by the strong offshore katabatic winds that descend from the Antarctic Ice Sheet. These winds open *polynyas* near the coast that expose open water to the strong, cold winds.

AABW contributes dense, cold bottom and deep water to all of the world's major ocean basins, especially the Indian and Pacific Oceans. In the Atlantic Ocean, AABW makes up the bottom water underneath the saltier but warmer and less dense NORTH ATLANTIC DEEP WATER (NADW). AABW gradually warms as it spreads northwards by mixing with overlying water. In many ways, AABW can be thought of as the dominant water mass in the global ocean; Johnson (2008) estimated the volume of AABW as 36% of the global ocean volume (excluding the Arctic Ocean, continental shelves and some other marginal seas such as the Mediterranean, Red, Caspian and Black Seas), with a volume of $0.468 \times 10^9 \, \mathrm{km}^3$. The equivalent depth of AABW exceeds 2000 m over much of the Pacific (north and south of the equator) and Indian Oceans, and even in the southern Atlantic; only in the North Atlantic does the outflow of NADW reduce the influence of AABW.   NSA

Reference

Johnson, G.C. (2008) Quantifying Antarctic Bottom Water and North Atlantic Deep Water volumes. *Journal of Geophysical Research*, **113**, C05027, doi: 10.1029/2007JC004477.

Antarctic Circumpolar Current (ACC) An

ocean current that flows clockwise from west to east around Antarctica, unhindered by any intervening landmasses, and thus connecting the Atlantic, Pacific and Indian Ocean basins. The current is driven by the strong westerly winds in the latitudes of the Southern Ocean, with a mean flow speed of $4 \, \mathrm{km} \, \mathrm{h}^{-1}$ but with considerable variability on seasonal, interannual and decadal time-scales. The water transport is estimated at 135 Sv (sverdrup) through the Drake Passage between South America and the Antarctic Peninsula; with a small addition of flow in the Indian Ocean, the transport south of Tasmania reaches ~147 Sv, making the ACC the world's largest ocean current.

The ACC's northern boundary is defined by the northern edge of the Subantarctic Front. Here, cold Antarctic waters predominantly sink beneath warmer Subantarctic waters (typically temperature shifts across this convergence are from 5 °C (N) to 2 °C (S)). Associated regions of mixing and upwelling across a seasonally and longitudinal variable zone some 32–48 km in width result in very high nutrient levels and marine productivity, especially for phytoplankton, copepods and Antarctic krill. Further south is the Polar Front, which is marked by a transition to very cold, relatively fresh, Antarctic Surface Water at the surface. Even closer to the Antarctic continent is the Southern Antarctic Circumpolar Current Front.

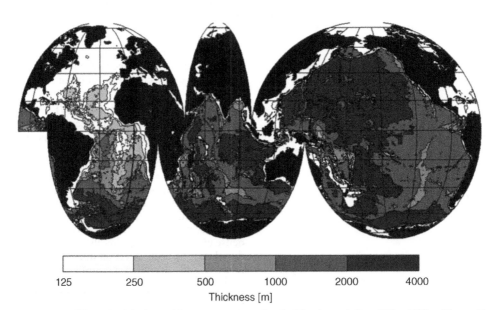

Antarctic Bottom Water depth in the world's oceans, contoured at doubling intervals from 125 to 4000 m. The small areas with values exceeding 4000 m are contoured, but not distinguished by a change in colour from those with values exceeding 2000 m.
Source: Johnson (2008). John Wiley & Sons.

Water associated with this front flows along the shelf break of the western Antarctic Peninsula and is the most southerly of the circumpolar flows.   TS

Reading
Orsi, A.H., Whitworth, T. and Nowling, W.D. (1995) On the meridional extent and fronts of the Antarctic Circumpolar Current. *Deep Sea Research, Series I*, **42**, 641–673.

Antarctic cold reversal (ACR)

A phase of colder climate revealed in Antarctic ice cores that correlates with the Late Glacial Interstadial warming recorded in Greenland ice cores. It spans the Bølling/ALLERØD and lasted until just before the onset of the YOUNGER DRYAS. The antiphase relationship between the south and the north is associated with changes in the Atlantic Meridional Overturning Circulation, often referred to as the bipolar seesaw. The ACR led, inter alia, to glacier advances in Patagonia and New Zealand.   ASG

Reading
Blunier, T., Chapellaz, J., Schwander, J., *et al.* (1998) Asynchrony of Antarctic and Greenland climate change during the last glacial period. *Nature*, **394**, 739–743.

Antarctic Ice Sheet – east and west

The large mass of ice, covering $13.5 \times 10^6 \, \text{km}^2$ (or 98%) of the surface of Antarctica. With a total volume of $25.4 \times 10^6 \, \text{km}^3$, it is the Earth's largest freshwater store. Were it to melt completely, global sea levels could rise by 60–70 m. The Antarctic Ice Sheet is divided into two unequal parts by the 3000 km long Transatlantic Mountains. The larger part, the East Antarctic Ice Sheet (EAIS), has a total grounded ice volume of $21.7 \times 10^6 \, \text{km}^3$ and a mean thickness of about 2.6 km, and the smaller part, the West Antarctic Ice Sheet (WAIS), which is less than five times the size of the EAIS, has a total grounded ice volume of $3 \times 10^6 \, \text{km}^3$ and a mean thickness of 1.8 km. Owing to a combination of atmospheric circulation patterns and high elevation, the South Pole Plateau (within the EAIS) is the coldest place on Earth, with a mean temperature at the Russian Vostok station of $-55\,^{\circ}\text{C}$.

Geological evidence suggests that Antarctica first became glaciated around 33 million years ago. It is thought that ice sheets of varying size existed during the Oligocene and into the Miocene, with some reaching or exceeding the volume of the current ice sheet. However, the presence of leaves and other organic matter in sediments from this time suggests that the climate was not as cold as at present, and that the ice sheets were transient features. The second main phase of sustained ice growth was around

13 million years ago, with another pulse of expansion around 7 to 5 million years ago. The extent and stability of the early Antarctic Ice Sheet remains the subject of debate. During the QUATERNARY, the ice sheet was considerably more extensive than at present during glacial maxima; a great deal of evidence suggests the ice sheet advanced onto what are now continental shelf areas as global eustatic sea levels fell. Most recently, the ice sheet likely reached its maximum extent around 18,000 years ago, retreating to its current extent around the last interglacial transition around 8000 years ago. Ice cores can be used to reconstruct an uninterrupted and detailed climate record over much of the Quaternary. The European Project for Ice Coring in Antarctica (EPICA) Dome C core contains the oldest ice yet recovered, going back about 800,000 years. Oxygen isotope analysis of this core shows evidence for a 'bipolar seesaw', the phenomenon where temperature changes in the northern and southern hemispheres are out of phase over timescales of thousands of years (Severinghaus, 2009).

Much of the bedrock beneath the EAIS currently lies below sea level, reaching 2.5 km below sea level in places. However, if the ice were removed, the bedrock would rebound, eventually to an elevation above sea level. As a result, the EAIS is considered a *continental ice sheet*, which appears to be relatively stable, having remained close to its present volume during several past INTERGLACIALS. In contrast, the WAIS is considered a *marine ice sheet*, as it resides on bedrock below sea level, which would otherwise be continental shelf seabed if the ice were not there. Therefore, the WAIS is thought to be more vulnerable to climate change, and in particular to ocean warming.

The central area of the EAIS is composed mainly of the Antarctic plateau, which extends over a diameter of 1000 km with an average elevation of 3 km. Ice flows radially outwards from this plateau. The largest outlet glacier of the EAIS is the Lambert Glacier, fed by its three tributaries (the Lambert, Mellor and Fisher Glaciers), which meet at the Amery Ice Shelf. Its grounded portion alone is over $1.48 \times 10^6 \, \text{km}^2$, or 16% of the total area of the EAIS (Yu *et al.*, 2010).

The topography of the WAIS is more complex than that of the EAIS, consisting of three major ice domes and numerous local domes and ridges. The *Admundsen Sea Sector* flows northwards via Pine Island Glacier, Thwaites Glacier and many smaller ice streams and outlet glaciers to reach small, fringing ice shelves at its seaward margin.

The *Weddell Sea Sector* drains northeastwards, around the Ellsworth Mountains and into the Ronne Ice Shelf (360,000 km²). Major ice streams in this sector include the Institute, Rutford, and Evans Ice Streams. The *Ross Sea Sector* drains westwards into the Ross Ice Shelf (850,000 km²), the largest floating ice mass on Earth, via the Ross or Siple Coast Ice Streams.

Extending northwards from the WAIS is the Antarctic Peninsula, a rugged mountain chain generally more than 2000 m high, which extends 1300 km from a line between Cape Adams (Weddell Sea) and a point on the mainland south of Eklund Islands. The ice caps and glaciers in the Antarctic Peninsula are nourished locally; thus, the Antarctic Peninsula is sometimes considered not to be part of the Antarctic Ice Sheet per se.

The average rate of ice loss from the Antarctic Ice Sheet as a whole has increased from 30 Gt a⁻¹ (sea level equivalent: 0.08 mm a⁻¹) during 1992–2001, to 147 Gt a⁻¹ (sea level equivalent: 0.40 mm a⁻¹) in 2002–2011 (Vaughan *et al.*, 2014), with air temperatures rising by 2.5 °C (1950–2000) compared with the global average rise of 0.6 °C per century. The most dramatic response has involved ice shelf collapse, with 28,000 km² being lost since 1960 from around the Antarctic Peninsula (e.g. Banwell *et al.*, 2013). The surface melt and runoff from the EAIS is negligible (apart from in the peak melt season on low-lying ice shelves), so any ice loss is due to increased ice discharge, which is also small. For the EAIS as a whole, a small rate of mass loss in Wilkes Land and a small rate of mass gain at the mouths of the Filchner and Ross ice shelves combine to give an average rate of mass loss of only 4 Gt a⁻¹ (Rignot *et al.*, 2008).   AB

Reading and References

Banwell, A.F., MacAyeal, D.R. and Sergienko, O.V. (2013) Breakup of the Larsen B Ice Shelf triggered by chain reaction drainage of supraglacial lakes. *Geophysical Research Letters*, **40**, 5872–5876, doi: 10.1002/2013GL057694. · Davies, B.J., Carrivick, J.L., Glasser, N.F., *et al.* (2012) Variable glacier response to atmospheric warming, northern Antarctic Peninsula, 1988–2009. *The Cryosphere*, **6**, 1031–1048. · Rignot, E., Bamber, J.L., van den Broeke, M.R., *et al.* (2008) Recent Antarctic ice mass loss from radar interferometry and regional climate modelling. *Nature Geoscience*, **1**, 106–110. · Severinghaus, J.P. 2009: Climate change: southern see-saw seen. *Nature*, **457**, 1093–1094, doi: 10.1038/4571093a. · Vaughan, D.G. (2008) West Antarctic Ice Sheet collapse – the fall and rise of a paradigm. *Climatic Change*, **91**, 65–79. · Vaughan, D.G., Comiso, J.C., Allison, I., *et al.* (2014) Observations: cryosphere. In T. F. Stocker, D. Qin, G.-K. Plattner, *et al.* (eds), *Climate*

Antarctic ice sheet: (a) mean surface mass balance for 1989–2004 from regional atmospheric modelling; (b) ice sheet velocity for 2007–2009 determined from satellite data; (c) changes in ice sheet surface elevation for 2003–2008 determined from ICEsat altimetry; temporal evolution of ice loss determined from GRACE for time periods (d) 2003–2012, (e) 2003–2006 and (f) 2006–2012.
Source: Vaughan *et al.* (2014: figure 4.14). Reproduced with permission of the IPCC.

Change 2013: The Physical Science Basis. Contribution of Working Group I to the Fifth Assessment Report of the Inter-governmental Panel on Climate Change. New York: Cambridge University Press. · Yu, J., Liu, H., Jezek, K.C., *et al.* (2010) Analysis of velocity field, mass balance, and basal melt of the Lambert Glacier–Amery Ice Shelf system by incorporating Radarsat SAR interferometry and ICESat laser altimetry measurements. *Journal of Geophysical Research*, 115, B11102, doi: 10.1029/2010JB007456.

antecedent drainage A drainage system that has maintained its general direction across an area of tectonic uplift. As uplift proceeds, the river may maintain its original drainage pattern and become incised into the emerging landscape. Large-scale examples include the Indus and Brahmaputra of the Himalayas and the Grand Canyon formed through incision of the Colorado River.   MEM

Reproduced with permission of Cory Mathews, iSearchNotes. com 2010.

antecedent moisture The soil moisture condition in an area before a rainfall, which moderates the area's run-off response to the rainfall. Antecedent moisture condition is usually expressed as an index and may be estimated by weighting past rainfall events to derive an antecedent precipitation index (API). The API is a weighted sum of preceding rainfall within given time units, and if a daily time base is used the API is often calculated as

$$API_t = API_{t-1}k + P_t$$

where API_t is the antecedent precipitation index for day t, P_t is the precipitation on day t, and k is a decay factor ($k < 1.0$ and usually $0.85 < k < 0.98$).

Alternatively, a water budget based upon the preceding rainfall and evapotranspiration rates can be used to provide an estimate of soil moisture storage (see WATER BALANCE) or base flow in the area may be used as an index of soil moisture or antecedent moisture condition.   AMG

Reading
Chow, V.T. (1964) *Handbook of applied hydrology.* New York: McGraw-Hill.

antecedent precipitation index An index of moisture conditions in a catchment area used to assess the amount of effective rainfall that will form direct surface run-off. If there has been no rain for several weeks less rainfall will get into the streams than if the ground surface is already saturated. The index is calculated on a daily basis and assumes that soil moisture declines exponentially when there is no rainfall. Thus, we have

$$API_t = kAPI_{t-1}$$

where API_t is the index t days after the starting point. The value of k will depend upon the potential loss of moisture, so has a seasonal variation between 0.85 and 0.98. An allowance is made for any precipitation input during the period.   PS

anteconsequent stream A stream that flows consequent on an early uplift but antecedent to later stages of the same tectonic uplift.

Reading
Shelley, D. (1989) Anteconsequent drainage: an unusual example formed during constructive volcanism. *Geomorphology*, 2, 363–367.

Anthropocene An as of yet informal term designating the current epoch of geologic time as the human era, variously considered as dating from either the onset of the Industrial Revolution (late eighteenth century) or the rise of agriculture (in which case the term would be closely synchronous with the Holocene). The designation is intended to reflect the substantial global impact of humans on the world's environment as being comparable to those of geological processes. Its formal recognition as a geological term is hotly contested.   DRM

anthropochore Plant introduced to an area by humans (see also ALIENS). This may take the form of an intentional introduction; for example, the introduction of various types of European and North American trees to the southwestern Cape of South Africa by colonial settlers. These trees were planted to augment dwindling supplies of locally available wood for fuel and shelter. In other situations, introduction may be entirely an accidental by-product of human activity; seeds of exotic plant species may be transported to a novel locality in, for example, aircraft tyre treads or in containers on board a ship. In some instances, plants introduced intentionally may disperse from their new localities and into places where they become ecologically problematic. For example, the ornamental garden rhododendron, *Rhododendron ponticum*, originally an Asian species, has become a significant invasive pest in the woodlands of Britain.   MEM

Table 2 Classification of anthropogenic landforming processes

1 *Direct anthropogenic processes*
  1.1 Constructional
      tipping: loose, compacted, molten
      graded: moulded, ploughed, terraced
  1.2 Excavational
      digging, cutting, mining, blasting of cohesive
      or noncohesive materials
      cratered
      trampled, churned
  1.3 Hydrological interference
      flooding, damming, canal construction
      dredging, channel modification
      draining
      coastal protection

2 *Indirect anthropogenic processes*
  2.1 Acceleration of erosion and sedimentation
      agricultural activity and clearances of
      vegetation
      engineering, especially road construction and
      urbanization
      incidental modifications of hydrological
      regime
  2.2 Subsidence: collapse, settling
      mining
      hydraulic
      thermokarst
  2.3 Slope failure: landslide, flow, accelerated creep
      loading
      undercutting
      shaking
      lubrication
  2.4 Earthquake generation
      loading (reservoirs)
      lubrication (fault plane)

Source: Haigh (1978).

anthropogeomorphology

anthropogeomorphology The study of the role of humans as a geomorphological agent. There are very few spheres of human activity that do not create landforms. There are those landforms produced by direct anthropogenic processes. These are relatively obvious in form and origin and are frequently created deliberately and knowingly (Table 2). Landforms produced by indirect anthropogenic processes are often less easy to recognize, not least because they do not so much involve the operation of a new process or processes as the acceleration of natural processes. They are the result of environmental changes brought about inadvertently by human technology. ASG

Reading and References
Brown, E.H. (1970) Man shapes the earth. *Geographical Journal*, **136**, 74–85. · Goudie, A.S. (2006) *The human impact*, 6th edition. Oxford/Cambridge, MS: Blackwell/MIT Press. · Haigh, M.J. (1978) Evolution of slopes on artificial landforms, Blaenavon, UK. *University of Chicago, Department of Geography Research Paper* 183. · Jennings, J.N. (1966) Man as a geological agent. *Australian Journal of Science*, **28**, 150–156. · Sherlock, R.L. (1922) *Man as a Geological Agent*. London: Witherby.

anthrosols Soils created or modified by human actions. There are many ways in which humans have modified soils, causing, for example, increased SALINIZATION, acidification, compaction and erosion. Traditional agriculturalists have often employed laborious techniques to augment soil fertility and to reduce such properties as undesirable acidity. In Britain, for example, the addition of chalk to light sandy land goes back at least to Roman times, and the 'marl pits' from which the chalk was dug are a striking feature of the Norfolk landscape. Similarly, in the sandy lowlands of Belgium, the Netherlands, Germany, and western Russia there are soils that at least as far back as the Bronze Age have been built up (often by over 50 cm) and fertilized with a mixture of manure, animal bedding, sods, litter or sand. Such humus-rich soils are called Plaggen soils (Blume and Leinweber, 2004). They also occur in Ireland and Scotland, where the addition of seasand to peat was carried out in pre-Christian times. Likewise, before European settlement in New Zealand, the Maoris used large quantities of gravel and sand, carried in flax baskets, to improve soil structure. Another type of soil that owes much to human influence are 'paddy soils'. Long-continued irrigation, levelling and manuring of terraced land in China and elsewhere have changed the nature of the pre-existing soils. Among the most important modifications (Zhang and Gong, 2003) are an increase in organic matter and in base saturation, and the translocation and reduction of iron and manganese.

In Amazonia, studies have revealed some distinctive 'Amazonian dark earths' under what was previously thought to be pristine rain forest. In pre-European times, before their numbers were decimated, there appear to have been shifting cultivator communities in the region who, over a few hundreds or thousands of years, deliberately or non-deliberately modified soils with various types of rubbish, such as food remains, ash, human excrement, and collapsed houses (Neves *et al.*, 2003).

Many urban soils have distinctive characteristics (Jim, 1998). Soils in cities, because of the limited vegetation cover, may suffer from a decline in organic content through time. In addition they may be compacted by traffic, churned up and eroded during construction, contaminated with large amount of rubble, and polluted with heavy metals such as cadmium, lead and mercury.   ASG

References

Blume, H.-P. and Leinweber, P. (2004) Plaggen soils: landscape history, properties, and classification. *Journal of Plant Nutrients and Soil Science*, **167**, 319–327. · Jim, C.Y. (1998) Urban soil characteristics and limitations for landscape planting in Hong Kong. *Landscape and Urban Planning*, **40**, 235–249. · Neves, E.G., Petersen, J.B., Bartone, R.N. and da Silva, C.A. (2003) Historical and socio-cultural origins of Amazonian dark earths. In J. Jehmann (ed.), *Amazonian dark earths: origin, properties, management*. Dordrecht: Kluwer; pp. 29–50. · Zhang, G. L. and Gong, Z.-T. (2003) Pedogenic evolution of paddy soils in different soil landscapes. *Geoderma*, **115**, 15–29.

antibiosis   A specific form of antagonism that involves the formation by one organism of a substance that is harmful to another organism.

anticentre   The point opposite the epicentre, above the focus, of an earthquake.

anticline   A type of geological FOLD that is convex-upwards in shape (i.e. the strata form an arch) which develops as a result of laterally applied compression resulting from tectonic activity.   DJN

anticyclone   An extensive region of relatively high atmospheric pressure, typically a few thousand kilometres across, in which the low-level winds spiral out clockwise in the northern hemisphere and counterclockwise in the southern hemisphere. Anticyclones are common features of surface weather maps and are generally associated with calm, dry weather.

They originate either from strong radiative cooling at the Earth's surface ('cold anticyclones') or from extensive subsidence through the depth of the troposphere ('warm anticyclones'). Cold anticyclones, or highs, form across the wintertime continents and are shallow features produced by cold, dense air that is confined to the lower troposphere. The mobile ridges that occur in the polar air between frontal systems are also cold highs. Warm anticyclones are semi-permanent features of the subtropical regions of the world. Here, the descending branch of the HADLEY CELL ensures the persistence of a large downward flow of mass to supply the outflowing surface winds. The compression of the subsiding air leads to a deep, anomalously warm troposphere within which the anticyclonic circulation persists with height. Although the subsidence and dry air in the highs tend to dampen convective activity, extensive and sometimes persistent low-layer cloud can occur in some regions.   RR

Reading

Marshall, J. and Plumb, R.A. (2008) *Atmosphere, ocean and climate dynamics: an introductory text*. London: Elsevier Academic Press.

antidune   A ripple on the bed of a stream or river similar in form to a sand dune but which migrates against the direction of flow (i.e. upstream).

antiforms   Upfolds of strata in the Earth's crust; *synforms* are downfolds. In both cases the precise stratigraphic relationships of the rocks are not known, whereas in the case of anticlines and synclines they would be.

antipleion   An area or a specific meteorological station where the mean annual temperature is lower than the average for the region.

antipodal bulge   The tidal effect occurring at the point on the Earth's surface opposite that where the pull of the Moon's gravity is strongest. Hence, it is the tidal effect at the point where lunar attraction is weakest.

antipodes   Any two points on the Earth's surface that are directly opposite each other so that a straight line joining them passes through the centre of the Earth.

antitrades   A deep layer of westerly winds in the TROPOSPHERE above the surface TRADE WINDS. In a simple way they represent the upper limits of the HADLEY CELL within which occurs the poleward transfer of heat, and momentum and water vapour.   BWA

antitriptic wind   See WIND.

aphanitic   Microcrystalline and cryptocrystalline rock textures. Pertaining to a texture of which the crystalline components are not visible with the naked eye.

aphelion   The point of the orbit of a planet, or other solar satellite, that is farthest from the sun.

aphotic zone   The portion of lakes, seas and oceans at a depth to which sunlight does not penetrate.

aphytic zone The portion of the floor of lakes, seas and oceans that, owing to their depth, is not colonized by plants.

apogee The point of the orbit of the Moon or a planet that is farthest from the Earth.

aposematic coloration The conspicuous and distinctive markings on a plant or animal which communicate that it is poisonous or distasteful to potential predators.

applied geomorphology The application of geomorphology to the solution of land stewardship problems, especially to the development of resources and the diminution of hazards (Hails, 1977). The great American geomorphologists of the second half of the nineteenth century – Powell, Dutton, McGee and Gilbert – were employed by the US Government to undertake surveys to enable the development of the West, and R.E. Horton, one of the founders of modern quantitative geomorphology, was active in the soil conservation movement generated by the 'Dust Bowl' conditions of the 1930s. It is notable that all these workers made fundamental contributions to theory, and thus to 'pure' geomorphology, even though much of their work was concerned with the solution of immediate environmental problems. In recent decades the role of the geomorphologist has developed, partly because with increasing population pressures and technological developments the impact of human activity on geomorphological processes has increased (Goudie, 1993) and partly because geomorphology as a discipline is now equipped with better analytical tools, including numerical models and geospatial data.

The role of the geomorphologist in environmental management (Cooke and Doornkamp, 1990) can be subdivided into six main categories. First of these is the mapping of geomorphological phenomena. Landforms, especially depositional ones, may be important resources of useful materials for construction, while maps of slope angle categories may help in the planning of land use, and maps of hazardous ground may facilitate the optimal location of engineering structures. Second, because landforms are relatively easily recognized on air photographs and remote-sensing imagery, they can be used as the basis for mapping other aspects of the environment, the distribution of which is related to their position on different landforms. The third category is the recognition and measurement of the speed at which geomorphological change is taking place. Such changes may be hazardous to humans. By using sequential photographs, maps, or remotely sensed imagery, and by monitoring processes with appropriate instrumentation (Goudie, 2006), areas at potential risk can be identified, and predictions can be made as to the amount and direction of change. For example, by calculating rates of soil erosion in different parts of a river catchment an estimate can be made of the likely life of a dam before it is silted up, and measures can be taken to reduce the rates of erosion in the areas where the erosion is highest. Indeed, the fourth category of applied geomorphology is to assess the causes of the observed changes and hazards, for, without a knowledge of cause, attempts at amelioration may have limited success. Fifth, having decided on the speed, location and causes of change, appropriate solutions can be made by employing engineering and other means. Sixth, because such means may themselves create a series of sequential changes in geomorphological systems, the applied geomorphologist may make certain recommendations as to the likely consequences of building, for example, a groyne to reduce coastal erosion. Examples of engineering solutions having unforeseen environmental consequences, sometimes to the extent that the original problem is heightened and intensified rather than reduced, are all too common, especially in many coastal situations (Bird, 1979). Applied geomorphology contributes extensively to the challenges of river management and river health, the erosional consequences of wildfires, road and highway routing and stability, and the management of mine site impacts and mine site rehabilitation.   ASG/DLD

Reading and References
Allison, R. (ed.) (2002) *Applied geomorphology: theory and practice*. Chichester: John Wiley & Sons, Ltd. · Bird, E.C.F. (1979) Coastal processes. In K. J. Gregory and D. E. Walling (eds), *Man and environmental processes*. Folkestone: Dawson; 81–101. · Cooke, R.U. and Doornkamp, J.C. (1990) *Geomorphology in environmental management*, 2nd edition. Oxford: Oxford University Press. · Goudie A.S. (1993) *The human impact on the environment*, 1st edn. Oxford: Blackwell. · Goudie, A.S., Lewin, J., Richards, K., *et al.* (eds) (1990) *Geomorphological techniques*, 2nd edition. London: Unwin Hyman. · Goudie, A.S. (2006) *The human impact on the natural environment*, 6th edition. Oxford/Cambridge, MA: Blackwell/MIT Press. · Hails, J.R. (ed.) (1977) *Applied geomorphology*. Amsterdam: Elsevier. · Verstappen, H. Th. (1983) *Applied geomorphology*. Amsterdam: Elsevier.

applied meteorology The study of the atmosphere and its behaviour through the use of archived and real-time atmospheric data, or forecasts, to address practical problems in a wide range of economic, social and environmental fields (Table 3). The need to analyse and apply

Table 3 Applied meteorology: sectors and activities where climate has significant social, economic and environmental significance

Primary sectors	General activities	Specific activities
Food	Agriculture	Land use, crop scheduling and operations, hazard control, productivity, livestock and irrigation, pests and diseases, soil tractionability
	Fisheries	Management, operations, yield
Water	Water disasters	Flood/droughts/pollution abatement
	Water resources	Engineering design, supply, operations
Health and community	Human biometeorology	Health, disease, morbidity and mortality
	Human comfort	Settlement design, heating and ventilation, clothing, acclimatization
	Air pollution	Potential, dispersion, control
	Tourism and recreation	Sites, facilities, equipment, marketing, sports activities
Energy	Fossil fuels	Distribution, utilization, conservation
	Renewable resources	Solar/wind/water power development
Industry and trade	Building and construction	Sites, design, performance, operations, safety
	Communications	Engineering design, construction
	Forestry	Regeneration, productivity, biological hazards, fire
	Transportation	Air, water and land facilities, scheduling, operations, safety
	Commerce	Plant operations, product design, storage of materials, sales planning, absenteeism, accidents
	Services	Finance, law, insurance, sales

Source: Thomas (1981).

atmospheric information for such purposes arises because weather and climate impinge directly on vital human concerns such as agriculture, water resources, energy, health and transportation. Applied climatology, the application of the study of Earth's climate to practical concerns, encompasses both climatology and meteorology. Different applications to various weather-sensitive activities require different kinds of meteorological data. For example, weather forecasts and atmospheric data can be usefully applied throughout the construction industry from the initial planning of location, through the design of buildings and the on-site construction phase to the control of energy and other running costs when the building is complete. Increasingly, spatially distributed rainfall data, such as PRISM (4 km resolution; Daly *et al.*, 2008) and TRMM (25 km resolution; e.g. Anders *et al.*, 2006), or real-time forecast models, such as MM5 (4 km resolution; e.g. Mass *et al.*, 2003), are being incorporated into flood forecasts and landslide warning systems as well as geomorphological studies.   KS/DRM

Reading and references

Anders, A.M., Roe, G.H., Hallet, B. *et al.* (2006) Spatial patterns of precipitation and topography in the Himalaya. In S. Willett, N. Hovius, M. T. Brandon and D. Fisher (eds), *Tectonics, climate, and landscape evolution*, Geological Society of America Special Papers, vol. 398. Boulder, CO: Geological Society of America; pp. 39–53. · Daly, C., Halbleib, M., Smith, J.I., *et al.* (2008) Physiographically-sensitive mapping of temperature and precipitation across the conterminous United States. *International Journal of Climatology*, **29**, 2031–2064. · Mass, C.F., Albright, M., Ovens, D., *et al.* (2003) Regional environmental prediction over the Pacific Northwest. *Bulletin of the American Meteorological Society*, **84**, 1353–1366. · Thomas, M.K. (1981) The nature and scope of climate applications. Canadian Climate Centre (unpublished).

aquaculture   The commercial cultivation of plants and animals in aquatic environments, both fresh water and marine (mariculture). Fish aquaculture includes the production of food crops and ornamental (i.e. aquarium) fish. Plant aquaculture includes hydroponic farming as well as open environment farming (e.g. seaweed production). Fish farming refers to closed-system and pen or cage production. Fish ranching refers to the cultivation of migratory species that are seeded, released and harvested upon their return to the 'ranch'. Most aquacultural production occurs in fresh water, and marine production is dominated by shellfish. The importance of aquaculture in the provision of food for the growing global population is reflected in annual production growth rates of 5–7% in recent years.   DJS/DLD

Reading

Bostock, J., McAndrew, B., Richards, R., *et al.* (2010) Aquaculture: global status and trends. *Philosophical Transactions of the Royal Society B*, **365**, 2897–2912.

aquiclude   See AQUIFUGE and GROUNDWATER.

aquifer Refers either to a permeable or porous subsurface rock that holds water, or to the body of water itself. This water is termed GROUNDWATER and is important both as a resource for human use and as a component of the hydrological cycle. A confined aquifer occurs between two imperme-able rocks (termed aquicludes), while the upper limit of an unconfined aquifer is marked by the water table. DSGT

aquifuge An impermeable rock incapable of absorbing or transmitting significant amounts of water. Unfissured, unweathered granite is an example. Certain rocks, such as clay and mud-stone, are very porous and absorb water, but when saturated are unable to transmit it in significant amounts under natural conditions. Such forma-tions are known as *aquicludes*. The term *aquitard* is also sometimes used to describe the hydrological characteristics of the less permeable bed in a strategic sequence that may be capable of trans-mitting some water, but not in economically sig-nificant quantities.

aquatic macrophyte The term is most com-monly used to describe freshwater plants defined on the basis of its large size (i.e. nominally large enough to be seen without the aid of a microscope but most frequently much larger). In marine con-ditions, in which case they are known as seaweeds, such plants may be extremely large; for example, kelp is a commonly occurring brown algae of coastal waters which may exceed 60 m in length. Seaweeds are most abundant in the intertidal and subtidal zone, where environmental conditions are such that specialized adaptations are required for survival. For example, twice a day the intertidal seaweeds are exposed to the atmosphere and must cope with a desiccating atmosphere and direct solar radiation, whereas during twice-daily tidal inunda-tion periods they must endure submersion by sea-water and the physical impact of waves in the surf zone. In order to deal with the rigours of survival under such circumstances, seaweeds have evolved unique metabolic and anatomical characteristics. For example, the body, or thallus, of the organism consists of a root-like holdfast and a stem-like stipe supporting leaf-like blades used in photosynthesis. Cell walls are composed of cellulose and gel-form-ing (e.g. agar) substances, accounting for the slimy and rubbery feel of many seaweeds, a feature that may cushion the thalli against wave impact and also help them resist drying during low tide. Especially in Asia, seaweeds, which are rich in nutrients such as iodine, are used as food. While the macrophytes of the oceans are usually algae, aquatic plants in fresh water – for example, the cosmopolitan

duckweeds (family *Lemnaceae*) – may belong to higher evolutionary groups. Under conditions of nutrient enrichment of lakes due to inputs of agricultural fertilizer, such plants may rapidly increase in population and contribute to the prob-lem of EUTROPHICATION. MEM

Reading
Kim, S.K. (2012) *Handbook of marine algae*. Oxford: John Wiley & Sons, Ltd.

aquitard See AQUIFUGE and GROUNDWATER.

arboreal Pertaining to trees.

arches, natural A bridge or arch of rock join-ing two rock outcrops that has been produced by natural processes of weathering and erosion.

Durdle Door, Dorset, England, is a classic sea arch, formed through differential erosion of Jurassic Portland limestone. Photograph by Alice Thomas.

archipelago A sea or lake containing numer-ous islands or a chain or cluster of islands.

arctic Variously defined. The popular definition is to include all areas north of the Arctic Circle (Lat. 66.5°N), which is the latitude at which the sun does not rise in mid-winter or set in mid-summer. A more useful natural definition includes land areas north of the tree line and oceans normally affected by Arctic water masses. DES

Reading
Nuttall, M. and Callaghan, T.V. (eds) (2000) *The arctic: environment, people, policy*. Amsterdam: OPA. · Sugden, D.E. (1982) *Arctic and Antarctic*. Oxford: Blackwell.

arctic–alpine flora A group of plants display-ing a disjunct geographical distribution that embra-ces both the lowland regions of the arctic and the high-altitude mountain areas of the temperate and even the tropical zones. The main mountain systems involved are the Rockies, the Alps, the Himalayas and high intertropical mountains, such as those of East Africa. Classic examples of the flora include *Anemone alpina* (Alps/Arctic), *Polygonum viviparum* (Alps, Altai, Himalayas/Arctic), *Ranunculus pygmaeus* (Alps, Rockies/Arctic), *Salix herbacea* (Alps, Urals, Rockies/Arctic) and *Saxifraga oppositifolia* (Alps/Arctic). PAS

Reading
Quinn, J.A. (2008) *Arctic and alpine biomes*. Westport, CT: Greenwood Press.

arctic haze A reddish-brown atmospheric haze, which is often observed in the Arctic, especially in the winter and spring, when atmospheric conditions are calm. It consists primarily of atmospheric pollutants derived from industrial sources in Europe and northern Asia. These include sooty and acidic particles. ASG

arctic smoke See FROST SMOKE.

areic Without streams or rivers.

arena A shallow, broadly circular basin hemmed in by a rim of higher land.

arenaceous Pertaining to, containing or composed of sand. Applied to sedimentary rocks composed of cemented sand, usually quartz sand.

areography The study of the geographical ranges of plant and animal taxa. It focuses on the form and size of taxonomic ranges, and differs in emphasis from BIOGEOGRAPHY (concerned with the delimitation of floral and faunal sets and the origins of their constituent elements) and ecogeography (concerned with the reasons for the form and size of taxonomic ranges). ASG

Reading
Rapoport, E.H. (1982) *Areography: geographical strategies of species*. Oxford: Pergamon.

arête A fretted, steep-sided rock ridge separating valley or cirque glaciers. The basic form is the result of undercutting or BASAL SAPPING by glaciers which evacuate any rock debris and thus maintain steep rock slopes. Arêtes are common whenever mountains rise above glaciers; for example, in mountain chains and as NUNATAKS protruding above ice sheets. DES

argillaceous Pertaining to, containing or composed of clay. Applied to rocks that contain clay-sized material and clay minerals.

aridisols The soils of dry climates, which are grouped as one of the 11 soil orders of the US system of soil taxonomy. Aridisols show limited differentiation into horizons (see HORIZONS, SOIL), owing to lack of water to break down and translocate materials, and of organic matter, which normally darkens the upper parts of other soils. Aridisols may have subsurface layers rich in salts, calcium carbonate, gypsum (calcium sulphate) or other materials that are normally removed by water from the soils of wetter regions. Part of the mineral fraction of many aridisols is composed of materials derived from elsewhere and delivered by long-distance wind transportation. DLD

Reading
Dunkerley, D.L. (2011) Desert soils. In D. S. G. Thomas (ed.), *Arid zone geomorphology: process, form and change in drylands*, 3rd edition. Chichester: John Wiley & Sons, Ltd; pp. 101–129. · Skujins J. (ed.) (1991) *Semiarid lands and deserts: soil resource and reclamation*. New York: Marcel Dekker.

arkose A sandstone containing more than 25% feldspar. Any feldspar-rich sandstone.

armoured mud balls (also called clay balls, pudding balls, mud pebbles and mud balls). Roughly spherical lumps of cohesive sediment, which generally have diameters of a few centimetres, though much larger examples have been reported. Many examples are lumps of clay or cohesive mud that have been gouged from stream beds or banks by vigorous currents. They often occur in badlands and along ephemeral streams, but can also be found on beaches, in tidal channels, etc. ASG

armouring A term used of heterogeneous river bed material when coarse grains are concentrated sufficiently at the bed surface to stabilize the bed and inhibit transportation of underlying finer material. An armour layer is coarser and better sorted than the substrate, and is typically only one or two grains thick. Genetically, a distinction exists between armoured and paved beds. An armoured bed is mobile during floods, but the coarser veneer reforms on or after the falling limb of a flood capable of disrupting and transporting the armouring grains as finer grains are winnowed and the protective layer is recreated. The substrate remains protected until the next event capable of entraining the armour grains. A paved surface, however, is more stable and is markedly coarser than the substrate. Whereas an armour layer is characteristic of an equilibrium channel in heterogeneous sediment, a paved layer often occurs in a channel experiencing degradation, and arises because of scour and removal of a substantial thickness of bed material such that the coarsest component is left as a lag deposit. The immobility of a paved surface is often indicated by discoloration and staining of the component grains; the occasional transport of armour grains keeps them clean.   KSR

Reading
Gomez, B. (1984) Typology of segregated (armoured paved) surfaces: some comments. *Earth Surface Processes and Landforms*, **9**, 19–24.

arroyo A trench with a roughly rectangular cross-section excavated in valley-bottom alluvium with a through-flowing stream channel on the floor of the trench (Graf, 1983). Although the term gully is used for similar features, a gully is V-shaped in cross-section instead of rectangular and is excavated in colluvium instead of alluvial fill. The term *arroyo* in the sense of a stream bed has been used in Spanish since at least the year 775, but its modern use in English physical geography dates from the exploration and survey of the American West in the 1860s.   WLG

Reading and References
Graf, W.L. (1983) The arroyo problem – paleohydrology and paleohydraulics in the short term. In K. J. Gregory (ed.), *Background to palaeohydrology*. Chichester: John Wiley & Sons, Ltd. · Harvey, J.E. and Pederson, J.L. (2011) Reconciling arroyo cycle and paleoflood approaches to late Holocene alluvial records in dryland streams. *Quaternary Science Reviews*, **30**, 855–866.

artesian A term referring to water existing under hydrostatic pressure in a confined aquifer. The water level in a borehole penetrating an artesian aquifer will usually rise well above the upper boundary of the water-bearing rocks and may even flow out of the borehole at the surface, in which case it is known as a flowing artesian well.

Water moves in artesian aquifers as a result of differences in fluid potential (see EQUIPOTENTIALS), water moving towards areas of lower HYDRAULIC HEAD. Natural outflow points are artesian springs, where water boils up under pressure. As a consequence, the surface of an artesian spring is usually domed upwards. Artesian waters are often highly mineralized as a result of a long residence time underground. Hence, artesian springs may sometimes build mounds of chemical precipitates deposited from emerging GROUNDWATER, especially if they are located in a tropical arid environment where evaporation is great.   PWW

artificial recharge See RECHARGE.

aspect The orientation of the face of a SLOPE. Aspect is an important control on slope MICROCLIMATE. In the northern hemisphere, for example, south-facing slopes receive more solar radiation than north-facing slopes. Therefore, south-facing slopes tend to be hotter and drier, support different vegetation assemblages and develop different soils. This is one cause of valley asymmetry (see Table 4). Aspect may also be a control on precipitation. In the mid-latitudes of the northern hemisphere, west-facing slopes tend to receive more precipitation than east-facing slopes, because the latter are in a RAIN SHADOW. In the European Alps a distinction is made between light and sunny *adret* slopes, and the shady *ubac* slopes, which have lower snowlines and timberlines.   DJS/ASG

Reading
Birkeland, P.W. (1999) *Soils and geomorphology*, 3rd edition. New York: Oxford University Press. · Kirkby, M.J. (2004) Aspect and geomorphology. In A. S. Goudie (ed.), *Encylopedia of geomorphology*. London: Routledge; pp. 34–36.

association, plant Basic unit of classification of plant COMMUNITY in which dominant or typically co-occurring species are used as the defining characteristic. For example, the temperate deciduous forests of large areas of the Appalachians of eastern North America are characterized by an association, in this instance

Table 4

Climatic regime	N-facing	S-facing	Geomorphic impact
Very cold (arctic or high altitude)	Permanently frozen	Some freeze–thaw	Greater solifluction and other activity on S-facing slopes
Moderately cold	Some freeze-thaw	Mainly unfrozen	Greater disturbance of vegetation and solifluction on N-facing slopes
Moist temperate	Cooler and moister	Warmer and drier	Where water is not limiting, differences due to aspect are weak
Warm semi-arid	Cooler and moister	Warmer and drier	S-facing slopes have sparser and more xeric vegetation, and greater runoff and erosion

Source: Kirkby (2004).

co-dominance, of oak and chestnut. The approach has been especially widely applied in Europe, where the BRAUN-BLANQUET SCALE system has been used to classify common plant associations at a range of spatial scales, although in North America the term is more generally applied to large-scale vegetation formations regarded as CLIMAX VEGETATION.

In essence, the concept of a plant association derives from the ecologist's desire to account for and, in so doing, classify the relationships between different plant species (i.e. PHYTOSOCIOLOGY) and to explain their spatial distribution within the environment. It is, of course, impossible to represent accurately the complex totality of plant distributions, so there exists the need to simplify the situation by identifying particularly conspicuous and commonly recurring species assemblages. Fundamentally, this is a floristic approach to community classification that attempts to compensate for the fact that, while the full species complement of communities better expresses their relationships to each other, some species provide a more sensitive expression of these relationships; that is, they typify a particular set of habitat conditions. The approach, therefore, seeks to identify patterns and emphasizes certain diagnostic species as indicators of given environments or relationships. Two kinds of diagnostic species may be identified within an association; namely, *character* species, which are centred upon, or have their dominant areas of distribution within, a particular association, as against their absence from other associations; *differential* species are present in one association but absent from most or all other associations (although they need not be centred upon the association which they define). This is illustrated in the figure, which shows how nine different plant species form associations along an environmental gradient, in this instance of moisture. Species 1 and 3 are character species for association B and, indeed, have their populations centred within that association. Species 4, on the other hand, while having its distribution centred in association A, is a differential species for subassociation Ba, and distinguishes the 'moist' group of species within association B as a whole. In this way, classification is a form of gradient analysis, since the pattern of plant associations is seen to mirror that of gradients in environmental characteristics.   MEM

Reading and Reference
Begon, M., Harper, J.L. and Townsend, C.R. (2006) *Ecology: from individuals to communities*, 4th edition. Oxford: Blackwell Science. · Westhoff, V. and van der Maarel, E. (1973) The Braun-Blanquet approach. In R. H. Whittaker (ed.), *Handbook of vegetation science 5: ordination and classification of communities*. The Hague: Junk; pp. 617–726. · Whittaker, R.H. (1975) *Communities and ecosystems*, 2nd edition. New York: Macmillan.

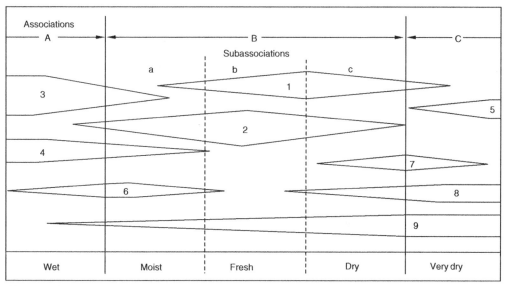

Diagrammatic representation of the distribution of six hypothetical plant species along a moisture gradient and the manner in which such distributions form associations. For further explanation, see text.
Source: Whittaker (1975), originating in Westhoff and van der Maarel (1973).

asthenosphere A zone within the earth's upper MANTLE, extending from 50 to 300 km from the surface to a depth of around 700 km, characterized by a lower mechanical strength and lower resistance to deformation than the regions above and below it. It is approximately, though not exactly, equivalent to the zone in the mantle that transmits seismic waves at a low velocity, due to its partially melted state. In the PLATE TECTONICS model the asthenosphere is regarded as the deformable zone over which the relatively rigid LITHOSPHERE moves.   MAS

Reading
Mörner, N.-A. (ed.) (1980) *Earth rheology, isostasy and eustasy.* Chichester: John Wiley & Sons, Ltd.

astrobleme A term put forward by Dietz (1961) meaning 'star wound' and referring to the erosional remnant or scar of a structure of extraterrestrial origin produced before the Pliocene by the impact of a meteorite on the Earth's surface.   ASG

Reference
Dietz, R.S. (1961) Astroblemes. *Scientific American*, **205**, 51–58.

asymmetric valley A river valley or glacial valley of which one side is inclined at a different angle to the other. Such valleys are a feature of periglacial areas where differences in aspect cause considerable differences in the strength of frost weathering and solifluction, but they can also be caused by structural circumstances (see UNICLINAL SHIFTING).   ASG

Reading
Churchill, R.R. (1982) Aspect-induced differences in hillslope processes. *Earth Surface Processes and Landforms*, 7, 171–182.

asymmetrical fold A fold in geological strata that has one side dipping more steeply than the other.

Atlantic coastlines Coastlines where the trend of the mountain ranges is at right angles or oblique to the coastline (e.g. southwest Ireland), whereas the Pacific (or concordant) type of coastline is parallel to the general trend-lines of relief (e.g. the Adriatic coast of Croatia).   ASG

Atlantic multidecadal oscillation (AMO)
See MERIDIONAL OVERTURNING CIRCULATION (MOC).

atmometer An instrument for measuring the rate of evaporation.

atmosphere The gaseous envelope of air surrounding the Earth and bound to it by gravitational attraction. Within the atmosphere several vertical regions based on temperature are distinguished, as shown in the diagram. In the lowest layer, the troposphere, temperature generally decreases with elevation. Most 'weather' occurs in the troposphere. Above that is the stratosphere, in which temperature increases with height, due to the absorption of solar radiation by ozone. This temperature structure creates a thermally stable situation, so that there is relatively little exchange between the troposphere and stratosphere. However, there is increasing evidence that processes in the stratosphere can affect weather phenomena. The composition of the atmosphere has changed over geologic time, with biological processes contribution greatly to its evolution. Atmospheric oxygen, for example, is produced almost exclusively by plant photosynthesis. Except for water vapour, the relative proportions of the major constituent gases (oxygen, nitrogen, and argon) are more or less constant to a height of about 100 km. The percentage volume of water vapour varies between less than 1% and more than 3% and decreases rapidly upwards from the surface. The trace gases methane, carbon dioxide and ozone vary locally, vertically and seasonally. Their distribution reflects their sources. Natural sources of carbon dioxide include plant and animal respiration, decay of organic matter, and release from sediments and volcanoes. Those for methane include microbes, termites, rice paddies and swamps, volcanoes and ruminant animals. Since their major sources are land based, both methane and carbon dioxide are more abundant in the northern hemisphere than in the southern hemisphere. Ozone forms from the dissociation of

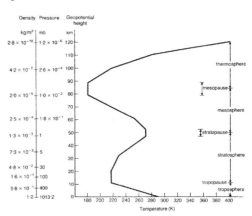

Average temperature structure of the atmosphere from 0 to 120 km.
Source: McIntosh (1972).

oxygen molecules via the absorption of ultraviolet radiation high in the stratosphere, but also from pollution in urban environments. The atmosphere also contains natural pollutants in the form of AEROSOLS, dust and smoke from volcanoes, forest fires, soil erosion, and so on, and man-made pollutants such as sulphur dioxide, nitrogen dioxide, nitric oxide, chlorofluorocarbons, lead and carbon monoxide. Many of these pollutants have been increasing rapidly over the last 100–200 years.   JET/SEN

Reading and Reference
Barry, R.G. and Chorley, R.J. (2010) *Atmosphere, weather and climate*. New York: Routledge. · McIntosh, D.H. (1972) *Meteorological glossary*. London: HMSO.

atmospheric composition   A unit mass of dry air is made up of 75.5% nitrogen, 23.2% oxygen (O_2), 1.3% argon, 0.01% carbon dioxide (CO_2) and smaller proportions of gases such as neon and helium (Table 5). At a height of about 100 km molecular diffusion becomes comparable with mixing due to air motion and the light gases tend to float upwards. Also at these heights diatomic molecules (especially O_2) become split into their atomic components. Atmospheric air contains water vapour from about 4% by mass at temperatures of 30 °C, such as found in the tropics, to 1% at temperatures of 0 °C, such as found in subpolar latitudes and in the middle TROPOSPHERE. Many molecules are spectacularly peculiar. Water has a large latent heat and changes state readily; consequently, it plays a major role as a store of latent energy. Its high solvency encourages chemical reactions in the sea and in living matter. CO_2 dissolves readily in seawater, where it can be utilized by living matter ultimately to form

carboniferous rocks. Otherwise it would clog up wavelengths through which terrestrial RADIATION escapes to space and lead to a hot 'runaway greenhouse' atmosphere like that of Venus (see GREENHOUSE EFFECT). Ozone (at 0.0001% by mass) removes virtually all solar radiation with wavelength less than 0.3 µm which would otherwise destroy living matter.   JSAG

Reading and Reference
Goody, R.M. and Walker, J.C.G. (1972) *Atmospheres*. Englewood Cliffs, NJ: Prentice-Hall. · McIntosh, D.H. (1972) *Meteorological glossary*. London: HMSO.

atmospheric energetics   Concerns the energy content of the atmosphere and how it is changed from one form to another. Almost all energy is heat, latent heat and POTENTIAL ENERGY. THERMODYNAMIC DIAGRAMS relate these. A small fraction of the total of sensible, latent and potential energy (known as available potential energy) can be converted into the KINETIC ENERGY of the WIND.   JSAG

atmospheric instability   In general, a system is unstable if an introduced disturbance increases in magnitude through time. Conversely, the system is stable if the introduced disturbance is damped out through time. In the atmosphere this idea is applied at two main scales: cyclone scale and cloud scale.

Cyclone-scale instability is exemplified by the growth of cyclones within the extratropical westerlies. They are a manifestation of baroclinic instability. Cloud-scale instability results in vertical displacement of air parcels in an atmosphere initially in hydrostatic equilibrium. Cumulus clouds frequently result from this type of instability. (See also VERTICAL STABILITY/INSTABILITY.)   BWA

Reading
Barry, R.G. and Chorley, R.J. (2003) *Atmosphere, weather and climate*, 9th edition. London: Routledge.

atmospheric layers   These are principally the TROPOSPHERE (or overturning layer) 0–10 km above the Earth's surface, the STRATOSPHERE (or layer of constant temperature) at 10–25 km, the ozonosphere (warmed through photochemistry involving oxygen) at 25–60 km, the MESOSPHERE (some similarity with troposphere) at 60–100 km, and between 100 and 500 km the thermosphere (molecular conductivity balancing energy input), the IONOSPHERE (electrical charge on particles significant) and the exosphere (molecules liable to escape into orbit). Similar layers occur at similar values of the pressure (which is related to height) in

Table 5   Composition of dry air

	Molecular weight ($^{12}C = 12.000$)	Volume (%)	Weight (%)
Dry air	28.966	100.0	100.0
Nitrogen	28.013	78.09	75.54
Oxygen	31.999	20.95	23.14
Argon	39.948	0.93	1.27
Carbon dioxide	44.010	0.03	0.05
Neon	20.183	0.0018	0.0012
Helium	4.003	5.2×10^{-4}	7.2×10^{-5}
Krypton	83.800	1.0×10^{-4}	3.0×10^{-4}
Hydrogen	2.016	5.0×10^{-5}	4.0×10^{-6}
Xenon	131.300	8.0×10^{-6}	3.6×10^{-5}
Ozone	47.998	1.0×10^{-6}	1.7×10^{-6}

Source: McIntosh (1972).

atmospheres of other planets. Near the ground, there is a hierarchy of boundary layers: convective (containing the active regions of cumulus-scale motion) about 1 km deep; mechanical or Ekman (mixing due to mechanical stirring) about 300 m deep; logarithmic or constant flux about 10 m deep; and finally an unnamed layer penetrated by material objects (such as trees, grass and waves) that interfere with the flow of air and transfer momentum, heat, moisture, salt, pollen, and so on into the atmosphere.   JSAG

Reading
Ahrens, C.D. (2012) *Essentials of meteorology: an invitation to the atmosphere*, 7th edition. Stanford, CA: Cengage. · Goody, R.M. and Walker, J.C.G. (1972) *Atmospheres*. Englewood Cliffs, NJ: Prentice-Hall.

atmospheric waves
An abstraction convenient for describing some phenomena closely related to the physical processes responsible for propagating characteristics through the atmosphere.

Elastic waves
Propagate at the speed of sound ($330 \, \mathrm{m \, s^{-1}}$) and transmit pressure pulses. They are usually of tiny amplitude but represent the practical limiting signal velocity (playing a role like that of the speed of light in classical physics).

Gravity (–inertia) waves
Represent the action of restoring forces (gravity and Coriolis) acting because the atmosphere is stably stratified. They are analogous to the waves on the interface between two immiscible liquids when the lower is slightly denser than the upper. Short waves (gravity waves) have a characteristic period of about 600 s, which is evident in many natural phenomena, such as LEE WAVES and overshoot of cumulus tops. Very long waves (gravity–inertia waves) are dominated by Coriolis accelerations (GEOSTROPHIC WIND) with an upper bound to the duration of the period of $2\pi/$Coriolis parameter ($\simeq 12 \, \mathrm{h}$); they are exemplified by cloud bands caused by mountain chains.

Cyclone waves
Eddies in the western circumpolar flow around a hemisphere with horizontal dimensions of at most a few thousand kilometres. Although frequently quasi-circular in plan view, vertical cross-sections reveal wave forms in the temperature and pressure distribution. These waves, known also as baroclinic waves, lie within and are inextricably linked to ROSSBY WAVES.

Rossby waves
These are very large perturbations (both wavelength and amplitude of several thousand kilometres) in the extratropical high atmosphere. They are usually two to six in number, encircle the globe from west to east in each hemisphere, contain the jet streams and are vital to the formation of extratropical cyclones and anticyclones, and hence extratropical climate.   JSAG

Reading
Ahrens, C.D. (2012) *Essentials of meteorology: an invitation to the atmosphere*, 7th edition. Stanford, CA: Cengage. · Atkinson, B.W. (1981) Atmospheric waves. In B. W. Atkinson (ed.), *Dynamical meteorology: an introductory selection*. London: Methuen; pp. 110–115.

atoll   An annular form of CORAL ALGAL REEF consisting of an irregular elliptical reef, often breached by channels, around a central lagoon. There are over 400 atolls recorded in the world, most of which are found in tropical waters of the Indo-Pacific. Atolls vary greatly in size and shape, as well as in the depth of the central lagoon. The largest atoll is Kwajalein in the Marshall Islands ($120 \, \mathrm{km} \times 32 \, \mathrm{km}$). As with all coral algal reefs, atolls are sensitive to fluctuations in relative sea level. Some atolls (e.g. Aldabra Atoll in the Seychelles archipelago) are now elevated by a few metres above present sea level; others have become drowned as their growth has failed to keep pace with changing sea level (e.g. Saya de Malha, Indian Ocean). Low, sandy islands (called cays) may form on the reef rim of atolls. Micro-atolls are rounded forms found often on reef flats, usually single colonies of massive corals less than 6 m in diameter with a flat or concave upper surface devoid of living coral. They grow preferentially where water is ponded at low tide.   HAV

Reading
Guilcher, A. (1988) *Coral reef geomorphology*. Chichester: John Wiley & Sons, Ltd. · Woodroffe, C.D. and McLean, R. (1990) Microatolls and recent sea level change on coral reefs. *Nature* **244**, pp. 531–4.

Atterberg limits   The results of tests (index tests) that, arbitrarily defined, show the properties of soils which have COHESION, in that they represent changes, in state or water content, from solid to plastic to liquid materials. The plastic limit (PL) is the minimum moisture content at which the soil can be rolled into a thread 3 mm in diameter without breaking. The liquid limit (LL) is the minimum moisture content at which the soil can flow under its own weight. These are the ones most commonly used, but the shrinkage limit (SL) is the moisture content at which further loss of moisture does not further decrease the volume of the sample. The PL and LL are often combined to give the plasticity index (PI) from $\mathrm{PI} = \mathrm{LL} - \mathrm{PL}$ and the liquidity index (LI) from

LI = $(100m - PL)/(LL - PL)$, where m is the natural moisture content of the soil. A chart of PI as ordinate, plotted against LL, is often used for comparing different types of soil and for classifying them.   WBW

Reading
Mitchell, J.K. (1976) *Fundamentals of soil behavior*. New York: John Wiley & Sons, Inc. · Whalley, W.B. (1976) *Properties of materials and geomorphological explanation*. Oxford: Oxford University Press.

aufeis   See ICING.

auge   A hot, dry wind that blows from the south of France to the Bay of Biscay.

aulacogens   Can be thought of as a continental rifting system in which sea-floor spreading proceeded for a while and then ceased. Such 'aborted pull-aparts' most commonly occur at places in continental lithosphere where three directions of sea-floor spreading are tending to occur at the same time. The common occurrence is for two of these directions to become predominant and for the third axis to become an aulacogen. ASG

Reading
Dewey, J.F. and Burke, K. (1974) Hot spots and continental break-up: implications for collisional orogeny. *Geology*, **2**, 57–60.

aureole, metamorphic   The zone of metamorphosed rock adjacent to an intrusion of igneous rock.

aurora borealis   The 'Northern Lights'. Flashing white and coloured luminescence in the ionized layers of the Earth's atmosphere about 400 km above the poles. The result of solar particles being trapped in the Earth's magnetic field. The term 'aurora australis' has been applied to the phenomenon in the southern hemisphere.

autecology   The ecology of individual organisms and of particular species. Originally used for the study of relationships between a single organism and its environment, the term is now equally widely used for the study of relationships between plants or animals of the same species, particularly species populations, and their environments (population ecology). Autecology provides the fundamental basis for understanding the distribution of organisms and their behaviour in communities, and involves the ecology of organisms at different stages of their life histories together with the environmental controls on growth and reproduction.   JAM/MEM

autochthonous   Matter that is formed or accumulates within a defined space (such as a lake or catchment) and which has not been subject to transport. The term is most frequently applied to sediments, and, for example, in the case of EVAPORITES refers to the fact they accumulate in situ from substances available in the immediate locality. In aquatic communities, the autochthonous input of organic matter is that which is derived from photosynthesis of the locally occurring plants, either the benthic AQUATIC MACROPHYTES and ALGAE associated with the shallower water nearer the shoreline or the planktonic, mainly algal, flora of the open water. A substantial proportion of the organic matter accumulation in such situations is derived, however, from dead material formed outside the lake and transported into the system either by run-off or wind. This is known as allochthonous organic matter or sediment, and its relative proportion depends on the dimensions of the water body and the kinds of terrestrial community associated with its catchment area. Some communities consist almost entirely of autochthonous material; for example, the nutrient-deficient and mainly organic sediments of raised bogs (OMBROTROPHIC mires) are derived almost entirely from local sources because the level of the surface is such that inputs via run-off are not possible. Generally, the relative proportion of autochthonous to allochthonous material in a catchment increases downstream. The open ocean, for example, derives the majority of its organic and inorganic sediments from autochthonous sources. In tectonics, the term 'autochthonous' is used to describe rock formations in Alpine structures that have not been displaced by major thrusting, although they have been folded and faulted.   MEM

autocorrelation   The property of persistence or dependence in sequences of values measured over time. It is usually measured by comparing each value in the sequence either with its immediately previous (lag = 1) value or with the value at a fixed previous time or distance (lag > 1). For a sequence $x_1 .. x_n .. x_N$ the comparisons for lag $r(\leq N)$ are between the $(N - r)$ pairs x_n and $x_{(n-r)}$ as n ranges from $r + 1$ to N. The pairs of values are then used to calculate a correlation coefficient, which measures the degree of dependence for the given lag. The correlation coefficient is calculated as for a normal least-squares correlation, although in this context it is called the coefficient of autocorrelation. A correlogram may then be constructed in which the successive values of the coefficient for different lags is plotted against the lag. Comparison between correlograms for different types of sequence gives some idea of the type and degree of dependence present, if any (for lag zero the coefficient is necessarily 1.0). A sequence of independent values clearly shows zero coefficients – that is, no autocorrelation. Many, if not most, time and space sequences show some degree of autocorrelation, so care must be exercised in obtaining valid independent samples or in applying parametric statistical tests.   MJK/MEM

autogenic stream A stream whose flow is sustained by a continuously positive water balance along its course. Most perennial streams of humid areas are of this kind. The distinction is made between these streams and ALLOGENIC STREAMS, which, for part or perhaps most of their course, traverse areas of negative water balance and which are instead sustained by a water source in wetter headwaters. Derivation: terms from the Greek *autos* (self) and *allos* (other). Examples of the latter include the Darling River system in Australia and the lower Nile in Egypt.   DLD

autogenic succession The process of community change (SUCCESSION) caused by the influence of organisms, particularly plants, on their own environment as classically described by Sir Arthur Tansley (1935). Through various mechanisms, organisms may so change their environment that other species obtain a competitive advantage so that the original organisms are eventually replaced, in effect having brought about their own destruction. The mechanism of ALLOGENIC SUCCESSION results from the action of external environmental factors independent of the organisms themselves. Consider, for example, a vegetation change in response to a change in soil pH. If this was caused by the in-situ accumulation of acidic plant litter the succession is autogenic, but if the cause was prolonged leaching due to heavy rainfall then it is described as allogenic, although the two concepts may be very difficult to distinguish. (See also COMPETITION and ECOSYSTEM.)   JAM/MEM

Reference
Tansley, A.G. (1935) The use and abuse of vegetational concepts and terms. *Ecology*, **16**, 284–307.

autotrophic A descriptive term for those organisms that synthesize (usually, but not exclusively, through PHOTOSYNTHESIS) organic substances from simple inorganic source materials. Autotrophs are exemplified by the primary producer green plants in the assimilation of inorganic carbon dioxide and water into organic carbohydrates using solar radiation as the energy source to drive the process. By comparison, heterotrophs, consumers, must rely on organic molecules already synthesized. Autotrophs can be divided into two groups. Photoautotrophs are photosynthesizing and contain chlorophyll, the catalyst that facilitates the entrapment of energy from sunlight. Chemoautotrophs, mostly bacteria, are organisms in which energy is obtained from the oxidation of inorganic compounds *without* the use of light. Among the more remarkable of these organisms are the bacteria found in great concentrations in the ocean depths feeding, literally, on the hydrogen sulphide thrown out by hot volcanic vents. It has now been realized that there is a very substantial bacterial biomass within the Earth's crust as a whole, somewhat humorously accorded the acronym SLIME (subsurface lithoautotrophic microbial ecosystems) by Stevens and McKinley (1995). An autotrophic bacterial flora may well dominate the Earth's biomass, and yet its existence was unheard of until the 1960s.   MEM

Reference
Stevens, T.O. and McKinley, J.P. (1995) Lithoautotrophic microbial ecosystems in deep basalt aquifers. *Science*, **270**, 450–454.

autovariation The variation within a geophysical system that is forced from within the system itself; that is, with no external forces acting. An example is autovariation of the atmosphere. While variability is frequently forced by processes at the lower or upper boundaries (e.g. sea-surface temperatures, solar radiation), interaction within the atmosphere itself can produce variability. Sources of autovariations include interactions between atmospheric waves, interactions between wave disturbances and the mean circulation, fluctuating zonal winds, and tropical/extratropical interactions. Autovariations include internally generated patterns of climate variability such as the El Niño/Southern Oscillation phenomenon or the North Atlantic Oscillation.   SEN

avalanche The sudden and rapid movement of ice, snow, earth or rock down a slope. Avalanches are an obvious and important mechanism of mass wasting in mountainous parts of the Earth. They are also highly significant on subaqueous continental margins and deltas as well as in extraterrestrial environments; for example, on Mars. Avalanches occur when the shear stresses on a potential surface of sliding exceed the shear strength on the same plane. Failure is sometimes associated with increased shear stress in response to slope steepening or loading (e.g. slope undercutting or snow or deltaic sediment accumulation), with reduced shear strength within the material (e.g. increased PORE WATER PRESSURE or the growth of weak snow crystals) and sometimes with a combination of the two, especially when associated with an external trigger such as an earthquake.

Avalanches are commonly subdivided according to the material involved. *Snow avalanches* occur in predictable locations in snowy mountains and create distinctive ground features as they plunge down the mountain side (Laute and Beylich, 2014). *Debris avalanches* involve the rapid downslope movement of sediment (Meyer *et al.*, 2014). On land they are commonly associated with saturated ground conditions. In subaqueous environments they reflect sediment overloading. One large example off the Spanish Sahara involved 18,000 km^2 of disturbance (Embley and Jacobi, 1977). *Rock avalanches* are very rapid downslope movements of bedrock that become shattered during movement. These avalanches sometimes achieve velocities as high as 400 km h^{-1} owing to the presence of trapped interstitial air; they can travel tens of kilometres from their source, sometimes with devastating effects on human life.   DES

Reading and References
Embley, R.W. and Jacobi, R.D. (1977) Distribution and morphology of large submarine sediment slides and slumps on Atlantic continental margins. *Marine Geotechnology*, **2**, 205–28. · Laute, K. and Beylich, A.A. (2014) Morphometric and meteorological controls on recent snow avalanche distribution and activity at hillslopes in steep mountain valleys in western Norway. *Geomorphology*, **218**, 16–34. · Meyer, N.K., Schwanghart, W., Korup, O., *et al.* (2014) Estimating the topographic predictability of debris flows. *Geomorphology*, **207**, 114–125. · Nicoletti, P.G. and Sorriso-Valvo, M. (1991) Geomorphic controls of the shape and mobility of rock avalanches. *Bulletin of the Geological Society of America*, **103**, 1365–1373.

avalanche tarns   Small water-filled depressions produced by repeated avalanche impact.

Reading
Fitzharris, B.B. and Owens, I.F. (1984) Avalanche tarns. *Journal of Glaciology* **30**, 308–312.

aven   A hole or shaft in the roof of a cave passage that may be either a rather large blind roof pocket or a tributary inlet into the cave system. It may connect the cave either with the surface or with overlying chambers and passages. Many avens close upwards to impenetrable fissures but may still be important hydrological routes. In parts of France, aven is equivalent to the British term *pothole*.   ASG

avulsion   The diversion of a river channel to a new course at a lower elevation on its floodplain as a result of floodplain aggradation. It causes established meander belts to become abandoned and new ones to form.

Reading
Smith, N.D., Cross, T.A., Dufficy, J.P. and Clough, S.R. (1989) Anatomy of an avulsion. *Sedimentology*, **36**, 1–23.

azimuth   The arc of the sky extending from the zenith to the point of the horizon where it intersects at 90°.

azoic   Without life. Pertaining to the period of Earth history before organic life evolved or to portions of the seas and oceans where organisms cannot exist.

azotobacter   The principal nitrogen-fixing bacteria. An aerobic bacteria that obtains energy from carbohydrates in the soil zone k.

B

backing wind   See WIND.

backshore   The backshore of the coastal zone lies between the highest point reached by marine action and the normal high-tide level. On a low coast, the backshore zone is often in the form of a berm, above the normal reach of the tide. The berm often slopes gently landwards. On shingle BEACHes, the berm crest can attain 13 m above normal high-tide level, as on Chesil Beach, Dorset. On sandy coasts, foredunes may form in the backshore zone, and washover fans are associated with barrier island backshore zones. On a steep coast, the backshore is that part of the platform and cliff foot affected by waves under storm conditions. A wave-cut notch is a common feature.   CAMK

Reading
Bird, E.C.F. (2008) *Coastal geomorphology: an introduction.* 2nd edition. Chichester: John Wiley & Sons, Ltd.

backswamp   Low-lying marshy or swampy area on a FLOODPLAIN where overbank flood or tributary drainage water may become ponded between river levées and valley sides or other relatively elevated alluvial sediments. Backswamps may be artificially drained as their organic-rich sediments may be good for agriculture.   JL/DSGT

backwall   The arcuate cliffed head of a cirque basin or a landslide.

backwash   The return flow of water down a BEACH after a wave has broken. It is the return to the sea of the swash or uprush of the wave. It plays an important part in determining the gradient of the swash slope in association with the size of the beach material. On a coarse pebble beach the backwash is reduced in volume through percolation, so that a steeper slope is necessary to maintain equilibrium between swash and backwash. On a fine sand or wet beach the backwash is a large proportion of the swash, so a flat beach can remain in equilibrium. Rhomboid ripple marks are sometimes formed by backwash. Long waves and steep waves enhance the backwash and are associated with flatter swash slope gradients.   CAMK

Reading
Bird, E.C.F. (2008) *Coastal geomorphology: an introduction,* 2nd edition. Chichester: John Wiley & Sons, Ltd. · Demarest, D.E. (1947) Rhomboid ripples marks and their relationship to beach slope. *Journal of Sedimentary Petrology,* **17**, 18–22.

backwearing   The parallel retreat of a slope without a change in overall form or inclination. The term may be applied to escarpments and to side slopes; it is commonly used to contrast with down-wearing of a slope in which material is lost from the upper segments of the slope with consequent decreases in inclination. Parallel retreat implies that the resistance of rock and soil of the slope is constant into the hill that is being eroded, or that resistance does not control the slope form.   MJS

badlands   Generally regarded as the archetypal example of the effects of vigorous water EROSION, badlands can resemble miniature desert landscapes with weirdly shaped HOODOOS, barren steep and rounded slopes scarred by RILLS and gullies (see GULLY), and a maze of winding channels. The name likely derives from the early French explorers' expression *mauvaises terres à traverser* (meaning bad lands to cross) for such terrain in North America's western plains. While usually associated with dryland areas, badlands can form wherever weak, unconsolidated materials are exposed to periodic high-intensity rainfall and rapid RUN-OFF. Badlands have formed on marine silts in Canada's high arctic and on deeply weathered granites and basaltic LAVA flows in the humid tropics (see WEATHERING); they occur either naturally on weak mudstones and poorly cemented sandstones, or on industrial spoilheaps and unwisely used agricultural land. The rapidity of erosion, which can remove several millimetres of surface material in a single storm, coupled with the often infertile character of the material, inhibits plant growth, leaving the surface unprotected against heavy rains and thereby encouraging faster

The Dictionary of Physical Geography, Fourth Edition. Edited by David S.G. Thomas.
© 2016 John Wiley & Sons, Ltd. Published 2016 by John Wiley & Sons, Ltd.

rates of water erosion. Badlands materials that contain swelling CLAY minerals develop DESICCATION cracks, allowing percolation of water that generates subsurface flow to create an often complex network of PIPES, tunnels, and sinkholes formed by collapse. Some of these features resemble those found in limestone (KARST) areas, but whereas limestone caves and so on form by chemical solution processes (see SOLUTION, LIMESTONE), in badlands they are the result of water erosion and are called PSEUDOKARST features to distinguish them from true karst landforms.   IC

Reading
Howard, A.D. (2009) Badlands. In A. D. Abrahams and A. J. Parsons (eds), *Geomorphology of arid environments*, 2nd edition. Berlin: Springer; pp. 265–299. · Torre, D., Calzolari, C. and Rodolfi, G. (2000) Badlands in changing environments: an introduction. *Catena*, **40**, 119–125. · Wainwright, J. and Brazier, R.E. (2011) Slope systems. In D. S. G. Thomas (ed.), *Arid zone geomorphology: process, form and change in drylands*, 3rd edition. Chichester: John Wiley & Sons, Ltd; pp. 209–233.

bajada   Confluent alluvial and pediment fans at the foot of mountain and hillslopes encircling desert basins.

ball lightning   See LIGHTNING.

bank erosion   Removal of material from the side of a river channel. This may be accomplished by several processes: particle by particle removal following surface wash, frost heave, groundwater sapping and the dislodgement and fall of material and the subsequent entrainment of particles by flowing water; abrasion by transported ice; and mass failure of the bank. Bank erosion is usually associated with high flows, but may be at a maximum as water levels fall after a high-flow period and the lateral support provided by the deep water and the apron of previously slumped material is removed. Rates of erosion depend on prior conditions affecting bank strength, and in particular water content, which may vary seasonally. Bankside vegetation and root systems are also an important control on bank stability.

In detail, bank erosion forms also reflect bank composition. In coarser materials, erosion may be by particle entrainment with steep banks that have talus slopes at the base as long as such material is not picked up by the flowing water. In cohesive sediment, bank failure may be in the form of shallow slips. In composite banks, with an upper cohesive fine layer above coarser materials, the process may be by removal of the underlying material accompanied by periodic collapse of overhanging 'cantilevers' of fine bank-top material. Until they are broken up and disaggregated, collapsed blocks may temporarily protect the bank base, especially where given additional strength by vegetation and a root mat.

Bank erosion rates are highest where flow in river channels is asymmetric, as on the outer bank of meander bends. The exact locations of such points may be a determinant in evolving channel patterns. For example, if located downstream of the bend apex, a meander loop may tend to translate downvalley rather than expand laterally. High rates of bank erosion may also relate to flow diverted by BARS in the channel, and the fact that non-cohesive banks may erode relatively rapidly may, among other factors, promote the formation of braided channel patterns (see BRAIDED RIVER).   JL

Reading
Charlton, R. (2007) *Fundamentals of fluvial geomorphology.* Abingdon: Routledge. · Richards, K. (1982) *Rivers: form and process in alluvial channels.* London: Methuen. · Schumm, S.A. (1977) *The fluvial system.* New York: John Wiley & Sons, Ltd.

bank storage   Water absorbed and stored in the void spaces in the bed and banks of a stream, lake or reservoir and returned in whole or in part as the water level falls. Drainage of such water by seepage into the channel may contribute to the maintenance of river flows at low water, while under dry conditions there may be a downstream decrease in river discharge because of percolation into bank storage. Losing streams where this is particularly important occur in arid and semi-arid environments, in allogenic streams (those fed by precipitation from outside the local area, such as the Nile, which is fed by lakes in central Africa) and on regulated rivers where low flow discharges are augmented by reservoir releases.   DRM

bankfull discharge   The river discharge that exactly fills the river channel to the bankfull level without spilling on to the floodplain and which depends upon the definition of CHANNEL CAPACITY. Bankfull discharge has often been assumed to be a significant or critical channel-forming flow that is important in determining the size and shape of the river channel and it has, therefore, sometimes been regarded as equivalent to DOMINANT DISCHARGE. The frequency of occurrence of bankfull discharge varies considerably and has been quoted as ranging from 4 months to 5 years in Scotland, 1.3 to 14 years in lowland England and 1 to 30 years in 28 basins in the western USA, but care should be taken in quoting such recurrence intervals because at least 16 different methods are

45

available (Williams 1978) to calculate bankfull discharge. KJG

Reading and References
Shibata, K. and Ito, M. (2014) Relationships of bankfull channel width and discharge parameters for modern fluvial systems in the Japanese Islands. *Geomorphology*, **214**, 97–113. · Williams, G.P. (1978) Bankfull discharge of rivers. *Water Resources Research*, **14**, 1141–1154.

banner cloud   Often seen to extend downward from isolated peaks rather like a flag. The classic example is that on the Matterhorn. The mechanism of the cloud is not yet understood.   BWA

bar, nearshore   A longitudinal sediment ridge developed in or adjacent to the surf zone as a result of wave action (see WAVES). Nearshore bars may occur individually or in sets of many bars (Levoy *et al.*, 2013). Troughs separate them from the foreshore and from each other. Where multiple bars are present, their height and spacing usually increase with distance offshore. The presence of a nearshore bar is often associated with the development of a STORM BEACH, and the form may be ephemeral. On other coasts, however, bars may be in long-term morphodynamic EQUILIBRIUM (Greenwood and Davidson-Arnott, 1979), changing position and form only moderately through time.

There are three commonly recognized bar forms: parallel (to the beach) bars that are linear to sinuous; crescentic bars that are convex shoreward; and transverse bars that are attached to the foreshore. Intertidal ridge and runnel systems are also frequently described as a fourth type of nearshore bar. Longshore currents are common in troughs, and RIP CURRENTS are associated with transverse bar systems. The location and geometry of subaqueous bars is dependent upon complex feedback between nearshore hydrodynamics, bar morphology and sediment characteristics.   DJS

References
Greenwood, B. and Davidson-Arnott, R.G.D. (1979) Sedimentation and equilibrium in wave-formed bars. *Canadian Journal of Earth Science*, **16**, 312–332. · Levoy, F., Anthony, E.J., Monfort, O., *et al.* (2013) Formation and migration of transverse bars along a tidal sandy coast deduced from multi-temporal lidar datasets. *Marine Geology*, **342**, 39–52.

barchan   A crescent-shaped sand dune, in which the horns point downwind in the direction of dune migration. Barchans are a type of TRANSVERSE DUNE that tends to form where either sand transport rates are high and/or sand supply is limited; thus, they frequently occur on the margins

of sand seas and migrate over nonsandy surfaces. Barchan dunes are not necessarily a very common type of dune, but the frequent simplicity of form has made them the source of investigations into aeolian sand transport on dune forms. (See DUNE.)   DSGT

Reading
Hastenrath, S. (1987) The barchan sand dunes of south Peru revisited. *Zeitschrift fur Geomorphologie, Neue Folge*, **31**, 167–178. · Moosavi, V., Moradi, H., Shamsi, S.R.F. and Shirmohammadi, B. (2014) Assessment of the planimetric morphology of barchan dunes. *Catena*, **120**, 12–19.

barchanoid ridge   A type of TRANSVERSE DUNE that has the appearance of a series of linked BARCHAN dunes, perpendicular to the resultant sand transport direction.   DSGT

baroclinicity   A measure of the fractional rate of change of atmospheric density in the horizontal. It is also frequently defined as a state where surfaces of equal pressure and surfaces of equal density intersect. In zones of large baroclinicity, spontaneous generation of new motion systems is likely: primary depressions (see CYCLONE) in the baroclinic zone of middle (30–60°) latitudes, secondary depressions in the frontal zones of the primary depressions and thunderstorms in the fronts of the secondary depressions.   JSAG

barometer   An instrument for measuring atmospheric pressure. Historically, there were two types: mercury and aneroid. Historically, mercury barometers worked on the principle that atmospheric pressure was sufficient to support a column of mercury in a glass tube. As the pressure varied, so did the height of the column. Mercury-free aneroid barometers comprised a series of vacuum chambers with springs inside to prevent total collapse. Atmospheric pressure changes were recorded by the compression or expansion of the chambers. This mechanism was frequently used in a barograph, which is a barometer that gives a graph of pressure changes through time. Modern atmospheric pressure data are generally collected using solid-state pressure transducers, which permit operating ranges of 500–1100 hPa, with a resolution often of <0.1 hPa. Results may be logged into solid-state memory or transmitted over private wireless networks or the cellular telephone network.   DLD

Reading
Meteorological Office (1956) *Handbook of meteorological instruments. Part I: instruments for surface observations.* London: HMSO.

barranca A steep-sided gully or ravine.

barrier island Barrier islands are elongate, shore-parallel structures of sand and/or gravel separated from the mainland by a lagoon. They are not attached at either end. Wave-dominated barriers in microtidal settings are often long and characterized by low dunes, punctuated by washover channels and with washover fans on their inner margins. With a mesotidal range, barriers often show a 'drumstick' planform and are separated by tidal inlets with ebb- and flood-tide deltas. Many barriers are undergoing long-term landward migration under moderate rates of sea-level rise by barrier 'rollover'; there is debate as to whether or not barriers can 'overstep' (a sudden shift of position landwards) under high rates of sea-level rise. Barrier islands are common on gently sloping coasts with wide shelves and coastal plains where sediment supply is good; locations include the eastern seaboard of the USA, West Africa; Atlantic coast of South America and the southern North Sea. TS

Reading
Davis, R.A. (ed.) (1994) *Geology of Holocene barrier island systems*. Berlin: Springer-Verlag. · Fitzgerald, D.M. and van Heteren, S. (1999) Classification of paraglacial barrier systems: coastal New England, USA. *Sedimentology*, **46**, 1083–1108. · Otvos, E.G. (2012) Coastal barriers – nomenclature, processes, and classification issues. *Geomorphology*, **139–140**, 39–52. · Stutz, M.I. and Pilkey, O.H. (2011) Open-ocean barrier islands. Global influence of climatic, oceanographic, and depositional settings. *Journal of Coastal Research*, **27**, 207–222.

barrier reef Set of coral reefs separated from a non-reefal landmass by a deep lagoon. Charles Darwin identified oceanic barrier reefs as occupying the mid position of a developmental sequence of reefs, caused by the continued upgrowth of coral on the subsiding basement of a central land mass, generally a volcanic island. Fringing reefs form the start of this sequence and atolls the end. Thermal cooling of oceanic lithosphere provides a modern explanation for the subsidence process. With lateral plate migration accompanying cooling, it might be expected that the Darwinian sequence will be seen in space, with fringing reefs around recently extinct volcanoes, leading to atolls at the distant, older ends of island chains. However, this succession is not seen in all cases, and the additional factors of varying rates of sea-level rise, the nature of surfaces for coral colonization following periods of sea-level fall and regional plate flexure all influence modern large-scale reef morphology. Coring of barrier reef margins has revealed the interplay between subsidence and sea-level variations. Classic oceanic barrier reefs characterize many of the island groups of the South Pacific Ocean. Non-oceanic, shelf settings also support extensive barrier reef systems, such as western Madagascar, the Red Sea, Southeast Asia (Philippines and Indonesia), eastern Papua New Guinea and New Caledonia. The Great Barrier Reef forms the largest stretch of reefs in the world, being almost 2000 km long, and is actually a compound form of many different reefs. The Meso-American barrier reef system characterizes the western Caribbean Sea. TS

Reading
Blanchon, P., Granados-Corea, M., Abbey, E., *et al.* (2014) Postglacial fringing-reef to barrier-reef conversion on Tahiti links Darwin's reef types. *Scientific Reports*, 4: 4997. doi: 10.1038/srep04997. · Cabioch, G., Montaggioni, L., Thouveny, N., *et al.* (2008) The chronology and structure of the western New Caledonian barrier reef tracts. *Palaeogeography, Palaeoclimatology, Palaeoecology*, **268**, 91–105. · Purdy, E.G. and Winterer, E.L. (2006) Contradicting barrier reef relationships for Darwin's evolution of reef types: *International Journal of Earth Sciences*, **95**, 143–167.

barrier spit The part of a barrier island system that is attached to an eroding headland. It is generally parallel to the coast and separated from it by a lagoon. An inlet, which is often tidal, separates a barrier split from a detached barrier island. One theory, first put forward by G. K. Gilbert (1890), suggests that barrier spits are the first stage of barrier island formation. They are fed by longshore drift of material derived from the eroding headland and are generally formed mainly of sand. Some barrier spits may be attached to older beach ridges, but one end of the elongated feature is always attached to the mainland while the other is free but associated with a barrier island. CAMK

Reading and Reference
Bird, E.C.F. (2008) *Coastal geomorphology: an introduction*, 2nd edition. Chichester: John Wiley & Sons, Ltd. · Gilbert, G.K. (1890) *Lake Bonneville*. US Geological Survey monograph 1. Washington, DC: US Government Printing Office; pp. 23–65. · Schwartz, M.L. (1971) The multiple causality of barrier islands. *Journal of Geology*, **79**, 91–94.

basal ice A relatively thin layer (with a vertical extent of up to tens of metres) of debris-rich ice present at the base of glaciers, which is produced at and interacts with the glacier bed. As a result of its different mode and environment of formation it may differ from 'normal' glacier ice in terms of its overall extent and properties (including its rheology), debris, solutes and gases. ASG

Reading
Hubbard, B. and Sharp, M. (1989) Basal ice formation and deformation: a review. *Progress in Physical Geography*, 13, 529–558.

basal sapping The process whereby a slope is undercut at the base. It can describe the retreat of a hillslope or escarpment by erosion along a spring line (mid-latitudes) by salt weathering (arid environments) or by glacial erosion (CIRQUE).

basal sliding The process by which a glacier slides over bedrock. It is distinguished from processes associated with the internal deformation of ice within the body of a glacier. Two important processes of sliding are *enhanced basal creep*, where pressures caused by irregularities on the bed increase stresses within the ice and allow it to deform round the obstacles, and *regelation*, where the pressures induced by the bed obstacles cause ice to melt on the upstream sides, flow round the obstacles as water, and refreeze on to the glacier on the downstream sides. Basal sliding is closely related to the presence and nature of water at the ice–rock interface. If no water is present, basal sliding is restricted to enhanced basal creep and the ice–rock interface remains immobile. Water allows movement at the ice–rock interface, and the higher the water pressure the quicker the rate of sliding. Glacier sliding takes place in a stick–slip fashion with periodic movements of a few centimetres, and probably relates to the impeding effect and sudden release of small frozen patches at the ice–rock interface. A recently recognized additional form of basal sliding incorporates the deformation of weak sediments beneath the weight of a glacier.   DES

Reading
Benn, D.I. and Evans, D.J.A. (2010) *Glaciers & glaciation*, 2nd edition. London: Routledge (especially chapter 4).

basalt A fine-grained and dark-coloured igneous rock. The lava extruded from volcanic and fissure eruptions. Rocks composed primarily of calcic plagioclase and pyroxene with or without olivine.

base exchange (or CATION EXCHANGE) A vital soil reaction whereby bases or cations such as calcium, magnesium, sodium and potassium are made available as plant nutrients. These cations are usually loosely bonded on the surface of clay and organic colloidal particles in the soil complex, and cation exchange takes place when hydrogen cations, derived from organic decomposition and atmospheric sources, replace the metal cations, releasing the latter into the soil water around the root hairs, where the plant may absorb them.   KEB

Reading
Pears, N. (1985) *Basic biogeography*, 2nd edition. London: Longman. · Trudgill, S.T. (1988) *Soil and vegetation systems*, 2nd edition. Oxford: Oxford University Press.

base flow A term often used to describe the reliable background river flow component from a drainage basin. Base flow, or delayed flow, was originally considered to be the result of effluent seepage from groundwater storage but in some drainage basins this sustained component of flow is also produced by interflow from zones above the water table and particularly by throughflow or lateral drainage through the soil. The DEPLETION CURVE or base flow recession curve describes the way in which base flow discharge recedes at a particular site during a period without rainfall.   AMG

base level The lower limit to the operation of subaerial erosion processes, usually defined with reference to the role of running water. Sea level acts as a general base level for the continents, although there can be a wide range of local base levels, some above and some below sea level. The term was first used by J. W. Powell (1875), who defined it in a very broad fashion to include: first, the concept of the sea as 'a grand base-level, below which the dry lands cannot be eroded'; second, the existence of local or temporary base levels 'which are the levels of the beds of the principal streams which carry away the products of erosion'; and, finally, as 'an imaginary surface, inclining slightly in all its parts towards the lower end of the principal stream draining the area'. W. M. Davis (1902) considered that the breadth of Powell's definition had led to variety of practice and confusion. He proposed that base level be restricted to 'simply . . . the level base with respect to which normal sub-aerial erosion proceeds'. This suggestion has largely been adopted.

Base levels can be identified on a number of different scales. A stream channel acts as the base level for processes on the adjacent slope. A major stream is the base level for the courses of its tributaries. A lake or a reservoir or a waterfall is the base level for the entire basin upstream. A band of resistant rock may act as a base level for a part of a river's long profile until such time as the band is breached. Clearly, base levels can and will alter over time, with implications for the operation of subaerial erosional processes. By and large, a fall in base level creates an increase in potential energy, by increasing the total relief, and may result in an acceleration of rates of erosion and downcutting. A rise in base level, on the other hand, reduces relief and potential energy and is frequently marked by aggradation. Over QUATERNARY time scales, sea-level changes associated with EUSTASY and ISOSTASY will influence the 'grand' base level, with complex repercussions in many drainage basins, even those where the effects of direct glaciation or tectonism are absent.   BAK

References
Davis W.M. (1902) Base-level, grade, and peneplain. *Journal of Geology*, **10**, 77–111. · Powell, J.W. (1875) *The exploration of the Colorado River and its canyons*. New York: Dover Press.

base saturation   The condition arising when the CATION-EXCHANGE CAPACITY of a soil is saturated with exchangeable bases – calcium, magnesium, sodium and potassium – when expressed as a percentage of the total cation-exchange capacity.

basement complex   A broad term for the older rocks, usually Archean igneous and metamorphic rocks, that underlie more recent sedimentary rocks in any region. Such rocks are a feature of the ancient shield areas of the Earth's surface.

basic rocks   Igneous rocks containing less than about 55% silica, basalt being a typical example (see ACID ROCKS).

basin-and-range   A type of terrain, as found in Utah and Nevada, where there are fault block mountains interspersed with basins. The basins often contain lakes, or sediments that are evidence of lake occurrence during times of wetter climates.

batholith   A large mass of intrusive igneous rock that extends to great depth and can cover or underlie very large areas.

bathymetry   The measurement of water depth.

bauxite   The main ore of aluminium. An impure aluminium hydroxide associated with the clay deposits of weathering zones, especially in tropical regions.

beach   A coastal accumulation of varied types of unconsolidated sediments, predominantly sand-sized but sometimes of gravel. Carbonate beaches are formed of biogenic sediments, such as coral fragments, coralline algae or shell debris.

The sediment is derived from rivers, coastal rock weathering or offshore sources and is moved onshore and alongshore by waves to form a beach. Beaches vary enormously in width and alongshore continuity. They can be 'swash aligned' (where waves arrive parallel to the shore and sediment volumes remain constant within bounded littoral cells) or 'drift aligned' (where waves break at an angle to the shore and drive longshore transport). A particular form of embayed beach can be described in planform by a logarithmic spiral. On a typical microtidal wave-dominated beach four zones are present: a zone of shoaling waves; a breaker zone; a surf zone; and, on the beach face itself, a swash zone. The top of the swash zone may be marked by a berm and is followed by the backshore, comprising a subaerial beach that can grade into dunes or a barrier of coarser sediments. Beach profiles are determined by wave characteristics (height and period) and the size and sorting of the sediments. Coastal morphodynamics identifies a series of characteristic beach states. The end members are described as low-angle 'dissipative' profiles with wide surf zones in fine sands at one extreme versus, at the other extreme, narrow, steep 'reflective profiles', perhaps supporting rhythmical alongshore forms such as beach cusps, in coarser materials. Many beaches, however, cycle through a series of more topographically complex 'intermediate' beach states, characterized by shore-normal ridges and runnels. On macrotidal beaches, beachface processes migrate rapidly up and down the beach with the tide, with varying rates of percolation and drainage on the beachface from lagged beach water-table responses. With relative sea-level fall, beaches can become isolated above sea level and their sediments weakly cemented, giving rise to fossil RAISED BEACHes.   TS

Reading and Reference
Davidson-Arnott, R. (2010) *Introduction to coastal processes and geomorphology*. Cambridge: Cambridge University Press. · Goudie, A.S. (2001) *The nature of the environment*, 4th edition. Oxford: Blackwell. · Komar, P.D. (1998) *Beach processes and sedimentation*, 2nd edition. Englewood Cliffs, NJ: Prentice-Hall. · Masselink, G., Hughes, M.G. and Knight, J. (2011) *An introduction to coastal processes and geomorphology*, 2nd edition. Abingdon: Hodder Education. · Woodroffe, C.D. (2002) *Coasts: form, process and evolution*. Cambridge: Cambridge University Press.

beach nourishment   The replenishment of sand volume on an eroding beach, which in Europe and North America is now often being preferred to the construction of a physical coastal defence structure. Nourishment does not solve the problem of erosion itself; it is

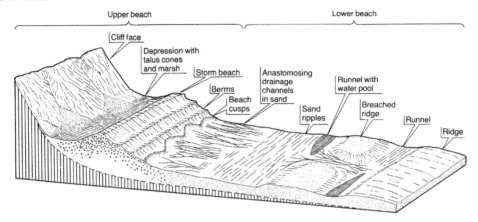

The idealized features of a sand and shingle beach.
Source: Goudie (2001: figure 8.4). Reproduced with permission of Andrew Goudie.

not a one-off solution, but rather leads to a commitment to future renourishment activity. The nourished profile is a disequilibrium profile that is subsequently reworked by shore-normal and shore-parallel wave processes in subsequent months. Strategies to slow sand loss rates include overnourishment and the introduction of coarser-than-natural renourishment sands. A major difficulty with renourishment schemes is finding an environmentally acceptable source of sediment that resembles the sediment lost as closely as possible. Sand can be sourced from offshore, back-barrier locations or by artificially reversing alongshore drift, trucking downdrift deposits to updrift locations. In some cases, non-beach sands have been used for nourishment.   TS

Reading
Hanson, H., Brampton, A., Copbianco, M., *et al.* (2002) Beach nourishment projects, practices, and objectives – a European overview. *Coastal Engineering*, 47, 81–111.

beach ridge Accumulations of sediment forming a prominent feature on many beaches. Beach ridges can take a wide variety of forms. Ridges may be formed near the top of shingle or sandy beaches (in sand these are called BERMS). Ridge features may also be formed in association with runnels near low-tide level on shallow-gradient beaches. Multiple ridge forms often occur when sediment has been left by successive storm events.   HAV

Reading
Davidson-Arnott, R. (2010) *Introduction to coastal processes and geomorphology*. Cambridge: Cambridge University Press. · Komar, P.D. (1976) *Beach processes and sedimentation*. Englewood Cliffs, NJ: Prentice Hall.

beach rock A sedimentary rock, or consolidated chemical sediment, that forms in the intertidal zone on beaches, most notably in the tropics. Beach rock may also develop along extratropical coastlines, such as around the Mediterranean and the Red Sea. It forms where a layer of beach sand and other material becomes consolidated by the secondary deposition of calcium carbonate at about the level of the water table. The cementing material may be aragonite or calcite and may come from groundwater or seawater (Stoddart and Cann, 1965). Biochemical precipitation of calcium carbonate by microorganisms may also be involved. Beach rock often forms relatively quickly, and is forming today in some areas as evidenced by the inclusion of recent human artefacts in the consolidated layer. Once formed, beach rock is relatively persistent, but often contains a suite of erosional features zoned according to their position relative to sea level.   HAV

Reference
Stoddart, D.R. and Cann, J.R. (1965) Nature and origin of beachrock, *Journal of Sedimentary Petrology*, 35, 243–247.

beaded drainage A series of small pools connected by streams. The pools result from the thawing of ground ice and may be 1–3 m deep and up to 30 m in diameter (see THERMOKARST).

bearing capacity The bearing capacity of a soil is the value of the average contact pressure between a foundation and the soil below it that will produce a (shear) failure in the soil. Strictly, this is the 'ultimate' bearing capacity; in soil mechanics, a 'maximum safe' value is used, which is the ultimate value divided by a factor of safety. Various methods are available to enable calculation of appropriate values, according to the soil propert-ies and the likely mode of failure.   wBW

Reading

Smith, G.N. (1974) *Elements of soil mechanics for civil mining engineers*. London: Crosby Lockwood Staples.

Beaufort scale Admiral Sir Francis Beaufort formalized a scale of WIND based on its effect on a man-of-war (Table 1). The scale was later adapted for use on land. It was still internationally useful in 1946, when it was argued that the anemometers in common use were unable to register the shorter term gust capable of destroying a ship.   JSAG

Reading and Reference

List, R.J. (ed.) (1951) *Smithsonian meteorological tables*, 6th revised edition. Washington, DC: Smithsonian Institu-tion; p. 119 (contains full listing of the scale). · Mather, J.R. (1987): Beaufort wind scale. In J. E. Oliver and R. W. Fairbridge (eds), *The encyclopedia of climatology*. New York: Van Nostrand Reinhold.

bed roughness The surface relief at the base of a fluid flow, as on the bed of a river channel. This may consist of several elements: particle roughness, commonly defined with ref-erence to size of larger particles in relation to the depth of fluid flow; form roughness pro-duced by bedforms and the distorting effects of channel bends and 'spill' resistance, as where rapid changes occur where flow spills around protruding boulders. Velocity formulae (see FLOW EQUATIONS), relating flow velocity to the hydraulic radius and slope of river channels for UNIFORM STEADY FLOWS, incorporate a rough-ness coefficient (Manning's n), or related Darcy–Wiesbach friction coefficient or CHÉZY EQUATION. Sometimes, roughness coefficients are estimated simply from the grain size on the bed, but it has to be remembered that this size may be variably effective at different depths and discharges, as may other forms of roughness or resistance.   JL

Table 1 Reasonable current specifications for the Beaufort wind force scale

Beaufort no.	Descriptive term	Windspeed (knots)		Effect on sea surface
		Mean	Limits	
0	Calm	0	<1	Sea like a mirror.
1	Light air	3	1–4	Ripples with the appearance of scales; no foam crests.
2	Light breeze	7	5–8	Small wavelets; crests have glassy appearance and do not break.
3	Gentle breeze	11	9–12	Large wavelets; crests begin to break; perhaps scattered white horses.
4	Moderate breeze	15	13–16	Small waves becoming longer; fairly frequent white horses.
5	Fresh breeze	19	17–21	Moderate waves; many white horses; chance of some spray.
6	Strong breeze	24	22–26	Large waves form; white foam crests extensive; probably some spray.
7	Near gale	29	27–31	Sea heaps up and white foam from breaking waves blown in streaks.
8	Gale	34	32–36	Moderately high waves of great length; edges of crests begin to break into the spin-drift; foam blown in streaks.
9	Strong gale	39	37–42	High waves; dense streaks of foam; crests of waves begin to topple, tumble and roll over.
10	Storm	45	43–48	Very high waves with long overhanging crests; sea surface takes white appearance; visibility affected by spray.
11	Violent storm	52	49–55	Exceptionally high waves; sea completely covered by long white patches of foam; everywhere edges of wave crests blown into froth.
12	Hurricane		>55	Air filled with foam and spray; sea completely white with driving spray; visibility seriously affected.

Source: Mather (1987: table 2, p. 162).

Reading
Charlton, R. (2007) *Fundamentals of fluvial geomorphology.* Abingdon: Routledge. · Graf, W.H. (1971) *Hydraulics of sediment transport.* New York: McGraw-Hill. · Richards, K. (1982) *Rivers: form and process in alluvial channels.* London: Methuen.

bedding plane The interface between two strata of sedimentary rocks; often a plane of weakness between two such strata.

bedform reconstitution The time that it takes for a bedform to move one wavelength in the direction of net transport (Allen, 1974). DSGT

Reference
Allen, J.R.L. (1974) Reaction, relaxation and lag in natural sediment systems: general principles, examples and lessons. *Earth Science Reviews,* **10,** 263–342.

bedforms A generic term for features developed by fluid flow over a deformable bed, as developed by wind on a bed of sand, streamflow over alluvial sediments, or ice over a sediment base. Bedforms may vary in size from small-scale RIPPLES to larger bars, DUNES or DRUMLINS. A hierarchy of forms may be present at any one place superimposed on one another and possibly related to different formative flows. The dimensions of bedforms may relate to flow magnitudes, as is particularly evident when comparing the giant 'ripples' tens of metres in length in gravel-sized material produced by the catastrophic draining of the Pleistocene ice-dammed Lake Missoula in North America with the ripples of dimensions in centimetres developed in sand in many shallow flows. Bedform development and interactions may be studied in the field, experimentally in a FLUME and increasingly through the use of COMPUTATIONAL FLUID DYNAMICS. JL/DSGT

bedload, bedload equation Fluid-transported sediment that moves along or in close proximity to the bed of the flow. This movement of the heavier and larger particles may be by rolling, sliding or saltation (the last being movement in a series of hops resulting from grains impacting on one another). For a given transporting flow the particles that move in this way are ones that are too heavy to be kept suspended in the flow itself since their fall velocity is greater than the upward velocity component of the turbulent fluid.

For rivers, bedload usually amounts to less than 10% of total sediment transport, though higher proportions have been reported for mountain streams. Measurements available, however, are relatively few and possibly unreliable. The problem is that any device inserted in flowing water to trap just that sediment moving at or near the bed itself interferes with the pattern of fluid flow and sediment movement. Techniques include permanent traps or slots in the bed, both ones which are periodically emptied and therefore assess just the total amount of sediment transport in the sampling period, and ones which include a conveyor system or weighing device that samples sediment continuously (e.g. Leopold and Emmett, 1976). Other portable basket-, bag- or pan-type samplers may be lowered onto the bed and then retrieved. These measurements are liable to varying and often unknown error (especially through faulty positioning on the bed) and trapping efficiency, but some long-term series of observations on European and North American rivers are now available. These show that, even for a given discharge, bedload transport rates are unsteady and uneven across streams; it is also well known that bed material (which may include an additional component of suspension load) may accrete and move discontinuously in the form of migratory bars. When bedload movement is in this form, it has been suggested that transport rates may be approximated by the volumetric transfer rate of such BEDFORMS over time.

Bedload movement has also been studied using tracers and the acoustic monitoring of interparticle collisions; a particular aim may be to ascertain the hydraulic conditions at which bed material starts to move. Difficulties in measuring bedload transport, both practical and conceptual, have led to the development of a series of empirically calibrated bedload equations. Given certain information (e.g. stream velocity, discharge or stream power and measures of bed material size and sorting), bedload transport rates may be calculated rather than directly measured. Some equations involve prediction of transport rates in terms of 'excess' shear or power above the threshold value at which transport starts. Some involve computation of total sediment transport rates rather than the bedload alone. From a practical point of view the problem is often that such equations may have to be used in conditions beyond those for which they have been designed and calibrated and, in practice, equations may give very different estimates from one another. Nonetheless, properly used, such estimates have proved very useful as an aid to the design of reservoirs and other engineering works on rivers where bedload movement may be a practical problem. JL

Reading and Reference

Charlton, R. (2007) *Fundamentals of fluvial geomorphology*. Abingdon: Routledge. · Graf, W.H. (1971) *Hydraulics of sediment transport*. New York: McGraw-Hill. · Leopold, L.B. and Emmett, W.W. (1976) Bed load measurements, East Fork River, Wyoming. *Proceedings of the National Academy of Sciences of the United States of America*, **73**, 1000–1004.

bedload pit trap An excavation made in the bed of a stream in order to intercept part or the whole of the bedload delivered to the site in order to be able to measure the size characteristics of the particles being carried and also the rate of sediment transport. The pit needs to be sufficiently long that grains cannot jump across it. Bedload traps may range from passive collecting pits lined with a durable material, and excavated during times of low flow to retrieve and weigh the trapped material, to recording traps that may contain at their base a bladder which develops increasing internal pressure as bedload falls in upon it (Reid *et al.*, 1980). This pressure can then be continuously recorded by running a small pressure tube to a transducer housed on the bank. Generally, pit traps only sample a width of channel of about 1 m, and the data derived from this are scaled up to estimate the sediment load being carried within the whole bank-to-bank width of the stream. DLD

Reference

Reid, I., Layman, J.T. and Frostick, L.E. (1980) The continuous measurement of bedload discharge. *Journal of Hydraulic Research*, **28**, 243–249.

bedload tracers Markings or other devices applied to bedload grains collected from a study stream, commonly pebbles and larger particles, which are then returned to the channel at known positions. Following a sediment transport event, the bed is searched for these marked grains, whose distance of travel can then provide useful information on stream competence and sediment transport mechanics (Laronne and Carson, 1976). Tracers have included durable coloured paints, fluorescent paints searched for under ultraviolet light, and ferrous metal or ceramic magnets attached to the stone, or embedded within it, that are relocated using a metal detector. Some studies have employed many thousands of tracer particles that have been tracked for periods of years. DLD

Reference

Laronne, J.B. and Carson, M.A. (1976) Interrelationships between bed material morphology and bed-material transport for a small, gravel-bed channel. *Sedimentology*, **23**, 67–85.

bedrock The consolidated, unweathered rock exposed at the landsurface or underlying the soil zone and unconsolidated surficial deposits.

Beer's law Establishes the linear relationship between the absorbency of light energy by a gas and the concentration of the gas, for a specific wavelength of energy. The equation is $A(\lambda) = a(\lambda)lc$, where $A(\lambda)$ is absorbency at a specific wavelength, $a(\lambda)$ is the absorptivity coefficient for the gas at the specified wavelength, l is the path length through the gas and c is the concentration of the gas. Beer's law is used by climatologists to estimate the transmittance of solar radiation through gases in the atmosphere for use in energy balance studies and climate modelling. NJS

benchmark A surveyor's mark, usually cut on rock, a kerbstone or other relatively permanent structure, indicating a point of reference for levelling or surveying. Two types of benchmark can be defined; temporary benchmarks are often established when surveying a small area, but the height of these temporary benchmarks relative to the national height datum can only be accurately determined with reference to a national benchmark. DJN

Benioff zone The zone or plane of earthquake foci beneath some continental margins. The inclined plane dips deeper on the island of the continental margin.

benthic The benthic zone is the ecological region at the lowest level of a body of water such as an ocean or a lake, including the sediment surface and some subsurface layers. Organisms living in this zone are called benthos; for example, the benthic invertebrate community, such as crustaceans and polychaetes (unlike PLANKTON, which are suspended in the water column). The organisms generally live in close relationship with the substrate bottom, and many are permanently attached to the bottom. The superficial layer of mobile sediment at the base, the benthic boundary layer, is an integral part of the benthic zone, influencing both physical and biological processes at this interface. TS

berg wind A type of FÖHN wind blowing, mainly in winter, off the interior plateau of South Africa, roughly at right angles to the coast.

Bergeron–Findeisen mechanism See CLOUD MICROPHYSICS.

berghlaup An Icelandic term (literally, rock-leaping) which is sometimes used to refer to a large fallen earth or rock mass. No genetic interpretation is usually intended, and it may be considered as an equivalent of bergsturz. WBW

Bergmann's rule Proposed by the German scientist C. Bergmann in 1847, the 'rule' states that, all other things being equal, an individual member of a species of warm-blooded creatures will have a greater body size in colder climates. Although not all species 'obey' this rule, it does appear to apply fairly widely amongst carnivores that have a wide geographical distribution (Klein, 1984). Where this has been demonstrated through investigation of modern living animals, it has proved a useful palaeoenvironmental tool for studying temperature changes in the late Quaternary, where the remains of a particular species occur through sediments of different ages (e.g. Avery, 1983). The rule has also been considered in studies examining postulated effects of current climate change on animal size (Teplitsky and Millien, 2014). DSGT

References
Avery, D.M. (1983) Palaeoenvironmental implications of the small Quaternary mammals of the fynbos region. In H. J. Deacon, Q. B. Hendry and J. J. N. Lambrechts (eds), *Fynbos palaeoecology: a preliminary synthesis*. SA National Scientific Progress Report no. 75. Pretoria: CSIR; pp. 139–55. · Klein, R.G. (1984) The large mammals of southern Africa. In R. G. Klein (ed.), *Southern African prehistory and palaeoenvironments*. Rotterdam: Balkema; pp. 107–46. · Teplitsky, C. and Millien, V. (2014) Climate warming and Bergmann's rule through time: is there any evidence? *Evolutionary Applications*, 7, 156–168.

bergschrund The crevasse occurring at the head of a cirque or valley glacier because of the movement of the glacier ice away from the rock wall.

berm A ridge of sand parallel to the coastline, commonly found on the landward side of steeply sloping beaches. It is a nearly horizontal feature formed by deposition at the upper limit of the swash zone. When steep beaches are transformed into more shallow gradient beaches, due to a change in wave regime, the berm is removed and a longshore bar deposited just below low-tide level. HV

Bernoulli's theorem and effect These describe the relationship between the pressure in a fluid flow in the direction of flow (dynamic pressure) and at right angles to it (static pressure).

The theorem states that the sum of these two pressures is equal to the local HYDROSTATIC PRESSURE. The dynamic pressure exceeds the hydrostatic pressure by an amount $\rho u^2/2$, where u is the local flow velocity and ρ is the density of the fluid. The static pressure is thus less than the hydrostatic pressure by the same amount. The Bernoulli effect is produced where flow velocities differ laterally. The difference in static pressures tends to push objects towards the streams of more rapid flow; for example, blowing between two oranges which are hanging so that they almost touch, causes them first to swing in to hit each other. MJK

biennial oscillation See MACROMETEOROLOGY.

bifurcation ratio $R_b = \sum n / \sum (n+1)$, the ratio of number of streams of a particular order $\sum n$ to number of streams of the next highest order $\sum (n+1)$. It is therefore dependent upon techniques of stream ordering and gives an expression of the rate at which a stream network bifurcates. It has been correlated with hydrograph parameters and sometimes with sediment delivery factors. In a Strahler fifth-order drainage basin four values of R_b could be calculated, and so a weighted mean bifurcation ratio WR_b was recommended by Schumm (1956):

$$WR_b = \frac{\sum [R_{b_{n:n+1}}(N_n + N_{n+1})]}{N}$$

(See also ORDER, STREAM.) KJG

Reading and Reference
Gregory, K.J. and Walling, D.E. (1976) *Drainage basin form and process*. London: Edward Arnold. · Schumm, S.A. (1956) The evolution of drainage systems and slopes in badlands at Perth Amboy, New Jersey. *Bulletin of the Geological Society of America*, **67**, 597–646.

bioaccumulation The build-up in concentration of a toxic substance in the body of an individual organism over time. Bioaccumulation occurs when an organism is unable to excrete the substance or break it down within the body, so that the substance remains in the body and its concentration increases each time an additional amount of the substance is taken in. Repeated exposure to toxic, conservative pollutants, such as heavy metals and pesticide residues, can result in numerous ill effects, including impaired reproductive capacity and birth defects, and sometimes ultimately in the organism's death. In many cases the effects may depend on the health of the animal. When chlorinated hydrocarbons such as pesticides and polychlorinated biphenyls are bioaccumulated they tend to be concentrated in fatty tissues and released to circulate in the body when fat reserves are used during times of poor feeding.

The diet of animals that feed on bioaccumulators is enriched in these conservative materials, and if, as is often the case, they, too, are unable to excrete them, they in turn acquire an even greater body concentration of the substance, a process known as biomagnification. Confusingly, some authors also use the term bioaccumulation for this process of biomagnification up a food chain.   NJM

Reading

McLachlan, M.S. (1996) Bioaccumulation of hydrophobic chemicals in agricultural food chains. *Environmental Science & Technology*, **30**, 252–259. · Newman, M.C. and Jagoe, R.H. (1996) Bioaccumulation models with time lags: dynamics and stability criteria. *Ecological Modelling*, **84**, 281–286. · Woodwell, G.M., Wurster, C.F. and Isaacson, P.A. (1967) DDT residues in an east coast estuary: a case of biological concentration of a persistent insecticide. *Science*, **156**, 821–824.

biochar   A form of manufactured charcoal produced through the high-temperature pyrolysis of organic materials in the absence of oxygen. Feedstock materials include wood and plant materials such as straw and crop residues. Volatile materials are removed during the pyrolysis, and this results in a very high carbon content in the residual biochar, which may be up to 90% carbon by weight. When added to soils, biochar exhibits resistance to decomposition, and the resulting long residence times provide a means to sequester carbon in soils for centuries to perhaps millennia. There may additionally be benefits for some soil properties significant for cultivation and agricultural production, though there is also the potential for detrimental effects (e.g. the blocking of pore spaces by fine particles of biochar, which would result in reduced aeration and drainage).   DLD

Reading

Verheijen, F., Jeffery, S., Bastos, A.C., *et al.* (2010) Biochar application to soils. A critical scientific review of effects on soil properties, processes and functions. European Commission, Joint Research Centre, JRC Scientific and Technical Reports. EUR 24099 EN. Luxembourg: Office for the Official Publications of the European Communities.

biochemical oxygen demand (BOD)

The amount of oxygen used for the biochemical degradation of organic compounds by a unit volume of water at a given temperature and for a given time. The measure is commonly used as an index of organic pollution because the amount of dissolved oxygen in water has a profound effect on the plants and animals living in it. BOD is measured by storing a test sample in darkness at 20 °C, for 3 days (to give BOD3) or 5 days (to give BOD5), and the amount of oxygen taken up by the microorganisms present is measured in milligrams per litre.

Water samples may also contain substances such as sulphides, which are oxidized by a purely chemical process, and in these cases the oxygen absorbed may form part of the BOD result. For this reason, the BOD test alone is no longer considered an adequate criterion for judging the presence or absence of organic pollutants, although it is still widely used, particularly to assess the polluting capacity of sewage. Other aquatic pollutants, such as pesticides, dissolved salts and heavy metals, cannot be assessed using the BOD test.   NJM

Reading

Clark, R.B. (1997) *Marine pollution*, 4th edition. Oxford: Clarendon Press. · Hoch, B., Berger, B., Kavka, G. and Herndl, G.J. (1995) Remineralization of organic matter and degradation of the organic fraction of suspended solids in the River Danube. *Aquatic Microbial Ecology*, **9**, 279–288.

biocides   See PESTICIDES.

bioclastic   Pertaining to rock composed of fragmented organic remains.

bioclimatology   'Deals with the relationships between climate and living organisms, including humans. It focuses upon the interactions between the biosphere and the Earth's atmosphere over seasons or years (as opposed to biometeorology, which looks at daily and monthly interactions) . . . A development in bioclimate studies is the building of relatively sophisticated bioclimatic models either to predict climatic changes on animals or plants or to infer past climates from the distribution and composition of past faunas and floras' (Huggett, 2010: 18).   ASG

Reference

Huggett, R. (2010) *Physical geography, the key concepts*. London: Routledge.

biodegradation   The decomposition of organic substances by microorganisms. The metabolism of aerobic (oxygen-utilizing) bacteria is primarily responsible for this breakdown.

The death and decay of organic matter is essential for the replenishment of raw materials in the biosphere. Reactions involved in the degradation of organic compounds are frequently those of their synthesis in reverse (Horne, 1978). Catalysts for the degradation reactions are produced by microorganisms, and are often identical or similar enzymes to those employed in synthesis (Bailey *et al.*, 1978). Natural toxins and some complex structures are resistant to breakdown. For example, parts of the intricate chlorophyll molecule can be found fossilized in ancient sedimentary rocks. Also, organic compounds synthesized by industrial chemistry (e.g. plastics and chlorinated hydrocarbons) are not readily biodegradable and may persist in the environment for a considerable time (see BIOLOGICAL MAGNIFICATION). The presence of biodegradable material in bodies of water can lead to a significant reduction in the dissolved oxygen content of the water, as oxygen is needed for microorganic activity to take place (BIOCHEMICAL OXYGEN DEMAND (BOD)).   RLJ

References

Bailey, R.A., Clark, H.M., Ferris, J.P., *et al.* (1978) *Chemistry of the environment*. New York: Academic Press. · Horne, R.A. (1978) *The chemistry of our environment*. New York: John Wiley & Sons, Inc.

biodiversity

The concept of biological diversity, or biodiversity, is an attempt to capture the extraordinary complexity of life on Earth. This has become a major global environmental issue because environments are being degraded at an accelerating rate and much diversity is being irreversibly lost through exploitation and the destruction of natural habitats. There are numerous definitions of biodiversity, but the most widely adopted is that developed by the 1992 United Nations Conference on Environment and Development in Rio de Janeiro, Brazil, at which the Biodiversity Convention was established. The definition reads as follows: *Biological diversity means the variability among living organisms from all sources, including terrestrial, marine and other aquatic ecosystems and the ecological complexes of which they are part; this includes diversity within species, between species and ecosystems.* In other words, biodiversity is the variety of life in all its manifestations and at all spatial scales. Although the concept embodies the full range of variability from the genetic to the ecosystem, the most commonly employed measure of biodiversity is species diversity; that is, the number of distinct biological species. Although this is a simplification of the concept it is a useful and commonly quick measure of how biologically diverse is a particular locality or region and it enables comparative analyses across geographical space and through time. For some types of organisms (e.g. terrestrial plants, mammals or birds) the patterns of species diversity are relatively well known, but it is still the case that our knowledge of total global biodiversity is very incomplete. The numbers of species formally recorded is somewhat more than 1.6 million, but the actual number will probably never be known. More than half the world's species are found within the tropical rainforest regions, many of which are remote and physically or practically inaccessible. Estimates of global total species numbers vary from somewhere between 3 million and 50 million, but the actual number will probably never be known – and certainly losses of species through extinction under the combined onslaughts of land use change for agriculture, urban development and industry, invasive alien species and anthropogenic climate change have accelerated.

Biodiversity is unevenly distributed, although there are several clear spatial patterns; for example, the latitudinal biodiversity gradient by which the diversity of most types of organisms is greatest in the tropics. Elevated species numbers within the humid tropics is a persistent and recurring pattern, and yet the factors underlying the evolution of such patterns are imperfectly understood. Some regions well beyond the tropics are also exceptionally rich, including several of the so-called Mediterranean-climate regions of the world, such as southwest Australia and the southwestern Cape region of South Africa. These form some of the 25 or so global biodiversity 'hot spots', a term coined by Norman Myers to draw attention to the need to focus global conservation efforts on the most biologically diverse regions of the world.   MEM

Reading

Gaston, K.J. and Spicer, J.I. (2004) *Biodiversity: an introduction*, 2nd edition. Oxford: Blackwell. · Myers, N., Mittermeier, R.A., Mittermeier, C.G., *et al.* (2000) Biodiversity hotspots for conservation priorities. *Nature*, **403**, 853–858.

bioerosion

Destruction and removal of consolidated mineral or lithic substrate by the direct action of organisms. It is often discussed in the context of the erosion of hard marine substrates, and particularly associated with coral reefs. Marine bioerosion includes both the removal of surface substrate by grazers (where the agents can be echinoids, gastropod molluscs and chitons, and parrotfishes) and the loss of substrate by boring organisms. The latter can be macroborers (such as polychaetes, bivalve molluscs, sponges) or microborers (including cyanobacteria, algae and fungi). Coral reef bioerosion not only generates considerable volumes of lagoonal and inter-reefal sediments, but also results in the generation of complex microhabitats which are colonized by a crypotofauna, critial for reef biodiversity and productivity. In terrestrial environments, bioerosion is typically performed by pioneer plants or plant-like organisms such as lichen, and is mostly chemical (e.g. by acidic secretions on limestone) or mechanical (e.g. by roots growing into cracks) in nature.   TS

Reading
Hutchings, P. (2011) Bioerosion. In: D. Hopley (ed.), *Encyclopedia of modern coral reefs; structure, form and process.* Dordrecht: Springer; pp. 139–156.

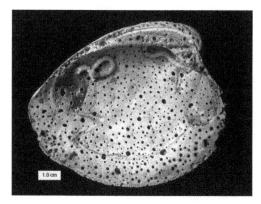

Clionid sponge borings, on a shell of the modern hard clam (*Mercenaria mercenaria*) from North Carolina, following the death of the host. Photograph by M. A. Wilson.

biofilm/extracellular polymeric substances (EPSs)
A biofilm is any group of microorganisms in which cells stick to each other on a living or nonliving surface. Much of the early research in this field concentrated upon the role of microalgae (microphytobenthos) and the embedding of these life forms within a self-produced matrix of EPS constituting 50–90% of a biofilm's total organic matter. EPSs are mostly composed of polysaccharides (exopolysaccharides) and proteins, but include other macromolecules such as DNA, lipids and humic substances. It was predicted that biostabilization would be more influential in marine habitats due to the high ion concentrations that help in the binding process, with research emphasizing the links between photosynthetic activity, migration and EPS production of microalgae during tidal cycles. More recently, the stabilizing role of heterotrophic bacteria in fine sediments has been recognized and the presence of high EPS concentrations in freshwater systems detected. Biofilms change the erosive response of fine sediments to hydrodynamic forcing. They provide important ecosystem services, being the base of the food chain, transferring carbon to higher trophic levels, controlling self-purification of organic compounds and biodegrading anthropogenic pollutants. See also BIOSTABILIZATION.   TS

Reading and Reference
Gerbersdorf, S.U. and Wieprecht, S. (2015) Biostabilization of cohesive sediments: revisiting the role of abiotic conditions, physiology and diversity of microbes, polymeric secretion, and biofilm architecture. *Geobiology*, 13, 68–97. · Le Hir, P., Monbet, Y. and Orvain, F. (2007) Sediment erodability in sediment transport modelling: can we account for biota effects? *Continental Shelf Research*, 27, 1116–1142. · Paterson, D.M. and Black, K.S. (1999) Water flow, sediment dynamics and benthic biology. *Advances in Ecological Research*, 29, 155–193. · Underwood, G.J.C. and Paterson, D.M. (2003) The importance of extracellular carbohydrate production by marine epipelic diatoms. *Advances in Botanical Research*, 40, 183–240.

biofuel
Liquid fuels, including ethanol and biodiesel, manufactured primarily from biological materials such as crops (e.g. corn, soybeans) grown for conversion to fuels, or from crop residues, or from other organic wastes. Though they are often seen as a more sustainable alternative to fossil fuels, the production of biofuels from crops and crop residues raises significant environmental issues. Removal of residues such as corn stover may expose soils to an increased risk of erosion, with the potential for off-site water quality and river health impacts. The removal of crops or residues additionally reduces the return of carbon to the soil. Depletion of this carbon store has the potential to offset some of the environmental gains sought through the use of biofuels. Thus, benefits such as fuel security have to be weighed against potential environmental impacts.   DLD

Reading
Dale, V.H., Kline, K.L., Wiens, J. and Fargione, J. (2010) *Biofuels: implications for land use and biodiversity.* Biofuel and Sustainability Reports, January). Ecological Society of America. http://www.esa.org/biofuelsreports (accessed 22 June 2015).

biogeochemical cycles
The cycling, at various scales, of minerals and compounds through the ecosystem. The cycles (see CARBON CYCLE and NITROGEN CYCLE) involve phases of weathering of inorganic material, uptake and storage by organisms, and return to the pool of the soil, the atmosphere or ocean sediments. An increasing amount of research has been focused on the working out of the details of such cycles during the past decade as a result of concern over global environmental change and nutrient budgets. Classic studies such as that of the Hubbard Brook Experimental Forest by Likens and associates are often cited (Begon *et al.*, 1990: 688–691) as examples of

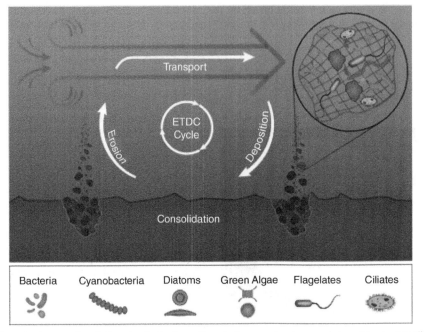

Biofilm formation impacts the whole cycle of cohesive sediments, from erosion, transport, deposition to consolidation. *Source:* Gerbersdorf and Wieprecht (2015). Reproduced with permission of John Wiley & Sons.

well-documented small-scale studies. Diagrammatic representations of major element cycles are common in textbooks, but there are fewer *quantified* studies on a global scale (Mannion, 1991). The biogeochemistry of carbon has attracted particular attention because of the concern over GLOBAL WARMING and the GREENHOUSE EFFECT.   KEB

Reading and References
Begon, M., Harper, J.L. and Townsend, C.R. (2005) *Ecology*, 4th edition. Oxford: Wiley-Blackwell. · Degens, E.T., Kempe, S. and Richey, J.E. (eds) (1991) *Biogeochemistry of major world rivers*. Chichester: John Wiley & Sons, Ltd. · Furley, P. and Newey, W. (1983) *Geography of the biosphere*. London: Butterworths. · Mannion, A.M. (1991) *Global environmental change*. Harlow: Longman Scientific and Technical.

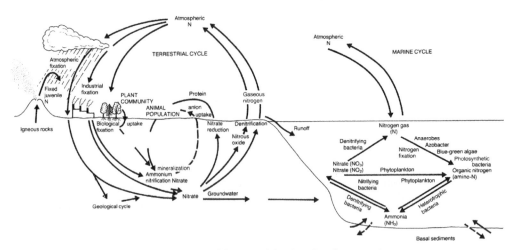

Biogeochemical cycles. Generalized diagrams of the terrestrial and marine nitrogen cycles. *Source*: Furley and Newey (1983: figure 3.7). Reproduced with permission of Peter Furley.

biogeocoenosis (biogeocenose) A combination on a specific area of the Earth's surface of a particular biocoenosis (biotic community) and its ECOTOPE (physical environment); for example, a forest, a peat bog or an oyster bank. Introduced by V. N. Sukachev in 1944, the term was widely used in the former USSR as an equivalent to the western term ECOSYSTEM; a biogeocoenosis type is equivalent to an abstract ecosystem type. A biogeocoenosis is generally considered to possess a degree of homogeneity in its structure and a certain coherence in its functioning. JAM

Reading
Sukachev, V.N. and Dylis, N. (1964) *Fundamentals of forest biogeocoenology*. Edinburgh: Oliver & Boyd. · Troll, C. (1971) Landscape ecology (geoecology) and biogeocenology – a terminological study. *Geoforum*, **8**, 43–46.

biogeography The science of biological distribution patterns, and a discipline examining the characteristic spatial and temporal occurrence of the Earth's organisms. As the name suggests, there are two components to this form of natural science. The biological component entails that its objects of study are biological entities, from species up through higher orders of taxonomic classification. The geographical component embodies the identification of distributional patterns and the search for an explanation as to the factors that underlie them. Depending on the level of taxonomy in question and the spatial scale of interest, there may be a strong degree of overlap between biogeography and ecology, since ecologists are also interested in organism distributions. The interdisciplinary science of biogeography has several elements that tend to distinguish it from ecology, however, especially its focus on generally larger scale spatial distributions (commonly regional, continental or even global) and also its frequent concern with higher taxonomic orders such as families. A further distinguishing factor is the biogeographical interest in evolutionary development over time and the dynamics of distribution patterns. Biogeographers also focus on the nature of the relationship between humans and organisms, although such an approach is not their sole preserve.

Biogeography has a distinguished history, although its identity as a named science is more recent, so that some of its earlier and more famous proponents would not have described themselves as biogeographers per se. Perhaps the first biogeographical studies are the inventories of plant species generated for particular geographical areas (one of the oldest emanates from fourth-century China). The German polymath Alexander von Humboldt is arguably the first biogeographer as, in his seventeenth-century explorations of South America, he collated enormous numbers of plant specimens together with observations on their distribution patterns. The realization soon followed that species maintained characteristic distributions constrained by their adaptations to the physical environmental characteristics (especially climate) in their respective ranges. Both Charles Darwin and Alfred Russell Wallace developed a keen interest in the distribution patterns of organisms and their adaptational relationships to the environment. Such information was the key to the establishment of the notion of natural selection, although it was the German A. W. F. Schimper who formalized ideas on the nature of vegetation–climate interactions in his 1903 *Plant geography on a physiological basis*. The twentieth-century history of biogeography reflects several diverse and distinctive developments in the discipline and is influenced by advances made in associated natural sciences such as botany, ecology and even geology.

Several distinctive forms of biogeography may be recognized in the contemporary literature. Prominent among them is a focus on distribution patterns, often of plant taxa (phytogeography), and their evolution over geological time. Few would argue that, ecologically, the rainforests of Africa and South America are functionally similar, but the taxonomic identities of the two regions are markedly distinct so that taxa in similar habitats are different because they have different histories. The subdiscipline of vicariance biogeography arose out of the recognition that history, by way of continental drift and other large-scale geological and climatic changes, has played the dominant role in determining taxonomic distribution patterns at the global scale. Cladistic biogeography goes one stage further in recognizing that the precise evolutionary relationships between taxa are a direct function of the particular sequence of events in their respective geographical areas of distribution; that is, the relationships between geographical areas can be reconstructed from knowledge of the taxa occurring there.

Ecological ideas, in particular the development of the ecosystem concept, have proved central, especially in the contemporary geographers' (as opposed to biologists') approach to biogeography. This spawned the consideration of major vegetation formations, or biomes, which could be seen as an expression of the environmental relationships at the global scale. Another expression of ecological ideas, in this case the branch of population ecology, arose out of

MacArthur and Wilson's (1967) ideas on island biogeography, which viewed species complements on islands, microcosms of larger scale ecosystems, as the product of an equilibrium between colonizing species and those becoming extinct. The potential for this idea to help resolve the growing conservation crisis was soon realized and, although direct application of the equilibrium approach has proved fruitless, it encouraged critical thinking as to the mechanisms underlying accelerated human-induced extinction. Indeed, biogeography exhibits a strong degree of involvement in that area of science dealing more broadly with human–environment relationships. Biogeographers, particularly those with training as geographers, have also made significant contributions in the field of Quaternary palaeoecology, where a dynamic approach to the rather more static large-scale biome studies has been fostered. MEM

Readings and Reference
Brown, J.H., Riddle, B.R. and Lomolino, M.V. (2010) *Biogeography*, 2nd edition. New York: Sinauer. · Cox, C.B. and Moore, P.D. (2010) *Biogeography: an ecological and evolutionary approach*, 8th edition. Oxford: Blackwell. · MacArthur, R.H. and Wilson, E.O. (1967) *The theory of island biogeography*. Princeton, NJ: Princeton University Press. · Whittaker R.J. and Fernandez-Palacios, J.M. (2007) *Island biogeography: ecology, evolution, and conservation*. Oxford: Oxford University Press.

biogeomorphology Encapsulates concisely a developing and previously much neglected approach to geomorphology that explicitly considers the role of organisms. Two main foci are (*a*) the influence of landforms/geomorphology on the distribution and development of plants, animals and microorganisms; and (*b*) the influence of plants, animals and microorganisms on Earth surface processes and the development of landforms. ASG

Reading
Viles, H.A. (ed.) (1988) *Biogeomorphology*. Oxford: Blackwell.

bioherm *a.* An ancient coral reef.
b. An organism that plays a role in reef formation.

biokarst A KARST landform, usually small in scale, produced mainly by organic action. Strictly speaking, the term PHYTOKARST should be restricted to phenomena produced by plants alone. Biokarst features can either be erosional (as where organisms bore into or abrade carbonate rock surfaces) or constructional (as in the case of certain tufas and reef forms). ASG

Reading
De Waele, J. and Furlani, S. (2013) Seawater and biokarst effects on coastal limestones. *Treatise on Geomorphology*, **6**, 341–350. · Viles, H.A. (1984) Biokarst: review and prospect. *Progress in Physical Geography*, **8**, 523–543.

bioleaching The process of using microorganisms to liberate metals from mineral ores. Direct leaching is when microbial metabolism changes the redox state of the metal being harvested, rendering it more soluble. Indirect leaching includes redox chemistry of other metal cations that are then coupled in chemical oxidation or reduction of the harvested metal ion and microbial attack upon and solubilization of the metal matrix in which the metal is physically embedded. Microbial mining of copper sulphide ores has been practised on an industrial scale since the late 1950s, and since then bioleaching has also become common at uranium and gold mines. Bioleaching technologies have been developed by the mining industry in response to the lower grades of accessible ores and lower operating costs. Microbial mining saves energy, creates minimal pollution and recovers resources from tailings that would otherwise be wasted. An allied development in mining biotechnology is the use of bacterial cells to detoxify waste cyanide solution from gold mining operations (Agate, 1996). Bioleaching is also used to extract metals from scrap products (Pant, 2014). NJM

References
Agate, A.D. (1996) Recent advances in microbial mining. *World Journal of Microbiology and Biotechnology*, **12**, 487–495. · Pant, D. (2014) A review of electronic waste management microbial participation: a green technology. *International Journal of Environment and Waste Management*, **13**, 23–36.

biological control (biocontrol) The control of pestilential organisms such as insects and fungi through biological means rather than the application of man-made chemicals. This can include breeding resistant crop strains, inducing fertility in the pest species, disruption of breeding patterns through the release of sterilized animals or spraying juvenile hormones to interrupt life cycles, breeding viruses that attack the pests, or the introduction or encouragement of natural or exotic predators to control pest outbreaks. ASG

biological crust See CRUST.

biological magnification The increased concentration of toxic material at consecutive

TROPHIC LEVELS in an ecosystem. Toxins, such as persistent pesticides and heavy metals, are incorporated into living tissue from the physical environment. Physical, chemical and biological processes operate to amplify the harmful substances in food chains by concentrating the quantities in individual organisms. The effect of these substances on organisms varies. In general, reproductive capacity is impaired at low concentrations and death occurs at high concentrations, sometimes via disease. Approximately 10% of food at one trophic level is transferred to the next, the remainder being removed by respiratory and executive activity; but as toxic materials are not so readily broken down as other components of organic tissue (BIODEGRADATION), their transfer efficiency is higher. Thus, a build-up of them occurs at successive trophic levels (Woodwell, 1967).

DDT (PESTICIDES) residues are an example of accumulation in the physical environment. They have a low biodegradability and are excreted slowly from organisms because they become dissolved in fatty tissues. Woodwell (1967) reports (in a study undertaken with Wurster and Isaacson) DDT concentrations of up to $36\,kg\,ha^{-1}$ after two decades of application of the pesticide to a New York marsh ecosystem. Marsh plankton contained 0.04 ppm, minnows 1 ppm and a carnivorous gull 75 ppm of DDT in their tissues.   RLJ

Reading and Reference

Jorgensen, S.E. and Johnsen, I. (1981) *Principles of environmental science and technology*. Amsterdam: Elsevier. · Odum, E.P. (1975) *Ecology: the link between the natural and social sciences*, 2nd edition. London: Holt, Reinhart & Winston. · Woodwell, G.M. (1967) Toxic substances and ecological cycles. *Scientific American*, **216**, 24–31.

biological productivity The rate at which organic matter accumulates over time within any given area. The term is usually applied at the ECOSYSTEM level, but may also be used with reference to an individual organism or population to account for growth, or the increase in BIOMASS over time. Biological productivity is a composite term and is used to refer generally to productivity that may be considered separately as NET PRIMARY PRODUCTIVITY (NPP), gross primary productivity and secondary productivity (of consumers), all of which involve the transfer of energy within an ecosystem. Biological productivity is measured in units of weight accumulating per unit area over time; for example, as grams per square metre per day or tonnes per

hectare per year. There are defined relationships between gross primary productivity (GPP) and NPP in an ecosystem. GPP represents the total rate at which green plants convert solar radiation into carbohydrates via the process of PHOTOSYNTHESIS. NPP is the net rate of organic matter accumulating after allowance is made for the fact that the green plants themselves need to utilize some of the assimilated energy in order to exist and that some energy is lost to the system by the death or herbivory of the photosynthesizing plants. Human populations are dependent for their existence on biological productivity, albeit often in an artificial and manipulated form as agricultural production. Humans, as consumers, rely on both NPP of agricultural crops and the secondary productivity of herbivores.   MEM

bioluminescence The light produced by some living organisms or the process of producing that light, characteristic of glow-worms and some marine fish.

biomantle A layer or zone in the upper part of a soil that has been produced primarily by BIOTURBATION. Where present, biomantles generally extend from the soil surface to the depth reached by burrowing invertebrates such as worms, ants and spiders, or small vertebrates (moles, gophers, etc.). Over extended periods of time, the excavation of fresh burrows by organisms can progressively affect all of the upper part of the soil column. This may do one of two things: either, it will overturn and homogenize the layer that is being worked, and so produce a distinctive soil fabric, or, if there are stones in the soil too large to be moved, these will gradually accumulate as a layer at the base of the worked layer. The texture of the biomantle layer can be affected as well as its structure. If the burrowing organisms build mounds of excavated spoils at the surface, then once again through extended periods of time all transportable materials in the worked layer will be brought to the surface before they are again mixed downward. While at the surface, the spoils are exposed to erosional processes (both wind and water), which may progressively winnow out finer, more readily moved grains. Thus, continued bioturbation can lead to a textural coarsening of the biomantle layer of the soil. Such processes can be triggered in other ways, as when tree-throw brings up subsoil attached to the roots and exposes this to erosional sorting. Progressively, this, too, could result in the development of a distinctive bio-mantle whose depth would correspond to the rooting depth of the trees. DLD

Reading
Johnson, D.L. (1990) Biomantle evolution and the redistribution of earth materials and artifacts. *Soil Science,* **149**, 84–102. · Nooren, C.A.M., van Breemen, N., Stoorvogel, J.J. and Jongmans, A.G. (1995) The role of earthworms in the formation of sandy surface soils in a tropical forest in Ivory Coast. *Geoderma,* **65**, 135–148.

biomass The total mass of biological material in any given area and is a result of NET PRIMARY PRODUCTIVITY (NPP). Total global biomass is difficult to estimate because of the high degree of spatial variation within and between BIOMES. Distribution of biomass between above- and below-ground components is a further problem for accurate estimation, as is the fact that forest vegetation types have complex three-dimensional structures. Earlier estimates (e.g. see that of Whittaker and Likens (1975)) placed the total continental phytomass (plant biomass) at 1837×10^9 t (dry weight) and that of the oceans and estuaries at only 3.9×10^9 t (dry weight). On a global scale, therefore, terrestrial biomass far exceeds that of the oceans. On the other hand, the balance between terrestrial and oceanic animal biomass, which is at much lower levels in general, is much more even. Developments in remote sensing have, more recently, facilitated more accurate assessments of global biomass and its distribution (see Table 2).

On land, more than 80% of the biomass is located in the world's forests, and more than half of this is in the tropical forests. Note that, while crop biomass is a very small proportion (<1%) of the world's total standing biomass, croplands represent a somewhat higher (~7%) proportion of NET PRIMARY PRODUCTIVITY (NPP) due to interventions such as irrigation, inputs of fertilizer, and so on. DW/MEM

References
Schlesinger, W.H. and Bernhardt, E.S. (2010) *Biogeochemistry: an analysis of global change.* Oxford: Academic Press. · Whittaker, R.H. and Likens, G.E. (1975) Net primary production and plant biomass for the earth. In H. Reith and R. H. Whittaker (eds), *The primary production of the biosphere.* New York: Springer-Verlag; pp. 305–328.

biome A mixed community of plants and animals (a biotic community) occupying a major geographical area on a continental scale, otherwise referred to as an ecoregion. Most commonly applied to terrestrial environments, each biome is characterized by characteristic similarity of vegetation structure or physiognomy rather than by species composition, and is strongly related to prevailing climate. Globally, the major biomes include tundra, taiga (boreal forest), temperate deciduous forest, temperate grassland, the so-called Mediterranean-type biome, desert biome (both hot and cold desert types), savanna and tropical rain forest. Needless to say, large areas of many biomes have undergone transformation by human societies, and small-scale maps that appear in texts and atlases may be misleading (Holzner *et al.*, 1983). JAM/MEM

Table 2 Biomass and net primary production in terrestrial ecosystems – adapted by Schlesinger and Bernhardt (2010) from other sources

Biome	Area (10^6 km²)	NPP (g C m⁻² a⁻²)	Total NPP (10^{15} g C a⁻¹)	Biomass (g C m⁻²)	Total plant C pool (10^{15} g C)
Tropical forests	17.5	1250	20.6	19,400	320
Temperate forests	10.4	775	7.6	13,350	130
Boreal forests	13.7	190	2.4	4150	54
Mediterranean shrublands	2.8	500	1.3	6000	16
Tropical savannas/ grasslands	27.6	540	14.0	2850	74
Temperate grasslands	15.0	375	5.3	375	6
Deserts	27.7	125	3.3	350	9
Arctic tundra	5.6	90	0.5	325	2
Crops	13.5	305	3.9	305	4
Ice	15.5				
Total	**149.3**		**58.9**		**615**

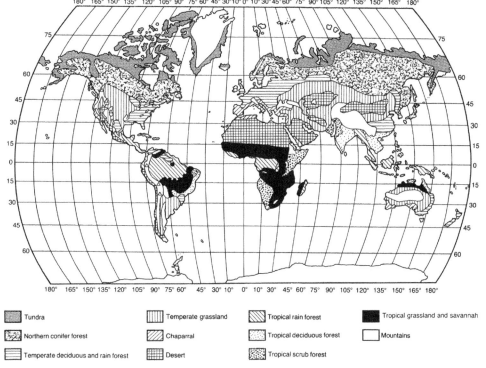

Tundra	Temperate grassland	Tropical rain forest	Tropical grassland and savannah
Northern conifer forest	Chaparral	Tropical deciduous forest	Mountains
Temperate deciduous and rain forest	Desert	Tropical scrub forest	

Major biomes of the world as if unaffected by human activity.
Source: Simmons (1979: figure 3.3).

Readings and Reference

Furley, P.A. and Newey, W.W. (1983) *Geography of the biosphere*. London: Butterworths. · Holzner, W., Ikusima, I. and Werger, M.J.A. (eds) (1983) *Man's impact on vegetation*. The Hague: W. Junk. · Lomolino, M.V., Riddle, B.R., Whittaker, R.J. and Brown, J.R. (2010) *Biogeography*, 4th edition. Sunderland, MA: Sinauer. · Simmons, I. (1979) *Biogeography, natural and cultural*. London: Edward Arnold. · Walter, H. (1979) *Vegetation of the Earth*, 2nd edition. Berlin: Springer-Verlag.

biometeorology

The study of the effects of weather and climate on plants, animals and humans. The International Society of Biometeorology, founded in 1956, has classified the subject into six main groups: phytological, zoological, human, cosmic, space and palaeo. Human biometeorology includes the study of the influence of weather and climate on healthy humans and on their diseases and the effect of microclimates in houses and cities on health. Although Hippocrates discussed some of these topics over 2000 years ago, it is only in the second half of the twentieth century that the main developments in this science have taken place. In some countries the old name of bioclimatology is still used.   DGT

Reading

Tromp, S.W. (1980) *Biometeorology*. London: Heyden.

biosphere

The zone, incorporating elements of the hydrosphere, lithosphere and atmosphere, in which life occurs on Earth. The term is occasionally used to refer only to the living component alone, although it is more commonly conceived as a zone of interaction between the other 'spheres'. This is appropriate because life is dependent upon energy, processes and materials that are located in all three of the Earth's other conceptual spheres, to the extent that the scheme is often represented as a series of overlapping circles with the biosphere in the nodal position. Used in this way, biosphere is synonymous with ecosphere. Implicit in the concept is the fact that life within the biosphere is organized hierarchically. The hierarchy has a fundamentally cellular basis and is structured through progressively larger and more complex agglomerations of organs, organisms and populations. By incorporating elements of the physical environment also, the higher levels of organization in the biosphere are represented by its ECOSYSTEMS and BIOMES.

In proportion to the dimensions of the Earth and its atmosphere as a whole, the biosphere occupies a remarkably thin and, arguably, fragile layer. It extends a maximum of a few hundred metres above the land surface and penetrates relatively little into the crust, although recent research suggests that the microbial biomass may be considerably larger (see ANAEROBIC bacteria). Certainly, the most conspicuous organisms are found within a few metres above and below the Earth's surface. Approximately two-thirds of the biosphere is characterized by water, as oceans.

Crucial in the evolution of the biosphere has been the emergence of photosynthesizing green plants and the consequent modification of the atmosphere that appears to have facilitated the familiar ecosystem processes of the contemporary Earth. Indeed, the prominence of atmospheric oxygen in its role in ecological processes is a direct result of the evolutionary development of photosynthesizing AUTOTROPHIC organisms. In essence, the Earth's contemporary atmosphere is a product of the biosphere.

Two fundamental kinds of processes characterize the biosphere: ENERGY FLOW and nutrient recycling; both involve functional linkages and exchanges between the various spheres in the Earth system. Most of the energy that activates the biosphere is derived from solar radiation and becomes ecologically important via its assimilation through photosynthesis. Subsequently, this energy is utilized by organisms in successive TROPHIC LEVELS as part of the FOOD CHAIN. Progressively smaller absolute quantities of energy are available along the length of the food chain so that consumer biomass (e.g. herbivores and primary and secondary carnivores) is less than for the autotrophs. The incoming energy driving the entire system is eventually dissipated as respirational or heat loss but is, of course, kept topped up by the constancy of solar inputs. Nutrients within the biosphere, however, are not renewed in the same way and, instead, are recycled through the active components of the Earth system over time. Hydrogen, carbon, oxygen and nitrogen, together with some 40 or so other essential mineral elements, recycle within the biosphere and maintain its integrity. Carbon, for instance, recycles through the atmosphere, biosphere, lithosphere and hydrosphere and, in an ideal, efficient biogeochemical cycle is consistently replenished so that circulating quantities remain stable over time (see BIOGEOCHEMICAL CYCLES). Owing to excessive human exploitation of resources in the biosphere, however, imbalances in some of the nutrient cycles are becoming increasingly apparent and express themselves, for example, as environmental problems such as EUTROPHICATION and GLOBAL WARMING.

Arising out of the biosphere concept, in which the Earth may be envisaged as a single, self-contained ecosystem, is the idea that actions on any scale anywhere on Earth may have a cumulative, by implication negative, effect on the whole system. James Lovelock in 1974 took this idea to what he considered to be its logical conclusion; that is, that the Earth is in effect a superorganism and that it strives to maintain a regulated atmospheric composition and temperature condition. The so-called GAIA HYPOTHESIS is controversial but has served to focus attention on the nature of the biosphere and the human interaction with it.   MEM

Reading
Bradbury, I.K. (1998) *The biosphere*, 2nd edition. Chichester: John Wiley & Sons, Ltd.

biostabilization fixation of sedimentary grains by biofilms and microbial mats. Biostabilization increases sediment stability or reduces the potential for erosion by tidal currents, waves and river flows by enhanced cohesion, binding by filaments/roots, surface protection/armouring, flow and wave attenuation by biota. Microbial activity 'glues' fine sediment particles together to impact sediment stability, floc characteristics of the entrained material, and thus, further sediment transport and deposition. During consolidation processes, the biological matrix undergoes chemical changes that further enhance the binding forces. See also BIOFILM/EXTRACELLULAR POLYMERIC SUBSTANCES (EPSS).   TS

Reading
Gerbersdorf, S.U. and Wieprecht, S. (2015) Biostabilization of cohesive sediments: revisiting the role of abiotic conditions, physiology and diversity of microbes, polymeric secretion, and biofilm architecture. *Geobiology*, **13**, 68–97. · Stal, L.J. (2010) Microphytobenthos as a biogeomorphological force in intertidal sediment stabilization. *Ecological Engineering*, **36**, 236–245.

biostasy A term that was applied by Erhart (1956) to periods of soil formation, with rhexistasy referring to phases of denudation. In periods of biostasy there is normal vegetation, while in phases of rhexistasy there is dying out or lack of vegetation as a result of soil erosion resulting from climatic changes, tectonic displacement, and so on. The period of rhexistasy is characterized by mechanical reworking, whereas biostasy is characterized by chemical decomposition.   ASG

Reference

Erhart, H. (1956) *La genèse des sols en tant que phénomène géologique*. Paris: Masson

biota   The entire complement of species of organisms, plants and animals found within a given region.

biotope   The HABITAT of a biocoenosis, or a microhabitat within a biocoenosis. In the first sense the word is synonymous with ECOTOPE, the effective physical environment of a biocoenosis or biotic community. In the second sense it refers to a small, relatively uniform habitat within the more complex community; for example, although a forest community occupies its own habitat, each layer or stratum within the forest may be regarded as a separate biotope. (See also NICHE.)   JAM

Reading

Daubenmire, R. (1968) *Plant communities: a textbook of plant synecology*. New York: Harper & Row. · Hirt, U., Mewes, M. and Meyer, B.C. (2011) A new approach to comprehensive quantification of linear landscape elements using biotope types on a regional scale. *Physics and Chemistry of the Earth*, **36**, 579–590.

bioturbation   The mixing of soil or surficial layers of terrestrial, marine or lacustrine sediments by the physical action of organisms. A familiar example would be the mulching of topsoil by the action of earthworms, termites or moles, which activity results in the organic and clastic soil particles being selectively reworked and sorted. Ploughing and mulching are forms of intentional human-induced bioturbation in soils aimed at mixing the topsoil and optimizing drainage and nutrient conditions. In the case of marine or lacustrine sediments, BENTHIC micro- and macrofauna, through their activities at the sediment surface and near surface, may overturn the accumulating deposits such that apparent inconsistencies in chronological order of sedimentation may result. This is potentially problematic if the sediments are the object of palaeolimnological or palaeoceanographic study, for the microfossils and other sediment characteristics of interest may become displaced. Such a situation is most serious under circumstances of very slow rates of sediment accumulation, where physical disturbance over a few centimetres can re-sort materials deposited over several hundreds or thousands of years, hence impacting on the level of temporal resolution possible in the palaeoenvironmental reconstruction. It is for this reason that deep ocean sediments are unable to resolve environmental events of less than 1000 years duration.   MEM

black box   The term given to any model or treatment of a system (see SYSTEMS) where the internal components of the system are ignored or excluded from direct treatment, by only considering the relationship between inputs to the system and outputs from the system. A good example might be a statistical treatment of the relationship between rainfall and runoff, which does not consider the operations of processes within the system (e.g. soil infiltration, initial soil moisture conditions, interception of rainfall by vegetation and subsequent evaporation). In practice, most models have an element of blackbox treatment, as there are normally always some processes operating at scales smaller than the model. These may be dealt with using statistical treatments that may have only a poor resemblance to the actual way in which they work.   SNL

blanket bog   A type of bog, often composed of peat, that drapes upland terrain and infills hollows in areas of high precipitation and low evapotranspiration.

blind valley   A steep-sided, river-cut valley that terminates abruptly in a precipitous cliff. Although blind valleys can be found in any terrain near the source of a river, they are usually produced when a river flows onto limestone bedrock and sinks into subterranean passages, enabling downcutting to occur in the active river valley and the progressive development of a steep cliff where the valley ends.   PAB

blizzard   A snow storm, either of falling snow or deflated snow, usually accompanied by low temperature and high winds.

block faulting   The process whereby large regions are tectonically disrupted to form complex systems of troughs and ridges or basins and block mountains. The result of tectonic uplift and subsidence of adjacent blocks of the Earth's crust following faulting and fracturing on a grid pattern.

block fields, block streams   Spreads or lines of boulders, generally angular, formed by in-situ shattering by frost of a bedrock surface. They may surround features such as tors and nunataks and other landforms subjected to severe periglacial processes.

blocking   An extreme state in which the tropospheric circulation takes the form of large-amplitude stationary waves. Such flow is frequently manifest in stationary anomalies in the weather that may have significant economic repercussions, such as the 1976 drought in the UK. Prediction of the initiation and persistence of blocks is difficult. Hydrology, transfer of energy by both solar and terrestrial radiation, interaction of synoptic-scale weather systems and the larger scale mean flow, and resonance of stationary waves are all possible relevant factors. JSAG

blow-hole Vertical shaft leading from a sea cave to the surface. Air and water may be forced through the hole with explosive force as a rising tide or large waves cause large changes in pressure in the underlying cave.

blowouts Erosional hollows, depressions, troughs or swales within a dune complex, often, but not exclusively, in coastal areas (Carter *et al.*, 1990). They form readily in vegetated dunes for a variety of reasons: shoreline erosion and/or wash-over, vegetation die-back and soil nutrient deficiency, destruction of vegetation by animals and fire, and human activities. However, blowout topography need not necessarily arise from erosional processes; it may develop as areas of non-deposition between mobile dune ridges or as gaps in incipient foredunes. Two basic blowout morphologies have been identified: saucer blowouts (shallow, ovoid and dish-shaped with a steep marginal rim) and trough blowouts (deep, narrow, steep-sided with more marked downwind depositional lobes, and marked deflation basins). ASG

Reading and Reference
Barchyn, T.E. and Hugenholtz, C.H. (2013) Dune field reactivation from blowouts: Sevier Desert, UT, USA. *Aeolian Research*, 11, 75–84. · Carter, R.W.G., Hesp, P.A. and Nordstrom, K.F. (1990) Erosional landforms in coastal dunes. In K. F. Nordstrom, N. P. Psuty and R. W. G. Carter (eds), *Coastal dunes: form and process*. Chichester: John Wiley & Sons, Ltd; pp. 217–250.

bluehole A circular, steep-sided hole that occurs in coral reefs. The classic examples come from the Bahamas (Dill, 1977), but other examples are known from Belize and the Great Barrier Reef of Australia (Backshall *et al.*, 1979). Although volcanicity and meteorite impact have both been proposed as mechanisms of formation, the most favoured view is that they are the product of karstic processes (e.g. collapse dolines) that acted at times of low sea level when the reefs were exposed to subaerial processes. ASG

References
Backshall, D.G., Barnett, J. and Davies, P.J. (1979) Drowned dolines – the blue holes of the Pompey Reefs, Great Barrier Reef. *BMR Journal of Australian Geology*

and Geophysics, 4, 99–109. · Dill, R.F. (1977) The blue holes – geologically significant sink holes and caves off British Honduras and Andros, Bahama Islands. In D. L. Taylor (ed.), *Proceedings of the Third International Coral Reef Symposium, Vol. 2: Geology*. Miami, FL: University of Miami; pp. 238–242.

Blytt–Sernander model Provides the classic terminology of the HOLOCENE (see Table 3). It was established by two Scandinavians, A. G. Blytt and R. Sernander, who, in the late nineteenth and early twentieth centuries, undertook various palaeobotanical investigations that revealed vegetational changes from which climatic changes were inferred. They introduced the terms Boreal, Atlantic, Sub-Boreal and Sub-Atlantic for the various environmental fluctuations that took place. Modern workers recognize that because factors other than climate affect vegetation change (e.g. human intervention, soil deterioration

Table 3 The classic European Holocene sequence

Period	Zone number	Blytt–Sernander zone name	Radiocarbon years BP
Post Glacial	IX	Sub-Atlantic	post 2450
	VIII	Sub-Boreal	2450–4450
	VII	Atlantic	4450–7450
	VI	Late Boreal	7450–8450
	V	Early Boreal	8450–9450
	IV	Pre-Boreal	9450–12,250
Late Glacial	III	Younger Dryas	10,250–11,350
	II	Allerød	11,350–12,150
	Ic	Older Dryas	12,150–12,350
	Ib	Bølling	12,350–12,750
	Ia	Oldest Dryas	

through time) the model may be simplistic. ASG

bog In the widest sense, the term bog may be taken to include any nutrient-poor, acidic peatland community with a distinctive plant community of sphagnum mosses, ericaceous sedges and coniferous trees. This broader use of the term includes poor fen. Bog is used in the restrictive sense to refer to those peatlands where the bog surface is irrigated more or less exclusively by precipitation inputs and is unaffected by minerotrophic groundwater. Another term for bog is ombrotrophic MIRE. The nutrient supply to the bog is low, although nitrogen may be supplemented by atmospheric enrichment derived from industrial, urban and agricultural sources. Bog biodiversity is low. Typically the pH is <4.5, compared with fens where the pH range is 4.5–7.5.   ALH

Reading
Mitsch, W.J. and Gosselink, J.G. (2011) *Wetlands*, 4th edition. Chichester: John Wiley & Sons, Ltd.

bog bursts   The sudden disruption of a bog so that there is a release of water and peat that may then flow over a considerable distance.

bogaz   Narrow, deep ravines and chasms in karst areas, the products of limestone solution along bedding planes, joints and fissures.

Bølling   A short-lived Late Glacial interstadial dated at about 12,350–12,750 BP (see INTERSTADIAL and LATE GLACIAL).

bolson   A low-lying trough or basin surrounded by high ground and having a playa at its lowest point to which all drainage trends.

bombardment   The striking of the Earth by bodies such as asteroids, comets and meteoroids (bolides). Impacts can create landforms, including craters (astroblemes), generate tsunamis, set off volcanic activity, send dust into the atmosphere and may be implicated in mass extinctions, such as that at the end of the Cretaceous period (Pope *et al.*, 1994). Noteworthy bolide impact craters include: vredefort crater in South Africa, the largest known impact crater on Earth (300 km diameter); Sudbury Basin in Ontario, Canada (250 km diameter); Chicxulub crater off the coast of Yucatan (170 km diameter); Manicouagan Reservoir in Quebec, Canada (100 km diameter); Popigai crater in Russia (100 km diameter); Acraman crater in South Australia (90 km diameter) and the Eocene Chesapeake Bay impact crater (90 km diameter).   ASG

Reference
Pope, K.O., Baines, K.H., Ocampo, A.C. and Ivanov, B.A. (1994) Impact winter and the Cretaceous/Tertiary extinctions: results of a Chicxulub asteroid impact model. *Earth and Planetary Science Letters*, **128**, 719–725.

Bond cycles   Groups of very rapid global temperature changes during the QUATERNARY, individually called DANSGAARD–OESCHGER (D–O) EVENTS, they are known as Bond cycles after Bond *et al.* (1992, 1993). Bond and coworkers demonstrated that the D–O events, identified

in ICE CORE records, were also present in the deep-sea core record of ocean sediments. This suggests a very strong likelihood of these events having a climatic causation, and also indicate complex ice–ocean–atmosphere linkages.   DSGT

Reading and References
Bond, G., Heinrich, H., Broecker, W., *et al.* (1992) Evidence for massive discharges of icebergs into the North Atlantic Ocean during the last glacial period. *Nature*, **360**, 245–249. · Bond, G., Broecker, W., Johnsen, S., *et al.* (1993) Correlation between climate records from North Atlantic sediments and Greenland ice. *Nature*, **365**, 143–147. · Williams M.A.J., Dunkerley, D., de Deckker, P., *et al.* (1998) *Quaternary environments*, 2nd edition. London: Arnold.

bora   From Latin *boreas*, the north wind. A dry, cold, gusty, northeast wind that affects the northern part of the Adriatic Sea and the Dalmatian coast (see Jurcec (1981)). Peak frequency occurs during the months from October to February and a maximum gust speed of $47.5\,\mathrm{m\,s^{-1}}$ has been recorded at Trieste. Two types of bora are recognized: anticyclonic and cyclonic, according to the atmospheric pressure field. The term bora is now applied to similar winds in other parts of the world; for example, Oroshi in central Japan (Yoshino, 1976).   AHP

References
Jurcec, V. (1981) On mesoscale characteristics of bora in Yugoslavia. In G. H. Liljequist (ed.), *Weather and weather maps*. Stuttgart: Birkauser Verlag. · Yoshino, M.M. (1976) *Local wind bora*. Tokyo: University of Tokyo Press.

bore   A tidal wave that propagates as a solitary wave with a steep leading edge up certain rivers. Formation is favoured in wedge-shaped shoaling estuaries at times of spring tides. Other local names include *egre* (England, River Trent), *pororoca* (Brazil), *mascaret* (France).   DTP

boreal climate   In Köppen's classification scheme a climate that is characterized by a snowy winter and warm summer, with a large annual range of temperature, such as occurs between 60° and 40° north.

boreal forest   The northern coniferous zone of the Holarctic. The most northerly section, transitional to the TUNDRA, is synonymous with TAIGA. The southerly sections are frequently made up of dense forest with closed canopies permitting little light to reach the floor, which possesses a variable cover of lichens, mosses and herbaceous plants. Typical hardy and undemanding trees include spruce, fir and hemlock, with pine providing more open formations and having a denser ground cover. The forest is bounded approximately to the north by the 10 °C July average isotherm and to the south by areas with more than 4 months above 10 °C.   PAF

Reading
Larsen, J.A. (1980) *The boreal ecosystem.* New York: Academic Press.

boreal (period) Period of time that, according to the BLYTT–SERNANDER MODEL division of European peat stratigraphy, occurred during the earlier part of the Holocene (approximately 8500 BP to 7500 BP) and was associated with somewhat warmer and drier conditions than occurred either immediately before or afterwards. It is inferred that the relatively high level of humification of peats associated with this period is indicative of both higher temperatures and lower precipitation levels. The interpretation of regionally contemporaneous climatic phases in the European Holocene is based on characteristic raised bog stratigraphy and associated pollen sequences, although the rigid zonation implied by the scheme has now been largely abandoned due to the recognition that relationships between peat stratigraphy, pollen zonation and climate change are more complex.   MEM

bornhardt A distinctive hill with bare rock surfaces, dome-like summit and steep sides, described from East Africa by German geologist Bornhardt (1900) as an INSELBERG that now often takes his name. Usually in granite or gneiss, and found on CRATONS and in deeply eroded OROGENS. Notable groups include the 'sugar loaves' of Rio de Janeiro and Idanre Hills of Nigeria; similar to exfoliation domes of Yosemite. Low porosity, from metasomatic infusion of granite or gneiss, may cause resistance to DENUDATION. Often surrounded by deeply weathered or less resistant rocks, they emerge as ground surface is lowered (Thomas, 1978). Bornhardt rock surfaces are diversified by etch pits and TAFONI.   MFT

Reading and References
Bornhardt, W. (1900) *Zur Oberflächengestaltung und Geologie Deutsch-Afrikas,* Berlin: Reimer. · King, L.C. (1975) Bornhardt landforms and what they teach. *Zeitschrift für Geomorphologie, Neue Folge,* **19,** 299–318. · Thomas, M.F. (1978) The study of inselbergs. In H. Bremer and J. N. Jennings (eds), *Inselbergs/Inselberge.* Zeitschrift für Geomorphologie, Supplementbände, vol. 31. Berlin: Borntraeger; pp. 1–41.

boss A small batholith or any dome-shaped intrusion of igneous rock, especially one exposed at the surface through erosion of the less resistant host rocks.

botryoidal Having a form resembling a bunch of grapes, often applied to aggregate minerals.

bottom-sets Beds of stratified sediment that are deposited on the bottom of the lake or sea in advance of a delta.

Bouguer anomaly A measure of the gravitational pull over an area of the Earth after the Bouguer correction to a level datum, usually sea level, has been applied.

boulder clay See TILL.

boulder train A stream of boulders derived by glacial transport from a specific and identifiable bedrock source, and carried laterally in a more or less straight line, thereby permitting former directions of ice movement to be inferred.

boulder-controlled slopes First described by Bryan (1923) from the Arizona deserts. Bryan recognized slopes formed on rock with a veneer of boulders and assumed that the angle of the bedrock surface had become adjusted to the angle of repose of the 'average-sized joint fragment'. Measurements have subsequently shown that boulder-covered slopes exist over the range of angles up to about 37° and that the existence of a boulder cover may be due to a number of causes. Boulders may also lie upon bedrock that has resulted from the development of a slope angle in equilibrium with the mass strength of the rock (Selby, 1982). The idea that the boulders control the angle of the bedrock slope on which they lie is, therefore, open to question.   MJS

Reading and References
Bryan, K. (1923) *Contributions to economic geology, 1921, Part II, mineral fuels – erosion and sedimentation in the Papago country, Arizona, with a sketch of the geology.* US Geological Survey Bulletin 730-B. Washington, DC: US Government Printing Office; pp. 19–90. · Melton, M.A. (1965) Debris-covered hillslopes of the southern Arizona Desert – consideration of their stability and sedimentation contribution. *Journal of Geology,* 73, 715–729. · Selby, M.J. (1982) Rock mass strength and the form of some inselbergs in the central Namib Desert. *Earth Surface Processes and Landforms,* 7, 489–497.

boundary conditions Many physical phenomena can be modelled by mathematical deduction leading to generalized equations; in order to obtain simplified specific solutions to these equations their applicability is deliberately constrained by the definition of particular circumstances known as 'boundary' or 'initial' conditions (Wilson and Kirkby, 1975: 206–207, 222–225). In particular, the solution of differential equations requires definition of boundary conditions so that expressions can be found for the arbitrary constants resulting from integration of the equations.

An example is the theoretical derivation of the logarithmic velocity profile in turbulent flow. The rate of change of velocity v with height above the bed y – the 'velocity shear' – itself decreases with height according to the differential equation

$$\frac{\mathrm{d}v}{\mathrm{d}y} = \frac{K}{y}$$

where K is a constant incorporating the bed shear stress. The variation of velocity with height is obtained by integrating this equation, which gives:

$$v = K \ln y + C$$

Here, C is a constant of integration. It can be evaluated by specifying the boundary condition that $v = 0$ when $y = y_0$, so that

$$0 = K \ln y_0 + C$$

Therefore:

$$C = -K \ln y_0$$

This is actually an initial condition, since the height is defined at which the velocity is zero, and negative velocities are assumed not to occur. The velocity profile equation can now be simplified by inserting the expression for C:

$$v = K \ln y - K \ln y_0$$
$$= K \ln(y/y_0)$$

This describes a curve plotting as a straight line on a graph with a logarithmic height axis and an arithmetic velocity axis; the intercept on the height axis where $v = 0$ is y_0, and the gradient of the line is K (Richards, 1982: 69–70).

Theoretical models may require multiple boundary conditions, some of which are necessarily dynamic in order to provide realistic solutions. The slope profile shape characteristic of different types of slope process is modelled by a partial differential equation in which the rate of change of local slope sediment transport with distance from the divide x equals the rate of change of local slope surface elevation y with

time t. To solve this equation and derive the characteristic form of profile as a graph of y against x, the initial conditions are: (a) the divide is fixed at $x = 0$ and sediment transport is zero at the divide; (b) an initial slope shape $(y = f(x))$ is defined from which the characteristic form evolves.

A boundary condition then defines the slope foot; this base level is fixed at x_1, and although a fixed base level elevation $(y = 0$ at $x = x_1)$ may be set, a dynamic boundary condition may be established in which $y(x_1) = f(t)$. Quite different characteristic profiles will emerge from the solution of the initial general equation according to the nature of this boundary condition, which models the behaviour of basal erosion at the slope foot.   KSR

Reading and References
Richards, K.S. (1982) *Rivers: form and process in alluvial channels*. London: Methuen. · Sumner, G.N. (1978) *Mathematics for physical geographers*. London: Edward Arnold. · Wilson, A.G. and Kirkby, M.J. (1975) *Mathematics for geographers and planners*. Oxford: Oxford University Press.

boundary layer   When a fluid and a solid are in relative motion the boundary layer is the zone in the fluid closest to the solid surface within which a velocity gradient develops because of the retarding frictional effect of contact with the solid. The fluid is at rest relative to the solid immediately adjacent to the surface, but with distance from the surface the frictional effect diminishes and velocity increases, at a rate dependent on the local flow characteristics. The velocity gradient of the boundary layer occurs in overland flow on hillslopes, river flow in channels, the swash and backwash of beaches, airflow over a desert dune, but also immediately adjacent to a sand grain falling through water or air, where it is the *relative* motion producing the boundary layer rather than the fluid motion over a static solid surface. The diagram illustrates the development of a boundary layer over a surface parallel to the direction of motion within deep water flow. A laminar boundary layer forms at the entry point where flow begins over the surface, but at some distance downstream this is replaced by a turbulent boundary layer if the flow conditions are appropriate; that is, if the REYNOLDS NUMBER exceeds about 2000. The boundary layer is 'fully developed' if the velocity profile extends to the surface of the flow, which is normal in rivers. In a deep fluid layer, however, the motion at some distance from the surface is unaffected by the boundary influence and the velocity is that of a free, or external, fluid stream.

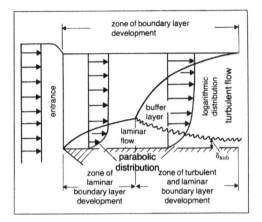

Within a laminar boundary layer viscous forces within the flow are pronounced, and adjacent fluid layers are affected by the molecular interference of the fluid viscosity. The velocity increases with distance from the solid in an approximately parabolic curve (Allen, 1970: 36–39). In a turbulent boundary layer, the pattern of velocity increase with distance from the bed is very complex. Close to the bed the fluid is sufficiently retarded for viscous effects to be pronounced and laminar flow occurs; this is the very thin 'laminar sublayer'. If grains on a sedimentary surface are smaller in diameter than the thickness of the laminar sublayer the flow is 'hydrodynamically smooth', and the grains are protected against entrainment. In the HJUL-STRÖM CURVE, threshold velocities are seen to be higher for silt and clay sizes than for sand sizes. Above the laminar sublayer is a buffer zone before the true turbulent velocity profile is reached. In the turbulent boundary layer, interference between fluid elements occurs at a scale controlled by the depth of eddy penetration, and measured velocity profiles indicate that velocity increases with the logarithm of distance from the surface (Richards, 1982: 68–72). Under the BOUNDARY CONDITIONS defined above it is shown that this relationship takes the form

$$v = K \ln(y/y_0)$$

where v is velocity, y is distance from the surface, y_0 is the height where the velocity is zero and K is a constant that is equal to v_*/K. Here, K is the von Karman constant (0.4) and v_* is the 'shear velocity', defined as $(\tau_0/\rho_w)^{1/2}$. This is a measure of the steepness of the velocity profile, which is dependent on bed shear stress and water density.

In hydrodynamically rough conditions where grains are large relative to the thickness of the laminar sublayer, y_0 equals 1/30th of the D_{65} grain diameter. If the above equation is converted to common logarithms and the expressions for K and y_0 are inserted, the equation for the logarithmic velocity profile in a hydrodynamically rough, turbulent boundary layer, becomes

$$\frac{v}{v_*} = 5.75 \log\left(\frac{y}{D_{65}} + 8.5\right)$$

This equation may be used to fit to velocity data from the lower 10–15% of the flow in order to project the curve to the bed. The local bed shear stress can then be estimated, as well as the velocity close to the bed sediment at heights where measurement is impractical, especially in the field. Note that both the turbulent fluctuations and the rapid increase of velocity above the bed material – causing strong lift forces, which occur under hydrodynamically rough bed conditions in turbulent flow – are important factors in the entrainment of sediment by the flow.

KSR

Reading and References

Allen, J.R.L. (1970) *Physical process of sedimentation.* London: Allen & Unwin. · Leeder, M.R. (1982) *Sedimentology: process and product.* London: Allen & Unwin; pp. 47–66. · Richards, K.S. (1982) *Rivers: form and process in alluvial channels.* London: Methuen. · Van Haren, H. (2015) Impressions of the turbulence variability in a weakly stratified, flat-bottom deep-sea 'boundary layer'. *Dynamics of Atmospheres and Oceans,* **69,** 12–25.

bounding surface A break between different primary sedimentary structures. Bounding surfaces may occur in a hierarchy, with major surfaces representing hiatuses in deposition due to factors that include climate change and others representing hiatuses in episodic deposition and/or the erosion of one unit prior to deposition of the next. The term is often used in describing aeolian deposits, where dating studies of sediments either side of bounding surfaces show that their significance may vary (Leighton *et al.,* 2013). DSGT

Reading

Brookfield, M.M. (1977) The origin of bounding surfaces in ancient aeolian sandstones. *Sedimentology,* **24,** 303–332. · Leighton, C.L., Thomas, D.S.G. and Bailey, R.M. (2013) Allostratigraphy and Quaternary dune sediments: not all bounding surfaces are the same. *Aeolian Research,* **11,** 55–60.

bourne A stream or stream channel on chalk terrain that flows after heavy rain.

Bowen ratio Named after American astrophysicist Ira S. Bowen (1898–1978). The ratio of heat energy used for sensible heating (conduction, convection) to the heat energy used for latent heating (evaporation of water, sublimation of snow). Applicable to any moist surface, the Bowen ratio ranges from near zero for ocean surfaces to greater than two for desert surfaces, with negative values possible. The Bowen ratio is often employed in the surface energy balance equation in order to estimate fluxes of latent heat in the 'Bowen ratio–energy balance' method.　　AWE

Bowen's reaction series A series of minerals that crystallize from molten rock of a specific chemical composition, wherein any mineral formed early in the chain will later react with the melt, forming a new mineral further down the series; the minerals formed under decreasing temperatures of crystallization are more stable in the weathering environment.　　ASG

brackish Pertaining to water that contains salt in solution, usually sodium chloride, but which is less saline than seawater.

braided river A river whose flow passes through a number of interlaced branches that divide and rejoin. The term has been applied both to short reaches where a river splits around an island and to very extensive river networks on valley bottoms or alluvial plains, the whole of which may be criss-crossed by rapidly shifting channels with freshly deposited sediment between them. Braiding may be more apparent at some low levels than at others. For example, single channels of low sinuosity at high flows may assume a braided pattern as channels thread their way between sets of emergent BEDFORMS at low flow. By contrast, single channels may become multiple channels at high flows as inactive channels are reoccupied and developed.

The term 'braided' is applied in a general sense to a whole family of multiple channel river patterns, some of which are given separate names, including 'anastomosing', ANABRANCHING and 'wandering' rivers. The term *braided river* is sometimes restricted in use to systems where channels are temporary and varying over time, anabranching and anastomosing to more stable multichannel systems developed under aggrading conditions with levées and backswamps (though Nanson and Knighton (1996) use anastomosing to refer to a subset of anabranching rivers). Wandering rivers may consist of alternate stable single channel reaches and unstable multichannel 'sedimentation zones'.

The term has also been used as an alternative for patterns that are transitional between meandering and braided.

Braided river patterns – in the general sense of multichannel systems – appear to be created in various ways. Mid-channel bar development may lead to division of the channel and enlargement of the bar by accretion, possibly with the development of a vegetated island. Alternatively, migratory bars, exposed only at low flows, may simply be exposed bedforms continually shifting at high flows by erosion and accretion. Scour at channel junctions may be important for the local entrainment of sediment that is then redeposited as flows diverge again downchannel. Overbank flood flows also scour out new chute channels.　　JL/DSGT

Reading and Reference
Nanson, G.C. and Knighton, D.A. (1996) Anabranching rivers: their cause, character and classification. *Earth Surface Processes and Landforms*, 21, 217–239. · North, C.P., Nanson, G.C. and Fagan, S.D. (2007) Recognition of the sedimentary architecture of dryland anabranching (anastomosing) rivers. *Journal of Sedimentary Research*, 7, 925–938. · Sambrook Smith, G.H., Best, J.L., Bristow, C.S. and Petts, G.E. (eds) (2006) *Braided rivers: processes, deposits ecology and management*. IAS Special Publication 36. Chichester: John Wiley & Sons, Ltd.

brash A mass of fractured rock that has been weathered in situ; also applied to a mixture of shattered rock or ice.

Braun-Blanquet scale Standard means of measuring plant cover in a sample quadrat. The phytosociological scheme was developed by Josias Braun-Blanquet, a Swiss ecologist (1884–1980), who devised this widely used method for the quantitative description of vegetation communities (see COMMUNITY) and published it in his 1928 text book *Pflanzensoziologie*. The method is based on a five-point scale accounting for plant species cover abundance as follows:

5	Cover over 75%
4	Cover 50–75%
3	Cover 25–50%
2	Cover 5–25%
1	Cover 1–5%
+	Species present but with negligible cover

When cover values are computed for all the species in a quadrat, and for a number of quadrats in a community, it is possible to arrange the data in such a way as to identify typically recurring groups of species, or plant associations (see

71

ASSOCIATION, PLANT). Although not widely utilized by English-speaking ecologists, the Braun-Blanquet scheme was adopted in mainland Europe and was influential in the development of the theoretical principles around the identification of plant associations and communities. Braun-Blanquet developed further five-point scales to account for, for example, the constancy (presence) of species within communities. Constancy is a measure of the evenness or otherwise of species distribution and may be expressed as the frequency of occurrence of particular species either in sample quadrats within a community or in stands between different communities. The scale is as follows:

5	Frequency 81–100%
4	Frequency 61–80%
3	Frequency 41–60%
2	Frequency 21–40%
1	Frequency 1–20%

Species classified as '1' under this scheme are regarded as 'rare', whereas those with a score of '5' are 'constant'. (See COVER, PLANT.)   MEM

breccia   A rock that has been greatly fractured into angular fragments, generally less than 2 mm in diameter, by tectonic activity, volcanism or transport over short distances.

brodel   A highly contorted and irregular structure in soils that have been subjected to churning by frost processes.

brousse tigrée   Vegetation banding, which may include grassland patterns but which generally consists of bands of more closely spaced trees alternating with bands of sparser vegetation. Its nature and origin have been well described thus by Mabbutt and Fanning (1987: 41): 'All are developed in arid or semi-arid areas, in open low woodlands or tall shrublands, with average annual rainfalls of between 100 and 450 mm; they occur on slopes of the order of 0.25%, too gentle for the development of drainage channels, but steep enough to maintain organized patterns of sheetflow . . . In drier areas the banding may be restricted to the better-watered depressions, but it is commonly best-developed on low interfluves, with the intervening depressions marked by uniformly dense tree cover. Such tracts of more concentrated sheetflow have been named "water lanes".'

Brousee tegrée are in fact one of several patterns of what are now termed self-organized vegetation systems found in drylands, where

moisture stresses create competition between vegetation types (Bailey, 2011).   ASG

References
Bailey, R.M. (2011) Spatial and temporal signatures of fragility and threshold proximity in modelled semi-arid vegetation. *Proceedings, Biological Sciences, The Royal Society*, **278**, 1064–1071. · Mabbutt, J.A. and Fanning, P.C. (1987) Vegetation banding in western Australia. *Journal of Arid Environments*, **12**, 41–59.

Brune curve   The empirically based formula published by Brune (1953) linking the size of a reservoir to its efficiency in trapping sediment carried into it by streamflow. Brune used data from 44 reservoirs of varying size, all in the USA. He established a curve linking sediment trap efficiency to the ratio of dam storage volume C to the annual inflow volume I. This showed that reservoirs having I/C ratios of around 0.1 have trap efficiencies approaching 90% for particles of silt size and larger. The ratio used by Brune is proportional to the average retention time of the impoundment, and thus reflects opportunities for sediment particles to settle.   DLD

Reference
Brune G.M. (1953) Trap efficiency of reservoirs. *Transactions of the American Geophysical Union*, **34**, 407–418.

Brunhes–Matuyama   The magnetic polarity epoch boundary (see PALAEOMAGNETISM) marking the major change or reversal in the Earth's magnetic field that occurred at around 730,000 years BP (determined by K–Ar dating) or 780,000 years BP (based upon dating using the oxygen-isotope record preserved within sea-floor sediments). The Brunhes–Matuyama boundary marks the change between the Matuyama epoch, when the magnetic field was reversed and aligned south to north, and the present-day Brunhes polarity epoch, where the magnetic field is aligned 'normally' (i.e. north to south).   DJN

brunizem   A prairie soil developed under grassland in temperate latitudes. Characteristically, a brown surface zone overlies a leached horizon that grades into a brown subsoil on non-calcareous bedrock.

Bruun rule   As sea level rises in response to global warming, many sandy beaches, especially those in closed bays (pocket beaches), may suffer from accelerated erosion. Bruun (1962) developed a widely cited and elegantly simple model of the response of a sandy beach to sea-level rise in a situation where the beach was initially in equilibrium, neither gaining nor losing sediment. The extent of recession was predicted by using a formula that translates into a 'rule of thumb' whereby the coastline retreats 50–100 times the dimensions of the rise in sea level: a 1 m rise would cause the beach to retreat by 50–100 m. The Bruun rule has been widely used. It can be stated mathematically as follows (Gornitz *et al.*, 2002: 68):

$$S = \frac{AB}{d}$$

where S is shoreline movement, A is sea level rise, d is maximum depth of beach profile, measured from the berm elevation for each project location to the estimated depth of closure, and B is the horizontal length of the profile, measured from the beginning of the berm to the intersection with the estimated depth of closure.

It is important to recognize, however, that there are some constraints on its applicability. The rule assumes that no sand is lost to long-shore transport and that an offshore 'closure depth' exists beyond which there is no sediment exchange. Moreover, the rule does not allow for shoreward transport of sediment as overwash or for those situations where the slope of the coastal plain is too gentle for sufficient sand to be available as a source for supplying the offshore. In addition, the rule was originally proposed for beaches that were initially in equilibrium. However, only a small proportion of the world's sandy beaches can in fact be considered to be in equilibrium. Beach erosion is widespread.

List *et al.* (1997) tested the validity of the rule for the barrier islands of Louisiana (USA). Using bathymetric surveys over about a century, they established that only a portion of their study profiles met the equilibrium criterion of the Bruun rule. They suggested that if the Bruun rule is inadequate for hindcasting it would also be inadequate for forecasting future rates of beach retreat. There are those who now suggest that use of the Bruun rule should be abandoned (Cooper and Pilkey, 2004). ASG

References

Bruun, P. (1962) Sea level rise as a cause of shore erosion. *American Society of Civil Engineers Proceedings: Journal of Waterways and Harbors Division*, **88**, 117–130. · Cooper, J.A.G. and Pilkey, O.H. (2004) Sea-level rise and shoreline retreat: time to abandon the Bruun rule. *Global and Planetary Change*, **43**, 157–171. · Gornitz, V., Couch, S. and Hartig, E.K. (2002) Impacts of sea level rise in the New York City metropolitan area. *Global and Planetary Change*, **32**, 61–88. · List, J.H., Sallenger, A.H., Hansen, M.E. and Jaffe, B.E. (1997) Accelerated sea level rise and rapid coastal erosion: testing a

causal relationship for the Louisiana barrier islands. *Marine Geology*, **140**, 437–465.

buffer A solution to which large amounts of acid or alkaline solutions may be added without markedly altering the original hydrogen ion concentration (pH).

buffer strip A belt of vegetated land, generally running continuously along the banks of a stream, and maintained with the intention of protecting the stream habitat from disturbance related to land use, perhaps agriculture or forest logging, occurring beyond the buffer strip. Buffer strips are commonly required by the legislation that governs timber harvesting in forests that are also important for water supply. The intention is that run-off water carrying eroded soil will be slowed, and some of the eroded particles trapped in the buffer strip, before reaching and polluting the stream. Nutrients in agricultural run-off may also be held in the soils of the buffer strip. Strips of preserved forest are commonly 20–100 m in width, and may be required along all perennial water courses in the area being logged. Buffer strips are also used to trap soil eroded from tilled fields, and may take the form of grassed or wooded zones at the hillslope foot (Daniels and Gilliam, 1996). Buffer strips may also serve as corridors of preserved habitat for wildlife, and by shading of the water course may limit any rise in water temperature that would occur following exposure to solar heating.

Much attention has been paid to the width required for efficient protection of the stream habitat. Experimental work generally suggests that a buffer of 30–50 m offers useful protection. DLD

Reference

Daniels R.B. and Gilliam J.W. (1996) Sediment and chemical load reduction by grass and riparian filters. *Soil Science Society of America Journal*, **60**, 246–251.

buffering capacity A buffer is a chemical compound that has the capacity to absorb or exchange hydrogen or hydroxide ions, and that allows the system to assimilate a limited amount of these ions without changing appreciably in pH. Buffering capacity is the quantitative ability of a solution to absorb hydrogen or hydroxide ions without undergoing a pronounced change in pH.

Soils are more strongly buffered in comparison with precipitation or freshwater. Different buffering systems come into play at particular ranges of soil pH. Carbonate minerals buffer soil pH within the range >8 to 6.2; silicates from 6.2 to 5.0; cation-exchange capacity from pH 5.0 to 4.2;

aluminium from pH 5.0 to 3.5; iron from pH 3.8 to 2.4; and humic acids from pH 6 to 3.   ALH

bulk density   The relationship of the mass of a soil or sediment to its volume, typically expressed in grams per cubic centimetre, using either a naturally damp or oven-dried sample (from which the wet and dry bulk density respectively can be calculated). Bulk density is measured in the laboratory from a sample that has been extracted using an open-ended metal cylinder of known volume. The cylinder is driven into the ground, carefully removed, has caps placed on either end and is then stored in a polythene bag. If the wet bulk density is required, the mass of the sample is determined by subtracting the weight of the cylinder plus caps from the total weight of the sample, cylinder and caps. For dry bulk density, the sample is carefully removed from the cylinder, dried at 105 °C and weighed. In both cases, the volume of the sample can be determined by measuring the radius and length of the sampling cylinder. The bulk density can then be determined by dividing the appropriate weight by the original volume of soil. Dry bulk density is a parameter used, along with the density of particles within the sample, in determining the porosity of a soil or sediment, with wet bulk density used when estimating soil moisture content. DJN

Reading
Rowell, D.L. (1994) *Soil science: methods and application.* Harlow: Longman; pp. 67–69.

bush encroachment   The process whereby shrubs come to dominate areas that previously were largely open grassland or mixed grass and woodland. It has been described from southern, eastern and Sahelian Africa, Australia and South America. The term is largely applied to SAVANNA environments and has widely been attributed to disturbances caused by the impact of marked grazing pressure by domestic livestock (e.g. Walker *et al.*, 1981). Bush encroachment is seen as a major threat to sustainable productivity in dryland pastoral systems, but its occurrence and persistence can prove difficult to assess given the inherent temporal and spatial variability of disequilibrium savanna ecosystems (Behnke *et al.*, 1993).

Several models have been expounded to explain the process of bush encroachment. Soil-based models (e.g. Walker and Noy-Meir, 1982) assume that grasses have a competitive advantage over shrubs and trees in tapping rainfall that infiltrates into the ground, the so-called 'two-layer' model. Changes within soil resources, in terms of the vertical distribution

of moisture and nutrients, may alter the balance in favour of shrub species. Ecological models (e.g. Westoby *et al.*, 1989) place an emphasis on changes in grazing strategies and in the occurrence of fire, the latter seen as a natural mechanism that suppresses the potential for shrubs to become dominant. Clearly, soil and ecological elements are interlinked and are both affected by grazing patterns and natural events such as droughts. Recent empirical research is suggesting that soil water and nutrient distribution changes are not evident in areas where bush encroachment has occurred in central southern Africa (Dougill *et al.*, 1999) and that interactions between fire regimes, grazing levels and rainfall distributions are the key to understanding the occurrence and potential persistence of bush encroachment.   DSGT

References
Behnke, R.H., Scoones, I. and Kerven, C. (1993) *Range ecology at disequilibrium: new models on natural variability and pastoral adaptation in African savannas.* London: ODI. · Dougill, A.J., Thomas, D.S.G. and Heathwaite, A.L. (1999) Environmental change in the Kalahari: integrated land degradation studies for nonequilibrium dryland environments. *Annals, Association of American Geographers,* **89**, 420–442. · Walker, B.H. and Noy-Meir, I. (1982) Aspects of the stability and resilience of savanna ecosystems. In B. J. Huntley and B. H. Walker (eds), *Ecology of tropical savannas.* Berlin: Springer-Verlag; pp. 556–590. · Walker, B.H., Ludwig, D., Holling, C.S. and Peterman, R.S. (1981) Stability of semiarid savanna grazing systems. *Journal of Ecology,* **69**, 473–498. · Westoby, M., Walker, B.H. and Noy-Meir, I. (1989) Opportunistic management for rangelands not at equilibrium. *Journal of Range Management,* **42**, 266–274.

bushveld   The savanna lands of sub-Saharan Africa, ranging from open grassland, through parkland with scattered trees to dense woodland.

butte   A small, flat-topped and often steep-sided hill standing isolated on a flat plain. Often attributed to erosion of an older land surface, the butte representing a remnant or outlier.

Buys Ballot's law   An observer in the northern hemisphere, standing with their back to the wind, will have low pressure to their left and high pressure to their right; the converse is true in the southern hemisphere. This law was formulated in 1857 by the Dutch meteorologist Buys Ballot. (See also CORIOLIS FORCE and GEOSTROPHIC WIND.)   BWA

bypass flow   The movement of water through the soil along a pathway other than that provided by the microscopic pore spaces within the soil matrix. Among the alternate pathways that may be available are shrinkage cracks, faunal burrows and voids left following the decay of plant roots. These may be classified as 'macropores', if >1 mm in diameter. In many cases, macropores occupy a few per cent of the total soil volume, and a somewhat larger fraction of the total void space present in the soil. The significance of bypass flow is that the larger conduits that carry the flow allow much faster flow than is possible in the laminar conditions arising in the ordinary network of soil micropores that are only micrometres in diameter. Consequently, water may travel downslope towards streams much more rapidly when the soil matrix is bypassed. This allows rain water to provide a greater contribution to channel flow than would be possible if all flow were through the soil matrix. Furthermore, there is less opportunity for rapidly delivered water to have its properties, such as acidity, moderated within the soil, since the area of contact with the soil is small and the period of residence short. In areas affected by acid rain, a major contribution to streams via bypass flow may result in a rapid drop in pH that is taxing for aquatic biota. Vertical bypass flow may carry water downward through the soil column rapidly, and result in very rapid water table fluctuations. If there is a significant volume of macropores that are not interconnected, the filling of these may delay the onset of runoff.

An important feature of bypass flow is that it can transmit water rapidly through soils whose matrix is not saturated. Normally, the hydraulic conductivity of soils declines markedly as they fall below saturation, since the largest pores drain first, leaving only small pores to transmit flow. The rapid transmission of water under unsaturated conditions can also be achieved in materials lacking macropores, through what has been termed *preferential flow* (e.g. Stagnitti *et al.*, 1995). This may involve concentrations of flow along restricted paths through a medium that is relatively homogeneous that arise from instabilities in the wetting front that may preferentially wet-up 'fingers' of soil extending to considerable depth.   DLD

Reading and Reference

Beven, K. and Germann, P. (1982) Macropores and water flow in soils. *Water Resources Research*, **18**, 1311–1325. · Chen, C. and Wagenet, R.J. (1992) Simulation of water and chemicals in macropore soils. Part 1. Representation of the equivalent macropore influence and its effect on soilwater flow. *Journal of Hydrology*, **130**, 105–126. · Dingman S.L. (1994) *Physical hydrology*. Englewood Cliffs, NJ: Prentice-Hall. · Stagnitti F., Parlange, J.-Y., Steenhuis T.S., *et al.* (1995) Transport of moisture and solutes in the unsaturated zone by preferential flow. In V. P. Singh (ed.), *Environmental hydrology*. Dordrecht: Kluwer; pp. 193–224.

C

caatinga A form of thorny woodland found in areas such as northeast Brazil, and characterized by many xerophytic species.

caballing The mixing of two water masses of identical in-situ densities but different in-situ temperatures and salinities, such that the resulting mixture is denser than its components. It sinks and forms a front. Caballing fronts are important in the hyperventilation of the Japan Sea and in deep water formation in both the Arctic and Antarctic. ASG

caesium-137 (^{137}Cs) analysis The use of spatial patterns of accumulation or depletion of the anthropogenic isotope ^{137}Cs (or radiocaesium) to determine rates of sedimentation and erosion. ^{137}Cs, a fallout product from thermonuclear explosions, does not occur naturally. It decays with a half-life of 30 years. ^{137}Cs analysis provides a means of rapidly assessing sediment loss or accumulation rates through the decades since 1954, when the isotope was first released from weapons testing.

^{137}Cs is delivered to the surface via precipitation and gravitational settling, where it is adsorbed strongly on clay and silt-sized particles. Thereafter, the isotope is redistributed as the soil particles are transported by erosion (Walling and Quine, 1991). Assuming that the initial distribution of ^{137}Cs across the landscape is relatively uniform, areas with higher ^{137}Cs contents indicate depositional sites, while those with low ^{137}Cs content are sites where erosion processes are active. Stable sites should have ^{137}Cs contents that reflect the initial input of the radioisotope, less the loss owing to radioactive decay, and are used as reference locations. Consequently, samples are collected for ^{137}Cs analysis, and their ^{137}Cs contents related to that of the reference site in order to examine whether erosion or deposition has taken place. KB

Reading and Reference
Walling, D.E. (1999) Linking land use, erosion and sediment yields in river basins. *Hydrobiologia*, **410**, 223–240.
· Walling, D.E. and Quine, T.A. (1991) Use of ^{137}Cs

measurements to investigate soil erosion on arable fields in the UK: potential applications and limitations. *Journal of Soil Science*, 42, 147–165.

Cainozoic (Cenozoic) A geological era spanning the Palaeocene, the Eocene, the Oligocene, the Miocene, the Pliocene and the Pleistocene (Table 1). See GEOLOGICAL TIMESCALE. ASG

Reference
Berggren, W.A. (1969) Cainozoic stratigraphy, planktonic foraminiferal zonation and the radiometric time-scale. *Nature*, **224**, 1072–1075.

calcicole A plant that thrives in lime-rich soil. Examples include wood sanicle (*Sanicula europaea*) and traveller's joy (*Clematis vitalba*). Plants that cannot tolerate such conditions are termed calcifuge; examples include common heather (*Calluna vulgaris*) and most other ericaceous plants. The effect of pH on mineral nutrition appears to be the operative factor. KEB/GF

Reading
Lee, J.A. (1998) The calcicole–calcifuge problem revisited. *Advances in Botanical Research*, **29**, 1–30.

calcifuge Any plant that grows best on acidic soils; for example, bracken.

calcite compensation depth The critical depth in the oceans below which the rate of solution of the calcite crystalline form of calcium carbonate exceeds the rate of deposition. Calcite is chemically more stable than aragonite so that the

Table 1 Subdivisions of the Cainozoic era

	Date of beginning, million years ago
Pleistocene	1.8
Pliocene	5.5
Miocene	22.5
Oligocene	36.0
Eocene	53.5
Palaeocene	65.0

Source: Berggren (1969).

The Dictionary of Physical Geography, Fourth Edition. Edited by David S.G. Thomas.
© 2016 John Wiley & Sons, Ltd. Published 2016 by John Wiley & Sons, Ltd.

latter is markedly more soluble at given temperatures and partial pressures of carbon dioxide. The two crystal forms, therefore, have differential solubilities in ocean water. Accordingly, the depth below which calcite dissolves, its compensation depth, is somewhat deeper than that of aragonite. (See CARBONATE COMPENSATION DEPTH.)   MEM

calcite saturation index Measure of the degree to which water contains dissolved calcium carbonate in the calcite crystal form in relation to the amount it contains at saturation level. Saturation is, in turn, defined as the calcite concentration above which point it comes out of solution and is deposited. The index is calculated as the absolute concentration of dissolved calcite in a sample of water, divided by the saturation concentration for a given condition of water temperature, pressure, pH and partial pressure of carbon dioxide.   MEM

calcrete A type of near-surface DURICRUST, predominantly composed of calcium carbonate, which occurs in a range of forms from powdery to nodular to highly indurated. It results from the cementation and displacive introduction of calcium carbonate into soil profiles, sediments and bedrock, in areas where vadose and shallow phreatic groundwater becomes saturated with respect to calcium carbonate. The term is synonymous with caliche and kunkur but distinct from other calcium carbonate cemented materials, including cave deposits (such as SPELEOTHEMS), lacustrine STROMATOLITES, spring deposits (such as TUFA or travertine), marine deposits (such as BEACH ROCK) or cemented dune sand (AEOLIANITE).

Calcretes are generally white, cream or grey in colour, though mottling and banding is common. They are widespread in semi-arid areas such as the High Plains of the USA, the Kalahari Desert, southern Africa, in North Africa, Rajasthan, India and in western Australia, where evapotranspiration is in excess of precipitation. They can form by either pedogenic (soil-forming) processes or by direct precipitation from groundwater. Pedogenic calcretes develop through a variety of stages from powdery varieties to fully indurated hardpans, and typically form at the land surface. In contrast, groundwater calcretes may form at depths of several metres, with sites of carbonate precipitation linked to the water table position and to the presence of landscape depressions such as ephemeral lakes or palaeochannels. Carbonate source materials can be distant or local, can be moved either laterally or vertically to site of precipitation, and include weathered bedrock, surface run-off, dust, groundwater and surface water. Important mechanisms leading to the precipitation of $CaCO_3$ include evapotranspiration, water uptake by soil organisms and shifts in environmental pH to above pH 9.0. The mean global chemical composition is 78% $CaCO_3$, 12% silica, 3% MgO, 2% Fe_2O_3 and 2% Al_2O_3, although variations in chemistry may occur dependent upon host sediment characteristics, cement composition, the presence of authigenic silica and silicates, the mode of origin and stage of development.   DJN

Reading
Wright, V.P. (2007) Calcrete. In D. J. Nash and S. McLaren (eds), *Geochemical sediments and landscapes*. Chichester: John Wiley & Sons, Ltd; pp. 10–45. · Scott, K. and Pain, C. (2009) *Regolith science*. Clayton: CSIRO Publishing.

caldera A large, roughly circular, volcanic depression. Calderas usually have a number of smaller vents and can also contain a large crater lake. The distinction between volcanic craters and calderas is essentially one of size, 1–2 km being the lower limit for the diameter of a caldera. Maximum diameters are in excess of 40 km. Calderas probably form in a variety of ways, but most proposed mechanisms attribute a primary role to collapse or subsidence, which may be related to explosive eruptions.   MAS

Reading
Francis, P. and Oppenheimer, C. (2003) *Volcanoes*. Oxford: Oxford University Press.

calms Winds with a velocity of less than one knot and which are represented by a force of zero on the BEAUFORT SCALE.

calving The breaking away of a mass of ice from a floating glacier or ice shelf to form an iceberg or brash ice (small fragments). Large tabular icebergs calve from ice shelves, while smaller icebergs and brash ice are commonly produced by valley or outlet glaciers. Most calving is induced by stresses set up within the floating ice mass by ocean swell. The incidence of calving from major ice sheets appears to be increasing as a consequence of global warming impacts on ice sheet melting and ice and sea temperatures.   DES/DSGT

Reading
Pritchard, H.D., Ligtenberg, S.R.M., Fricker, H.A., *et al.* (2012) Antarctic ice-sheet loss driven by basal melting of ice shelves. *Nature*, **484**, 502–505.

cambering The result of warping and sagging of rock strata that overlie beds of clay. The plastic nature of the clays causes the overlying rocks to flow towards adjacent valleys, producing a convex outline to the hill tops. Classic examples of cambering occurred in the Pleistocene when, under periglacial conditions, great rafts of limestone or sandstone subsided over lias and other clays along the escarpments of southern England. Cambering is often associated with the development of VALLEY BULGES.   ASG

canopy   Usually taken to be the uppermost stratum of woodland vegetation, the tree-top layer, though the term may also be used for any extensive above-ground leaf-bearing parts of plants. Despite the obvious importance of this zone in the interception of light and precipitation, and in the production of flowers, relatively little work has been reported, presumably due to practical difficulties. The role of the canopy in the woodland light climate, and therefore in tree regeneration and ground flora, is a vital one: some 80% of incoming radiation may be intercepted in this zone and 10% reflected from the upper leaves and twigs.   KEB

capacity/non-capacity load   A classification of the load of sediment being carried by a stream, according to whether the stream has excess (unsatisfied) capacity, in which case it carries a *non-capacity load*, or is moving as much material as available stream energy permits (a *capacity load*). The figure shows that, in general, sediment transport capacity declines for increasingly large (and therefore heavy) particles. This is because a stone resting on the bed experiences an entraining drag force exerted by the water that is proportional to the area of the stone facing into the current. But the same stone experiences a retaining force (its weight) that is proportional to its volume. Given that for increasingly large stones, area increases as the square of the diameter while weight increases as the cube of the diameter, retaining forces outstrip drag. The figure also shows a notional curve of sediment supply that suggests a similar decline in the supply of materials of increasingly large size. The bulk of the particles fed to most streams come from soil erosion in the catchment and along the channel banks, and are mostly fine. Fine particles are light and can easily be carried in turbulent streamflow. The enormous capacity for the transport of fine sediment is exemplified by rivers carrying HYPERCONCENTRATED FLOWS. Thus, most streams carry a non-capacity load of fine sediments; this is clearly because, despite their available capacity, they are only fed limited quantities of fine materials by run-off from the catchment. Thus, the supply curve crosses the capacity curve at some diameter D_{crit}. For particles larger than this, which are harder to transport, capacity generally lies below supply. The size corresponding to this diameter is approximately the silt–sand boundary (0.063 mm).

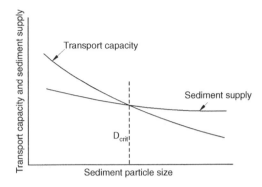

Many streams flow on beds of pebbles, cobbles and other coarse particles. The way in which these materials move generates a number of sedimentological features that further contribute to their resistance to motion, and to their involvement in non-capacity transportation. These include ARMOURING and cluster bedforms, in which smaller pebbles come to rest in the quieter water just downstream of larger, stationary, ones. These lee-side clusters of particles may be matched by upstream or stoss-side accumulations that come to rest as the flow passes up and over the obstacle clast that has triggered the accumulations. Both lee-side and stoss-side accumulations are protected from the force of the water and thus are not entrained. Imbrication, the geometric packing of platy particles so that they rest one upon the other, like a series of books that has been pushed over, also contributes stability to the particles concerned.   DLD

Reading
Shen, H.W. (1971) Wash load and bed-load. In H. W. Shen (ed.), *River mechanics*. Fort Collins: H. W. Shen; chapter 11, pp. 11–1 to 11–30.

capillary forces   Essentially SURFACE TENSION and adsorptive forces. Water will rise up a narrow (capillary) tube as a result of adsorptive forces between the water and the tube surface and tension forces at the water surface. These forces bind soil moisture to the soil particles so that it is held in an unsaturated soil at less than atmospheric pressure. This is often called a SUCTION or tension, and its strength may be determined using a TENSIOMETER.   AMG

Reading
Baver, L.D., Gardner, W.H. and Gardner, W.R. (1991) *Soil physics*, 5th edition. New York: John Wiley & Sons, Inc. · Smedema, L.K. and Rycroft, D.W. (1983) *Land drainage*. London: Batsford.

cap-rock A stratum of hard, resistant rock that overlies less competent strata and protects them from erosion.

capture (or river capture) The capture of part of one drainage system by another system during the course of drainage pattern evolution. Interpretation of drainage networks in terms of river capture was an integral feature of the Davisian CYCLE OF EROSION, and the distance to base level or sea level, the exposure of easily eroded rocks, or the effects of discharge increase following climatic change could all be reasons why one river was able to erode more rapidly and so capture the headwaters of another. The beheaded stream becomes a misfit stream as it is now too small for the valley. River capture has certainly featured prominently in the evolution of world river systems; for example, the easternmost tributary of the Indus was captured by the Ganges in geologically recent times, transferring drainage from a large area of the Himalayas from Pakistan to India. Knowledge of the sequence of river capture is sometimes necessary in the location of placer deposits, which are alluvial deposits containing valuable minerals. Placer deposits from ore deposits may occur in an area no longer directly connected to the drainage system with ores outcropping in the headwaters (Schumm, 1977).   KJG

Reference
Schumm, S.A. (1977) *The fluvial system*. Chichester: John Wiley & Sons, Ltd.

carapace *a.* The upper normal limb of a recumbent fold.
b. A soil crust that is exposed at the surface, especially a surficial calcrete.

carbon cycle The 'life' cycle, carbon being one of the three basic elements (with hydrogen and oxygen) making up most living matter. Over 99% of the Earth's carbon is locked up in calcium carbonate rocks and organic deposits such as coal and oil, both being the result of millions of years of carbon fixation by living organisms on land and in the oceans. The biotic cycle is similarly split into terrestrial and oceanic subsystems. Photosynthesis by pigmented plants fixes the carbon dioxide from air and water. Almost half is returned by plant respiration; the rest builds up as plant materials. The carbon is then returned to the atmosphere via animal respiration or plant decomposition. Fossil fuel consumption has increased atmospheric carbon dioxide fairly dramatically in the past few decades and is the basis of current concern over a GREENHOUSE EFFECT. KEB

Reading
Bach, W., Crane, A.J., Berger, A.L. and Longhetto, A. (eds) (1983) *Carbon dioxide*. Dordrecht: D. Reidel. · Bradbury, I.K. (1991) *The biosphere*. London: Belhaven Press. · Goudie, A.S. (1993) *The human impact*. 4th edition. Oxford: Blackwell.

carbon dating See RADIOCARBON DATING.

carbon sequestration The process of capture and long-term storage of atmospheric carbon dioxide (CO_2). It can refer to: (1) Natural biogeochemical cycling of carbon between the atmosphere and reservoirs, such as by chemical weathering of rocks. (2) The process of removing carbon from the atmosphere and depositing it in a reservoir; when carried out deliberately, this may also be referred to as CO_2 removal which is a form of geoengineering. (3) The process of carbon capture and storage, where CO_2 is removed from fuel gases, such as at power stations, before being stored in underground reservoirs. Interest in carbon sequestration has increased in recent years in the context of slowing the atmospheric and marine accumulation of greenhouse gases released by burning fossil fuels.   TS

Reading
Metz, B., Davidson, O., de Coninck, H., *et al.* (eds) (2005) *Carbon dioxide capture and storage. Special report*. Intergovernmental Panel on Climate Change. Cambridge: Cambridge University Press.

carbonate budget, of coral reefs A quantitative measure, generally from some form of census-based approach, to establish the net rate of carbonate production within a given reef or carbonate sedimentary environment. Carbonate is added to the reef structure by corals, calcareous algae and other encrusters and by the precipitation of marine cements. These constructive processes are offset by destructive biological (BIOEROSION) and physical processes; the eroded carbonate may be reincorporated into the reef framework or exported from the reef system. A summation of the rates at which these different processes operate allows the net rate of reef carbonate production to be quantified, typically expressed in terms of kilograms of $CaCO_3$ per square metre per year. Although methodologically difficult to construct, such budgets are more meaningful than assessments of reef health from simple coral cover statistics. They are useful in mapping intra- and inter-reef variations in carbonate production and in tracking reef performance through time, including the identification of state changes (i.e. production-dominated to erosion-dominated states) consequent upon either internal or external

79

(a) Production-dominated

(b) Import-dominated

(c) Export-dominated

(d) Bioerosion-dominated

Production   Import   Bioerosion
                 Export
Cementation   Dissolution   Imported material

Budgetary state models for different types of reefs, in terms of the relative importance of different carbonate-producing processes and pathways.
Source: Kleypas *et al.* (2001). Reproduced with permission of Springer Science and Business Media.

forcing. See also BIOFILM/EXTRACELLULAR POLYMERIC SUBSTANCES (EPSS).   TS

Reading
Eakin, C. (1996) Where have all the carbonates gone? A model comparison of calcium carbonate budgets before and after the 1982–1983 El Nino at Uva Island in the eastern Pacific. *Coral Reefs*, **15**, 109–119. · Kleypas, J.A., Buddemeier, R.W. and Gattuso, J.-P. (2001) The future of coral reefs in an age of global change. *International Journal of Earth Sciences*, **90**, 426–437. · Perry, C.T., Spencer, T. and Kench, P.S. (2008) Carbonate budgets and reef production states: a geomorphic perspective on the ecological phase-shift concept. *Coral Reefs*, **27**, 853–866.

carbonate compensation depth The critical depth in the oceans below which the rate of calcium carbonate solution exceeds that of deposition. It has long been observed that calcium carbonate remains of planktonic organisms are restricted to ocean floors above a certain depth. Below around 4000–5000 m in the Pacific, and somewhat deeper in the Atlantic and Indian Oceans, the calcium carbonate tests of plankton are dissolved. Controlling factors include pressure, partial pressure of dissolved carbon dioxide, temperature and pH. The solubility of calcium carbonate increases as the water temperature drops; since ocean bottom water is relatively cold (2–3 °C) and has higher pressures and carbon dioxide concentrations and correspondingly lower pH, this means that, in general, calcium carbonate goes into solution at such depths. (See also CALCITE COMPENSATION DEPTH.)   MEM

carbonation The reaction of minerals with dissolved carbon dioxide in water. The process is dominant in the weathering of limestone, since rainwater contains a small proportion of carbon dioxide (0.03% by weight) and thus acts as weak acid dissolving limestone rock. The conventional chemical reaction is

$$CO_2 + H_2O + CaCO_3 \rightleftharpoons Ca(HCO_3)_2$$

The $Ca(HCO_3)_2$ molecules have never been detected in solution and, while the product of carbonation is well known, the chemical process is not fully explained by this conventional equation (Picknett *et al.*, 1976).   PAB

Reference
Picknett, R.G., Bray, L.G. and Stenner, R.D. (1976) The chemistry of cave water. In T. D. Ford and C. H. D. Cullingford (eds), *The science of speleology*. London: Academic Press; pp. 213–266.

carnivore An animal-eating mammal of the order Carnivora, which depends solely on other carnivores or HERBIVORES for its food and which is located in the higher TROPHIC LEVELS of ecological systems. Carnivores may be predators (e.g. the lion or wolf among the large land animals, many species of beetles, molluscs, centipedes and mites among the smaller), scavengers (e.g. jackals and seagulls) or animal parasites, including a wide range of bacteria, protozoa, nematodes and winged insects. Excepting the parasites, most carnivores are not restricted to a single species for their food supply; their ranges accordingly tend to be larger than those of the animals on which they depend. DW

carrying capacity Represents the population size that the resources of an environment can just maintain without a tendency to decrease or increase. Begon *et al.* (1986: 209) explain it thus: 'As population density increases, the per capita birth rate eventually falls and the per capita death rate eventually rises. There must, therefore, be a density at which these curves cross. At densities below this point, the birth rate exceeds the death rate and the population increases in size. At densities above the cross-over point, the death rate exceeds the birth rate and the population declines. At the cross-over density itself, the two rates are equal and there is no net change in population size. This density therefore represents a *stable equilibrium*, in that all other densities will tend to approach it. In other words, intraspecific competition, by acting on birth rates and death rates, can *regulate* populations at a stable density at which the birth rate equals the death rate. This density is known as the *carrying capacity* of the population and is usually denoted by *K*.' ASG

Reference
Begon, M., Townsend, C.R. and Harper, J.L. (2006) *Ecology: from individuals to ecosystems*, 4th edition. Malden, MA: Blackwell.

carse A flat area of alluvium adjacent to an estuary.

cascading systems See SYSTEMS.

case hardening The feature or process of formation of a hard, resilient crust on the surfaces of boulders and outcrops of soft, porous rock through the filling of voids with natural cement. The cement may consist of a range of different materials, including iron and manganese oxides, silica and calcium carbonate. Beneath the hard surface the rock may be weakened, so that cavernous weathering may occur if the crust is breached. ASG

cataclasis The process of rock deformation accomplished by the fracture and rotation of mineral grains, as in the production of a crush breccia.

cataclinal Pertaining to a stream or river that trends in the same direction as the dip of the rocks over which it flows.

catastrophe/catastrophism Any major, normally short-lived and sudden event leading to widespread undesirable or disruptive change, as might be triggered by storm surges, floods and hurricanes. The theory of catastrophism sought to explain major events of Earth history, and the resulting geological structures and surface features,

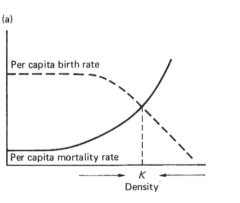

(a)

Per capita birth rate

Per capita mortality rate

K

Density

(b)

K

(c)

K

Carrying capacity. Density-dependent birth and mortality rates lead to the regulation of population size. When both are density dependent (a), or when either of them is (b and c), their two curves cross. The density at which they do so is called the carrying capacity *K*. Below this the population increases, and above it the population decreases; *K* is a stable equilibrium. But this figure is a mere caricature of real populations.
Source: Begon *et al.* (2006: figure 5.7). Reproduced with permission of John Wiley & Sons.

in terms of catastrophic events. The origins of catastrophism have often been traced to Baron Georges Cuvier (1769–1832), who believed that Earth history had to be subsumed within biblical history. He argued that, given the shortness of the creation, only catastrophic events such as the Deluge could explain the observed geological structures, fossils and features of the living world.

The theory of catastrophism is often placed in opposition to the doctrine of UNIFORMITARIANISM, which basically explains Earth history in terms of the kinds of gradual processes we see acting today, a view that was championed by James Hutton (1726–1797) and Charles Lyell (1797–1875) – see Gould (1984) for a more comprehensive description of the varied meanings of uniformitarianism. As a result of the work of Lyell and others, uniformitarian views have dominated the study of geology and the Earth sciences since the latter part of the nineteenth century. Darwin's acceptance of uniformitarianism, for example, had a major influence on his ideas about the possibility of evolution by natural selection.

In recent years catastrophism has once again become accepted as an explanatory tool (see Marriner *et al.* (2010)). The term *neo-catastrophism* is used to describe this modern mode of thinking. In palaeontology and biology, for example, the idea of 'punctuated equilibria' (i.e. discrete, sudden changes in species as opposed to gradual evolution) has gained support (Gould and Eldridge, 1977), and there have been many explanations of past mass extinctions invoking catastrophic events. It is now accepted that many geomorphological features can, in fact, be explained by recourse to catastrophic ideas. A classic example of this is provided by the work of Bretz on the channelled scablands of eastern Washington. Bretz suggested in 1923 that these scablands could best be explained by the action of a single gigantic flood over a period of only a few days. Bretz's views did not achieve much recognition when they were published, but they have since been shown to be broadly correct. Huggett (1990) provides a useful review of the history of catastrophism and its importance to biology, geology and geomorphology. It is clear that most environmental systems are affected by both catastrophic and gradual changes and the concept of a NONLINEAR SYSTEM is often used to describe systems that can demonstrate these different types of dynamics.   RH–Y

References and Reading
Benson, R.H. (1984) Perfection, continuity and common sense in historical geology. In W. A. Berggren and J. A. van Couvering (eds), *Catastrophes and Earth history*. Princeton, NJ: Princeton University Press; pp. 35–76. · Gould, S.J. (1984) Toward the vindication of punctuational change. In W. A. Berggren and J. A. van Couvering (eds), *Catastrophes and Earth history*. Princeton, NJ: Princeton University Press; pp. 9–34. · Gould, S.J. and Eldridge, N. (1977) Punctuated equilibria: the tempo and mode of evolution reconsidered. *Paleobiology*, **3**, 115–151. · Huggett, R.J. (1990) *Catastrophism. Systems of Earth history*. London: Edward Arnold. · Marriner, N., Morhange, C. and Skrimshire, S. (2010) Geoscience meets the four horsemen? Tracking the rise of neocatastrophism. *Global and Planetary Change*, 74, 43–48.

catchment   The area draining into a body of water (stream, river, or lake). See also DRAINAGE and WATERSHED.

catchment control   The nature and distribution of land use and land cover in a catchment affect the quality and quantity of water yield. Urban growth provides an instance of how yield and quality can be affected. Catchment control involves the adjustment and arrangement of land use in a catchment, often on the basis of scenarios of future change, so that as far as possible an appropriate quality and quantity of water suitable for distribution throughout the year can be ensured at minimum cost to the community.

catena   A sequence of contrasting soils formed along a topographic slope, which have acquired their different characteristics from differences in soil drainage, leaching, MICROCLIMATE erosion and depositional processes, and other factors which vary with slope position. The term was introduced by Milne (1936). In a catena, soils on steep upper sites may be well drained and better aerated, and may lose easily erodible particles by water transport to flatter sites downslope where they come to rest. Footslope conditions may be generally wetter, less well drained and aerated, and the soils perhaps deeper owing to deposition of eroded materials, though very fine particles may leave the slope altogether. Soil differences of this kind can be further accentuated by associated differences in plant cover, soil biota, biological mixing and in the supply of organic materials to the soil. The differentiated soils that make up a catenary sequence are often derived from the same parent materials and developed under the same climatic conditions. There may be differences in features like stone content, and in the sizes of rock particles found within the soils, if these materials are sorted during downslope migration from an outcrop near the crest. Along the catena, the upper layers of soils are most mixed by biota, and most susceptible to slope wash. Thus, along lower parts of the catena, the soils may be composed of relatively immobile lower layers derived primarily from the underlying bedrock together with mobile upper layers in which the material is actually undergoing progressive downslope movement. DLD

Reading and Reference
Gerrard, J. (1992) *Soil geomorphology*. London: Chapman & Hall; chapter 3, pp. 29–50. · Milne, G. (1936) Normal erosion as a factor in soil profile development. *Nature*, **138**, 548–549.

cation exchange
The interchange of cations between ADSORPTION sites on CLAY particles or organic matter, and the soil water. Clay particles generally carry a negative surface charge, which holds cations adjacent to the surface; many ions are also held in the interlayer spaces between the sheets that make up clay minerals. There are similar sites on organic macromolecules where cations can be held. Since most soils contain both clays and organic materials, cation exchange is often thought of as arising on the 'clay–humus' complex. The enormous surface area of clays, reaching hundreds of square metres per gram, is the other reason that clays are so important in retaining cations in the soil. The most important ions held are Na^+, K^+, Ca^{2+} and Mg^{2+}, these being important plant nutrients, and the amounts held are expressed by the CATION-EXCHANGE CAPACITY. In the soil, the roots, root hairs and MYCORRHIZAL FUNGI extract ions from the soil water. This increases the concentration gradient near the soil exchange complexes and additional ions are exchanged into solution, often with H^+ taking their place. In soils affected by acidic precipitation, cation exchange is involved in the neutralization process, but this depletes the store of macronutrient cations available in the exchange complex. DLD

Reading
Brady, N.C. and Weil, R.R. (1996) *The nature and properties of soils*, 11th edition. Englewood Cliffs, NJ: Prentice-Hall. · White, R.E. (2006) *Principles and practice of soil science: the soil as a natural resource*, 4th edition. Malden, MA: Blackwell.

cation-exchange capacity
A measure of the ability of the soil exchange complex, which consists of clays and organic matter (see CATION EXCHANGE), to supply the ions K^+, Na^+, Ca^{2+} and Mg^{2+} to the soil water. Cation exchange capacity (CEC) is determined in the laboratory by chemically extracting all of the available cations. It normally increases with the clay content of the soil, and is also influenced by the soil pH, not least because, in acid soils, H^+ occupies some exchange sites. Commonly, amounts of the main cations

held in exchange sites amount to 10–100 me (milli-equivalents) per 100 g of dry soil. Frequently, somewhat larger amounts are additionally held on the soil organic matter. Soils where the CEC is dominated by sodium are termed sodic soils and can be dispersible and unstable when wet. High CEC is associated with elevated soil mechanical strength. DLD

Reading
Brady, N.C. and Weil, R.R. (1996) *The nature and properties of soils*, 11th edition. Englewood Cliffs, NJ: Prentice-Hall. · White, R.E. (2006) *Principles and practice of soil science: the soil as a natural resource*, 4th edition. Malden, MA: Blackwell.

cation-ratio dating
Biogenic ROCK VARNISH coatings or patinas formed on surface boulders, rock engravings and surface artefacts provide an opportunity to obtain chronological control on the timing of exposure of underlying surfaces. Such surface varnishes are common in DRYLANDS and may exhibit a detailed microstratigraphy when examined in thin section. As the ages thus established relate to the timing of COLONIZATION of exposed substrates they represent minimum ages of surface formation. The method is based on the observation that the ratio of certain cations – (potassium and calcium)/titanium – in varnish decreases with age. It is generally accepted that this is the result of preferential leaching of the more mobile potassium and calcium. The minimum age range of the method relates to the time required for initiation of a visible rock varnish. Visible varnishes form over periods of the order of a few thousand years, while incipient varnish development may be observed microscopically over as little as 100 years.

While the cation-ratio method requires calibration to other techniques to derive absolute age estimates, the relatively low costs (compared with RADIOCARBON DATING or COSMOGENIC ISOTOPE methods), its high speed of analysis and its ability to generate relative or absolute chronologies in varnishes that do not contain sub-varnish organic matter mean that the method has considerable utility despite its numerous limitations. SS

Reading
Dorn, R.I. (1994) Surface exposure dating with rock varnish. In C. Beck (ed.), *Dating in exposed and surface contexts*. Albuquerque, NM: University of New Mexico Press; pp. 77–114. · Walker, M. (2005) *Quaternary dating methods*. Chichester: John Wiley & Sons, Ltd.

causality
The relationship between events in which a second event or configuration B can be

seen as the product of a prior event A; in other words, A is cause and B is effect. A simple causal relationship is one where B is only and always the result of A. An obvious example is the reaction of litmus paper to the application of an acid solution.

In physical geography (and in historical science in general) causality can rarely be established in a simple experimental fashion, but has to be inferred by repeated observations of A and B. Several problems arise. First, the joint occurrences of A and B may be fortuitous and there may be no physical connection between them. Second, both A and B may be responses to some other, truly causal event or variable C, and the apparently direct causal link between them misleading. Third, A may be a necessary but not a sufficient cause of B; that is, some further agency or group of agencies is involved.

It is particularly difficult to infer causality with certainty when observations are spatially contiguous or coincident, although similar problems arise with temporal sequences.

causse A term synonymous with karst, derived from the name of the limestone landscape of the Central Massif of France.

cave A natural hole or fissure in a rock, large enough for a human to enter. Although caves can be found in any type of rock, they are most common in limestone regions and are formed by solutional processes of joint enlargement. Caves can be either horizontal or vertical in general form; the latter are usually termed potholes. Those produced by solutional processes are normally initiated (i.e. by joint enlargement) in the saturated or phreatic zone. Lowering of the water table allows normal stream or vadose conditions to cut canyons in the more circular phreatic cave tubes. Thus, compound cave cross-sections can result; in this specific case a keyhole-shaped passage is produced. (Indeed, the 20 km cave Agen Allwedd in South Wales is named from the Welsh: Keyhole Cave.) Solutional processes alone do not account for all limestone cave systems; often, when the water table lowers, the overburden of rock, now no longer supported by a water-filled cavity, collapses, producing extensive boulder falls in cave passages.

The general pattern of a cave system depends not only on the processes that have led to its formation, but also on the regional jointing, folding and faulting. Caves develop along lines of weakness, and the structural geology of the area will dictate the plan and depth of a cave almost as much as fluctuations in the water table. Solutional caves can also form in rock salt, although such cavities usually form as isolated chambers rather than integrated cave passage networks. Ice, too, can provide solutional cave systems; some systems can be very long lasting (Bull, 1983).

Although caves can also be produced by tectonic activity (which is regularly referred to in textbooks as a viable mechanism of cave formation), in practice they are few and far between. They form as cavities on the limbs and crests of tightly folded rocks, but normally only very small recesses are formed, never long cave systems.

The largest of the cave systems formed in non-karstic rocks (PSEUDOKARST) are found in lava. Well-documented, long cave systems exist in Hawaiian lava flows (Wood, 1976), sometimes exceeding 10 km in passage length. They are not, of course, the results of solutional processes, but rather are products of heat loss at the edges of lava flows, with corresponding continual flowing of molten lava in the core of the flow. Repeated eruptions utilize the same passages to transport their lava along these gently dipping tubes, perpetuating the lava cave system.   PAB

References
Bull, P.A. (1983) Chemical sedimentation in caves. In A. S. Goudie and K. Pye (eds), *Chemical sediments in geomorphology*. London: Academic Press; pp. 301–319.
· Wood, C. (1976) Caves in rocks of volcanic origin. In T. D. Ford and C. H. D. Cullingford (eds), *The science of speleology*. London: Academic Press; pp. 127–150.

cavern See CAVE.

cavitation Occurs in high-velocity water (above $8–16 \, \mathrm{m \, s^{-1}}$) in irregular channels, when local acceleration causes pressure to decrease to the vapour pressure of water and airless bubbles form. Subsequent local deceleration and increased pressure result in bubble collapse. This process is a manifestation of the conservation of energy: increased kinetic energy during flow acceleration is balanced by decreased pressure energy (see GRADUALLY VARIED FLOW). The bubble implosion generates shock waves that erode adjacent solid surfaces like hammer blows. Cavitation erosion occurs in waterfalls, rapids and especially in subglacial channels where velocities of $50 \, \mathrm{m \, s^{-1}}$ have been observed (Barnes, 1956). Typical erosional products are potholes and crescent-shaped depressions called *sichelwannen*.   KSR

Reference
Barnes, H.L. (1956) Cavitation as a geological agent. *American Journal of Science*, **254**, 493–505.

celerity The speed at which a flood wave moves along a river channel. In general, in the absence of major longitudinal changes in channel form, the celerity exceeds the flow speed of the water for flows within the channel. At high river stage, when shallow overbank flow is present, the flow speed may exceed the celerity. These differences allow flood stage to peak at a downstream station earlier or later than the peak in sediment concentration arrives. The term celerity is sometimes applied to related phenomena, such as the celerity of the peak in riverine suspended sediment concentration, the celerity of a lahar or mudflow, or the upstream celerity of tsunami propagation into estuaries and rivers. In all cases, the term refers to the speed of movement of the physical feature or characteristic, rather than the flow speed of the viscous material itself.   DLD

Reading

Bull, L.J. (1977) Relative velocities of discharge and sediment waves for the River Severn, UK. *Hydrological Sciences Journal*, 42, 649–660. · Marcus, W.A. (1989) Lagtime routing of suspended sediment concentrations during unsteady flow. *Geological Society of America Bulletin*, 101, 644–651.

cell circulation, beaches Multiscale features on sedimentary coasts. At the small scale,

sediment cells on individual beaches are driven by variations in wave breaker height, with return flow taking place through rip currents at locations of lowest breaker heights. At the largest scale, entire coastlines can be characterized by a series of alongshore littoral cells, such as the five major cells along the coastline of southern California between Santa Barbara and Tijuana. Cell boundaries may be either 'fixed' between topographic features (such as pocket beach cells between rocky headlands) or 'free', with exchanges of sediment into, and out of, the cell; transfers can take place alongshore or in an onshore–offshore direction. As well as rocky headlands and submarine canyons, cell boundaries can be associated with river and estuary mouths. Human interventions at the coast can influence cell budgets, either positively through BEACH NOURISHMENT or negatively through interference with riverine sediment supply, interruption of longshore drift or direct sand mining.   TS

Reading

Davidson-Arnott, R. (2010) *Introduction to coastal processes and geomorphology*. Cambridge: Cambridge University Press.

Cell circulation at La Jolla, California: a classic example of cell circulation at the small scale caused by wave convergence between the two submarine canyons and wave divergence over the canyon heads. S.I.O.: Scripps Institution of Oceanography.

cellular automata An approach to distributed numerical modelling in physical geography that explicitly recognizes that, in many SYSTEMS, feedbacks between components of a system are spatially distributed. This follows from basic conservation laws, and notably the CONTINUITY EQUATION. For example, if one area of a bed of a stream erodes then the eroded material may well be available for deposition further downstream. Cellular automata have a number of key aspects: (1) process rules that apply to each cell; (2) strong dependence of these cell properties; (3) feedback between cell properties and process operations at the cell level; and (4) process rules that connect cells to one another. Despite widespread application in the physical and biological sciences in general, applications in physical geography are more unusual. A good example is provided by Murray and Paola (1994), who developed a cellular automata model for a BRAIDED RIVER. Their model divides the river into cells in the cross-stream and downstream directions and has (1) a discharge-based bedload equation (see BEDLOAD, BEDLOAD EQUATION) driven by (2) the DISCHARGE passing through each cell, which causes (3) the bed elevation of the cell to increase (where more sediment is supplied to the cell than can be transported) or decrease (where less sediment is supplied to the cell than can be transported), and where the discharge and sediment leaving any one cell is routed downstream whilst being distributed laterally according to prevailing bed slope. The model produced sensible behaviour when tested using a range of specially developed quantitative methods. Cellular automata reflect one of the basic principles of any numerical model: they seem to have just enough process representation to capture the fundamentals of system behaviour. However, some view them as just another form of distributed numerical modelling, whilst others are concerned about the simplicity of process representation and recognize the possibility that these models produce reasonable representations of real systems, but not necessarily for the right reasons. Nevertheless, these models may have much potential in any dynamic environmental system with strong spatial connectivity between system components. SNL

Reading and Reference
Eastwood, E., Nield, J., Baas, A., and Kocurek, G. (2011) Modelling controls on aeolian dune-field pattern evolution. *Sedimentology*, 58, 1391–1406. · Murray, A.B. and Paola, C. (1994) A cellular model of braided rivers. *Nature*, 371, 54–57. · Murray, A.B. and Paola, C. (1998) Properties of a braided stream model. *Earth Surface Processes and Landforms*, 22, 1001–1025.

Cenozoic See CAINOZOIC.

cerrado A form of savanna vegetation, comprising grasses, small trees and tangled undergrowth, found in Brazil.

channel capacity The ability of the channel to convey discharge, which is a function of channel geometry (width and depth), as well as flow velocity (and therefore bed roughness). Channel size at a particular location is related to a range of channel-forming discharges that have often been approximated by a single BANKFULL DISCHARGE or DOMINANT DISCHARGE value. Most recent research has shown how a range of flows acting upon the locally available bed and bank sediment determine channel size and capacity.

Channel capacity can be defined for different recurrence interval flows, but is commonly equated with the bankfull flow. The recurrence interval for bankfull flows varies widely between regions, from a typical 1- to 2-year recurrence interval in temperate regions to >20 years in arid regions (Williams, 1978).

In New South Wales it has been shown that a range of flows (recurrence interval 1.0–1.4 years on the annual series) affects the bedforms in the channel and a less frequently occurring range of flows (recurrence interval 1.6–4 years) is responsible for the channel capacity (Pickup and Warner, 1976). Andrews (1980) has defined effective discharge as the discharge that transports the largest fraction of the annual sediment load over a period of years, and from 15 gauging stations showed such effective discharges to be equalled or exceeded between 1.5 and 11 days per year and to have recurrence intervals on the ANNUAL SERIES ranging from 1.2 to 3.3 years.

Channel capacity can change with flow variations, such as those arising from urbanization (which can greatly increase channel capacity; e.g. Trimble, 1997) or by sediment infilling due to an increased in the sediment delivery to channels, which can dramatically reduce channel capacity and result in overbank flooding at stages formerly contained within the channel (e.g. Stover and Montgomery, 2001). DRM

Reading and References
Andrews, E.D. (1980) Effective and bankfull discharges of streams in the Yampa river basin, Colorado and Wyoming. *Journal of Hydrology*, 46, 311–330. · Pickup, G. and Warner, R.F. (1976) Effects of hydrologic regime on magnitude and frequency of dominant discharge. *Journal of Hydrology*, 29, 51–75. · Richards, K.S. (1982) *Rivers: form and process in alluvial channels*. London: Methuen. · Riley, S. (1972) A comparison of morphometric measures of bankfull. *Journal of Hydrology*, 17, 23–31. · Williams,

G.P. (1978) Bankfull discharge of rivers. *Water Resources Research*, **14**, 1141–1154. · Stover, S.C. and Montgomery, D.R. (2001) Channel change and flooding, Skokomish River, Washington. *Journal of Hydrology*, **243**, 272–286. · Trimble, S.W. (1997) Contribution of stream channel erosion to sediment yield from an urbanizing watershed. *Science*, **278**, 1442–1444.

channel classification An approach to the description of river channels based upon characteristic morphological or sedimentological parameters. The most basic form of channel classification is based upon planform morphology and typically involves four types of channel pattern distinguished based on sinuosity and style of branching: straight, MEANDERING, braided and anastomosing (see BRAIDED RIVER and ANASTOMOSING RIVER). However, actual observation of the range of fluvial forms seen in a variety of different environments has resulted in a number of additional classificatory variables.

Descriptions of channel patterns are largely qualitative, but attempts have been made to discriminate channel patterns quantitatively, both in terms of morphological properties and possible explanatory variables. In terms of quantification of channel patterns, this may involve indices of sinuosity and braiding (e.g. Church, 1992). Relatively more attention has been given to identification of possible variables that might explain different channel planform types. Leopold and Wolman (1957) recognized a continuum of river channel patterns from braiding through meandering to straight, and the role of other environmental controls. They distinguished between braided and meandering planforms on the basis of two variables, BANKFULL DISCHARGE Q_b and channel slope S, via the relation

$$S = 0.013Q_b^{-0.44} \qquad (1)$$

where braided channels had slopes steeper than this criterion and meandering channels had lower slopes. Ferguson (1987) shows how it is possible to obtain theoretical support for the model developed by Leopold and Wolman (1957), and that the threshold identified in Equation (1) is actually a threshold of specific STREAM POWER. This simple threshold has been modified to recognize that discharge and channel slope are rarely the only controls on river channel morphology. For instance, Carson (1984) argued that sand-bed rivers braid at lower slopes than gravel-bed rivers for a similar discharge regime, and this will follow from the differences in the dominant modes of transport and also differences in bank stability. Sand-bedded channels have more erodible banks, and hence braid more easily. This ties in with the observation of Murray and Paola (1994) that braiding is the fundamental instability of laterally unconstrained free-surface flows over COHESIONless beds and that meandering results from the partial suppression of braiding due to lateral channel constraints or transport of sediment in suspension. Tal and Paola (2007) subsequently demonstrated the role of bank cohesion from vegetation in maintaining dynamic single-thread, meandering channels.

Channels can undergo changes in planform over a wide variety of temporal and spatial scales (Richards, 1982). These may be long term, in response to secular change in external factors (e.g. tectonic-related uplift, climate). They may also be short term, due to perturbation of the discharge or sediment regime. Harvey (1987), for instance, describes how a hillslope failure event in the Howgill Fells, northern England, caused a change in river channel pattern from meandering to braided. However, simple divisions between timescales in this manner ignore the role of a particular river channel planform in determining the way in which it responds to secular changes in external factors. For instance, it has been argued that the braiding process may help to maintain a channel in its braided state because of the effects of braiding upon the redistribution of sediment, whereby local deposition within the channel creates the topographic complexity that drives the braiding process (e.g. Lane and Richards, 1997). If a channel is dependent upon braiding to maintain it in its braided state, then it may be more sensitive to changes in external factors that reduce or eliminate the braiding process. Nevertheless, the basic discriminatory functions may still provide a useful indicator of the probability of change between system states: the closer to the threshold or discriminating condition, the more likely that a river will change its pattern (Ferguson, 1987). Church (1992) systematized a conceptual model for the controls on the morphological types of large rivers through the planform pattern and bed material.

Many other channel classifications address various aspects of channel morphology, processes and behaviour. Frissell *et al.* (1986) proposed classifying channels through a nested set of hierarchical scales of valley segments, channel reaches with similar bedform morphology and individual channel units (features like pools or habitat types). Several systems of channel classification for upland streams have focused on differences in transport processes (Whiting and Bradley,

1993) and in the relative transport capacity as expressed in the bedform morphology (Montgomery and Buffington, 1997).   SNL/DRM

References

Carson, M.A. (1984) The meandering–braided river threshold: a reappraisal. *Journal of Hydrology*, 73, 315–334. · Church, M. (1992) *Channel morphology and typology*. In P. Calow and G. E. Petts (eds), *The river handbook*. Oxford: Blackwell; pp. 126–143. · Ferguson, R.I. (1987) Hydraulic and sedimentary controls of channel pattern. In K. S. Richards (ed.), *River channels: environment and process*. Oxford: Blackwell; pp. 129–158. · Frissell, C.A., Liss, W.J., Warren, C.E. and Hurley, M.D. (1986) A hierarchical framework for stream habitat classification: viewing streams in a watershed context. *Environmental Management*, 10, 199–214. · Harvey, A. (1987) Sediment supply to upland streams: influence on channel adjustment. In C. R. Thorne, J. C. Bathurst and R. D. Hey (eds), *Sediment transport in gravel-bed rivers*. Chichester: John Wiley & Sons, Ltd; pp. 121–150. · Lane, S.N. and Richards, K.S. (1997) Linking river channel form and process: time, space and causality revisited. *Earth Surface Processes and Landforms*, 22, 249–260. · Leopold, L.B. and Wolman, M.G. (1957) *River channel patterns: braided, meandering, and straight*. US Geological Survey Professional Paper 282-B. Washington, DC: US Government Printing Office. · Murray, A.B. and Paola, C. (1994) A cellular model of braided rivers. *Nature*, 371, 54–57. · Richards, K.S. (1982) *Rivers: form and process in alluvial channels*. London: Methuen. · Montgomery, D.R. and Buffington, J.M. (1997) Channel reach morphology in mountain drainage basins. *Geological Society of America Bulletin*, 109, 596–589. · Whiting, P.J. and Bradley, J.B. (1993) A process-based classification for headwater streams. *Earth Surface Processes and Landforms*, 18, 603–612. · Tal, M. and Paola, C. (2007) Dynamic single-thread channels maintained by the interaction of flow and vegetation. *Geology*, 35, 347–350.

channel resistance   Water flowing in a river channel encounters various sources of resistance that oppose downstream motion and result in energy loss. The potential energy of the water is converted to kinetic energy, and thence to work in overcoming frictional resistance and generating heat, as well as in transporting sediment. Channel resistance, or roughness, includes grain resistance controlled by bed material size, internal distortion resistance, which encompasses the form resistance of bedforms and flow separation at bends, and spill resistance caused by local acceleration at obstacles. Irregularity of channel form and bank vegetation add to flow resistance. The combined effect of these resistances is summarized by the composite roughness coefficients in the Manning and Chézy equations. Channel resistance is stage dependent, and often reaches a local minimum at about bankfull stage, when relative roughness is least. Resistance is also affected by high suspended sediment concentrations, by water temperature, and by riparian and in-channel vegetation, both rigid and flexible.   DLD

Reading

Robert, A. (2011) Flow resistance in alluvial channels. *Progress in Physical Geography*, 35, 765–781.

channel storage   The volume of water that can be stored along a river channel because of the variations in channel morphology. As a flood travels along a river channel the shape of the hydrograph will change as a result of the storage of water in the channel (see UNIT HYDROGRAPH). Prediction of the character of the hydrograph along the channel is called FLOOD ROUTING.   KJG

channelization (or river channelization)   The modification of river channels for the purposes of flood control, land drainage, navigation and the reduction or prevention of erosion. River channels may be modified by engineering works, including realignment, or by maintenance measures by clearing the channel. Channelization can influence the downstream morphological and ecological characteristics of river channels through channel erosion to give larger channels, deposition of the sediment released and change in the river ecology. Because of these consequences downstream from channelization schemes and because of the effects that channelization measures can have on the landscape by aesthetic degradation, alternative methods of stream restoration or stream renovation have been suggested.   KJG

Reading

Brookes, A. (1988) *Channelized rivers – perspectives for environmental management*. Chichester: John Wiley & Sons, Ltd. · Brookes, A., Gregory, K.J. and Dawson, F.H. (1983) An assessment of river channelization in England and Wales. *Science of the Total Environment*, 27, 97–111. · Keller, E.A. (1976) Channelization: environmental, geomorphic and engineering aspects. In D. R. Coates (ed.), *Geomorphology and engineering*. Binghampton, NY: State University of New York Press; pp. 115–140.

chapada   A wooded ridge or elevated plateau in the savanna areas of South America, especially Brazil.

chaparral   A vegetation type encountered in areas experiencing Mediterranean climates, characterized by evergreen shrubs with small leathery leaves. (See also MATORRAL.)

char   The solid residue left after the combustion of organic materials in fires. Char is one of many forms of pyrogenic carbon compound whose formation depends upon the temperature reached during burning, and other factors. Most chars contain a structure involving many aromatic carbon rings. There are other residues from burning, such as soot, which forms from condensation of volatile compounds from the vapour phase. After forest wildfire, char is known to accumulate at loadings of tonnes per hectare. Analysis of char abundance is required for an analysis of the net effect of forest fire on carbon release and sequestration.   DLD

Reading
Worrall, F., Clay, G.D. and May, R. (2013) Controls upon biomass losses and char production from prescribed burning on UK moorland. *Journal of Environmental Management*, **120**, 27–36.

charcoal Consisting mostly of carbon, with some ash, the black residue of incomplete combustion of wood. It is widely manufactured through heating wood in the absence of oxygen to produce an easily transportable fuel source that is black, soft, brittle and lightweight. Charcoal is also consistently a product of vegetation wildfires in fire-prone ecosystems, such as those in the savanna and Mediterranean-type climate regions. Charcoal manufacture has been a historical practice in many forested regions globally, although the mass-production of electricity using other energy sources such as coal and nuclear has largely replaced it in the developed world. In developing world countries, however, particularly in rainforest, savanna and semi-arid environments, charcoal production, both at commercial and subsistence levels, remains very common and is responsible for significant deforestation and a range of associated environmental problems (Chidumayo and Gumbo, 2013). When preserved in sediments, charcoal is a potential powerful palaeoenvironmental tool, especially when used in conjunction with other proxies, such as POLLEN ANALYSIS.   MEM

Reading and Reference
Chidumayo, E.N. and Gumbo, D.J. (2013) The environmental impacts of charcoal production in tropical ecosystems of the world: a synthesis. *Energy for Sustainable Development*, **17**, 86–94. · Scott, A.C. (2010) Charcoal recognition, taphonomy and uses in palaeoenvironmental analysis. *Palaeogeography, Palaeoclimatology, Palaeoecology*, **291**, 11–39.

chattermarks Crescent-shaped gouges found on the surfaces of rocks and rock particles (even sand grains) either as individual features or as trails. They can be produced either by the grinding of rock-armoured basal ice riding over a rock outcrop

to produce crescentic gouge trails – Chamberlain (1888) on rock; Gravenor (1979) on sand grains – or by impaction of subrounded grains on other grains in wind or water environments. These latter forms are termed Hertzian cracks in engineering science. Chattermarks on sand grains may also be produced by chemical etching, particularly in a beach environment (Bull *et al.*, 1980).   PAB

References
Bull, P.A., Culver, S.J. and Gardner, R. (1980) Chattermark trails as palaeoenvironmental indicators. *Geology*, **8**, 318–322. · Chamberlain, T.C. (1888) The rock scorings of the great ice invasions. In *Seventh annual report of the US Geological Survey to the Secretary of the Interior 1885–'86 by J. W. Powell Director*. Washington, DC: US Government Printing Office; pp. 147–248. · Gravenor, C.P. (1979) The nature of the Late Paleozoic glaciation in Gondwana. *Canadian Journal of Earth Sciences*, **16**, 1137–1153.

cheiorographic coast The characteristic coastline of areas that have experienced complex tectonic uplift and subsidence, being made up of alternating deep bays and promontories.

chelation The chemical removal of metallic ions in a rock or mineral by biological weathering. The term derives from the Greek *chela* meaning claw and reflects the process by which the metallic ion is sequestered, held between a pincher-like arrangement of two atoms (a ligand). These ligands most frequently attach themselves to the metal ion through nitrogen, sulphur or oxygen atoms. Ligands are produced by organic molecules of plant, animal and microbial origin and are important, and much neglected, processes of rock disintegration.   PAB

Reading
Ehrlich, H.L. (1981) *Geomicrobiology*. New York: Marcel Dekker.

cheluviation Results when water containing organic extracts combines with soil cations to form a chelate. This solution then moves downwards in the soil profile by a process of eluviation, transferring aluminium and iron sesquioxides into lower horizons.   ASG

chemosphere A term sometimes applied to the region of the atmosphere, mainly between 40 and 80 km in altitude, in which photochemical processes are important.

chenier A shore-parallel beach ridge, of sand, gravel or shell debris, surrounded by low-lying swamp deposits. They are generally slightly curved, with smooth seaward margins but ragged landward margins due to washovers. Conditions conducive to their formation include low wave energy, low tidal range, effective longshore currents and a variable supply of predominantly fine-grained sediment. Chenier ridge/swamp sequences are called chenier plains. The name is derived from the area to the west of the Mississippi delta, where the individual chenier ridges are up to 3 m high, 1000 m wide and 50 km long. Here, and in the Huanghe delta, China, alternation between mud deposition and ridge formation has been linked to delta lobe switching, the ridges being formed by winnowing of coarse materials on shoreline retreat at times of low mud supply. Elsewhere, however, it has been argued that these sequences are triggered by sea-level change or variations in storminess.   ASG/TS

Reading
Augustinus, P.G.E.F. (ed.) (1989) Cheniers and chenier plains. *Marine Geology*, **90**, 219–351. · Otvos, E.G. (2005) Cheniers. In M. L. Schwartz (ed.), *Encyclopedia of coastal sciences*. Dordrecht: Springer; pp. 233–235. · Saito, Y., Wei, H., Zhou, Y., *et al.* (2000) Delta progradation and chenier formation in the Huanghe (Yellow River) delta. *Journal of Asian Earth Sciences*, **18**, 489–497.

chernozem   A black soil rich in humus and containing abundant calcium carbonate in its lower horizons. A soil type characteristic of temperate grasslands, notably the Russian steppes.

chert   A cryptocrystalline variety of silica (e.g. flint), or more specifically a limestone rock in which the calcium carbonate has been replaced by silica.

Chézy equation   A flow equation developed by Antoine Chézy in 1769 and experimentally tested using data from the River Seine (see FLOW EQUATIONS). Its derivation assumes UNIFORM STEADY FLOW in which no acceleration or deceleration occurs along a reach, and the resistance to flow must therefore balance the component of the gravity force acting in the direction of the flow (Sellin, 1969):

$$v = C\sqrt{Rs}$$

where v is mean velocity, R is the HYDRAULIC RADIUS (often taken to be the mean depth in wide, shallow channels) and s is energy or bed slope. C, the Chézy coefficient, is essentially a measure of 'smoothness' or the inverse of channel resistance, and is therefore inversely related to the coefficient in the similar MANNING EQUATION.   KSR

Reference
Sellin, R.H.J. (1969) *Flow in channels*. London: Macmillan.

chine   A small ravine or canyon that reaches down to the coast, especially in southern England. Chines are well developed in sandstones near Bournemouth.

chinook   A warm, dry wind that blows down the eastern slopes of the Rocky Mountains of North America. It is warmed adiabatically during its descent and produces marked increases in temperatures, especially in the spring months. It has some similarity to the föhn winds of Europe.

chlorofluorocarbons (CFCs)   Organic compounds of human origin derived from hydrocarbons. Large-scale production of CFCs did not occur until the 1950s, though they were invented in the 1920s. CFCs contain chlorine, fluorine and carbon atoms arranged in a chemically stable (inert) structure. This compound is non-flammable, nontoxic, noncorrosive and unreactive with most other substances. This allowed for the widespread use of CFCs as coolants, insulators, foaming agents, solvents and aerosols. CFCs are extremely stable in the lower atmosphere. The molecules do not dissolve in water or break down in biological processes, so that CFC molecules eventually drift up into the stratosphere, where the sun's electromagnetic radiation in the upper atmosphere breaks apart the CFC molecules, freeing chlorine atoms. The excess chlorine atoms in the stratosphere react with ozone molecules, producing chlorine monoxide and molecular oxygen – in essence destroying the ozone molecule. The chlorine monoxide molecule can combine with an oxygen atom to produce an oxygen molecule and free the chlorine atom to begin the process all over again. One chlorine atom could destroy up to 100,000 ozone molecules. Another feature of chlorine in the upper atmosphere is the long residence time of the chlorine atoms (40–100 years). Because of the relationship between ozone destruction and chlorine the Montreal Protocol was passed in 1987 (revisions in 1989, 1990, 1992 and 1995) which diminished and eventually eliminated the production of CFCs. However, even with the Montreal Protocol amendments and adjustments, the levels of CFCs in the stratosphere will not return to the pre-1980 levels until the year 2050.   JAS

Reading
Prather, M.J., McElroy, M.B. and Wofsy, S.C. (1984) Reductions in ozone at high concentrations of stratospheric halogens. *Nature*, **312**, 227–231. · Ramanathan, V., Cicerone, R.J., Singh, H.B. and Kiehl, J.T. (1985) Trace gas trends and their potential role in climate change. *Journal of Geophysical Research*, **90**, 5547–5566.

chott   A seasonal lake, often very saline, flooded only during the winter months. Applied especially to the tectonically formed lake basins of North Africa.

chronosequence   A sequence of soils, each having undergone weathering and soil development (pedogenesis) for a different period of time. If they are located on similar parent materials, and have experienced the same climatic and other influences, the soils of a chronosequence can be used to study the rates and mechanisms of soil formation. Various sites may host chronosequences, including flights of river terraces of varying age, coastal terraces produced by tectonic processes, multiple glacial till sheets, and successive lava flows, as well as sites disturbed by human activity such as rehabilitated mine sites of varying age. Chronosequences have been used to document rates of horizon development, accumulation of organic carbon, soil carbonates, clay enrichment, development of hardpans, and many other soil features.   DLD

Reading
Eash, N.S. and Sandor, J.A. (1995) Soil chronosequence and geomorphology in a semi-arid valley in the Andes of southern Peru. *Geoderma*, **65**, 59–79.

chute   A narrow channel with a swift current, applied both to rivers and to the straits between the mainland and islands.

circadian rhythm   The approximately 24-hour rhythm of activity exhibited by most living organisms: humans, higher animals, insects and plants. The cycle is to some degree independent of day and night cycles and seems to be an important organizing principle in animal and plant physiology. Organisms isolated from external stimuli will continue to display circadian rhythms of temperature, respiration, hormone levels, and so on for some time, but may get 'out of phase' and need a diurnal cycle to reset their 'internal clocks'.   KEB

Reading
Luce, G.G. (1972) *Body time*. London: Temple Smith.

circulation index   A numerical measure of properties or processes of the large-scale atmospheric circulation. Indices have been devised to measure the strength of the east–west component of the circulation in middle latitudes (i.e. the zonal index) and the north–south component (i.e. the meridional index). The indices are usually in terms of differences in the mean pressures of two specified latitudes. The mean pressures are calculated along each of the latitudes. Lamb (1966) suggests that indices can usefully express circulation vigour if measured at points where the main air streams are most regularly developed. Indices were first used in statistical investigation connected with long-range forecasting (e.g. Walker's North Atlantic oscillation; see Forsdyke (1951)) but are now used more widely in studies of climatic change and the general circulation.   AHP

References
Forsdyke, A.G. (1951) Zonal and other indices. *Meteorological Magazine*, **80**, 156–160. · Lamb, H.H. (1966) *The changing climate*. London: Methuen.

cirque (also corrie, cwm)   A hollow, open downstream but bounded upstream by an arcuate, cliffed headwall, with a gently sloping floor or rock basin. The cirque floor is eroded by glacier sliding, while the backwall is attacked by BASAL SAPPING and subaerial rock weathering. Cirques are common in formerly glaciated uplands and have long caught the imagination of physical geographers. They were originally thought to have been formed during the waxing and waning of ice sheet glaciations, but few were occupied by active glaciers during ice sheet withdrawal. Instead, it seems likely that they represent many stages during the past few million years when marginal glaciation affected mid-latitude uplands. Most mid-latitude cirques show a preferred orientation towards the northeast in the northern hemisphere and towards the southeast in the southern hemisphere, reflecting mainly the effect of shade in protecting the glacier from the sun but also the effect of wind-drifted snow accumulated by predominantly westerly winds. Preferred orientation is less important in polar and tropical mountains. Cirque altitude is an indication of former snow lines, and it is common for basin altitudes to increase with distance from a coast. Cirques have attracted much morphometric analysis and, although the main controls on their morphology remain unclear, it seems that they tend to become more enclosed and deeper with time.   DES

Reading
Derbyshire, E. and Evans, I.S. (1976) The climatic factor in cirque variation. In E. Derbyshire (ed.), *Geomorphology and climate*. New York: John Wiley & Sons, Inc. · Gordon, J.E. (1977) Morphometry of cirques in the Kintail–Affric–Cannich area of northwest Scotland. *Geografiska Annaler, Series A: Physical Geography*, **59**(3–4), 177–194.

cirrocumulus   See CLOUDS.

cirrostratus   See CLOUDS.

cirrus   See CLOUDS.

cladistics (or phylogenetic systematics)

The elucidation of the evolutionary history of groups of organisms. Hennig (1966) was responsible for its initial development. Subsequently, it has been refined into a method with more general properties than those stated by Hennig and having a much wider application than he intended (Humphries and Parenti, 1986).

Biological systematics seeks to describe and to classify the variation between organisms. Such classifications are often hierarchical and reveal patterns of association (TAXONOMY). Resemblances among the intrinsic properties of organisms (encompassing everything from their chemistry to conduct) can be classified to reveal phylogenetic patterns. As Eldredge and Cracraft (1980) point out, the pattern of similarity of features in organisms has resulted from either evolution by descent or special creation (CREATIONISM). If evolution by ancestry and descent is accepted, together with the fact that novel characteristics appear in organisms at different times during its course, a hierarchical phylogenetic pattern of clustered intrinsic resemblances (akin to similarities in taxa) can be formulated.

According to Hennig (1966), it is possible to distinguish two principal categories of resemblance in a monophyletic taxon (one having two or more species and including an ancestor and descendants). First, true evolutionary similarities. These may be of two kinds, either those acquired from a distant common ancestor and retained by one or more of its descendants, or those possessed solely by a particular group of organisms that acquired them from a recent common ancestor. The second category is that of false or misleading resemblance and derives from adaptations made during parallel and convergent EVOLUTION. Cladists contend that the intermittent retention of resemblances acquired from an ancient source renders such traits only useful as clustering agents where they can claim novel status. Their methodology also indicates that because ancestors do not appear to have special assemblages of novelties their definition is difficult.

Accordingly, phylogenetic systematists concentrate upon recently acquired evolutionary resemblances in an attempt to ascertain patterns of novelties and establish proximate relationships between taxa. The intrinsic characteristics of organisms are examined in detail. An attempt is made to establish whether these characteristics are primitive or derived. Widespread traits within groups are thought to have originated there, while characters occurring in only a few instances are considered as derived. Significant patterns of phylogenetic resemblance have also emerged after the analysis of homologous characteristics at various stages in the life cycle of organisms. The taxon being investigated is also compared in detail with its nearest relative (sister), as part of the search for character distributions. The aim is to find a universal set (one or more) of similarities in order that a group in which they are all present may be defined. Resemblances that are common to only some of the group can also be used to categorize subsets. Inferred inherited similarities that define subsets at some level within a universal set are known as homologies. Thus, homology is synonymous with synapomorphy and refers to novel, shared resemblance endowed by a recent common ancestor (Eldredge and Cracraft, 1980).

Data from such analyses are presented as branching diagrams or cladograms, upon which are illustrated clusters of common resemblances regarded as evolutionary innovations. It is often possible to map more than one cladogram from one data set. When compared, these exhibit different patterns that require further investigation. An example provided by Eldredge and Cracraft (1980) will illustrate the principles and problems of cladistic analysis.

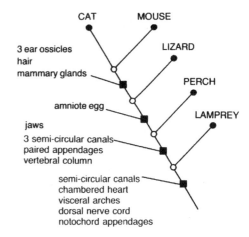

A cladogram for five kinds of vertebrates. Each level of the hierarchy (denoted by branch points) is defined by one or more similarities interpreted as evolutionary novelties.
Source: Eldredge and Cracraft (1980: figure 2.1). Reproduced with permission of Joel Cracraft.

The cat, mouse, lizard, perch and lamprey are vertebrates with a universal set of resemblances that include semicircular canals in their heads and a chambered heart. Subgroups can be made of the perch and lamprey as they lack an amniote egg, and the lizard, perch and lamprey which do not possess hair or mammary glands. The apparent conflict in the relationships between these organisms can be clarified with reference to the stages in their life cycles. Each has a substantial amount of cartilage in at least one stage of their ontogeny. Cartilage is superseded by bone in the advanced stages of development of the cat, mouse, lizard and perch. Thus, bone is diagnostic of a subgroup (each member of which is also part of a bigger group that includes the lamprey) whose members possess a vertebral column, paired appendages, jaws and semicircular canals. A further subgrouping emerges with reference to an amniote egg, which is possessed by the cat, mouse and lizard. Consideration of earlier developmental stages of the organisms indicates that all five then had a general vertebrate egg, an amniotic membrane subsequently emerging in the cat, mouse and lizard. The cat and mouse can be combined in isolation from the lizard as they are endowed with hair, mammary glands and three ear ossicles, these again being adaptations from more widespread vertebrate traits. Such relationships can be expressed in the cladogram shown here.

The next stage in the cladistic analysis of these data is to look outside the group of five organisms for others with equivalent shared resemblances. In this context, two other chordates, the lancelet amphioxus and the tunicates, are relevant. They, however, do not have the characteristics of the cat, mouse, lizard and perch grouping, nor of the large group that includes the lamprey. Thus, the amphioxus, tunicates and lamprey comprise outgroups and the cat, mouse, lizard and perch a subgroup. The latter can be further subgrouped in that cats, mice and lizards possess amniote eggs but perch do not, their eggs being analogous to those of the outgroups. Finally, cats and mice are the exclusive possessors of hair, mammary glands and three ear ossicles, qualifying as a subgroup by virtue of these evolutionary novelties.

A progression in this type of analysis, whereby more complex and accurate hypotheses concerning ancestor–descendant relationships can be formulated, comes via additional information and takes the form of a phylogenetic tree, which can also be represented diagrammatically. At the end of this sequence is the phylogenetic scenario which consists of a tree with an overlay of adaptational narrative (Eldredge, 1979).

Cladistics has caused considerable controversy and acrimony among systematists. Some cladistic viewpoints conflict, often sharply, with certain of those of the other major biological taxonomists – the evolutionary systematists and

the numerical taxonomists. Eldredge and Cracraft (1980) state that cladistic strategy involves investigation of the structure of nature, and the formulation of hypotheses concerning this which contain the basis for their own evaluation. This approach differs from that of the Darwinian school of evolutionary taxonomy, which, they contend, invokes acknowledged processes in the explanation of form, such that its conclusions are not amenable to disproval. The cladists do not claim to prove anything, but to carry out pattern analysis without fixed ideas concerning causal processes, and within a framework that accepts that evolutionary patterns and processes are related. They also accept that morphological and taxonomic diversity are related but not synonymous, stating that evolutionary systematists believe that they are equivalent. The principles of cladistics also have a bearing on speciation. Cladists recognize species as distinctive entities to the extent that the emergence of a new one interferes with the pattern of ancestry and descent. Darwinists explain long-term intrinsic modifications in species populations by means of natural selection and adaptation. Cladism leads to the notion that such gradual speciation is unlikely. Cladists believe that species remain unmodified for a period, and are quite suddenly replaced by others. Speciation occurs in areas of various dimensions. When these areas are defined, they can be related to the phylogenetic histories of the relevant species as depicted by cladograms (VICARIANCE BIOGEOGRAPHY).

Cladists disagree with numerical taxonomists who, they contend (mainly on the basis of computerized cluster analysis), feel that evolutionary history defies reason. They are at variance with both evolutionary systematists and numerical taxonomists over a fundamental tenet of their method. This relates to cladistic assemblages that comprise novelties rather than retentions, the latter being the usual characteristics employed by alternative approaches. In consequence, the cladists' interpretation of the evolutionary status of certain taxa differs from that of their fellow systematists.   RLJ

Reading and References
Eldredge, N. (1979) Cladism and common sense. In J. Cracraft and N. Eldredge (eds), *Phylogenetic analysis and paleontology*. New York: Columbia University Press; pp. 165–197. · Eldredge, N. and Cracraft, J. (1980) *Phylogenetic patterns and the evolutionary process*. New York: Columbia University Press. · Hennig, W. (1966) *Phylogenetic systematics*. Urbana, IL: University of Illinois Press. · Humphries, C.J. and Parenti, L.R. (1986) *Cladistic biogeography*. Oxford: Clarendon Press.

clast A coarse sediment particle, usually larger than 4 mm in diameter; clast sizes include pebbles, cobbles and boulders.

clastic Composed of or containing fragments of rock or other debris that have originated elsewhere.

clay Term applied in the textural classification of soils and sediments to particles having diameters of less than 2 µm, as well as in soil chemistry to a wide range of crystalline hydrous layer silicate minerals (the 'clay minerals'). The clay minerals, of which there are many varieties, are derived from the weathering of primary silicate minerals, and are made up of stacked layers of tetrahedrally and octahedrally structured sheets of oxygen atoms bound to cations, notably Si^{4+}, Al^{3+}, Fe^{3+}, Fe^{2+} and Mg^{2+}. These materials ordinarily occur as small crystalline fragments less than a few micrometres in size. Whilst many clay particles, *texturally* defined, do consist of such clay mineral fragments, clay-sized particles of other materials, such as quartz, also occur. Furthermore, clay mineral particles, which carry a negative surface charge because of ionic substitution within their crystal lattice (say, Si^{4+} replaced by Al^{3+}, leaving an unbalanced negative charge), are often held together in clumps or flocs, and when held in this way the composite particle composed of many clay particles can itself be of much larger size. (See also PARTICLE SIZE.)   DLD

Reading
Brady, N.C. and Weil, R.R. (1996) *The nature and properties of soils*, 11th edition. Englewood Cliffs, NJ: Prentice-Hall.

clay dune A dune (often LUNETTE DUNE) where a significant proportion of the constituent sediments (up to 30%) comprises clay minerals. These are likely to have been transported to the dune, via saltation and creep, as CLAY PELLETS, which may have become broken down, or further aggregated, after deposition. Clay dunes usually border lacustrine basins and result from the deflation of basin floor sediments during PLAYA or PAN stages. Clay dunes occur in parts of the Kalahari, Australia, North Africa and Texas, USA.   DSGT

Reading
Bowler, J.M. (1973) Clay dunes: their occurrence, formulation and environmental significance. *Earth Science Reviews*, 9, 315–338.

clay–humus complex Consists of a mixture of clay particles and decaying organic material that attracts and holds the cations of soluble salts within the soil profile.

clay pellet A particle of sediment that, although SAND or SILT sized, is in fact composed of aggregated CLAY-size particles. Pellet formation may occur through the process of FLOCCULATION or through the breakdown of clay layers through salt efflorescence. Sand-size clay pellets will be transported through the processes of SALTATION, CREEP and REPTATION. On deposition, pellets may become deflocculated. If a particle size analysis of such a sediment is conducted, the resulting distribution may overrepresent clay in respect of the transport processes and mechanisms through which the deposit was achieved. One environment where clay pellet formation may be important is on the floor of PANS and PLAYAS, with pellets transported by the wind to an adjacent LUNETTE DUNE.   DSGT

claypan A stratum of compact but not cemented clayey material found within the soil zone.

clay-with-flints An admixed deposit of clay and gravel, predominantly flint, occurring locally in depressions in the chalk uplands of southern England. Probably the insoluble components of the chalk and/or reworked Tertiary deposits.

clear-water erosion The erosion caused by rivers whose sediment load has been removed by the construction of a dam and reservoir. With a reduced sediment load, incision occurs rather than aggradation. This can create serious problems for bridges and other man-made structures downstream.

cleavage In minerals, the plane along which a crystal can be split owing to its internal molecular arrangement. Also those planes of weakness in fissile rocks that are not related to jointing or bedding.

CLIMAP In 1971 a consortium of scientists from many institutions was formed to study the history of global climate over the past million years, particularly the elements of that history recorded in deep-sea sediments. This study is known as the CLIMAP (Climate, Long-range Investigation, Mapping and Prediction) project. One of CLIMAP's goals is to reconstruct the Earth's surface at particular times in the past, and a good example is contained in CLIMAP Project Members (1976).   JGL

Reading and Reference
CLIMAP Project Members (1976) The surface of the ice-age earth. *Science*, **191**, 1131–1137. · Gates, W.C. (1976) Modelling the ice-age climate. *Science*, **191**, 1138–1144.

climate The long-term atmospheric characteristics of a specified area. Contrasts with weather. These characteristics are usually represented by numerical data on meteorological elements, such as temperature, pressure, wind, rainfall and humidity. These data are frequently used to calculate daily, monthly, seasonal and annual averages, together with measures of dispersion and frequency.

Climatic statistics have been published for a huge number of stations, and there are some useful publications that have summarized tables of major climatic phenomena. Among the most useful are Landsberg (1969–1984), Wernstedt (1972) and NOAA (2004).   ASG

Reading and References
Landsberg, H.E. (ed. in chief) (1969–1984) *World survey of climatology*, 15 vols. Amsterdam: Elsevier. · Müller, M.J. (1982) *Selected climatic data for a global set of standard stations for vegetation science*. The Hague: Junk. · NOAA (2004) *Climatic data records from environmental satellites*. Washington DC: National Academies Press. · Wernstedt, F.L. (1972) *World climatic data*. Lemont, PA: Climatic Data Press.

climate change Defined by the Intergovernmental Panel for Climate Change (IPCC) as any alteration in climate over time, whether due to natural variability or as a result of human activity (Stocker *et al.*, 2013). This differs from that of the United Nations Framework Convention on Climate Change (UNFCCC), which defines it as a change of climate attributed directly or indirectly to human activity that alters the composition of the global atmosphere that is in addition to natural climate variability observed over comparable time periods (UNFCCC, n. d.). It can also be considered to be a change in the state of the climate that can be identified by statistically significant changes in the mean and/or the variability of its properties and that persists for an extended period, typically decades or longer. The study of climate change can be classified into a number of different areas of investigation, including climate reconstruction, climate change theory and climate modelling.

Climate reconstruction, or PALAEOCLIMATOLOGY, relates to the study of the Earth's past climates and determination of those past climates. Records of actual measures of climate, such as temperature and precipitation, are often only available for limited places and for limited times of the past. Modern weather service observations, for example, have only been available for the last century and half, and then primarily only in western Europe and North America. Consequently, other measures of climate variability, termed climate proxies, are needed to extend our understanding of past climates both geographically and temporally.

For the time associated with human civilizations, many historical records, diaries, planting records and even clothing and building materials can supply useful information on climate conditions for a given civilization. More detailed and geographically diverse information, however, can be obtained from various geological and biological indicators, termed physical climate proxies. Analyses of such biologic indicators as tree rings, lake varves, vegetation growth boundaries and pollen distributions have provided detailed climate information such as the precipitation and temperature regimes of relatively remote regions of the Earth from several thousand to millions of years into the past.

Other techniques can also be used to evaluate the climates of the Earth's past. Isotope analysis can be applied to glacial ice cores or to deep-sea marine sediment cores to gain data on past temperature and global ice volumes (see OXYGEN ISOTOPE). The biogenic material found in a marine core may also be analysed for proxy climate data, since certain species display identifiable temperature-dependent characteristics. For example, the coiling direction of some kinds of foraminfera is dependent on water temperature, which in turn is influenced by atmospheric temperatures.

Abiotic proxy evidence can be garnered from such sources as the analysis of isotopic concentrations in calcium carbonate (such as the well-dated core taken from Devil's Hole, Nevada), the analysis of the extent of ancient desert sediments and the analysis of fossil pollen accumulated within sediments. In all cases, however, consideration has to be given to the relative influence of climatic and other environmental factors that have affected the magnitude and distribution of living species and the operation of ecological, sedimentary and geomorphological processes.) Lowe and Walker (2014) give an excellent summary of various means of climate reconstruction.

A second major component in the field of climate change is the formulation and evaluation of climate change theories. Such theories are constructed to explain either the various shifts in climate revealed through climate reconstruction and study of secular climate records or to account for anticipated changes in global or regional climate. Although hundreds of climate change theories have been hypothesized in the last century or so, two have reached great prominence and acceptance among climatologists.

95

The astronomical theory of climate change, often referenced by the name of one of its leading early researchers, Multin Milankovitch, states that periodic changes in the Earth's climate are linked to cyclic orbital variations of the Earth. In particular, the climatic changes of the last 2 million years have been statistically linked to variations in three of the Earth's orbital mechanisms. Obliquity, changes in the axial tilt of the Earth, operates with a periodicity of 41,000 years. Precession of the longitude of the equinoxes, the wobble in the Earth's orbit that changes the timing in which perihelion (the time at which the Earth is closest to the sun) occurs in relation to the equinoxes, occurs with variability of 21,000 years. The third mechanism, eccentricity, the variability in the shape of the Earth's orbit, has a periodicity of 100,000 years. The cycles associated with these mechanisms have been identified in the analyses of deep-sea marine sediment cores across the globe.

A second climate change theory, the GREEN-HOUSE EFFECT, has received significant scientific and public attention in recent decades. In brief, carbon dioxide and other gases, including water vapour, act to absorb terrestrial radiation in the lower atmosphere. Such absorption acts as an extra source of heat to the surface. Consequently, it is hypothesized, when all other factors are held constant, that increasing amounts of various greenhouse gases will lead to general global warming. This theory has been used to explain the abnormal global warmth revealed in reconstruction evidence from the age of the dinosaurs (the Cretaceous). Greenhouse theory has also been used to produce future scenarios of the Earth's climate given the increasing amount of human-produced greenhouse gases emitted into the atmosphere. Two current concerns associated with this theory are the identification of potential feedback mechanisms that may serve to mitigate the global warming, and the identification of associated climate impacts beyond global warming, such as changes in regional precipitation.

A third subdiscipline in climate change studies is the creation, testing and application of numerical climate models. Climate models have become a major tool of investigation for climate change theories, as well as a means for evaluating regional differences in climate derived from reconstruction proxies and secular climate records. A hierarchy of climate models exists based on the complexity of their physical relationships and the number of their spatial dimensions.

Radiative–convective models are generally one-dimensional models that focus on mathematical expression of the radiative processes in the atmosphere, particularly as influenced by atmospheric water vapour and aerosols. Energy balance models (EBMs) have been formulated using a single point determination (zero-dimensional), with latitudinal dependence (one-dimensional), with latitude and vertical components (two-dimensional) and with all three spatial dimensions (three-dimensional). EBMs are constructed around a primary equation that addresses the various energy transfers (such as by advection of sensible heat, latent heat or by ocean currents) that occur in the Earth's climate system.

The most complex of numerical climate models are those known as general circulation models (GCMs –see GENERAL CIRCULATION MODELLING), which incorporate many of the features of both radiative–convective models and EBMs. GCMs are mathematical computer models describing the primary controls (such as radiative fluxes), energy transfers, circulations and feedbacks existing in the Earth–ocean–climate system. GCMs have their roots in the early numerical weather models developed by J. G. Charney and by J. von Neumann, the so-called 'father' of modern computing, in the late 1940s. Most GCMs are based on seven fundamental (or 'primitive') equations of the atmosphere and they 'parameterize' various physical processes.

Climate models have been used to study the interrelationships between climate processes (such as albedo and temperature), evaluate and explain reconstruction proxy data (such as the GCM simulations of the Cretaceous and Pleistocene climates) and to address possible effects of increased atmospheric carbon dioxide and other greenhouse gases. Over the past decade considerable effort has been invested into quantifying and understanding the uncertainties of climate model simulations Washington et al. (2009)) give an excellent review of both fundamentals of computer modelling and the applications to which they have been applied.

Climate change study has become one of the most important sciences in recent years. Appreciating the variability inherent in our climate system is becoming increasingly critical to both understanding the growth and decline of past human civilizations and, indeed, other species on this planet, as well as the formulation of future public policy regarding all elements of human life.

The IPCC is the leading international body for the assessment of climate change. It was established by the United Nations Environment Programme (UNEP) and the World Meteorological

Organization (WMO) in 1988, and endorsed by the UN General Assembly in the same year. It aims to provide a clear scientific view on the current state of knowledge in relation to climate change and its potential environmental and socio-economic impacts. Currently, 195 countries are members of the IPCC. Governments participate in the review process and the plenary sessions, where main decisions about the IPCC work pro-gramme are taken and reports are accepted, adopted and approved. Its Fifth Assessment Report (AR5; Stocker *et al.*, 2013) is the most recent. It confirms that warming of the climate system is unequivocal, with each of the last three decades being successively warmer at the Earth's surface than any preceding decade since 1850. Compared with its earlier assessments, the more detailed and longer observations and improved climate models now enable the human contribu-tion to be detected with more certainty. AR5 concludes that it is extremely likely that human influence has been the dominant cause of the observed warming since the mid-twentieth cen-tury.   RSC, RH-Y, SG, SJN, GS

Reading and References
Berger, A. and Loutre, M.F. (2006) Milankovitch theory and paleoclimate. In S. E Elias (ed. in chief), *Encyclopedia of Quaternary science*. Amsterdam: Elsevier; pp. 1017–1022. · Bradley, R.S. (2014) *Paleoclimatology: reconstructing cli-mates of the Quaternary*, 3rd edition. San Diego, CA: Academic Press. · Fletcher, C. (2012) *Climate change*. Chi-chester: John Wiley & Sons, Ltd. · Lowe, J. and Walker, M. (2014) *Reconstructing Quaternary environments*, 3rd edition. Routledge: London. · McGuffie, K. and Henderson-Sell-ers, A. (2014) *The climate modelling primer*, 4th edition. Oxford: Wiley-Blackwell. · Stocker, T.F., Qin, D., Platt-ner, G.-K., *et al.* (eds) (2013) *Climate change 2013: the physical science basis*. Cambridge: Cambridge University Press. · UNFCCC (n.d.) First steps to a safer future: Introducing The United Nations Framework Convention on Climate Change. http://unfccc.int/essential_background/convention/items/6036.php (accessed 27 July 2015). · Washington, W.M., Buja, L. and Craig, A. (2009) The computational future for climate and Earth system models: on the path to petaflop and beyond. *Philo-sophical Transactions of the Royal Society A: Mathematical, Physical and Engineering Sciences*, **367**, 833–846.

climate modelling   See GENERAL CIRCULATION MODELLING.

climate   modification See   WEATHER MODIFICATION.

climate sensitivity parameter A numeri-cal index of the amount by which the mean surface temperature of the Earth would change in response to a unit change in radiative forcing.

The radiative forcing is primarily the solar radia-tion received by the Earth, but also includes energy trapped within the Earth–atmosphere system, and resulting fluxes of longwave radiation. For exam-ple, greenhouse gases in the atmosphere trap heat energy, and the resulting radiant heat flux is esti-mated to increase radiative forcing by about $4\,\mathrm{W\,m^{-2}}$ for a doubling of carbon dioxide. The climate sensitivity parameter λ links this change in radiative forcing ΔQ to the rise in mean surface temperature ΔT_s as follows:

$$\Delta T_s = \lambda \Delta Q$$

The value of λ is $<1\,K\mathrm{m^2\,W^{-1}}$, but its value is difficult to determine exactly. This is because many feedback processes, such as changes in snow, ice and cloud cover, and the water vapour content of the atmosphere, are triggered by a change in T_s, and so affect the value of λ.   DLD

Reading
Houghton, J.T., Jenkins, G.J. and Ephraums, J.J. (eds) (1990) *Climate change: the IPCC scientific assessment*. Cam-bridge: Cambridge University Press.

climatic classification A systematic parti-tioning of the Earth's climates based on the aver-ages of weather elements, the variations of weather elements and the effects of weather elements on the land surface environment.

The first attempt at climate classification was made by early Greek civilizations, who used only latitude to subdivide the Earth's global climate into three distinct zones: torrid, temperate and frigid. Further attempts at classifying the Earth's climates were not made until the end of the nineteenth and the beginning of the twentieth centuries. In the late 1800s, German plant phys-iologist Wladimir Köppen (1846–1940) began to recognize that a plant or plant communities at a particular place reflected a synthesis of the weather elements of that location. Comparing the global distribution of vegetation with his own data on world distributions of temperature and precipitation, Köppen identified a correlation between the atmosphere and the biosphere, and he subsequently began to distinguish cli-mates based on those interrelations. The first version of the Köppen climate classification sys-tem, based solely on temperature, was published in 1884. Numerous modifications have been made by Köppen himself and by other scientists, including his contemporary Rudolf Geiger.

The Köppen climate classification system uses six major climate groups that are subdivided once, and in some cases twice, based on specific temperature   and   moisture   characteristics.

Relative magnitudes of precipitation and its predominant timing within the year, as well as data on the warmest and coldest months of the year, provide a finer definition of the climate groups. Largely based on the distribution patterns of the Earth's vegetation, many of the climate groups derived from the Köppen system bear names associated with the type of vegetation that they support, such as the tropical rain forest, desert and tundra climates.

Several scientists of the early to mid twentieth century opposed Köppen's system, the most notable of whom was C. Warren Thornthwaite (1899–1963). Thornthwaite (1948) argued that the classification of the Earth's climates required a greater understanding of spatial variations in the climatic water budget, not just simply variations in temperature and moisture. Accordingly, the 1948 climate classification system published by Thornthwaite is based on the concept of potential evapotranspiration, or the amount of evapotranspiration that would occur from a well-watered land area. This maximum amount of water loss through evapotranspiration is a function of air temperature and vegetation cover. Thornthwaite's method differs from that of Köppen in that it expresses the effectiveness of precipitation and the efficiency of air temperature in the process of evapotranspiration, both of which indicate the degree to which vegetation can be supported. Similar to the method of Thornthwaite, Mikhail Budyko later developed an approach in which, instead of temperature, the net radiation available for evaporation from a moist surface is related to the amount necessary to evaporate the mean annual precipitation (Budyko, 1974). As a reflection of the general climate, Budyko used that ratio to classify vegetation zones. Arthur Strahler (1978) followed Thornthwaite's water balance approach, but delineated climate zones on the basis of soil moisture and its seasonality. Others, such as Peveril Meigs (1953), developed schemes specifically for arid regions.

These schemes suffer from several weaknesses. For one, they are empirical. For another, they do not incorporate climatic changes, so that the boundary between one climate type and the next may differ from year to year. This can be the case when 30-year mean data are used. Recognizing these weaknesses, Flohn (1957) and Trewartha and Horn (1980) proposed alternative 'genetic' approaches, focusing on the systems generating the specific climate type. These approaches thus link climate types to the general atmospheric circulation, but they have the disadvantage of being nonquantitative.

Classification was once a major focus of climatology. Today, there has been a shift towards understanding the processes creating climate (i.e. climate dynamics). In this case, the 'genetic' approaches have more pedagogical value, but the water balance approaches have more meaning for agricultural or environmental planning. AWE/SEN

Reading and References

Budyko, M.I. (1974) *Climate and life*. New York: Academic Press. · Flohn, H. (1957) Zur Frage der Einteilung der Klimazonen. *Erdkunde*, **11**, 161–175. · Köppen, W. (1900) Versuch einer Klassifikation der Klimatre, Vorzugweise nach ihren Beziehungen zur Pflanzenwelt. *Geographische Zeitschrift*, **6**, 593–611. · Meigs, P. (1953) The distribution of arid and semiarid homoclimates. In *UNESCO, arid zone research series 1, review of research on arid zone hydrology*. Paris: UNESCO; pp. 203–210. · Strahler, A.N. (1978) *Physical geography*. New York: John Wiley & Sons, Inc. · Thornthwaite, C.W. (1948) An approach towards a rational classification of climate. *Geographical Review*, **38**, 55–94. · Trewartha, G.W. and Horn, L.H. (1980) *An introduction to climate*. New York: McGraw-Hill.

climatic geomorphology

This subject developed during the period of European colonial expansion and exploration at the end of the nineteenth century, when unusual and spectacular landforms were encountered in newly discovered environments like deserts and the humid tropics. In the USA, W. M. Davis recognized 'accidents', whereby non-temperate climatic regions were seen as deviants from his normal cycle and introduced, for example, his arid cycle (Davis, 1905). Some regard Davis as one of the founders of climatic geomorphology (see Derbyshire (1973)), though the French (e.g. Tricart and Cailleux, 1972) have criticized him for his neglect of the climatic factor in landform development. Much important work on dividing the world map into climatic zones (morphoclimatic regions) with distinctive landform assemblages was attempted both in France (e.g. Tricart and Cailleux, 1972) and in Germany (e.g. Büdel, 1982). This geomorphology described itself as geographical (Holzner and Weaver, 1965).

In recent years, certain limitations have become apparent:

1 Much climatic geomorphology has been based on inadequate knowledge of rates of processes and on inadequate measurement of process and form.
2 Many of the climatic parameters used for morphoclimatic regionalization have been seen to be meaningless or crude.
3 The impact of climatic changes in Quaternary times has disguised the climate–landform relationship.

4 Climate is one step removed from process.

5 Many supposedly diagnostic landforms are either relict features or have a form that gives an ambiguous guide to origin.

6 Climate is but one factor in many that affect landform development.

7 Climatic geomorphology tends to concentrate on the macroscale rather than investigating the details of process.

8 Macroscale regionalization has little inherent merit. ASG

Reading and References
Birot, P. (1968) *The cycle of erosion in different climates.* London: Batsford. · Büdel, J. (1982) *Climatic geomorphology.* Princeton, NJ: Princeton University Press. · Davis, W.M. (1905) The geographical cycle in an arid climate. *Journal of Geology,* **13**, 381–407. · Derbyshire, E. (ed.) (1973) *Climatic geomorphology.* London: Macmillan. · Derbyshire, E. (ed.) (1976) *Geomorphology and climate.* Chichester: John Wiley & Sons, Ltd. · Holzner, L. and Weaver, G.D. (1965) Geographical evaluation of climatic and climato-genetic geomorphology. *Annals of the Association of American Geographers,* **55**, 592–602. · Stoddart, D.R. (1969) Climatic geomorphology. In R. J. Chorley (ed.), *Water, earth and man.* London: Methuen; pp. 473–485. · Tricart, J. and Cailleux, A. (1972) *Introduction to climatic geomorphology.* London: Longman.

climatic hinge An imaginary line in the tropical zone that separates areas receiving the bulk of precipitation in the northern summer from those affected by summer rainfall in the southern hemisphere. Although this currently equates to the position and movement of the INTERTROPICAL CONVERGENCE ZONE (ITCZ), the concept has been used in palaeoclimatic studies, particularly in Africa, to identify the precipitation equator once it became apparent that long-term changes in the two hemispheres were neither identical nor synchronous. PSH

climatic optimum See ALTITHERMAL.

climato-genetic geomorphology An attempt to explain landforms in terms of fossil as well as contemporary climatic influences. Büdel (1982) recognized that landscapes were composed of various relief generations and saw the task of climato-genetic geomorphology as being to recognize, order and distinguish these relief generations, so as to analyse today's highly complex relief.

Reference
Büdel, J. (1982) *Climatic geomorphology.* Princeton, NJ: Princeton University Press.

climatology Traditionally concerned with the collection and study of data that express the prevailing state of the atmosphere. The word *study* here implies more than simple averaging. Various methods are used to represent climate; for example, both average and extreme values, frequencies of values within stated ranges and frequencies of weather types with associated values of elements. Modern climatology is concerned with explaining, often in terms of mathematical physics, the causes of both present and past climates. JGL

Reading
Barry, R.G. and Chorley, R.J. (1992) *Atmosphere, weather and climate,* 6th edition. London: Routledge. · Lockwood, J.G. (1979) *Causes of climate.* London: Edward Arnold.

climax vegetation Vegetation COMMUNITY regarded as existing in a state of equilibrium with given climatic and edaphic conditions and capable of self-perpetuation. The idea was initially developed by the American biologist F. E. Clements in 1916, who hypothesized that, following colonization of a new surface (e.g. a lava flow or bare sand dune), ecological SUCCESSION would proceed until a final stage was reached in which the community was in harmony with prevailing environmental conditions. Climate was considered to be the dominant ecological determinant, and the end product of succession was therefore referred to as the climatic climax community; for example, deciduous woodland in the cool temperate climates of northwest Europe and eastern North America. The climax community is perceived as having greater species diversity than earlier successional stages and is a correspondingly more stable community under conditions of higher levels of complexity and interdependence. Intuitively attractive though the notion is, it is now realized that the dynamics of ecosystems are such that equilibrium, implying stasis, is unlikely to be achieved due to spatial and temporal variability in environmental determinants, imposing ecological stresses of varying degrees of magnitude upon communities. Change, as opposed to stability, in community composition is probably the norm, and the term *mature community* is a more appropriate one to describe ecosystems eventually developing when the degree of stress is low. MEM

Reading
Burrows, C.J. (1990) *Processes of vegetation change.* London: Unwin Hyman.

climbing dune A TOPOGRAPHIC DUNE that has developed on the windward side of a topographic obstacle, as airflow and sand transport is disrupted by the barrier. Where the hillslope is less than 30° sand transport is not disrupted, but at 30–50° a dune develops, up to twice as long as the height of the hill (Cooke *et al.*, 1993). If the barrier is a discrete hill rather than an escarpment, the dune may eventually wrap around the obstacle. See also ECHO DUNE and SAND RAMP.   DSGT

Reference
Cooke, U., Warren, A. and Goudie, A.S.G. (1993) *Desert geomorphology*. London: UCL Press; pp. 353–354.

climogram   Usually refers to two types of climatic diagram, a climograph and a hythergraph, that were introduced by T. Griffith Taylor in 1915. The diagrams depict graphically the annual range of climatic elements at a particular location, emphasis being placed on the way in which these elements affect the comfort of humans. In a climograph, mean monthly WET-BULB TEMPERATURE is plotted as the ordinate and mean monthly relative humidity as the abscissa. The points are then joined to form a 12-sided polygon. In a hythergraph, mean monthly temperature is plotted against mean monthly precipitation. Data for several stations, representing different climatic zones, are usually plotted on a single climogram, for comparative purposes. Many other types of climogram have been devised (Monkhouse and Wilkinson, 1971).   DGT

Reference
Monkhouse, F.J. and Wilkinson, H.R. (1971) *Maps and diagrams*. London: Methuen; pp. 246–250.

cline   A gradation in the range of physiological differences within a species that result from the adaptations to different environmental conditions.

clinometer   An instrument for measuring angles in the vertical plane, particularly those of dipping rocks and hillslopes.

clinosequence   A group of related soils that differ in character as a result of the effect of slope angle and position.

clint   The ridge or block of limestone between the runnels (grikes) on a rock outcrop. Clints form on remnant features caused by the solution of the limestone by water under soil or drift cover, etching weaknesses or joint patterns in the rock. Clint is an English term (helk is also used), but the features are called KARREN in German. The most famous clints and grikes can be found at Malham Tarn, Malham, Yorkshire.   PAB

clisere   The series of climax plant communities that replace one another in a particular area when CLIMAX VEGETATION is subjected to a major change in climate.

clitter   Massive granite boulders, especially those found on Dartmoor. Clitter probably represents a type of blockstream or blockfield that may have resulted from the fashioning of tors.

cloud computing   A term typically used to identify a type of Internet-based computing in which key resources are shared, possibly over a large number of devices. The word *cloud* is used as a metaphor for the Internet, arising from the way the Internet is represented in computer network diagrams.

In essence, cloud computing involves the outsourcing of computer services over the Internet. The computer user simply employs the services as and when required without having to have any knowledge, responsibility or control over the technologies or resources providing the service. A variety of types of cloud service are available: platform as a service, software as a service and infrastructure as a service. In each case the service is typically sold on demand, allows elastic usage and is managed by the provider so that the user needs only a computer with Internet access.   GF

Reading
Pallis, G. (2010) Cloud computing. The new frontier of Internet computing. *IEEE Internet Computing*, **14**, 70–73.

cloud dynamics   The role played by air motions and cloud microphysics in the development, evolution and form of clouds. There exists a cascade of dynamics on various time and space scales, all of which influence the motions leading to cloud development and within clouds. Clouds generally result from vertical motions forced at the surface via topography, convergence of air, thermal instability and lifting in the vicinity of fronts. Extensive sheets or layers of stratiform cloud are an expression of the gentle, uniform and widespread ascent of deep layers of moist air. In contrast, localized convective or cumuliform cloud bears witness to the presence of more vigorous but much smaller scale ascent of moist bubbles of air. Gravity leads to the descent of particulates, rain droplets and ice crystals, which in turn create currents within clouds. Sometimes the currents can be powerful (see DOWNBURSTS). Cloud microphysical processes, such as latent heat release upon cloud formation, drop coalescence and break-up, deposition and sublimation, and melting, also influence cloud dynamics. Entrainment of air also influences motion within the cloud.

Widespread ascent is associated with the extensive low-level convergence and frontal upgliding that characterize middle-latitude depressions, and involves vertical velocities of a few centimetres per second, which can reach $10\,cm\,s^{-1}$ in the vicinity of the fronts. Cirrostratus, altostratus and stratus are products of this kind of ascent, although they can be modified if they occur in thin layers and are not shielded by any higher level cloud. A stratus layer cools at its top by radiation to space and is warmed at its base by absorbing infrared radiation emitted from below. This process destabilizes the layer, leading to overturning and a dappled appearance as the stratus is gradually transformed into stratocumulus. Cirrocumulus and altocumulus can also form in this way.

Localized ascent is associated generally with strong surface heating in an unstable atmosphere, such as occurs occasionally over summertime continents or in polar air outbreaks across oceans. The rising convective bubbles transport heat upwards in a central core of air, which becomes visible as a cumulus cloud once condensation occurs. When these clouds occur on settled days in the form of fair-weather cumulus they tend to be only 1–2 km deep and display updraughts of up to $5\,m\,s^{-1}$. In unsettled weather, convection can extend to the tropopause, and these cumulonimbus clouds exhibit updraughts often stronger than $5\,m\,s^{-1}$ and exceptionally up to $30\,m\,s^{-1}$. These convective clouds are surrounded by clear sky in which the air is sinking gently.

Cumulonimbus clouds go through a unique dynamical life cycle that is most developed at maturity with an organized up- and downdraught, the latter of which flows out at the surface as a gust front. The dissipating stage is dominated by downdraughts associated with extensive rainshafts and evaporative cooling of the raindrops. RR/SEN

Reading
Cotton, W.R., Bryan, G.H. and van den Heever, S.C. (2010) *Storm and cloud dynamics*, 2nd edition. San Diego, CA: Elsevier. · Houze, R.A. Jr (2014) *Cloud dynamics*. San Diego, CA: Elsevier.

cloud forest Occupies those zones in mountainous terrain where clouds occur sufficiently regularly to provide moisture to support the growth of forest, which is often of broad-leaved evergreen type.

cloud microphysics Deals with the physical processes associated with the formation of cloud droplets and their growth into precipitation particles. Liquid water droplets and ice crystals in clouds all possess a nucleus; the former have hygroscopic nuclei (e.g. dust, smoke, sulphur dioxide and sodium chloride) and the latter freezing nuclei that form most commonly between $-15\,°C$ and $-25\,°C$ (e.g. very fine soil particles). The hygroscopic nuclei vary in diameter between $4 \times 10^{-7}\,m$ and $2 \times 10^{-5}\,m$ and exhibit concentrations of about $10^{8}\,m^{-3}$ in oceanic air and $10^{9}\,m^{-3}$ in continental air. Microphysics is concerned with explaining how cloud droplets that have diameters from less than $2 \times 10^{-6}\,m$ up to $10^{-4}\,m$ grow to precipitation particle diameters of up to $5 \times 10^{-4}\,m$ (drizzle) and above (rain).

It is observed that at a given temperature below $0\,°C$ the relative humidity above an ice surface is greater than over liquid water and that saturation vapour pressure over water is greater than over ice, notably between $-5\,°C$ and $-25\,°C$. The result is that in clouds where ice crystals and supercooled water droplets coexist the crystals grow preferentially by SUBLIMATION at the expense of the droplets. This Bergeron–Findeisen mechanism is important in precipitation production especially in the extratropics, where a good deal of clouds have substantial layers colder than $0\,°C$. This means that most rain and drizzle in these areas starts off as ice crystals or snowflakes high in the parent cloud.

In natural clouds the droplet size varies, and those that have grown on giant nuclei are large and exhibit the greatest fall speeds. These droplets grow by colliding with slower moving smaller droplets, and this process of coalescence is efficient enough to produce drizzle-size or sometimes raindrop-size particles. It is most favoured in deep clouds with prolonged updraughts and is fairly common in clouds of tropical maritime origin. Cloud electrification increases the efficiency of coalescence. RR

Reading
Mason, B.J. (1962) *Clouds, rain and rainmaking*. Cambridge: Cambridge University Press.

cloud seeding See WEATHER MODIFICATION.

cloud streets Rows of cumulus or stratocumulus clouds lying approximately along the direction of the mean wind in the layer they occupy. A single cloud street may form downwind of a persistent source of thermals, but extensive areas of parallel streets are common, being associated with a curvature of the vertical profile of the horizontal wind. The axes of adjacent rows are separated by distances that are approximately three times the depth of the layer of convection. They are especially frequent over the oceans during outbreaks of cold arctic or polar air. The air motion within the cloud streets is one of longitudinal roll vortices, with air undergoing spiralling motions.   KJW

Reading

Agee, E.M. and Asai, T. (eds) (1982) *Cloud dynamics*. Advances in Earth and Planetary Science, vol. **13**. Tokyo/Dordrecht: Terra Scientific/D. Reidel. · Atkinson, B.W. (1981) *Mesoscale atmospheric circulations*. London: Academic Press. · Stull, R.B. (1988) *An introduction to boundary layer meteorology*. Amsterdam: Kluwer.

clouds   Both the most distinctive feature of the Earth viewed from space (see SATELLITE METEOROLOGY) and the most transient element of the climate system. At any time about half of the globe is covered by clouds composed of water droplets or ice crystals and occurring at altitudes throughout the TROPOSPHERE. The dynamic nature of cloud processes (see CLOUD DYNAMICS) is most readily viewed on a bright summer day, when vertical development is vigorous. Clouds are formed when air is cooled below its saturation point (see CLOUD MICROPHYSICS). This cooling is usually the result of ascent accompanied by ADIABATIC expansion. Different forms of VERTICAL MOTION give rise to different cloud types: vigorous local CONVECTION causes convective clouds; forced ascent of stable air (usually over an adjacent AIR MASS) produces layer clouds and forced lifting over a topographic feature gives rise to orographic clouds, which may be convective or stable in

character. Clouds can also form as a result of other processes, such as the cooling of the lowest layers of the atmosphere in contact with a colder surface in, for instance, the formation of radiative FOG as a result of radiative cooling of the surface at night and advective fog resulting from movement of warm, moist air across a cold surface. The International Classification scheme (WMO, 1956) groups clouds into four basic categories and uses the latin names given to them by the English chemist Luke Howard in 1803: *cumulus*, a heap or pile; *stratus*, a layer; *nimbus*, rain; and *cirrus*, a filament of hair. Additionally, the height of the cloud can be indirectly identified by the use of the following terms: *strato*, low level; *alto*, middle level; and *cirro*, high. Thus, middle-level cumuliform cloud, often known as a 'mackerel sky', is called altocumulus. Clearly, many different combinations of character and height descriptors are possible. The passage of a warm-sector depression system often includes rain from nimbostratus clouds, while conditions in summer frequently lead to cumuliform development, perhaps culminating in a fully fledged cumulonimbus cloud and a THUNDERSTORM.   AH-S

References

Harvey, J.G. (1976) *Atmosphere and ocean*. London: Artemis. · WMO (1956) *International cloud atlas*. Geneva: World Meteorological Organization.

cluse   Originally a French term, now in general use to describe specifically a steep-sided valley that

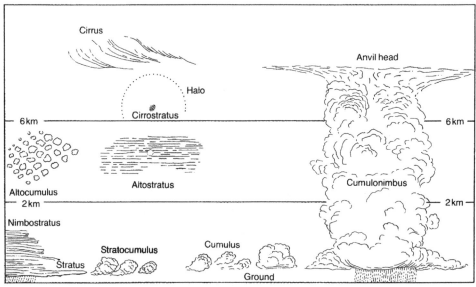

Cloud genera showing typical heights in middle latitudes.
Source: Harvey (1976: figure 3.5).

cuts through a mountain ridge in the Jura mountains.

cluster bedform   A particular type of BED-FORM associated with gravel-bed rivers, and comprising an aggregation of individual grains, often of varying sizes. Cluster bedforms can be classified (e.g. Richards and Clifford, 1991) by their orientation with respect to the flow: either transverse or parallel. Transverse bedforms can be classified into three scales of feature: (1) TRANSVERSE RIBS; (2) transverse CLAST dams; and (3) STEP–POOL SYSTEMS. Transverse ribs are the smallest features, comprising regularly spaced pebble, cobble or boulder ridges formed when rolling or sliding grains stall against already stationary grains. Transverse clast dams are larger in scale, often associated with a wider particle size distribution, with larger clasts forming the building blocks of the dams. It is not uncommon to observe infilling of fines between these dams. Step–pool systems are associated with particularly steep, coarse-bedded channels, with a low depth to grain size ratio. They are stable at most discharges, and the timescales at which these three types of bedform adjust to flow and sediment transport are one of their major characteristic features. Ribs and dams are more dynamic than step–pool systems because they generally involve smaller clasts and, therefore, are moved at a much wider range of discharges. The main type of bedform that is parallel to the flow is the pebble cluster. This comprises a large core clast, with larger particles that have stalled on its STOSS side, and smaller particles that accumulate in the flow separation zone in its lee. These are exceptionally common in all gravel-bed channels and are also very mobile. As a group, cluster bedforms are now recognized to play a critical part of the dynamics of gravel-bed rivers. The can significantly enhance ROUGHNESS and create important flow structures, both of which have implications for the bedload transport process (see BEDLOAD, BEDLOAD EQUATION). In general, they increase roughness, reduce sediment transport rates and may be a critical part of the sediment sorting process.   SNL

Reading and Reference
Bluck, B.J. (1987) Bedforms and clast size changes in gravel-bed rivers. In K. S. Richards (ed.), *River channels: environment and process.* Oxford: Blackwell. · Richards, K.S. and Clifford, N.J. (1991) Fluvial geomorphology: structured beds in gravelly rivers. *Progress in Physical Geography*, **15**, 407–422.

CMIP   The Coupled Model Intercomparison Project. This developed under the auspices of the World Climate Research Programme as a systematic way to compare the performance of models by establishing a standard protocol for examining the output of coupled atmospheric–ocean general circulation models. Such models simulate climate by adjusting the forcings, such as carbon dioxide and solar radiation. 'Coupled' refers to the fact that the ocean responds to the atmosphere and vice versa. Part of the project is to have the participating models simulate the same scenarios and compare outputs among models and with observations. The current phase, CMIP5, includes experiments on (1) decadal hindcasts and predictions and (2) 'long-term' simulations and (3) 'atmosphere-only' simulations (i.e., with fixed sea-surface temperatures) for certain extremely complex models.   SEN

Reading
Meehl, G.A., Covery, C., McAvaney, B., *et al.* (2005) Overview of the Coupled Model Intercomparison Project. *Bulletin of the American Meteorological Society*, **86**, 89–93.

coastal dunes   Deposits of AEOLIAN sand adjacent to large bodies of water. Coastal dunes may form in marine or lacustrine environments, with forms dominated by transverse ridges, parabolics and blow outs, though other types can occur too. Large dune complexes are most common along low-relief coasts, adjacent to rivers, on BARRIER ISLANDS and spits, or along arid coastlines. Coastal dunes are also distinguished from dunes in other environments by the morphological signatures of hydrodynamic erosion, especially dune scarping, and by the importance of vegetation in controlling dune development. The sediment source for most coastal dunes is an adjacent beach, where aeolian erosion and transport provide the linkage of the systems. Along many coasts, sand stored in the dunes represents a significant fraction of the material in the sediment budget. In drier environments, coastal dunes may provide the source of sediment for more extensive dunefields, since vegetation growth does not inhibit dune migration and expansion.

In populated areas, coastal dune systems often require careful management. In many environments, these dunes provide critical habitat for plants and animals, and they may serve as important, local sources of fresh water. The dunes also offer important protection against coastal erosion and flooding. They act as a sediment reservoir to feed the nearshore system during erosive events, and linear foredune systems act as barriers to flooding when water levels are elevated. For these reasons, coastal dunes are often socially valuable landforms.   DJS/DSGT

Reading
Masselink, G., Hughes, M.G. and Knight, J. (2011) *Introduction to coastal processes and geomorphology*, 2nd edition. London: Hodder Education. · Nordstrom, K.F. (2014), *Coastal dunes*. In G. Masselink and R. Gehrels (eds), *Coastal environments and global change*. Chichester: Wiley-Blackwell; p. 178–192. · Sherman, D.J. and Bauer, B.O. (1993) Dynamics of beach–dune systems. *Progress in Physical Geography*, 17, 413–447.

coccoliths The mineralized (calcite) components of single-celled green ALGAE called coccolithophores. They are generally spherical or oval in shape and are usually smaller than 20 μm and are often referred to as nanofossils. Although a few species are adapted to either fresh or brackish water, coccolithophores are superabundant in the world's oceans, and are one of the principal components of modern deep-sea sediments. Studies have shown that many species are associated with distinct thermal characteristics, and they provide an important means of determining ocean palaeo-temperatures, and hence past climates.   SLO

cockpit karst The scenery produced by the solution of limestone resulting in a hummocky terrain of conical residual hills surrounded by DOLINES or SINKHOLES. They are very characteristic of Jamaican limestone scenery and are often associated with humid tropical karst landscapes. The hills are also called KEGELKARST or MOGOTES. PAB

Reading
Sweeting, M.M. (1972) *Karst landforms*. London: Macmillan.

coefficient of permeability See PERMEABILITY.

coefficient of variation (CV) A relative measure of the variability within a group of observations or other numerical data. It is defined as the standard deviation divided by the mean and is expressed as a percentage:

$$CV = \frac{s}{x} \times 100$$

For example, the numbers $0, 1, \ldots, 10$ have a mean of 5 and a standard deviation of 3.3, and hence a CV of 66.3%. A useful property of the CV is that it is not affected by the units of measurement. Thus, using the CV, the variability in yearly rainfalls over a catchment, scaled in millimetres, can be compared directly with the variability in resulting river flows measured in megalitres, since the variability is expressed as a dimensionless percentage of the mean. For the numbers $0, 1, \ldots, 100$, both the mean and the standard deviation are 10 times larger than for the numbers $0, 1, \ldots, 10$ (50 and 33.2 respectively) but the CV is the same (66.3%).   DLD

Reading
Clark, W.A.V. and Hosking, P.L. (1986) *Statistical methods for geographers*. New York: John Wiley & Sons, Inc.

coevolution The evolutionary interaction between two species so that over a long period of time they become co-adapted to each other through a series of selective influences. The predator–prey relationship in animals provides some of the clearest examples. A predator such as a cheetah has evolved to outrun prey such as springbok, whose defence is to run faster themselves, and to evolve more sensitive detection mechanisms, forcing the cheetah to stalk more quietly and unobtrusively. Coevolution between insects and plants has been the subject of much recent work. Chemical defences may be used by plants against particular insects (e.g. oak produces leaves with more tannin as a defence against defoliating caterpillars), or else mimicry may be employed. An example of this is the extraordinary way in which the leaves of *Passiflora* species in central America mimic the leaves of other rainforest species in an attempt to deceive heliconiid butterflies who otherwise may lay their eggs on the leaves. Some *Passiflora* species even produce small swellings on the leaves to mimic butterfly eggs and hence escape predation! Coevolution must also have occurred between larger herbivores and grazed vegetation: the production of spines, of tough unpalatable leaves and of basal shoots as in grasses are all responses to grazing pressure.

Coevolution can also lead to symbiosis or mutualism, as in lichens, and other relationships that may be looked upon as controlled or 'beneficial' parasitism, such as that between mycorrhizal fungi and their host plants. The whole field of pollination ecology also throws up many examples of coevolution between insects and plants.   KEB

Reading
Thompson, J.D. (2014) *Interaction and coevolution*. Chicago, IL: University of Chicago Press.

cohesion The attraction of particles to each other (usually clay minerals in soils) that is not governed directly by a FRICTION law (i.e. it is independent of STRESS) but does provide a measure of the strength of a material. Thus, sands do not exhibit cohesion but what is termed (intergranular) friction, while clays, or soils that contain clays, show cohesion. The strength is supplied by the structure of the clay minerals and the way in which chemical bonding is produced in these structures. It can be measured, as in soil mechanics, by the MOHR–COULOMB EQUATION.   WBW

col   A pass or saddle between two mountain peaks, or a narrow belt of low pressure separating two areas of high pressure.

cold front   A frontal zone in the atmosphere where, from its direction of movement, rising warm air is being replaced by cold air. It has an average slope of about 1 in 60 over the lowest several hundred metres, rather like the forward bulge of a density bore. Above this level both warm and cold fronts have similar slopes. A cold front is often found to the rear of an EXTRATROPICAL CYCLONE, marking a sudden change from warm, humid and cloudy weather to clear, brighter weather. Satellite images usually show cold fronts clearly, as a thick band of cloud spiralling away from the depression centre. Where uplift within the warm air is rapid, rainfall may be heavy along the cold frontal zone. As it passes, temperatures fall suddenly together with a rapid veering of the wind.   PS

cold pole   Can be defined as the location of lowest mean monthly temperature, or of lowest mean annual temperature, or of the coldest air in the TROPOSPHERE. The last case is usually indicated by the area of lowest THICKNESS on the chart of 1000–500 mb thickness: in January this is found in the northern hemisphere over northeast Siberia, and in July over the North Pole. In the southern hemisphere, the lowest 1000–500 mb thickness is found over the Antarctic continent. The lowest surface air temperatures in the northern hemisphere are found in northeast Siberia at Verhojansk and Oymjakon, where they have reached −68 °C. The lowest surface air temperatures in the southern hemisphere are recorded in Antarctica, where they have reached −88 °C.   JGL

Coleoptera   Beetles, one of several orders of insects. In Quaternary studies, the remains of Coleoptera have proved especially valuable for the following reasons: (1) unlike other orders, their hard chitinous exoskeletons, especially wing cases, withstand burial and preservation in sediments; (2) the order is one of the largest in the insect kingdom; (3) remains are well documented, facilitating relatively easy identification to the species level; (4) many species are stenotypic (i.e. have distinct environmental controls on their distribution); and (5) they respond relatively rapidly in their distribution in response to environmental changes. The use of Coleoptera remains in Quaternary studies has especially been developed by G. R. Coope, who noted that, for the

aforementioned reasons 'Coleoptera [are] one of the most climatically significant components of the whole terrestrial biota' (Coope, 1977). MEM

Reference
Coope, G.R. (1977). Quaternary Coleoptera as aids in the interpretation of environmental history. In F. W. Shotton (ed.), *British Quaternary studies: recent advances*. Oxford: Oxford University Press.

colloid   Any noncrystalline, partially solid substance. Often applied to individual, lens-shaped particles of clay.

colluvium   Material that is transported across and deposited on slopes as a result of wash and mass movement processes. It is frequently derived from the erosion of weathered bedrock (ELUVIUM) and its deposition on low-angle slopes, and can be differentiated from ALLUVIUM, which is deposited primarily by fluvial agency. Colluvium may be many metres thick, often contains fossil soil layers that represent halts in deposition, shows some crude bedding downslope, and is generally made of a large range of grain sizes. Cut-and-fill structures may represent phases when stream incision has been more important than colluvial deposition.   ASG

colmation   Used to describe the processes that contribute to surface sediments (usually in a stream bed) becoming clogged with fine particles. This reduces POROSITY, HYDRAULIC CONDUCTIVITY and INFILTRATION capacity.   DSGT

colonization   The occupation by an organism of new areas or habitats, thereby extending either its geographical or its ecological range. For colonization to succeed, three main phases are usually necessary: effective dispersal or migration, germination or breeding and establishment, and, finally, survival in the new site through time. Most colonizing species are subject to R-SELECTION. Species that are able to increase their numbers exponentially on arrival in a new area have strong advantage (e.g. annual and biennial plants), while animals exhibiting behavioural adaptability and a marked ability to learn new behaviour tend to be natural colonizing species. Humans have proved a potent aid to many colonizers, spreading them to new areas where they have subsequently flourished (e.g. *Opuntia stricta* in Australia) or opening up new habitats ripe for colonization. (See also ALIENS, DISPERSAL and R- AND K-SELECTION.)   PW

Reading
Elton, C.S. (1958) *The ecology of invasions by animals and plants*. London: Methuen. · Krebs, C.J. (2014) *Ecology:*

the experimental analysis of distribution and abundance, 6th edition. London: Pearson. · MacArthur, R.H. (1972) *Geographical ecology*. New York: Harper & Row.

combe, coombe A small, often narrow valley. Frequently applied to the dry or seasonal stream valleys of chalk country.

comfort zone The range of meteorological conditions within which the majority of the population, when not engaged in strenuous activity, will feel comfortable. It is often expressed in terms of effective temperature (ET), which combines dry-bulb and WET-BULB TEMPERATURES and air movement into a single biometeorological index. The comfort zone varies with climatic zone and season of the year. Personal factors such as age, clothing, occupation and degree of acclimatization are also important. In the tropics, the comfort zone is often taken to range from 19 to 24.5 °C ET. In the UK, comparable values are 15.5–19 °C ET in summer and 14–17 °C ET in winter (Air Ministry, 1959). The effects of direct exposure to solar radiation are not normally included in a consideration of the comfort zone.   DGT

Reading and Reference
Air Ministry (1959) *Handbook of preventive medicine*. London: HMSO; p. 164. · Terjung, W.H. (1966) Physiologic climates of the conterminous United States: a bioclimatic classification based on man. *Annals of the Association of American Geographers*, 56, 141–179.

commensalism The weakest type of association of species living together in symbiosis. One or more 'guest' species benefit from living in association with a 'host' species, but the latter neither benefits nor is harmed by the presence of the guests. Mites that live in human hair represent a case in point, as also does the pitcher plant (*Nepenthes*), which can host up to 19 species.

comminution The reduction of a rock or other substance to a fine powder, often as a result of abrasion.

community Any assemblage of populations of living organisms in a prescribed area or habitat. The term *community* is an ecological unit used in a more general, broad, collective sense to include groups of various sizes and degrees of integration. The community comprises a typical species composition that has resulted from the interaction of populations over time.

Botanists and zoologists have defined the term *community* in widely differing ways. Three main ideas are involved in community definitions, which claim to have one or more of the following attributes:

- co-occurrence of species;
- recurrence of groups of the same species;
- homeostasis or self-regulation.

Two opposing schools have developed in ecology over the question of the nature of the community. The *Organismic school* suggests that communities are integrated units with discrete boundaries. The *Individualistic school* suggests that communities are not integrated units but collections of populations that require the same environmental conditions.   AP

Reading
Krebs, C.J. (2014) *Ecology: the experimental analysis of distribution and abundance*, 6th edition. London: Pearson.

compaction In engineering terms, the expulsion of air from the voids of a soil. This is usually achieved by artificial means; for example, with various types of roller in road and dam construction. It differs from CONSOLIDATION, which is usually a natural process, although the latter may involve some air expulsion. The aim of compaction is to achieve a high bulk density or lower VOID RATIO, so improving the overall strength of the soil.   WBW

compensation flows Designated river discharges that must continue to flow in a river below a direct supply reservoir or abstraction point to allow for other riparian activities and interests downstream. Such flows are legally established in different ways in different countries and at different times, and may be necessary to maintain water quality, to satisfy the needs of wildlife or recreation, or to provide water for other abstractors or users.   JL

competence The competence of flowing water is the maximum particle size transportable by the flow. A related concept is the threshold flow, which is the minimum flow intensity capable of initiating the movement of a given grain size. When the largest grains on a stream bed are just mobile the flow is the threshold (critical) flow for those grains, and their size represents the competence of the flow. It is possible to predict theoretically the threshold flow conditions for non-cohesive grains coarser than 0.5–0.7 mm, by balancing moments caused by fluid drag and particle immersed weight (Carson, 1971: 26; Richards, 1982: 79–84). The threshold shear stress increases linearly with the grain diameter, while threshold velocity increases with the square root of grain diameter or the one-sixth power of its weight (the so-called 'sixth-power law'). These relationships can be used to define the competence of a given shear stress or velocity (see HJULSTRÖM CURVE). However, the competence to *maintain* particle motion is usually greater than that to *initiate* motion, so large grains may be carried *through* a reach in which they cannot be entrained.

In flowing water, grain motion is normally caused by fluid stresses. In air, however, an impact threshold also occurs. The kinetic energy of a sand grain travelling by SALTATION in air allows it to move grains up to six times its size. The concept of fluid competence is therefore less applicable to aeolian transport.   KSR

References
Carson, M.A. (1971) *The mechanics of erosion*. London: Pion. · Richards, K.S. (1982) *Rivers: form and process in alluvial channels*. London: Methuen.

competition
The inevitable result of interactions between members of the same and different species attempting to secure limited resources from the environment, and leading to increase and decrease in abundance of more and less successful organisms respectively. Competition takes many forms: between plants for light, water, space and nutrients; between animals for mates, food, shelter and for social position within a hierarchy. While many of the principles are the same for both plants and animals, there are several obvious practical differences, such as the ability of plants to spread and reproduce asexually. Competition is also a very complex phenomenon, with many of the precise mechanisms by which it occurs being poorly understood. The literature tends, therefore, to be overreliant upon a relatively small number of classic case studies. More recently, there has been increasing recognition of the role of other factors in the distribution of species and the structure of plant and animal communities, particularly herbivory, predation and habitat disturbance.

The classic experiments concluded that species grown together in the same environment did not do as well as when grown separately, for one species tended to triumph and the other became extinct. This led to the 'competitive exclusion principle' and quantitative expressions based on the logistic curve of POPULATION DYNAMICS. All discussion of competition also involves consideration of the concept of the NICHE, the position of an organism within the community and its interrelationships with the organisms around it. KEB/MEM

Reading
Begon, M., Harper, J.L. and Townsend, C.R. (2006) *Ecology: from individuals to communities*, 4th edition. Oxford: Blackwell. · Townsend, A.P. (2011) *Ecological niches and geographic distributions*. Princeton, NJ: Princeton University Press.

complex response
The term introduced by Schumm (1973) to describe the variety of linked changes that may occur within a drainage basin in response to the single passage of a geomorphological threshold (see THRESHOLD, GEOMORPHOLOGICAL). These changes may well be both spatially and temporally separated, and it may, in consequence, be difficult *ex post facto* to establish that all are ultimately due to a single event.   BAK

Reference
Schumm, S.A. (1973) Geomorphic thresholds and the complex response of drainage systems. In M. Morisawa (ed.), *Fluvial geomorphology*. Publications in Geomorphology. Binghamton, NY: State University of New York; pp. 299–310.

compressing flow
Compressing and extending flows describe longitudinal variations in velocity in a moving medium. The concept is widely used in glacial geomorphology to describe longitudinal variations in flow along the length of a glacier (Nye, 1952). Compressing flow refers to a reduction in the length of a unit of glacier ice in a downstream direction, and extending flow refers to an increase in length in the same direction. Such changes must be accompanied by a corresponding variation in glacier width or height. As valley walls often prevent lateral expansion, the variations in a glacier are commonly accompanied by a change in depth, with compressing flow associated with thickening and extending flow with thinning. The association of compressing and extending flows with changing ice thickness leads to predictable spatial associations. Extending flow is characteristic of the accumulation zone, where the glacier mass increases downstream, while compressing flow is common in the ABLATION zone, where the glacier thins downstream, and especially near the snout where thinning is most marked. Extending flow is also associated with a bed convexity, which steepens downstream, as in the case of an ICE FALL, while compressing flow occurs as the glacier crosses a bed concavity. Variations in the rate of BASAL SLIDING along the length of a glacier can also induce compressing and extending flows. For example, a change from a warm- to cold-based thermal regime or a thinning of the basal water layer in a downstream direction favours compressing flow, while the opposite conditions favour extending flow. DES

107

Reading and Reference

Benn, D.I. and Evans, D.J.A. (2010) *Glaciers & glaciation.* London: Hodder. · Nye, J.F. (1952) The mechanics of glacier flow. *Journal of Glaciology*, 2(12), 82–93.

compressional Pertaining to descending atmospheric air masses, which as a result of pressure increases are warm and dry. Also pertaining to geological faults and fractures that are the product of lateral increases of pressure and to earthquake waves, specifically P-waves.

computational fluid dynamics A subdiscipline of fluid mechanics that uses numerical methods to analyse problems involving fluid flows. The basis for most applications of computational fluid dynamics is the Navier–Stokes equations, which are derived based on applying Newton's second law ($F = ma$) to fluid motion along with the assumption that the stress in the fluid is the sum of viscous and pressure terms (conservation of momentum, mass and energy). Computational methods may rely on finite-volume, finite-element or finite-difference methods. DRM

conchoidal fracture A rock or mineral fracture that is shaped like a shell; that is, concave down. The characteristic fracture of siliceous rocks and minerals.

concordant Within the same plane. Applied particularly to the mountain summits of a region, the concordance of summits attesting to the existence of a plateau prior to incision and dissection of the land surface.

concordant coast A coastline where the general orientation is parallel with, and controlled by, geological structures. The western coastlines of the Americas are usually concordant, exhibiting control by adjacent faults. The Dalmatian coast of Croatia and Montenegro is concordant (see DALMATIAN COAST). The east coast of the USA is discordant, with Cape Cod and Long Island representing orientations perpendicular to the general coastal trend. DJS

concretion A solid lump or mass of a substance, or aggregates of these, incorporated within a less competent host material.

condensation The process of the formation of liquid water droplets from atmospheric water vapour mainly onto 'large' hygroscopic nuclei with diameters between 2×10^{-7} m and 10^{-6} m. The cooling necessary to produce condensation in the atmosphere is effected either by ADIABATIC expansion during ascent, by contact with a colder surface or by mixing with colder air. The LATENT HEAT released during this process can be a significant source of heat in weather systems. RR

conductance, specific A measure of the ability of an aqueous solution to conduct an electrical current. It is reported in units of microsiemens per centimetre at $t\,°C$ ($\mu S\,cm^{-1}\,t\,°C$). Pure water has a very low specific conductance (SC), but the conductance will increase with an increasing concentration of charged ions in solution. Individual ions exhibit different conductance values, and the level of conductance recorded for a given total ion concentration will therefore vary according to its ionic composition. For most dilute natural waters an increase in temperature of 1 °C will increase the conductance by approximately 2%, and values of SC are normally corrected to a reference temperature of 20 or 25 °C.

Because SC measurements are simple and rapid to make, and reflect the total ion concentration, they have been widely used as a means of estimating the total dissolved solids (TDS) concentration of water samples. The relationship between SC and TDS for waters of different chemical composition is generally determined empirically and commonly takes the form:

$$TDS = KSC$$

where K varies between 0.55 and 0.75. This relationship may be complicated by changing ionic composition and by the presence of nonionized material in solution (e.g. SiO_2 and dissolved organic matter), which will contribute to TDS but not SC. DEW

Reading

Foster, I.D.L., Grieve, I.C. and Christmas, A.D. (1982) The use of specific conductance in studies of natural waters and solutions. *Hydrological Sciences Bulletin*, 26, 257–269.

cone of depression The shape of the depression in the WATER TABLE surface around a well that is being actively pumped for GROUNDWATER. Pumping results in a lowering (or DRAW DOWN) of the water table that is greatest in the well itself, but reduces radially with distance from the well. In a rock with horizontal PERMEABILITY that is uniform in every direction, the cone of depression will be symmetrical about the well. But where groundwater flow occurs more readily in one direction than in another (perhaps because of the influence of dominant joints), the cone of depression will be asymmetrical and elongated in the direction of greatest permeability. Cones of depression can develop in wells pumping both unconfined and confined aquifers. PWW

Reading
Fitts, C.R. (2013) *Groundwater science*, 2nd edition. Waltham, MA: Academic Press. · Linsley, R.K., Kohler, M.A. and Paulhus, J.L.H. (1988) *Hydrology for engineers*, 3rd edition. New York: McGraw-Hill.

confined groundwater   See GROUNDWATER.

conglomerate   A rock that is composed of or contains rounded or water-worn pebbles and cobbles more than about 2 mm in diameter.

connate water   Water trapped in the interstices of sedimentary rocks at the time of their deposition. It is usually highly mineralized and is not involved in active GROUNDWATER circulation, although connate waters may be expelled from their original location by compaction pressure and migrate, accumulating in more permeable formations. (See also JUVENILE WATER and METEORIC WATER/GROUNDWATER.) PWW

connectivity   In hydrology, the extent to which run-off and sediment-contributing areas within a landscape are connected to stream channels via continuous flow pathways. Only when source areas are linked or connected is there capacity for the delivery of water and sediment from hillslopes to the slope foot, stream or basin outlet. The extent of connectivity can change dynamically in time, varying with antecedent soil moisture content, storm duration, with the nature of available run-off pathways (surface, subsurface, channelized, nonchannelized, etc.) and with catchment characteristics (forested, grassed, agricultural, burnt). For instance, in short rainfall events, there may be insufficient time for overland flow draining from upslope source areas to reach the slope foot. During run-off events, flow pathways that allow rapid movement of water and sediment, such as pipe and tunnel flow, may yield connectivity whereas slower flow paths (e.g. overland flow through grass or litter) may not. These conditions may all vary seasonally, with variation in rainfall climate, soil moisture levels and the phenologic condition of the vegetation. DLD

Flow pathways on a schematic hillslope.
Source: Freeman *et al.* (2007). Reproduced with permission of John Wiley & Sons.

Reference
Bracken, L.J., Wainright, J., Ali, G.A., *et al.* (2013) Concepts of hydrological connectivity: research approaches, pathways and future agendas. *Earth-Science Reviews*, **119**, 17–34. · Freeman, M.C., Pringle, C.M. and Jackson, C.R. (2007) Hydrologic connectivity and the contribution of stream headwaters to ecological integrity at regional scales. *Journal of the American Water Resources Association*, **43**, 5–14.

consequent stream A stream that flows in the direction of the original slope of the land surface.

conservation Used in its more general sense to mean environmental conservation, the term relates to the sustainable utilization, through intervention and management, of the entire range of environmental resources and processes. Conservation is distinguished from preservation insofar as it incorporates the element of sustainable utilization, whereas preservation implies protection from any form of use. Thus, conservation refers to the active maintenance of elements of the environment; for example, soil, water, biological diversity and other natural resources in association with the recognition that exploitation of such resources is essential to human existence. The broader use of the term, however, is relatively recent; historically, conservation has been synonymous with the establishment of set-aside areas, so-called nature reserves, for the preservation of, in the main, wild animals. More recently, the need for conservation of the spectrum of environmental processes, both within and beyond reserve boundaries, has been recognized. Many politicians have jumped on the 'conservation bandwagon', and the increase in environmental awareness that marked the 1990s, following the Rio Earth Summit and subsequent major international events, has led to the wider application of conservation (i.e. sustainable utilization) principles. From being a rather negative concept dealing with the preservation of existing environmental processes, conservation now invokes active and positive intervention as an integral component of economic development. The traditional 'conservation versus development' debates are placed in perspective by the acceptance of an expanded view of conservation.

The most common context for conservation, nevertheless, remains the goal of maintaining the Earth's biological diversity, a branch of science that is referred to as conservation biology. Different levels of biological conservation are apparent, ranging from entire ecosystem conservation to the conservation of communities, habitats, species, populations or even gene pools. Biological conservation is based on sound scientific principles emanating from the fields of, in particular, genetics and population ecology. With respect to conservation of endangered plant and animal species within reserve areas, significant advances were made in the 1960s following the development of ISLAND BIOGEOGRAPHY, and nowadays theory and practice around the management of so-called minimum viable populations represents a key element of conservation biology. Conservation using these principles is both in situ (i.e. in the natural environment) and ex situ (e.g. in botanical and zoological gardens).

Much of the debate around conservation remains focused on the protected area network. The World Conservation Strategy, established in 1980, formally introduced international policy aimed at conserving the environment through the maintenance of essential ecological processes and the preservation of genetic and species diversity. The call for each country to conserve a minimum of 10% of its geographical area was made, although there is no scientific basis for such an arbitrary value and it by no means guarantees the meaningful protection of most of the plant and animal species within any country. In any case, very few nations would be able to boast a conservation estate of 10% of their areas. In practice, politics and economics, as opposed to science, have tended to dictate which areas are managed for conservation purposes. Numerous criteria can be put forward to evaluate an area, such as total species diversity, the presence of rare species and the naturalness or degree to which a site has been modified by human activities. By employing these and other conservation criteria it is hoped that land-use planning decisions that minimize the environmental impact of developments will emerge.

To most, including politicians, biological conservation has an undeniable value, although it has proved remarkably difficult to convert such value into the kind of economic entity that influences hard decisions about where to spend taxpayers' money. Of late, attempts to answer the question 'why conserve nature?' have taken on a less abstract form with the development of ecological economics (Costanza, 1991). Although it is a difficult process, it is possible to sum the various ethical, aesthetic and utility values of conserving; for example, biological diversity to construct a more persuasive, economically based argument in favour of conservation.

Increasing concern with the degree of SOIL EROSION, especially in seasonal rainfall climates, has resulted in better soil conservation practices with the intention of maintaining soil structure and fertility. Soil management actually has a long history in the early agricultural areas around the Mediterranean and the Middle and Far East, as reflected in the adoption of, for example, terraces and mulching techniques. Since 1934 in the USA, the Soil Conservation Service has offered advice to farmers about methods of soil tillage and management aimed at minimizing soil loss by wind and water, a movement that has since been emulated in many other countries.   MEM

Readings and Reference
Costanza, R. (ed.) (1991) *Ecological economics: the science and management of sustainability*. New York: Columbia University Press. · Macdonald, D.W. (ed.) (2007) *Key topics in conservation biology*. Malden, MA: Blackwell. · Shmelev, S.E. (ed.) (2012) *Ecological economics: sustainability in practice*. Dordrecht: Springer. · Sutherland, W.J. (2000) *The conservation handbook: research, management and policy*. Malden, MA: Blackwell.

consociation   A natural vegetation community which is dominated by one species.

consolidation   In engineering terms, the expulsion of water from the void spaces (see VOID RATIO) of a soil. This is achieved by natural burial processes as sediment accumulates or when a structure is erected on a soil. A normally consolidated soil (usually clay) is one that has never been subjected to an overburden pressure (i.e. load on top) greater than that which it currently has. An over-consolidated soil is one that has had a greater overburden pressure. This last effect is usually achieved by denudation of the overlying material: lodgement TILL is overconsolidated because of the pressures previously applied by the glacier above. Overconsolidated clays have higher strengths than an otherwise equivalent normally consolidated clay, measured by an overconsolidation ratio.   WBW

constant slope   A term used by Wood (1942) to define the straight part of a hillside surface, lying below the free face, and having an inclination determined by the angle of repose of the TALUS material forming it. The constant slope extends upwards until it buries, or replaces, the free face and so eliminates the supply of fresh talus debris. The WANING SLOPE of lower inclination is formed of weathered talus and eventually extends upwards and replaces the constant slope. The idea of a 'constant' slope is a theoretical construct with constancy of slope length and slope angle being of limited duration in a geological time span.   MJS

Reading and Reference
Wood, A. (1942) The development of hillside slopes. *Proceedings of the Geologists' Association*, **53**, 128–40. · Young, A. (1972) *Slopes*. Edinburgh: Oliver & Boyd.

continental climate   A climate type characteristic of the interior of large landmasses in the mid-latitudes. Being far removed from the modulating influence of the oceans, these areas experience large variations in temperature from season to season and year to year. One measure of continentality is the annual temperature range. The most extreme continentality is in the interior of Asia, where the absolute annual range of temperature can exceed 100 °C and monthly mean temperatures can fall to −40 °C in winter. Although many areas of continental climate are desert or semi-arid, the continentality is generally not the reason for the aridity. The continental climates of the central US and Asia, for example, are produced largely by orographic effects; that is, the rain shadows in the lee of mountains.   SN

Reading
Nicholson, S.E. (2011) *Dryland climatology*. Cambridge: Cambridge University Press.

continental drift   The movement of continents relative to each other across the Earth's surface. Although it was a subject of speculation by numerous early workers, the first comprehensive case for continental drift was presented by Alfred Wegener in 1912. Wegener cited various lines of evidence in support of his notion of a supercontinent (Pangaea) that gradually separated into a northern (Laurasia) and a southern (Gondwanaland) landmass before finally splitting into the continents of the present day. The evidence included the matching configuration of opposing continental coastlines, the similarity of geological structures on separate continental masses, the anomalous location of ancient deposits, indicating specific climatic conditions, and the distribution of fossil species through time. The theory was rejected at first by most geologists and geophysicists, largely because of a lack of a viable mechanism, but further support for continental drift was provided during the 1950s and 1960s through evidence from PALAEOMAGNETISM. In the late 1960s the idea was incorporated into the PLATE TECTONICS model.   MAS

Reading
Du Toit, A.L. (1937) *Our wandering continents*. Edinburgh/New York: Oliver & Boyd/Hafner. · Frisch, W., Meschede, M. and Blakey, R. (2011) *Plate tectonics, continental drift and mountain building*. Berlin: Springer-Verlag. · Hallam, A. (1973) *A revolution in the Earth sciences: from continental drift to plate tectonics*. Oxford: Clarendon Press. · McElhinny, M.W. (ed.) (1977) *Past distribution of the continents*. Amsterdam: Elsevier. · Wegener, A. (1966) *The origin of continents and oceans*. Translated by John Biram from the fourth revised German edition. New York: Dover.

continental freeboard The average elevation of the continents relative to MEAN SEA LEVEL. Continental freeboard changes as a result of expansion and contraction of the volumes of the ocean basins relative to changes in continental volumes. It is an indicator of the ratio of the volume of ocean basins (or the volume of water in the basins) to the volume of the continents. If the ratio increases, there is eustatic fall of sea level and freeboard rises (see EUSTASY). Conversely, if the ratio decreases, sea level rises and freeboard falls. The present freeboard is approximately 750 m. Cogley (1984), including large, submerged regions of continental crust, estimated freeboard to be 126 m.

According to Wise (1974), the average freeboard has been relatively stable over the last 2.5 billion years, at a level about 20 m higher than present. Since the Precambrian, variation in freeboard has been within a range of ± 60 m of the mean for more than 80% of the time. He attributed this to isostatic equilibria in the LITHOSPHERE and described the feedback mechanisms responsible (see ISOSTASY). Galer and Mezger (1998) have formalized this concept as

$$\Delta f = \left(\frac{\rho_o}{\rho_m} - 1\right)\Delta d_r + \left(\frac{\rho_b}{\rho_m} - 1\right)\Delta d_b$$
$$- \left(\frac{\rho_c}{\rho_m} - 1\right)\Delta d_c$$

where f is freeboard, ρ is density and the subscripts o, m, b and c refer to ocean, mantle, oceanic crust and continental crust, d_r is reference water depth, and d_b and d_c are the thicknesses of oceanic and continental crusts.

Changes in freeboard are estimated using palaeogeographic maps of the area of continents, where inundation is interpreted from marine deposits in the geological record. The areal estimates are combined with HYPSOMETRIC CURVES to reconstruct the continental elevation distributions.   DJS

References
Cogley, J.G. (1984) Continental margins and the extent and number of the continents. *Reviews of Geophysics and Space Physics*, **22**, 101–122. · Galer, S.J.G. and Mezger, K. (1998) Metamorphism, denudation and sea level in the Archean and cooling of the Earth. *Precambrian Research*, **92**, 389–412. · Wise, D.U. (1974) Continental margins, freeboard and the volumes of continents and oceans through time. In C. A. Burk and C. L. Drake (eds), *The geology of continental margins*. New York: Springer-Verlag; pp. 45–58.

continental islands These occur in close proximity to a continent, to which they are also geologically related, and which are detached from the mainland by a relatively narrow, shallow expanse of sea. They were formerly united to the mainland. Oceanic ISLANDS, on the other hand, are and have been geographically isolated and rise from the floors of the deep ocean basins.

continental shelf A portion of the continental crust below sea level, consisting of a very gently sloping, rather featureless, surface forming an extension of the adjacent coastal plain and separated from the deep ocean by a much more steeply inclined continental slope. The gradient of most continental shelves is between 1 and 3 m km^{-1}. Some shelves reach a depth of over 500 m, but 200 m is often conveniently used as a depth limit. Their mean width is around 70 km, but there is a marked variation between different coasts. It is estimated that continental shelves cover approximately 5% of the Earth's surface.   MAS

Reading
Kennett, J.P. (1982) *Marine geology*. Englewood Cliffs, NJ: Prentice-Hall. · Shephard, F.P. (1973) *Submarine geology*. 3rd edition. New York: Harper & Row.

continental slope Lies to the seaward of the CONTINENTAL SHELF and slopes down to the deep sea floor of the abyssal zone.

continuity equation A statement that certain quantities, such as mass, energy or momentum, are conserved in a system, so that for any part of the system the net increase in storage is equal to the excess of inflow over outflow of the quantity conserved.

The most familiar example of a continuity equation is perhaps the STORAGE equation in hydrology that describes the conservation of mass for water. Continuity equations are equally applicable to the conservation of total mass of earth materials, or of mass for particular chemical elements or ions. In these contexts the continuity equation is one fundamental basis for models of hillslope or soil evolution (e.g. Kirkby, 1971). The equation is usually used in these models as a partial differential equation, which formally connects rates of change at a site over time to rates of change over space at a given time. One simple form of the continuity equation is

$$\frac{\partial s}{dt} + \frac{\partial Q}{dx} = a$$

where s is the amount stored (e.g. as elevation of rock and soil materials), Q is the rate of flow (e.g. as total transport of earth materials downslope), a is the net rate of accumulation (e.g. as wind-deposited dust), and x and t are respectively distance downslope and time elapsed.

An equation of this type cannot normally be solved without specifying the relevant processes that control the rates of flow, accumulation, and so on. Continuity equations are equally relevant to the conservation of energy in micro-meteorological studies and in many other physical and chemical systems. The concept of continuity dates back to Leonardo da Vinci and remains a fundamental principle underlying our understanding of the physical world.   MJK

Reading and Reference
Davidson, D.A. (1978) *Science for physical geographers*. London: Edward Arnold. · Kirkby, M.J. (1971) Hillslope process-response models based on the continuity equation. In D. Brunsden (ed.), *Slopes: form and process*. Transactions, Institute of British Geographers, Special Publication No. 3. London: Institute of British Geographers; pp. 15–30.

contour
A line on a map that joins areas of equal height or equal depth. The diagram illustrates a hypothetical island with two hills and the contour heights shown in the three-dimensional version on top and the two-dimensional representation of these below.

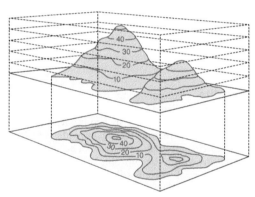

contrails
Aircraft, both civil and military, discharge some water vapour from their engines into the atmosphere as linear wispy clouds called contrails (short for 'condensation trails'). In Britain these are often called vapour trails. At present, the water content of the stratosphere is very low, as is the exchange of air between the lower stratosphere and other regions. Consequently, comparatively modest amounts of water vapour discharged by aircraft could have a significant effect on the natural balance. It is possible that contrails and the development of thin cirrus clouds could lead to warming of the Earth's surface. Observations of cirrus cloud cover over the last 50 years near to major USA flight corridors show a clear upward trend that seems to correlate with US domestic jet fuel trends over the same period (Yang *et al.*, 2010). However, the Intergovernmental Panel on Climate Change (Solomon *et al.*, 2007: 30) concludes 'No best estimates are available for the net forcing from spreading contrails. Their effects on cirrus cloudiness and the global effect of aviation aerosol on background cloudiness remain unknown'. The whole role of aircraft impacts on the atmosphere and climate is discussed by Lee *et al.* (2010). ASG

References
Lee, D.S., Pitari, G., Grewe, V., *et al.* (2010) Transport impacts on atmosphere and climate: aviation. *Atmospheric Environment*, 44, 4678–4734. · Solomon, S., Qin, D., Manning, M., *et al.* (2007) *Climate change 2007: the physical science basis*. Cambridge: Cambridge University Press. · Yang, P., Hong, G., Dessler, A.E., *et al.* (2010) Contrails and induced cirrus. *Bulletin of the American Meteorological Society*, 91, 473–478.

contributing area
The area of a catchment that is, or appears to be, providing water for storm run-off. The term was first used by Betson (1964) and may be calculated as the stream discharge divided by the rainfall (or net rainfall) intensity, usually expressed as a proportion or percentage of total catchment area. This ratio is meaningless when it is not raining, and is usually calculated over a total storm period, in the form total storm run-off (normally calculated as QUICKFLOW) divided by total storm rainfall. For many well-vegetated small catchments, the contributing area is less than 5% of the catchment, but is commonly many times greater where vegetation is sparse (within the area of storm rainfall).   MJK

Reference
Betson, R.P. (1964) What is watershed runoff? *Journal of Geophysical Research*, 69, 1541–1552.

convection
In general, mass movement within a fluid resulting in transport and mixing of properties of that fluid. In the atmosphere, a class of fluid motion in which warmer air goes up while colder air goes down. Unfortunately, fluid dynamicists sometimes use the word in place of ADVECTION. In the case of fair-weather convection, air at low levels is slowly warmed by sunshine absorbed at the ground until it can lift away from the surface to be replaced suddenly by comparatively cool air from above. If the air is sufficiently moist, cumulus cloud may form above the updraught as an indicator of the process. *Cumulonimbus* convection is more complex (see CLOUDS). Air rising from the surface is usually slightly cooler than its immediate environs but it is very moist, so that when it has risen above cloud base and made use of the latent energy of the water vapour by condensing to cloud droplets it becomes very buoyant. This air then accelerates vigorously, dragging the reluctant air near the surface after it. Slantwise convection occurs when the warm and cold air are initially side by side. This mechanism accounts for sea breezes, valley winds and weather systems (see MESOME-TEOROLOGY).   JSAG

Reading

Ludlam, F.H. (1980) *Clouds and storms.* University Park, PA: Pennsylvania State University Press.

convective precipitation Precipitation that arises from vigorous uplift from warm land or water surfaces, with associated convergence and entrainment of moist air. There are diverse mechanisms leading to convective precipitation, including isolated thunderstorm cells over land or widespread convective uplift over warm tropical oceans. Owing to the importance of surface heating, there can be marked diurnal cycles in the arrival of convective precipitation, such as the late afternoon maximum frequently seen over land. Convective rainfall is typically more intense than stratiform rainfall, and may exhibit a larger median drop diameter. Consequently, convective precipitation may deliver more erosive energy to the soil surface than stratiform precipitation that occurs when the atmosphere exhibits greater stability. Rainfall rates of 5–$10\,\mathrm{mm\,h}^{-1}$ or more (i.e. heavy rainfall) are frequently associated with convective precipitation. Vigorous convective conditions may also be associated with hail and lightning, as well as strong downdrafts and gust fronts flanking the column of heavy rain. In drylands, these downdrafts may generate dust storms in the area surrounding the convective rainfall cell. DLD

Reference

Tremblay A. (2005) The stratiform and convective components of surface precipitation. *Journal of the Atmospheric Sciences*, **62**, 1513–1528.

Simplified model depicting the life cycle of an ordinary thunderstorm that is nearly stationary. (Arrows show vertical air currents. Dashed line represents freezing level, 0°C isotherm.)

The life cycle of a thunderstorm, showing updrafts and downdrafts. Dashed line shows the freezing level.
http://www.atmo.arizona.edu/students/courselinks/fall12/atmo336/lectures/sec2/sevthun.html

convergence (see also DIVERGENCE) The meeting of air masses in the lower atmosphere, contributing to CONVECTION (see also INTERTROPICAL CONVERGENCE ZONE (ITCZ)) or the meeting of tectonic plates (see PLATE TECTONICS), leading to uplift.

coordination number Refers to the packing of objects around a given body. The packing of sand particles can be represented by a coordination number that refers to the number of contacts a given particle makes with other particles. The higher the coordination number the greater the interparticle friction and interlocking and the higher the SHEAR STRENGTH of the material. The value also affects (but is not a substitute for) POROSITY and PERMEABILITY of a material. In chemical structures the coordination is the number of contacts a central atom in a molecule has with its neighbours and is important in determining the properties of materials, including clay minerals. WBW

Reading

Whalley, W.B. (1976) *Properties of materials and geomorphological explanation.* Oxford: Oxford University Press.

coprolite Animal dung or excrement preserved or fossilized over time. Such deposits, particularly if preserved in sediments suitable for absolute or relative dating, provide valuable palaeoenvironmental evidence. The chemical conditions within the coprolite may facilitate the preservation of other fossils; for example, plant remains, including pollen grains, fish scales or parts of other devoured animals. Subsequent analysis may then allow for the reconstruction either of the diet of the animal that produced the excrement or of the surrounding vegetation types at the time of its production. Several types of coprolite have proved useful in this way; for example, hyena dung deposited in caves may preserve pollen blown into the cave. Excreta of the packrat (in North America) or hyrax (in Africa), both of which develop dung middens over extended periods of time, have been utilized for POLLEN ANALYSIS, since the sticky surface of fresh deposits acts as an excellent pollen trap. Hyrax middens in particular may reveal long, continuous and high-resolution records for arid and semi-arid localities that may otherwise have proved difficult to study from the palaeoenvironmental perspective (Chase *et al.*, 2012). MEM

Reference

Chase, B.M., Scott, L., Meadows, M.E., *et al.* (2012) Rock hyrax middens: a palaeoenvironmental archive for southern African drylands. *Quaternary Science Reviews*, **56**, 107–125.

coquina   A carbonate rock that consists largely or wholly of mechanically sorted, weakly to moderately cemented fossil debris (especially shell debris) in which the interstices are not necessarily filled with a matrix of other material.   ASG

coral–algal reef   See CORAL REEFS AND CORAL–ALGAL REEFS.

coral bleaching   Corals respond to thermal stress, and coupled increases in solar irradiance, by whitening or 'bleaching'. Bleaching is the visible sign of the degeneration and/or loss of large populations (hundreds of thousands) of 'zooxanthellae' (single-celled dinoflagellate algae) that live within coral tissues. Photosynthesis by the algal symbiont in the protective environment of the coral host produces a surplus of energy-rich compounds that are used by the coral for 'light-enhanced' calcification and skeletal growth. The expulsion of the zooxanthellae thus leads to the cessation of these processes. Zooxanthellae populations are naturally dynamic, and even healthy corals periodically bleach – generally when the water temperature exceed $1\,°C$ over the mean monthly temperature in the warmest month of the year – before recovery. However, greater temperature anomalies, particularly if prolonged, can lead to coral mass mortality (>90% loss of living coral cover) and the replacement of hard coral substrates by 'soft reefs' of algal turf or fleshy algae. This process has been described by the term 'phase shift'; determination of this threshold is important because, whereas accreting hard reefs can track sea-level rise, soft reefs lay down no hard substrate. Furthermore, the establishment of new coral recruits is difficult on soft, mobile surfaces.   TS

Reading
Baker, A.C., Glynn, P.W. and Riegel, B. (2008) Climate change and coral reef bleaching: an ecological assessment of long-term impacts, recovery trends and future outlook. *Estuarine, Coastal and Shelf Science*, **80**, 435–471. · Brown, B.E. (1997) Coral bleaching: causes and consequences. *Coral Reefs*, **16**, S129–S138. · Done, T.J. (1992) Phase shifts in coral reef communities and their ecological significance. *Hydrobiologia* **247**, 121–132. · Fitt, W.K., McFarland, F.K., Warner, M.E. and Chilcoat, G.C. (2000) Seasonal patterns of tissue biomass and densities of symbiotic dinoflagellates in reef corals in relation to coral bleaching. *Limnology and Oceanography*, **45**, 677–685. · Van Oppen, M.J.H. and Lough, J.M. (eds) (2009) *Coral bleaching: patterns, processes, causes and consequences.* Berlin: Springer.

coral disease   First described from the 1970s and thought to be a factor in the degradation and loss of coral communities worldwide. Although there have been field descriptions of over a dozen different coral diseases, only three diseases are well known: 'black band' disease, which affects non-acroporid corals and has a global extent; 'white band' disease, affecting acroporid corals and which has effectively restructured the composition of Caribbean reefs at a regional scale; and 'plague', which impacts massive and platy corals. All three diseases are characterized by a sharp boundary between apparently healthy coral tissue and freshly exposed coral skeleton; progress of tissue destruction has been measured at rates of $3–10\,mm\,day^{-1}$, and small coral colonies can be killed within weeks or even days. For black band disease, the primary causative pathogen has been identified as a cyanobacterium (*Phormidium corallyticum*), part of a microbial consortium also containing sulphate-reducing and sulphate-oxidizing bacteria. A fourth disease, aspergillosis, is caused by the terrestrial fungus *Aspergillus sydowii* and has been responsible for the mass mortality of sea fans on Caribbean reefs. The exact relations between reef degradation, the presence of coral diseases and anthropogenic influences are unclear. Whilst in some areas the incidence of black band disease appears correlated with high nutrient levels (perhaps derived from terrestrial runoff), this and other diseases are also known at locations remote from strong human pressures. TS

Reading
Edmunds, P. (1991) Extent and effect of black band disease on Caribbean reefs, *Coral Reefs*, **10**, 161–165. · Gladfelter, W. (1982) Whiteband disease in *Acropora palmata*: implications for the structure and growth of shallow reefs. *Bulletin of Marine Science*, **32**, 639–643. · Richardson, L.L. (1998) Coral diseases: what is really known? *Trends in Ecology and Evolution*, **13**, 438–443.

coral reefs and coral–algal reefs   A rigid, wave-resistant marine structure, usually covering several square kilometres, containing colonies of scleractinian corals and (coral–algal forms) crustose coralline algae as the main framework builders. In the Indo-Pacific reef province, coralline algae often build a prominent 'algal ridge' at the seaward margin of the reef system. Mechanical and biological breakdown of framework provides abundant within-basin sediments for lagoonal infill and reef rim coral island formation. Not to be confused with cold-water coral reefs, which do not build wave-resistant structures, or with 'coral community', a local assemblage of individual coral colonies on a non-reefal substrate. The distribution of coral reefs is controlled by environmental factors, notably water temperature, clarity and salinity. Most reef-forming corals prefer sea temperatures between 17 and $33\,°C$, salinities of between 30 and 38‰ and clear water. Because of

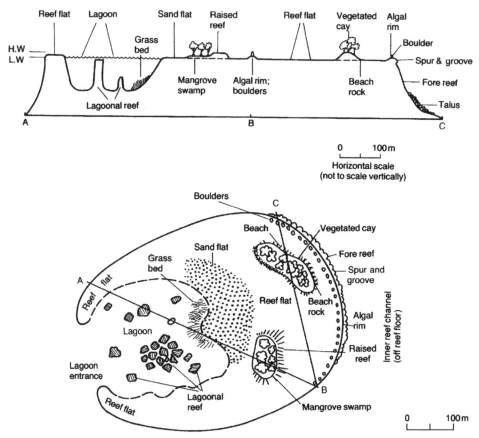

Coral–algal reefs. Plan and cross-section of an ideal coral reef, showing the major environments.

these factors, coral reefs are found mainly between latitudes 30°N and 30°S, particularly on the western margins of the Pacific, Indian and Atlantic Ocean basins. Light is also important. Although individual corals can be found at water depths in excess of 100 m, reef framework construction is usually restricted to the upper 40 m. In very turbid waters the depth limit may be as low as 5 m. Contemporary coral reefs are a living surface veneer on older reef materials. In oceanic settings, shallow water limestones may exceed over 1000 m in thickness and represent the interaction of long-term plate subsidence with sea-level fluctuations; elsewhere, Holocene reefs may be as little as a few metres in thickness.

Darwin's classification of oceanic reef types envisages a temporal sequence of fringing reefs, BARRIER REEFS and ATOLLS on subsiding, mid-plate volcanic islands. Bank reefs, bank barrier reefs, platform reefs and ridge reefs (found in the Red Sea) have also been described. Reefs formed

on continental coastlines are often complex, multiple features that are difficult to categorize and which have been shown to evolve through a development series. Patch reefs can also be found within larger reef structures. The geomorphology of coral reefs is controlled by the interplay of growth and erosion, producing a reef front (often with coral spur and sand-filled groove topography) on the ocean side, a reef crest and a back reef zone. Where sea level has recently been attained, as in the Caribbean Sea, the back reef environment is generally a shallow lagoon. Where sea level has been at, or slightly above, present sea level over the last 6000 years, the back reef is often a cemented plate, or 'reef flat', drying at low tide and, in places, extending for several kilometres in width.

The coral–algal reef ecosystem has very different environmental conditions from the surrounding sea, encouraging a flourishing of life. Scleractinian corals form reef framework though a process known as 'light-enhanced calcification'. This results from the presence in coral tissues of photosynthesizing unicellular algae known as zooxanthellae (genus *Symbiodinium*). Open-structured branching corals grow by branch tip extension at 30 cm a^{-1}, whereas solid hemispherical corals grow by radial expansion, typically 1 cm a^{-1}. Calcareous, encrusting organisms (such as coralline algae, corals, bryozoans, gastropods and serpulid worms) attach themselves to cavities within the primary reef framework, forming a secondary structure.

Information on past rates of reef growth in the geological record suggests that modern reefs have the potential to keep up with even high estimates of near-future accelerated sea-level rise. However, there remains considerable uncertainty as to whether such potential growth can be realized, either as a result of increased storminess removing growth increments or because of physiological constraints on reef growth rates and framework strength. Whilst robust at moderate to high energy levels, reefs are vulnerable to catastrophic events, such as tsunamis, hurricanes and cyclones. Corals are susceptible to both high solar irradiance and high seawater temperatures (when they show CORAL BLEACHING), low seawater temperatures, low salinities and high sedimentation rates. They are outcompeted by fleshy algae at high nutrient loadings. Human stresses on reefs and reef-related ecosystems, such as pollution and overfishing, have degraded reefs in many areas and made them less resilient to global environmental change. A new emerging concern is that coral colony growth and framework maintenance are likely to be susceptible to ocean acidification. TS

Reading
Hallock, P. (2001) Coral reefs, carbonate sediments, nutrients and global change. In G. D. Stanley, Jr (ed.), *The history and sedimentology of ancient reef systems*. New York: Kluwer; pp. 387–427. · Hoegh-Guldberg, O., Mumby, P.J., Hooten, A.J., *et al.* (2007) Coral reefs under rapid climate change and ocean acidification. *Science*, **318**, 1737–1742. · Hopley, D., Smithers, S.G. and Parnell, K.E. (2007) *The geomorphology of the Great Barrier reef: development, diversity and change*. Cambridge: Cambridge University Press. · Kench, P., Perry, C. and Spencer, T. (2009) Coral reefs. In O. Slaymaker, T. Spencer and C. Embleton-Hamann (eds), *Geomorphology and global environmental change*. Cambridge: Cambridge University Press; pp. 180–213. · Massel, S.R. and Done, T.J. (1993) Effects of cyclone waves on massive coral assemblages on the Great Barrier reef; meteorology, hydrodynamics and demography. *Coral Reefs*, **12**, 153–166. · Montaggioni, L.F. (2005) History of Indo-Pacific coral reef systems since the last glaciation: development patterns and controlling factors. *Earth-Science Reviews*, **71**, 1–75.

core The intensely hot (2700 K) inner part of the Earth. It begins at around 2900 km from the surface at the Gutenberg discontinuity. The outer portions of the core may be liquid and the inner solid. The geomagnetic field arises from the liquid outer core, and both the inner and outer cores rotate. DLD

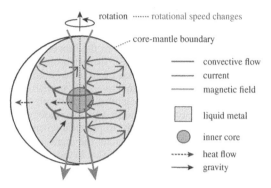

Schematic illustration of the geodynamo problem. Three factors, electrically conducting fluids, convection, and rotation, are necessary for the MHD geodynamo. We focus on the effect of rotational speed change. *Source:* Miyagoshi and Hamano (2013). Reproduced with permission of American Physical Society.

Reference
Deguen, R. (2012) Structure and dynamics of Earth's inner core. *Earth and Planetary Science Letters*, **333–334**, 211–225. · Miyagoshi T, Hamano Y. 2013. Magnetic field variation caused by rotational speed change in a magnetohydrodynamic dynamo. *Physical Review Letters* **111**, **124501**.

corallith A free-living, spheroidal growth of coral, 5–25 cm in diameter, typically of the massive coral genera *Porites*, *Pavona* and *Siderastrea*. They are found in the shallow back reef environments of Caribbean reefs and on Indo-Pacific reef flats. Movement is necessary to maintain a complete envelope of living tissue. This may be as a result of traction under waves, strong currents (either directly or indirectly through tidal currents periodically flushing sediment grains at corallith contact points with the substrate) or biological disturbance (e.g. by grazing reef fish). Once anchored, large coralliths can develop into stable coral micro-atolls in tidal lagoons. TS

Reading
Scoffin, T.P., Stoddart, D.R., Tudhope. A.W. and Woodroffe, C. (1985) Rhodoliths and coralliths of Muri Lagoon, Raroptionga, Cook Islands. *Coral Reefs*, **4**, 71–80.

corestone A cobble or boulder of relatively unweathered rock that is or has been incorporated within the weathered rock that surrounds it.

Coriolis force Also known as the geostrophic force. An apparent force on moving particles in a frame of reference that itself is moving, usually rotating. Such a force is required if Newton's laws of motion (see EQUATIONS OF MOTION) are to be applied in the rotating framework. The Coriolis force is of major importance to the movement of both oceanic waters and air. In meteorology the Coriolis force per unit mass of air arises from the Earth's rotation and is equal to $-2\Omega \times \mathbf{v}$, where Ω and $\mathbf{v}$ are vectors representing respectively the angular velocity of the Earth and the velocity of the air relative to the Earth. In practical terms the force 'deflects' air particles to the right in the northern hemisphere and to the left in the southern hemisphere. It affects only the direction, not the speed of the wind. (See also GEOSTROPHIC WIND and WIND.) BWA

Coriolis parameter Twice the component of the Earth's angular velocity Ω about the local vertical, $2\Omega \sin \phi$, where ϕ is latitude.

cornice An overhanging accumulation of wind-blown snow and ice found on a ridge or a clifftop, usually on the lee side.

corniche An organic protrusion growing out from steep rock surfaces at about sea level and providing a narrow pavement or sidewalk-like path at the foot of sea cliffs. Comparable rock ledges caused by erosional processes and coated with organic material are termed trottoirs. Corniches are often formed of calcareous algae. They are largely intertidal, being best developed in the inlets of exposed coasts, and generally protrude about 0.2–2.0 m. Vermetids and serpulids may contribute to their development. ASG

corrasion Mechanical erosion of rocks by material being transported by water, wind and ice over and around them. Also see ABRASION.

corrie See CIRQUE.

corrosion The process of solution by chemical agencies of a rock in water as distinct from the mechanical wearing away of rock by water or its bedload (CORRASION).

cosmogenic isotope Radiogenic and stable isotopes may be created by the interaction of extraterrestrial cosmic rays and terrestrial atoms and are cosmogenic isotopes. The best known of such isotopes is radiocarbon (^{14}C), but other important isotopes include ^{10}Be and ^{26}Al. Stable cosmogenic nuclides include ^{3}He and ^{21}Ne. Their formation may be atmospheric (of meteoric origin), as is the case for ^{14}C, or result from interactions with terrestrial rocks in the upper few metres of Earth regolith (of hypogene origin). The advent of accelerator mass spectroscopy (see also RADIOCARBON DATING) has allowed accurate analysis of cosmogenic isotope concentrations, which are typically low (at best a few atoms per gram per year) in whole rock samples and individual minerals and precipitates. As the isotopes exhibit a wide range of half-lives, they provide a potential for establishing ages over a considerable range of time: from a few thousand to in excess of 5 million years.

Central to dating via cosmogenic isotopes is the assumption of a constant flux of cosmic rays and their secondary products for any given area, and therefore a constant (and known) rate of cosmogenic isotope production at source through time. In-situ (hypogene) accumulation of cosmogenic isotopes has been described for exposed country rock, alluvial fan surfaces, flood and shoreline deposits and lava flows. Much of the atmospheric cosmogenic ^{36}Cl production is transported by the hydrological system to the oceans, while some remains trapped in evaporitic deposits associated with closed basins.

Key applications of cosmogenic isotope analysis include the estimation of exposure ages, the history of exposure and burial, and the calculation of long-term erosion rates. Cosmogenic isotope analyses have shed light on geomorphic questions not previously capable of solution. SS/DLD

Reading
Cockburn, H.A.P. and Summerfield, M.A. (2004) Geomorphological applications of cosmogenic isotope analysis. *Progress in Physical Geography*, **28**, 1–42. · Lal, D. (1998) In situ-produced cosmogenic isotopes in terrestrial rocks. *Annual Reviews of Earth and Planetary Sciences*, **16**, 355–388. · Walker, M. (2005) *Quaternary dating methods*. Chichester: John Wiley & Sons, Ltd.

coulée A flow of volcanic lava that has cooled and solidified.

couloir A deep gorge or ravine on the side of a mountain, especially in the Alps.

Coulomb equation See MOHR–COULOMB EQUATION.

coupled GCMs General circulation models (GCMs) are global models of the atmosphere or the ocean that describe their circulations at the planetary scale. At the time of the initial development of numerical models of climate, GCMs for the global atmosphere and the oceans were developed in isolation. These reductionist-modelling systems were forced by the sea surface temperature for the atmospheric GCMs and with atmospheric fluxes for the oceanic GCMs, which were obtained from other alternative sources. While these forced GCMs have been extensively used in research and operational predictions of atmospheric or oceanic climate, there was a growing consensus that such numerical models are inherently limited in not resolving the coupled air–sea interactions (Barsugli and Battisti, 1998; Bretherton and Battisti, 2000). Moreover, these reduced modelling systems were incapable of simulating or predicting El Niño and the Southern Oscillation events, which is the most well-known naturally occurring coupled ocean–atmosphere phenomenon on our planet that influences global climate variability. Such limitations of the reduced modelling systems motivated the development of coupled GCMs, which now not only resolve the interactions between the atmosphere and the ocean, but also include the interactions amongst other components of the Earth's climate system, which include the biosphere and the cryosphere. These newer coupled GCMs are now commonly referred to as Earth system models. However, reference to coupled GCMs sometimes makes a subtle difference from the Earth system models by referring to only resolving the air–sea interactions in the modelling system while ignoring the other interactions. VM

References

Barsugli, J.J. and Battisti, D.S. (1998) The basic effects of atmosphere–ocean thermal coupling on midlatitude variability. *Journal of Atmospheric Science*, **55**, 477–493. · Bretherton, C.S. and Battisti, D.S. (2000) An interpretation of the results from atmospheric general circulation models forced by the time history of the observed sea surface temperature distribution. *Geophysical Research Letters*, **27**, 767–770.

cover, plant The proportion, or percentage, of ground occupied by the aerial parts of a plant species or group of species. With the overlapping of species in most plant communities, the combined percentage will nearly always exceed 100%, except in very open vegetation. Special scales have been devised for estimating both the degree and character of plant cover in quadrats of a chosen size, such as the DOMIN SCALE and the BRAUN-BLANQUET SCALE. Some methods involve sampling, like the number of points touching a given species, whereas other systems are simply estimates by eye. PAS

Reading

Kershaw, K.A. (1973) *Quantitative and dynamic plant ecology*, 2nd edition. London: Edward Arnold. · Willis, A.J. (1973) *Introduction to plant ecology*. London: Allen & Unwin; especially chapter 11.

coversand A term originally applied to any aeolian sand that masked older sediments, but this generic definition has led to its indiscriminate use. Coversands are typically thin sandy deposits of cold-climate aeolian origin formed into a flat spatially continuous relief of uniform thickness. Only in valleys and against topographic barriers do they increase in thickness. DUNES directly associated with coversands are relatively rare and often indistinct. Whilst coversands are predominantly aeolian they do not preclude incorporation of sand or subsequent reworking by other processes. Use of the orientation coversand of dune morphology, bedding inclination and unit thickness has enabled the reconstruction of regional palaeowind directions (Bateman, 1998).

Coversands are widely found on the lowlands of Europe, with occurrences in Britain, Netherlands, Germany, Denmark and Poland. The northerly coversand limits coincide with the maximal position of the late PLEISTOCENE ice sheet. Coversands occupy similar ice maximal positions in North America.

European coversands are all relict features with two main phases of coversand deposition reported: one around 18,000–15,000 years ago (Older Coversand) and another between 13,000 and 11,000 years ago (Younger Coversand) (Koster, 1988; Bateman, 1998). On the basis of their stratification they have been classified into two main FACIES. One, associated with the Younger Coversand phase, is typically unimodal, well sorted, with parallel laminations and a large sand component derived from local sources. This type has little evidence of PERMAFROST and indicates increasingly dry conditions. The other facies, associated with the Older Coversand phase, is characterized by an alternation of well-sorted parallel-laminated beds of loam and sand with evidence of deposition under wet/moist conditions. Cryogenic deformation and FROST WEDGE casts are found only in the lower part of the facies (Older Coversand I). MDB

Reading and References

Bateman, M.D. (1998) The origin and age of coversand in N. Lincolnshire, UK. *Permafrost and Periglacial Processes*, **9**, 313–325. · Koster, E.A. (1988) Ancient and modern cold-climate aeolian sand deposition: a review. *Journal* *of Quaternary Science* **3**, 69–83. · Kasse, C. (1997) Cold-climate aeolian sand-sheet formation in north-western Europe (*c.*14–12.4 ka): a response to permafrost degradation and increased aridity. *Permafrost and Periglacial Processes*, **8**, 295–311.

crab-holes Small abrupt depressions in the ground surface that vary in diameter from a few centimetres to more than a metre, and in depth from around 5–60 cm. They occur in sediments that are prone to vertical cracking and horizontal piping.   ASG

Reading
Upton, G. (1983) Genesis of crabhole microrelief at Fowlers Gap, western New South Wales. *Catena*, **10**, 383–392.

crag and tail A landform consisting of a small rock hill and tapering ridge that is produced by selective erosion and deposition beneath an ICE SHEET. They range in scale from tens of metres to kilometres in length, with the tail pointing in the down-ice direction. The hill, or crag, is usually of strong rock that has resisted glacial erosion and forms an obstruction to the ice, producing a 'pressure shadow' in its lee. This extends in a down-ice direction in proportion to the ice velocity and thickness and creates a gradually tapering zone of minimal erosion or even a cavity. Although often similar in appearance, there are two types of crag and tail, dependent upon the composition of the tail and processes that led to its formation. Erosional crag and tails consist of a highly resistant rock crag that protected less resistant bedrock in its lee from the full force of glacial erosion. The tail in

this type consists of bedrock. Depositional crag and tails were formed by the inflow of glacial sediments into a cavity produced in the lee of the rock obstruction, and hence have tails composed of unconsolidated sediments. These tend to be smaller in scale. In practice, it is hard to differentiate between these two types as they may both have glacial sediments at the surface of the tail. The significance of these landforms is that, in common with SUBGLACIAL BEDFORMS, they record former directions of ice flow and indicate that the ice was at its pressure melting point, which permitted sliding at the bed.   CDC

Reference
Benn, D.I. and Evans, D.J.A. (2010) *Glaciers & glaciation*, 2nd edition. London: Arnold.

crater A depression at the crest or on the flanks of a volcanic cone where a pipe or vent carrying gases and lava reaches the surface. Also the scar left by a meteorite when it impacts on the surface of a planet or moon.

craton A continental area that has experienced little internal deformation since the Precambrian (about 570 million years ago). These areas can be divided into a very stable core, known as a SHIELD, and a marginal platform zone where gently tilted or flat sedimentary rocks bury the Precambrian basement. Crustal movement in cratons is largely vertical and results in the formation of broad domes and basins. The term *craton* is also used in an alternative sense to refer to the very ancient core of shield.   MAS

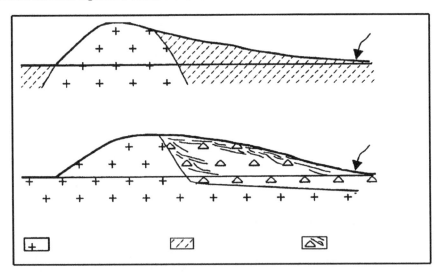

Crag and tail forms: erosional (top) and depositional (bottom); ice flow from left to right.
Source: Benn and Evans (2010). Reproduced with permission of David Evans.

Reading
Spencer, E.W. (1977) *Introduction to the structure of the Earth*, 2nd edition. New York: McGraw-Hill. · Windley, B.F. (1984) *The evolving continents*, 2nd edition. New York: John Wiley & Sons, Inc.

creationism   The belief that attributes the origin of all species of organisms (and indeed all matter) to special creation as opposed to evolution. Creationists maintain that plants and animals were brought into existence, in their present form, by the direct intervention of divine power. Creationists usually suggest that all organisms were created at the same time, or over a very short period (as in the Genesis accounts), although Louis Agussiz (1807–1873) entertained the possibility of 'multiple creations'.

'Creation science' or 'scientific creationism' attempts to demonstrate the literal truth of accounts of the creation, such as that in the book of Genesis, using the techniques of modern science. Bone beds, such as that of the Rhaetic in Britain, are attributed to a catastrophic destruction of organisms caused by the deluge. Creation scientists have drawn attention to imperfections in the fossil record, and the rarity within it of links between the major groups of organisms, arguing that these invalidate the theory of evolution. They have also taken certain recent scientific doubts about natural selection as a *principal* mechanism of evolution as calling into question the whole of evolutionary doctrine.

Unquestionably, creationists have caused textbook writers to be much more careful in their wording, but most evolutionists would argue that many of the creationist ideas are based upon the misunderstanding, or in some cases, the deliberate distortion, of scientific evidence. See EVOLUTION.   PHA

Reading
Dawkins, R. (2009) *The greatest show on earth: the evidence for evolution*. London: Random House. · Ruse, M. (1982) Creation science: the ultimate fraud. In J. Cherfas (ed.), *Darwin up to date*. London: New Science Publications; chapter 2, pp. 7–11.

creep   The gradual downslope movement of soil and rock debris on hillsides or glacier ice under the influence of gravity.

crevasse   A chasm or deep fissure in the surface of an ice sheet or glacier. Also a breach in a levée along the bank of a river.

critical load   A concept in pollution studies that involves the idea there is a certain pollution load level above which harmful effects on biological systems, such as decline and disappearance of fish populations, will occur.   ASG

Reading
Brodin, Y.-W. (ed.) (1992) The critical-load concept: an instrument to combat acidification and nutrient enrichment. *Ambio*, **21**, 332–387.

critical velocity (or critical erosion velocity)   The flow velocity required to initiate the ENTRAINMENT of sediment particles by wind or water; for example, in the study of fluvial sediment transport, the critical velocity of streamflow is that velocity which, if increased slightly, would trigger the beginning of motion in the bed materials. Thus, it indicates the threshold velocity below which the bed is stable and above which particles on the bed would begin to move along in the flow. The critical erosion velocity is primarily a function of the mean particle size of the sediment in question. Generally, larger particles have a higher entrainment threshold. Particles $<0.06\,mm$ (silt/clay), however, may also require faster fluid flows for entrainment because they tend to have additional molecular and electrostatic forces of cohesion, often become protected from erosion by larger particles or lie below the ROUGHNESS LENGTH in the zero-velocity layer. In aeolian systems, small particles such as these also have an affinity for the retention of moisture, thus increasing the forces resistant to entrainment.

There are two critical velocity thresholds: fluid (or static) and dynamic. The fluid threshold refers to the critical velocity where sediment is entrained only by the drag and lift forces of the moving fluid. However, one of the most significant influences on the entrainment of sediment is the existence of grains already in motion. In particular, the impact of grains in SALTATION on a surface often leads to the splashing of other grains into the fluid flow, which may, in turn, also be entrained by the flow and undergo saltation. Such a saltating system carries inertia and often continues even when the flow velocity is reduced to below that of the fluid critical erosion velocity. Great efforts have been made to develop both empirical and theoretical formulae to enable the critical velocity to be calculated for various kinds of stream bed sediment. In general, these have greatest success in sands and when the material concerned is well sorted. The values of critical erosion velocity for grains of known size for both wind and water are commonly determined by wind tunnel or flume experiments. Thresholds of motion on natural sediment beds are also influenced by factors such as sediment size mixtures, surface crusting, surface slope, moisture and vegetation.

A diversity of factors influences the value of critical velocity, including flow properties (depth, water temperature) and particle properties (density, size, shape, interparticle cohesion). In addition, aspects of the sedimentology of the stream bed are very important, especially in gravel streams and where the bed material involves poorly sorted sediments (see CAPACITY/NON-CAPACITY LOAD). Under these conditions, particle exposure and friction between particles lodged tightly together can greatly affect the critical velocity. Various forms of critical velocity relation have been developed. One difficulty that arises in the use of velocity is that this parameter varies continuously from the bed to the water surface. Some critical velocity equations employ the *bottom velocity*, while others are based on the *mean velocity* of the flow. An empirical curve developed by Hjulström relates particle diameter to critical velocity expressed as the mean. However, the critical velocity is known to vary with water depth and with the slope of the channel, so that empirical relations provide only a general guide to entrainment conditions. DLD/GFSW

Reading
Graf, W.H. (1971) *Hydraulics of sediment transport.* New York: McGraw-Hill. · Knighton, D. (1998) *Fluvial forms and processes: a new perspective.* London: Arnold. · Wiggs, G.F.S. (2011) Sediment mobilisation by the wind. In D. S. G. Thomas (ed.), *Arid zone geomorphology: process, form and change in drylands,* 3rd edition. Chichester: John Wiley & Sons, Ltd; pp. 455–486. · Yalin, M.S. (1972) *Mechanics of sediment transport.* Oxford: Pergamon.

critical zone That part of the Earth extending from the vegetation canopy to the water table. The concept of the critical zone provides a framework within which to explore the complex interactions of the atmosphere, biota, water, soil, organic matter, nutrients and rock that are influential in many ecosystem processes. Processes within the critical zone are subject to development or perturbation from land-use change, fire, climate variability, shifts in base level, and mass movements and other erosional processes. The term 'critical zone' in large measure corresponds to the long-established term 'regolith', which has been defined as the zone of the Earth's near-surface environments extending from 'fresh rock to fresh air' (Eggleton, 2001). DLD

Concept and key components of the critical zone.
Source: Lin (2010). This work is distributed under the Creative Commons Attribution 3.0 License.

Reading and Reference

Akob, D.M. and Küsel, K. (2011) Where microorganisms meet rocks in the Earth's critical zone. *Biogeosciences*, **8**, 3531–3543. · Eggleton, R.A. (ed.) (2001) *The regolith glossary*. Canberra: The Cooperative Research Centre for Landscape Evolution and Mineral Exploration. · Lin, H. (2010) Earth's critical zone and hydropedology: concepts, characteristics, and advances. *Hydrology and Earth System Science*, **14**, 25–45.

cross-bedding

The arrangement of laminae and beds in sedimentary strata at different angles from the principal planes of stratification. The pattern of cross-bedding provides evidence of the environment and modes of deposition.

cross-lamination

Thin layers of sediment, often only a few millimetres thick, that dip obliquely to the main bedding plane. The cross-laminae represent individual sedimentation units resulting from small-scale fluctuations in velocity and in rates of supply and deposition of silts and sands forming the lee, stoss and trough laminae of migrating ripples.   JM

cross-profile, valley/river channel

A profile may be surveyed at right angles to the river flow direction across a river channel or across the valley in which the river channel occurs. Information from contours or topographic maps is often sufficiently detailed to draw valley cross-profiles. (See also CHANNEL CAPACITY.)   KJG

crumb structure

Pertains to those soils that have their fine particles accumulated in the form of aggregates or crumbs, so that they have a more open, coarser and workable texture.

crust

Used in various ways in physical geography, but often refers to the uppermost layer of a soil or sediment, where properties including hardness and porosity differ from the underlying material. This may be due to the addition of fine material that enters pore spaces from subaerial sources, the precipitation of fines and salts through capillary rise, compaction through rain impact or animal trampling, the growth of biological (algal, bacterial) communities, and so on (see SOIL CRUST). Crusts may be micrometre to millimetre thick and are perhaps more common in dryland environments than elsewhere. Various crust terminologies exist; for example, for biological crust, physical crust, clay crust, silica crust. Features known as DURICRUSTS may form at the surface or in subsurface locations, and are therefore not necessarily true crusts.   DSGT

Reading

Belnap, J. and Lange, O.L. (eds) (2003) *Biological soil crusts: structure, function and management*. Dordrecht: Springer. · Van Breeman, N. and Buurman, P. (2002) *Soil formation*. Dordrecht: Springer.

cryergic

Periglacial in its broadest sense. Pertaining to the periglacial features and processes that occur in those areas not immediately adjacent to glaciated areas.

cryosphere

All regions on and beneath the surface of the Earth and ocean where water is in solid form, including snow cover, sea ice, lake ice, river ice, glaciers and ice sheets, and frozen ground (which includes permafrost). Snow is sensitive to individual weather events, river lake and sea ice to seasonal fluctuations and longer trends, permafrost and glaciers to decadal and longer trends, and ice sheets to millennial and geological timescale changes. Seasonal frozen ground has the largest areal component $(5.2 \times 10^7 \, \text{km}^2)$, about 50% of the northern hemisphere landmass. Ice sheets contain the largest proportion of the volume of the cryosphere $(3.3 \times 10^7 \, \text{km}^3)$.   TS

Reference

Slaymaker, O. and Kelly, R.E.J. (2007) *The cryosphere and global environmental change*. Malden: Blackwell.

cryoplanation

The flattening and lowering of a landscape by processes related to the action of frost. (See also ALTIPLANATION.)   ASG

Reading

Priesnitz, K. 1988: Cryoplanation. In Clarke, M.J., ed., *Advances in periglacial geomorphology*. Chichester: Wiley. Pp. 49–67.

cryostatic pressures

Freezing-induced pressures thought to develop in the ACTIVE LAYER in pockets of unfrozen material that are trapped between the downward-migrating freezing plane and the perennially frozen ground beneath. Although recorded in experimental studies, the existence of substantial cryostatic pressures in the field has yet to be convincingly demonstrated. Generally, the presence of voids in the soil, the occurrence of frost cracks in winter and the weakness of the confining soil layers lying above prevent pressures of any magnitude from forming. Nevertheless, cryostatic pressures are often invoked to explain various forms of patterned ground and mass displacements (cryoturbations) in the active layer.   HMF

Reading

French, H.M. (1976) *The periglacial environment*. London: Longman; especiaaly pp. 40–44. · Mackay, J.R. and

Mackay, D.K. (1976) Cryostatic pressures in non-sorted circles (mud hummocks), Inuvik. *Canadian Journal of Earth Sciences*, **13**, 889–897. · Washburn, A.L. (1979) *Geocryology: a survey of periglacial processes and environments.* New York: John Wiley & Sons, Inc.; especially p. 167.

cryoturbation The process whereby soils, rock and sediments are churned up by frost processes to produce convolutions or involutions. The process is especially active in the zone above permafrost that is subject to seasonal freezing and thawing – the active layer. (See also FROST HEAVE.) ASG

Reading
Vandenberghe, J. (1988) Cryoturbations. In M. J. Clarke (ed.), *Advances in periglacial geomorphology.* Chichester: John Wiley & Sons, Ltd; pp. 179–198.

cryovegetation Consists of plant communities comprising such types as algae, lichens and mosses that have adapted to life in environments where there is permanent snow and ice.

cryptovolcano A small, roughly circular area of greatly disturbed strata and sediments that, though suggestive of volcanism, does not contain any true volcanic materials.

cuesta A ridge that possesses both scarp and dip slopes.

cuirass An indurated soil crust that mantles the land surface, protecting the underlying unconsolidated sediments from erosion.

cultural geomorphology This term encapsulates the scientific study of human engagement with geomorphological landscapes and encompasses cultural reactions to and perceptions of landscape, and how these should be considered by geomorphologists, especially in terms of improving environmental management and landscape conservation (Goudie and Viles, 2010: 99; Panizza and Piacente, 2003). Geomorphology is seen as part of our cultural heritage in the same way that archaeological and historical sites are. In addition, humans are playing an increasingly important role in modifying landforms and land-forming processes (ANTHROPOGEOMORPHOLOGY), and there is an increasing concern about the impacts of geomorphological hazards on human societies. ASG

References
Goudie, A.S. and Viles, H.A. (2010) *Landscapes and geomorphology.* Oxford: Oxford University Press. · Panizza, M. and Piacente, S. (2003) *Geomorfologia culturale.* Bologna: Pitagora Editrice.

cumec A measure of discharge, being an abbreviation for cubic metre per second.

cumulative soil profiles These receive influxes of parent material while soil formation is still going on; that is, soil formation and deposition are concomitant at the same site. Their features are thus partly sedimentological and partly pedogenic. Among topographic sites favourably situated for their formation are areas receiving increments of loess, river floodplains, and colluvial and fan deposits at the base of hillslopes. ASG

cumulonimbus See CLOUDS.

cumulus See CLOUDS.

cupola A dome-shaped mass of igneous rock that projects from the surface of a batholith.

current bedding Layering produced in accumulating sediment by fluid flow that is oblique to the general stratification. Laminae (less than 1 cm) or thicker strata in tabular or wedge-shaped units may comprise near-parallel sets bounded by plane, curved or irregular surfaces, with boundaries representing accretionary limits or erosional truncations. Such features are produced through the development of fluid BEDFORMS varying in size from CURRENT RIPPLES to bars, and their form may give indications of the direction of current flow, flow regime and sediment supply. Also called cross-bedding (a term now preferred by most sedimentologists and subdivided, into cross-lamination and cross-stratification) or false bedding, this phenomenon may help in the identification of sedimentary environments. For example, large-scale current bedding may relate to delta growth, the progradation of aeolian dune slip faces or to river channel point bar sedimentation. Smaller scale features derive from ripple development, and different kinds of ripples (e.g. straight-crested or linguoid) may produce contrasted current bedding patterns. JL

Reading
Allen, J.R.L. (1970) *Physical processes of sedimentation.* London: Allen & Unwin. · Allen, J.R.L. (1982) *Sedimentary structures*, vol. 1. Amsterdam: Elsevier.

current meter An instrument for measuring the velocity of flowing water in freshwater and marine environments. Many principles have been employed, and available types include rotating, electromagnetic, optical and pendulum current meters. The rotating current meter is the most widely used for river measurements. It consists of a propeller (horizontal axis type) or a rotor formed by a series of cups (vertical axis type) that rotates at a speed proportional to the flow velocity. The revolutions are counted over a fixed period of time, and velocity is computed from calibration data. The meter body may be mounted on a wading rod or suspended on a cable. Current approaches include laser Doppler flow meters, acoustic Doppler flowmeters, and noncontact radar-based and acoutstic flowmeters that can be mounted above the moving liquid.   DEW/DLD

Reading
Nezu, I. and Rodi, W. (1986) Open channel flow measurements with a Laser Doppler anemometer. *Journal of Hydraulic Engineering*, **112**, 335–355.

current ripples   Small-scale wave-like undulations developed by fluid flow over a sandy or coarse silty bed, in water or in air. Their spacing or wavelength is usually less than 50 cm and the height difference between trough and crest is seldom more than 3 cm. Larger sand features may be called dunes or sandwaves and larger dynamically related features in coarser sediment may be termed bars.

Ripples developed on the upper windward slope of a small barchan sand dune. Photograph by David Thomas.

Current ripples may be described in terms of their plan and profile characteristics. They may be straight, sinuous or indented (linguoid, cuspate, lunate) in plan and may have peaked, rounded and asymmetrical crests. Ripple development occurs through flow separation from the bed; dimensions may be related to applied fluid shear stress.   JL

Reading
Allen, J.R.L. (1968) *Current ripples*. Amsterdam: North-Holland.

cusp, beach   A three-dimensional, scallop-shaped BEACH form occurring in rhythmic sets along the shore. Beach cusps can form through erosion and accretion and are most common on REFLECTIVE BEACHes. The regular spacing of beach cusps has made them a focus of considerable research attention. The EDGE WAVES hypothesis and the system self-organization hypothesis are two competing models. The former requires that beach cusp spacing be tied to edge wavelengths. The latter requires that spacing be a function of run-up height. Attempts to discern the appropriate model remain unsuccessful because of a lack of appropriate field data.   DJS

Reading
Coco, G., Huntley, D.A. and O'Hare, T.J. (1999) Beach cusp formation: analysis of a self-organization model. In N. C. Kraus and W. G. McDougal (eds), *Coastal sediments '99*. Reston, VA: American Society of Civil Engineers; pp. 2190–2205.

cutan   A thin coating of clay on soil particles or lining the walls of a void in the soil zone.

cut-off   An abandoned reach of river channel, produced particularly where a meander loop has become detached from the active river channel because the neck of the loop has been breached. The abandoned reach may be occupied by an OXBOW lake that gradually fills with sediment. Different cut-off processes are possible: with highly sinuous channels adjacent reaches may impinge on each other directly (neck cut-offs), while in other cases the scouring out of longer short-circuiting channels across the inside of meander bends during floods may produce cut-offs at much lower sinuosities (chute cut-offs). The accumulation of sediment in bars may also lead to the detachment of channel reaches. The term is sometimes applied to the new channel itself as well as to the abandoned one or to the process in general.   JL

cutter   Linear slots cut in bedrock by solution along a guiding structural element; they are equivalent to the British term *grike* (White, 1988). ASG

Reference
White, W.B. (1988) *Geomophology and hydrology of karst terrain*. New York: Oxford University Press.

cyanobacteria   Photosynthetic, single-celled organisms lacking an enclosed nucleus or other specialized cell structures. Often referred to as blue–green algae, they are prokaryotes and more appropriately classified as bacteria. Cyanobacteria contain chlorophyll, although it is distributed throughout the cell rather than confined in cell organelles as in the green plants. Other pigments in many species impart a bluish or reddish tinge. In fresh waters subject to EUTROPHICATION, cyanobacteria may form toxic blooms. In the tropics, mats of cyanobacteria filaments grow into humps called STROMATOLITES.   MEM

cycle, biochemical See BIOGEOCHEMICAL CYCLES.

cycle, climatic A recurrent climatic phenomenon. The term is best reserved for changes of strictly periodic origin, such as the annual temperature cycle, but it is often used loosely to describe many changes of climate that occur at approximately fixed intervals.

The cycle of major glaciations with a peak about every 10^5 years is a good example. Various short-period cycles have been described by numerous authors, but they are often the subject of some controversy. Examples are the supposed climatic cycles related to solar activity in general and SUNSPOTS in particular. (See MILANKOVITCH HYPOTHESIS.)   JGL

cycle of erosion The sequence of denudational processes and forms that, in theory, exist between the initial uplift of a block of land and its reduction to a gently undulating surface or PENEPLAIN close to BASE LEVEL. William Morris Davis was the leading American geomorphologist of the late nineteenth and early twentieth centuries, and his great contribution was to produce in the 1880s and 1890s this deductive model of landscape evolution, which he called the cycle of erosion or the geographical cycle. He believed that landscapes were the product of three factors: structure (geological setting, rock character, etc.), process (weathering, erosion, etc.) and time (stage) in an evolutionary sequence. Stage was what most interested him. He suggested that the starting point of the cycle was the uplift of a broadly flat, low-lying surface. This was followed by a phase he termed *youth*, when streams become established and started to cut down and to develop networks. Much of the original flat surface remained. In the phase he termed *maturity* the valleys widened so that the original flat surface had been largely eroded away and streams drained the entire landscape. The streams began to meander across wide floodplains and the hillslopes become gradually less steep. In *old age* the landscape became so denuded that a low-relief surface close to sea level (a peneplain) developed, with only low hills (monadnocks) rising above it.

Initially, the Davisian model was postulated in the context of humid temperate ('normal')

conditions, but it was then extended by Davis and successors to other environments, including arid, glacial, coastal, savanna, limestone and periglacial landscapes. Chorley *et al.* (1973) argued that Davis's cyclic model was not very successful in Germany. Here, W. Penck's model of slope evolution, often seen as the antithesis of Davis, involved more complex tectonic changes than that of Davis, and he regarded slopes as evolving in a different manner (by slope replacement rather than slope decline) through time. An alternative model of slope development by parallel retreat leading to pediplanation was put forward by L. C. King in southern Africa. His model represents an amalgam of the views of Davis and Penck; episodic uplift resulting in both downwearing and backwearing, with the parallel retreat of slopes leading to the formation of low-angle rock cut surfaces (pediments) that coalesced to form pediplains. By the mid-twentieth century the Davisian model was becoming less dominant, partly because there was a growing awareness of crustal mobility that could not sustain notions of initial uplift followed by prolonged structural quiescence. In addition, the frequent and intense changes in global climate and sea level that appear to have occurred during the Quaternary render it improbable that any area will contain the simple sequence of forms postulated in the model; and even if these occur, it is difficult to see how they can be the products of any simple, unidirectional set of processes. ASG

Reading and Reference
Chorley, R.J. (1965) A re-evaluation of the geomorphic system of W.M. Davis. In R. J. Chorley and P. Haggett (eds), *Frontiers in geographical teaching*. London: Methuen. · Chorley, R.J., Beckinsale, R.P. and Dunn, A.J. (1973) *History of the study of landforms*, vol. 2. London: Methuen. · Davis, W.M. (1899) The geographical cycle. *Geographical Journal*, **14**, 481–504. · Kennedy, B.A. (2006) *Inventing the Earth: ideas on landscape development since 1740*. Oxford: Blackwell.

cyclone (or depression) A region of relatively low atmospheric pressure, typically 1000–2000 km across, in which the low-level winds spiral counterclockwise in the northern hemisphere and clockwise in the southern hemisphere. Cyclones are common features of surface weather maps and are frequently associated with windy, cloudy and wet weather.

Cyclogenesis (or the formation of cyclones) occurs in preferred areas and is usually most vigorous in wintertime; such areas in the northern hemisphere are the western North Atlantic, western North Pacific and Mediterranean Sea. The birth and subsequent movement of cyclones is closely linked to the presence of the planetary scale ROSSBY WAVES in the atmosphere. They form frequently in the downstream or eastern limb of these large-scale troughs and move as features embedded in the deep, generally poleward, flow.

The inward-spiralling air ascends within the system and flows out in the upper troposphere. In any atmospheric column, if more mass is exported aloft than is imported at low levels, the surface pressure will fall, while surface pressure will rise and the low will fill if there is a net gain of mass in the column.

Cyclones are highly transient features that are associated with disturbed weather and, if frontal, with strong horizontal gradients of TEMPERATURE and HUMIDITY, and sharp changes in cloud cover and type. Across the extratropical ocean basins they carry out very important heat transport in a meridional direction, which offsets to some extent the persistent equator to pole imbalance in the RADIATION budget. In frontal cyclones, warm air is transported poleward and cooled while cold air moves towards the equator and is warmed. RR

Reading
Palmén, E. and Newton, C.W. (1969) *Atmospheric circulation systems*. New York: Academic Press.

cyclostrophic A term that relates to the balance of forces in atmospheric systems in which the flow is tightly curved; for example, near the centre of a hurricane. In this case the centrifugal force is substantially larger than the CORIOLIS FORCE, and the cyclostrophic wind V is

$$V = \frac{P_n^{1/2}}{R_T}$$

where P_n is the horizontal pressure gradient force and R_T the local radius of curvature of the isobars. RR

cymatogeny The warping of the Earth's crust over horizontal distances of tens to hundreds of kilometres with minimal rock deformation, producing vertical movements of up to thousands of metres. The term, introduced by L. C. King (1959), describes crustal movements intermediate between EPEIROGENY and OROGENY and applies not only to the formation of broad domal uplifts but also to the linear vertical movements represented by mountain ranges such as the Andes. Uplift is assumed to be induced by vertical movements associated with processes active within the Earth's MANTLE and not to arise from the large-scale horizontal movements proposed in the PLATE TECTONICS model. MAS

Reading and Reference
King, L.C. (1959) Denudational and tectonic relief in southeastern Australia. *Transactions of the Geological Society of South Africa*, **62**, 113–138. · King, L.C. (1967) *The morphology of the Earth*, 2nd edition. Edinburgh/New York: Oliver & Boyd/Hafner.

D

dalmatian coast A coastline characterized by chains of islands running parallel to the mainland, deep bays and steep shorelines, being the product of subsidence of an area of land with mountain ridges running parallel to the coast.

Daly level Named after R. A. Daly, who first put forward the theory that coral reefs were affected by cold periods during the Pleistocene when colder temperatures prevailed and there were lower sea levels. Daly suggested that during these phases the coral would be destroyed and wave-cut platforms, formed by marine planation, would develop on the dead coral. These platforms are, according to Daly, evidenced by present-day lagoon floors. Subsequent sea-level rise would encourage regrowth of corals at the edges of these platforms. If Daly's theory is correct, lagoon floors should be flat and occur at similar levels throughout the world.   HAV

Reading
Stoddart, D.R. (1973) Coral reefs: the last two million years. *Geography*, **58**, 313–323.

dambo Shallow, seasonally flooded treeless depressions in catchment headwaters that lack a permanent channel (Boast, 1990) are distinctive and prominent features of savanna landscapes across substantial regions of the tropical and subtropical latitudes. The flanking slopes are gentle and they are associated with land surfaces of very low relief. Dambos were first described from present-day Zambia by Ackermann (1936), and the term now embraces equivalents elsewhere (e.g. *vlei* in South Africa, *fadama* in Nigeria). Dambos are typical valley forms of the SAVANNA of Central and south Central Africa but are described from other continents where they develop similarly on gently undulating land surfaces under tropical seasonal climates. They are typically associated with a characteristic CATENA whereby upper slopes are underlain by ferralsols (ultisols) and the valley floors are associated with VERTISOLS.

Many dambos broaden out upstream, becoming bottle or hammer shaped. This suggests that they may drain from former enclosed hollows or pans, and/or that they become widened by chemical sapping of surrounding slopes. The importance of chemical DENUDATION processes (etchplanation) is indicated by the occurrence of core boulders, and the formation of 2 : 1 smectite clays in the central seepage zone where divalent ions (Ca^{2+}, Mg^{2+}, Fe^{2+}) liberated by rock weathering become fixed. However, most dambos are also erosional forms and show evidence of buried stream channels, infilled with sands and clays, and are products of landscape evolution over longer periods, including the late Quaternary (Thomas and Goudie, 1985). Drainage is an important factor determining their characteristic vegetation patters, since where water tables rise seasonally close to the surface many tree species are excluded, and this produces the distinctive pattern of *miombo* woodland (dominated by *Brachystegia* spp.) interspersed with linear grasslands in the lower lying *dambos* found across much of south Central Africa (Goudie, 1996).

Dambo margins are prized by farmers for their moist, productive soils, and dambo floors are valued for dry-season grazing. These wetlands are delicate ecosystems, deriving their distinctive features from a combination of geomorphic setting, environmental history and a seasonal climate with an intense dry season.   MFT/MEM

References
Ackermann, E. (1936) Dambos in Nordrhodesien. Wissenschaft Veröff. *Deutschen Museums für Länderkunde zu Leipzig*, Neue Folge, 4, 147–157. · Boast, R. (1990) Dambos: a review. *Progress in Physical Geography*, 14, 153–177. · Goudie, A.S. (1996) The geomorphology of the seasonal tropics. In A. S. Goudie, W. M. Adams and A. Orme (eds), *The physical geography of Africa*. Oxford: Oxford University Press. · Thomas, M.F. and Goudie, A.S. (eds) (1985) *Dambos, small channelless valleys in the tropics: characteristics, formation, utilisation*. Zeitschrift für Geomorphologie, Supplementbände, vol. **52**. Berlin: Borntraeger.

dams The damming of rivers by artificial structures to create reservoirs has been one of the most dramatic and widespread deliberate impacts that humans have had on the natural environment. Such structures change river hydrology, sediment loads, riparian vegetation, patterns of aggradation

---

The Dictionary of Physical Geography, Fourth Edition. Edited by David S.G. Thomas.
© 2016 John Wiley & Sons, Ltd. Published 2016 by John Wiley & Sons, Ltd.

and erosion, the migration of organisms, seismic activity, and so on. ASG

Reading

Petts, G.E. (1984) *Impounded rivers – perspectives for ecological management.* Chichester: John Wiley & Sons, Ltd.

Dansgaard–Oeschger (D–O) events

The high resolution of climatic change proxies preserved in ICE CORE evidence, particularly in the form of OXYGEN ISOTOPE records, has allowed identification of rapid CLIMATE CHANGE during the QUATERNARY period, particularly in the last GLACIAL cycle. For example, Dansgaard *et al.* (1989) identified evidence of very rapid warming, of up to 7 °C in ~50 years at the end of the YOUNGER DRYAS. Other evidence has shown that INTERSTADIALS during the last glacial began equally rapidly. Such rapid temperature changes are known as Dansgaard–Oeschger events, 25 of which have been identified with temperature elevations of 10–15 °C. Their cause is not fully understood. DSGT

Reference

Dansgaard, W., White, J.W.C. and Johnsen, S.J. (1989) The adrupt termination of the Younger Dryas event. *Nature*, **339**, 532–534.

Darcy's law

Defines the relationship between the discharge of a fluid through a saturated porous medium and the gradient of HYDRAULIC HEAD. It is the most important law of GROUNDWATER hydrology.

Henri Darcy was a French engineer. In 1856 he published the results of an experiment undertaken to determine the nature of water flow through sand. He found the outflow discharge from the sand to be directly proportional to the loss in hydraulic head *h*

Darcy's law. Apparatus showing the relationships expressed in Darcy's law: A, constant-head device; B, manometers; C, porous medium being measured; D, outflow with discharge *Q*. Note the change in head *h* with distance *l*.

Source: Smith *et al.* (1976: figure 6.2). Reproduced with permission of Elsevier.

after a given length *l* of flow through the sand. This relationship may be written as

$$\nu = -K\frac{dh}{dl}$$

where ν is the specific discharge, dh/dl is the hydraulic gradient and K is a constant of proportionality known as the hydraulic CONDUCTIVITY. Thus, if dh/dl is held constant, $\nu \propto K$.

The specific discharge is sometimes termed the macroscopic velocity, the Darcy velocity or Darcy flux. It is the volume rate of flow through any cross-sectional area perpendicular to the flow direction (Freeze and Cherry, 1979). It has the dimension of a velocity, but it should be clearly distinguished from the microscopic velocities of water passing through individual intergranular spaces during flow through the porous medium.

Darcy's law is valid for groundwater flow in any direction, including when it is being forced upwards against gravity in circuitous groundwater flow paths. It may also be used to describe the flow of moisture in soil. However, there are limits to its validity. Freeze and Cherry (1979) point out that, if it were universally valid, a plot of specific discharge ν against hydraulic gradient dh/dl would reveal a straight-line relationship for all gradients between zero and infinity. This is not the case. There appear to be both upper and lower limits to its validity.

Darcy's law is only applicable under conditions of laminar flow. But at the upper limits to laminar flow, where the REYNOLDS NUMBER is in the range 1–10, the law breaks down. Hence, Darcy's law is not valid in the nonlinear laminar flow regime. Evidence is less conclusive at the lower limit, but some work suggests that there may be a threshold hydraulic gradient below which flow does not occur.

The above equation describing Darcy's law is one-dimensional in form, but it may also be developed to describe three-dimensional flow (Freeze and Cherry, 1979). In three dimensions, specific discharge ν is a vector with components ν_x, ν_y and ν_z. Hydraulic conductivity K may not be the same in each direction. Thus, a three-dimensional generalization of Darcy's law may be written as

$$\nu_x = -K_x\frac{dh}{dx}$$

$$\nu_y = -K_y\frac{dh}{dy}$$

$$\nu_z = -K_z\frac{dh}{dz}$$

PWW

Reading and References

Castany, G. 1982: *Principes et méthodes de l'hydrogéologie.* Paris: Dunod. · Fitts, C.R. 2013. *Groundwater Science,*

2nd edition. Waltham, MA: Academic Press. · Freeze, R.A. and Cherry, J.A. (1979) *Groundwater*. Englewood Cliffs, NJ: Prentice-Hall. · Smith, D.I., Atkinson, T.C. and Drew, D.P. (1976) In T. D. Ford and G. H. D Cullingford (eds), *The science of speleology*. London: Academic Press. · Todd, D.K. 1980: *Groundwater hydrology*. 2nd edition. New York and Chichester: John Wiley & Sons, Ltd.

Darwinism   The biological theory of evolution by natural selection as propounded by Charles Darwin (1809–1882) and set out in *On the Origin of Species* (1859) and *The Descent of Man* (1871). In demonstrating the mutability of species Darwin succeeded in setting the whole idea of scientific enquiry free from theological constraints. Thus, the biblical notion of successive deluges modifying the landscape (CREATIONISM) was superseded by a theory of 'evolution' whereby random variations in fauna and flora would be selectively preserved and inherited by subsequent generations. Darwinism is sometimes taken to refer to any kind of evolution (the term did not appear in *On the Origin of Species* until the fifth edition), indeed to any kind of evolutionary theory that relies on the natural selection principles but rejects the doctrine of inheritance of acquired characteristics as suggested by Lamarck. Specifically, however, the idea of adaptation to the environment and selective change at the species level revolutionized the scientific community of the nineteenth century so much that the development of geography as a science became possible. Darwin's theory allowed the disciplines geomorphology, pedology, ecology and natural history more generally to calculate more *time* in which the sequential progression of development occurred; dispelled were the time-limited ideas of Bishop Ussher and the Anglican Church that the world was a divine creation formed 4004 BC. The pre-Darwinian ideas of landform studies in the Linnaeus taxonomic form produced only confusion in the early nineteenth-century scientific community. This was due in the main to the failure of taxonomic organization to provide any unifying principle that would allow scientists to order the myriad of landform types that had been recognized.

W. M. Davis, utilizing the Darwinian principle of *evolution* through time, provided a cycle of landscape evolution from youth to maturity and finally to old age. This was a direct analogue of the Darwinian idea of a plant or animal undergoing sequential change through time. Such change embodies a second important quality of Darwinism: the idea of organization in change. The importance of this in nineteenth-century scientific thinking was the rejection of the preconception of Platonic thinking of immutability of form. Dominant in the rationale of geologists at this time was the idea of changeless ideas or forms. Indeed, it was generally held that change and variation were no more than illusions and that genuine reality was of fixed types permanently distinguished from one another. Such ideas delayed the recognition of evolution in nature and specifically prevented the science of ecology from developing. Indeed, the idea of the interrelationship between fauna and flora and their environment is a basic tenet of ecological understanding. Darwinism became the underlying principle of the subject.

Darwinism also provides geography with the idea of struggle and selection. Cause–effect relationships preoccupied pre-Darwinian thought at the expense of *process*. Subsequent Darwinian disciples have stressed the importance of the environmental influence. In a broad sense these 'Darwinian impacts', which include Social Darwinism, Darwinism and its influence upon politics, theology, philosophy, psychology, anthropology, literature and even music, all affect our general viewpoint within the specific field of physical geography. All these facets of the 'Darwinian revolution', as it is now called, relied initially (as did geology and geomorphology) on the idea of evolution. Darwin's important message, which lay relatively unheeded until the 1930s, was the idea of randomness or chance variation in nature. Indeed, natural selection of species was effected (according to Darwin) by this mechanism of chance.   PAB/TS

Reading and References

Darwin, C.R. (1859) *On the origin of species by means of natural selection; or, the preservation of favoured races in the struggle for life.* London: John Murray. · Darwin, C.R. (1871) *The descent of man and selection in relation to sex.* London: John Murray. · Green, J.C. (1980) The Kuhnian paradigm and the Darwinian revolution in natural history. In G. Gutting (ed.), *Paradigms and revolutions*. Notre Dame, IN: University of Notre Dame Press. · Oldroyd, D.R. (1983) *Darwinian impacts*. Milton Keynes: Open University. · Stoddart, D.R. (1966) Darwin's impact on geography. *Annals of the Association of American Geographers*, **56**, 683–698. · Wallace, A.R. (1895) *Natural selection and tropical nature*. London: Macmillan.

data logger   An instrument that records environmental data (e.g. temperature, wind speed, river stage) from external monitoring equipment, so allowing an observer to be absent from the apparatus during data collection.

Data loggers are particularly useful where data have to be collected over long time periods in harsh terrain (e.g. deserts or arctic environments) or where data are required at different places at the same time over spatially extensive areas. A good example of the historic use of data loggers is in the recording of air pressure on barometers using an ink line drawn on a chart attached to a continuously rotating and clock-work drum (see PRESSURE, AIR and BAROMETER).

Digital data loggers are electronic and can be programmed by computer to accept data from a wide range of measuring probes recording environmental variables from wind speed and humidity to river velocity and bank erosion. With the introduction of solar power and satellite technology it is now possible for data loggers set up in remote locations to record and then transfer data to base stations perhaps many thousands of kilometres away. Such an approach means that the data logger and measuring equipment may only have to be visited on an annual basis for routine maintenance.

Data loggers are also now able to accept very high frequency measurements of environmental data (>10 Hz). Whilst this can prove very useful in certain experimental situations, it can also be problematic in terms of the amount of data that can be collected in a short space of time, which then needs to be stored and analysed.

Important variants of data logging apparatus include event loggers (for tipping-bucket rain gauges or other switch closure logging) and voltage loggers that convert data to engineering units, process high-frequency data and log only the average value, loggers with alarm channels that can be used to trigger other devices, and very small data loggers that can be packed with samples to record temperature and other conditions during transport. Many data loggers are weatherproof and communicate using wireless or optical data links.   GFSW/DLD

Reading
Whalley, W.B. (1990) Measuring and recording devices. In A. Goudie (ed.), *Geomorphological techniques*, 2nd edition. London: Unwin Hyman; pp. 186–191.

daya   A small, silt-filled solutional depression found on limestone surfaces in some arid areas of the Middle East and North Africa.

Reading
Mitchell, C.W. and Willimott, S.G. (1974) Dayas of the Moroccan Sahara and other arid regions. *Geographical Journal*, **140**, 441–453.

dead ice topography   See STAGNANT ICE TOPOGRAPHY.

débâcle   The breaking up of ice in rivers in spring.

debris flows   A form of rapid mass movement that involves the viscous flow of a slurry-like mixture that is 60–80% by weight solids (rock and weathered material) with a limited amount of water. Frequently, the bore-like front of a debris flow carries the largest rocks or boulders, with the later parts of the flow showing diminishing grain sizes. Debris flows exhibit properties that are distinctively different from fluvial flows, including rheological properties. The flows may be convex-upward in cross profile, may cause substantial scour erosion when confined within steep channels, and typically involve a depositional zone where the flow reaches low-gradient footslopes. Debris flows may occur once a threshold of debris water content is exceeded, and their occurrence is often linked to periods of heavy rain. Debris flows are frequently recorded in the period after forest fires, when debris in the landscape is more mobile owing to the absence of vegetation and the relative ease with which unconsolidated materials can become saturated. In many locations, debris flows pose a significant hazard to settlement and infrastructure.   DLD

Reading
Iverson, R.M. (1997) The physics of debris flows. *Reviews of Geophysics*, **35**, 245–296. · Iverson, R.M. (2014) Debris flows: behaviour and hazard assessment. *Geology Today*, **30**, 15–20.

deciduous forest   An area in which the dominant life form is trees whose leaves are shed at a particular time, season or growth stage. Deciduousness is a protective mechanism against excessive transpiration and represents an alternative strategy to being evergreen. Leaf shedding occurs most commonly in either cold or dry conditions, when the availability of soil water to roots is reduced. If transpiration were to continue unchecked during these periods, deciduous trees with a full leaf canopy would suffer serious water deficiencies. During the resting season, buds are enclosed in tough, protective scales, but the timing of bud formation varies for each different tree species.   PAF

decomposer   An organism that helps to break down dead or decaying organic material, and so aid the recycling of essential nutrients to plant producers. Bacteria and small fungi are the major decomposer groups. Digestive enzymes are released from the bacterial cells or fungal filaments, which turn the organic material into a soluble form capable of being ingested, and at the same time carbon dioxide and water are released back into the environment. In natural ecological systems, many thousands of decomposer organisms ensure the efficient operation of the detritus food chain (see FOOD CHAIN, FOOD WEB). DW

deductive (science) Also known as critical rationalism, this employs a research method based on logical structures and systematic explanations. Scientific explanations are based on judgements derived from the analysis and assessment of observations and measurements (Popper, 1972). In a formal sense, a HYPOTHESIS is established and tested, though MULTIPLE WORKING HYPOTHESES may be established and tested against each other. The hypothesis is accepted or refuted on the basis of the balance of the evidence obtained by data analysis. A key component of deductive science is that nothing is ever proved, since further data collection and analysis could always provide evidence that contradicts the accepted hypothesis; thus, acceptance and rejection are asymmetrical alternatives. The deductive approach is widely used in physical geography, even if not in a rigid formal sense. Its use and basis are analysed in a geographical context by Haines-Young and Petch (1986). (See also INDUCTIVE.) DSGT

References

Haines-Young, R.H. and Petch, J.R. (1986) *Physical geography: its nature and methods*. London: Paul Chapman. · Popper, K.R. (1972) *The logic of scientific discovery*. London: Hutchinson.

deep weathering A term widely used to denote the existence of a chemically weathered layer or mantle (see also WEATHERING PROFILE) exceeding the depth of the soil profile. It describes the deep penetration of chemical decay into susceptible rocks, often producing a *saprolite* with significant clay content. In extraglacial areas deep weathering is widespread in feldspathic rocks (especially granite, but also gneiss and many volcanic rocks), in the humid climates of temperate and tropical regions. Often seen as typical of tropical cratons (shields) such as West and Central Africa, western Australia and South America, where profiles can exceed 100 m, but thick saprolites are found in Palaeozoic mountains of temperate areas such as Appalachia, western and central Europe and eastern Australia, and in Mesozoic rocks within the humid tropics.

Deep weathering is associated with the concentration of oxides, and saprolites are mined for economic minerals (e.g. gold, aluminium, copper, nickel). In geomorphology, the stripping of weathering covers by erosion is thought to reveal bedrock forms (see BORNHARDT and TOR), and some glaciated landscapes may be little more than exposed portions of this bedrock relief.

Spatial patterns of deep weathering indicate thickening beneath summits in humid areas, deep troughs following shatter belts and fault zones, and a strong influence from fracture (joint) patterns in the bedrock (Thomas, 1966). Described by Gerrard (1988), Ollier (1984) and Thomas (1994). MFT

References

Ollier, C.D. (1984) *Weathering*, 2nd edition. London: Longman. · Gerrard, A.J. (1988) *Rocks and landforms*. London: Unwin Hyman. · Thomas, M.F. (1966) Some geomorphological implications of deep weathering patterns in crystalline rocks in Nigeria. *Transactions of the Institute of British Geographers*, **40**, 173–193. · Thomas, M.F. (1994) *Geomorphology in the tropics*. Chichester: John Wiley & Sons, Ltd.

deflation The process whereby the wind removes fine material from the surface of a beach or a desert. It is a process that contributes to the development of LOESS, DUST storms and some STONE PAVEMENTS. ASG

deforestation The removal of trees from a locality. This removal may be either temporary or permanent, leading to partial or complete eradication of the tree cover. It can be a gradual or rapid process, and may occur by means of natural or human agencies, or a combination of both.

Spurr and Barnes (1980) enumerate the major causes of deforestation. Natural tree removal is of relatively little significance on a global scale. Its mechanisms often lead to partial and temporary clearance that is followed by secondary succession, as a result of which forest develops again. The major natural cause of tree removal is fire resulting from lightning strike. Such burns are an essential part of certain forest ecosystems (e.g. in some types of pine forest). Gales may cause trees to be broken or uprooted in what is termed windthrow. Disease can also lead to the elimination of forest trees. Native animals (e.g. elephants in savanna woodland) can also damage trees by removing foliage and bark, and by trampling and uprooting them. Temporary severe weather (such as cold or drought) can lead to tree death (DIE-BACK), while secular modifications of climate may contribute to deforestation. In the latter instance, several millennia of reduced temperatures and increased precipitation may, for example, accelerate the accumulation of soil organic matter and inhibit forest growth by preventing tree regeneration.

The principal cause of deforestation is human activity. Forests are often permanently cleared on a large scale for a variety of agricultural and urban–industrial purposes using cutting and burning techniques. Some 33% of the biosphere ($\sim 4 \times 10^9$ ha) is at present forested, compared with approximately 42% about a century ago. Although this represents a pronounced recent decline, it is merely part of a process that has been operational in certain parts of the world for millennia.

As Lamb (1979) notes, while some trees continue to be removed, major deforestation is no longer a feature of most developed countries. There are some exceptions (e.g. parts of the boreal coniferous forests), but, in general, these areas have already been effectively deforested, and their emphasis is now upon woodland preservation. In northwest Europe, for instance, postglacial pollen records indicate the presence of an extensive mixed deciduous forest from around $(7-5) \times 10^3$ years BP, after which time it was progressively cleared by human activities. The forest soils, also found as fossil soils (see PALAEOSOLS), were fertile, and hence enticing to prehistoric and later agriculturalists.

Some developed countries have a considerably shorter history of major deforestation. These (such as the USA, Australia and New Zealand) were settled by Europeans, who, together with their introduced animals, brought about substantial tree losses in hundreds rather than thousands of years. This is illustrated by New Zealand, which was over two-thirds forested at the time of European colonization and now has less than a 20% tree cover.

The most significant deforestation at present is in the less-developed countries. Tropical tree cover has been removed for millennia by native inhabitants as an aid to hunting, for fuel, itinerant agriculture and settlement. As population increases, the cleared area is getting larger, and secondary succession to forest rarer, with lasting changes being brought about in the vegetation. Tropical rain forest is also being subjected to the constant, organized, commercial removal of its hardwood timber for export to developed countries. If, as is increasingly the case, such felling is entire (clear), regeneration of similar forest vegetation is unlikely.

Forests are complex ecological structures. They represent the optimum sites for photosynthesis within the biosphere, and thus contain a substantial proportion of the Earth's biomass. Moreover, they possess considerable biotic diversity, which makes them important gene pools (see GENECOLOGY). Their photosynthetic activity means that they assimilate a considerable amount of atmospheric carbon dioxide (CO_2), the concentration of which is increasing as a result of fossil fuel burning. Increased CO_2 concentration may be a contributor to a global warming of climate (GREENHOUSE EFFECT). Forest removal could therefore amplify this trend, because less CO_2 will be taken up by the trees. Additionally, the burning of wood releases CO_2 to the atmosphere. Tree burning also depletes atmospheric oxygen, and destroys an important source of oxygen (see PHOTOSYNTHESIS).

As well as destroying trees, deforestation eliminates dependent animal habitats in the forest ecosystem. Deforestation involving fire may also kill animal as well as plant life. As Tivy and O'Hare (1981) observe, tree removal causes changes in the light, temperature, wind and moisture regimes of an area. A greater quantity of light is able to reach the ground, where temperatures will also be increased. A consequence of this is an accelerated rate of organic matter decomposition. Both temperature ranges and wind speeds are greater after forest clearance. There is also a higher intensity of rainfall and a lower evapotranspiration rate. Hence, more run-off occurs, as does enhanced leaching and soil erosion. Studies at the Hubbard Brook Experimental Catchment in the USA (Likens *et al.*, 1977) have revealed details of the biogeochemistry (BIOGEOCHEMICAL CYCLES) resulting from deforestation. For example, compared with a forested area, run-off from one that was deforested increased and had a higher concentration of dissolved matter. The increased run-off was principally the result of the diminution in evapotranspiration, while the greater quantity of dissolved matter was derived mainly from that not taken up by plants as nutrients, and from substances released by the higher rate of organic matter breakdown on the former forest floor.   RLJ

Reading and References
Lamb, R. (1979) *World without trees*. London: Wildwood House. · Likens, G.E., Bormann, F.H., Pierce, R.S., *et al.* (1977) *Biogeochemistry of a forested ecosystem*. New York: Springer-Verlag. · Richards, J.F. and Tucker, R.P. (eds) (1988) *World deforestation in the twentieth century*. Durham, NC: Duke University Press. · Spurr, S.H. and Barnes, B.V. (1980) *Forest ecology*, 3rd editionn. New York: John Wiley & Sons, Inc. · Tivy, J. and O'Hare, G. (1981) *Human impact on the ecosystem*. Edinburgh: Oliver & Boyd.

deformable beds As Hart (1995: 173) remarked, 'When glaciers move over an unconsolidated bed, it has been shown by numerous workers that there is a coupling between the glacier and the underlying bed. This leads to an increase in the velocity and/or decrease in the slope angle of the glacier, and the deformation of the sediments below.' In this relatively new paradigm, subglacial deformation is seen to be of critical importance as opposed to simply subglacial deposition and sliding (Hart and Rose, 2001), for many glaciers and ice masses overlie soft, often saturated sediments, rather than hard bedrock, as has been recognized for former continental ice sheets in Europe and North America. This has implications for rates of glacier flow, for surging and the formation of such landforms as flutes and drumlins.   ASG

References
Hart, J.K. (1995) Subglacial erosion, deposition and deformation associated with deformable beds. *Progress in Physical Geography*, **19**, 173–191. · Hart J. and Rose, J. (2001) Approaches to the study of glacier bed deformation. *Quaternary International*, **86**, 45–58.

deformation   A general geological term for the disruption of rock strata, folding, faulting and other tectonic processes.

De Geer moraine   Morainic ridges a few metres high lying transverse to the direction of former ice flow. Also called *cross-valley* and *washboard moraine*. There are almost as many theories about transverse moraines as there are researchers, but one common characteristic is that they are often associated with the presence or former presence of lakes. They may be a subaqueous example of a push moraine (perhaps annual). Others may relate to subglacial thrusting as proposed for rogen moraines, with material thrust up into basal cavities.   DES

Reading
Larsen, E., Longva, O. and Follestad, B.A. (1991) Formation of De Geer moraines and implications for deglaciation dynamics. *Journal of Quaternary Science*, **6**, 263–277.

deglaciation   The process by which glaciers thin and withdraw from an area. The usual cause is climatic amelioration, which reduces snow accumulation or increases ABLATION, but sea-level rise relative to the land can also increase calving, and thus the rate of ablation. The literature is full of arguments about whether the dominant process of deglaciation is thinning (downwasting) or snout retreat (back-wasting). Since both must occur together, it is probably helpful to regard the relative dominance of one or other as varying according to glacier type, location and the nature of the climatic or sea-level change. (See also STAGNANT ICE TOPOGRAPHY.)   DES

degradation   The destruction or reduction of quality of a physical phenomenon, as in land degradation, soil degradation or vegetation degradation. Often used in the context of DESERTIFICATION. The term has also sometimes been used in the context of the lowering, and often flattening, of a land surface by erosion.   DSGT

degree day   For a given day, the difference between the mean temperature and a given threshold, normally 15.5 °C for a heating degree day:

$$\text{degree days} = 15.5 - \frac{\text{maximum temperature} + \text{minimum temperature}}{2}$$

If the mean temperature is greater than 15.5 °C, negative degree days can be used to estimate air conditioning needs. For space heating, two other formulae have been devised for the days when the maximum temperature is above 15.5 °C.

If the daily maximum temperature is above 15.5 °C by a lesser amount than the daily minimum temperature is below 15.5 °C, then

$$\text{degree days} = \frac{1}{2}(15.5 - \text{minimum temperature})$$
$$-\frac{1}{4}(\text{maximum temperature} - 15.5).$$

If the daily maximum temperature is above 15.5 °C by a greater amount than the daily minimum temperature is below 15.5 °C, then

$$\text{degree days} = \frac{1}{4}(15.5 - \text{minimum temperature})$$

Generally speaking, degree days are used for heating purposes and ACCUMULATED TEMPERATURE is used for agricultural purposes.   JET

delayed flow   The part of the streamflow that lies below an arbitrary cut-off line drawn on the hydrograph, representing the more slowly responding parts of the catchment.

The division between QUICKFLOW and delayed flow is usually made by a line that rises from the start of the hydrograph rise at a gradient of $0.55\,\text{L}\,\text{s}^{-1}\,\text{km}^{-2}\,\text{h}^{-1}$ until the line meets the falling limb of the hydrograph. This procedure was suggested by Hewlett (1961) as an arbitrary but objective basis of separating the hydrograph peaks associated with each storm. It was earlier proposed as a replacement for older methods of hydrograph separation (e.g. Linsley *et al.*, 1949) that were considered to have only a spurious physical basis (see BASE FLOW).   MJK

References
Hewlett, J.D. (1961) *Soil moisture as a source of base flow from steep mountain watersheds*. Southeastern Forest Experiment Station, Station paper 132. Asheville, NC: US Department of Agriculture Forest Service. · Linsley,

R.K., Kohler, M.A. and Paulhus, J.L.H. (1949) *Applied hydrology*. New York: McGraw-Hill.

delta An accumulations of river-derived sediment deposited at the coast when a stream enters a receiving body of water, which may be an ocean, gulf, lagoon, estuary or lake. Deltas result from the interaction of fluvial and marine (or lacustrine) forces. Sediment accumulations at the mouths of rivers are subject to reworking by waves and tidal currents. Owing to long-term sediment loading and shorter term sediment compaction, deltaic environments are naturally subsiding systems, and it is only for part of an individual delta's history that sedimentation outpaces land subsidence and builds a delta lobe. In the aggradational phase, delta planform typically reflects the interaction between river, wave and tide regimes (Galloway, 1975). Figure 1 shows a tripartite classification using selected major deltas to illustrate variations

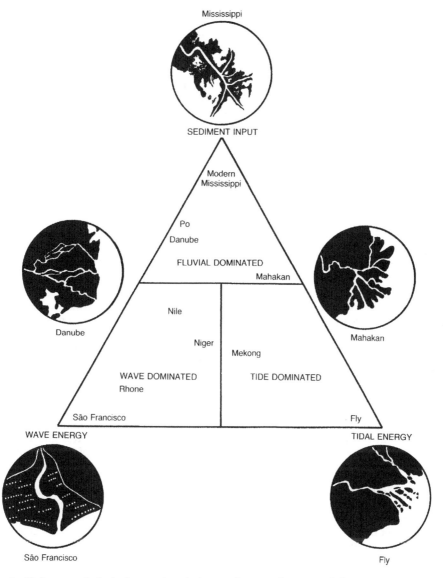

Figure 1 Delta morphological types in relation to three environmental factors.
Source: Galloway (1975). Houston Geological Society.

135

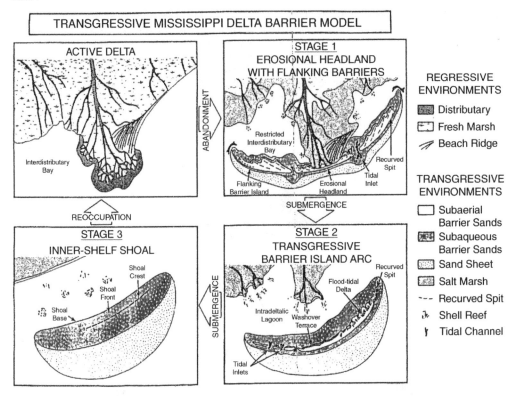

Figure 2 The delta cycle, as illustrated by the growth and decay (stages 1–3) of a Mississippi delta complex. *Source*: Penland *et al.* (1988). Reproduced with permission of the Society for Sedimentary Geology.

in deltaic characteristics associated with the relative contribution of each process. The building phase is only the first stage in the 'delta cycle'. In time, the river, seeking a shorter path to the sea, abandons its previous course and the associated delta. The now subsiding delta headland erodes, with the released sediments building flanking barriers. With continued subsidence, the barriers become detached from the former headland, forming an arc of barrier islands and, in time, these barriers are then overwashed and converted to sand shoals (Figure 2). Detailed studies from the Mississippi deltaic plain have shown the duration of the complete delta cycle in this setting to take 1000–1500 years; there have been six major deltaic complexes formed, and abandoned, over the last 7500 years in the Mississippi delta plain.

Deltas consist of subaerial and subaqueous components. The framework of the delta is constructed from narrow sand levees which are formed by the rapid ($100\,\mathrm{m\,a^{-1}}$) progradation of lateral distributary bars when the delta lobe is actively extending seawards. Breaks in the levee divert water and sediment into the shallow interdistributary bays where sand is deposited in fan-shaped micro-deltas known as 'crevasse splay deposits'. These sediments become overlain with silt/clay sediments and marsh vegetation. As sediment supply is reduced and inundation increases with subsidence, freshwater marshes are replaced by brackish and then saline vegetation communities; ultimately, increased physiological stress and wave action lead to the break-up of marshes and their replacement by open water bays. The subaqueous delta front and the prodelta to seaward have traditionally been seen as environments of 'quiet suspension sedimentation' of silts and clays respectively. However, underwater mapping now shows this region to be characterized by highly dynamic mass-movement processes, moving large quantities of loaded and underconsolidated sediments (often with high pore-water pressures) along well-defined transport pathways in a range of modes.   BGT/TS

Reading and References
Galloway, W.E. (1975) Process framework for describing the morphologic and stratigraphic evolution of deltaic depositional systems. In M. L. Broussard (ed.), *Deltas: models for exploration*. Houston, TX: Houston Geological Society. · Penland, S., Boyd, R. and Suter, J.R. (1988) Transgressive depositional systems of the Mississippi delta plain. *Journal of Sedimentary Petrology* **58**, 932–949. · Reed, D.J. (2002) Sea-level rise and coastal marsh sustainability: geological and ecological factors in the Mississippi delta plain. *Geomorphology*, **48**, 233–243. · Roberts, H.H. (1997) Dynamic changes of the Holocene Mississippi River delta plain: the delta cycle. *Journal of Coastal Research*, **13**, 606–627. · Woodroffe, C.D. (2002) *Coasts: form, processes and evolution*. Cambridge: Cambridge University Press; chapter 7. · Wright, L.D. (1978) River deltas. In R. A. Davis Jr (ed.), *Coastal sedimentary environments*. New York: Springer-Verlag.

demoiselle A pillar of earth or other unconsolidated material that is protected from erosion by a capping boulder.

denitrification The removal of nitrogen from a reservoir, commonly groundwater enriched in nitrogen from agricultural land use, or industrial flue gases containing NO_x. Nitrogen is one of the components of organic matter (e.g. in amino acids and proteins) and occurs naturally in the environment as well as at elevated levels in contaminated soils and soil water. Within the soil, especially in waterlogged reducing environments, bacteria are able to convert nitrates (NO_3^-) into nitrogen gas, which then escapes from the system. Rates of gas release of $>50\,\mathrm{mg\,N_2\,m^{-2}\,day^{-1}}$ have been observed from riparian soils carrying forest cover, and the ability of the soil zone to retain and process nitrogen compounds before they reach the stream environment is an important goal in the preservation of streamside BUFFER STRIPS. Soil denitrification is exploited in the treatment of wastewater (e.g. from tertiary sewage treatment plants) that is used in irrigation. Industrial denitrification is the removal of NO_x compounds that are involved in the acid rain phenomenon from the flue gases released by industries using fossil fuels. Denitrification of this kind employs a catalytic reduction process that yields nitrogen gas.   DLD

Reading
Pinay, G. and Decamps, H. (1988) The role of riparian woods in regulating nitrogen fluxes between the alluvial aquifer and surface water: a conceptual model. *Regulated Rivers: Research and Management*, **2**, 507–516.

density current A descending body of air or water with a high suspended sediment load. Turbidity currents on lake-bed and sea-floor slopes, clouds of falling volcanic ash and dust storms are all examples of density currents.

density dependence The action of environmental factors controlling the growth of populations of organisms, which vary with the density of the population. In contrast, factors that affect population size independently of the number of individuals present are described as density independent. Climatic and physical factors, such as floods or earthquakes, are usually classed as density-independent factors. Resources (food, shelter) are often considered to be density dependent, as are biotic factors such as competition and parasites.

The relative importance of density-dependent and -independent factors in regulating population numbers has been a subject of considerable debate. Andrewartha and Birch (1954) argued that natural populations are controlled primarily by density-independent factors and emphasized the ultimate role of severe climatic conditions. An alternative view stressed the importance of density-dependent factors involving competition (Lack, 1954). These differences in viewpoint may also be influenced by contrasts in the ecosystems analysed, which range from small organisms, such as insects, in arid areas in the case of Andrewartha and Birch, to larger organisms, such as birds, in temperate landscapes in the case of Lack.

The regulation of populations may often involve the interaction of density-independent and -dependent factors. Moreover, density-dependent factors may sometimes vary in their effectiveness from one population density to another. In some cases predation may control population size at low prey densities but become ineffective at high densities.   ARH

Reading and References
Andrewartha, H.G. and Birch, L.C. (1954) *The distribution and abundance of animals*. Chicago, IL: University of Chicago Press. · Lack, D. (1954) *The natural regulation of animal numbers*. Oxford: Oxford University Press. · Odum, E.P. (1971) *Fundamentals of ecology*, 3rd edition. Philadelphia, PA: W.B. Saunders; chapter 7, pp. 162–233.

denudation Literally, the laying bare of underlying rocks or strata by the removal of overlying material. It is usually defined as a broader term than 'erosion', to include weathering and all processes that can wear down the surface of the Earth. While some denudational processes may continue to operate below sea level, most discussions are concerned with subaerial denudation.

The principal agent of subaerial denudation on the Earth is water, as Playfair (1802) clearly recognized long ago. Other important agents are related to stresses generated by pressure (atmospheric, gravitational and crustal) and the actions of organisms, including humans. The impact of bodies such as meteorites, although of limited importance on the modern Earth, assumes great significance on Mercury and the Moon.

The rate at which denudation proceeds is fundamentally dependent upon the intensity with which the different agents operate, singly or collectively, and the ability of the ground surface and the underlying materials to withstand the stresses generated. It is this intimate connection between the nature of the applied force and the actual resistance of Earth materials that makes it difficult to produce realistic generalizations about denudation rates, either over space or through time. Surface geometry, the chemical constituents of rocks, tectonic setting and climate are key factors in determining the denudation environment.

Denudation can be classified as *chemical* denudation or *mechanical* denudation.   BAK/DLD

Reading and Reference

Fournier, F. (1960) *Climat et érosion: la relation entre l'érosion du sol par l'eau et les précipitations atmosphériques.* Paris: Presses Universitaires de France. · Francis, P. (1981) *The planets.* London: Penguin Books. · Langbein, W.B. and Schumm, S.A. (1958) Yield of sediment in relation to mean annual precipitation. *Transactions of the American Geophysical Union*, **39**, 1076–1084. · Ollier, C.D. (1981) *Tectonics and landforms.* London: Longman. · Playfair, J. (1802) *Illustrations of the Huttonian theory of the Earth.* London: Cadell & Davies. · Willenbring, J.K., Codilean, A.T. and McElroy, B. (2014) Earth is (mostly) flat: apportionment of the flux of continental sediment over millennial time scales. *Geology*, **41**, 343–346.

denudation chronology An attempt by geomorphologists to reconstruct the erosional history of the Earth's surface. The original definition of geomorphology that emerged in the US Geological Survey in the 1880s saw the new science of geomorphology as being in all essentials equivalent to what is now termed denudation chronology.

In the mid-nineteenth century planed-off surfaces had been identified by British geomorphologists in areas of complex structure and lithology, such as mid-Wales, while, with the exploration of the walls of the Colorado Canyon by Major Powell and his colleagues, great unconformities were recognized, leading to the concepts of BASE LEVEL and peneplain. In Europe, Suess postulated that planation surfaces might be susceptible to correlation on a worldwide basis as a result of worldwide (eustatic) changes of sea level in the geological past. Denudation chronology, therefore, arose as a prime focus of geomorphology, the aim of which was to use the study of erosional remnants to reconstruct the history of the Earth where the stratigraphic record was interrupted or unclear. Techniques were developed to help in the identification of such erosional remnants, including superimposed contours, altimetric frequency curves, and so on (Richards, 1981), and particular energy was expended on trying to fathom out whether surfaces were the product of marine or subaerial denudation. Crucial in such an analysis was the degree of adjustment of streams to structure: streams on subaerial peneplains were thought to be better adjusted than those developed on marine planation surfaces. Much of the evidence for denudation chronology was morphological, with all the implications that this has for its reliability (Rich, 1938). The tectonic warping of small remnants rendered height correlation difficult. Supposedly accordant summits might have been greatly lowered by erosion, areas of flat ground might be susceptible to a whole range of different interpretations of their origin, and adjustments of streams to structure might be affected by a variety of tectonic factors, including antecedence (Jones, 1980). The more successful attempts at denudation chronology were able to supplement the morphological evidence with information gained from deposits resting on the planation surfaces. In southern England, for example, Wooldridge and Linton (1955) were able to use the Lenham Beds to establish the presence of the supposed marine Calabrian Transgression of Plio-Pleistocene times. Other notable studies include those of Baulig (1935) in France, D. W. Johnson (1931) in the USA and E. H. Brown (1960) in Wales. In the 1960s, as geomorphology became less concerned with evolution and more concerned with process studies, morphometry and systems, considerable dissatisfaction was expressed about denudation chronology as a basis for the discipline (Chorley, 1965), but with developments in plate tectonics, in our knowledge of the importance of Pleistocene events, in the amount of information that can be gained from a study of submarine deposits in basins like the North Sea, and with improvements in dating techniques, it remains a viable branch of study.   ASG

References

Baulig, H. (1935) *The changing sea level.* London: Philip. · Brown, E.H. (1960) *The relief and drainage of Wales: a study in geomorphological development.* Cardiff: University of Wales Press. · Chorley, R.J. (1965) The application of quantitative methods to geomorphology. In R. J. Chorley and P. Haggett (eds), *Frontiers in geographical teaching.* London: Methuen; pp. 148–163. · Johnson, D.W. (1931) *Stream sculpture on the Atlantic Slope: a study in the evolution of Appalachian rivers.* New York: Columbia University Press. · Jones, D.K.C. (ed.) (1980) *The shaping of southern England.* London: Academic Press. · Rich, J.L.

(1938) Recognition and significance of multiple erosion surfaces. *Bulletin of the Geological Society of America*, **49**, 1695–1722. · Richards, K.S. (1981) Geomorphometry and geochronology. In A. S. Goudie (ed.), *Geomorphological techniques*. London: Allen & Unwin; pp. 38–41. · Wooldridge, S.W. and Linton, D. (1955) *Structure, surface and drainage in south-east England*, 2nd edition. London: Philip.

denudation rates

Provide a measure of the rate of lowering of the land surface by erosion processes per unit time and are expressed in millimetres per 1000 years or in the direct equivalent of cubic metres per square kilometre per year. These rates are commonly calculated using information on sediment (physical denudation) and solute yields (chemical denudation) from drainage basins, coupled with an estimate of soil or rock density. As such, they are an index of the rate of denudation in the upstream catchment area. Maximum reported denudation rates are probably those for the island of Taiwan, which exceed $10,000 \, m^3 \, km^{-2} \, a^{-1}$ (Li, 1976).

There are, however, several important problems in the derivation and interpretation of denudation rates based on measurements of river loads. For example, in the case of suspended sediment yields, it must be recognized that only a proportion of the eroded sediment will be transported to the basin outlet and that the associated denudation rate may be an underestimate. With the dissolved load, however, a large proportion may reflect non-denudational sources and should not be included in the calculation (Janda, 1971). Furthermore, it may be unrealistic to convert values of river load to a uniform rate of surface lowering over the entire catchment and to assume that current river loads are representative of past conditions (Meade, 1969). DEW

References

Janda, R.J. (1971) An evaluation of procedures used in computing chemical denudation rates. *Bulletin of the* *Geological Society of America*, **82**, 67–80. · Li, Y.H. (1976) Denudation of Taiwan Island since the Pliocene epoch. *Geology*, **4**, 105–107. · Meade, R.H. (1969) Errors in using modern stream-load data to estimate natural rates of denudation. *Bulletin of the Geological Society of America*, **80**, 1265–1274.

deoxygenation

The depletion of the amount of oxygen dissolved in a reservoir such as a surface water body (stream, lake or artificial impoundment). Deoxygenation can result from the oxidation of organic material, as may happen in the lower water column of a large reservoir where oxidation of the remains of the former land vegetation can progressively consume available oxygen until reducing conditions set in. Deoxygenation in surface water can also result from EUTROPHICATION or excessive enrichment of the available nutrient supply, which may result in seasonal blooms of algae. Oxidation of the remains of these microorganisms may again lead to progressive deoxygenation of the waterbody. DLD

depletion curve (or recession curve or base-flow recession curve)

Represents gradual drainage of water from storage in a drainage basin, and it is often possible to construct a master depletion curve for a particular site on a river system. This depletion curve will usually provide a very reliable means of predicting the decline of base-flow discharge at a site during dry conditions, although some drainage basins may exhibit seasonal variations in the form of the curve as a result of differences in loss of stored water through evapotranspiration (Federer, 1973).

The depletion curve may be derived by producing a composite curve from the recession limbs of storm HYDROGRAPHS at a gauging site (see diagram) and one of a number of functions may be fitted to the depletion curve to describe it and to allow quantitative comparison of curves from different sites within a drainage basin,

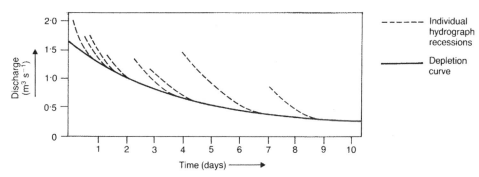

Construction of a depletion curve.

different drainage basins or to check the consistency of curves produced from different hydrograph recession limbs at the same site. One widely applied depletion function is

$$Q_t = Q_o e^{-at}$$

where Q_o is flow at any time in the period of depletion or base flow, Q_t is flow after time t from flow Q_o, a is the recession coefficient, and e is the base of natural logarithms.

This function will plot as a straight line on semi-logarithmic graph paper, and a modification of the method of depletion curve construction shown in the diagram is to plot the recession limbs on semi-logarithmic graph paper so that they define a straight line. However, in practice a perfect straight line is rarely found, and this is probably partly a result of the fact that a number of stores are contributing to the depletion curve, all of which produce a recession flow at different rates.

The sources of discharge contributing to the depletion curve are likely to be very different in different drainage basins. Traditionally, it has been assumed that depletion flow is derived from effluent seepage from an aquifer and so the recession coefficient a has often been called the aquifer coefficient, implying that it is a simple parameter of the drainage characteristics of a single aquifer. However, in practice such a simple situation is highly unlikely and the number of stores contributing to base flow will vary with the structure and size of the catchment. Many drainage basins are not underlain by efficient aquifers and yet they exhibit depletion flow that is largely generated from soil moisture storage. Higher discharges on a depletion curve are almost certainly produced by drainage from soil moisture and from the unsaturated zone of aquifers as well as from the saturated zone. Drainage basins underlain by more than one aquifer may experience effluent seepage from different aquifers at different points on the drainage network and may even lose flow by influent seepage at some locations. As a result, the depletion curve should not be expected to have a simple form because every store contributing to flow should have its own depletion curve, which should produce a complex composite curve (see RECESSION LIMB OF HYDROGRAPH, RECESSION CURVE). AMG

Reading and References

Bako, M.D. and Owoade, A. (1988) Field application of a numerical method for the derivation of baseflow recession constant. *Hydrological Processes*, **2**, 331–336. · Federer, C.A. (1973) Forest transpiration greatly speeds streamflow recession. *Water Resources Research*, **9**, 1599–1605. · Hall, F.R. (1968) Base-flow recessions – a review. *Water Resources Research*, **4**, 973–983. · Linsley, R.K., Kohler, M.A. and Paulhus, J.L.H. (1988) *Hydrology for engineers*, 3rd edition. New York: McGraw-Hill.

depression See CYCLONE.

depression storage Consists of water trapped in small surface depressions or hollows during a rainfall event. Depression storage must be filled before overland flow can occur, and this component of the total volume of rainfall will eventually be evaporated or will infiltrate the soil. (See also SURFACE DETENTION and SURFACE STORAGE.) AMG

Reading

Carvajal, F., Aguilar, M.A., Aguera, F., *et al.* (2006) Maximum depression storage and surface drainage network in uneven agricultural landforms. *Biosystems Engineering*, **95**, 281–293.

depth–duration curve Relates the magnitude of rainfall to its duration. This type of curve is usually constructed to relate the magnitude of extreme rainfall events to their duration at a single site or over an area. AMG

Reading

Niemczynowicz, J. (1982) Areal intensity–duration–frequency curves for short term rainfall events in Lund. *Nordic Hydrology*, **13**, 193–204. · Shaw, E.M. (1988) *Hydrology in practice*, 2nd edition. New York: Van Nostrand Reinhold.

desalinization (or desalination) The production of fresh water from saline brines, especially seawater, by distillation or any other process.

desert Scientific definitions of deserts have been based on a range of criteria, including the nature and development of drainage systems, the types of rock weathering processes that operate, ecological communities and the potential for crop growth. Most definitions of deserts, however, are related to moisture deficiency, and the term is generally synonymous with DRYLANDS. True desert conditions are most clearly represented by the HYPER-ARID and arid components of drylands. Deserts do experience significant rainfall events, but these are either unreliable, irregular or confined to only part of the year. Some deserts also have perennial rivers, with sources in wetter regions, flowing through them; for example, the Nile, which waters extremely low rainfall areas of Egypt and the Sudan. The location of deserts worldwide is determined by four factors that lead to low rainfall: subtropical zones of high atmospheric pressure; continentality (distance from the ocean); rain shadow effects; and the effect

Table 1 Arid zone landscapes in different regions (expressed as a percentage of area[a])

	SW USA	Sahara	Libya	Arabia	Australia[b]
Mountains	38.1	43	39	47	16
Low-angle bedrock surfaces	0.7	10	6	1	14
Alluvial fans	31.4	1	1	4	
River plains	1.2	1	3	1	13
Dry watercourses	3.6	1	1	1	
Badlands	2.6	2	8	1	—
Playas	1.1	1	1	1	1
Sand seas	0.6	28	22	26	38
Desert flats[c]	20.5	10	18	16	18
Recent volcanic deposits	0.2	3	1	2	—

[a] Percentages given are only approximate, with the degree of accuracy differing between areas.
[b] From Mabbutt (1977). Categories used by Mabbutt do not necessarily coincide with those used in other areas; included for comparison only. The remaining data are from a study by Clements et al. (1957) for the US Army.
[c] Undifferentiated: includes areas bordering playas.

of cold ocean currents on atmospheric convection.

Deserts are rarely totally devoid of plant and animal life. Desert species have adaptive strategies that permit the accumulation and retention of available moisture, and/or physiologies that allow biological functions to be slowed or shut down at times of acute moisture stress. Deserts embrace a wide range of landscape systems (see Table 1) that reflects their tectonic and continental settings and the interplay between different geomorphological processes. SAND SEAS are not dominant at the global scale despite popular images of deserts and early views that suggested aeolian processes were the dominant influence on desert landscapes (see AEOLATION). The major climatic changes that have affected the Earth in the QUATERNARY period are known on theoretical and empirical grounds to have impacted on the distribution and extent of desert conditions. The rock record has also been used to identify the sporadic existence of desert sand seas and dune deposits as far back as the Proterozoic.

At the scale of Quaternary glacial–interglacial cycles, changes in the extent of deserts and drylands can be expected to have occurred due to changes within the partitioning of moisture in the global hydrological cycle and alterations in the positioning of major climatic systems. Human actions may also influence the extent of desert-like conditions today through the impact of DESERTIFICATION.   DSGT

Reading and References

Blume, H.-P. and Berkowicz, S.M. (1995) *Arid ecosystems*. Advances in Geoecology 28. Cremingen: Catena Verlag. · Clements, T., Merriam, R.H., Stone, R.O., *et al*. (1957) *A study of desert surface conditions*. Headquarters Quartermaster Research and Development Command, Technical report EP53. Natick, MA: US Army Environmental Protection Research Division. · Thomas, D.S.G. (ed.) (2011) *Arid zone geomorphology: process, form and change in drylands*, 3rd edition. Chichester: John Wiley & Sons, Ltd. · Mabbutt, J.A. (1977) *Desert landforms*. Canberra: ANU Press.

desert pavement See STONE PAVEMENT.

desert varnish See ROCK VARNISH.

desertification (also **desertization**) A term coined by the French forester Aubreville in 1949 to describe land degradation. Since then there have been over 100 published definitions of this controversial term, which gained prominence after the 1977 UN Conference on Desertification, itself prompted by social and environmental concerns in the sub-Saharan Sahel region of Africa. The term has been confused with DROUGHT in some circles (though there are clear links with drought), and its social impacts have sometimes been associated with famine that may in fact have nonenvironmental causes (Ollson, 1993). Desertification has consequently been heavily critiqued in recent years in both the environmental and social sciences (e.g. Thomas and Middleton, 1994; Stiles, 1995). In 1995, The UN Convention to Combat Desertification (CCD) was signed; it has since been ratified by the governments of over 150 countries. Coincident with the CCD have been attempts to clarify the scientific and social dimensions of this environmental issue, and to establish an agreed definition.

In the CCD, desertification is defined as land degradation in arid, semi-arid and dry-subhumid areas resulting from various factors, including climatic variations and human activities. The problem is therefore confined to the susceptible DRYLANDS, with land degradation regarded as soil erosion, internal soil changes, depletion of groundwater reserves and irreversible changes to vegetation communities (see also SOIL EROSION, SALINIZATION and BUSH ENCROACHMENT). In many respects, desertification is no different than land degradation occurring worldwide, except that it specifically refers to its occurrence within dryland areas. Whether there is anything special about desertification, in environmental process terms, is a moot point that has been considered by writers such as Mainguet (1991). It is perhaps the political and social dimensions of the problem (including its apparent severe occurrence in some of the world's poorest nations, in Africa and Asia) that have given credence to desertification as an environmental issue worthy of special and urgent consideration – see Stiles (1995) and Thomas (1997).

In environmental terms, the soil degradation component of desertification comprises water and wind erosion, and physical and chemical changes within the soil. These factors reduce the potential productivity of the land. A systematic global survey of soil degradation, commissioned by the United Nations Environment Programme and conducted in the early 1990s, estimated that soil degradation and erosion affects up to 1035 million hectares of drylands worldwide, though gained reliable and agreed assessments of the full extent of desertification have been difficult, and sometimes controversial (Thomas and Middleton, 1994). Overall, the survey suggested that water erosion was the dominant form of soil desertification in 48% of affected areas, wind erosion in 39%, chemical changes (that include nutrient depletion and salinization) in 10% and physical changes (including soil crusting and compaction) in 4%. The full data set is explored in Middleton and Thomas (1997); it can, however, be additionally noted that the level of desertification was only extreme or severe in 4% of all affected lands. Even given the precision limitations of the global survey, it serves to identify the diversity and insidiousness of desertification, and adds further weight to dispelling images of sand dunes advancing over productive land, a common media misrepresentation, as the common face of the problem. As Toulmin (1995: 5) has concisely noted, 'dryland degradation does not involve moving sand dunes. Rather, it concerns the gradual impoverishment of agricultural and pastoral systems, which makes them less productive and more vulnerable to drought'. The status of vegetation degradation as a form of desertification is even more difficult to assess given the natural dynamics of dryland ecosystems and debates over how permanent vegetation changes really are (e.g. see BUSH ENCROACHMENT).

Though the CCD definition includes possible multiple causes for desertification, it is undoubtedly the case that the principal agent of degradation is human actions. Though desertification has often been viewed as a particular problem of the twentieth century, the ability of humans to detrimentally alter drylands is not new. Salinization in Mesopotamia around 2500 BC has been attributed to agriculture, while pre-Hispanic societies in central Mexico caused severe soil erosion (O'Hara *et al.*, 1993). In the twentieth century it is the growth and changing distributions of human populations in drylands, and the spread of technological advances, that have increased the propensity for dryland degradation. Several issues make drylands especially susceptible to degradation by humans. These do not necessarily relate to the sometimes supposed fragility of their ecosystems – which, as Behnke *et al.* (1993) show, is disputable for many dryland areas – but to the nature of environmental systems and human activities.

1 The inherent natural variability of dryland climates and ecosystems tends to be overridden today when humans use technologies and methods imported from more consistent temperate environments.
2 Geomorphic processes in drylands tend to be characterized by significant periods of quiescence, punctuated by abrupt episodes of activity. Thus, areas cleared of natural vegetation by grazing pressures, or bare in the period immediately following harvesting, are particularly susceptible to rapid erosion from high-intensity rain storms or windy conditions.
3 The rapid growth of urban areas in deserts and drylands since the Second World War has placed significant pressures on limited water resources and options for waste disposal.
4 In dryland areas of the developing world, declining rural populations, in response to migration to urban centres, can lead to the failure of traditional soil and vegetation conservation techniques. Tiffen *et al.* (1994) have shown how more people on the land can in certain circumstances halt and reverse environmental degradation.   DSGT

References

Behnke, R.H., Scoones, I. and Kerven, C. (1993) *Range ecology at disequilibrium: new models on natural variability and pastoral adaptation in African savannas*. London: ODI. · Mainguet, M. (1991) *Desertification*. Berlin: Springer Verlag. · Middleton, N.J. and Thomas, D.S.G. (1997) *World atlas of desertification*, 2nd edition. London: UNEP/Edward Arnold. · O'Hara, S.L., Street-Perrot, F.A. and Burt, T.P. (1993) Accelerated soil erosion around a Mexico highland lake caused by prehispanic agriculture. *Nature*, **362**, 48–51. · Ollson, L. (1993) On the causes of famine: drought, desertification and market failure in the Sudan. *Ambio*, **22**, 395–403. · Stiles, D. (ed.) (1995) *Social aspects of sustainable dryland management*. Chichester: John Wiley & Sons, Ltd. · Thomas, D.S.G. (1997) Science and the desertification debate. *Journal of Arid Environments*, **37**, 599–608. · Thomas, D.S.G. and Middleton, N.J. (1994) *Desertification: exploding the myth*. Chichester: John Wiley & Sons, Ltd. · Tiffen, M., Mortimore, M. and Gichuki, F. (1994) *More people, less erosion: environmental recovery in Kenya*. Chichester: John Wiley & Sons, Ltd. · Toulmin, C. (1995) *The convention to combat desertification: guidelines for NGO activity*. IIED paper no. 56. London: IIED.

desiccation This is both a concept and a process. The concept of progressive desiccation has existed since at least the mid-nineteenth century, and is the belief that parts of the world are getting drier, through reduced rainfall levels and depleted groundwater reserves, giving rise to the spread of deserts. It has been attributed to both climatic change and human mismanagement. Warren and Khogali (1992) have used the term to describe the reduction of moisture in drylands that results from a dry event at the scale of decades; that is, longer than a DROUGHT. Desiccation is also used to describe the process of the drying out of an individual water body (e.g. a lake) during a drought period or at the end of the rainy season. (See also DESERTIFICATION and DRYLANDS.) DSGT

Reference
Warren, A. and Khogali, M. (1992) *Assessment of desertification and drought in the Sudano-Sahelian region*. New York: UNSO/UNDP.

Desiccation cracks in the clay sediments on the floor of a dry lake.
Photograph by David Thomas.

design discharge The discharge that a structure or development is designed to resist or to cope with. A dam across a river must be designed to retain a flood of a particular size, and a flood prevention scheme will also be designed to convey a flood of a particular magnitude. Although very large recent events may provide the experience against which schemes may be designed, a design discharge is usually selected by reference to a specific RECURRENCE INTERVAL, and this has to be such that it will provide a reasonable expectation of protection and yet not be too costly. Therefore, for land drainage works or for flood prevention schemes, estimated recurrence intervals in the range 50–75 years, or occasionally 100 years, may be used. KJG

desquamation Onion-weathering. The disintegration of rocks, especially those in desert areas, by peeling of the surface layers.

desulphurization The removal of sulphur compounds, primarily sulphur dioxide from the flue gases of coal-fired power stations and other industrial sources in order to prevent their escape into the atmosphere, where they may contribute to the acid rain phenomenon. DLD

Reading
Department of the Environment (1991) *Manual of acidic emission abatement technologies. Volume 1: coal-fired systems*. London: HMSO.

Devensian See LATE GLACIAL.

dew, dewpoint Dew is formed as a condensate from moist air that is cooled by contact with a surface that loses heat by RADIATION. It occurs once the chilling has lowered the air's TEMPERATURE to its dewpoint, which is the temperature at which an air sample becomes saturated by cooling at constant pressure (see PRESSURE, AIR) and absolute HUMIDITY. The numerical value of dewpoint is obtained by using humidity tables in conjunction with measurements of dry-bulb and wet-bulb temperatures. RR

Reading
Monteith, J.L. (1957) Dew. *Quarterly Journal of the Royal Meteorological Society*, **83**, 322–341.

diabatic See ADIABATIC.

diachronous Pertaining to a sedimentary unit of a single facies that belongs to two or more units of geological time.

diaclinal Describes those rivers whose courses cross the strike of a geological structure at right angles.

143

diagenesis Post-depositional changes that have altered a sediment, particularly cementation and compaction.

diamictite, diamicton Terms proposed by the American geologist R. F. Flint for non-sorted terrigenous deposits and rocks containing a wide range of particle sizes, regardless of genesis. Examples of diamictites include till and mudflow deposits.

diapir An anticlinal fold that has resulted from the upward movement of mobile rocks, such as halite, lying beneath more competent strata. Sometimes a surface dome produced by such movements.

diastrophism Tectonic processes that produce dramatic changes in the shape of the Earth's surface, such as orogenies, faulting and folding.

diatoms These are microscopic, unicellular ALGAE (Bacillariophyceae) with a shell, called a frustule, that is made of silica. The frustule, which is etched by rows of tiny holes, is composed of two valves that are held together by siliceous belts called girdle bands. Round *et al.* (1990) recognized three principal types of diatom pattern: (1) centric, (2) simple pennates and (3) raphid pennates. Centric diatoms are radially symmetrical and are generally planktonic (floating), while the pennates are bilaterally symmetrical. The shape, size and style of ornamentation on the shell vary from one species to another and are used to help identify the range of diatoms present in a sample. The frustule of all except the most fragile of diatoms is generally well preserved in sediments and allows the identifications of diatoms down to species and in exceptional cases subspecies level. Diatoms are very small, ranging in size from about 5 μm to 2 mm, although most species fall into the size range 20–200 μm. A very high powered light microscope is required to identify and count the different species, with the diatom magnified between 200 and 1000 times.

Diatoms can be found in almost every aquatic environment where there is sufficient light for photosynthesis, and distinct diatom communities exist within a wide range of microhabitats. Diatoms are commonly preserved in LAKE sediments and are useful palaeoenvironmental indicators for several reasons. If well preserved and abundant the frustules may be readily identified and counted. Furthermore, many species live in very defined ecological conditions, thus providing important information on pH, salinity and mineral conditions. Unlike POLLEN, which gives a regional picture of environmental change, the results from diatom analysis generally relate to the lake being studied, providing a more detailed view of change on a local scale.   SLO

Reference
Round, F.E., Crawford, R.M. and Main, D.G. (1990) *The diatoms.* Cambridge: Cambridge University Press.

diatreme The general term for vents and pipes that have been forced through sedimentary strata by the forces of underlying volcanism. Kimberlite pipes are examples of diatremes.

die-back Mortality beginning at the extremities of plants. One or more phenomena lead to stress, which manifests itself in the decline and death of leaves, shoots and roots. Plants susceptible to die-back are often woody and possess restricted environmental tolerances. In deciduous trees, a characteristic first sign of it is the premature discoloration and fall of leaves on outermost branches, with no subsequent leaf regrowth in that area.

There appear to be a number of possible causes of die-back. Indeed, regional-scale die-back of forests has emerged as a global concern that is expected to escalate under model projections of CLIMATE CHANGE (Michaelian *et al.*, 2011). Pollution, especially ACID PRECIPITATION, is a significant factor (Driscoll *et al.*, 2001). Other, natural, causes include waterlogging; for example, reducing conditions in salt-marsh sediments in southern England result in the accumulation of sulphide, the toxic effect of which may lead to die-back in several species (Lamers *et al.*, 2013).

Progressive die-back lowers the resistance of plants to other environmental factors (e.g. insect attack) unfavourable to their survival, and thus can combine with such factors to bring about mortality.   RLJ/MEM

References
Driscoll, C.T., Lawrence, G.B., Bulger, T.J., *et al.* (2001) Acidic deposition in the northeastern United States: sources and inputs, ecosystems effects, and management strategies. *BioScience*, **51**, 180–198. · Lamers, L.P.M., Govers, L.L., Janssen, I.C.J.M., *et al.* (2013) Sulfide as a soil phytotoxin – a review. *Frontiers in Plant Science*, **4**, 268. · Michaelian, M., Hogg, E.H., Hall, R.J. and Arsenault, E. (2011) Massive mortality of aspen following severe drought along the southern edge of the Canadian boreal forest. *Global Change Biology*, **17**, 2084–2094.

diffluence The rate of change of the direction of a fluid flow in the direction transverse to the motion. It is by definition positive if the stream lines spread apart downstream and negative (i.e.

confluent) if they converge downstream. The term is most commonly applied to atmospheric flows. The term has other geophysical applications, such as the case of a glacier overflowing its valley into an adjacent one.   SN

diffusion equation

diffusion equation   A flow equation for *transient* flow through a saturated homogeneous, porous rock in which hydraulic conductivity is the same in each direction (see HYDRAULIC CONDUCTIVITY). *Steady* flow through a porous medium requires the rate of fluid flow into a given volume of the rock to be equal to the rate of flow out of it. This is expressed mathematically by the equation of continuity:

$$-\frac{dv_x}{dx} - \frac{dv_y}{dy} - \frac{dv_z}{dz} = 0$$

where v_x, v_y and v_z are specific discharges in directions x, y and z. If the porous rock is homogeneous and hydraulic conductivity is the same in each direction, then another equation incorporating DARCY'S LAW can be written to express the steady-state saturated flow through it:

$$\frac{d^2h}{dx^2} + \frac{d^2h}{dy^2} + \frac{d^2h}{dz^2} = 0$$

where h is the HYDRAULIC HEAD and x, y and z are a coordinate system defining position. This partial differential equation is known as the *Laplace equation*. It is incorporated into another equation termed the *diffusion equation*, in order to describe *transient* saturated flow through a porous medium with similar homogeneous properties, as follows:

$$\frac{d^2h}{dx^2} + \frac{d^2h}{dy^2} + \frac{d^2h}{dz^2} = \frac{pg(\alpha + n\beta)}{K}\frac{dh}{dt}$$

where p is density, g is gravitational acceleration, α is the vertical compressibility of the aquifer, β is the compressibility of water, n is porosity and K is hydraulic conductivity.

Since $pg(\alpha + n\beta) = S_s$, the specific storage, the right-hand side of the equation may be simplified to $(S_s/K)dh/dt$.

The specific storage is the volume of water that a unit volume of aquifer releases from storage per unit decline in h. The solution $h(x, y, z, t)$ describes the value of the hydraulic head at any point in a flow field at any time (Freeze and Cherry, 1979). PWW

Reference

Freeze, R.A. and Cherry, J.A. (1979) *Groundwater*. Englewood Cliffs, NJ: Prentice-Hall.

diffusive processes

diffusive processes   Processes for which sediment transport is dominantly slope dependent and thus diffusion-like in terms of rates being dominated by a gradient (in this case the topographic gradient or slope). Sediment transport Q_s by diffusive processes is often cast in landscape evolution models as $Q_s = DS^n$, where D is a rate parameter (hillslope diffusivity), S is the slope, and $n = 1$ for linear diffusion, $n > 1$ for nonlinear diffusion and $n < 1$ for viscoplastic creep. Diffusive processes tend to reduce relief and infill valleys.   DRM

digital elevation model (DEM)

digital elevation model (DEM)   A digital model or representation of the land surface. A DEM can be either raster based (composed of a regular grid of squares, or cells) or vector based (composed of a triangular irregular network). DEMs can be made from a variety of data sources (field surveys, photogrammetry, lidar data or satellite data) See also DIGITAL TERRAIN MODEL (DTM).   DRM

Reading

Moore, I., Grayson, R.B. and Ladson, A.R. (1991) Digital terrain modelling: a review of hydrological, geomorphological, and biological applications. *Hydrological Processes*, 5, 3–30.

digital image processing

digital image processing   A REMOTE SENSING technique involving the handling and modification of images that are held as discrete units; for example, the sampling, correction and enhancement of a Landsat/MSS image can be achieved by this method.

A discrete image comprises a number of individual picture elements known as pixels, each one of which has an intensity value and an address in two-dimensional image space. The intensity value of a pixel, which is recorded by a digital number (DN), is dependent upon the level of ELECTROMAGNETIC RADIATION received by the sensor from the Earth's surface and the number of intensity levels that have been used to describe the intensity range of the image. There are three stages in the processing of a discrete image. First, the image data are computer manipulated, then the results of these manipulations are displayed. These three stages can be performed using a range of computers; workstations and the larger personal computers are currently the most popular among physical geographers.

There are many techniques for the processing of digital images, and physical geographers tend to concentrate on six of them: image restoration and correction, image enhancement, data compression, colour display, image classification and

145

the development of geographic information systems.

1 Image restoration and correction form the first stage in any image-processing sequence and include, first, the restoration of the image by the removal of effects whose magnitudes are known, like the nonlinear response of a detector or the curvature of the Earth and, second, the correction of the image by the suppression of effects whose magnitudes can only be estimated, such as atmospheric scatter or sensor wobble.

2 Image enhancement involves the 'improvement' of an image in the context of a particular application. The most popular image enhancements are the selective increase in image contrast (stretching), ratioing wavebands against each other to display differences between wavebands, and digital filtering to smooth or sharpen edges within an image.

3 Data compression involves the reduction of many images into one image for ease of interpretation.

4 Colour display involves the combination of images with colour, again for ease of interpretation.

5 Image classification can be achieved by several techniques, notably the density slicing of one image or the supervised classification of several images. Density slicing involves the grouping of image regions with similar DN, either automatically or interactively. Supervised classification involves the careful choice of wavebands, the location of small but representative training areas, the determination of the relationship between object type and DN in the chosen wavebands, the extrapolation of these relationships to the whole image data set and the display and accuracy assessment of the resultant images.

6 Geographic information systems involve the combination and use of any spatial data that can be referenced by geographic coordinates. The three processing steps for this operation are, first, data encoding, where spatial data are broken into polygons or grids; second, data management, where these data are spatially filed; and third, data manipulation, where these data are retrieved, transformed, analysed, measured, composited or modelled.   PJC

Reading
Curran, P.J. (1985) *Principles of remote sensing.* Harlow: Longman Scientific and Technical. · Mather, P.M.

(1987) *Computer processing of remotely-sensed images: an introduction.* Chichester: John Wiley & Sons, Ltd. · Moik, J.G. (1980) *Digital processing of remotely-sensed images.* Washington, DC: National Aeronautics and Space Administration. · Richards, J. (1986) *Remote sensing digital image analysis: an introduction.* Berlin: Springer-Verlag.

digital terrain model (DTM) A model of the land surface used in a GEOGRAPHIC INFORMATION SYSTEM (GIS). A digital elevation model is a type of DTM that simply models height, whereas a DTM also models terrain shape. In practice, however, the terms are often used interchangeably. The commonest form of DTM consists of point estimates of elevation located on a regular grid with the resolution of the model set by the grid dimensions or grid size. This can be used in a standard RASTER GIS, and DTMs in this form are available for many regions. The other common format is the triangulated irregular network, where points of known height are connected into a series of triangles covering the land surface. Typical applications of DTMs include: (1) calculation of basic terrain parameters, such as slope steepness; (2) hydrological analyses, including the upslope contributing drainage area, or catchment area for any point on a stream; (3) determining the visibility between points (viewshed analysis), used in telecommunications, military planning and visual impact assessment.   SMW/DRM

Reading
Goodchild, M.F., Parks, B.O. and Steyaert, L.T. (1993) *Environmental modeling with GIS.* New York: Oxford University Press. · Moore, I., Grayson, R.B. and Ladson, A.R. (1991) Digital terrain modelling: a review of hydrological, geomorphological, and biological applications. *Hydrological Processes,* 5, 3–30. · Weibel R. and Heller M. (1991) Digital terrain modelling. In D. J. Maguire, M. F. Goodchild and D. W. Rhind (eds), *Geographical Information Systems,* vol. 1. Harlow: Longman; pp. 269–297.

dikaka Accumulation of dune and sand covered by scrub or grass vegetation, extended to include plant-root cavities in dune sediments (calcified root tubules).

dilation (or dilatation) Describes the action of PRESSURE RELEASE in a rock mass by the removal of overlying material by erosional processes. Severe glaciation may cause dilation joints to open in glaciated terrain, while in granite areas the opened joints on many INSELBERGS may be a result of pressure release following the removal of the overlying sedimentary or metamorphosed rocks. The joints often approximately parallel the ground surface configuration.   ASG

dilution effect A term used to describe the behaviour of those solute concentrations in a stream that decrease during a storm run-off event. This decrease is ascribed to the dilution of solute-rich base flow by additional inputs of storm run-off which, in view of its shorter residence time within the drainage basin, possesses lower solute concentrations. Some solute concentrations in a stream, however, may increase during periods of storm run-off.   DEW

dilution gauging A method of measuring river discharge by introducing a tracer into the river channel and timing the passage of the tracer over a known length of channel. (See also DISCHARGE.)

Reading
Water Research Association (1970) *River flow measurement by dilution gauging*. Water Research Association technical paper TP74. Medmenham: Water Research Association. · White, K.E. (1978) Dilution methods. In R. W. Herschy (ed.), *Hydrometry: principles and practices*. Chichester: John Wiley & Sons, Ltd.

diluvialism The belief in the role of Noah's flood, as reported in the book of Genesis, in shaping the landscape. Before the true origin of glacial drift was recognized, such materials were ascribed to a great deluge, when 'waves of translation' covered the face of the Earth. The heterogeneous and unsorted character of the drift seemed ample proof that it had been laid down in the turbulent waters of a universal flood, and Dean Buckland of Oxford termed such material 'diluvium' to distinguish it from the 'alluvium' formed by rivers. By the 1830s, recognition of the often complex stratigraphy of the drift, and the Ice Age, greatly weakened the diluvial viewpoint. As the catastrophic interpretation of landscape and geological history gave way to UNIFORMITARIANISM, diluvialism became obsolete.   ASG

Reading
Davies, G.L. (1969) *The Earth in decay*. London: Macdonald.

dimensionless number A dimensionless number is one that is scaled by a parameter that (1) has identical dimensions to it and (2) has basic theoretical or empirical bases for being used in such a way. The need for dimensionless numbers arises from the fact that the magnitude of a variable will tend to change as the scale of the investigation changes. Thus, by scaling the variable's magnitude, with respect to the experiment, it becomes possible to compare information acquired from a range of different experiments, even where the experiments are conducted at different scales. For instance, in BOUNDARY LAYER flows we tend to express the height above the boundary (units of length) as a proportion of the total boundary depth (also units of length). Use of depth means that identical dimensions are being used. The theoretical basis of this scaling is also strong, as we know that boundary layer characteristics do change as a function of depth, so by scaling using depth we can compare situations with different depths. It follows that dimensionless numbers should be central to the representation of most empirical evidence in PHYSICAL GEOGRAPHY.   SNL

dimethylsulphide (DMS) The most abundant volatile sulphur compound in seawater is produced by planktonic algae and bacterial decay. It oxidizes in the atmosphere to form a sulphate aerosol that is a major source of cloud-condensation nuclei. Because of this, DMS may have an important climatic impact through its impact on cloud albedo and the Earth's radiation budget (Charlson *et al.*, 1987). Increasing cloud production and albedo caused by increased planktonic productivity resulting from global warming could act as a negative feedback in the climate system. This is a hypothesis that merits careful consideration.   ASG

Reference
Charlson, R.J., Lovelock, J.E., Andreae, M.O. and Warren, S.G. (1987) Oceanic phytoplankton, atmospheric sulphur, cloud albedo and climate. *Nature*, **326**, 655–661.

dip The angle between the inclination of sedimentary strata and horizontal.

dipslope The more gentle slope of a cuesta; the slope of the land surface that approximates the dip of the underlying sedimentary rocks.

dipwell A device designed to measure the position of the water table below the ground surface. The water table position is measured at atmospheric pressure. Thus, a dipwell differs from a PIEZOMETER, which measures pressure head in the saturated zone, and a TENSIOMETER, which measures matric potential in the unsaturated zone. A dipwell may be constructed and installed very simply using plastic drain pipe ~40–60 mm internal diameter drilled (2–3 mm holes) along the length of the pipe at regular intervals and sealed at the base with a rubber bung. A loose cap should be fitted to the top of the dipwell above the ground surface to stop direct precipitation input. The length of the dipwell depends on the estimated water table depth over the hydrological year. Continuous water table measurements can be obtained by installing a calibrated pressure transducer inside the dipwell and linking this to a data logger. ALH

dirt cone   A conical hill or dome of ice that is completely mantled by till or rock fragments. It owes its existence and form to the mantle of debris that retards the rate of surface lowering resulting from the melting of the ice.

discharge   The volume of flow of water or fluid per unit time. It is usually expressed in cumecs, which are cubic metres per second, but for small discharges it may be more conveniently expressed as litres per second. In Imperial units, the cusec (cubic feet per second) was originally used, and it is still employed in the USA. It is necessary to measure discharge and to obtain continuous records of discharge variation for the investigation of the HYDROLOGICAL CYCLE.

Discharge may be measured in a number of different ways, and the method adopted at a particular gauging station will depend upon the size of the river, the stability of the channel, the variability of the flow and of the sediment transported, and the length and accuracy of the record required. The major methods of discharge measurement are as follows.

1 *Volumetric gauging* involves collection of the total volume of flow over a specified period of time. It is the most accurate method but can only be used where it is easy to collect the discharge in a large container and to time the increases in water level. It can therefore be used to measure flow from small plots or experimental areas.

2 *Control structures* are structures installed in the cross-section of the stream channel that includes weirs and flumes. Both types of control structure have a formula that relates depth of water to discharge. A weir is a structure placed across the channel and may be sharp crested, in which case the plate inserted in the channel cross-section has a sharp edge on the V notch and the angle at the centre of the V may be 90° or 120° or other angles. Alternatively, the weir may be broad crested; this is preferred for large basins. It has various forms, including flat-V and Crump types, and often needs a rating curve established to relate the depth of water above the weir and the velocity of water. Where the gauging station needs to measure a range of flows it may be necessary

to construct a compound weir where a V notch may occur in the centre of a rectangular cross-section, for example. The flume is an artificial channel constructed by raising the channel bed into a hump or by contracting the sides of the channel, or by combining both. The cross-section of the flume is adapted to suit the range and magnitude of discharge, and it may be rectangular, triangular or trapezoidal. Flumes have the advantage that silt and debris are easily carried through, whereas it could collect upstream of a weir. Several types of flume exist, and in a standing-wave flume or critical-depth flume there is a direct relationship between depth of water upstream of the throat of the flume and discharge. In other cases (Parshall and Venturi flumes) the head is measured within and upstream of the throat and the difference between the two values is directly related to discharge.

3 *Velocity–area technique* is the most frequently used method for discharge measurement and depends upon the fact that discharge is

$$Q = Va$$

where V is velocity and a is water cross-sectional area. The velocity is usually measured by current meter; this is used in each of several verticals across the channel, and the spacing between the verticals should not exceed 5% of the channel width. At many gauging stations the velocity, and hence discharge, is measured at a range of flows and a rating curve is constructed relating discharge to depth of water or stage. This rating curve can then be the basis for converting continuous records of river stage into discharge values.

4 *Dilution gauging* is a method of discharge measurement depending upon calculation of the degree of dilution by the flowing water of an added tracer solution that may be sodium chloride or sodium dichromate. The tracer may be injected either at a constant rate or by gulp injection. In the latter case an amount of the tracer solution is introduced instantaneously into the stream and the passage of the 'ionic wave' or slug past a downstream site at a known distance is measured usually using a conductivity meter.

5 *The slope–area method* of estimating discharge is effected by using a flow equation whereby velocity V can be estimated by

surveying water surface slope S, hydraulic radius R and estimating roughness n using an equation such as the Manning equation:

$$V = \frac{R^{2/3}S^{1/2}}{n}$$

6 *Electromagnetic gauging* is particularly useful where there is no stable-discharge relationship or where weed growth impedes flow. An electric current passed through a large coil buried beneath the river bed induces an electromotive force in the water and the force recorded by probes at each side of the river is directly proportional to the average velocity through the cross-section.

7 *Ultrasonic gauging* can be used where water flow is not hampered by vegetation or sediment. The time taken for acoustic pulses beamed from transmitters on one side of the river to travel to sensors on the other side is recorded and gives mean velocity at a specified depth.

Most of the above methods of discharge measurement will provide a value for a single moment. To obtain continuous records of discharge it is usual to employ a stage recorder, which will give continuous records of water depth that can subsequently be converted to discharge values.

The discharge record obtained from a gauging station has to be expressed in a form that can be analysed in relation to controlling parameters. This can be done by establishing the general character of the discharge record and by calculating daily, monthly or annual flows for a specific period, usually a year. A further way is to calculate the total run-off volume for a specified period; this is usually expressed as depth of run-off from the entire catchment area and calculated by dividing the total volume of water that passes the gauging station by the surface area of the drainage basin. The run-off R can then be compared directly with precipitation P over the basin during the same time period and the ratio gives the run-off percentage, $(R/P) \times 100$. If the average flow is plotted for the year, a diagram can be drawn to show the river regime. Such regime diagrams reflect the broad influence of climate, and whereas some river regimes will show a major concentration during a short period (e.g. in an area affected by snowmelt), other areas will have a fairly uniform distribution of discharge throughout the year.

A discharge record may also be analysed for a short period and individual HYDROGRAPH events may be identified either by extracting specific parameters from a single hydrograph or by generalizing the hydrographs as UNIT HYDROGRAPHS.

Variation in river discharge reflects the pattern of climate over the basin, particularly the incidence and intensity of precipitation, the drainage basin characteristics and also the effects of change in the drainage basin including LAND USE changes.   KJG

Reading
Gregory, K.J. and Walling, D.E. (1976) *Drainage basin form and process*. London: Edward Arnold. · Herschy, R.W. (ed.) (1978) *Hydrometry, principles and practices*. Chichester: John Wiley & Sons, Ltd.

disclimax A stable plant community resulting from the disturbance of CLIMAX VEGETATION. The word is normally used for communities disturbed to such an extent by the activities of humans or domesticated animals that the former climax has been largely replaced by new species. The species may even be introduced, as in the case of the prickly pear cactus, which has formed a disclimax over wide areas in Australia. It is similar to the term plagioclimax, which has been used for many plant communities of the English landscape produced by grazing, cutting or burning over many centuries. (See also SERE, SUBCLIMAX and SUCCESSION.)   JAM

Reading
Oosting, H.J. (1956) *The study of plant communities*. San Francisco, CA: W.H. Freeman. · Tansley, A.G. (1949) *The British Islands and their vegetation*, vol. I. Cambridge: Cambridge University Press. · Vogl, R.J. (1980) The ecological factors that produce perturbation-dependent ecosystems. In J. Cairns Jr (ed.), *The recovery process in damaged ecosystems*. Ann Arbor, MI: Ann Arbor Science.

disconformity An unconformity in a geological sequence that is not represented by a difference in the inclination of the strata above and below.

discordance An unconformity. A difference in the inclination of two strata that are contiguous.

disjunct distribution A geographical distribution pattern in which two or more populations of an organism are exceptionally widely separated for the organism concerned, thus creating a major discontinuity, and may involve now isolated relict populations, exceptional long-range migration or the separation of populations through continental movement. The southern hemisphere beeches (*Nothofagus*), for example, are thought to have been formerly linked on one continental landmass, Gondwanaland, but today they occur widely disjunct in South America, New Zealand, Australia and New Guinea. PAS

Reading
Stott, P.A. (1981) *Historical plant geography: an introduction.* London: Allen & Unwin.

dispersal The mechanism of migration by which plants and animals are disseminated over the surface of the Earth. In plants, dispersal involves the transport of any spore, seed, fruit or vegetative portion that is capable of producing a new plant in a new locality. These propagules or diaspores may be carried by air or water, on or inside animals, by the movement of soil or rock, or they may be exploded from the parent plant. In animals, dispersal depends on the mechanisms for movement, which may range from facilities for flying to swimming and running. For both plants and animals, humans are a potent agent of dispersal. (See also ALIENS and MIGRATION.) PAS

Reading
Seddon, B. (1971) *Introduction to biogeography.* London: Duckworth; especially chapter 8.

dissection The destruction of a relatively flat landscape through incision and erosion by streams.

dissipative beach Beaches can be classified into two basic types: dissipative and reflective. A third type, termed intermediate, represents those beach states that contain elements of the two basic types (Wright *et al.*, 1984).

Under dissipative conditions, incident waves break and lose much of their energy before reaching the beach face. Broken waves or dissipative bores form the resulting surf zone with bore height decreasing in amplitude towards the shore. Depending on incident wave height, the surf zone may be as wide as 500 m under fully dissipative conditions.

Dissipative beaches are characterized by a wide, low-gradient beach face extending from the foot of the foredune into the surf zone. Multiple bars or breaker zones may be present across the surf zone. Water circulation in the surf zone is dominated by strong onshore flow in the upper water column (alongshore if incident wave approach is oblique to the shoreline) and offshore towards the bed (Wright *et al.*, 1982). BGT

References
Wright, L.D., Guza, R.T. and Short, A.D. (1982) Dynamics of a higher energy dissipative surf zone. *Marine Geology*, 45, 41–62. · Wright, L.D. and Short, A.D. (1984) Morphodynamic variability of surf zones and beaches: a synthesis. *Marine Geology*, 56, 93–118.

dissolved load Material in solution transported by a river and including both inorganic and organic substances. There is no clear boundary between a true solution and the presence of material as fine colloidal particles, and all material contained in a water sample that has been passed through a 0.45 μm filter is conventionally regarded as dissolved. Sources of the dissolved load include rock weathering, atmospheric fallout of aerosols of both oceanic and terrestrial origin, atmospheric gases, and decomposition and mineralization of organic material. In general, the total concentration ($mg\,L^{-1}$) of material in solution in a river will decline as discharge increases due to a DILUTION EFFECT, but the load transported ($kg\,s^{-1}$) will increase.

On a global basis, the dissolved loads of perennial rivers and streams range from $<1.0\,t\,km^{-2}\,a^{-1}$ to a maximum of about $500\,t\,km^{-2}\,a^{-1}$, although even higher levels may exist in streams draining saline deposits. Meybeck (1979) has estimated the mean dissolved load transport by rivers from the land surface of the globe to the oceans at $37.2\,t\,km^{-2}\,a^{-1}$, and Walling and Webb (1983) have pointed out the importance of mean annual run-off and lithology in controlling the global pattern of dissolved load transport. DEW

References
Meybeck, M. (1979) Concentration des eaux fluviales en éléments majeurs et apports en solution aux océans. *Revue de Géologie Dynamique et de Géographie Physique*, 21, 215–246. · Walling, D.E. and Webb, B.W. (1983) The dissolved loads of rivers: a global overview. In B.W. Webb (ed.), *Dissolved loads of rivers and surface water quantity/quality relationships*. IAHS publication no. 141. Wallingford: IAHS Press; pp. 3–20.

dissolved oxygen This will be present in most natural waters, but interest in this water quality parameter has focused largely on rivers because of its importance for fish and other aquatic life and the potential for significant spatial and temporal variation. The dissolved oxygen content of streamflow primarily reflects interaction with the overlying air, since oxygen from the atmosphere is dissolved in the water. Assuming equilibrium conditions, the dissolved oxygen concentration is essentially a function of the water temperature, which influences its solubility, and of the atmospheric pressure, which reflects the partial pressure of the gas. The solubility of oxygen at 0 °C and 760 mm atmospheric pressure is $14.6\,mg\,L^{-1}$ and this will decrease with increasing temperature to $7.6\,mg\,L^{-1}$ at 30 °C. Deviations from the equilibrium may occur in both a positive and a negative direction. Supersaturation can result from the production of oxygen within the

Table 2 Average composition of world river water

	Concentration (mg L^{-1})								
	Ca^{2+}	Mg^{2+}	Na$^+$	K$^+$	Cl$^-$	SO$_4$$^{2-}$	HCO$_3$$^-$	SiO$_2$	Total
Livingstone (1963)	15.0	4.1	6.3	2.3	7.8	11.2	58.4	13.1	118.2
Meybeck (1979)	13.4	3.35	5.15	1.3	5.75	8.25	52.0	10.4	99.6

water body through photosynthesis by macrophytes and algae during daylight hours. Oxygen will be consumed by the respiration of aquatic organisms and by the biochemical oxidation of organic material and pollutants, and reduction in dissolved oxygen concentration will occur if this consumption exceeds the rate of atmospheric reaeration. The potential oxygen consumption associated with an organic pollutant is expressed by its biochemical oxygen demand value.   DEW

dissolved solids The total concentration of dissolved material in water is frequently expressed as a total dissolved solids (TDS) concentration. This parameter can be determined by evaporating to dryness a known volume of water. The evaporation is normally carried out at a temperature of 103–105 °C, but some standard procedures specify a temperature of 180 °C. The value obtained should be viewed as only approximate, as there is a possibility of loss of dissolved material by volatilization or of incomplete dehydration of the residue. The residue will contain both inorganic and organic material. An alternative procedure involves summation of the results obtained from analysis of individual constituents, although these need to be expressed in terms of an anhydrous residue (e.g. bicarbonate will exist as carbonate) in order to ensure comparability. There may be poor correspondence between results from the two methods, through the problems outlined above and the general lack of data for dissolved organic material.

Because of their generalized nature, measurements of TDS concentration are of limited value in water quality assessment and pollution studies, but they are frequently employed in investigations of DISSOLVED LOAD transport by rivers and in associated assessments of chemical DENUDATION RATES.

On a global scale, discharge-weighted TDS concentrations encountered in rivers range between minima of approximately 5–8 mg L^{-1} recorded in several tributaries of the Amazon, and maxima of 5000 mg L^{-1} or more associated with streams draining areas of saline deposits. Values, however, are typically in the range 30–300 mg L^{-1}, and Meybeck (1979) cites a mean inorganic TDS concentration of 99.6 mg L^{-1}

for world river water. Dissolved organic matter generally constitutes only a small proportion of the TDS, and concentrations commonly fall in the range 2–40 mg L^{-1}, with a mean of approximately 10 mg L^{-1}. The inorganic component is dominated by a limited number of major elements, and Ca^{2+}, Mg^{2+}, Na$^+$, K$^+$, Cl$^-$, HCO$_3$$^-$, SO$_4$$^{2-}$ and SiO$_2$ generally account for 99% of the material, with a number of lesser constituents comprising the remainder. Several workers have attempted to define a world average river composition in terms of major inorganic constituents, and those provided by Livingstone (1963) and Meybeck (1979) are listed in Table 2. In the latter case, an attempt has been made to deduct anthropogenic contributions.   DEW

References

Livingstone, D.A. (1963) Chapter G. Chemical composition of rivers and lakes. In M. Fleischer (ed.), *Data of geochemistry*, 6th edition. US Geological Survey Professional Paper 440-G. Washington, DC: US Government Printing Office. · Meybeck, M. (1979) Concentrations des eaux fluviales en éléments majeurs et apports en solution aux océans. *Revue de Géologie Dynamique et de Géographic Physique*, **21**, 215–246.

distributary A stream channel that divides from the main channel of a river. One of the channels a river subdivides into when it becomes braided or when it reaches its delta.

diurnal tides Occur in a limited number of parts of the world and consist of only one high and one low tide in each 24-h period. They only occur where the coastline has the correct configuration.

divergence The phenomenon of air flowing outwards from an air mass being replaced by air descending from above; also the splitting of oceanic currents, often as a result of offshore winds, allowing upwelling of cold water from the depths.

Written as div V, it is given by

$$\text{div } V = \frac{\partial u}{\partial x} + \frac{\partial v}{\partial y} + \frac{\partial w}{\partial z}$$

where V is the three-dimensional velocity vector, having x, y and z components u, v and w; units are reciprocal time. Often in meteorology the term is used to mean horizontal divergence, being the instantaneous fractional rate of change of an infinitesimal horizontal area within the fluid, given by

$$\text{div } V_H = \frac{\partial u}{\partial x} + \frac{\partial v}{\partial y}$$

Typical values of large-scale horizontal divergence in the atmosphere are about $10^{-5}\,\text{s}^{-1}$. Negative divergence is called convergence.

Local variations of density in the atmosphere are relatively small, so that horizontal divergence and convergence usually result in VERTICAL MOTION. KJW

Reading
Atkinson, B.W. (ed.) (1981) *Dynamical meteorology: an introductory selection*. London: Methuen.

doab A term used in the Indian subcontinent to describe the low alluvial plain between two converging rivers.

doldrums A zone of light, variable winds, low atmospheric pressure, high humidity and temperature, and frequent cloudy and unsettled weather located near or slightly north of the equator. The doldrums are bounded to the north by the northeast trade winds of the northern hemisphere and to the south by the southeast trade winds of the southern hemisphere (see INTERTROPICAL CONVERGENCE ZONE (ITCZ)). They shift north and south with the seasons, being farthest north in June to October. In a broader sense, the term *doldrums* refers to the lethargic, monotonous, warm, humid weather of summer and to the listless, often despondent, human response to it. WDS

doline An enclosed depression found in KARST landscapes. They are typically sub-circular in plan, tens to hundreds of metres in diameter, from a few metres to a kilometre in width, and a few metres to tens of metres (and sometimes more) in depth. They may have vertical or gentle sides and they range from being saucer shaped to conical or even cylindrical (Ford and Williams, 2007). They occur in carbonate rocks, EVAPORITES and even in siliceous rocks such as quartzites. Some dolines are the result of concentrated dissolution, some result from collapse into underlying cavities and some are the result of subsidence or suffusion of overlying materials into the underlying bedrock. ASG

Reference
Ford, D.C. and Williams, P.W. (2007) *Karst geomorphology and hydrology*. Chichester: John Wiley & Sons, Ltd.

dolocrete A form of calcrete in which magnesium carbonate is a major component, and which probably forms as a groundwater precipitate near the water table of a brackish water body. Dolocretes, which may form lower down in a profile than standard calcrete, have been described from a number of current dryland areas, including Australia (Dixon, 2010) and Kuwait (Khalaf and Abullah, 2013). ASG

References
Dixon, J.C. (2010) Origin of calcrete and dolocrete in the carbonate mantle of St Vincent basin, southern South Australia. *Cadernos do Laboratorio Xeolóxico de Laxe*, **35**, 109–122. · Khalaf, F.I. and Abdullah, F.A. (2013) Petrography and diagenesis of cavity-fill dolocretes, Kuwait. *Geoderma*, **207**, 58–65.

dome dune See ZIBAR.

domin scale Plant abundance scale that is used to rate the relative vigour and cover of species within a sample quadrat. It is similar in principle to the BRAUN-BLANQUET SCALE, although has 10 divisions, and was also developed to facilitate the quantitative analysis of co-occurrences of plant species and is a phyto-sociological tool used to identify plant associations (see ASSOCIATIONS, PLANT and COVER, PLANT). MEM

dominant discharge The discharge to which the average form of river channels is related. The dominant discharge that determines the size of a river channel cross-section or the size of river channel pattern at a particular location will also depend on the character and quantity of sediment transported and also on the composition of the bed and bank materials. The dominant discharge will not be a single value but a range of flows. (See also CHANNEL CAPACITY.) KJG

dominant organism An organism of principal importance in either the whole or part of a community. Dominance may be physiognomic (of a particular form, such as a tree – see LIFE FORM), taxonomic (of an evolutionary category, such as a species) or ecological (of quantity and function, such as the amount of standing crop (see BIOMASS), the competitive ability of a producer or consumer). An organism can exhibit more than one type of dominance. This is especially so in the case of physiognomy and taxonomy (Shimwell, 1971).

Both plants and animals may dominate. However, plants comprise most of the Earth's biomass and are essential for the survival of animals (see ECOSYSTEM). Thus, strictly speaking, plants always dominate communities (Odum, 1975). Nevertheless, as Daubenmire (1968) points out, there are instances (e.g. in agricultural ecosystems) in which animals assume an important role. Similar regard may be given to the current ascendancy of humans in the majority of world ecosystems.   RLJ

Reading and References
Daubenmire, R.F. (1968) *Plant communities: a textbook of plant synecology*. New York: Harper & Row. · Odum, E.P. (1975) *Ecology: the link between the natural and social sciences*, 2nd edition. London: Holt Reinhart & Winston. · Shimwell, D.W. (1971) *Description and classification of vegetation*. London: Sidgwick & Jackson. · Willis, A.J. (1973) *Introduction to plant ecology*. London: Allen & Unwin.

dominant wind The wind that plays the most significant role in a particular local situation, in contrast to the PREVAILING WIND.

donga Derived from the Nguni word *Udonga*, meaning a wall, the term used in southern Africa to describe a gully or badland area caused by severe erosion. Dongas are especially prevalent in colluvium and in weathered bedrock in areas where the mean annual rainfall lies between 600 and 800 mm. Where the materials in which they are developed have high exchangeable sodium

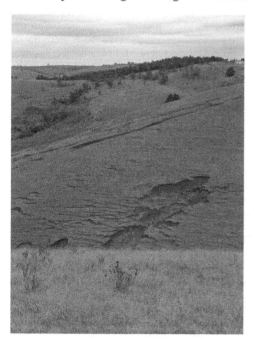

Donga development on steep slopes in Swaziland. Land-use pressures, particularly grazing, are likely causes of enhanced soil erosion and gully (donga) development. Photograph by David Thomas.

contents, they may have highly fluted 'organ pipe' sides (Stocking, 1978). At Rorke's Drift they provided a snare for unwary troops in the battle between the British and Zulus.   ASG

Reference
Stocking, M.A. (1978) Interpretation of stone lines. *Southern African Geographical Journal* **60**, 121–134.

dormant volcano A volcano that, though not currently or perhaps even recently active, is not extinct since it is likely to erupt in the future.

double mass analysis A plot of cumulative values of one variable, or values from one site against cumulative values of another variable or from another site. It has often been used to plot the values of precipitation recorded at one station against the records of another station or against the average of several other stations. Unless the double mass curve that is plotted has a slope of 1.0 there is a difference between the sites, and this may be interpreted in terms of instrument siting or of temporal variations that affect one variable but not the other.   KJG

Reading
Dunne, T. and Leopold, L.B. (1978) *Water in environmental planning*. San Francisco, CA: W.H. Freeman.

downbursts Parcels of cold air descending from the middle and upper levels of thunderstorms. When the air hits the surface, it flows out from a central point in a straight line in all directions. The winds can reach over 150 miles per hour and can be more damaging than a tornado. The small bursts, less than 2.5 miles in extent, are termed microbursts and can last 5–15 min. The larger macrobursts can last up to 30 min. Downbursts occur much more frequently than tornadoes and create severe hazards for aircraft.   SN

downscaling The name given to the procedure of taking information at large spatial and temporal scales to local or regional scales. In the context of weather or climate, downscaling can either be done by numerical models or by statistical methods. The former case is often referred as dynamical downscaling. It involves a numerical model of weather or climate, which is run at a relatively high resolution over a regional domain. It is forced at the lateral boundaries by a corresponding weather or climate model of a coarser resolution with either global or a regional coverage of the planet (but which circumscribes the spatial domain of the high-resolution model). The statistical downscaling method is based on developing statistical relationships between the local scale and the large scale from a limited period of data available at both scales. This statistical relationship is then used to generate the local-scale information from the given large-scale information for periods outside of the period when both local- and large-scale information were available. The dynamical downscale procedure is computationally intensive but uses physically and dynamically consistent numerical models to generate local climate or weather information. The statistical method, on the other hand, is relatively inexpensive computationally but requires the data to be stationary so that the derived statistical relationships are sustained. This condition of stationarity may be harder to find in the context of global climate change, where the planet is continuing to warm in many regions and may have an influence on changing the statistical relationships between the local and large scales (Milly *et al.*, 2008).   VM

Reference

Milly, P.C.D., Betancourt, J., Falkenmark, M., *et al.* (2008) Stationarity is dead: whither water management? *Science*, **319**, 573–574.

downwelling   The piling up and pushing downwards of surface ocean waters. It occurs at locations where surface currents meet, such as in the subpolar gyre of the North Atlantic Ocean and in the Southern Ocean where cold Antarctic water sinks below warmer South Pacific and South Atlantic waters. There is also downwelling on coastlines where the wind blows in such a direction that it causes the Ekman transport to move water towards the coast. Regions of downwelling have low productivity because the nutrients in the water column are utilized but are not continuously resupplied by nutrient-rich water from below. However, downwelling is beneficial in promoting deep-ocean ventilation, bringing dissolved oxygen down from the surface to help facilitate aerobic respiration in organisms throughout the water column. Without this renewal, the dissolved oxygen in the sediment and within the water column would be quickly used up by biological processes. Under these conditions, anaerobic bacteria would take over decomposition, leading to a build-up of hydrogen sulphide. In these toxic conditions, few benthic animals survive. In the most extreme cases, a lack of downwelling could lead to mass extinction. Palaeontologists have suggested that, 250 million years ago, deep-ocean ventilation slowed nearly to a halt and the ocean became stagnant. Low oxygen, sulphide and methane-rich waters filled the deep ocean and progressed onto the continental shelves, wiping out 95% of all marine species in the great Permian extinction. TS

Reference

Pond, S. and Pickard, G. (1978) *Introductory dynamic oceanography*. Oxford: Pergamon Press.

draa   A large-scale accumulation of aeolian sand dunes geographically distinct from other sand fields but smaller than a sand sea.

drainage   May refer either to the natural drainage of the land surface or to the system of land drainage introduced by human activity. Natural drainage of the land surface is organized in drainage basins, which are those areas in which water is concentrated and flows into the DRAINAGE NETWORK. The drainage basin is usually defined by reference to information on surface elevation (e.g. from contours on topographic maps), although the position of the WATERSHED on the ground surface, which is the line separating flow to one basin from that to the next, may not correspond to the PHREATIC DIVIDE beneath the surface. The pattern of natural drainage has been studied in relation to (i) the DRAINAGE DENSITY and drainage basin characteristics, which can be quantified and used in rainfall–run-off modelling and in the interpretation of river discharge; (ii) the nature of the drainage network including the pattern of the drainage and also the stream order (see ORDER, STREAM); and (iii) the evolution of the drainage pattern. In the course of drainage evolution the details of the several patterns, such as trellis or rectangular (see diagram for DRAINAGE NETWORK), may be related to geological structure such as the alternation of hard and soft rocks or to the presence of joints or faults. Where drainage patterns are discordant with the structure and cross folds or faults, for example, it has been suggested that either the drainage has been superimposed from a cover rock that originally occurred above the rocks at present exposed in the landscape, or the drainage was antecedent and the drainage pattern was maintained as the structures were developed by endogenetic uplift giving the folded and/or faulted structures.   KJG

drainage density   Calculated by dividing the total length of stream channels in a basin ΣL by the drainage basin area A_d as

$$D_d = \frac{\Sigma L}{A_d}$$

Drainage density is a very significant measure of drainage basin character because the values of D_d reflect the climate over the basin and the influence of other drainage basin characteristics, including rock type, soil, vegetation and land use, and topographic characteristics. Drainage density also has an important influence upon streamflow because water flow in channels is faster than water flow over or through slopes: the higher the drainage density is, the faster is the hydrograph rise and the greater the peak discharge. Although drainage density is a very significant parameter it must be used with careful attention given to the extent of the drainage network – which can be determined from topographic maps of different scales, from remote-sensing sources or from field survey – and also to the composition of the DRAINAGE NETWORK, because networks are composed of channels that have flows for different periods of the year. KJG

Reading

Gregory, K.J. (1967) Drainage networks and climate. In E. Derbyshire (ed.), *Geomorphology and climate*. Chichester: John Wiley & Sons, Ltd; pp. 289–318.

drainage network   The system of river and stream channels in a specific basin or area (see diagram). Whether a particular headwater stream is included in the drainage network depends upon channel type, and networks have been analysed qualitatively and quantitatively. Qualitative classifications of drainage networks depend upon the way in which drainage patterns reflect either underlying geological structure (rectangular, parallel, dendritic, trellised), prevailing regional slope (radial, centripetal), geomorphological history (deranged), or some combination of these (e.g. annular). Quantitative analysis has focused upon DRAINAGE NETWORK structure founded upon stream order (see ORDER, STREAM) and upon DRAIN-AGE DENSITY. When viewed according to channel process, the drainage network can be regarded as composed of PERENNIAL STREAMS, INTERMITTENT STREAMS and EPHEMERAL STREAMS.   KJG

Reference

Garner, H.F. (1974) *The origin of landscapes*. New York: Oxford University Press; pp. 60–61.

draw down   The extent to which the WATER TABLE is reduced in elevation as a result of pumping water from a well. The amount of draw down diminishes logarithmically with distance from the site of pumping, and this determines the shape of the CONE OF DEPRESSION in the water table. Under ARTESIAN conditions, draw down may also occur in the potentiometric surface should heavy pumping occur at a bored well. In a coastal AQUIFER, the reduction in HYDRAULIC HEAD caused by draw down will encourage saltwater intrusion beneath the well (see GHYBEN–HERZBERG PRINCIPLE). PWW

Drainage network. Types of drainage pattern.
Source: Garner 1974. Reproduced with permission of Oxford University Press.

Reading

Fitts, C.R. (2013) *Groundwater science*, 2nd edition. Waltham, MA: Academic Press. · Kohler, M.A. and Paulhus, J.L.H. (1988) *Hydrology for engineers*, 3rd edition. New York: McGraw-Hill.

dreikanter A pebble found on the surface in desert regions; it has three distinct facets on its upper surface as a result of wind abrasion.

drift potential A term used to describe the *potential sand transport* due to aeolian processes in a particular wind environment. Three different types of drift can be calculated when a SAND ROSE is produced: the potential drift from each compass direction for which wind data are available, the total drift potential, which is the sum of all individual directional potentials, and the resultant or net drift potential, which is the resolved value in the resultant drift direction   DSGT

dripstone Any accumulation of water-soluble salts on the roofs, walls and floors of caves. A general term for stalactites and stalagmites.

drop size (and the distribution of drop sizes) This forms one of the major determinants of the effects of rain on exposed soils. Similarly, beneath plant canopies, the sizes of water drops that fall from the leaves and branches (termed *gravity drops*) are equally important to soil splash. In general, larger drops have a higher TERMINAL VELOCITY and deliver more energy to the soil surface, where it may be expended in breaking down soil aggregates or splashing grains. In rain, drop sizes are no smaller than about 0.1 mm diameter (smaller drops are kept aloft by atmospheric turbulence) and no larger than about 6 mm (larger drops are unstable and break up during their fall). The modal diameter is often about 2–2.5 mm. It tends to be larger in intense thunderstorms (when only larger and heavier drops are able to fall through the strong updrafts feeding moisture into the convective clouds) and smaller in low-intensity rain.   DLD

Reading

Bubenzer, G.D. and Jones, B.A. (1971) Drop size and impact velocity effects on the detachment of soils under simulated rain. *Transactions of the American Society of Agricultural Engineers*, **14**, 625–628.

drought This is 'a rather imprecise term with both popular and technical usage' (Kemp, 1994: 41). Droughts can occur in any environment, and the perception, and therefore definition, of a drought varies with the climate conditions that generally pertain in the environment concerned.

While a drought linked to a rainfall deficit is strictly a meteorological phenomenon, drought may also be assessed relative to economic and social needs. To this end, an assessment solely linked to rainfall deficits might more appropriately be called a meteorological drought, to distinguish it from considerations related to other factors. In particular, an assessment of an even broader term might be: 'There are a plethora of purpose-specific subtypes of drought that have been proposed [see Agnew and Anderson (1992) for a review] which include water-supply drought, climatic drought, hydrological drought, socio-economic drought and economic drought, which means a water shortage that negatively affects the established economy of the country or region affected' (Sandford, 1987). Some of these subtypes are not especially helpful in the analysis of water deficits, but the term AGRICULTURAL DROUGHT has relatively wide usage and may be helpful in the assessment of impacts upon crop production. In order to take non-crop-producing areas into account, Rasmussen (1987) provides a useful wider socioeconomic definition that refers to drought as an extended and significant negative departure of rainfall from the regime that society has adjusted to. The remainder of this entry focuses upon the meteorological focus of drought.

Meteorological drought is the condition of dryness resulting from a lack of PRECIPITATION. The deficit is commonly sufficiently persistent for EVAPORATION to lead to a substantial decrease in the moisture content of soils and in other hydrological parameters, such as groundwater flow and streamflow. Official definitions exist in many countries. In the former USSR a drought was defined as a period of 10 days with total rainfall not exceeding 5 mm; in India it is when rainfall is less than 80% of normal levels (both from Agnew and Anderson (1992)). A number of indices exist for the assessment of drought from meteorological data, of which the PALMER DROUGHT SEVERITY INDEX is one widely used example.

In the UK, an *absolute drought* is a period of at least 15 consecutive days during which no day reports more than 0.2 mm of rain. A *partial drought* is a spell of at least 29 days during which there may be some days that experience slight rain but for which the mean daily rainfall does not exceed 0.2 mm. Absolute droughts occur about once a year on average in the lowlands of southeast England. In contrast, more prolonged drought is in a sense a regular annual feature of some tropical climates where one rainy season is separated from the next by a long dry season, found characteristically across areas at the limit of the poleward excursion of the INTER-TROPICAL CONVERGENCE ZONE (ITCZ).

Droughts in the mid-latitudes are associated with the unusual persistence of anticyclonic conditions and especially with the presence of a BLOCKING anticyclone. Under such a regime the rain-bearing frontal systems are steered around the flanks of the stationary high-pressure area so that anomalous dryness in one place is linked to anomalous wetness in others. For example, May 1975–August 1976 was a period of extreme drought stretching from Scandinavia to western France, with southern England recording only 50% of the long-term mean precipitation for a 16-month period. In August 1976 the same area recorded less than 50% of the mean precipitation, while at the same time Iceland to the northwest and the northern Mediterranean to the south of the blocking high reported over 150% of the respective means. The Sahelian drought of subtropical Africa that commenced in 1969 and which still persisted through the 1970s is believed to be the result of various factors, including the anomalous southward expansion of the Azores anticyclone and, through the operation of TELE-CONNECTIONS, the effects of EL NIÑO. This long period of rainfall depression may also be called a DESICCATION event. A detailed analysis of meteorological data by Agnew (1990) has shown that, whilst the effects of the Sahel drought have undoubtedly been severe, its spatial distribution was markedly uneven from year to year, reflecting the spottiness of rainfall in DRYLANDS. Consequently, some of the environmental and social consequences attributed to drought impacts may be due to other factors (including human actions) that form part of DESERTIFICATION.   DSGT/RR

References

Agnew, C. (1990) Spatial aspects of drought in the Sahel. *Journal of Arid Environments*, **18**, 279–293. · Agnew, C. and Anderson, E. (1992) *Water resources in the arid realm*. London: Routledge. · Kemp, D.D. (1994) *Global environmental issues: a climatological approach*. London: Routledge. · Rasmussen, E.M. (1987) Global climate change and variability: effects on drought and desertification in Africa. In M. Glantz (ed.), *Drought and hunger in Africa*. Cambridge: Cambridge University Press. · Sandford, S. (1987) Towards a definition of drought. In M. T. Hinchey (ed.), *Proceedings of symposium on drought in Botswana, Gaborone*. Hanover, NH: Clark University Press; pp. 33–40.

drought index

There are a range of indices that measure, in slightly different ways and for differing purposes, the severity of individual droughts compared with normal rainfall. The advantage of the use of an index in categorizing a drought is that it is a quantified approach, allowing comparisons between different droughts and from place to place. Index values are usually calculated on a monthly or annual basis from meteorological data, but seasonal calculations can occur too. Indices in use by drought planners include the Palmer drought severity index (PDSI) and the standardized precipitation index.

For example, the PDSI (Palmer, 1965) is a single numerical value that takes into account PRECIPITATION, potential EVAPOTRANSPIRATION, RUN-OFF and SOIL MOISTURE DEFICIT to depict extended periods (months, years) of abnormal dryness or wetness. The index value is 'standardized' so that comparisons can be made between geographic locations and over months. The index uses an arbitrarily selected scale ranging from below −6.0 (extreme drought) to above +6.0 (extreme moist spell). The index has been used to delineate drought and wetness disaster areas, for monitoring the status of aquifer, stream and reservoir water supplies, and to indicate the potential intensity of forest fires. The index has a number of limitations (Alley, 1984): it does not perform as well over short time periods (days, few weeks) which are used for monitoring the impact on crops and agricultural operations.   DSGT/LN

Reading and References

Alley, W. (1984) The Palmer drought severity index: limitations and assumptions. *Journal of Climate and Applied Meteorology*, **23**, 1100–1109. · Hayes, M.J. (n.d.) *Comparison of major drought indices: introduction*. http://drought.unl.edu/Planning/Monitoring/ComparisonofIndicesIntro.aspx (accessed 26 June 2015). · Palmer, W. (1965) *Meteorological drought*. US Department of Commerce Weather Bureau, Research paper 45.

drumlin

An oval-shaped hill, largely composed of glacial drift, formed beneath a GLACIER or ICE SHEET and aligned in the direction of ice flow. There are no strict definitions relating to their size, but they tend to be up to a kilometre long and 50 m in relief. They are widespread in formerly glaciated areas and are especially numerous in Canada, Ireland, Sweden and Finland. Drumlins are considered to be part of a family of related landforms including FLUTES, mega-scale glacial lineations and ROGEN MORAINES, which are collectively referred to as SUBGLACIAL BEDFORMS. Their formation remains controversial (see below), but in spite of this they are extremely useful for reconstructing former ice sheets. The word drumlin is a derivation of a Gaelic word for a rounded hill.

Whilst there are many variations in shape, the 'classic' drumlin is a smooth, streamlined hill that resembles an egg half buried along its long-axis. They tend to exist as fields or swarms of landforms rather than as isolated individuals, with a typical swarm comprising tens to thousands of drumlins. Viewed en masse, drumlins within a swarm display a similar long-axis orientation and morphology to their neighbours, and are closely packed, usually within two to three times the dimensions of their drumlin length. The majority of drumlins in a swarm have their highest elevation and blunter end pointing in an upstream direction, with the more gently sloping and pointed end, or tail, facing down-ice. The upstream blunt end is called the stoss end and the downstream end is called the lee. A common measure of their shape is the elongation ratio, which is the maximum drumlin length divided by maximum width. Typical elongation ratios are 2 : 1 to 7 : 1. Variations in drumlin shape include spindle-like forms and two-tailed forms that resemble a BARCHAN dune in plan view, and they also exist as perfect circular hills with an elongation ratio of 1 : 1. There is a whole branch of investigation, called drumlin morphometry, that uses measures of shape, size and spacing to try to develop or test theories for their formation.

The timing or synchroneity of drumlin formation within a field remains unknown. Some researchers deem drumlin formation as occurring close to ice margins and believe fields are built up incrementally as the margin retreats, whereas others believe that extensive patterns of drumlins may have formed approximately synchronously under wide swathes of an ice sheet. Drumlin patterns have been found to lie cross-cutting each other, with some superimposed upon others. This demonstrates that older landforms can be preserved beneath ice flow and that more than one flow direction is recorded.

The internal composition of drumlins reveals a perplexing array of different sediment types and structures. Some have rock cores surrounded by a concentric sheath of TILL, but they are mostly filled with unconsolidated sediments that are poorly sorted, and may contain silts, sands, gravel and boulders. They may, however, also be found with fluvially sorted sediments at their core or in a lee-side position. Some interpret this as demonstrating that fluvial deposition was part of the drumlin formation process, but others favour a two-stage process of fluvially deposited material that was later shaped into a drumlin. Tectonic structures such as thrusts and folds have also been found and have been taken to imply the sediments have been deformed during the drumlinization process.

Owing to the inaccessibility of glacier beds, active drumlin formation has not been observed

first hand, so it is perhaps no surprise that their formation remains something of a mystery. Part of the difficulty is that a good theory must be capable of explaining the full range of observed drumlins and other subglacial bedforms and their wide variation in shapes, scales and internal composition. There have been many hypotheses and theories that attempt to explain their formation. Menzies (1979) and Patterson and Hooke (1995) provide good overviews of the 'drumlin problem'. Put simply, drumlins may have formed by a successive build of sediment to create the hill (i.e. deposition or accretion) or pre-existing sediments may have been depleted in places leaving residual hills (i.e. erosion), or possibly a process that blurs these distinctions. Hypotheses have been proposed for all these cases, but most common have been those involving some form of sediment accretion. These, however, have difficulty explaining drumlins with cores of pre-existing fluvially sorted sediments.

Observations of the nature of the bed of contemporary ice sheets have revealed that the forward motion of ice can, in part, be accomplished by deformation of the soft sedimentary bed. This has led to the deforming bed model of glacier flow, which has become the most widely accepted, but still unproven, mechanism for drumlin formation. If the sediments of the bed are weak they may deform as a result of the shear stress imparted by the overlying ice. If parts of this deforming till layer vary in relative strength, then the stronger, stiffer portions will deform less and remain static, whilst the intervening weaker portions will deform more readily and become mobile. The relative strength is thought to be controlled by grain size, with coarse-grained sediments (e.g. gravels) remaining strong as they do not allow a build-up of pore water pressures, and fine-grained sediments as easily deformable. So a till layer with spatially variable strength will have static or slow-moving strong patches around which the weaker, more deformable till will flow. This can explain the cores of drumlins (strong patches; rock-cored, coarse-grained or with preserved fluvially sorted sediments) surrounded by more easily deformed till that is responsible for the streamlining. It also explains the occurrence of folds and thrusts commonly observed in drumlins. In this deforming bed model of drumlin formation (Boulton, 1987), the position of each drumlin is controlled by sediment inhomogeneities and the streamlined shape by deformation. As deformation continues, drumlins may be uprooted and become mobile.

An alternative to the above is a model developed by Shaw *et al.* (1989) that views drumlins and other subglacial bedforms to be the result of meltwater erosion and deposition as a consequence of large floods beneath the ice. In this meltwater model, regional-scale outburst floods from the central regions of the ice sheet produce sheet flows of water tens to hundreds of kilometres wide and deep enough to separate the ice from its bed. Turbulent water during the flood stage erodes giant drumlin-shaped scours in the base of the ice, which are then infilled with sediment as the flood wanes and as the ice presses down onto its bed. This is the cavity-fill drumlin and explains how fluvially derived sediments may appear in drumlins. A related mechanism is also thought to operate whereby vortices in the flood water erode down into the till bed leaving intervening ridges of the original material, which are the second type of meltwater drumlins. These could, therefore, contain tills or almost any material, as the composition is unrelated to the shaping event.

The difficulty in evaluating these theories arises from the fact that the deforming bed and meltwater models are each so comprehensive as to be able to predict the wide variety of observed drumlin characteristics. This makes it hard to use geomorphological observations to test between them. Also, both are still at the stage of qualitative theories rather than physically based models. Deforming beds have been observed to exist, and so have subglacial floods, but the question remains as to which are capable of producing drumlin forms and over the widespread patterns for which they are observed. The answers must surely lie in numerical modelling to examine plausible mechanisms tested against large-scale drumlin patterns.   CDC

Reading and References

Benn, D.I. and Evans, D.J.A. (2010) *Glaciers & glaciation*, 2nd edition. London: Arnold. · Boulton, G.S. (1987) A theory of drumlin formation by subglacial sediment deformation. In J. Menzies and J. Rose (eds), *Drumlin symposium*. Rotterdam: Balkema; pp. 25–80. · Menzies, J. (1979) A review of the literature on the formation and location of drumlins. *Earth Science Reviews*, **14**, 315–359. · Patterson, C.J. and Hooke, R.LeB. (1995) Physical environment of drumlin formation. *Journal of Glaciology*, **41**, 30–38. · Shaw, J., Kvill, D. and Rains, B. (1989) Drumlins and catastrophic subglacial floods. *Sedimentary Geology*, **62**, 177–202.

dry deposition   The process by which pollutant gases or particles are transferred directly from the atmosphere on to liquid and solid surfaces. This is particularly important for the deposition of

SO_2, but deposition of NO_2 is slow and dry deposition is not a major removal process for atmospheric NO_x. Deposition is primarily through turbulent transfer, and the velocity of deposition depends upon relative humidity, pH and, especially in urban areas, aerodynamic resistance, whereby surfaces exposed to the wind experience greatest deposition. Particulate deposition tends to concentrate near to source areas, and particulates deposited between rains can be concentrated in the early stages of surface run-off, creating an acid surge. Dry deposition can be increased by land-use changes (e.g. afforestation) that increase aerodynamic resistance and/or surface area.   BJS/SN

Reading
Review Group on Acid Rain (1987) *Acid deposition in the United Kingdom*, 2nd report. Warren Spring Laboratory. London: HMSO.

dry valley   A valley that is seldom, if ever at the present time, occupied by a stream channel. These valleys are widespread on a variety of rock types, including sandstones, chalk and limestone in southern England, but they are also known from many other parts of the world, including the coral reefs of Barbados. An enormous range of hypotheses has been put forward to explain why they are generally dry.

Hypotheses of dry valley formation
Uniformitarian
1 Superimposition from a cover of impermeable rocks or sediments.
2 Joint enlargement by solution through time.
3 Cutting down of major through-flowing streams.
4 Reduction in catchment area and groundwater lowering through scarp retreat.
5 Cavern collapse.
6 River capture.
7 Rare events of extreme magnitude.

Marine
1 Non-adjustment of streams to a falling Pleistocene sea level and associated fall of groundwater levels.
2 Tidal scour in association with former estuarine conditions.

Palaeoclimatic
1 Overflow from proglacial lakes
2 Glacial scour
3 Erosion by glacial meltwater
4 Reduced evaporation caused by lower temperatures

159

5 Spring snowmelt under periglacial conditions

6 Run-off from impermeable permafrost.

The uniformitarian hypotheses require no major changes of climate or base level, merely the operation of normal processes through time: the marine hypotheses are related to base-level changes, and the palaeoclimatic hypotheses are associated primarily with the major climatic changes of the Pleistocene. Dry valleys show a considerable range of shapes and sizes, from mere indentations in escarpments, to great winding chasms like Cheddar Gorge in the Mendips. ASG

Reading

Goudie, A.S. (1993) *The nature of the environment*, 3rd edition. Oxford: Blackwell; section 4.15.

dry weather flow A term used to describe low flow in a river as an alternative term to BASE FLOW. It is also used to refer to the total daily rate of flow of domestic and trade waste sewage in a sewer in dry weather. The daily total of sewage discharge is used to represent dry weather flow because both domestic and trade waste sewage vary greatly in quantity with 24-h periods. The dry weather flow is the background flow in the sewer and may be 'measured after a period of seven consecutive days during which rainfall has not exceeded 0.25 mm' (Bartlett, 1970).

Reading and Reference

Bartlett, R.E. (1970) *Public health engineering-design in metric sewerage*. Oxford: Elsevier. · Linsley, R.K., Kohler, M.A. and Paulhus, J.H.L. (1982) *Hydrology for engineers*, 3rd edition. New York: McGraw-Hill.

drylands These cover approximately 47% of the Earth's land surface and embrace HYPER-ARID, arid, SEMI-ARID and dry subhumid environments (Middleton and Thomas, 1997). Drylands are closely allied to the notion of *aridity*, which has four main causal factors: tropical and subtropical atmospheric stability; continentality; topographically induced rain shadows; and in some coastal situations, such as the southwestern coasts of South America and southern Africa, cold ocean currents that reduce evaporation from the sea surface.

Attempts to quantify aridity have focused on the balance between PRECIPITATION inputs and moisture losses through EVAPOTRANSPIRATION. Recent attempts to establish the extent of drylands on the basis of moisture availability have used an aridity or moisture index in the form P/PET, where P is annual precipitation and PET is potential evapotranspiration. Meigs' (1953) moisture index used aggregated monthly to annual moisture surplus and deficit data instead of P, with PET calculated by the Thornthwaite method. Because of limitations in the availability of meteorological data from many desert and dryland areas, the simpler Penman method of

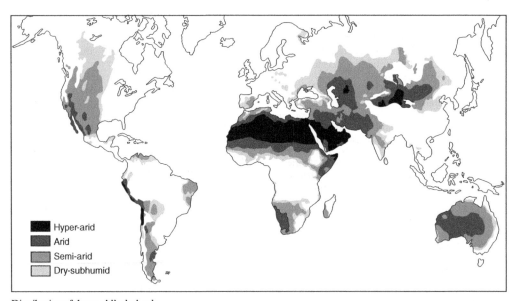

Distribution of the world's drylands.
Source: Thomas (2011). Reproduced with permission of John Wiley & Sons.

calculating PET is often used. To further rationalize values, Hulme also used meteorological data for a defined time period (1951–1980) to calculate P/PET, rather than simply taking mean values from each station supplying data. This overcomes the problem that in some developing parts of the world the mean values calculated from data runs of a few decades would be treated as equivalent to those produced from a century or more of data from parts of Europe. This approach also allows account to be given to a major climatological characteristic of drylands: high interannual and interdecadal climatic variability. This permits the construction of dryland climate surfaces for different decades, allowing spatial changes in their extent to be determined (Hulme, 1996).

In terms of P/PET values, hyper-arid areas have values <0.05; arid have values of 0.05 to <0.20; semi-arid have values of 0.20–0.50 and dry-subhumid have values of 0.50–0.65 (Hulme, 1996). The world's drylands support approximately 17% of the human population; in Africa, this rises to nearly 50% of that continent's population. The significance of this is that these people are both susceptible to the natural climatic variability inherent in drylands and are potential agents of environmental change, including DESERTIFICATION. Arid, semi-arid and dry-subhumid areas have together been termed the susceptible drylands, because of their potential to be degraded by human actions. Indirect human impacts, through possible anthropogenic global warming, may also impact significantly on drylands. It has been estimated that drylands provide a net contribution to enhanced greenhouse gas levels of only 5–10% of the global total (Williams and Balling, 1995), but many dryland areas are predicted to be amongst the biomes that respond most rapidly to global warming, through decreased soil moisture levels, a greater incidence of drought and higher mean temperatures. Significantly, however, some view drylands as having the potential to play a critical role on the mitigation of global warming, through carbon sequestration. Opportunities exist for enhancing dryland soil and biomass carbon storage, not least because drylands are a major reserve (~75%) of global soil carbonate carbon, which participates less in global carbon fluxes than organic carbon does. If anti-desertification land restoration measures proposed by the United Nations were to be implemented, the sequestered carbon would be equivalent to 15% of current annual carbon dioxide emissions. This demonstrates clearly the global significance of dryland environments.   DSGT

Reading and References

Hulme, M. (1996) Recent changes in the world's drylands. *Geophysical Research Letters*, **23**, 61–64. · Meigs, P. (1953) World distribution of arid and semi-arid homoclimates. In *Arid zone hydrology*. UNESCO Arid zone research series 1. Paris: UNESCO; pp. 203–209. · Middleton, N.J. and Thomas, D.S.G. (1997) *World atlas of desertification*, 2nd edition. London: UNEP/Edward Arnold. · Squires, V.R. and Glenn, E.P. (1997) Carbon sequestration in drylands. In N. J. Middleton and D. S. G. Thomas (eds), *World atlas of desertification*, 2nd edition. London: UNEP/Edward Arnold; pp 140–143. · Thomas, D.S.G. (ed.) (2011). *Arid zone geomorphology: process, form and change in drylands*. Chichester: John Wiley & Sons, Ltd. · Williams, M.A.J. and Balling, R.C. (1995) *Interactions of desertification and climate*. London: Edward Arnold.

du Boys equation   Paul du Boys (1879) developed one of the earliest bedload transport equations from consideration of the TRACTIVE FORCE of flowing water. The bedload transport weight per unit width of channel per unit time g_b is expressed as a function of the mean bed shear stress τ_0 in excess of the threshold shear stress τ_{oc} required to initiate particle motion:

$$g_b = A(\tau_0 - \tau_{oc})\tau_0$$

The constant A and the threshold mean bed shear stress τ_{oc} are both regarded as functions of sediment particle size alone. Derivation of the equation is simply based on the assumption that bed sediment moves in discrete layers whose velocity decreases linearly from the bed surface to zero at a depth dependent on the fluid shear stress at the bed (Embleton and Thornes, 1979: 238–239).
KSR

References

Du Boys, P.F.D. (1879) Études du régime du Rhône et l'action exercée par les eaux sur un lit à fond de graviers indéfiniment affouillable. *Annales des Ponts et Chaussées, Séries* 5, **18**, 141–195. · Embleton, C. and Thornes, J.B. (eds) (1979) *Process in geomorphology*. London: Edward Arnold.

dune   A dune is a subaerial accumulation of sediment, with particles usually sand sized (2 mm or less in diameter). Dunes are formed by the wind, in aeolian environments, and in water, generated by turbulent flow structures in fluvial environments. Active AEOLIAN dunes are aerodynamically shaped, with different forms being the result of the overall sand-transporting wind environment, the supply of sediment and the characteristics of the terrain in which they occur. Dunes may vary in size from less than a metre to over 200 m high, and they may be ephemeral forms or the result of sediment accumulation over many millennia. Dunes usually comprise quartz sand, but can also consist of carbonate sediments, gypsum, CLAY PELLETS (see also CLAY DUNE) and volcanic ash. The constituent particles of active dunes remain loose, but gypsum and carbonate sands may readily become cemented. Dunes most commonly occur in dryland and coastal environments, where suitable sediment is available for entrainment by the wind and where surface vegetation does not impede transport by SALTATION and CREEP. Approximately 20% of the world's drylands are covered by aeolian deposits, including SAND SHEETS and dunes. Extensive areas of dunes and sand sheets are called SAND SEAS. Active dunes are defined as those that are presently undergoing aeolian processes, though rates of surface change may be slow and total dune mobility does not necessarily occur. Dunes that are stabilized, fixed and/or degraded, sometimes termed relict/relic dunes, occur when the total package of environmental factors favouring aeolian processes does not now exist. Relict dunes occur on the margins of many active sand seas and are testimony to expanded dryland conditions at various times during the Holocene and late Quaternary.

A TOPOGRAPHIC DUNE occurs where airflow and sediment transport is disrupted by hills, valleys or other structural features. Most dunes, however, are not influenced by topography and occur in relatively low-relief situations, where the initiation of dune forms is probably largely a function of changes in surface ROUGHNESS, which leads to sediment deposition and the formation of a sand patch. Following initiation, dune development results from the transport of sediment up the windward or *stoss* slope of the patch towards the *crest*. The growing dune increasingly modifies its own BOUNDARY LAYER, such that flow separation on the lee side allows a SLIP FACE to develop. Desert dunes commonly occur not as isolated forms but in dunefields, where many individual dunes exist. Following McKee (1979), *simple dunes* occur where individual dunes are discrete forms, *compound dunes* occur when adjacent dunes of the same basic type coalesce or merge with each other and *complex dunes* comprise superimposed forms of different dune types.

Different types of simple dune result primarily from variations in formative wind regimes, with marked morphological variability occurring within all dune types. TRANSVERSE DUNES, LINEAR DUNES and STAR DUNES form respectively in unimodal, bimodal and multimodal wind regimes. A SAND ROSE may be constructed from wind data to assess the wind environments for different dune environments. The terminology used to describe dunes is wide ranging and reflects local usages; for example, see BARCHAN, LONGITUDINAL DUNE, SEIF and RHOURD. Other types of dune result from

the controlling influences of vegetation covers (see NEBKHA, PARABOLIC DUNES and BLOWOUTS), sediment type (see ZIBAR) and sediment supply (see AKLÉ and LUNETTE DUNE). Not all dune types are mobile and migrate, which is a function of the wind environments in which they occur (Thomas, 1992). Transverse dunes are migratory forms, moving downwind in the overall direction of sand transport; linear dunes tend to extend downwind, but they may also experience slow lateral movement if one transport component of the bimodal regime is stronger than the other;, and star dunes are accumulatory forms that may also experience a slow migration in one direction. With the exception of blowouts and parabolic forms, most coastal dunes – especially in temperate environments – are stationary, accumulating sand bodies, anchored by the vegetation that traps sand deflated from the beach zone.   DSGT

Reading and References

Lancaster, N. (2011) Desert dune processes and dynamics. In D. S. G. Thomas (ed.), *Arid zone geomorphology: process, form and change in drylands*, 3rd edition. Chichester: John Wiley & Sons, Ltd; pp. 487–515. · McKee, E.D. (ed.) (1979) *A study of global sand seas*. United States Geological Survey Professional Paper 1052. Washington, DC: US Government Printing Office. · Thomas, D.S.G. (1992) Dune activity: concepts and significance. *Journal of Arid Environments*, **22**, 31–38. · Warren, A. (2013) *Dunes: dynamics, morphology, history*. Chichester: Wiley-Blackwell.

dune memory See MEMORY CAPACITY (LANDFORM).

dune network Sand dunes in coastal and continental settings do not always produce discrete forms, but may overlap or encroach upon each other to produce *compound* or *complex* forms. In extreme cases, dune networks, sometimes called AKLÉ or reticulate dunes, may result, especially where the supply of sand or the thickness of the sand body is large. Dune networks may be distinguished from compound and complex dunes by the fact that the interdune areas are completely enclosed by dune bodies, creating interdune cells that are usually a few tens of metres across. Dune networks are usually formed by overlapping TRANSVERSE DUNE forms, in which elements with different orientations respond to sand-transporting winds deriving from different directions. Dune networks may form relatively stable patterns, especially where the ridges are large and respond to different seasonal wind directions, or may comprise unstable, ephemeral patterns that respond to short-term wind directional changes. (See DUNE.)   DSGT

Reading
Warren, A. and Kay, A.W. (1987) The dynamics of dune networks. In L. E. Frostick and I. Reid (eds), *Desert sediments, ancient and modern*. Geological Society of London, special publication 35. Oxford: Blackwell; pp. 205–212.

durability A term generally used with reference to building materials, which gives an indication of the service life of the material (e.g. a building stone). It is not an absolute quality; nor can it be easily quantified, as it must be related to tests appropriate to the type of deterioration that the structure is likely to undergo (e.g. various weathering phenomena). Such tests can be standardized. The term can also be applied to the durability of sediment in relation to downstream comminution during fluvial transport. Recently, the concept has been extended to include the relative erosion resistance, or erodibility, of bedrock channel beds, which Sklar and Dietrich (2001) showed to be a function of rock tensile strength.   DRM

Reading
Frohnsdorff, G. and Masters, L.W. (1980) The meaning of durability and durability prediction. In P. J. Sereda and G. G. Litvan (eds), *Durability of building materials and component*. American Society for Testing and Materials special technical publication 691. Philadelphia, PA: ASTM; pp. 17–30. · Sklar, L.S. and Dietrich, W.E. (2001) Sediment and rock strength controls on river incision into bedrock. *Geology*, **29**, 1087–1090.

duricrust A hard crust or nodular layer formed at or near the land surface by the processes of weathering. The crust may be composed dominantly of iron oxides (LATERITE, FERRICRETE), aluminium oxides (alcrete, bauxite), silica (SILCRETE) or calcium carbonate (CALCRETE). Crusts may also form from more soluble minerals such as gypsum and halite, but are generally not classified as duricrusts as they do not persist in the landscape beyond rainfall events.

The formation of duricrusts requires stable, low-gradient landscapes and high temperatures, which promote intense chemical activity. The crusts themselves form in a range of climatic regimes. Where rainfall is high and leaching intense, residual accumulations of insoluble ions (Al–Fe–Ti) form the bulk of the material, whereas environments in which evaporation exceeds precipitation are dominated by absolute accumulation, as in the case of calcrete. Thus, on a global scale, alcretes and ferricretes are associated with humid tropical regions and calcretes with deserts. However, local conditions frequently determine the duricrust type, particularly

in semi-arid regions, and inter-grade crusts of two or more components are common. Silica mobility in particular is responsive to changes in pH, and silcrete occupies a wide climatic range. The persistence of duricrusts in the landscape may also lead to survival of both climatic change and continental wandering. The bauxites of Le Baux (France) and the sarsen stones (silcretes) of southern England and the Paris Basin are relicts of warmer, wetter climates of the geological past.

Models of duricrust formation require movement of the ions through the near-surface zone, either downwards from the atmosphere, upwards from weathering rock, or laterally through the regolith. Where translocation and precipitation occur in the soil profile the resulting duricrust can be described as pedogenic. Non-pedogenic crusts, sometimes termed groundwater duricrusts, have subsurface water as a transporting agent and frequently occur on the margins of valleys and ephemeral lakes. Precipitation may occur directly to form pisoliths and nodules, or may indurate pre-existing rock, sediment or soil, thereby preserving the host fabric.

Regional duricrust suites in ancient shield landscapes, such as India, Australia and southern Africa, have originated in the Tertiary era. As they are resistant to erosion they may form cap rocks on mesas and escarpments, suggesting relief inversion in the long term. Not all duricrusts, however, are of great antiquity; examples have been reported of ferricretes and calcretes containing twentieth-century artefacts, and of silcretes formed by bacteria in the space of a few years.   PSH

Reading
Nash, D.J. and McLaren, S. (2007) *Geochemical sediments and landscapes*. Chichester: John Wiley & Sons, Ltd. · Scott, K. and Pain, C. (2009) *Regolith science*. CSIRO Publishing.

duripan A synonym for silcrete. An indulated and cemented silica-rich accumulation within the soil zone or zone of weathering.

dust This consists of fine particles (commonly silica) of less than about 0.07 mm diameter (SILT size) that have undergone ENTRAINMENT by wind and are suspended in the atmosphere. Whilst dust sources may include fine-grained glacial deposits or even industrial or domestic AEROSOLS, the greatest sources are the DESERT or semi-desert areas of the world where the lack of stable vegetated surfaces allows the EROSION of available fine sediment by strong winds. Where visibility is reduced to less than 1 km by suspended sediment a dust storm condition is said to exist; with visibility up to 2 km it is termed a dust HAZE. Turbulent eddies within the atmosphere are capable of keeping fine sediment entrained in the atmosphere for many days, and airborne dust can be transported thousands of kilometres from its source. Evidence for such long-range transport comes from reports of 'red rain' in the UK caused by Saharan dust and red dust settling on snow in New Zealand originating from the Channel Country of central Australia.

The dustiest places on Earth tend to be in the SEMI-ARID regions with between 100 and 200 mm of annual precipitation, rather than the hyper-arid deserts. This is because the WEATHERING and FLUVIAL activity in these semi-arid areas helps to produce and concentrate fine sediment in dry lake beds or floodplains, which may then be available for entrainment into the atmosphere. Pye (1987) also notes that human cultivation and activity in the semi-arid zone may disrupt stable sediment surfaces that may then be eroded. The Seistan Basin in Iran records 80 dust storms a year and is considered to be one of the dustiest regions on the globe (Washington and Wiggs, 2011).

The importance of dust has been generally underestimated both in its geomorphological significance and also its effects on human health and activities. More recently, the significance of airborne dust has gained ground in both academic and applied research. It has been estimated that the Niger river, the most significant active river draining the Sahara, carries 15×10^6 t a^{-1} of sediment, whilst airborne dust accounts for between 60×10^6 and 200×10^6 t a^{-1} (Cooke et al., 1993). The vast LOESS deposits of the world are now also thought to be partly aeolian in origin, with the deposition of airborne dust contributing significantly to both the 'glacial' and 'desert' loess theories.

Environmental hazards caused by dust are associated with the DEFLATION, transportation and deposition of fine sediment. In arid and semi-arid areas in particular, the deflation of silty sediment from soils is very damaging as it is in the top few centimetres that most soil nutrients are accumulated. The removal of these soil nutrients by high winds can also cause severe problems downwind, where dust deposition may bury plants or contaminate food and water supplies. It is, however, the transportation of dust in the atmosphere that has received the most attention in terms of effects on human health and activity. Dust entrained in a strong wind may be highly abrasive, and major crop and live-stock damage can occur. A lot of attention has also focused on fine sediment particles less than

10 µm in diameter (called PM10s). Such PM10s are easily inhaled into the deepest parts of the human lung and may contribute to lung damage and cancer. Problems that include bronchitis, emphysema and conjunctivitis have also been linked to airborne dust. Visibility reduction on roads and runways during severe dust storms has also caused loss of life (Washington and Wiggs, 2011), whilst at a larger scale there is mounting concern as to the reflective properties of airborne dust reducing the input of solar radiation to the Earth's surface and hence contributing to global cooling.

The Dust Bowl on the Great Plains of the USA in the mid-1930s is perhaps the best known example of how disruptive human activities and agricultural mismanagement may result in severe and life-threatening dust storms (Worster, 1979). The environmental hazards associated with dust transport are beginning to be understood, and a recognition of the consequences of poor management techniques on dust generation in hazardous areas is leading to improved management practices aimed at reducing the ERODIBILITY of agricultural surfaces and also reducing the EROSIVITY of the wind. There are still many areas, however, where airborne dust generation is a serious problem. One problem that is very difficult to control is the use of off-road vehicles, which can lead to the loss of 8–80 kg km^{-1} of silt (Pye, 1987). Urban activities around localized water supplies, building sites and military manoeuvres can all seriously disrupt naturally wind-resistant surfaces and lead to an increase in dust storm potential (Pye, 1987). Dust blown from the exposed former sea-bed of the Aral Sea (Gill, 1996), which is rapidly desiccating due to water mismanagement in the surrounding basin, has been called one of the worst environmental disasters to date.   GFSW

References

Cooke, R.U., Warren, A. and Goudie, A. (1993) *Desert geomorphology*. London: UCL Press. · Gill, T. (1996) Eolian sediments generated by anthropogenic disturbance of playas: human impacts on the geomorphic system and geomorphic impacts on the human system. *Geomorphology*, 17, 207–228. · Pye, K. (1987) *Aeolian dust and dust deposits*. London: Academic Press. · Washington, R. and Wiggs, G.F.S. (2011) Desert dust. In D. S. G. Thomas (ed.), *Arid zone geomorphology: process, form and change in drylands*, 3rd edition. Chichester: John Wiley & Sons, Ltd; pp. 517–537. · Worster, D. (1979) *Dust bowl*. Oxford: Oxford University Press.

dust devil A small WHIRLWIND (up to ~0.5–10 m diameter, and up to ~1 km high), but often smaller and often short lived (minutes rather than hours). They rarely grow to do damage, but may lift dust and dry vegetation. Various local names exist; for example, whoosher whoosher in parts of southern Africa, willy willy in Australia. DSGT

dust storm A large volume of dust-sized sediment blown into the atmosphere by a strong wind. The ENTRAINMENT of dust from the ground surface is controlled by the nature of the wind, the nature of the soil or sediment itself and the presence of any surface obstacles to wind flow. Dust storms can occur in any environment but are most common in drylands. Meteorologists define a dust storm as a dust-raising event that reduces horizontal visibility to 1000 m or less, as distinct from a 'dust haze'. The other types of dust event are 'blowing dust', which is material being entrained within sight of the observer but not obscuring visibility to less than 1000 m, and the 'dust devil', which is a localized column of dust that neither travels far nor lasts long. McTainsh and Pitblado (1987) have further refined these definitions by including the relevant international meteorological observers' 'SYNOP' present weather (or 'ww') codes for each event type.

Different terrain types vary greatly in their susceptibility to dust storm occurrence. Although silts and clays occur widely in soils and sediments, a number of factors influence their susceptibility to the deflation of dust. Among these factors are the ratios of clay-, silt- and sand-sized particles, the moisture content of the soil, the presence of particle cements, such as salts or organic breakdown products, the compaction of sediments and the presence of crusts or armoured surfaces. The most favourable dust-producing surfaces are areas of bare, loose and mobile sediments containing substantial amounts of sand and silt but little clay. Terrains that satisfy these conditions are most commonly found in geomorphologically active landscapes where tectonic movements, climatic changes and/or human disturbance are responsible for rapid exposure, incision and reworking of sediment formations containing dust (Pye, 1987). NJM

Reading and References
McTainsh, G.H. and Pitblado, J.R. (1987) Dust storms and related phenomena measured from meteorological records in Australia. *Earth Surface Processes and Landforms*, 12, 415–424. · Middleton, N.J. (1997) Desert dust. In D. S. G. Thomas (ed.), *Arid zone geomorphology: process, form and change in drylands*, 2nd edition. Chichester: John Wiley & Sons, Ltd; pp. 413–436. · Pewe, T.L. (ed.) (1981) *Desert dust: origins, characteristics and effects on man*. Geological Society of America Special Papers, vol. 186. Boulder, CO: Geological Society of America. · Pye, K. (1987) *Aeolian dust and dust deposits*. London: Academic Press. · Goudie, A.S. and Middleton, N.J. (1992) The changing frequency of dust storms through time. *Climatic Change*, 20, 197–225.

dust veil index A quantitative method developed by Lamb (1970) for comparing the magnitude of volcanic eruptions. The formulae use observations either of the depletion of the solar beam, temperature lowering in the middle latitudes or the quantity of solid matter dispersed as dust. The reference dust veil index is 1000, assigned to the Krakatoa 1883 eruption, and the index is calculated using all three methods, where the information is available, for statistical comparison purposes. ASG

Reference
Lamb, H.H. (1970) Volcanic dust in the atmosphere with a chronology and assessment of its meteorological significance. *Philosophical Transactions of the Royal Society of London, Series A*, 266, 425–533.

dyke A sheet-like intrusion of igneous rock, usually orientated vertically, which cuts across the structural planes of the host rocks. The wall or trough formed by differential weathering of such an intrusion when exposed at the land surface.

dynamic equilibrium A term that denotes a system fluctuating about some apparent average state, where that average state itself is also changing through time. The condition of a time-varying mean makes establishing that a system is in dynamic equilibrium difficult. Huggett (1980), however, argues that dynamic equilibrium in physical geography generally 'is taken either as synonymous with "steady state" or with some false equilibrium in which the system appears to be in equilibrium (*qua* steady state) but in reality is changing very, very slowly with time'. (See also STEADY STATE.) BAK

Reference
Huggett, R. (1980) *Systems analysis in geography*. Oxford: Clarendon Press.

dynamic source area The changing part of a catchment that is physically contributing overland flow at any time. The SOURCE AREA of a catchment is envisaged as changing in response to subsurface flow conditions before and during storms. This concept is important to an understanding of hillslope hydrological response. MJK

Reading
Kirkby, M.J. (ed.) (1978) *Hillslope hydrology*. Chichester: John Wiley & Sons, Ltd.

dynamical meteorology Study of the forces acting on, and the subsequent motion of air in, weather systems; compare with SYNOPTIC METEOROLOGY and SATELLITE METEOROLOGY. The control of most of the important physical processes in the atmosphere, such as condensation, depends critically on air motion (see WIND). Hence, dynamical meteorology is seen by many to represent the core of meteorology.   JSAG

Reading
Aktinson, B.W. (ed.) (1981) *Dynamical meteorology: an introductory selection*. London: Methuen.

dzud A NATURAL HAZARD, particularly used as a term in Mongolia, Inner Mongolia in China and adjacent areas. It refers to an extremely cold winter with thick snow lying on the ground, preventing the livestock of commonly nomadic pastoralists from gaining access to fodder, and leading to widespread livestock deaths. The term is more complex, in that official dzud occurrences in Mongolia are frequently preceded by high-drought summers that weaken livestock before winter. Further, several varieties of dzud are identified, such as 'iron dzud', where the ground is frozen solid.   DSGT

Reading
Fernández-Giménez, M.E., Batkhishig, B. and Batbuyan, B. (2012) Cross-boundary and cross-level dynamics increase vulnerability to severe winter disasters (dzud) in Mongolia. *Global Environmental Change*, **22**, 836–851.

E

Earth system science (ESS)   Developed in the 1980s by NASA to promote use of their satellite products, which for the first time provided data of a scale, frequency and detail not previously known. ESS has been widely adopted in the scientific community as a means to view the Earth system holistically, in some definitions including physical and social components (Pitman, 2005). The focus should be on the linkages between the atmosphere, biosphere, hydrosphere and lithosphere, as this is where important dynamical changes and feedbacks occur, rather than solely on specific system parts, and in doing so gives significance to the complexity of the whole. ESS relates not simply to processes, but to change (including the impact of one change upon others); it is obviously very relevant, therefore, to many aspects of physical geography.

The significance, novelty and terminology of ESS has been questioned by some, however (e.g. Clifford and Richards, 2005). They argue, amongst other things, that ESS is not a new framework, that it is in fact the fundamental of the discipline of geography, and that it is not possible to make scientific advancements without detailed knowledge of the individual components. They strongly argue that ESS has philosophical shortfalls and limitations in is certainly not the only approach to use in the quest for answers to critical environmental questions.   DSGT

References
Clifford, N. and Richards, K. (2005) Earth system science: an oxymoron? *Earth Surface Processes and Landforms*, **30**, 379–383. · Pitman, A.J. (2005) On the role of geography in Earth system science. *Geoforum*, **36**, 137–148.

Earth observation   A term used widely for Earth-orientated REMOTE SENSING.

earth pillar   A pinnacle of soil or other unconsolidated material that is protected from erosion by the presence of a stone or boulder at the top.

earthflow   See MASS MOVEMENT TYPES.

earthquake   A series of shocks and tremors resulting from the sudden release of pressure along active faults and in areas of volcanic activity. The shaking and trembling of the Earth's surface associated with subterranean crustal movements. There are several ways of measuring earthquake intensity, including the RICHTER SCALE, MERCALLI SCALE and MOMENT MAGNITUDE SCALE.   DSGT

easterly wave   A westward-propagating disturbance in the pressure field of the low latitudes. The pattern of motion produces regions of divergence and convergence, with the latter generally associated with precipitation. Easterly waves occur in the Atlantic, Pacific and Indian Oceans, as well as over continental locations. The characteristics of the waves are significantly different in these various locations. The most intensely studied are the African easterly waves (AEWs) that propagate across North Africa. These have typical wavelengths on the order of 2000–5000 km and they occur in two distinct frequency bands of 3–5 days and 6–9 days. Some 25–35% of the rainfall over the Sahel and adjacent regions is associated with the waves (Mekonnen *et al.*, 2006). AEWs occur along two east–west tracks. Those in the northerly track (~18–20°N) seldom produce rainfall. Waves on the southern track are usually, but not always, associated with convection and precipitation. The classic understanding of the origin of AEWs is combined barotropic/baroclinic instability associated with horizontal and vertical shear of the wind. Recent studies suggest that an additional 'trigger' is needed for wave development, such as deep convection or orographic influence.   SEN

Reading and Reference
Nicholson, S.E. (2013) The West African Sahel: a review of recent studies on the rainfall regime and its interannual variability. *ISRN Meteorology*, Article ID 4534521. · Mekonnen, A., Thorncroft, C.D. and Aiyyer, A.R. (2006) Analysis of convection and its association with African easterly waves. *Journal of Climate*, **19**, 5405–5421. · Webster, P.J. and Curry, J.A. (2008) *Tropical meteorology, oceanography and climate*. Academic Press.

---

eccentricity Literally non-concentric, it is widely used to describe an orbit that does not have its axis located centrally; that is, the moving body does not always remain equidistant from the axial point. The Earth has an eccentric orbit around the Sun, and since the degree of eccentricity varies through time, on a cycle of approximately 96,000 years, this is one component of Earth-orbital theories of the causes of CLIMATE CHANGE. (See also MILANKOVITCH HYPOTHESIS.) DSGT

echo dune A TOPOGRAPHIC DUNE that is detached from the obstacle that induces its formation, due to the development of a reverse flow eddy that sweeps the area adjacent to the obstacle clear of sand. The reverse flow eddy develops when the upwind hillslope is steeper than 50°. An echo dune may form either upwind or downwind of the topographic barrier. DSGT

ecological energetics The study of energy fixation, transformation and movement within ecological systems. The three most important types of energy found are solar energy, chemically stored energy and heat energy. Shortwave energy from the sun, in the visible light spectrum of from 0.36 to 0.76 μm, is fixed within the cells of green plants, blue–green and other algae, and phytoplankton by transformation, through the process of PHOTOSYNTHESIS, into chemical energy, following which it is passed down either the grazing or the decomposer food chain (see FOOD CHAIN, FOOD WEB). The longwave heat energy associated with plant and animal RESPIRATION may leave ecological systems at any time. Heat is not directly available for reuse as an energy source by organisms. Both plant and animal communities at high altitudes (above 2000 m) may also receive small but significant amounts of ultraviolet radiation, with wavelengths of less than 0.36 μm, which is potentially damaging to cell tissue, and against which chemical and physical defences are required.

To avoid confusion, and since all forms of ecologically useful energy can be converted into heat, the calorie or kilocalorie is used as the standard unit of measurement in ecological energetics. One calorie is equivalent to the amount of heat required to raise the temperature of 1 g of water by 1 °C; and one kilocalorie (kcal) is 10^3 calories. The joule (J), a measure of mechanical or work energy, is now increasingly utilized: the amount of work needed to raise 1 g weight against gravity to a height of 1 cm $= 981$ erg; 10^7 erg $= 1$ J; 4.2 J $= 1$ cal. The conversion factor between mechanical and heat energy was not discovered until the mid-nineteenth century.

Studies in ecological energetics are inevitably complex, not least because of the large numbers of different organisms found within most biological systems, the uncertainty as to the precise roles of some of these in respect of energy transfer, and the often complicated technology required to unravel the intricacies of energy flow and their changing patterns in time and space. Even so, some understanding of the essential energy controls of individual land-based organisms had been gained from A. L. Lavoisier's studies in France on respiration as early as 1777. And from the 1920s, investigations into the energy flow of biologically simple lake systems were under way, particularly in North America.

But the major theoretical step forward was provided by R. L. Lindemann in 1942. It was he who first formalized the concept of TROPHIC LEVELS and the energy relationships between them. He suggested that the standing crop (BIOMASS) of each trophic level might be measured and the result then translated into energy equivalents. He hypothesized that the mean energy flow between different levels could be represented by simple equations, which took into account the ultimate controls of the first and second laws of thermodynamics. If the energy content of any trophic level is given as Λ along with a subscript to indicate the food-chain role of the organisms within it (Λ_1: producers; Λ_2: herbivores; Λ_3: carnivores; etc.), and the passage of energy between any two levels is designated by λ, then that which is transferred from Λ_n to Λ_{n+1} will be λ_{n+1}. Further, the heat loss in respiration may be termed R_1, R_2, and so on, according to the particular trophic level under review. The heat loss R coupled with the quantity of energy in transfer between any two trophic levels may jointly be described as λ_n and is the amount that, at the lower trophic level, does not go into biomass production. From this, the rate of energy exchange in any trophic level may be expressed as

$$\frac{\Delta \Lambda_n}{\Delta_t} = \lambda_n + \lambda'_n$$

or in other words, the rate at which energy is taken up by that level λ_n to form new biomass minus the rate at which energy is lost from it λ'_n.

In many respects Lindemann's theory marked a turning point in the study of ecological energetics, for it provided a framework for all later research. It encouraged Slobodkin (1959) to undertake a series of laboratory experiments on *Daphnia* populations, from which he deduced

168

that the mean transfer of energy from one trophic level to another would be ~10% (it is now known that it can vary widely from this; see TROPHIC LEVELS). Others (e.g. Odum, 1971) have accumulated a vast amount of data on the patterns of energy transfer in both natural and agricultural systems. Still others (e.g. Kozlovsky, 1968) have suggested alternative methods of examining the efficiency of energy transfer between trophic levels, as for example that related to consumption efficiency:

$$\text{Consumption efficiency} = \frac{\text{intake of food at trophic level } n}{\text{net productivity at trophic level } n-1} \times 100$$

This may increase slightly from the lowermost trophic levels, but seems in general to fall between 20 and 25%; and the majority of the remainder goes into the decomposer food chain.

Overall, ecological energetics has brought home the importance of small and microscopic organisms in the patterns of energy transfer everywhere, and the need to maintain the presence of these in balanced systems. A good deal is now known about the fixation of solar energy into plant cells at the primary producer level, but much still remains to be learnt about energy transfer and fixation further down the food chain (see BIOLOGICAL PRODUCTIVITY and NET PRIMARY PRODUCTIVITY (NPP)).   DW

Reading and References
Kozlovsky, D.G. (1968) A critical evaluation of the trophic level concept. I. Ecological efficiencies. *Ecology*, 49, 48–60. · Lindemann, R.L. (1942) The trophic–dynamic aspect of ecology. *Ecology*, 23, 399–418. · Odum, E.P. (1971) *Fundamentals of ecology*, 3rd edition. Philadelphia, PA: W.B. Saunders. · Phillipson, J. (1966) *Ecological energetics*. London: Edward Arnold. · Slobodkin, L.B. (1959) Energetics in *Daphnia pulex* populations. *Ecology*, 40, 232–243.

ecological footprint   A measure of human demand on the Earth's ecosystems. It is a standardized measure of demand for natural capital that may be contrasted with the planet's ecological capacity to regenerate. It represents the amount of biologically productive land and sea area necessary to supply the resources a human population consumes, and to assimilate its associated waste. For 2007, humanity's total ecological footprint was estimated at 1.5 planet Earths – the world's population uses ecological services 1.5 times as quickly as the Earth can renew them (http://www.footprintnetwork.org/en/index.php/GFN/page/world_footprint/).   ts

Reading
Rees, W.E. (1992) Ecological footprints and appropriated carrying capacity: what urban economics leaves out. *Environment and Urbanization*, 4, 121–130.

ecology   The scientific study of the interactions between organisms and their environment. Ecology examines the relationships between organisms belonging to both the same and different taxonomic groups, and between those organisms and their physical environmental circumstances. The word was first used by Ernst Haeckel in 1869, but many of its concepts are much more recent. The ECOSYSTEM concept is central to an understanding of the nature of ecological relationships and dates from the 1930s' syntheses of Arthur Tansley. Still other associated concepts, such as those developed in relation to population ecology, came much later, so that ecology can be considered as a relatively new scientific discipline. Essentially, according to Begon *et al.* (1996), ecology operates at three levels: the organism, the population and the community. At the organism and population levels (autecology) ecology is concerned with the way in which individuals are influenced by the physical environment and the presence or absence of particular species, while at the community level (synecology) ecology is concerned with the composition of assemblages of different species. A number of concepts integral to ecology have emerged owing to the focus on the ecosystem as the fundamental unit of study, in particular the ideas of NICHE, BIOLOGICAL PRODUCTIVITY, SUCCESSION and CLIMAX VEGETATION. Ecology has had a profound influence on the understanding of relationships between humans and their environment. The ongoing search for a more complete understanding of the role of humans in the BIOSPHERE is really a function of intellectual developments within the science of ecology.   MEM

Reading and Reference
Begon, M., Harper, J.L. and Townsend, C.R. (1996) *Ecology: individuals, populations and communities*, 3rd edition. Oxford: Blackwell. · Bush, M.B. (1997) *Ecology of a changing planet*. Englewood Cliffs, NJ: Prentice-Hall.

ecosystem   The North American ecologist E. P. Odum (1969) defined an ecosystem as 'any unit that includes all of the organisms in a given area interacting with the physical environment so that a flow of energy leads to . . . exchange of materials between living and non-living parts of the system'. This amplifies the earlier definition given by A. G. Tansley, who coined the term in 1935, and confirms the concept of the ecosystem in an aggregative hierarchy (see Tansley (1946)). Individuals aggregate into populations, populations come together in communities, and a community plus its physical environment comprise an ecosystem. In many ways the concept is independent of scale, for the definition is valid for a drop of water with a few microorganisms in it or for the whole of planet Earth, but in usual practice the term is used for units below the scale of the major world BIOMES. Although an ecosystem may be characterized in synoptic terms (i.e. by an inventory of its components, both biotic and physical), the essential features of the term are (*a*) that it implies a functional and dynamic relation between the components, going beyond a frozen mosaic of species distribution, and (*b*) that it is holistic, implying that the whole possesses emergent qualities that are not predictable from our knowledge of the constituent parts. The study of functional relationships in ecosystems has usually concentrated on phenomena that can be accurately measured and which are common to both biotic and abiotic parts of the system: energy, water and mineral nutrients are frequent examples. Energy flow through the various TROPHIC LEVELS of a system and its dissipation into heat can be used to see how the system has in its evolution partitioned the energy, and how efficiently it is passed from level to level. Studies of nutrients have often revealed mechanisms for keeping them within the ecosystem: under natural conditions relatively little of the nutrient capital of the system is lost in run-off or animal migration. In arid areas, the use of water by the system may be similarly conserved by a variety of adjustments within the ecosystem, as well as the physiological responses of individual plants and animals. The temporal dimensions of the system are also amenable to study; for example, population numbers through time are often collected. For most species, each ecosystem has a CARRYING CAPACITY, an optimum level for a particular population, which may be a simple number or subject to fluctuations of various kinds. Again, the changes in species composition and physiognomy of an ecosystem through time may be studied, as in the SUCCESSION from bare ground left by a glacier through various types of vegetation to a stable, self-reproducing forest. When succession has apparently terminated at an ecosystem type that sustains itself and gives way, under natural conditions, to no other, then this is said to be a mature or climax ecosystem. The applied side of the concept is evident in the idea of BIOLOGICAL PRODUCTIVITY, which is the rate of organic matter production per unit area per unit time, a rate that can be used to compare natural ecosystems with those affected by human activity or indeed totally human-made. The concept of STABILITY is important in the human–biophysical interface because it relates to the resilience of an ecosystem to human-induced perturbation. If we perform a particular act of environmental manipulation, will an ecosystem recover its former state (given the cessation

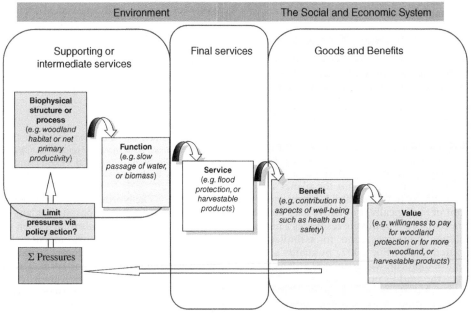

The cascade model of ecosystem services

of the impact) or will it perhaps break down irreversibly? The concept itself may also apply to ecosystems with a large human-directed component, such as agriculture (the term agro-ecosystem is sometimes used), pastoralism, fisheries, forestry and even cities themselves, as in the work on Hong Kong by K. Newcombe and colleagues (Boyden *et al.*, 1981), in which the urban area is seen as a functional ecosystem with inputs and outputs of energy and matter. IGS

Reading and References
Boyden, S., Miller, S., Newcombe, K. and O'Neill, B. (1981) *The ecology of a city and its people: the case of Hong Kong*. Canberra: ANU Press. · Odum, E.P. (1969) The strategy of ecosystem development. *Science*, **164**, 262–270. · Putman, R.J. and Wratten, S.D. (1984) *Principles of ecology*. London: Croom Helm. · Tansley, A.G. (1946) *Introduction to plant ecology*. London: Allen & Unwin.

ecosystem goods and services
These have been variously defined as the benefits ecosystems make to human well-being (see MILLENNIUM ECOSYSTEM ASSESSMENT), or the direct and indirect contributions that ecosystems make to human well-being (The Economics of Ecosystems and Biodiversity, TEEB; UK National Ecosystem Assessment, UK NEA). The concept seeks to capture the way in which people and ecosystems are linked and how different kinds of value can be assigned to different types of ecosystem output. Services can be grouped into *provisioning services*, which represent the material and energetic outputs of ecosystems (e.g. food and fibre); *regulating services*, which mediate the quality of environmental conditions that affect the well-being of people (e.g. air and water quality, climate); and *cultural services* that represent the non-material, intellectual and spiritual aspects of living systems (sense of place, aesthetic quality). The terms *supporting or intermediate services* are used to refer to the ecological structures, processes and functions that underpin the 'final' provisioning, regulating and cultural services that are directly consumed or used by people. Although some sources such as the Millennium Ecosystem Assessment regard 'goods' and 'services' as synonymous, others (e.g. UK NEA) prefer to use the term goods alongside benefits to represent the things that can be valued. RH-Y

Reading
Fisher B., Turner R.K., and Morling P. (2009) Defining and classifying ecosystem services for decision making. *Ecological Economics*, **68**, 643–653. · Millennium Ecosystem Assessment (2005) *Ecosystems and Human Well-Being: Synthesis*. Washington, DC: Island Press. http://www.unep.org/maweb/en/Index.aspx (accessed 27 June

2015). · Potschin, M.B. and Haines-Young, R.H. (2011 Ecosystem services: exploring a geographical perspective. *Progress in Physical Geography*, **35**, 575–594. · The Economics of Ecosystems and Biodiversity, TEEB: http://www.teebweb.org/ (accessed 27 June 2015). · The UK National Ecosystem Assessment. http://uknea.unep-wcmc.org/Home/tabid/38/Default.aspx (accessed 27 June 2015).

ecotone
The transition on the ground between two plant communities. It may be a broad zone and reflect a gradual blending of two communities or it may be approximated by a sharp boundary line. It may coincide with changes in physical environmental conditions or be dependent on plant interactions, especially COMPETITION, which can produce sharp community boundaries even where environmental gradients are gentle. It is also used to denote a mosaic or interdigitating zone between two more homogeneous vegetation units. They have special significance for mobile animals through edge effects (such as the availability of more than one set of HABITATs within a short distance). JAM

Reading
Allee, W.C., Emerson, A.E., Park, O., *et al.* (1951) *Principles of animal ecology*. Philadelphia, PA: W.B. Saunders. · Daubenmire, R. (1968) *Plant communities: a textbook of plant synecology*. New York: Harper & Row.

ecotope
The physical environment of a biotic community (biocoenosis). It includes those aspects of the physical environment that are influences on or are influenced by a biocoenosis. Together with its biocoenosis, the ecotope forms an integral part of a BIOGEOCOENOSIS. There are two major component parts of the ecotope: the effective atmospheric environment (climatope) and the soil (edaphotope). (See also BIOTOPE and HABITAT.) JAM

ecotoxicology
'The science which includes all studies carried out with the intention of providing information to further our understanding of the effects that chemicals and radiations (that become bio-available as a direct or indirect result of man's activities) exert on organisms in their natural habitats' (Depledge 1990: 251). ASG

Reading and Reference
Depledge, M.H. (1990) New approaches in ecotoxicology: can individual physiological variability be used as a tool to investigate pollution effects? *Ambio*, **19**, 251–252. · Walker, C.H., Sibly, R.M., Hopkin, S.P. and Peakall, D.B. (2012) *Principles of ecotoxicology*, 4th edition. Boca Raton, FL: CRC Press.

ecotype
Coined by Turesson (1922) to describe populations of organisms within a single

species that exhibit genetically produced differences in morphology or physiology which have adapted to a particular habitat, but can interbreed with other ecotypes (ecospecies) of the same species without loss of fertility. Well-known examples come from plants that, because of their low mobility, exhibit evolutionary isolation of subpopulations on small geographic scales. In many cases, these differences amongst habitats result from developmental responses; the phenotypes of populations are fixed, but vary among individuals from place to place, which may limit the exchange of genes. Such region and habitat differences in adaptations broaden the ecological tolerance ranges of many species by dividing them into smaller subpopulations, each differently adapted to consistent local environmental conditions. AP

Reading and Reference
Krebs, C. (2014) *Ecology: the experimental analysis of distribution and abundance*. 6th edition. New York: Pearson. · Turesson, G. (1922) The genotypical response of the plant species to the habitat. *Hereditas*, **6**, 147–236.

ecozone Large-scale ecological unit of classification with broadly consistent landscape and vegetation characteristics and where organisms and their physical environments endure as a system. The scale is such that an ecozone defines an area of at least $100,000 \text{ km}^2$ and which is, therefore, mappable at a scale of $1 : 3,000,000$. Diagnostic features include landforms, soils, water features, vegetation and climate; in essence, an ecozone is a large-scale ECOSYSTEM. The term is somewhat loosely and variously applied as the approximate equivalent to BIOME, zonobiome or ecoregion. Schulze (1995) recognizes nine terrestrial ecozones on the global scale as follows: polar and sub-polar zone, boreal zone, humid mid-latitude zone, arid mid-latitude zone, tropical and sub-tropical arid zone, Mediterranean zone, seasonal tropical zone, humid subtropical zone and humid tropical zone. The correspondence of this scheme with numerous biome-scale global ecological treatments is striking. Some countries (e.g. Canada and the USA) use the ecozone concept as a mapping unit in the development of large-scale land use and ecological planning frameworks. MEM

Reading and Reference
Bailey, R.G. (2009) *Ecosystem geography*, 2nd edition. New York: Springer. · Schulze, J. (1995) *The ecozones of the world: the ecological divisions of the geosphere*. Hamburg: Springer.

edaphic A term referring to environmental conditions that are determined by soil characteristics. Edaphic factors include the physical, chemical and biological properties of soils such as pH, particle size distribution and organic matter content. Plants and animals may be influenced by soil characteristics, although in many instances these edaphic factors interact with other aspects of the HABITAT.

Areas underlain by serpentine rock provide a well-defined example of the influence of edaphic factors on plant distribution. Serpentine soils are low in major nutrients but contain very high levels of chromium, nickel and magnesium. These sites are occupied by distinctive plant communities adapted to these unusual soil conditions. ARH

Reading
Watts, D. (1971) *Principles of biogeography*. New York: McGraw-Hill; chapter 4, pp. 175–184.

eddies In any model or data set we tend to see a slowly varying signal (the 'drift' or the 'climate') plus resolved fluctuations (eddies) and unresolved fluctuations (TURBULENCE and perhaps noise). In a useful record or model these signals and fluctuations define distinct physical processes. When visualizing the nature of the eddy flow it is desirable to subtract the mean flow by plotting the flow relative to the eddy. (Imagine trying to describe the mechanism of a motor-car engine relative to coordinates fixed in a pedestrian.) When this is done there are usually some closed STREAMLINES that are taken by some authors to define an eddy. There is also a tendency to notice eddies only in flow plotted relative to the ground. This makes the definition more pictorial and less mechanistic. Many eddies are noticed because they are the finite-amplitude form of mathematically well-defined solutions to a linearized equation; for example, Kelvin–Helmholtz. A popular class of eddies is found to the lee of, or propagating downwind of, bluff bodies like steep mountains or islands, such as Jan Mayen or Madeira, where accompanying cloud features can be seen on satellite pictures. JSAG

Reading
Atkinson, B.W. (1981) *Dynamical meteorology; an introductory selection*. London: Methuen. · Holton, J.R. and Hakim, G.J. (2013) *An introduction to dynamic meteorology*. Waltham, MA: Academic Press.

eddy diffusivity (eddy viscosity) A conceptual device designed to represent the mixing effect of eddies. By analogy with the kinetic theory of gases, the flux of a quantity is supposed to be proportional to the mean gradient of the quantity, the distance over which the air moves and the speed with which it does so. Molecular mean free path of 10^{-7} m multiplied by molecular speed of $300\,\mathrm{m\,s^{-1}}$ gives the kinematic molecular diffusivity of $3 \times 10^{-5}\,\mathrm{m^2\,s^{-1}}$. Eddy radius of 10 m multiplied by a fluid velocity of $1\,\mathrm{m\,s^{-1}}$ gives kinematic eddy diffusivity of $10\,\mathrm{m^2\,s^{-1}}$. JSAG

Reading
Holton, J.R. and Hakim, G.J. (2013) *An introduction to dynamic meteorology*. Waltham, MA: Academic Press.

edge waves Gravity or infragravity WAVES with crests oriented parallel to the shoreline, generated by reflection of incident waves and subsequent refraction of the reflected wave (see REFRACTION, WAVE). Their amplitude is a maximum at the water line. The wavelength λ is controlled by its period T_e, modal number n and beach slope β:

$$\lambda = \frac{g}{2\pi} T_e^2 (2n + 1)\tan \beta$$

where g is gravitational acceleration. T_e may be synchronous or subharmonic with incident wave period. The mode describes the offshore structure of the edge wave relative to the mean water level. Standing edge waves may be the mechanism for the formation of rhythmic BEACH topography including beach cusps (see CUSP, BEACH), rip current channels and nearshore bars (see BAR, NEARSHORE). DJS

Reading
Holman, R. (1983) Edge waves and the configuration of the shoreline. In P. D. Komar (ed.), *Handbook of coastal processes and erosion*. Boca Raton, FL: CRC Press; pp. 21–33.

EDM See ELECTROMAGNETIC DISTANCE MEASUREMENT.

effective precipitation Actual PRECIPITATION minus the amount lost back to the atmosphere by EVAPORATION or EVAPOTRANSPIRATION. It is therefore the amount that enters a soil plus runoff. It is usually calculated on an annual or monthly basis. It can also be defined with reference to a specific purpose. An agroclimatologist (see AGROMETEOROLOGY), for instance, calculating the availability of water for plant growth, may simply assume that it equals the amount reaching the soil surface and that intercepted (see INTERCEPTION) by vegetation:

$$ER = S + ITC$$

or they may take into account RUN-OFF and drainage losses; that is:

$$ER = S + ITC - DRA - RO$$

A hydrologist examining discharge may consider it necessary to include run-off and GROUNDWATER contributions; that is:

$$ER = RO + GWflow$$

whereas a hydrogeologist may be interested only in the amounts of water remaining in aquifers:

$$ER = GWstor$$ CTA/DSGT

effective stress A concept related to total stress in the MOHR–COULOMB EQUATION, which takes the effects of pore water pressure into account when the strength of a soil is calculated. The pressure exerted by water in soil pore spaces acts upon the grains, tending to force them apart when saturated (positive pore water pressure) or pull them together when the pore spaces have only a small amount of water in them (negative pore water pressure). The normal stress σ is modified by the subtraction of the pore water pressure u so that $\sigma' = (\sigma - u)$, where σ' is the effective normal stress. The Mohr–Coulomb equation or failure criterion then becomes

$$s = c' + \sigma' \tan \phi$$

where the primes denote that effective stresses have been considered. Positive values of u thus tend to decrease the soil strength but negative values (suction) increase it. WBW

Reading
Mitchell, J.K. and Soga, K. (2005) *Fundamentals of soil behavior*, 2nd edition. New York: John Wiley & Sons, Inc. · Whalley, W.B. (1976) *Properties of materials and geomorphological explanation*. Oxford: Oxford University Press.

effluent The polluted water or waste discharged from industrial plants.

effluent stream A stream that flows from a lake or the small distributary of a river.

egre A tidal bore.

E_h See REDOX POTENTIAL.

Ekman layer The transition layer in the atmosphere above the surface ($\sim$10 m deep) layer where the friction of the Earth's surface causes winds to blow across the isobars at an angle; and below the geostrophic wind level (about 1 km above the surface) where there is no friction and the pressure gradient force is balanced by the CORIOLIS FORCE, causing winds to blow parallel to the isobars. In the Ekman layer, eddy viscosity and density are assumed to be constant. This layer was originally theorized by V. W. Ekman in 1902 to describe wind-driven ocean currents. It is also known as the Ekman spiral due to the spiralling directional shift of the winds with altitude clockwise (counterclockwise) in the northern (southern) hemisphere.

In the ocean, the Ekman spiral describes the currents generated by surface winds. At the surface the frictional effect is maximum and the currents move at about 45° from the wind's direction. As depth increases, current speed decreases and direction becomes increasingly opposed to the surface wind direction. The depth at which the current direction and surface wind direction are 180° apart is called the depth of fractional influence. NJS

Ekman spiral

An idealized mathematical description of wind distribution in the atmospheric BOUNDARY LAYER. It is an equiangular spiral that forms the locus of the end points of the wind vectors as a function of height, all having a common origin. The GEOSTROPHIC WIND is its limit point. The assumed conditions under which this description is valid are that EDDY DIFFUSIVITY and density are constant within the layer, and the geostrophic wind is constant and unidirectional.

If the x-direction is taken along the geostrophic wind direction, the equations for the components of the wind vectors u and v, in the x- and y-directions respectively, at any level z are

$$u = v_g(l - e^{-z/\beta}\cos^{z/\beta}); \quad v = v_g e^{-z/\beta}\sin^{z/\beta}$$

where v_g is the geostrophic wind speed and $\beta = (2K_M/f)^{1/2}$, where K_M is the eddy viscosity and f is the CORIOLIS PARAMETER.

Through most of the layer the wind blows across the isobars towards lower pressure at an angle that is maximum at the surface at 45°.

The theory of the spiral was developed by Ekman in 1902 for the variation of wind-driven ocean current with depth below the surface. KJW

Reading
Holton, J.R. and Hakim, G.J. (2013) *An introduction to dynamic meteorology*. Waltham, MA: Academic Press.

elastic rebound theory

The contention that the faulting of rocks results from the sudden release, through movement, of the elastic energy that has accumulated owing to pressure and tension in the Earth's crust. Sudden movement dissipates the energy.

electromagnetic distance measurement

A highly accurate method for determining the distance between two points based on measuring the transit time of electromagnetic waves between an emitting instrument (commonly mounted on a theodolite), a reflector and back again. Most often using infrared radiation over distances of the order of 1 km, instruments based on microwave emission can operate over much longer distances. GFSW

Reading
Wilson, J.P. (1983) *Land surveying*. Plymouth: MacDonald & Evans.

electromagnetic radiation

Energy that is propagated through space or through a material in the form of an interaction between electric and magnetic wave fields. This is the link between the Earth's surface and the majority of sensors used in REMOTE SENSING. The three measurements used to describe these waves are: wavelength λ (μm), which is the distance between successive wave peaks; frequency γ (Hz), which is the number of wave peaks passing a fixed point in space per unit time; and velocity c (m s^{-1}), which within a given medium is constant at the speed of light. As wavelength has a direct and inverse relationship to frequency, an electromagnetic wave can be characterized either by its wavelength or by its frequency.

Electromagnetic radiation occurs as a continuum of wavelengths and frequencies from short wavelength, high-frequency cosmic waves to long wavelength, low-frequency radio waves.

The wavelengths that are of greatest interest in remote sensing are visible and near-infrared radiation in the waveband 0.4–1 μm, infrared radiation in the waveband 3–14 μm and microwave radiation in the waveband 5–500 mm. PJC

Reading
Campbell, J.B. and Wynne, R.H. (2011) *Introduction to remote sensing*. New York: Guildford Press. · Drury, S. (1990) *A guide to remote sensing: interpreting images of the Earth*. Oxford: Oxford University Press. · Monteith, J.C. and Unsworth, M.H. (1990) *Principles of environmental physics*, 4th edition. Oxford: Academic Press.

electron spin resonance (ESR)

A phenomenon, relating to the presence of a paramagnetic effect in insulating minerals, that is exploited to determine ages for bones, teeth, precipitated carbonates, shells, corals and volcanic materials. The effect is measured by placing a sample within an external magnetic field and exposing it to microwaves. The age produced relates to the time of precipitation of the specific mineral used. Materials may be dated back to millions of years. The ESR signal measured relates to a population of trapped charges within the crystalline lattice of some minerals. Hydroxyapatite is the most common mineral exploited when measurements are made on bones and tooth enamel. The trapped charges come about from the interaction of electrons within the crystal and energy deposited following natural radioactive decay of uranium, thorium and potassium in the environment and a small contribution from cosmic rays. Additionally, uranium incorporated into enamel and bone may also make an internal dose contribution. Displaced electrons are transferred to the conduction band and after a short period of diffusion are lodged in traps between the conduction and valence bands. These traps come about due to ionic defects and impurities within the atomic structure. The trapped electron population grows larger at a rate that is directly related to the level of environmental radiation, and for paramagnetic centres. When a sample containing such populated paramagnetic centres is brought into an external magnetic field and exposed to microwaves the magnetic moment of the centre changes relative to the magnetic field. This occurs only for specific magnetic field strengths and microwave frequencies. At such times the centre is said to undergo resonance and its frequency is characteristic for a given trap type. By convention, ESR spectra are plotted as the first derivative of ESR intensity versus g-value (the ratio of the microwave frequency to the magnetic field strength). The suitability of a signal for dating relates to the following criteria:

1 There must have been a zeroing effect or process that deletes all previously accumulated populated traps.
2 The ESR signal must grow in proportion with radiation dose.
3 The signal must be stable over geological timescales.
4 The number of trapping sites is fixed or changes systematically.
5 The ESR signal is not influenced by sample preparation methods, anomalous fading or secondary influences (such as superimposed signals from organic radicals).

As in LUMINESCENCE DATING, the age of a sample is calculated by establishing the total amount of radiation damage, the sensitivity of the sample to radiation damage and the rate of radiation dose. The age equation is

$$Age(ka) = \frac{Equivalent\ dose(Gy)}{Dose\ rate(Gy\ ka^{-1})}$$

The palaeodose is frequently also termed the accumulated dose or equivalent dose. Additional radiation dose is applied to samples in the laboratory to ascertain the sample sensitivity to radiation, and hence the palaeodose. Unlike luminescence-based techniques, the trapped charge responsible for producing the ESR signal is not destroyed during the measurement process. As such, ESR measurements are repeatable and may be measured to a high degree of precision. To simplify the determination of the radiation dose level during antiquity (the dose rate), it is common to remove the outer 2 mm of material from a specimen, in doing so removing the effects of beta irradiation from the surrounding environment. A complication, however, is the need to model the timing of incorporation of any uranium that is present within the sample. For this reason, sites that produce samples with low natural uranium levels are preferred.

ESR, in combination with THERMO-LUMINESCENCE, has played a key role in providing dates for archaeological sites in the Levant that are considered to be the earliest known localities of *Homo sapiens sapiens*. A recent application of the method relates to dating fault movements.   SS

Reading
Grun, R. and Stringer, C.B. (1991) Electron spin resonance dating and the evolution of modern humans. *Archaeometry* **31** (2), 153–199. · Walker, M. (2005) *Quaternary dating methods*. Chichester: John Wiley & Sons, Ltd.

ellipsoid Closed surface, symmetrical about three mutually perpendicular axes, in which all plane sections are either ellipses or circles. The name is also given to the solid contained by such a surface. The ellipsoid is used to represent the Earth's shape (see GEOID) in geodetic calculations. The ellipsoidal Earth is slightly flatter at the poles, the difference between polar and equatorial radii being approximately 21 km. In a geological context, the triaxial ellipsoid may be used to represent various optical properties of crystals or to denote values of stress or strain within materials.   MEM

El Niño The name translates from Spanish as 'the boy child'. Peruvian fishermen originally used the term – a reference to the Christ child – to describe the appearance, around Christmas, of a warm ocean current off the equatorial South American coast of Peru, Ecuador and Chile. This is normally a region of upwelling of cold, nutrient-rich water brought up from the lower depths of the ocean to replace surface water driven westward by the TRADE WINDS. During typical episodes, from late December to March, the

(a)

(b)

The El Niño–Southern Oscillation. (a) Typical pattern of sea-level pressure anomalies associated with the Southern Oscillation.
Source: Enfield (1989). Reproduced with permission of John Wiley & Sons. (b) Sea-surface temperature anomalies (°C) during the mature phase of El Niño.
Source: Rasmusson and Carpenter (1982). Reproduced with permission of the American Meteorological Society.

upwelling usually weakens and is replaced to some extent by warm water moving in from the west and north.

Changes to the atmosphere and ocean circulation during El Niño events include: (1) warmer than normal ocean temperatures across the central and eastern tropical Pacific Ocean; (2) increased convection or cloudiness in the central tropical Pacific Ocean – the focus of convection migrates from the Australian/Indonesian region eastward towards the central tropical Pacific Ocean; (3) weaker than normal easterly trade winds; and (4) low (negative) values of the Southern Oscillation index.

El Niño events occur irregularly at intervals of 2–7 years, although the average has been, until recently, about once every 3–4 years and lasting 12–18 months. El Niño events generally have

repercussions over far distant parts of the globe. They often result in flooding in California and parts of the midwestern USA, while the southern half of the USA experiences cooler than normal winters. Winters are generally warmer than normal in the northern half of the USA. During El Niño years, there are fewer hurricanes in the Atlantic. El Niño is also associated with extremely wet conditions in eastern equatorial Africa and drought in southern Africa.

The global impacts of an El Niño can be quite different for various episodes, depending on the episodes' characteristics. Recently, a second type of El Niño has been recognized. Termed the El Niño Modoki, it is characterized not by warming in the eastern Pacific, but instead in the central Pacific. Abnormally cold temperatures prevail in the western and eastern Pacific.

El Niño is frequently associated with warming in the Atlantic and Indian Oceans. Consequently, it is sometimes difficult to ascertain whether the impacts of El Niño are direct or linked to an associated phenomenon. For example, many of the impacts of Pacific decadal variability or the Indian Ocean zonal mode are similar to those of El Niño.

Until the 1997–1998 El Niño, the 1982–1983 episode had been the strongest and most devastating of the twentieth century. During that period the trade winds not only collapsed, they also reversed and caused weather-related disasters on nearly every continent. Australia, Africa and Indonesia suffered droughts, while Peru received the heaviest rainfall on record. The 1982–1983 El Niño was blamed for thousands of deaths and billions of dollars in damage to property and livelihoods worldwide. The damage from the 1997–1998 event was even greater, possibly totally over $30 billion.   MLH/SEN

Reading
Ashok, K., Behera, S.K., Rao, S.A., *et al.* (2007) El Niño Modoki and its possible teleconnection. *Journal of Geophysical Research: Oceans*, **112**, C11007. · Battisti, D.S. and Sarachik, E.S. (1995) Understanding and predicting ENSO. *Reviews of Geophysics*, **33**, 1367–1376. · Enfield, D.B. (1989) El Niño, past and present. *Review of Geophysics*, **27**, 159–187. · Rasmusson, E.M. and Carpenter, T.H. (1982) Variations in tropical sea-surface temperature and surface wind fields associated with the Southern Oscillation/El Niño. *Monthly Weather Review*, **110**, 354–384. · Wang, H., Kumar, A., Wang, W., *et al.* (2012) US summer precipitation and temperature patterns following the peak phase of El Niño. *Journal of Climate*, **25**, 7204–7215.

eluviation The downwards transportation by percolating water of certain soil materials from upper layers of the soil towards lower layers, where they may be set down. Both materials in solution and fine solid particles may undergo eluviation. (The corresponding term for the arrival of such materials and their accumulation in deeper layers is ILLUVIATION.) Also used to refer to the lateral transport of soil materials that takes place in seepage water moving down the topographic slope (*lateral eluviation*). Evidence includes loss of fine clays and soluble materials from the upper soil horizons (though this can be explained in other ways) and in the presence of accumulations of these materials in the subsoil (see HORIZON, SOIL). Depositional clay skins (CUTANS) surrounding larger grains, or lining the walls of void spaces, and more abundant in the lower soil than in the upper parts, provide one line of evidence for the eluviation of materials.   DLD

Reading
Brady, N.C. and Weil, R.R. (2008) *The nature and properties of soils*, 14th edition. Englewood Cliffs, NJ: Prentice-Hall.

eluvium The material that is produced through the rotting and weathering of rock in one place. In-situ weathered bedrock.

Emerson test A test that determines the stability of soil aggregates when in contact with water. Small air-dry soil aggregates, 3–5 mm in diameter, are placed in beakers of distilled water and their condition observed for up to 12 h (Emerson, 1967). The sample may slake (slowly or rapidly disintegrate into separate grains that lie on the bottom of the beaker, but leaving the water clear), partially disperse (indicated by a turbid cloud of fine particles surrounding the test aggregate) or remain stable. The rate and extent of these processes are scored, and aggregates that remain stable can be subjected to mechanical stirring to inspect for dispersion, such as might result from raindrop impact or soil tillage.   DLD

Reference
Emerson, W.W. (1967) A classification of soil aggregates based on their coherence in water. *Australian Journal of Soil Research*, **5**, 47–57.

emissions scenario Plausible representation of the future development of emissions of substances that are potentially radiatively active (e.g. greenhouse gases, aerosols) based on assumptions about driving forces (such as demographic and socioeconomic development, technological change) and their key relationships. Concentration scenarios, derived from emission scenarios, are used as input to a climate model to compute climate projections. The various assessment reports of the Intergovernmental Panel on Climate Change (IPCC) have evolved a series of emissions scenarios. The IS92 scenarios were used as a basis for the climate projections in the IPCC Second Assessment Report (1996). The IPCC produced the Special Report on Emissions Scenarios(Nakićenović and Swart, 2000) that were used, among others, as a basis for the climate projections presented in the Third Assessment Report (2001) and the Fourth Assessment Report (2007). New emission scenarios for climate change, the four Representative Concentration Pathways, were developed for, but independently of, the Fifth IPCC Assessment Report (2014).   TS

References
Meinshausen, M., Smith, S., Calvin, K., *et al.* (2011) The RCP greenhouse gas concentrations and their extensions

from 1765 to 2300. *Climatic Change*, **109**, 213–241. · Nakićenović, N. and Swart, R. (eds) (2000) *Emissions scenarios*. Special Report, Intergovernmental Panel on Climate Change. Cambridge: Cambridge University Press.

endangered species One of a group of terms used to describe the status of wildlife. The following definitions are those employed by the Species Survival Commission of the World Conservation Union (IUCN) and are accepted for use by international bodies such as the Convention on Trade in Endangered Species of Flora and Fauna.

extinct:	species not definitely located in the wild during the past 50 years;
endangered:	species in danger of extinction, whose survival is unlikely if the causal factors continue operating;
vulnerable:	species believed likely to move into the 'endangered' category in the near future if the causal factors continue operating;
rare:	species with small world populations that are not at present 'endangered' or 'vulnerable';
indeterminate:	species known to be 'endangered', 'vulnerable' or 'rare', but not enough information is available to determine which category is appropriate;
insufficiency known:	species that are suspected of belonging to one of the above categories but are not definitely known to be, due to lack of information;
threatened:	a general term used to denote species that are in any of the above categories.

ASG

endemism The confinement of plant and animal distributions to one particular continent, country or natural region. Thus, the white spruce (*Picea glauca*) is endemic to North America, the coast redwood (*Sequoia sempervirens*) to coastal California and southern Oregon and the genus *Dasynotus* to a few square kilometres of Idaho. Organisms confined to one small island, one mountain range or just a few restricted localities are termed 'local' or 'narrow' endemics, whereas those with more substantial distributions are called 'broad' endemics.

The number of endemics in the northern hemisphere is lower than that in the southern hemisphere, although their count is reduced in both hemispheres where the lands were occupied by the cap of the continental glacier during the Pleistocene. Endemism is most marked on islands, such as Darwin's famous Galápagos Islands, particularly in the warmer regions of the world. No less than 90% of the native plants of the Hawaiian Islands are endemic to the island group. Endemism is also marked on isolated mountain tops, especially in the tropics. There is usually a close relationship between the number and type of endemics and the geological age of the habitats they occupy. Naturally, the study of endemism is one of the main ways of characterizing the different faunal and floristic regions of the world. (See also ISLAND BIOGEOGRAPHY, REFUGIA and VICARIANCE BIOGEOGRAPHY.) PAS

Reading
Richardson, I.B.K. (1978) Endemic taxa and the taxonomist. In H. E. Street (ed.), *Essays in plant taxonomy*. London: Academic Press; pp. 245–262. · Stott, P. (1981) *Historical plant geography*. London: Allen & Unwin. · Williamson, M. (1981) *Island populations*. Oxford: Oxford University Press.

endogenetic Pertaining to the forces of tectonic uplift and disruption originating within the Earth and to the landforms produced by such processes (see EXOGENETIC).

endoreic Drainage systems that do not reach the oceans, terminating in inland locations. They are therefore a feature of some DRYLANDS, where, due to insufficient flow from wetter (allogenic) headwater areas and the excess of EVAPOTRANSPIRATION over local desert (endogenic) precipitation, rivers may terminate before draining to the coast. De Martonne and Aufrere (1928) classified the world's drylands on the basis of endoreism and areicism (no drainage at all).

The Okavango in central southern Africa is an example of such a river, terminating in an extensive inland delta or fan. Tectonic and structural factors may also contribute to endoreism. The Okavango is the last remaining major river from a more extensive endoreic system that developed with the division of Gondwanaland and the creation of a tectonic rim (or hingeline) around southern Africa. The endoreic system has now largely been captured by more aggressive allogenic rivers that have cut back through the hingeline and captured endoreic channels (Thomas and Shaw, 1988). DSGT

References
De Martonne, E. and Aufrere, L. (1928) Map of internal basin drainage. *Geographical Review*, 17, 414. · Thomas, D.S.G. and Shaw, P. (1988) Late Cainozoic drainage evolution in the Zambesi basin: geomorphological evidence from the Kalahari rim. *Journal of African Earth Sciences*, 7, 611–618.

energy flow May be loosely defined as the energy transformations that occur within the planet Earth system. Energy is the capacity to do work and, for present purposes, includes mechanical, chemical, radiant and heat energy. Mechanical energy may be further subdivided into kinetic and potential energy. Kinetic energy, or free 'useful' energy, is possessed by a body by virtue of motion and is measured by the amount of work required to bring the body to rest. Potential energy is stored and becomes useful only when converted into the free form and can do work. This includes movement, friction and the expenditure of heat. On Earth, energy sources are solar radiation, rotational energy of the solar system and radiogenic heat involving geothermal heat flow. It is no exaggeration to claim that physical geography is the science of energy flow from these sources through the atmospheric, oceanic, aeolian, fluvial (hydrological cycle), glacial, biological, human, tectonic and geothermal systems as described by the laws of thermodynamics and as defined by complex circulation patterns and thermal gradients. The main circulations are atmospheric circulation, heat and moisture balance, photosynthesis and ecological energetics, the hydrological cycle and tectonics. The important gradients are latitudinal, altitudinal, seasonal, daily, heat gradients and across-system contrasts (e.g. land–sea). DB

Reading
Bloom, A.L. (1969) *The surface of the Earth*. Englewood Cliffs, NJ: Prentice-Hall. · Caine, N. (1976) A uniform measure of sub aerial erosion. *Bulletin of the Geological Society of America*, **87**, 137–140. · Chapman, D.S. and Mach, A.N. (1975) Global heat flow: a new look. *Earth and Planetary Science Letters*, **28**, 23–32.

energy grade line Water flowing down a channel loses energy because of the work done in overcoming FRICTION, both internally and with the channel perimeter, and in transporting sediment. The rate of energy loss per unit length of channel is measured by the energy gradient, which is the slope of the energy grade line. This line may be plotted above the water surface at a distance equal to the velocity head ($v^2/2g$), and therefore measures the variation of the total energy (potential and kinetic) of the flow (see GRADUALLY VARIED FLOW). In UNIFORM STEADY FLOW the energy grade line, water surface and channel bed are parallel. In gradually varied flow, the three slopes differ but the energy grade line always slopes downwards in the direction of flow. KSR

engineering geomorphology 'The application of geomorphic theory and techniques to solve earth-related engineering problems' (Giardino and Marston, 1999: 1). It is one type of applied geomorphology and is very important for understanding and mitigating geomorphological hazards, including those triggered by human activities. The contributions that geomorphologists can make to understanding and reducing the impact of geomorphic hazards include (Goudie and Viles, 2010): the mapping of hazard-prone areas; constructing the history of past hazardous events; establishing their frequency and magnitude; predicting the occurrence and location of future events (such as slope failures); monitoring geomorphological change; and using knowledge of geomorphological processes to advise on appropriate mitigation strategies. A good review of the role of geomorphologists in environmental management is provided by Cooke and Doornkamp (1990). ASG

References
Cooke, R.U. and Doornkamp, J.C. (1990) *Geomorphology in Environmental Management*, 2nd edition. Oxford: Oxford University Press. · Giardino, J.R. and Marston, R.A. (1999) Engineering geomorphology: an overview of changing the face of the Earth. *Geomorphology*, **31**, 1–11. · Goudie, A.S. and Viles, H.A. (2010) *Landscapes and geomorphology*. Oxford: Oxford University Press.

englacial Describes conditions within the body of a glacier and is therefore to be distinguished from the *subglacial* environment that is beneath a glacier, the *supraglacial* environment on the glacier surface and the *proglacial* environment in front of the glacier margin.

enhanced thematic mapper See THEMATIC MAPPER.

ensemble climate modelling An atmospheric climate state that is dictated by the changes in the boundary conditions (e.g. sea-surface temperature, soil moisture, snow cover, greenhouse gas concentrations) is considered to be far more predictable than an atmospheric climate state that is sensitive to the initial conditions. This is unlike atmospheric weather, where predictability is significantly dependent on the initial conditions. In other words, it is possible that a given climate state could be simulated or predicted by a numerical model far beyond the initial time and the 5–7-day predictability limit of synoptic weather systems (Lorenz, 1963; Arpe *et al.*, 1985; Shukla, 1998). Therefore, in order to diagnose the predictable component of the atmospheric climate simulated or predicted by a numerical model, it is necessary to conduct several realizations of the model integrations with slightly different (perturbed) initial conditions from one another. The collection of these model integrations is called ensemble climate modelling, which is practised rigidly for seasonal climate prediction and simulations. In the last decade or more, there is the advent of using multiple climate models in addition to using multiple initial conditions to obtain quantitative measures of model uncertainty in addition to initial condition uncertainty in the simulated or predicted climate (Palmer *et al.*, 2004; Kirtman *et al.*, 2014). Such a collection of multiple model integrations is also sometimes referred to as ensemble climate modelling.   VM

References
Arpe, K., Hollingsworth, A., Tracton, M.S., *et al.* (1985) The response of numerical weather prediction systems to FGGE level IIb data. II: forecast verifications and implications for predictability. *Quarterly Journal of the Royal Meteorological Society*, **111**, 67–101. · Lorenz, E.N. (1963) Deterministic non-periodic flow. *Journal of Atmospheric Science*, **20**, 130–141. · Kirtman, B., Min, D, Infanti, J.M., *et al.* (2014) The North American multimodel ensemble: phase-1 seasonal to interannual prediction; phase-2 toward developing intraseasonal prediction. *Bulletin of the American Meteorological Society*, **95**, 585–601. · Palmer, T.N., Doblas-Reyes, F.J., Hagedorn, R., *et al.* (2004) Development of a European multimodel ensemble system for seasonal-to-interannual prediction (DEMETER). *Bulletin of the American Meteorological Society*, **85**, 853–872. · Shukla, J. (1998) Predictability in the midst of chaos: a scientific basis for climate forecasting. *Science*, **282**, 728–731.

ENSO An abbreviation for the EL NIÑO– SOUTHERN OSCILLATION phenomenon. This phenomenon is a natural variation in the ocean– atmosphere system that occurs every 3–7 years. Though it takes place in the tropical south Pacific, ENSO affects the weather worldwide, and next to the seasons themselves it is the most important variation in the global climate system.

Under 'normal' conditions, atmospheric pressure is higher in the western equatorial Pacific than in the eastern equatorial Pacific; as a result, surface flow is from east to west. This easterly flow piles up warm surface water in the western Pacific, where sea-surface heights are about 0.5 m higher than in the east. The easterly flow also causes an upwelling of cold water along the coast of South America, and thus sea-surface temperatures there are about 8 °C cooler than in the west. There is also a tendency for rising motion in the western equatorial Pacific, where the pressure is low, and subsidence in the eastern equatorial Pacific, where the pressure is high.

The Southern Oscillation is a shift in this east– west pressure pattern. This gradient is quantified by the Southern Oscillation index, generally calculated as a normalized pressure difference between Darwin (Australia) and Tahiti. When the pressure gradient weakens (low-index years), the subsidence in the east is weaker, as is the ascent in the west. This redistributes the rainfall, with generally lower than normal rainfall occurring in the western sector and higher than normal rainfall in the eastern sector. The opposite occurs when the gradient intensifies (high-index years).

The high-index events are generally coupled with LA NIÑA episodes and the low-index events with El Niño episodes. This association is what led to the term ENSO. However, El Niño and the Southern Oscillation can be independent phenomena.

A typical ENSO event evolves over the course of an 18-month period. Atmospheric 'mass' seesaws from the eastern to the western Pacific, and the easterly trade winds weaken and recede eastward. This recession reduces the coastal upwelling and allows the warm waters amassed in the western Pacific to 'slosh' back towards the east. Precipitation follows the warm water eastward, and the typically arid regions of coastal South America receive heavy rainfall. Indonesia and Australia, in contrast, experience drought. The eastward displacement of the warmest water also has TELECONNECTIONS outside the tropical Pacific; drought in northeastern Brazil and southeastern Africa, for example, usually accompany ENSO, while the western USA experiences wet winters.   RSW/SEN

entrainment The process by which surface sediment is incorporated into a fluid flow (such as air or water) as part of the operation of EROSION. Sediment is entrained into a flow when forces acting to move a stationary particle overcome the forces resisting movement.

In water and air, sediment is subjected to three major forces of erosion: lift, surface drag and form drag. Lift is a result of the fluid flowing directly over the particle forming a region of low pressure (the Bernoulli effect), in contrast to the relatively high pressure beneath the particle. The particle then gets 'sucked' into the flow. Surface drag is the shear stress (see SHEAR STRESSES) on the particle provided by the flow velocity, and the form drag relates to the pressure differences upstream and downstream of the particle. When these movement forces overcome particle cohesion, packing and weight, the particle tends to shake in position and then lift off, spinning into the flow. The velocity required to initiate this entrainment is termed the critical erosion velocity (see CRITICAL VELOCITY (OR CRITICAL EROSION VELOCITY)) and is primarily a function of mean particle size.

Once entrained into the fluid, sediment may move in a number of ways. Coarse particles tend to slide or roll along the bed, whilst smaller particles may undergo SALTATION. Very fine particles (silt or clay) tend to be transported by suspension, where the turbulent flow structures within the fluid carry the particles without impacting the surface.

In GLACIAL environments sediment may become entrained into moving ice by falling onto the ice surface from above or by joint block removal, which involves the 'plucking' of joint-separated rock by glacial ice.   GFSW

Reading

Thorne, C.R., Bathurst, J.C. and Hey, R.D. (eds) (1987) *Sediment transport in gravel-bed rivers*. Chichester: John Wiley & Sons, Ltd. · Wiggs, G.F.S. (2011) Sediment mobilisation by the wind. In D. S. G. Thomas (ed.), *Arid zone geomorphology: process, form and change in drylands*, 3rd edition. Chichester: John Wiley & Sons, Ltd; pp. 456–486.

An entrenched meander: Horseshoe Bend, Arizona.
Photograph by David Thomas.

entrenched meander   A meander or bend in

a river channel that has become incised into the surrounding landscape, often as a result of local tectonic uplift (see MEANDERING).

entropy   A concept in thermodynamics that

describes the quantity of heat supplied at a given temperature. The entropy of a unit mass of substance remains constant in the process of adiabatic expansion; that is, when no additional heat is supplied to the mass. The term has been transferred by analogy to other fields (notably information theory) and introduced into geomorphology by Chorley (1962) and Leopold and Langbein (1963) as a measure of energy unavailable to perform work (high entropy equals greater unavailability of energy). This analogue use has been widely criticized as inappropriate.   BAK

References

Chorley, R.J. (1962) *Geomorphology and general systems theory*. United States Geological Survey Professional Paper 500-B. Washington, DC: US Government Printing Office. · Leopold, L.B. and Langbein, W.B. (1963) *The concept of entropy in landscape evolution*. United States Geological Survey Professional Paper 500-A. Washington, DC: US Government Printing Office.

envelope curve   A curve drawn on an X–Y

scatterplot that delimits the part of the graph within which points representing a particular phenomenon occur. When the relationship between the two variables involves a great deal of scatter, techniques like the fitting of a least-squares regression line become inappropriate. It is under these conditions that an envelope curve may prove to be a useful analytical tool. The figure shows an example in which the sizes of floods are plotted in relation to the river catchment area. The envelope curve delimits the largest floods occurring for any basin area, and its form suggests that there is an approach toward an upper limit on flood sizes for even the largest basins.   DLD

Reference

Enzel, Y., Ely, L.L., House, P.K., *et al.* (1993) Paleoflood evidence for a natural upper bound to flood magnitudes in the Colorado River basin. *Water Resources Research*, **29**, 2287–2297.

environmental assessment   An objective

attempt to evaluate the quality of the environment in terms of biophysical attributes and/or aesthetic value. It is used to give the natural environment credibility and comparability with socioeconomic data in planning decisions. The concept of environmental assessment has been important since the 1960s with the gradual recognition that problems such as loss of habitat and genetic variety, pollution, population growth and an increasing reliance on nonrenewable energy and mineral resources cannot be solved by economic growth or technology and that finite limits exist within the environment. Increasingly, factors such as environmental beauty and the range of noneconomic values that can be ascribed to environments are included in assessments.   SMP

environmental economics   The aim to con-

serve, maintain, use and reuse natural resources so that the quality of life is retained without excessive waste. It recognizes that the major cause of environmental problems in market systems is failure of the incentives generated by these markets to lead towards efficient use of resources.

At the beginning of the twentieth century Alfred Marshall suggested that the basis of orthodox economics – the market – presented an oversimplified idea of reality and introduced the concept of external costs and benefits. These describe instances where one fiscally independent economic unit directly affects another, without intervention of the market. On the cost side, the instance of a locomotive igniting an adjacent field is commonly quoted; a good

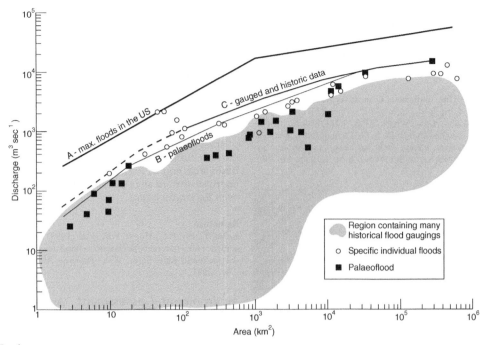

Envelope curve.
Source: Enzel *et al.* (1993).

example of a beneficial externality is that of bees pollinating an orchard owner's apples.

The 1960s and 1970s witnessed the rapid development of these ideas owing to an increasing awareness of environmental problems, particularly pollution, and the notion that both free enterprise and Marxist economics encourage the squandering of natural resources. A profound asymmetry had developed in the effectiveness and efficiency of the market system that works well in stimulating the exploitation, processing and distribution of basic resources, but almost completely fails in the efficient disposal of residuals to common property assets.

The natural environment has clearly been largely ignored in the conventional economic account, and commonly the Earth is regarded as a bottomless rubbish dump. Boulding (1966, 1970) describes this as the 'cowboy economy', in which success is measured in terms of the amount of material turned over by the factors of production. He compares this with the 'spaceship economy', in which maintenance of existing capital stocks within limits is the criterion for success.

The practice of economics is certainly still the single most important feature governing the relationship between humans and the environment. However, environmental economics acknowledges

its own limitations for regulating current human resource requirements. Price cannot always be equated with value; for instance, unspoilt countryside may have intrinsic value but cannot have an accurate price fixed upon it. The interest of the individual in the market is often not the same as the general interest (Hardin, 1968); indeed, some resources (e.g. clean air) are not within the market and so are not subject to the choices normally available. Furthermore, market economies are not well suited to respond to problems that suddenly force themselves upon resource managers or the public (e.g. disease causing a harvest to collapse) nor to cope with the long time lags within which complex technology needs to develop substitutes.

The ultimate goal of environmental economics is to reach the steady state with little or no economic growth in the industrialized nations (Boulding, 1966). In so doing it attempts to investigate that which now governs price and supply of materials and, in particular, the role of energy, the economic relations of rich and poor countries and the construction of new measures of human welfare in terms of the whole resource process, rather than just a portion of it, which has for so long been the sole concern of economists. SMP

Reading and References

Boulding, K.E. (1966) Ecology and economics. In F. Fraser Darling and J. P. Milton (eds), *Future environments of North America*. New York: Natural History Press. · Boulding, K.E. (1970) The economics of the coming spaceship Earth. In H. Jarrett (ed.), *Environmental quality in a growing economy*. Baltimore, MD: Johns Hopkins University Press. · Cottrell, A. (1978) *Environmental economics*. London/New York: Edward Arnold/Halsted Press. · Hardin, G. (1968) The tragedy of the commons. *Science*, **162**, 1243–1248. · Knese, A.V. (1977) *Economics and the environment*. London: Penguin. · Marshall, A. (1930) *Principles of economics*, 8th edition. London: Macmillan.

environmental impact

A net change, either positive or negative, in human health and well-being and in the stability of the ecosystem on which human survival depends. The change may result from an accidental or a planned action and can affect the change in balance either directly or indirectly.

Direct impacts are generally premeditated and planned and are commonly felt soon after environmental modification. The effects are often long term, but normally reversible, and include alterations such as land-use changes, various constructional and excavational activities, the direct ecological impact of agricultural practices and the direct effects of weather modification programmes.

In contrast, indirect effects are normally unplanned and are often socially, if not economically, undesirable. Effects are often delayed until well after the original impact and depend upon the sensitivity of the system to change, the existence of threshold conditions, and interaction between different side effects of the initial impact. Many impacts are long term, cumulative and irreversible, difficult to identify and almost impossible to predict, and include the introduction of DDT and other toxic elements into the environment and the subsequent accumulation of those into food chains over a wide area, triggering long-term and possibly long-range climatic modifications by particulate and gaseous pollution and indirect local climatic effects associated with changed land surface configuration or material composition.

Many impacts are caused by pressures related to the rapid increase in population growth, especially on physical resources – land, food, water, forests, metals – and on the biological environment, whose ability to remove and recycle human waste and provide an important set of functions, such as pest control or fish production, is being severely strained. In addition, there are pressures on society's ability to dispense services – education, medicine and law or administration – and in personal values such as privacy, freedom from restrictive regulations and the opportunity to chase a lifestyle.

It is not only population growth that causes these pressures. The consumption of materials and energy per person have grown simultaneously, linked with the type of technology that is being employed to facilitate consumption, and the economic, political and social forces influencing decision-making are contributory factors. The ability of both individuals and governments to react has not kept pace.

The environmental impact may be the final stage of a basic resource process, whereby humankind takes a resource from the environment, uses it in some fashion and then returns it to the environment. Human actions, whether they be legislative proposals, policies, programmes or operational procedures, may well set in motion or accelerate environmental effects – which are, for example, dispersal of pollutants, soil erosion or displacement of persons – unless some form of preventative management is initiated. If management at this stage is ineffectual or avoided, however, an environmental impact will probably occur, and any subsequent management may be too late, too expensive or merely palliative. Unfortunately, most knowledge is still concerned with what is put into the system rather than with how the environment responds to it, and it is this response that defines the nature and magnitude of the environmental impact.

If, at a later stage, humans complete their action, the response may be a lessening of the environmental impact (Figure 1) owing to natural homeostasis in the system. A good example is a colliery spoil heap that gradually revegetates sometime after mining has ceased. It is rare, however, for the new equilibrium even to approach the original environmental quality. There are two main approaches to the assessment of environmental impacts. In the first the problem of resources receives particular attention, and some of the more subjective ecological and aesthetic aspects of the environment, although recognized as important, are not included. One attempt (Ehrlich *et al.*, 1977) tried to relate the problem in terms of population and consumption:

environmental impact = population ×

consumption of goods per person ×

environmental impact per quantity of goods

consumed

or

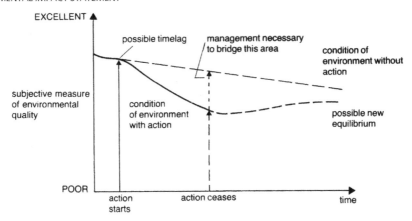

Figure 1 Environmental impact: changing environmental impact resulting from human activity.

$$\text{environmental impact} = \text{population} \times$$

$$\text{affluence} \times$$

$$\text{technology}$$

However, most forms of consumption give rise to many forms of environmental impact; changes in technology might reduce some impacts but increase others, and different impacts associated with alternative technologies (e.g. oil spills and coal mining) are difficult to compare, with the result that the above measure is almost impossible to quantify.

The second approach considers the probable consequences of human intervention in the natural environment with the goal of minimizing environmental damage, while developing resources. Known as an environmental impact assessment (EIA), it was developed by ecologists and economists in the USA and became a legal requirement under S102(2)(C) of the Natural Environmental Policy Act 1969 for all major federal actions affecting the human environment.

In Europe, an EIA is used early in existing planning procedure and does not replace it. It identifies adverse effects, suggests alternatives to proposals and considers short-term local environmental use in relation to long-term productivity. It imposes limitations, therefore, on the magnitude and impact of resource processes.

The flowchart (Figure 2) outlines a general impact evaluation methodology. The critical part is stage 5 – preliminary appraisal – as the value of any assessment of impact will depend heavily on the thoroughness with which the appraisal work is undertaken. Both the specific characteristics and the potential interactions between the existing local situation and that in the development

proposal are evaluated and are normally identified using an impact matrix, which acts as a framework and checklist.

Most commonly used is the Leopold matrix, which lists all possible environmental, social and economic parameters and attempts to evaluate the importance and magnitude of each impact identified. Others include the component interaction matrix, normally used for biological, physical and climatic dependencies, the disruption matrix, which measures environmental disturbance for each alternative proposal, and the Sorenson matrix, a three-stage computer analysis that may also suggest corrective action or control mechanisms.

Baseline studies (stage 6) are used to provide information for the matrix, and both are considered to appraise the potential impacts, and produce an ENVIRONMENTAL IMPACT STATEMENT. SMP

Reading and Reference
Clark, B. (1976) Evaluating environmental impacts. In T. O'Riordan and R. Hey (eds), *Environmental impact assessment*. Farnborough: Saxon House. · Ehrlich, P.R., Ehrlich, A.H. and Holdren, J.P. (1977) *Ecoscience*. San Francisco, CA: W.H. Freeman. · O'Riordan, T. (1981) *Environmentalism*, 2nd edition. London: Pion. · Glasson, J., Therivel, R. and Chadwick, A. (2005) *Introduction to environmental impact assessment*. Routledge: Abingdon. · Simmons, I.G. (1981) *The ecology of natural resources*, 2nd edition. London: Edward Arnold.

environmental impact statement The summary of all the information gathered on each potential ENVIRONMENTAL IMPACT that might be realized for a given development proposal. It is an integral part of the procedure of environmental impact analysis.

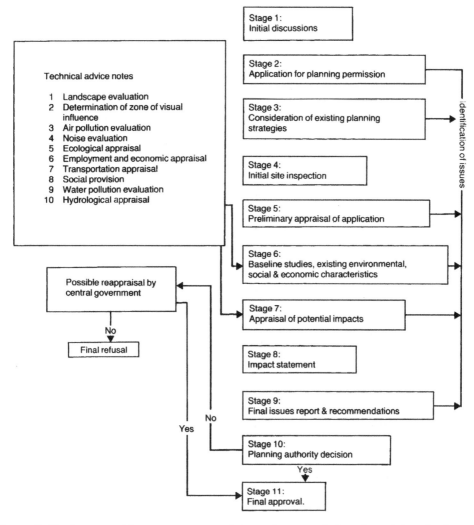

Figure 2 Environmental impact: an approach to project appraisal under existing development control procedures. *Source:* Clark (1976).

The environmental impact statement (EIS) discusses as succinctly as possible (1) a brief description of a proposed action, (2) the likely impact of the proposed action on the environment, (3) any predicted adverse or beneficial effects as a result of the proposal, (4) whether the impacts are likely to be long term or short term, (5) whether the impacts will be reversible or irreversible, (6) the range of direct and indirect impacts associated with the proposed action and (7) whether the impacts are likely to be of local and/or national strategic significance. In addition, the prospects for the area

under consideration if the development does not take place are outlined so that decision-makers can compare the potential effects of approving the application with the implications of the 'no change' alternative.

Most EISs do not need nonenvironmental questions (such as economic impact) to be included, although many agencies voluntarily include such information. The aim of the EIS is to identify and develop methods and procedures and use empirical information to ensure that unquantified environmental amenities and values are given appropriate consideration in decision-making along with economic and technical considerations. An EIS may, therefore, include rankings or hierarchies to differentiate the degree of importance or magnitude of impacts, although this procedure can lead to a false sense of objectivity. SMP

Reading
Glasson, J., Therivel, R. and Chadwick, A. (2005) *Introduction to environmental impact assessment*. Routledge: Abingdon. · Wathern, P. (ed.) (1990) *Environmental impact assessment: theory and practice*. London: Routledge.

environmental issue A concern that has arisen as a result of the human impact on the natural environment and/or the ways in which the natural environment affects human society. Such issues include DEFORESTATION, DESERTIFICATION, SOIL EROSION, POLLUTION, CLIMATE CHANGE, the OZONE HOLE and ACID RAIN. Research into, and teaching of, environmental issues is not the domain of the physical geographer alone since the word 'issue' inherently implies that human values play a key role in defining the subject matter (Middleton, 2013); consequently, some issues (e.g. the carbon dioxide problem) are highly controversial. A physical geographer may approach these applied issues from a pure stance, bringing a knowledge of physical environmental processes and their rates of change, while a human geographer's approach may involve perceptions, cultural contexts and political viewpoints. Both approaches are important. A full understanding of desertification, for example, requires both an appreciation of the physical processes operating in the dryland environment and a grasp of the social, economic and political factors that affect the everyday lives of desert inhabitants (Thomas and Middleton, 1994). In the study of soil erosion it has become clear that we are well aware of how soil erosion works as a set of processes, and physical scientists have developed many methods for combating the soil erosion problem. Nevertheless, soil erosion problems still plague many world regions, suggesting that the emphasis on the technical is not sufficient. The answers to the question 'why is soil erosion still a problem?' lie in an appreciation of the way local inhabitants view the problem (some may not view it as such at all) and the factors that force or encourage people to abuse their soil resource (Blaikie, 1985). Hence, a full appreciation of any environmental issue must consider it as a set of physical processes viewed in a social context.

This understanding of what makes an issue an issue reflects the need for a holistic approach to the subject matter, and environmental issues have been advocated as a rallying point for physical geography to offset the subject's tendency towards fragmentation through the increasing development of specialisms (e.g. Stoddart, 1987). According to this view, physical geography should focus on both the way the natural environment works and how this knowledge can be used to improve our understanding of human–environment relations, including the synergy that occurs between the two sets of forces, to concentrate on applied problems. Some see this role of applied environmental management as requiring a broader base for physical geography, involving further integration with the social science aspects of geography as a whole, although a division between such 'environmentalists' and the pure school of 'reductionist' physical geography continues to be apparent (Newson, 1992). NJM

References
Blaikie, P. (1985) *The political economy of soil erosion*. London: Longman. · Middleton, N.J. (2013) *The global casino: an introduction to environmental issues*, 5th edition. London: Arnold. · Newson, M. (1992) Twenty years of systematic physical geography: issues for a 'New Environmental Age'. *Progress in Physical Geography*, **16**, 209–221. · Stoddart, D.R. (1987) To claim the high ground: geography for the end of the century. *Transactions of the Institute of British Geographers, New Series*, **12**, 226–327. · Thomas, D.S.G. and Middleton, N.J. (1994) *Desertification: exploding the myth*. Chichester: John Wiley & Sons, Ltd.

environmental lapse rate See LAPSE RATE.

environmental magnetism Thompson and Oldfield (1986), who produced an accessible synthesis of magnetic theory, measuring techniques and the applications of magnetic measurements in a wide range of environmental contexts, first coined the term *environmental magnetism*. Earlier studies of the natural magnetic properties of crustal rocks showed how certain rock minerals enabled the recording of the Earth's palaeomagnetic field, which eventually led to the confirmation of plate tectonics as the mechanism of continental drift (Vine and Mathews, 1963). And the French geophysicist Le Borgne (1955) carried out simple magnetic studies of soil and deduced that strongly magnetic minerals formed as a result of pedogenic processes and fire. These studies subsequently triggered many applications of magnetic measurements in geomorphology, pedology, climate change and pollution studies. The dichotomy of palaeomagnetism and mineral magnetism, though far from being mutually exclusive, still represents the main magnetic framework today. Both divisions exploit the magnetic properties of iron-bearing minerals, particularly those that exhibit ferrimagnetic (e.g. magnetite) and imperfect antiferromagnetic (e.g. haematite) behaviour. These minerals retain a measurable magnetization after a magnetic field has been removed, termed remanent magnetization.

Natural remanent magnetization (NRM) is induced by the Earth's magnetic field as the minerals cool in igneous rocks from molten lava, precipitate from solution as structured crystals or physically align with the field in the surface of lake and marine sediments. The NRM of a sample can be analysed to reconstruct the original inclination (vertical) and declination (horizontal) angle components, and often the intensity of the magnetic field. When viewed over geological time-scales, the data provide a magnetostratigraphic record showing periods when the polarity of the Earth's magnetic field was either the same as today's (normal) or when it was reversed. Comparison of the palaeomagnetic pole positions between rocks of different ages allows reconstruction of polar wander paths and the direction of continental drift. Over the past few million years, igneous rock records are well dated using potassium–argon ages, and the record shows three major polarity zones in the last 3.4 million years: normal Brunhes (0–0.78 Ma), reversed Matuyama (0.78–2.60 Ma), and normal Gauss (2.60–3.40 Ma), with a larger number of shorter but nevertheless worldwide polarity intervals. The polarity record provides the means for dating samples, correlating different sediment and rock sequences and examining the mechanisms of the Earth's interior. In sediments deposited over the past few thousand years, the polarity remains normal, but there are significant fluctuations in the magnetic field caused by secular variation in the position of the geomagnetic pole. Records of secular variation in recent lake sediment records compare well with the laboratory records of declination and inclination kept since the sixteenth century in Europe and USA, and are used as a sediment dating technique. The NRM preserved in fired artefacts such as tiles and hearths is often used to help date archaeological contexts. Recently scientists have even succeeded in matching the NRM of red paint pigments from museum paintings to historical records of secular variation!

In mineral magnetic studies, measurements such as remanent magnetization induced in the laboratory and magnetic susceptibility are used to identify iron-bearing minerals, the processes of their formation and their origins (see also MAGNETIC SUSCEPTIBILITY). Primary minerals, such as titano-magnetite, are found in igneous rocks and derived sediments, while secondary ferrimagnetic minerals (SFMs) are formed as a result of biogeochemical processes in soils and sediments, fire and

combustion of fossil fuels. The formation of SFMs in soil often leads to magnetic enhancement of surface or subsurface horizons, a phenomenon used in geomorphology to help trace the erosion and transport of surface soil from slopes to river. In some landscapes, the distribution of both primary and secondary minerals enables a variety of sources to be classified and their sediments to be traced through stream networks. Several studies have extended sediment-source tracing to lake and estuarine sediment records to enable reconstruction of the history of erosion and have shown clearly the long-term impact of human activities. In contrast, SFMs formed in thick loess deposits, such as those on the Chinese Loess Plateau, are sensitive to climatic conditions, particularly rainfall, and have been used to reconstruct glacial–interglacial sequences of climate for the region. These long terrestrial records parallel magnetic-based climate records from marine sediments – see Maher and Thompson (1999). In the North Atlantic, the presence of primary minerals in the sediments has been linked to the southward movement of icebergs and the release of minerogenic detritus following their melting. In warmer oceans, the magnetic records respond variously to the glacial–interglacial shifts in wind-blown dusts, biological productivity and the deposition of carbonate, and the presence of magnetite formed by bacteria. A major source of human magnetite is fossil fuel combustion, especially fly ash from coal. Magnetic measurements also help to discriminate between metallurgical dusts, rust particles and vehicle emissions. Measurements of recent lake sediments and peat profiles record the timing and trends of magnetite pollution from many countries. Magnetic properties of atmospheric dusts, road dusts and river sediments in urban areas help to identify sources of pollution, and in some studies have provided an estimation of pollution by heavy metals and polycyclic aromatic hydrocarbons. During the 1990s the subject of environmental magnetism expanded greatly. As a result, there is now a better appreciation of mineral dissolution, iron sulphide formation and magnetotactic bacteria, a stronger marriage between theory and measurement, and a more comprehensive use of both room-temperature and thermal magnetization and remanent properties.   JAD

Reading and References

Dearing, J.A. (1999) *Environmental magnetic susceptibility.* Kenilworth: Chi Publishing. · Evans, M.E. and Heller, F. (2003) *Environmental magnetism. Principles and applications of enviromagnetics.* San Diego, CA: Academic Press. · Le Borgne, E. (1955) Susceptibilité magnétique anormale du sol superficiel. *Annales de Geophysique,* **11**, 399–419. ·

Maher, B.A. and Thompson, R. (1999) *Quaternary climates and magnetism*. Cambridge: Cambridge University Press. · Thompson, R. and Oldfield, F. (1986) *Environmental magnetism*. London: Allen & Unwin. · Vine, F.J. and Mathews, D.H. (1963) Magnetic anomalies over ocean ridges. *Nature*, **199**, 947–949.

environmental management Provides resources from the bioenvironmental systems of the planet but simultaneously tries to retain sanative, life-supporting ecosystems. It is therefore an attempt to harmonize and balance the various enterprises that humans have imposed on natural environments for their own benefit. To achieve this, long-term strategies are evolved, based on reducing stress on ecosystems from contamination or overuse. In addition, environmental management pursues short-term strategies that are sufficiently flexible to preserve the long-term options; in other words, no resource process that brings about irreversible environmental changes should be allowed to develop.

This temporal scale is of major importance in environmental management; for example, engineering solutions may be necessary in the short term to check localized coastal erosion, but in the long term conservation of the entire coast with acceptance of slow erosion, accretion or movement may be required. Extended time perspectives may also have to accommodate extreme events.

Different approaches are similarly evident with the spatial element. Using the coastal example again, management may consider a single beach profile or a complete sedimentary cell involving supply of material from a river mouth, movement, storage in beaches and abstraction to dunes or marine deeps. In a predominantly agricultural area, the maintenance of a wild population of predatory birds for scientific interest or pest control is not possible with an isolated nature reserve, but requires a whole network of protected areas.

Environmental management, therefore, is dealing with the rationalization of the resource process – the flows of material (and energy) from natural states through a period of contact with humans to their ultimate disposal. It has much in common with environmental planning, and the two are sometimes used interchangeably. Strictly, planning approaches the natural resource problem with a cultural or demand bias, whereas in environmental management the emphasis is on the resources themselves. It is a dynamic discipline that does not urge the preservation of resources at all costs but attempts to identify or specify major groups of resources, consider the way each changes, evaluate and resolve conflicting demands upon them and finally conserve the resources. It believes that the very process by which renewable resources are produced can be manipulated, but that the production of nonrenewable resources is virtually unmanageable because of the timescale under which they develop.

Traditionally, environmental management has considered that, generally speaking, environmental problems need more adjustment to socioeconomic systems and has been more concerned with the maintenance of the ecological and geomorphological balance – for instance, the study of and control of movements of pollutants and pesticides in food chains, overtrampling of ecologically interesting swards – leading to the preparation of conservation strategies. However, it is becoming more socially aware as renewable resource systems become so thoroughly altered by humans that they can seldom be left to produce or even function in a stable condition without interdisciplinary management involving biological and physical, economic and political, and scientific and aesthetic approaches.

In terms of values, environmental management is rather ambivalent towards economic growth, recognizing that there is an absolute limit to materials and the surface area of the planet, but seeing little reason to prevent resource use unless ecological stability is threatened. In any natural ecosystem the overall limiting factor must be the amount of incident solar energy, but within this context other boundaries may operate. Humans may alleviate a critical unit (e.g. by using chemical fertilizers) or may introduce a new lower limit (e.g. untreated sewage in coastal waters may decrease the light reaching littoral and sublittoral vegetation, limit productivity and so reduce the recreational potential of the system).

Different societies have differing attitudes to environmental management determined by their own order of priorities. In the USA and Europe the dominant purpose of environmental management in the past has been to obtain useful materials, an emphasis that is decreasing in favour of more concern about the life-supporting role of the ecosystem and the aesthetic value of the environment given greater impacts upon it. In contrast, the struggle to obtain food has always dominated environment management in countries such as Egypt, where the gathering of useful materials has taken a secondary role and is largely regarded only as a basic development aimed at export markets, and the care of wildlife and aesthetic preservation of the environment is of peripheral interest with little value.

It is unrealistic to pretend that totally successful environmental management is currently much more than a concept, except perhaps in the relatively simple situation of Antarctica, where resource processes are readily identifiable and human intrusion is limited. Some process response reactions (e.g. the avoidance of flood hazard) might be regarded as environmental management at a local scale, but there are critical problems within the concept. One of the most important is the dualism between ecology and economics, which have completely different resource and time perspectives. It is also quite difficult in many cases to distinguish between changes that have been brought about exclusively by humans and those which are at least partly natural. A further difficulty is that problems of environmental management are not the same worldwide on account of differing attitudes of wealth and its distribution, sociopolitical systems, population growth rates, and the implementation of western 'developed' culture leading to rapid urbanization and industrialization that takes no account of natural environmental processes. Indeed, in many instances, environmental problems are caused by unbalanced or overrapid development rather than a complete disregard for environmental management. SMP

Reading
Hyde, P. and Reeve, P. (2011) *Essentials of environmental management*. London: IOSH Services. · Simmons, I.G. (1981) *The ecology of natural resources*, 2nd edition. London: Edward Arnold.

epeiric sea   A shallow body of marine water on the continental shelf that is connected with an ocean.

epeirogeny   The warping of large areas of the Earth's crust without significant deformation. It can be contrasted with OROGENY, which is associated with linear zones of uplift. Epeirogenic uplift can affect regions thousands of kilometres across and is the major form of uplift in most CRATONS. The causes of the predominantly vertical movements associated with epeirogeny are uncertain but may be related to expansion resulting from localized heating within the crust or at the base of the LITHOSPHERE, possibly in conjunction with phase changes in the MANTLE.   MAS

Reading
Crough, S.T. (1979) Hot spot epeirogeny. *Tectonophysics*, **61**, 321–333. · Ollier, C.D. (1981) *Tectonics and landforms*. London: Longman.

ephemeral plant   A plant, generally found in arid and semi-arid regions, in which the life cycle is completed within a very short period, perhaps only a few weeks. According to RAUNKIAER'S LIFE FORMS classification, these are therophytes. Seed germination in such species is triggered by a particular combination of environmental conditions, usually involving substantial rainfall and, as a consequence, although they are nominally 'annuals', they do not necessarily reappear annually. These species have a large reservoir of seeds capable of surviving many years in the dormant state before germination is triggered by moisture inputs. Seed germination and growth is then exceptionally rapid, followed by flowering and seed production. Because the life cycle is completed in such a brief period following rain, it is unusual for these species to have other adaptations to drought, although succulence is a feature of a significant proportion of ephemerals in the Namib Desert. The diversity of ephemeral species is especially great in the winter-rainfall desert regions of the world, such as in the western South American and southwestern African coastal arid zones. In the semi-arid parts of the southwestern Cape of South Africa, spectacular floral displays of ephemeral plants, normally following spring rains, regularly occur, in which there is mass flowering of plant species, many of which belong to the Asteraceae (daisy) family. Most of the plants are insect pollinated, and the annual display of brightly coloured flowers represents an important component of the ecological dynamics of these regions in supporting an impressive diversity of wasps, beetles and other pollinating insects whose populations are correspondingly ephemeral.   MEM

Reading
Inouye, R.S. (1991) Population biology of desert annual plants. In G. A. Polis (ed.), *The ecology of desert organisms*. Tucson, AZ: University of Arizona Press; pp. 25–54.

ephemeral stream   A stream that is often one of the outer links of a DRAINAGE NETWORK and which contains flowing water only during and immediately after a rainstorm that may be fairly intense. As the water flows along the ephemeral stream channel it may infiltrate into the channel bed as a transmission loss by influent seepage; therefore, the peak discharge may decrease downstream along the ephemeral channel by as much as 5% per kilometre of channel. In arid and semi-arid areas of the world ephemeral streams are very extensive and represent the major channel type.   KJG

Reading
Renard, K.G. and Laursen, E.M. (1975) Dynamic behaviour model of ephemeral streams. *Journal of the Hydraulic Division of the American Society of Civil Engineers*, **101**, 511–528. · Thornes, J.B. (1977) Channel changes in ephemeral streams: observations, problems and models. In K. J. Gregory (ed.), *River channel changes*. Chichester: John Wiley & Sons, Ltd; pp. 317–335.

The highly ephemeral Aoub River, in southern-central Namibia. This river may flow only once every decade or more, subject to heavy rainfall in its arid headwaters. Heavy rain in 2011 led to exceptional flows and to a verdant green landscape in this usually arid region. Photograph by David Thomas.

epicentre The point on the Earth's surface that lies directly above the focus of an earthquake.

epiclimate (or nanoclimate) The climate on the surface of leaves, in the air cavities in litter, along the slopes of an ant hill or in the fissures in rocks. Epiclimate extends vertically a few centimetres, or perhaps a decimetre, and extends horizontally for centimetres. Studies of epiclimate are important for ecophysiology and population ecology of very small organisms.   RCB

epilimnion The surface layer of water of a lake or sea. The water that lies between the surface and the thermocline.

epipedon A diagnostic surface horizon that includes the upper part of the soil that is darkened by organic matter, or the upper eluvial horizons, or both.

epiphyte A plant that grows upon the surface of another plant but does not obtain food from the host plant.

epoch A unit of geologic time equivalent to a series, a division of a period. See GEOLOGICAL TIMESCALE.

equation of state A relationship between properties of a material, and/or forces acting on it, in a state of equilibrium. The most familiar examples of equations of state in physical geography refer to the balance of forces acting on a body or within a material in equilibrium. Stability analysis for landslides is, for instance, based on such a balance. Equations of state may also refer to other properties, and the term has a special significance in thermodynamics as the unique relationship between temperature, pressure and volume for a body of fluid.   MJK

equations of motion Expressions governing the motion of a body or a material under the action of a force or forces. The equations most commonly take the form

$$\text{force} = \text{mass} \times \text{acceleration}$$

in either its linear form or as moments (torques) about a centre. Since force is a vector, the equation is a vector equation, and may be resolved to give up to three component equations in directions that are mutually at right angles. Equations of motion are one example of expressions that control the rate of a process, usually subject to the constraints of the CONTINUITY EQUATION. The term was originally used in the context of solid bodies, but may also be used to describe the motion of a fluid such as water or air, either travelling with a physical body of fluid or describing motion at a fixed point.   MJK

equatorial rain forest A lowland evergreen TROPICAL FOREST lying approximately 5° north and south of the equator in near-continuous rainfall climates, over 2000 mm a^{-1}, and not limited by low temperatures. The forests are multilayered, over 30 m tall, shallow rooted and often buttressed, containing a profusion of climbing plants and epiphytes and the greatest diversity and abundance of plants and animals of any terrestrial biome. Their BIOLOGICAL PRODUCTIVITY also heads the league for terrestrial biomes. Although the main global formations are comparable in structure, life forms and animal adaptations, the biological evolution and species composition is profoundly different in each area.   PAF

Reading
Golley, F.B., Lieth, H. and Werger, M.J.A. (eds) (1982) *Tropical rain forest ecosystems*. Amsterdam: Elsevier. · Longman, K.A. and Jenik, N. (1987) *Tropical forest and its environment*, 2nd edition. London: Longman. · Turner, I.M. (2004) *The ecology of trees in the tropical rain forest*. Cambridge: Cambridge University Press.

equatorial trough A narrow, fluctuating belt of unsteady, light, variable winds, low atmospheric pressure and frequent small-scale disturbances. It is located near the equator between the trade wind belts of the two hemispheres. However, its position, breadth and intensity are constantly changing. From time to time it disappears completely, especially over the continents. Along its meandering position occur most of the frequent, heavy showers for which the tropics are so well known. The equatorial trough includes the prevailing calms of the DOLDRUMS and is frequently referred to as the INTERTROPICAL CONVERGENCE ZONE (ITCZ).   WDS

Reading
Jury, M.R. (2010) Ethiopian decadal climate variability. *Theoretical and Applied Climatology*, **101**, 29–40.

equifinality Arises when a particular morphology (e.g. a landform) can be generated by a number of alternative processes, process assemblages or process histories. Under such circumstances the morphology alone cannot be used as a basis for reconstructing the process of origin of a feature. For example, a central assumption in CLIMATIC GEOMORPHOLOGY is that landforms differ significantly between climatic zones because of variation in the climatic factors that control weathering, run-off, erosion and deposition. However, specific landforms may originate in different ways and, therefore, are not restricted to single climate zones. U-shaped valleys are characteristic of glaciated highlands, but also occur in high-relief subtropical areas where basal sapping maintains steep valley sides after intensive chemical weathering at the water table (Wentworth, 1928). Tors are also features produced by quite distinct sets of processes in different areas; both deep chemical weathering with subsequent stripping of the weathered mantle and frost-shattering with mass movement generate similar morphological features. Thus, the supposed characteristic forms of particular process assemblages and climatic regimes may in fact have diverse origins that display equifinality, and a simple correlation between form, process and climate cannot be demonstrated.   KSR

Reading and Reference
Nicholas, A.P. and Quine, T.A. (2010) Quantitative assessment of landform equifinality and palaeoenvironmental reconstruction using geomorphic models. *Geomorphology*, **121**, 167–183. · Wentworth, C.K. (1928) Principles of stream erosion in Hawaii. *Journal of Geology*, **36**, 385–410.

equilibrium A concept commonly applied to environmental open systems; that is, systems in which the quantities of stored energy or matter are adjusted so that input, throughput and output of energy or matter are balanced (e.g. Ohara *et al.*, 2014). For example, the Earth receives shortwave solar radiation at the top of the atmosphere. Of the total receipt ($263 \, \text{kcal} \, \text{cm}^{-2} \, \text{a}^{-1}$), 31% is reflected and 69% ($181 \, \text{kcal} \, \text{cm}^{-2} \, \text{a}^{-1}$) are absorbed. Input

and output are balanced by the Earth maintaining an equilibrium mean temperature such that it emits $181 \, \text{kcal} \, \text{cm}^{-2} \, \text{a}^{-1}$ of longwave radiation. In a river system, equilibrium is often defined as a balance of erosion and deposition. This is achieved by morphological adjustments that maintain sediment transport continuity. If a short river reach experiences more bedload input from upstream than output downstream, the excess is deposited and the slope is steepened and the cross-section shallowed. The bedload input can then be transported through the reach more effectively, and the output is increased to balance the input. Any local particle detachment (bed erosion) is balanced by deposition. This equilibrium is maintained by negative feedback: the deposition of excess load changes reach morphology so that the transport capacity increases and further deposition is prevented. The inputs to an open system vary through time (e.g. on a seasonal basis), but so long as average annual input is constant the system state is constant, and the equilibrium is one in which the relationship between form and process is stationary. This is a STEADY STATE (Chorley and Kennedy, 1971: 201–203). If the annual average input is changing through time sufficiently slowly for the system to adjust, the condition is a DYNAMIC EQUILIBRIUM. Technically, however, there is always a lag between the change in the process input variable and the internal morphological adjustment of the system, so the term *quasi-equilibrium* is sometimes used in this case.   KSR

References
Chorley, R.J. and Kennedy, B.A. (1971) *Physical geography: a systems approach*. London: Prentice-Hall. · Ohara, N., Jang, S.H., Kure, S., *et al.* (2014) Modeling of interannual snow and ice storage in high-altitude regions by dynamic equilibrium concept. *Journal of Hydrologic Engineering*, **19** (12), art. no. 04014031.

equilibrium line A notional line describing some sort of balance between process and form. The notion can be applied widely (e.g. to the profile of a slope, plan of a beach, long profile of a glacier) and is closely bound up with concepts of EQUILIBRIUM in natural systems. (See DYNAMIC EQUILIBRIUM and EQUILIBRIUM SHORELINE.)

equilibrium line of glaciers A notional altitudinal line on a glacier where ABLATION balances accumulation. In most situations this means that the summer is just warm enough to melt the snow and ice that has accumulated during the previous winter. Above the equilibrium line on a glacier is the accumulation zone where accumulation exceeds ablation each year, while below the equilibrium line is the ablation zone where ablation exceeds accumulation each year. The amount of snow and ice melted at the equilibrium line each year is a measure of the activity of a glacier, with high values implying high velocities (Andrews, 1972). Glaciers are most active in mid-latitude, temperate areas and become less active towards continental interiors and the poles. The equilibrium line altitude varies in a similar way. Where there is high winter accumulation the summer temperature must be high in order to melt the snow and ice. Thus, equilibrium line altitudes tend to be low near maritime coasts and to rise towards continental interiors in response to precipitation gradients. This pattern is brought out by the distribution of both present-day mountain glaciers and abandoned glacial CIRQUES.

In temperate environments where the glacier ice is at the pressure melting point the equilibrium line may coincide with the FIRN line, which marks the line separating bare ice from snow at the end of the ablation season. But on cold glaciers the snowmelt may percolate down and freeze onto the glacier as SUPERIMPOSED ICE. Under these circumstances the positions of the firn line and equilibrium line on a glacier may differ.   DES

Reading and Reference

Andrews, J.T. (1972) Glacier power, mass balances, velocities and erosion potential. *Zeitschrift für Geomorphologie, Neue Folge*, **13**, 1–17. · Benn, D.I. and Evans, D.J.A. (2010) *Glaciers & glaciation*, 2nd edition. London: Routledge.

equilibrium shoreline

A hypothetical state that actual shorelines may or may not approximate. It is a dynamic state in which the geometry of the beach reflects a balance between materials, processes and energy levels (climate). The ideal EQUILIBRIUM beach has curvature and sand prism characteristics that are adjusted so closely that the energy available transports the detritus supplied over a period to be measured in years rather than months, days or seconds.   ASG

Reading

Castelle, B., Marieu, V., Bujan, S., *et al.* (2014) Equilibrium shoreline modelling of a high-energy meso-macrotidal multiple-barred beach. *Marine Geology*, **347**, 85–94. · Tanner, W.F. (1958) The equilibrium beach. *Transactions of the American Geophysical Union* **39**, 889–891.

equipotentials

Lines on a GROUNDWATER map joining points of equal fluid potential (or HYDRAULIC POTENTIAL). Fluid potential at any point is the product of HYDRAULIC HEAD and acceleration due to gravity. Consequently, since gravitational acceleration is practically constant, a WATER TABLE contour map with equal contour intervals is a potentiometric map in the horizontal plane. The distance between the contours

(equipotentials) depicts the gradient of the potential. Hence, hydraulic gradient varies inversely with contour spacing (Ward and Robinson, 1990). In accordance with DARCY'S LAW, water always flows in a down-gradient direction perpendicular to the equipotentials, the path followed by a particle of water being known as a streamline. A mesh formed by a series of equipotentials and corresponding streamlines is known as a flow net.

Equipotentials may also be constructed in the vertical plane. If fluid potential increases with depth, groundwater flow will be towards the surface, but if it decreases vertically the flow will be downward (Hubbert, 1940).   PWW

Reading and References

Fitts, C.R. (2013) *Groundwater science*, 2nd edition. Waltham, MA: Academic Press. · Hubbert, M.K. (1940) The theory of ground-water motion. *Journal of Geology*, **48**, 785–944. · Ward, R.C. and Robinson, M. (2000) *Principles of hydrology*, 4th edition. Maidenhead: McGraw-Hill.

era

The largest unit of geological time, being a span of one or more periods.

erg

A sand desert. A desert area characterized by sand sheets and dunes. A SAND SEA.

ergodic hypothesis

As used in geomorphology, suggests that under certain circumstances sampling in space can be equivalent to sampling through time. Geomorphologists have sometimes sought an understanding of landform evolution by placing such forms as regional valley-side slope profiles and drainage networks, or sequences of landforms (Vincent and Kattan, 2006) in assumed time sequences. The concept of the cycle of erosion was based to a large extent on ergodic assumptions, as was Darwin's model of coral reef development. Chorley *et al.* (1984: 33) point to certain dangers in ergodic reasoning: landforms may be assembled into assumed time sequences simply to fit preconceived theories of denudation; there is always a risk of circular argument; and form variations may result from factors other than their position in time.   ASG

References

Chorley, R.J., Schumm, S.A. and Sugden, D.E. (1984) *Geomorphology*. London: Methuen. · Vincent, P. and Kattan, F. (2006) Yardangs on the Cambro-Ordovician Saq sandstones, north-west Saudi Arabia. *Zeitschrift fur Geomorphologie*, **50**, 305–320.

erodibility

The susceptibility of a surface on sediment to EROSION. Erodibility is a function of factors such as rock hardness and strength, the particle size distribution of a consolidated or unconsolidated material, the degree and nature of organic content and plant cover, and so on. See also EROSIVITY.   DSGT

erosion   The process through which soil and rock material are removed from the land surface by wind, water, or ice and transported to other locations. Erosion may occur by mechanical or chemical means through a wide range of specific erosional processes. The erodibility, or erosion resistance of Earth materials varies widely, and is greatly increased by weathering processes that loosen or dissolve material, facilitating its removal from and transport across Earth's surface. Rates of terrestrial erosion vary from less than 10^{-4} mm a^{-1} in flat-lying areas in arid continental interiors to upwards of 10 mm a^{-1} in the most rapidly incising Himalayan rivers.   DRM

erosion pin   Any rod, usually metal and ranging from a stake to a nail, that is fixed into the ground surface and used to monitor surface level changes (both losses and gains). Pins may initially be applied flush to the ground surface, to minimize intrusion into sediment movement processes, but as such are susceptible to burial. Normally they are left protruding and surface change measured against the initial known length of the protruding part. Pins should be as thin as possible, but need to be sufficiently robust, and sufficiently embedded into the ground surface, to resist disturbance. Pins have been used to monitor, for example, slope erosion and surface change on sand dunes (e.g. Wiggs *et al.*, 1995).   DSGT

Reference
Wiggs, G.F.S., Thomas, D.S.G., Bullard, J.E. and Livingstone, I. P. (1995) Dune mobility and vegetation cover in the southwest Kalahari Desert. *Earth Surface Processes and Landforms*, 20, 515–529.

erosion surface   A term commonly used in Britain to describe a flattish plain resulting from erosion. Since, strictly speaking, erosion surfaces may be far from flat, it is probably more helpful to use the term *planation* surface instead.

Planation surfaces assume a central role in a geomorphological approach concerned with the evolution of landscape since they are generally regarded as the end product of either a cycle of erosion – the peneplain in the Davisian sense (see CYCLE OF EROSION) – or of a particular blend of surface processes; for example, the pediplain in a semi-arid environment (see PEDIMENT), the etch-plain in a humid tropical environment or the wave-cut platform in the coastal environment.

In the first half of the twentieth century, when the study of landscape evolution was the prime

goal of geomorphology, much attention was devoted to the identification of present and relict planation surfaces as the key to understanding DENUDATION CHRONOLOGY (King, 1950; Wooldridge and Linton, 1955). In many parts of Britain, relict planation surfaces (summit planes) are common and comprise master features of the total landscape (e.g. Brown, 1960; Sissons, 1967). A major difficulty was the problem of dating, and the evolutionary approach in geomorphology became unfashionable in the 1960s and 1970s. Now, with new forms of radiometric dating and stratigraphic evidence available from offshore sediments, there are signs of a new lease of life for studies of landscape evolution and real prospects of understanding the significance of relict planation surfaces (e.g. Lidmar-Bergström, 1982). (See also BASE LEVEL.)   DES

Reading and References
Brown, E.H. (1960) *The relief and drainage of Wales.* Cardiff: University of Wales Press. · Brown, E.H. and Clayton, K. (eds) (n.d.) *The geomorphology of the British Isles.* London: Methuen (a series of regional volumes; varying dates). · King, L.C. (1950) The study of the world's plain lands. *Quarterly Journal of the Geological Society*, **106**, 101–131. · Lidmar-Bergström, K. (1982) *Pre-Quaternary geomorphological evolution in southern Fennoscandia.* Sveriges Geologiska Undersökning C, 785. Uppsalla: Sveriges Geologiska Undersökning. · Ollier, C.D. (1979) Evolutionary geomorphology of Australia and Papua-New Guinea. *Transactions of the Institute of British Geographers, New Series*, 4 (4), 516–539. · Sissons, J.B. (1967) *Evolution of Scotland's scenery.* Edinburgh: Oliver & Boyd. · Wooldridge, S.W. and Linton, D.L. (1955) *Structure, surface and drainage in south-east England*, 2nd edition. London: George Philip.

erosivity   A measure of the potential ability of a soil or land surface to be eroded by a given geomorphological agency. For given soil and vegetation conditions the effects of a storm, for example, can be compared with another storm quantitatively and a scale of erosivity values can be produced. The ERODIBILITY of a surface is its vulnerability to erosion. Thus, EROSION can be considered as a function of both erosivity and erodibility and they are related in the universal soil loss equation, details of which are given in Morgan (1986).   WBW/DSGT

Reading and Reference
Morgan, R.P.C. (1986) *Soil erosion and conservation.* London: Longman. · Vrieling, A., Hoedjes, J.C.B. and van der Velde, M. (2014) Towards large-scale monitoring of soil erosion in Africa: accounting for the dynamics of rainfall erosivity. *Global and Planetary Change*, 115, 33–43.

erratic (glacial)   A rock or boulder that has been carried to its present location by the action of a glacier. In the English-speaking world in the first half of the nineteenth century erratics were commonly attributed to ice rafting in the universal flood, although in Switzerland, Germany and Norway their true origin had already been appreciated. The significance of erratics in demonstrating former widespread glaciers in Britain was shown by Agassiz (1840).   DES

Reference
Agassiz, J.L.R. (1840) Glaciers, and the evidence of their having once existed in Scotland, Ireland and England. *Proceedings of the Geological Society*, 3, 321–322.

eruption   A discharge of volcanic material, either gaseous, liquid or solid, at the Earth's surface.

escarpment   The steeper slope of a cuesta. Often used as a synonym for a cuesta.

esker   A sinuous ridge of coarse gravel representing the deposits of a MELTWATER stream normally flowing subglacially. Eskers may be hundreds of kilometres in length and 100 m in height (Storrar *et al.*, 2014). In many situations they are beaded, which means that mounds occur

along their length, particularly at points where they change direction. It is common to find that eskers form complex patterns of tributaries and distributaries and that sometimes ridges are discontinuous or linked by rock-cut meltwater channels. Most eskers are the channel deposits of subglacial meltwater rivers, and their orientation is usually parallel to that of overall ice flow (Shreve, 1972). The deposits are related to both closed channel flow and open channel flow within a conduit (Saunderson, 1977).   DES

References
Saunderson, H.C. (1977) The sliding bed facies in esker sands and gravels: a criterion for full-pipe (tunnel) flow? *Sedimentology*, 24, 623–638. · Shreve, R.L. (1972) Movement of water in glaciers. *Journal of Glaciology*, 11, 205–214. · Storrar, R.D., Stokes, C.R. and Evans, D.J.A. (2014) Morphometry and pattern of a large sample (>20,000) of Canadian eskers and implications for subglacial drainage beneath ice sheets. *Quaternary Science Reviews*, 105, 1–25.

ESP   See EXCHANGEABLE SODIUM PERCENTAGE.

ESR   See ELECTRON SPIN RESONANCE.

	Head	Upper reaches	Middle reaches	Lower reaches	Mouth
Salinity (p.p.t.)	0.5–5	5–18	18–25	25–30	30–40
Deposit	Muds	Muds	Mud and sand	Sand	Sand and rock
Vegetation	Reeds Sedges Rushes	Abundant organic detritus Salt-marsh flora Mainly perennial halophytes with mud algae and phytoplankton			Marine algae
Fauna	Oligohaline partly tolerant	Species tolerant of estuarine conditions (euryhaline)			Stenohaline marine forms

The estuarine environment. Zones of transition separate the different environments shown.
Source: Furley and Newey (1983: figure 13.12). Reproduced with permission of Peter Furley.

estuary The section of a river that flows into the sea and is influenced by tidal currents. Estuaries form transition zones between freshwater rivers and saltwater oceans, with fluctuations in water level, salinity, temperature and velocity. They are constantly modified by erosion and deposition, resulting in tidal flats and salt marshes, deltas, spits and lagoons. A funnelling of tidal currents may produce powerful periodic waves or estuarine bores. Shallow sedimentary estuaries are rich in nutrients and very high in BIOLOGICAL PRODUCTIVITY, providing nurseries for fish and other animals. Deep estuaries, such as fiords, are colder, less productive and less biologically diverse.

Estuaries can be classified into a number of types on the basis of chemical characteristics (Dyer, 1986): salt wedge estuaries in tideless seas, partially mixed estuaries where there are appreciable tidal movements, and well-mixed estuaries where the strength of the tidal currents is strong relative to river flow. They can also be classified on the basis of their tidal range. This determines the tidal current and residual current velocities, and therefore the amount and source of sediments. *Microtidal estuaries* occur where tidal range is less than 2 m and so are dominated by freshwater inflow upstream of the mouth and by wind-driven waves seaward of the mouth. They often contain a fluvial delta and spits and bars at the seaward margin. In *mesotidal estuaries* (tidal range ~2–4 m) tidal currents are of greater importance, but because of the still somewhat modest tidal range the tidal flow does not extend very far upstream. Thus, most mesotidal estuaries are relatively stubby. In the case of *macrotidal estuaries*, tidal ranges in excess of 4 m produce a situation where tidal influences extend far inland. Such estuaries have long, linear sand bars parallel to the tidal flow, but their most distinguishing characteristic is their trumpet-shaped flare. The Severn Estuary in Britain, the Delaware Estuary in the USA and the Plate Estuary in Latin America are prime examples of this type. PAF/ASG

Reading and References

Dyer, K.R. (1986) *Coastal and estuarine sediment dynamics.* Chichester: John Wiley & Sons, Ltd. · Furley, P. and Newey, W. (1983) *Geography of the biosphere.* London: Butterworth. · Ketchum, B.H. (ed.) (1983) *Estuaries and enclosed seas.* Amsterdam: Elsevier. · Wolanski, E. (2007) *Esturine ecohydrology.* Amsterdam: Elsevier.

eulittoral zone The portion of the coastal zone that extends seawards from high-water mark down to the limit of attached plants (generally at a depth of 40–60 m).

The nature of global sea-level change over the last 250,000 years.
Source: A.S. Goudie (1984). *The nature of the environment.* Oxford: Blackwell. Reproduced by permission of Andrew Goudie.

euphotic zone The surface layer of a body of water in which photosynthesis can take place because of the availability of light.

eustasy A term that embraces sea-level changes of a worldwide nature. Local changes of sea level complicate the eustatic pattern and are caused by local factors such as ISOSTASY, OROGENY and EPEIROGENY. During the first decades of the twentieth century, it was proposed that most sea-level oscillations and strandlines of the Quaternary were glacio-eustatic– see Guilcher (1969). It was believed, correctly, that sea level oscillated in response to the quantity of water stored in ice caps during repeated phases of glaciation and deglaciation.

The transgressions of the interglacial were succeeded by regressions during glacials, and the height of the various stages declined during the course of the Pleistocene. Total melting of the two main current ice caps – Greenland and Antarctica – would raise sea level a further 66 m if they both melted. Deep-sea core evidence, however, does not suggest that in previous interglacials of the Pleistocene these two ice caps did disappear, and without a general melting of them the sea level would only have been a few metres higher than now in the interglacials. This fact suggests that a simple glacio-eustatic theory of progressive sea-level decline during the Pleistocene cannot be upheld, and factors other than glacio-eustasy must be responsible for the proposed high sea levels of early Pleistocene times. Because other, sometimes local, factors have played a role, age correlation of shorelines over wide areas through measurement of elevation alone is not possible.

Nevertheless, low Quaternary sea levels brought about by the ponding up of water in the ice caps were quantitatively extremely important. Minimum eustatic sea levels around the time of the Last Glacial Maximum were some

195

130 m lower than at present, although they may have been even lower earlier in the Pleistocene.

Although orogeny is normally regarded as being an essentially local factor of sea-level change, and eustasy as being worldwide, there is one class of process, orogenic eustasy, whereby a local change can have worldwide effects. It therefore acts as some sort of a link between these two main types of change, and is a worldwide change of sea level produced by changes in the volumes of the ocean basins resulting from orogeny (mountain building) and tectonically induced changes in the volume of the ocean basins (Worsley *et al.*, 1984).

The importance of a third type of eustatic change was emphasized by Mörner (1980). This is termed geoidal eustasy. The shape of the Earth is not regular, and at present the GEOID (caused by the Earth's irregular distribution of mass) has a difference between lows and highs of as much as 180 m. The ocean surface reflects this irregularity in the geoid surface, which varies according to forces of attraction (gravity) and rotation (centrifugal), and will respond by deformation to a change in these controlling forces. The possible nature of such changes is still imperfectly understood, but they include fundamental geophysical changes within the Earth, changes in tilt in response to the asymmetry of the ice caps, changes in the rate of rotation of the Earth and redistribution of the Earth's mass caused by ice-cap waxing and waning.

Although glacio-eustasy is the most important of the eustatic factors that have affected world sea levels during the course of the Quaternary, it is worth looking at some of the other minor eustatic factors that have played a role, especially over the long term. The infilling of the ocean basins by terrestrial sediment, for example, would lead to sea-level rise. Higgins (1965) estimated that this could lead to a rise of 4 mm per 100 years, equivalent to a rise of 40 m in a million years. Two additional factors are the addition of juvenile water from the Earth's interior and the variation of water level according to temperature. The latter could raise the level of the sea by about 60 cm for each 1 °C rise in temperature of the seawater. The former could probably add about 1 m of water in a million years. Much has been written in recent years about the effect of CLIMATE CHANGE on global eustatic sea levels as higher temperatures induce ice-sheet melt and an increase in the volume of ocean water itself. This is discussed in greater detail elsewhere (see SEA LEVEL).   ASG/MEM

Reading and References

Goudie, A.S. (1984). *The nature of the environment.* Oxford: Blackwell. · Guilcher, A. (1969) Pleistocene and Holocene sea level changes. *Earth Science Reviews*, 5, 69–98. · Higgins, C.G. (1965) Causes of relative sea-level changes. *American Scientist*, 53, 464–476. · Mörner, N.A. (1980) *Earth rheology, isostasy and eustasy.* New York: John Wiley & Sons, Inc. · Worsley T.R., Nance, D. and Moody, J.B. (1984) Global tectonics and eustasy for the past 2 billion years. *Marine Geology*, 58, 373–400. · Yokoyama, Y., Lambeck, K., de Decker, P., *et al.* (2000) Timing of the Last Glacial Maximum from observed sea-level minima. *Nature*, **406**, 713–716.

eutrophic   Pertaining to lakes and other freshwater bodies that abound in plant nutrients and which are therefore highly productive. Lakes tend to become more eutrophic as they become older, and eutrophication can also result from the addition of nutrients as a result of pollution. This can cause phenomena such as algal blooms. (See also NUTRIENT STATUS.)

eutrophication   The addition of mineral nutrients to an ECOSYSTEM, generally raising the NET PRIMARY PRODUCTIVITY (NPP). It is usually used of human-induced additions of elements such as nitrogen and phosphorus to saltwater and freshwater, which are naturally low in those elements, but it also occurs in terrestrial systems and may be a natural phenomenon. In current usage, it very often relates to the loads of nitrogen and phosphorus in fresh and offshore waters heated by such effluents as sewage, fertilizer run-off and detergents. The effects are often algal blooms, deoxygenation of water through consequent bacterial activity and, in the sea, rapid growth of small organisms called dinoflagellates, which are implicated in 'red tides'. Eutrophication is usually a

Eutrophication is a problem in parts of Lake Victoria, leading to algal bloom growth, and is a function of nitrogen and phosphorus loading from run-off affected by soil erosion, and possibly from atmospheric dust sources. This picture is from Kisumo, Kenya, on the lake's eastern shoreline. Photograph by David Thomas.

local or regional problem at most, though enclosed seas like the Mediterranean may be more vulnerable than open oceans. Its main drawback is the loss of expensively gained nutrients that then have to be replaced, since they cannot be economically retrieved from the water in which they have become diluted.   IGS

Reading

Whitton, B.A. (ed.) (1975) *River ecology*. Berkeley, CA: University of California Press.

evaporation   The change of state between water and water vapour. Evaporation of surface water is the prime source of atmospheric moisture. Total evaporation is the flux of vapour from free-standing water surfaces, bare soil, vegetation surfaces, snow and ice. Evapotranspiration includes these sources plus the water transpired by plants. Three processes simultaneously contribute to the net evapotranspiration: absorption of radiation, advection of heat, and turbulent and molecular exchange. These processes are controlled respectively by available energy, wind and water vapour. The processes are represented by net radiation or temperature, wind speed or turbulence, and the atmospheric saturation deficit or soil moisture content. The wind acts to maintain high vapour pressure gradients between the surface and the ground. Over bare soil, evaporation is controlled by the atmospheric variables when surface water is sufficiently available. This situation is sometimes referred to as radiation limited, but the term *atmosphere controlled* is more appropriate, because wind also plays a determining role. When the surface is relatively dry, soil moisture is the major control on evaporation. This is a water-limited or soil-controlled situation.

The assessment of evapotranspiration is different for the atmosphere-controlled case and the soil-controlled case. In the former case, this is essentially the assessment of potential evapotranspiration or water demand. In the latter case actual evapotranspiration or water use must be assessed. Evaporation over a water surface is predominantly atmosphere controlled. Over land the process is generally atmosphere controlled when the soil moisture is close to or above field capacity and actual evapotranspiration equals potential to a first approximation. Numerous methods are applied to estimating actual and potential evapotranspiration. The choice of method depends on the definition of potential evapotranspiration utilized, the available input data, the purpose of the calculation and surface characteristics.

Potential evapotranspiration E_p is commonly defined as the rate at which evapotranspiration

can be sustained from a complete and uniform vegetation cover when water availability is unlimited. This definition is conceptually appropriate, but it is problematic in practice because the rate depends not only on the atmosphere, but also on the surface conditions of soil and vegetation type. Recognizing this, Penman (1956) modified the definition to 'the amount of water transpired . . . by a short green crop, completely shading the ground, of uniform height and never short of water'. This is one of many definitions and is what Federer *et al.* (1996) term 'reference surface potential evapotranspiration'. With so many operative definitions, there is considerable ambiguity to the term 'potential evapotranspiration' and a variety of assessment methods. Thus, that reference to this concept should be accompanied by information regarding the reference surface and method of calculation.

Potential evapotranspiration is assessed via evaporation pans or empirical formulae. The term *combination methods* refers to methods that are based on a combination of mass transfer and energy balance considerations. In the direct measurement approach, the volume of water lost from the pans is recorded. Although this sounds like a simple and robust calculation, in reality the value of E_p obtained depends a great deal on the characteristics of the pan.

The most common empirical approaches to estimating potential evapotranspiration are those of Thornthwaite, Priestley–Taylor and Penman–Monteith. The Priestley–Taylor approach uses only radiation and is an adaptation of the Penman formula for evaporation from a free water surface. Thornthwaite's calculation of potential evapotranspiration requires only temperature information. The Penman–Monteith 'combination' method utilizes concepts of mass transfer and energy balance. It thus requires several data types: net radiation, air temperature, wind speed and relative humidity.

Early approaches to calculating evaporation from a free water surface included mainly the mass-transfer concept developed by Dalton in 1802 and the energy balance. Dalton reasoned that evaporation would be proportional to the difference between actual and saturation vapour pressure and would occur at a rate that is dependent on wind speed. Equations based on this concept are in the form

$$E_p = (b_1 + b_2 u)(e_{sat} - e_a)$$

where u is wind speed, e_a is the atmospheric vapour pressure, e_{sat} is the saturation vapour pressure, and b_1 and b_2 are empirical constants. Penman's (1948) formula for potential evapotranspiration, also called the Penman combination equation, added net radiation as a factor, so that E_p is calculated as

$$E = \frac{R_N \Delta + \gamma E_a}{\Delta + \gamma}$$

where R_N is the net radiation, Δ is the slope of the curve relating saturation vapour pressure to temperature and γ is the psychrometric constant, equal to 0.622. E_a is a term describing the contribution of mass transfer to evaporation and is in the form shown in the earlier equation. Monteith (1965) revised the Penman equation so that it would be valid in the water-limited case and could incorporate the effects of vegetation on evapotranspiration. The development of this approach was based on biophysical concepts related to water passage through vegetation. It incorporated a series of resistances to water passages offered by the plants and the atmosphere. Dingman (1994) provides an excellent discussion of the development and suggests that Penman–Monteith is the best approach to estimating potential evapotranspiration over vegetated surfaces.

Actual evapotranspiration E is likewise a difficult quantity to determine. Because of the complexity of the process and the number of degrees of freedom (numerous characteristics of the wind, soil, soil moisture and vegetation cover all play a role), a tremendous number of methods exist. The most common methods of assessing actual evaporation are based on turbulent transfer, potential evapotranspiration, and water balance, but direct measurements can be made via instruments such as lysimeters.

The water balance methods are most useful in calculating evapotranspiration for large catchments. They require a tally of inputs and outputs, calculating evapotranspiration as the difference between precipitation, run-off and changes in soil moisture storage w:

$$E = P - N - \frac{dw}{dt}$$

The turbulent-transfer methods include the Bowen ratio energy-balance approach, eddy correlation and what is often termed aerodynamic methods (the terms *profile* and *mass transfer* methods are also used). These generally require measurements of the gradients and fluxes of quantities such as momentum, sensible heat and water vapour. In the aerodynamic approach, the rate of exchange of water vapour is assumed equivalent to that of heat or momentum and E is assessed by measuring vertical profiles of temperature or wind. The Bowen ratio approach also

relies on near-surface profiles and E is calculated as a residual in the energy balance equation, such that

$$LE = \frac{R_{net} - S}{1 + B}$$

where L is the latent heat of condensation, B is the Bowen ratio of sensible to latent heat transfer to the atmosphere (H/LE) and S, sensible heat transfer, is H plus the heat transfer to the ground. The eddy-correlation methods involve direct measurement of turbulent fluxes. The evaporative flux of water vapour E can be expressed as

$$E = \rho_a \omega' q'$$

where ω' and q' are the instantaneous departures of vertical velocity and specific humidity from their mean values, and ρ_a is the density of air.

Several approaches to calculating actual evaporation are based on potential evapotranspiration and express the actual rate in terms of the proportion of E_p that is realized. These are commonly based on the degree of soil saturation and are expressed in the form

$$E_o = f(\theta)E_p$$

where θ is water in the root zone of plants. A linear relationship is often assumed, so that

$$\frac{E_o}{E_p} = \frac{w}{w^*}$$

where w and w^* are the available soil moisture and the soil water storage capacity. The latter is equivalent to field capacity. This is the so-called 'bucket method' used in many climate models. However, the relationship varies greatly with soil type and it tends be linear only for clay soils.   CTA/SEN

Reading and References

Dingman, S.L. (1994) *Physical hydrology*. New York: Macmillan. · Federer, C.A., Vörösmarty, C. and Fekete, B. (1996) Intercomparison of methods for calculating potential evaporation in regional and global water balance models. *Water Resources Research*, **32**, 2315–2321. · Mengelkamp, H.-T. (2006) Evaporation over a heterogeneous land surface: the EVA-GRIPS Project. *Bulletin of the American Meteorological Society*, **87**, 775–786. · Monteith, J.L. (1965) Evaporation and environment. In G. E. Fogg (ed.), *The state and movement of water in living organism*. 19th Symposia of the Society for Experimental Biology, no. XIX. Cambridge: Cambridge University Press; pp. 205–234. · Nicholson, S.E. (2011) *Dryland climatology*. Cambridge: Cambridge University Press. · Penman, H.L. (1948) Natural evaporation from open water, bare soil, and grass. *Proceedings, Royal Society of London, Series A*, **193**, 120–145. · Penman, H.L. (1956) Evaporation: an introductory survey. *Netherlands Journal of Agricultural Science*, **4**, 7–29. · Shuttleworth, W.J. (1991) Evaporation models in hydrology. In T. J. Schmugge and J. André (eds), *Land surface evaporation.* New York: Springer; pp. 93–120. · Thornthwaite, C.W. (1948) An approach towards a rational classification of climate. *Geographical Review,* **38**, 55–94. · Willmott, C.J., Rowe, C.N. and Mintz, Y. (1985) Climatology of the terrestrial seasonal water cycle. *Journal of Climatology,* 5, 589–606.

evaporite   A water-soluble mineral (or rock composed of such minerals) that has been deposited by precipitation from saline water as a result of evaporation, especially in coastal sabkhas or in salt lakes of desert areas. Among the most common minerals are sodium chloride (halite) and calcium sulphate (gypsum and anhydrite). Evaporites can be subject to solution and so can give rise to karstic forms and ground subsidence, but they also contribute to salt weathering, and may produce various types of patterned ground.   ASG

Reading
Warren, J.K. (1989) *Evaporite sedimentology.* Englewood Cliffs, NJ: Prentice-Hall.

evapotranspiration   The diffusion of water vapour into the atmosphere from a vegetated surface. It has five components: standing water surfaces, water on the surfaces of plants and soil (interception and detention water respectively), soil moisture and water transpired from plants via passage from the roots through the stomata. For details of processes, see EVAPORATION.

evolution   Cumulative development in the characteristics of species over time. Classically, evolution is regarded as the progressive change in features of populations occurring through the course of sequential generations brought about by the process of natural selection. In fact, evolution must also explain the diversification of organisms of presumed common ancestry through geological time and must, therefore, account for speciation and extinction as well as progressive change as a result of natural selection. Convergent evolution is the process by which unrelated or distantly related groups evolve similar morphologies or adaptational traits; for example, the reduced limbs of whales and penguins. COEVOLUTION is the contemporaneous evolution of two ecologically linked taxonomic groups, such as flowering plants and their pollinators.

Charles Darwin is regarded as the father of evolutionary theory, although natural scientists had been grappling with the problem of explaining the diversity of life well before his publication, in 1859, of the first edition of *On the Origin of Species by Natural Selection.* In 1832, the British geologist Charles Lyell applied the term evolution to the notion of organic transmutation over

time. Indeed, Charles Darwin himself did not use the word extensively at all, preferring instead the phrase 'descent with modification'. Darwin's famous voyage, as ship's naturalist, aboard the British brig HMS *Beagle,* undertaken between 1831 and 1836, was to be instrumental in formulating his ideas about evolution. His short visit to the Galapagos Islands, 800 km off the west coast of South America, appears to have been especially influential. It was here that Darwin observed and reported on morphological variations between individuals belonging to the same species and was also struck by the degree of adaptation to the environment displayed by several apparently closely related different species of 'finch', all confined to different habitats or to completely different islands. These ideas were to prove crucial in the development of his idea of modification between generations, and of common ancestry – in short, of evolution.

It transpires that Darwin and a contemporary British zoologist, Alfred Russel Wallace, developed the idea of evolution by natural selection independently. Wallace penned a letter to Darwin in 1858 in which he outlined his own hypotheses concerning the development of species through time and which appear to have persuaded Darwin that it was time to go public with the theory. On 1 July 1858 a joint paper was presented, *in absentia,* to the Linnean Society in Piccadilly and *Origin* soon followed. Furore greeted its publication, fuelled by the logical conclusion of common ancestry that humans were, according to Darwin and Wallace's ideas, descended from the apes, an idea that clearly affronted the dominant conservative Christian philosophy of the Victorian age. But acceptance did eventually come; indeed, the theory profoundly influenced the development of thinking on a number of other aspects of physical geography, such as the CYCLE OF EROSION and SUCCESSION. It appears to have frustrated Darwin that the crux, for him at least, of the idea of evolution, that of natural selection (or 'survival of the fittest' as it has been commonly perceived), was relatively ignored in the heat of the debate that centred on common ancestry.

The essential mechanism of natural selection may be summarized as follows:

1 In most species, far more young are produced than can possibly survive to reproductive age.
2 Organisms vary: individuals in a population differ from each other in numerous minor morphological and behavioural ways. Some of these variations are heritable and may be passed on to the succeeding generation.

199

3 There is a 'struggle' for existence, competition between organisms, in which those better suited to the prevailing environment are more likely to survive to reproductive age.

4 Individuals with characteristics best suited to survival are more likely to reproduce successfully and will, ultimately, predominate in a population. Adaptive change is therefore a result of natural selection operating over long periods based on the gradual accumulation of small, favourable variations present in populations.

Acceptance of these ideas was slow in coming, not least because the mid-nineteenth-century scientific understanding of the processes of 'heritability' was incomplete. Gregor Mendel's theory of 'particulate inheritance' put forward in the 1860s certainly provided a mechanism, although it was only 80 years later that the 'modern synthesis' emerged (Huxley, 1942). According to this integration, evolution was interpreted as a combination of speciation operating in concert with natural selection. Processes other than natural selection (e.g. GENETIC DRIFT) may also play a role.

More recently, Darwin's and Wallace's ideas have come under scrutiny, especially in regard to the pace of evolutionary change. Their view of the process of natural selection was one of incremental, gradual accumulation of 'successful' adaptations within a species over time. The fossil record, on the other hand (despite its almost notorious incompleteness and arguable unreliability), reveals no such gradual changes in morphology of species over time, but rather a history of relative stability interrupted by relatively rapid changes. In 1977, two American biologists, Stephen Jay Gould and Niles Eldredge, published their explanation of this observation. They argued that evolution is not a gradual process, but is instead characterized by relatively long periods of negligible or limited change (stasis) punctuated by short, intense periods of rapid development based on speciation events. Gould (1996) has further argued that apparently progressive adaptation to environmental conditions over time, as implied by natural selection, is probably an artefact, rather than fundamental, of the way the process operates.

The most recent interpretations of evolution, described as a 'post-modern' synthesis (Bennett, 1997), assess the relative importance of natural selection and speciation events in the evolutionary process. Bennett argues, on the basis of evidence from the Quaternary period, that the incremental changes brought about by natural selection are relatively unimportant in macro-evolution and that the processes of migration, extinction and speciation prompted by environmental changes on 10^3-to10^4-year timescales are the real pacemakers of evolution.   MEM

Reading and References

Bennett, K.D. (1997) Evolution and ecology: the pace of life. Cambridge: Cambridge University Press. · Darwin, C. (1859) *On the origin of species by natural selection, or the preservation of favoured races in the struggle for life.* London: John Murray. · Desmond, A. and Moore, J. (1991) *Darwin.* London: Michael Joseph. · Gould, S.J. (1996) *Life's grandeur: the spread of excellence from Plato to Darwin.* London: Jonathan Cape. · Gould, S.J. and Eldredge, N. (1977) Punctuated equilibria: the tempo and mode of evolution reconsidered. *Paleobiology,* **3,** 115–151. · Huxley, J. (1942) *Evolution: the modern synthesis.* London: Allen & Unwin.

evorsion   The erosion of rock or sediment in a river bed through the action of eddies and vortices.

exaration   The process by which glaciers pluck or quarry bedrock. Abrasion is not involved.

exchangeable sodium percentage (ESP)

This measures the proportion of sodium among the major cations present in the soil exchange complex (see CATION EXCHANGE). It is given by

$$ESP = \frac{[Na^+]}{[Na^+ + K^+ + Ca^{2+} + Mg^{2+}]} \times 100$$

ESP is significant because the abundance of sodium in the soil exchange complex strongly influences the behaviour of the soil when in contact with water. The sodium ion is small and carries only a single electrical charge. It is therefore very mobile in the soil environment. When soil with a high ESP is wetted by dilute rainwater, sodium rapidly moves into solution in the immediate vicinity of the soil clays where it ordinarily resides. This sets up strong concentration gradients between the soil water and the rainwater, and osmosis drives water molecules from the rainwater into the more saline water among the soil clays. This flow of water requires that there be expansion of the soil matrix, and this may disrupt the soil aggregates, leading to dispersive behaviour and the release of small, readily erodible particles. This unstable behaviour that is triggered by large amounts of sodium can be reduced if the soil solution carries large concentrations of other salts, whose presence restricts the entry of additional sodium. Alternatively, if very soluble materials are applied to the soil surface, so that infiltrating rainwater rapidly collects a load of dissolved ions, the damaging effects of osmotic breakdown can be curtailed.   DLD

Reading
Brady, N.C. and Weil R.R. (2008) *The nature and properties of soils*, 14th edition. Englewood Cliffs, NJ: Pearson Prentice-Hall.

exfoliation Onion-weathering or desquamation. The weathering of a rock by peeling off of the surface layers.

exhaustion effects Encountered frequently in detailed studies of the variation of suspended sediment concentrations in rivers and streams during storm run-off events, and attributable to a progressive reduction or exhaustion in the availability of sediment for mobilization and transport. They may occur during an individual event, where they will be reflected in the occurrence of maximum suspended sediment concentrations and loads before the hydrograph peak, or during a closely spaced sequence of hydrographs, where they will give rise to a progressive decrease in the sediment concentrations associated with similar levels of water discharge.   DEW

exhumation The exposure of a subsurface feature through the removal by erosion of the overlying materials.

exogenetic Pertaining to processes occurring at or near the surface of the Earth and to the landforms produced by such processes. By contrast, ENDOGENETIC processes originate within the Earth.

exotic A term normally used to describe a plant or animal that is kept, usually in a semi-natural or artificial manner, in a region outside its natural provenance. A wide range of garden plants and aviary birds are exotics, having been introduced by horticulturalists and bird fanciers from widely differing regions of the world. The term tends to be used in a rather restricted sense to describe especially spectacular species collected from markedly different climatic regimes. Exotic should be contrasted with the term ALIENS, which is the commoner designation for introductions that are more generally naturalized in their new locations.   PAS

expanding Earth The idea that the physical size of the planet may have increased through geological time has a long history, having been raised periodically during the eighteenth, nineteenth and twentieth centuries. Various lines of evidence have been put forward in support of the idea. Carey (1975, 1976) performed exacting map reconstructions of the supercontinent of Pangaea in the configuration of about 200 Ma BP but found that there were always gaps, which he termed *gaping gores*, that were difficult to account for. He found that these could all be closed at once if the continental reassembly were carried out on a globe of smaller diameter than the present Earth (70–80% of its present size). Others (e.g. Steiner, 1977) tallied the areas of new crust created by sea-floor spreading and consumed along SUBDUCTION ZONES, and argued that more crust had been created than consumed, therefore requiring the surface area of the globe to increase. Other geometric arguments have been raised, including the difficulty of reconciling the simultaneous expansion, since the break-up of Pangaea, of the Pacific and Atlantic Oceans, both of which have evolved since that time. The length of the day has increased through geological time, as revealed by the number of diurnal growth bands per annual cycle in fossil corals. Earth expansion has been raised as a mechanism to account for the slowing rotation of the planet that is the cause of this day-length variation, but it rather appears to be related to the steady retreat of the Moon.

Diverse mechanisms have been proffered to account for the expansion of the planet. Ongoing planetary cooling, recrystallization and change in rock fabrics and overall density provide one category. Another involves the very slow temporal change in what are thought of as 'constants', like the gravitational constant G. Decline in parameters like this, involved fundamentally in determining the forces that bind matter together, are envisaged in a number of cosmological hypotheses, and may play a part in the observed expansion of the universe.   DLD

References
Carey, S.W. (1975) The expanding Earth: an essay review. *Earth Science Reviews*, 11, 105–143. · Carey, S.W. (1976) *The expanding Earth*. Amsterdam: Elsevier. · Steiner J. (1977) An expanding Earth on the basis of sea-floor spreading and subduction rates. *Geology*, 5, 313–318.

expansive soils Soils whose volume can be increased substantially as they become wet, and which shrink as they dry out, leading to a range of what are termed 'shrink–swell phenomena'. The wetting and drying cycles are mostly seasonal. Because of the requirement that there be a large surface area across which water can be exchanged, expansive soils commonly have at least 30% clay, often fine clay, and the most important swelling minerals are 2 : 1 layered smectite clays such as montmorillonite. Water and exchangeable cations enter the spaces between structural sheets within these minerals, causing a volume increase. As determined using a ribbon of soil whose length is measured when dry and when wet, expansion may amount to 20% in some cases. Volume increase accounts for a number of distinctive features in expansive soils, including marked local variation in the soil caused by subsoil material being forced towards the surface (associated shearing of the soil may produce SLICKENSIDES), and by various kinds of buckling of the soil surface (see GILGAI). Drying of these soils may be associated with the opening of deep cracks in the soil, which may later conduct water and facilitate rewetting of the subsoil. Expansive soils form the Vertisol order in the US soil taxonomy. They are associated with many management problems, including damage to foundations and roads, cracking of pipelines and leakage from impoundments via tension cracks. DLD

Reading
Coulombe, C.E., Wilding, L.P. and Dixon, J.B. (1996) Overview of vertisols: characteristics and impacts on society. *Advances in Agronomy*, 57, 289–375.

experimental catchment A small drainage basin used for detailed investigations of hydrological and geomorphological processes. During the UNESCO International Hydrological Decade (1965–1974) the term was frequently used in a more limited context to refer to small ($<4\,km^2$) catchment studies, where the vegetation cover or land use was deliberately modified in order to study the hydrological impact of such changes. These studies were seen as experiments. Various strategies involving single, paired and multiple watershed experiments have been employed in order to decipher the nature and magnitude of changes in catchment behaviour. DEW

Reading
Rodda, J.C. (1976) Basin studies. In J. C. Rodda (ed.), *Facets of hydrology*. Chichester: John Wiley & Sons, Ltd; pp. 257–297. · Toebes, C. and Ouryvaev, V. (eds) (1970) *Representative and experimental basins: an international guide for research and practice.* Paris: UNESCO.

experimental design See also DEDUCTIVE. The prior planning, in terms of methods, sampling, statistical replication and validation, of a programme of data collection, experimentation or testing. This forms a vital initial phase of HYPOTHESIS testing in geographical work as well as across the disciplines generally. It is necessary to consider, in designing an experiment, the variability (temporal and/or spatial) of many phenomena, because this will determine where and when sampling would be required in order to derive representative values. Some phenomena, like distributions of soil moisture, vary diurnally, or seasonally, or are dependent on antecedent weather conditions. Often, there will be marked spatial variation also. In sampling such phenomena, the researcher has to consider the number of samples that would be required in order to estimate the value of the variable with an acceptable level of error, and whether the time and budget available would permit this. In studying a process like soil erosion where there are many influential variables, a decision has to be made about which will be measured. The ones selected need to be physically informative, perhaps statistically independent, and adequate for resolution of the research problem being studied. In addition, the particular means by which a variable will be measured requires consideration. For instance, in describing plant cover on an experimental plot, a worker might employ projected canopy cover, or biomass, or might decide to tally separately the leaf area of individual species growing at different heights above the soil surface. Each description has some value, but the worker must evaluate the various options in advance of fieldwork, and select the means of description (termed *parameterization*) that seems likely to yield the most relevant data.

Anderson and McLean (1974) identified 11 stages in the conduct of an experiment:

1 recognition that a problem exists;
2 formulation of the problem;
3 selecting factors and treatments to be used in the experiment;
4 specifying the variables to be measured;
5 definition of the 'inference space' for the problem – that is, the range of conditions to which it is hoped that the results will apply;
6 random selection of subjects or sites;
7 assignment of treatments to the subjects or sites;
8 deciding on the kinds of statistical analysis that will be used with the data;
9 collection of the data;
10 analysis of the data;
11 drawing up of conclusions.

Not all of these will be required in all geographical studies, but many will. Jumping from step 1 or 2 to step 9 is all too common, but can only result in mistakes being made and inefficiency in the collection and later attempted analysis of the results. Often, it is desirable to run a small pilot investigation after step 2 in order to become more familiar with the issues and locations involved, and to test data-gathering equipment or methods. Especially if working in a remote area, or one that is costly to visit, a 'shakedown' trial will often prove helpful.   DLD

Reading and Reference
Anderson, V.L. and McLean, R.A. (1974) *Design of experiments*. New York: Marcel Dekker. · Montgomery, D.C. (2012) *Design and analysis of experiments*, 8th edition. New York: John Wiley & Sons, Inc.

extending flow   See COMPRESSING FLOW.

external forcing   Forces acting on a system are categorized as external or internal. Internal forcings are interactions between components of a system. External forcings are factors that are not part of a system, but exert influence on the system. The atmosphere provides a good illustration of the difference. Internal forces are interactions between parts of the atmosphere, such as the interactions between waves in the general atmospheric circulation or between waves and the mean circulation. External forcings are imposed at the upper and lower boundaries of the atmosphere: solar radiation at the top, fluxes from the Earth's surface at the bottom. These include gases, particulates, energy, water and solutes. What is external versus internal depends on how a system is defined. If the system is not the atmosphere, but the atmosphere and Earth system, the external forcing is that from the Sun.   SEN

Reading
Shukla, J., (1981) Dynamical predictability of monthly means. *Journal of Atmospheric Science* **38**, 2547–2572.

extinction   The disappearance of an individual organism, a group of organisms or a local population from existence. Although many recorded extinctions are directly attributable to the actions of humans, the fossil record clearly reveals that nearly all lineages have become extinct without leaving descendants and that evolutionary conservatism will frequently lead to extinction in the face of significant environmental changes, such as climatic change, the development of new competitors, predators or diseases, or the loss of a major food supply. Many classic mass extinctions, such as that of the dinosaurs, remain enigmas unexplained by modern science, although there is never a dearth of speculation, scientific or otherwise, on the causes of their demise. The process of extinction continues, however, and over the past 200 years or so it is estimated that some 53 birds and 77 mammals have become extinct.

One particularly relevant principle in the discussion of extinction is that of 'competitive exclusion' or the 'ecological replacement principle'. When one group disappears, another appears to take its place in a corresponding or identical habitat. This theory, therefore, asserts that when an ecological resource is used simultaneously by more than one kind of organism and when this resource is insufficient to furnish all their needs, then the better or best adapted organism will eventually eliminate the competitors. It has even been argued that new forms must always replace old ones and that for every species that evolves another becomes extinct. It is clear, however, that competition is not always present in the process of extinction, which may simply reflect evolutionary conservatism in relation to the speed and direction of environmental change. Indeed, competition may resolve itself by producing diverging adaptations, leading to overall diversification and not to extinction. Other factors involved in extinctions include overspecialization, reduced mutation, loss of behavioural versatility and changes in community patterns.

Whatever the causes, the palaeontological record clearly indicates that extinction has paralleled evolution throughout the entire history of the plant and animal kingdoms, a fact that has yet to be faced by the modern CONSERVATION movement, which so often aims to maintain all forms, whatever the cost and against all odds, in the battle between birth and death. (See also ISLAND BIOGEOGRAPHY.)   PAS

Reading
Cox, C.B. and Moore, P.D. (1993) *Biogeography: an ecological and evolutionary approach*, 5th edition. Oxford: Blackwell. · Krebs, C.J. (1978) *Ecology: the experimental analysis of distribution and abundance*. New York: Harper & Row. · MacArthur, R.H. (1972) *Geographical ecology*. New York: Harper & Row. · Stanley, S.M. (ed.) (1987) *Extinction*. New York: Scientific American Library.

extratropical cyclone   An area of low pressure that develops in the westerly wind belts of middle latitudes and is associated with characteristic fronts and weather patterns. Also known as a depression or low. At the surface the extratropical cyclone consists of an area of low pressure surrounded by winds blowing in an anticlockwise direction in the northern hemisphere and in a clockwise direction in the southern hemisphere. Central pressure values vary greatly from about 930 hPa to 1015 hPa. The most intense cyclones usually occur in winter and, on average, those of the southern hemisphere are deeper than those in the northern hemisphere. The diameter of the

system may range from about 500 km to over 3000 km. The associated weather is unsettled with cloud, strong winds, periods of rain and sudden changes of temperature as the frontal zones pass. Satellite photographs show the beauty of the cloud patterns around the cyclones. They usually consist of a large 'comma-shaped' mass of cloud near the centre of the low with bands of frontal cloud near the warm and cold fronts streaming away towards the equator on a curved path (see Figure 1). Many differences of detail are found between cyclones but almost all follow this basic pattern. Over time the cyclones tend to move eastwards or northeastwards, but some will remain almost stationary or follow unusual tracks depending upon air circulation in the middle atmosphere.

Cyclones are dynamic features of our atmosphere undergoing a sequence of changes from initiation to decay. The start of cyclone development involves a weak surface trough or front and a wave in the upper level westerlies. Intensification of the surface low takes place as a result of divergence in the upper westerly wave that leads to the formation of a cyclonic circulation along the surface front. Establishment of the cyclonic circulation allows the flow of warm air to occur to the east of the cyclone centre and cold air transfer to the west of the centre. This temperature modification changes the distribution of uplift from one in which there is a general area of ascent and low-level convergence to one in which ascent takes place mainly east and poleward of the surface low and descent west and equatorward of the low. The circulation associated with the cyclone causes the cold front to move eastward and equatorward and the warm front to move eastward and poleward. This stage of development is shown in Figure 2(a). Further advection of warm and cold air together with changes in the upper airflow produce a gradual (or sometimes rapid) change in the appearance of the surface low. By stage (d) the storm has reached its maximum intensity and started to occlude. Cold air is present at all levels over the cyclone centre. As the system decays, advection over the centre almost ceases. Surface friction and internal dissipation ensure that this eddy in the westerly circulation loses its identity.

In the northern hemisphere, troughs in the westerlies have favoured positions of location because of the mountain ranges and the distribution of land and ocean. As a result, cyclones have preferred areas of formation: off Newfoundland and northeast of Japan for example. In the southern hemisphere, the westerly flow is

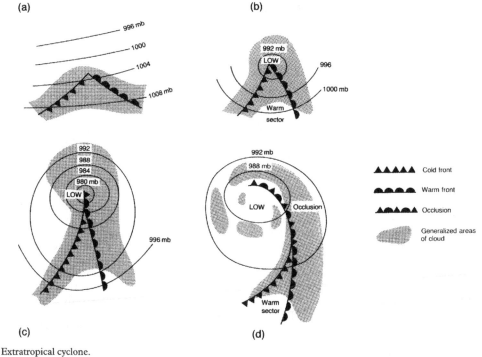

Extratropical cyclone.

more symmetrical about the pole and longitudinal variations in cyclone formation are less pronounced. (See also CYCLONE.) PS

Reading
Carlson, T.N. (1998) *Mid-latitude weather systems*. Boston, MA: American Meteorological Society. · Martin, J.E. (2006) *Mid-latitude atmospheric dynamics*. Hoboken, NJ: John Wiley & Sons, Inc.

extreme value theory A group of statistical methods that is widely applied in the study of natural phenomena, such as floods. Extreme values represent the outliers among populations, such as the shortest individuals, or the most intense storm events, or the largest flood experienced in each year of a period of record. Sets of such extreme values often follow a form of statistical distribution that is called an *extreme value distribution*. Many different mathematical functions have been employed to describe and analyse data for which some form of extreme value distribution is appropriate. These include the Gumbel, log Gumbel, Weibull and others (Kinnison, 1985). Two approaches exist for practical extreme value analysis. The first method relies on deriving block maxima (minima) series as a preliminary step. In many situations it is usual to extract the annual maxima (minima), generating an 'annual maxima series'. The second method relies on extracting, from a continuous record, the peak values reached for any period during which values exceed a certain threshold (falls below a certain threshold). This method is generally referred to as the 'peak-over-threshold' method and can lead to several or no values being extracted in any given year. Different kinds of distributions apply to each form of investigation. If an appropriate extreme value function can be fitted to some observational flood size data collected over 30 years, for example, then it may be used to estimate the magnitude of the larger events that might have been encountered had a longer period of record, say 500 or 1000 years, been available. DLD/TS

Reference
Kinnison, B.R. (1985) *Applied extreme value statistics*. Columbus, OH: Battelle Press.

extrusion flow of glaciers A view that the lower layers of a glacier are squeezed out by the weight of the overlying ice and move faster than it. The concept has been abandoned as physically impossible.

extrusion, volcanic A feature produced by rocks that have been deposited at the Earth's surface after eruption, in a molten or solid state, from volcanic vents and fissures.

exudation basin A depression occurring at the head of glaciers emanating from the Greenland ice cap.

eye The centre of a cyclone where air descends from the upper atmosphere, filling the low-pressure zone. A calm area around which cyclonic winds blow. Also the source of a river/spring.

eyot An islet in a river or lake.

F

fabric The three-dimensional arrangement of particles of a sediment. It is a bulk property and can be specified at a variety of scales and in a number of ways, while the volume of sediment analysed impacts on the measurement and interpretation of fabric data. For example, for the finest sediments, clays, it is possible to refer to the 'clay fabric' over a volume of a few cubic millimetres, while for coarser sediments fabric may require measurement over a volume of a cubic metre or so.

Most usually, clay fabrics are related to the engineering properties of the soil, especially to the response to STRAIN; this is similar to the analysis of strain and fabrics of metamorphic rocks. The micro-fabric of the clay component may differ significantly from the overall fabric of the sediment, as this takes much larger particles into account and covers a much greater volume. The fabrics of stones (clasts) are often used as indicators of palaeocurrent direction or ice sheet movement (TILL FABRIC ANALYSIS). Fabric can also be used to help to identify a specific geomorphological mechanism (e.g. imbrication of pebbles in a stream), as well as to help explain the mass behaviour of a sediment (e.g. preferred permeability direction).

Fabrics are sometimes determined in two dimensions, but ideally three dimensions should be used. Where stones are employed, a preferred orientation is determined from analysis of a number (more than 50) of individual measurements that are statistically analysed to give a measure of the fabric for that position. A clast long axis a needs to be determined and its orientation, or azimuth (with respect to, say, true north) and its plunge (with respect to an angle below the horizontal) determined. These two values can be plotted on an equal-area projection (or net) and contoured for all the clasts taken. Alternatively, the maximum projection (a/b) plane can be determined and the orthogonal to this (the c-axis) measured with respect to azimuth and plunge. This again can be plotted. Values of azimuth and plunge (of a line) or dip (of a plane) can be used to calculate mean orientations directly by various statistical measures and tested against several types of spherical distribution. WBW

Reading
Goudie, A.S. (ed.) (2005) *Geomorphological techniques*, 2nd edition. London: Unwin Hyman. · Whalley, W.B. (1976) *Properties of materials and geomorphological explanation.* Oxford: Oxford University Press.

facet A flat surface on a rock or pebble produced by abrasion.

facies The characteristics of a rock or sediment that are indicative of the environment under which it was deposited. A distinct stratigraphic body that can be distinguished from others on the basis of appearance and composition. Lateral variations in the nature of a stratigraphic unit.

factor of safety The ratio of RESISTANCE to FORCE; whenever this falls below 1.0, acceleration will take place. Most often this term is used in the context of slope stability; LANDSLIDE movement takes place when the sum of resisting forces falls below the sum of driving forces. Since the least resistant material is unlikely to have been sampled, and there is an error margin on the estimation of force, factors of safety below 1.3 are generally considered unstable; higher or lower threshold values may be appropriate. Since both resistance and force vary over time, the worst case expected should be used; for example, zero cohesion or maximum pore-water pressure. Calculation of a factor below 1.0 means that some measurements or assumptions are inaccurate, otherwise movement would have taken place. IE

The Dictionary of Physical Geography, Fourth Edition. Edited by David S.G. Thomas.
© 2016 John Wiley & Sons, Ltd. Published 2016 by John Wiley & Sons, Ltd.

Multiple fairy circles, 2–3 m in diameter, in southern Namibia. Photograph by David Thomas.

fairy circles A type of intriguing, but little understood, vegetation pattern that has been likened to an ostrich skin (Picker, 2012). These occur in their thousands in the pro-Namib in Namibia (Albrecht *et al.*, 2001) in areas with sandy soils and where the mean annual rainfall is 50–150 mm. They consist of bare areas ~2–12 m in diameter surrounded by perennial grasses such as *Stipagrostis*. In general, they seem to decrease in size from north to south, with the largest occurring in the far north of the country. The circles become larger as aridity increases. Fairy circles are widespread between the Orange River and southern Angola. It is possible that their origin owes something to the foraging action of termites (Juergens, 2013) or ants (Picker, 2012), or to growth inhibition as a result of allelopathic compounds released by dead *Euphorbia damarana* plants, but as yet there is no entirely satisfactory explanation for their origin. It has also been proposed that they may be the result of micro-seepage of gases and hydrocarbons (Naudé *et al.*, 2011). They are ephemeral features that have a lifespan on the order of decades (Tschinkel, 2012).   ASG

References
Albrecht, C.F., Joubert, J.J. and De Rycke, P.H. (2001) Origin of the enigmatic, circular, barren patches ('fairy rings') of the pro-Namib. *South African Journal of Science*, **97**, 23–27. · Juergens, N. (2013) The biological underpinnings of Namib Desert fairy circles. *Science*, **339**, 1618–1621. · Naudé, Y., van Rooyen, M.W. and Rohwer, E.R. (2011) Evidence for a geochemical origin of the mysterious circles in the Pro-Namib desert. *Journal of Arid Environments*, **75**, 446–456. · Picker, M. (2012) Little landscapers: social insects transform landscapes by recycling and releasing nutrients and increasing floral diversity.

Veld & Flora, **98**, 174–177. · Tschinkel, W.R. (2012) The life cycle and life span of Namibian fairy circles. *PLoS ONE*, 7, e38056.

falling dune A TOPOGRAPHIC DUNE that has developed on the lee side of a topographic obstacle, as airflow and sand transport is disrupted by the barrier. A falling dune usually develops when a CLIMBING DUNE has reached the top of an obstacle, allowing sand to be transported over the barrier, permitting deposition on the lee side. (See also ECHO DUNE and LEE DUNE.)   DSGT

fallout The descent of solid particles from the Earth's atmosphere and/or the particles themselves. Finer, lighter particles tend to fall out further from the source than coarser, heavier particles do. The particles may be of natural origin, such as mineral dust from a dust storm, or from anthropogenic sources, such as sulphur particles emitted by an oil-fired power plant. Both can be construed as part of the flow of material through biogeochemical cycles. The measurement of caesium-137 deposited on soils after radioactive fallout from atmospheric weapons testing has been widely practised as a means to estimate erosion and deposition (e.g. Bajracharya *et al.*, 1998).   NJM

Reference
Bajracharya, R.M., Lal, R. and Kimble, J.M. (1998) Use of radioactive fallout cesium-137 to estimate soil erosion on three farms in west central Ohio. *Soil Science*, **163**, 133–142.

false-bedding The stratification of sediments in several units inclined to the general stratification. The product of fluvial, littoral and aeolian sedimentation.

fan delta   An alluvial fan extending into a body of water. The subaerial fan results from fluvial and slope processes, and shares many of the characteristics of alluvial FANS. The subaqueous fan is affected also by lake or marine processes, like a DELTA.   IE

Reading
Schumm, S.A., Mosley, M.P. and Weaver, W.E. (1987) *Experimental fluvial geomorphology*. New York: John Wiley & Sons, Inc.; pp. 351–365, 372–374

fanglomerate   Indurated alluvial fan gravel.

fans   Depositional landforms whose surface forms a segment of a cone that radiates downslope from the point where the stream leaves the source area. The coalescing of many fans forms a depositional piedmont that is commonly called a *bajada*. Each fan is derived from a source area with a drainage net that transports the erosional products of the source area to the fan apex in a single trunk stream. The plan view of the cone-shaped deposit is broadly fan-shaped with the contours bowing downslope. Overall, radial profiles are concave and cross-fan profiles are convex. They vary greatly in size from less than 10 m in length to more than 20 km, and many large fans are thicker than 300 m. The debris that makes up fans decreases in size downfan, but is frequently coarse, and much of it has been transported by mudflow activity. Deposition is caused by decreases in depth and velocity where streamflow spreads out on a fan, and by infiltration of water into permeable superficial deposits.

Alluvial fans are widespread, especially in arid, mountainous and periglacial areas, but are especially notable in particular tectonic environments where there is a marked contrast between mountain front and depositional area. Uplift creates mountainous areas that provide debris and increased stream competence.   ASG

Reading
Bull, W.B. (1977) The alluvial-fan environment. *Progress in Physical Geography* **1**, 222–270. · Harvey, A.M., Mather, A.E. and Stokes, M. (eds) (2005) *Alluvial fans: geomorphology, sedimentology, dynamics*. Geological Society Special Publication 251. London: Geological Society.

fatigue failure   Fracture as a result of repeatedly applied cyclic stresses, at levels far below the instantaneously determined strength of a material; it is widely recognized as a major contributor to failure in metals. In aircraft, regular vibrations have led to disastrous fracturing. Experimental studies of cyclic stresses in rocks have shown that fatigue fracturing occurs at 80–60% of the ultimate strength and after a number of cycles ranging from 10^3 to 10^6. The importance of fatigue failure in rocks exposed to repeated wetting and drying, thermal expansion and contraction, and salt crystallization and dissolution is not known. It may, however, be a significant cause of several forms of rock fracturing that have not yet been explained satisfactorily.   MJS

Reading
Selby, M.J. (1993) *Hillslope materials and processes*, 2nd edition. Oxford: Oxford University Press; chapter 8.

Fault plane

Block before faulting

Normal fault

Reverse fault

Strike-slip fault

Oblique-slip fault

Fault types.
Source: Press & Siever (1978).

fault A crack or fissure in rock, the product of fracturing as a result of tectonic movement. The line along which displacement of formerly adjacent rocks has taken place as a result of earth movements. Normal faults develop under a pattern of stress that is predominantly tensional. A down-faulted block between a pair of more or less normal faults is known as a graben, and the up-faulted block is called a horst. By contrast, reverse faults are normally associated with zones of compression. Where the angle of dip is low the term *thrust fault* is used. When the mean compressive stress is vertical, strike-slip faults are formed (also called wrench faults and trans-current faults). Where both horizontal and vertical movements are significant, the term *oblique-slip* fault is applied.   ASG

Reference
Press, F. and Siever, R. (1978) *Earth*. San Francisco, CA: W.H. Freeman.

faunal realms The largest divisions into which the faunas of the world are customarily grouped; also called faunal kingdoms or empires. The earliest enduring classification of world faunas was drawn up by Sclater in 1858 and dealt with the distribution of birds, a group already well known by that date. Alfred Russel Wallace was responsible for revising this system in his early classic on zoogeography, *The Geographical Distribution of Animals*, in 1876 (see WALLACE'S REALMS). In this, he subdivided each of Sclater's original six regions into four subregions. The Sclater–Wallace scheme is still in use today, despite attracting continuous criticism.

feather edge The thin edge of a wedge-shaped sedimentary rock that tapers out and eventually disappears as it abuts an area where no deposition has taken place.

feedback See SYSTEMS.

feedback loops or homeostasis See SYSTEMS.

feldspars Aluminosilicates of K, Na and Ca, and as such are important components of many rocks and rock-weathering products. Feldspars have alkali and plagioclase types and in total frequency are the second most abundant mineral (after quartz) in SAND.

felsenmeer See BLOCK FIELDS, BLOCK STREAMS.

fen A type of inland wetland, fed by surface flows or groundwater, which is pH neutral or alkaline and exhibits relatively high dissolved minerals but few other nutrients. Fens are usually dominated by grasses, reeds and sedges, and in some cases develop into carr woodland (typical species being alder and willow). Fens have been damaged in the past by land drainage, and associated peat shrinkage and by peat cutting. Some are now being carefully restored through the reintroduction of more natural water flow regimes, improvements in water quality and removal of invasive woody plants.   TS

Reading
Friday, L.E. (ed.) (1997) *Wicken Fen: the making of a wetland nature reserve*. Colchester: Harley Books. · Gore, A.J.P. (ed.) (1983) *Mires: swamp, bog, fen and moor – regional studies*. Ecosystems of the World 4B. Amsterdam: Elsevier.

ferrallitization A combination of intense weathering and efficient removal of the more soluble weathering products under warm, wet conditions. There are three basic aspects to the process: intensive and continuous weathering of parent material involving the release of iron and aluminium oxides and silica, as well as of bases; translation of soluble bases and silica; and formation of 1 : 1 kaolin-type clays.

Ferrel cell A thermally indirect, weak, MERIDIONAL CIRCULATION in middle latitudes of the atmosphere, in which air rises in the colder regions around 60° latitude and descends in the warmer regions around 30° latitude (subtropic high-pressure belts). This cell, in which the surface winds are predominantly westerly, forms part of a three-cell (in each hemisphere) mean meridional circulation pattern, the other two cells being the Hadley and polar cells.   KJW

Reading
Holton, J.R. and Haskim, G.J. (2013) *An introduction to dynamic meteorology*, 5th edition Waltham, MA: Academic Press.

Ferrel's law States that all moving masses of air move to the right (as they proceed) in the northern hemisphere and to the left in the southern hemisphere. This is because air, in moving from higher to lower pressure, is subject to the effect of the Earth's rotation (see CORIOLIS FORCE).   KJW

ferricrete An iron-pan or a near-surface zone of iron oxide cementation. An accumulation of iron oxides and hydroxides within the soil zone as a result of weathering or soil-forming processes such as laterization. It is a type of duricrust and tends to be associated with deep weathering profiles.

fetch or fetch length The distance over which the wind has blown across the sea surface.

Along with wind speed and wind duration, the fetch determines the size of waves produced. The longer the fetch (and the higher the wind speed and the longer its duration), the more wind energy is imparted to the water surface and the larger the resulting sea state will be, ultimately producing a 'fully arisen sea'. It follows, therefore, that wave heights can be 'fetch limited'.   TS

Reading
Young, I.R. (1999) *Wind generated ocean waves*. New York: Elsevier.

fiard   A river valley that has been drowned by the sea owing to a rise in sea level or subsidence of the land.

field capacity   The condition reached when a soil holds the maximum possible amount of water in its voids and pores after excess moisture has drained away. A measure of the volume of water a soil can hold under these conditions.

field drainage   The process of artificially accelerating the movement of water through soil. This is undertaken to lower levels of saturation, to reduce the extent of surface water ponding or run-off and to minimize the periods during which saturation or surface water exists on agricultural land. This process is alternatively called underdrainage, in contrast to the surface and arterial drainage that involve ditching and the modification of existing water courses to accelerate channelized run-off.

Field drainage may be implemented by excavating a network of trenches and inserting plastic piping or tubing, earthenware pipes or permeable rock fill, and then backfilling the trenches again. In heavy cohesive soils an extra network of mole drainage may be added. This is undertaken by drawing a former (the 'mole') through the soil at the required depth, the hole remaining open for a number of years afterwards. Deep ploughing ('subsoiling') may also accelerate field drainage.   JL

Reading
Green, F.H.W. (1978) Field drainage in Europe. *Geographical Journal*, **114**, 171–174.

Finger Lakes   A group of semi-parallel lakes in New York State. The name originated in the Indian legend of the Great Spirit placing his hand on the Earth causing the finger-like indentations. The accepted explanation is that they were cut by glacial erosion.   DES

Reading
Mullins, H.T. and Hinchey, E.J. (1989) Erosion and infill of New York Finger Lakes: implications for Laurentide ice sheet deglaciation. *Geology*, 17, 622–625.

fiord (fjord)   A glacial trough whose floor is occupied by the sea. Fiords are common in uplifted mid-latitude coasts; for example, in Norway, East Greenland, eastern and western Canada and Chile. Historically, the origin of fiords has generated a great deal of interest; for example, an extreme view championed by Gregory (1913) that they are essentially tectonic forms. Today, it seems easiest to think of them as essentially glacial with such features as steep sides, deepened rock basins and shallow thresholds at the coast as characteristic of erosion by ice streams or valley glaciers exploiting either pre-existing river valleys or underlying weaknesses in the bedrock (Holtedahl, 1967).   DES

Reading and References
Gregory, J.W. (1913) *The nature and origin of fiords*. London: John Murray. · Holtedahl, H. (1967) Notes on the formation of fjords and fjord valleys. *Geografiska Annaler, Series A: Physical Geography*, **49**, 188–203. · Syvitski, J.P.M., Burrell, D.C. and Skei, J.M. (1987) *Fiords: processes and products*. New York: Springer-Verlag.

fire   This is a very important ecological factor, especially in regions with seasonally dry climates. Fire has played a significant role in the evolution of plant species in many different biomes, including boreal forests, Mediterranean-climate shrublands, savanna and temperate grasslands and is a major determinant of species diversity. For plant species adapted to burning, so-called pyrophytes, fire is an integral and essential component of their life cycles, provided that the fire regime is appropriate. Fire-adapted species may survive fire either by resprouting, usually from underground parts, or by regeneration from soil or canopy-stored seed bank, in which case the adult plant is actually killed by the fire. Fire regime is accounted for by variations in the type, intensity, timing and frequency of fire, all of which are dependent on climate, availability of fuel and occurrence of ignition events. Fire has many positive implications for plant species with appropriate adaptations, since the combination of open space, increased nutrient availability and temporary reduction in numbers of potential seed predators is favourable for seedling establishment in the immediate post-fire environment. Fire may stimulate flowering, seed release or seed germination in such instances.

People have altered and manipulated the fire regime for a very long time, perhaps for more than 1 million years (Gowlett *et al.*, 1981). Fire is utilized by people in a range of environments for a variety of reasons; for example, in clearing vegetation, attracting game, to deter pests and disease, to improve grazing for domestic stock and to make pottery. Given the virtual ubiquity of current human populations, it is difficult to know what would constitute a 'natural' fire regime. Most fire-prone vegetation types are now subject to anthropogenic burning regimes, either intentionally or otherwise; indeed, it has been argued that some vegetation types (e.g. the savannas) may owe their distribution and characteristics to long-term human-induced fires.

Land stewardship and climate change are thought to be increasing the likelihood of very large fires, the 'mega-fires' of more than 10^4 ha. Indeed, there is an increasing challenge in attempting to understand and manage fire regimes for the protection of biodiversity, particularly as settlement and infrastructure occupy areas of flammable vegetation.   MEM/DLD

Reading and Reference
Attiwil, P. and Binkley, D. (2013) Exploring the mega-fire reality: a 'Forest Ecology and Management' conference. *Forest Ecology and Management*, **294**, 1–3. · Bond, W.J. and van Wilgen, B.W. (1996) *Fire and plants*. London: Chapman and Hall. · Gowlett, J.A.J., Hairns, J.W.K., Walton, D.A. and Wood, B.A. (1981) Early archaeological sites, hominid remains and traces of fire from Chesowanja, Kenya. *Nature*, **294**, 125–129. · Stephens, S.L., Burrows, N., Buyantuyev, A., *et al.* (2014) Temperate and boreal forest mega-fires: characteristics and challenges. *Frontiers in Ecology and the Environment*, **12**, 115–122.

firn   The term generally applied to snow that has survived a summer melt season and which has not yet become glacier ice.

firn line   See EQUILIBRIUM LINE.

fission track dating   Based on the principle that traces of an isotope, ^{238}U, occur in minerals and glasses of volcanic rocks, and that this isotope decays by spontaneous fission over time, causing intense trails of damage called tracks. These narrow tracks, between 5 and 20 µm in length, vary in their number according to the age of the sample. By measuring the numbers of tracks an estimate can be gained of the age of the volcanic minerals.   ASG

Reading
Walker, M. (2005) *Quaternary dating methods*. Chichester: John Wiley & Sons, Ltd.

fissure eruption   The eruption of volcanic gases, lavas and rocks through a large crack or chasm in the Earth's surface rather than through a pipe or vent.

flash flood   Flash floods are local or rapid increases in DISCHARGE commonly, but not exclusively, in previously dry valleys. They are normally associated with rapid transport of large amounts of water and/or sediment. They may be either natural, such as due to a sudden rainstorm (e.g. Schick and Lekach, 1987) or involve anthropogenically created hazards such as a dam break (e.g. Jarrett and Costa, 1985).   SNL

References
Jarrett, R.D. and Costa, J.E. (1985) Hydrology, geomorphology, and dam break modeling of the July 15, 1982, Lawn Lake Dam and Cascade Lake Dam failures, Larimer County, Colorado. US Geological Survey, Professional Paper 1369. Washington, DC: US Government Printing Office. · Schick, A.P. and Lekach, J. (1987) A high magnitude flood in the Sinai desert. In L. Mayer and D. Nash (eds), *Catastrophic flooding*. Boston, MA: Allen & Unwin; chapter 18, pp. 381–410.

flashiness   The rapidity with which the stage discharge increases at a stream cross-section. The hydrograph (a graphical plot of water discharge versus time) of a flashy stream shows a rapid increase in discharge over a short period, with a quickly developed high peak in relationship to normal flow (Ward, 1978: 10). On the other hand, a sluggish stream develops peaks more slowly, and the peaks are relatively lower than those developed on flashy streams.   WLG

Reading and Reference
Schick, A.P. (1970) Desert floods: interim results of observations in the Nahal Yael research watershed, southern Israel, 1965–1970. In Symposium on the results of research on representative and experimental basins: proceedings of the Wellington symposium, vol. 1. International Association of Scientific Hydrology publication. No. 96. Paris: IASH–UNESCO; pp. 478–493. · Ward, R. (1978) *Floods: a geographical perspective*. London/New York: Macmillan/John Wiley & Sons, Inc.

flatiron   Steep triangular cliff facets resulting from the presence of a capping of rock resistant to erosion which protects the underlying more readily eroded rocks.   ASG

Reading
Schmidt, K.-H. (1989) Talus and pediment flatirons – erosional and depositional features on dryland cuesta scarps. *Catena Supplement*, **14**, 107–118.

float recorder   A device once in widespread use for recording the water level in a river channel, lake, well or similar situation. A flexible cable or steel tape connecting a float and counterweight passed over a pulley that transmitted vertical movements of the float to the recording mechanism. In modern river stage recording, float recorders are rarely used, having been superseded by digital pressure transducers, ultrasonic level gauges and other technologies. These are free from the mechanical problems of floats, and interface directly with modern data-logging equipment. DLD

floating bogs Refers to any lake-fill bog or kettle-water wetland with a quaking *Sphagnum* mat. Such bogs, which are common in parts of North America, are generally referred to as 'schwingmoor' or 'kesselmoor' in Europe. ASG

Reading
Warner, B.G. (1993) Palaeoecology of floating bogs and landscape change in the Great Lakes drainage basins of North America. In F. M. Chambers (ed.), *Climate change and human impact on the landscape*. London: Chapman & Hall; pp. 237–245.

flocculation The process of aggregating into small lumps, applied especially to clays.

flood A high water level along a river channel or on a coast that leads to inundation of land that is not normally submerged. River floods that involve inundation of the FLOODPLAIN may be caused by:

- *Precipitation* when storm precipitation is very intense; when precipitation is very prolonged and follows a period of wetter than average conditions; when the snowpack melts and snowmelt floods are an annual feature of many river regimes; when rain falls on snow and accelerates snowmelt; or when ice and snowmelt are combined and melting river ice produces break-up floods.
- *Collapse of dams*, which may be natural where a landslide temporarily blocks a river or ice or log jams obstruct the river channel; or which may occur when a manmade dam bursts and the impounded water flows down the river valley as a flood wave or when a landslide into a reservoir leads to a wave overtopping the dam wall.
- *Drainage of ice-dammed lakes*, which can lead to the release of great volumes of water (in Iceland JÖKULHLAUPS are flood waves that roar down-valley as a result of the failure of a lake dammed by, or contained within, a glacier).
- *Loss of channel capacity resulting from sedimentation* resulting from accelerated erosion in the catchment area.
- *Breaching of levees* designed to contain floodwaters within the alluvial ridge.

In addition to being effected by the above causes, coastal floods may also be produced by:

- *High tides* – especially in combination with river floods.
- *Storm surges* – wind-driven inundation of coastal land, especially when low pressure associated with cyclones is coincident with high tides.
- *Tsunami* – large waves produced by submarine earthquakes, volcanic eruptions, landsliding or slumping.

The FLOOD FREQUENCY in a particular area may vary over time, especially as a result of human activity, and in some areas the damage produced by floods has increased despite expenditure on flood protection simply because there is a tendency to assume that flood protection measures have completely eliminated the flood hazard. Responses to flood hazard include:

- *Flood protection* by structural measures along the river or coast to reduce the effects of flooding; these include walls and embankments, river diversion schemes, flood barriers.
- *Flood reduction* by taking action in the drainage basin by afforestation, controlled vegetation or agricultural changes; or by the construction of small or large dams.
- *Flood adjustment* by adjusting to the hazard by accepting it; by land-use zoning; by taking emergency measures when floods occur; or by flood-proofing so that flooding will damage structures and buildings as little as possible.

Research on palaeoflood hydrology uses slackwater deposits and other kinds of field evidence to suggest that, in some areas, floods larger than those seen in the historic period have occurred during the Holocene. KJG/DLD

Reading
Baker, V.R. (1987) Paleoflood hydrology and extraordinary flood events. *Journal of Hydrology*, **96**, 79–99. · Baker, V.R., Cochel, R.C. and Patton, P.C. (1988) *Flood geomorphology*. Chichester: John Wiley & Sons, Ltd. · Di Baldassarre, G., Viglione, A., Carr, G., *et al.* (2013) Socio-hydrology: conceptualizing human–flood interactions. *Hydrology and Earth System Sciences*, **17**, 3295–3303. · Ward, R.C. (1978) *Floods: a geographical perspective*. London: Macmillan.

flood frequency An analysis using data from the period of hydrologic records to establish a relationship between discharge and return period or probability of occurrence as a basis for estimating discharges of specific recurrence intervals. Several methods are available for calculating the recurrence intervals or return periods T, but one most frequently employed ranks the N years of record from the highest (rank $m = 1$) to the lowest (rank $m = N$) and uses

$$T = \frac{N+1}{m}$$

To obtain the N years of record, use of the highest discharge from each year of record is the basis for the ANNUAL SERIES, whereas using the N highest independent discharges provides the annual exceedance series. The recurrence intervals may be plotted against the discharges for a specific gauging station and the curve fitted by eye or by fitting an appropriate theoretical probability distribution. A range of distributions are available, and the Gumbel extreme value type 1 based on GUMBEL EXTREME VALUE THEORY is one frequently used. The flood frequency plot of discharge against recurrence interval can be the basis for estimating the recurrence interval of discharge including the BANKFULL DISCHARGE and for giving the DESIGN DISCHARGE for a particular return period when a structure is being planned. When a regional flood frequency analysis is required using the data from a series of gauging stations within the region it is necessary to undertake a homogeneity test that identifies a homogeneous region according to the 10-year flood estimated from the probability curve for each gauging station.   KJG

Reading

Newson, M.D. (1975) *Flooding and flood hazard in the United Kingdom*. Oxford: Oxford University Press. · Ward, R.C. (1978) *Floods: a geographical perspective*. London: Macmillan.

floodout   The part of an ephemeral desert stream system, located in the lowest reaches of the channel, where organized channel flow gives way to multiple distributary channels and ultimately to shallow nonchannelized flow. Progressive transmission losses taking place along the channel from its headwater source areas progressively reduce the sediment-transporting capacity of the stream. Resulting deposition in the channel, together with a diminished ability to flush away wind-blown sands that may accumulate there, lead to falling channel capacity (Mabbutt, 1977). Many floodouts are located where EPHEMERAL STREAMS first contact dunefields lying across their path. When large floods arrive at such a location, overbank flows spill from the channel and cut many small channels leading away from the trunk stream, perhaps draining along interdune corridors. Floodout locations appear to become semi-permanent in this way, and frequently channels collapse at floodouts before reaching saline lake basins that would otherwise be their flow terminus.   DLD

Reading and Reference

Mabbutt, J.A. (1977) *Desert landforms*. Canberra: Australian National University Press. · Tooth, S. (1999) Floodouts in central Australia. In A. J. Miller and A. Gupta (eds), *Varieties of fluvial form*. Chichester: John Wiley & Sons, Ltd; pp. 219–247.

floodplain   The *hydraulic floodplain* is an engineering concept for any surface subject to river flooding within a given return period (say, every 100 years). Such surfaces may be formed of bedrock or Pleistocene sediment and only slightly modified by the contemporary river. The *genetic floodplain* is a geomorphological term for a relatively flat alluvial landform, constructed largely by the flow regime of the present river and subject to flooding. Commonly, floodplains flank a clearly defined river channel, but some occur in valleys without channels, while others form downstream of channels (a *floodout*). A genetic floodplain shows strong affinity to the river that has formed it, with sedimentology, stratigraphy and morphology reflecting flow and catchment conditions. With changes in flow regime, rivers may incise, leaving abandoned floodplains as RIVER TERRACES.

There are three main processes of floodplain formation: LATERAL ACCRETION, OVERBANK DEPOSIT and BRAIDED RIVER accretion. Three less common processes are oblique accretion (the onlapping of relatively steeply dipping convex-bank deposits, as the channel migrates), counterpoint accretion (sediment deposited against the concave bank of tightly curving bends) and abandoned channel (OXBOW) accretion. These processes are determined in individual reaches by variations in stream power, sediment character and local vegetation, making it possible to classify floodplains into a wide variety of types, but essentially three broad classes: (1) high-energy, noncohesive sediment, (2) medium-energy, noncohesive sediment and (3) low-energy cohesive sediment floodplains. In detail, floodplain surfaces consist of LEVÉES, scrolls (prior levees), abandoned channels and BACKSWAMPS that greatly influence the passage of overbank flow.

Floodplains generally support fertile agricultural soils, but can be hazardous, rapidly changing geomorphological environments for human occupation. Flood risk management requires the accurate prediction of flood water levels, flow paths, velocities and inundation extents. This can be progressed either through a raster-based approach, which uses a relatively simple process representation, with channel flows being resolved separately from the floodplain using either a kinematic or diffusive wave approximation, or with a data-intensive finite-element hydraulic model aiming to solve the full two-dimensional shallow water equations. Data availability, model parametrization and model validation remain challenging for such studies.   GCN/TS

Reading

Horritt, M.R. and Bates, P.D. (2001) Predicting floodplain inundation: raster-based modelling versus the finite-element approach. *Hydrological Processes* 15, 825–842. ·

Knighton, D. (1998) *Fluvial forms and processes*. London: Arnold. · Lewin, J. (1978) Floodplain geomorphology. *Progress in Physical Geography*, **2**, 408–437. · Nanson, G.C. and Croke, J.C. (1992) A genetic classification of floodplains. *Geomorphology*, 4, 459–486.

flood routing The technique of determining the timing and shape of the flood wave at successive points along a river channel. Techniques for flood peak estimation applicable to small areas are not appropriate for larger basins because as a flood wave is routed along the channel it will change in shape as a result of the storage offered by the channel and the characteristics of the channel morphology. Flood routing may be accomplished either by a theoretical method or by an empirical method, of which the MUSKINGUM METHOD is well known.   KJG

Reading
Lawler, E.A. (1964) Flood routing. In V. T. Chow (ed.), *Handbook of applied hydrology*. New York: McGraw-Hill; section. 25-II.

floristic realms The largest divisions into which the floras of the world are customarily grouped; also called floristic kingdoms or empires. In turn, these may be subdivided into floristic regions, provinces and districts. Floristic realms are mainly characterized by the plant families that are endemic to that particular portion of the globe. The main realms are constantly under review (see Moreira-Muñoz (2007)), but those normally recognized are: the realms within the Holarctic floral kingdom, which embraces North America, Greenland, Europe and the northern part of Asia; the realms of the Palaeotropical kingdom; the Neotropical kingdom; the Austral or Australian kingdom, which includes southern South America, southern Africa, Australia and New Zealand. The Cape floral realm or kingdom is a distinct case with extreme diversity and species uniqueness, at the southern extremity of Africa. The boundary between the Holarctic and the tropical realm is, with a few exceptions, marked by the southward limit of the northern hemisphere coniferous family (Pinaceae) and the northward limit of the palms (Palmae or Arecaceae).

The evolution of these different realms reflects above all the geological history of the Earth and the movement of continents brought about by the processes of plate tectonics and sea-floor spreading. (See also faunal realms; WALLACE'S REALMS.)   PAS/MEM

Reading and References
Good, R. (1974) *The geography of the flowering plants*, 4th edition. London: Longman. · Goudie, A.S. (2001) *The nature of the environment*, 4th edition. Oxford: Blackwell. · Moreira-Muñoz, A. (2007) The austral floristic realm revisited. *Journal of Biogeography*, **34**, 1649–1660.

flow duration curve (FDC) A plot that shows the percentage of time that flow in a stream is likely to equal or exceed some specified value of interest. For example, it can be used to show the percentage of time streamflow can be expected to exceed a design flow of some specified value (e.g. $20 \, \mathrm{m}^3 \, \mathrm{s}^{-1}$), or to show the discharge of the stream

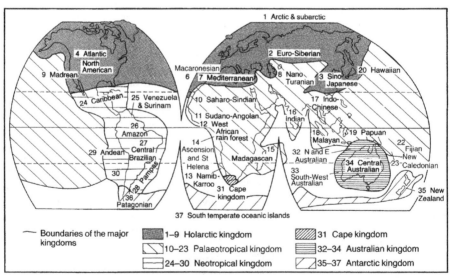

Floristic realms.
Source: Goudie (2001). Reproduced with permission of Andrew Goudie.

that occurs or is exceeded some percentage of the time (e.g. 75% of the time). The basic time unit used in preparing an FDC will markedly affect its appearance. In most studies, mean daily discharges are used, and these will give a steep curve. However, when the mean flow over a longer period is used (such as mean monthly flow), the resulting curve will tend to be flatter due to averaging of short-term peaks with intervening smaller flows during a month. Extreme values are averaged out more and more as the time period gets longer (e.g. for an FDC based on annual flows at a long-record station). FDCs that plot steeply throughout denote highly variable flows with a large stormflow component, whereas gently sloping curves indicate a large baseflow component, as for example in groundwater-fed streams. An FDC characterizes the ability of the basin to provide flows of various magnitudes. Information concerning the relative amount of time that flows past a site are likely to equal or exceed a specified value of interest is very useful for the design of structures.   ASG

Reading
Ward, R.C. and Robinson M. (2000) *Principles of hydrology.* London: McGraw Hill.

flow equations   These define the interrelationships between velocity, depth, slope and boundary roughness or energy losses in open-channel flow. A variety of flow conditions can be defined, including UNIFORM STEADY FLOW and GRADUALLY VARIED FLOW. In the former the water surface is parallel to the bed, and the width, depth and velocity are constant along a reach. Under such constrained circumstances the velocity formulae developed by Chézy and Manning (see CHÉZY EQUATION and MANNING EQUATION) relate mean velocity to depth, slope and a roughness coefficient. These equations may be used to estimate flood velocities if flood depth and slope are measured and a suitable value of the roughness coefficient is identified. The equations that define point velocity variation with height above the channel bed – that is, the velocity profile shape within the BOUNDARY LAYER – also relate to uniform flow conditions. In gradually varied flow, which is more common in natural river channels, the depth and velocity vary from section to section, and bed and water surface slopes are not parallel. Under these conditions the flow properties must be modelled by an energy-balance equation based on the principle of conservation of energy between closely spaced sections. This is the Bernoulli equation (see GRADUALLY VARIED FLOW), which defines the total energy of the flow as the sum of potential energy, pressure energy and kinetic energy.   KSR

flow regimes   Four flow regimes are defined, in terms of the depth and velocity of open channel flow, by the combination of criteria based on the Froude and Reynolds numbers. These describe the relationships between inertial, viscous and gravitational forces acting on the flow. The Froude number distinguishes subcritical (Fr < 1) from supercritical (Fr > 1) flow, and the Reynolds number identifies laminar (Re < 500) and turbulent (Re > 2000) flow with a transitional state between these limits. The flow regimes are laminar subcritical, laminar supercritical, turbulent subcritical and turbulent supercritical. In river channels, turbulent subcritical conditions are normal, while overland flow on hillslopes is often laminar except when disturbed by rainsplash.

In SANDBED CHANNELS, the progressive change of flow intensity during the passage of a flood along a reach is associated with a systematic sequence of changing bedforms. In the subcritical regime, as Froude and Reynolds numbers increase, small sand ripples up to 4 cm high and 60 cm long are replaced first by dunes with superimposed ripples and then by dunes up to 10 m high and 250 m long, depending on river size. At Froude numbers approaching unity, as the flow passes into the upper regime supercritical state, dunes wash out to a plane bed with intense sediment transport, and then antidunes and standing waves form. Thus, the flow regimes are closely related to bedform changes, which are in turn associated with changes of sediment transport rate and ROUGHNESS (Simons *et al.*, 1961).   KSR

Reading and Reference
Richards, K.S. (1982) *Rivers: form and process in alluvial channels.* London: Methuen. · Simons, D.B., Richardson, E.V. and Albertson, M.L. (1961) *Flume studies using medium sand (0.45 mm).* United States Geological Survey water supply paper 1498A. Washington, DC: US Government Printing Office.

flow resources   These may be defined as NATURAL RESOURCES that are 'naturally renewed within a sufficiently short timespan to be of relevance to human beings' (Rees, 1990: 15). They are also known as *renewable resources.* Two categories of flow resource can be discerned: those where flows are dependent on human activity, and whose future availability may be compromised by excessive use (see also SUSTAINED YIELD), and those where human usage has no impact on future availability. To this effect, Rees (1990) gave solar energy, air, water (at the global scale), wind and tidal energy as examples of the latter, termed *noncritical zone* flow resources, and fish, forests, soil and water in aquifers as examples of those in danger of losing their renewability through human actions. These can be termed *critical zone flow resources*, and in effect they can become STOCK RESOURCES if their ability to regenerate is compromised. DSGT

Reference
Rees, J. (1990) *Natural resources: allocation, economics and policy*, 2nd edition. London: Routledge.

flow till See TILL.

flow visualization This provides tools that have led to major advances in our understanding of fluid dynamics and boundary layers and allows the tracking and following of individual turbulent flow structures as they develop. Many flow visualization techniques employ foreign tracers in the flow, such as dyes, smoke and air bubbles. ASG

Reading
Buffin-Bélanger, T. and Kirkbride, A.D. (2004) Flow visualization. In A. S. Goudie (ed.), *Encyclopedia of geomorphology*. London: Routledge; pp. 387–388.

fluid mechanics A branch of applied mathematics dealing with the motion of fluids and the

Building a saltmarsh in the Large Wave Flume (Grosser Wellenkanal, GWK) of Forschungszentrum Küste (FZK) in Hannover, Germany, to test over-marsh wave energy dissipation under storm waves. The flume is 330 m long, 7 m deep and 5 m wide. Photograph by Tom Spencer.

conditions governing such motion. Mechanics generally includes kinematics and dynamics, which respectively cover the geometry of motion and the forces involved in motion. Thus, the continuity equation that defines channel discharge as the product of cross-section area and velocity is a kinematic statement, while the theoretical velocity profiles of laminar and turbulent boundary layers are the product of dynamical analysis of forces acting upon and within moving fluids. HYDRAULICS deals specifically with the mechanics of *liquid flow*, particularly water. KSR

flume When used to measure the flow of water in open channels, a flume is a specially shaped, fixed hydraulic structure. Under free-flow conditions, it forces flow to accelerate in such a manner that the flow rate through the flume can be characterized by a level-to-flow relationship as applied to a single head measurement within the flume. Acceleration is accomplished through a convergence of the sidewalls, a change in floor elevation or some combination of the two; there is a wide variety of specialized designs that have been produced for specific channel and flow conditions. Calibration may involve formulae or field rating. The term *flume* is also used to describe a laboratory facility for the physical modelling of either flows and/or waves (generated by a mechanical wavemaker), to study their effects on structures, sediment transport and bed deposition/erosion. Often, the side walls contain windows or are completely transparent, allowing for a clear visual observation of the experiment and the easy deployment of optical instrumentation. The very largest flumes – like the Deltares' Delta Flume at 240 m long, 7 m deep and 5 m wide – allow for physical models on a scale close to prototype.

flute Also known as fluted moraine, these are elongated streamlined ridges of sediment produced beneath a GLACIER or ICE SHEET and which are aligned in the direction of ice flow. They are the smallest landform type in the family of SUBGLACIAL BEDFORMS, and typically occur as numerous parallel ridges, tens of centimetres to a few metres high and wide and tens of metres in length. These rather delicate landforms are best observed close to recently retreated glacier margins, and owing to their poor preservation potential are much less commonly found in relation to former ice sheets. CDC

Reading
Benn, D.I., and Evans, D.J.A. (2010) *Glaciers & glaciation*, 2nd edition. London: Hodder Education.

fluting The production of flute marks on soft, cohesive sediment surfaces by flow-induced scour, often associated with TURBIDITY CURRENTS. Flute marks are usually streamlined, and characteristic shapes include bulbous, conical and linguiform. Turbulent vortices cause the EROSION from either direct fluid stresses on the bed or by ABRASION from entrained particles. A bed defect acts as a locus for flow separation and vortex generation. The resulting erosion potential is greatest near the upstream end of the flute. Flute marks are often preserved as casts on the bottom of a subsequently deposited sand bed, where they are termed sole marks.   DJS

Reading
Allen, J.R.L. (1982) *Sedimentary structures: their character and physical basis*, vol. II. Developments in Sedimentology, 30B. Amsterdam: Elsevier.

fluvial This is the common usage of the term and refers to anything that is of or referring to a river, including things that grow or live in a river or that are produced by the action of a river. It is commonly combined as follows: (1) fluvial geomorphology, which refers to the study of the morphology of environments worked by rivers; (2) fluvial processes, which refer to flow processes and solute and sediment transport in rivers; (3) fluvial deposits or sediments, which refer to material worked or deposited in the landscape by rivers; and (4) fluvial erosion or denudation, which refers to erosion of a land surface by rivers.   SNL

fluvioglacial See GLACIOFLUVIAL.

fluviokarst A type of landform developed in limestone areas by a combination of river action and of true KARST processes. Included in this category are gorges and DRY VALLEYS.

flux The rate of flow of some quantity. It is perhaps most easily perceived in the context of fluids (air, water), but it also applies to many other elements of the natural environment, such as mass and energy. Many apparently unchanging forms in the natural world ensue from different fluxes into and out of the system. As such, fluxes are a primary element in natural environmental processes.   BWA

flyggberg An asymmetrical hill 1–3 km across and 100–300 m high shaped by the action of overrunning ice. Swedish derivation: *flygg*, a steep rock face; *berg*, mountain.

flysch Deposits of marine sandstones, shales, marls and clays produced during the initial uplift of the Alps by sedimentation and later deformation of

materials eroded from the uplifted rocks. The word now refers to any thick succession of alternations of the rocks mentioned above, interpreted as having been deposited by turbidity currents or mass-flow in a deep water environment within a geosynclinal belt.

fog If the air near the ground is cooled enough and its moisture content is high, CONDENSATION may occur and a fog will develop. By definition, a fog exists if the visibility in cloudy air is 1 km or less. There are several ways by which moist air may be cooled enough to form fog: loss of RADIATION at night (radiation fog), warm air passing over a cold surface (advection fog) and air flow up an incline (upslope fog). Fog may also form when warm rain falls through cold, but saturated, air (frontal fog) and when cold air passes over warm water (steam fog).   WDS

Reading
Strangeways, I. (2006) *Precipitation: theory, measurement and distribution*. Cambridge: Cambridge University Press.

föhn (sometimes foehn; chinook in North America) The name given to a strong and gusty downslope (or 'fall') wind that occurs on the lee side of a mountain range when stable air is forced to flow over the mountains by the large-scale pressure gradient. The air at the foot of the mountains is characteristically dry and warm as a result of the ADIABATIC compression during its descent. In some cases the air may ascend over the range, with cloud and precipitation occurring on the windward side, but in other cases a stable layer near the summit level blocks the cross-mountain flow and air descends from the summit level. Föhn onset can give rise to dramatic increases in temperature, and such winds are important in accelerating snowmelt in spring along the northern flanks of the Alps and Caucasus and the eastern flanks of the Rocky Mountains.   RGB

Reading
Gohm, A. and Mayr, G.J. (2004) Hydraulic aspects of föhn winds in an Alpine valley. *Quarterly Journal of the Royal Meteorological Society*, **130**, 449–480.

fold A bend in formerly planar strata of rock resulting from movement of the crustal rocks. If the strata are flexured in just one plane they are termed a monocline. The arching up of strata produces an anticline, while the depression of strata produces a syncline. In a recumbent fold the strata are overturned and both limbs of the folds are nearly horizontal. The horizontal compression that creates recumbent folds may lead to the shearing of the upper part of the fold along a

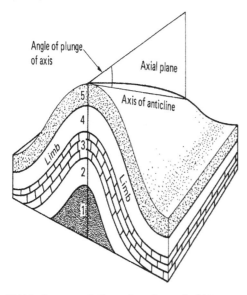

Fold 1: diagrammatic illustration of parts of a fold.

Fold 2: diagrammatic illustration of symmetrical, asymmetrical, overturned and recumbent folds.
Source: Press and Siever (1978: figures 20.23 and 20.24).

thrust fault. The strata moved forward over a thrust fault form a nappe.   ASG

Reference
Press, F. and Siever, R. (1978) *Earth*. San Francisco, CA: W.H. Freeman.

foliation   Fine layering of rocks, usually metamorphic and igneous rock, as a result of parallel orientation of minerals. Slates and schists are typical foliated rocks.

food chain, food web   Food chains represent the transfer of food (and, therefore, energy) from one type of organism to another, in sequence and in a linear relationship; for instance, in a terrestrial environment: lion → springbok → grass. Normally, however, this would represent an extreme oversimplification of the feeding relationships present in any community, which are much better portrayed as a food web (see diagram).

There are two major types of food web, termed grazing and detrital. At the base of the grazing food web are producer organisms that are autotrophs; that is, they are able to fix incident light energy and so manufacture food from subsequent chemical reactions (see PHOTOSYNTHESIS). These may be green plants, blue–green or other algae, or phytoplankton. All other organisms within it are dependent heterotrophs; that is, they eat or rearrange existing organic matter. In the overall sequence of food consumption patterns, they may be classed as primary consumers (HERBIVORES), secondary consumers (CARNIVORES, which eat herbivores), tertiary consumers (top carnivores, which eat other carnivores), and so on. Although essentially similar, detrital food chains begin at producer level with the breakdown of dead or decaying organic material by DECOMPOSER organisms. Consumers then take up the nutrients released by the decomposers, and some of the decomposers themselves; and they in turn are eaten by a range of carnivores.   DW/MEM

Reading
McCann, K.S. (2012) *Food webs*. Princeton, NJ: Princeton University Press.

foraminifera   Often abbreviated to forams, these comprise a soft body enclosed within a shell that often resembles a gastropod or cephalopod in form. Depending on the species, the shell may be made of organic compounds, sand grains and other particles cemented together, or crystalline calcite. Fully grown individuals range in size from

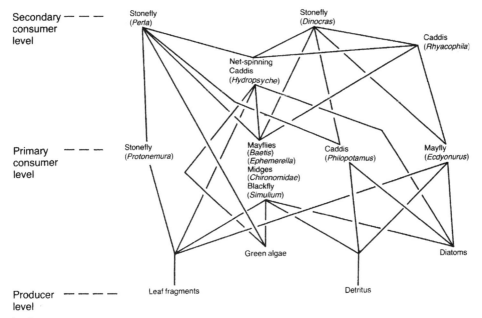

Food web.

about 0.4 mm to almost 20 cm in length. They are found in all marine environments, from the intertidal zone to the deepest ocean trenches, and from the tropics to the poles, but species can be very particular about the environment where they live. Because different species are found in different environments, their fossils can be used to determine past environments.　　　　　　　　SLO

force　An agency changing the velocity of an object. Force is mass times acceleration; hence, it is measured in dimensions of mass times length per the square of time: $1 \, \text{kg} \, \text{m} \, \text{s}^{-2}$ is equivalent to $1 \, \text{N}$ (newton). Like velocity and acceleration, force is a vector quantity: it has direction as well as magnitude. It is common to resolve force into components in two or three mutually perpendicular directions; for example, normal to the surface and tangential to the surface. The latter is a shear force and the former contributes to RESISTANCE. Body forces (gravitational, magnetic) are produced without physical contact; surface forces result from physical contact with other bodies, leading to STRESS. All mechanical processes involve force, for which the energy is derived from a combination of gravity, climate and geothermal heat. For geomorphological work to be done, forces must overcome resistances.　　IE

Reading
Carson, M.A. and Kirkby, M.J. (1972) *Hillslope form and process*. Cambridge: Cambridge University Press; chapter

3. · Chapman, R.E. (1995) *Physics for geologists*. London: UCL Press; chapters 1 and 2

forebulge　A flexural bulge in front of a load on the lithosphere. The load causes the lithosphere to flex by depressing the plate beneath it. Because of the flexural rigidity of the lithosphere, the area around the load is uplifted by a height that is 4% of that of the depression under the load. Mantle flow continues until the load is in isostatic equilibrium, a process that takes on the order of 10,000–20,000 years. One cause for forebulge formation is loading of continental lithosphere by ice sheets during continental glaciations; it is estimated that the central North Sea basin was uplifted by ~170 m during the Pleistocene as a result of forebulge dynamics. As ice sheets retreat and thin, so isostatic rebound in the former ice-covered areas is accompanied by forebulge collapse; under certain conditions the forebulge also migrates towards the former ice centres. Areas in regions characterized by forebulge collapse show higher rates of postglacial sea-level rise than neighbouring regions not so affected.　　TS

foredune　A distinctive coastal dune, with geometry controlled by wind, nearshore processes, sand and vegetation (see COASTAL DUNES). Foredunes develop in locations at least partly protected from EROSION by nearshore processes. Vigorous plant colonization causes the dune to migrate towards the water, but hydrodynamic erosion moves the dune front landward. As a result, the location of the dune represents an EQUILIBRIUM between the two systems. The characteristic morphology of mature foredunes is a near-continuous ridge with an erosion scarp on the windward side and dune grass cover on the crest and lee slope. PARABOLIC DUNES may develop from BLOWOUTS through the foredune.   DJS

Reading
Bauer, B.O. and Sherman, D.J. (1999) Coastal dune dynamics: problems and prospects. In A. S. Goudie, I. Livingstone, and S. Stokes (eds), *Aeolian environments, sediments, and landforms*. Chichester: John Wiley & Sons, Ltd.

foreset beds   Layers of sediment that have been laid down on the inclined surface of an advancing deltaic deposit or sand dune.

forest   See BOREAL FOREST, DECIDUOUS FOREST, MONSOON FOREST and TROPICAL FOREST.

forest hydrology   Concerned with the 'study of how forests affect the hydrologic cycle, with particular reference to the regulation of streamflow, water supplies, and erosion control' (Storey *et al.*, 1964). The relationships between water and forests have been the subject of scientific and political interest for several centuries. In 1215, King Louis VI of France set out a decree for the management of forests and related waters, and in Switzerland a law of 1342 protected forests from overcutting as an avalanche protection measure (Storey *et al.*, 1964). The impacts of deforestation on accelerated soil erosion have been a subject of study and debate in the Mediterranean area for 500 years. In the New World the US Government issued its first report on the hydrological impacts of deforestation in 1849, and in 1891 began setting aside vast areas of forest for the purpose of watershed protection.

From the human perspective, forests are considered to have beneficial effects on the hydrological cycle (Kittredge, 1948). The forest foliage breaks the fall of raindrops, lessening the erosion caused by drop splash on the surface below. Large quantities of moisture are intercepted by the foliage, temporarily stored on leaves and branches, and then allowed slowly to drop to the surface. This process ensures that precipitation is more likely to be absorbed into the surface instead of being fed to the surface too rapidly for effective infiltration.

The overhead canopy of vegetation in the forest also reduces water loss from the surface by means of evaporation, although this saving is at some cost because the trees transpire large amounts of moisture through their leaves. The forest root system acts as an important soil binder and retards erosion. The litter on the forest floor, organic material that has fallen from the actively growing vegetation, absorbs large amounts of moisture that otherwise would quickly run off into channels. Finally, the forest protects snow cover with shade, preventing rapid melt and flash flooding in downstream areas. Melting takes place over a relatively longer period of time, releasing water later in the melt season, when it is more useful for irrigation.   WLG

Reading and References
Chang, M. (2013) *Forest hydrology: an introduction to water and forest*, 3rd edition. Boca Raton, FL: CRC Press. · Kittredge, J. (1948) *Forest influences*. New York: McGraw-Hill. · Storey, H.C., Hobba, R.L. and Rosa, J.M. (1964) Hydrology of forest lands and rangelands. In V.T. Chow (ed.), *Handbook of hydrology*. New York: McGraw-Hill.

form line   A contour line on a map, the precise position of which has not been accurately surveyed but interpolated.

form ratio   The ratio of the depth of a stream or river to its width.

Fourier analysis   The process of decomposing a periodic function into its constituent simple sine or cosine waves. The analysis is named after the French mathematician Jean Baptiste Joseph Fourier, who showed that any continuous function can be produced from a set of basic sine and cosine waves.

Fourier analysis is used to decompose a function into simpler components and, if desired, the function can be reconstructed from the components via a Fourier synthesis. The decomposition process is often referred to as a Fourier transform, and the synthesis achieved via an inverse transform. Fourier analysis is commonly encountered in a range of numerical analyses, including signal and image processing. In the latter, for example, a Fourier transform decomposes an image into its sine and cosine components with the output being a representation of the image in the frequency domain. This conversion from spatial to frequency domains can be helpful in identifying particular components and forms the basis of a range of image analysis tasks, such as filtering and data compression.   GF

Reading
Millman, A.S. (1999) Mathematical principles of remote sensing. Making inferences from noisy data. Chelsea, MI: Ann Arbor Press.

fracking Also known as hydraulic fracturing, this is a relatively new technology for obtaining hydrocarbons, and especially gas, from shales. It is a technique that involves drilling into suitable gas-bearing bedrock. Water is mixed with sand and chemicals, and the mixture is injected at high pressure into a borehole to create small fractures (typically less than 1 mm), along which fluids such as gas may migrate to the well. It has been widely adopted in parts of the USA and is currently being appraised in the UK. There has been support for fracking on the basis that it can provide cheap energy and security of supply. On the other hand, it has been opposed on the grounds that it can cause seismic activity, lead to groundwater pollution and, in areas of water deficit, cause water shortages.   ASG

Reading
Rahm, D. (2011) Regulating hydraulic fracturing in shale gas plays: the case of Texas. *Energy Policy*, **39**, 2974–2981.

fractal A geometric form that is infinitely repeated at all scales. This mathematical concept has been applied in geography (physical and human) by including repetition that is statistical rather than exact, by accepting that there are upper and lower limits to the scales involved, and sometimes by allowing the type of variability to change with scale. Thus, geographers and Earth scientists deal with *statistical pseudo-fractals* or *pre-fractals*, which are irregular over a broad range of scales. Interesting experiments on whether mapped surfaces are scale free in this way, or scale bound, can be performed if extraneous clues such as contour labels or anthropogenic features are excluded: Can we estimate map scale and contour interval?

Fractal concepts have been applied with varying degrees of success to time series of processes, to turbulence-related features in the atmosphere and oceans, to fractured and other features in geology, to PARTICLE FORM and to REMOTE SENSING images. One of the earliest contributions was the work of H. E. Hurst on the long-term record of DISCHARGE for the River Nile, which shows *persistence* of deviations above or below average. Here, we concentrate on fractal aspects of three distinct features in geomorphology: the coastline, the land surface and the drainage network.

Indices of coastal irregularity have a long history, but the scaling properties of coastal length were noted by L. F. Richardson and developed by B. Mandelbrot. Many coastlines are *self-similar*: one part, magnified, statistically resembles the whole. Hence, as measurements are made at higher resolution (e.g. shorter straight-line segments), the estimated length of a coast increases without limit. This increase may be linear on a log–log plot, for segments from around 100 m to 100 km. The rate of increase gives the fractal dimension D usually between 1.05 and 1.40. Thus, the irregularity of coastlines gives them a fractal dimension greater than their conceptual Euclidean dimension (1.0), but less than that of Brownian motion (random increments: 1.5).

Similar relationships are found for topographic contours and for profiles, but the fractal dimensions estimated tend to differ. Fractal concepts can be applied to the land surface by scaling the vertical dimension by a different factor than the horizontal; this means it can be *self-affine* rather than *self-similar*. A smooth, differentiable surface has a dimension (Euclidean) of 2.0. The land surface tends to show fractal dimensions between 2.05 and 2.40; that is, 1.0 more than coastlines. However, estimation is more difficult for surfaces, and there is little agreement between the results of different techniques, among which box counting and *variogram* analysis are to be preferred. In addition to sampling problems, these differences indicate that real land surfaces deviate from the assumptions (e.g. self-affinity, invertibility and anisotropy) underlying the techniques: even impossible values of D are sometimes calculated. Surfaces simulated by approximations to *fractional Brownian motion* are useful as null hypotheses for comparison with real surfaces, or as initial rough surfaces on which geomorphologic processes can be simulated.

Most published results for land surfaces estimate fractal dimension over a limited scale range, between 10-fold and 100-fold (one to two cycles of decimal logarithms). The concept of scale independence is more valuable if it applies over at least three cycles of length (e.g. 100 m to 100 km), which is very demanding in terms of data. However, curvature or changes of slope are commonly observed leading to estimation of two or even three different dimensions, each over a limited scale range. More recent work has suggested that land surfaces exhibit *multifractality*; that is, continuous variation of fractal dimension with scale. Even so, there are many characteristics that are incompatible with either unifractality or multifractality.

The situation is different for DRAINAGE NETWORKS. Rain falls over the whole land surface and run-off produces a space-filling network of drainage. The fractal dimension of river networks in map view approaches the theoretical maximum value of 2.0. Models based on fractal-related processes, such as self-avoiding invasive percolation, can simulate realistic drainage basins. Realism is increased by use of multi-fractal models. Note that river channels have D around 1.16, but river networks have D around 1.9.   IE

Reading
Evans, I.S. and McClean, C.J. (1995) The land surface is not unifractal: variograms, cirque scale and allometry. In R.J. Pike and R. Dikau (eds), *Advances in geomorphometry – proceedings of the Walter F. Wood memorial symposium.* Zeitschrift für Geomorphologie, Supplementband, vol. 101 Berlin: Gebruder Borntraeger; pp. 127–147. · Gao, J. and Xia, Z. (1996) Fractals in physical geography. *Progress in Physical Geography* 20, 178–191. · Lavallée, D., Lovejoy, S., Schertzer, D. and Ladoy, P. (1993) Nonlinear variability of landscape topography: multifractal analysis and simulation. In N. S. N. Lam and L. De Cola (eds), *Fractals in geography.* Englewood Cliffs, NJ: Prentice-Hall; pp. 158–92. · Xu, T.-B., Moore, I.D. and Gallant, J.C. (1993) Fractals, fractal dimensions and landscapes: a review. *Geomorphology* 8, 245–262.

fractal dimension (or fractal)

A concept put forward by Mandelbrot (1982) that refers to the space filling of curves such that the dimension is between 1 and 2; the trail of Brownian motion is a fractal. The length of a coastline can be measured in different ways that will give values greater than the linear distance between the end points. In extreme cases, the length is very large but it can be characterized by fractals. Complex lines can be analysed by using the ideas of fractal dimension. For a closed loop (such as the outline of a particle) with perimeter P, measured by stepping off a constant length S, a plot of log P versus log S tends to give a constant slope b. The fractal dimension D is equal to the slope term plus one ($D = b + 1$) and $\log P \propto (D - 1)\log S$.   WBW

Reading and Reference
Mandelbrot, B.B. (1982) *The fractal geometry of nature.* San Francisco, CA: W.H. Freeman. · Orford, J.D. and Whalley, W.B. (1983) The use of the fractal dimension to quantify the morphology of irregular particles. *Sedimentology*, 30, 655–668.

fracture

The term given to the splitting of a material into two or more parts; the material is said to have failed. It usually refers to brittle fracture where the stressed body ruptures rapidly with the release of energy to form new surfaces by crack propagation after little or no plastic deformation. Brittle fracture may take place through crystals, along cleavage planes or between grains to give 'intergranular' fracture. Brittle fracture occurs normal to the maximum applied tensile stress component. This type of failure is usual with hard rocks. Ductile failure is fracture that occurs after extensive plastic deformation and with slow crack formation; this is typical of clays.   WBW

Reading
Whalley, W.B. (1976) *Properties of materials and geomorphological explanation.* Oxford: Oxford University Press. · Whalley, W.B., Douglas, G.R. and McGreevy, J.P. (1982) Crack propagation and associated weathering in igneous rocks. *Zeitschrift für Geomorphologie*, 26, 33–54.

fractus

Cloud having a broken or shattered appearance, perhaps a convective cloud growing in a place where the wind shear tears apart the incipient cloud, as in fractocumulus. It is a temporary phase of development in which buoyancy and shear are not yet reconciled.   JSAG

fragipan

An acidic, cemented horizon between the base of the soil zone and the underlying bedrock or parent material. Fragipans are normally compact and brittle, and may often be bonded by clays. Other agents may also be involved in the bonding, including silica (Bridges and Bull, 1983), iron, aluminium and organic matter. Many are the result of periglacial processes, and they are widespread in Europe and North America.   ASG

Reading and Reference
Bridges, E.M. and Bull, P.A. (1983) The role of silica in the formation of compact and indurated horizons in the soils of South Wales. In P. Bullock and C. P. Murphy (eds), *Soil micromorphology. Volume 2: soil genesis.* Berkhamsted: A.B. Academic; pp. 605–613. · Grossmann, R.B. and Carlisle, F.J. (1969) Fragipan soils of the eastern United States. *Advances in Agronomy*, 21, 237–279.

frazil ice

Fine spicules of ice in suspension in water, commonly associated with the freezing of seawater.

free face

The wall of a rock outcrop that is too steep for debris to rest upon it. It is the portion of a cliff that lies above a scree or talus, and from which, through rockfall and other processes, scree formation may take place.

freeze–thaw cycle

A cycle in which temperature fluctuates both above and below 0 °C. The amplitude of the temperature change and the period of time over which the fluctuation occurs are important considerations since freezing does not occur instantaneously, nor does it always occur at 0 °C. A typical diurnal freeze–thaw cycle is one in which the temperature ranges from +0.5 °C to −0.5 °C within a 24-h period.

Because of their supposed significance with respect to frost shattering of rock, the frequency and efficacy of freeze–thaw cycles have been the subject of both field and laboratory investigations. Field studies indicate that most freeze–thaw cycles per year (~40–60) occur in subarctic alpine regions that experience diurnal temperature rhythms. High latitudes experience few cycles on account of the seasonal temperature regimes. In all areas, most cycles occur in the upper 0–5 cm of the ground, and only the annual cycle occurs at depths in excess of 20 cm. Laboratory studies, in which rock samples are subject to repeated freeze–thaw cycles of varying amplitude and intensity, suggest that the number of freeze–thaw cycles is more important than their intensity as regards rock shattering. If correct, the low frequency of freeze–thaw cycles recorded in present-day periglacial environments suggests that frost shattering may be overemphasized as a physical weathering process. Hydration shattering and cryogenic (i.e. frost) weathering in general may be equally, if not more, important.   HMF

Reading
French, H.M. (2007) *The periglacial environment*, 3rd edition. Chichester: John Wiley & Sons, Ltd.

freezing front The boundary between frozen or partially frozen ground and nonfrozen ground. During freezing in permafrost regions the freezing fronts move downwards from the ground surface and upwards from the permafrost table. Only a downward moving freezing front exists in seasonally frozen ground. The freezing front is sometimes equated with the cryofront, the position of the 0 °C isotherm in the subsurface, forming the boundary between cryotic (i.e. temperature <0 °C) and non-cryotic (i.e. temperature >0 °C) ground. The permafrost base, the permafrost table, and the top and base of the cryotic portion of the active layer all constitute cryofronts, or freezing fronts. (See also PERMAFROST.)   HMF

Reading
Van Everdingen, R.O. (1976) Geocryological terminology. *Canadian Journal of Earth Sciences*, **13**, 862–867.

freezing index A measure of the combined duration and magnitude of the below-freezing temperatures that occur in a freezing season. It is expressed in DEGREE DAYS.

friction A force resisting relative motion between two solids, two fluids or between a solid and a fluid. It is a fundamental property in studies of sediment transfer or transport, since frictional resistance to motion must be overcome before masses or particles of sediment can be moved, and sedimentary landforms modified as a result.

The force required to move a solid block on a plane surface is its weight W multiplied by the static coefficient of friction μ between the block and the underlying surface. If the surface is sloping at an angle $\beta°$, the normal force on the surface is $W \cos \beta$, and the force required to move the block is $\mu W \cos \beta$. The block will slide under its own weight when $\tan \beta = \mu$ (Statham, 1977: 12–14). The interlocking friction in a sediment between individual grains is measured by the angle of internal friction ϕ, which is, however, difficult to define and measure since it is dependent on density, water content and test conditions (Statham, 1977: 41–49). Dry sediment on a slope will slide if the slope angle equals or exceeds the angle of internal friction, which is therefore a limiting or threshold slope angle in the landscape.

Friction also occurs between a solid bed surface and a fluid passing over it. This slows the immediately adjacent fluid to a standstill, but shear in the fluid above allows the development of a velocity gradient within the BOUNDARY LAYER, in which the frictional resistance is successively less effective with distance from the bed. In UNIFORM STEADY FLOW the friction exerted by the bed on the flow is equal and opposite to the drag of the fluid on the bed, and in loose, mobile sediment, this drag (the TRACTIVE FORCE) may overcome the resistance to motion of non-cohesive sediment grains, which is dependent on their weight and the frictional contact between them.   KSR

Reference
Statham, I. (1977) *Earth surface sediment transport*. Oxford: Clarendon Press.

fringing reef See REEF.

front A sharp transition zone separating air of different temperatures and origins. The term was introduced by the Bergen School of Meteorology in 1918 as part of their work on EXTRATROPICAL CYCLONE structure. The front has a three-dimensional form. It extends into the atmosphere as a gently sloping surface of about 1 in 100, so that the cold, denser air appears as a wedge shape beneath the warmer air. The front lies in a trough of lower pressure accompanied by changes in wind velocity, pressure and temperature as the front passes a site. The intensity of change varies greatly from one front to another.

Horizontal convergence and associated vertical motion are a necessary feature of a well-defined front. According to early ideas, the rising of the warm air at a front took place along the frontal surface itself, but recent work, based on Doppler radar, research aircraft, satellite images and rain gauges, has shown a much more complex structure of air movements near fronts. Basic to the uplift of air at the warm front is a well-defined flow of low-level, moist, warm air within the warm sector that moves parallel to the cold front and then ascends above the main warm frontal surface, eventually running parallel to the surface warm front but at a higher level (see figure (b)). Interaction between this flow and that at even higher levels (600 hPa) may trigger off potential instability that produces linear areas of heavier precipitation. Similar rain bands of subsynoptic scale, known as mesoscale precipitation areas (MPAs), have been found associated with cold and occluded fronts. Five types of frontal rain bands have been identified: (1) warm frontal; (2) warm sector; (3) cold frontal – wide; (4) cold frontal – narrow; and (5) post frontal (see figure (a)). Aircraft observations have confirmed that clouds developing in frontal MPAs do have different liquid water contents and ice particle concentrations from other clouds associated with the fronts. Radar provides a continuous picture of the movement and patterns of the MPAs, but

forecasting their initiation is difficult because of the lack of detailed information about atmospheric structure near fronts, and because the horizontal resolution of the present numerical weather prediction models is insufficiently fine to represent the features precisely.   PS

Reading
Atkinson, B.W. (1981) *Meso-scale atmospheric circulations*. London: Academic Press. · Browning, K.A. (1982) *Nowcasting*. London: Academic Press. · Carlson, T.N. (1991) *Mid-latitude weather systems*. London: Harper-Collins.

frontogenesis The process of intensification of the thermal gradient at a frontal zone. It takes place mainly by horizontal confluence and convergence when the isotherms are suitably aligned. Since uplift must follow surface convergence, frontogenesis is helped if the upper atmospheric air movements favour the continuation of rising air. The reverse process is known as frontolysis. In this situation the thermal gradient becomes progressively less distinct and the cloud band eventually disperses.   PAS

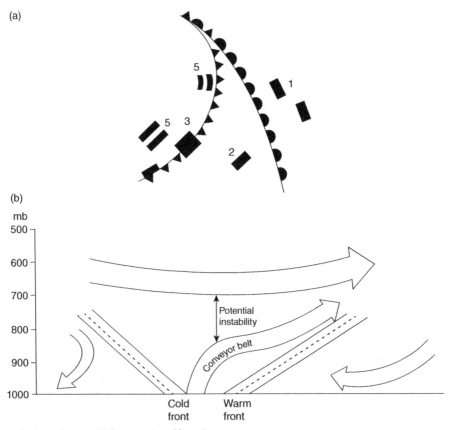

(a) Frontal rain band types. (b) Cross-section of frontal system.

frost action The mechanical weathering process caused by alternative or repeated cycles of freezing and thawing of water in pores, cracks and other openings, usually at the ground surface. The expansion of water upon freezing (approximately 9% by volume) forces material, commonly rock, apart. Termed frost wedging, its efficacy largely depends upon the frequency of FREEZE–THAW CYCLES, the availability of moisture and the lithological/strength characteristics of the material. Other terms for frost wedging include frost shattering, gelifraction and frost riving. The term 'frost action' is sometimes used to include a wider range of frost-related processes, such as frost heaving, frost creep, thermal contraction cracking and frost weathering. (See also FROST WEATHERING.)   HMF

frost creep The downslope movement of the particles as a result of the frost heaving of the ground and subsequent settling upon thawing, the heaving being predominantly normal to the slope and the settling more nearly vertical. Although frost creep is commonly associated with GELIFLUCTION (and is usually included within it in rate measurements because of the difficulty of distinguishing between their contributions to total movement) it is a separate process. Movement associated with frost creep decreases from the surface downwards and depends upon frequency of FREEZE–THAW CYCLES, angle of slope, moisture available for heave and frost susceptibility of soil. Studies in East Greenland indicate that frost creep exceeds gelifluction by not more, and probably less than, 3 : 1 in most years, and frost creep usually resulted in 30–50% of total annual movement on slopes. In the Colorado Rockies, measurements indicate that solifluction is a more effective process than frost creep in the saturated axial areas of lobes, but less effective than frost creep at their edges.   HMF

Reading
Benedict, J.B. (1970) Downslope soil movement in a Colorado alpine region: rates, processes and climatic significance. *Arctic and Alpine Research*, **2**, 165–226. · Washburn, A.L. (1979) *Geocryology: a survey of periglacial processes and environments*. New York: John Wiley & Sons, Inc.

frost heave The predominantly upward movement of mineral soil during freezing caused by the migration of water to the freezing plane and its subsequent expansion upon freezing. Frost heaving is usually associated with the active layer above permafrost or with seasonally frozen ground. As such, ICE SEGREGATION is an essential component of frost heave. Field studies indicate that heave occurs not only during the autumn freeze-back period, but also during winter when ground temperatures are below 0 °C. Frost heaving processes include the upheaving of bedrock blocks, upfreezing of objects, tilting of stones, formation of NEEDLE ICE, and the sorting and migration of soil particles. Frost heaving presents important geotechnical problems in the construction of roads, buildings, pipelines and airfields in cold environments.   HMF

Reading
French, H.M. (2007) *The periglacial environment*, 3rd edition. Chichester: John Wiley & Sons, Ltd. · Mackay, J.R. (1983) Downward water movement into frozen ground, western Arctic coast. *Canadian Journal of Earth Sciences*, **20**, 120–134.

frost smoke (also **arctic smoke**) A fog produced by the contact of cold air with relatively warm water. It is commonly associated with leads of open water that open up in a sea-ice cover but may also occur at the ice edge or over water that is beginning to freeze.

frost weathering A general term used to describe the complex of weathering processes, both physical and chemical, that operate, either independently or in combination, in cold non-glacial environments. The most important physical weathering process is frost wedging, which characteristically produces angular fragments of varying sizes. The predominant size to which rocks can be ultimately reduced by frost wedging is generally thought to be silt. Porous and well-bedded sedimentary rocks, such as shales, sandstones and limestones, are especially susceptible to frost weathering. Features attributed to frost weathering include extensive areas of angular bedrock fragments (blockfields and blockslopes) and irregular bedrock outcrops termed tors. It has been estimated that frost weathering (i.e. rockfalls induced by frost wedging) over a 50-year period caused steep rock faces in Longyeardalen, Spitsbergen, to retreat at a rate of 0.3 mm a^{-1}.

Many aspects of cold climate weathering are not fully understood. For example, it has been suggested that hydration shattering may be responsible for the large blockfields and blockslopes characteristic of low-temperature alpine and polar environments, but this has yet to be proved. Equally, experimental studies in the former USSR indicate that, under cold conditions, the ultimate size reduction of quartz (0.05–0.01 mm) is smaller than for feldspar (0.1–0.5 mm), a reversal of what is normally assumed for temperate environments. Finally, the current emphasis upon frost wedging in periglacial environments should not obscure the fact that chemical weathering can be significant. The dominance of physical weathering tends to mask chemical effects. In places, physical and chemical effects combine, as in salt wedging. Like frost wedging, this process breaks up rock into silt-size particles and is particularly effective in cold, arid regions, such as the ice-free areas of Antarctica. Solutional effects in limestone terrain may also be present, and karst terrain exists in permafrost regions, further illustrating the inadequacy of a simplistic view of frost weathering.   HMF

Reading
French, H.M. (2007) *The periglacial environment*, 3rd edition. Chichester: John Wiley & Sons, Ltd. · Van Everdingen, R.O. (1981) Morphology, hydrology and hydrochemistry of karst in permafrost terrain near Great Bear Lake, NWT. National Hydrology Research Institute, paper 11, Calgary, AB: Environment Canada. · Jahn, A. (1976) Contemporaneous geomorphological processes in Longyeardalen, Vestspitsbergen (Svalbard). *Biuletyn Peryglacjalny* **26**, 253–268. · White, S.E. (1976) Is frost action really only hydration shattering? *A review. Arctic and Alpine Research*, **8**, 1–6.

frost wedge   See ICE WEDGE.

Froude number   The dimensionless ratio Fr of inertial to gravity forces in flowing water:

$$\mathrm{Fr} = \frac{v}{(gd)^{0.5}}$$

where v is velocity, g is gravitational acceleration and d is depth. The term $(gd)^{0.5}$ is the velocity of a small gravity wave (a surface ripple), and if Fr < 1 the flow is subcritical or tranquil and ripples formed by a pebble dropped into the water travel upstream because their velocity exceeds that of the stream. If Fr > 1 the flow is supercritical, and when Fr = 1 the flow is critical. Sudden spatial changes from supercritical to subcritical are HYDRAULIC JUMPS. In sandbed channels, temporal changes of flow regime (see FLOW REGIMES) during floods cause a consistent sequence of bedform changes, with sand dunes washing out to form antidunes at a local Froude number of approximately unity. However, *mean* Froude numbers in natural channels rarely exceed 0.4–0.5, because the associated rapid energy losses cause bank erosion, channel enlargement and a reduction of flow velocity and Froude number – an example of negative feedback (see SYSTEMS). KSR

fulgurite   A tube in sand or rock produced by the fusing effects of a lightning strike (Withering, 1790). Sand fulgurites are especially common in areas of dry, loose, quartz sand typical of deserts.   ASG

Reading and Reference
Pasek, M.A., Block, K. and Pasek, V. (2012) Fulgurite morphology: a classification scheme and clues to formation. *Contributions to Mineralogy and Petrology*, **164**, 477–492. · Withering, W. (1790) An account of some extraordinary effects of lightning. *Philosophical Transactions of the Royal Society of London, Series D*, **80**, 293–295.

fulje   A depression between barchans or barchanoid sand ridges, especially where dunes are pressing closely on one another. In Australia the term may be used to describe a blowout or small parabolic dune.   ASG

fumarole   A small volcanic vent through which hot gases are emitted.

fungi   A diverse grouping of organisms involved in nutrient cycling, lacking the ability to photosynthesize. An important group occurs as mycorrhizae (see MYCORRHIZAL FUNGI), a mutualistic (symbiotic) association with plant roots occurring in up to 80% of terrestrial plant species (van der Heijden *et al.*, 1998). Fungi are capable of breaking down a wide range of organic materials, including cellulose, both living and dead. They inhabit a broad range of environments, marine as well as terrestrial. Fungi display branching systems of filaments called *hyphae* that can extend in all directions from the location where a fungal spore germinates, and these can generate very large osmotic gradients that result in the transfer of nutrients from the medium in which the fungus grows. In association with an alga, fungi form the *lichens*, another very important group of nonvascular plants, especially in the soils of dry environments, where they are able to survive severe moisture stress.   DLD

Reference
Van der Heijden, M.G.A., Klironomos, J.N., Ursic, M., *et al.* (1998) Mycorrhizal fungal diversity determines plant biodiversity, ecosystem variability and productivity. *Nature*, **396**, 69–72.

funnelling ratio   A means of describing the volume of STEM FLOW that is carried to the soil along the trunk of a tree or other large plant. It was defined by Herwitz (1982) as the ratio of the observed volume of stem flow to the volume that would be expected if the water was intercepted solely on parts of the plant directly above the tree base, and equal in area to the basal area of the trunk. A ratio >1 indicates that water is being funnelled into the trunk from peripheral parts of the canopy. Using canopy dimensions of rain forest trees, and stem flow volumes measured using stem-flow collars, Herwitz found funnelling

ratios as high as 156. Funnelling of this intensity delivered sufficient stem flow volume to exceed the infiltration capacity of the soil around the tree, and to create local sources of Hortonian surface run-off. DLD

Reference

Herwitz, S.R. (1982) The redistribution of rainfall by tropical rainforest canopy tree species. In E. M. O'Loughlin and L. J. Bren (eds), *The first national symposium on forest hydrology, Melbourne, 11–13 May 1982*: preprints of papers. National Conference Publication 82/6. Barton, ACT: Institution of Engineers, Australia; pp. 26–29.

fynbos Heathlands found in the Mediterranean-climate region of southwestern South Africa. Like the plant communities that characterize other parts of the world with similar fire-prone and temperate winter-rainfall climates (e.g. macchia or maquis of the Mediterranean basin, and chaparral of southwestern USA), fynbos is dominated by sclerophyllous (i.e. evergreen, hard-leaved) shrubs. In the case of fynbos, plant diversity is exceptionally high, and there may be as many as 120 different plant species within a 10 m × 10 m sample plot. Taxa typically include shrubs belonging to the families Proteaceae and Ericaceae, and a group of reed-like plants of the Restionaceae family. MEM

Reading

Cowling, R.M. (ed.) (1992) *The ecology of fynbos: nutrients, fire, diversity*. Cape Town: Oxford University Press.

G

gabbro A basic igneous rock composed of calcic plagioclase and clinopyroxene with or without orthopyroxene and olivine. Usually coarse grained and dark grey to black in colour.

Gaia hypothesis In 1979 James Lovelock published his so-called Gaia hypothesis, which proposed that the Earth system was in some respects a self-regulating (via FEEDBACKS) single system (i.e. a 'living planet'). Gaia is named after the Greek goddess of the Earth.

Reference
Lovelock, J.E. (1979) *Gaia: a new look at life on Earth*. Oxford: Oxford University Press.

gallery forest Forest that lines the banks of a river in an area where away from the river's favourable hydrological circumstances such forest does not occur.

garrigue Xerophytic, evergreen scrubland occurring on thin soils in areas with a dry 'Mediterranean type' of climate. Much of it may result from anthropogenic landscape degradation. It consists of low thorny shrubs and stunted evergreen oaks.

gauging stations Sites at which river flow is determined. The gauging sites may only be equipped to provide point measurements in time or they may provide continuous measurements. The accuracy of the flow estimates will vary according to the gauging technique employed. (See also DISCHARGE and HYDROMETRY.) AMG

GCM See GENERAL CIRCULATION MODELLING.

geest Ancient alluvial sediments that still mantle the land surfaces on which they were originally deposited.

gelifluction A type of solifluction occurring in periglacial environments underlain by permafrost. Suitable conditions for gelifluction occur in areas where downward percolation of water through the soil is limited by the permafrost table

and where the melt of segregated ice lenses provides excess water that reduces internal friction and cohesion in the soil. Particularly favoured sites include areas beneath or below late-lying snowbanks. Rates of movement, which generally vary between 0.5 and $10.0\,\mathrm{cm\,a^{-1}}$, usually decrease with depth. Frost creep is usually measured as a component of gelifluction. As with solifluction, features related to gelifluction include sheets, stripes and lobes. HMF

Reading
French, H.M. (2007) *The periglacial environment*, 3rd edition. Chichester: John Wiley & Sons, Ltd.

genecology The study of the genetics of populations in relation to habitat; the study of species (and other taxa) through a combination of the methods and concepts of both ecology and genetics. Some species display a range of ECOTYPES – genetic varieties existing in different environments; the investigation of these ecotypes constitutes a part of genecology. PHA

general atmospheric circulation The planetary system of major wind and pressure features. Its components have space scales on the order of 5×10^3 km and they are associated with processes acting on timescales of months. Examples of its components are the westerly winds of the mid-latitudes and the trade winds of the subtropics, and pressure systems such as the subtropical highs and the equatorial zone of low pressure. This planetary-scale circulation spawns the development of smaller-scale features, such as the synoptic-scale highs and lows and tropical storms depicted on weather maps.

The atmospheric general circulation is shaped largely by five factors. The two most important are the latitudinal variation of solar radiation (i.e. the equator-to-pole temperature gradient) and the Earth's rotation. The former factor is manifested as the pressure gradient force, which pulls air parcels towards low pressure. The latter is manifested as the Coriolis force, an apparent force directed perpendicular to the motion of an air parcel. Acting together, these two forces

The Dictionary of Physical Geography, Fourth Edition. Edited by David S.G. Thomas.
© 2016 John Wiley & Sons, Ltd. Published 2016 by John Wiley & Sons, Ltd.

produce air motion that is parallel to the contours of atmospheric pressure (i.e. isobars). This is referred to as geostrophic motion and occurs at elevations high enough that the friction from the Earth's surface is negligible. Note that, in the lower latitudes, the Coriolis force is so weak that geostrophic flow, such as the mid-latitude westerlies, does not occur.

The fundamental importance of these two factors was shown in a series of laboratory experiments at the University of Chicago in the 1950s. These were commonly termed the 'dish pan' experiments, a reference to the laboratory apparatus, which was a circular pan that could be rotated and/or heated and cooled. The pan contained a dense fluid and a tracer dye.

When the pan was merely rotated, and no heating occurred, the fluid rotated in a flat plane in the direction in which the pan was turning. This circulation is analogous to the westerly winds and is termed a 'Rossby-type' circulation. To simulate the latitudinal distribution of solar radiation, the pan was heated along the edges and cooled in the centre. Under some circumstances the east–west circulation developed waves that produce north–south variations in the flow. The former is referred to as 'zonal' flow and the latter is referred to as 'meridional' flow. In the presence of this heating, but with no rotation, a circulation resulted in which the warm fluid rose along the edges and the colder fluid in the centre sank. This overturning is termed a 'Hadley-type' circulation. When the 'dish pan' was subjected to both rotation and differential heating, the Rossby circulation prevailed in the centre (i.e. at higher latitudes) and the vertical Hadley overturning was limited to outer portions (i.e. at lower latitudes). This is precisely the situation that prevails, to a first approximation, in the actual atmosphere.

The strength of the westerlies increases with the magnitude of the mid-latitude temperature gradient. The reason for this is the density contrast between warm and cold air. This enhances the latitudinal pressure gradient, which, at a given latitude, is proportional to the magnitude of the geostrophic wind. The density contrast also causes pressure to decrease with height more rapidly in the cold air, so that the pressure gradient increases with elevation when a horizontal thermal gradient exists. This relationship, termed the thermal wind, prescribes an increase in westerlies/easterlies when the cold air is poleward/equatorward of the warm air. Thus, easterly winds increase with height over Asia, where the Tibetan plateau (average elevation over 4500 m) provides an elevated heat source that

lies adjacent to a relatively cool air mass over the Indian Ocean to the south. In general, easterly winds prevail throughout most of the atmospheric column in the tropics, while westerlies prevail aloft in the mid and high latitudes.

The rotation and latitudinal heating gradients create the essence of the atmospheric circulation, but they prescribe features that are solely a function of latitude. This is an unrealistic situation. Three secondary factors act to produce variations within latitudinal belts, regional detail and the easterly winds that prevail in the tropics. The foremost of these is the thermal contrast associated with the distribution of land and ocean. Because cold is denser than warm, high pressure tends to overlie cooler areas and low pressure tends to overlie warmer areas. Therefore, the seasonal change of the land–water thermal contrast (land warmer in summer, colder in winter) produces much of the seasonal change of the general atmospheric circulation. The pressure patterns associated with the distribution of continents also produce large-scale north–south undulations in the flow.

The additional secondary factors are large-scale topography and the need to conserve angular momentum of the Earth–atmosphere system. This last factor requires the development of easterly winds to balance the angular momentum of the westerlies. Large-scale topographic features, such as major mountain ranges, create waves that transport air in a north–south direction. Mountain ranges also produce 'rain shadows' in their lee, areas of general subsidence where the climate is generally arid or semi-arid.

The surface features of the general atmospheric circulation are illustrated in the figure. Dominant features include three major wind systems, associated with particular latitudinal zones, and four major pressure systems that give rise to the wind systems. The dominant pressure systems are the subtropical highs and the mid-latitude lows. The polar highs and the general belt of low pressure in the equatorial regions complete the picture. The subtropical highs are remarkably stable features, but the mid-latitude lows are the statistical mean of migrating cyclones in the mid-latitudes with life times of generally a few days.

Air circulates counterclockwise around lows and clockwise around highs in the northern hemisphere (the reverse in the southern hemisphere). Consequently, the major wind systems are flanked by these pressure systems. Thus, westerlies prevail in mid-latitudes, easterlies in the subtropical and polar latitudes. These easterlies are called the trade winds, and they are

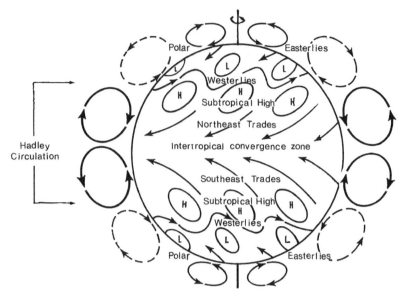

Schematic diagram of the general atmospheric circulation.
Source: Nicholson (2011) Reproduced with permission of Cambridge University Press.

northeasterly in the northern hemisphere and southeasterly in the southern hemisphere. Over the oceans the trades converge near the equator in a region referred to as the equatorial trough or the intertropical convergence zone (ITCZ). Because of the stability of the subtropical highs, the trade winds are remarkably persistent in their occurrence and direction. In the vicinity of the subtropical highs, winds are generally weak and variable.

The global pattern of pressure also gives rise to a series of three vertical circulation cells. During the course of the year the cells vary greatly in magnitude and location. These vertical cells result because air converges into the centre of lows, forcing it to rise, but diverges in the vicinity of high-pressure cells, resulting in sinking motion (Figure 2). These cells influence the general global pattern of precipitation because rising motion promotes precipitation while the sinking motion suppresses its formation. The largest and strongest of these vertical cells is the tropical Hadley cell, which is the dominant mean air movement in the low latitudes (Webster, 2004). Its ascending branch is associated with the convergence of the trade winds in the ITCZ and equatorial heating; its descending branch is associated with the centres of the subtropical highs. In the mid-latitudes, the rising air in the prevailing cyclones is coupled with sinking motion in both subtropical and polar highs. This produces two other vertical cells in

each hemisphere, but they are relatively weak, especially in comparison with the prevailing wind systems.

The above picture of two Hadley cells on both sides of the equator is an oversimplification. It represents the circulation as a function of latitude, averaged over the Earth as a whole. The two-celled pattern exists only in the transition seasons, and the pattern is interrupted in many longitudinal sectors. At other times there is one dominant cell of overturning, that of the winter hemisphere with an extensive area of rising motion in the equatorial latitudes and well into the summer hemisphere and a region of subsidence in the subtropical latitudes of the winter hemisphere.

In addition to the north–south-oriented Hadley cells, a vertical circulation oriented east to west also plays a major role in producing the Earth's patterns of climate. This cell, termed the Walker circulation, owes its existence to two factors. The first is that, for complex dynamic reasons, the eastern flank of the subtropical highs is more stable than the western. Consequentially, a circulation results with sinking motion in the east but a tendency for rising motion in the west. Orographic effects, which produce ascent over major mountain ranges, further modify the east–west circulations.

The classic picture of the Walker circulation is depicted in Figure 3, but in reality the situation is more complex. Although the tendency for rising and sinking motion exists in the areas depicted, there is not necessarily a closed circulation cell connecting them (Hastenrath, 2007). The Walker circulation is best developed over the Pacific Ocean, while the vertical cell over the Indian Ocean is apparent only in northern-hemisphere autumn. Minor cells are evident over the continents in some seasons. The western branch of each of the major cells is linked to areas of intense convection over the Amazon, Central Africa and the Pacific maritime continent.

The major consequence of the Walker circulation is a tendency for the west coasts of continents in the subtropical and mid-latitudes to be considerably drier than the eastern portions of the continent. This is clearly apparent over Africa, South America, the USA and, to a lesser extent, Australia.

The seasonal changes in the general atmospheric circulation are produced by the land–water contrast and by the change from strong latitudinal temperature gradients in the winter to weak latitudinal gradients in summer. In winter, highs tend to be located over the relatively cold continents ans lows over the relatively warm oceans. In summer, lows tend to be located over the warmer continents, with strong high-pressure systems over the oceans. The circulation features are displaced poleward with respect to their winter positions and generally weaker.

The land–water contrast and latitudinal temperature gradients result in very different characteristics of the general atmospheric circulation in the two hemispheres. Because the southern hemisphere is primarily water, the flow is primarily east–west, the seasonal changes are minimal and the westerlies are very strong. In the northern hemisphere the meridional elements (i.e. waves in the flow) are very prominent. The intensity and location of circulation features change significantly with the seasons, as do the number of waves, which are very strongly influenced by the latitudinal temperature gradient. Because of the absence of warm landmasses and the presence of the extremely frigid Antarctic continent, the southern hemisphere is colder than the northern hemisphere. As a result, the meteorological equator, the belt of low pressure, has a mean location in the northern hemisphere.

The prevailing patterns of the general atmospheric circulation are the main factors in the global patterns of aridity and rainfall. Areas where low pressure prevails tend to be humid; those where high pressure prevails tend to be dry. This pattern is very pronounced in the low latitudes, where the subtropical highs represent the transition between the tropics and extratropics.

The highs promote aridity not only because the prevailing subsidence inhibits the production of precipitation, but also because the subsidence produces a temperature inversion in the vicinity of the trade winds. The thermal stability associated with the inversion further inhibits the production of precipitation. Thus, deserts are common in the subtropics. The semi-arid regions lie in regions where there is a seasonal change in the prevailing high- and low-pressure systems. For example, Mediterranean climates with dry summers and wet winters lie along the northern flank of the subtropical highs; rainfall occurs when the highs are displaced southward and mid-latitude storm systems can then traverse the region. On the southern flank of the highs, summer rainfall and winter aridity prevail as a shift occurs between the dominant of the equatorial low-pressure regime and the subtropical high pressure. Year-round precipitation occurs in most of the equatorial region and in the mid-latitudes. SEN

Reading and References

Hastenrath, S. (2007) Equatorial zonal circulations: historical perspectives. *Dynamics of Atmospheres and Oceans*, **43**, 16–24. · Nicholson, S.E. (2011) *Dryland climatology*. Cambridge: Cambridge University Press (especially chapter 4). · Webster, P.J. (2004) The elementary Hadley circulation. In H. Diaz and R. Bradley (eds), *The Hadley circulation: past, present and future*. Advances in Global Change Research, vol. **21**. Dordrecht: Kluwer; pp. 9–60.

general circulation modelling

One of the primary tools of investigation now used by climate researchers is the general circulation model (GCM). GCMs are mathematical computer models describing the primary controls (such as radiative fluxes), energy transfers, circulations and feedbacks existing in the Earth–ocean–climate system. GCMs have their roots in the early numerical weather models developed by J. G. Charney and the so-called 'father' of modern computing, J. von Neumann, in the late 1940s. Most GCMs are based on seven fundamental (or 'primitive') equations. These equations are:

- The equation of state (or ideal gas law). This basic physics principle mathematically describes the relationship between pressure, density and temperature.
- The three equations of motion (or, when expressed in vector notation, often called 'the Navier–Stokes equation'). These equations describe the forces producing airflow in the three spatial dimensions. The forces mathematically included in those equations include gravity, the pressure gradient force,

the Coriolis effect and various acceleration terms.

- The first law of thermodynamics. This principle links temperature changes to changes in energy inputs into the system and energy changes within the system.
- An equation addressing conservation of mass. This mathematical relationship ensures that mass is neither created nor destroyed throughout the climate domain during the course of the simulation.
- An equation addressing moisture distribution and phase through the model domain. Because water is of critical importance to the climate system, separate accounting of moisture (and its various phases of ice, snow, liquid water and water vapour) must be conducted within the GCM.

To ensure mathematical closure, these seven equations must be associated with seven field variables (variables expressed as functions of time and space). Consequently, the seven fundamental variables in a GCM are pressure p, temperature T, density ρ, zonal (E–W) wind u, meridional (N–S) wind v, vertical wind w, and specific humidity q. In general, the mathematical expressions are solved such as the change in one of the variables over time is expressed as the result of the given variable's spatial variability. For example, all other factors being constant, a net convergence of moisture into a given area will lead to an increasing level of moisture at that location over time.

GCMs are constructed over spatial domains that can be expressed by Cartesian or spatial coordinates (termed gridpoint models) or by Fourier transform functions (referred to as spectral models). The spatial resolution of these models continues to be refined. Early GCMs of the 1970s and 1980s commonly had spatial dimensions of 5° latitude × 5° longitude or even 10° × 10°. Today, many GCMs operate with resolutions of 2.5° latitude × 2.5° longitude or have higher resolution submodels operating within larger spatial domains.

Temporally, the equations of a given GCM are solved over a fairly short time basis (on the order of minutes or hours), and the results are subsequently used as the new input to compute additional changes in the state variables over time. This means that an incredible number of calculations (over both space and time) must be made to produce a single GCM simulation. Normally, because of these heavy computational demands, GCM simulations are run only for a limited number of simulated years.

The first GCMs only addressed the fundamental principles of atmospheric circulation with very limited or extremely simplified inputs and outputs. Today, GCMs are much more complex and include mathematical expressions of ocean circulation and interaction with the atmosphere, with land surface processes such as vegetation or cryosphere (snow and ice surfaces) and atmospheric chemistry (including radiative processes associated with ozone and carbon dioxide). Although inclusion of these processes and others has greatly improved GCMs, these models remain imperfect representations of the actual Earth–ocean–climate system. Consequently, the results from climate simulations by GCMs should not be taken as true predictions of climate change but rather as either information used to improve our understanding of the climate system or as potential outlooks of future climate based only on our current understanding of climate.

Because GCMs have become such critical tools of climatic investigation, they have been used to study a number of important climate concerns. GCMs have been used as primary tools in the investigation of (1) the global climatic effects of increased atmospheric carbon dioxide; (2) the effects of deforestation on global climate; (3) the reconstruction of past climates (such as the Cretaceous time of the dinosaurs); and (4) studies of the global effects of the South Pacific phenomena known as El Niño–South Oscillation.

Advances in numerical programming continue to focus on the inclusion of high-resolution spatial domains and of more complex processes into GCMs. However, because of the heavy numerical demands made by these inclusions, such advancement in GCMs is often limited by the size, speed and availability of modern computers.   RSC

general system theory This was largely developed by the biologist von Bertalanffy, whose basic statement appeared in 1950 and was substantially extended in 1962. The fundamental proposition of the theory is that systems (defined as structured sets of objects and their attributes) may be identified in all studies of phenomena; and that more may be learnt by the comparison of the ways in which similar (isomorphic) systems function than by the standard academic concentration on the distinctiveness of their component parts. Thus, what is of interest in a comparison of (say) towns, drainage basins and bathtubs is that they may be considered to share fundamental attributes that relate to their physical boundedness and to the transfers of energy and mass across those boundaries. These transfers serve to integrate the components of each system, and this interrelatedness of parts is a diagnostic and important feature.

Since its introduction, general system theory (GST) has become a field of study in its own right. The first explicit use of von Bertalanffy's ideas in geomorphology was made by Strahler (1950: 676), who stated: 'A graded drainage system is perhaps best described as an open system in a steady state . . . which differs from a closed system in equilibrium in that the open system has import and export of components'. This and subsequent studies by Strahler and others were then taken up by Chorley (1960) in a major paper that focused on the ideas of GST, rather than simply on concepts of physical systems. Chorley's discussion introduces a series of the key concepts: notably, open and closed systems, entropy, steady state, self-regulation, equilibrium, hierarchial differentiation and organization. While it may be argued that the *framework* provided by GST has been widely adopted in physical geography, actual explicit use has been somewhat limited and succeeded by the wider use of systems in a more general sense.   BAK

Reading and References
Bennett, R.J. and Chorley, R.J. (1978) *Environmental systems: philosophy, analysis and control.* London: Methuen. · Chorley, R.J. (1960) *Geomorphology and general systems theory.* United States Geological Survey Professional Paper 500-B. Washington, DC: US Government Printing Office. · Chorley, R.J. and Kennedy, B.A. (1971) *Physical geography: a systems approach.* London: Prentice-Hall. · Strahler, A.N. (1950) Equilibrium theory of erosional slopes approached by frequency distribution analysis. *American Journal of Science*, **248**, 673–696, 800–814. · Von Bertalanffy, L. (1950) The theory of open systems in physics and biology. *Science*, **111**, 23–29. · Von Bertalanffy, L. (1962) General systems theory, a critical review. *General Systems*, 7, 1–20.

genetic drift   The effect of sampling error in causing random changes in the relative frequency of genes in a gene pool. Random change in gene frequency occurs in all populations, and it is maintained that genetic drift provides a mechanism for evolution. The smaller the population size, the greater the possibility of gene loss or fixation: a gene normally present in 1 in 10,000 individuals may not be present in a population of 100. Particularly important is the 'founder effect': when a small sample of an organism's population is isolated – for example, on an island or mountain peak – it may have a different gene frequency from the parent group. This is one reason why organisms on remote islands are frequently endemic species or subspecies (see ENDEMISM). An animal that has been reduced to a very small population by climatic catastrophe, disease or overhunting is likely to have an impoverished gene pool for some time after recovery.   PHA

Reading
Bonnel, M.L. and Selander, R.K. (1974) Elephant seals: genetic variation and near-extinction. *Science*, **184**, 908–909.

geo   A deep, narrow cleft or ravine along a rocky sea coast that is flooded by the sea.

geochronology   A term used both to describe the chronologies of long-term environmental change, usually in the context of the QUATERNARY period, derived from the use of numerical age dating techniques (e.g. LUMINESCENCE DATING, RADIOCARBON DATING, URANIUM SERIES DATING), and to describe the techniques used to produce a chronology of change (i.e. a *geochronometric technique*).   DSGT

geoconservation   Action taken with the intent of conserving and enhancing geological and geomorphological features, processes, sites and specimens. As successful conservation often depends on understanding and valuing the feature, process, site or specimens to be conserved, the actions taken often also include promotional and awareness raising activities. Associated with the UNESCO World Heritage List and Geoparks, and at more regional and local levels with the development of Regionally Important Geological/Geomorphological Sites (RIGS) and Local Geodiversity Action Plans (LGAPs) respectively.   TS

Reading
Burek, C.V. and Prosser, C.D. (eds) (2008) *The history of geoconservation.* The Geological Society of London, Special Publication No. 300. London: The Geological Society.

geocryology   The study of frozen, freezing and thawing terrain (but not glaciers) is known as permafrost science or more generally termed geocryology. This widespread term is usually associated with earth materials having a temperature below 0 °C. Geocryology seeks to promote an understanding of the dynamics of such environments, especially the study of the origins and history of permafrost.   AP

Reading
Washburn, A.L. (1979) *Geocryology: a survey of periglacial processes and environments.* London: Edward Arnold.

geode   A roughly spherical or globular inclusion within a mass of rock. Geodes are hollow and frequently exhibit mineral crystals growing into the central void.

geodesy The determination of the size and shape of the Earth by mathematical means and surveys.

geodiversity The variety of earth materials, forms and processes that constitute and shape of the Earth, either the whole or a specific part of it. Studying geodiversity involves the processes of recognizing and assessing the value of this variety; such study, therefore, has cultural and societal aspects. Relevant materials include minerals, rocks, sediments, fossils, soils and water. Forms may comprise folds, faults, landforms and other expressions of morphology or relations between units of earth material. Any natural process that continues to act upon, maintain or modify either material or form (e.g. tectonics, sediment transport, pedogenesis) represents a further aspect of geodiversity.   TS

Reference

Gray, M. (2004) *Geodiversity: valuing and conserving abiotic nature.* Chichester: John Wiley & Sons, Ltd.

geoengineering The term *geoengineering* is often applied to ambitious attempts at global climatic modification (Lovelock, 2008). In particular, such techniques may be developed to combat global warming. Geoengineering methods can be divided into two basic 'classes' (Royal Society, 2009). The first of these is a group of carbon dioxide (CO_2) removal techniques that aim to address the root cause of climate change by removing greenhouse gases from the atmosphere: land-use management to protect or enhance land carbon sinks; the use of biomass for carbon sequestration as well as a carbon-neutral energy source; enhancement of natural rock-weathering processes to remove CO_2 from the atmosphere; direct engineered capture of CO_2 from ambient air; and the enhancement of oceanic uptake of CO_2 by fertilization of the oceans with naturally scarce nutrients, or by increasing upwelling of ocean water. The second class of techniques are called solar radiation management techniques. These aim to offset the effects of increased greenhouse gas concentrations by causing the Earth to absorb less solar radiation. Among the methods put forward are: placing shields or reflectors in space to reduce the amount of solar energy reaching the Earth; brightening the surface of the Earth (e.g. by covering deserts with mirrors or other reflective material) so that more incoming radiation is reflected back to space; and injecting sulphate aerosols into the low stratosphere to mimic the cooling effect achieved by volcanic eruptions.   ASG

References

Lovelock, J. (2008) A geophysiologist's thoughts on geoengineering. *Philosophical Transactions of the Royal Society A: Mathematical, Physical and Engineering Sciences,* **336**, 3883–3890. · Royal Society (2009) *Geoengineering the climate.* London: Royal Society.

geographical information science (GISci) In essence this is the science that supports research with geographical information systems (GISs). Although a difficult and contentious term to define, GISci can be regarded as a field of research that studies the theory, concepts and methods that underpin GISs, such as those connected with the acquisition, management, analysis, storage and visualization of geographical data. Through GISci, geographical data may be transformed into geographical information and contribute to furthering knowledge and understanding of the world.   GF

Reading

Goodchild, M.F. (1992) Geographical information science. *International Journal of Geographical Information Science,* **6**, 31–45. · Goodchild, M.F. (2010) Twenty years of progress: GIScience in 2010. *Journal of Spatial Information Science,* **1**, 3–20. · Mark, D.M. (2003) Geographic information science: defining the field. In M. Duckham, M. F. Goodchild and M. F. Worboys (eds), *Foundations of geographic information science.* New York: Taylor and Francis; pp. 1–18.

geographic information system (GIS) Computer system for the storage, analysis and display of spatial data. Spatial data relates to features on the surface of the Earth, at any scale from the global to the local, and consists of two elements: location (where things are) and attributes (what things or places are like). There are two main ways of storing this information on the computer, giving two main types of GIS: (1) In VECTOR GIS, location is stored by classifying objects as points (e.g. spot heights, buildings), lines (e.g. rivers, roads) and areas (e.g. woodlands, towns). Each object has attributes associated with it, normally stored in a database file. (2) In RASTER GIS, location is stored by imposing a grid of square PIXELS over the area of interest – since the pixels are of fixed size, if the location of one point on the grid is known, the location of all other points can be calculated. Attributes are stored by recording a value in each pixel – these are usually presence/absence of an object, value in a classification (e.g. land use) or value of a geographical variable (e.g. elevation).

The input of data is an important and time-consuming element of using a GIS. The commonest sources of data are maps, remote sensing, surveys and address-based data. Maps are normally captured by digitizing. The paper map is attached to a digitizing table, which can record the location of points, lines and areas on the map under operator control. This produces vector data, but this can be converted to raster by software if needed. REMOTE SENSING data are already in a digital, raster format, but raw satellite images must often be processed to generate useful information from the radiation values. Many large surveys, such as population censuses, provide data for small areas. If the boundaries of the areas are available in digital format, the survey information can be attached as attributes. Personal information often contains some form of postal address or postal code (e.g. postcode in the UK, zip code in the USA). This provides a point location that can be stored in a GIS, allowing the mapping of items such as crime incidence or cases of disease. Personal data can also be aggregated to provide a picture of the typical make-up of small areas. Such geodemographic information can then be used to target products and services more closely at potential customers.

Some GISs have developed from systems for computer cartography, and the production of paper maps from digital map data is still an important application of GISs. Advantages of digital methods include the ability to map only selected features, to experiment with the symbolism used and to produce maps unconstrained by traditional map sheet boundaries ('seamless mapping').

The major advantage of GISs lies in their ability to analyse data, which takes two main forms: (1) Analysis that uses simple queries of existing data. An important example is facilities management – the use of a GIS for managing infrastructure such as cables and pipes. This is best suited to a vector GIS, and is commonly used by utility companies. (2) Analysis that modifies or combines existing sets of data. This can be done using both vector and raster, although the types of data and analysis involved are often different.

A vector GIS is well suited to analysis involving discrete objects, such as the use of digitized road networks to estimate travel times between points. A raster GIS is well suited to the analysis of environmental data, where phenomena vary continuously across space. Such analysis commonly makes use of a DIGITAL TERRAIN MODEL (DTM) of the land surface. Both vector and raster can be used for cartographic modelling in which operations are used to modify or combine the original layers of data to answer queries. For instance, to find potential sites for an agricultural crop, layers indicating where individual factors such as soil type, temperature and precipitation are suitable might be combined (using an overlay operation) to find those areas where all the factors are suitable.

In the past, GIS software took the form of general-purpose packages that required considerable training to use. Increasingly, some elements of GIS functionality are finding their way into more widely available software and onto the World Wide Web. For instance, route-finding facilities are provided in simple PC packages, and tourist information services may make use of a map interface, with a radial search to locate all attractions near to a given location.   SMW

Reading
Bernhardsen, T. (2002) *Geographic information systems: an introduction*. New York: John Wiley & Sons, Inc. · Burrough, P.A., McDonnell, R. and Lloyd, C.D. (2013) *Principles of geographical information systems*. Oxford: Oxford University Press. · Longley, P., Maguire, D., Goodchild, M.F. and Rhind, D. (2011) *Geographical information systems: principles, techniques, management and applications*, 3rd edition. Chichester: John Wiley & Sons, Ltd.

geohazard   A geological or environmental situation in which there is a risk of significant damage; for example, in areas of high relief, slopes may be prone to failure through various mass movement processes (see MASS MOVEMENT TYPES). See also NATURAL DISASTER and NATURAL HAZARD. MEM

geoheritage   Elements of the Earth's physical environment that may be considered to have significant scientific, educational, cultural, religious or aesthetic value. Geoconservation represents the measures taken to preserve geodiversity, which is the variety of rocks, minerals, fossils, landforms, sediments, water and soils, together with the natural processes that develop them. Many geoheritage sites, or geosites, can be tourist destinations and may provide significant local and regional socio-economic benefits (Dowling, 2011). MEM

Reference
Dowling, R.K. (2011) Geotourism's global growth. *Geoheritage*, **3**, 1–13.

geoid   The equipotential surface that would be assumed by the sea surface in the absence of tides, water-density variations, currents and atmospheric effects. It varies above and below the geometrical ellipsoid of revolution by as much as 100 m due to the uneven distribution of mass within the Earth. The mean sea-level surface varies about the geoid by typically decimetres, but in some cases by more than a metre.   DTP

Table 1 Divisions of geological time

Era	Sub-era/period/subperiod/ epoch		Age (Ma) begins
Cainozoic	Quaternary	Holocene	0.01
		Pleistocene	2.6
	Neogene	Pliocene	5.3
		Miocene	14.2
			23.0
	Palaeocene	Oligocene	33.9
		Eocene	56.0
		Palaeocene	66.0
Mesozoic	Cretaceous	Late	100.0
		Early	145.0
	Jurassic	Late	164.0
		Middle	174.0
		Early	201.0
	Triassic		252.0
Palaeozoic	Permian		299.0
	Carboniferous		359.0
	Devonian		419.0
	Silurian		444.0
	Ordovician		485.0
	Cambrian		541.0
Precambrian			>4000

geological timescale The divisions of geo-logical time are listed in Table 1. The scheme is revised from time to time and some temporal boundaries are presented differently in various sources. The table below represents the scheme as presented by the International Commission on Stratigraphy and Geological Society of America in 2012. Note that the term Tertiary (Neogene plus Palaeogene), present in older schemes (e.g. Harland *et al.* 1990) is not now officially recog-nized. There is currently debate as to whether the ANTHROPOCENE should be added as a formal epoch.

Reading

Harland, W.B., Armstrong, R.L., Cox, A.V., Craig, L.E., Smith, A.G. and Smith, D.G. (1990) *A geologic time scale: 1989*. Cambridge: Cambridge University Press. · Walker, J.D., Geissman, J.W., Bowring, S.A., and Babcock, L.E. (2013) The Geological Society of America geologic time scale. *GSA Bulletin*, 125, 259–272.

geomorphology A term that arose in the Geological Survey in the USA in the 1880s, possi-bly coined by J. W. Powell and W. J. McGee. In 1891 McGee wrote: 'The phenomena of degra-dation form the subject of geomorphology, the novel branch of geology'. He plainly regarded geomorphology as being that part of geology which enabled the practitioner to reconstruct Earth his-tory by looking at the evidence for past erosion, writing: 'A new period in the development of geologic science has dawned within a decade.

In at least two American centres and one abroad it has come to be recognised that the later history of world growth may be read from the configuration of the hills as well as from the sediments and fossils of ancient oceans . . . The field of science is thereby broadened by the addition of a coordinate province – by the birth of a new geology which is destined to rank with the old. This is geomorphic geology, or geomorphology'.

Many scientists had studied the development of erosional landforms – see Chorley *et al.* (1964) – before the term was thus defined, and since that time its meaning has become broader. Many geo-morphologists believe that the purpose of geo-morphology goes beyond reconstructing Earth history (see DENUDATION CHRONOLOGY), and that the core of the subject is the comprehension of the form of the ground surface and the processes that mould it. In recent years there has been a tendency for geomorphologists to become more deeply involved with understanding the processes of ero-sion, weathering, transport and deposition, with measuring the rates at which such processes oper-ate, and with quantitative analysis of the forms of the ground surface (morphometry) and of the materials of which they are composed. Geo-morphology now has many component branches (e.g. ANTHROPOGEOMORPHOLOGY and APPLIED GEO-MORPHOLOGY). ASG

Reading and References

Anderson, R.S. and Anderson, S.P. (2010) *Geo-morphology: the mechanics and chemistry of landscapes*. Cam-bridge: Cambridge University Press. · Chorley, R.J., Dunn, A.J. and Beckinsale, R.P. (1964) *The history of the study of landforms*, vol. 1. London: Methuen. · McGee, W.J. (1891) *The Pleistocene history of northeastern Iowa*. Eleventh Annual Report of the US Geological Survey. Washington, DC: US Government Printing Office; pp. 189–577. · Huggett, R.J. (2003) *Fundamentals of geo-morphology*. Abingdon: Routledge. · Tinkler, K.J. (1985) *A short history of geomorphology*. London: Croom Helm.

geoparks Conservation or heritage areas, akin to national parks, where the primary defining crite-ria are geological or geomorphological character-istics. UNESCO promotes a Global Geoparks Network, established in 1998, which has the slightly different goal of supporting the gazetting of geo-graphical areas where geological heritage is the focus of local protection, education and develop-ment. As of January 2015 the network included 110 geoparks, including 31 in China, 10 in Spain, 9 in Italy, 7 in Japan, 6 in the UK and 27 countries with five or less. There are none in the USA, but many United States National and other parks can be considered as equivalents because of the geological or geomorphic basis of their designation. DSGT

Reading
Global Network of National Geoparks. www.globalgeopark.org (accessed 3 July 2015).

geophyte A herbaceous plant that has parts beneath the ground surface which survive when the parts above ground die back.

geoproxy A landform that is not forming today but which has characteristics that allow formation in the past under different environmental conditions to be deduced. Such features can include river terraces, moraines, sand dunes, shorelines around now-dry lake basins, and so on.   DSGT

Reading
Thomas, D.S.G. (2013) Reconstructing paleoenvironments and palaeoclimates in drylands: What can landform analysis contribute? *Earth Surface Processes and Landforms*, 38, 3–16.

geostatistics A subfield of statistics that is focused on the analysis of spatial data sets. Early developments in geostatistics were associated with geological applications, especially in relation to mining, and focused on solving problems encountered in the analysis of spatially correlated data. Geostatistics is based on the theory of random variables and allows the uncertainty associated with spatial estimation and simulation to be modelled in a manner that avoids problems such as spatial AUTOCORRELATION. Geostatistics is widely used to describe the spatial features of a data set and in the interpolation of data. For example, a semi-variogram may be used to describe spatial correlation and used with a kriging analysis to derive optimal estimates for the value of a variable at a location from a set of measurements acquired at other locations. The analysis essentially aims to estimate local values of the variable of interest through the use of a model based on weighted linear combinations of available data.   GF

Reading
Cressie, N.A.C. (1993) *Statistics for spatial data*. New York: John Wiley & Sons, Inc.

geostrophic wind The geostrophic wind is a wind with a velocity determined by an exact balance of the CORIOLIS FORCE and the horizontal pressure gradient force. This balance results in a configuration of velocity and pressure as described by BUYS BALLOT'S LAW; that is, in the northern hemisphere, when one has one's back to the wind, low pressure lies on the left and high pressure on the right. The converse is true in the southern hemisphere. The geostrophic wind thus blows *along* the isobars and its magnitude is a direct function of the horizontal pressure gradient force, and an inverse function of height and latitude. It is not defined at the equator. At other latitudes the geostrophic wind is a reasonable approximation to the real wind.   BWA

Reading
Atkinson, B.W. (1972) The atmosphere. In D. Q. Bowen (ed.), *A concise physical geography*. Amersham: Hulton Educational; pp. 1–76. especially pp. 33–48. · Hess, S.L. (1959) *Introduction to theoretical meteorology*. New York: Henry Holt.

geosyncline A very large depression, perhaps several hundred kilometres across and up to 10 km deep, the terrestrial or marine floor of which is built up by sedimentation.

geothermal energy Gravitational collapse and radioactive breakdown generate heat in the Earth's interior, which may be 'extracted' for human use by pumping water into the Earth and extracting it as steam.   DSGT

Gerlach trough A sediment trap designed to catch a sample of OVERLAND FLOW and the sediment it carries down a hillside. Troughs with a variety of shapes and sizes have been used to collect slopewash sediment. They have an upslope lip that is either flush with the surface or inserted beneath the uppermost organic soil layers; disturbance from installing this lip is a major source of error. Sediment settles out and/or is filtered out of water that eventually overflows from the trough, sometimes through a total water or flow rate meter. The trough is protected by a lid. A second major source of error is the accurate delimitation of the area that yields sediment to the trough. This type of installation is an alternative to direct measurements of slope lowering by erosion pins, and so on.   MJK

Reading
Goudie, A. (ed.) (2003) *Geomorphological techniques*, 2nd edition. London: Unwin Hyman.

geyser A spring or fountain of geothermally heated water that erupts intermittently with explosive force as a result of increases in pressure beneath the surface. Significant numbers occur in Wyoming (USA), Iceland, New Zealand, and in the pre-Andes of Argentina.

Ghyben–Herzberg principle This refers to the relationship between fresh water and saltwater in a coastal aquifer. Ghyben and Herzberg were two European scientists who independently investigated this relationship around the turn of the

Lady Knox, a geothermal geyser in Waiotapu, New Zealand.
Photograph by Mair Thomas.

century. They found that since fresh water is less dense than seawater, it rises above underlying intruding saltwater. In unconfined aquifers beneath small islands, a lens of fresh water floats on seawater that surrounds and underlies it, whereas at the edges of larger landmasses there is a sloping interface with fresh water extending to the coast near the surface and seawater penetrating inland at depth (see diagram). The Ghyben–Herzberg principle can be expressed by

$$Z_s = \frac{\rho_f}{\rho_s - \rho_f} Z_w$$

where Z_s is the depth below sea level to the interface between fresh water and saltwater; ρ_f and ρ_s are the density of fresh water and saltwater respectively; and Z_w is the elevation of the water table above sea level. Hence, if the density of the fresh water is 1 and that of seawater is 1.025, then under hydrostatic equilibrium the depth Z_s to the interface is 40 times the height of the water table above sea level. Consequently, if pumping from a well in a coastal aquifer results in a draw down of the water table by 1 m, then saltwater will intrude upwards beneath the well by a distance of 40 m (see diagram).

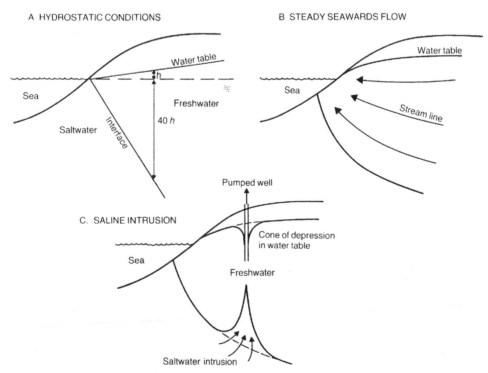

Ghyben-Herzberg principle: Stages of saltwater intrusion.

The Ghyben–Herzberg principle simplifies the relationship usually found in nature, because groundwater conditions are usually dynamic rather than static. As a result, the equation usually underestimates the depth to the interface with saltwater, which is commonly seaward of the calculated position.   PWW

Reading

Hubbert, M.K. (1940) The theory of groundwater motion. *Journal of Geology*, **48**, 785–944. · Pool, M. and Carrera, J. (2011) A correction factor to account for mixing in Ghyben–Herzberg and critical pumping rate approximations of seawater intrusion in coastal aquifers. *Water Resources Research*, **47**, art. no. W05506. · Ward, R.C. and Robinson, M. (1990) *Principles of hydrology*, 3rd edition. Maidenhead: McGraw-Hill.

gibber   The Australian term for the resistant pebbles, often covered by ROCK VARNISH, that form surface veneers across large areas of gibber desert (or stony desert). The term was taken into English from the Dharuk indigenous language of the Sydney area in the late eighteenth century. The original meaning was a stone or boulder. Gibber deserts exhibit extensive desert pavements in which the rock material is frequently highly resistant SILCRETE.   DLD

Reading

Taylor, G. and Eggleton, R.A. (2001) *Regolith geology and geomorphology*. Chichester: John Wiley & Sons, Ltd.

Gibbsian distribution   A distribution used in statistical mechanics that, for a system capable of many different configurations, links the probability of occurrence of any one state to the total energy associated with that state. This distribution has been employed in the study of the structure of stream networks, which topology shows can exist in principle in an extremely large number of different forms but which in nature display certain kinds of regularity (as exhibited in some of the empirically based 'laws' of drainage basin composition). The analysis of system energy using the Gibbsian distribution seeks to develop explanations for these observed regularities in the form of stream networks.   DLD

Reading

Troutman, B.M. and Karlinger, M.R. (1994) Inference for a generalized Gibbsian distribution on channel networks. *Water Resources Research*, **30**, 2325–2338.

gilgai   A class of soil surface forms, including various kinds of undulations and closed depressions. Derived from an Australian aboriginal word meaning 'waterhole'. Gilgai features include up to three components: the *depression* or low-lying part, the *mound* or elevated part, and sometimes a *shelf* or planar area lying at an elevation between that of depression and mound. The relative spatial extent of each component leads to four basic gilgai types: (1) mound and depression equally developed; (2) mound of greater extent than depression; (3) depression of greater extent than mound; (4) gilgai with mound, shelf, and depression (Paton, 1974). The distance separating adjacent mounds or depressions is typically 1–20 m, and the amplitude from depression to mound 10–50 cm, but reaching 1–2 m in rare instances. The development of gilgai is ill understood but appears to be related to shrink–swell behaviour in the subsoil materials of EXPANSIVE SOILS. In some cases, the mound overlies a diapiric structure involving intense deformation, in which subsoil material approaches or reaches the soil surface.   DLD

Reference

Paton, T.R. (1974) Origin and terminology for gilgai in Australia. *Geoderma*, **11**, 221–242.

gipfelflur   A plane within which uniform summit levels occur in a mountainous region, especially where the uniformity is neither structural nor the residual portion of a peneplain.

glacial   *a.* (adjective). Describes a landscape occupied by glaciers. The term *glaciated* describes a landscape that has been covered by glaciers, but normally in the past. The term is also used to identify sediments of glacial affinity or origin.

b. (noun). Those eras of time when ice sheets were expanded and average global climates were colder and drier than during the intervening INTERGLACIALS, such as exists at present. During many of the 20 or so PLEISTOCENE glacials, ice sheets covered Canada and the northern USA, northern Europe, Britain north of the environs of London and northwestern Siberia. In addition, the existing ice sheets of Greenland and Antarctica expanded offshore, while mountain glaciers throughout the world extended into lower altitudes. Sea ice extended further towards the equator as global ocean temperatures fell. Atmospheric and oceanic circulation was modified, extending subtropical deserts and restricting equatorial rain forests to discrete islands by the spread of savanna conditions. Sea levels were lower due to more water being tied up in ice bodies. Mid-latitude areas in the northern hemisphere saw extensive deposition of loess in Europe, China and North America.

Reading
Benn, D. and Evans, D.J.A., (2010) *Glaciers & glaciation*. London: Hodder Arnold. · Gillespie, A.R., Porter, S.C. and Atwater, B.F. (2004) *The Quaternary period in the United States*. Amsterdam: Elsevier. · Lowe, J. and Walker, M. (2015) *Reconstructing Quaternary environments*, 3rd edition. London: Routledge.

glaciation   *a.* The covering of land by year-round ice, leading to the development of GLACIERS, ICE CAPS and ICE SHEETS.
b. In a temporal sense, the processes associated with the occurrence of a GLACIAL period.

Reading
Benn, D., and Evans, D.J.A. (2010) *Glaciers & glaciation*. London: Hodder Arnold. · Gillespie, A.R., Porter, S.C. and Atwater, B.F. (2004) *The Quaternary period in the United States*. Amsterdam: Elsevier. · Lowe, J. and Walker, M. (2015) *Reconstructing Quaternary environments*, 3rd edition. London: Routledge.

glaciation level (also called glacial limit)
The altitude above which mountain glaciers occur. Since glacier location is also influenced by topography, in particular by the need for sufficiently gentle slopes on which to form, the commonly used method of fixing the glaciation level is to take the altitude midway between the highest topographically suitable mountain without a glacier and the lowest topographically suitable mountain carrying a glacier (Østrem, 1966). Delimited in this way, the glaciation level varies predictably over the globe. It rises from near sea level in high latitudes towards the equator in response to temperature changes, but superimposed on this trend are variations reflecting depression in humid areas, such as the mid-latitude and equatorial regions. The glaciation level also rises from maritime coastal locations towards continental interiors.   DES

Reference
Østrem, G. (1966) The height of the glacial limit in southern British Columbia and Alberta. *Geografiska Annaler, Series A: Physical Geography*, **48**, 126–138.

glacier   A mass of snow and ice that, if it accumulates to sufficient thickness, deforms under its own weight and flows. If there is insufficient snow to maintain flow, as occurs in some dry polar areas, the glacier may be essentially stagnant and termed a *glacier réservoir* (Llibboutry, 1965). If the snowfall is sufficient the snow is transformed to ice and flows from the accumulation zone to the ABLATION zone as a *glacier évacateur*.
There are three main types of glacier:

1 *Ice sheet or ice cap*, where the ice builds up as a dome over the underlying topography. Such domes are often drained radially by outlet glaciers.
2 *Ice shelf*, where the ice forms a floating sheet in a topographic embayment and flows towards the open sea.
3 *Mountain glaciers*, which are constrained by the underlying topography of the mountains and form a wide variety of types; for example, cirque, valley, piedmont glaciers.   DES

Reading and Reference
Benn, D.I., and Evans, D.J.A. (2010) *Glaciers & glaciation*, 2nd edition. London: Hodder Arnold. · Llibboutry, L. (1965) *Traité de glaciologie*, 2 vols. Paris: Masson.

glacier mass balance (GMB)   The difference between accumulation and ablation (melting and sublimation) of a glacier. Mass balance is now calculated by several means. The simplest 'mass budget' method, long used, determines the amount of snow accumulated during winter and the amount of snow and ice removed (ablated) by melting in the summer. GMB is the difference between these two parameters. If the amount of snow accumulated during the winter is larger than the amount of snow and ice melted during the summer, then the mass balance is positive and the glacier has increased in volume. On the other hand, if the melting of snow and ice during the summer is larger than the supply of snow in the winter, the GMB is negative and the glacier volume decreases.
More advanced means are now available, illustrated by approaches used in studies of the GREENLAND ICE SHEET. First, the 'mass budget' approach has been advanced to use climate models to calculate snow accumulation and surface run-off. These are combined with calculations of ice discharge into the ocean, which requires knowledge of grounding lines, ice thickness and velocity. Ice thickness is measured using airborne radio-echo sounding, and ice velocity is calculated from satellite radar interferometry or repeat-pass tracking of ice-surface features. This technique indicates that ice loss is approximately equally partitioned between surface run-off and ice flux to the ocean (van den Broeke *et al.*, 2009).
Second, mass balance may be determined using satellite radar or laser altimetry. Altimetry is used to determine surface height or volume changes across the ice body, which can be converted to changes in mass using estimates of the average densities of snow, firn and ice. Third, mass balance can be determined through the use of the Gravity Recovery and Climate Experiment (GRACE). By using variations in the distance

between two identical polar-orbiting satellites to detect small changes in the Earth's gravitational field, GRACE can provide valuable information about short-term (usually monthly) changes in ice-sheet mass.

A glacier with a sustained negative balance is out of equilibrium and will retreat, while one with a sustained positive balance is out of equilibrium and will advance. GMB is reported in meters of water equivalent. This represents the average thickness gained (positive balance) or lost (negative balance) from the glacier during that particular year). Changes in precipitation and temperature are major controls of GMB, and there is great interest in how changes in GMB relate to ongoing climatic change. Globally, the World Glacier Monitoring Service (2008) suggests that the average annual mass loss of glaciers during the decade from 1996 to 2005 was twice that of the previous decade (1986–1995) and over four times that of the decade from 1976 to 1985. The positive mass balance (and advance) of some Scandinavian glaciers in recent decades, notwithstanding rising temperatures, has been attributed to increased storm activity and precipitation inputs coincident with a high index of the North Atlantic Oscillation (NAO) in winter months since 1980 (Zeeberg and Forman, 2001). A positive GMB phase in the Austrian Alps between 1965 and 1981 has been correlated with a negative NAO index (Schöner et al., 2000). Indeed, the GMB of glaciers in the north and south of Europe have been inversely correlated (Six et al., 2001).　　　　　　　　　　　　　　　AB/ASG

Reading and References

Benn, D.I., and Evans, D.J.A. (2010) *Glaciers & glaciation*. 2nd edition. London: Hodder Arnold. · Schöner, W., Auer, I. and Böhm, R. (2000) Climate variability and glacier reaction in the Austrian eastern Alps. *Annals of Glaciology*, 23, 31–38. · Six, D., Reynaud, L. and Letreguilly, A. (2001) Bilans de masse des glaciers alpines et scandinaves, leurs relations avec l'oscillation due climat de l'Atlantique nord. *Comptes Rendus Academie des Sciences, Science de la Terre et des Planètes*, 333, 693–698. · Van den Broeke, M., Bamber, J., Ettema, J., et al. (2009) Partitioning recent Greenland mass loss. *Science*, **326**, 984–986. · World Glacier Monitoring Service (2008) Global glacier changes: facts and figures. www.grid.unep.ch/glaciers/ (accessed 4 July 2015). · Zeeberg, J. and Forman, S.L. (2001) Changes in glacier extent of north Novaya Zemlya in the twentieth century. *The Holocene*, **11**, 161–175.

glacier milk   A popular name given to glacial meltwater with sufficient suspended sediment load to give it a milky-green colour.

glacier table   A stone resting on a pillar of ice that protrudes above a glacier surface. The ice has

been protected from melting by the presence of the overlying stone.

glacieret   A small glacier, such as may develop from a snow patch.

glacierization   The process whereby a landscape is progressively covered by glacier ice.

glacioeustasy   See EUSTASY.

glaciofluvial   The activity of rivers that are fed by glacial MELTWATER. The main characteristics of such streams are the highly variable discharge and the high sediment loads. Discharge varies markedly on a wide variety of timescales. Variations over a matter of seconds or minutes relate to the sudden release or closure of basal water pockets as a result of glacier sliding. Diurnal fluctuations respond to high rates of melting by day and produce high flows in the evenings. Fluctuations over a matter of days reflect prevailing weather patterns, whereas a strong seasonal summer flow reflects the effect of a glacier in storing winter precipitation only to release it in the ABLATION season. One particularly sudden seasonal peak in discharge may accompany the rapid emptying of a marginal or subglacial lake (see JÖKULHLAUP). The muddy colour of meltwater streams reflects their high suspended sediment loads, and measurements as high as $3800\,mg\,l^{-1}$ have been measured. In addition, the bedload is high and may amount to 90% of the suspended sediment load. Not surprisingly, glaciofluvial landforms may reflect prodigious feats of erosion and sedimentation (Table 2). Formerly glaciated areas, particularly in mid-latitudes, contain abundant erosional evidence in the form of deeply incised meltwater channels and giant pot-holes, subglacial channel courses such as ESKERS and KAMES, and extensive areas of proglacial OUTWASH and lake deposits (glaciolacustrine).　　　　　　DES

Reading and Reference

Benn, D.I. and Evans, D.J.A. (2010) *Glaciers & glaciation*, 2nd edition. London: Hodder Arnold. · Elliston, G.R. (1973) Water movement through the Gornergletscher. *Symposium on the Hydrology of Glaciers, Cambridge 9–13 September, 1969*. International Association of Scientific Hydrology Publication 95. Paris: IASH; pp. 79–84. · Østrem, G., Bridge, C.W. and Rannie, W.F. (1967) Glaciohydrology, discharge and sediment transport in the Decade Glacier area, Baffin Island, NWT. *Geografiska Annaler, Series A: Physical Geography*, 49, 268–82. · Price, R.J. (1973) *Glacial and fluvioglacial landforms*. Edinburgh: Oliver & Boyd.

glacioisostasy   The state of balance that the Earth's crust will attempt to achieve when loaded

Table 2 A classification of glaciofluvial deposits

Dominant sediment	Environment	General form	Relationship to ice	Genetic term
Ice-contact deposits				
Sand and gravel	Fluvial	Ridge	Marginal, subglacial, englacial,	Esker
		Mound	supraglacial	Kame
				Kame complex
		Spread with depressions	Marginal	Kettled sandur
Proglacial deposits				
Sand and gravel	Fluvial	Spread	Proglacial	Sandur
Silt and clay	Lacustrine		Proglacial/marginal	Lake plain
Sand and gravel		Terraces, ridges		Beach
Clay, sand and gravel		Terrace		Kame delta
Silt and clay	Marine	Spread		Raised mud flat
Sand and gravel		Terraces, ridges		Raised beach
Clay, sand and gravel		Terrace		Raised delta

Source: Price (1973: table 3, p. 138).

or unloaded by an ICE SHEET. ISOSTASY describes the equilibrium between a less dense and rigid lithosphere 'floating' on the dense and more mobile mantle. The addition of an ice sheet provides greater loading, which, given enough time, will depress the lithosphere by an amount related to the thickness of ice and in proportion to the ratio of the densities of ice and the mantle.

Since the density of glacier ice is about one-third that of the crust/mantle, it follows that depression is about 0.3 times the ice thickness. This is a simplification because depression will be greatest under the central portion of the ice sheet, diminishing radially away from this. As the lithosphere possesses rigidity, some of the load is transmitted beyond its margins, which can lead to the depression extending as far away as 250 km.

The lithosphere can respond to a change in loading on a 1000 year timescale but may take over 10,000 years to reach an equilibrium. As the ice sheets of the last glacial disappeared between 10,000 and 7000 years ago the land surface in these areas is still adjusting to the unloading. This is often called glacioisostatic rebound, and is as much as 1 cm per year for the central portion of the LAURENTIDE ICE SHEET with an estimated 150 m of uplift still to take place. In areas of close proximity to former ice sheets, RAISED BEACHes provide a partial record of the glacioisostatic rebound, although the effects of GLACIOEUSTASY and HYDROISOSTASY also need to be taken into account. CDC

Reading
Benn, D.I. and Evans, D.J.A. (2010) *Glaciers & glaciation*, 2nd edition. London: Hodder Arnold. · Sabadini, R.,

Lambeck, K. and Boschi, E. (eds) (1991) *Glacial isostasy, sea level, and mantle rheology*. Dordrecht: Kluwer.

glaciotectonism Those structures and landforms (e.g. displaced megablocks) produced by deformation and dislocation of pre-existing soft bedrock and drift as a direct consequence of glacier ice movement. ASG

Reading
Aber, J.S. (1985) The character of glaciotectonism. *Geologie en Mijnbouw*, **64**, 389–395.

glacis A gentle pediment slope, especially in arid and semi-arid regions.

glei, gley A clayey soil rich in organic material that usually develops in areas where the soil is waterlogged for long periods. Various component processes are involved: the reduction of ferric compounds, the translocation of iron as ferrous compounds or complexes, and precipitation of iron as mottles and minor indurations.

Glen's law The relationship between the deformation of ice over time and SHEAR STRESS discovered by J. W. Glen (1955). It has the form

$$\dot{\varepsilon} = A\tau^n$$

where $\dot{\varepsilon}$ is the strain rate (deformation rate), τ is the shear stress, A is a constant depending on ice temperature, crystal size and orientation, impurity content and perhaps other factors, and n is a constant whose mean value is normally taken as equal to 3. The relationship models the secondary creep of ice, which involves several separate processes, such as the movement of dislocations within crystals, crystal growth, the migration of crystal boundaries and recrystallization.

Glen's law is of fundamental importance in understanding glacier motion. It shows how sensitive glacier ice is to increasing shear stress; for example, when the shear stress is doubled, the rate of deformation increases eight times. This inherent sensitivity helps explain the characteristic shallow surface profile of glaciers. It also explains why most internal deformation occurs at the bottom of glaciers, and it shows how glaciers move by internal deformation in the absence of BASAL SLIDING.   DES

Reference
Glen, J.W. (1955) The creep of polycrystalline ice. *Proceedings of the Royal Society of London, Series A*, **228**, 519–538.

glint line   The escarpment of Palaeozoic rocks that borders the Scandinavian and Laurentide shields and is associated with a line of lakes. Infrequently used today.

global dimming   The gradual reduction in the amount of global direct irradiance at the Earth's surface that was observed for several decades after the start of systematic measurements in the 1950s. It is thought to have been caused by an increase in particulates such as sulphate aerosols in the atmosphere due to human action. The effect varies by location, but worldwide it has been estimated to have been of the order of a 4% reduction over the three decades from 1960 to 1990. Global dimming creates a cooling effect that may have partially counteracted the effect of greenhouse gases on global warming. It has also interfered with the hydrological cycle by reducing evaporation and may have reduced rainfall in some areas. It has been estimated that solar dimming due to rising aerosol concentrations in the atmosphere around 1980 led to an increase in river runoff by up to 25% in the most heavily polluted regions in Europe. Conversely, these regions may experience reduced fresh water availability in the future, as air quality improvements are set to lower aerosol loading and associated solar dimming.   TS

Reading
Gedney, N., Huntingford, C., Weedon, G.P., *et al.* (2014) Detection of solar dimming and brightening effects on northern hemisphere river flow. *Nature Geoscience*, **7**, 796–800.

global environmental change   There are two components to this (Turner *et al.*, 1990): systemic global change and cumulative global change. In the systemic meaning, 'global' refers to the spatial scale of operation and comprises such issues as global changes in climate brought about by atmospheric pollution (e.g. GLOBAL WARMING). In the cumulative meaning, 'global' refers to the areal or substantive accumulation of localized change, and a change is seen to be 'global' if it occurs on a worldwide scale or represents a significant fraction of the total environmental phenomenon or global resource. Both types of change are closely intertwined. For example, the burning of vegetation can lead to systemic change through such mechanisms as carbon dioxide release and albedo change, and to cumulative change through its impacts on soil and biotic diversity.

Because of the magnitude of the HUMAN IMPACT in causing global environmental change, Crutzen and colleagues have introduced the term *Anthropocene* (e.g. Crutzen, 2002; Steffen *et al.*, 2007) as a name for a new epoch in Earth's history when human activities have 'become so profound and pervasive that they rival, or exceed the great forces of Nature in influencing the functioning of the Earth System' (Steffen, 2010). In the last 300 years, they suggest, we have moved from the Holocene into the Anthropocene. They identify three stages in the Anthropocene. Stage 1, which lasted from *c.* 1800 to 1945 they call 'The Industrial Era'. Stage 2, which extends from 1945 to *c.* 2015, they call 'The Great Acceleration', and Stage 3, which may perhaps now be starting, is a stage when people have become aware of the extent of the human impact and may thus start stewardship of the Earth system.   ASG

References
Crutzen, P.J. (2002) Geology of mankind. *Nature*, **415**, 23. · Steffen, W. (2010) Observed trends in Earth system behaviour. *Interdisciplinary Reviews, Climate Change*, **1**, 428–449. · Steffen, W., Crutzen, P.J. and McNeill, J.R. (2007) The Anthropocene: are humans now overwhelming the great forces of nature? *Ambio*, **36**, 614–621. · Turner II, B.L., Kasperson, R.E., Meyer, W.B., *et al.* (1990) Two types of global environmental change. Definitional and spatial-scale issues in their human dimensions. *Global Environmental Change*, **1**, 14–22.

Global Positioning System (GPS)   A satellite-based system that provides civilian and military users with positioning, navigation and timing (PNT) services. The most widely used system is that owned by the USA. It consists of a series of 24 satellites, whose 12-h orbits are designed so that at any point on the Earth, between five and eight satellites are always overhead. The satellites carry high-precision atomic clocks and constantly transmit time signals using a unique identifying code. These signals can be used by a receiver to estimate how far away each satellite is. Given information on the distance (pseudorange) to at least four satellites whose position is known, the receiver can then estimate its own position by trigonometry. The accuracy depends on the quality and type of signal from the satellites, which can be affected by atmospheric conditions and sky blockage, as well as processing methods and types of receiver. The positional accuracy with a single receiver for civilian users is currently between 5 and 10 m, 95% of the time, and the height accuracy is generally 15–20 m, 95% of the time. Military users receive a more accurate coded signal from the satellites. Greater positional accuracy of up to 10 cm can be achieved using differential GPS (DGPS). DGPS uses a network of fixed, ground-based reference stations to broadcast the difference between the positions indicated by the satellite systems and the known fixed positions. These stations also broadcast the difference between the measured satellite pseudoranges and actual (internally computed) pseudoranges, and the receiver can then correct their pseudoranges by the same amount to improve accuracy.

The use of GPS has grown rapidly and the technology advanced such that many inexpensive and highly mobile devices (e.g. smartphones, cameras) can use it. This has proved important in crowdsourcing, as members of the general public now have the ability to locate themselves with high accuracy and provide georeferenced data. For example, citizen sensing allows the public to acquire georeferenced photography that may be shared over social media or Internet projects that can be used to support scientific studies of the Earth and its environment.   RH-Y/GF

Reading

GPS.gov. http://www.gps.gov/ (accessed 5 July 2015). · Kaplan, E.D. and Hegarty, C.J. (eds) (2005) *Understanding GPS: principles and applications*. London: Artech House. · Raper, J., Gartner, G., Karimi, H., and Rizos, C. (2007) Applications of location-based services: a selected review. *Journal of Location Based Services*, 1, 89–111. · Shoval, N., Kwan, M.P., Reinau, K.H., and Harder, H. (2014) The shoemaker's son always goes barefoot: implementations of GPS and other tracking technologies for geographic research. *Geoforum*, 51, 1–5.

global warming   The warming of the atmosphere and oceans due to increasing concentrations of certain gases in the atmosphere (see GREENHOUSE EFFECT). The major gases contributing to the warming are the so-called 'greenhouse gases': carbon dioxide, methane and nitrous oxide. These are also termed 'trace gases' because they are very small in concentration. Excluding water vapour, these gases together comprise much less than 1% of the atmospheric mass, but they produce up to 35% of the radiative heating of the troposphere (Kiehl and Trenberth, 1997). The remainder of the radiative heating is via cloud and water vapour absorption. However, both cloud amount and water vapour content of the atmosphere are influenced by temperature. Hence, there is feedback from the increase in greenhouse gases.

This problem has been of concern since the mid-twentieth century, when an analysis of measurements established in 1958 at the Mauna Loa Observatory in Hawaii demonstrated the steady rise in carbon dioxide. Subsequently, it was realized that the other greenhouse gases, such as methane and nitrous oxide, were similarly increasing, heightening the concern about global warming. By 2010 the three major greenhouse gases had reached concentrations of 391 ppm for carbon dioxide, 1803 ppb for methane and 324 ppb for nitrous oxide (Stocker *et al.*, 2014). These levels represent increases with respect to pre-industrial levels of 40%, 150% and 20% respectively. Ice core measurements show that these levels are unprecedented over the last 800,000 years. The heating potential of methane is 21 times greater than that of carbon dioxide, and that of nitrous oxide is 310 times greater (Houghton *et al.*, 1996).

Completing the picture are man-made substances such as chlorofluorocarbons, which are low in concentration but have huge potential as warming agents. These anthropogenic substances have heating potential thousands of times greater than carbon dioxide, and their combined contributions to global warming have been roughly 15% of that due to carbon dioxide (Houghton *et al.*, 1996).

The reason for the concern about global warming is its potential impacts, including cryospheric melting, sea-level rise, shifts in vegetation and precipitation belts, changes in hurricane characteristics, modifications to land use and human affairs, and changes in disease distribution. Recognizing the global dimensions of the problem, in 1988 the World Meteorological Organization and the United Nations Environmental Program established the Intergovernmental Panel on Climate Change (IPCC). It was tasked with preparing a comprehensive review on global climate change. Since that time, five major assessments have been produced, the fifth being published in several parts

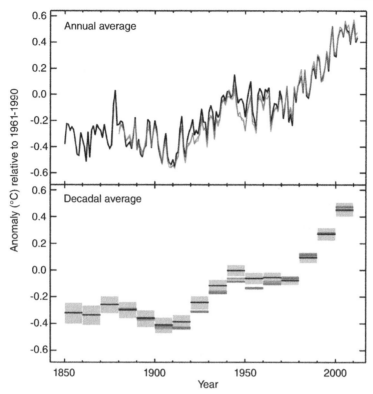

Global warming: Annual average observed combined land and ocean temperature anomalies, 1850–2012.

during 2013 and 2014. Contributors include hundreds of scientists in diverse disciplines.

In the fifth assessment the panel concluded that the evidence of the warming of the climate system is unequivocal and that many of the observed changes are unprecedented over at least decades, and probably millennia. Each of the last three decades has been successively warmer than all proceeding decades since 1850. Since 1900, warming has taken place over nearly the entire globe, both atmosphere and ocean. The warming has been particularly large in central Canada, eastern South America and Asia. At the same time, increasing trends in precipitation have occurred over the eastern USA and Canada, Europe, western Asia, parts of Australia and southern South America. Increases of greater magnitude and larger spatial extent have occurred since 1950, while precipitation has decreased over much of Africa, the Mediterranean, eastern Asia and southern Australia. Over the last two decades, numerous changes have been observed in the cryosphere as well: decreases in the Greenland and Antarctic ice sheets, shrinking glaciers nearly worldwide,

reduced Arctic sea ice cover and spring snow cover in the northern hemisphere. Commensurate with this, global sea level has increased on average some 50 mm or more; since 1900 the increases has been nearly 200 mm.

The fifth IPCC assessment also concludes that the human influence is clear in the warming of atmosphere and ocean, changes in global water cycle, reductions in snow and ice cover, and sea level rise. The evidence for this has grown since the fourth assessment and indicates that that anthropogenic factors account for more than half of the global temperature increase over the last half century. Between 1951 and 2010 the overall greenhouse gas contribution to global mean surface warming was on the order of 0.5 to 1.3 °C.

SEN

References
Kiehl, J.T. and Trenberth, K.E. (1997) Earth's annual global mean energy budget. *Bulletin of the American Meteorological Society*, **78**, 197–208. · Houghton, J.T., Meira Filho, L.G., Callander, B.A., *et al.* (eds) (1996) *The science of climate change. Report of Working Group I to the second IPCC assessment report.* Cambridge: Cambridge University Press. · Stocker, T.F., Qin, D., Plattner, G.-K., *et al.*

245

(eds) (2014). *Climate change 2013: the physical science basis. Contribution of Working Group I to the fifth assessment report of the IPCC.* Cambridge: Cambridge University Press.

GLOSS (Global Sea-Level Observing System)

A worldwide network of sea-level gauges, defined and developed under the auspices of the Intergovernmental Oceanographic Commission. Its purpose is to monitor long-term variations in the global level of the sea surface by reporting the observations to the Permanent Service for Mean Sea Level. The levels of the gauges are fixed by a satellite-based global positioning system, capable of accurately locating points in a three-dimensional geometric framework.   DTP

gnammas

Holes produced in rock surfaces, especially igneous rocks and sandstones, by weathering processes. They are a type of rock basin.

gneiss

A coarse-grained igneous rock, often a granite, that has been metamorphosed, producing a banded or foliated structure.

goletz terrace (also known as cryoplanation terrace or altiplanation terrace)

A hillside or summit bench that is cut in bedrock and transects lithology and structure. It is confined to cold climates.

Reading
Washburn, A.L. (1979) *Geocryology.* London: Edward Arnold.

Gondwanaland   See SUPERCONTINENT.

gorge

A deep and narrow section of a river valley, usually with near-vertical rock walls. More generally a narrow valley between hills or mountains.

gouffres

Large pipes or vertical shafts that occur in limestone areas.

graben

A valley or trough produced by faulting and subsidence or uplift of adjacent blocks (horsts).

grade, concept of

One of the most confusing concepts in geomorphology, partly because of its inextricable relationship with gradient. Introduced by G. K. Gilbert (1876), it relates the gradient of a channel to the balance between corrasion (erosion), resistance and transportation. The idea was adopted, adapted and debated by W. M. Davis, J. E. Kesseli and J. H. Mackin in particular and resurfaces in the influential paper by S. A. Schumm and R. W. Lichty (1965) as the 'graded' time span. It is now generally assumed to be roughly equivalent to DYNAMIC EQUILIBRIUM and, in practical terms, should be viewed – by extension from work on REGIME THEORY in alluvial canals – as a state in which channel form is relatively constant despite variations in flow (usually 2–10 years). The application to hillslope form and materials is distinctly problematic.   BAK

References
Gilbert, G.K. (1876) The Colorado Plateau province as a field for geological study. *American Journal of Science and Arts (Third Series),* **12,** 16–24, 85–103. · Schumm, S.A. and Lichty, R.W. (1965) Time, space and causality in geomorphology. *American Journal of Science,* **263,** 110–119.

graded bedding

Comprises sedimentary units that exhibit a vertical gradation in mean grain size. Normal grading is where a fining-upward sequence is present, and may result from deposition in a waning current, from a decrease in sediment supply or from the progressive sorting or settling out of different size fractions. Inverse grading exhibits an upward-coarsening sequence, and may result from deposition in rising flow conditions or from an increase in sediment supply.   JM

graded slopes

Those possessing a continuous regolith cover without rock outcrops. The concept of grade was used by G. K. Gilbert (1876) to indicate a condition of balance between erosion and deposition, brought about by adjustments between the capacity of a stream to do work and the quantity of work that the stream has to do. This definition was formally introduced by W. M. Davis (1899, 1954: 267) and applied to hillslopes in the words: 'a graded waste sheet . . . is one in which the ability of the transporting forces to do work is equal to the work they have to do. This is the condition that obtains on those evenly slanting, waste-covered mountain sides which have been reduced to a slope that engineers call *the angle of repose*'. Rocky outcrops are not graded because waste can be removed from them faster than it is supplied by weathering. On slopes from which outcrops have been eliminated, the 'agencies of removal are just able to cope with the waste that is there weathered plus that which comes from farther uphill' (Davis, 1954: 268). Graded waste slopes decline in angle as the waste becomes finer in texture as a result of weathering 'so that some of its particles may be moved even on faint slopes' (Davis, 1954: 269).

Because of the difficulty of determining the volumetric relationships between weathering and removal, and texture and removal processes,

Young (1972: 100) has suggested that the term graded slope be used to indicate those lacking outcrops; this definition would make it equivalent to a soil or regolith-covered slope.   MJS

References

Davis, W.M. (1899) The geographical cycle. *Geographical Journal*, 14, 481–504. · Davis, W.M. (1954) The geographical cycle. In D. W. Johnson (ed.), *Geographical essays*. New York/London: Dover/Constable. · Gilbert, G.K. (1876) The Colorado Plateau province as a field for geological study. *American Journal of Science and Arts (Third Series)*, 12, 16–24, 85–103. · Young, A. (1972) *Slopes*. Edinburgh: Oliver & Boyd.

graded time A time span intermediate between the longer interval of 'cyclic time' and the shorter period of 'steady time' (Schumm and Lichty, 1965). It is defined as 'a short span of cyclic time during which a graded condition or dynamic equilibrium exists' (Schumm and Lichty, 1965: 114), with respect to the landforms and, by reference to Mackin's (1948) discussion of grade (see GRADE, CONCEPT OF), it is implied that this 'short span' will be a 'period of years'. In Schumm and Lichty's view, the chief practical considerations governing studies on the graded timescale are that time and initial relief become irrelevant to the enquiry, while the morphology of drainage networks and hillslopes and the hydrologic outputs of drainage basins are dependent variables, contingent upon the independent controls of geology, climate, vegetation, disposition of relief above base level and the manner in which run-off and sediment are generated within the landscape.

The concept is discussed again by Schumm (1977: 10–13), and rather confusingly it is there redefined (figures 1–5) as equivalent to steady-state equilibrium time, with a time span of 100–1000 years. It seems clear that Schumm intends the term to be used to imply an intermediate timescale in which the focus of investigations is on fluctuations in hydrologic outputs, channel morphology and hillslope form viewed as responses to spatial or temporal patterns of variation in the 'independent' variables listed in the 1965 paper.   BAK

References

Mackin, J.H. (1948) Concept of the graded river. *Bulletin of the Geological Society of America*, 59, 463–512. · Schumm, S.A. (1977) *The fluvial system*. New York: John Wiley & Sons, Inc. · Schumm, S.A. and Lichty, R.W. (1965) Time, space and causality in geomorphology. *American Journal of Science*, 263, 110–119.

gradient wind Results from a balance of horizontal pressure gradient force, CORIOLIS FORCE and

the centripetal acceleration (or centrifugal force) that exists when air moves in a curved path, such as occurs in a cyclone and an anticyclone. BUYS BALLOT'S LAW applies to this wind just as it does to the geostrophic wind, but because more forces are involved, the velocity of the gradient wind is different to that of the GEOSTROPHIC WIND. The one exception to this occurs when airflow is straight, giving a zero centripetal acceleration and, hence, geostrophic balance and wind.   BWA

Reading

Hess, S.L. (1959) *Introduction to theoretical meteorology*. New York: Henry Holt.

gradually varied flow In most natural river channels the flow is gradually varied because the cross-section and bed slope change downstream and the water surface is not parallel to the bed. Under these conditions the FLOW EQUATIONS for UNIFORM STEADY FLOW are not strictly applicable except locally, and a more detailed analysis of the flow energy is required. So long as the streamlines in a short reach are approximately parallel, and pressure within the flow is therefore hydrostatic, the total energy of a unit mass of water at the bed is the sum of its potential energy, pressure energy and kinetic energy:

$$E = \rho_w g z_1 + \rho_w g d_1 + \frac{\rho_w v_1^2}{2}$$

or

$$E = \rho_w g \left(z_1 + d_1 + \frac{v_1^2}{2g} \right)$$

where the term in parentheses is the 'total head' H_1, and

$$H_1 = z_1 + d_1 + \frac{v_1^2}{2g}$$

is the Bernoulli equation (see BERNOULLI'S THEOREM AND EFFECT). This can be used to define the conservation of energy between two adjacent sections. An energy balance equation between sections 1 and 2 states that

$$z_1 + d_1 + \frac{v_1^2}{2g} = z_2 + d_2 + \frac{v_1^2}{2g} + h_L$$

where h_L is the head or energy loss between the sections.

It is clear that in gradually varied flow the water surface, bed slope and energy grade line are not parallel. The energy grade line always slopes downwards in the direction of flow and measures the rate of dissipation of energy (the energy loss) caused by flow resistance and sediment transport. If the water surface slope is gentler than the energy slope the kinetic energy term decreases downstream as the flow decelerates, whereas a steeper water surface slope would reflect accelerating flow and conversion of energy from potential to kinetic forms (Richards, 1982: 72–76). In a POOL AND RIFFLE stream the variations of velocity and depth from section to section reflect this continual conversion of energy from potential to kinetic forms, and vice versa, in response to the changing bed slope.   KSR

Reference

Richards, K.S. (1982) *Rivers: form and process in alluvial channels*. London: Methuen.

granite   A coarsely crystalline igneous rock composed predominantly of quartz and alkali feldspars. Additional constituents are commonly mica and hornblende.

granulometry   The process or method of PARTICLE SIZE determination of sediments, usually referring to those with a diameter of 2 mm or less. A number of granulometric methods exist, including the use of settling tubes, by sieving or by laser defraction. Owing to the different principles involved, data obtained by one method may not be directly comparable to those from another.   DSGT

grasslands   Regions in which the natural or the plagioclimax vegetation is dominated by grasses or grass-like plants and non-grass-like herbs. They include temperate grasslands of the steppes, prairies, pampas and veld, tropical grasslands or SAVANNAS, and smaller zones on mountains, in high latitudes and as patches within other plant formations resulting from fire, soil or drainage controls. Before human modification of the natural plant and animal cover, grasslands probably occupied around 40–45% of the land surface, a figure increased by the maintenance of grazing land and decreased by conversion to other forms of land use to around 25% today.   PAF

Reading

Coupland, R.T. (ed.) (1979) *Grassland ecosystems of the world*. Cambridge: Cambridge University Press.

gravel   Though frequently used to describe small, rounded, unconsolidated rock fragments in fluvial systems, gravel can also refer to any sediment comprising uncemented rock fragments coarser than SAND (2 mm diameter) and smaller than pebbles, with the upper size limit in the 10–60 mm range. Gravel particles may be angular or rounded, spheroid or rod shaped, and may

result from in-situ weathering or rock breakdown during transport.   DSGT

gravimetric method   A means of soil moisture determination involving taking, weighing, oven drying and reweighing a soil sample and expressing the moisture content (or sample loss in weight) as a percentage of the oven dry weight of the sample.   AMG

Reading

Reynolds, S.G. (1970a) The gravimetric method of soil moisture determination, I: a study of equipment, and methodological problems. *Journal of Hydrology*, **11**, 258–273. · Reynolds, S.G. (1970b) The gravimetric method of soil moisture determination, II: typical required sample sizes and methods of reducing variability. *Journal of Hydrology*, **11**, 274–287. · Reynolds, S.G. (1970c) The gravimetric method of soil moisture determination, III: an examination of factors influencing soil moisture variability. *Journal of Hydrology*, **11**, 288–300.

gravity   The force imparted by the Earth to a mass that is at rest relative to the Earth. All masses are attracted to each other according to Newton's law of universal gravitation, but the Earth is also rotating, and therefore a centrifugal force is also exerted on the mass in question. Hence, the force observed, and commonly called gravity, is the combination of the true gravitational force and the centrifugal force. The standard acceleration of gravity at sea level at 45° latitude is $9.80665 \ \mathrm{m \ s^{-2}}$.
BWA

gravity faulting   A process that operates in mountainous areas; high available relief enables major rock movements to occur under the influence of gravity, creating hilltop valleys and depressions, and sometimes double summits (*doppelgrate*).   ASG

Reading

Beck, A.C. (1968) Gravity faulting as a mechanism of topographic adjustment. *New Zealand Journal of Geology and Geophysics*, **11**, 191–199. · Paschinger, V. (1928) Untersuchungen über Doppelgrate. *Zeitschrift für Geomorphologie*, **3**, 204–236.

gravity wave   A wave disturbance in which buoyancy acts as a restoring force on fluid parcels displaced from an equilibrium state. The restoring force acts only in the vertical, frequently producing simple harmonic motion around the equilibrium level. The wave ensues because this simple harmonic motion occurs in a horizontal flow. The resultant of the two components of velocity (vertical and horizontal) at any instant gives the velocity of the parcel of air under consideration. A sequence of such velocities throughout one cycle of simple harmonic motion describes the gravity wave form. Examples of gravity waves are lee waves in the atmosphere and water waves. Gravity waves are also known as buoyancy waves for obvious reasons.   BWA

great escarpment An escarpment formed by the uplift associated with rifting that contributed to the break-up of Gondwanaland. Great escarpments are found, with various degrees of dissection and erosion, around the margins of southern hemisphere continents. The great escarpment of southern Africa extends more or less continuously, 150–200 km inland of the modern coastline, from Angola to southern Tanzania.   DSGT/MEM

Reading
Ollier, C. and Rabassa, J.J. (eds) (2014) *Gondwana landscapes in southern South America*. Dordrecht: Springer.

great interglacial A phase of Pleistocene history identified in the classic four-glacial model developed by A. Penck and E. Brückner (see PENCK AND BRÜCKNER MODEL), who believed that the interglacial between the Mindel and the Riss glacials lasted a longer time than any others. Studies of the Pleistocene record in ocean cores tend not to support this view.   ASG

greenhouse effect The greenhouse effect is a natural phenomenon that occurs when short-wave solar radiation from the sun passes largely unattenuated through the Earth's atmosphere, is absorbed at the planetary surface, is reradiated upward as longer wavelength thermal radiation, and is absorbed by various atmospheric constituents and again reradiated. Since some of this latter radiation flux is directed downward, it results in a surface warming that would not occur in the absence of an atmosphere. And this extra warming is what is commonly called the greenhouse effect.

The most important 'greenhouse gases' are water vapour and carbon dioxide (CO_2). Their presence in the atmosphere allows the Earth to maintain an average temperature of approximately 15 °C. Without them, the surface temperature of the planet would be about −19 °C, and the Earth could not support life. Consequently, it is clear that we owe our very existence to the greenhouse effect. With the onset of the Industrial Revolution, however, the CO_2 content of the air began to rise, and people began to worry that this phenomenon might have a 'dark side' we had not anticipated (see GLOBAL WARMING).

Spurred on by this concern, scientists are attempting to determine whether increasing concentrations of atmospheric greenhouse gases will have any effect on global temperature or climate patterns. Scientists who use computer models (see GENERAL CIRCULATION MODELLING) to study these questions commonly report that a temperature rise of 1.5–4.5 °C will result from a doubling of the atmosphere's CO_2 content, leading to catastrophic events such as the melting of Earth's polar ice caps, which could produce widespread flooding and famine. Some scientists suggest these estimates are too extreme, pointing out the shortcomings of computer models and the difficulties of emulating the many interacting complexities of the natural world. However, global temperature has already risen roughly 1 °C during the last half century and 1.5 °C since pre-industrial times, indicating that the more extreme estimates could be realistic.   SEN

Reading
Blindoff, N.L., Stott, P.A., Achuta Rao, K.M., *et al.* (2014) Detection and attribution of climate change: from global to regional. In T. F. Stocker, D. Qin, G.-K. Plattner, *et al.* (eds), *Climate change 2013: the physical science basis. Contribution of Working Group I to the fifth assessment report of the IPCC*. Cambridge: Cambridge University Press. · Collins, M., Knutti, R., Arblaster, J.M., *et al.* (2014) Long-term climate change: projections, commitments and irreversibility. In T. F. Stocker, D. Qin, G.-K. Plattner, *et al.* (eds), *Climate Change 2013: the physical science basis. Contribution of Working Group I to the fifth assessment report of the IPCC*. Cambridge: Cambridge University Press. · Karl, T.R., Nicholls, N. and Gregory, J. (1997) The coming climate. *Scientific American*, 276 (5), 78–83.

Greenland Ice Sheet Covers 1.71×10^6 km^2, or about 80%, of the surface of Greenland. It is the largest ice mass in the northern hemisphere, and globally is second in size only to the ANTARCTIC ICE SHEET. The geographical position of the ice sheet ranges from 60° to 83°N, and 73° to 110°W. The average ice thickness is ~1.8 km, but in places is almost twice that. With a volume of around 2.85×10^6 km^3, the Greenland Ice Sheet contains 10% of the Earth's total fresh water; were it to melt completely the global sea level would rise by about 7 m. The weight of the ice sheet has depressed the underlying bedrock to elevations close to sea level. The ice-sheet periphery is bounded mostly by coastal mountains, through which land- and marine-terminating outlet glaciers drain.

The bulk of the Greenland Ice Sheet formed during the Late Pliocene (around 3 million years ago), when glaciation intensified due to cooling in the northern hemisphere. Since then, Greenland's climate has been highly variable, and the ice sheet has expanded and contracted many

Greenland ice sheet. (a) Bed elevation and bathymetry. Bathymetry is plotted on an opaque scale. The red contour is at 0 m above sea level. (b) Ice thickness as determined from the difference between surface and bed digital elevation models, with contours at 500, 2000 and 3000 m.
Source: Bamber *et al.* (2013). This work is distributed under the Creative Commons Attribution 3.0 License.

times during the glacial and interglacial periods of the QUATERNARY. Several ice cores taken from the central part of the ice sheet reveal important clues about the magnitude, timing and causes of climate change during the last glacial period; for example, oxygen isotope analysis reveals 25 rapid warming events (called DANSGAARD–OESCHGER (DO) EVENTS).

Calculations indicate that the ice sheet currently has a negative mass balance (see GLACIER MASS BALANCE), and that mass is being lost at an accelerating rate. Since 1992–2001, the average rate of mass loss has increased sixfold, from $34\,\mathrm{Gt\,a^{-1}}$ (sea level equivalent, $0.08\,\mathrm{mm\,a^{-1}}$) to $215\,\mathrm{Gt\,a^{-1}}$ over the period 2002 to 2011 (sea level equivalent, $0.59\,\mathrm{mm\,a^{-1}}$) (e.g. Vaughan *et al.*, 2014). This rapid mass loss since the early 1990s coincides with increasing atmospheric and ocean temperatures in the Arctic, which have risen at almost twice the global average rate in the past 100 years. Albedo feedback due to atmospheric soot additions to the ice surface may also contribute to ice warming and melting (Box *et al.*, 2012). Under the most extreme Intergovernmental Panel on Climate Change greenhouse gas emissions scenario, Greenland's shrinking ice sheet is expected to raise sea levels by 12 cm by 2100.

The seasonal extent of surface meltwater on the Greenland Ice Sheet, which has been monitored by satellites since 1979, with the eight highest melt years on record all occurring since 2002. To date, 2012 is the highest melt year, when, on 12 July, satellite observations revealed that melting had occurred across 98.6% of its entire extent (Nghiem *et al.*, 2012).

The response of the Greenland Ice Sheet to a changing climate is complex, depending upon multiple interactions between the ice sheet, the atmosphere and the ocean. Only recently have complex mathematical models been developed (e.g. Gagliardini *et al.*, 2013) that can simulate this system as a whole. Though the ultimate fate of the Greenland Ice Sheet remains unknown, such models predict that it will be highly susceptible to the substantial increases in air and ocean temperatures that are predicted over the next century.   AB

References
Bamber, J.L., Griggs, J.A., Hurkmans, R.T.W.L., *et al.* (2013) A new bed elevation dataset for Greenland. *The Cryosphere*, **7**, 499–510. · Gagliardini, O., Zwinger, T., Gilet-Chaulet, F., *et al.* (2013) Capabilities and performance of Elmer/Ice, a new-generation ice sheet model. *Geoscience Model Development*, **6**, 1299–1318. doi: 10.5194/gmd-6-1299-2013. · Nghiem, S.V., Hall, D.K.,

Mote, T.L., *et al.* (2012) The extreme melt across the Greenland ice sheet in 2012. *Geophysical Research Letters*, **39**, L20502. doi: 10.1029/2012GL053611 · Vaughan, D.G., Comiso, J.C., Allison, I., *et al.* (2014) Observations: cryosphere. In: T. F. Stocker, D. Qin, G.-K. Plattner, *et al.* (eds), *Climate change 2013: the physical science basis. Contribution of Working Group I to the fifth assessment report of the Intergovernmental Panel on Climate Change.* Cambridge: Cambridge University Press.

grey wether   See SARSEN.

greywacke   A sediment composed of coarse fragments of quartz and feldspar, usually poorly sorted.

grike (gryke)   The cleft or runnel in bare limestone pavements that separates the CLINTS. Grikes are formed when limestone is dissolved by water, probably under a soil cover (Trudgill, 1972) normally along joint pattern weaknesses. Grikes are called *Kluftkarren* in German.   PAB

Reference
Trudgill, S.T. (1972) The influence of drifts and soils on limestone weathering in NW Clare. *Proceedings of the University of Bristol Speleological Society*, **13**, 113–118.

GRIP   Acronym for the Greenland ICE CORE Project.

ground frost   Ground, but not necessarily air, below freezing. On cool, dry nights there is not enough moisture in the air to stop terrestrial (heat) RADIATION escaping to space. When, as is usual, the earth is a good radiator but a poor conductor of heat, the ground may cool sufficiently to give CONDENSATION (dew). If the air is dry enough the moisture may sublimate into crystalline form to make frost. Such conditions are hazardous for plants, whose cells are disrupted by ice crystals. Whether crystals form depends on the amount of 'antifreeze' (sugar and starch) in their tissues.   JSAG

Reading
Monteith, J.L. (1973) *Principles of environmental physics.* London: Edward Arnold.

ground ice   A body of more or less clear ice within frozen ground. It takes many forms, some of the more common being PORE ICE, SEGREGATED ICE, ice veins and ICE WEDGES, PINGO ice and massive icy beds. Buried glacier ice, buried ICING ice and buried snowbank ice are sometimes regarded as forms of ground ice even though they are of surface origin. In places, ground ice may constitute more than 50% by volume of the upper 2–3 m of permafrost. Generally speaking, ground

ice amounts decrease with increasing depth. Aggradational landforms associated with the formation of ground ice include open- and closed-system pingos, ice wedge polygons, palsas and peat plateaux, and seasonal frost mounds. The degradation of ice-rich permafrost causes a THERMOKARST and results in thaw slumping, thaw depressions and lakes and hummocky unstable topography.   HMF

Reading
Mackay, J.R. (1972) The world of underground ice. *Annals of the Association of American Geographers*, **62**, 1–22.

ground moraine   See TILL.

ground-penetrating radar (GPR)   A form of radar, using radio-frequency electromagnetic radiation commonly in the frequency range 100 MHz to 1 GHz, that can be used to make images of subsurface structures within the soil and regolith to a depth of some metres. Reflected signals detected by GPR may be able to indicate the position of soil horizons, the water table, the soil–bedrock interface, cavities or buried objects. GPR can also be used to map soil moisture. GPR uses an antenna that is placed in contact with the ground surface, which both transmits very short pulses into the ground and detects reflected signals. Signal propagation and reflection are affected by the conductivity of the materials beneath the antenna, and their dielectric properties. Wet clays attenuate the signal very strongly, but dry sandy materials cause reduced attenuation and yield better GPR performance. GPR is usually applied in the field by collecting data along transects with the apparatus carried along on a sled or trolley, with the position of the antenna tracked using GPS.   DLD

Reading
Doolittle, J.A. and Collins, M.E. (1995) Use of soil information to determine application of ground penetrating radar. *Journal of Applied Geophysics*, **33**, 101–108. · Minet, J., Bogaert, P., Vanclooster, M. and Lambot, S. (2012) Validation of ground penetrating radar full-waveform inversion for field scale soil moisture mapping. *Journal of Hydrology*, **424–425**, 112–123.

grounding line (grounding zone)   The grounding line of a GLACIER or ice stream marks the boundary between grounded ice (ice on land) and floating ice (ice over water). Where the ice surface gradient is small, this boundary may consist of a transition region of lightly grounded and barely floating ice, rather than a well-defined line, in which case, the term 'grounding zone' may be more appropriate. The areas of floating ice between the grounding line and the calving front are called ice shelves or tongues, and these fringe Antarctica and occupy many fjords and bays along the coast of Greenland.

By definition, ice thickness at the grounding line is equal to the critical thickness required to enable the ice to float, which varies depending on ice and ocean water density (Seroussi et al., 2014).

Determining the flux of ice across a grounding line is often important as it represents a mass loss from the grounded ice part of an ice sheet or glacier, and therefore a contribution to sea level. The flux of sediment across a grounding line is also important as sedimentary depocentres, including grounding-zone wedges, moraine ridges and ice proximal fans, may build up during still-stands or re-advances of the grounding line (Powell and Alley, 1997).

The grounding line of a glacier or ice stream is often identifiable from surface elevation data by a significant down-glacier decrease in surface slope. This indicates the transition from grounded ice, where all resistance to flow is due to basal drag, to the floating ice, where basal drag is zero (Rignot et al., 1997). Ice flow models often struggle to model this transition zone because one set of assumptions must be entirely replaced by another (e.g. Goldberg et al., 2012). AB

References

Goldberg D.N., Little, C.M., Sergienko, O.V., et al. (2012) Investigation of land ice–ocean interaction with a fully coupled ice–ocean model, part 1: model description and behavior. *Journal of Geophysical Research*, 117, F02037, doi: 10.1029/2011JF002246. · Powell, R.D. and Alley, R.B. (1997) Grounding-line systems: processes, glaciological inferences and the stratigraphic record. In P. F. Barker and A. C. Cooper (eds), *Geology and seismic stratigraphy of the Antarctic margin II*. Antarctic Research Series 71. Washington, DC: American Geophysical Union; pp. 169–187. · Rignot, E.J., Gogineni, S.P., Krabill, W.B. and Ekholm, S. (1997) North and northeast Greenland ice discharge from satellite radar interferometry. *Science*, 276, 934–937. · Seroussi, H., Morlighem, M., Larour, E., et al. (2014) Hydrostatic grounding line parameterization in ice sheet models. *The Cryosphere*, 8, 2075–2087, doi: 10.5194/tc-8-2075-2014.

groundwater Strictly defined, groundwater includes all subsurface water in a solid, liquid or gaseous state, other than that which is chemically combined with any minerals present. In practice, it is usual to include all subsurface water that is part of the HYDROLOGICAL CYCLE, which encompasses all groundwater except CONNATE WATER, which is prevented from active circulation, and some frozen groundwaters, which may have a very long turnover time. This definition is wide enough to encompass soil water components such as saturated THROUGHFLOW, but it is more common to distinguish groundwater as a larger, deeper, more slowly moving body located in sediments or bedrock (AQUIFERS), as opposed to soil water.

Groundwater occurs within two vertically separated zones: a VADOSE (or aerated or unsaturated) zone beneath the soil which is situated above a PHREATIC (or saturated) zone. The boundary between the two is termed the WATER TABLE or PIEZOMETRIC SURFACE and marks the upper limit of saturation in an unconfined aquifer. The shape of the water table usually mimics that of the topography, but in a much smoother, more subdued manner, with its depth below the surface varying spatially with climate

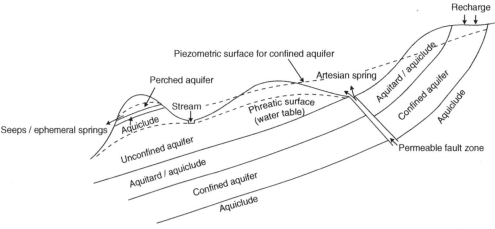

Forms of groundwater storage.

and rock type and temporally due to seasonal variations in groundwater RECHARGE.

The upper vadose zone is often simply regarded as a zone in which water is in transit from the surface to the phreatic zone. However, it holds around 1.5% of global nonfrozen fresh water, and recent research in areas of nonporous fractured bedrock and karst terrain indicates that it can be an important store of water, sustaining base flow in subterranean streams above the water table. Groundwater in the phreatic zone accounts for around 95% of the global fresh, nonfrozen water reserve. It can be considered as being either unconfined if it is stored near the surface within an aquifer that is situated above an impermeable *aquiclude* (a rock which holds and transmits little water) or semipermeable *aquitard*, or confined if it is held within an aquifer between two aquicludes. Confined groundwater is often referred to as ARTESIAN, and may emerge at the surface at an *artesian spring* or well. An artesian spring is one in which pressure within the confined groundwater body forces the water upwards towards the *potentiometric surface* (i.e. the level which the phreatic surface would reach were the groundwater not confined, which would lie above the ground surface). Perched water tables are a variety of unconfined groundwater aquifer of relatively limited extent that stand above the main groundwater body and are formed either above an aquiclude of limited area or where percolation rates exceed the hydraulic conductivity of an aquitard (see PERCHED GROUNDWATER). Groundwater has been encountered at depths of up 4000 m during drilling and mining, but, in general, it diminishes below a few hundred metres. The lower limit of groundwater penetration is determined by a variety of factors, including the availability of voids in the rock and the hydrostatic pressure. At great depths there may be sufficient lithostatic pressure exerted by surrounding rocks that all voids are closed and the rock is effectively impervious.

Groundwater is an important contribution to base flow in some streams, especially those situated upon permeable aquifers, with water entering the channel by effluent seepage. This water may have taken a considerable time from falling as precipitation and contributing to recharge before exfiltrating, with phreatic groundwater flow-through times varying from 2 weeks to 10,000 years depending upon location. Flow-through times are also significant in terms of human utilization of groundwater resources. Areas where groundwater is extracted at a higher rate than recharge is occurring are likely to experience falling water tables and, in exceptional cases, surface subsidence due to aquifer collapse. Excessive groundwater extraction in coastal zones may lead to the intrusion of saline water into the aquifer.

In addition to forming a major part of the hydrological system, groundwater can also act as an important geomorphological agent. Subsurface groundwater flow through fissures and joints in limestone bedrock can lead to the development of KARST landforms, such as caves and subterranean channels by solutional processes. Emerging groundwater can contribute to the erosion of rock or sediment faces, and the development of scarps, DRY VALLEYS and some canyons, by spring SAPPING or seepage erosion. The height of the water can also act as an important control on the operation of specific processes, such as where the maximum extent of aeolian DEFLATION is limited by the depth to the regional water table and in PLAYA systems where fluctuations in water table height influence water chemistry and surface morphology. Groundwater can also act as an important medium for transporting minerals in solution, which is important in the development of many near-surface duricrusts such as CALCRETES, SILCRETES and LATERITE.   DJN

Reading
Brown, A.G. (ed.) (1995) *Geomorphology and groundwater*. Chichester: John Wiley & Sons, Ltd. · Fitts, C.R. (2013) *Groundwater science*, 2nd edition. Waltham: Academic Press. · Jones, J.A.A. (1997) *Global hydrology: processes, resources and environmental management*. Harlow: Longman.

growan Decomposed granite or related rock as found on Dartmoor, southwest England, and its environs. The growan may be produced by chemical weathering or metamorphic processes, and it has been suggested that the stripping of growan from corestones of sounder rock may play a role in the formation of tors.

groyne A man-made barrier running across a beach and into the sea that has been constructed to reduce the erosion of the beach by longshore currents. Although groynes may serve to reduce erosion at the place where they are constructed, by reducing the movement of material along the coastline by longshore drift they may cause beach starvation and erosion elsewhere.

grumusol Under the classification of the Seventh Approximation this may be regarded as a vertisol and is in effect a modern American term for a black cotton soil.

grus An accumulation of poorly sorted angular quartz grains and clayey material derived locally from weathered granite.

guano Thick accumulations of bird excrement, usually found on islands where the birds nest. The material is used as a fertilizer as it is rich in phosphates. In some caves (and belfries) there may be a substantial accumulation of bat guano.

gull A fissure or crack, sometimes sediment filled, that opens up on escarpments as a result of the tensions produced by CAMBERING.

gully A steep-sided trench or channel, often deep (several metres), that is cut into poorly consolidated bedrock, weathered sediment or soil. The agent of gullying is ephemeral flow of running water. A gully is deeper and longer than a RILL. (See also ARROYO, CHANNEL CLASSIFICATION, FLUVIAL and WADI.)

gully erosion The pronounced erosion, by ephemeral streams, of soils and other poorly consolidated sediments, producing networks of steep-sided channels.

Gumbel extreme value theory A theory appropriate for the analysis of extreme values; it has been applied particularly to flood frequency analysis where the Gumbel extreme value 1 distribution may be used as the theoretical distribution to fit to the distribution of flood frequency values for a particular gauging station. KJG

Reading
Gumbel, E.V. (1958) *Statistics of extremes*. New York: Columbia University Press.

gumbo An area of clayey soil that turns to sticky mud when wet. Any damp, sticky clay.

gustiness factor An index of the variations in wind speed. It is calculated from the ratio of the total range of wind speed between gusts and lulls to the mean wind speed in the given period. Gustiness factors are highest in urban areas where the surface roughness is high, and lowest in coastal sites or exposed upland locations where surface friction is small. A gustiness factor may also be defined in terms of wind direction, whereby the angular width in radians is the measure of lateral gustiness. For small values it is nearly equivalent to the speed ratio. PS

guyot A flat-topped mountain on the sea floor, especially in the Pacific, which does not reach the sea surface. A sea mount or drowned island, which is a truncated volcano, formed as a result of subsidence associated with sea-floor spreading. ASG

Reading
Watts, A.B. (1984) The origin and evolution of seamounts. *Journal of Geophysical Research*, **89**, 1106–1286.

gypcrete Gypsum crusts, as found in deserts, and comprising loose, powdery or cemented crystalline accumulations dominated by calcium sulphate dihydrate at or near the ground surface. ASG

Reading
Watson, A. (1983) Gypsum crusts. In A. S. Goudie and K. Pye (eds), *Chemical sediments and geomorphology*. London: Academic Press; pp. 133–161.

gypsum The evaporite mineral calcium sulphate dihydrate ($CaSO_4 \cdot 2H_2O$) formed as an evaporite deposit in enclosed basins and also widespread as a cementing agent in surface or near-surface crusts (such as GYPCRETE) in some arid and hyperarid environments. DJN

gyre A large-scale ocean circulation system generated by atmospheric circulation. Gyres play a fundamental role in general ocean circulation, and in maintaining the global heat balance. Each gyre comprises four major currents: a western and eastern boundary current, and two transverse currents. Prevailing winds drive surface waters east or west across the ocean basins, and the CORIOLIS FORCE deflects the resulting currents to an angle of about 45° relative to the wind direction. The continents bounding the basins also contribute to the deflection.

The largest systems are the five subtropical gyres that are each centred at about 30° north or south latitude. There are north and south pairs in the Atlantic and Pacific Oceans, and a southern gyre only in the Indian Ocean. The flow in these gyres mimics that of the subtropical anticyclones that sit over them. The western boundary currents include five of the six largest currents on Earth, including the Gulf Stream and the Kuroshio. The western boundary currents are substantially narrower and faster than the eastern boundary currents. There are also smaller, subpolar gyres in the North Atlantic and North Pacific, centred around 50°. These exhibit counterclockwise circulation similar to that of the semi-permanent, mid-latitude cyclones. Subpolar gyres are absent in the southern hemisphere because there are no continents to block and redirect the circumpolar circulation. DJS

Reading

Pedlosky, J. (1990) The dynamics of the oceanic subtropical gyres. *Science*, **248**, 316–322. · Stewart, R.W. (1969) The atmosphere and the ocean. *Scientific American*, **221**, 76–105. · Williams, R.G. and Fellows, M.J. (2003) Physical transport of nutrients and the maintenance of biological production. In M. J. R. Fasham (ed.), *Ocean biogeochemistry*. Berlin: Springer; pp. 19–50.

gyttja A nutrient-rich peat or organic mud that contains much plankton.

H

habitat The overall environment, but more often specifically the physical environment, in which organisms live. It may be examined in a range of scales that extend from the macroscale (continental, subcontinental) to the mesoscale (regional, local) and microscale, the latter being of particular significance to the numerous small animal species of submicroscopic and microscopic size. All organisms must be morphologically and genetically adapted to the habitats in which they reside for any length of time. Such adaptation usually begins with the acquisition of tolerance to the conditions therein.

For plants on land, the major influential factors of habitat that can limit growth may be grouped under four headings: climatic, topographic, edaphic and biotic. Of these, *climatic* factors, especially those related to cold tolerance and the provision of adequate amounts of heat and moisture in the growing season, are normally regarded as being most important, certainly on a continental, subcontinental or regional scale, but they are frequently of much less significance at a local and a microscale level. Climatic influences are those of light (energy), heat, moisture availability and wind.

Those temperature conditions that influence plant growth and physiology are in most cases a response to latitude, altitude and distance from the sea. In general, most of the higher plants stop growth activity once external temperatures fall to 5 °C, so as to enable cold-hardiness to form. There are very few trees in places where mean annual external temperatures are lower than 10 °C; and little activity is exhibited by plants generally when the immediate temperature around them exceeds 45 °C.

Habitat moisture circumstances (EFFECTIVE PRECIPITATION plus the effect of the soil moisture reserve) also vary widely. In all but the wettest terrestrial biological systems (e.g. bog and marsh land), efficient means of gaining moisture are required. Within the overall evapotranspirational limits, which are set by energy availability (Penman, 1963), the flow of water into plants generally rises with an increase in TRANSPIRATION, which in turn is usually augmented by the maximum opening of stomata under strong light conditions, high temperatures and above-minimum wind speeds. Most plants can withstand some moisture stress, though if this becomes too great, or continues for too long, they may wilt and die. In arid areas, in which such stresses are severe and prolonged, xerophytes (i.e. drought-tolerant plants) have an inbuilt range of defences against them.

On a more local scale, climatic influences of habitat may be overshadowed by factors of topography and edaphic and biotic controls. Altitude effects and slope angle affect habitat. Among *edaphic controls* are those of soil, soil chemistry and soil water: any material deficiencies or excesses of soil nutrients or soil water compared with the mean are likely to restrain plant growth. *Biotic controls* include especially the influence of grazing (see HERBIVORES), of fire and of humans.

These external habitat factors may influence land-based animals too, both directly (temperature, water availability, etc.) and also indirectly, in that they determine to a large extent the nature of local and regional food chains (see FOOD CHAIN, FOOD WEB). In consequence of this, the NICHE role of animals is a further important habitat constraint, as also is their precise relationship with other animals (see COMMENSALISM, COMPETITION and PARASITE). For water-based organisms the main habitat conditions to which they respond are differences in water chemistry, temperature, light penetration and the general state of the food-chain web within the water body.   DW

Reading and Reference
Cox, C.B. and Moore, P.D. (2010) *Biogeography: an ecological and evolutionary approach*, 8th edition. New York: John Wiley & Sons, Inc. · Jorgensen, S.E. (ed.) (2009) *Ecosystem ecology*. Amsterdam: Elsevier. · Penman, H.L. (1963) *Vegetation and hydrology*. Farnham Royal: Commonwealth Agricultural Bureau. · Prentice, T.C. (1992) A global biome model based on plant physiology and dominance, soil properties and climate. *Journal of Biogeography*, **19**, 117–134. · Whitfield, P., Moore, P.D.

and Cox B. (2002) *Biomes and habitats*. New York: MacMillan.

haboob From the Arabic *habb*, to blow. A haboob is a DUST STORM generated by the evaporative outflow of a cumulonimbus cloud, which may exceed 80 km in extreme cases. Extremely turbulent conditions along the boundary of the cool, dense outflow contribute to dust raising from dry surfaces. ASG

Reference
Miller, S.D., Kuciauskas, A.P., Liu, M., *et al.* (2008) Haboob dust storms of the southern Arabian Peninsula. *Journal of Geophysical Research: Atmospheres*, **113**, art. no. D01202.

hadal zone Pertaining to the greatest depths (more than 6000 m) of the oceans. ASG

Hadley cell The large-scale thermally driven circulations existing in tropical latitudes and most prominent over the Atlantic and Pacific Oceans. There is one Hadley cell in each hemisphere, heated air rising near the equator in the INTERTROPICAL CONVERGENCE ZONE (ITCZ), flowing poleward aloft, descending at a latitude of 30–40°, especially in the eastern half of the very intense subtropical high-pressure areas at these latitudes, and then flowing either poleward or equatorward near the Earth's surface. Because of the Earth's rotation, the equatorward-moving currents are deflected towards the west and become the northeast and southeast trade winds of the northern and southern hemispheres respectively. The poleward-moving currents are deflected towards the east and become the middle-latitude westerlies. The upper poleward-moving branch of the Hadley cell rapidly gains westerly momentum, which is concentrated in the subtropical JET STREAM at a height of 12–15 km above the tropical highs. WDS

haff A coastal lagoon separated from the open seas by a sand pit formed by longshore drifting of sediments.

hagg A channel that separates hummocks in a peat bog.

hail Solid precipitation that falls in the form of ice particles from cumulonimbus clouds. The high concentration of liquid water in these clouds provides an environment that is favourable for the quick growth of ice particles by both coalescence and collision with supercooled water droplets.

Hail is commonly spherical, and the larger stones are composed of concentric shells of clear and opaque ice with a diameter as large as 10 cm in severe convective storms. Hailstones grow larger by being transported in the strong up- and downdraughts that characterize cumulonimbus clouds. RR

haldenhang A degrading rock slope that underlies an accumulation of talus or scree.

half-life A parameter used to describe aspects of the radioactive decay process. It is the time required for one half of a given number of a radionuclide to decay to a protégé product. Alternatively, this may be expressed in the quantum-mechanical view as the interval of time within which an unstable atomic nucleus has a probability of decay of 0.5. When large numbers of atoms are involved, the two definitions are in effect equivalent, but it is important to realize that a half-life is only exhibited because some atoms remain stable for much longer than others of the same radioactive element.

Each radioisotope has a fixed half-life, which is typically determined by mass spectrometry measurements. If it is possible to assume that at the time of formation of a deposit that no protégé isotopes were present, then measurement of the relative abundance of parent radioisotopes and protégé products, coupled with the knowledge of the half-life provide the basis for radiometric dating methods. The half-lives of some widely used radioisotopes are listed in Table 1 (Geyh and Schleicher, 1990).

Radiocarbon (^{14}C) and radioactive potassium (^{40}K) are both key RADIOISOTOPES for dating purposes. See also RADIOCARBON DATING,

Table 1 Half-lives of some widely used radioisotopes

Element	Symbol and radiation type[a]	Half-life (a)
Hydrogen (tritium)	$^{3}H(\beta^{-})$	12.43
Lead	$^{210}Pb(\beta^{-})$	22.3
Caesium	$^{137}Cs(\beta^{-})$	30.17
Carbon	$^{14}C(\beta^{-})$	5568
Chlorine	$^{36}Cl(\beta^{-})$	301×10^{3}
Uranium	$^{238}U(\alpha)$	4.47×10^{9}
Potassium	^{40}K (electron capture)	12.50×10^{9}
Thorium	$^{232}Th(\alpha)$	14.05×10^{9}

Source: Geyh and Schleicher (1990).
α: alpha particle emission; β⁻: electron emission.

GEOCHRONOLOGY, POTASSIUM–ARGON (K/Ar) DAT-
ING and URANIUM SERIES DATING.   SS/DLD

Reading and Reference
Faure, H. (1986) *Principles of isotope geology*, 2nd edition.
Chichester: John Wiley & Sons, Ltd. · Geyh, M.A. and
Schleicher, H. (1990) *Absolute age determination*, trans. R.
C. Newcomb. Berlin: Springer-Verlag.

halocarbons   Carbon compounds containing
fluorine, chlorine, bromine or iodine. The group
includes halogenated fluorocarbons based on
ethane and methane, such as CHLOROFLUOROCAR-
BONS (CFCs) and hydrochlorofluorocarbons
(HCFCs), and halons containing bromine and
perfluorocarbons (PFCs). Along with associated
compounds such as sulphur hexafluoride, these
halocarbons are effective greenhouse gases whose
presence is solely due to anthropogenic activity.
They have both a direct radiative effect and an
impact through depletion of ozone.
   Although the emission of CFCs and HCFCs has
been controlled by the Montreal Protocol and atmospheric concentrations are expected to decline over
the next century, PFCs and some of the halon group
have residence times exceeding 1000 years and may
have an influence on future climate.   PSH

haloclasty   The disintegration of rock as a
result of the action of salts, which may result
from salt crystallization, salt hydration or the thermal expansion of salts. It is especially important in
arid areas, but may also play a role in building
decay in cities. (See also SALT WEATHERING.)   ASG

halons   Members of the halogenated fluoro-
carbon (HF) group of ethane- or methane-based
compounds in which H^+ ions are partially or
completely replaced by chlorine, fluorine and/or
bromine. This group also includes chlorofluorocarbons. Halons are HFs that contain
bromine; for example, halon 1211 (CF_2BrCl)
and halon 1301 (CF_3Br). They are long-lived
and have been implicated in stratospheric
ozone depletion, where their damage potential
has been estimated at 3–10 times that of equivalent chlorofluorocarbon molecules.   BJS

haloturbation   The disturbance of soils and
surficial sediments by the growth of salt crystals
by a process akin to cryoturbation. Common in
salt lakes and playas, haloturbation leads to the
sorting of coarser particles to form patterned
ground.   PSH

hamada, hammada   A desert region that
does not have any surficial materials other than
boulders and exposed bedrock.

hamra   A red, sandy soil that also contains clay.

hanging valley   A tributary valley whose floor
is discordant with the floor of the main valley.
Hanging valleys are a hallmark of glacial erosion
in mountains. The discordance of the valley floors
was one of the arguments used in the early nineteenth century to suggest that rivers did not cut
valleys, and the objection was only removed when
the role of glacier activity was appreciated. The
valley cross-sections were adjusted to the glaciers
they held, and it is likely that the glacier surfaces
met concordantly (Penck, 1905).   DES

Reference
Penck, A. (1905) Glacial features in the surface of the Alps.
Journal of Geology, **13**, 1–17.

hardness   The resistance of a material to
scratching. Moh's scale, which is not quantitative,
depends upon the ability of one mineral in a series
(usually 1 to 10, but also 1 to 8 or 1 to 12) to scratch
others below it but not those above, e.g. quartz (7)
will scratch feldspar (6) but not topaz (8). More
quantitative measures (e.g. those of Brinell, Rock-
well and Vickers) depend on the load applied to a
standard indenter applied for a given time to
produce a given pattern on the material under
test. The SCHMIDT HAMMER test used in concrete
research gives an approximate value of hardness
via the 'rebound value' (in reality a measure of the
coefficient of restitution of the material). This can
be used to determine the compressive strength of
the material.   WBW

hardpan   A compacted or cemented subsurface horizon within the soil zone, sometimes
termed a duripan.

harmattan   A dry, northeasterly wind of West
Africa that blows in the winter months. Blowing
out of the Sahara, it frequently carries much sand
and dust.

harmonic analysis   See FOURIER ANALYSIS.

Hawaiian eruption   A type of volcanic eruption characterized by the emission of large quantities of highly fluid, basic lava issuing from vents
and fissures. Explosive eruptions of material are
rare.

hazard   See NATURAL HAZARD.

haze   A term used to describe a reduction in
visibility due to the suspension of fine aerosols in
the atmosphere. Caused by DUST particles of 1 μm
(10^{-6} m) or less entrained into the atmosphere by

wind or human activity (commonly off-road vehicle driving), a haze results in a reduction in visibility to less than 2 km (but more than 1 km, otherwise it is termed a DUST STORM). A haze may last many hours or days owing to the long-term suspension of fine particles by the turbulent nature of the wind (see TURBULENCE). Haze may also result from smoke from fires. GFSW

head See SOLIFLUCTION.

headcut The upslope limit of a gully system, characterized by a steep wall that is cut back, migrating upslope, as further erosion occurs.

headwall A steep, arcuate slope around the head of a CIRQUE or LANDSLIDE. Gradients exceed the ANGLE OF REPOSE (ANGLE OF INITIAL YIELD) of granular materials (around 34°). Sometimes a headwall is divided into a backwall and two side-walls. IE

headward erosion The processes involved in the upslope migration of the head of a gully or source of a stream.

headwater A stream that forms the source and upper limit of a river, especially a large one.

heat budget The magnitude of terms related to the heating and cooling of the surface or atmosphere. Although most commonly referring to surface heating, a heat budget can be calculated for any level of the atmosphere.

The transfer of heat to or from a substance is effected by one or more of the processes of conduction, CONVECTION or RADIATION. The common effect of such a transfer is to alter either the temperature or the state of the substance or both. A heated body may acquire a higher temperature (sensible heat) or change to a higher state (liquid to gas, or solid to liquid) and therefore acquire latent or hidden heat. Conduction is the process of heat transfer through matter by molecular impact from regions of high temperature to regions of low temperature without the transfer of the matter itself. It is the process by which heat passes through solids, but its effects in fluids (liquids and gases) are usually negligible in comparison with those of convection. In contrast, convection is a mode of heat transfer in a fluid, involving the movement and mixing of substantial volumes of the substance concerned. Conduction is the main method of heat transfer in solid rocks and the soil, while the convection process frequently operates in the atmosphere and oceans. Radiation is heat transfer via electromagnetic radiation and requires no medium for transfer to occur. The Earth and atmosphere receive radiation from the Sun via what is termed longwave radiation (primarily ultraviolet and visible) and emit radiation in the longer wavelengths (primarily infrared). Although the atmosphere absorbs some solar radiation, its main source of heating is via long-wave radiation and latent heat exchange from the surface.

The net heat input to the surface, the radiant energy available to drive the climate system, is termed the net radiation R_N. This is the difference between the solar radiation absorbed and the net longwave radiation. The latter is the difference between the longwave radiation emitted by the surface emission of and its heat gain via longwave radiation from the atmosphere. R_N can be expressed as

$$R_N = R_{sw}\downarrow - R_{sw}\uparrow + R_{lw}\downarrow - R_{lw}\uparrow$$
$$= R_{sw}\uparrow(1 - a_s) + R_{lw}\downarrow - R_{lw}\uparrow$$

where the subscript sw refers to shortwave or solar radiation reaching the surface and the subscript lw refers to the longwave radiation emitted by the Earth, atmosphere and clouds. The upward arrows represent radiative loss and the downward arrows represent radiative gain. $R_{lw}\uparrow$ is radiation emitted spaceward from the Earth's surface and it is equal to $\varepsilon\sigma T^4$, where ε is surface emissivity, σ is the Boltzmann constant and T is temperature. $R_{lw}\downarrow$ is that emitted Earthward by clouds and the atmosphere. $R_{sw}\downarrow$ is the solar radiation received at the Earth's surface; $R_{sw}\uparrow$ is that reflected back towards space and is equal to the product of the solar radiation received and the surface albedo a_s.

Over time a heat balance must exist at the Earth's surface. For that reason, the net radiation must be balanced by other forms of heat transfer, namely sensible and latent heat exchanges described in the first paragraph. Thus:

$$R_N = R_{sw}\downarrow(1 - a_s) + R_{lw}\downarrow - \varepsilon\sigma T^4 = H + LE_a$$

where H is the sensible heat flux via conduction into/out of the soil and via convection to the atmosphere or surface water bodies. The evaporative heat flux is LE_a, where L is the latent heat of vaporization and E_a is the actual evaporation.

The values of the surface heat fluxes are dependent upon surface characteristics. For a soil, surface soil moisture is the dominant control on albedo, emissivity, evaporation and temperature. Vegetative cover similarly influences the albedo and, by affecting wind, the sensible and latent heat transfer to the atmosphere.

The surface provides most of the heat to the tropospheric atmosphere. Sensible and latent heat transfer have different effects on the atmosphere. Sensible (i.e. convective) heat transfer S generally impacts the lower boundary layer, which is also termed the mixed layer. Latent heat transfer generally occurs much higher in the atmosphere, as it is released to the air when moisture condenses in the form of clouds. Thus, latent heat has a larger scale effect. For this reason, it is important to know the relative importance of these two forms of heat transfer. This is expressed via the Bowen ratio β, which can be written as

$$\beta = \frac{\text{sensible heat loss to atmosphere } S}{\text{latent heat loss to atmosphere } LE_t}$$

In the absence of atmosphere advection, β can vary between $+\infty$ for a dry surface with no evaporation to zero for an evaporating wet surface with no sensible heat loss. Under special circumstances, such as dew formation or advection, it can become negative. JGL/SJN

Reading
Nicholson, S.E. (2011) *Dryland climatology*. Cambridge: Cambridge University Press; especially chapter 6. · Wallace, J.M. and Hobbs, P.V. (2006) *Atmospheric science: an introductory survey*. New York: Academic Press.

heat island   See URBAN METEOROLOGY.

heathlands   These are mineral-based and usually distinguished by acidophilous vegetation. In Britain, heathlands are commonly associated with the sandstones of southern and eastern England. Heathlands exhibit a wide range of moisture conditions; the vegetation is usually dominated by heather (*Calluna*) communities.

Heathland in Britain mainly formed as a secondary habitat resulting from tree clearance in areas such as Breckland in East Anglia. This habitat is now relatively rare. For example, Breckland covered an area of 22,000 ha in 1880 and was reduced to just 7500 ha in 100 years as a result of agricultural improvement. ALH

Reading
Evans, D. (1997) *A history of nature conservation in Britain*, 2nd edition. London: Routledge. · Rodwell, J.S. (ed.) (1991) *British plant communities, vol. 2: mires and heaths*. Cambridge: Cambridge University Press.

Heinrich events   Layers of rock, debris-rich but FORAMINIFERA-poor sediment, identified in the deep sea core record from the North Atlantic were reported by Heinrich (1988). These have been attributed to deposition from the bottom of iceberg armadas resulting from the catastrophic break-up of the margins of the polar ice cap and adjacent ice sheets. These resulted from relatively rapid ice decay due to global warming events, and have been interpreted as important evidence of the link between oceans and the atmosphere. Heinrich events occurred irregularly during the last glacial cycle, with the five most recent dated to 14.5–13.5 ka, 22–19 ka, 27 ka, 35.5 ka and 52 ka. Heinrich events are probably linked to DANSGAARD–OESCHGER (D–O) EVENTS. See also BOND CYCLES. Their impacts on climate may have been global but this is debatable. DSGT

Reference
Heinrich, H. (1988) Origin and consequences of cyclic ice-rafting in the northeast Atlantic Ocean during the past 130,000 years. *Quaternary Research*, **29**, 142–152.

helical flow   Spiral (roll) airflow vortices in the lower atmosphere that develop in unstable thermal conditions. Parallel helical vortices were once thought to lead to the formation of linear sand dunes. DSGT

helictite   A small calcium carbonate SPELEOTHEM that grows in curved or spiralling forms in limestone caves. It grows by the slow accretion of calcium carbonate crystals from the top of the structure, fed by an internal canal. When the crystal-growth forces exceed that of the normal hydraulic force of the flowing water, the crystals can grow away from the vertical. Helping this erratic crystal growth, there may often be clay particle blockages in the feeder canal or even draughts within the cave passage. Helictites are quite common in limestone caves of all environments and can often produce intricate tangled structures. Famous ones can be found in the Kango Caves of South Africa. PAB

hemera   A zone or any other period of geological time as determined from an assemblage of fossils.

herbivore   A plant-eating organism; an animal that feeds directly on photosynthetic species. Other than the plant itself, it is the organism that lies at the base of the food chain. Most herbivores are very small; for example, the aphid species that live off cultivated plants. The largest are the grazing and browsing mammals of Africa (such as the giraffe or elephant). The herbivore group also includes humans' domesticated animals. On land, the smaller the size of the herbivore the more likely it is to be linked with one plant species alone.

Overall, in terrestrial environments, the ecological effect of herbivores under normal conditions is greatest within grasslands, where up to 35% of production may be consumed; within forests, the equivalent figure is customarily 4–7% (Whittaker, 1975). Overstocking of larger herbivores may result in the unbalanced and excessive removal of vegetative resources in all land habitats. In aquatic systems, herbivores are relatively unimportant, with most lighter plants being eaten by decomposers, moulds and bacteria.   DW

Reference
Whittaker, R.H. (1975) *Communities and ecosystems*, 2nd edition. New York: Macmillan.

herbivore use intensity (HUI)   This measures the impact of grazing and browsing animals on the environment, reflecting the density of animals per unit area. This can show marked spatial variations in some managed environments, particularly when activities are focused on single water sources. (See also PIOSPHERE.)   DSGT

Reading
Danell, K., Bergstrom, R., Duncn, P. and Pastor, J. (eds) (2006) *Large herbivore ecology: ecosystem dynamics and conservation*. Cambridge: Cambridge University Press.

heterosphere   The outer portion of the Earth's atmosphere beyond the hemosphere.

heterotrophs   Organisms that depend upon organic foods in order to obtain energy. Most bacteria, fungi and animals fall into this category.

hiatus   A gap in the stratigraphic record or in geological time that is not represented by any sediments.

hierarchical rhythmic topography   Nesting of repetitive alongshore morphological features on beaches, such as beach cusps being nested within outer rhythmic bars (where variants include crescentic bars, shore-welded bars and transverse bars) associated with the presence of rip currents and edge waves. This topography is likely to be particularly associated with coarse-grained reflective beaches. Such features may be migratory, with small features moving more slowly than the encasing large forms.   TS

Reading
Woodroffe, C.D. (2002) *Coasts: form, process and evolution*. Cambridge: Cambridge University Press.

high-energy window   The suggestion, first introduced by Neumann (1972), that in the mid-Holocene on tropical coasts there was a period when wave energy was higher than at present. This occurred during the phase when the present sea level was being first approached by the Flandrian (Holocene) transgression and prior to the protective development of coral reefs. The 'window' may have operated on a more local scale on individual reefs, with waves breaking not on margins of an extensive reef flat as at the present time but more extensively over a shallowly submerged reef top prior to the development of the reef flat (Hopley, 1984).   ASG

References
Hopley, D. (1984) The Holocene 'high energy window' on the central Great Barrier Reef. In B. G. Thom (ed.), *Coastal geomorphology in Australia*. Sydney: Academic Press; pp. 135–150. · Neumann, A.C. (1972) Quaternary sea level history of Bermuda and the Bahamas. *American Quaternary Association Second National Conference Abstracts*; pp. 41–44.

hillslope flow processes   The mechanisms and routes of flow followed by precipitation down hillsides to a stream channel.

Some precipitation is intercepted before reaching the ground, and if not evaporated will reach it directly or by stemflow, with some delay and with some spatial redistribution, particularly concentrating round tree trunks, for example. If rainfall intensity exceeds the INFILTRATION capacity, or if the soil is saturated, surface depressions will fill and overflow to give OVERLAND FLOW, which normally reaches the stream channels but may infiltrate into the soil downslope, especially if a rainstorm is very brief or very local, or if the soil thickens appreciably. Infiltration occurs most readily into soil crumbs or peds, but also overflows down soil cracks and other structural voids, thereby bypassing the surface soil layers and usually infiltrating into peds farther down, but in some cases connecting directly to the groundwater table.

Within the soil the flow is mainly vertical in the unsaturated zone, flowing under the predominant influence of hydraulic potential gradients. Soil porosity normally decreases with depth, though there are many exceptions from this general rule. If downward flow is impeded then a saturated zone may develop, either in contact with the groundwater body as a whole or perched above it. In such a layer, hydraulic potential gradients are negligible and saturated subsurface flow or THROUGHFLOW predominates. As throughflow accumulates downslope, the level of saturation in the soil rises towards the surface, or the SATURATION DEFICIT falls, and flow may re-emerge on the surface as return flow. The

261

area of zero saturation deficit defines the SOURCE AREA for a rapid streamflow response to subsequent rain by overland flow production as described above. MJK

Reading
Kirkby, M.J. (ed.) (1976) *Hillslope hydrology.* Chichester: John Wiley & Sons, Ltd. · Lu, N. and Godt, J.W. (2013) *Hillslope hydrology and stability.* Cambridge: Cambridge University Press.

hindcasting A synonym for retrospective forecasting. This refers to conducting forecasts of past weather or climate in a set-up that mimics operational real-time forecasting. This is done for at least three well-known reasons: (1) To test the viability of the new forecast system for operational purposes in terms of producing the forecast on time. (2) To test the forecast skill of the system (which can be either statistical or numerical model) with available observations to determine that the forecast skill is better than climatology, persistence or the skill of any other existing forecast tools that it is trying to replace. (3) This exercise of hindcasting is also conducted to estimate the errors of the forecasts so that an appropriate correction may be applied to the forecasts posteriorly. The hindcasts are now routinely conducted when a new weather or climate forecast system is introduced (Saha *et al.*, 2006; Kirtman *et al.*, 2014). VM

References
Kirtman, B., Min, D., Infanti, J.M. *et al.* (2014) The North American Multimodel Ensemble: phase-1 seasonal to interannual prediction; phase-2 toward developing intraseasonal prediction. *Bulletin of the American Meteorology Society,* **95**, 585–601. · Saha, S., Nadiga, S., Thiaw, C., *et al.* (2006) The NCEP climate forecast system. *Journal of Climate,* **19**, 3483–3517.

Hjulström curve In 1935 the Swedish geomorphologist Filip Hjulström presented an empirical curve defining the threshold flow velocities required to initiate motion of grains of different sizes on a stream bed. By displaying this curve in conjunction with curves of the settling velocity or depositional velocity, the conditions required for transport, traction and deposition of sediment can be defined (see diagram). The threshold flow velocity is at a minimum for well-graded sand particles in the 0.2–0.5 mm size range, and higher velocities are necessary to entrain both finer and coarser sediments. Finer sediment includes cohesive silt and clay, whose entrainment is inhibited by particle aggregation and by submergence in the laminar sublayer (see BOUNDARY LAYER).

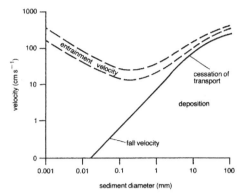

Hjulström curve.

Coarser sediments (gravel, pebbles and cobbles) require higher threshold velocities simply because they are heavier. However, there is no unique mean velocity at which grains of given size first move, because the bed velocity is a more critical control of entrainment by flowing water. For a particular threshold bed velocity the mean velocity is greater in deeper water. Also, the particle size distribution affects the threshold velocity, since coarse grains in a poorly sorted sediment may shelter finer grains and prevent their entrainment. A similar shape of curve defines the fluid threshold in air (Bagnold, 1941), but the different fluid densities result in different entrainment velocities in the two media. KSR

References
Bagnold, R.A. (1941) *The physics of blown sand and desert dunes.* London: Chapman & Hall. · Hjulström, F. (1935) Studies of the morphological activities of rivers as illustrated by the River Fyris. In *Bulletin of the Geological Institute, University of Uppsala,* vol. XXV. Uppsala: Almqvist and Wiksell; pp. 221–527.

hoar frost The deposition of ice crystals usually by the direct sublimation of water molecules, but also by the freezing of dew, on a solid surface exposed to the atmosphere. The surface, if cooled at night by loss of radiation, in turn cools the water vapour immediately above it to its frost point and sublimation begins. The crystals are usually white. JET

hodograph A scheme for representing wind speeds and directions at different heights at a given time by plotting them on a polar coordinate diagram. Successive winds are plotted as vectors so that they all 'blow' from the respective directions towards the central point, which represents the station at which the observations were made. The

A hodograph. The blank diagram comprises compass directions and a series of concentric circles, the radii of which represent wind speed. The centre of all the circles represents the observing station. Three winds are shown (dashed lines) for the levels 400, 500 and 700 mb. The wind at 700 mb, for example, blows from a direction of about 70° with a speed of about 8 m s⁻¹. The solid arrows show the size and direction of the vector difference of the winds at the top and bottom of any layer. This vector difference is known as the thermal wind.

hodograph is then constructed by drawing a series of vectors that join the successive end points of the plotted winds.

For example, the three winds plotted in the diagram for 700, 500 and 400 mb (broken lines) are shown with the hodograph construction (solid lines) that represents the shear vectors for the 700–500 mb and 500–400 mb layers. Assuming the observed winds to be geostrophic, the shear vectors can be considered to be THERMAL WINDS and, therefore, to provide information about the nature of temperature ADVECTION.

Between 700 and 500 mb the wind backs with height and within this layer the air is blowing from a relatively cold region (cold air lies to the left of thermal wind). So winds that back with height signify cold advection and those that veer (500–400 mb here) mean warm advection. The thermal winds can be measured directly from the scale on the polar diagram.   RR

Reading
Atkinson, B.W. (1968) *The weather business*. London: Aldus Books.

hogsback   A long ridge of rock dipping steeply on both sides that is the exposure of a stratum of rock that has been tilted until the originally horizontal beds are almost vertical.

Holocene   The second series/epoch of the Quaternary period, following the Pleistocene.

Often called the postglacial, it has extended from *c*. 11,700 calendar years BP until the present and has been characterized by interglacial conditions. It has been marked by various climatic fluctuations (e.g. see ALTITHERMAL, BLYTT–SERNANDER MODEL, HYPSITHERMAL, LITTLE CLIMATIC OPTIMUM, LITTLE ICE AGE, MEDIEVAL WARM PERIOD AND NEOGLACIAL) and by rapid sea-level rise (e.g. Flandrian transgression – see TRANSGRESSION). The term was introduced during the 1860s and means 'entirely modern'. In recent years there has been a growth in the use of the term ANTHROPOCENE to describe the part of the Holocene (and sometimes earlier, e.g. the last *c*. 12,000 years) in which humans became a dominant environmentally shaping force; this has yet to be adopted as a formal subdivision of the Quaternary.   ASG/DSGT

Reading
Walker, M., Johnsen, S., Olander Rasmussen, S., *et al.* (2009) Formal definition and dating of the GSSP (Global Stratotype Section and Point) for the base of the Holocene using the Greenland NGRIP ice core, and selected auxiliary records. *Journal of Quaternary Science*, 24, 3–17.

holokarst   A term coined to describe the Dinaric karst belt of the countries to the east of the Adriatic Sea, where the full suite of karst landforms is found and limestone solution processes dominate the landscape. Cvijic contrasted this landscape with the less well developed karst landscape further inland, which he termed merokarst. The term has been generally used to describe any limestone landscape with a fully developed range of karst features.   HAV

Reading
Cvijic, J. (1893) Das Karstphänomen: Versuch einer morphologischen Monographie. In A. Penck (ed.), *Geographische Abhandlung*, vol. V. Vienna: E. Hölze; pp. 217–329. · Cvijic, J. (1925) Types morphologiques des terrains calcaires. Le holokarst. *Comptes Rendus de l'Académie des Sciences*, 180, 592–594.

homoclines   Locations that experience the same types of climatic regimes.

hoodoo   An unusually shaped pillar or outcrop of rock, a few to tens of metres high, produced by erosion. A North American term, hoodoos are usually found in dryland regions and are recognized as having a variable shape (like a totem pole) that might be due to the inclusion of material from different lithologies or due to differential erosion.   DSGT

Hoodoos on the skyline, Bryce Canyon National Park, USA. Photograph by David Thomas.

horizon, soil

A recognizable layer within a soil, running essentially parallel to the soil surface, that has been differentiated from the materials above and below it by the operation of various weathering and soil-forming processes, such as ELUVIATION. Horizons may be differentiated on the basis of properties like colour, texture, structure, stoniness or accumulations or deficits of iron, manganese, carbonates or other materials. The principal horizon designations O, A, B, C and R are used to indicate characteristic kinds of modifications to the original parent material of the soil, and indicate something of the genetic linkages between the horizons found in the soil 'profile' (the characteristic series of layers from the surface downwards). The O horizons are accumulations of organic matter lying on top of the mineral soil; the A horizons form the uppermost layer of the mineral soil and display accumulations of humus and often a relative deficit of clay, iron and other materials, resulting in a sandier texture than the B horizon below, where the mobile materials have accumulated. The C horizons represent parent materials altered in limited ways, generally unaffected by biota or organic materials, and

form a transition to the R horizon of underlying unweathered rock. These major horizon types are further subdivided both on the basis of depth and distinctive properties; for example, into A1, where the major feature is accumulated organic matter, and A2, where loss of mobile materials has been most influential on the soil character. Suffixes are also used; for example, Bca to indicate a B horizon with clear carbonate accumulation. Horizonation may reflect in part the operation of processes of hillslope transport, so that the notion of layers genetically linked only through the *vertical* flux of material cannot be applied in all situations. In the US *Soil Taxonomy*, an important diagnostic role is given to various additional kinds of horizons that are recognized in tandem with the O, A, B, C, R system. These include the *epipedon*, a horizon formed at the surface. If darkened by organic matter, for example, and the colouring affects the top of the B horizon as well as the A, then the epipedon includes part of both the A and the B horizons. Similar subsoil horizons include *argillic horizons* (where illuviated clays have accumulated), *natric* and *calcic* horizons of sodium and carbonate accumulation, *oxic* horizons and *duripans*, hardpans that are cemented by silica.   DLD

Reading
Brady, N.C. and Weil, R.R. (1996) *The nature and properties of soils*, 11th edition. Englewood Cliffs, NJ: Prentice-Hall. · Schoeneberger, P.J., Wysocki, D.A., Benham, E.C., *et al.* (2012) *Field book for describing and sampling soils*, Version 3. Lincoln, NE: Natural Resources Conservation Service, National Soil Survey Center. · United States Department of Agriculture, Soil Conservation Service (1988) *Soil taxonomy*. Malabar, FL: Krieger Publishing Co.

horn, glacial

A pyramidal peak with three or more distinct faces steepened by glacial undercutting. The classic situation occurs when cirque glaciers encroach on a mountain from all sides.

horse latitudes

The latitude belts over the oceans at approximately 30–35° north and south where winds are predominantly calm or very light and the weather often hot and dry. These latitudes represent the normal axes of the subtropical high-pressure belts. The origin of the name is uncertain, but it is suggested that the crews of sailing ships carrying horses to the West Indies had occasionally when becalmed either to jettison their live cargo as fodder ran out or eat it. The term 'subtropical high-pressure areas' is now frequently used in preference to 'horse latitude'.   AHP

horst

An upstanding block of the Earth's crust that is bounded by faults and has been uplifted by tectonic processes. The down-faulted areas that bound horsts are called grabens.

Horton overland flow

Rainfall in excess of the infiltration capacity of the soil runs off as Horton overland flow, named after the hydrologist Robert E. Horton who first studied and described the process. In Horton's original version of the model (1933), infiltration capacity was calculated as an exponential decay of the form

$$f = f_c + (f_o - f_c)\exp(-ct)$$

where f is the infiltration capacity at time t, and f_c, f_o and c are constants that depend on the soil and its moisture distribution at the start of infiltration. At the start of a rainstorm all rainfall is considered to infiltrate until the infiltration capacity falls to the current rainfall intensity. Subsequently, the infiltration curve determines how much water enters the soil, and any excess runs off as OVERLAND FLOW. Total overland flow discharge is obtained by routing this flow downslope, combined with the overland flow produced at each site. This overland flow is then considered to supply the sharply rising and falling peak of stream hydrograph response to rainstorms. The model was first developed for agricultural soils in an area of very intense rainfalls, and it works very well under these conditions. At the simplest level the model appears to imply a uniform rate of overland flow production so that flow discharge increases linearly downslope. Allowance for wetter soils downslope, however, forecasts increasing production downslope, so that Horton's model is able to perform well as a forecasting tool even in humid forested environments, where its physical basis is undermined by the rarity of observable overland flow, and where PARTIAL AREA MODELS are now preferred by many researchers.

Attempts to improve the Horton model have focused on improving the infiltration equation used, or by using an equation that takes into account the reduced rate at which infiltration capacity falls if rainfall intensity is at less than the maximum capacity. While rainfall–run-off models based on Horton overland flow have been applied widely, run-off by Horton overland flow most commonly occurs in arid and semi-arid regions, and on disturbed lands and agricultural fields. The lack of observable overland flow on well-vegetated hillsides in humid and tropical catchments reflects the dominance of run-off generation in these areas by THROUGHFLOW within the soil as a direct source of stormflow and overland flow production in saturated SOURCE AREAS.   DRM

Reading and Reference
Horton, R.E. (1933) The role of infiltration in the hydrological cycle. *Transactions of the American Geophysical Union*, **14**, 446–460. · Kirkby, M.J. (ed.) (1978) *Hillslope hydrology*. Chichester: John Wiley & Sons, Ltd.

Horton's laws

Two laws of drainage composition were suggested by Robert E. Horton (1945): the law of stream *numbers*, whereby an inverse geometric series related the number of streams of a particular order (see ORDER, STREAM), and the law of stream *lengths*, which was based on a geometric series between mean lengths of streams of each order. These two laws were subsequently complemented by three others giving an inverse geometric series relation between mean slope of streams of a particular order, a geometric series relating mean basin area of streams of a particular order and the law of contributing areas, which was the logarithmic relation of drainage areas of each order and the total stream lengths that they contained and supported. Although modifications to these five laws were made with the advent of different systems of stream ordering and some useful comparisons were made using them, it was realized by 1970 that the 'laws' were at least in part a consequence of the definition of stream ordering and, therefore, largely statistical relationships. Kirchner (1993) subsequently showed that Horton's original laws describe virtually all possible channel networks and, therefore, support no particular conclusion about the origin or structure of stream networks.   KJG/DRM

Reading and Reference
Horton, R.E. (1945) Erosional development of streams and their drainage basins; hydrophysical approach to quantitative morphology. *Bulletin of the Geological Society of America*, **56**, 275–370. · Kirchner, J.W. (1993) Statistical inevitability of Horton's laws and the apparent randomness of stream channel networks. *Geology*, **21**, 591–594.

hot spot

A small area of the Earth's crust where an unusually high heat flow is associated with volcanic activity. Of approximately 125 hot spots thought to have been active over the past 10 million years, most are located well away from plate boundaries (see PLATE TECTONICS). The major theory of hot spot formation involves the effects of a PLUME of hot MANTLE rising to the surface. Some researchers consider that hot spots are sufficiently stationary with respect to the mantle to provide a reference frame for determining plate motions and CONTINENTAL DRIFT.

In the context of BIODIVERSITY, hot spots are defined as regions of elevated biological diversity with a large proportion of endemic species (see

ENDEMISM) and which is experiencing exceptional rates of habitat loss (Myers *et al.*, 2000). MAS/MEM

Reading and Reference
Anderson, D.L. (2005) Scoring hotspots: the plume and plate paradigms. In G. R. Foulger, J. H. Natland, D. C. Presnall and D. L. Anderson (eds), *Plates, plumes, and paradigms*. Geological Society of America Special Papers, vol. 388 Boulder, CO: Geological Society of America; pp. 31–54. · Courtillot, V., Davaille, A., Besse, J. and Stock, J. (2003) Three distinct types of hotspots in the Earth's mantle. *Earth and Planetary Science Letters*, 205, 295–330. · Myers, N., Mittermeier, R.A., Mittermeier, C.G., *et al.* (2000) Biodiversity hotspots for conservation priorities. *Nature*, 403, 853–858. · Possingham, H.P. and Wilson, K.A. (2005) Biodiversity: turning up the heat on hotspots. *Nature*, 436, 919–920.

hot spring An emission of hot, usually geothermally heated, water at the land surface.

hum A residual hill in limestone country formed through surface lowering of the surrounding land surface.

human impact Recognition that humans have had a major suite of impacts on their environment has a long history, as Glacken (1967) pointed that out at length in a scholarly monograph. An important stimulus to such ideas arose in the seventeenth and eighteenth centuries as western Europeans became aware of the ravages inflicted in the tropics by their overseas expansion. In the nineteenth century, Marsh (1864) wrote the first full-length study of how humans were transforming the Earth's surface by means of deforestation and other processes. Subsequently, many historical geographers were concerned with such activities as the use of fire, the clearing of woodland and the drainage of wetlands, and in 1956 many of these issues were considered in a great symposium on 'Man's Role in Changing the Face of the Earth' (Thomas, 1956). In the 1970s, the use of the term GLOBAL ENVIRONMENTAL CHANGE became prevalent, while in the following decades a suite of textbooks started to appear (e.g. Goudie, 2013) that assessed the human impact on the environment. Steffen *et al.* (2004) sought to review the ways in which biogeochemical systems interact at a global scale, and the term 'Earth system science' started to be widely employed.

Early humans had greater power to cause landscape transformation than is often accepted. Among the earliest impacts were those associated with the deliberate adoption and use of fire in the Palaeolithic. Another important human impact was a spasm of animal extinction, often called 'Pleistocene overkill', achieved as humans entered Australasia and the Americas. At around the start of the Holocene, about 10,000 years ago, many crops and animals were domesticated in various parts of the world, and this led to the 'agricultural revolution' which led to wholesale modification of land cover (Barker, 2006). In the Old World, the secondary applications of domesticated animals were explored. This has been termed 'The Secondary Products Revolution'. The plough was particularly important in this process – the first application of animal power to the mechanization of agriculture – as was the adoption of the wheeled cart, the development of textiles from animal fibres and the use of animal milk (which provided a means whereby large herds could use marginal or exhausted land). This encouraged the development of the pastoral sector with transhumance or nomadism. One further development in human cultural and technological life that was to increase human power was the mining of ores and the smelting of metals. Evidence for copper smelting occurs at Catal Hüyük in Turkey from the sixth millennium BC, while the smelting of iron ores may date back to the late third millennium BC.

The current impact of humans is enormous. In the last 300 years the global areas of cropland and pasture have increased by around five- to sixfold (Goldewijk, 2001). One measure of the extent of the human impact is the human appropriation of net primary production (HANPP) as a percentage of the total amount of NET PRIMARY PRODUCTIVITY (NPP) generated on land. There is a considerable range of estimates of HANPP as a percentage of NPP, but Imhoff *et al.* (2004) give a global figure of around 20%. As McNeill (2003) has pointed out, the twentieth century was a time of extraordinary change. The human population increased from 1.5 billion to 6 billion, the world's economy increased 15-fold, the world's energy use increased 13- to 14-fold, freshwater use increased 9-fold, and the irrigated area by 5-fold. In the hundred centuries from the dawn of agriculture to 1900, McNeill calculates that humanity only used about two-thirds as much energy (most of it from biomass) that it used in the twentieth century. Indeed, he argued that humankind used more energy in the twentieth century than in all preceding human history put together.

Humans have had a massive impact on vegetation (Ellis, 2011), and through that on animals, soil quality and erosion rates, land surface albedo, evapotranspiration and geomorphological processes. Plant communities have been transformed by the introduction and spread of INVASIVE SPECIES, and we have now appropriated a large amount of the world's BIOMASS for our own use. Smil (2013) has estimated that, through harvesting, deforestation and conversion of grasslands and wetlands, humans have reduced the stock of global terrestrial plant mass by up to 45% in the last 2000 years, with a third of this being achieved in the twentieth century. We now have the greatest biomass or anthropomass of any species, and it is an order of magnitude greater than the mass of all terrestrial mammals alive today. We are also currently in the throes of the most severe mass extinction event for 65 million years. A good review of global threats to animals, including hunting, pollution, ecological invasions and loss of habitat, is provided by Baillie *et al.* (2010).

Humans have deliberately modified river systems by the creation of dams, reservoirs, embankments and CHANNELIZATION. Urbanization has also had an impact on river flows and can cause an increase in flood hazard. Lake levels have also been modified, not least through interbasin water transfers, as was the case of the Aral Sea, but also in the case of the Owens and Mono lakes in California. There may also have been widespread cultural eutrophication of lakes. Groundwaters have also been exploited extensively (Wada et al., 2010) for municipal supply and irrigation. Water pollution is also a major concern and is an increasing issue in marginal seas and in the oceans leading to cultural eutrophication and hypoxia (see MARINE POLLUTION).

Humans create a whole suite of landforms deliberately through excavation, construction and dumping (see ANTHROPOGEOMORPHOLOGY). However, there are also many incidental consequences of human activities, including accelerated erosion and sedimentation, ground subsidence, changes in the weathering environment, acceleration and triggering of mass movements and coastal erosion and deposition.

At a global scale, humans have started to modify climate by increasing the loading of greenhouse gases (such as carbon dioxide, methane and nitrous oxide) in the atmosphere, modifying surface ALBEDO (reflectivity) through land cover change, and adding aerosol particles (e.g. dust and soot) to the atmosphere. Air quality is also being modified by urban pollution, photochemical smogs, the creation of ACID PRECIPITATION and stratospheric OZONE depletion.

It is now clear that humans will continue to modify their environment in coming decades. New technologies will be developed and applied (e.g. genetic modification), and increasing human population levels will lead to further changes in land cover and in the exploitation of natural resources. The effects of land cover

changes, as Slaymaker *et al.* (2009) point out, may be at least as important as the changes that will be caused by future global climatic change. However, GLOBAL WARMING will greatly modify biomes, lead to massive changes in the cryosphere and cause sea levels to rise.   ASG

References
Baillie, J.E.M., Griffiths, J. Turvey, S.T., *et al.* (2010) *Evolution lost. Status and trends of the world's vertebrates.* London: Zoological Society of London. · Barker, G. (2006) *The agricultural revolution in prehistory. Why did foragers become farmers?* Cambridge: Cambridge University Press. · Ellis, E.C. (2011) Anthropogenic transformation of the terrestrial biosphere. *Philosophical Transactions of the Royal Society A*, **369**, 1010–1035. · Glacken, C. (1967) *Traces on the Rhodian shore: nature and culture in western thought from ancient times to the end of the eighteenth century.* Berkeley, CA: University of California Press. · Goldewijk, K.K. (2001) Estimating global land use change over the past 300 years: the HYDE database. *Global Biogeochemical Cycles*, **15**, 417–433. · Goudie, A.S. (2013) *The human impact on the natural environment*, 7th edition. Oxford: Wiley-Blackwell. · Imhoff, M.L., Bounoua, L., Ricketts, T., *et al.* (2004) Global patterns in human consumption of net primary productivity. *Nature*, **429**, 870–873. · Marsh, G.P. (1864) *Man and nature*. New York: Scribner. · McNeill, J. (2003) Resource exploitation and over-exploitation: a look at the 20th century. In B. Benzing and B. Herrmann (eds), *Exploitation and overexploitation in societies past and present.* Münster: LIT Verlag; pp. 51–60. · Slaymaker, O., Spencer, T. and Embleton-Hamann, C. (eds) (2009) *Geomorphology and global environmental change.* Cambridge: Cambridge University Press. · Smil, V. (2013) *Harvesting the biosphere. What we have taken from nature.* Cambridge, MA: The MIT Press. · Steffen, W., Sanderson, A., Tyson, P. *et al.* (2004) *Global change and the Earth system. A planet under pressure.* Berlin: Springer. · Thomas, W.L. (ed.) (1956) *Man's role in changing the face of the Earth.* Chicago, IL: University of Chicago Press. · Wada, Y., van Beek, L.P.H., van Kempen, C.M., *et al.* (2010) Global depletion of groundwater resources. *Geophysical Research Letters*, **37**, L20402, doi: 10.1029/2010GL044571.

humate A collective term for the dark brown to black gel-like humic substances formed as a result of the decomposition of organic matter in soils and sediments. It may be translocated down the profile by vadose water. Subsurface accumulations of this material occur most commonly in podzols. If the humate dries out it may harden sufficiently to create a humicrete.   ASG

Reading
Pye, K. (1982) Characteristics and significance of some humate-cemented sands (humicretes) at Cape Flattery, Queensland, Australia. *Geological Magazine*, 119, 229–242.

humic acid Humic colloids may be chemically fractionated into fulvic acid, humic acid and humin, although the divisions are not clear cut. Humic colloids are built up of carbon, oxygen, hydrogen, nitrogen, sulphur and phosphorus, and contain polysaccharides, proteins and other organic materials, including waxes and asphalts. Humic colloids are negatively charged and may be bound to clay through polyvalent cations (e.g. Ca^{2+} and Al^{3+}) or to positively charged surfaces of iron or aluminium hydrated oxides. ALH

Reading
Russell, E.W. (1973) *Soil conditions and plant growth.* Harlow: Longman.

humidity A term that relates to the water vapour content of the atmosphere and is expressed in a variety of ways.

Thus, the relative humidity of an air sample is expressed as a percentage found by relating the observed VAPOUR PRESSURE at a given temperature to the saturation value of vapour pressure at that temperature. It is evaluated by using tables or a special slide rule in conjunction with dry-bulb and wet-bulb temperature readings. Relative humidity commonly displays a daily cycle that has a phase opposite to that of temperature, so that the highest values are observed near dawn and the lowest during the afternoon.

Absolute humidity indicates the actual amount of water vapour present in a sample of air. For example, the humidity mixing ratio of an air parcel is defined to be the ratio of the mass of water vapour to the mass of dry air with which the water vapour is associated. At the surface it may range between near zero in frigid polar areas to $25\,g\,kg^{-1}$ in very humid tropical air. Absolute humidities can be deduced from wet-bulb temperature readings. RR

humus Partially decomposed organic matter that accumulates on and within the soil zone.

hurricane See TROPICAL CYCLONES.

hydration The process whereby an anhydrous mineral, one not containing water molecules within its crystal structure, takes up water to form a crystallographically distinct mineral. The partial decomposition of rocks by water.

hydraulic conductivity The term given to the parameter K in the equation defining DARCY'S LAW. It is concerned with the physical properties of both the fluid and the material through which it flows, reflecting the ease with which a liquid flows and the ease with which a porous medium permits it to pass through it. Hydraulic conductivity has

the dimensions of a velocity and is usually expressed in metres per second. Since hydraulic conductivity may vary according to direction, K_x, K_y and K_z can be used to represent the hydraulic conductivity values in the x, y and z directions. Hydraulic conductivity should be distinguished from the term PERMEABILITY (or INTRINSIC PERMEABILITY) k, which refers only to the characteristics of the porous medium and not to the fluid that passes through it.

The saturated hydraulic conductivity of soils and other sediments may be measured in the laboratory with a PERMEAMETER. If samples are essentially undisturbed, the results are point values representative of the field conditions. However, the hydraulic conductivity of an aquifer (see GROUNDWATER) is best determined by field methods. Piezometer tests can be used to determine in-situ K values in a porous material around a piezometer tip. Pumping tests at a well provide measurements representative of a much larger aquifer volume (Freeze and Cherry, 1979). PWW

Reading and Reference
Fitts, C.R. (2013) *Groundwater science*, 2nd edition. Waltham, MA: Academic Press; especially chapter 3. · Freeze, R.A. and Cherry, J.A. (1979) *Groundwater*. Englewood Cliffs, NJ: Prentice-Hall.

hydraulic diffusivity *(D)* The ratio of the hydraulic conductivity K to the specific storage S_s or of transmissivity T to storativity S, as follows:

$$D = \frac{K}{S_s} = \frac{T}{S}$$

The square root of hydraulic diffusivity is proportional to the velocity of HYDRAULIC HEAD transmission in the aquifer.

The specific storage is the volume of water that a unit volume of aquifer releases from storage per unit decline in hydraulic head, whereas the dimensionless parameter storativity is the volume of water it releases per unit surface area per unit decline in hydraulic head above the surface. Transmissivity (or transmissibility) is the product of the hydraulic conductivity and aquifer thickness.

The hydraulic diffusivity is therefore an aquifer parameter that combines the transmission characteristics and the storage properties. PWW

Reading
Fitts, C.R. (2012) *Groundwater science*, 2nd edition. Waltham: Academic Press.

hydraulic force The water flowing in a channel reach exerts a hydraulic force on the bed and may move grains of unconsolidated bed material. This is usually known as the TRACTIVE FORCE and is the effective component of the weight of water acting parallel to the bed in the direction of flow. It is usually expressed in relation to the bed area over which it acts and is then called the unit tractive force, or the bed shear stress (force per unit area). The roughness of the channel bed imposes a frictional drag on the flow (see BOUNDARY LAYER), and in a reach where no flow acceleration occurs this is equal and opposite to the hydraulic force of the flow on the bed.   KSR

hydraulic geometry

The hydraulic geometry of alluvial channels was defined by Leopold and Maddock (1953) as: (1) the at-a-station adjustment of flow characteristics such as width w, depth d and velocity v within a section as discharge Q varies; and (2) the general downstream adjustment of these flow properties as discharge increases at a constant flow frequency. Both at-a-station and downstream, these adjustments are commonly described by the following power functions:

$$w = aQ^b; \; d = cQ^f; \; v = kQ^m$$

And since $w\, dv = Q$, it follows that $ack = 1$ and $b + f + m = 1$. The overall flow geometry of a river system is represented by relationships for an upstream and a downstream section at both a high- and a low-frequency flow (see diagram). However, the downstream trends are only approximate because factors other than discharge, such

as bed and bank sediment properties, control the changing shape of the channel within which the flow then adjusts at discharges that are not competent to change the section.

At-a-station flow geometry varies between rectangular sections with steep cohesive banks in humid areas and parabolic sections in sandy sediments in semi-arid regions. In the former, $b = 0.05, f = 0.45$ and $m = 0.50$, and in the latter $b = f = m = 0.33$. Downstream, the adjustment of width and depth accommodates more of the increase in discharge, and $b = 0.50, f = 0.40$ and $m = 0.1$ (Richards, 1982: 148–159). Downstream exponents vary according to the trends in ROUGHNESS and long-profile slope, but their general similarity reflects the tendency of all rivers to create forms that allow equal dissipation of energy and minimize total work. At the local reach scale, hydraulic autogeometry describes a further pattern in flow and channel form.   KSR

Reading and References

Ferguson, R.I. (1986) Hydraulics and hydraulic geometry. *Progress in Physical Geography*, **10**, 1–31. · Leopold, L.B. and Maddock, T. (1953) *The hydraulic geometry of stream channels and some physiographic implications*. United States Geological Survey Professional Paper 252. Washington, DC: US Government Printing House. · Richards, K.S. (1982) *Rivers: form and process in alluvial channels*. London: Methuen.

hydraulic gradient

If piezometric tubes are inserted into flowing water (soil throughflow, groundwater flow, pipe flow or open channel

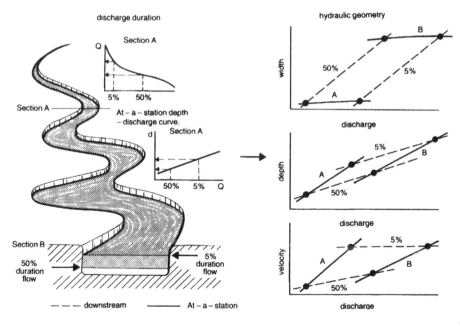

269

flow) water will rise to the hydraulic grade line. The slope of this line is the hydraulic gradient. So long as conditions remain hydrostatic, which requires that no extreme flow curvature occurs and that slopes are less than 1 in 10, the hydraulic gradient is the water table or water surface slope, or the piezometric surface slope in the case of confined groundwater flow or pipe flow.   KSR

hydraulic head *(h)*   The sum of the elevation head z and the pressure head h_p. Each of these parameters has the dimension of length and is measured in metres above a convenient datum, usually taken as sea level. The change of hydraulic head with distance l is known as the hydraulic gradient dh/dl and is an important term in the equation defining DARCY'S LAW.   PWW

Reading
Fitts, C.R. (2012) *Groundwater science*, 2nd edition. Waltham: Academic Press.

hydraulic jump   A given discharge in an open channel of constant width can be conveyed either at a high velocity and shallow depth or at a low velocity in deep water; a sudden change between these states along a channel is a hydraulic jump. The FROUDE NUMBER Fr passes from Fr > 1 (supercritical flow) to Fr < 1 (subscritical flow) during this transition. A hydraulic jump may be caused naturally by a sudden reduction of bed slope, or a large submerged obstacle (a boulder or bank slump), and is deliberately generated by

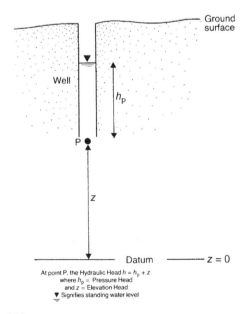

At point P, the Hydraulic Head $h = h_p + z$
where h_p = Pressure Head
and z = Elevation Head
▼ Signifies standing water level

hydraulic engineers to dissipate energy over weirs and dams, to ensure that upstream flow is critical in measuring installations such as flumes, and below sluices to ensure that they are not drowned (Sellin, 1969: 52–68).   KSR

Reference
Sellin, R.J.H. (1969) *Flow in channels*. London: Macmillan.

hydraulic potential (or fluid potential)
At any point in a porous substance, this is the product of the HYDRAULIC HEAD and the acceleration due to gravity. It is the mechanical energy of water (in this case) per unit mass. Since gravitational acceleration is almost constant near the Earth's surface, hydraulic potential is very closely correlated with hydraulic head (Freeze and Cherry, 1979). In accordance with DARCY'S LAW, subsurface water flows from points of high hydraulic potential towards points of low hydraulic potential. This occurs down the hydraulic gradient perpendicular to the EQUIPOTENTIALS.   PWW

Reading and Reference
Fitts, C.R. (2012) *Groundwater science*, 2nd edition. Waltham: Academic Press. · Freeze, R.A. and Cherry, J.A. (1979) *Groundwater*. Englewood Cliffs, NJ: Prentice-Hall.

hydraulic radius   The hydraulic radius R, or the hydraulic mean depth, is the ratio of wetted perimeter P to the cross-sectional area A of flow in a channel:

$$R = \frac{A}{P}$$

It measures the efficiency of a section in conveying flow, as it is the area of flow per unit length of water–solid contact. In wide, shallow channels the mean depth is a convenient approximation for R; when the width/depth ratio of a rectangular channel is 50, this approximation results in a 4% error, but when it is 10 the error increases to 17%.   KSR

hydraulics   The science in which the principles of FLUID MECHANICS are applied to the behaviour of water flowing in pipes or open channels over rigid or loose boundaries. Basic hydraulic theories relating to water flow deal with the development and refinement of FLOW EQUATIONS that define the flow velocity in terms of depth, slope and boundary roughness, with assessment of the forces, momentum and energy losses within the flow, and with analysis of velocity variations in two- and three-dimensional BOUNDARY LAYERS. Some of the most difficult problems in hydraulics occur in the area where it impinges most obviously on physical geography – namely, in the analysis of flow over loose boundaries, when the flow transports sediment at the bed, creates bedforms and, therefore, affects its own flow resistance in a complex feedback process.

Applied hydraulics is concerned with the economic development of water supply and, therefore, deals with the storage, conveyance and control of water. Storage problems include assessment of the ability of the physical supply within an area to cope with estimated and forecast demand, as well as the difficulties of maintenance of supply – for example, because of overexploitation of groundwater, or reservoir sedimentation and loss of storage capacity. Hydraulic design is concerned with dams, spillways and conveyance and diversion systems such as canals and irrigation ditches. There are close links between REGIME THEORY, which considers the design of stable sediment-bearing artificial channels, and the geomorphological study of natural river morphology. The design of control and measurement systems is a further important aspect of hydraulic analysis, in which spatially varied flow including free overfalls and HYDRAULIC JUMPS is deliberately manipulated to provide control conditions for the purpose of flow measurement in weirs and flumes.   KSR

Reading
Chow, V.T. (1959) *Open-channel hydraulics*. Tokyo: McGraw-Hill. · Sellin, R.H.J. (1969) *Flow in open channels*. London: Macmillan.

hydrodynamic levelling The transfer of survey datum levels by comparing mean sea level at two sites, and adjusting them to allow for gradients on the sea surface due to currents, water density, winds and atmospheric pressures.   DTP

hydrofluorocarbons (HFCs) A group of manufactured chemicals containing the elements carbon, hydrogen and fluorine. They are colourless, odourless and unreactive gases used mainly in refrigeration and air conditioning equipment and as propellants in industrial aerosols (replacing the formerly used chlorofluorocarbons and hydrochlorofluorocarbons that have been shown to damage the ozone layer in the upper atmosphere). HFCs are also used for foam blowing, solvent cleaning and in fire extinguishers. The main sources of HFCs include the manufacture of, leakage from and end-of-life disposal of refrigeration and air conditioning equipment and aerosols. Although these chemicals do not impact on ozone,

their chemical stability results in long-term accumulation in the atmosphere, where they can have significant CLIMATE CHANGE effects, for they are potent greenhouse gases (see GREENHOUSE EFFECT).   MEM

Reading
Montzka, S.A., Dlugokencky, E.J. and Butler, J.H. (2011) Non-CO_2 greenhouse gases and climate change. *Nature*, **476**, 43–50.

hydrofracturing The rupture of rock caused when water enters cracks in rock that may be little wider than the combined diameters of a few molecules of water. Under cold conditions, if ice seals off the end of such a crack, liquid water may be forced towards the tip of the crack and so extend it.   ASG

hydrogeological map A cartographic representation of subterranean water resources information with associated surface water and geological data. Hydrogeological maps may be prepared from continental to local scales, depending on their purpose. Continental- and regional-scale maps commonly depict the major aquifers and identify water-bearing lithologies and GROUNDWATER basins. Major SPRINGS and their discharges are also frequently depicted. Water balance information is sometimes contoured. Larger scale local maps, in addition, usually show WATER TABLE (or piezometric) contours (or EQUIPOTENTIALS in the case of ARTESIAN aquifers) and may provide information on the direction of groundwater flow, aquifer thickness, borehole locations, recharge areas and saltwater intrusions. Water quality data are also sometimes presented.   PWW

hydrographs (see UNIT HYDROGRAPH) These show discharge plotted against time for a point in a drainage basin. The discharge is usually monitored at a site on the stream channel network (stream hydrograph), but it may be measured at a section on a hillslope (hillslope hydrograph).

Stream hydrographs represent the pattern of total discharge through time and are made up of flow from different stores within the drainage basin (see HYDROLOGICAL CYCLE). During dry periods, discharge gradually recedes and the hydrograph at a particular site may often be represented by a standard DEPLETION CURVE or base-flow recession curve. During rainfall events, the stream discharge responds in the form of a storm hydrograph (see diagram). The shape of the storm hydrograph is affected by both the spatial and temporal pattern of the rainfall, ANTECEDENT MOISTURE conditions and by

271

Storm hydrograph.

the drainage basin characteristics, including catchment geometry, rock and soil types, vegetation and land-use pattern and drainage network structure.

Hydrograph analysis often begins with a separation of the hydrograph into components. Traditional hydrograph separation subdivides the hydrograph into surface run-off or direct run-off, base flow and sometimes interflow. The subdivision is essentially empirical, with straight lines joining either the beginning of the storm hydrograph rise or a point beneath the hydrograph peak to one or more points on the falling limb of the hydrograph. These points on the falling limb may be identified from the curvature of the falling limb, from the point of divergence of the falling limb from the depletion curve or by identifying breaks of slope on semi-logarithmic plots of the hydrograph (Barnes, 1939; Gregory and Walling, 1976). Hibbert and Cunningham (1967) avoided the generic terms for describing the components of the separated storm hydrograph by suggesting a separation into quick flow and delayed flow by a straight line from the point of hydrograph rise to the falling limb of the hydrograph with a slope of $0.551 \, s^{-1} \, km^{-2} h^{-1}$. Hydrograph separation based upon water quality variations (e.g. Pinder and Jones, 1969; Pilgrim et al., 1979) show that straight-line separations do not accurately reflect the way in which storm hydrographs are composed of water from different source areas.

Whichever separation technique is used, parameters of the storm hydrograph may be analysed to represent hydrograph size and shape, and some widely used parameters include the time of rise, base time, run-off volume, peak discharge, percentage run-off and hydrograph peakedness. The unit hydrograph technique (Sherman, 1932; see UNIT RESPONSE GRAPH) is often applied to separated storm hydrographs since it provides a simple means

of summarizing and predicting the storm hydrograph at a site.   AMG

Reading and References
Barnes, B.S. (1939) The structure of discharge recession curves. *Transactions of the American Geophysical Union*, **20**, 721–725. · Gregory, K.J. and Walling, D.E. (1976) *Drainage basin form and process*. London: Edward Arnold. · Hibbert, A.R. and Cunningham, G.B. (1967) Streamflow data processing opportunities and applications. In W. E. Sopper and H. W. Lull (eds), *International symposium on forest hydrology*. Oxford: Pergamon. · Linsley, R.K., Kohler, M.A. and Paulhus, J.L.H. (1988) *Hydrology for engineers*, 3rd edition. New York: McGraw-Hill. · Pilgrim, D.H., Huff, D.D. and Steele, D.M. (1979) Use of specific conductance and contact time relations for separating flow components in storm runoff. *Water Resources Research*, **15**, 329–339. · Pinder, G.F. and Jones, J.F. (1969) Determination of the groundwater component of peak discharge from the chemistry of total runoff. *Water Resources Research*, **5**, 438–445. · Sherman, L.K. (1932) Stream flow from rainfall by the unit-graph method. *Engineering News Record*, **108**, 501–505.

hydroisostasy The reaction of the Earth's crust to the application and removal of a mass of water, as when eustatic sea-level changes have affected the depth of water over the continental shelves causing the crust to be depressed at times of high sea level and to be elevated at times of low sea level. The same process can operate in lake basins like Lake Bonneville in response to climatically induced fluctuations in lake level, and may create warped shorelines.   ASG

Reading
Bloom, A.L. (1967) Pleistocene shorelines: a new test of isostasy. *Bulletin of the Geological Society of America*, **78**, 477–494.

hydrolaccolith A mound of ice formed by frost heaving of frozen underground water, resembling a LACCOLITH in section. To some, the term hydrolaccolith is synonymous with the terms ice laccolith and PINGO. However, others believe that they differ from pingos in that they are seasonal forms (whereas pingos are perennial), and differ from ice laccoliths in that they do not form within the active layer of permafrost ground. Hydrolaccoliths range in size between 1 and 10 m diameter, and are usually less than 2 m in height.   ASG

hydrological cycle The hydrological cycle describes the continuous movement of all forms of water (vapour, liquid and solid) on, in and above the Earth's surface, and it is the central concept of HYDROLOGY.

The hydrological cycle includes the condensation and freezing of water vapour in the

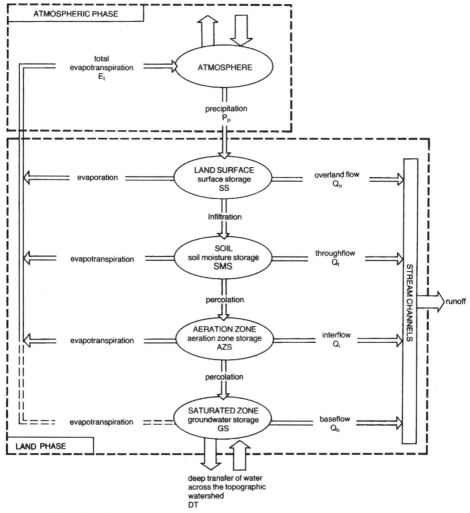

Components of the hydrological cycle associated with a drainage basin.

atmosphere to form liquid or solid precipitation, the movement of water from precipitation through one or more of a range of conceptual stores, including SURFACE STORAGE, soil moisture storage, groundwater storage, stream channels and the oceans until at some stage the water returns to the atmosphere as water vapour through the processes of evaporation and transpiration. The figure schematically presents the drainage basin hydrological cycle or the possible routes that water may follow within a drainage basin to form part of stream discharge at the basin outlet or to be lost to the basin through evapotranspiration. The time taken for water to pass to and from a store, the capacity of the store and the volume of water contained in each store at a particular time will strongly influence the discharge response of the drainage basin to a rainfall event.   AMG

Reading
Kirkby, M.J. (ed.) (1978) *Hillslope hydrology.* Chichester: John Wiley & Sons, Ltd.

hydrology   The science concerned with the study of the different forms of water as they exist in the natural environment. Its central focus is the circulation and distribution of water as it is expressed by the WATER BALANCE and HYDROLOGICAL CYCLE. Hydrology embraces not only the study of water quantity and movement, but also the

273

degree to which these are affected by human activities, including deliberate management of water resources and the inadvertent effects of humans on hydrological processes. It is often subdivided into physical and applied hydrology. Physical hydrology includes the detailed measurement and analysis of information on hydrological processes to improve understanding of the functioning of the hydrological system and also the refinement of statistical and mathematical methods of predicting and modelling these physical processes. Applied hydrology is concerned with the application of the understanding of hydrological processes to their modification and management. 'Water resources and pollution on the one hand and flooding and erosion on the other, are the chief concerns of the hydrologist' (Rodda *et al.*, 1976).

Hydrology is not only a long-established science that has been given major impetus at the international level, it is also an interdisciplinary subject bringing together specialists from an enormous range of disciplines, including biology, chemistry, civil engineering, environmental planning, forestry, geology, geomorphology, mathematics and physics.   AMG

Reading and Reference
Dunne, T. and Leopold, L.B. (1978) *Water in environmental planning*. San Francisco, CA: W.H. Freeman. · Gregory, K.J. and Walling, D.E. (1976) *Drainage basin form and process*. London: Edward Arnold. · Kirkby, M.J. (1978) *Hillslope hydrology*. Chichester: John Wiley & Sons, Ltd. · Linsley, R.K., Kohler, M.A. and Paulhus, J.L. (1982) *Hydrology for engineers*, 3rd edition. New York: McGraw-Hill. · Linsley, R.K., Downing, R.A. and Law, F.M. (1976) *Systematic hydrology*. London: Butterworth. · Rodda, J.C. (1976) *Facets of hydrology*. Chichester: John Wiley & Sons, Ltd. · Ward, R.C. and Robinson, M. (1990) *Principles of hydrology*, 3rd edition. Maidenhead: McGraw-Hill.

hydrolysis   *a*. The disintegration of organic compounds through their reaction with water.
b. The formation of both an acid and a base by a salt when it dissociates with water. This is an important mechanism of chemical weathering of rock.

hydrometeorology   The application of meteorology to hydrological problems. Hydrometeorology is concerned particularly with the atmospheric part of the HYDROLOGICAL CYCLE. This involves the input of water to the surface through PRECIPITATION and outputs by EVAPORATION and TRANSPIRATION. Precipitation input to a catchment area depends upon atmospheric factors that can be examined at both short and long timescales. Rainfall from an individual cloud will

be determined by *microphysical processes* within the CLOUD as well as by the synoptic controls on cloud development. The extent to which suitable atmospheric conditions for cloud growth and precipitation formation prevail will be influenced by atmospheric circulation changes not only on a daily scale, but also at the seasonal and annual timescales. One of the main applications of hydrometeorology is for flood prediction. Statistical techniques are used to analyse existing precipitation data in order to calculate the return of flood events (i.e. how often a particular flood level is likely to occur), to estimate the PROBABLE MAXIMUM PRECIPITATION over an area or even to simulate river-level variations. Some engineering design problems need similar information to provide cost-effective solutions when building reservoirs, bridges and sewers and planning irrigation schemes.   PS

Reading
Shaw, E.M. (1988) *Hydrology in practice*, 2nd edition. London: Van Nostrand Reinhold.

hydrometry   The measurement of water flow in channels. Accurate flow measurements are difficult to achieve, particularly when the cross-section of the flowing water is large or when the channel is shifting, irregular or subject to heavy weed growth. The techniques of flow or discharge measurement fall into three groups. Since discharge at a cross-section is equal to the product of the mean velocity of flow and the cross-sectional area of the flowing water, methods of gauging the flow are based upon either the direct measurement of discharge or on the product of an indicator of water cross-sectional area (usually through water depth or stage at a fixed site) and a means of estimating velocity of flow. In addition, measurements of the discharge may be just point measurements in time or they may be continuous. Measurements of stage, velocity or discharge are usually undertaken at fixed sites known as gauging stations. Table 2 summarizes the main methods of flow measurements, and developments of these techniques are discussed in Cole (1982) (see DISCHARGE).   AMG

Reference
Cole, J.A. (ed.) (1982) *Advances in hydrometry*. Proceedings of the Exeter symposium. International Association of Hydrological Sciences publication 134. Wallingford: IAHS Press.

hydromorphy   A process that occurs in soils that produces gleying and mottling as a result of the intermittent or permanent presence of excess water.

Table 2 Hydrometry: methods for river flow measurement

Variable	Point measurement	Continuous measurement	Comments
Stage	1 Stage board 2 Float and counterweight 3 Tape and electrical contact 4 Pressure bulb 5 Crest stage gauge (for peak flow only)	1 Float and counterweight or pressure bulb attached to continuous recorder 2 Bubble gauge and recorder	Float and counterweight or tape and electrical contact both require a level water surface and so are usually used with a stilling well
Velocity	1 Current meter 2 Floats 3 Tracer (e.g. NaCl or $NaCr_2O_3$) velocity 4 Electromagnetic flowmeter 5 Ultrasonic flowmeter	Electromagnetic flowmeter and recorder (average velocity in cross-section) Ultrasonic flowmeter and recorder (average velocity at one or more depths in the cross-section)	
Discharge	1 Current meter (mean or midsection method) 2 Tracer dilution (gulp or continuous injection) 3 Electromagenetic gauging 4 Ultrasonic gauging 5 Structures (weirs and flumes) 6 Volumetric 7 Slope area method (e.g. Manning equation)	Continuously monitored stage with established stage–discharge relationship Continuous measurement of velocity and stage Continuous measurement of velocity and stage Continuous stage measurement coupled with a stage–discharge relationship	Choice of wading or cableway or moving boat methods Smooth, straight section required Suitable for turbulent sections or where high velocities or irregular bed make the use of current meter difficult Suitable for sections with aquatic growth and unstable bed Suitable for very wide rivers and estuaries Suitable for small rivers Suitable for measuring very low flows Suitable for estimating flows at an ungauged site

hydrophobic soils Soils that are non-wettable or water repellent due to their characteristic physical and chemical properties. Hydrophobicity may be caused by fats (lipids) released by certain plants, which coat particles of the soil. Following vegetation wildfires, especially those of very high temperatures, hydrocarbon residue occupies pore spaces in the soils, making them impervious to water, thereby restricting INFILTRATION, which may in turn induce run-off and promote SOIL EROSION.   MEM/DLD

Reading
Debano, L.F., Mann, L.D. and Hamilton, D.A. (1970) Translocation of hydrophobic substances into soil by burning organic matter. *Soil Science Society of America Proceedings*, **34**, 130–133. · Dekker, L.W., Ritsema, C.J., Oostindie, K., *et al.* (2009) Methods for determining soil water repellency on field-moist samples. *Water Resources Research*, **45** (4), W00D33, doi: 10.1029/2008WR007070. · Doerr, S.H. (1998) On standardizing the 'water drop penetration time' and the 'molarity of an ethanol droplet' techniques to classify soil hydrophobicity: a case study using medium-textured soils. *Earth Surface Processes and Landforms*, **23**, 663–668.

hydrophyte A herbaceous plant that has parts beneath water that survive when the parts above water die back.

hydrosphere The Earth's water, which exists in both fresh and saline form and may occur in a liquid, solid or gaseous state. Land, sea and air each contribute to the total volume of water, which is conveyed between various locations and transformed from one state to another (HYDROLOGICAL CYCLE). The overall quantity of water in the hydrosphere remains more or less constant. An insignificant increment accrues from volcanic water vapour, while dissociation of water vapour in the upper atmosphere (photodissociation) represents a minor loss.

About 70% of the Earth's surface is occupied by water. Some 97.3% of its volume is currently in the oceans $(1350 \times 10^{15}\,\mathrm{m}^3)$, the maximum extent of which is in the southern hemisphere. Of the 2.7% of terrestrial water, most is polar snow and ice $(29 \times 10^{15}\,\mathrm{m}^3)$. Groundwater (the majority below soil level) accounts for $8.4 \times 10^{15}\,\mathrm{m}^3$, lakes (its main superficial terrestrial location) and rivers $0.2 \times 10^{15}\,\mathrm{m}^3$, and water in living organisms $0.0006 \times 10^{15}\,\mathrm{m}^3$. Water vapour is the most important variable constituent of the atmosphere, but it only accounts for $0.013 \times 10^{15}\,\mathrm{m}^3$ of the total amount of water in the hydrosphere (Peixoto and Ali Kettani, 1973).   RLJ

Reading and Reference
Peixoto, J.P. and Kettani, M.A. (1973) The control of the water cycle. In F. Press and R. Siever (eds), *Planet Earth (readings from Scientific American)* San Francisco, CA: W.H. Freeman.

hydrostatic equation
This equation is as follows:

$$\frac{\partial p}{\partial z} r = \rho g$$

where p is pressure, z is height, ρ is air density and g is the acceleration due to gravity. Alternatively, it may be written as

$$-\frac{1}{\rho}\frac{\partial p}{\partial z} = g$$

The latter equation reveals that the upward pressure gradient force (left-hand side) is balanced by the downward force of gravity (right-hand side). Such a balance is known as hydrostatic equilibrium. The assumption that the atmosphere is a hydrostatic equilibrium has proved very useful in the analysis of large-scale motion because, at that scale, buoyancy forces may safely be ignored. At smaller scales (e.g. motion within clouds) we cannot validly assume that hydrostatic equilibrium exists.   BWA

Reading
Hess, S.L. (1959) *Introduction to theoretical meteorology.* New York: Henry Holt.

hydrostatic pressure
The fluid pressure p exerted by the underlying column of water in a body of water when at rest. It is given by Pascal's law as

$$p = \rho g \psi + p_0$$

where ρ is the mass density of the water, g is the acceleration due to gravity, ψ is the pressure head (or vertical depth of water) and p_0 is atmospheric pressure at the surface of the water. In groundwater hydrology practice, the latter is usually set equal to zero and p is calculated in pressure above atmospheric and expressed in newtons per square metre or pascals.   PWW

Reading
Fitts, C.R. (2012) *Groundwater science*, 2nd edition. Waltham: Academic Press.

hydrothermal alteration
Alteration of rocks due to earth pressures and temperatures, which are milder than those which produce metamorphism. It may give rise to chemical changes that are similar to weathering. Generally, weathering effects decrease from the surface downwards, while hydrothermal alteration increases downwards towards the fluids that give rise to the activity. Pneumatolysis is similar but relates to gases (including steam) at still higher temperatures and pressures.   WBW

hyetograph
A self-recording instrument for measuring rainfall continuously. A chart showing the distribution of rainfall over a region.

hygric weathering
Wetting and drying (hygric weathering) of rocks such as shales and marbles (Koch and Siegesmund, 2004) can cause them to disintegrate. Ravina and Zaslavsky (1974) argued that cracking could be produced by pressures arising from the very high electrical gradients in the water of the electrical double layers at rock–water interfaces. The pressures are said to vary with relative humidity and diurnal temperature ranges, both of which are often high in drylands. However, also of importance is the disintegration of susceptible rocks or loosely consolidated material following either the introduction of water or exposure to the atmosphere. This can cause shrinkage and swelling (Weiss *et al.*, 2004), and is called slaking. Clays (especially smectites) and mudrocks are prone to this form of failure, especially in the presence of saline waters. In spite of their aridity, desert surfaces may be wetted by rainfall, fog, dew or fluctuating groundwater on more occasions than one might intuitively think. Fog is a potentially very important source of moisture for weathering processes, particularly in coastal deserts, while in some desert areas

appreciable amounts of dewfall may occur on a surprisingly large number of nights in the year and may, therefore, provide conditions for salts to hydrate and for hygroscopic salts to take up moisture (Davila *et al.*, 2008). In addition to fog, rain and dew, another important source of moisture can be groundwater discharge.   ASG

References

Davila, A.F., Gómez-Silva, B., de los Rios, A., *et al.* (2008) Facilitation of endolithic microbial survival in the hyperarid core of the Atacama Desert by mineral deliquescence. *Journal of Geophysical Research*, **113**, G01028, doi: 10.1029/2007JG000561. · Koch, A. and Siegesmund, S. (2004) The combined effect of moisture and temperature on the anomalous expansion behaviour of marble. *Environmental Geology*, **46**, 350–363. · Ravina, I. and Zaslavsky, D. (1974) The electrical double layer as a possible factor in desert weathering. *Zeitschrift für Geomorphologie*, **21**, 13–8. · Weiss, T., Siegesmund, S., Kirchner, D. and Sippel, J. (2004) Insolation weathering and hygric dilatation: two competitive factors in stone degradation. *Environmental Geology*, **46**, 402–413.

hygrograph   A self-recording instrument for measuring atmospheric humidity.

hygrometer   A device for measuring the relative humidity of the atmosphere. Those which give a continuous record are called hygrographs. ASG

hypabyssal rock   Igneous rock that is intrusive but has consolidated in a zone above the base of the Earth's crust and hence has distinct structural characteristics.

hyper-arid   These areas are the most arid DRY-LANDS and cover around 7.5% of the world's land surface. They represent the true DESERTS where biological productivity is lowest and rainfall most sporadic and variable. Hyper-arid areas have P/PET values of less than 0.05 (see DRYLANDS for methodology). The world's hyper-arid areas include much of the Sahara, Arabia, parts of central Asia and the coastal Namib and Atacama deserts in Namibia and Peru respectively. In most hyper-arid areas, periods in excess of 12 months without any rainfall have been recorded.   DSGT

hyperconcentrated flow   A river flow transporting an exceptionally large concentration of suspended particulates. Desert streams in flood commonly carry high concentrations of $5–50 \, \mathrm{g \, l^{-1}}$, but concentrations of hundreds of grams per litre have been reported, and in rare cases more than half of the weight of a water sample has been composed of sediment. These extraordinary concentrations fall into the range termed hyperconcentrated, and clearly grade progressively into mud flows, one form of mass movement (see MASS MOVEMENT TYPES). Beverage and Culbertson (1964) suggested that any streamflow carrying in excess of 40% by weight sediment be classed as hyperconcentrated. For ordinary sediment densities, this corresponds to flows composed of <80% water by volume.   DLD

Reference

Beverage, J.P. and Culbertson, J.K. (1964) Hyperconcentrations of suspended sediment. *Proceedings of the American Society of Civil Engineers, Journal of the Hydraulics Division*, **90**, 117–128.

hyperspectral imagery   Refers to the images provided by hyperspectral remote sensing, also known as imaging spectrometry, which have many (often hundreds) contiguous, registered, spectral bands of calibrated radiance units so that the spectrum for each pixel in the image of a scene can be derived. The spectral signal recorded through hyperspectral remote sensing should be able to reveal individual relevant features (absorption, reflectance, transmittance and emittance) with at least three or more contiguous spectral bands at a spectral resolution smaller than the spectral width of the feature itself. The ability to detect features in the spectra makes this technology ideal for finding objects, identifying materials or detecting processes in the environment. The field of geology has benefitted immensely from hyperspectral imagery since different minerals have a unique spectral response and can be located in the hyperspectral imagery accordingly. Indeed, the production of spectral libraries allows matching of spectral signatures of known environmental properties with that found in imagery to afford rapid mapping of the scene of interest and its monitoring over time. Other application areas include ecology, which exploits the known spectral features of the foliar chemistry of vegetation, as indeed does precision agriculture. Gases that are emitted from industries such as coal and oil-fired power plants, hazardous waste incinerators and cement plants can also be detected using hyperspectral imaging at the longer wavelengths of 3–5 and 7–11 µm. The early remote-sensing instruments providing hyperspectral imagery were airborne sensors (e.g. the NASA JPL Airborne Imaging Spectrometer (AIS) superseded by the Airborne Visible/Infrared Imaging Spectrometer (AVIRIS)). Satellite-based hyperspectral sensors are still rare; EOA-1 Hyperion and CHRIS/PROBA are examples that are operational, and the German Environmental Mapping and Analysis Program (EnMAP) hyperspectral satellite mission is one in the production phase.   DSB

Reading
Goetz, A.F.H. (2009) Three decades of hyperspectral remote sensing of the Earth: a personal view. *Remote Sensing of Environment*, 113, S5–S16. · Schaepman, M.E. (2009) Imaging spectrometers. In T. A. Warner, M. D. Nellis, and G. M. Foody (eds), *The SAGE handbook of remote sensing*. SAGE: London; pp. 166–178.

hypogene Term applied to mineral and ore deposits formed by water rising towards the surface, in contrast to supergene.

hypolimnion The lowest layers of cold water that occur at the bottom of an ocean, sea or lake, whereas the EPILIMNION consists of the warmer, less dense upper layers. The intermediate zone is called the thermocline.   ASG

hypothesis A set of beliefs or assumptions that we hold about the structure of the world or the way it behaves. Hypotheses are the basis of explanation in science, in that explanations are the logical consequence of the hypotheses we have accepted (Haines-Young and Petch, 1986). For example, explanations of many contemporary global biogeographical patterns are based on accepted hypotheses about the movement of the continents over geological time and the theory of evolution.

But why do we accept one hypothesis and not another? The choice is based on the judgements we make about a hypothesis, given how well it has stood up to testing. Hypotheses can be mistaken because they are initially conjectures. Thus, we demand that scientific hypotheses should be *testable* (Popper, 1972). In principle, we should be able to refute hypotheses by making some measurement or observation that can potentially contradict them. We also demand that hypotheses should be consistent with the other hypotheses we have accepted. As scientists we cannot hold inconsistent ideas. Thus, conflicts between hypotheses have to be resolved by eliminating those that are mistaken or resolving the difference by developing other testable hypotheses that account for the conflict. In the end, the hypotheses we accept and use are those that have withstood our best attempts to refute them.

Baker (1996) has described the role of hypothesis testing in the context of Quaternary geology and geomorphology, while Llambi (1998) has considered such ideas in an ecological context. Ayala (2009) provides a succinct account of the scientific method and the role of hypothesis testing, illustrating the issues by reference to the work of Darwin. Inkpen and Wilson (2013) provide a more detailed treatment of the nature of hypotheses in physical geography.   RH-Y

References
Ayala, F.J. (2009) Darwin and the scientific method. *Proceedings of the National Academy of Sciences*, 106 (Supplement 1) 10033–10039. · Baker, V.R. (1996) The pragmatic roots of American Quaternary geology and geomorphology. *Geomorphology*, 16, 197–215. · Haines-Young, R.H. and Petch, J.R. (1986) *Physical geography: its nature and methods*. London: Paul Chapman. · Inkpen, R. and Wilson, G. (2013) *Science, philosophy and physical geography*, 2nd edition. Routledge. · Llambi, L.D. (1998) An epistemological debate in ecology: Popper and the hypothesis test. *Interciencia*, 23 (5), 286–294. · Popper, K.R. (1972) *The logic of scientific discovery*. London: Hutchinson.

hypsithermal A term introduced by Deevey and Flint (1957) for a warm HOLOCENE phase covering four of the traditional Blytt–Sernander pollen zones, embracing the Boreal through to the Sub-Boreal (8950–2550 BP). It is broadly equivalent in meaning to ALTITHERMAL or climatic optimum.   ASG

Reference
Deevey, F.S. and Flint, R.F. (1957) Post-glacial hypsithermal interval. *Science*, 125, 182–184.

hypsographic curve A generalized profile of the surface of the Earth and the ocean floors. A curve or graph that represents the proportions of the area of the surface at various altitudes above or below a datum.

hypsometric curve See HYPSOGRAPHIC CURVE.

hypsometry The measurement of the elevation of the land surface or sea floor above or below a given datum, usually mean sea level.   ASG

Reading
Cogley, J.G. (1985) *Hypsometry of the continents*. Zeitschrift für Geomorphologie, Supplementbände, vol. 53 Berlin: Borntraeger.

hysteresis A term borrowed from the study of magnetism and used to describe a bivariate plot that evidences a looped form and, therefore, a different value of the dependent variable according to whether the independent variable is increasing or decreasing. Examples include relationships between matric potential and soil moisture content, river discharge and water stage, and suspended sediment and solute concentrations and river discharge. In the latter case, different concentrations are associated with a given level of discharge on the rising and falling limbs of a hydrograph, and this hysteresis effect may be ascribed either to a difference in timing or lag between the response of the two parameters or to asymmetry in either relative behaviour. Both clockwise and anticlockwise hysteresis loops have been reported for relationships between suspended sediment and solute concentrations and discharge. From a study of the River Rother in the UK, Wood (1977) suggested that the precise form of the hysteresis loop associated with the relationship between suspended sediment concentration and discharge for specific events could be related to the period between successive flow events and to the duration and intensity of each event.

DEW

Reference

Wood, P.A. (1977) Controls of variation in suspended sediment concentration in the River Rother, West Sussex, England. *Sedimentology*, **24**, 437–445.

I

ice Normally the solid form of water formed by: (a) the freezing of water; (b) the condensation of atmospheric water vapour directly into ice crystals; (c) the sublimation of solid ice crystals directly from water vapour in the air; or (d) the compaction of snow. Each form of ice has important implications for the physical geography of the Earth.

a. Ice derived from the freezing of water involves SEA ICE and lake ice, which contribute a characteristic morphology to water surfaces and shorelines in high and mid-latitudes; also GROUND ICE, which forms a significant component of permafrost landscapes (see also ICE WEDGE and PINGO) and the freezing of water within a snowpack to form ice lenses or, in the case of a glacier, SUPERIMPOSED ICE. Repeated freezing and thawing is also an efficient means of WEATHERING and can cause rapid rock breakdown.

b. Ice crystals in the atmosphere begin as minute ice particles that form around condensation nuclei. These nuclei may be dust particles with favourable molecular structure or even crystals of sea salt. They will grow so long as they exist in an atmosphere with an excess supply of water vapour and will form an ice cloud or, if conditions are suitable near the ground surface, an ice fog. If the ice crystals reach a critical size they begin to fall as snow. RIME ICE can contribute to the growth of the falling ice crystal and also directly to the ground surface whenever supercooled water droplets freeze on impact with a solid object.

c. The sublimation of ice crystals directly from water vapour in the air can take place within a snowpack in response to strong temperature gradients and causes the growth of fragile ice crystals known as *depth hoar*. The process also produces hoar frost, the ice equivalent of dew.

d. The compaction of snow to form GLACIER ice involves a number of metamorphic processes whose overall effect is to increase the crystal size and eliminate air passage. Snow that has survived a summer melt season and begun this process of transformation is known as FIRN. When consolidation has proceeded sufficiently far to isolate the air into separate bubbles then the firn becomes glacier ice. Carbon dioxide ice occurs extraterrestrially and is of geomorphological importance; for example, on Mars (Lucchitta, 1981). DES

Reading and Reference
LaChapelle, E.R. (1973) *Field guide to snow crystals*, 3rd edition. Seattle, WA: University of Washington Press. · Lucchitta, B.K. (1981) Mars and Earth: comparison of cold-climate features. *Icarus*, **45**, 264–303. · Paterson, W.S.B. (1981) *The physics of glaciers*. Oxford: Pergamon.

ice age A period in the Earth's history when ice sheets were extensive in mid and high latitudes. Such conditions were often accompanied by the widespread occurrence of sea ice, permafrost, mountain glaciers in all latitudes and sea-level fluctuations.

Ice ages have affected the Earth on many occasions. There are records of major glaciations in the Precambrian (on several occasions), the Eocambrian (650–700 million years ago), the Ordovician (*c.* 450 million years ago), the Permo-Carboniferous (250–300 million years ago) and during the Cainozoic (the past 15 million years or so). On occasions the evidence of even the ancient ice ages is splendidly clear: ice directions can be inferred from striations and grooves, basal temperatures and basal water conditions from the type of till, and sea-level fluctuations from rhythmic marine sediments. The freshness of the striations cut by the Ordovician ice sheet in what is now the Sahara is quite stunning to the visitor.

Within an ice age there are sharp fluctuations of ice extent. During the Cainozoic, mid-latitude ice sheets built up on many occasions (see GLACIAL) only to disappear during interglacials. The fluctuations within the ice age can be related to cyclic variations in the amount of solar radiation received by the Earth. Cycles of varying radiation receipt were postulated by Milankovitch (see MILANKOVITCH HYPOTHESIS) in 1924 and related to variation in the Earth's orbital eccentricity, tilt and precession. Evidence from deep sea cores reveals a similar cyclic variability of ice build-up and decay (Hays *et al.*, 1976).

The Dictionary of Physical Geography, Fourth Edition. Edited by David S.G. Thomas.
© 2016 John Wiley & Sons, Ltd. Published 2016 by John Wiley & Sons, Ltd.

It is not clear why ice ages have occurred during part of the Earth's history only. Perhaps an ice age needs a continent to be positioned in an approximately polar position so that ice can build up permanently and chill the Earth as a whole.

The behaviour of the Quaternary ice ages has been investigated in considerable detail. The chronology of ice advance and retreat is clearly linked to the patterns of high-latitude insolation, as set out in the Milankovitch hypothesis. However, the dominant 100,000-year periodicity of the last million years involves the slow isostatic adjustment of the crust an asthenosphere to the ice loading. Feedbacks from changes in the concentration of atmospheric carbon dioxide and from iron fertilization of the oceans by terrestrial dust are also involved in the climate changes from glacial to interglacial.   DES/DLD

Reading and Reference
Abe-Ouchi, A., Saito, F., Kawamura, K., *et al.* (2013) Insolation-driven 100,000-year glacial cycles and hysteresis of ice-sheet volume. *Nature*, **500**, 190–194. · Bar-Or, R., Erlick, C. and Gildor, H. (2008) The role of dust in glacial–interglacial cycles. *Quaternary Science Reviews*, **27**, 201–208. · Denton, G.H., Andeerson, R.F., Toggweiler, J.R., *et al.* (2010) The last glacial termination. *Science*, **328**, 1652–1656. · Hays, J.D., Imbrie, J. and Shackleton, N.J. (1976) Variations in the Earth's orbit; pacemaker of the ice ages. *Science*, **194** (4270), 1121–1132. · Imbrie, J. and Imbrie, K.P. (1979) *Ice ages*. London: Macmillan. · John, B. (ed.) (1979) *The winters of the world*. Newton Abbot: David & Charles.

ice blink   A mariner's term for the white glare on the underneath of clouds indicating the presence of pack ice or glacier ice.

ice cap   A dome-shaped GLACIER with a generally outward and radial flow of ice. The difference between an ice cap and an ice sheet is normally taken to be one of scale, with the former being less than 50,000 km^2 in area and the latter larger. The marginal regions of the ice cap may be drained by OUTLET GLACIERS, which flow beyond the ice cap in rock-walled valleys.   DES

ice contact slope   A slope, formerly banked up against an ice mass, that has experienced slumping as a result of ice melting. Such slopes are commonly associated with KAME TERRACES and ESKERS.   DES

ice core   A column of ice extracted by drilling into an ice sheet. An ice core is typically of the order of 10 cm in diameter and up to 3000 m long. Since the 1960s, many have been taken from parts of the Greenland and Antarctic ice sheets for the purpose of deriving information about past climates. Glacier ice accumulates in annual layers, and so increasing depth down a core corresponds to increasing age. By measuring properties of ice of different ages throughout the core, it is possible to reconstruct changes over timescales from annual/decadal up to a complete glacial–interglacial cycle (i.e. 100,000 years). Analysis of the data obtained from ice cores has provided an uninterrupted record of change that has revolutionized our understanding of rates of climate change and the interactions between components of the climate system.

Air bubbles within the ice provide samples of the atmosphere at the time they were formed and so can be used for measurements of past concentrations of greenhouse gases, such as carbon dioxide and methane. The relative abundance of small traces from atmospheric fallout, such as wind-blown dust, sea salt, volcanic ash, and pollen, can be used to provide evidence of particular events, such as volcanic eruptions or aeolian activity, and can also inform about past wind directions and strength. The OXYGEN ISOTOPE ratio ($^{18}O/^{16}O$) of ice can be used as an estimate of global ice volumes and temperatures in a similar way to the methods used in ocean cores. Measured or observed annual ice increments can be used to count down from the surface to derive ages and information about relative changes in snow accumulation and ablation.   CDC

Reading
Lowe, J.J., and Walker, M.J.C. (2015) *Reconstructing quaternary environments*, 3rd edition. London: Routledge.

ice cored moraine   A ridge or spread of glacial rock debris that contains a buried ice core. Protected from melting by the overlying debris, the ice core can survive for considerable periods of time.   DES

ice dam   A blockage of drainage caused by ice that leads to periodic and/or rapid fluctuations in meltwater drainage. Large ice dams occur subglacially and are particularly catastrophic. The drainage outbursts are usually triggered when the glacier internal drainage network developing during the ablation season taps the subglacial lake. An initially small outflow melts open a very large passage in a matter of hours. The Icelandic word JÖKULHLAUP is often used to describe the increased discharge associated with the breaching of a subglacial ice dam. Increased discharge can be extreme: the burst of ice-dammed Lake Agassiz, associated with the retreat of the Laurentide Ice Sheet, is regarded to have had a major impact on climate and parts of the American landscape (e.g. Teller *et al.*, 2002).

In High Arctic regions, streamflow occurs in late spring when valleys are choked with snow. Channel development begins with a saturation of the valley snowpack and water movement within or through the snow. The ponding and subsequent release of water behind snow dams formed by drifts is a common occurrence in a range of environments. The term *ice jam* is sometimes used.   HMF/DSGT

Reading and reference

Teller, J.T., Leverington, D.W. and Mann, J.D. (2002) Freshwater outbursts to the ocean from glacial Lake Agassiz and their role in climate change during the last deglaciation. *Quaternary Science Reviews*, **21**, 879–887. · Wilcox, A.C., Wade, A.A. and Evans, E.G. (2014) Drainage events from a glacier-dammed lake, Bear Glacier, *Alaska: remote sensing and field observations. Geomorphology*, **220**, 1–49.

ice dome   A term used to describe the main form of an ice sheet or ice cap and to distinguish it from streaming flow associated with OUTLET GLACIERS.   DES

ice fall   A heavily crevassed area of a glacier associated with flow down a steep rock slope. The zone is one of EXTENDING FLOW and is marked by arcuate rotational slumps.   DES

ice field   An approximately level area of ice that is distinguished from an ice cap because its surface does not achieve a dome-like shape and because flow is not radial outwards.   DES

ice floe   A piece of floating sea or lake ice that is not attached to the land. In the Arctic and Antarctic, ice floes are commonly from tens of metres to several kilometres across and 2–3 m thick. Some ice floes form in one winter and melt the following summer. Where they survive from one year to the next they are known as multiyear ice and tend to be tougher and, in places where they have been rafted on top of one another or crushed together, thicker. (See also SEA ICE.)   DES

Reading

Nansen, F. (1897) *Farthest north*, **2** vols. London: Constable.

ice flow   The movement of ice by internal deformation or BASAL SLIDING. Most studies of ice flow concern glaciers, but relatively little is understood, for example, about the flow of debris-rich ground ice in permafrost areas.

A glacier flows in response to shear stresses set up in the ice mass by the force of gravity. These vary according to the thickness of the glacier and its surface slope and can be calculated from the equation: $\tau = \rho g h \sin \alpha$, where τ is the shear stress, $\rho g h$ is the weight of overlying ice and α is the slope of the ice surface. Internal deformation of the glacier takes place mainly through the action of CREEP, which is modelled for glaciers by GLEN'S LAW. This shows that glacier flow is highly sensitive to an increase in shear stress, and this is why most internal deformation occurs at the bottom of a glacier. Near the glacier bottom, bedrock obstacles set up locally high stresses and enhanced basal creep is the mechanism by which they are passed. In situations where the base of the glacier is at the pressure melting point and a film of water exists at much of the ice–rock interface, basal sliding contributes to glacier flow and rates may equal or exceed the rate of internal deformation for the glacier as a whole. Where glaciers overlie saturated soft sediments it has been discovered that deformation of the sediment may contribute to the forward movement of the glacier (Boulton, 1979).

Most glaciers flow at the rate of 10–$100 \, \mathrm{m \, a^{-1}}$, but where rates of basal sliding are high, as in the case of SURGING GLACIERS, flow rates may exceed several kilometres per year.   DES

Reference

Boulton, G.S. (1979) Processes of glacier erosion on different substrata. *Journal of Glaciology*, **23**, 15–38.

ice fog   A suspension of minute ice crystals in the air reducing visibility at the Earth's surface. The optimum conditions for ice fog build-up are temperatures below $-30 \, °C$ and a supply of water vapour. Such conditions are common in and around Arctic settlements in winter, where an inversion causes low air temperatures and vehicles and heating plants contribute more water vapour than can be absorbed without condensing. Such fogs are associated with severe pollution.   DES

Reading

Benson, C.S. (1969) The role of air pollution in Arctic planning and development. *Polar Record* **14**, 783–790.

ice front   The vertical cliff forming the seaward edge of an ice shelf or floating glacier.

ice jam   A blockage caused by the accumulation of pieces of river ice or sea ice in a narrow channel.

ice rind   A stage in the growth of sea ice when the accumulating ice crystals coagulate to form a brittle skin.

ice segregation Sometimes called ice lensing, ice segregation is the process by which water freezes, thereby causing heave of the ground surface. Primary (i.e. capillary) and secondary heave can be distinguished. In primary heave, the critical conditions for the growth of segregated ice are

$$P_i - P_w = \frac{2\sigma}{r_{iw}} < \frac{2\sigma}{r}$$

where P_i is the pressure of ice, P_w is the pressure of water, σ is the surface tension of ice to water, r_{iw} is the radius of the ice–water interface and r is the radius of the largest continuous pore openings.

Secondary heave is not clearly understood but may occur at temperatures below $0\,°C$ and at some distance behind the freezing front. Pore-water expulsion from an advancing freezing front is another mechanism for ice segregation, especially massive ice bodies, provided that pore-water pressures are adequate to replenish groundwater that is transformed into ice. HMF

Reading
Miller, R.D. (1972) Freezing and heaving of saturated and unsaturated soils. In *Frost action in soils*. Highway Research Record no. 393. Washington, DC: Highway Research Board; pp. 1–11. · Washburn, A.L. (1979) *Geocryology: a survey of periglacial processes and environments*. New York: John Wiley & Sons, Inc.; especially pp. 68–70. · Williams, P.J. (1977) General properties of freezing soil. In P.J. Williams and M. Fremond (eds), *Soil freezing and highway construction*. Ottawa: Paterson Centre, Carleton University.

ice sheet A large dome-shaped glacier (over $50,000\,km^2$ in area) with a generally outward and radial flow of ice. On a continental scale such ice sheets can exceed a thickness of 4 km, as they do in Antarctica. A simple model can be used to approximate the surface slope of an ice sheet, assuming the ice is perfectly plastic. The profile equation is

$$h = 3.4(L - x)^{1/2}$$

where L is the distance from the ice sheet centre to edge and h is the ice thickness at a distance $L - x$ from the edge and both quantities are in metres (Cuffey and Paterson, 2010). Today, the Greenland, East Antarctic and the smaller West Antarctic ice sheets are subject to concern about the potential impacts on sea-level change of significant melting caused by global warming. During glacial times, other extensive ice sheets, including the Laurentide (North America), Patagonian (South America) and Weichselian (Europe) existed. DES

Reading and Reference
Cuffey, K.M. and Paterson, W.S.B. (2010) *The physics of glaciers*, 4th edition. London: Academic Press. · Khan, S.A., Kjær, K.H., Bevis, M., *et al.* (2014) Sustained mass loss of the northeast Greenland ice sheet triggered by regional warming. *Nature Climate Change*, 4: 292–299. · Steig, E.J., Schneider, D.P., Rutherford, S.D., *et al.* (2009) Warming of the Antarctic ice-sheet surface since the 1957 International Geophysical Year. *Nature*, 457: 459–462.

ice shelf A floating sheet of ice attached to an embayment in the coast. It is nourished by snow falling onto its surface and by land-based glaciers discharging into it. The seaward edge is a sheer cliff rising some 30 m above sea level. The shelf surface is virtually flat, although it rises slightly inland with an increase in ice thickness in the same direction. Freed of basal friction associated with land-based glaciers, ice velocities are high and commonly $1-3\,km\,a^{-1}$. Periodically, calving removes huge tabular ICEBERGS from the front. Some 30% of the Antarctic coastline is fringed by ice shelves. The Ross Ice Shelf extends 900 km inland and is 800 km across. DES

Reading
Robin, G. de Q. (1975) Ice shelves and ice flow. *Nature*, 253, 168–172. · Thomas, R.H. (1974) Ice shelves: a review. *Journal of Glaciology*, 24, 273–286.

ice stream A relatively narrow zone of swiftly moving ice within an ice sheet or ice cap, often bordered by spectacular crevasses. The high velocities probably reflect high sliding velocities associated with a basal water film. Ice streams often form the heads of OUTLET GLACIERS. DES

Reading
Drewry, D.J. (1983) Antarctic ice sheet: aspects of current configuration and flow. In R. Gardner and H. Scoging (eds), *Megageomorphology*. Oxford: Oxford University Press.

ice wedge A massive, generally wedge-shaped, ground ice body composed of foliated or layered, vertically oriented ice that extends below the permafrost table. Large ice wedges may be $1-2\,m$ wide near the top and extend downwards for $8-10\,m$. They form in cracks in polygonal patterns originating in winter by thermal contraction of the ground into which water from melting snow penetrates in the spring. Repeated annual contraction and subsequent cracking of the ice in the wedge, followed by freezing of water in the crack, lead to an increase in width and depth of the wedge. Ice wedges require PERMAFROST for their formation and existence. They often give a distinct polygonal micro-relief to the tundra surface. HMF

Reading
Mackay, J.R. (1974) Ice-wedge cracks. Garry Island. NWT. *Canadian Journal of Earth Sciences*, **11**, 1336–1383.

iceberg A floating mass of ice that has been either calved or torn off from the snout of a glacier or the floating margin or ice shelf at the edge of an ice sheet. The shape and size of an iceberg are related to the dynamics and morphology of the glacier or ice sheet, particularly the distribution of crevasses. Heavily crevassed surging glaciers typically produce small icebergs (termed 'brash ice' if <2 m across or 'bergy bits' if <10 m across), whilst large outlet glaciers and ice shelves produce large 'tabular icebergs' (>500 m across). Exceptionally large icebergs are termed 'ice islands'.

Icebergs are most common in polar regions but, if either exceptionally large or carried by strong ocean currents, may drift as far south as Newfoundland or as far north as the Falkland Islands. Icebergs are important media for the transportation and deposition of glacially eroded debris from polar regions, with sediment deposited when icebergs melt, overturn or are grounded in shallow water. Grounded icebergs are also erosional agents, creating flat-bottomed troughs or furrows where they are dragged along lake or sea beds. Increased calving of icebergs from Arctic and Antarctic ice shelves in recent years is widely considered to be a direct result of global climatic change.   DJN

icehouse A time at which continental ice sheets are present because of low temperatures associated with low levels of greenhouse gases within the atmosphere. It is the reverse of greenhouse conditions. Earth's most recent transition from a relatively warm greenhouse climate (with no ice sheets on the planet) to a relatively cold icehouse climate (with ice sheets and glaciers) occurred as the Cainozoic climate cooled and culminated as the Tertiary gave way to the Quaternary Period. A key change took place at the Eocene–Oligocene transition at *c.* 33–34 million years ago (Wade *et al.*, 2012).   ASG

Reference
Wade, B.S., Houben, A.J., Quaijtaal, W., *et al.* (2012) Multiproxy record of abrupt sea-surface cooling across the Eocene–Oligocene transition in the Gulf of Mexico. *Geology*, **40**, 159–162.

icing A mass of surface ice formed during the winter by successive freezing of sheets of water that may seep from the ground, from a river or from a spring. Icings are widespread in periglacial areas.   DES

igneous rock Rock formed when molten material, magma, solidifies, either within the Earth's crust or at the surface.

illuviation The precipitation and accumulation of material within the B horizon of a soil after the material has been leached from the surface or overlying soil horizons.

imbrication A regular overlapping, or shingling, of nonspherical sedimentary particles as a result of their deposition by fluids. Imbrication, one form of an anisotrophic sedimentary FABRIC, is most commonly associated with pebble and cobble deposits, but may occur in sand and boulder deposits. The plane described by the long *A* and intermediate *B* axes of an imbricate clast tends to DIP towards the flow at angles less than 20°, and this affords potential palaeoflow evidence in the stratigraphic record. Imbricated deposits represent a relatively stable bed configuration and may result in either bed ARMOURING or bed pavement.

Preferential orientations of the long axes may also be palaeoflow indicators. The *A* axis tends to be aligned perpendicular to the flow when the FROUDE NUMBER is low and transport is mainly as bedload (see BEDLOAD, BEDLOAD EQUATION). The *A* axis tends to assume a flow-parallel orientation with higher Froude numbers or when there is substantial transport via SALTATION. Imbrication is common on gravel beaches, on bars and outwash fans in BRAIDED RIVERS, and glacial TILLS, and is often a key to interpreting FACIES. The term *imbrication* is also used to describe overlapping tabular THRUST sheets.   DJS

Reading
Leeder, M.R. (1982) *Sedimentology: process and product.* London: George Allen & Unwin. · Reineck, H.-E. and Singh, I.B. (1986) *Depositional sedimentary environments*, 2nd edition. Berlin: Springer-Verlag. · Rust, B.R. (1972) Pebble orientation in fluvial sediments. *Journal of Sedimentary Petrology*, **42**, 384–388.

impermeable Having a structure or texture that does not allow the movement of water through a rock or soil material under the natural conditions in the groundwater zone.

impervious Impermeable. Having a texture that does not allow the movement of water, oil or gas through a rock or soil material. Under certain conditions a rock may have an impervious texture though a stratum of the rock is permeable owing to joints and fractures.

in and out channel The name given to a small, discontinuous channel produced by meltwater flow from a glacier onto the adjacent hillside.

Indian Ocean zonal mode (Indian Ocean dipole)
An east–west mode of sea-surface temperature (SST) variability in the tropical Indian Ocean. It is a coupled ocean–atmosphere phenomenon that is probably forced by the internal dynamics of the Indian Ocean, but some maintain that El Niño Southern Oscillation (ENSO) provides a trigger. The Indian Ocean zonal mode (IOZM) is in some ways analogous to the Pacific ENSO, in that it involves irregular oscillations of SSTs along the equator and associated shifts in the wind regime. In its positive phase, SSTs are anomalously warm in the western Indian Ocean and anomalously cold in the eastern Indian Ocean. The pattern of anomalies is reversed in the negative phase. Wind shifts occur in conjunction with the SST changes, notably decreased/increased equatorial westerlies in its positive/negative phase. The IOZM has a strong impact on precipitation in several areas, including India, East Africa and Australia. The positive phase is associated with droughts in Australia but above average rainfall in East Africa. SEN

Reading
Ihara, C., Kushnir, Y. and Cane, M.A. (2008) Warming trend of the Indian Ocean SST and Indian Ocean dipole from 1880 to 2004. *Journal of Climate*, **21**, 2035–2046. · Saji, N.H., Goswami, B.N., Vinayachandran, P.N. and Yamagata, T. (1999) A dipole mode in the tropical Indian Ocean. *Nature*, **401**, 360–363. · Shi, L., Hendon, H.H., Alves, O., *et al.* (2012) How predictable is the Indian Ocean dipole? *Monthly Weather Review*, **140**, 3867–3884.

incised meander
See MEANDERING.

inconsequent stream
A stream not apparently related to land-surface features or major geological controls, but following minor surface features without being developed into an organized pattern overall. The term was used by G. K. Gilbert (1877). It became largely superseded by the synonymous 'insequent' of W. M. Davis (1894) and is now used hardly at all. JL

Reading and References
Davis, W.M. (1894) Physical geography as a university study. *Journal of Geology*, **2**, 66–100. · Davis, W.M. (1899) The geographical cycle. *Geographical Journal*, **14**, 481–504. · Gilbert, G.K. (1877) *Report on the geology of the Henry Mountains*. Washington, DC: Government Printing Office.

inductive
Science based on the principle that accumulated knowledge allows generalizations to be developed that provide theories or laws. In its purest form, data would be collected without the collector (scientist) prejudging what that data might demonstrate in relation to the phenomena being assessed. This may be believed to lead to clear objectivity in analysis, but in reality the distinction between observation and theory that this approach requires is not clear cut (Haynes-Young and Petch, 1986). A DEDUCTIVE approach to scientific research is more widely employed in response to limitations with the inductive method (Popper, 1972). DSGT

References
Haines-Young, R.H. and Petch, J.R. (1986) *Physical geography: its nature and methods*. London: Paul Chapman. · Popper, K.R. (1972) *The logic of scientific discovery*. London: Hutchinson.

induration
The process of hardening through cementation, desiccation, pressure or other causes, applied particularly to sedimentary materials. Examples of indurated surface materials include DURICRUSTS, BEACH ROCK, AEOLIANITE and case-hardened outcrops of limestones, sandstones and granites. Induration of alluvium can have a significant effect on river channel morphology (Nanson *et al.*, 2005), while CASE HARDENING of rocks can contribute to the formation of cavernous weathering forms, caprocks and some types of tropical karst. ASG

Reference
Nanson, G.C., Jones, B.G., Price, D.M. and Pietsch, T.J. (2005) Rivers turned to rock: Late Quaternary alluvial induration influencing the behaviour and morphology of an anabranching river in the Australian monsoon tropics. *Geomorphology*, **70**, 398–420.

infiltration
The process by which water percolates into the soil surface. Two main zones can be observed in the soil when infiltration is proceeding at its maximum rate; that is, from a surface that is saturated with a thin layer of standing water. There is an upper transmission zone with an almost constant moisture content close to saturation. Below this is a sharp wetting front in which the moisture content declines rapidly towards its pre-infiltration value. Within the transmission zone, flow is driven mainly by gravitational forces. Across the wetting front there is a strong hydraulic or tension gradient, tending to push the water into the drier soil in the way that water is drawn into a fine capillary tube. This hydraulic gradient advances the wetting front down into the soil and so allows additional water to infiltrate at the surface. The rate at which water can infiltrate under these ideal circumstances is called the infiltration capacity, and it decreases as the wetting front advances deeper into the soil.

Infiltration capacity may be expressed either in terms of time since the process began, or in terms of current moisture storage. The main advantage of the storage expressions is that they may remain valid during infiltration at less than the capacity rate, as commonly occurs at the start of a rainstorm, when infiltration capacity tends to be very high. One example of a widely used empirical infiltration equation is that used in the HORTON OVERLAND FLOW MODEL. Another equation with a better theoretical basis was put forward by Philip (1957a–e, 1958a,b):

$$f = A + Bt^{1/2}$$

where f is the infiltration capacity at time t and A and B are constants that depend on the soil and its initial moisture distribution.

In this expression, the constant term A mainly represents the steady rate of infiltration under gravitational potential (i.e. the weight of the water), and the time-dependent term is due to the hydraulic potential gradient at the advancing wetting front. It may be seen that the infiltration capacity is initially very high and decreases steadily towards a constant rate, which is usually achieved within 1–2 h.

An example of a storage-based infiltration equation is the Green and Ampt (1911) equation:

$$f = A + \frac{C}{S}$$

where f and A are as above, S is a soil water storage value and C is a constant of the soil and its initial moisture.

In the original formulation of this expression the storage S was the total amount of water infiltrated since the start, but an alternative is to budget S as representing a store of infiltrated water that leaks at steady rate A. This version has the advantage that, if converted to a time-dependent form under conditions of surface ponding, it is exactly equivalent to the Philip equation (with $B = (2C)^{1/2}$). During a rainstorm of constant or varying intensity, this kind of storage-based model allows estimation of the infiltration capacity at any time in terms of the water that has actually entered the soil previously, which may have been at any rate less than (or equal to) the current infiltration capacity.

Infiltration may be likened to the process of pouring water into a bottle: it may fail to get in either because it is being poured in too fast or because the bottle is already full. On a hillside, saturated THROUGHFLOW may increase downslope or in areas of flow concentration until the SATURATION DEFICIT is zero; in other words, the bottle may be full or almost full. A second criterion for infiltration may therefore also be expressed in storage terms, in that ponding will occur when soil water storage reaches a critical level S_c.

In areas of high rainfall intensities and low infiltration capacity, infiltration capacity is commonly exceeded and the Horton overland flow model is generally applicable to the estimation of streamflow. This includes areas that are naturally or artificially clear of dense vegetation; that is to say, semi-arid areas and (seasonally) cultivated fields, as will be seen below. In areas of low rainfall intensity and/or dense vegetation cover, including forested areas and much of the humid-temperate zone, infiltration capacities are seldom exceeded, and PARTIAL AREA MODELS, which estimate the areas of saturation and the volumes of THROUGHFLOW, are generally more appropriate for estimating streamflow volumes and HILLSLOPE FLOW PROCESSES.

Rates of infiltration are usually compared in terms of the steady long-term rate (A in the equations above). This rate responds to some extent to soil texture, typically ranging from $0–4\,mm\,h^{-1}$ on clays to $3–12\,mm\,h^{-1}$ on sands where the soils are initially wet and unvegetated. Vegetation cover and protection by coarse particles shields the surface from raindrop impact, which is otherwise liable to break down the top layer of soil peds and pack the resulting soil grains down into the next layer as a thin, impermeable crust. On a crusted surface the infiltration capacity is commonly very low except in extremely intense storms, which may break the crust; where crusting is prevented, capacities are much higher. Within the soil, structure has a greater influence on infiltration capacity than texture does, and vegetation and its associated organic soil have a strong influence on soil structure. Thus, vegetation cover increases infiltration capacity in two ways, so that steady rates may be $50–100\,mm\,h^{-1}$ under a good cover, compared with less than $10\,mm\,h^{-1}$ on bare crusted soil.

Where soil structural voids are marked and soil textural pores fine, as for a cracked clay soil for example, soil water may bypass much of the soil mass. At the surface, soil peds allow infiltration at maximum capacity, and excess rainfall overflows down the structural voids. In the largest voids the water flows as a film down each wall and infiltrates into additional peds below the surface. The advance of water down each void is limited by this infiltration into the walls. In most cases rainfall enters the soil peds within a few tenths of a metre of the surface, but along a highly convoluted wetting front that follows the geometry of the largest voids. The resulting pattern may then be conceived either as a greater average depth of penetration of infiltrated water or as shallow penetration to match the current rainfall intensity. In extreme cases water may bypass the soil as a whole and make direct contact with groundwater via bedrock fissures, and so on. It should be pointed out that almost all real soils show some structural voids, but that local bypassing is only important where there is a very marked contrast between the textural and structural pore size distributions.

The importance of infiltration in physical geography lies in its role within the catchment and hillslopes' hydrological cycle in partly determining the flow routes of precipitation to the streams and so determining the timing of stream-flow response. Infiltration also plays an important part in separating groups of hillslope processes. Water that travels as overland flow takes little part in supplying soil moisture for plant growth and, because it comes into little and rather brief physical contact with mineral soil, picks up little solute load except from the litter layer. OVERLAND FLOW, therefore, tends to dilute stream solute concentrations, which therefore tend to be lower during intense storms when overland flow is greatest. Overland flow also erodes and transports all surface wash/soil erosion. The infiltrated water flows through the soil as throughflow, which is able to carry negligible amounts of suspended material, but is in intimate contact with mineral soil grains from which it is effective in leaching solutes and promoting chemical weathering. The infiltrated water is also responsible for providing water for plant growth and for establishing patterns of hydraulic potential that have a powerful influence on slope stability in the context of mass movement. In other words, the process of infiltration is a critical regulator of the landscape system in both the short, hydrological term and in longer erosional time spans.   MJK

Reading and References

Goudie, A. (ed.) (2003) *Geomorphological techniques*, 3rd edition. London: Taylor and Francis. · Green, W.H. and Ampt, G.A. (1911) Studies on soil physics I: the flow of air and water through soils. *Journal of Agricultural Science*, 4, 1–24. · Horton, R.E. (1945) Erosional development of streams and their drainage basins: hydrophysical approach to quantitative morphology. *Bulletin of the Geological Society of America*, 56, 275–370. · Knapp, B.J. (1978) Infiltration and storage of soil water. In M.J. Kirkby (ed.), *Hillslope hydrology*, Chichester: John Wiley & Sons, Ltd; pp. 43–72. · Philip, J.R. (1957a) The theory of infiltration: 1. The infiltration equation and its solution. *Soil Science*, 83, 345–358. · Philip, J.R. (1957b) The theory of infiltration: 2. The profile at infinity. *Soil Science*, 83, 435–448. · Philip, J.R. (1957c) The theory of infiltration: 3. Moisture profiles and relation to experiment. *Soil Science*, 84, 163–178. · Philip, J.R. (1957d) The theory of infiltration: 4. Sorptivity and algebraic infiltration equations. *Soil Science*, 84, 257–264. · Philip, J.R. (1957e) The theory of infiltration: 5. The influence of the initial moisture content. *Soil Science*, 84, 329–340. · Philip, J.R. (1958a) The theory of infiltration: 6. Effect of water depth over soil. *Soil Science*, 85, 278–286. · Philip, J.R. (1958b) The theory of infiltration: 7. *Soil Science*, 85, 333–337.

inflorescence   The flowering shoot of a plant.

influent   Either a tributary stream or river, or a term applied to a stream that supplies water to the groundwater zone.

infrared imagery   See THERMAL INFRARED LINESCANNER.

infrared thermometer/thermometry
Bodies reflect, emit and sometimes transmit radiation falling upon them. The efficiency of the surface in generating contributions is known, respectively, as reflectivity, emissivity and transmissivity. If the latter is zero (i.e. an opaque material) then reflectivity + emissivity = 1. The wavelengths near, but longer than, the visible spectrum are termed infrared and constitute most of the radiation from a 'hot' body. This radiation can be detected by an instrument that does not require contact with the body; when allowance is made for the emissivity of the surface, its temperature can be measured. A THERMAL INFRARED LINESCANNER is a more complex instrument used in remote sensing.   WBW

infragravity wave   Waves of periods 20–200 s. Unlike gravity waves, they are too long to be dissipated in the shoaling zone and hence cause significant variations in water level in the breaker and surf zones, often in the form of standing waves. Infragravity waves in the surf zone have been shown to relate to the generation of rip currents, cusps and crescentic bars.   TS

ingrown meander   See MEANDERING.

inheritance   See DARWINISM and EVOLUTION. In the geomorphological context, landforms or entire landscapes can be inherited from a past geological period when environmental or tectonic conditions were different. These features retain their characteristics because subsequent processes and events were insufficiently potent or long-lasting to obliterate them. Classic examples include the inselberg-dominated landscapes, particularly common in Africa but widespread elsewhere, which are usually traced back to an origin in the early Tertiary. MEM

Reading
Thomas, M.F. (1978) The study of inselbergs. In H. Bremer and J.N. Jennings (eds), *Inselbergs/Inselberge*. Zeitschrift für Geomorphologie, Supplementbände, vol. 31 Berlin: Borntraeger; pp. 1–41.

inlier An outcrop of rock that is completely surrounded by younger formations, frequently the result of erosion of the crest of an anticline.

inselberg (German for island hill) A general class of large residual hill that usually surmounts an eroded plain. Small residual rock masses tend to be called TORS; large, domed residuals tend to be called domed inselbergs or bornhardts, while large accumulations of boulders in the form of a hill are called koppies.

Inselbergs occur in a wide range of rock types, but the most common lithologies appear to be sandstones and conglomerates (e.g. Ayers Rock in Australia or Meteora in Greece) or gneisses and granites, especially those that have widely spaced joints and a high potassium content. ASG

Reading
Bremer, H. and Jennings, J. (eds) (1978) *Inselbergs/Inselberge*. Zeitschrift für Geomorphologie, Supplementbände vol. **31** Berlin: Borntraeger. · Pye, K., Goudie, A.S. and Thomas, D.S.G. (1984) A test of petrological control in the development of bornhardts and koppies on the Matopos Batholith, *Zimbabwe. Earth Surface Processes and Landforms*, **9**, 455–467. · Migoń, P. (2006) *Granite landscapes of the world*. Oxford: Oxford University Press.

insequent stream A drainage network that has developed as a result of factors which are not determinable.

An inselberg in southern Namibia. Photograph by David Thomas.

Inselbergs of resistance are those that are left as prominent landforms as a result of their superior resistance (brought about by the jointing density of the rock or its mineralogical composition), while inselbergs of position remain as prominent features because they are on divides farthest from lines of active erosion.

There has been considerable debate as to inselberg origin, and three main mechanisms have been proposed: that they are produced by scarp retreat across bedrock; that they are a result of scrap retreat across deeply weathered rocks; or that they result from differential weathering followed by stripping of the regolith.

insolation a term used in two senses:

1 The intensity at a specified time, or the amount in a specified period, of direct solar radiation incident on unit area of a horizontal surface on or above the Earth's surface.
2 The intensity at a specified time, or the amount in a specified period, of total (direct and diffuse) solar radiation incident on unit area of a specified surface of arbitrary slope and aspect.

In general, insolation depends on the solar constant, calendar date, latitude, slope and aspect of surface, and degree of transparency of the atmosphere. JGL

Rocks that may have been split by insolation weathering, Namibia. Photograph by David Thomas.

insolation weathering

The disintegration of rock in response to temperature changes setting up stresses. Early travellers heard, or reported they heard, sounds like pistol shots as rocks cooled in the evening, and thus arose a classic process envisaged in desert geomorphology. Experimental work in the twentieth century and recognition that weathering appears to be far more effective in the presence of moisture resulted in pure insolation weathering being related to a position of relatively lowly importance (Schattner, 1961). However, thermal fatigue has been cited as a process generating regolith on asteroids (Molaro and Byrne, 2012; Delbo *et al.*, 2014). Modern analytical methods have provided evidence that the precipitation of calcium carbonate may enlarge proto-fissures, with the entry of dust, especially containing swelling clays, further driving crack expansion in warm desert environments (Dorn, 2011). DLD

Reading and References

Delbo, M., Libourel, G., Wilkerson, J., *et al.* (2014) Thermal fatigue as the origin of regolith on small asteroids. *Nature*, **508**, 233–236, doi: 10.1038/nature13153. · Dorn,

R.I. (2011) Revisiting dirt cracking as a physical weathering process in warm deserts. *Geomorphology*, **135**, 129–142. · Eppes, M.C., McFadden, L.D., Wegmann, K.W. and Scuden, L.A. (2010) Cracks in desert pavement rocks: further insights into mechanical weathering by directional insolation. *Geomorphology*, **123**, 97–108. · Molaro, J. and Byrne, S. (2011) Rates of temperature change of airless landscapes and implications for thermal stress weathering. *Journal of Geophysical Research*, **117**, E10011, doi: 10.1029/2012JE004138, 2012. · Schattner, I. (1961) Weathering phenomena in the crystalline of the Sinai, in the light of current notions. *Bulletin of the Research Council of Israel, Section G: Geo-Sciences*, **10G**, 247–265.

instantaneous unit hydrograph (IUH)

See UNIT HYDROGRAPH.

intact strength

The strength of a rock sample that is free of large-scale structural discontinuities, such as joints, fissures or foliation partings. It is usually expressed as a measure of unconfined compressive strength. MJS

integrity of rivers

This has been defined by Graf (2001: 6) as referring 'to a set of active fluvial processes and landforms wherein the channel, flood plains, sediments and overall spatial configuration maintain a DYNAMIC EQUILIBRIUM, with adjustments not exceeding limits of change defined by societal values. Rivers possess physical integrity when their processes and forms maintain active connections with each other in the present hydrologic regime'. Geomorphological integrity is basic to river biodiversity and ecosystem functioning since the channel pattern provides habitat for biota and the physical framework for ecosystem processes. There are many cases where human activities, such as dam construction and channelization, have changed river dynamics and reduced river integrity. River restoration seeks to recover this integrity (Elosegi *et al.*, 2010). ASG

References

Elosegi, A., Diez, J. and Mutz, M. (2010) Effects of hydromorphological integrity on biodiversity and functioning of river systems. *Hydrobiologia*, **657**, 199–215. · Graf, W.L. (2001) Damage control: restoring the physical integrity of America's rivers. *Annals of the Association of American Geographers*, **91**, 1–27.

intensity of rainfall

See RAINFALL INTENSITY.

interannual variability

Variations of climatic parameters from year to year. This may refer specifically to variations of timescales of 2–6 years, in contrast to lower frequency variations on interdecadal timescales. However, interannual variability may also be used with reference to the measurement of the likely probable departure in any one year from the annual mean value in a long data set (i.e. for rainfall) and is usually expressed as a percentage. Variations on timescales shorter than 1 year are termed 'intra-seasonal'. DSGT/SJN

interbasin transfers A method of water supply whereby the natural or regulated flow from one river system is transferred, usually by pumping, to another river system. This method of water supply is likely to increase in future decades, and many schemes have already been evaluated, such as the possibility of the southward diversion of flow from Siberian rivers such as the Lena and the Ob and the transfer of water from the Chiang Jiang (Yangtze) to the Huang He (Yellow River) basin in China. Developments have been approached cautiously because of the possible ENVIRONMENTAL IMPACTS of such interbasin transfers, and it is known that in addition to changes of the water balance, especially in the evaporation term, there may also be substantial changes in the morphology and ecology of the river channels themselves because of the increase or decrease in streamflows. KJG

interception The process by which precipitation is trapped on vegetation and other surfaces before reaching the ground. Interception loss is the component of intercepted precipitation that is subsequently evaporated, although this is also frequently described as interception. The character of the intercepting surfaces has a major impact on the amount of precipitation that is intercepted and then lost through evaporation. (See also STEM FLOW and THROUGHFLOW.) AMG

Reading
Courtney, F.M. (1981) Developments in forest hydrology. *Progress in Physical Geography*, 5, 217–241. · Crockford, R.H. and Richardson, D.P. (1990) Partitioning of rainfall in a eucalyptus forest and pine plantation in southeastern Australia (four papers). *Hydrological Processes*, 4, 131–188. · Durocher, M.G. (1990) Monitoring spatial variability of forest interception. *Hydrological Processes*, 4, 215–229.

interception capacity The maximum volume or depth of water that can be retained on a plant canopy exposed to rain, generally under windless conditions. Interception capacity is determined in various ways, often by noting the weight gain on specimens that are uprooted for the purpose. It may also be estimated by measuring rainfall above the canopy, STEM FLOW, and THROUGHFALL beneath the canopy, and finding the intercepted amount from [rainfall − (stem flow + throughfall)]. For many plants, interception capacity amounts to a few millimetres of water depth over the projected area of the canopy. The value of interception capacity is influenced by leaf shape, leaf surface texture, leaf geometry and the rigidity or flexibility of the foliage under the imposed weight of water. Interception capacity may be reached in large storms but not in small; overall, canopy interception, followed by evaporative return of the water to the atmosphere, consumes 10–20% of the rainfall that falls over a dense forest. DLD

Reading
Aston, R.R. (1979) Rainfall interception by eight small trees. *Journal of Hydrology*, **42**, 383–396.

interflow A component of streamflow that responds to rainfall more slowly than surface run-off and more rapidly than BASE FLOW. In the HORTON OVERLAND FLOW MODEL streamflow was initially separated into overland flow and groundwater flow, and various procedures were used for partitioning the stream hydrograph into these components. Interflow was originally introduced as an intermediate hydrograph component that fell between the other two. It was considered to represent groundwater that re-emerged as overland flow; or else it was considered to be shallow groundwater flow. Some literature now uses the term interflow interchangeably with subsurface soil flow or THROUGHFLOW, but its original physical identification was rather tenuous. MJK

Reading
Linsley, R.K., Kohler, M.A. and Paulhus, J.L.H. (1949) *Applied hydrology*. New York: McGraw-Hill.

interfluve The area of high ground that separates two adjacent river valleys.

interglacial A phase of warmth between glacials when the great ice sheets retreated and decayed, and tundra conditions were replaced by forest over the now temperate lands of the northern hemisphere. The Holocene is an interglacial, but some of the Quaternary interglacials may have been slightly warmer than today. Iversen (1958) identified various stages in a glacial–interglacial cycle in northwestern Europe. Development begins with a *protocratic* phase, characterized by rising temperature, raw, basic or neutral mineral soils, favourable light condition, and a pioneer vegetation of small plants, with exacting requirements for both nutrients and light. In the following *mesocratic* phase, comprising the climax of the interglacial, there are maximum temperatures, brown forest soils, mull plants, dense climax forest and a vegetation that, while still demanding nutrients, is tolerant of shade. The *oligocratic* phase arises as a result of soil development and involves more acid soils and more open vegetation. The *telocratic* phase marks the end of interglacial forest development. Heaths expand and bogs develop in response to climatic deterioration. The climatic deterioration culminates in a *cryocratic* phase when cold conditions and soil instability are hostile to tree growth.   ASG

Reference
Iversen, J. (1958) The bearing of glacial and interglacial epochs on the formation and extinction of plant taxa. In O. Hedberg (ed.), *Systematics of today*. Acta Universitets Upsaliensis, vol. **6** Uppsala: Lundequistska Bokhandeln; pp. 210–215.

intermittent spring   A natural outflow point of underground water that sometimes dries up completely (see SPRINGS). Normally, such a spring is intermittent because the WATER TABLE upstream of the spring has fallen to or below the elevation of the spring, with the result that the hydraulic gradient leading to the spring is zero.   PWW

intermittent stream   A stream is classified as intermittent if flow occurs only seasonally when the water table is at the maximum level. The drainage network is composed of ephemeral, intermittent and perennial streams, and the network expands during rainstorms and extends to limits affected by antecedent conditions, especially antecedent moisture. Flow may occur along intermittent streams for several months each year but will seldom occur when the water table is lowered during the dry season.   KJG

interpluvial   A relatively dry phase interspersed with the wetter phases (pluvials) of the Pleistocene and Holocene. In many parts of the tropics the period at the end of the Late-Glacial Maximum (between *c.* 18,000 and 13,000 years ago) was dry enough to cause lake levels to fall and dune fields to expand.   ASG

interrill flow   The overland flow that moves as a thin layer, perhaps with some deeper and faster flow filaments within it, but which is not yet organized into small channels (RILLS) that are excavated by the erosive power that results from channelization. The zones of a soil surface carrying interrill flow expand and contract dynamically with run-off rate, as rills, too, enlarge or contract. Thus, the division cannot be applied rigorously in the spatial sense and is rather a distinction based on erosional processes. Interrill areas are the primary sediment sources, where raindrop splash is the major detachment agency. Rills, in contrast, are principally zones of transport and receive their sediment load from the influx of interrill flow.   DLD

Reading
Sharma, P.P. (1996) Interrill erosion. In M. Agassi (ed.) *Soil erosion, conservation, and rehabilitation*. New York: Marcel Dekker; chapter 7, pp. 125–152.

interrill processes   Those surface processes, particularly relating to the dislodgment and transport of soil and sediment particles, from areas of the land surface where overland flow has not yet been transformed to channelized flow in the form of rills or other small flow concentrations. In the interrill zone, the overland flow is generally shallower and slower moving than channelized flow. As a result, the forces exerted by the impact of rain droplets are more important for sediment movement than are drag forces exerted by the shallow fluid sheet flow. These latter forces, however, become dominant in deeper and faster channelized flow of the rill zone, while drop impacts have little or no effect. The physical processes that are active in the interrill zone during rainfall remain incompletely understood, but involve multiple interactions among drop impacts, shallow flow, and the range of particle sizes that is available for entrainment and redeposition. These processes exhibit marked spatial and temporal variability as a function of factors including rainfall intensity, drop size distribution, kinetic energy, the depth of surface ponding or overland flow and its speed of movement, the soil aggregate stability and shear strength, and the range of particle sizes present at the soil surface.   DLD

References
Kinnell, P.I.A. (2005) Raindrop-impact-induced erosion processes and prediction: a review. *Hydrological Processes*, **19**, 2815–2844. · Kinnell, P.I.A. (2012) Modeling of the effect of flow depth on sediment discharged by rain-impacted flows from sheet and interrill erosion areas: a review. *Hydrological Processes*, **27**, 2567–2578, doi: 10.1002/hyp.9363.

interstadial   There is as yet no universally accepted definition that differentiates an interstadial from an interglacial. However, it may be defined as a relatively short-lived period of lesser glaciation and relatively greater warmth and thermal improvement during the course of a major glacial phase. During such phases conditions were not of sufficient magnitude and/or duration to permit the development of temperate deciduous forest of the full interglacial type. Information about interstadial environment conditions have been obtained and assessed using faunal and floral evidence, the timing of which has been obtained mainly from radiocarbon dating where this technique permits. AP

Reading
Goudie, A.S. (1992) *Environmental change*, 3rd edition. Oxford: Clarendon Press. · Lowe, J.J. and Walker, M.J.C. (2015) *Reconstructing Quaternary environments*, 3rd edition. London: Routledge.

interstices Voids such as pores and fissures that occur within a rock. They can be classified according to their origin, shape and size. Primary interstices are those formed when the rock was created, such as intergranular pores in a sandstone, whereas secondary interstices are the result of later tectonic activity or weathering, such as fault planes and voids left by the differential corrosion of minerals. Interstices are most often small and interconnected, although large isolated interstices termed *vugs* also sometimes occur. Both small primary and large secondary interstices can be present simultaneously in a rock. For example, well-jointed and bedded sedimentary rocks, especially if rich in carbonate, may have a geometrical lattice of large interstices produced by solution along the fissures system and a fine intergranular POROSITY within the main body of the rock. PWW

intertropical convergence zone (ITCZ)

An equatorial zone of low pressure into which the trade winds converge. In an idealized picture of the atmosphere, the wind convergence corresponds to a pressure minimum and a maximum in cloudiness and rainfall. However, this picture is seldom realized, and the various minima and maxima are seldom collocated. Because of the ease of viewing cloudiness and rainfall from satellites, the ITCZ is often tracked via the cloudiness maximum. On satellite photographs, especially over the oceans, it appears as a narrow, well-defined cloud band. Occasionally, two ITCZs will be visible, especially in the eastern tropical Pacific Ocean shortly after the equinoxes. The split appears to be related to cool, upwelled water at the equator. The ITCZ meanders with the season. The extreme positions occur in February and August, when temperatures are highest in the respective summer hemisphere.

The concept of the ITCZ and accompanying rainfall is most valid over the oceans because the trade winds are largely confined to ocean regions. Although the ITCZ and its seasonal migration have been used to explain the seasonality of rainfall over continents, such as Africa and Asia, the origin of rainfall over continental areas of the tropics is quite different. Over North Africa, for example, a clear wind convergence zone exists and migrates latitudinally with the

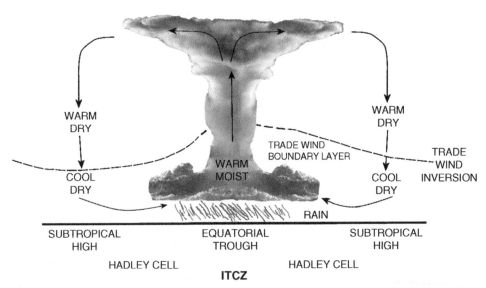

(a) Classic view of the ICTZ as produced by the convergence of the trade winds into the equatorial trough, creating rising motion and precipitation. The ITCZ is viewed as the rising branch of the Hadley cell, with the sinking branch coinciding with the subtropical highs of both hemispheres.

(b) Photograph showing cloud development associated with the ITCZ over the Atlantic Ocean.

seasons. However, maximum rainfall occurs well to the south, and rainfall variations are not strongly linked to changes in the ITCZ. For this reason, some now use terms such as 'tropical rain belt' (e.g. Nicholson, 2009) or rain band (e.g. Zhang *et al.*, 2006) to refer to what was once called the ITCZ. Over equatorial Africa, a well-developed ITCZ associated with precipitation is not evident.   SEN

References and reading
Barry, R.G. and Chorley, R.J. (2003) *Atmosphere, weather and climate*, 8th edition. Routledge. · Nicholson, S.E. (2009) A new picture of the structure of the 'monsoon' and land ITCZ over West Africa. *Climate Dynamics*, **32**, 1155–1171. · Zhang, C. (2001) Double ITCZs. *Journal of Geophysical Research*, **106**, 11785–11792. · Zhang, C., Woodworth, P. and Gu, G. (2006) The seasonal cycle in the lower troposphere over West Africa from sounding observations. *Quarterly Journal of the Royal Meteorological Society*, **132**, 2559–2582.

intrazonal soil   A soil group comprising well-developed soils, the main characteristics of which can be attributed more to local factors such as relief, drainage or parent material than to climate factors.

intrenched meander   A meander or bend in a river channel that has become incised into the surrounding landscape as a result of local tectonic uplift (see MEANDERING).

intrinsic permeability (or specific permeability)   A measure of the capacity of a rock or soil under natural conditions to transmit fluids. It depends upon the physical properties of the porous medium, such as pore size, shape and distribution. It is measured in square metres or darcy (d) units ($1\,\mathrm{d} \approx 10^{-8}\,\mathrm{cm}^2$). Intrinsic permeability k is related to HYDRAULIC CONDUCTIVITY K (which takes into account the physical properties of the liquid as well as the rock) as follows:

$$K = \frac{k\rho g}{\mu}$$

where ρ, the mass density, and μ, the dynamic viscosity, are functions of the fluid alone, and g is acceleration due to gravity (Freeze and Cherry, 1979).   PWW

Reference
Freeze, R.A. and Cherry, J.A. (1979) *Groundwater*. Englewood Cliffs, NJ: Prentice-Hall.

introductions, ecological   Introductions, usually deliberate, of an organism into new regions lying outside the range of its natural occurrence, by which it is hoped to bring about some specific ecological condition or control in the receiving areas or habitats. Although any introduction of a species, accidental or otherwise, into a new environment is likely to have considerable ecological repercussions, the concept of ecological introductions is mostly used to describe carefully planned introductions made with specific ecological intentions in mind. For example, where aquatic weeds are an increasing problem, water authorities may decide to introduce the herbivorous grass carp (*Ctenopharyngdon idella*) to bring them under control. In many instances, the aim is to create or recreate some particularly effective food chain or ecological relationship.

A famous example of such an introduction is afforded by the deliberate spread of an Australian ladybird, *Novius (Vedalia) cardinalis*, to California. The fluted or cottony cushion scale insect (*Icerya purchasi*), another native of Australia, appeared in California to threaten the famous citrus groves of the area. Within a few years this threat had been overcome by the numerous descendants of the 139 specimens of the ladybird introduced from Australia, where it was a natural enemy of the fluted scale insect and from where it had been introduced to California to control the alien pest. As Charles Elton (1958) has written, this was a 'miracle of ecological healing' in which 'Australia administered the poison, but it also supplied the antidote'. The success was repeated in many other countries where the fluted scale became a problem; for example, New Zealand and Egypt. It is a perfect example of an ecological introduction in which one alien species is controlled by a natural predator from its own ecosystem, thus using and re-establishing a natural ecological chain. Another pest, this time from the plant kingdom, required the introduction of the cinnabar moth, which cleared vast areas of the prickly pear (*Opuntia* species) in Australia.

In most cases of predator control, the number of introductions need not be great, partly because of the speed of breeding, but also because of the fundamental principle of the ecological pyramid, in which predators are virtually always scarcer than their prey. Many ecological introductions fail, however, and some may even go badly wrong, leading to new and more serious problems, including ecological explosions, the cure proving worse than the disease. Finally, there have been attempts to reintroduce formerly native animals and plants into their old habitats in order to try to establish the past ecological order; for example, the wolf, a top carnivore, in certain forest areas in the west of Germany. (See also ALIENS and BIOLOGICAL CONTROL.) PAS

Reading and Reference

Elton, C.S. (1958) *The ecology of invasions by animals and plants*. London: Methuen. · Samways, M.J. 1981: *Biological control of pests and weeds*. London: Edward Arnold. · Simmons, I.G. (1979) *Biogeography: natural and cultural*. London: Edward Arnold. · Van Emden, H.F. (1974) *Pest control and its ecology*. London: Edward Arnold.

intrusion A mass of igneous rock that has penetrated older rocks through cracks and faults before cooling. The process of emplacement of such a mass of rock.

invasive species The introduction and spread of species from one area to another by humans sometimes leads to the introduced species spreading explosively and becoming invasive. This may be because in the new area they do not have some of the organisms that control them in their native environment. In Britain, for instance, many elm trees died in the 1970s because of the accidental introduction of the Dutch elm disease fungus that arrived on imported timber. The American chestnut *Castanea dentata* was, following the introduction of the chestnut blight fungus in ornamental nursery material from Asia late in the 1890s, almost eliminated throughout its natural range in less than 50 years. In Western Australia, the great jarrah forests have been invaded and decimated by an introduced root fungus, *Phytophthora cinnamomi*. The spread of the disease within the forests was facilitated by road building, logging and mining activities that involved movement of soil or gravel containing the fungus. More than 3×10^6 ha of forest have been affected. In South Africa, the unique and species-rich *fynbos* heathland has been threatened by the spread of introduced pines and various types of Australian acacia. Significant declines have taken place in native species richness (Gaertner *et al.*, 2009). In the USA, kudzu (*Pueraria montana*) has been a particularly difficult invader. It was introduced into the USA from Japan in 1876 as an ornamental and forage crop plant. From 1935 to the mid-1950s it was recommended as an effective means of reducing soil erosion. Unfortunately, this large vine, with massive tap roots and the ability to spread at about 30 cm per day, now smothers large expanses of the southeastern USA. No less dangerous has been the grass *Imperata cylindrica*. Originating from East Asia, it was introduced to Alabama in the mid-1900s as a forage crop and soil stabilizer. It has since invaded nearly 500,000 ha within the USA and over 500×10^6 ha worldwide (Holzmueller and Jose, 2011). Introduced animals can also broaden their ranges and increase in numbers explosively, as the example of the rabbit in Australia has shown. In terms of the number of countries where they are considered to be invasive, the Global Native Species Information Network (2011) (http:///www.niss.org) listed rats (*Rattus rattus* and *Rattus norvegicus*), cats (*Felis catus*), goats (*Capra hircus*), American mink

(*Mustela vison*) and the house mouse (*Mus musculus*) as the most widespread invasive animals.

Oceanic islands have often been particularly vulnerable to plant invasions (Sax and Gaines, 2008). Their simple ecosystems inevitably have diminished stability, and introduced species often find that the relative lack of competition enables them to broaden their ecological range more easily than on the continents. Many successful invasive plants have escaped the natural enemies that hold them in check, freeing them to utilize their full competitive potential. Moreover, because the natural species inhabiting remote islands have been selected primarily for their dispersal capacity, they have not necessarily been dominant or even very successful in their original continental setting. Therefore, introduced species may prove more vigorous and effective. There may also be a lack of indigenous species to adapt to conditions such as bare ground caused by humans.

Invasive plants pose a number of threats to natural ecosystems. These have been discussed by Cronk and Fuller (1995) and Holzmueller and Jose (2011):

• Replacement of diverse systems with single-species stands of aliens, leading to a reduction in biodiversity; for example, as where Australian acacias have invaded the fynbos in South Africa.
• Direct threats to native faunas by producing a change of habitat.
• Alteration of soil chemistry. For example, the African *Mesembryanthemum crystallinum* accumulates large quantities of salt. Some plants exude allelopathic chemicals (weapons) that are inhibitory to native species.
• Modification of geomorphological processes, especially rates of sedimentation and movement of mobile land forms (e.g. dunes and salt marshes).
• Plant extinction by competition for light, nutrients, and so on.
• Change in fire regimes.
• Alteration of hydrological conditions (e.g. reduction in groundwater levels caused by some species having high rates of transpiration).

There are many examples of ecological explosions (bioinvasions) caused by the creation of new habitats by humans. Some of the most striking are associated with the establishment of artificial lakes in place of rivers. Riverine species that cannot cope with the changed conditions tend to disappear, while others that can exploit the new courses of food, and reproduce themselves under the new conditions, multiply rapidly in the absence of competition. Vegetation on land flooded as the lake waters rise decomposes to provide a rich supply of nutrients that then allow explosive outgrowth of organisms. In particular, floating plants may form dense mats of vegetation, which in turn support large populations of invertebrate animals, may cause fish deaths by deoxygenating the water and can create a serious nuisance for turbines, navigators and fishermen.   ASG

Reading and References
Catford, J.A., Vesk, P.A., Richardson, D.M. and Pyšek, P. (2012) Quantifying levels of biological invasion: towards the objective classification of invaded and invasible ecosystems. *Global Change Biology*, **18**, 44–62. · Cronk, Q.C.B. and Fuller, J.L. (1995) *Plant invaders*. London: Chapman & Hall. · Gaertner, M., Breeyen, A.D., Hui, C. and Richardson, D.M. (2009) Impact of alien plant invasions on species richness in Mediterranean-type ecosystems: a meta-analysis. *Progress in Physical Geography*, **33**, 319–338. · Holzmueller, E.J. and Jose, S. (2011) Invasion success of cogongrass, an alien C4 perennial grass, in the southeastern United States: exploration of the ecological basis. *Biological Invasions*, **13**, 435–442, doi: 10/1007/s10530-010-9837-1. · Sax, F.F. and Gaines, S.D. (2008) Species invasions and extinction: the future of native biodiversity on islands. *Proceedings of the National Academy of Sciences of the United States of America*, **105**, 11490–11497.

inversion of temperature   An increase of temperature with height, the inverse of the normal decrease of temperature with height that occurs in the TROPOSPHERE. Temperature inversion layers are very stable and greatly restrict the vertical dispersion of atmospheric pollutants. They can form in several different ways. (1) Radiative cooling of the air near the ground at night. These inversions are very common on clear nights, but dissipate rapidly after sunrise. (2) Advective cooling of warm air passing over a cold surface. These persistent inversions may be accompanied by thick fog if the air is moist. (3) A cold air mass undercutting a warm air mass along a FRONT. These frontal inversions act as an invisible barrier between the two AIR MASSES. (4) Radiative heating in the upper atmosphere. The STRATOSPHERE and thermosphere are examples of this type of inversion. Thunderstorm clouds and pollutants rarely penetrate far into the stratosphere because of its great stability. (5) Descent and ADIABATIC heating of air from the upper troposphere. These subsidence inversions are most common near and to the east of ANTICYCLONES. They may be very intense and persist for days, trapping noxious pollution in a thin air layer near the ground. WDS

Reading
Battan, L.J. (1979) *Fundamentals of meteorology.* Englewood Cliffs, NJ: Prentice-Hall.

inverted relief The condition resulting from the erosion of areas of high relief, such as anticlines, to produce low-lying areas, such as valleys, which simultaneously results in the originally low-lying inclines becoming hills. Equally, the deposition of resistant lag gravels, lava streams or duricrusts in river valleys may cause them to be left upstanding in a subsequent phase of erosion.

involution The refolding of large nappes, producing complex structures of more recent nappes within old nappes. Also a term synonymous with cryoturbation.

ion concentrations Studies of the DISSOLVED SOLIDS content of precipitation, run-off and water from other phases of the hydrological cycle will frequently consider the concentrations of individual constituents. With the exception of dissolved silica and small quantities of dissolved organic matter, the dissolved material is largely dissociated into charged particles or ions, and water analyses are therefore generally expressed in terms of concentrations of individual ions (milligrams per litre or milliequivalents per litre). The major cations (positively charged ions) in natural waters are Ca^{2+}, Mg^{2+}, Na^+ and K^+; and the major anions (negatively charged ions) are HCO_3^{2-} and CO_3^{2-}, SO_4^{2-}, Cl^-, F^-, and NO_3^-. DEW

ionic wave technique See DILUTION EFFECT and DISCHARGE.

ionosphere Region above a height of about 50 km in which the gas density is so low and the temperature so high that positive and negative ions can move with some degree of independence. Electric currents so generated cause daily fluctuations in the Earth's magnetic field, affect the propagation of radio waves and respond to solar flares. Other planets seem to have similar layers at similar values of pressure. JSAG

Reading
Goody, R.M. and Walker, J.C.G. (1972) *Atmospheres.* Englewood Cliffs, NJ: Prentice-Hall.

IPCC The Intergovernmental Panel on Climate Change (IPCC) was established in 1988 by the World Meteorological Organization (WMO) and the United Nations Environment Programme (UNEP) to assess the available scientific, technical and socio-economic information in the field of climate change. The IPCC is organized into three interrelated working groups: Working Group I concentrates on the science of the climate system, Working Group II focuses on impacts of climate variability and response options, and Working Group III deals with economic and social dimensions of climate change. Approximately 2500 scientists from throughout the world are directly involved with the IPCC as editors, lead authors, contributing authors and reviewers. The IPCC has released five assessment reports, the first in 1990 and the most recent in 2014 in three parts (The physical science basis; Impacts, adaptation and vulnerability; Mitigation of climate change); the IPCC continues to produce technical papers and develop methodologies (e.g. national greenhouse gas inventories) for use by parties to the Climate Change Convention. RCB/DSGT

iridium layer Iridium (Ir) is an element (atomic number, 77; atomic weight, 192.2) that occurs in concentrations averaging about 0.5 parts per billion (ppb) by weight in the Earth's crust, and about 2 ppb in the universe. In carbonaceous meteorites, the abundance averages 550 ppb. The iridium layer is a clay deposit found at the Cretaceous–Tertiary boundary, where concentrations are commonly about 50 ppb. There are two plausible sources for the excess iridium: large-scale volcanic eruption, associated with the formation of the Deccan Traps, for example, or a massive extraterrestrial impact. It is widely assumed that the Chicxulub impact is the source of the iridium layer. DJS

Reading
Alvarez, L.W., Alvarez, W., Asaro, F., and Michel, H.V. (1980) Extraterrestrial cause for the Cretaceous/Tertiary extinction. *Science*, **208**, 1095–1108.

irrigation This is 'the practice of applying water to the soil to supplement the natural rainfall and provide moisture for plant growth' (Wiesner, 1972: 23). The frequency and method of water delivery to the area to be irrigated can vary significantly. It may occur only once during a crop's growth cycle (e.g. to stimulate germination) or every day – more than once in some cases if, for example, the crop is being grown in glass houses. Barrow (1987) notes that there are three basic irrigation strategies:

1 complete irrigation, in low rainfall areas and most crop water requirements must be met;

2 supplementary irrigation, when rainfall is adequate but improved crop yields can be gained by irrigating;

3 protective irrigation, where there is a risk of crop quality being damaged by a rainfall shortage.

Irrigation has probably existed for thousands of years, and many qanats (underground tunnel transfer) systems in the Middle East may be very old. Many different irrigation methods exist. Water may be derived from perennial rivers and transported by channels and canals to areas of need. Groundwater may be pumped to the surface and used locally by an individual farmer or collectively. Water may also be harvested, either from slopes surrounding a farmed area by constructing bunds to catch and divert run-off, or by building dams on seasonal water courses. In many situations it is common for as little as 20% of the water involved to reach the crop that it is intended for. To this end, sub-surface and trickle irrigation systems (Agnew and Anderson, 1992) may offer a number of advantages.

The most appropriate method of irrigation for a particular situation depends on a range of factors, including water availability, crop type, soil type and the seasonality of climate. Whatever method is used, environmental damage can result from excessive watering. SALINIZATION is a particular effect in this respect, especially in DRYLANDS areas or where centre pivot irrigation is employed. Water tables may be lowered, too, by excessive pumping of groundwater. Large-scale irrigation schemes may cause especial problems, including in the context of DESERTIFICATION. Especial concern on the direct and indirect consequences of irrigation have been associated with schemes in the Aral Sea basin and on the High Plains of the USA. It is estimated that all of Egypt's agriculture is irrigation dependent, 60% of Japan's and 50% of India's.   DSGT

References

Agnew, C. and Anderson, E. (1992) *Water resources in the arid realm.* London: Routledge. · Barrow, C.J. (1987) *Water resources and agricultural development in the tropics.* London: Longman. · Wiesner, A. (1972) *The role of water in development: an analysis of principles of comprehensive planning.* New York: McGraw-Hill.

island arc   A chain of islands, mostly volcanic in origin, with a characteristic arcuate plan-form, rising from the deep ocean and associated with an ocean trench. Island arcs, such as those of the southwestern Pacific, are generally located fairly near to continental masses and their curvature is typically convex towards the open ocean. Some island arcs comprise an inner arc of active volcanoes and an outer arc of non-volcanic origin formed from sediments thrust up from the ocean floor. According to the PLATE TECTONICS model, island arcs form as a result of the volcanism induced by subduction of oceanic LITHOSPHERE.   MAS

Reading

Karig, D.E. (1974) Evolution of arc systems in the western Pacific. *Annual Review of Earth and Planetary Sciences,* 2, 51–75. · Sigimura, A. and Uyeda, S. (1973) *Island arcs: Japan and its environs.* Amsterdam: Elsevier.

island biogeography   In general terms, the study of the distribution and evolution of organisms on islands; more narrowly, the examination of MacArthur and Wilson's equilibrium theory of island biogeography. Island floras and faunas have always fascinated biogeographers and biologists, and all the great nineteenth-century biologists were intrigued by the highly distinctive plants and animals and geographically isolated environments they found on these 'natural laboratories'. In a famous lecture to the British Association given in 1866, J. D. Hooker outlined the main characteristics and origins of island floras, while Darwin wrote of his famous Galápagos finches (the Geospizinae) and in 1880 A. R. Wallace published his classic book entitled *Island life or the phenomena and causes of insular faunas and floras.*

In the main, island biota are more polar in character than their neighbouring mainland counterparts and there are usually fewer species than on a similar-sized continental area. Moreover, the species mix tends to be disharmonic, being different from that on the mainland and often seeming out of harmony ecologically. For example, there are sometimes no top carnivores. The mix is usually an assemblage of taxa noted for their capacity to accomplish long-range dispersal and migration. In many instances the island progeny of these able dispersers have lost their dispersal ability, a phenomenon well exemplified by the ill-fated dodo (*Raphus cucullatus*) of Mauritius, a large flightless bird related to the pigeon. The small populations on islands are subject to a range of distinctive ecological pressures, and under these conditions evolution is accelerated, with new forms developing through adaptive radiation and hybridization.

Islands are therefore noted for their endemic organisms and for the large and abnormal percentage of their floras and faunas that are endemic. For example, some 45% of the birds of the Canaries are endemic and no less than 90% of the flora of the Hawaiian Islands, the most isolated of all floristic regions. Islands are also the homes of relict organisms that have survived there but have become extinct elsewhere. This is especially the case on islands that were once part of a continental system, as with Crete, a former remnant of an old mountain system that connected the Balkans with southern Anatolia. Extinction, however, is also known in island biota, and it is the balance between immigration and extinction that is at the heart of the now much discussed equilibrium theory of island biogeography.

This theory was first published by R. H. MacArthur and E. O. Wilson in 1963 as the equilibrium theory of island zoogeography, but was widened to include plants in their book of 1967, simply entitled *The theory of island biogeography*. The theory was mainly stimulated by thoughts on the distributions found across oceanic islands in the Pacific. The core of the theory is simple. It is argued that the number of species on an island is determined by a balance between immigration (which is a function of the distance of the island from the mainland) and extinction (which is a function of island area). The theory assumes a pool of species P, which can immigrate to the island and which is the number of species on the neighbouring landmass.

The theory can be presented in the form of an equation:

$$S_t = S_t + I + V - E$$

where S_t is the number of species at time t, I is the number of immigrants by time $t + 1$, V is the number of new species evolving in situ on the island and E is the number of extinctions.

This simple and seemingly logical theory has come under a wide range of criticism. First, it deals only with the number of species and not the number of individuals of the species on the island. In other words, it ignores population numbers. Second, it does not really deal with evolution, although in chapter 7 of the book the theory is tentatively extended to include this process. Third, it appears to ignore historical factors that might, for example, mean that many organisms are relicts, subject only to extinction and with no potential for immigration. Fourth, the theory lumps together all species and treats them as functioning in a similar manner. Fifth, there are serious problems in defining both immigration and extinction, and many argue that it is unacceptable to make immigration solely a function of distance. Finally, of course, it must not be assumed that all islands are in

'equilibrium', for in many this has yet to be reached, even in terms of the theory. Yet, despite all these criticisms, the theory has stimulated a new and invigorated interest in island biogeography.

It should also be noted that the theory has received a much wider application than its use on oceanic islands and that it has been related to the fauna and flora of biological 'islands' on continental areas, such as the remaining relicts of tropical rain forest. It is now being used to help determine the minimum size of viable conservation areas in which the local populations will be able to maintain themselves in some form of equilibrium. In general, it is true to say that the main tenets of island biogeography also apply to a wide range of such biological islands, including isolated mountain tops, ponds and lakes, and tracts of woodland surrounded by agriculture.

One thing is certain, namely that the arrival of humans on many isolated islands has seriously disrupted and altered their ecosystems. Extinction rates have increased markedly with their presence, especially where they have created ecological disharmony by the introduction of alien species (see ALIENS). The Atlantic island of St Helena, for example, has seen the demise of its endemic St Helena ebony (*Trochetia melanoxylon*), which was destroyed by goats, first introduced in 1513, and by the deforestation of the island for fuel. All that is left is a barren landscape with a few relict fragments of the original biota persisting on cliffs and ridges. On the other hand, many species introduced by humans have themselves begun to change and form distinctive island races. It is believed that the special forms of the long-tailed field mouse (*Apodemus sylvaticus*) found on the Scottish and Scandinavian islands have developed from ancestors brought to these scattered locations by the Vikings. The house mouse (*Mus musculus*) on the island of Skokholm off Wales is some 30% different in form from the mainland populations in Pembrokeshire. Yet it was probably only introduced to the island around 1900 by rabbit catchers and farmers. Thus, islands continue to be wonderful laboratories for the study of immigration, extinction and evolution, and they will remain at the centre of biogeography for a long time to come. (See also ALIENS, ENDEMISM, EXTINCTION, REFUGIA and SPECIES–AREA CURVE.)   PAS

Reading
Gilbert, F.S. (1980) The equilibrium theory of island biogeography: fact or fiction? *Journal of Biogeography*, 7, 209–235. · MacArthur, R.H. and Wilson, E.O. (1967) *The theory of island biogeography*. Princeton, NJ: Princeton

University Press. · Whittaker, R.J. and Fernández-Palacios, J.M. (2007) *Island biogeography: ecology, evolution, and conservation*. Oxford: Oxford University Press.

islands Land surfaces totally surrounded by water and smaller in size than the smallest continent (Australia). Oceanic islands are built up from the ocean floor and are part of the basal structure, not attached to continents, as in the example of the Hawaiian group. Continental islands are part of the neighbouring continental geological structure, as exemplified by the British Isles. The dispersal and colonization of plants and animals to islands is related to the distance from the species source. Hence, oceanic islands tend to be occupied by a smaller number of species, highly adapted to the available HABITAT or NICHE and frequently endemic.   PAF

Reading
Gorman, M. (1979) *Island ecology*. London: Chapman & Hall. · MacArthur, R.H. and Wilson, E.O. (1967) *The theory of island biogeography*. Princeton, NJ: Princeton University Press. · Menard, H.W. (1986) *Islands*. New York: Scientific American Books. · Nunn, P. (1993) *Oceanic islands*. Oxford: Blackwell.

isochrones Lines joining points on the Earth's surface at which the time is the same. Lines joining points that experienced a seismic wave at the same time.

isocline A fold that is so pronounced that the strata forming the limbs of the fold dip in the same direction at the same angle.

isolation, ecological The ecological or habitat separation of one population from another so that interbreeding is normally prevented, even though the organisms involved may have overlapping geographical ranges (see SYMPATRY). Thus, although two closely related organisms live in the same region, interbreeding does not take place because they occupy different habitats. A classic example is afforded by two sympatric African species of *Anopheles* mosquito, the one, *A. melas*, confined to brackish water habitats, the other, *A. gambiae*, to freshwater.   PAS

Reading
Ross, H.H. (1974) *Biological systematics*. Reading, MA: Addison-Wesley.

isopleths Lines drawn on maps connecting points that are assumed to be of equal value (e.g. contours on a topographical map). Among specific types of isopleth are those shown in the Table 3.

Table 3 Some isopleth types

Type	Connects up points of equal
Isobar	Barometric pressure
Isobase	Uplift or subsidence during a specified time period
Isobath	Distance beneath the surface of a water body
Isobathytherm	Temperature at a given depth below sea level
Isocheim	Mean winter temperature
Isoflor	Floral character
Isoglacihypse	Altitude of the firn line
Isohaline	Salinity in the oceans
Isohel	Recorded sunshine hours
Isohyet	Rainfall amount
Isomer	Mean monthly rainfall expressed as a percentage of the mean annual rainfall
Isoneph	Cloudiness
Isonif	Snow depth
Isopach	Rock-stratum thickness
Isoryme	Frost intensity
Isotach	Wind or sound velocity
Isothere	Mean summer temperature
Isotherm	Temperature
Isothermobath	Seawater temperature at a given depth

isostasy The condition of hydrostatic equilibrium between sections of the LITHOSPHERE with respect to the underlying MANTLE. Units of the comparatively rigid outer layer of the Earth in effect 'float' in the more mobile and denser material at greater depth. Isostatic adjustment was originally thought to occur by vertical movements of the crust with respect to the underlying MANTLE, but some isostatic models now assume that adjustment occurs through the movement of the lithosphere, which comprises not only the crust but also an underlying zone of comparatively rigid mantle.

Two models of isostasy were proposed during the nineteenth century. G. B. Airy noted that the gravitational attraction of the Himalayan mountains was less than could be explained if the range was simply above a radially homogeneous crust and mantle. The gravity anomaly was explained by Airy as a result of crustal blocks of equal density but different thicknesses.

The thickest blocks form the highest topography and are supported by roots of light crust, which have displaced the denser underlying mantle. At a depth equal to, or greater than, the thickness of the crust, the pressure in the mantle is constant and hydrostatic equilibrium is attained. An alternative model proposed by J.H. Pratt attempted to explain isostasy by variations in the density rather than the thickness of crustal blocks. In this model the crust is assumed to be of equal thickness and areas of high elevation are associated with low-density crust, which is more buoyant with respect to the underlying mantle than adjacent areas of denser crust. Although these models make unrealistic assumptions about the fluid nature of the mantle and the ability of crustal blocks to move independently, they describe adequately the gross variations in gravitational attraction over the Earth's surface. The crustal thickness model is particularly applicable to most continental mountain systems, whereas the density model provides a more adequate explanation of the relief of the mid-ocean ridges.

While the lithosphere has a tendency to attain isostatic equilibrium, several factors may prevent this from occurring. For example, temperature and density variations associated with convection in the mantle can lead to marked gravity anomalies indicative of a lack of isostatic equilibrium. Another factor is the rigidity of the lithosphere, which means that variations in surface loading over a small area may not promote isostatic compensation, while compensations to a really extensive change in loading may produce vertical movements well beyond the zone of loading itself. This is especially well illustrated by glacial isostasy, the response of the lithosphere to the loading and unloading of the surface by ice. Rates of uplift estimated from raised shorelines and other features may exceed $20 \, mm \, a^{-1}$. Much slower rates of isostatic compensation are associated with erosional unloading of the continents, although these rates are sustained over much longer periods and are consequently an important factor in continental uplift. (See GLACIOISOSTASY and RHEOLOGY.)   MAS

Reading

Andrews, J.T. (ed.) (1974) *Glacial isostasy*. Stroudsburg, PA: Dowden, Hutchinson & Ross. · Lyustikh, E.N. (1960) *Isostasy and isostatic hypotheses*. New York: American Geophysical Union Consultants Bureau. · Mörner, N.-A. (ed.) (1980) *Earth rheology, isostasy and eustasy*.

Chichester: John Wiley & Sons, Ltd. · Smith, D.E. and Dawson, A.G. (1983) *Shorelines and isostasy*. London: Academic Press.

isotope   A form of an element that, while always having the same number of protons in the nucleus, has another form or forms with differing numbers of neutrons. There are three main types of isotope.

Radiogenic or radioactive isotopes (see also RADIOISOTOPES) are unstable and decay to a more stable isotopic form over predictable (typically very long) periods of time. These isotopes are utilized extensively in geological dating.

Stable isotopes, also called environmental isotopes, are chemically stable and do not undergo transformation to other isotopic forms. Though chemically the same, their physical properties are slightly different. These differences result in heavier isotopes having lower mobility (which means that heavier molecules have lower diffusion velocities and a lower collision frequency with other molecules) and generally (although not always) having higher binding energies.

During natural cycles of evaporation, condensation, photosynthesis and diagenesis a natural fractionation takes place that is related to temperature and other factors. The value of the ratio $^{18}O/^{16}O$, for example, is used in ice and marine core determinations of palaeotemperature (see also OXYGEN ISOTOPE). Hydrogen/deuterium (dD) variations are also used for a similar purpose.

Cosmogenic isotopes are produced by the interaction of extraterrestrial cosmic rays and other atoms, in the atmosphere and in rocks (see COSMOGENIC ISOTOPE).   SS/DLD

References

Allegre, C.J. (2008). *Isotope geology*. Cambridge University Press. · Dickin, A.P. (2005) *Radiogenic isotope geology*, 2nd edition. Cambridge University Press. · Michener, R. and Lajtha, K. (2007) *Stable isotopes in ecology and environmental science*, 2nd edition. Blackwell.

J

jet stream   A distinct wind maximum in the atmospheric circulation that is characterized by a core of high winds and strong wind shear surrounding the core. The speed and degree of shear required to merit the term 'jet' depend on the spatial scale of the flow, but there is no universal agreement on the criteria required.

The largest jet streams reside in the upper troposphere. They result from large-scale, latitudinal temperature gradients. The association with temperature gradients is explained by a concept termed the 'thermal wind', an acceleration of geostrophic wind speed with height that results from horizontal temperature gradients. Because pressure diminishes with height more rapidly in cold air than warm air, a temperature gradient at one level is associated with a pressure gradient further up. The pressure gradient controls the wind speed.

For this reason, the major jet streams are collocated with regions of strong latitudinal temperature gradients. If cold air lies poleward of warm air, as is generally the case, the westerly speed increases with height. Thus, the equator-to-pole temperature contrast produces the well-known westerly jet streams of the higher latitudes. When the direction of temperature contrast is reversed, easterly speed increases with height. This is the case over Asia, where the Tibetan Plateau provides a heating source poleward of the comparatively cool Indian Ocean. This gradient results in the Tropical Easterly Jet (TEJ), centred over the western equatorial Indian Ocean.

Two upper tropospheric westerly jets exist in each hemisphere, one in the higher latitudes and one in the subtropics. The higher latitude jet is associated with the strong temperature gradient in the vicinity of the Polar Front and hence is often called the Polar Front Jet (PFJ). Like the Polar Front, it migrates around the hemisphere. In both hemispheres the core of the PFJ is generally around 250 mb and average core speeds are $\sim$22 m s^{-1}. Instantaneous speeds can reach at least $\sim$112 m s^{-1}.

The Subtropical Jet (STJ) tends to merge with the PFJ in summer, but they are distinct in winter, when the latitudinal gradient has two maxima. One maximum lies near the Polar Front and the other lies near the area of continent–ocean contrast in the subtropics. In the northern hemisphere the STJ has three geographical maxima, corresponding to areas of strong thermal contrast between land and water. These areas are geographically fixed, so that the position of the STJ is relatively stable, in contrast to the migrating PFJ. Other factors contribute as well to the development of the STJ, such as outflow aloft from regions of convection and from the Hadley circulation cells. The STJ is roughly collocated with the surface position of the descending branch of the Hadley cells. The core of the northern hemisphere STJ is at about 200 mb, with speeds of about 40–50 m s^{-1}. The STJ of the southern hemisphere is evident over Australia, where its core lies at roughly 27°S in winter and its mean speed exceeds 50 m s^{-1} over South America.

The TEJ is a prominent summer feature with maximum winds of about 30–40 m s^{-1} near 150 mb. A similar feature exists in the southern hemisphere, in the austral summer. The TEJ extends from eastern Asia to West Africa, centred at about 0° of latitude.

Numerous other jet streams, smaller in scale and magnitude, lie in the mid and lower troposphere. They strongly affect regional climates in that they variously transport moisture, intensify aridity, create mesoscale circulations, modify the diurnal cycle of precipitation, or trigger or promote convective disturbances.

Mid-tropospheric easterly jet streams exist in the lower latitudes, but only the African Easterly Jet (AEJ) has been well studied. It is a consequence of the temperature contrast between the warm Saharan air and the cooler air to the south over the coastal forests and the Gulf of Guinea in the Atlantic. A southern hemisphere counterpart appears in the southern subtropics of Africa during some seasons, a consequence of the dry-season temperature contrast between the woodland to the south and the equatorial rain forest. Mid-level easterly jets exist also near the equator

The Dictionary of Physical Geography, Fourth Edition. Edited by David S.G. Thomas.
© 2016 John Wiley & Sons, Ltd. Published 2016 by John Wiley & Sons, Ltd.

over western South America and the far eastern Atlantic and over northwestern Australia.

Far more numerous are the low-level jets. Although both weaker in magnitude and smaller in spatial scale than the mid- and upper-level jets, these features can be locally very important. Some, particularly coastal jets, result mainly from thermal gradients. Others are influenced by orographic features, such as wind being channelled through a narrow gap in the terrain or constrained by high mountains. These generally lie in the interior of continents.

Low-level jets with cores at elevations ranging from 500 to 2000 m, are apparent along several coasts and in some continental interiors. Coastal jets are common along cold-water coasts where upwelling occurs: Peru, Somalia, California, north central Chile, the Benguela coast, the equatorial eastern Pacific, the Caribbean and the Gulf of California. The most well-known example of an interior jet is that in the Great Plains of the central USA. Other interior jets include the Turkana Jet of East Africa and the South American Low Level Jet east of the Andes. Similar low-level jets are found in the Iranian desert, over Australia and near the Bodélé Depression of North Africa.

Jet streams play important roles in the global climate system. The upper level jets of the subtropical and mid-latitudes are important for the development and steering of mid-latitude cyclones. The subtropical jets further play a role in transmitting the influence of the tropics to the mid-latitudes. The large-scale tropical jet streams influence the intensity and location of convection and provide steering for tropical disturbances. The TEJ, for example, shows considerable variation within the Indian monsoon, and its fluctuations are highly correlated with monsoon convection. Both the low-level Bodélé Jet and the AEJ of the northern hemisphere transport massive amounts of dust from the African continent to the Atlantic and even into the Caribbean. Other low-level jets, especially the coastal jets, enhance aridity. Areas of the Benguela and Peruvian coasts where the jets are best developed are virtually rainless.　SEN

Reading
Nicholson, S.E. (2011) *Dryland climatology*. Cambridge: Cambridge University Press; chapter 4.

joint probability estimates　The joint probability indicates the chances of two or more events occurring concurrently. The magnitude of the joint probability is a function of the probability with which each event occurs and on whether or not the event occurrences are dependent on each other. If there are two events and the occurrence of one has no effect on the probability that the other will occur, the events are independent and the joint probability is calculated as the product of the individual event occurrence probabilities. If the events are not independent the calculation of the joint probability needs to include the conditional probability of one event occurring given the occurrence of the other event.

Joint probability estimates are widely used in support of activities such as flood management, which focus on a variable such as the overall water level that is a function of multiple source variables such as wave height, sea level and river flow. The joint probability that the set of source variables may simultaneously take on high values, and hence act to form a situation in which flooding may occur, is a key method to inform flood risk management.　GF

Reading
Hawkes, P.J. (2005) Use of joint probability methods in flood management. A guide to best practice. R&D Technical Report FD2308/TR2. London: Defra.

jökulhlaup　An Icelandic term for catastrophic drainage of a subglacial or ice-dammed lake. The lake may build up seasonally or over several years only to drain in a matter of hours when conditions are suitable for meltwater to open up tunnels in the glacier, mainly by frictional heating. In Iceland, some jökulhlaups may be triggered by volcanic activity.　DES

Reading
Björnsson, H. (1992) Jökulhlaups in Iceland: prediction, characteristics and simulation. *Annals of Glaciology* **16**, 95–106. · Björnsson, H. (2011) Understanding jökulhlaups: from tale to theory. *Journal of Glaciology*, **56**, 1002–1010. · Larsen, M., Tøttrup, C., Mätzler, E., *et al.* (2013) A satellite perspective on jökulhlaups in Greenland. *Hydrology Research*, **44**, 68–77.

juvenile water　Water that originates from the interior of the Earth and has not previously existed as water in any state. Consequently, it has not previously participated in the hydrological cycle. The term was coined by Meinzer (1923), who contrasted juvenile with meteoric water (see METEORIC WATER/GROUNDWATER).　PWW

Reference
Meinzer, O.E. (1923) *Outline of ground-water hydrology: with definitions*. US Geological Survey water-supply paper 494. Washington, DC: US Government Printing Office.

K

K- and r-selection   See r- AND K-SELECTION.

K-cycle   The name given to a concept of landscape development involving the cyclic erosion of soils on upper hillslopes during unstable climatic phases and soil development during stable phases. The term is much used in Australia.

kame   An irregular mound of stratified sediment associated with GLACIOFLUVIAL activity during ice stagnation. It is a Scottish term for a landform much prized for the variety it adds to golf courses.   DES

kame terrace   A terrace formed between a hillside and a glacier by glaciofluvial activity. The landform is commonly associated with the former presence of stagnant ice down-wasting in valleys.

kamenitza   A generally small solutional basin developed on the surfaces of soluble rocks such as limestones. They are a type of LAPIÉ.   ASG

kaolin   A clay mineral, mainly hydrated aluminium silicate, or any rock or deposit composed predominantly of kaolinite. China clay or other material from which porcelain can be manufactured.

karren (singular karre)   A collective name describing small limestone ridges and pool structures that have developed as a result of the solution of rock by running or standing water. There are many types of karren, all differentiated by morphology (Bögli, 1960). The commonest are rillenkarren (sharp ridges between rounded channels). In Britain the best examples are on Hutton Roof Crag, Kirkby Lonsdale (Cumbria), while spectacular karren scenery can be found at Lluc in Mallorca. The term is German in origin; the French equivalent is *lapiés*.   PAB

Reference
Bögli, A. (1960) Kalklösung und Karrenbildung. In H. Lehmann and H. Bremer (eds), *Internationale Beiträge zur Karstmorphologie*. Zeitschrift für Geomorphologie, Supplementbände, vol. 2. Berlin: Borntraeger; pp. 4–21.

karst   Generally, the term given to limestone areas that contain topographically distinct scenery, including CAVES, SPRINGS, BLIND VALLEYS, KARREN and DOLINES (see Table 1). Specifically, Karst is a region of limestone country between Carniola and the Adriatic coast, which is characterized by typical limestone topography.

Karst regions are typified by the dominant erosional process of solution, the lack of surface water and the development of stream sinks (dolines), cave systems and resurgences or springs. Indeed, the process of stream sinking is known as karstification. All the resultant landforms associated with karst scenery depend upon this phenomenon of stream sinking.

The most abundant rock type that exhibits karst features is limestone, and the best karst scenery can be found when that limestone is pure, very thick in areas of upstanding relief, in an environment that provides enough water for solutional processes. Other calcareous rocks, such as chalk, fit many but not all of the above prerequisites. Chalk is often too soft to give rise to distinctive karst scenery.   PAB

Reading and References
Ford, D. and Williams, P. (2007) *Karst geomorphology and hydrology*, 2nd edition. Chichester: John Wiley & Sons, Ltd. · Ford, T.D. and Cullingford, C.H.D. (eds) (1976) *The science of speleology*. London: Academic Press. · Jennings, J.N. (1985) *Karst geomorphology*. Oxford: Blackwell; pp. 73–82. · Summerfield, M.A. (1991) *Global geomorphology*. Harlow/New York: Longman/John Wiley & Sons, Inc.

katabatic flows   Downslope winds, often coupled with, or induced by, the large-scale atmospheric circulation. These flows may reach surrounding lowlands as dry warm or cold winds, blowing at speeds in excess of $50\,\mathrm{m\,s^{-1}}$ for several days. Examples of warm katabatic wind are the FÖHN on the north slopes of the Alps in Europe and the Chinook on the east slopes of the Rockies in the USA. These strong winds derive their warmth either from ADIABATIC compression during

---

Table 1 Classification of solutional microforms developed on limestone

Form[a]	Typical dimensions	Comments
Forms developed on bare limestone		
Developed through areal wetting		
Rainpit	<30 mm across, <20 mm deep	Produced by rain falling on bare rock. Occurs in fields on gentle rather than steep slopes. Can coalesce to give irregular, carious appearance
Solution ripples	20–30 mm high; may extend horizontally for >100 mm	Wave-like form transverse to downward water movement under gravity. Rhythmic form implies that periodic flows or chemical reactions are important in their development
Solution flutes (rillenkarren)	20–40 mm across, 10–20 mm deep	Develop due to channelled flow down steep slopes. Cross-sectional form ranges from semicircular to V-shaped but is constant along flute
Solution bevels	0.2–1 m long, 30–50 mm high	Flat, smooth elements usually found below flutes; flow over them occurs as a thin sheet
Solution runnels (rinnenkarren)	400–500 mm across, 300–400 mm deep, 10–20 m long	Down-runnel increase in water flow leads to increase in cross-sectional area. May have meandering form. Ribs between runnels may be covered with solution flutes
Developed through concentration of run-off		
Grikes (kluftkarren)	500 mm across, up to several metres deep	Formed through the solutional widening of joints or, if bedding is nearly vertical, of bedding planes
Clints (flackkarren)	Up to several metres across	Tabular blocks detached through the concentration of solution along near-surface bedding planes in horizontally bedded limestone
Solution spikes (spitzkarren)	Up to several metres	Sharply pointed projections between grikes
Forms developed on partly covered limestone		
Solution pans	10–500 mm deep, 0.03–3 m wide	Dish-shaped depressions usually floored by a thin layer of soil, vegetation or algal remains. Carbon dioxide contributed to water from organic decay enhances dissolution
Undercut solution runnels (hohlkarren)	400–500 mm across, 300–400 mm deep, 10–20 m long	Like runnels but become larger with depth. Recession at depth probably associated with accumulation of humus or soil, which keeps sides at base constantly wet
Solution notches (korrosionkehlen)	1 m high and wide, 10 m long	Produced by active solution where soil abuts against projecting rock giving rise to curved incuts
Forms developed on covered limestone		
Rounded solution runnels (rundkarren)	400–500 mm across, 300–400 mm deep, 10–20 m long	Runnels developed beneath a soil cover that become smoothed by the more active corrosion associated with acid soil waters
Solution pipes	1 m across, 2–5 m deep	Usually become narrower with depth. Found on soft limestones, such as chalk, as well as mechanically stronger and less permeable varieties

Source: Summerfield (1991: table 6.6), based largely on discussion in Jennings (1985).
[a] The commonly encountered German terms are given in parentheses.

descent or from heat released by condensation on the windward slopes of the mountains or from both mechanisms together. This heat can increase the temperature of the air by 20 °C or more. Warm katabatic winds occur most frequently during the cooler months and when there is a rapid fall in sea-level pressure from the highlands to the lowlands.

Cold katabatic winds occur when a large pool of cold air, forming perhaps over a mountain glacier or on ice caps, becomes so deep that it spills over into the highlands. Heat release by condensation is not involved here, so the air remains cold. The glacier winds of Greenland and Antarctica, the BORA along the Adriatic coast, and the MISTRAL along the French Mediterranean coast are good examples of cold katabatic winds. WDS

Reading

Atkinson, B.W. (1981) *Meso-scale atmospheric circulations.* New York: Academic Press. · Gedzelman, S.D. (1980) *The science and wonders of the atmosphere.* New York: John Wiley & Sons, Inc.

kata-front

kata-front Any front at which the warm air is subsiding relative to the cold air. As a result, frontal activity is weak with only a belt of shallow stratiform cloud marking its presence. The change from an ana- to a kata-frontal structure can be seen sometimes on satellite images by their different cloud characteristics. Kata-fronts tend to develop at some distance from the cyclone centre where uplift in the warm air is less marked. PAS

kavir Iranian term for PLAYA.

kegelkarst Groups of residual conical-shaped limestone hills produced by limestone solution in adjoining DOLINES or SHAKEHOLES. The remnant limestone blocks are steep-sided and often heavily vegetated. They are also called cone-karst, COCKPIT KARST or MOGOTES. PAB

Reading

Ford, D. and Williams, P. (2007) *Karst geomorphology and hydrology,* 2nd edition. Chichester: John Wiley & Sons, Ltd.

Kelvin wave A long wave in the oceans whose characteristics are altered by the rotation of the Earth. In the northern hemisphere the amplitude of the wave decreases from right to left along the crest, viewed in the direction of wave travel. DTP

kettle, kettle hole An enclosed depression resulting from the melting of buried ice. Kettle holes are characteristic features of STAGNANT ICE TOPOGRAPHY. DES

keystone species A species whose removal from the ecosystem of which it forms a part leads to a series of adverse effects (including extinctions) in that system. An example of this is provided by elephants in savannas (Waithaka, 1996). They diversify the ecosystems that they occupy and create a mosaic of habitats by browsing, trampling and knocking over of bushes and trees. They also disperse seeds through their eating and defecating habits and maintain or create water holes by wallowing. All these roles are beneficial to other species. Conversely, where human interference prevents elephants from moving freely within their habitats and leads to their numbers exceeding the carrying capacity of the savanna, their effect can be environmentally catastrophic. Equally, if humans reduce elephant numbers in a particular piece of savanna, it may become less diverse and less open, and its water holes may silt up. This will be to the detriment of other species. ASG

Reading and Reference

Waithaka, J.M. (1996) Elephants: a keystone species. In T. R. McClanahan and T. P. Young (eds), *East African ecosystems and their conservation.* New York: Oxford University Press; pp. 284–285.

khamsin A hot, dry wind that blows from the desert to the south across the North African coast. Also called the sirocco and ghibli.

kinematic wave Consists of zones of high and low density that travel through a medium at a velocity which is generally different from that of the medium as a whole.

One set of solutions of the CONTINUITY EQUATION for material in a medium is generally dominant under certain circumstances. The theory was originally developed by Lighthill and Whitham (1955) and has been applied in a number of contexts within physical geography. The continuity equation in its differential form may be rewritten, using the same notation, as

$$c\frac{\partial s}{\partial x} + \frac{\partial s}{\partial t} = a$$

In this formulation, c is defined as the kinematic wave velocity, equal to dQ/ds evaluated at x, and not necessarily or usually a constant. For the simplest case where c is constant and $a = 0$, the complete solution to the above equation is

$$s = f(x - ct)$$

for an arbitrary function f. This represents a wave travelling without change of form at velocity c. In more complex cases different parts of the wave travel at different velocities, but the concept of a wave remains.

Kinematic waves have been applied to glacier response to climatic changes (Nye, 1960), to the movement of riffles in stream beds (Langbein and Leopold, 1968), to the movement of stream knickpoints and valley-side terraces, and to the routing of river and overland and flow flood peaks. MJK

Reading and References

Freeze, R.A. (1978) Mathematical models of hillslope hydrology. In M. J. Kirkby (ed.), *Hillslope hydrology.* Chichester: John Wiley & Sons, Ltd. · Langbein, W.B. and Leopold, L.B. (1968) *River channel bars and dunes – theory of kinematic waves.* United States Geological Survey Professional Paper 122L. Washington, DC: US Government

Printing Office. · Lighthill, M.J. and Whitham, G.B. (1955) On kinematic waves I: flood movements in long rivers. *Proceedings of the Royal Society Series A*. **229**, 281–316. · Nye, J.F. (1960) The response of glaciers and ice sheets to seasonal and climatic changes. *Proceedings of the Royal Society A: Mathematical, Physical and Engineering Science*, **256**, 559–584

kinematics The branch of mechanics dealing with the description of the motion of bodies without reference to the force producing the motion. BWA

Reading
Petterssen, S. (1956) *Weather analysis and forecasting*. New York: McGraw-Hill; especially chapters 2 and 3.

kinetic energy Energy due to the translational movement of a body. It is not so definitive as it looks. It depends on the frame of reference. For example, an object lightly tossed from a rapidly moving railway train has potentially lethal energy for a bystander, and vice versa. And similarly for scale; for example, 'temperature' of a gas represents the kinetic energy of individual molecules of the gas.   JSAG

kingdoms of animals and plants The simple classic division of all living things (except the noncellular, problematical viruses) into two categories of plants and animals (*plantae* and *animalia*) is no longer found entirely satisfactory and has been abandoned by most life scientists. The principal problem with the traditional twofold classification is that it groups organisms that are very unlike one another under the same heading. A few organisms lie uneasily in either category.

No classification of kingdoms is entirely satisfactory; the following is in widespread use.

- **Monera** Not having a well-defined nucleus: bacteria and blue–green algae. Acellular organisms; that is, those lacking clear division into cells.
- **Protista** A cellular organisms mostly lacking chlorophyll: flagellates (some of which do possess chlorophyll), amoebae, foraminifera, sporozoans, ciliates. Some of these form colonies, being incipiently multicelled.
- **Fungi** All kinds of fungi, including slime moulds. This group has long been included with the plants, but its members have a long, quite distinct evolutionary history. Non-photosynthetic organisms, mostly with definite cell walls.
- **Plantae** Six phyla (major groups) of photosynthetic organisms with cell walls, ranging from acellular forms (some algae) to much more complex organisms with highly developed organs and systems of organs (ferns, flowering plants).
- **Animalia** Multicelled, non-photosynthetic organisms without cell walls. Some classifications recognize over 300 phyla, varying from simple forms with few cells (the mesozoa), through sponges (which have partly differentiated tissues), to a great diversity of complex metazoans with well-developed organs and organ systems (e.g. worms, insects, molluscs, vertebrates).

See also FAUNAL REALMS and FLORISTIC REALMS. PHA

klippe An outcrop of rock that is separated from the rocks upon which it rests by a fault. It may represent an erosional remnant of a nappe or may have been emplaced by gravity sliding.

knickpoint A break in profile, generally in the long profile of a river. This was especially thought of as the product of REJUVENATION, where a steeper gradient lower reach is receding headward as a result of local or general lowering of base level, and this steeper profile intersects with a gentler upper one. Some knickpoints may be sharply defined, as in a waterfall, or they may be much less distinguished and only apparent after detailed field survey of stream profiles.

The term is also applied to any profile irregularity, as at tributary confluences or associated with lithological or structural controls, and not just those produced following rejuvenation. Furthermore, rejuvenation may itself be generated in alternative ways – extensively by eustatic sea-level change, or by isostatic and tectonic movements, or by changes in river discharge or sediment load. The early geomorphologists W. M. Davis and W. Penck (writing of *Knikpunkte* in German) were among the foremost in developing studies of such phenomena, though they used the word differently and in more restricted senses than are now generally adhered to. Emphasis is now placed on the interaction of many factors in stream system development, so that several possibilities for breaks in long profile would need to be explored.   JL

Reading
Bressan, F., Papanicolaou, A.N. and Abban, B. (2014) A model for knickpoint migration in first- and second-order streams. *Geophysical Research Letters*, **41**, 4987–4996.

knock-and-lochan topography A landscape of ice-moulded rock knobs with intervening lochans that have been eroded along lines of structural weakness. The type site is in the northwestern highlands of Scotland (Linton, 1963), but it is also characteristic of much of the Canadian and Scandinavian shields (Sugden, 1978).   DES

References
Linton, D.L. (1963) The forms of glacial erosion. *Transactions of the Institute of British Geographers*, **33**, 1–28. · Sugden, D.E. (1978) Glacial erosion by the Laurentide ice sheet. *Journal of Glaciology*, **20**, 367–391.

kolk   A form of macroturbulence, or large-scale vortex structure, that arises in the turbulent flow of water in rivers. The term was introduced by Matthes (1947). A kolk involves a rapidly rotating cylinder of flow, or vortex, that is initially attached at its lower end to the bed surface and rises away from it. The kolk subsequently breaks away and rises towards the water surface, where its presence is marked by a 'boil' on the water surface that dissipates after a short time (Jackson, 1976). Kolks may entrain sediment particles from the bed and release them some distance higher, so fostering the downstream motion of the bed material. Kolk development in turbulent flow over a dune was reported from numerical simulations by Grigoriadis *et al.* (2009), who presented three-dimensioanl visualizations of the phenomenon.   DLD

References
Grigoriadis, D.G.E., Balaras, E. and Dimas, A.A. (2009) Large-eddy simulations of unidirectional water flow over dunes. *Journal of Geophysical Research*, **114**, F02022, doi: 10.1029/2008JF001014. · Jackson, R.G. (1976) Sedimentological and fluid-dynamic implications of the turbulent bursting phenomenon in geophysical flows. *Journal of Fluid Mechanics*, **77**, 531–560. · Matthes, G.H. (1947) Macroturbulence in natural stream flow. *Transactions of the American Geophysical Union*, **28**, 255–265.

koniology (coniology)   The scientific study of atmospheric DUST together with its solid pollutants, such as soot, pollen, microbial spores, and so on.

A small granite kopje, southern Namibia.
Photograph by David Thomas.

kopje   A hillock or rock outcrop, applied especially in South Africa.

Köppen climatic classification   A system of climatic differentiation based upon temperature and precipitation linked to vegetation zones. It is one of the most widely used methods of classification but has undergone numerous modifications since it was devised about 1900 by Russian climatologist W. Köppen. The scheme is sometimes called the Köppen–Geiger classification as it was later updated by Rudolf Geiger. Five major categories subdivide the Earth's climates: tropical (A), dry (B), temperate (C), continental (D) and polar and alpine (E). Further subdivisions are made on the basis of the rainfall regime, temperature characteristics and any other special features. Each region can then be identified on the basis of a sequence of letters; for example, Csb indicates a coastal Mediterranean climate with a mild winter and a dry but warm summer.

Updates to the Köppen classification are made periodically as new data become available (e.g. Kottek *et al.*, 2006; Peel *et al.*, 2007). Climate change is also modifying the classification appropriate to some parts of the Earth's surface (Chen and Chen, 2013).   DLD

Reading and References
Barry, R.G. and Chorley, R.J. (2003) *Atmosphere, weather and climate*, 8th edition. London: Psychology Press. · Chen, D. and Chen, H.W. (2013) Using the Köppen classification to quantify climate variation and change: an example for 1901–2010. *Environmental Development*, **6**, 69–79. · Kottek, M., Grieser, J., Beck, C., *et al.* (2006) World map of the Köppen–Geiger climate classification updated. *Meteorologische Zeitschrift*, **15**, 259–263. · Peel, M.C., Finlayson, B.L. and McMahon, T.A. (2007) Updated world map of the Köppen–Geiger climate classification. *Hydrology and Earth System Sciences*, **11**, 1633–1644.

krotovina   Infilled animal burrows or filaments found in soils and sediments such as loess.

krummholz   From the German, meaning crooked wood, refers to the stunted and gnarled woodlands characteristic of forest margins at high altitudes and high latitudes. The dwarfing, distortion and, in extreme conditions, the prostrate habit of trees is a result of the combined effects of wind and cold. Such features are common in the transition zone between the subalpine forest and alpine tundra in high latitudes, or in the elfin

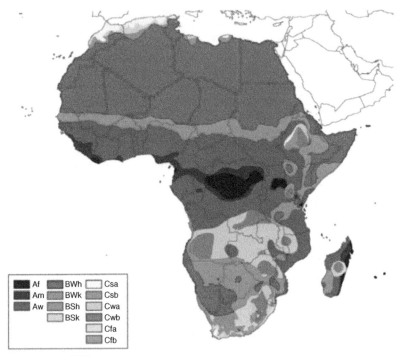

New Köppen climate map of Africa.
Source: Peel *et al.* (2007). This work is distributed under the Creative Commons Attribution 3.0 License.

woods of high tropical elevations, bearing a heavy cover of epiphytes and a ground layer cushioned by mosses, herbaceous plants and grasses. The word is also used to describe dense, tangled thickets in tropical forest.   PAF

kumatology   A neglected term developed by the great British amateur geographer Vaughan Cornish for the study of wave-like forms encountered in nature.   ASG

kunkar   See CALCRETE.

kurtosis   A descriptor of the shape of a distribution. Kurtosis provides a measure of the peakedness of a distribution relative to a normal distribution and can be quantified in a variety of ways. Typically, a high kurtosis is associated with a data set that has a very distinct peak near the mean value (leptokurtic distribution), while a low kurtosis arises when the distribution is relatively flat and does not have a sharp peak (platykurtic distribution).

Kurtosis is often used in the description of sediments, where it typically relates to both sorting (standard deviation) and differences from a normal distribution. A flat distribution would be found for a poorly sorted sediment, such as a till; a peaked distribution would be found in a well-sorted sediment, such as a wind-blown sand.   GF

Kyoto Protocol   The Kyoto Protocol is an international agreement linked to the United Nations Framework Convention on Climate Change, which commits its parties to binding greenhouse gas (GHG) emission reduction targets. It was adopted in Kyoto, Japan, on 11 December 1997, and entered into force on 16 February 2005. The basis for its implementation was set out in the Conference of Parties meeting in Marrakesh, Morocco, in 2001 and is referred to as the 'Marrakesh Accords'. The first commitment period started in 2008 and ended in 2012, and during this time 37 industrialized countries and the European Community committed to reduce GHG emissions by an average of 5%

compared with 1990 levels. The period since 2012 is covered by the 'Doha Amendment to the Kyoto Protocol', which commits signatories to reduce GHG emissions by at least 18% below 1990 levels between 2013 and 2020; the amendment will enter into force when at least three-quarters of the 192 parties to the Kyoto Protocol have agreed to the new commitments.   RH-Y

Reading

Hoffmann, M.J. (2011) *Climate governance at the crossroads: experimenting with a global response after Kyoto*. Oxford: Oxford University Press. · United Nations Framework Convention on Climate Change (2014) *Kyoto protocol*. http://unfccc.int/kyoto_protocol/items/2830.php (accessed 10 July 2015) · Yamin, F. (ed.) (2012) *Climate change and carbon markets: a handbook of emissions reduction mechanisms*. Taylor & Francis.

L

laccolith A mass of intrusive rock that, though concordant with the host rocks, has domed up the overlying strata. The base of the laccolith is either horizontal or convex downward.

Reading
Corry, C.E. (1988) *Laccoliths: mechanisms of emplacement and growth*. Geological Society of America Special Papers, vol. 220. Boulder, CO: Geological Society of America.

lacustral See PLUVIAL.

lacustrine Of or pertaining to lakes.

lag gravel An accumulation of coarse rock fragments at the land surface that has been produced by the removal of finer particles, generally by deflation.

lag time The period elapsing between the occurrence of a causative phenomenon and its resulting effect, as in the time difference between peak storm rainfall and the later peak in stream discharge that results from it. This may be an important consideration in many physical processes. For example, BEDFORMS in rivers may be related to river flows, but such flows vary over time and it also takes a finite time for the bedforms themselves to develop. Thus, there may be delayed response between development and a change in flow.   JL

Reading
Allen, J.R.L. (1974) Reaction, relaxation and lag in natural sedimentary systems: general principles, examples and lessons. *Earth Science Reviews*, **10**, 263–342.

lagoon A generally shallow, often elongated body of water separated from a larger body of water by sand shoals, barrier islands or coral reefs. Lagoons are commonly divided into coastal lagoons and atoll lagoons. Coastal lagoons can be subdivided into three geomorphic types: 'choked', 'restricted' and 'leaky', reflecting the degree of exchange of water with the coastal sea (Kjerfve, 1994). Similarly, salinity can range from nearly fresh to hypersaline waters. The maximum depth of coastal lagoons in the Mediterranean Sea is 30 m, although the mean depth is rarely higher than 2 m (Perez-Ruzafa *et al.*, 2010). Atoll lagoons can be deep (often with depths of >50 m with the deepest reaching >90 m) and are generally circular in form. They can be very large; for example, $120 \times 32 \, km^2$ at Kwajalein (Marshall Islands) and $79 \times 34 \, km$ at Rangiroa (Tuamotu Archipelago).   TS

Reading and References
Guilcher, A. (1988) *Coral reef geomorphology*. New York: John Wiley & Sons, Inc. · Kjerfve, B. (ed.) (1994) *Coastal lagoon processes*. Amsterdam: Elsevier. · Perez-Ruzafa, A., Marco, C. and Perez-Ruzafa, I.M. (2011) Mediterranean coastal lagoons in an ecosystem and aquatic resources management context. *Physics and Chemistry of the Earth*, **36**, 160–166.

lahar A mass movement feature on the flank of a volcanic cone. The volcanic ash may move either as a mudslide when saturated with water or as a dry landslide as a result of Earth tremors.

lake A lake is a body of water contained within continental boundaries. Worldwide, almost all natural lakes were formed by tectonic, volcanic or glacial activity, with the majority being glacial in origin. Others may occupy depressions formed by DEFLATION (see PAN and PLAYA). The relatively still waters of lakes have led to them being termed lentic environments. The size, depth and shape of lakes vary enormously from shallow, ephemeral systems, playa lakes, to huge deep lakes, such as the Siberian Lake Baikal. This 25-million-year-old lake is over 1600 m in depth and contains about 20% of the Earth's unfrozen fresh water, more than the combined total of the five Great Lakes of North America. By far the largest lake in terms of area and volume is the vast saline Caspian Sea that covers an area of nearly 370,000 km^2 and contains over 78,000 km^3 of water. Most of the world's 1 million plus lakes are, however, very small freshwater bodies.

Once a lake basin has formed, physical, biological and chemical factors interact to produce discernible structures in the water. Thermal stratification is one of the most important physical events in a lake's annual cycle, and it dominates most aspects of lake structure. If a lake is deep enough, heating of its surface waters by the sun will cause a warm, light layer to form, the epilimnion, above a colder dense layer, the hypolimnion. A third layer, the metalimnion, forms a transitional zone between the two. The temperature gradient between the layers is termed the thermocline. During the summer the volume of the epilimnion increases at the expense of the hypolimnion. In autumn, heat loss exceeds the input of solar heat and destratification will occur in all but the deepest lakes. If the waters become fully mixed the lake is termed holomictic. Meromictic lakes are deep water bodies, such as Lake Tanganyika, that remain stratified throughout the year. The majority of lakes are dimictic, with their waters mixing twice a year during the spring and autumn. Shallow lakes that are mixed throughout the year are termed polymictic. Not only is the heat transmitted by light responsible for various kinds of thermal stratification in water bodies, but it also regulates the rate of chemical reactions and biological processes. An increase in heat during the summer, for example, will increase metabolic activity and the rate of recycling of organic and mineral components.

Lakes can also be classified based on their primary biological productivity or trophic status. Those that are nutrient poor and experience low productivity are termed oligotrophic. Oligotrophic lakes are often deep with steep slopes and relatively small drainage areas. With few nutrients, oligotrophic lakes support relatively low levels of algae and, consequently, are characterized by clear blue waters that are highly transparent. Lakes that are nutrient rich and display high levels of primary productivity are termed eutrophic (see EUTROPHICATION). In many cases these are shallow water bodies, less than 10 m deep, with gentle sloping edges and a large drainage-area-to-lake-surface ratio. Characterized by high nutrient levels, the waters of eutrophic lakes are characterized by low clarity, and in extreme cases unsightly blooms of blue–green algae can form a crust on the surface of the lake. The sediments of eutrophic lakes become enriched with nutrients as organic matter accumulates. Initially, this tends to increase the biomass of rooted macrophytes, but phytoplankton growth can become so dense that it can shade out submerged plants. The trophic status of a lake will often change over time, with waters usually becoming more nutrient rich as they get older. In recent decades, however, there has been growing concern about cultural eutrophication as a result of human activities. Cultural eutrophication is not a new phenomenon. For example, Lake Patzcuaro, in the highlands of central Mexico, became eutrophic following widespread deforestation and increased catchment erosion some 900 years ago (Metcalfe et al., 1989). Elsewhere, discharge of domestic waste into lakes and rivers has resulted in widespread eutrophication, and in recent years problems have been associated with the intensification of agriculture and the increased use of organic fertilizer. Invariably, some of the fertilizer applied is lost from the field and washes into lakes and rivers. Given the higher productivity of eutrophic lakes, eutrophication is not always undesirable. And there are many naturally eutrophic lakes where large quantities of fish are harvested for food, and in some cases water bodies are managed to maximize productivity. (See also LIMNOLOGY.)   SLO

Reading and Reference
Heinonen, P. (ed.) (2000) *Hydrological and limnological aspects of lake monitoring*. New York: John Wiley & Sons, Inc. · Metcalfe, S.E., Street-Perrott, F.A, Brown, R.B., et al. (1989) Late Holocene human impact on lake basins in central Mexico. *Geoarchaeology*, 4, 119–141. · Wetzel, R.G. (2001) *Limnology*, 3rd edition. San Diego, CA: Academic Press.

lake breeze   See SEA/LAND BREEZE.

laminar flow   A type of flow in which the movement of each fluid element is along a specific path with uniform velocity, with no diffusion between adjacent 'layers' of fluid. Injected dye maintains a straight, coherent thread. The shear stress between adjacent layers increases from zero at the surface to a maximum at the fluid–solid contact (e.g. the river bed), and the flow velocity increases parabolically with height above the bed. Fluid motion is laminar if viscous forces are so strong relative to inertial force that the fluid viscosity significantly influences flow behaviour. The viscosities of air and water are so low that laminar flow is rare in these fluids. Laminar flow is characterized by a REYNOLDS NUMBER of below 500; for example, in the very shallow water of overland flow on hillslopes, and then only when the water is undisturbed by raindrop impact.   KSR

La Niña   The cold counterpart of El Niño, a period of anomalously cold temperatures in the equatorial Pacific. La Niña (the girl child) can last up to 2 years or longer and its effects are generally opposite those of El Niño. It frequently occurs just before or just after an El Niño episode. Like El Niño, it is associated with east–west pressure contrasts over the Pacific, and with an intensification of the trade winds over the Pacific. La Niña brings abnormally wet conditions to the western equatorial Pacific and abnormally dry conditions to the eastern equatorial Pacific.

La Niña is often accompanied by significant changes in the atmospheric circulation in the mid and low latitudes of both hemispheres, evoking a global pattern of climate anomalies. La Niña is often the time of extreme drought in Peru. It is generally characterized by warm winters in the southeastern USA, colder and wetter than normal winters in the Pacific Northwest, and drier than normal conditions in the American Southwest, central Plains, and Southeast. La Niña is also linked to drought/wet conditions in equatorial/southern Africa.   SJN

land breeze   See SEA/LAND BREEZE.

land capability   A measure of the value of land for agricultural purposes. The capability unit is described as a group of soils that are nearly alike, based on an interpretation of soil data. The soils in each will: (1) produce similar kinds of cultivated crops and pasture plants with similar management practices; (2) require a similar conservation treatment and management under the same kind and condition of vegetative growth; and (3) have comparable potential productivity.

Subclasses are defined according to their limitations for agricultural use and hazards to which they are exposed. In the USA, four general limitations are recognized: erosion hazard, wetness, rooting zone limitations and climate, of which all but the last are closely related to geomorphology. In the UK, another subdivision – gradient or soil pattern – is also added.   SMP

Reading
Bibby, J.C. and Mackney, D. (1969) *Land use capability classification*. London: Soil Survey.

land systems   Subdivisions of a region into areas having within them common physical attributes that are different from those of adjacent areas. Any one land system normally has a recurring pattern of topography, soils and vegetation, reflecting the underlying rocks (geology), erosional and depositional processes (geomorphology) and the climate under which these processes operate. A land unit, the detailed component of a land system, is particularly useful in evaluating land for agricultural and engineering purposes and in devising problem-orientated classifications. The resultant land systems maps are easily interpreted, and both rapid and economical to produce.   SMP

Reading
Veatch, J.O. (1983) *Agricultural land classification and land types of Michigan*. Michigan Agricultural and Experimental Station special bulletin 544. East Lansing, MI: Michigan State College, Agricultural Experiment Station.

land-bridge   An isthmus or other connection between two land masses across which animals and plants move to colonize a new environment.

Landsat   The Landsat program consists of a series of REMOTE SENSING satellite missions that provides a long-term moderate spatial resolution monitoring capability. The first Landsat satellite, originally named the Earth Resources Technology Satellite before being renamed Landsat 1, was launched in 1972. Since then, Landsat satellites have been in continuous operation through a series of missions, with Landsat 8 launched in 2013 and plans for future missions established. The satellite systems have evolved over time, with the key sensors typically acquiring imagery in more wavebands and at a finer spatial resolution, but the changes have been made in way that facilitates data continuity so as to aid long-term monitoring studies. Key sensors include the THEMATIC MAPPER (TM) and Enhanced Thematic Mapper+. The latter acquires imagery in visible to thermal infrared wavelengths for almost any location on the Earth's surface every 16 days. The latest, Landsat 8, satellite includes the OPERATIONAL LAND IMAGER system. The spatial resolution of the imagery acquired ranges from 15 m in the visible to 60 m for thermal infrared. All Landsat data are available free from the United States Geological Survey.   GF

Reading
Lillesand, T.M., Kiefer, R.W. and Chipman, J.W. (2008) *Remote sensing and image interpretation*, 6th edition. New York: John Wiley & Sons, Inc. · National Research Council (2013) *Landsat and beyond. Sustaining and enhancing the nation's land imaging program*. Washington, DC: The National Academies Press.

landscape ecology   A term introduced by the German geographer Carl Troll, who also later used the term geo-ecology. It has two components (Vink, 1983): an approach to the study of the landscape that interprets it as supporting natural and cultural ecosystems; and the science that investigates the relationships between the biosphere and anthroposphere and either the Earth's surface or the abiotic components.   ASG

Reading and Reference
Turner, M.G., Gardner, R.H. and O'Neill, R.V. (2001) *Landscape ecology in theory and practice*. New York:

Springer-Verlag. · Vink, A.P.A. (1983) *Landscape ecology and land use*. London: Longman.

landslide A landslip. The movement down-slope under the influence of gravity of a mass of rock or earth. A mass of rock or earth that has moved in such a way. (See MASS MOVEMENT TYPES).

Reading
Evans, S.G. and DeGraff, J.V. (eds) (2002) *Catastrophic landslides: effects, occurrence, and mechanisms*. Boulder, CO: Geological Society of America. · Selby, M.J. (1982) *Hillslope materials and processes*. Oxford: Oxford University Press.

lapié See KARREN.

lapse rate The rate of *decrease* of a quantity with height, usually applied to temperature but sometimes also to the mixing ratio (see HUMIDITY) of water vapour to air. The typical rate of temperature change in the TROPOSPHERE is $6.5\,\mathrm{K\,km^{-1}}$, whereas for the temperature of dry air displaced adiabatically it is $10\,\mathrm{K\,km^{-1}}$ (dry adiabatic lapse rate). It follows that the troposphere is usually stable to dry adiabatic processes. JSAG

Type of
movement

Falls

Rockfall

Slab failure

Toppling
failure

Rock
avalanche

Slides

Slump

Planar slide

Wedge failure

Block slide

Creep

Cambering

Curving of strata

A classification of landslides in rock.
Source: Selby 1982. Reproduced with permission of Oxford University Press.

Reading
Ackerman, S.A. and Knox, J.A. (2014) *Meteorology: understanding the atmosphere*, 4th edition. Sudbury, MA: Jones and Bartlett.

Late Glacial A term used for the span of time between the Last Glacial Maximum (in Britain; the Devensian, in North America, *c.* cal. 19,000–23,000 BP) and the beginning of the HOLOCENE (*c.* cal. 11,700 BP). It was marked by various minor stadials and interstadials (e.g. see ALLERØD). ASG

latent heat That part of the thermal energy involved in a change of state, like the 2.4×10^6 J kg^{-1} of energy released when water vapour condenses to liquid. This process makes rising cloudy air cool less rapidly than does the ambient air with height and hence the cloudy air becomes buoyant. EVAPORATION of liquid water at the ground 'saves up' solar energy in latent form until it can be released in cloudy convection. Typical English thunderstorms rain French water. JSAG

Reading
Ackerman, S.A. and Knox, J.A. (2014) *Meteorology: understanding the atmosphere*, 4th edition. Sudbury, MA: Jones and Bartlett.

lateral accretion The process by which bed sediments accumulate at the side of a channel as it shifts laterally. The term applies notably to the sediment accumulating in POINT BAR DEPOSITS, but lateral accretion can also occur in BRAIDED RIVERS as channels shift and bars enlarge. Such deposits may contrast with the vertical accretion of sediments deposited from suspension, which are usually finer in size and may accrete beyond the confines of channels, and they may be particularly important volumetrically among the near-surface sediments of many FLOODPLAINS. Such sediments may possess a distinctive type of cross-bedding, called epsilon cross-bedding by J. R. L. Allen, in which the dipping or sigmoid-shaped beds represent successive increments of accretion developed at right-angles to the general flow direction. JL

Reading
Allen, J.R.L. (1970) *Physical processes of sedimentation*. London: Allen & Unwin. · Collinson, J.D. and Thompson, D.B. (1982) *Sedimentary structures*. London: Allen & Unwin. · Reading, H.G. (ed.) (1986) *Sedimentary environments and facies*, 2nd edition. Oxford: Blackwell Scientific.

lateral flow Applied particularly to subsurface near-horizontal or ground slope-aligned water movement, in contradistinction to vertical water movement. Water flow along permeable soil horizons may be an important mechanism in the transfer of water from soils and hillslopes into streams without such flow taking place over the surface of the ground or through deep percolation to groundwater. (See also INTERFLOW and THROUGHFLOW.) JL

lateral migration The movement of stream channels across valley floors through bank erosion and accompanying deposition on the opposite bank. This may proceed at varying rates on different rivers, on some amounting to several metres a year and on others none at all. JL

Reading
Osborn, G. and Du Toit, C. (1991) Lateral planation of rivers as a geomorphic agent. *Geomorphology*, 4, 249–260.

lateral moraine See MORAINE.

laterite A surface accumulation of the products of rigorous chemical selection, developing where conditions favour greater mobility of alkalis, alkaline earths and silica than of iron and aluminium. Bauxite is a laterite rich in aluminium. Laterite was originally defined by its ability to harden rapidly and irreversibly on exposure to the air, a property that led to its use as building bricks in southern India (Buchanan, 1807). The term has been extended to include related materials that were hard or contained hard parts, even though this induration may be an original result of iron segregation rather than of exposure.

Laterite profiles vary enormously in scale. The laterite may be a few centimetres to tens of metres thick, and below this the saprolite, leached or unleached of iron, varies from a few centimetres to over 100 m. Thick profiles develop on low-angle slopes (Goudie, 1973), and laterites can be divided into those that result from relative accumulation of iron and aluminium sesquioxides and those that result from absolute accumulation.

Relative accumulations owe their concentration to the removal of more mobile components, and absolute accumulations to the physical addition of materials. ASG

Reading and References
Buchanan, F. (1807) *A journey from Madras through the countries of Mysore, Kanara and Malabar*. London: East India Company. · Goudie, A.S. (1973) *Duricrusts of tropical and subtropical landscapes*. Oxford: Clarendon Press. · Widdowson, M. (2007) Laterite and ferricrete. In D. J. Nash and S. McLaren (eds), *Geochemical sediments and landscapes*. Chichester: John Wiley & Sons; pp. 45–94.

latosol A deep, red, yellow and brown, highly weathered, residual iron-rich soil typically found

Late Holocene lava flows in the Markagunt Plateau volcanic field, Utah, USA.
Photograph by David Thomas.

in the humid tropics. The term is used to embrace a range of residual iron soils, including lateritic soils, highly weathered ferralitic soils and moderately weathered ferruginous soils, and is broadly synonymous with the United States Department of Agriculture (USDA) term oxisol. In addition to being iron-rich, latosols tend to have a moderate to high clay content, dominated by kaolinite, goethite, haematite and gibbsite, and may merge at depth into saprolite and unweathered bedrock (see also LATERITE and SAPROLITE). DJN

Laurasia The northern part of Pangaea, a supercontinent thought to have been broken up by continental drift. The southern continent, Gondwanaland, was separated from it by the Tethys Ocean.

Laurentide Ice Sheet The ICE SHEET that covered much of North America during the last glaciation. At its maximum extent it covered most of Canada, ranging from the Queen Elizabeth Islands in the north to just south of the US border, and from beyond the current coastline in the east to where it met the Cordilleran Ice Sheet covering the Rocky Mountains in the west. It is thought to be the largest ice sheet to grow and decay during glaciations of the QUATERNARY period, and at its maximum is estimated to have made up 35% of the world's ice volume, containing sufficient water to lower global sea level by around 50 m. It occupied an area of around $11 \times 10^6 \, \text{km}^2$ and reached a probable thickness of 2–3 km. The ice sheet comprised three main ice dispersal centres, called the Keewatin, Baffin and Labrador sectors. In its strict sense the name Laurentide Ice Sheet only applies to the ice mass during the last (Wisconsinan) glacial period, with its maximum extent at 21,000 years BP, although we know that similar ice sheets existed during earlier glacials.

The significance of this ice sheet lies in its interaction with the climate system. It was large enough to alter atmospheric circulation patterns, and during its decay we now know that abrupt releases of meltwater and icebergs were of great enough magnitude to modify ocean circulation, thus forcing abrupt flips in the climate of the north Atlantic region. CDC

Reading
Benn, D., and Evans, D.J.A. (2010) *Glaciers & glaciation*, 2nd edition. London: Hodder Arnold. · Fulton, R.J. and Prest, V.K. (1987) The Laurentide Ice Sheet and its significance. *Geographie Physique et Quaternaire*, **41**, 181–186. · Clark, P.U., Licciardi, J.M., MacAyeal, D.R. and Jenson, J.W. (1996) Numerical reconstruction of a soft-bedded Laurentide Ice Sheet during the last glacial maximum. *Geology*, **24**, 679–682.

lava Molten rock material that is extruded from volcanoes and volcanic fissures. Consolidated lava flows can form distinct landscape features in areas of present or past volcanic activity.

law of the wall A relationship describing the semi-logarithmic increase in fluid (normally wind) velocity at an increasing distance from a rough bed under conditions of TURBULENCE. Under normal conditions on flat, unvegetated

315

surfaces and in the absence of thermal effects a turbulent velocity profile will plot as a straight line on a semi-logarithmic chart with velocity increasing away from the bed. The gradient of the semi-logarithmic profile is proportional to the shear velocity u^* and is a result of the surface ROUGHNESS producing a drag on the overlying fluid flow. The relationships between the principal controlling parameters on the profile can be described by the Karman–Prandtl velocity distribution:

$$\frac{u}{u^*} = \frac{1}{k}\ln\frac{z}{z_0}$$

where u is the fluid velocity; u^* is shear velocity, k is the von Karman constant (0.4), z is height and z_0 is ROUGHNESS LENGTH.

The law of the wall provides a particularly useful technique by which shear velocity can be calculated if fluid velocity is known at a number of heights from the bed. Such an approach is commonly used when determining sediment ENTRAINMENT and the critical erosion velocity (see CRITICAL VELOCITY (OR CRITICAL EROSION VELOCITY)) in fluid flows because the shear velocity represented by the velocity profile is proportional to the SHEAR STRESS at the surface.   GFSW

Reading
Livingstone, I. and Warren, A. (1996) *Aeolian geomorphology: an introduction.* Longman: Harlow. · Wiggs, G.F.S. (2011) Sediment mobilisation by the wind. In D. S. G. Thomas (ed.), *Arid zone geomorphology: process, form and change in drylands,* 3rd edition. Chichester: John Wiley & Sons, Ltd; pp. 351–372.

leachate   Liquid that passes through a soil, sediment or other permeable material, carrying away substances taken into solution (leached) from it. Applied to the water that passes through litter and the upper parts of the soil, which may be made more potent as a solutional agency by organic acids derived from the plant residues. Leachates also commonly require management at landfill and mine tailings sites, where percolating rainwater is capable of creating a contaminated leachate that must be prevented from escaping into the environment (Sophocleous *et al.*, 1996).   DLD

Reference
Sophocleous, M., Stadnyk, N.G. and Stotts, M. (1996) Modelling impact of small Kansas landfills on underlying aquifers. *Journal of Environmental Engineering,* **122,** 1067–1077.

leaching   The downward movement of water through the soil zone that results in the removal of water-soluble minerals from the upper horizons

and their accumulation in the lower soil zone or groundwater.

leaching requirement   The fraction of irrigation water that must be leached through the root zone to maintain the soluble salt level in the soil at an acceptable level in relation to the salt tolerance of the proposed crop.

leaf area index (LAI)   A measure of the abundance of foliar cover as projected above a unit area of the ground surface. LAI is usually defined as the one-sided leaf area, and is employed to scale estimates of gas exchange, the attenuation of solar radiation, the interception of rainwater, and similar ecosystem processes. LAI can be found by direct measurement, such as mechanical defoliation and the direct measurement of total leaf area, or non-destructively through the use of optical instruments (including wide-angle or fisheye lens photography) that are based on the extinction of the solar beam by leaves. LAI can also be estimated from remote sensing data such as surface reflectance data. LAI is dimensionless, since it is measured in metres squared per metres squared. Values of LAI exceed 40 in some kinds of forest, and 20 in some crop plants. More commonly, LAI is <10.   DLD

Reading
Asner, G.P., Scurlock, J.M.O. and Hicke, J.A. (2003) Global synthesis of leaf area index observations: implications for ecological and remote sensing studies. *Global Ecology & Biogeography,* **12,** 191–205. · Ganguly, S., Nemani, R.R., Zhang, G., *et al.* (2012) Generating global leaf area index from Landsat: algorithm formulation and demonstration. *Remote Sensing of Environment,* **122,** 185–202, doi: 10.1016/j.rse.2011.10.032.

Le Chatelier principle   Named after the French inorganic chemist, H. L. Le Chatelier (1850–1936), it defines a condition of a system in STABLE EQUILIBRIUM in which a change in any one of the governing forces will cause the equilibrium to shift so that the original condition is restored. In other words, the initial change sets up an internal reaction that is equal and opposite to that change and there is no net alteration in the system.

As used by physical geographers, this concept is generally conflated with ideas of homeostasis, negative feedback and self-regulation.

In its original form, and related to the thermodynamics of (strictly, isolated) systems, it states: 'If the temperature of a system in equilibrium be raised, the equilibrium will shift in favour of the reaction in which heat is absorbed (endothermic reaction); the converse also applies'. This shows how weathering of rocks, more or less in equilibrium in the lithosphere, tends towards exothermic (heat-evolving) reactions. Oxidation, hydration and carbonation are volume-increasing reactions in weathering typically of this kind.   BAK/WBW

lee depression
Region of low pressure found downwind of a mountain chain, representing the large-scale part of the drag of the ground on the air. The cyclonic circulation is due to the interplay of the vorticity and divergence of the air to the lee of the obstacle. Occasionally such features appear to develop and move away, as in cyclogenesis in the lee of the European Alps, which results in a depression over northern Italy that forces water up the Adriatic Sea. This occasionally results in the flooding of Venice.   JSAG

Reading
Ackerman, S.A. and Knox, J.A. (2014) *Meteorology: understanding the atmosphere*, 4th edition. Sudbury, MA: Jones and Bartlett. · McIntosh, D.H. and Thom, A.S. (1969) *Essentials of meteorology*. London: Wykeham Publications.

lee dune
When winds transporting sand encounter a topographic obstacle, air and sand flow may be accelerated around the obstacle to accumulate in a zone of slack (or separated) flow on the lee side. Aeolian accumulation may then extend downwind parallel to the direction of sand transport, so that lee dunes are usually linear forms. (See also TOPOGRAPHIC DUNE.)   DSGT

lee eddy
A closed circulation (primarily in the vertical plane) often found on the downward side of steep obstacles. On the smaller scale, it defines a good place for scenic picnics; on the larger scale, it identifies places prone to recirculation of pollutants.   JSAG

lee waves
Waves in the atmosphere of about 6 km wavelength (see GRAVITY) extending downwind of an obstacle in trains that may be up to 400 km long. Often made visible (remarkably so on satellite pictures) by alternating bands of clear and cloudy air (lenticular CLOUDS), they are conceptually important as identifying some aspects of the irreversibility of atmospheric processes. JSAG

Reading
Atkinson, B.W. (1981) *Meso-scale atmospheric circulations*. London: Academic Press. · Ludlam, F.H. (1980) *Clouds and storms*. University Park, PA: Pennsylvania State University Press.

lessivage
Involves the translocation of silicate clays in colloidal suspension in a soil profile without any change in their chemical composition. It contrasts with podzolization, in which the clay materials decompose and the hydrous oxides of iron and aluminium are mobilized.

levée
A visible ridge running alongside a FLOODPLAIN stream, DELTA distributary or intertidal creek, composed generally of coarse sand to silt-grade suspended sediment deposited by floodwaters as they overtop channel banks. The levée may slope gently away from the river and consist of progressively finer sediment as distance from the channel increases. Rather different features of the same name may also be created on steeper slopes by debris flows; here, they may comprise boulders or coarse material. Levées may be artificially created or raised for flood protection but add a risk of failure.   TS

Reading
Reading, H.G. (ed.) (1986) *Sedimentary environments and facies*, 2nd edition. Oxford: Blackwell Scientific.

lichenometry
A technique for dating Holocene events that was developed in the 1950s. It is especially useful for dating glacial fluctuations over the past 5000 years or so. It is believed that most glacial deposits are largely free of lichens when they are formed, but that once they become stable, lichens colonize their surfaces. The lichens become progressively larger through time. By measurement of the largest lichen thallus of one or more common species, such as *Rhizocarpon geographicum*, an indication of the date when the deposit became stable can be obtained.   ASG

Reading
Innes, J.L. (1985) Lichenometry. *Progress in Physical Geography*, 9, 187–254. · Worsley, P. (1990) Lichenometry. In A. S. Goudie (ed.), *Geomorphological techniques*, 2nd edition. London: Unwin Hyman; pp. 422–428.

lidar (the acronym for light detection and ranging)
An active REMOTE SENSING system based on laser beams as the source of illumination. Ordinarily, these are emitted from the sensor as a series of pulses of a very narrow beam of coherent electromagnetic radiation at a certain wavelength towards the target of interest. The time difference between pulse emission and the backscattering of that same pulse (known as the range) recorded by the lidar is dependent on the location and properties of that target. Early lidar systems were pure ranging systems, which only recorded the time for one backscattered pulse (the first return) from the sensor to the target. More recently, lidar systems have been able to record the last return pulse in addition to the first

return, and more often than not the full waveform of the backscattered pulses. In various degrees of sophistication, this backscattered data (as a three-dimensional (3D) point cloud) provides information principally on the 3D structure of the target and secondarily on the properties of that target. This can be further enhanced by using the intensity data that are also collected, which is a measure of the strength of the backscattered pulse. Intensity is primarily determined by the reflectivity of the object struck by the laser pulse, and this in turn is a function of the wavelength used (most commonly in the near infrared). Intensity is used as an aid in feature detection and extraction, in lidar pulse classification, and as a substitute for aerial imagery when none is available.

Lidars are often mounted on an airborne platform (fixed-wing or otherwise) and this is known as airborne laser scanning (ALS), and increasingly they are found on satellite and terrestrial platforms. ALS is supported by measurements made by a GLOBAL POSITIONING SYSTEM (GPS) and an inertial measurement unit that measure the position and rotation of the lidar sensor, affording accurate measurements of targets. Common applications for these data including the profiling of water depths (known as bathymetry), terrestrial elevation/terrain modelling, canopy height measurement of crops (for crop yield management) and of forests, and the 3D structure retrieval of urban areas (known as 3D city modelling). These applications exploit the differences in the backscattered pulses to the lidar. In the case of water, the strong first return is from the water surface, with the bottom of the water body giving rise to a weaker return. Urban areas give rise to a strong first return from which a digital surface model can be computed, and if the terrain is known this provides the height of all features in the urban environment (heights of buildings, street furniture, etc.). In forested environments it is often found that the emitted pulse can pass through the canopy, giving rise to multiple return echoes that can be usefully recorded with the full waveform systems. This means a wealth of information can be retrieved about forests in addition to their canopy heights, including forest biomass, structure and species. A similar set of data can be retrieved from satellite-mounted lidars, which are terrain focused (e.g. ICESat/GLAS), but these have a different set of sampling protocols and spatial resolutions (having a large footprint). Satellite lidars that are atmosphere focused are deployed to measure properties of clouds and aerosols. Terrestrial lidars are commonly deployed as a means of acquiring detailed and highly accurate 3D data rapidly and efficiently. Applications are wide ranging from a physical geography point of view, including geomorphological mapping and forest biomass estimation. In each case the areal coverage is limited, although using a kinematic system improves capture rates. There are many advantages to using lidar to capture information about the physical environment, with the biggest one being that as they are part of active remote sensing systems they do not rely on solar radiation as its source; they can be deployed any time of day or night in any season. Recent years has seen the costs of deploying lidar systems (particularly airborne ones) reduce dramatically, so that it has become an economic source of data even for regional or country-wide surveys.　　　　　DSB

Reading

Dassot, M., Constant, T. and Fournier, M. (2011) The use of terrestrial lidar technology in forest science: application fields, benefits and challenges. *Annals of Forest Science*, **68**, 959–974. · Fuiji, T. and Fukuchi, T. (2005) *Laser remote sensing*. Boca Raton, FL: CRC Press. · Hyyppä, J., Wagner, W., Hollaus, M. and Hyyppä, H. (2009) Airborne laser scanning. In T. A. Warner, M. D. Nellis and G. M. Foody (eds), *The SAGE handbook of remote sensing*. London: SAGE; pp. 199–211. · Mallet, C. and Bretar, F. (2009) Full-waveform topographic lidar: state-of-the-art. *ISPRS Journal of Photogrammetry and Remote Sensing*, **64**, 1–16.

life form The body shape of an organism at maturity, most commonly used in reference to plants. RAUNKIAER'S LIFE FORMS classification, for example, is based mainly on the nature of perennating buds and their position and protection. This distinguishes plants over or below 25 cm from the ground, at soil level, below ground or lying in mud or water. Other life-form features of plants include the length of shoots or the nature and density of the root system. Animal classifications also employ life-form attributes that result from morphological adaptations to the environment.　　　　PAF

lightning A luminous discharge associated with a thunderstorm. Several types can be distinguished, principally:

1 cloud discharges (also called sheet lightning) occur between different parts of a thunderstorm, giving a diffuse illumination;
2 ground discharges (also called forked lightning) occur between cloud and ground along a tortuous path with side branches from a main channel;

3 air discharges occur between a part of the cloud and the adjacent air, but otherwise are similar in appearance to a ground discharge; and

4 ball lightning, reported to have the appearance of a moving, luminous globe-discharge about 20 cm in diameter, sometimes disappearing in a violent explosion.

In addition to being a significant natural hazard, lightning is an important agent in fire ecology and may also have miscellaneous geomorphological effects.   KJW

Reading
Ackerman, S.A. and Knox, J.A. (2014) *Meteorology: understanding the atmosphere*, 4th edition. Sudbury, MA: Jones and Bartlett. · Norin, J. (1986) Geomorphological effects of lightning. *Zeitschrift für Geomorphologie*, **30**, 141–150.

limiting angles (of slopes)
Describe the upper and lower angles at which distinct processes or forms may occur either in a given locality or under particular environmental conditions. The upper (maximum) limiting angles for continuous regolith and vegetation cover are commonly regarded as being in the range 40–45° in western Europe; in Papua New Guinea on mudstones under rain forest the limiting angle is in the range 70–80°. Lower limiting angles for the occurrence of landslides have been quoted for a few areas: in thin regolith under temperate climates this may be in the range 18–40°, but in periglacial environments the angles are much lower and in the range 1–8°. Further examples are given by Young (1972: 165).   MJS

Reference
Young, A. (1972) *Slopes*. Edinburgh: Oliver & Boyd.

limiting factors
Those factors in ecosystems that are in short supply and can thus inhibit efficient and productive ecological development. The concept of limiting factors was introduced in the 1840s by the German chemist Justus von Liebig, who found that the yield of a crop could be increased only be supplying the plants with more of the nutrient that was present in the smallest quantities.   ASG

Reading
Blackman, F. (1905) Optimal and limiting factors. *Annals of Botany*, **19**, 281–295. · Park, C.C. (1980) *Ecology and environmental management*. Folkestone: Dawson; pp. 94–99.

limnology
Derived from the Greek *limnos* (pool, lake, swamp), limnology is the study of the physical, chemical and biological processes of fresh or saline waters surrounded by land. The actual term limnology was first used by Forel in his 1892 book on the limnology of Lake Leman, although biological limnology began in 1674 when Leeuwenhoek provided the first microscopic description of the green ALGAE, *Spirogyra*. His pioneering study not only outlined the seasonal cycle of algae in lakes, but gave the first insights about food chain dynamics. By the late 1700s Saussure had devised a way of measuring temperature in deep water and discovered that lakes are thermally stratified. An enormous amount of information on lakes is now available, and limnologists are not only concerned with investigating the physical, chemical and biological properties of lake systems but also their preservation and enhancement. Researchers are not only interested in the present and future condition of lakes, but also the past, with the reconstruction of past lake environments being the domain of palaeolimnologists. (See also LAKES.)   SLO

limon
French term used for fine silty sediments laid down by the wind (or possibly water) and therefore equivalent to LOESS.

line squall
See SQUALL LINE.

lineament
A large-scale linear feature on the land surface, such as a trough or ridge, that is the product of the structural geology of a region.

linear dune
Linear dunes can be over 50 m high and may extend for tens of kilometres in length. They are the most common type of desert sand dune (Fryberger and Goudie, 1981) and usually occur in extensive dune fields in which they are the dominant or only major dune type that is present; for example, in the Kalahari, Australian continental dune fields, and the western Saharan ergs. Despite this, the mechanisms that lead to the formation of linear dunes have proved controversial; for example, one unsubstantiated theory was that they formed under the influence of parallel HELICAL FLOW vortices in the lower atmosphere. It is now known, and substantiated by field investigations by Tsoar (1978) and Livingstone (1986), that linear dunes form in acute bimodal wind regimes, with overall dune orientation parallel/subparallel to the resultant direction of sand transport. In most situations it is seasonal differences in wind regime that make up the components of the formative regime; if one of these results in greater net transport than the other, a slow lateral movement of the dune may be caused. Overall, however, linear dunes do not migrate, but extend slowly in the net transport direction.

The SLIP FACE on a linear dune may be relatively weakly developed, as is the case on a SAND RIDGE, or it may alternate seasonally from one side of the dune to the other. This may give rise to the dune developing a sinuous crestal profile, termed a SEIF dune. Overall, the lack of migratory behaviour, and occurrence of dune extension at the downwind end, means that linear dunes can be regarded as 'sand-passing' forms. This type of activity can mean that the plinths of linear dunes are relatively stable, to the extent that they may support comparatively dense vegetation covers, with only upper slopes experiencing sand movement. Livingstone and Thomas (1993) have noted that the activity of linear dunes may be extremely episodic, making it difficult in some situations to distinguish between relict forms, inherited from drier climatic periods, and episodically active dunes with limited surface activity.   DSGT

References
Fryberger, S.G. and Goudie, A.S. (1981) Arid geomorphology. *Progress in Physical Geography*, 5, 420–428. · Livingstone, I. (1986) Geomorphological significance of wind flow patterns over a Namib linear dune. In W. G. Nickling (ed.), *Aeolian geomorphology*. Boston, MA: Allen and Unwin; pp. 97–112. · Livingstone, I. and Thomas, D.S.G. (1993) Modes of linear dune activity and their palaeoenvironmental significance: an evaluation with reference to southern African examples. In K. Pye (ed.), *The dynamics and context of aeolian sedimentary systems*. The Geological Society of London, Special Publication No. 72. London: The Geological Society; pp. 91–101. · Tsoar, H. (1978) *The dynamics of longitudinal dunes: final technical report*. London: European Research Office, US Army.

liquid limit The maximum amount of water an unconsolidated sediment or material can hold before it becomes a turbid liquid.

lithification The process of the formation of a consolidated rock from originally unconsolidated sediments through cementation or other diagenetic processes.

lithology The macroscopic physical characteristics of a rock.

lithosol Surficial deposits that do not exhibit soil horizons.

lithosphere The Earth's crust and a portion of the upper MANTLE, which together constitute a layer of strength, relative to the more easily deformable ASTHENOSPHERE below. On the basis of worldwide heat flow measurements, it has been estimated that the lithosphere varies in thickness from only a few kilometres along the crest of mid-ocean ridges where, according to the PLATE TECTONICS model, new lithosphere is being created, to over 300 km beneath some continental areas. Oceanic lithosphere capped by continental crust tends to be thinner but more dense than continental lithosphere, which is capped by continental crust.   MAS

Reading
Pollack, H.N. and Chapman, D.S. (1977) On the regional variations of heat flow, geotherms and lithospheric thickness. *Tectonophysics*, 38, 279–296. · Walcott, R.I. (1970) Flexural rigidity, thickness and viscosity of the lithosphere. *Journal of Geophysical Research*, 75, 3941–3954. · Wilson, J.T. (ed.) (1976) *Continents adrift and continents aground*. San Francisco, CA: W.H. Freeman.

litter The remains of dead vegetation material, especially tree leaves that are present on the ground surface. They are broken down into essential nutrients by a wide range of DECOMPOSER organisms (bacteria, fungi) and other associates, among which are earthworms, springtails, mites and millipedes. The rate of breakdown and, conversely, the degree of accumulation of litter varies with climate. The amount of litter present in tropical rain forests, where breakdown is fast, may be only $20 \, kg \, ha^{-1}$, or less than 1% of the above-ground BIOMASS, whereas in the cold, dry climate of high-latitude boreal forests litter accumulation is substantial due to slow rates of breakdown, reaching $300 \, kg \, ha^{-1}$, or ~30% of the above-ground biomass (Rodin and Basilevich, 1967).   DW

Reference
Rodin, L.E. and Basilevich, N.I. (1967) *Production and mineral cycling in terrestrial vegetation*. Edinburgh: Oliver and Boyd.

litter dam A small barrier made up of particles of plant litter, especially leaves, twigs and flower parts. Litter dams are built up when these particles, floating on surface run-off, periodically come to rest and then trap additional particles delivered by the flow. In this way dams may grow to heights of 5 cm or more. The dams run approximately along the contour, and often occur in tiers. They are important, especially in small run-off events, because they retard or hold water and so promote infiltration. They also trap eroded soil, which settles in the impoundment. In dry climates especially, seeds also lodge within litter dams, and the water and nutrients available there may support germination.   DLD

Little Climatic Optimum A phase in early medieval times (*c.* AD 750–1200) when conditions were relatively clement in Europe and North

America, allowing settlement in inhospitable parts of Greenland, reducing the problems of ice on the coast of Iceland, and allowing widespread cultivation of the vine in Britain. Now more commonly referred to as the MEDIEVAL WARM PERIOD.   ASG

Reading
Lamb, H.H. (1995) *Climate, history, and the modern world*, 2nd edition. London: Routledge.

Little Ice Age (LIA)   See also NEOGLACIAL.
Occurring after the MEDIEVAL WARM PERIOD, the LIA occurred within the period between AD 1500s and 1800s, with duration and extent varying globally. While glaciers in some parts of the world extended at this time (e.g. Harrison *et al.*, 2014), it was no true 'ice age'. In Europe, winters became colder, snowfall was higher and, consequently, the growing season and crop production declined, and has been linked to periods of famine. Environmental records from, for example, Patagonia, New Zealand and Malawi suggest that the LIA also impacted in the southern hemisphere. Attribution is unclear, with various analyses suggesting changes in solar output, volcanic activity, ocean circulation or orbital forcing as the primary cause, or even an increase in forested areas leading to atmospheric carbon draw-down (Ruddiman, 2003).   DSGT

Reading and References
Araneda, A., Torrejón, F., Aguayo, M., *et al.* (2007) Historical records of San Rafael glacier advances (North Patagonian Icefield): another clue to 'Little Ice Age' timing in southern Chile? *The Holocene*, 17, 987–998. · Broecker W.S. (2000) Was a change in thermohaline circulation responsible for the Little Ice Age? *Proceedings of the National Academy of Sciences of the United States of America*, 97, 1339–1342. · Harrison, S., Rowan, A.V., Glasser, N.F., *et al.* (2014) Little Ice Age glaciers in Britain: Glacier-climate modelling in the Cairngorm Mountains. *The Holocene*, 24, 135–140. · Ruddiman, W.F. (2003) The anthropogenic greenhouse era began thousands of years ago. *Climatic Change*, 61, 261–293.

littoral, littoral zone   The part of a sea, lake or river that is close to the shore. In coastal environments the littoral zone extends from the high-water mark, which is rarely inundated, to shoreline areas that are permanently submerged. Three zones are recognized: (1) the supralittoral zone (also called the splash, spray or supratidal zone), the area above the spring high-tide line that is regularly splashed, but not submerged by seawater; (2) the eulittoral zone (midlittoral or mediolittoral zone), the intertidal zone between the upper and lower limits of high and low spring tides respectively; and (3) the sublittoral zone, which is permanently covered with seawater and can extend as far as the edge of the continental shelf (~200 m water depth). In freshwater systems, the littoral zone may form a narrow or broad fringing wetland. Typically, four zones are recognized, from higher to lower on the shore: wooded wetland, wet meadow, freshwater marsh and submerged aquatic vegetation.   TS

load, stream   The total mass of material transported by a stream. The units employed vary according to the time-base considered. Short-term loads may be expressed in kilograms or tonnes per day, whereas annual loads are expressed in tonnes per year. The total includes both organic and inorganic material and comprises three major components: first, material carried in solution; second, material transported in suspension; and third, material moving on the bed of the stream as bedload. The magnitude of the load and the relative importance of the three load components may vary markedly in both time and space. (See also BEDLOAD, BEDLOAD EQUATION, DISSOLVED LOAD and SUSPENDED LOAD.)   DEW

load structures   Irregular contortions found in fine-grained deposits where sands have been deposited on water-saturated hydroplastic silts or muds. Differences in density, compaction and pore-fluid pressures cause lobes of sand to sink into the underlying silts and/or tongues of mud to rise up into the sand horizons. The resulting load structures exhibit contorted and deformed laminae, often folded, festooned or detached.   JM

Reading
Rodin, L.E. and Basilevich, N.I. (1967) *Production and mineral cycling in terrestrial vegetation*. Edinburgh: Oliver & Boyd.

local climate   See MESOCLIMATE and MICROCLIMATE.

local winds   Those winds that differ from the general winds expected from the pressure pattern due to topographical or urban or other effects. Four main types of local wind have been identified: (1) those winds intensified by topographical features, such as a narrow mountain gap or urban canyon; (2) winds that blow along the pressure gradient, such as land and sea breezes, mountain and valley winds and, on larger scales, föhn, chinook, bora and mistral winds; (3) winds associated with vertical instability, such as those accompanying thunderstorms; (4) strong winds due to flow over a level surface with a strong pressure gradient, such as uninterrupted flow over a level surface with

a strong pressure gradient, such as sirocco and blizzard. JET

lodgement till   See TILL.

loess

The original German word *Löss* was simply a name for a particular form of loose, crumbly earth. In due course, definitions became more constricted, and that of Flint (1957: 181) received wide currency: 'a sediment, commonly nonstratified and commonly unconsolidated, composed predominantly of silt-sized materials, ordinarily with accessory clay and sand, and deposited primarily by wind'. Particularly thick and extensive deposits of loess occur in Asia (e.g. the Chinese Loess Plateau), but it is also common in central Europe, parts of North and South America, and on some desert margins. The depositional processes of loess have been debated in the literature. Today, though, it is widely accepted that loess is primarily wind deposited, although the *formation* of silt-sized quartz fragments may be attributed to a wide range of mechanisms, particularly, in the context of higher and mid-latitude occurrences, glacial grinding (e.g. Smalley and Vita-Finzi, 1968), with the very widespread development of loess deposits around the borders of Pleistocene ice sheets supporting this view. Rock weathering (Wright, 2007) and aeolian attrition may also be important.

The fossil soils and faunal and floral remains in thick, dated loess sections provide an environmental record that is equalled only by that preserved in the deep sea cores (Kukla, 1977). The application of LUMINESCENCE DATING methods to loess has allowed a direct record of accumulation histories to be developed that, when combined with other new proxy analyses, has challenged some older views on the controls on accumulation (Stevens *et al.*, 2007) ASG/DSGT

References

Flint, R.F. (1957) *Glacial and Pleistocene geology.* New York: John Wiley & Sons, Inc. · Kukla, G.J. (1977) Pleistocene land–sea correlations: I. Europe. *Earth Science Reviews*, **13**, 307–374. · Smalley, I.J. and Vita-Finzi, I.J. (1968) The formation of fine particles in sandy deserts and the nature of 'desert' loess. *Journal of Sedimentary Petrology*, **38**, 766–774. · Stevens, T., Thomas, D.S.G., Armitage, S.J., *et al.* (2007) Reinterpreting climate proxy records from late Quaternary Chinese loess: a detailed OSL investigation. *Earth-Science Reviews*, **80**, 111–136. · Wright, J.S. (2007) An overview of the role of weathering in the production of quartz silt. *Sedimentary Geology*, **202**, 337–351.

logan stone

Any large boulder that is so balanced that it readily rocks.

long profile, river

The graph representing the relation between altitude and distance along the course of the river. The profile is usually concave upwards, is graded to a local or regional BASE LEVEL, and may be punctuated by KNICKPOINTS where the river cuts through former valley floors or river terraces. The profile may be plotted for the bed of the river channel or for the bankfull or channel capacity stage where analysis is to be related to contemporary processes, but it will be plotted for the floodplain or valley floor when related to valley development. The long profile of an entire river or valley may be approximated by one of several equations, but in detail over short distances the river long profile is punctuated by pools and riffles. KJG

Reading
Richards, K.S. (1982) *Rivers: form and process in alluvial channels.* London: Methuen; pp. 222–251.

longitudinal dune

Term sometimes used in place of LINEAR DUNE.

longshore drift

Longshore drift is the transport of beach material along the coast. There are two main processes involved. Beach drifting is caused by the oblique upward transport of material by the swash of short, little-refracted WAVES, and its return straight down the swash slope by the BACKWASH, thus moving it alongshore. The process takes place only in the swash zone. Longshore currents in the surf zone, generated by waves approaching the coast at an angle, also move material alongshore. The transport rate depends upon the wave energy and angle of approach mainly. Longshore transport with steep waves is usually at a maximum along the submarine bar crest. The process is of great importance in accounting for long-term coastal erosion and deposition. CAMK

Reading
Hardisty, J. (1990) *Beaches: form and process.* London: Unwin Hyman. · Komar, P.D. (1971) The mechanism of sand transport on beaches. *Journal of Geophysical Research*, **76**, 713–721.

lopolith

An igneous intrusion similar to a laccolith but saucer-shaped on both its upper and lower surfaces.

löss   See LOESS.

louderback

A lava flow on the surface of the dip slope of a faulted block, the presence of which proves that the topography is the product of faulting rather than erosion. This is because the lava flow is separated from its source location by the faulting process. The term is named after the geologist G. D. Louderback who first described such flows in the western USA.   DSGT

low flow analysis
An analysis of the frequency, magnitude and persistence of low discharge and its relationship with climatic and catchment characteristics.   AMG

Reading
Institute of Hydrology (1980) *Low flow studies.* Wallingford: Institute of Hydrology. · Task Committee on Low Flows (1981) Characteristics of low flows. *American Society of Civil Engineers: Journal of the Hydraulics Division,* **106**, 717–732.

luminescence dating
Provides a means of directly determining the age of mineral grains in numerous sedimentary environments. They are based on estimating the time elapsed since a sediment was last exposed to daylight. They are radiation damage (or sometimes termed 'trapped charge') GEOCHRONOLOGY methods involving quantifying the accumulation of a radiation-related signal. Initial developments in the field, resulting in THERMOLUMINESCENCE (TL) dating, focused on the use of heat to stimulate a time-dependent signal from mineral grains that had been 'zeroed' by exposure to heat during formation or usage (e.g. pottery, burnt flint). In the early 1980s it was realized that exposure to daylight was also capable of 'zeroing' previously accumulated radiation damage, and the method was applied to sediment. The date so established was the time elapsed since a sediment was last exposed to daylight (a depositional age). A significant drawback to TL methods was the difficulty in establishing the completeness of the resetting (bleaching) process at deposition. A significant advantage of the related development optical dating methods is the effectiveness of signal resetting at deposition.

Optically stimulated luminescence (OSL) dating involves the generation of a signal from a sample by optical stimulation Convention now is to further classify the form of signal according to the wavelength of optical stimulation (e.g. green light or infrared stimulated luminescence). Unlike TL, the eviction of charge from traps is observed only as a rapidly depleting signal when expressed as a function of light exposure. Minerals are variously sensitive to differing wavelengths of light; quartz, for example, is sensitive to green ($\lambda \approx 500$ nm) and shorter wavelengths; feldspars exhibit a wide range of sensitivities to visible, red and near-infrared photons.

The age of a sample is derived from the general age equation:

$$\text{Age(ka)} = \frac{\text{Equivalent dose(Gy)}}{\text{Dose rate(Gy ka}^{-1})}$$

where the palaeodose P (sometimes referred to as equivalent dose De or accumulated dose AD) is the accumulated radiation damage (measured in grays), and the dose rate is the rate at which the sample absorbs energy from its immediate proximity; this comprises alpha, beta, and gamma radiation from ^{40}K, ^{238}U, ^{235}U and ^{232}Th and their protégé products, together with a typically small cosmic ray contribution (Aitken, 1985, 1998).

There are a variety of protocols that have been employed for evaluation of De. Most utilize measurements of the response of single or multiple aliquots of refined samples of quartz, feldspar or mineral mixtures to laboratory radiation to quantify the radiation sensitivity of the mineral used, and all require some form of thermal pretreatment to remove unstable luminescence components from those that are stable over geological time periods. Multiple aliquot approaches are little used now, and are a significant source of error, while single-grain dating (whereby final De values are the result of specifically targeting multiple, but individually identifiable, grains, thereby allowing variance ion behaviour to be better understood) is growing in use.

Palaeodoses are typically calculated either for progressive temperature (TL) or exposure time (OSL) intervals as a means of testing the stability and reproducibility of the estimate. Dose rate evaluation may be achieved by a wide variety of field or laboratory-based spectroscopic, nuclear and chemical methods (Aitken, 1985). Dose rate is generally less well investigated than the equivalent dose and may be an insufficiently recognized source of error in ages.

The maximum age range of TL and OSL methods is controlled both by the quantity of radiation damage that a given mineral species can accommodate and the rate at which the damage accumulates. At typical environmental dose rates quartz saturates at between 100 and 150 ka, although under certain circumstances ages of over 700 ka may be obtainable. Feldspars demonstrate a considerably greater capacity to accumulate charge and correspondingly exhibit a maximum age range up to 800 ka, though many feldspars exhibit anomalies that may result in age underestimation. Fine-grained polymineral mixtures from loess have also been dated up to 800 ka.

The minimum age range of the methods is determined by both the completeness of resetting of previously accumulated luminescence at deposition (including the degree to which the solar spectrum is filtered in subaqueous depositional environments), sample sensitivity to ionizing radiation and subtle charge reorganization effects that may occur during deposition, or which are the result of laboratory pretreatment procedures.

Luminescence dating has grown in use since the 1980s and has proved especially valuable in desert dune environments where other dating techniques cannot be applied. Luminescence is also applied to fluvial, glacial and other sediments, and has been widely used in archaeological contexts.   SS/DSGT

Reading and References
Aitken, M.J. (1985) *Thermoluminescence dating*. London: Academic Press. · Aitken, M.J. (1998) *Optical dating*. Oxford: Oxford University Press. · Liritzis, I., Singhvi, A.K., Feathers, J.K., *et al.* (2013) *Luminescence dating in archaeology, anthropology and geoarchaeology: an overview*. Heidelberg: Springer.

lunette dune   A crescent-shaped sand dune found on the downwind margin of some PAN or PLAYA basins, with the term first used by Hills (1940). Lunettes are widely identified in Australia, southern Africa (Telfer and Thomas, 2006), Tunisia and the US High Plains (Bowen and Johnson, 2012). Lunettes are in fact a form of TRANSVERSE DUNE; while not present on the margins of all pans, they are widespread in environments whether the pan depression occurs in a sandy substrate or where processes on the pan floor lead to the release of sand-size sediment particles. Lunette dunes may form directly by the deflation of sediments from dry pan surfaces or from deflation from beach sediments that result from wave transport, so that their development could occur during dry or wet phases. While this may influence the palaeoenvironmental significance of lunettes dated to the late Quaternary period, those containing CLAY PELLETS (also called CLAY DUNES) are almost certainly a result of deflation directly from dry floors.   DSGT

Reading and References
Bowen, M.W., and Johnson, W.C. (2012) Late Quaternary environmental reconstructions of playa–lunette system evolution on the central high plains of Kansas, United States. *Bulletin of the Geological Society of America*, **124**, 146–161. · Bowler, J.M. (1986) Spatial variability and hydrological evolution of Australian lake basins: analogue for Pleistocene hydrological change and evaporite formation. *Palaeogeography, Palaeoclimatology,Palaeoecology*, **54**,

21–41. · Hills, E.S. (1940) The lunette: a new landform of aeolian origin. *Australian Geographer*, **3**: 1–7. · Telfer, M.W. and Thomas, D.S.G. (2006) Complex Holocene lunette dune development, South Africa: implications for palaeoclimate and models of pan development in arid regions. *Geology*, **34**: 853–856.

lynchet   A terrace on a hillside, generally held to be human-made and produced by ploughing. Lynchets are widespread in southern England and northern France.

lysimeter   An instrument for assessing evapotranspiration losses from a vegetated soil column. Lysimeters may be used to assess either actual or POTENTIAL EVAPOTRANSPIRATION losses and the estimates are derived using a WATER BALANCE approach. A column of soil and vegetation is placed in a container and replaced in the soil so that the vegetation and soil conditions are as similar as possible to their surroundings. The container should be as large as possible to allow free growth of the vegetation and to reduce the significance of boundary effects. There are two main types of lysimeter, the drainage type and the weighing type, although some weighing lysimeters also allow drainage. Input of water to the lysimeter is assessed using rain gauges, output is measured as drainage from the base of the container enclosing the soil column, and changes in soil moisture storage are estimated by repeatedly weighing the soil column. These measurements allow the estimation of losses of water through evapotranspiration. If estimates of potential evapotranspiration are required, the lysimeter and a surrounding area are irrigated to ensure that the soil moisture is maintained at field capacity.

There is no standard size for a lysimeter. Larger instruments are less influenced by boundary effects but present problems if soil moisture changes are to be determined accurately by changes in weight of the soil column. Occasionally, special environmental circumstances allow the construction of very large drainage or weighing lysimeters (Schrader *et al.*, 2013).   AMG/MEM

Reading and Reference
Reyenga, W., Dunin, F.X., Bautovich, B.C., *et al.* (1988) A weighing lysimeter in a regenerating eucalyptus forest: design, construction and performance. *Hydrological Processes*, **2**, 301–314. · Schrader, F., Durner, W., Fank, J., *et al.* (2013) Estimating precipitation and actual evapotranspiration from precision lysimeter measurements. *Procedia Environmental Sciences*, **19**, 543–552.

M

maar   An old volcanic crater, usually shallow. A pond or lake formed in such a depression. The largest maars, up to ~8000 m in diameter, occur in northwest Alaska and up to no more than 300 m deep. The term derives from the Latin *mare*, or sea.

macroecology   The study of the relationships between organisms and their environment over broad geographical areas and/or time periods. It is a subfield of ecology that typically aims to understand and explain the patterns of life, notably those relating to the abundance, distribution and diversity of species.

In macroecology, attention is focused upon broad-scale patterns in the ecological characteristics of organisms and ecosystems and is approached from a 'top down' perspective in which understanding arises from the studies of the system as a whole rather than from knowledge of the component parts. At such broad scales manipulative experiments are typically not feasible, and research tends to use statistical and/or mathematical methods applied to observed or simulated data sets. Examples of typical subjects for macroecological research are the latitudinal gradients in species richness, species–area relationships and variations in range size distributions, with the broad patterns observed used to uncover processes and further understanding.   GF

Reading
Gaston K.J. and Blackburn, T.M. (2000) *Pattern and process in macroecology.* Malden, MA: Blackwell. · Keith, S.A., Webb, T.J., Böhning-Gaese, K., *et al.* (2012) What is macroecology? *Biology Letters,* **8,** 904–906.

macrofossils   Animal or plant fossil remains visible with the naked eye but which usually require microscopic examination for identification. The commonest macrofossils are those of the genus *Sphagnum,* which occur in and may comprise the bulk of many PEAT deposits; other common macrofossils include seeds and fruits (often abundant in lake sediments), insect remains and molluscs. Many can be identified down to species level. The study of all three groups has been well developed in Quaternary palaeoecology to demonstrate local vegetational changes and hydroseral development, and to investigate wider phenomena such as climatic change (insects, molluscs and peat macrofossils) and marine transgressions (molluscs).   KEB

Reading
Birks, H.J.B. and Birks, H.H. (1980) *Quaternary palaeoecology.* London: Edward Arnold. · Godwin, H. (1975) *History of the British flora,* 2nd edition. Cambridge: Cambridge University Press.

macrometeorology   Three spatial scales of atmospheric motion are distinguished: micro, meso and macro or planetary scale. The lower limit of the macroscale is variously defined in the range of a hundred to a few thousand kilometres. Macroscale systems include the major features of the GENERAL ATMOSPHERIC CIRCULATION, such as upper-level JET STREAMS, waves in the westerlies (ROSSBY WAVES), MONSOONS and subtropical highs. Various phenomena that characterize climatic variability are also macroscale. Examples are the SOUTHERN OSCILLATION and the Atlantic meridional overturning (see MERIDIONAL OVERTURNING CIRCULATION).   SJM

macrophyte   Although in the strict sense of the word this is a plant large enough to be seen without the aid of a microscope, in practice it refers to the larger photosynthesizing plants in a community. More typically used in relation to aquatic or marine habitats (see AQUATIC MACROPHYTE), in the context of terrestrial environments the term most likely refers to plants of more substantial physical stature; for example, trees and shrubs as opposed to grasses and herbs. In the terrestrial sense, then, the term is relative and would not be used to describe all those plants large enough to be seen by the naked eye. Thus, in a temperate deciduous forest, the imposing oak and beech trees, reaching perhaps 30 m in height, would be described as macrophytes, whereas the smaller herbaceous flowering plants in the understorey would not.   MEM

macropore Soil pore spaces range in size from the very small voids or 'micropores' between the clay, silt and sand particles making up the soil matrix, to larger openings, 'macropores', created by shrink–swell processes, the roots of plants or burrowing organisms. The micropores, often defined as those <30 μm in diameter, retain water within the soil matrix because, within them, water molecules are strongly attracted to the mineral surfaces. Within the larger macropores, faster turbulent flow may occur. Various attempts to define macropores according to size have been made, including suggested minimum equivalent diameters of 75 μm or 100 μm. However, macropores may be planar cracks or voids of irregular shape. They facilitate the aeration and drainage of soils and result in a much larger hydraulic conductivity than would arise from the microporosity alone. Field evidence shows that water can rapidly pass through soil by following preferential flow paths along interconnected macropores, in this way speedily bypassing the much less permeable soil matrix.   DSGT

Reading
Beven, K. and Germann, P. (2013) Macropores and water flow in soils revisited. *Water Resources Research*, **49**, 3071–3092.

Madden–Julian Oscillation (MJO) Intra-
aseasonal variations in tropical precipitation on timescales of roughly 30–60 days. The MJO affects global circulation in the tropics and subtropics, but also the wintertime jet streams that influence weather over the North Pacific and western North America. The cloudiness and precipitation associated with the MJO propagate around the globe in the low latitudes.   SJN

Reading
Wheeler, M. and Hendon, H. (2004) An all-season real-time multivariate MJO index: development of an index for monitoring and prediction. *Monthly Weather Review*, **132**, 1917–1932.

maelstrom A powerful tidal current or whirlpool.

magma Fused, molten rock material found beneath the Earth's crust from which igneous rocks are formed. Magma may contain gases and some solid mineral particles.

magnetic anomaly A fluctuation in the strength of the magnetic field mapped over large areas of the ocean floor. Magnetic anomalies were so named when they had been mapped but not yet explained. Their origin lies in sea-floor spreading,

which results when new basaltic rock wells up along the mid-oceanic fracture system and, cooling as it does so, is magnetized in the direction of the Earth's own magnetic field. The geomagnetic field periodically reverses, however, with the southern magnetic pole exchanging locations with the northern magnetic pole. New basalts intruded into the sea floor during a time of REVERSED POLARITY then have a magnetic orientation opposite to that of the present-day field. Thus, when the apparent strength of the field in this vicinity is measured, it appears anomalously weak, the field in the basalts opposing the Earth's field. Likewise, above rocks magnetized normally, the field appears anomalously strong since the field of the rocks reinforces the Earth's field. DLD

Reading
McElhinny M.W. (1973) *Palaeomagnetism and plate tectonics*. Cambridge: Cambridge University Press.

magnetic declination The angle separating true north (the direction to the northern geographic pole of rotation) from magnetic north, the direction to the northern magnetic pole. Since it is related to fluid processes in the outer core, the geomagnetic field is not completely stable in its orientation and, indeed, drifts westward, lagging the Earth's rotation.   DLD

Reading
Merrill, R.T., McElhinny, M.W. and McFadden, P.L. (1996) *The magnetic field of the Earth: paleomagnetism, the core, and the deep mantle*. San Diego, CA: Academic Press.

magnetic storm A high level of magnetic disturbance produced by particles of solar origin, causing rapid field variations over the Earth. Such storms disturb the ionosphere, causing anomalous radio propagation and adversely affecting cable telegraphy. During storms, aurora, arcs and rays of coloured light appearing in the sky are visible much further towards the equator than their usual position.   KJW

magnetic susceptibility Iron-bearing minerals are widespread in the natural environment and their presence may be diagnostic of geology, soil processes, sediment pathways, pollution and biological conditions. Large differences in the way iron is organized at the atomic level allow these minerals to be studied and classified by their magnetic properties. The simplest and most fundamental property is the ease by which a mineral is magnetized, which is termed *magnetic susceptibility*. Measurements of environmental samples provide first-order information about the type of iron-bearing minerals and their concentrations.

Routine measurements are rapid, nondestructive and may be made in the laboratory or field using portable probes and sensors. In the natural environment, metals, iron-bearing minerals and materials can be divided according to their magnetic properties: ferromagnetic (e.g. iron, nickel), ferrimagnetic (e.g. magnetite, maghemite, titanomagnetite, greigite), imperfect antiferromagnetic (e.g. goethite, haematite) and paramagnetic (e.g. ferrihydrite, lepidocrocite, ilmenite, iron silicates). These all show a positive susceptibility, but it is the strongly magnetic ferrimagnetic group that dominates a bulk magnetic susceptibility value in the majority of samples. The susceptibility of magnetite is about 1000 times higher than a typical paramagnetic mineral. Materials that are generally considered as nonmagnetic (e.g. quartz, water, organic matter, $CaCO_3$) are termed diamagnetic and show a very weak and negative susceptibility. Within the ferrimagnetic group especially, minerals may also be divided according to the type of internal organization into magnetic domains controlled by crystal dimensions. Thus, for magnetite there are multidomain, pseudo-single domain, stable single domain ($<0.2\,\mu m$) and superparamagnetic domain ($<0.03\,\mu m$) states. Identification of the domain state provides important information about the origins and formation of ferrimagnetic minerals. Primary minerals derived from igneous rock are predominantly multidomain, pseudo-single domain and stable single domain, and secondary minerals produced bio-geochemically in soil and sediments, or through fire, are mainly stable single domain and superparamagnetic. (See also ENVIRONMENTAL MAGNETISM.)

Volume susceptibility κ is defined as the ratio of magnetization M to an applied magnetic field H: $\kappa = \mathbf{M}/\mathbf{H}$. *Mass specific susceptibility* χ is defined as the volume susceptibility divided by sample density ρ: $\chi = \kappa/\rho$. Preferred measurement units are SI, where magnetization and field are expressed in amperes per metre and density is expressed as kilograms per cubic metre. Thus, κ is dimensionless and χ is expressed in units of cubic metre per kilogram. For magnetite, $\kappa \approx 3.1$ (SI) and $\chi \approx 600 \times 10^6 m^3 kg^{-1}$. Most routine measurements are made in low magnetic fields (typically $80\,A\,m^{-1}$) at room temperature, but there are three additional sets of measurements: (i) in high magnetic fields (typically $>10^6\,A\,m^{-1}$); (ii) in low fields using two AC frequencies set a decade apart; and (iii) at low or high temperatures. *High field susceptibility* is controlled by nonferrimagnetic behaviour. In practice, it is often a good estimate of the concentration of total iron contained in the paramagnetic iron oxides, iron hydroxides and iron-bearing silicates, which

in terms of mass contribution may outweigh the ferrimagnetic minerals by orders of magnitude.

Frequency-dependent susceptibility χ_{FD} detects very fine grained ferrimagnetic minerals, lying within the superparamagnetic domain range. Expressions of χ_{FD} are in mass specific ($\chi_{LF} - \chi_{HF}$) or percentage terms ($[(\chi_{LF} - \chi_{HF})/\chi_{LF}] \times 100$), where LF and HF denote low and high frequency (typically 470 Hz and 4700 Hz respectively). Percentage χ_{FD} values range from zero in samples containing no superparamagnetic grains to $\sim 15\%$ in samples where superparamagnetic grains completely dominate the magnetic mineral assemblage. The measurements are most useful in studies of modern and old soils where superparamagnetic grains are an important component of secondary ferrimagnetic mineral formation.

Thermal susceptibility measurements χ_T from $-196\,°C$ (liquid nitrogen) to room temperature, and from room temperature to $\sim 700\,°C$, may provide valuable information about mineral and domain type. Two types of diagnostic behaviour are common. Thermal disordering causes the susceptibility of paramagnetic minerals to decline with increasing temperature (Curie–Weiss law) and in other minerals and domains occurs at specific temperatures known as Curie points, Néel points or blocking temperatures. For instance, the ordered ferrimagnetic behaviour of magnetite becomes disordered and paramagnetic at $\sim 580\,°C$ with an almost total loss of susceptibility. Transitions in magneto-crystalline properties cause changes in susceptibility at specific temperatures, notably the Verwey transition in multidomain magnetite ($-155\,°C$) and Morin transition in multidomain haematite ($-10\,°C$). Measurements made at low temperature are essentially nondestructive with reversible χ_T changes, while those at high temperature often lead to irreversible changes, mineral destruction and the formation of new minerals. Magnetic susceptibility measurements are based on the singular property of in-field magnetization; further room-temperature and thermal measurements, especially of remanent magnetization in a zero field, are needed in many studies in order to identify mineral types, domains and their concentrations using their magnetic properties alone.   JAD

Reading
Dearing, J.A. (1999) *Environmental magnetic susceptibility.* Kenilworth: Chi Publishing. · Dunlop, D.J. and Özdemir, Ö. (1998) *Rock magnetism: fundamentals and frontiers.* Cambridge: Cambridge University Press. · O'Reilly, W. (1984) *Rock and mineral magnetism.* Glasgow: Blackie. · Thompson, R. and Oldfield, F. (1986) *Environmental magnetism.* London: George Allen and Unwin.

Makatea island topography, Mangaia, southern Cook Islands, South Pacific.
Photograph by from Tom Spencer.

makatea An elevated rim of karst-eroded reef limestone of Tertiary age, sometimes surrounded on its seaward margin by Pleistocene limestone reef terraces, on islands in the South Pacific Ocean. Makatea islands exhibit three elements from island centres outwards: a central volcanic core; surrounding marginal swamps; and an encircling limestone rim. Although superficially resembling elevated barrier reefs, detailed field studies suggest erosional origins, with aggressive drainage waters from the volcanic core creating a karst marginal plain. The solution and seaward retreat of the original island-fringing reefs results in a vertical inner wall to the limestone rim and creates the low-lying swamps between the volcano and the limestone rim. The repeated rejuvenation of limestone solution at the volcano–makatea contact is promoted by island uplift, which itself is the result of crustal flexure at ~200 km from lithospheric loading by young (<2 million years) volcanoes. Thus, in the southern Cook Islands, the makatea islands are associated with loading from the volcano of Rarotonga. Confusingly, the island of Makatea, Tuamotu Archipelago, does not show the characteristics of a makatea island; rather, it appears to be an uplifted atoll.   TS

Reading
Stoddart, D.R., Scoffin, T.P. and Spencer, T. (1985) Reef growth and karst erosion on Mangaia, Cook Islands: a reinterpretation. In E. C. F. Bird (ed.), *Geomorphology of changing coastlines*, Zeitschrift für Geomorphologie, Supplementbände, vol. 57. Berlin: Borntraeger; pp. 121–140.

mallee Eucalyptus scrub characteristic of the semi-arid areas of Australia.

mammilated surface The surface of a rock outcrop that has been smoothed and rounded by erosional processes.

mangrove, mangrove forest Mangroves are plants (mostly tree species) that show a series of physiological, morphological and life history adaptations to the difficulties of both establishment in mobile substrates subject to current flows and wave activity and subsequent survival over a wide range of salinities (0–90‰) in waterlogged, anoxic sediments. These remarkable adaptations include prop, 'knee' and buttress roots, oxygen-capturing pneumatophores, salt exclusion and secretion mechanisms and viviparous seedlings. In addition, highly evolved plant–soil–microbe relations allow the efficient use of nutrients, rapidly cycled between trees and large below-ground reservoirs. As forests, they can densely vegetate the upper intertidal areas (typically at levels from just below mean sea level to ~2 m above this level) of tropical coastlines. In recent decades, mangroves have been seen to provide a wide range of ecological services. It has been claimed that mangroves provide important nursery grounds for juvenile fish and shellfish; through the outwelling of nutrients, critically support estuarine and nearshore productivity; protect against coastal flooding and erosion; act as nutrient and pollutant filters; and play a key role in global biogeochemical cycling through the sequestration of carbon. However, in spite of these critical roles, degradation and loss of mangrove ecosystems over the past three decades has been intense and appears to be increasing. The immediate concern, therefore, is

Mallee vegetation, south central Australia.
Photograph by David Thomas.

to determine what benefits are lost as sizeable areas of wetland are converted to other uses as a result of coastal development and population growth. These issues were brought into sharp relief by the impact of the Asian TSUNAMI in December 2004, where it was widely claimed that the presence of mangroves significantly reduced loss of life and property behind them. This role has, however, been challenged. Coastal scientists have identified a need for '"detangling the effect" of the physical and vegetative properties that determine whether mangroves . . . are effective coastal barriers against periodic, economically damaging storm events' (Barbier *et al.*, 2008) and proposed a research agenda to provide an 'in-depth investigation of the protection function of various mangrove formations and coast-geomorphological settings, various root types and various species compositions' (Koedam and Dahdouh-Guebas, 2008). TS

Reading and References

Alongi, D.M. (2009) Paradigm shifts in mangrove biology. In G. M. E. Perillo, E. Wolanski, D. R. Cahoon and M. M. Brinson (eds), *Coastal wetlands: an integrated ecosystem approach*. Amsterdam: Elsevier. · Barbier, E.B., Koch, E.W., Silliman, B.R., *et al.* (2008) Coastal ecosystem-based management with nonlinear ecological functions and values. *Science*, **319**, 321–323. · Ellison, J.C. (2009) Geomorphology and sedimentology of mangroves. In G. M. E. Perillo, E. Wolanski, D. R. Cahoon and M. M. Brinson (eds), *Coastal wetlands: an integrated ecosystem approach*. Amsterdam; Elsevier; pp. 565–591. · Koedam, N. and Dahdouh-Gerbas, F. (2008) Ecological quality changes precede changes in quantity in mangrove forests. http://www.ulb.ac.be/sciences/biocomplexity/pub/Koedam&Dahdouh-Guebas_2008_Science.pdf (accessed 12 July 2015). · McIvor, A.L., Spencer, T., Möller, I. and Spalding, M. (2012) *Storm surge reduction by mangroves*. Natural Coastal Protection Series: Report 2. Cambridge Coastal Research Unit Working Paper 41. The Nature Conservancy and Wetlands International.

manned Earth resources satellites

Satellites carrying a human crew plus photographic and other REMOTE SENSING devices to acquire images of the Earth's surface. The US manned satellites of Mercury, Gemini and Apollo were designed and operated by the National Aeronautics and Space Administration (NASA). A complementary series of manned earth resources satellites were launched by the Commonwealth of Independent States (e.g. through the Vostok programme), and China is the only other country to send a manned satellite into space (the Shenzhou 5 in 2003). It is only the image products available from the NASA satellites that are readily available, with these being so successful that were the stimulus for NASA to develop the unmanned satellite, LANDSAT.

Space stations/spacelabs have also been used for remote sensing purposes. The Space Shuttle was a series of NASA spacecraft designed to shuttle backwards and forwards between Earth and space. They started flying in April 1981 with the last mission ending on 21 July 2011. The orbiter vehicle on the Shuttle carried astronauts who took photographs of the Earth's surface on a regular basis and had two remote sensing packages named OSTA and Spacelab. The OSTA package is a collection of remote sensing experiments, including three environmentally useful

Table 1 Values of Manning's roughness coefficient for various types of natural channel

Channel type	Normal value	Range
Small channels (width <30 m)		
Low-gradient streams		
Unvegetated straight channels at bankfull stage	0.030	0.025–0.033
Unvegetated winding channels with some pools and shallows	0.040	0.033–0.045
Winding vegetated channels with stones on bed	0.050	0.045–0.060
Sluggish vegetated channels with deep pools	0.070	0.050–0.080
Heavily vegetated channels with deep pools	0.100	0.075–0.150
Mountain streams (with steep unvegetated banks)		
Few boulders on channel bed	0.040	0.030–0.050
Abundant cobbles and large boulders on channel bed	0.050	0.040–0.070
Large channels (width >30 m)		
Regular channel lacking boulders or vegetation	—	0.025–0.060
Irregular channel	—	0.035–0.100

Source: Based on data in Chow (1964).

sensors, a SIDEWAYS LOOKING AIRBORNE RADAR (SLAR – synthetic aperture radar (SAR) type), a pair of television cameras and an optical imager. The European Space Agency (ESA) had the Spacelab, which included an aerial camera and SLAR (SAR type). Since November 2000 it has been the International Space Station programme, a joint project among five participating space agencies (NASA, Roskosmos, Japan Aerospace Exploration Agency, ESA, and the Canadian Space Agency) that has been the principal vehicle for remote sensing from a manned spacecraft. The ISS is funded until 2024 and via its Crew Earth Observation programme will provide science results for physical geographers.   DSB

Reading
Amsbury, D.L. (1989) United States manned observations of earth before the space shuttle. *Geocarto International*, 4, 7–14. · Crew Earth Observations (CEO). http://www.nasa.gov/mission_pages/station/research/experiments/86.html (accesse 12 July 2015).

Manning equation
In 1891 the Irish engineer Robert Manning summarized uniform flow data in the popular, widely used flow equation (see FLOW EQUATIONS)

$$\nu = \frac{kR^{2/3}s^{1/2}}{n}$$

where ν (m^3 s^{-1}) is velocity, R (m) is the HYDRAULIC RADIUS (mean depth in wide channels) and s (m m^{-1}) is the slope. The Manning ROUGHNESS coefficient n is a general measure of channel resistance that is numerically constant regardless of the measurement units being used. The coefficient k, therefore, accommodates the variation of measurement system, and is 1 for SI units and 1.49 for Imperial units. In straight natural

streams, n ranges from 0.03 for smooth sections to 0.10 for rocky or heavily vegetated sections (Chow, 1959: 98–123; Barnes, 1967) (see Table 1). It is related to the coefficient C in the CHÉZY EQUATION by

$$C = \frac{R^{1/6}}{n}$$

KSR

Reading and References
Barnes, H.H. (1967) Roughness characteristics of natural channels. *US Geological Survey water supply paper* 1849. · Chow, V.T. (1959) *Open channel hydraulics*. London: McGraw-Hill. · Chow, V.T. (ed.) (1964) *Handbook of applied hydrology*. New York: McGraw-Hill. · Manning, R. (1891) On the flow of water in open channels and pipes. *Transactions of the Institution of Civil Engineers of Ireland*, **20**, 161–207.

mantle
The zone within the Earth's interior extending from 25 to 70 km below the surface to a depth of 2900 km and lying between the partially molten core and the thin surface crust. The uppermost rigid section of the mantle forms the lower part of the LITHOSPHERE. Extending a few hundred kilometres below this region is the ASTHENOSPHERE, a layer in which the magnesium and iron-rich silicate rocks of the mantle are probably partially molten. Most of the convection within the mantle, viewed by some as a crucial mechanism of PLATE TECTONICS, is generally thought to occur in this zone. MAS

Reading
Davies, P.A. and Runcorn, S.K. (eds) (1980) *Mechanisms of continental drift and plate tectonics*. London: Academic Press. · Ringwood, A.E. (1975) *Composition and petrology of the Earth's mantle*. New York: McGraw-Hill. · Wyllie, P.J. (1976) The earth's mantle. In J. T. Wilson (ed.), *Continents adrift and continents aground*. San Francisco, CA: W.H. Freeman; pp. 46–57.

mantle plume   A convectional flow of hot rock that rises through the MANTLE to the base of the LITHOSPHERE and which gives rise to a HOT SPOT on the surface. In the oceans, mantle plumes create lines of large volcanoes (such as the Hawaiian chain) if the overlying plate is moving with respect to the plume. On the continents, mantle plumes have probably been responsible for generating voluminous and extensive accumulations of BASALT flows, such as those of the Karoo (South Africa) and the Deccan (India), many of which are located at PASSIVE MARGINS and were thus originally close to sites of continental break-up. They are also probably capable of causing surface uplift over areas in excess of 1000 km across, both by the direct effects of heating of the lithosphere and by the associated thickening of the crust through the addition of enormous quantities of igneous material.   MAS

Reading
Summerfield, M.A. (1991) *Global geomorphology*. London/New York: Longman/John Wiley & Sons, Inc. · White, R.S. and McKenzie, D. (1989) Magmatism at rift zones: the generation of volcanic continental margins and flood basalts. *Journal of Geophysical Research*, **94**, 7685–7729.

maquis   Scrub vegetation of evergreen shrubs characteristic of the western Mediterranean, and broadly equivalent to chaparral. Many areas of maquis probably represent human-induced degradation of the natural forest vegetation.

margalitic   Soil A horizons that are dark coloured with a high base status, Ca and Mg being the predominant exchangeable cations.

marginal channel   A meltwater stream flowing along the edge of a glacier, most often in a glacier's ablation area.

marine isotope stage (MIS)   Some parts of ocean floors provide excellent long term (multimillion-year) depositories of sediments that are undisturbed by erosion, movement or reworking. The incorporation of marine organism remains, and the successful retrieval of sedimentary cores from these deposits from the 1950s onwards, allowed long records of environmental change to be determined through analysis of the OXYGEN ISOTOPE records in these sediments. Alternating warm and cold periods of history, embracing the QUATERNARY period and extending to *c.* 6 million years ago, have been determined, called marine isotope stages. Warm stages, with relatively low ^{18}O concentrations, have odd numbers; cold stages have even numbers. MIS 1 continues to the present and commences 14 thousand years ago, MIS 2 covers the last glacial, including the Last Glacial Maximum, back to 29

thousand years ago, and so on. Some stages are subdivided (e.g. MIS 5, into 5a–5e, representing warmer and cooler periods within an overall warm stage), and to date over 100 stages are numbered and dated. The scheme is subject to periodic revision.   DSGT

Reading
Lisiecki, L.E. and Raymo, M.E. (2005) A Pliocene–Pleistocene stack of 57 globally distributed benthic δ ^{18}O records. *Paleoceanography*, **20**, 1–17.

marine pollution   The pollution of the oceans and seas. At first sight, as Jickells *et al.* (1991: 313) point out, two contradictory thoughts may cross our minds on this issue:

The first is the observation of ocean explorers, such as Thor Heyerdahl, of lumps of tar, flotsam and jetsam, and other products of human society thousands of kilometres from inhabited land. An alternative, vaguer feeling is that given the vastness of the oceans (more than 1,000 billion billion litres of water!), how can man have significantly polluted them?

What is the answer to this conundrum? Jickells *et al.* (1991: 330) draw a clear distinction between the open oceans and regional seas and in part come up with an answer:

The physical and chemical environment of the open oceans has not been greatly affected by events over the past 300 years, principally because of their large diluting capacity.

Coastal areas (including estuaries) in close proximity to large concentrations of people show evidence of increasing concentrations of various substances that are almost certainly linked to human activities (Howarth *et al.*, 2011). This has produced what have been termed 'dead zones'. Thus, the partially enclosed North and Baltic Seas (Conley *et al.*, 2011) show increases in phosphate concentrations as a result of discharges of sewage and agricultural waste. The same is true of the more enclosed Black Sea. The Mediterranean also suffers from many threats, including increased sedimentation and pollution from growing urban centres and from intensive agriculture (Claudet and Fraschetti, 2010).

On a global basis, organic point-source pollution has been identified as one of the greatest threats. This can cause accelerated or cultural EUTROPHICATION that can lead to excessive growths of algae and to hypoxia (Zhang *et al.*, 2010). Coastal and estuary water are sometimes affected by algal foam and scum, often called 'red tides'. Some of these blooms are so toxic that consumers of seafood that have been exposed to them can be affected by diarrhoea, sometimes fatally. They may also be sufficiently toxic to kill marine animals. Long-term studies in many parts of the world suggest that these red tides are increasing in extent and frequency as coastal pollution worsens and nutrient enrichment occurs more often.

The increasing exploitation of offshore oil reserves, a growing number of them located in deep waters, has created a new dimension with regard to oil pollution. This was brought out starkly in April 2010, when the rig the *Deepwater Horizon* exploded in the Gulf of Mexico and released huge amounts of oil, some of which reached the coastlines of states bordering the Gulf. This, however, was not a unique event (Jernelöv, 2010).

A further threat that is attracting increasing attention is the pollution of the marine environment by plastic debris (Barnes *et al.*, 2009). This threatens marine life due to ingestion and due to entanglement (especially in discarded fishing gear). Some plastics contain polychlorinated biphenyls, which can lead to reproductive disorders and death. Drifting plastics can also provide a platform for invasive species. Another topic of current debate is the extent to which new technologies and pressures, including ocean acidification brought about by increasing carbon dioxide levels, may begin to impact on the 'Last Great Wilderness', the deep recesses of the oceans (Ramirez-Llodra *et al.*, 2011).   ASG

Reading and References

Barnes, D.K.A., Galgani, F., Thompson, R.C. and Barlaz, M. (2009) Accumulation and fragmentation of plastic debris in global environments. *Philosophical Transactions of the Royal Society*, **364**, 1985–1998. · Clark, R.B. (2001) *Marine pollution*, 5th edition. Oxford: Clarendon Press. · Claudet, J. and Fraschetti, S. (2010) Human-driven impacts on marine habitats: a regional meta-analysis in the Mediterranean Sea. *Biological Conservation*, **143**, 2195–2206. · Conley, D.J., Carstensen, J., Aigars, J., *et al.* (2011) Hypoxia is increasing in the coastal zone of the Baltic Sea. *Environmental Science and Technology*, **45**, 6777–6783. · Howarth, R., Chan, F., Conley, D.J., *et al.* (2011) Coupled biogeochemical cycles: eutrophication and hypoxia in temperate estuaries and coastal marine ecosystems. *Frontiers in Ecology and Environment*, **9**, 18–26. · Jernelöv, A. (2010) The threats from oil spills: now, then, and in the future. *Ambio*, **39**, 353–366. · Jickells, T.D., Carpenter, R. and Liss, P.S. (1991) Marine environment. In B. L. Turner, W. C. Clark, R. W. Kates, *et al.* (eds.) *The Earth as transformed by human action.* Cambridge: Cambridge University Press; pp. 313–334. · Ramirez-Llodra, E., Tyler, P.A., Baker, M.C., *et al.* (2011) Man and the last great wilderness: human impact on the deep sea. *PLoS ONE*, **6** (8), e22588, doi: 10.1037/journal.pone.0022588. · Zhang, J., Gilbert, G., Gooday, A.J., *et al.* (2010) Natural and human-induced hypoxia and consequences for coastal areas: synthesis and future development. *Biogeosciences*, 7, 1443–1467.

marker horizon   A distinctive sedimentary layer or stratigraphic unit that is found in different locations but is of the same age in each location. The concept needs to be used with extreme caution, as many sediments can appear to be the same but have different ages. The concept works well with TEPHRA, as deposits from different volcanic eruptions have distinct geochemical signatures, allowing clear identification.   DSGT

Markov process   A statistical process in which the probability of an event, in a sequence of random events, is influenced by the outcome of the preceding event. Any system in which an outcome depends in part on the previous outcome may be described as exhibiting a Markov property. Any system in which successive events are partially dependent is said to have a 'memory' of the preceding event. The methodology is formalized using a transition matrix in which entries represent the probabilities of transition from one state to another. When the matrix is powered, the probabilities represent transitions from one state to another in two steps. Repeated powering will develop a matrix in which all rows are the same. States may also be defined that are absorptive, when the realization of the process terminates, or reflexive, when a process enters a state and then returns to the previous state. The utility of the process depends on the length of the record on which the probabilities are established, whether the series exhibits the Markov property, is stationary and whether the transition probabilities are invariant with time.   DB

Reading

Harbaugh, J.W. and Bonham-Carter, G. (1970) *Computer simulation in geology.* New York: John Wiley & Sons, Inc. · Thornes, J.B. and Brunsden, D. (1977) *Geomorphology and time.* London: Methuen.

marl   An argillaceous, non-indurated calcium carbonate sediment, frequently grey to blue–grey in colour, formed in freshwater environments. PSH

mass balance   Changes in the mass of a sedimentary landform over time. The concept is fundamental to the understanding of a glacier, where a zero mass balance describes a situation where accumulation balances ABLATION, a positive mass balance describes an increase in glacier mass and a negative mass balance a reduction in glacier mass. Recent years have witnessed increased levels of interest in glacier and ice sheet mass balance studies owing to the possible impact of temperature rise on ice volume. Examples include those on the major ice caps of Greenland and Antarctica (Shepherd *et al.*,

2012) and the glaciers of the Himalayas (Kääb *et al.*, 2012). The concept can also be applied to other sedimentary landforms; for example, a beach, talus or sand dune.   DES/MEM

References
Kääb, A., Berthier, E., Nuth, C., *et al.* (2012) Contrasting patterns of early twenty-first century glacier mass change in the Himalayas. *Nature*, **488**, 495–498. · Shepherd, A., Ivins, E.R., Geruo, A., *et al.* (2012) A reconciled estimate of ice-sheet mass balance. *Science*, **338**, 1183–1189.

mass movement types   The modes of hillslope failure. Various methods have been used to classify mass movements, including reference to parent material, the causes and mechanics, slope and landslide geometry, shape of failure surface, post-failure debris distribution, soil moisture and physical properties. The most generally accepted method is the complex variable classification of Varnes (1958). The most common general groupings are linked to the processes of fall, slide, flow and creep.

Falls of rock or soil are the free movement downslope of slope-forming materials. Failures may be in circular, plane, wedge, toppling or settling modes. Slides of rock, mud and soil take place by movement on a sliding surface that may be circular, noncircular or planar in form. Circular features may be single, multiple or successive in character. Noncircular failures are usually controlled by lithology or rock structures such as bedding planes, which impart a

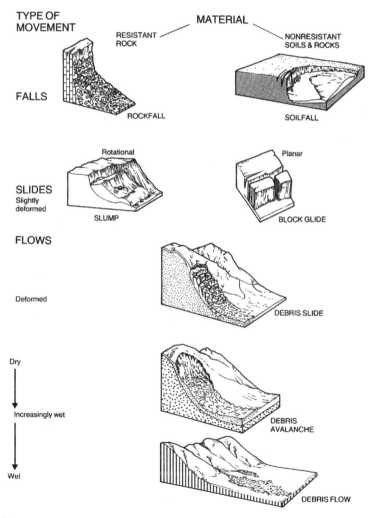

Mass movement types.
Source: Selby 1982. Reproduced with permission of Oxford University Press.

translational component to the slide. A graben form at the head or retrogressive failure may also be evident. Planar failures are usually shallow and translational, including debris slides and mudslides, but may be of immense size, such as the major rock slides along discontinuities in mountain regions.

All slides may disintegrate upon failure to develop into an 'avalanche' or 'flow'. These involve complex processes such as fluidization, liquefaction, cohesionless grain flow, remoulding and possibly air lubrication. Flows represent a transitional set of processes lying between streamflow or mass transport and mass movement. The term *debris flow* is used for the mass movement of a wet mixture of granular slides, clay, water and air under the influence of gravity with intergranular shear distributed evenly through the mass. They usually occur in three modes, as simple flows on hillslides, as valley-confined flows debouching onto fans or as catastrophic flows that effectively destroy everything in their path and override topography. This is a convenient division based on scale, but the processes are probably similar. Creep includes the continuous, gravity or mass creep of hillslopes affected by the force of gravity in which low-continued stresses deform the slope materials at a fairly constant rate but to a considerable depth. The forms that result are gravity slides or *sackung*, cambering and valley bulging (also produced by periglacial conditions, which accelerate the process), outcrop curvature and botanic deformation. Creep is more commonly regarded as a seasonal process in which surface materials are continuously affected by the expansion and contraction of the soil and by soil heave caused by changes of temperature, moisture, freezing, crystal growth or chemical changes. Varieties occur as soils, scree or talus, and intense activity develops into faster solifluction processes under periglacial conditions (gelifluction, congelifluction, congeliturbation). DB

Reading and References
Clague, J.J. and Stead, D. (eds) (2012) *Landslides: types, mechanisms and modeling*. Cambridge: Cambridge University Press. · Selby, M.J. (1993) *Hillslope materials and processes*. Oxford: Oxford University Press. · Sharpe, C.F.S. (1938) *Landslides and related phenomena*. New York: Columbia University Press. · Varnes, D.J. (1958) Landslide types and processes. In E. B. Eckel (ed.), *Landslides and engineering practice*. Highway Research Board special report 29. Washington, DC: HRB, National Research Council; pp. 20–47.

mass strength (of rock) A measure of the resistance to erosion and instability of an entire rock mass inclusive of its discontinuities, contained water and weathering products. For geomorphic purposes, mass strength has been assessed quantitatively by taking into account the following parameters: (1) strength of intact rock; (2) state of weathering of the rock; (3) spacing of joints, bedding planes, foliations or other discontinuities within the rock mass; (4) orientation of discontinuities with respect to a cut slope; (5) width of the discontinuities; (6) lateral or vertical continuity of the discontinuities; (7) infill of the discontinuities; and (8) movement of water within or out of the rock mass. MJS

Reading
Dackombe, R.V. and Gardiner, V. (1983) *Geomorphological field manual*. London: Allen & Unwin. · Selby, M.J. (1980) A rock mass strength classification for geomorphic purposes: with tests from Antarctica and New Zealand. *Zeitschrift für Geomorphologie, Neue Folge*, 24, 31–51.

matorral The Chilean equivalent of the evergreen, xerophilous, woody plants of Mediterranean environments. (See also CHAPARRAL and MAQUIS).

Reading
Miller, P.C. (1981) *Resource use by chaparral and matorral*. Berlin: Springer Verlag.

matrix flow The flow of water through the ordinary microscopic pore spaces that make up perhaps 50% of the volume of many soils. There is significant attraction between the water molecules and the walls of the micropores ($<10\,\mu m$ diameter) and mesopores (10–$1000\,\mu m$ diameter); in addition, the pathway that must be followed by water in order to traverse a path through the soil is very tortuous and, therefore, longer than the actual straight-line length would suggest. Both these factors, together with others like obstruction of some pores and dead-end passages, retard water traversing the micropores. The flow travels slowly and in the laminar flow state, at speeds of millimetres to centimetres per day, in accordance with DARCY'S LAW. Much higher speeds are achieved if there are larger conduits, such as faunal burrows, within the soil (see BYPASS FLOW) that can permit turbulent flow. Speeds may then reach millimetres to centimetres per second. DLD

Reading
Dingman, S.L. (2008) *Physical hydrology*, 2nd edition. Long Grove, IL: Waveland Press.

maturity The stage of development of a river system or landscape at which processes are most efficient and vigorous and are tending towards maximum development. In landscape evolution, the period of maximum diversity of form.

Maunder minimum   A term that was introduced in 1976 by J. A. Eddy, and named after two solar astronomers, Annie and Walter Maunder, to describe a state of low sunspot numbers (and thus of the output of solar radiation) between 1645 and 1715 AD. This coincided with one of the coldest portions of the Little Ice Age.   ASG

Reading
Eddy J.A. (1976) The Maunder minimum. *Science*, **192**, 1189–1202. · Shindell, D.T., Schmidt, G.A, Mann, M.E., *et al.* (2001) Solar forcing of regional climate change during the Maunder minimum. *Science*, **294**, 2149–2152.

mean sea level   An average elevation of the sea surface based upon elevation data that are referenced to an established datum. Mean SEA LEVEL may be estimated using temporal data for a specific location, or from temporal and spatial data to represent the sea surface. These estimates are used to establish baseline conditions for the measurement of sea-level changes. Global mean sea level closely approximates the GEOID, with maximum differences of the order of 1 m, as a result of, for example, ocean currents, atmospheric pressure systems and different water densities (Pugh, 1987). Compared with the Earth's ellipsoid, there is more than 150 m of relief across the time-averaged sea surface. For example, there is a depression averaging about −100 m off the coast of India and a bulge of about 70 m near New Guinea (Tooley, 1993).

Contemporary data are usually obtained from tide gauges or satellite altimetry. The latter has been especially valuable in providing open-ocean data. Time and space averages of mean sea level are useful for the assessment of seasonal changes in water surface elevation, and changes caused by other atmospheric phenomena, such as EL NIÑO. The 1997–1998 El Niño was characterized by sea levels that were, relative to the annual mean, higher than normal in the eastern Pacific and lower than normal in the western Pacific.

Carefully obtained estimates of mean sea level are also valuable in the detection and quantification of sea-level trends. Recent estimates of the rate of global sea-level rise (Houghton *et al.*, 1996) are between 1.0 and 2.5 mm a^{-1}. The range in these estimates derives largely from uncertainty over the quality of the tide gauge records (especially vis-à-vis the vertical stability of the gauge location) used to obtain local mean sea-level data.   DJS

References
Houghton, J.T., Meiro Filho, L.G., Callander, B.A., *et al.* (1996) *Climate change 1995: the science of climate change.* Cambridge: Cambridge University Press. · Pugh, D.T.

(1987) *Tides, surges and mean sea level.* Chichester: John Wiley & Sons, Ltd. · Tooley, M.J. (1993) Long term changes in eustatic sea level. In R. A. Warrick, E. M. Barrow and T. M. L. Wigley (eds), *Climate and sea level change: observations, projections and implications.* Cambridge: Cambridge University Press; pp. 81–107.

meandering   The sinuous winding of a river, as in the River Menderes in southwest Turkey. Various, more restricted definitions have been applied; for example, that a river may be arbitrarily considered as meandering if it flows in a single channel that is more than one and a half times the direct downvalley length of the reach in question. Other definitions have suggested that meandering must involve the development of regular bends of repeated geometry; these bends may have characteristic amplitude, wavelength and curvature, and they may be approximated by geometrical curves such as circular arcs or so-called sine-generated curves, or recently by migration-rate models. Irregular and compound bend patterns are also possible.

The term is also applied to analogous flow patterns (as in the Gulf Stream or atmospheric jet streams) and to the dynamic processes by which such planforms may develop. For rivers, these include outer-bank erosion and inner-bank deposition on channel bends, and these may be related to the hydraulics of flow and sediment flux in curving channels.

Different kinds of meandering may be distinguished; these include free (developing in alluvial materials without hindrance), forced or confined (where barriers such as bedrock valley sides affect meander generation) and valley (rather than just channel) meandering. Meanders in bedrock may be incised, either as they develop (ingrown) or vertically entrenched. (See also CHANNEL CLASSIFICATION.)   JL

Reading
Collinson, J.D. and Lewin, J. (eds) (1983) *Modern and ancient fluvial systems.* International Association of Sedimentologists special publication 6. Oxford: Blackwell Scientific. · Gregory, K.J. (ed.) (1977) *River channel changes.* Chichester: John Wiley & Sons, Ltd. · Richards, K. (1982) *Rivers.* London: Methuen.

Medieval Warm Period   A warm phase well represented in Europe and parts of North America, but not necessarily globally, in early medieval times. It is often called the Little Optimum, and lasted from *c.* 800–1300 AD. It was followed by the cooling phase of the Little Ice Age, and was a time of Norse settlement in southern Greenland and of extensive farming in Iceland. Grapes were grown in northern England, figs and olives grew in regions of Europe well north of their current range, and many European glaciers retreated (Grove and Switsur, 1994). It may have been a time of drought and aeolian activity in the southwest USA. ASG

Reading and Reference
Esper, J. and Frank, D. (2009) The IPCC on a heterogeneous Medieval Warm Period. *Climatic Change*, **94**, 267–73. · Grove, J.M. and Switsur, R. (1994) Glacial geologic evidence for the Medieval Warm Period. *Climatic Change*, **26**, 143–169.

Mediterranean climate
Characterized by hot, dry, sunny summers and a winter rainy season. In the KÖPPEN CLIMATIC CLASSIFICATION this is the Cs climate, while in Thornthwaite's (1948) classification it is designated as a subhumid mesothermal climate. Circulation in summer is dominated by expansions of the subtropical high-pressure cells and in winter by travelling mid-latitude depressions (Perry, 1981).

Typical monthly mean temperatures are between 21–27 °C in summer and 4–13 °C in winter with annual rainfall totals between 38 and 76 cm. A number of distinctive LOCAL WINDS are associated with Mediterranean climates; for example, mistral, bora and Santa Ana (Meteorological Office, 1964).

This climatic type is found around the Mediterranean Sea, in southern California, the central Chilean coast, the South African coast near Cape Town and in western Australia. AHP

References
Meteorological Office (1964) *Weather in the Mediterranean*, 2 vols. London: HMSO. · Perry, A. (1981) Mediterranean climate – a synoptic reappraisal. *Progress in Physical Geography*, **5**, 105–113. · Thornthwaite, C.W. (1948) An approach to a rational classification of climate. *Geographical Review*, **38**, 55–94.

mega dune (or DRAA)
The largest aeolian bedform in Wilson's (1972) hierarchy, with a wavelength of up to 5500 m and a height of 20–450 m. Given their size, mega dunes are in many cases compound or complex DUNE forms. DSGT

Reference
Wilson, I.G. (1972) Aeolian bedforms: their development and origins. *Sedimentology*, **19**, 173–210.

megafan
As Horton and DeCelles (2001: 44) observed:

Fluvial megafans constitute volumetrically significant depositional elements of sedimentary basins adjacent to mountain belts. A fluvial megafan is a large (10^3–10^5 km^2), fan-shaped (in plan-view) mass of clastic sediment deposited by a laterally mobile river system

that emanates from the outlet point of a large mountainous drainage network.

Alternative names include megacone, inland delta, wet alluvial fan and braided stream fan. Although many of the best examples occur on the north side of the Ganga plain in India at the edge of the Himalayas (e.g. the Kosi Fan), some, such as the Okavango Fan of the Kalahari in Botswana, have developed in low-relief grabens. ASG

Reference
Horton, B.K. and DeCelles, P.G. (2001) Modern and ancient fluvial megafans 1. The foreland basin system of the Central Andes, southern Bolivia: implications for drainage network evolution in fold-thrust belts. *Basin Research*, **13**, 43–63.

megashear
A term introduced by S. W. Carey (1976) to describe a strike-slip fault with a large lateral displacement greatly exceeding the thickness of the crust. In a strike-slip fault a section of the crust moves horizontally with respect to an adjoining section, as on the San Andreas fault system. Megashears form a component of Carey's model of global TECTONICS based on the assumption of an expanding Earth. MAS

Reading and Reference
Carey, S.W. (1976) *The expanding Earth*. Amsterdam: Elsevier. · Neev, D., Hall, J.K. and Saul, J.M. (1982) The Plesium megashear system, across Africa and associated lineament swarms. *Journal of Geophysical Research*, **87**, 1015–1030.

megathermal climate
A climate in which no month has a mean temperature below 18 °C. Such conditions are found in the humid tropics and subtropics.

mekgacha
Dry or fossil valley systems from the Kalahari of southern Africa. They occur in a largely featureless terrain where at present surface run-off is of very limited occurrence, while they often commence abruptly (in contrast to a BLIND VALLEY, which terminates abruptly). The origins of mekgacha have been variously ascribed to fluvial activity during wetter episodes, ephemeral flow during high rainfall events, river capture and groundwater sapping along geological lineaments. ASG/DSGT

Reference
Shaw, P.A., Thomas, D.S.G. and Nash, D.J. (1992) Late Quaternary fluvial activity in the dry valleys (mekgacha) of the Middle and southern Kalahari, southern Africa. *Journal of Quaternary Science*, 7, 273–281.

meltout till   See TILL.

meltwater   Water produced by the melting of snow or ice. In polar periglacial areas the melting of the winter snow cover and ice in the river channels occurs in the spring and early summer. The resulting flood may last from a few days to a few weeks and account for up to 90% of the total discharge. Sediment loads of such floods are commonly suppressed by the still-frozen state of the ground and river channel.

Glacier meltwater flow peaks somewhat later in the season and is less sudden in its build up and decline, but sediment loads are high. Meltwater in a glacier is derived mainly from summer melting of the glacier surface, but a small though significant amount is derived from bottom melting of the glacier by geothermal heat and by frictional heating associated with glacier flow (so long as the glacier base is at the pressure melting point).

On glaciers that are below the pressure melting point the meltwater is normally restricted to surface flow. Where the ice is at the pressure melting point, meltwater penetrates the ice via MOULINS until it flows at the glacier bed. It can be distinguished from basally derived meltwater by its lower solute content (Collins, 1979). Two hydrological systems exist at the glacier bed: that of the conduits and that of the intervening areas where a film of water exists between glacier ice and bedrock. In certain bedrock depressions water can be trapped to form subglacial lakes up to many kilometres across (Oswald and Robin, 1973).   DES

Reading and References
Collins, D.N. (1979) Quantitative determination of the subglacial hydrology of two alpine glaciers. *Journal of Glaciology*, **23**, 347–362. · Lliboutry, L. (1983) Modifications to the theory of intraglacial water-ways for the case of subglacial ones. *Journal of Glaciology*, **29**, 216–226. · Oswald, G.K.A. and Robin, G.de Q. (1973) Lakes beneath the Antarctic ice sheet. *Nature*, **245**, 251–254. · Shreve, R.L. (1972) Movement of water in glaciers. *Journal of Glaciology*, **11**, 205–214.

memory capacity (landform)   The ability of a constructional landform to retain a form resulting from development in one regime after environmental conditions have changed. Applied to sand dunes, the term 'dune memory' was coined by Warren and Kay (1987). All other things being equal, a dune containing a low volume of sand has a short memory, perhaps of only a few hours or days, while a large dune has a long memory that may extend for thousands of years. The concept can apply to other landforms, such as beaches, and is related to the principle of BEDFORM RECONSTITUTION.   DSGT

Reference
Warren, A. and Kay, S. (1987) Dune networks. In L. E. Frostick and I. Reid (eds), *Desert sediments: ancient and modern*. Geological Society, London, Special Publications, vol. **35**. London: The Geological Society; pp. 205–212.

Mercalli scale   A scale between 1 and 12 for measuring the intensity of an EARTHQUAKE, first developed in 1902 by Giuseppe Mercalli (Table 2). It is subjective, and measures qualitatively the experience of affected people and damage to structures.   DSGT

Reference
US Geological Survey (1989) *The severity of an earthquake*. General Interest Publications of the US Geological Survey. Washington, DC: US Government Printing Office; http://pubs.usgs.gov/gip/earthq4/severitygip.html (accessed 12 July 2015).

mere   A lake, especially in Cheshire and East Anglia, where meres are developed in glacial outwash deposits.

meridional circulation   Usually defined as the average, over all longitudes, of the flow in the meridional plane. Though very much weaker than the zonal (i.e. along latitudes) component, the convergence–divergence of the meridional component defines the main climatic zones. Spatial unrepresentivity of observations makes the arithmetic unreliable compared with the values of wind speeds of 0.1–$1\,\mathrm{m\,s^{-1}}$ observed. Annual averages confirm the picture of a direct (upward-moving air warmer than downward-moving air) tropical cell, bounded on the equatorial side by the INTERTROPICAL CONVERGENCE ZONE (ITCZ) and by an indirect cell (upward-moving air cooler than downward-moving air) between 30° and 60° latitude. There is some suggestion of a second direct cell between 60° and 90°. Downward velocities bringing dry air towards the surface at 30° (and possibly 90°) coincide with the desert belts at these latitudes. Fluxes of sensible, gravitational potential and latent energy in the meridional circulation are each much greater than their sum. Fluxes of ANGULAR MOMENTUM at high and low levels are each large compared with their sum. Thus, the circulation acts largely to convert one type of thermal energy into another and to adjust the vertical distribution of zonal wind. To some extent the circulation is an arithmetic artefact, in that the actual flow takes place in specific geographical areas (such as Indonesia) often on occasions of intense tropical cumulonimbus (see CLOUDS) rather than being spread uniformly in space and time.   JSAG

Table 2 Modified Mercalli scale of earthquake intensity

I	Not felt except by a very few under especially favourable circumstances.
II	Felt only by a few persons at rest, especially on upper floors of buildings. Delicately suspended objects may swing.
III	Felt quite noticeably indoors, especially on upper floors, but many people do not recognize it as an earthquake. Standing motor cars may rock slightly. Vibration like passing truck.
IV	During the day felt indoors by many, outdoors by few. At night some awakened. Dishes, windows, doors disturbed; walls make creaking sound. Sensation like heavy truck striking building. Standing motor cars rocked noticeably.
V	Felt by nearly everyone; many awakened. Some dishes, windows, etc. broken; a few instances of cracked plaster; unstable objects overturned. Disturbances of trees, poles, and other tall objects sometimes noticed. Pendulum clocks may stop.
VI	Felt by all, many frightened and run outdoors. Some heavy furniture moved; a few instances of fallen plaster or damaged chimneys. Damage slight.
VII	Everybody runs outdoors. Damage negligible in buildings of good design and construction; slight to moderate in well-built ordinary structures; considerable in poorly or badly designed structures; some chimneys broken. Noticed by persons driving motor cars.
VIII	Damage slight in specially designed structures; considerable in ordinary substantial buildings, with partial collapse; great in poorly built structures. Panel walls thrown out of frame structures. Fall of chimneys, factory stacks, columns, monuments, walls. Heavy furniture overturned. Sand and mud ejected in small amounts. Changes in well-water levels. Disturbs persons driving motor cars.
IX	Damage considerable in specially designed structures, well-designed frame structures thrown out of plumb; great in substantial buildings, with partial collapse. Buildings shifted off foundations. Ground cracked conspicuously. Underground pipes broken.
X	Some well-built wooden structures destroyed; most masonry and frame structures destroyed with foundations; ground badly cracked. Rails bent. Landslides considerable from river banks and steep slopes. Shifted sand and mud. Water splashed over banks.
XI	Few, if any, masonry structures remain standing. Bridges destroyed. Broad fissures in ground. Underground pipelines completely out of service. Earth slumps and land slips in soft ground. Rails bent greatly.
XII	Damage total. Waves seen on ground surfaces. Lines of sight and level distorted. Objects thrown upward into the air.

Source: Modified from US Geological Survey (1989).

Reading

Lorenz, E.N. (1967) *The nature and theory of the general circulation of the atmosphere*. Geneva: World Meteorological Organization.

meridional overturning circulation (MOC)

A global ocean circulation cell in which surface waters in the high latitudes cool and thus become denser; this dense water sinks and flows towards the lower latitudes. This colder water eventually mixes with other water in tropical and subtropical regions, becoming less dense. As a result of the density change, this water returns to the sea surface and ultimately flows back towards the higher latitudes to complete the cell. Several components of this system in both the northern and southern hemispheres are individually identified. The best known is The Atlantic Meridional Overturning Circulation (AMOC), a major current in the Atlantic Ocean in which warm, salty water flows northwards in the upper layers of the Atlantic and colder water flows southwards deep in the Atlantic. The AMOC exhibits multidecadal swings in sea-surface temperatures. These swings are termed the Atlantic Multidecadal Oscillation (AMO).

Palaeoclimatic evidence suggests that there have been abrupt changes in the AMOC, particularly during glacial times, and that these changes have had a profound impact on climate in the Atlantic and around the world. SJN

mesa A steep-sided plateau of rock, often in horizontally bedded rocks, surrounded by a plain.

mesoclimate The intermediate scale of climatic conditions between the MICROCLIMATE and the macroclimate. Although there is not a clear demarcation between these classes, the mesoscale is often taken to include the horizontal scale between 100 m and 200 km (Oke, 1987). Some features of the atmosphere, such as sea breezes, are the product of MESOMETEOROLOGY at this scale. At the lower end of the scale, the term *local climate* is often used. The distinguishing features of a local climate are usually a product of topographic features or proximity to water bodies. The urban environment also creates a distinct mesoclimate that includes the urban heat island and characteristic air circulation patterns resulting from the heat island effect and the urban structures obstructing the wind field. Being static, these factors provide a more permanent foundation for the generation of surface climate. An extensive, south-facing slope would be an example of a favourable local climate in the northern hemisphere as more solar energy would be received than on a horizontal surface and so it is likely to be warmer than flat terrain or terrain sloped in other directions.

Examples of mesoscale features are the MOUNTAIN/VALLEY WINDS generated over complex terrain and the SEA/LAND BREEZES characteristic of many coastal locations. The Great Lakes also produce notable mesoclimates. Thermal belts around the lakes are favourable for vineyards, but the lakes also create the well-known snowbelts on their eastern sides. Mesoscale winds can be important in assisting the dispersion of air pollution as they do blow when the prevailing winds are light. Unfortunately, they do not always produce a clearance of the air. In Los Angeles and Milwaukee, pollution carried inland during the daytime sea breeze can be returned across the city during the night as the land breeze reverses the flow without being strong or unstable enough to mix clean with polluted air.

As computational technology and meteorological understanding have improved, it is now possible to produce realistic mesoscale models to predict the patterns of climatic elements that should be experienced. The areas of greatest progress have been on mesoscale dynamical models of boundary layer flow over complex terrain and models of lake or sea breezes. Such mesoscale models are now incorporated into general circulation models (see GENERAL CIRCULATION MODELLING) to provide regional detail within the large-scale model.   PS/SJN

Reading and Reference

Barry, R.G. (2008) *Mountain weather and climate*. Cambridge: Cambridge University Press. · Oke, T.R. (1987) *Boundary layer climates*. London: Routledge.

mesometeorology
A subdiscipline of meteorology that focuses on phenomena with horizontal scales of 20–250 km and lifetimes of a few to a few tens of hours. Mesoscale circulations are smaller than synoptic-scale features, such as the transient highs and lows that dominate weather maps.

Mesoscale circulations may be conveniently categorized into those that are topographically induced and those that are intrinsically products of the free atmosphere. Each of these categories may be divided into two parts. Thus, topographically induced systems may result from mechanical and thermal processes and the free-atmosphere

systems may result from convective and nonconvective processes.

Mechanically forced circulations include LEE WAVES, downslope winds (see BORA and FÖHN) and circulations within wakes. Lee waves are just that – waves in the airflow to the lee of the hill that triggers them. Their typical wavelength is about 5–15 km and their crest-to-trough amplitude is about 0.5 km. Such waves can produce extremely high winds at the surface, resulting in damage as severe as that produced by a tornado. An exceptionally strong lee wave in Boulder, Colorado, produced $10 million worth of property damage in 1982. These waves result primarily from vertical oscillations induced in a stable airstream by the hill. A wind with strong vertical shear and a direction almost perpendicular to the main axis of the hill is also favourable to the development of lee waves. Circulations in wakes usually take the form of vortices, frequently in trains known as vortex streets. These streets develop behind an obstacle, usually an island, when the flow in the wake region does not mix with that in the surrounding region.

Thermally induced mesoscale circulations include SEA/LAND BREEZES and slope and VALLEY WINDS. Both are the result of a thermally direct, diurnally varying, vertical circulation. In the case of the sea/land breeze, the land–sea thermal contrast generates low-level flow from the sea by day and low-level flow from the land by night. In hilly areas, the daytime mechanism generates upslope and up-valley airflow near the surface, overlain by a 'return current' aloft that flows from the hills to the plains. At night the reverse occurs, resulting in katabatic flows downslope and downvalley.

Moving gravity waves are nonconvective, free-atmosphere systems. Little is currently known about this type of wave, but we do know that they have long wavelengths, up to 500 km and that they are frequently associated with strong vertical shear of the horizontal wind. Convective, free-atmosphere circulations comprise severe local storms (see THUNDERSTORM), shallow cellular circulations and circulations in CYCLONES. Severe local storms have been studied for decades yet remain a challenge. It now appears that, in addition to ATMOSPHERIC INSTABILITY, a strong vertical shear of horizontal wind is vital for long-lived (several hours), large (up to 100 km across) storms to exist.

Shallow cellular circulations were first fully documented only in the satellite era. The cells are usually hexagonal, may be open (i.e. cloud-free area surrounded by cloud) or closed (i.e. cloud area surrounded by a hexagonal ribbon of clear air), have diameters of about 30 km and depths of about 2 km. Their full explanation is as yet unknown. Circulations within cyclones are also recently observed features. Radar and autographic rain-gauge records have revealed systems about 50 km across, with lifetimes of a few hours, that form, move and dissipate within synoptic-scale frontal areas. At present they are thought to result from CONVECTION that is released in layers of potentially unstable air over the frontal surfaces.

Much remains to be discovered and explained within the realms of mesometeorology. The sub-discipline is taking an increasingly important role within meteorology as a whole. The next decade holds exciting prospects for both observational and theoretical studies of mesoscale atmospheric circulations. BWA/SJN

Reading
Markowski, P., and Y. Richardson Y. (2010) *Mesoscale meteorology in midlatitudes*. Hoboken, NJ: Wiley-Blackwell.

mesophyte A plant that flourishes under conditions which are neither very wet nor very dry.

mesoscale cellular convection Over millions of square kilometres of ocean, heat transfer to the atmosphere takes the form of mesoscale cellular convection (MCC). The convection is visible in the particular cloud forms produced, and satellite imagery first noted this form of convection about 1970. The convection is shallow (~2 km deep) and forms distinctive horizontal patterns. In planform the convection takes two main forms: parallel lines of clouds, known as cloud streets, and hexagonal cells. The former is essentially two-dimensional and the latter three-dimensional convection.

The cloud streets may extend for a few hundred kilometres and they lie a few kilometres apart. The hexagonal cells are typically 30 km across and are of two types: closed cells and open cells. The former have uplift in their centre (hence are cloudy and 'closed') and compensatory subsidence on their sides, giving clear air. The open cells have a reversed configuration of vertical motion: uplift on the 'walls' and subsidence in the middle. These different forms are related to vertical stability and vertical shear of horizontal winds, particularly the latter. The details of these relationships are not yet fully clear.

Both types of MCC tend to occur over warm ocean surfaces adjacent to cold land surfaces, such as the Sea of Japan and the Atlantic east of Greenland. In such areas, very cold air frequently flows over warm ocean surfaces, thus creating the instability required to drive this kind of convection. BWA

mesosphere A portion of the Earth's atmosphere. Either that between the stratosphere (at about 40 km above the surface) and the thermosphere (80 km), or that between the ionosphere (400 km) and the exosphere (1000 km).

mesothermal climate Characterized by moderate temperatures, with the mean temperature of the coldest month between $-3\,°C$ and $+18\,°C$. Climates of this type are found mainly between latitudes 30° and 45° but may extend up to latitude 60° on the windward side of continents.

mesotrophic See NUTRIENT STATUS.

Schematic of a mesoscale convective system (MCS) and a photograph of such a system over the African Sahel. Vertical axis in cloud schematic is altitude in kilometres. Vertical axis above is atmospheric temperature estimated from 85 GHz radar (polarization-corrected temperature – PCT), a metric used to define an MCS.
Photograph by Sharon Nicholson.

Messinian salinity crisis An event that took place around 5.96–5.33 million years ago, at the end of the Miocene, during which the ancient Mediterranean Sea underwent desiccation. The growth of the Antarctic Ice Sheet at that time contributed to a marked sea-level reduction so that the western Tethys Ocean lost its connection with the Atlantic (see TETHYS SEA). Tectonic activity may have been a contributory factor. Isolation of this ancient Mediterranean from the world ocean caused it to become highly saline, so that extensive gypsum and halite deposits were precipitated. Subsequent uplift of residual deposits provides sedimentary evidence of the event at various locations around today's Mediterranean. Several cycles of opening and closure to the world ocean system are evidenced within the sedimentary deposits. The drying up of so large a body of water over a brief geological period would have had major consequences, facilitating, for example, the incision of rivers and the movements of fauna and flora. The final opening of the Straits of Gibraltar 5.33 million years ago may have been associated with a catastrophic flood event (Garcia-Castellanos *et al.*, 2009). ASG/DSGT

Reference

Garcia-Castellanos, D., Estrada, S.F., Jiménez-Munt, I., *et al.* (2009) Catastrophic flood of the Mediterranean after the Messinian salinity crisis. *Nature*, **462**, 778–781.

meta-analysis A study that integrates the results of many other previous studies into the same phenomenon. Linked to the concept of big data, and the volume of data that science today generates, meta-analytical approaches are increasingly used as a means to gain a 'big picture' of trends and differences in scientific studies of the same phenomenon. Widely used in medical science, the approach is becoming increasingly popular in studies of climate and ecological systems. Meta-analyses should employ statistical approaches to analysis; some, however, are more textural and qualitative in approach. DSGT

Reading

Borenstein, M., Hedges, L.V., Higgins, J.P.T and Rothstein, H.R. (2009) *Introduction to meta-analysis*. Chichester: Wiley-Blackwell.

metamorphism The processes by which the composition, structure and texture of consolidated rocks are significantly altered through the action of heat and pressure greater than that produced normally by burial, or by the introduction of additional minerals as a result of a marked change in the thermodynamic environment; for example, limestone may be converted to marble, mudstone to slate, and granite to gneiss.

The intensity of metamorphism is referred to as its grade, and high-grade metamorphic rocks include gneiss, granulite, blueschist, amphibolites and eclogites.

There are three classes of metamorphism: regional metamorphism, which occurs in orogenic belts; contact metamorphism around the boundaries of igneous intrusions; and dislocation metamorphism, resulting from friction along fault or thrust planes. ASG

metasomatism The processes by which a rock is physically or chemically altered, partially or wholly, as a result of the introduction of new materials. Contact metamorphism is the alteration of rocks adjacent to an igneous intrusion by metasomatism and high temperatures.

meteoric water/groundwater Water that is derived by precipitation in any form from the atmosphere. Originally defined by Meinzer (1923) and contrasted with JUVENILE WATER, meteoric groundwater is water with an atmospheric or surface origin rather than with a source in the interior of the Earth. (See also CONNATE WATER.) PWW

Reference

Meinzer, O.E. (1923) *Outline of ground-water hydrology: with definitions*. Geological Survey Water Supply Paper 494. Washington, DC: United States Government Printing Office.

meteorological satellite A type of satellite that is principally used to monitor the weather and climate of the Earth (carrying out the function of satellite meteorology). Typically, these satellites are polar orbiting, covering the entire Earth asynchronously, or are geostationary, hovering over the same spot on the equator. Observations from meteorological satellites have commonly been made by sensor systems designed to provide cloud imagery, radiation budget data and vertical profiles through the atmosphere. The first purpose-built meteorological satellite was the American Tiros-I, launched on 1 April 1960. This, a 'television and infrared observation satellite', carried a vidicon camera for visible radiation imaging and an infrared radiometer to provide target temperature data in the invisible thermal region of the electromagnetic spectrum. The data were transmitted to Earth by radio, a procedure followed by all subsequent meteorological satellites. Since their inception it has been evident that that they provided more uniform and more complete views of global weather than were available from conventional ('in situ') sources. Recently launched satellites for European weather forecasting include Meteosat-8 (3.5°W) and Meteosat-9 (0°) for Atlantic Ocean sensing and Meteosat-6 (63°E) and Meteosat-7 (57.5°E) over the Indian Ocean. A wide range of satellite products and derived products are now available for weather and climate research. Increasingly, in-situ observing networks are being planned and operated in full conjunction with meteorological satellites, which are now recognized as indispensable parts of any effective broad-scale atmospheric monitoring system. Other application areas that have benefited from these satellites include snowpack measurement and monitoring and detection of burn-off in gas and oil fields using imagery acquired at night.   DSB

Reading

Bowman, K. and Yang, P. (2014) *Satellite meteorology and atmospheric remote sensing.* Wiley-VCH Verlag GmbH. · Davis, G. (2007) History of the NOAA satellite program. *Journal of Applied Remote Sensing*, **1** (1), 012504, doi: 10.1117/1.2642347. · EUMETSAT (n.d.) Meteosat. http://www.eumetsat.int/website/home/Satellites/Current-Satellites/Meteosat/ (accessed 12 July 2015).

mesoscale The intermediate class in the common grouping of phenomena into micro-, meso-, and macroscale. These terms may be used to group phenomena on the basis of their lineal or areal spatial extent. The divisions are arbitrary and depend on the context and purpose of the work. Within the landscape, macroscale spatial features might be those of >250 m in linear dimension, such as hillslopes; mesoscale features might be those from 1 to 250 m in linear dimension, such as gullies; and microscale features are those <1 m, such as rills, splash pedestals, or other small erosional features. Within the soil, macropores are those >1 cm in diameter, and micropores are those <1 mm in diameter. Mesopores are therefore those in the diameter range 1 mm to 1 cm. In climate modelling, 'mesoscale' commonly refers to those models that have a spatial resolution in the range 1–100 km.   DLD

Reading

De Boer, D.H. (1992) Hierarchies and spatial scale in process geomorphology: a review. *Geomorphology*, **4**, 303–318. · Church, M. and Mark, D.M. (1980) On size and scale in geomorphology. *Progress in Physical Geography*, **4**, 342–390.

mesoscale convective systems Intermediate-scale weather systems that are larger than individual cumulus clouds but smaller than synoptic systems. Most convective precipitation is coupled with systems on this scale. This is contrary to the once-held belief that convective rainfall is generally associated with relatively small and isolated cumulus and cumulonimbus clouds. Mesoscale systems have various names, such as linear disturbances, line squalls, disturbance lines, sumatras, cloud clusters, mesoscale convective clusters or complexes (MCCs), and mesoscale convective systems (MCSs). Here, the term MCS will be utilized and it applies only to 'cold' systems; that is, those in which ice is present in the upper levels of the cloud (Nesbitt and Zipser, 2003). This distinction is important because the presence of ice enhances the efficiency of the precipitation process, generally producing more intense precipitation.

An MCS is essentially a large, continuous area of deep cloud. Definitions vary, but most require a contiguous area of cloud greater than $2000\,km^2$. The typical MCS is topped by a large anvil of stratiform cloud. Despite the intense convection associated with these systems, stratiform precipitation accounts for 73% of the rain area and contributes roughly 40% of the total rainfall for the tropics as a whole (Schumacher and Houze, 2003). MCSs may be at least partially generated by the aggregation of smaller systems as they grow during the course of the day, but they appear to be linked to larger scale systems as well. In the tropics, wave disturbances appear to organize the convection into mesoscale systems. Many systems are associated with squall lines.

Mesoscale convective systems were initially studied in the central USA, where they produce 30–70% of the rain falling during the months of April to September (Fritsch *et al.*, 1986). The mesoscale systems were initially identified with visible and infrared satellite imagery that provides little information about the precipitation produced by the system. More advanced satellite techniques allow for better descriptions of features, such as the presence of ice, that are more closely correlated with precipitation. Over the tropical land areas MCSs average roughly $10,000\,km^2$ in size and comprise only about 2% of the precipitation-bearing features. However, they provide about 50% of total rainfall tropics-wide and up to 90% of the rainfall in areas over select land areas (e.g. Nesbitt *et al.*, 2006). By comparison, small cold-cloud precipitation features provide some 30% of the rainfall over land.

Table 3 Definitions of climate scales

Climate	Horizontal distribution (m)	Climatic example	Lifetime of phenomenon (s)
Microclimate	10^{-2}–10^{2}	Small-field climate	10^{-1}–10^{1}
Local climate	10^{2}–10^{4}	Slope climate	10^{1} to 10^{4}
Mesoclimate	10^{3} to 2×10^{5}	Urban climate	10^{4} to 10^{5} s
Macroclimate	2×10^{5} to 5×10^{7}	Regional climate	10^{5} to 10^{6}

MCSs are in a constant state of evolution. These systems produce intense convective rainfall mainly during the afternoon, with most convective events lasting 3 h or less over land (Ricciardulli and Sardeshmukh, 2002). As the system evolves in the later hours of the day and into the night, the cloud anvil that tops the system spreads and produces a large area of stratiform cloud (Nesbitt and Zipser, 2003). Thus, the stratiform rain typically occurs at night and for a longer period of time, but the rain rate is roughly one-fourth the rain rate associated with convective clouds. SEN

References

Fritsch, J.M., Kane, R.J. and Chelius, C.R. (1986) The contribution of mesoscale convective systems to the warm season precipitation in the United States. *Journal of Climate and Applied Meteorology*, 25, 1333–1345. · Nesbitt, S.W. and Zipser, E.J. (2003) The diurnal cycle of rainfall and convective intensity according to three years of TRMM measurements. *Journal of Climate*, 16, 1456–1475. · Nesbitt, S.W., Cipelli, R. and Rutledge, S.A. (2006) Storm morphology and rainfall characteristics of TRMM precipitation features. *Monthly Weather Review*, 134, 2702–2721. · Ricciardulli, L. and Sardeshmukh, P.D. (2002) Local time- and space scales of organized tropical deep convection. *Journal of Climate*, 15, 2775–2790. · Schumacher, C. and Houze Jr, R.A. (2003) Stratiform rain in the tropics as seen by the TRMM precipitation radar. *Journal of Climate*, 16, 1739–1756.

meteorology The science concerned with understanding atmospheres. In recent years, since satellites have probed the atmospheres of planets other than Earth, it has become acceptable to use the term 'atmospheric science' and to apply the term to any atmosphere, not just that on Earth. BWA

microclimate The climate near the ground surface in which plants and animals live. There is no distinct and internationally agreed separation between the study of climate at the smallest or microscale and that of larger scales. This is not surprising as the processes operating at the ground surface form part of a continuum that is not readily separated into discrete subdivisions. As a result, the term is used frequently without a precise meaning. Yoshino (1975) suggests a horizontal scale of

0.01–100 m with anything larger belonging to the local-, meso- or macroscales of climate, whilst Oke (1987) takes 1000 m as his upper limit of the microscale. Averaging of temporal scales also produces variety. Yoshino (1975) suggests 10 s for the microscale in contrast to 10^{6} s (approximately 1 day) for the macroscale (see Table 3). Both spatial and temporal definitions indicate a highly variable environment in contrast to that of the macroscale.

Microclimate is inextricably linked to the processes taking place in the immediate Earth–air interface. Hence, the nature of the surface is extremely important in determining the type of microclimate likely to be experienced. The prime factor is the treatment of solar energy input. Surfaces will respond differently to absorption through their different ALBEDOS, or reflective capacities. How this energy is converted into heat to warm up the surface will depend upon the nature of the vegetation canopy and the degree of moisture availability, which in turn influences how much energy is used for LATENT HEAT and SENSIBLE HEAT. Soil factors may be significant, as dry soils conduct heat away from the surface more slowly than wet soils, and the colour of the soil will affect how much solar energy will be absorbed. Aspect and shadowing may affect the amount of energy a surface receives; in the northern hemisphere, south-facing slopes receive more energy and therefore are potentially warmer than north-facing slopes. At night, cold air may drain into topographic basins to produce a TEMPERATURE INVERSION. The sky-view factor (or proportion of the sky visible for longwave radiational cooling) can be important in determining the degree of cooling at night. If the sky view is unobstructed by buildings, trees or slopes, the loss of longwave radiation will be much greater than if the site is very sheltered, as in a deep, narrow valley. As a result, temperatures may fall to lower values. In winter, the consequence of these processes may be seen from the distribution of surface frost.

Of equal importance is the influence of wind speed and turbulence in mixing the air. If wind speeds are light, then the surface properties mentioned above can dominate. If winds are stronger, then the surface-generated thermal microclimates can become mixed to give a relatively uniform temperature where microscale differences are limited. Normally, wind speeds do decrease as the ground surface is approached, as a result of friction and shelter. This gives a logarithmic wind-speed profile above the surface.

As a result of this potential shelter, the thermal climate near the ground can become much more extreme than that observed at a standard climatological site about 1.5 m above the ground with instruments cased in a wooden screen. As the ground surface is the main area of absorption and emission of energy, it tends to become warmer by day and cooler by night than standard observation sites. Only if the atmosphere is cloudy with a strong wind is the effect nullified. For example, in a desert environment, the daily range of temperature in the air a few centimetres above the surface may be 50 °C greater than the range measured at a standard height. In practice, the majority of this difference occurs within the lowest 50 cm. Grass minimum temperatures are the only routine observations at climate stations that give any indications of conditions near the ground surface, and these are only read once per day. Under clear skies and light winds, minimum temperatures recorded here can be up to 10 °C cooler than in a Stevenson screen.

The greater range of temperatures is important for the growth and survival of plants and animals living in this environment. Seedlings emerging from the soil have to withstand greater daytime heating and night-time cooling than might be expected from normal climatological values. Ground frost can cause serious damage to newly emerging plants. Animals (including humans) can select appropriate microclimates for comfort and survival, and several useful schemes have been developed to relate possible microclimates to known physiological responses, thereby establishing the microclimate space in which an animal can exist. PS

Reading and References
Geiger, R., Aron, R.H. and Todhunter, P. (1995) *Climate near the ground*. Wiesbaden: Vieweg. · Oke, T.R. (1987) *Boundary layer climates*. London: Routledge. · Yoshino, M.M. (1975) *Climate in a small area*. Tokyo: Tokyo University Press.

microcracks These can occur in rocks of all kinds. They are generally short (a few centimetres long) and narrow (<1 mm), but they are important because they provide one of the main means by which rocks break down. Both physical and chemical weathering can take place within them as they provide a much enlarged surface area for chemical action to take place, and the crack tip is an important stress

concentrator in brittle fracture. A general classification of such cracks has been put forward by Farran and Thenoz (1965):

- microfissures – less than 1 µm in width and about the length of a crystal;
- microfractures – about 0.1 mm or less wide;
- macrofractures – greater than 0.1 mm wide and may be several metres long. WBW

Reading and Reference
Farran, J. and Thenoz, B. (1965) L'altérabilité des roches, ses facteurs, sa prévision. *Annales de l'Institut Technique du Bâtiment et des Travaux Publics*, supplément no. 215, 1534–1548. · Whalley, W.B., Douglas, G.R. and McGreevy, J.P. (1982) Crack propagation and associated weathering in igneous rocks. *Zeitschrift für Geomorphologie*, **26**, 33–54.

micro-erosion meter An instrument used to measure accurately the erosion or weathering of a rock from direct measurement on the rock surface. It comprises a spring-loaded probe connected to an engineer's dial gauge. In order to measure exactly the same area each time, the instrument is mounted on three legs that are placed onto studs drilled in the rock surface. Measurements have been achieved to the nearest 0.00001 mm. PAB

microfossils These are normally less than 2.00 mm in size and include the mineralized shell remains of animals (e.g. COCCOLITHS, DIATOMS, FORAMINIFERA, OSTRACODS) and the organic-walled remains of plants (see POLLEN ANALYSIS). Because of similar processing and study methods, several microfossil groups are studied under the umbrella term 'micropalaeontology'. Their small size means that they can only be studied with the aid of microscopes, and in the case of extremely small microfossils scanning electron microscopes. SLO

micrometeorology The science concerned with the study of physical phenomena taking place on a microscale in the atmosphere close to the surface. The development of micrometeorology has involved not only the examination and analysis of highly accurate observations made in the air layers adjacent to surfaces, but also detailed study of processes at surfaces that determine the properties of air in such layers. A principal feature of air near the ground is its turbulence, the complex mixing that is produced by eddies with sizes ranging from microscopic to tens of metres. The turbulence is generated by friction between moving air and the surface, and may be modified by temperature differences between air and ground. Understanding of the structure of turbulence has been a major theme of micrometeorology, of fundamental importance and with practical applications such as the dispersal of spores and the diffusion of air pollution. Exchange between the surface and the atmosphere leads to vertical gradients of temperature, wind speed and gas concentrations. Much progress in micrometeorology has been made by the detailed analysis of mean gradients (over periods of about 30 min) at sites where there is extensive uniform horizontal ground cover. Based on such analysis, rates of exchange at the surface can be deduced from measurements made entirely in the atmosphere. For example, rates of evaporation or of carbon dioxide uptake by growing crops can be studied in relation to diurnal changes in the weather. Gradient methods in micrometeorology have a sound physical basis but must rely on empirical corrections to allow for effects of temperature gradients on turbulence. Recent developments in instrumentation and computers suggest that, in future, methods of measuring turbulent transfer more directly by sensing the motion and composition of individual eddies – an eddy correlation approach – will become more readily available for applied research.

The surface processes most directly affecting the structure of the lower atmosphere begin with the absorption of radiant energy at the ground. Micrometeorology is concerned with the interception and fate of that energy, with the way it is used to heat vegetation, soil and water, and with the conversion of some of the energy to latent heat when water is evaporated (see EVAPORATION). Heat stored in the soil may be of great importance as an energy source at night, and the understanding of factors determining the temperature structure of the lower atmosphere at night is a challenging problem for micrometeorologists.

Although this topic can be pursued as a fundamental aspect of meteorological physics, much of micrometeorology has practical applications. Examples are particularly common in AGROMETEOROLOGY, including the study of forest meteorology. Particular progress has been made in understanding atmospheric and surface controls of evaporation, diffusion of pollution from point and area sources, uptake of air pollutants by plants, and the spread of animal and plant disease.   MHU

Reading
Arya, S.P. (1988) *Introduction to micrometeorology*. San Diego, CA: Academic Press. · Munn, R.E. (1966) *Descriptive micrometeorology*. New York: Academic Press. · Oke, T.R. (1987) *Boundary layer climates*, 2nd edition. London: Routledge. · De Vries, D.A. and Afgan, N.H. (eds) (1975) *Heat and mass transfer in the biosphere. I:*

transfer processes in plant environment. Washington, DC: Scripta Press.

microphytes Small organisms such as mosses, lichens, liverworts, algae and bacteria. They may be found in terrestrial or aquatic environments. Terrestrial microphytes are best known from the soil surfaces of the drylands, where they frequently occupy the exposed spaces between the scattered vascular plants such as grasses or shrubs. Microphytes alter the albedo, mechanical properties, nutrient status and water relations of dryland soils. One of the most widespread groups is the cyanobacteria, whose metabolic exudates and filamentous polysaccharide sheaths can bind fine materials such as silts into quite stable crusts. These may be called 'biological soil crusts' or 'microphytic crusts'. Frequently, only the top few millimetres of the soil is structurally affected by the microphyte colonies, but this is sufficient to limit erosion of the bulk soil by wind and water, except where the microphytic crust is damaged by vehicles, hard-hoofed grazing animals, or people. In severe droughts, microphytic crusts can be degraded because of abrasion by blowing sand, and this contributes to the release and mobilization of the underlying soil.   DLD

Reading
Belnap, J. and Gillette, D.A. (1998) Vulnerability of desert biological soil crusts to wind erosion: the influences of crust development, soil texture, and disturbance. *Journal of Arid Environments*, **39**, 133–142. · Eldridge, D.J. and Leys, J.F. (2003) Exploring some relationships between biological soil crusts, soil aggregation and wind erosion. *Journal of Arid Environments*, **53**, 457–466.

microtopography A term used to describe small-scale irregularities of the ground surface. Traditionally, geomorphologists have regarded microtopography as recognizable elements superimposed on the overall hillslope form, which occur in a variety of settings and environments. More recent interest in the application of FRACTALS has led to speculation that hillslope forms contain a spectrum of features at all scales. On slopes, microtopography is most frequently encountered in environments where soils are generally thin and bedrock is at, or close to, the surface. In KARST environments, water flow paths across rock surfaces generate a variety of microtopographic features, generally referred to as KARREN. In semi-arid environments, the swelling and heaving of clays, associated with wetting and drying cycles, can lead to the sorting of stone layers and the formation of stepped hillslope microtopography, termed desert GILGAI. The banded pattern of treads and risers is often reinforced by vegetation (Dunkerley and Brown, 1995). Similar sorting of fine and coarse particles is observed in periglacial environments due to FROST HEAVE and SOLIFLUCTION processes. The most conspicuous form of microtopography on humid temperate hillslopes is the TERRACETTE, a stepped sequence of treads and risers. Their origin remains unclear, but they are thought to be relict periglacial features accentuated by the action of trampling animals (Vincent and Clarke, 1976). Fluvial microtopography includes river BEDFORMS and clusters. A precise definition of where micro-topography begins and ceases is difficult. Reviewing previous studies of microtopography (or microrelief), Parsons (1988) notes its application to features with elevation differences between 2.5 cm and 3 m. More recent innovations in measuring techniques, such as in-situ laser reflection, permit irregularities to be characterized much more precisely. The statistical description of surfaces is fundamental to tribology, the science of wear. DH

References

Dunkerley, D.L. and Brown, K.J. (1995) Runoff and runon areas in patterned chenopod shrubland, arid western New South Wales, Australia. *Journal of Arid Environments*, **30**, 41–55. · Parsons, A.J. (1988) *Hillslope form*. London: Routledge. · Vincent, P.J. and Clarke, J.V. (1976) The terracette enigma. *Biuletyn Peryglacjalny*, **25**, 65–77.

microwave remote sensing

Refers to REMOTE SENSING using microwave wavelength radiation. The radiation measured may be either that emitted from the Earth in a passive system or that produced and backscattered to a sensor such as a synthetic aperture radar in an active system.

In the context of EARTH OBSERVATION, microwave radiation may be characterized as having a relatively long wavelength. While most Earth-orientated remote sensing is undertaken using visible to thermal infrared radiation with wavelengths that range from roughly 0.4 to 14.0 μm, microwave radiation has a wavelength usually measured in centimetres (e.g. C- and L-band RADAR systems would typically operate at a wavelength in the order of 5 cm and 20 cm respectively). The key attraction for the use of such long wavelength radiation is that it can pass through the atmosphere without significant interference, providing, in essence, an ability to see through the clouds; the atmospheric windows in which visible and infrared remote sensing are undertaken can be closed by cloud. The nature of the radiation interactions with the ground surface is also different to that at shorter wavelengths,

providing especially information on the roughness of the target. As such, microwave remote sensing provides data that are different from but complementary to those acquired at shorter wavelengths. For example, synthetic aperture radar systems may be used to derive information on forest biophysical variables such as biomass, while visible and near-infrared sensors could provide information on biochemical variables such as chlorophyll content. Because of the weather independence and sensitivity to biophysical variables, active microwave remote sensing has considerable potential for applications such as the estimation of tropical forest biomass and its dynamics. GF

Reading

Woodhouse, I.H. (2005) *Introduction to microwave remote sensing*. Boca Raton, FL: CRC Press.

midden

An accumulation of biological waste that is a result of human or other zoological activity. Shell middens, and so on, are important in archaeological research as they contain material that can both be dated and analysed geochemically, in terms of species, and so on. The middens of hyrax in southern Africa, and of other rodents in other continents, can serve a similar purpose. DSGT

mid-ocean ridge

Part of a coherent global system of interconnected underwater mountain chains (with a total length of 80,000 km) throughout the world's ocean basins. The ridges typically having a valley, or rift, running along their spine. These rifts are the sites of mantle upwelling between two separating tectonic plates (a constructive, or divergent, plate boundary) where emergent lava creates new crust on cooling, a process known by the general term 'sea-floor spreading'. Slow-spreading ridges ($\sim$25 mm a^{-1}), like the Mid-Atlantic Ridge (MAR), generally have large, wide rift valleys, sometimes as wide as 10–20 km, and very rugged terrain at the ridge crest (relief of up to 1000 m). Fast-spreading ridges (80–120 mm a^{-1}), like the East Pacific Rise (EPR), are narrow, sharp incisions surrounded by a generally flat topography sloping away from the ridge.

These shapes result from isostacy: close to the ridge axis there is hot, low-density mantle supporting the oceanic crust. As the oceanic plates cool, away from the ridge axes the oceanic mantle lithosphere (the colder, denser part of the mantle that, together with the crust, comprises the oceanic plates) thickens and the density increases. Thus, older sea floor is underlain by denser material and 'sits' lower. The width of the ridge is hence a function of spreading rate; slow-spreading ridges like the MAR have not spread as far as faster spreading ridges like the EPR for the same amount of cooling and consequent bathymetric drop-off.

There are two processes, ridge-push and slab-pull, thought to be responsible for the spreading seen at mid-ocean ridges, and there is some uncertainty as to which is dominant. Ridge-push occurs when the growing bulk of the ridge pushes the rest of the tectonic plate away from the ridge, often towards a subduction zone. At the subduction zone, 'slab-pull' comes into effect. This is simply the weight of the tectonic plate being subducted (pulled) below the overlying plate dragging the rest of the plate along behind it. The relatively shallow depth from which the upwelling mantle rises below ridges (upper 400 km only) is more consistent with the 'slab-pull' process. Nevertheless, other studies have suggested that the upper mantle (asthenosphere) is too plastic (flexible) to generate enough friction to pull the tectonic plate along.   TS

Reading
Parsons, B. and Sclater, J.G. (1977) An analysis of the variation of ocean floor bathymetry and heat flow with age. *Journal of Geophysical Research*, **82**, 803–827.

migration   A term used in many ways in physical geography. For example, in geomorphology, rivers can be described as migrating in the landscape over time, or their channels migrating within a FLOODPLAIN. Some types of sand dune have migration as their principle dynamical behaviour (e.g. BARCHAN dunes).

In a biogeographical/ecological sense, any movement of animals or plants from one region or habitat to another (see DISPERSAL) is also termed migration. More specifically, migration refers to the seasonal or cyclical movement of animals between two or more areas, a phenomenon common in mammals, birds, fishes and insects. Exactly how animals migrate still remains largely unanswered, although it is clear that no one theory is universally applicable throughout the animal kingdom. Navigation appears to be possible through the use of the sun as a compass, a similar use of the stars, the use of the Earth's magnetic field and the Coriolis effect, the recognition of the lie of the land and distinctive landmarks, and the use of smell. Environmental factors may drive migration, as animals seek food or suitable habitats for certain activities, such as breeding. Stable isotopes can now be used to track migration pathways, providing new data on animal behaviour. At a different timescale completely, in Quaternary studies, climate and environmental change or variability is one theory that may explain the evolution and migration of early human ancestors out of Africa, *c.* 120–70 ka ago.   PAS/DSGT

Reading
Hobson, K.A. and Wassanaar, L.I. (eds) (2008) *Tracking animal migration with stable isotopes*. London: Academic Press · Milner-Gulland, E.J. and Fryxell, J.M. (2011) *Animal migration: a synthesis*. Oxford: Oxford University Press. · Drake, N.A., Breeze, P. and Parker, A.G. (2013) Palaeoclimate in the Sahara and Arabian Deserts during the Middle Palaeolithic and the potential for hominin dispersals. *Quaternary International*, **300**, 48–61.

Milankovitch hypothesis   One of the most significant models of Earth history, which is seen by some (e.g. Imbrie and Imbrie, 1979) as the key to understanding the causes of the climatic fluctuations of the Pleistocene. The hypothesis is based on the fact that the position and configuration of the Earth as a planet in relation to the Sun is subject to change, thereby affecting the receipt of insolation at the Earth's surface. There are three such changes that have been identified, all three occurring in a cyclic manner: changes in the eccentricity of the Earth's orbit (a 96,000-year cycle), the precession of the equinoxes (with a periodicity of 21,000 years) and changes in the obliquity of the ecliptic – the angle between the plane of the Earth's orbit and the plane of its

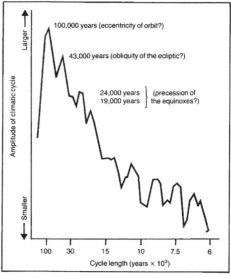

Amplitude of climate cycles and links to orbital changes.

rotational equator – which has a periodicity of about 40,000 years.

The Earth's orbit around the Sun is not a perfect circle but an ellipse. If the orbit were a perfect circle then the summer and winter parts of the year would be equal in their length. With greater eccentricity there will be a greater difference in the length of the seasons. Over a period of about 96,000 years the Earth's orbit can stretch by departing much further from a circle and then reverting to almost true circularity.

The precession of the equinoxes simply means that the time of year at which the Earth is nearest the Sun (perihelion) varies. The reason is that the Earth wobbles like a top and swivels its axis round. At the moment, the perihelion comes in January. In 10,500 years it will occur in July.

The third cyclic perturbation, change in the obliquity of the ecliptic, involves the variability of the tilt of the axis about which the Earth rotates. The values vary from 21°39′ to 24°36′. This movement has been likened to the roll of a ship. The greater the tilt, the more pronounced is the difference between winter and summer (Calder, 1974). Appreciation of the possible significance of these three astronomical fluctuations of the Earth goes back to at least 1842, when J. F. Adhemar made the suggestion that climate might be affected by them. His views were developed by Croll in the 1860s and by Milankovitch in the 1920s (Mitchell, 1965).

The major attraction of these ideas is that while the extent of temperature change caused by them may well only be of the order of 1 or 2 °C, the periodicity of these fluctuations seems to be largely comparable to the periodicity of the ice advances and retreats of the Pleistocene. Recent isotopic dating has shown that the record of sea-level changes preserved in the coral terraces of Barbados and elsewhere, and the record of heating and cooling in deep-sea cores, correlates well with theoretical insolation curves based on those of Milankovitch (Mesollela et al., 1969).

The variations in the Earth's orbit have been seen as 'the pacemaker of the ice ages' (Imbrie and Imbrie, 1979), for detailed statistical analysis of ocean cores shows that they possess statistically significant wavelike fluctuations with amplitudes of the order of around 100,000 years, 43,000 years and 19,000–24,000 years (see diagram). The most important of these cycles is the longest one, corresponding to variations in eccentricity (Hays et al., 1976). This applies back to 900,000 years ago, but probably not further (Pisias and Moore, 1981).

Thus, there is some substantial evidence to suggest that astronomical theories may be valid as an explanation of the longer scale of environmental changes.

Orbital changes also help to explain the very marked expansion of lakes that took place in tropical and subtropical areas around 9000 years BP. Recent analyses of theoretical insolation levels by Kutzbach (1981) indicate that radiation receipts in July at that time were larger than now (by about 7%) and that this led to an intensification of monsoonal circulation and associated precipitation.

Recent modelling suggests that the periodicity of climatic variations during the glacial–interglacial cycles is partly influenced by the response of the crust and asthenosphere to changing ice and water loads. Other feedbacks involved in climate change triggered by the orbital changes include variations in the concentration of carbon dioxide in the atmosphere, and the fluctuations of terrestrial dust loadings to the oceans, where iron fertilization of phytoplankton occurs (see ICE AGE). ASG/DLD

Reading and References

Andrews, J.T. (1975) *Glacial systems: an approach to glaciers and their environments*. North Scituate, MA: Duxbury. · Berger, A., Imbrie, J., Hays, J., *et al.* (eds) (1984) *Milankovitch and climate*, 2 vols. Dordrecht: Reidel. · Calder, N. (1974) *The weather machine*. London: BBC. · Hays, J.D., Imbrie, J. and Shackleton, N.J. (1976) Variations in the Earth's orbit: pacemaker of the ice ages. *Science*, **194**, 1121–1132. · Imbrie, J. and Imbrie, K.P. (1979) *Ice ages: solving the mystery*. London: Macmillan. · Kutzbach, J.E. (1981) Monsoon climate of the early Holocene: climate experiment with the Earth's orbital parameters for 9000 years ago. *Science*, **214**, 59–61. · Mesollela, K.J., Matthews, R.K., Broecker, W.S. and Thurber, D.L. (1969) The astronomical theory of climatic change: Barbados data. *Journal of Geology*, 77, 250–274. · Mitchell, J.M. (1965) Theoretical paleoclimatology. In H. E. Wright and D. G. Frey (eds), *The Quaternary of the USA*. Princeton, NJ: Princeton University Press; pp. 881–901. · Pisias, N.G. and Moore, T.C. (1981) The evolution of Pleistocene climate: a time-series approach. *Earth and Planetary Science Letters*, **52**, 450–458.

Millennium Ecosystem Assessment (MA)

An international assessment of the consequences of ecosystem change for human well-being, initiated by a call from the UN Secretary-General in 2000 (Reid *et al.*, 2005). The assessment involved more than 1300 experts worldwide, who used available scientific evidence to appraise of the state and trends of a range of ecosystem services (see ECOSYSTEM GOODS AND SERVICES), such as clean water, food, forest products, flood control and cultural services such as recreation (see figure). The results, which were published in 2005, also considered options to restore, conserve or

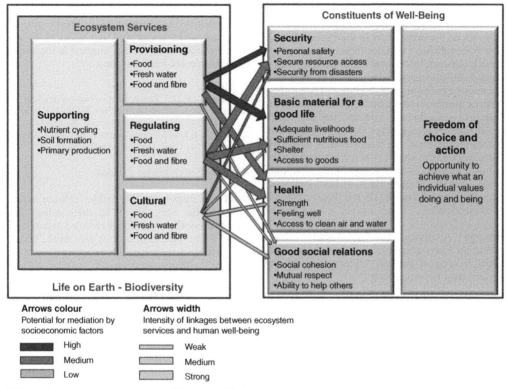

Linkages between ecosystem services and human well-being.
Reproduced with permission of Roy Haines-Young.

enhance the sustainable use of ecosystems. The MA found (Hassan *et al.*, 2005) that in the latter part of the twentieth century people have changed ecosystems more rapidly and extensively than in any comparable period of time in human history, and that while there have been gains in human well-being and economic development, many ecosystem services have been degraded. The assessment concluded that such degradation could worsen in the first half of this century and could prevent the UN's Millennium Development Goals from being achieved. RH-Y

Reading and References
Hassan, R., Scholes, R. and Ash, N. (eds) (2005) *Ecosystems and human well-being: current state and trends*, volume 1. Washington, DC: Island Press. http://www.unep.org/maweb/documents/document.766.aspx.pdf (accessed 12 July 2015). · Millennium Ecosystem Assessment (n.d.) Guide to the Millennium Assessment reports. http://www.unep.org/maweb/en/Index.aspx (accessed 12 July 2015). · Reid, W.V., Mooney, H.A., Cropper, A., *et al.* (2005) *Ecosystems and human well-being: synthesis*. Washington, DC: Island Press. http://www.unep.org/maweb/documents/document.356.aspx.pdf (accessed 12 July 2015).

mima mounds Quasi-evenly spaced ('self-organizing) mounds of earth, up to 1 m high, found in open landscapes in all environments from polar to desert, but not in Antarctica. Their origins have fascinated scientists for a long time, and the quest for a singular explanation for their development has not been forthcoming. This is unsurprising, since different processes may apply in different contexts. Biological origins predominate (termites, burrowing rodents), and spacing may reflect reduced competition among animals that may be responsible for development. Siesmic, geochemical, aeolian and other theories exist. DSGT

Reading
Gabet, E.J., Perron, J.T. and Johnson, D.L. (2014) Biotic origin for mima mounds supported by numerical modeling. *Geomorphology*, **206**, 58–66. · Naudé, Y., van Rooyen, M.W. and Rohwer, E.R. (2011) Evidence for a geochemical origin of the mysterious circles in the Pro-Namib desert. *Journal of Arid Environments*, **75**, 446–456.

minimum acceptable flow This concept dates from the Water Resources Act 1963, which requires the UK River Authorities to determine and to keep under review minimum acceptable flows for all the main rivers in their areas. Such a flow is very difficult to determine because it necessarily depends upon local circumstances, but it should take account of the character of the inland water and its surroundings and should be 'not less than the minimum which in the opinion of the river authority is needed for safeguarding the public health and for meeting (in respect of both quantity and quality of water) the requirements of existing lawful uses of the inland water, whether for agriculture, industry, water supply and other purposes, and the requirements of land drainage, navigation and fisheries' (Water Resources Act 1963, section 19).   AMG

mirage   An optical illusion produced by refraction in the lower atmosphere as a result of differential heating of the air. Objects beyond the horizon may become visible.

mire   A peat-accumulating wetland. This is a general term applied to peat-producing ecosystems that develop in sites of abundant water supply. Mires are made up of microtopes, mesotopes and macrotopes. Mire microtopes are small-scale topographical features such as a regular arrangement of ridges or hummocks and hollows associated with the mire surface. Mesotopes are mire systems developed from one original centre of peat formation; they may join to form a mire macrotope or mire complex. (See also RAISED MIRE and PEAT.)   ALH

Reading
Heathwaite, A.L. and Gottlich, Kh. (1993) *Mires: process, exploitation and conservation.* Chichester: John Wiley & Sons, Ltd.

miombo woodland   Savanna woodlands found in form of a broad belt (covering $2.7 \times 10^6\,km^2$) across south-central Africa, from Angola in the west to Tanzania to the east, and dominated by trees of the subfamily Caesalpinioideae, particularly Miombo (*Brachystegia*), Julbernardia and Isoberlinia. Miombo trees have been variously described along a spectrum from deciduous to evergreen; a useful division is that between wet and dry miombo based on the 1000 mm isohyet. In dry miombo, aboveground woody biomass averages around 55 t dry matter per hectare, whilst in wet miombo 90 t dry matter per hectare is typical. Characteristically, the trees shed their leaves for a short period in the dry season to reduce water loss and produce a flush of new leaves just before the onset of the rainy season. Fire burning is prevalent and maintains the open canopy. Miombo soils are generally nutrient poor,

acidic and low in organic matter. The dominant trees utilize ectomycorrhizae to obtain phosphorus and have large lateral and tap roots to access nutrients and water. Grazer biomass is low compared with more fertile savanna.   TS

Reading
Chidumayo, E.N. (2001) Climate and phenology of savanna vegetation in southern Africa. *Journal of Vegetation Science* 12, 347–354. · Coates Palgrave, K., Drummond, R.B., Moll, E.J. and Coates Palgrave, M. (2002) *Trees of southern Africa.* Cape Town: Struik Publishers.

misfit   See UNDERFIT STREAM.

mist   A suspension of visible, minute water drops in the atmosphere. In meteorology the term is used when visibility is reduced to between 1 and 2 km. The term haze is also used. Water contents of mists are typically a very small fraction of $1\,g\,m^{-3}$ and average drop radii are typically less than $1\,\mu m$. In an atmosphere containing many particulates, mists can occur with relative humidities as low as 80%.   KJW

mistral   A cold, dry north or northwest wind affecting the Rhône valley, particularly to the south of Valence. It is typically strong and squally and most violent in winter and spring. A depression over the Tyrrhenian Sea or Gulf of Genoa with high pressure advancing from the west towards Spain provides the necessary synoptic gradient. Market gardens and orchards require protection from the mistral by wind-breaks. Topographic channelling causes local strengthening of wind speeds in the Rhône valley (Barsch, 1965), while marked diurnal variations are common. In the area of maximum frequency of mistral in the Rhône delta an average of over 100 days per year are recorded.   AHP

Reading and Reference
Barsch, D. (1965) Les arbres et le vent dans la vallée meridionale du Rhone. *Revue de Géographie de Lyons* 40, 35–45. · Boyer, F., Orieux, A. and Powger, E. (1970) *Le Mistral en Provence occidentale.* Monographs de la météorologie 79. Paris: Nationale.

mixing corrosion   The increased degree of solutional corrosion that occurs when two saturated karst waters of different composition mix. (See also TROMBE'S CURVES.)

mixing models   Models used to explain or predict temporal variations in the solute concentrations of streamflow by taking account of the mixing of water from different sources or the mixing of water within a store. Gregory and

Walling (1976) report several studies where simple mass balance models, based on the mixing of individual run-off components, have been developed. Johnson *et al.* (1969) also describe a model involving the mixing of incoming precipitation with water stored within the basin, which in turn supplies the streamflow output.   DEW

References
Gregory, K.J. and Walling, D.E. (1976) *Drainage basin form and process*. London: Edward Arnold; pp. 222–223. · Johnson, N.M., Likens, G.E., Bormann, F.H., *et al.* (1969) A working model for the variation in stream water chemistry at the Hubbard Brook Experimental Forest, New Hampshire. *Water Resources Research*, 5, 1353–1363.

mobile belt   A linear crustal zone characterized by tectonic activity. The term is applied to both contemporary zones of OROGENY, where mountain ranges are actively being formed, and to ancient zones of intense crustal activity indicated by the effects of metamorphism, granite emplacement and faulting. In the latter sense, mobile belts are contrasted with the stable crustal regions of continents or CRATONS.   MAS

Reading
Spencer, E.W. (1977) *Introduction to the structure of the Earth*, 2nd edition. New York: McGraw-Hill. · Windley, B.F. (1984) *The evolving continents*, 2nd edition. New York: John Wiley & Sons, Inc.

moder   One of the three main forms of organic matter in soils. See also MOR and MULL. Between the neutral or slightly acid conditions in which mull develops and the very acid conditions of mor, there is an intergrade known as moder. Although much of the organic fraction is well decomposed and incorporated into the soil's mineral profile, the binding between the two remains weak. There is, therefore, no strong structural development, and a thin layer of litter and fermented material accumulates at the surface.

mogote (haystack hill)   A large residual limestone hill that is a remnant of limestone solution and erosional processes. It is roughly circular in shape, with steep sides that terminate abruptly in a flat alluvial plain. Mogotes are usually found in large numbers and are common in south China (where they inspired many classical landscape pictures), North Vietnam, France (Massif Central), the Philippines, Croatia and Java. In each area they have different local names. (See also COCKPIT KARST and KEGELKARST.)   PAB

Reading
Klimaszewski, M. (1964) The karst relief of the Kuelin area (South China). *Geographia Polonica*, 1, 187–212.

Mohorovičić discontinuity   Sometimes referred to as the 'moho', lies between the crust of the Earth and the underlying MANTLE. Occurring at around 30–40 km below the continents and at about 10 km beneath the oceans, it is a zone where seismic waves are significantly modified. ASG

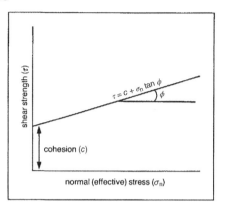

Relationship between normal stress and shear strength.

Mohr–Coulomb equation   A widely used relationship to describe the strength of soils in terms of COHESION and FRICTION of materials:

$$s = c + \sigma \tan \phi$$

where s is the shear strength at failure (sometimes denoted by T), c is the cohesion component and ϕ the friction of the soil; σ (sometimes σ_n) is the normal stress. Either cohesive or frictional components may be absent, but a plot of normal stress and shear stress for a soil with both is of the form shown above. The strength parameters c and ϕ are usually obtained by SHEAR BOX or TRIAXIAL APPARATUS tests. The equation is modified if pore water pressures are taken into account to give an EFFECTIVE STRESS analysis.   WBW

Reading
Mitchell, J.K. (1976) *Fundamentals of soil behavior*. New York: John Wiley & Sons, Inc. · Statham, I. (1977) *Earth surface sediment transport*. Oxford: Clarendon Press. · Whalley, W.B. (1976) *Properties of materials and geomorphological explanation*. Oxford: Oxford University Press.

molard   A term from the French Alps for 'conical mounds of broken slide rock deposited along the typically lobate margins of rock avalanche spoil debris – a debris cone' (Cassie *et al.*, 1988). They may be up to 20 m high.   ASG

Reference
Cassie, J.W., Van Gassen, W. and Coudess, D.M. (1988) Laboratory analogue of the formation of molards, cones of rock-avalanche debris. *Geology*, 16, 735–778.

molasse A term applied to any thick succession of continental deposits consisting in part of sandstones and conglomerates that were formed as a result of mountain building. The molasse facies is the main diagnostic feature of orogeny (mountain building).

moment magnitude scale A dimensionless scale for measuring the intensity of an EARTH-QUAKE. Abbreviated as M_w (meaning 'mechanical work'), the scale is based on the total moment release of an event, which is the product of the distance a fault moves and the energy required to move it. It has the advantage over the Richter scale (for which values for small and medium earthquakes are similar) of more accurately measuring earthquakes of greater energy.   DSGT

Reading
Hanks, T.C. and Kanamori, H. (1979) Moment magnitude scale. *Journal of Geophysical Research*, **84**, 2348–2350.

momentum budget Except for the small effects of tidal friction, the absolute ANGULAR MOMENTUM of the Earth and atmosphere combined must remain constant in time; and because the annual average rotation rate of the Earth is observed to remain almost constant, the atmosphere alone must also on average conserve its angular momentum. Thus, it is possible to consider the fluxes of angular momentum in an angular momentum budget. Such a momentum budget is applied to the time-averaged flow of the atmosphere; that is, the general circulation (see GENERAL ATMOSPHERIC CIRCULATION).

Angular momentum per unit mass of a westerly current is the product of angular velocity with the distance about the axis of rotation. In the tropics, where surface winds are easterly, the torque, or twisting moment, about the Earth's axis exerted by the Earth on the atmosphere is such that the atmosphere gains angular momentum from the Earth; while in the middle latitudes, where surface winds are westerly, the atmosphere gives up angular momentum to the Earth. Therefore, there must be a net poleward flux of angular momentum in the atmosphere between the tropics and middle latitudes, otherwise surface friction would decelerate both the easterlies and westerlies. This poleward flux of angular momentum must increase with latitude in the regions of the easterlies and decrease with latitude in the westerlies. In the northern hemisphere it is observed to reach a maximum at about 30° latitude. Poleward of the mid-latitude westerlies, the surface area of the Earth is relatively small, so that angular momentum

exchange is relatively unimportant. If the total angular momentum of the atmosphere is to remain constant, the exchanges between atmosphere and Earth in the easterlies and westerlies must be equal but of opposite sense.

In low latitudes, the poleward momentum transport is effected partly by the mean meridional flow and partly by eddies. In middle latitudes, the mean meridional flow is much too weak to effect a significant flux, so that the poleward transport is accomplished mainly by eddies. For eddies to effect a meridional transport of angular momentum they must be asymmetric, with the axes of troughs and ridges tilting from southwest to northeast. In such a trough–ridge system, zonal flow is larger in those portions where the meridional flow is poleward, so that a poleward angular momentum flux results. Observations show that eddies account for most of the poleward transport except within about 10° latitude of the equator.

The torque exerted by the Earth on the atmosphere is due partly to surface friction and partly to pressure differences across mountains. For instance, in the mid-latitude westerlies, observations show that surface pressures on the western slopes of mountains tend to exceed those on the eastern slopes at the same heights. In mid-latitudes, this mountain pressure torque is estimated to be about as large as the torque due to surface friction.   KJW

Reading
Holton, J.R. and Hakim, G.J. (2013) *An introduction to dynamic meteorology*. Waltham, MA: Academic Press.

monadnock Any residual hill or mountain that is isolated on a flat plain produced by erosion. It is a product of the late stages of the Davisian cycle of erosion, and rises above a peneplain.

monoclimax A theory of vegetation requiring that all the SERES (community sequences) in an area converge on a uniform and stable plant community, the composition of which depends solely on regional climate. Given sufficient time, it is argued, the processes of SUCCESSION overcome any major effects on vegetation of differences in other environmental factors, such as topography and soils.

It is now generally replaced by other theories of CLIMAX VEGETATION. Many vegetational terms had their origin in monoclimax theory. (See also POLYCLIMAX.)   JAM

Reading
Clements, F.E. (1916) *Plant succession*. Publication 242. Washington, DC: Carnegie Institute. · Matthews, J.A.

(1979) Refutation of convergence in a vegetation succession. *Naturwissenschaften*, **66**, 47–49. · Walker, D. (1970) Direction and rate in some British postglacial hydroseres. In D. Walker and R. G. West (eds), *Studies in the vegetation history of the British Isles*. Cambridge: Cambridge University Press; pp. 117–139.

monocline   A zone of steeply dipping strata in an area of horizontally bedded rocks. A zone of rocks that dip steeply to great depth.

monsoon   A seasonally reversing wind system. The term also refers to the rainfall that accompanies the seasonal shift. Derived from the Arabic word *mausim*, meaning season, the term originally referred to the winds of the Arabian Sea, which blow for about 6 months from the northeast and for 6 months from the southwest. In his classic text, Ramage (1971) puts forth three criteria that define a monsoon: a seasonal wind shift of at least 120° between January and July, a certain minimum intensity of wind, a strong directional persistence of the wind. By this definition, monsoons exist in Asia (Wang, 2006), South America (Vera *et al.*, 2006), West Africa (Nicholson, 2009), northern Australia (Hendon and Liebmann, 1990), and the southwest USA and northern Mexico (Higgins *et al.*, 2003). The most strongly monsoonal climate is that of the Indian subcontinent. There, the southwest monsoon prevails from roughly June to September and that is the season of intense rainfall. Rainfall does not occur every day; breaks in the monsoon lasting 10–30 days regularly occur (Webster and Hoyos, 2004). The northeast monsoon prevails during the remainder of the year, marking the dry season over the Indian subcontinent. The development of the monsoon is related to the strong summer heating of the Tibetan Plateau, but the system also reflects steady, coupled interactions between the land and ocean (Fasullo and Webster, 2003). Over West Africa, a similar seasonal shift occurs between prevailing northeasterly flow during the dry season and southwesterly flow during the summer rainy season. The southwestern USA, particularly Arizona, is also considered to have a seasonal monsoon, with rainfall peaking in August when the prevailing flow is southeasterly. Over the tip of northern Australia and the Indonesian maritime continent the monsoonal flow shifts from northwesterly during the summer rainy season to southeasterly during the dry winter.   SEN

References
Fasullo, J. and Webster, P.J. (2003) A hydrological definition of Indian Monsoon onset and withdrawal. *Journal of Climate*, **16**, 3200–3211. · Hendon, H.H. and Liebmann,

B. (1990) A composite study of onset of the Australian summer monsoon. *Journal of Atmospheric Science*, **47**, 2227–2240. · Higgins, R.W., Douglas, A., Hahmann, A., *et al.* (2003) Progress in Pan American CLIVAR research: the North American monsoon system. *Atmosfera*, **16**, 29–65. · Nicholson, S.E. (2009) A new picture of the structure of the 'monsoon' and land ITCZ over West Africa. *Climate Dynamics*, **32**, 1155–1171. · Ramage, C.S. (1971) *Monsoon meteorology*. Academic Press. · Vera, C., Higgins, W., Amador, J., *et al.* (2006) Toward a unified view of the American monsoon systems. *Journal of Climate*, **19**, 4977–5000. · Wang, B. (2006) *The Asian monsoon*. Springer. · Webster, P.J. and Hoyos, C.D. (2004) Prediction of monsoon rainfall and river discharge on 15–30 day time scales. *Bulletin of the American Meteorological Society*, **85**, 1745–1765.

monsoon forest   A lowland and montane division of TROPICAL FOREST characterized by seasonal climates with pressure and wind reversal. Monsoon forests occur when the total rainfall during the monsoon is sufficient to maintain forest, but where there is also a marked dry season. The resulting trees may be predominantly evergreen, depending upon the intensity of the wet season. They are, however, more characteristically semi-deciduous, particularly in the upper canopy, as a result of the seasonal water deficiency. The structure is more open than lowland evergreen forest, with greater light penetration to the ground and denser undergrowth, with locally abundant bamboos and lianas.   PAF

Reading
Golley, F.B., Lieth, H. and Werger, M.J.A. (eds) (1982) *Tropical rain forest ecosystems*. Amsterdam: Elsevier. · Longman, K.A. and Jenik, N. (1987) *Tropical forest and its environment*, 2nd edition. London: Longman.

monumented sections   River channel cross-sections that are precisely surveyed in relation to at least two fixed points and which can be resurveyed on a number of occasions to indicate the amount of channel change. Such sections were advocated as part of the Vigil Network of REPRESENTATIVE AND EXPERIMENTAL BASINS.   KJG

mor   Raw humus that is not admixed with mineral material.

moraine   A distinct landform fashioned by the direct action of a glacier (see Table 4 for contexts and moraine types). In the past the term was also used to describe glacial sediments (e.g. ground moraine), but now it is accepted that *moraine* should refer to the landforms and the word *till* to the sediment.

Classification of moraines.
Source: Chorley *et al.* (1984: figure 17.23).

Lateral morianes, Columbia icefield, Canada.
Photograph by David Thomas.

One group of moraines exists on the surfaces of glaciers and includes *lateral* moraines (see photograph), which form through the accumulation of valley-side material on either side of the glaciers, and *medial* moraines which form from the junction of lateral moraines as two glaciers meet. In the ABLATION areas of glaciers in particular, such moraines can form prominent upstanding ridges where the debris has protected the underlying ice from melting. The material in lateral and medial moraines is characteristically angular rockfall debris and undergoes minimum modification during transport.

A second group of moraines occurs at the edge of existing glaciers or in areas formerly covered by glaciers. One classification scheme,

Table 4 Classification of the major types of moraine

Parallel to ice flow	Transverse to ice flow	Lacking consistent orientation
Subglacial forms with streamlining	*Subglacial forms*	*Subglacial forms*
Fluted and drumlinized ground moraine	Rogen or ribbed moraine	Low-relief ground moraine
	De Geer or washboard moraine	Hummocky ground moraine
Drumlins and drumlinoid ridges	Subglacial thrust moraine	
Crag-and-tail ridges	Sublacustrine moraine	
Ice-pressed forms	*Ice-pressed forms*	*Ice-pressed forms*
Longitudinal squeezed ridges	Minor transverse squeezed ridges	Random or rectilinear squeezed ridges
Ice marginal forms	*Ice front forms*	*Ice surface forms*
Lateral and medial moraines	End moraines	Disintegration moraines
Some interlobate and kame moraines	Push moraines	
	Ice thrust/shear moraines	
	Some kame and delta moraines	

Source: Modified from Sugden and John (1976: table 12.1, p. 236), after Prest (1968).

based on glacier dynamics and position with regard to a glacier, is shown in the diagram. Subglacial forms constructed beneath moving glacier ice include uniform till sheets, as for example occur widely in Iowa and Illinois (Kemmis, 1981), and streamlined and transverse features. Glacial flutes are small-scale examples of ridges streamlined parallel to the direction of ice flow and are generally a few tens or hundreds of metres long and up to 2 m high. They build up in the lee of boulders that become lodged on the bed and thereby create a cavity or low-pressure zone in their lee. Drumlins are larger forms streamlined parallel to the direction of ice flow (see DRUMLIN). A ROGEN MORAINE forms transverse to ice flow.

Active glaciers that end on land build up moraines at the ice margin. The size of the moraine depends on the period that the margin lies in the same location and also on the amount of rock debris transported to the edge by the glacier. The common 'ice-dumped' moraine is typically a complex landform reflecting many different processes of debris accumulation, including slumping or flow off the glacier surface, and lodgement or deformation of basal till. Another common moraine is associated with deformation of sediments by pushing. Glaciers tend to advance in winter when ablation rates at the snout are low, and this advance may physically push sediments into a ridge up to 2 m high. Sometimes it may push a frontal snow bank or sheet of lake ice that itself deforms sediments (Birnie, 1977). If a glacier is in overall retreat, these annual advances are marked by a succession of small ridges whose spacing can be directly related to the amount of annual ablation. At a larger scale, push moraines may involve much larger areas of sediments in front of a glacier; for example, as can occur in the case of a SURGING GLACIER.

A disintegration moraine is the complex remains of a process whereby debris-bearing ice stagnates and melts in situ. An irregular landscape of hummocks and kettles, with differing combinations of subglacial and slumped surface debris is the end product. Such moraines form best where compressing flow brings large quantities of debris up to the ice surface.

Moraines have been widely used to delimit the former extent of glaciers. When they are dated unambiguously, a relatively sophisticated reconstruction of past climate can be made. Problems have arisen in recent years, however, because many moraines are found to have been built up by successive glacier advances over long time spans. DES

Reading and References
Benn, D.I. and Evans, D.J.A. (2010) *Glaciers and glaciation*, 2nd edition. London: Hodder Arnold. · Birnie, R.V. (1977) A snow-bank push mechanism for the formation of some 'annual' moraine ridges. *Journal of Glaciology*, **18**, 77–85. · Chorley, R.J., Schumm, S.A. and Sugden, D.E. (1984): *Geomorphology*. London: Methuen. · Kemmis, T.J. (1981) Importance of the regelation process to certain properties of basal tills deposited by the Laurentide ice sheet in Iowa and Illinois. *Annals of Glaciology*, **2**, 147–152. · Prest, V.K. (1968) *Nomenclature of moraines and ice-flow features as applied to the glacial map of Canada*. Geological Survey of Canada, paper 67-57. Ottawa: Geological Survey of Canada, Department of Energy, Mines and Resources. · Sugden, D.E. and John, B.S. (1976) *Glaciers and landscape*. London: Edward Arnold.

morphogenetic regions
Those regions in which it is claimed that certain geomorphological processes result from a particular set of climatic conditions, thereby giving distinctive regional landscapes.

morphological mapping
A means of mapping landforms. In the strict sense, morphological maps display only the shape of the ground with breaks of slope and gradients indicated. Such maps have been found useful by, for example, South African engineers. In a wider sense, morphological mapping has been used to map landforms in terms of their origin. This form of morphogenetic mapping, popular in central Europe, has found wide applications in terms of engineering geomorphology, resource surveys and pure research. DES

Reading
Smith, M.J., Paron, P., and Griffiths, J.S. (eds) (2011) *Geomorphological mapping: methods and applications*. Amsterdam: Elsevier.

morphodynamics
The mutual co-adjustment of form and process that is mediated through sediment transport, a set of relationships that have been found to be particularly useful in coastal studies. The presence of these interactions explains why, on the one hand, physically based models perform well at small spatial scales and over a limited number of time steps but, on the other hand, why model predictions often break down at larger 'event' and particularly 'engineering' space–time scales. TS

Reading
Cowell, P.J. and Thom, B.G. (1994) Morphodynamics of coastal evolution. In R. W. G. Carter and C. D. Woodroffe

(eds), *Late Quaternary shoreline morphodynamics*. Cambridge: Cambridge University Press; pp. 33–86.

morphometry The quantitative description of forms; in physical geography it refers to the Earth's surface (strictly geomorphometry), but in other sciences it is an approach that can relate to fossils or crystals, for example.

Evans (1972, 2005) has distinguished *general geomorphometry*, which is based upon an analysis of the entire land surface as a continuous, rough surface described by the attributes at a sample of points or from arbitrary areas, and *specific geomorphometry*, which relates to specific landforms and to the measurement of their size, shape and relationships. In both approaches, definitions should be made to allow relationships to process indices. Morphometry is necessary to characterize areas and landforms quantitatively, to allow areas studied by different scientists to be compared easily, to demonstrate how aspects of the land surface are interrelated and to provide parameters that can relate to processes in relationships from which processes may be estimated where only morphometric parameters are available.

In general geomorphometry, a specific part of the land surface could be described by an equation, but this would have so many terms and be so complex that it has not been used except for specific areas of landforms. Therefore, altitude at a point and the derivatives of slope and curvature have often been used as the basis for general geomorphometry, and Evans (2005) and others have identified five fundamental attributes that are altitude, gradient, aspect, profile convexity and plan convexity. Profile convexity is the rate of change along a line of maximum gradient, and plan convexity is the rate of change of aspect along a contour. These five attributes relate to a point or small area, and systems of general geomorphometry that have been put forward are based upon analysis of spatial patterns of some or all of the attributes. Variations between methods that have been suggested depend upon the relative significance accorded to the five attributes, but several schemes have been devised because of the relevance to trafficability, drainage, suitability for different types of land use and susceptibility to erosion hazard.

Specific geomorphometry has been used as a more realistic way of simplifying the task of Earth surface description, and morphometric methods have been devised for the description of coral atolls, karst depressions, glacial cirques, sand dunes, lake basins and many other landforms. Because the morphometry of drainage basins has attracted much attention, morphometry has sometimes been associated mainly with drainage networks and drainage basin morphometry. The earliest developments in drainage basin morphometry were based upon stream order (see ORDER, STREAM) and HORTON'S LAWS, and these could provide the basis for comparisons between areas but have been less valuable because of the statistical nature of these so-called 'laws'. The morphometry of drainage basins has focused (Gregory and Walling, 1976) upon the area, length, shape and relief attributes, and these have parallels at the level of the drainage NETWORK when DRAINAGE DENSITY is an important measure of relative length. A wide range of morphometric measures has been devised for drainage basins, and these have usually been defined as either ratio measures (such as ratio of maximum width to breadth to give an index of drainage basin shape) or as measures that depend upon comparison with an ideal shape, and drainage basins have been compared with a circle or with a lemniscate (pear-shaped) loop, for example. (See also CIRQUE, DRAINAGE DENSITY, DUNE and ORDER, STREAM.) KJG

Reading and References
Chorley, R.J. (ed.) (1972) *Spatial analysis in geomorphology*. London: Methuen. · Chorley, R.J. and Kennedy, B.A. (1971) *Physical geography: a systems approach*. London: Prentice-Hall. · Evans, I.S. (1972) General geomorphometry, derivatives of altitude and description statistics. In R. J. Chorley (ed.), *Spatial analysis in geomorphology*. London: Methuen; pp. 17–90. · Evans, I.S. (2005) General geomorphometry. In A Goudie (ed.), *Geomorphological techniques*, 2nd edition. London: Routledge; pp. 49–62. · Goudie, A. (ed.) (2005) *Geomorphological techniques*, 2nd edition. London: Routledge. · Gregory, K.J. and Walling, D.E. (1976) *Drainage basin form and process*. London: Edward Arnold; pp. 37–60.

morphotectonics (or tectonic geomorphology) The study of the interaction of tectonics and geomorphology. ASG

Reading
Ollier, C.D. (ed.) (1991) *Geomorphology and geoecology: morphotectonics and structural geomorphology*. Zeitschrift für Geomorphologie, Supplementbände vol. 82. Berlin: Borntraeger.

mosaic vegetation Dryland plant communities often exhibit marked patchiness that can include bare, stony, and vegetated areas, and the vegetated patches can be actively growing or senescent. This spatial structure, primarily developed on gently sloping landscapes, can be described as a vegetation mosaic. Critical aspects of the ecohydrologic role of vegetation mosaics include the compartmentalization of the landscape into run-off source zones (the less vegetated components of the mosaic) and run-on sink zones (the more densely vegetated components of the mosaic, where soil properties are altered by more abundant organic matter, soil fauna, etc.). This systematically delivers more plant-available moisture to the vegetated patches than could be sourced from precipitation alone, and this in turn allows a higher biomass to be maintained. There is a wide diversity of views on the mechanisms that underlie or account for mosaic development, derived primarily from modelling studies, but little agreement on many aspects of the phenomenon. However, feedback processes are clearly involved, since the denser the vegetation in the groves of plants, the more abundant is organic litter and associated decomposer organisms. These characteristics support greater soil infiltrability, and hence the maintenance of vascular plant cover.   DLD

Reading

Deblauwe, V., Couteron, P., Bogaert, J. and Barbier, N. (2012) Determinants and dynamics of banded vegetation pattern migration in arid climates. *Ecological Monographs*, 82, 3–21. · Tongway, D., Valentin, C. and Seghieri, J. (2001) *Banded vegetation patterning in arid and semiarid environments.* Ecological Studies vol. 149. New York: Springer Science+Business Media.

mottled zone   The portion of a soil zone or weathering profile immediately beneath a ferricrete or silcrete horizon, in which bleached kaolinitic material occurs with patches of iron staining.

moulin   A vertical cylindrical shaft by which surface meltwater flows into a glacier. Moulins tend to form at lines of structural weakness in the glacier and are usually 0.5–1.0 m in diameter and up to 25–30 m deep.   DES

Reading

Stenborg, T. (1969) Studies of the internal drainage of glaciers. *Geografiska Annaler: Series A, Physical Geography*, 51, 13–41.

mountain meteorology   Mountains exert an influence on the atmosphere both *mechanically* (by blocking the airflow, deflecting it over and around the barrier and through frictional drag) and *thermodynamically* (by acting as a direct, elevated heat source, as an indirect heat source through latent heat release in clouds formed over the mountains and as a moisture sink through

precipitation). The scales of mountain effects on the atmospheric circulation include:

- the planetary wave scale, with upper-air low-pressure troughs located over eastern North America and eastern Asia related, respectively, to the Rocky Mountains and the Tibetan Plateau upwind;
- the regional-synoptic scale, with the modification of frontal systems as they move across major mountain ranges and the formation of lee cyclones;
- the mesoscale of mountain-induced lee wave clouds and fall winds (föhn, bora);
- the local scale of mountain/valley and slope winds systems resulting from topoclimatic contrasts in diurnal heating patterns.

The characteristics of weather and climate in mountain areas are most closely related to the last three categories of meteorological phenomena.

A mountain climate can be considered to exist whenever the relief creates an altitudinal zonation of climatic elements (temperature, precipitation) sufficient to change the local vegetation characteristics. Exceptions to this criterion may occur, however, where vegetation is absent on hyperarid subtropical or polar mountains. The effect of altitude causes a temperature decrease (environmental lapse rate) of about 5–$6\,°C\,km^{-1}$, on average, although a temperature increase with height often occurs in mountain valleys and basins, with wintertime and/or nocturnal temperature inversions. There is also a general altitudinal decrease of water vapour content, a decrease of pressure (approximately $100\,mb\,km^{-1}$ in the lower troposphere), and an increase of incoming solar radiation (about 5–$15\%\,km^{-1}$ under cloudless skies). Orography redistributes and in many cases augments the precipitation that would otherwise have occurred through cyclonic or convective processes. The altitudinal enhancement mostly occurs as a result of increased amounts rather than greater frequency of precipitation. On windward mountain slopes, the zone of maximum precipitation, in a climatic sense, typically occurs at low elevations in equatorial zones, about 700–$1200\,m$ in the tropical (trade wind) zones, and at higher levels (up to $3000\,m$ and above) in mid-latitudes. In the lee of many mountain ranges, with respect to the prevailing wind direction, there is a reduction in average precipitation giving rise to a so-called 'rain shadow'.   RGB

Reading

Barry, R.G. (1992) *Mountain weather and climate.*, 2nd edition. London: Routledge. · Browning, K.A. and Hill,

F.F. (1981) Orographic rain. *Weather*, **36**, 326–329. · Smith, R.B. (1979) The influence of mountains on the atmosphere. *Advances in Geophysics*, **21**, 87–230. · Yoshino, M.M. (1975) *Climate in a small area: an introduction to local meteorology*. Tokyo: University of Tokyo Press.

mountain/valley wind

A local wind system produced in mountainous regions as a result of temperature differences. The circulation is best developed in summer, when the skies are clear and large-scale motions are weak, and in deep, straight valleys with a north–south axis. During the day the air above the slopes and floor of the valley is heated to a temperature well above that over the centre of the valley. Shallow upslope flow (see ANABATIC FLOWS) results in, and is compensated for, by air sinking in the valley centre. If the ascending air is moist enough, convective clouds may form along the valley ridges. Superimposed on this cross-valley flow is a VALLEY WIND blowing up-valley at low levels from the adjacent plains.

At night the valley surface and the overlying air cool by the emission of infrared radiation, causing the air to flow downslope under the influence of gravity. The convergence of these slope winds near the valley centre produces both a weak ascending motion and a low-level down-valley or mountain wind that flows out of the mountains onto the adjacent plains. At higher elevations a counter flow occurs from the plains to the valleys. WDS

Reading
Atkinson, B.W. (1981) *Meso-scale atmospheric circulations*. London: Academic Press. · Oke, T.R. (1987) *Boundary layer climates*, 2nd edition. London: Routledge.

mountains

Substantial elevations of the Earth's crust above sea level, which result in localized disruptions to climate, drainage, soils, plants and animals. Increases in altitude tend to repeat the bioclimatic patterns associated with a move towards higher latitudes, although cloudiness, day length and seasonal variations differ from the latitudinal progression. Temperature drops at a rate of approximately 0.5 °C per 100 m, and rainfall is often heaviest where moisture-laden winds are forced to rise (over 1500 m in the tropics). Vertical ZONATION of plants and animals is most clearly illustrated where mountains rise from tropical forest to tundra environments. PAF

Reading
Gerrard, A.J. (1990) *Mountain environments*. London: Belhaven Press. · Moore, D.M. (ed.) (1983) *Green planet: the story of plant life on Earth*. Cambridge: Cambridge University Press. · Price, L.W. (1981) *Mountains and man: a study of process and environment*. Berkeley, CA: University of California Press.

mudflats (or mud flats; also known as tidal flats)

Deposits of silt and clay found in sheltered intertidal areas. They range from soft muds in the most sheltered back-barrier and estuarine environments, to firm sands in more wave- and current-exposed areas. TS

mud lumps

Small-scale landforms found from the Mississippi delta region of the USA. Rapid forward growth of distributary channels deposits deltaic sand, mud and organic sediment on top of unstable prodelta clay. This causes loading that in turn causes diapiric intrusions of plastic clays through the overlying sands. Updoming or extrusion occurs, producing the mud lumps. ASG

Reading
Morgan, J.P., Coleman, J.M. and Gagliano, S.M. (1968) Mudlumps: diapiric structures in Mississippi delta sediments. In J. Brownstein and G. D. O'Brien (eds), *Diapirism and diapirs*. American Association of Petroleum Geologists Memoir 8. Tulsa, OK: American Association of Petroleum Geologists; pp. 145–161.

mud volcano

A mount built up of mud carried to the surface by geysers or gap eruptions in volcanically active regions.

mull

Humus admixed with mineral material in the surface horizons of the soil zone.

multiple working hypotheses

These represent a method used to test alternative explanations against each other (Haines-Young and Petch, 1986). When trying to explain aspects of the world we can often think of several plausible HYPOTHESES. Using this method we can test the logical consequences of each hypothesis and attempt to eliminate those ideas that are wrong. The availability of competing hypotheses helps us focus on just what observations or experiments are needed to examine ideas critically.

The method was initially discussed by Chamberlin (1995[1890]) and Gilbert (1896). Baker (1996) describes his ideas in a wider context. Examples of its use include Batterbee *et al.* (1985) and Turner (1997). RH-Y

References
Baker, V.R. (1996) The pragmatic roots of American Quaternary geology and geomorphology. *Geomorphology*, **16**, 197–215. · Batterbee, R.W, Flower, R.J., Stevenson, J. and Rippy, B. (1985) Lake acidification in Galloway: a palaeological test of competing hypotheses. *Nature*, **314**, 350–352. · Chamberlin, T.C. (1995[1890]) The method of multiple working hypotheses. *Journal of Geology*, **103**, 349–354 (originally printed in *Science*). · Gilbert, G.K. (1896) The origin of hypotheses, illustrated by the discussion of a topographical problem. *Science*, **3**, 1–3. · Haines-

Young, R.H. and Petch, J.R. (1986) *Physical geography: its nature and methods*. London: Paul Chapman. · Turner, R.E. (1997) Wetland loss in the northern Gulf of Mexico: multiple working hypotheses. *Estuaries*, **20**, 1–13.

multispectral scanner An optical remote sensing device that collects incident electromagnetic radiation (between visible and thermal infrared wavelengths) and converts it to a stored representation (usually an image) for remote sensing analysis. All these systems have in common the following basic components of optical imaging, spatial scanning, spectral selection, detection and electronics calibration and processing. The optical imaging component uses reflective optics to gather enough light for accurate measurement that is sampled by the spatial scanning component. There are three basic spatial scanning techniques: line scanners (along-track); pushbroom (across-track) scanners and frame scanners, all of which create a two-dimensional image spectrally filtered by the spectral selection component. In multispectral scanning systems, separate co-aligned optical apertures and focal planes with a different pass band filter on each focal plane are used to spectrally sample the electromagnetic radiation for each pixel in the image according to the wavebands of the sensing device. More commonly, a single aperture with beam splitters can be used to feed multiple focal planes with filters. These multispectral scanners typically collect imagery in a limited number (3–10) of irregularly and often widely spaced spectral bands, which is in contrast to hyperspectral scanners collecting imagery in hundreds of bands. After spectral selection, detectors convert the electromagnetic radiation to an electronic signal that is then processed to a final image by the electronics and calibration algorithms to an accurate representation of the scene with proper radiometric units and geometric orientations. The scanning systems that are most known to physical geography include the multispectral scanning system, THEMATIC MAPPER (TM) and enhanced thematic mapper+ that operated on board the Landsat satellites.   DSB

Reading
Campbell, J.B. and Wynne, R.H. (2011) *Introduction to remote sensing*, 5th edition. The Guildford Press. · Kerekes, J.P. (2009) Optical sensor technology. In T. A. Warner, M. D. Nellis and G. M. Foody (eds), *Handbook of remote sensing*. London: SAGE; pp. 95–107. · NASA (2015) Landsat science. http://landsat.gsfc.nasa.gov/ (accessed 13 July 2015).

muskeg A Canadian-Indian term for waterlogged depressions in the subarctic zone of Canada and Alaska. There are some 500,000 square miles of such marsh in Canada alone. These depressions are largely filled with peat and characterized by *Sphagnum* moss. It is a region of marshy depressions with scattered lakes, stagnant mosquito-infested pools and slow meandering streams.   AP

Reading
Radforth, N.W. (1969) Environmental and structural differentials in peatland development. In E. C. Dapples and M. E. Hophins (eds), *Environments of coal deposition*. Geological Society of America Special Papers, vol. 114. Boulder, CO: Geological Society of America; pp. 87–104. · Radforth, N.W. and Branner, C.O. (eds) (1977) *Muskeg and the northern environment in Canada*. Toronto: University of Toronto Press.

Muskingum method A method of FLOOD ROUTING that assumes that along any channel reach the difference between the inflow hydrograph I to the reach and the outflow hydrograph O from the reach is equal to the stored or depleted water in a specified time interval. Two simultaneous equations can be solved; namely, the water balance equation expressing change in storage:

$$\Delta S = I - O$$

and an equation for storage:

$$S = K[xI + (1 - x)O]$$

where x is a dimensionless constant for the channel reach and K is a storage constant that is obtained from hydrographs of I and O at each end of the reach.   KJG

Reading
Lawler, E.A. (1964) Flood routing. In V. T. Chow, *Handbook of applied hydrology*. New York: McGraw-Hill; section 25-II.

mutation A change in the structure or amount of DNA in the chromosomes in the cells of an organism, or the resulting change in the organism's characteristics. A mutation occurring in the gametes (reproductive cells) is inherited; if the mutation occurs elsewhere (in somatic or non-reproductive cells) it is not inherited. Inherited mutations caused by a change in the structure of the DNA molecule are known as gene mutations; those produced by a change in the amount of DNA are known as chromosomal mutations. These errors in the coding of inherited information occur at a low frequency, apparently spontaneously.

Most mutations are deleterious because of the long period of testing (by natural selection) the genome (the package of genes within the gamete or germ-cell) has undergone. They will be

eliminated. The extremely rare beneficial mutation will be incorporated into the genome by the
process of natural selection.   PHA

mutualism An interaction that benefits the
species involved. The most widespread mutual-
isms are between plants and animals, to use the
animals to improve the efficiency of plants' reproduction and to provide food for the animals in
return. For example, birds are attracted to fruits
and eat them. The fruits pass through the gut and a
portion is defecated shortly afterward. Mean-
while, the bird has flown from the site of the parent
plant and the seed is left in a supply of fertilizer,
ready to germinate.   ASG

mycorrhizal fungi These infect the roots of
host plants and exist in a mutualistic relationship
that is beneficial to both symbionts. The term
'mycorrhiza' was coined by Frank in 1885 from
the Greek meaning 'fungus-root' (*myko – rhiza*).
Mycorrhizae are very widespread, ranging from
arctic to desert to equatorial plant communities.
They also occur through a considerable range of
soil depths, spanning near-surface locations to
depths of some metres. The fungi produce filaments, or hyphae (about 3 µm in diameter) that
individually may extend up to a few centimetres
into the soil surrounding the root, but which may
reach total lengths of 50 m per gram of soil. They
reach further into the soil than do root hairs, which
may only be 1 mm in length. The hyphae provide
additional area exposed to the soil solution, and
provide an important source of nutrients like
phosphorus, nitrogen and potassium, especially
when these are only available at low levels in the
soil. Mycorrhizal fungi are capable of transporting
nitrogen in the form of ammonium (NH_4^+) as well
as NITRATE, so further enhancing the supply of this
nutrient. Fungal remains also contribute organic
matter to the soil in substantial amounts, and may
enhance the flow of water to the host plant. The
hyphae bind soil particles together physically, and
also by producing polysaccharides that act as
binding agents.   DLD

Reading
Allen, M.F. (1991) *The ecology of mycorrhizae*. Cambridge:
Cambridge University Press. · Jungk, A. and Claassen, N.
(1997) Ion diffusion in the soil–root system. *Advances in
Agronomy*, **61**, 53–110.

N

nanism (or microsomia) The condition of being dwarfed, often implying stunted, and refers to both plants and animals. Small size is implicit in the expression nanophyllous (small-leaved), or nanoplankton (the smallest plankton), or nanophanerophytes (shrubs under 2 m in height), although in SI terminology the prefix nano- strictly signifies a unit $\times 10^{-9}$. Artificial breeding of dwarf animals and plants is sometimes referred to as nanization. PAF

nappe A mass of rock that is thrust over other rocks by thrust faulting or a recumbent fold or both.

natural disaster A NATURAL HAZARD or GEOHAZARD that actually happens. An event that has a dramatically negative effect on humans but which is due to the occurrence of a natural event or process. For example, a famine may be due to the occurrence of a to-be-expected DROUGHT event in a DRYLANDS region, or large-scale death and destruction may result from the passage of a TORNADO through a populated area. If blame has to be ascribed in the context of the disaster, it lies not with the natural event but with the location of human activities or settlements in the locality or pathway of that event. DSGT

Reading
Abbot, P.L. (2008) *Natural disasters*. New York: McGraw-Hill.

natural hazard Any aspect of the physical environment's natural functions that may adversely affect human society to cause social disruption, material damage and/or loss of life, in which case the impact is referred to as a natural disaster or catastrophe. Physical geographers may classify natural hazards according to their sphere of occurrence into geological (e.g. earthquake), hydrological (e.g. flood), atmospheric (e.g. snowstorm) – all of which would be known as GEOHAZARDS – and biological (e.g. disease). Another approach divides hazards into rapid-onset, intensive events (short, sharp shocks such as earthquakes or debris flows) and slow-onset

pervasive events, which often affect larger areas over longer periods of time (such as droughts).

Although such classifications provide useful summaries of hazard types, many hazards can have a variety of effects. Hence, an earthquake may cause a tsunami wave at sea, landslides or avalanches on slopes, building damage and fires in urban areas, and flooding due to the failure of dams, as well as ground shaking and displacement along faults. Similarly, many natural disasters cause disruption to public hygiene and, consequently, result in heightened risks of disease transmission.

In practice, it is often difficult to distinguish between purely 'natural' events and human-induced events. In one sense, all 'natural' disasters can be thought of as human-induced since it is the presence of people that defines whether a hazard creates a disaster or not, and many of the natural physical processes that cause disasters can also be triggered or exacerbated by human action. The composite nature of many disasters also blurs the distinction: the major cause of death due to earthquakes is usually crushing beneath buildings, so is the disaster natural or human induced? Some researchers believe that the difficulty of making this distinction has rendered the division pointless, and so prefer to talk of 'environmental' hazards that refer to a spectrum with purely natural events at one end and distinctly human-induced events at the other (e.g. Smith, 2004). Increasing levels of concern over CLIMATE CHANGE has also led to much research on the possible implications of this for the frequency and magnitude of environmental hazards (McGuire *et al.*, 2002). NJM

Reading and References
Abbot, P.L. (2008) *Natural disasters*. New York: McGraw-Hill. · McGuire, B., Mason, I. and Kilburn, K. (2002) *Natural hazards and environmental change*. London: Arnold. · Smith, K. (2004) *Environmental hazards: assessing risk and reducing disaster*, 4th edition. London: Routledge.

natural resources Components of the natural environment that have a utility to humankind;

and following Zimmerman (1933), it is sometimes said that resources 'are not, they become'. This indicates that the natural environment contains components that are both useful and not useful to people, the utility being derived from societal needs and the technical ability to extract usage, rather than the mere existence of the elements in the environment. To this end, elements without utility were termed 'neutral stuff' by Zimmerman. The natural resource/neutral stuff dichotomy is not a static one, since over time different societies have found use for elements that were previously unused, and vice versa. Thus, at the end of the twentieth century, flints, so vital for stone-age cultures, had very limited practical usage (except perhaps as a luxury building material), while stone-age people had no use for uranium ore; indeed, they did not know what it was.

It is clear, from the perspective of geography, that natural resources can be considered from both their physical and human dimensions. In the latter case, demand, supply, differences in resource perceptions based on cultural development, wealth, and so on are all relevant issues for consideration. From a physical perspective, the natural distribution of resources in the biosphere and lithosphere, resource types, and the impacts of resource use on the wider environment are all currently topics of interest. In practice, however, the physical and human dimensions coexist and impinge upon each other, as is well illustrated, for example, in the works of Ian Simmons (1974, 1991).

Natural resources can be classified in many ways. A useful division can be made between STOCK RESOURCES and FLOW RESOURCES, with division between the two based on the time it takes for a resource to form relative to human life-spans. For stock resources, availability is a function of the natural abundance of the resource in the Earth system, knowledge of the distribution of reserves, having the means to extract reserves of that resource (which may be defined by technological developments; for example, sinking an oil well on land is generally easier and cheaper than tapping oil fields that lie beneath the North Sea), and the rate of usage, itself affected by demand. Over time, the status of a particular component of the base can change; for example, from being hypothetical, to conditional upon technology allowing exploitation, to proven and awaiting use, to used. The future availability of a resource may be revised on the basis of factors including new reserves being identified and exploited, and changes in demand that may relate, for example, to substitutes being developed or recycling rates increasing.

Many environmental issues and problems relate to the use of natural resources. Overuse of soil, excessive cultivation and the expansion of agriculture to marginal lands may lead to SOIL EROSION and in drylands to DESERTIFICATION, turning a flow resource into one that is being used unsustainably. The growth in use of fossil fuels in the twentieth century has been seen as a key cause of the carbon dioxide problem and GLOBAL WARMING. Damage to the OZONE layer may result from the release of certain chemicals into the atmosphere. Spatial inequalities and mismatches between the occurrence of particular natural resources and human demands contribute to the transport of resources over long distances, adding to various forms of POLLUTION and the consumption of energy resources. Overall therefore, natural resources are a fundamental component of human use and occupation of the Earth; their use and temporal changes in demand relate to cultural developments and advancements, as well as to the rise of environmental issues and concerns.   DSGT

Reading and References
Simmons, I.G. (1974) *The ecology of natural resources.* London: Edward Arnold. · Simmons, I.G. (1991) *Earth, air and water: resources and environment in the late 20th century.* London: Edward Arnold. · Walther, J.V. (2013) *Earth's natural resources.* Boston, MA: Jones and Bartlett. · Zimmerman, E.W. (1933) *World resources and industries.* New York: Harper.

natural selection See DARWINISM and EVOLUTION.

natural vegetation A general term for the total sum of plants in an area, grouped by communities but not as part of a taxonomic system. 'Natural' signifies the sum total of inheritance or genotype, but the term natural vegetation is also associated with environmental factors that encourage or constrain plant growth after an equilibrium between plants and their surroundings has been established. The larger groupings of plants illustrate a ZONATION that has a combined biological and environmental basis. Perhaps the most common usage is to denote the plant cover of any area prior to its modification by humans.   PAF

Reading
Walter, H. (1973) *The vegetation of the earth.* London: English Universities Press.

neap tide See TIDES.

nebkha (also nabkha) A small sand dune, from ~10 cm to ~2–3 m high, formed when wind-blown sand is trapped within or accumulates around a plant. Nebkhas are discrete forms, occurring individually or in nebkha fields, that develop in dryland areas where the vegetation cover is discontinuous. DSGT

neck A narrow isthmus or channel. A mass of lava that has solidified in the pipe or vent of a volcano.

neck cut-off The process in which a tight meander loop on an alluvial river is abandoned by the incision of a new linking channel that bypasses the loop. The other common mode of abandonment is by chute cut-off, when the flow exploits the low-lying part of an old meander scroll system to bypass part of a meander loop. (See also CUT-OFF.) DLD

needle ice A small-scale heave phenomenon produced by freezing and associated ice segregation at or just beneath the ground surface. Cooling at the ground surface results in ice crystals that grow upwards in the direction of heat loss. The needles, which can range in length from a few millimetres to several centimetres, may lift small pebbles or soil particles. The growth of needle ice is usually associated with diurnal freezing and thawing. It is widespread and particularly common in alpine locations in mid-latitudes, where the frequency of freeze–thaw cycles is at its greatest. Wet, silty, frost-susceptible soils are the sites of most intense needle ice activity. Needle ice frequently occurs in orientated stripes, and both wind direction and sunlight have been suggested as explanations for the pattern; it is not clear whether orientated needle ice patterns are primarily a shadow effect developed by thawing or a freezing effect.

Thawing and collapse of needle ice is thought significant for frost sorting, frost creep, the differential downslope movement of fine and coarse material, and the origin of certain micropatterned ground forms. The importance of needle ice as a disruptive agent has probably been underestimated, especially in exposing soil to wind and water in periglacial regions. In other areas it may be responsible for damage to plant materials when freezing causes vertical mechanical stress within the root zone. HMF

Reading
Lawler, D.M. (1988) Environmental limits of needle ice: a global survey. *Arctic and Alpine Research*, **20**, 137–159. · Mackay, J.R. and Mathews, W.H. (1974) Needle ice striped ground. *Arctic and Alpine Research*, **6**, 79–84. · Washburn, A.L. (1979) *Geocryology: a survey of periglacial processes and environments*. New York: John Wiley & Sons, Inc.; especially pp. 91–93.

negative feedback See SYSTEMS.

nehrung A sand or shingle spit that separates a HAFF from the open sea. A bar that isolates an estuary or lagoon from the sea.

neocatastrophism A term introduced into geoscience in the mid-twentieth century by palaeontologists concerned with sudden and massive extinctions of life forms, such as that which afflicted the great mammals at the end of the Pleistocene. It has been extended into geomorphology by those dealing with rapid outputs from, and rapid inputs to, interfluve systems (Dury, 1980). Much modern geomorphology is concerned with events of great magnitude and low frequency, and problems have been encountered with accommodating such an approach within the context of UNIFORMITARIANISM. ASG

Reference
Dury, G.H. (1980) Neocatastrophism? A further look. *Progress in Physical Geography*, **4**, 391–413.

neo-Darwinism An evolutionary theory that combines DARWINISM with modern genetics. It regards the gene pool of a population as the fundamental unit in evolution and takes into account larger mutations as well as the small heritable variations of Darwin. ASG

Reading
Berry, R.J. (1982) *Neo-Darwinism*. London: Edward Arnold.

neoglacial A small-scale glacial advance that occurred in the Holocene, after the time of maximum HYPSITHERMAL glacier shrinkage (Denton and Porter, 1970). Fluctuations appear to have been frequent and to have shown sparse temporal correlation between different areas (Grove, 1979), though the latest advance, the so-called Little Ice Age, was widespread between c. 1550 and 1850 AD. ASG

Reading and References
Denton, G.H. and Porter, S.C. (1970) *Neo-glaciation*. *Scientific American*, **222**, 101–110. · Grove, J.M. (1979) The glacial history of the Holocene. *Progress in Physical Geography*, **3**, 1–54. · Margreth, A., Dyke, A.S., Gosse, J.C. and Telka, A.M. (2014) Neoglacial ice expansion and late Holocene cold-based ice cap dynamics on Cumberland Peninsula, Baffin Island, Arctic Canada. *Quaternary Science Reviews*, **91**, 242–256.

neotectonics The study of the processes and effects of movements of the Earth's crust that have occurred during the Late Cainozoic (Neogene). Some investigators use the term in a more restricted temporal sense to refer to post-Miocene or even just Quaternary movements, while others regard neotectonics as involving all tectonic activity that has been instrumental in forming present-day topography. Neotectonic activity that has been directly monitored by geodetic relevelling or other measurement techniques during the present century is commonly referred to as recent crustal movements.

Several lines of evidence have been employed to establish the nature of neotectonic activity, depending on the size of area and time period being considered. Regional subsidence and uplift over several millions of years are largely investigated by the usual methods of structural geology. As the temporal and spatial scale contracts, geomorphological and sedimentological data become more important. For instance, coastal movements can be monitored by raised or down-warped shorelines, while inland the mapping of fluvial features and erosion surfaces in conjunction with detailed study (and especially dating) of associated deposits can provide valuable information. Rapid uplift in mountainous areas may be indicated by geomorphological evidence of the onset of glaciation, while both horizontal and vertical movements along faults can in many cases be related, respectively, to offset drainage and knickpoints.

Maximum rates of uplift estimated from geomorphological and other evidence, or measured directly by relevelling, vary by several orders of magnitude over the Earth's surface. Average rates of uplift over several millions of years rarely exceed $300 \, \text{mm} \, \text{ka}^{-1}$. However, rates of post-glacial isostatic uplift (see ISOSTASY) may exceed $20 \, \text{m} \, \text{ka}^{-1}$, while contemporary crustal movements in currently highly active tectonic zones average up to $10 \, \text{m} \, \text{ka}^{-1}$ or more. Such high short-term rates are clearly not sustained for more than a very limited period in geological terms. MAS

Reading
Fairbridge, R.W. (ed.) (1981) *Neotectonics*. Zeitschrift für Geomorphologie, Supplementbände vol. 40. Berlin: Borntraeger. · Vita-Finzi, C. (1986) *Recent earth movements – an introduction to neotectonics*. London: Academic Press. · Vyskocil, P., Green, R. and Maelzer, H. (eds) (1981) *Recent crustal movements, 1979*. Amsterdam: Elsevier. · Whitney, B.B. and Hengesh, J.V. (2015) Geomorphological evidence of neotectonic deformation in the Carnarvon Basin, Western Australia. *Geomorphology*, **228**, 579–596.

nephanalysis The term used to cover the analysis and interpretation of spatially organized cloud data. Coined in presatellite days, the term originally applied to 'the study of synoptic charts on which only clouds and weather are plotted' (Berry et al., 1945). The observations plotted were of cloud type, cloud amount, precipitation, weather, cloud ceilings and cloud-top heights. With the advent of METEOROLOGICAL SATELLITES it soon became clear that the contents of the satellite imagery were so rich and complex that many users would prefer to be provided with simpler cloud charts instead. Such charts are known as *satellite nephanalyses*. The earliest type was designed for use in the US Weather Bureau (see Godshall (1968)). Similar manual schemes were implemented in the late 1960s and early 1970s in other major meteorological centres. The term is little used today. ECB

References
Berry, F.A., Bollay, E. and Beers, N.R. (1945) *Handbook of meteorology*. New York: McGraw-Hill. · Godshall, F.A. (1968): Intertropical convergence zone and mean cloud amount in the tropical Pacific Ocean. *Monthly Weather Review*, **96**, 172–175.

nephoscope An instrument for measuring the height, direction of movement and velocity of clouds from a point on the ground.

neptunism The belief that a large proportion of the Earth's rocks are precipitates laid down in some chaotic fluid, a theory that was devised and popularized by the German geologist. A. G. Werner in the late eighteenth century, and imported into Britain in the early nineteenth century by R. Jameson. It contrasts with PLUTONISM. ASG

Reading
Davies, G.L. (1969) *The Earth in decay*. London: Macdonald.

neritic Pertaining to the part of the seas and oceans above the continental shelf.

ness A promontory or headland, especially in Scotland, but also in eastern and southern England.

net primary productivity (NPP) The net augmentation of green plant material per unit area per unit time on land, and of blue–green and other algae, phytoplankton, and higher plants in water bodies; or, the amount of stored cell energy produced by PHOTOSYNTHESIS. It may be expressed by the equation

$$NPP = \text{gross production} - \text{respiration}$$

(see BIOLOGICAL PRODUCTIVITY) and is measured in dry weight grams per square metre per day or year, dry weight tonnes per hectare per year, or in assimilated carbon equivalents, or energy equivalents. Although photosynthetic energy fixation is its main determinant, NPP may also be constrained by limiting factors that restrict growth, in particular cold and/or drought, and nutrient availability inadequacies.

NPP on land can be estimated by the harvest method, in which all parts of living plants (including roots if possible) are cut and weighed at the end of a set period of time. This is especially useful for crops, grasslands and forest plantings. For natural forests and woodlands, some form of forest dimension analysis is normally adopted (Clark et al., 2001). The NPP of plankton communities is more often ascertained by the technique of measuring uptake rates of carbon dioxide labelled with radiocarbon (^{14}C), which provide accurate values over 1 or 2 days.

The lowest NPPs on land occur in deserts, semi-deserts and tundras, at $0–250\,g\,m^{-2}\,a^{-1}$. Systems whose growth is restricted by cold or drought (boreal forest, semi-desert shrublands, tropical savannas, steppe grasslands) give values of $250–1000\,g\,m^{-2}\,a^{-1}$; temperate-latitude forests, in which NPP is restrained by seasonal cold, lie within $1000–2000\,g\,m^{-2}\,a^{-1}$; and tropical rain forest and some of the most nutrient-rich marshland may attain $2000–3000\,g\,m^{-2}\,a^{-1}$ and occasionally as high as $5000\,g\,m^{-2}\,a^{-1}$. NPP in agricultural systems also varies widely, from $\sim250–500\,g\,m^{-2}\,a^{-1}$ in many tropical subsistence crops, and for most nonintensive temperate-latitude farmland, to $750–1500\,g\,m^{-2}\,a^{-1}$ under modern intensive and frequently irrigated agriculture. Sugar cane is perhaps worthy of special mention, since it is one of the most productive of crops: its mean NPP is $\sim1725\,g\,m^{-2}\,a^{-1}$, but this can be augmented to $6700\,g\,m^{-2}\,a^{-1}$, or higher. NPP in oceans can range from $\sim2\,g\,m^{-2}\,a^{-1}$ in Arctic waters under the ice cap, to almost $5000\,g\,m^{-2}\,a^{-1}$ in some coral reefs, mangrove swamps and tropical estuaries. Most open oceans, which are poor in nutrients, have values that are equivalent to semi-desert on land, namely $40–200\,g\,m^{-2}\,a^{-1}$. Greater NPPs are found in regions of upwelling and on continental shelves, both of which are nutrient rich $(400–600\,g\,m^{-2}\,a^{-1})$, and when these spatially coincide they can reach $1000\,g\,m^{-2}\,a^{-1}$. A good deal of variation is also found in freshwater systems: OLIGOTROPHIC streams normally have very low NPPs, while those of EUTROPHIC marshes can reach $2500\,g\,m^{-2}\,a^{-1}$, with $5600\,g\,m^{-2}\,a^{-1}$ attained in swamps artificially enriched with sewage (Woodwell, 1970). Climate change may have a significant effect on NPP and its spatial distributions (e.g. Reeves et al., 2014).

DW

Reading and References
Clark, D.A., Brown, S., Kicklighter, D.W., et al. (2001) Measuring net primary productivity in forests: concepts and methods. *Ecological Applications*, **11**, 356–370. · Reeves, M.C., Moreno, A.L., Bagne, K.E. and Running, S.W. (2014) Estimating climate change effects on net primary productivity of rangelands in the United States. *Climatic Change*, **126**, 429–442, doi: 10.1007/s10584. · Whittaker, R.H. (1975) *Communities and ecosystems*, 2nd edition. New York: Macmillan. · Woodwell, G.M. (1970) The energy cycle of the biosphere. *Scientific American*, **223**, 64–74.

net radiation   The resultant flux of the solar and terrestrial radiation through a horizontal surface. The downwards (positive) flux of radiation consists of shortwave solar radiation plus infrared atmospheric counter-radiation. The upward (negative) flux consists of reflected shortwave radiation and infrared radiation from the ground surface. The net radiation is considered positive if the flux downwards exceeds that upwards, and in this case will add energy to the surface. Net radiation is also known as radiation balance. Typically, it is positive during the day and negative at night.

JGL

network   The structure composed of links and nodes that are the junctions of at least three links. Networks can be identified for any linear form or process, and Haggett and Chorley (1969) recognized two fundamental types of networks in geography: open and closed. Much emphasis has been placed upon drainage networks, where network delimitation and density are important considerations (see DRAINAGE DENSITY and DRAINAGE NETWORK). The last two decades has seen a growing interest in the development and use of so-called neural networks in physical geography (Hewitson and Crane, 1994). These have various applications, including rainfall-runoff and streamflow modelling (Abrahart et al., 2012).

KJG/MEM

References
Abrahart, R.J., Anctil, F., Coulibay, P. and Dawson, C.W. (2012) Two decades of anarchy? Emerging themes and outstanding challenges for neural network river forecasting. *Progress in Physical Geography*, **36**, 480–513. · Haggett, P. and Chorley, R.J. (1969) *Network analysis in geography*. London: Edward Arnold. · Hewitson, B.W. and Crane, R.G. (1994) *Neural nets: applications in geography*. Dordrecht: Kluwer.

neutron probe   An instrument for determining soil moisture content. It consists of a radioactive source of fast (or high-energy) neutrons, a slow-neutron detector and a counter unit. The method is based on the principle that fast neutrons emitted into the soil collide with the nuclei of atoms in the soil, notably the hydrogen nuclei of soil water, and as a result lose energy and slow down. A proportion of the resulting cloud of slow neutrons is scattered towards the probe, where it is sensed by the slow-neutron detector and translated into an estimate of soil moisture content using the mean count rate displayed on the counter unit and a soil moisture calibration curve. The neutron probe is introduced into the soil using a permanently sited access tube, and so repeated measurements of soil moisture content can be made at the same site without destruction of the site.   AMG

Reading
Schmugge, T.J., Jackson, T.J. and McKim, H.L. (1980) Survey of methods for soil moisture determination. *Water Resources Research*, **16**, 961–979.

neve   Another word for FIRN.

niche   This refers to the precise way a species relates to its environment and to the other species with which it interacts. While the HABITAT of a species describes its physical environmental circumstances, the word *niche* deals in addition with the functioning of the species in the community as a whole. In this sense, habitat is the species' 'address', while the niche is its 'profession'. Clearly, the occurrence or otherwise of a particular species depends on the availability of suitable habitat, but is also determined by the ability of that species to occupy its own niche within that space. No two species occupy precisely the same niche, an idea referred to as Gause's competitive exclusion principle, although possibly even a relatively large proportion of their habitat and resource requirements may overlap. Different species manage to coexist by partitioning their resources, perhaps the most well known example of this being provided by the herbivorous megafauna of the African savannas. The grasslands of the Serengeti plains of Tanzania, for example, support numerous large herbivores that appear to occupy identical niches (i.e. they are all grazers and all live within the same habitat). However, on closer inspection, it can be observed that owing to migration and other smaller spatial rearrangements of populations, different species rarely come into direct competition. Even when occupying the same locality, different species may be selectively grazing distinctive grass species, or different parts of the same species. G. Evelyn

Hutchinson further defined the niche as a multi-dimensional volume that must take into consideration a species's resource and habitat requirements, as well as its physical environment tolerances. Generalist species have correspondingly wide niches, while those with more exacting requirements by definition occupy narrower niche spaces.   MEM

Reading
Ricklefs, R.E. (1990) *Ecology*, 3rd edition. New York: Freeman; pp. 728–747.

nick point   A break in the long profile of a river. (See also KNICKPOINT.)

nimbostratus   See CLOUDS.

nimbus   See CLOUDS.

nitrate   An oxidized form of nitrogen, in combination with three atoms of oxygen. The chemical symbol for the nitrate ion is NO_3^-. Commonly produced in the soil by the bacterial conversion of ammonium (NH_4^+), one of the breakdown products produced by the decay of organic compounds. This process is known as nitrification. NO_3^- is also a common anion in aqueous (water) solutions. It is delivered in precipitation, having resulted from processes involving free nitrogen gas in the atmosphere, and is also derived from natural and artificial agricultural fertilizer. Nitrates are also generated as a by-product when fossil fuels are burned, and when in the atmosphere contribute to the acidification of rain. Consequently, NO_x compounds (a mixture of nitric oxide, NO, and nitrogen dioxide, NO_2) are increasingly removed from flue gases by various DENITRIFICATION processes.   DLD

nitrification and denitrification   Two essential components of the nitrogen cycle. The former involves the capture of nitrogen from the air, a process carried out by bacteria that live symbiotically in association with leguminous plants. They fix nitrogen in the form of highly soluble nitrates, which are taken up by plants, which convert them into useful organic compounds such as proteins. These, in turn, may be consumed by animals. The process of denitrification takes place when bacteria and fungi decompose plant or animal waste and convert it back again into nitrogen gases.   ASG

nitrogen cycle   Nitrogen (N) and phosphorus (P) are the rate-limiting nutrients in freshwaters; thus, they control the productivity of freshwater ecosystems and are implicated in the accelerated EUTROPHICATION of many rivers and lakes, particularly in the developed world. N in freshwaters takes several forms. Excluding molecular N_2, the dominant combined N fractions in freshwaters are dissolved inorganic N (NH_4^+, NO_2^-, NO_3^-), dissolved organic N, and particulate N, which is usually organic but may contain inorganic N. Organic N usually exists as an integral part of protein molecules or in the partial breakdown of these molecules (e.g. peptides, urea, amino acids). Ammoniacal N is usually present in freshwaters as a result of the biological decomposition of organic N.

Terrestrial N is an important source for the aquatic N system. Up to 40% of total N flux reaches the aquatic system through direct surface run-off or subsurface flow from the catchment. As a consequence, patterns of use, particularly land use, in the terrestrial system determine the magnitude and form of N inputs to the aquatic system.

There are five main reactions in N cycling. *Fixation* converts molecular N to ammonia through bacterial mediation. Some cyanobacteria can fix N directly and often form dense planktonic mats at the surface of lake waters. N fixation is primarily important in eutrophic lakes with large populations of cyanobacteria. Photosynthetic bacteria may fix N in the anoxic (no oxygen) zone. *Assimilation* of nitrate to organic N is mediated primarily by plankton in freshwaters. The rate of assimilation varies with nitrate concentration, and ammonia may be assimilated if available. *Mineralization* (ammonification) may occur both in the water column of lakes and rivers and in their sediments but is relatively more important in lake sediments. Mineralization may be rapid when plankton biomass dominates lake waters because this creates a low carbon to N ratio. Ammonification converts organic N in sediments to ammonium via microbial decomposition processes. The process is oxygen demanding and regenerates N in a form that is available for re-assimilation by primary producers. The rate of release of N from decomposing organic matter can be an important factor in determining nutrient limitation in freshwaters.

Nitrification is a two-stage oxidation process mediated by the chemoautotrophic bacteria, nitrosomonas (ammonia to nitrite) and nitrobacter (nitrite to nitrate). The oxidation of ammonium to nitrite by nitrosomonas is usually rate limiting, so nitrite is rarely present in appreciable concentrations in freshwaters. Nitrate, the end-product, is highly oxidized, soluble and biologically available. A high rate of nitrification is essential for efficient N cycling in fresh waters, particularly as nitrate is an important substrate for denitrification. *Denitrification* is a loss process for nitrate from rivers and lakes. Loss may also occur via dissimilatory nitrate reduction, but this process is less important than denitrification in fresh waters. Denitrification is controlled by the oxygen supply and available energy provided by organic matter. To function, this N cycling pathway requires anaerobic conditions and a fixed bacterial carbon supply. ALH

Reading
Burt, T.P., Heathwaite, A.L. and Trudgill, S.T. (1993) *Nitrate: process, patterns and management.* Chichester: John Wiley & Sons, Ltd.

nivation The localized erosion of a hillside by frost action, mass wasting and the sheet flow or rill work of meltwater at the edges of, and beneath, lingering snow patches. The term was introduced by Matthes (1900). The main effect of nivation is to produce nivation hollows, which, as they grow in depth, trap more snow and thereby enhance the process of deepening. Given adequate time and suitable conditions a nivation hollow may evolve into a cirque. Topographic and climatic controls strongly influence the distribution and orientation of nivation hollows. The most favoured locations are on hillsides protected from the sun and with an ample supply of drifted snow. In mid-latitudes these factors favour a northeastern orientation in the northern hemisphere and southeastern orientation in the southern hemisphere. DES

Reading and Reference
Embleton, C. and King, C.A.M. (1975) *Periglacial geomorphology.* London: Edward Arnold. · Matthes, F.E. (1900) Glacial sculpture of the Bighorn Mountains, Wyoming. In Twenty-first annual report of the United States Geological Survey to the Secretary of the Interior 1899–1900. Part II. General geology, economic geology, Alaska. Washington, DC: US Government Printing Office; pp. 167–190. · Thorn, C.E. (1988) Nivation: a geomorphic chimera. In M. J. Clark (ed.), *Advances in periglacial geomorphology.* Chichester: John Wiley & Sons, Ltd; pp. 3–31.

nivometric coefficient An index of snowfall efficacy, being the ratio of snowfall (water equivalent) to total annual precipitation. A coefficient of 1 implies precipitation entirely of snow.

Reading
Tricart, J. (1969) *Geomorphology of cold environments.* London: Macmillan.

noctilucent clouds Found near a height of 80 km where there is a minimum in the temperature. They are so extremely tenuous that they can be seen only against the light scattered by air molecules on summer nights, between latitudes of 50° and 70°, at least 1 h after the sun has set at the ground. While the cloud particles are likely to be ice, their origin and the mechanism for the formation of the clouds is under considerable debate.   JSAG

Reading
Ludlum, F.H. (1980) *Clouds and storms*. University Park, PA: Pennsylvania State Press.

nonconformity An angular unconformity or other discontinuity between strata where the older rocks are of plutonic origin.

nonlinear system In nature, most systems are nonlinear; they are viewed for simplicity as being linear in operation over a restricted range of action; for example, extension of a spiral spring (Hooke's law). More complicated systems can be modelled by approximating or simplifying to the linear case. True nonlinear systems, however, do not have an easily derived solution by formula, although they can be computed. Nonlinear dynamic systems may give rise to responses that are described as chaotic – where the system does not settle down to a fixed equilibrium condition or value. The earliest 'natural' chaotic systems studied were simple meteorological (Edward Lorenz) and the logistic equation in ecology (Robert May). For a steady-state condition, a 'control parameter' gives a single value (e.g. population in the logistic equation), but at a further value of the control parameter two solutions may be given and at still higher values four, eight, and so on, and so increasingly rapidly up to the chaotic regime. This is related to the Feigenbaum number.   WBW

Reading
Gleick, J. (1987) *Chaos*. Harmondsworth: Sphere/Cardinal/Penguin.

nonrenewable   resource See   STOCK RESOURCES.

normal cycle See CYCLE OF EROSION.

normal fault A fault with the fault plane inclined towards the side that has been downthrown.

normal stress The stress or load applied to the surface of an object (as either compression or tension). It is perpendicular to the SHEAR

STRESSES that act parallel to the surface. A normal stress produces a strain or deformation in the material.   WBW

normalized difference vegetation index (NDVI) A mathematical combination of the remotely sensed response measured in red (R) and near-infrared (nIR) wavelengths. It is widely used in studies of vegetated environments by REMOTE SENSING as the magnitude of the index is strongly related to important biochemical and biophysical properties of vegetation. The index should be calculated using radiometrically calibrated data with the equation NDVI = (nIR − R)/(nIR + R); values derived are constrained to lie on a scale from −1 to +1. Because healthy vegetation has a low reflectance in red wavelengths, due to absorption by chlorophyll for photosynthesis, and a high reflectance in the near-infrared, due especially to the internal structure of plant leaves, the index tends to be positively related to variables such as vegetation amount or biomass. Because of limitations associated with factors such as soil background or atmospheric effects, a wide range of other spectral indices have been developed, but the popularity of the NDVI is indicated in its common availability as a standard data product from major satellite remote sensing missions.   GF

Reading
Huete, A.R., Liu, H.Q., Batchily, K. and van Leeuwen, W. (1997) A comparison of vegetation indices over a global set of TM images for EOS-MODIS. *Remote Sensing of Environment*, **59**, 440–451. · Jones, H.G. and Vaughan, R.A. (2010) *Remote sensing of vegetation: principles, techniques and applications*. Oxford University Press, Oxford.

North Atlantic Deep Water (NADW)
NADW is dense, characterized by relatively cool temperatures (~2–4 °C), high salinities (above 34.9) and a high oxygen content. It occupies the bulk of the Atlantic basin between depths of around 1000 and 4000 m in the north, tapering to 1700–3000 m in the southern Atlantic. It is underlain by the colder, denser ANTARCTIC BOTTOM WATER (AABW) over much of the Atlantic Basin, including in the northern hemisphere to around 40°N, and it is overlain by Antarctic Intermediate Water (AAIW), which occupies mid-depths below the thermocline. NADW is a significant water mass in the global ocean; Johnson (2008) estimates it as making up 21% of the global ocean volume (excluding the Arctic Ocean, continental shelves and some other marginal seas such as the Mediterranean, Red, Caspian and Black Seas), a volume of $0.268 \times 10^9 \, \text{km}^3$.

NADW meridional salinity transect at 20–25°W for the Atlantic Ocean showing the extent of NADW and the main water masses that form NADW. UNADW: Upper NADW; LSW: Labrador Sea Water; SPMW: Subpolar Mode Water; MW: Mediterranean Water; ISOW: Iceland–Scotland Overflow Water; AAIW: Antarctic Intermediate Water; AABW: Antarctic Bottom Water.

Source: Talley *et al.* (2011). Reproduced with permission of Elsevier.

NADW is a complex water mass with several main sources (Talley *et al.*, 2011). Aside from the water entering the North Atlantic as AABW and AAIW, there are three main local sources of NADW: dense water formed in the Nordic Seas north of Iceland; Labrador Sea Water; and warmer but salty water entering the Atlantic at depth through the Straits of Gibraltar, which is formed by strong evaporation and convection in the western Mediterranean Sea.

NADW plays a vital role in the operation of the global THERMOHALINE CIRCULATION (THC), with variations in its formation implicated in several important aspects of millennial-scale climate variability during the last glacial period and during the early part of the Holocene (e.g. Broeker and Denton, 1990; Barber *et al.*, 1999). Variations in the production of NADW also seem to be linked with present-day, short-term climate variability (e.g. Bryden *et al.*, 2005), with NADW formation likely vulnerable to increased freshwater flux into the Arctic Ocean and North Atlantic region due to climate

warming in the twenty-first century (e.g. Stouffer *et al.*, 2006).

NSA

References

Barber, D.C.A., Dyke, A., Hillaire-Marcel, C., *et al.* (1999) Forcing of the cold event of 8200 years ago by catastrophic drainage of Laurentide lakes. *Nature*, **400**, 344–348. · Broeker, W.S. and Denton, G.H. (1990) The role of ocean–atmosphere reorganisations in glacial cycles. *Quaternary Science Reviews*, **9**, 305–341. · Bryden, H.L., Longworth, H.R. and Cunningham, S.A. (2005) Slowing of the Atlantic meridional overturning circulation at 25°N. *Nature*, **438**, 655–657. · Johnson, G.C. (2008) Quantifying Antarctic Bottom Water and North Atlantic Deep Water volumes. *Journal of Geophysiocal Research*, **113**, C05027, doi: 10.1029/2007JC004477. · Stouffer, R.J., Yin, J., Gregory, J.M., *et al.* (2006) Investigating the causes of the response of the thermocline circulation to past and future climate changes. *Journal of Climate*, **19**, 1365–1387. · Talley, L.D., Pickard, G.L., Emery, W.J. and Swift, J.H. (2011) *Descriptive physical oceanography: an introduction*, 6th edition. Boston, MA: Elsevier.

North Atlantic Oscillation (NAO)

Quasi-regular fluctuations in the pressure gradient over

the North Atlantic; that is, large-scale movements of atmospheric mass between the Azores or subtropical high and the Iceland or polar low. The NAO is assessed via the seasonal pressure difference between these systems. Several indices of the NAO have been developed, each using sea-level pressure at Reykjavik, Iceland, to represent the Icelandic low and a subtropical station such as Lisbon, Portugal, or Gibraltar to represent the Azores high. A more technical definition that is frequently used is the leading empirical orthogonal function of sea-level pressure anomalies over the North Atlantic (Hurrell, 1995).

The NAO is the dominant mode of winter climate variability in the North Atlantic, and its influence ranges from central North America to Europe and even into northern Asia. It controls the strength and flow pattern of the westerly winds and storm tracks across the North Atlantic, particularly during the period November to April. It may also influence the path of Atlantic tropical storms and hurricanes (Elsner *et al.*, 2000). When the NAO is positive, both the Icelandic low and the Azores high are stronger than usual, producing stronger and more frequent winter storms that progress along a more northerly track. The result is warm and wet winters in Europe, cold and dry winters in Greenland and northern Canada, but generally mild and wet winter conditions in the eastern USA. When the NAO is negative, storms are fewer, weaker and passing along a more southerly track. The flow tends to be directly west to east, with few north–south undulations. This brings moist air into the Mediterranean and cold air into northern Europe. The eastern USA tends to experience more cold outbreaks and abnormally snowy weather (Scaife *et al.*, 2014). SJN

References

Elsner, J.B., Liu, K.-B. and Kocher, B. (2000) Spatial variations in major U.S. hurricane activity: statistics and a physical mechanism. *Journal of Climate*, **13**, 2293–2305. · Hurrell, J.W. (1995) Decadal trends in the North-Atlantic Oscillation – regional temperatures and precipitation. *Science*, **269**, 676–679. · Scaife, A.A., Arribas, A., Blockley, E., *et al.* (2014) Skillful long-range prediction of European and North American winters. *Geophysical Research Letters*, **41**, 2514–2519.

notch Landform that develops at the base of a cliff, platform or reef flat, especially in limestone and on tropical coasts. Deep, narrow notches are characteristic of areas with a low tidal range. Their positions are related to lithological and structural controls, tidal characteristics and sea-level history. In general, the higher the amplitude of the waves and the higher the tidal range, the greater is the difference in elevation between the notch roof and the floor. Notches in the humid tropics may be 1–5 m in depth. Although mechanical action of waves may contribute to their development, most investigators now believe that chemical or biochemical corrosion, or biological boring and grazing activities, are important. ASG

Reading

Woodroffe, C.D., Stoddart, D.R., Harmon, R.S. and Spencer, T. (1983) Coastal morphology and Late Quaternary history, Cayman Islands, West Indies. *Quaternary Research*, **19**, 64–84.

nubbins Small lumps of earth produced by heaving due to the growth of needle ice.

nuclear waste (radwaste) Waste produced during the operation of nuclear facilities and as a result of decommissioning. Much comes from nuclear power stations, but other sources include hospitals and research institutions. Radwaste can be classified according to its volume, level of activity (high, medium, low) and its form (liquid, solid or gas).

High-volume low-activity solid wastes result from mining and uranium ore processing, from reactor operations, from final plant dismantling (decommissioning) and from soiled clothing, and so on. Generally speaking, low-activity wastes are characterized by radionuclides with short half-lives. Commonly, this type of waste is buried in designated shallow trenches, but in the past sea disposal has been used to remove much of this waste.

The disposal of *low-activity liquid waste* from nuclear power plants and fuel reprocessing factories depends upon the siting of the works. Those with a coastal location, or on a large river or lake, remove sufficient radionuclides from liquid streams by distillation or floc precipitation to produce effluents of 'acceptable' purity prior to discharge into the adjacent water body.

Medium-volume medium-activity wastes are produced by both reactor operation and fuel reprocessing (e.g. ion exchange resins, sludges and precipitates) and may include some plutonium-contaminated material. *Solid low-volume high-activity waste* comprises mainly fuel element cladding and solidified material from reprocessing. *High-activity liquid waste* is produced entirely in fuel reprocessing operations.

Disposal of high-activity wastes currently presents problems because of their potential as biological hazards. Not only are they highly active, but they also contain some very long-lived activity. The optional sequence of events in the management of high-level waste is:

1 Storage of fuel elements in ponds for months to several years.
2 Storage of highly active liquor produced in reprocessing fuel for not more than two decades.
3 Solidification into borosilicate glass, after which the glass blocks will be artificially cooled, for between 10 and 20 years.
4 The encapsulation of the blocks and their emplacement in a final deep geological repository. ASG

Reading

Ojovan, M.I. and Lee, W.E. (2013) *An introduction to nuclear waste immobilisation*. Oxford: Newnes.

nudation See SUCCESSION.

nuée ardente (glowing cloud) A cloud of superheated gas-charged ash produced by certain acidic volcanic eruptions (e.g. the eruption of Mt Pelée on Martinique in 1902 and 1903). The deposits produced by nuées ardentes are termed ash-flow tuffs, welded tuffs or ignim-brites. ASG

numerical modelling A method for obtaining particular solutions or deductions for a model that is expressed in mathematical or logical form, and for which general mathematical solutions are not appropriate and/or not available.

A numerical solution is often the only one available for any but the simplest model, but has the disadvantage that it lacks the generality of an analytical solution. In numerical modelling, all parameters must be given definite values and all variables assigned initial values. A model run is then a single realization of the model constrained by these particular values. A large number of trials is therefore needed satisfactorily to explore all the possibilities inherent in any model.

Models vary considerably in style and complexity. In 'black-box' models either the whole of a system or parts of it are considered solely in terms of empirical relationships between the input and output of the system. Most models in use in physical geography contain at least elements of this type. The commonest source of material for this type of model is a regression equation based on field observations. As understanding advances the black-box components within the total system become less significant as more components are based on established scientific principles. The level of empirical relationships in a useful forecasting model is partly constrained by our state of knowledge and partly by the level of detail that it is appropriate to represent. Such practical models are commonly developed drawing on the methods of systems analysis.

A stochastic element is often found in numerical models. For example, rainfall inputs to a hydrological model may be drawn at random from a known distribution in order to generate a probability distribution of high and low flows for hydraulic engineering design. Stochastic elements are usually included either to represent a model input of which a direct model is not needed, or to cover variability at scales below the level of resolution of the model. In the example above, a stochastic rainfall model may well be more appropriate and economic as an input than an independent model of the general atmospheric circulation! As another example, microtopography might be included within a model of long-term hillslope evolution as a random variation in process rate. The microtopography is below the general level of resolution of the model but could cause some of the observed variability in the forms of neighbouring hillslope profiles apparently subject to identical processes.

Numerical models require an underlying formulation in mathematical terms, and the range of models reflects the range of mathematical possibilities, which is immense. Perhaps the most common type of numerical model in physical geography is a solution to one or a family of differential equations, which are ultimately based on the CONTINUITY EQUATION with the substitution of suitable EQUATIONS OF MOTION or other expressions for the rates of the relevant processes of material or energy transport. In some cases, particular classes of solution are sought; for example, EQUILIBRIUM or KINEMATIC WAVE solutions. In all cases, the numerical model can only be run when boundary and initial conditions are specified. Initial conditions must specify the relationships that inputs and outputs must satisfy where they enter/leave the system of interest.

An example of a differential equation type is the model of hillslope evolution based on continuity of downslope sediment transport. Ignoring wind-blown dust, the continuity equation is

$$\frac{\partial Q}{\mathrm{d}x} + \frac{\partial z}{\mathrm{d}t} = 0$$

where Q is the rate of downslope sediment transport at distance x from the divide, and z is the elevation at x and time t.

For hillslope processes such as soil creep or soil EROSION, which are largely transport-limited removal, the sediment transport may be expressed in the form $Q = f(x)s^n$ for some function of slope distance on slope gradient s. Where $f(x)$ is constant, an analytical solution is available (Culling, 1963), but otherwise numerical modelling is the best method. In this case, any initial conditions may be used to describe the original slope profile form. The simplest boundary conditions are a fixed divide ($Q = 0$ at $x = 0$) and basal removal at a fixed base level ($z = 0$ at $x = x_1$), although others may be used. Runs of this model may be carried out on small microcomputers to follow long-term profile development.   MJK

Reading

Wainwright, J. and Mulligan, M. (2013) *Environmental modelling: finding simplicity in complexity*. Chichester: John Wiley & Sons, Ltd.

nunatak An Inuit-derived word describing a mountain completely surrounded by glacier ice, normally an ICE CAP or ICE SHEET. The nunatak hypothesis is the idea that plant and animal (but especially the former) communities have been isolated as refugia on nunataks. Sometimes these refugia may be as mountains, as with the normal usage of nunatak, but the nunatak hypothesis may also relate to much larger areas of isolated, ice-free land.   WBW

Reference

Gjaerveroll, O. (1963) Survival of plants on nunataks in Norway during the Pleistocene glaciation. In A. Löve and D. Löve (eds), *North Atlantic biota and their history*. Oxford: Oxford University Press; pp. 261–283.

nutrient A biologically essential chemical element and one that is required for the maintenance of life processes. In practice, the nutrients are classified into macronutrients, which are required in proportionally large quantities, and micronutrients, only trace concentrations of which are needed to sustain life but which are, nevertheless, essential. Macronutrients include nitrogen, phosphorus, sulphur, potassium, calcium, magnesium and iron, whereas the micronutrients, of which there are approximately 30, include manganese, zinc, copper, iodine, molybdenum and sodium. Because of adaptational differences, not all organisms require the same amounts of nutrients, nor even the same nutrients; for example, silicon is an essential nutrient for diatoms but is of no consequence to the higher plants. Humans and other vertebrates require relatively large quantities of calcium and phosphorus for the construction and maintenance of bone and need iodine in the manufacture of the thyroid hormone that controls the metabolic rate. Nutrients have uneven distributions within the environment; for example, carbon and oxygen elements are obtained directly from the atmosphere via PHOTOSYNTHESIS of green plants and can be made available to other organisms thereafter. Other nutrient elements enter the food chain via the soil or sediments (see FOOD CHAIN, FOOD WEB). Complex linkages between the atmosphere, lithosphere, hydrosphere and biosphere exist such that nutrients are constantly circulated in BIOGEOCHEMICAL CYCLES.   MEM

nutrient status A collective term usually applied to soils, peatlands and lakes whose nutrient status may be EUTROPHIC (rich in nutrients), *oligotrophic* (poor) or *mesotrophic* (transitional). Mires are also referred to as OMBROTROPHIC (rain feeding) or *rheotrophic* (flow feeding). In this context the use of the suffix 'trophic' is not to be confused with the trophic levels of animal communities feeding off each other and off the autotrophs or primary producers, the plants.   KEB

References

Trudgill, S.T. (1988) *Soil and vegetation systems*, 2nd edition. Oxford: Oxford University Press.

O

oasis An area within a desert region where there is sufficient water to sustain animal and plant life throughout the year.

obliquity A change in the tilt of the Earth's axis relative to the plane in which the bodies of the Solar System lie (the ecliptic), which varies between 21° and 25, with a periodicity of *c.* 41,000 years. Together with changes in the eccentricity of the Earth's orbit (a *c.* 100,000-year cycle) and the precession of the equinoxes (with cycles of *c.* 19,000 and 23,000 years) it is one of the components of the MILANKOVITCH HYPOTHESIS, which proposes that changes in the position and configuration of the Earth as a planet in relation to the Sun may determine climate as a result of changes in the amount of insolation received.   ASG

Reading
Anderson, D.E., Goudie, A.S. and Parker, A.G. (2013) *Global environments through the Quaternary*, 2nd edition. Oxford: Oxford University Press; section 9.4

obsequent stream A stream or river that is the tributary of a subsequent stream and flows in a direction opposite to the regional dip of the land surface.

obsidian hydration dating (OHD) Used by archaeologists and geologists to date events ranging in age from a few hundred years to several million years. The principle is based on the dating of obsidian (volcanic glass), which when a fresh surface is formed and exposed to the atmosphere will proceed to form a hydration layer due to the diffusion of ambient water. The hydration layers are firmly adherent to the parent glass and resistant to chemical dissolution. The thickness of the hydration layer, which varies from 1 μm to more than 50 μm, depends on the time of exposure. OHD requires that a measurement of hydration thickness or the depth of penetration of water into obsidian be measured and a rate of hydration be known. The measurement can be made optically or by using particle accelerators.   AP

Reading
Trembour, F. and Friedman, I. (1984) The present status of obsidian hydration dating. In W. C. Mahaney (ed.), *Quaternary dating methods: developments in palaeontology and stratigraphy 7*. Amsterdam: Elsevier; pp. 141–151. · Walker, M. (2005) *Quaternary dating methods*. Chichester: John Wiley & Sons, Ltd.

obstacle dune See TOPOGRAPHIC DUNE.

occlusion A complex frontal zone associated with the later stages of the life cycle of an extra-tropical cyclone. The name is derived from the associated occluding or uplifting of the warm sector air from the Earth's surface. Cold fronts tend to travel more quickly than warm fronts, so the area of warm air between narrows. Eventually the warm air is entirely aloft, with the occluded front marking the surface juxtaposition of the two cool or cold air masses. The detailed structure of the occlusion will be determined by the temperature difference of these two air masses. Where the cold air behind the cold front is cooler (warmer) than that ahead of the original warm front it is known as a cold (warm) occlusion. In practice, the temperature difference may be quite small and the occlusion difficult to classify.   PS

occult deposition The wet deposition of acidic pollutants, particularly sulphur dioxide and NO_x, onto surfaces by the impaction of fog and cloud droplets. Patterns of deposition are influenced by climatic factors that encourage fog and mist and by local variations in wind direction and intensity. Concentration of pollutants in occult deposition can be considerably higher (up to ×20) than those in wet deposition by rainfall. The term is particularly used in connection with acid deposition in urban environments, where it precipitates and concentrates pollutants on surfaces protected from rainwash. The process is important in, for example, coastal areas of low rainfall but high relative humidity and fog frequency and deposition can be increased by land-use changes; for example, afforestation, which increase surface area.   BJS

Reference
Building Effects Review Group (1989) *The effects of acid deposition on buildings and building materials in the United Kingdom*. London: HMSO.

ocean The general name for large bodies of saltwater making up around 70% of the Earth's surface. Open oceans or oceanic zones are those parts deeper than 200 m, whereas shallow coastal waters or neritic zones lie over continental shelves and are usually less than 200 m deep. There is only one ocean basin; the geographical subdivisions are made for convenience, because they are all interconnected. Shallow waters are more affected by changes in temperature, salinity, sedimentation and water movements. They are reached by sunlight and are richer in nutrients, and hence in plant and animal life, than the deeper, more constant oceans. In general, the nearer the land the higher the NET PRIMARY PRODUCTIVITY (NPP). PAF

Reading and Reference
Furley, P. and Newey, W. (1984) *The geography of the biosphere*. London: Butterworth. · Noone, K.J., Sumaila, U.R. and Diaz, R.J. (2013) *Managing ocean environments in a changing climate*. Amsterdam: Elsevier.

ocean acidification A reduction in the pH of the ocean over an extended period, typically decades or longer, which is caused primarily by uptake of carbon dioxide from the atmosphere. Between 1751 and 1994 the surface ocean pH is estimated to have decreased from approximately 8.25 to 8.14, representing an increase of almost 30% in H^+ ion concentration in the world's oceans. Ocean acidification is expected to impact ocean species to varying degrees. Photosynthetic algae and seagrasses may benefit from higher carbon dioxide conditions in the ocean. On the other hand, studies have shown that a more acidic environment has a dramatic effect on some calcifying species, including oysters, clams and calcareous plankton. Increasing ocean acidification has been shown to significantly reduce the ability of reef-building corals to produce their skeletons. TS

Reading
Orr, J.C., Fabry, V.J., Aumont, O., *et al.* (2005) Anthropogenic ocean acidification over the twenty first century and its impact on calcifying organisms. *Nature*, **437**, 681–686. · Pandolfi, J.M., Connolly, S.R., Marshall, D.J. and Cohen, A.L. (2011) Projecting coral reef futures under global warming and ocean acidification. *Science*, **333**, 418–422. · Provoost, P., van Heuven, S., Soetaert, K., *et al.* (2010) Seasonal and long-term changes in pH in the Dutch coastal zone. *Biogeosciences*, **7**, 3869–3878.

Ocean Drilling Programme (or Program – ODP) An international cooperative science venture focused on fundamental research into the history of the ocean basins and the nature of the crust beneath the ocean floor. The programme is funded principally by the US National Science Foundation with substantial contributions from several international partners. Deep-ocean drilling began initially with the Deep Sea Drilling

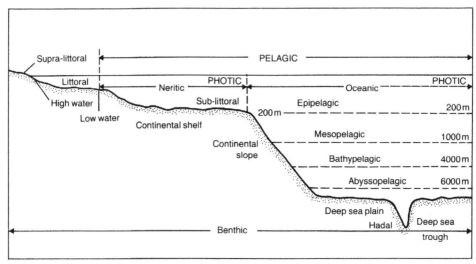

Ocean: classification of marine biome-zones and divisions.
Source: Furley and Newey (1984: figure 13.1). Reproduced with permission of Peter Furley.

Project (DSDP), headquartered at Scripps Institution of Oceanography. In 1964, the Joint Oceanographic Institutions for Deep Earth Sampling (JOIDES) was formed in the USA and in 1968 drilling operations began under their leadership with DSDP and its drilling vessel, *Glomar Challenger*. In 1983, Texas A&M University was designated as the principal science operator and the name Ocean Drilling Program was adopted. The drilling vessel currently in use by ODP is *JOIDES Resolution*, and it is this vessel that is responsible for the major ocean sediment coring activities of the programme.   MEM

ocean sediments   These have accumulated over millennia, particularly in deep ocean locations. On continental shelves (see CONTINENTAL SHELF) ocean sediments may preserve materials derived from terrestrial erosion that have been transported to the oceans by fluvial and aeolian processes. In ocean deeps, erosion or removal of sediments may be virtually absent such that long records of accumulation are preserved. Sediments may predominantly comprise the detritus derived from the accumulated remains of ocean-dwelling organisms, though in high latitudes ice-rafted glacial sediments may be a significant component of the total sediment. From carbonate sediments, long records of OXYGEN ISOTOPE changes may be determined, reflecting global environmental changes and glacial–interglacial cycles. Long deep sea core records, provided by the OCEAN DRILLING PROGRAMME, have yielded records of environmental change that have revolutionized understanding of Quaternary and late Tertiary period environments and climates.   DSGT

Reading
Einsele, G. (2000) *Sedimentary basins: evolution, facies, and sediment budget*. Berlin: Springer-Verlag; chapter 5, pp. 183–248.

oceanic crust   A portion of the Earth's crust underlying most ocean basins. There are two major crustal types: continental and oceanic. Oceanic crust comprises about 59% of the crust. The basaltic oceanic crust is sometimes called SIMA because of the abundance of silicon (50%) and magnesium (8%). It is denser (3000 kg m^{-3}) than the granitic continental crust (2700 kg m^{-3}), and thinner (~5 km in thickness versus ~30 km). Both oceanic and continental crusts float on the even denser mantle (4500 kg m^{-3}). The crust and mantle are separated by the MOHOROVIČIĆ DISCONTINUITY.

Studies of oceanic crust provided several key pieces of evidence supporting the hypothesis of PLATE TECTONICS. Oceanic crust is not more than

about 200 million years old, remarkably young compared with some continental crust that is several billion years old. This suggests that some mechanism of destruction and generation of oceanic crust must exist, and this is presumed to be sea-floor spreading. When collisions occur between continental and oceanic plates, the former tend to ride up over the latter because continental crust is less dense than oceanic crust. Sea-floor spreading forces oceanic crust beneath the continents in SUBDUCTION ZONES, where the rock is remelted. In the Pacific basin, the length of time required for crust to spread from the ridge to a subduction zone is about 200 million years, thus setting the maximum ages for oceanic crust.

As a result of sea-floor spreading, distinct, paired bands of palaeomagnetic signatures are found symmetrical about mid-ocean ridges. When new crust is formed, the axes of magnetite crystals within the molten rock align with the Earth's magnetic field. The rapid cooling of the sea floor basalts freezes the axes of those crystals into place. A record of periodic reversals of the Earth's polarity is thus preserved in the basalt.   DJS

Reading
Condie, K.C. (1989) *Plate tectonics and crustal evolution*, 3rd edition. Oxford: Pergamon. · Kearey, P., Klepeis, K.A. and Vine, F.J. (2009) *Global tectonics*. Chichester: John Wiley & Sons, Ltd.

oceanography   The study or description of the oceans encompassing the sea floor, the physics and chemistry of the seas and all aspects of marine biology.

Reading
Knauss, J.A. (2005) *Introduction to physical oceanography*. Long Grove, IL: Waveland Press. · Pickard, G.L. and Emery, W.J. (1990) *Descriptive physical oceanography: an introduction*, 5th edition. Oxford: Pergamon.

ogives (also known as Forbes bands) Alternating bands of light and dark ice that extend across the surface of some glaciers below ice falls. They are arcuate in response to the normal pattern of ice flow across a glacier. The combined width of a dark and light band corresponds to the distance the glacier moves in a year. The dark ice corresponds to the ice that traverses the ice fall in summer and is thus exposed to melting, while the light ice reflects the incorporation of snow as the ice traverses the ice falls in winter.   DES

Reading
Benn, D.I. and Evans, D.J.A. (2010) *Glaciers & glaciation*. London: Hodder Arnold. · Paterson, W.S.B. (1981) *The physics of glaciers*. Oxford: Pergamon.

oil-shale A shale that contains sufficient quantities of hydrocarbons to yield oil or petroleum gas when distilled. FRACKING may be used to extract the hydrocarbons.

okta Measurement of the amount of cloud cover is one of the two standard scales used by surface meteorological observers worldwide (the other scale being tenths). The observer reports the number of eighths or oktas of the celestial dome that is covered by clouds. Total cloud cover and layer cloud amount are reported in this fashion. Care must be taken to give equal weight to all areas of the sky, especially in the case of *cumuliform* clouds when cloud sides as well as cloud bases may be viewed.   AH-S

oligotrophic See NUTRIENT STATUS.

ombrotrophic Description of plants or plant communities that are associated with a rain-fed substrate that is poor in nutrients.

onion-weathering EXFOLIATION. The destruction of a rock or outcrop through the peeling off of the surface layers.

ontogeny (ontogenesis) The sequence of development during the whole life history of an organism. The term is also applied to the life history of lakes and other systems.   ASG

ooids Small (typically 2 mm in diameter), spheroidal, concentrically layered sedimentary grains, usually composed of calcium carbonate, but sometimes made up of iron- or phosphate-based minerals. Ooids usually form on the sea floor, most commonly in shallow tropical seas (around the Bahamas, for example, or in the Persian Gulf). After being buried under additional sediment, these ooid grains can be cemented together to form a sedimentary rock called an oolite.   TS

Reference
Tucker, M.E. and Wright, V.P. (1990) *Carbonate sedimentology*. Oxford: Blackwell Science.

oolite A sedimentary rock, usually calcareous but sometimes dolomitic or siliceous, that is composed of concentrically layered spheres (ooliths) that have formed by accretion on the surface of a grain. The Jurassic rocks of southern England contain the Greater and the Inferior Oolite, and originally obtained their name because of the supposed resemblance of their fabric to fish roe.

ooze Fine-grained organic-rich sediments on the floors of lakes and oceans.

open channel flow See FLOW REGIMES.

open system See SYSTEMS.

Operational Land Imager The Landsat 8 satellite (see LANDSAT), launched in 2013, does not carry THEMATIC MAPPER instrumentation, instead carrying a new system for monitoring the Earth's surface that combines THERMAL INFRARED IMAGER and the Operational Land Imager sensors that together measure reflected and emitted radiation across 11 wave bands. Four bands are in the visible portion of the electromagnetic spectrum: band 1, 0.435–0.451 μm (new, for coastal observations); band 2, 0.452–0.512 μm (blue); band 3, 0.533–0.590 μm (green); and band 4, 0.636–0.673 μm (red). Five bands are in the infrared: band 5, 0.851–0.8798 μm (near infrared); band 6, 1.566–1.651 μm (shortwave infrared); band 7, 2.107–2.294 μm (shortwave infrared); band 10, 10.60–11.19 μm (thermal infrared-1); band 11, 11.50–12.51 μm (thermal infrared-2). Band 8 is panchromatic at 0.503–0.676 μm, while band 9 at 1.363–1.384 μm is especially set to monitor cirrus clouds. While band 8 has a 15 m pixel resolution and bands 10 and 11 are at 100 m, the others remain at 30 m, as with the original Thermal Mapper.   DSGT

Reading
Irons, J.R., Dwyer, J.L. and Barsi, J.A. (2012) The next Landsat satellite: the Landsat Data Continuity Mission. *Remote Sensing of Environment*, **122**, 11–21.

opisometer An instrument for measuring distances on a map.

optical remote sensing Conducted using remote sensing devices that typically record the electromagnetic radiation in the visible (0.4–0.7 μm), near infrared (0.7–1.5 μm) and short/middle infrared (1.5–5.0 μm) that have been reflected by a target in a scene of interest. This region of the electromagnetic spectrum corresponds to the maximum spectral exitance of the Sun and is characterized by absorptions due to electronic transitions and changes in molecular rotation and vibration within molecules. Depending on the temperature of the scene of interest, there may be radiation that is emitted that is recorded by optical remote sensing devices (as a function of Wien's displacement law), but usually this radiation is emitted at longer wavelengths (5.0–15.0 μm), and this is known as thermal remote sensing. Aerial photography is a form of optical remote sensing, but recently there has been a proliferation of imaging devices available for optical remote sensing. These can be classified based on the number of spectral wavebands they have: panchromatic imaging systems have a single-band sensor and are used to detect radiation within a broad wavelength range (e.g., SPOT HRV-PAN and IKONOS PAN); multispectral imaging systems use a multichannel detector and record radiation within a narrow range of wavelength (e.g. LANDSAT OLI and WorldView 2); superspectral imaging systems consist of more than 10 narrow spectral bands (e.g. Terra MODIS and the Sentinel-2A/B sensors); and finally, hyperspectral imaging systems record hundreds of spectral bands to provide HYPERSPECTRAL IMAGERY).   DSB

Reading
Lillesand, T., Kiefer, R.J. and Chipman, J. (2009) *Remote sensing and image interpretation*, sixth edition. New York: John Wiley and Sons, Inc. · Warner, T.A., Nellis, M.D. and Foody, G.M. (2009) *The SAGE handbook of remote sensing*. London: SAGE.

order, stream   Identification of the links in a stream NETWORK is by the process of stream ordering. At least 11 methods of stream ordering have been devised. The first method to be adopted was proposed by R. E. Horton and led to the development of HORTON'S LAWS. In Horton ordering, all unbranched streams were designated first order, two first combined to make a second order and so on, and then the highest order streams were projected to the headwaters. Subsequently, Strahler (1952) suggested that the second stage should not be effected and Strahler ordering became widely used. Later developments have endeavoured to propose ordering methods that are mathematically consistent and which allow all components of the network to be differentiated according to their position. Thus, in the Horton and Strahler systems of ordering the order of a stream of order *n* is unaffected by the entry of a tributary of order less than *n*. Later ordering systems, such as those proposed by Shreve (1967) and by Scheidegger

Four methods of stream and segment ordering.

(1965), have overcome this fundamental objection. The objective of all ordering systems is to be able to describe a link in a drainage network anywhere in the world in an unambiguous manner, and also to provide an ordering system that can readily provide an indication of the discharge from a network.   KJG

Reading and References

Charlton, R. (2007) *Fundamentals of fluvial geomorphology*. London: Routledge. · Scheidegger, A.E. (1965) The algebra of stream-order numbers. In Geological survey research 1965: chapter B. United States Geological Society Professional Paper 525-B. Washington, DC: US Government Printing Office; pp. 187–189. · Shreve, R.L. (1967) Infinite topologically random channel networks. *Journal of Geology*, **75**, 178–186. · Strahler, A.N. (1952) Hypsometric (area–attitude) analysis of erosional topography. *Bulletin of the Geological Society of America*, **63**, 923–938.

organic weathering   The disintegration or destruction of a rock by living organisms or organic processes. It is a much neglected cause of weathering.

orocline   A structural arc that has formed by horizontal displacement subsequent to the development of the main structural features of an area.

orogens   Total masses of rock deformed during an orogeny (mountain-building episode).

orogeny   The event, or mechanism, of the construction of characteristically linear or arcuate mountain chains formed on continents. Such mountain chains, of which the Andes and the Himalayas are contemporary examples, are known as orogenic belts or simply orogens. Investigation of the results of recent orogeny shows it to produce crustal thickening, deformation and associated volcanic activity, although in some cases, such as the Himalayas, the latter is relatively less important. Many earlier theories of orogeny emphasized the accumulation of a thick wedge of sediments forming a geosyncline before uplift, but these ideas have now been largely subsumed within the more comprehensive model of PLATE TECTONICS.

A view to the Kali Gandaki gorge in Nepal. This is one of the steepest places on Earth and is created by a combination of mountain building (orogeny) and erosion.

Within this scheme, orogeny is of two fundamental types. Where a continent margin, such as that of western South America, is underridden by oceanic LITHOSPHERE being reabsorbed or subducted into the Earth's interior, largely thermal effects generate uplift and volcanic activity. In contrast, mainly mechanical effects are responsible for the Himalayan orogeny, which is associated with the collision of the northward-moving Indian continent and the Eurasian continent.   MAS

Reading
Frisch, W., Meschede, M. and Blakey, R. (2011) *Plate tectonics, continental drift and mountain building.* Berlin: Springer-Verlag. · Hsü, K.J. (ed.) (1982) *Mountain building processes.* London: Academic Press.

orographic precipitation Precipitation caused by the forced ascent of air over high ground. Uplift of air leads to cooling that, if the air is moist, may lead to condensation and eventually precipitation. The warm sector of an intense extratropical cyclone is the synoptic situation that demonstrates the orographic effect most clearly. Even where rain of convectional or cyclonic origins is falling, the orographic influence can still be seen in larger and sometimes longer precipitation events over the hills. The extra uplift will ensure that the precipitation processes in the clouds operate more effectively.   PS

osmosis Three forces act on water in soils: (1) gravity, (2) suction and (3) osmosis. Water moves by osmosis from regions of low solute concentration to regions of higher concentration separated by a semi-permeable membrane. Osmosis is rarely involved in large-scale water movement in soils owing to the lack of semipermeable membranes, which means that difference in salt concentration are quickly balanced by diffusion of dissolved salts. The exception is dry soils, where water moves as vapour and the air–water interface acts as a membrane allowing the passage of water but not solutes. At the small scale, osmosis results in localized movement of water into interlayer spaces of clay particles, resulting in swelling. Osmosis is important in plants, where the uptake of water traverses plant membranes.   ALH

ostracods Small crustaceans 0.3–30 mm in length. Their body and limbs are surrounded by two valves that are hinged dorsally to form the shell. When the animal dies the shell may be preserved as a fossil and it is possible to identify species based on the shell structure and morphology. Ostracods can be found in almost any global environment where water is present, and a wide range of factors, including water temperature, salinity and the nature of the substrate, govern their distribution. The analysis of ostracod shells, particularly their geochemistry, is proving a valuable tool in reconstructing past climates (e.g. Holmes, 1992).   SLO

Reference
Holmes, J.A. (1992) Non-marine ostracods as Quaternary palaeoenvironmental indicators. *Progress in Physical Geography*, **16**, 405–431.

outlet glacier A type of glacier that radiates out from an ICE DOME and often occupies significant depressions. Within an ice dome they can frequently be distinguished by a zone of high-velocity ice termed an ICE STREAM.

outlier An isolated hill lying beyond the scarp slope of a cuesta. A rock outcrop that is surrounded by rocks that are of an older age.

outwash Comprises stratified GLACIOFLUVIAL sands and gravels deposited at or beyond the ice margin. Outwash usually forms fan, valley bottom (valley train) or plain (SANDUR) deposits, often hundreds of metres thick, built up by aggrading braided or anastomosing meltwater channels that migrate laterally across the outwash surface. Periodic high-energy flood events are marked by high rates of transport of sediment reworked from older outwash and till. Where outwash has accumulated on the glacier margin itself, differential ice melt may produce pitted outwash. Outwash is often steeply graded close to the glacier, comprising coarse-grained, imbricated, noncohesive sediments deposited rapidly during unidirectional, high-flow regime conditions, both within channels and on longitudinal bars. Bar deposits exhibit crude horizontal bedding truncated by erosional contacts and scour-and-fill structures representing successive discrete flood events. The bar surfaces are often characterized by SAND LENSES, SILT drapes, TRANSVERSE RIBS and coarse gravel lags. CLASTS are typically poorly sorted, subangular to subrounded, and comprise heterogeneous lithologies. Outwash is more gently graded farther from the glacier, with finer grained, more cohesive, sandy facies types forming transverse and linguoid bars characterized by dune and ripple forms and by bar avalanche face sediments. These distal sediments often exhibit planar and trough crossbedding, and ripple and ripple drift CROSS-LAMINATION; LOAD STRUCTURES may also be present. Clasts tend to be better sorted and more rounded, while fewer, more resistant lithologies are represented in both the clast and heavy mineral populations. Coarsening or fining-upward sequences may reflect local flood events or periods of channel scour and infill, or longer term periods of ice advance or recession.   JM

Reading

Boothroyd, J.C. and Ashly, G.M. (1975) Process, bar morphology and sedimentary structures on braided outwash fans, northeastern gulf of Alaska. In A. V. Jopling and B. C. McDonald (eds), *Glaciofluvial and glaciolacustrine sedimentation*. Society of Economic Paleontologists and Mineralogists Special Publication 23. Tulsa, OK: SEPM Society for Sedimentary Geology; pp. 193–222. · Church, M. (1972) *Baffin Island sandurs: a study of arctic fluvial processes*. Geological Survey of Canada Bulletin 216. Ottawa: Department of Energy, Mines and Resources. · Church, M. and Gilbert, R. (1975) Proglacial fluvial and lacustrine environments. In A. V. Jopling and B. C. McDonald (eds), *Glaciofluvial and glaciolacustrine sedimentation*. Society of Economic Paleontologists and Mineralogists Special Publication 23. Tulsa, OK: SEPM Society for Sedimentary Geology; pp. 22–100. · Rust, B.R. (1978) Depositional models for braided alluvium. In A. D. Miall (ed.), *Fluvial sedimentology*. Canadian Society of Petroleum Geologists, Memoir 5. Calgary: Canadian Society of Petroleum Geologists; pp. 605–625.

outwash terrace   An outwash deposit that has been incised by meltwater to form a terrace. Incision of the outwash deposit is a response by meltwater streams to an increase in channel slope and/or an increase in the discharge/sediment load balance such that a period of proglacial aggradation is followed by a period of increased meltwater flow capacity and stream degradation. This change is likely to occur during glacial retreat when an overall decrease in meltwater and sediment supply occurs. A stepped sequence of outwash terraces may relate to successive periods of ice advance and retreat, or possibly to a more complex response to a single glacial event. Many outwash terraces can be traced up-valley to an associated terminal moraine; the terrace surfaces often exhibit traces of former braided channel networks and are pitted with kettle holes.   JM

overbank deposit   Flood sediment laid down by a river beyond its normal-flow channel; generally fine materials deposited from suspension in floodwaters, but it is possible for sheets of coarser bed material to be deposited overbank as well. Finer material may be deposited considerable distances from the channel.   JL

overbank flow   River discharge that has escaped from a channel when its banks were overtopped. Overbank flow inundates adjoining land, often the surface of a floodplain. Overbank flow is a normal occurrence that is expected every few years, because the geometry and capacity of channels is adjusted to conditions existing during the much more common lower flows. Overbank flows are sometimes erosive, but are frequently shallower and, therefore, slower moving than the stream itself. They also exert a drag on the channel flow, as momentum is exchanged between the relatively fast and slower moving flows. Deleterious consequences have sometimes followed from the control of overbank flows with artificial levée banks, since floodplain ecosystems that had evolved in the presence of periodic inundation no longer experienced this, or experienced only much less frequent flooding.   DLD

overdeepening, glacial   Often regarded as a prime characteristic of glacial erosion, overdeepening refers to the long profile of glacial troughs that tend to have a 'down-at-heel' profile with a steep gradient near or at their heads and a gentler slope, sometimes a reverse slope, towards their mouths (Linton, 1963). They are 'overdeepened' only when compared with river-long profiles; indeed, from a glacial viewpoint river valleys can be regarded as underdeepened! One explanation of the 'overdeepened' glacial long profile is that, unlike most river valleys, a glacier discharge is greatest at the midway EQUILIBRIUM LINE and decreases towards the snout. One might expect most erosion at the point of maximum discharge. Overdeepening at a different scale involves the excavation of rock basins by glaciers. This is feasible where rock conditions are favourable because glaciers can flow up a bed slope so long as the overall ice surface gradient is in the contrary direction.   DES

Reference

Linton, D.L. (1963) The forms of glacial erosion. *Transactions of the Institute of British Geographers*, **33**, 1–28.

overflow channel   A channel incised into the landscape by a lake overflow. In the glacial landscape overflows were considered a form of meltwater channel, where water from an englacial lake cut across low cols, through rock or surficial sediment, to release water. By the 1950s they had been identified in large numbers throughout the formerly glaciated areas of Britain, though little consideration had been given to the generation of the hydraulic energy necessary to incise through rock in a short period of time, or the lack of shoreline features associated with them. Subsequent research has established that many overflows –

for example, those associated with the deglaciation of the Canadian prairies – have been cut by sudden and catastrophic subglacial flow. Overflow channels have also been identified in lake basins subject to past changes in level in tropical and arid regions, such as East Africa and the southwest USA, where overflow has taken place at the lowest point on the catchment divide. PSH

overland flow A visible flow of water over the ground surface, however produced. Surface flow may be generated through a number of HILLSLOPE FLOW PROCESSES, including excess of rainfall intensity over infiltration capacity, excess of rainfall amount over soil storage capacity and the seepage of return flow. Where storms are brief or local, flow may run overland for some distance and then infiltrate. Flow velocities are typically of 3–$150\,\mathrm{mm\,s^{-1}}$, and flow depths normally of a few millimetres. Flow rarely consists of a uniform sheet of water, and commonly concentrates in threads of higher velocity. Overland flow is very important for the removal and transport of debris in SOIL EROSION by water. MJK

Reading
Emmett, W.W. (1978) Overland flow. In M. J. Kirkby (ed.), *Hillslope hydrology*. Chichester: John Wiley & Sons, Ltd; pp. 145–176.

overthrust A thrust fault with a fault plane dipping at a low angle and considerable horizontal displacement.

overtopping The process by which coastal barrier crests are built up as swash flows (of insufficient magnitude to reach across the crest and cause overwash) terminate on top of the crest. Sediment carried in overtopping flows is deposited at the swash limit and hence increments the barrier crest height. There should be a relationship between barrier crest height and the return period of storms generating the swash flows. Any increase in storm return period would be associated with a reduction in barrier crest height as increasing wave height would generate flows capable of forming overwash down the back of the barrier, eroding material from the crest in the process. Therefore, overtopping is part of a continuum of cross-barrier flow types, merging into overwash as the magnitude of the flow increases (Orford and Carter, 1982). The balance between overtopping and OVERWASHING controls the instability and migration of gravel barriers, whereas on sand barriers aeolian deposition is a more effective control on barrier crestal elevation than wave-generated overtopping. JO

Reference
Orford, J.D. and Carter, R.W.G. (1982) The structure and origins of recent sandy gravel overtopping and overwashing features at Carnsore Point, southeast Ireland. *Journal of Sedimentary Petrology*, **52**, 265–278.

overwashing The process by which storm-generated swash flows transport beach-face sediment over the top of the beach ridge and deposit it on the back slope. Overwash is a major contributor to the creation and renewal of the back slope of a beach. Overwash is regarded as a dominant process by which sand BARRIER ISLANDS are generated (with aeolian deposition as a co-dominant process). Transport is often constrained sufficiently to cut through the crest via an overwash throat. Backslope deposition often occurs in the form of discrete washover fans, but washover can occur along a broad front parallel to the barrier crest. Note that 'overwash' is the process and 'washover' the sedimentary result. Fan volume generally correlates with the height of the breaking wave (controlled for sediment size). The reflective nature of gravel-dominated barriers tends to increase the likelihood of periodic fans, whereas the dissipative nature of low-angle sand barriers militates against periodic overwash flows. Continual overwashing erodes a barrier's seaward side and extends the landward side. Over time, a barrier appears to be rolling onshore as sediment on the backslope is buried by subsequent overwash and eventually exhumed on the seaward barrier slope as the form of the barrier gradually retreats over its washover foundation. Overwash works towards a reduction of barrier height. The counter process that builds up the barrier crest is known as barrier OVERTOPPING. JO

Reading
Masselink, G., Hughes, M.G. and Knight, J. (2011) *Introduction to coastal processes and geomorphology*, 2nd edition. London: Hodder Education. · Orford, J.D. and Carter, R.W.G. (1982) The structure and origins of recent sandy gravel overtopping and overwashing features at Carnsore Point, southeast Ireland. *Journal of Sedimentary Petrology*, **52**, 265–278. · Orford, J.D. and Carter, R.W.G. (1984) Mechanisms to account for the longshore spacing of overwash throats on a coarse clastic barrier in southeast Ireland. *Marine Geology*, **56**, 207–226.

oxbow A lake, usually curved in plan, occupying a CUT-OFF channel reach that has been abandoned. The term may be applied also to an extremely curved active channel meander with only a narrow neck between adjacent reaches or even to the land within such a reach. The term derives from the U-shaped piece of wood fitted around the neck of a harnessed ox. Lakes of this type may become plugged with sediment where they adjoin the channel and then progressively fill in.   JL

oxidation In general terms the loss of electrons from an atom, but specifically the loss of oxygen from, or addition of hydrogen to, a substance. It is an important weathering mechanism, and the oxidation of iron compounds in rocks and soil can cause reddening to occur.

oxygen isotope Oxygen is the most abundant element on Earth. It occurs in three stable isotopic forms (see also ISOTOPE): ^{16}O (99.759%), ^{17}O (0.0374%), and ^{18}O (0.2039%). It is common to evaluate the ratio of these isotopes, and they are found to exhibit systematic variations that reflect physical and chemical processes. As ^{17}O is the least abundant it is generally not used or described, and the abundance of the more easily measured ^{18}O is usually compared with ^{16}O. Measurements are normally made in a mass spectrometer. This ratio is normally expressed by the $\delta^{18}O$ value (the Greek letter δ stands for 'difference'), which is a measure of how different a measured ratio is from a standard of modern seawater (standard mean ocean water, SMOW) or Pee Dee Belemnite (PDB). These differences are extremely small so are expressed as parts per thousand (per mil) rather than parts per hundred (per cent). Owing to fractionation processes, the ratio of the oxygen isotopes varies. As a result, observing the ratio of oxygen isotopes in a variety of deposits is the most widely employed proxy record of past climates. The principal deposits that have been used in oxygen isotope research are marine shells and ice preserved in high-latitude and high-altitude ice cores.

In the case of oxygen isotopic records preserved in ice, water (H_2O) molecules consisting of the most abundant ^{16}O isotope are slightly lighter than water formed from either ^{17}O or ^{18}O. This lighter water will preferentially undergo evaporation from oceans into the atmosphere and is more likely to remain there during condensation in comparison with water molecules that include the heavier (^{18}O) oxygen isotope. The latter process is controlled by temperature, so that the oxygen isotope ratio of past rainfall or snow events can be used to retrospectively estimate palaeotemperatures.

Similar isotopic fractionation effects occur when calcareous marine organisms incorporate oxygen from seawater to form their calcium carbonate ($CaCO_3$) exoskeletons. The most important group of organisms used for this are the bottom (benthonic) and surface (planktonic)

dwelling microorganisms called *foraminifera*. Measuring the isotopic ratio of the oxygen preserved in fossils was initially thought to provide a direct record of past ocean temperatures (Emiliani, 1955). It was then realized that the fractionation of oxygen isotopes from the seawater into the skeletal structure was also dependent on the isotopic composition of the water (Shackleton, 1967). The composition of the water varied dramatically during the changes in global climate between glacial and interglacial periods. In glacial periods the continents store vast amounts of water in the form of ice that is enriched in ^{16}O. As a result, global sea level falls by around 100 m and the water becomes enriched in ^{18}O. The systematic shift in the oxygen isotopic ratios of shells collected from deep ocean sediments has been used as a means of constructing a named marine $\delta^{18}O$ sequence of stages that is now the most widely used basis to subdivide portions of time during the past 2 million years.   SS

Reading and References
Emiliani, C. (1955) Pleistocene temperatures. *Journal of Geology* **63**, 538–578. · Lowe, J. and Walker, M. (2015) *Reconstructing Quaternary environments*, 3rd edition. Abingdon: Routledge. · Shackleton, N.J. (1967) Oxygen isotope analyses and Pleistocene temperatures re-assessed. *Nature*, **215**, 15–17.

ozone A form of oxygen, but whereas the molecules of ordinary oxygen each contain two atoms, the ozone molecule has three. The ozone layer is a relatively high ozone concentration zone that occurs at a height of 16–18 km in polar latitudes and at about 25 km over the equator. This layer helps to control the temperature gradient of the atmosphere and also, through absorption, controls the amount of ultraviolet (UV-B) radiation reaching the ground.

Ozone is constantly created and destroyed through natural chemical reactions. However, human actions are increasing the concentrations of certain substances that may accelerate the rate of ozone destruction in the stratosphere: oxides of nitrogen, hydrogen, bromide and chlorine. The oxides of nitrogen might be generated by high-flying supersonic aircraft emissions, while the offending chlorine comes from such sources as the chlorofluorocarbons and carbon tetrachloride. In recent years it has become apparent that the concentrations of ozone in the stratosphere have declined, most notably over the Antarctic, where an 'OZONE HOLE' has been detected by a combination of ground monitoring and satellite observations. The reasons for the development of this zone of depletion in the south polar area include the presence of a well-established polar vortex and the extreme cold that generate ice particles that play a role in the crucial chemical reactions. However, stratospheric ozone depletion is increasingly being recognized in other geographical regions. The most obvious cause for concern is that this depletion will reduce the effectiveness of the ozone layer as a filter for incoming UV-B radiation.

At lower levels in the atmosphere, ozone levels may, paradoxically, increase as a result of anthropogenic pollutants. Photochemical reactions can produce ozone. High ozone concentrations can adversely affect human health and damage vegetation.   ASG

Reading

Elsom, D. (1992) *Atmospheric pollution: a global problem*, 2nd edition. Oxford: Blackwell. · Wayne, R.P. (1991) *Chemistry of atmospheres*. Oxford: Clarendon Press.

ozone hole A region in the stratosphere 19–48 km above the Earth's surface has a high concentration of ozone. This layer of ozone absorbs most of the ultraviolet radiation from the Sun, preventing the damaging radiation from reaching the Earth's surface. In 1985, researchers reported finding a hole in the ozone layer occurring during several spring seasons since the late 1970s over Antarctica. The cause of the ozone hole was attributed to CHLOROFLUOROCARBONS (CFCs). Normally, chlorine (Cl) atoms are largely locked into 'safe'

compounds that cannot harm ozone. The stratosphere above Antarctica has one of the highest ozone concentrations. During the polar night (winter) the temperatures drop below $-80\,°C$ ($-112\,°F$). This frigid air allows the formation of polar stratospheric clouds. These clouds activate inert Cl into Cl that can break down ozone by sunlight. This Cl builds during winter and early spring. During late September and October (spring), a belt of stratospheric winds called the polar vortex encircles Antarctica. Once the sunlight reaches the Cl it activates it, destroying the ozone molecules. As the vortex weakens during the summer, ozone-depleted air spills out into the midlatitudes. The same process occurs in the Arctic, but to a lesser extent. The Arctic polar vortex is not as strong and the temperatures are warmer than the Antarctic. Depletion of ozone over the Arctic exceeds 35%. As a consequence of ozone depletion, the 1987 Montreal Protocol established a timetable for diminishing CFC emissions. The treaty was revised in 1989, 1990, 1992 and 1995 and called for a complete phase-out of CFCs and other ozone-destroying chlorine compounds by the year 2000, though full compliance has not been achieved.   JAS

Reading

Jacobson, M.Z. (2012) *Air pollution and global warming: history, science and solutions*, 2nd edition. Cambridge: Cambridge University Press.

P

Pacific decadal oscillation (PDO) A long-lived spatial pattern of sea-surface temperature anomalies in the North Pacific that acts on time scales of 15–25 and 50–70 years (Mantua *et al.*, 1997; Mantua and Hare, 2002). It is defined as the leading principal component of North Pacific monthly sea-surface temperature variability (poleward of 20°N) A similar pattern of variability is the interdecadal Pacific oscillation (IPO). It is the dominant pattern of variability throughout the North and South Pacific, once warming trends have been removed (Dai, 2013). The PDO is its North Pacific component. The spatial patterns of both the PDO and IPO strongly resembled El Niño Southern Oscillation (ENSO) and may represent multidecadal variations of ENSO. The impacts tend to be broader than those of ENSO, which are concentrated in the low latitudes. The IPO has a large influence on precipitation in the west and central USA, but particularly on the southwest (Meehl *et al.*, 2010). SEN

References
Dai, A. (2013) The influence of the inter-decadal Pacific oscillation on US precipitation during 1923–2010. *Climate Dynamics*, 41, 633–646. · Mantua, N.J. and Hare, S.R. (2002) The Pacific decadal oscillation. *Journal of Oceanography*, 58, 35–44. · Mantua, N.J., Hare, S.R., Zhang, Y., *et al.* (1997) A Pacific interdecadal climate oscillation with impacts on salmon production. *Bulletin of the American Meteorological Society*, 78, 1069–1079. · Meehl, G.A., Hu, A.X. and Tebaldi, C. (2010) Decadal prediction in the Pacific region. *Journal of Climate*, 23, 2959–2973.

pacific-type coast A longitudinal coast where folded belts trend parallel to the coastline.

pack ice See SEA ICE.

padang A vegetation type, developed on poor soils in southeast Asia, consisting of grass and shrubs.

palaeobotany The study of ancient or fossil plants and plant communities.

palaeochannel A river or stream channel (or, more generally, river valleys) that no longer conveys discharge and is no longer part of the contemporary fluvial system. Palaeochannels may be preserved as abandoned surface channels on river floodplains or may have been infilled by fluvial or other sediment and are exposed as isolated sediment sections. It is rare that complete channel systems are preserved unless there has either been rapid aggradation in the system or a relatively abrupt cessation of flow. The extent of many larger buried palaeochannel networks – for example, in Australia and northern Africa – have only been identified through the use of remote sensing techniques. Many palaeochannels are abandoned as a result of changes in the fluvial regime, the most common of which is reduction in flow due to climate change. Other causes of abandonment include STREAM CAPTURE, which may divert flow away from a section of the system, and RIVER METAMORPHOSIS, which may lead to the development of a new river system. Data derived from well-preserved and unmodified palaeochannels, such as the channel dimensions, cross-sectional shape, planform, sedimentary infill and terrace materials, can be used to estimate palaeodischarge values and can be compared with contemporary channels (see PALAEOHYDROLOGY). DJN

Reading
Drake, N.A., Blench, R.M., Armitage, S.J., *et al.* (2011) Ancient watercourses and biogeography of the Sahara explain the peopling of the desert. *Proceedings of the National Academy of Sciences of the United States of America*, 108, 458–462. · Maizels, J. (1990) Raised channel systems as indicators of palaeohydrologic change: a case study from Oman. *Palaeogeography, Palaeoclimatology, Palaeoecology*, 76, 241–277.

palaeoclimatology The study of climate prior to the period of instrumental measurements, few of which pre-date the nineteenth century. Indeed, instrumental records only span a tiny fraction (less than 10^{-7}) of the Earth's climatic history and so provide a record that is both an inadequate perspective on climatic variation and a very limited view of the evolution of climates. The foundation of palaeoclimatology is the use of climate-dependent proxy data. The principal sources of proxy data are ice cores, ocean cores, various types of terrestrial sediment (e.g. glacial deposits, periglacial features, loess, relict sand dunes, speleothems, tufas), biological evidence (e.g. tree rings, pollen, plant microfossils, diatoms, insect fossils) and historical records (e.g. writings, paintings, tax returns, phenological records). ASG

Reading
Bradley, R.S. (2013) *Quaternary paleoclimatology: reconstructing climates of the Quaternary*, 3rd edition. Boston, MA: Academic Press.

palaeoecology The term widely used for the research field that involves the study of fossils in order to infer past ecological processes, past biological environments and, hence, past biogeographical patterns. Fossil plant remains commonly studied are pollen and spores, diatoms, phytoliths and the so-called macrofossil-remains of plants, seeds, fruits, leaves and wood (see MACROFOSSILS). Fossil animal remains commonly studied are cladocera, molluscs, coleoptera (beetles) and skeletal fragments. Using the language of biology and geology, palaeoecologists describe the changing fossil composition of the sedimentary sequence. From these descriptions they infer changing biotic compositions, and hence changing environments, and deduce changes in environmental variables such as climate, soil and human influence. Palaeoecology is not restricted to the QUATERNARY, but it is for this period that its use is most widespread, as the good preservation of material increasingly allows quantified approaches to palaeoecological analyses and the application of sophisticated geochemical methods as well as more traditional forms of analysis. RHS

Reading
Birks, H.J.B. and Birks, H.H. (1980) *Quaternary palaeoecology*. London: Edward Arnold (reprinted 2004 by Blackburn Press). · Cappers, R.T.J. and Neef, R. (2012) *Handbook of plant palaeoecology*. Groningen: Barkhuis.

palaeoenvironment A past environment, with the term usually applied to environments beyond the time of historical records or at the scale of 10^4 years or longer. Palaeoenvironments and the occurrence of environmental changes at a range of spatial scales are investigated through the analysis of a diverse range of PALAEOENVIRONMENTAL INDICATORS, with the

timing of palaeoenvironmental changes established through the use of GEOCHRONOLOGY methods. The QUATERNARY period is especially investigated within physical geography and allied disciplines. DSGT

palaeoenvironmental indicators Lines of PROXY evidence for past environmental conditions. Palaeoenvironmental indicators may be derived through the analysis of sedimentary deposits on land or in oceans, including their biological and geochemical content, and/or geoproxies (see GEOPROXY), and include, for example, fossil pollen (see POLLEN ANALYSIS), preserved bone or shell remains and deep ocean cores whose isotopic content may be analysed, and RELIC/RELICT landforms. The diversity of palaeoenvironmental indicators used in the study of QUATERNARY environmental conditions is immense, with their utility enhanced through the application of one of many GEOCHRONOLOGY techniques that are available. Since the availability of palaeoenvironmental indicators is dependent upon their preservation and lack of subsequent disturbance or alteration, the availability and utility may decrease with age, as may the existence of a suitable dating method. The interpretation of palaeoenvironmental indicators may be somewhat imprecise or subject to revision in the light of interpretative and chronometric developments. DSGT

Palaeogene A subdivision of the TERTIARY period of the CAINOZOIC era, encompassing the Palaeocene, Eocene and Oligocene epochs. It is followed by the Neogene, containing the Miocene and Pliocene epochs. PSH

palaeogeography The geography of a former time, especially a specific geological epoch.

palaeohydrology The science of the waters of the Earth, their composition, distribution and movement on ancient landscapes from the occurrence of the first rainfall to the beginning of continuous hydrological records. The advance of palaeohydrology has been possible with greater understanding of contemporary HYDROLOGY, so that morphological evidence, including that from PALAEOCHANNELS, sedimentological evidence including characteristics of PALAEOMAGNETISM and information from organic deposits, especially by palynology and diatom analysis, and knowledge of the mechanisms of the hydrological cycle, can be employed to make quantitative estimates of hydrological conditions in the past. Most emphasis has been placed upon Quaternary palaeohydrology, but it has also been possible to indicate the palaeohydrological features of earlier geological periods. The advance of palaeohydrological investigation depends upon refinement of the relationships between parameters such as precipitation, run-off and temperature and also upon a closer liaison between palaeohydrology and PALAEOCLIMATOLOGY and PALAEOECOLOGY.   KJG

Reading

Benito, G., Baker, V.R. and Gregory, K.J. (1998) *Palaeohydrology and environmental change*. Chichester: John Wiley & Sons, Ltd. · Starkel, L., Gregory, K.J. and Thornes, J.B. (1991) Temperate palaeohydrology: fluvial processes in the temperate zone during the last 15,000 years. Chichester: John Wiley & Sons, Ltd.

palaeomagnetism   The intensity, direction and polarity of the Earth's magnetic field throughout geological time. Palaeomagnetic studies are possible because certain iron-rich rocks containing magnetic minerals become more or less permanently magnetized at the time they are formed. This can occur when a rock cools (thermal remanent magnetism), when an iron mineral is chemically altered to another form (chemical remanent magnetism) or when magnetic particles are deposited in calm water (detrital remanent magnetism). Systematic deviations in the magnetic orientation of rocks of different ages have enabled the previous latitudes of continents to be determined, while periodic polarity reversals of the magnetic field now provide both a record of the creation of new oceanic LITHOSPHERE and a basis for dating suitable deposits.   MAS

Reading

Cox, A. (ed.) (1973) *Plate tectonics and geomagnetic reversals*. San Francisco, CA: W.H. Freeman. · Frisch, W., Meschede, M. and Blakey, R. (2011) *Plate tectonics, continental drift and mountain building*. Berlin: Springer-Verlag. · Kennett, J.P. (ed.) (1980) *Magnetic stratigraphy of sediments*. Stroudsburg, PA: Dowden, Hutchinson and Ross.

palaeosol   A soil of an environment of the past, formed either by burial under later geological materials or because of a change in the climate or topographical conditions of soil formation. They are identifiable by any evidence that indicates the presence of a former land surface that has undergone some form of alteration in response to in-situ surface processes. Palaeosols represent periods of geomorphic stability.   ASG

Reading

Fenwick, I. (1985) Palaeosols – problems of recognition and interpretation. In J. Boardman (ed.), *Soils and Quaternary landscape evolution*. Chichester: John Wiley & Sons, Ltd; pp. 3–21. · Marković, S.B., Hambach, U., Stevens, T., *et al.* (2011) The last million years recorded at the Stari Slankamen (northern Serbia) loess–palaeosol sequence: revised chronostratigraphy and long-term environmental trends. *Quaternary Science Reviews*, **30**, 1142–1154.

Near-horizontal units in the foreground of this stabilized sand dune in the Strezlecki desert, Australia, have been interpreted as palaeosols.
Photograph by David Thomas.

palaeovelocity The velocity of a past flood event that was not directly observed, but for which an estimate is made using various traces of the flow that remain in the channel or the landscape. One method involves determining the slope and roughness of the former channel bed, and from these estimating the palaeovelocity from one of many standard flow equations. Various relations can be used to estimate the roughness from values of D_{50} or D_{84}, grain size percentiles for which the nominated fraction (50% or 84%) of the bed sediment is finer. These measures are used to indicate the size of roughness elements that would have been present on the bed, but they neglect other sources of roughness such as the channel banks, vegetation, and so on. Deriving estimates of D_{84} or other grain size percentiles from field measurement can be very difficult, and many workers have used approximations to D_{84} by measuring only the largest 10–20 stones carried by the former flows. The final piece of information required to estimate the palaeovelocity is the former depth of the flow, and this again must be estimated from limited evidence available in the field.   DLD

Reading
Church, M., Wolcott, J. and Maizels, J. (1990) Palae-ovelocity: a parsimonious proposal. *Earth Surface Processes and Landforms*, 15, 475–480.

pali ridge A sharply pointed ridge between two stream valleys on deeply dissected volcanic domes, especially in Hawaii.

Palmer drought severity index (PDSI) See DROUGHT INDEX

palsa A peat mound associated with the development of an ice lens. Genetically, the feature is similar to a PINGO, but is restricted to peat bogs. DES

Reading
Seppälä, M. (1972) The term 'palsa'. *Zeitschrift für Geomorphologie, Neue Folge*, 16, 463.

paludal sediments Deposits of marshes and swamps formed in areas of low and irregular topography, as along the banks of lakes, river floodplains, deltas, and so on.

paludification The expansion of a bog caused by the gradual rising of the water table as accumulation of peat impedes water drainage.   ASG

palynology See POLLEN ANALYSIS.

pan A closed (i.e. usually with no inlet or outlet) dryland depression that may hold an ephemeral shallow water body or which may have been occupied by a lake under past positive water balance conditions. The term is widely used in southern Africa and is equivalent to PLAYA and deflation basin in other localities, though pans are usually no more that 10^{1}–10^{4} m in diameter. Although aeolian deflation, from a susceptible surface, has often been regarded as the principal mechanism of formation, with the presence of a fringing LUNETTE

Salt-covered pan depressions, in the southwest Kalahari desert, Northern Cape, South Africa.
Photograph by Mike Meadows.

DUNE or CLAY DUNE taken as evidence for this, other factors, notably groundwater fluctuations and the concentration of salts, are now regarded as vital prerequisites for their development. The term *pan* is also applied to very small closed depressions that result from the drinking and wallowing activities of large mammals. DSGT

Reading
Shaw, P.A. and Bryant, R.G. (2011) Pans, playas and salt lakes. In D. S. G. Thomas (ed.), *Arid zone geomorphology: process, form and change in drylands*, 3rd edition. Chichester: John Wiley & Sons, Ltd.

panarchy Metaphor designed to describe systems of ecosystems developed by the ecologist C. S. 'Buzz' Holling. Holling suggests that complex systems are driven through adaptive cycles that exist at a range of spatial scales. The cycles are defined as consisting of four phases: exploitation (the environmental disturbance regime); conservation (the response); collapse (threshold exceedance and unpredictable behaviour of the system); and reorganization (recovery). A geomorphic analogue would be a glacial–interglacial cycle characterized by both orderly evolution and system collapse. The duration of these phases of adaptation in ecological systems depends on three factors: intrinsic wealth, connectivity and resilience of the system. Slaymaker *et al.* (2009) give a geomorphic analogue: the intensity and extent of a glaciation (intrinsic wealth), the connectivity of the erosion–sedimentation cycles (connectivity) and the time required for the recovery of the system during interglacials (resilience). TS

Reading and References
Gunderson, L.H. and Holling, C.S. (eds) (2002) *Panarchy: understanding transformations in human and natural systems*. Washington, DC: Island Press. · Slaymaker, O., Spencer, T. and Dadson, S. (2009) Landscape and landscape-scale processes as the unfilled niche in the global environmental change debate: an introduction. In: O. Slaymaker, T. Spencer, and C. Embleton-Hamann (eds), *Geomorphology and Global Environmental Change*. Cambridge: Cambridge University Press; pp. 1–34. · Walker, B.H., Gunderson, L.H., Kinzig, A.P., *et al.* (2006) A handful of heuristics and some propositions for understanding resilience in social–ecological systems. *Ecology and Society*, **11**, art. no. 13. http://www. ecologyandsociety.org/vol11/iss1/art13/ (accessed 13 July 2015).

panbiogeography See VICARIANCE BIOGEOGRAPHY.

panfan The surface produced when a hill or mountain is completely eroded so that the peripheral fans coalesce, as in the end stage of landscape evolution in an arid region.

Pangaea A postulated continental landmass, existing *c.* 260–180 Ma ago, that comprised all the present continental landmasses. This split to form Gondwana and Laurasia (see SUPERCONTINENT) under the influence of PLATE TECTONICS.

panplain A flat or almost flat landscape that has been produced by lateral erosion by rivers and lowering of divides and interfluves.

pantanal A type of savanna area along the sides of some Brazilian rivers. The land is seasonally flooded by river water but is very dry for most of the year.

parabolic dunes Crescentic sand accumulations in which the horns point away from the direction of dune movement (the opposite of BARCHANS). They may occur as rake-like clusters and can develop from blowouts in transverse ridges. Parabolic dune formation may be linked to contexts where vegetation provides some impedance to sand flow, as in some coastal dune fields and on desert margins. Modelling has enhanced understanding of their development (Nield and Baas, 2008) ASG

Reference
Nield, J.M. and Baas A.C.W. (2008) Investigating parabolic and nebkha dune formation using a cellular automaton modelling approach. *Earth Surface Processes and Landforms*, **33**, 724–740.

paraglacial Unstable conditions caused by a significant relaxation time in processes and geomorphic patterns following glacial climates. The withdrawal of glacier ice exposes landscapes that are in an unstable or metastable state, and consequently liable to modification, erosion and sediment release at rates greatly exceeding background denudation rates. These changes are accompanied by major shifts in ecological communities. Relaxation of landscape elements to nonglacial conditions operates over timescales of 10^1 to $>10^4$ years. The primary reworking of glacigenic sediment declines exponentially through time; alluvial and coastal systems may exhibit complex responses due to the reworking of secondary paraglacial sediment stores. The long relaxation time of paraglacial systems implies that many areas deglaciated in the Late Pleistocene or Early Holocene have still not fully adjusted (in terms of sediment supply) to nonglacial conditions. TS

(a)

(b)

(c)

The panarchy framework. Modified from Gunderson and Holling (2002): (a) the adaptive cycle; (b) components of the adaptive cycle; (c) panarchy in time and space.

Source: Slaymaker *et al.* (2009). Reproduced with permission of Cambridge University Press.

Reading

Ballantyne, C.K. (2002) Paraglacial geomorphology. *Quaternary Science Reviews*, **21**, 1935–2017. · Church, M. and Ryder, J.M. (1972) Paraglacial sedimentation: a consideration of fluvial processes conditioned by glaciation. *Bulletin of the Geological Society of America*, **83**, 3059–3072.

parallel retreat   The phenomenon of denudation of a landscape by lateral erosion of scarp slopes and hills that maintain their slope angle as erosion progresses. It is one of three classic models of slope evolution, the other two being slope downwearing (as in the Davisian cycle) and slope replacement (as in the model by W. Penck). L. C. King explained the great escarpments and inselbergs of Africa through this process.

parameterization   A process undertaken during development of numerical simulation models in which unknown values describing the strength (and structure) of relationships are

assigned to particular model components. This is undertaken in three main ways: (1) direct derivation of parameter values from specially designed laboratory or field experiments; (2) indirect derivation through calibration, where parameter values within a numerical model of a system are changed until optimal agreement with previously measured observations of the real system are obtained (optimization); and (3) sensitivity analysis, where effects of parameter changes upon model response are quantified to define probable parameter values. An example of this is the HORTON OVERLAND FLOW MODEL, which requires parameters that depend upon soil and soil moisture characteristics, and which may be determined directly by field experiments (i.e. 1); indirectly by changing these parameters with respect to measured overland flow (i.e. 2); or quantifying the effects of different parameter combinations upon model predictions (i.e. 3). Model parameterization raises difficult issues (Beven, 1989): (a) parameters may have poor physical meaning or are not readily measured such that direct derivation from experimentation is not possible; (b) if a large number of parameter values need to be specified, there may be many different possible parameter combinations for testing against real-world data; (c) different combinations of parameter values may produce similar results, such that the right results may be obtained for the wrong reasons; and (d) both (b) and (c) assume that high-quality test data are available. If they are not available, or are of poor quality, model parameterization may result in wrong results for the wrong reasons! Thus, modellers now give explicit attention to the effects of parameter uncertainty upon model predictions by undertaking many model simulations with different combinations of parameter values to assign uncertainty bounds to those predictions.   KSR

Reading and Reference
Beven, K.J. (1989) Changing ideas in hydrology: the case of physically-based models. *Journal of Hydrology*, **105**, 157–172. · Howes, S. and Anderson, M.G. (1988) Computer simulation in geomorphology. In M. G. Anderson (ed.), *Modelling geomorphological systems*. Chichester: John Wiley & Sons, Ltd; pp. 421–440.

páramo   Refers to a variety of alpine tundra ecosystem, above the continuous forest line yet below the permanent snowline. Some ecologists describe the páramo broadly as 'all high, tropical, montane vegetation above the continuous timberline', although a narrower definition classifies the páramo according to its regional placement in the northern Andes of South America and adjacent southern Central America. This high mountain biome exhibits a vegetation composed mainly of giant rosette plants, shrubs and grasses.   TS

Reading
Luteyn, J.L. (1999) *Páramos: a checklist of plant diversity, geographical distribution, and botanical literature*. New York: New York Botanical Garden Press.

parasite   An organism that exists in a somewhat imbalanced symbiosis with another in which, although it is able to extract energy (food) from its host for its own preservation, it is unlikely to develop this ability to the point where the host is killed. Some human parasites are, however, exceptions to this general rule; and there are others (e.g. the larvae of some Diptera and Hymenoptera). Since parasites do not require their own energy-production systems, they are often structurally simple; for example, mistletoe, which lives through tapping the phloem tissue of certain deciduous trees, often has no chlorophyll.

Parasites are usually much smaller than their hosts, and they may coexist with them externally or internally.   DW

parna   A word coined by Butler (1956) for aeolian clay deposits found in Australia. Parna deposits occur either as discrete dunes or as thin, discontinuous, widespread sheets. They contain some 'companion sand', in addition to the clay pellets that make up the greater portion of the material. They may be derived from the deflation of material from unvegetated, saline lake floors, or from other soil or alluvial surfaces. Parna is in effect a loessic clay. (See also CLAY DUNE and LUNETTE DUNE.)

Reading and Reference
Butler, B.E. (1956) Parna – an aeolian clay. *Australian Journal of Science*, **18**, 145–51. · Butler, B.E. (1974) A contribution to the better specification of parna and other aeolian clays in Australia. *Zeitschrift für Geomorphologie*, **20**, 106–116.

partial area model   Forecasts the saturated area around streams and channel heads as a basis for flood hydrograph prediction. This type of model is in direct contrast to the HORTON OVERLAND FLOW MODEL. THROUGHFLOW is generally routed downslope to estimate the areas where SATURATION DEFICIT is zero. This DYNAMIC SOURCE AREA is then assumed to generate OVERLAND FLOW, which provides the most rapidly responding part of the stream hydrograph.   MJK

Reading
Kirkby, M.J. (ed.) (1978) *Hillslope hydrology*. Chichester: John Wiley & Sons, Ltd; especially chapters 6–9.

partial duration series   A series of events analysed in flood frequency analysis and consisting of all flood peak discharges above a specified threshold discharge. The series is particularly useful when it is needed to know the frequency with which a particular flood discharge is exceeded. (See also FLOOD FREQUENCY.)   KJG

particle form   Has been used as a composite term to cover several properties of the morphology of sedimentary particles. Thus, in a functional expression:

$$F = f(\mathrm{Sh}, A, R, T, F, \mathrm{Sp})$$

where Sh denotes the shape of the particle, A its angularity, R its roundness, T is the surface texture and Sp is the sphericity. Thus, a complete description of the form would entail observations of these component parts. However, the ease with which this may be done depends upon factors such as the size of the particles to be examined. Particle size itself does not come into the expression, but it is relatively easy to measure axial ratio lengths on gravel size particles, much less so with sands. Both shape and sphericity of particles can be determined by measurement of *a*, *b* and *c* axes. Shape has been defined as 'a measure of the relation between the three axial dimensions of an object'; this is related to, but not the same as, sphericity, 'a measure of the approach of a particle to the shape of a sphere'. Angularity and roundness are at different ends of a continuum and are usually measured by comparing the particle with a 'standard' chart. This method has problems with operator variance, and it is likely that computer-based methods using Fourier analysis will supersede it. Analysis of surface texture has been a much-used technique on its own with the advent of the scanning electron microscopy, but it, too, is largely based on qualitative assessments and comparisons.　　　WBW

Reading

Bull, P.A. (1981) Environmental reconstruction by electron microscopy. *Progress in Physical Geography*, **5**, 368–397. · Clark, M.W. (1981) Quantitative shape analysis: a review. *Mathematical Geology*, **13**, 303–320. · Goudie, A.S. (ed.) (2005) *Geomorphological techniques*, 2nd edition. London: Taylor and Francis. · Whalley, W.B. (1972) The description and measurement of sedimentary particles and the concept of form. *Journal of Sedimentary Petrology*, **42**, 961–965.

particle size

Can refer either to an individual particle or to a mass of particles. It is useful for characterizing or describing one aspect of a particle's morphology, form being another important component. Various names have been given to size ranges (e.g. clay, silt, sand), but these are useful as rough, ordinal-scale descriptors only, and the actual size range covered by each name varies from scale to scale. For particles that can be measured by picking up an individual and measuring lengths with a rule or calipers, three orthogonal axes give a good measure of size. For smaller particles, especially those that can only be seen easily with a microscope or hand lens, it is more usual to measure a single (long) axis. Because the size of particles may range from microscopic aerosols (<1 µm) through clay size (<2 µm), silt, sand and cobbles to even larger blocks, no single measurement technique can be applied. This means that the definition of size varies according to the technique used. This restriction applies even to the normal range of particles seen in a soil.

For instance, if the long *a* axes of cobbles (above 60 mm) are measured with calipers but smaller particles are measured by sieving, the sizes obtained by these methods are not strictly comparable. This difficulty of making measurements on a continuous scale also occurs when sieving becomes difficult and precipitation methods are employed. Sieves give 'nominal diameters' that roughly correspond to the intermediate or *b* axis of a grain. When, as is usual, a sample is taken containing perhaps many thousands of grains, size fractions are determined by sieving through progressively finer sieves. The weights held on each sieve are recorded and expressed as a 'percentage passing' or 'percentage finer' than a given mesh. Such data may be graphed as histograms or as a cumulative size distribution. Although millimetre and micrometre sizes can be recorded, it is often convenient to transform the sizes into the ϕ scale by

$$\phi = -\log_2 d$$

where *d* (mm) is the particle size. This transform takes into account the generally log-normal size distribution of most sediments. The usual statistical treatments of distributions can be applied to the data obtained by sieving, for example. Thus, mean, SKEWNESS, SORTING and KURTOSIS of a size range can be derived by various methods (most often graphically). Such size parameters can be used to compare sediments, to determine their possible origin or to use in some other way (e.g. to suggest their geotechnical behaviour).　　　WBW

Reading and Reference

Briggs, D. (1977) *Sediments*. London: Butterworth. · Goudie, A.S. (ed. 2005) *Geomorphological techniques*, 2nd edition. London: Taylor and Francis. · Pettijohn, F.J., Potter, P.E. and Siever, R. 1972: *Sand and sandstone*. New York: Springer-Verlag. · Syvitski, J.P.M. (ed. 1991) *Principles, methods and applications of particle size analysis*. Cambridge: Cambridge University Press.

passive margin

The margin of a continent formed as a result of the break-up of a large continental mass, or supercontinent. Passive margins are so called since, in comparison with the active continental margins that coincide with zones of plate convergence, they are characterized by low levels of tectonic activity once continental rupture has been completed. The fragmentation of the supercontinent of Pangaea, which began about 180 Ma ago, created the many new passive continental margins that form a substantial proportion of the perimeters of the present-day continents. There are two types of passive margin: rifted margins, formed by divergent plate motion, and sheared margins, where the movement between two adjacent continental blocks has been essentially transform.　　　MAS

Table 1 Major class intervals used in description of sediment sizes

mm	φ	Descriptive size classes	
		Boulder	
256	−8		
64	−6	Cobble	
32	−5		Gravel
16	−4	Pebble	
4	−2		
2.83	−1.5	Granule	
2.00	−1.0		
1.41	−0.5	Very coarse sand	
1.00	0.0		
0.71	−0.5	Coarse sand	
0.50	1.0		
0.35	1.5	Medium sand	Sand
0.25	2.0		
0.177	2.5	Fine sand	
0.125	3.0		
0.088	3.5	Very fine sand	
0.0625	4.0		
0.031	5.0	Coarse silt	
0.0156	6.0		Silt
0.0078	7.0	Fine silt	
0.0039	8.0		
		Clay	Clay

Source: Based on Pettijohn *et al.* (1972: table 3-2, p. 71).

Reading
Frisch, W., Meschede, M. and Blakey, R. (2011) *Plate tectonics, continental drift and mountain building*. Berlin: Springer-Verlag. · Summerfield, M.A. (1991) *Global geomorphology*. London/New York: Longman/John Wiley & Sons, Inc. (republished 2013 by Routledge).

pastoralism System of land use based on the rearing of livestock. Pastoralism may be nomadic, whereby the herders move their animals from place to place, either seasonally or more frequently to reflect grazing opportunities and availability, or sedentary, as in ranch-based systems. Nomadic pastoralism is particularly well suited to areas with highly seasonal and/or low rainfall levels (e.g. in semi-arid areas or in SAVANNA ecosystems) or in areas where marked seasonal temperature regimes make higher altitude areas usable only in summer months. In African savanna areas, nomadic pastoralism has often, and wrongly (Livingstone, 1991; Mace, 1991), been blamed as a major cause of DESERTIFICATION, contributing to its demise and replacement by sedentary ranch systems that are largely unsuited to highly variable environments and which may be susceptible to DROUGHT impacts.   DSGT

References
Livingstone, I. (1991) Livestock management and 'overgrazing' among pastoralists. *Ambio*, **20**, 80–85. · Mace, R. (1991) Overgrazing overstated. *Nature*, **349**, 280–281.

paternoster lake One of a series of lakes occupying basins in a glacial trough. The name is taken from the term for rosary prayer-beads.

patterned ground A general term for the more or less symmetrical forms, such as circles, polygons, nets and stripes, that are characteristic of, but not necessarily confined to, soils subject to intense frost action. Patterned forms, such as MIMA MOUNDS, are also found in dry regions. Description is usually based upon geometric form and the degree of uniformity in grain size (i.e. sorting). The most common type of patterned ground is the non-sorted circle, sometimes termed earth hummock or mud-boil. The processes responsible for the formation of patterned ground are not clear; some forms may be polygenetic and others may be combination products in a continuous system having different processes as end members. In some instances thermal or desiccation cracking is clearly essential. In others, cryoturbation, the lateral and vertical displacement of particles associated with repeated freezing and thawing, combined with site-specific factors such as moisture availability, lithology, and slope appear essential. HMF

Reading
French, H.M. (2007) *The periglacial environment*, 3rd edition. Chichester: John Wiley & Sons, Ltd.

pavement See STONE PAVEMENT.

peak discharge The highest discharge achieved at a stream gauging site within a specific time period, usually during the passage of a storm hydrograph (see HYDROGRAPHS). Peak discharge is a widely used parameter of the flood characteristics of a site and is the usual parameter to be analysed in FLOOD FREQUENCY analysis. (See also DISCHARGE.) AMG

peat A deposit of partially decomposed or undecomposed organic material derived from plants. Peat deposits form where the rate of accumulation of dead plant remains exceeds the rate at which they are decomposed. This imbalance results where microbial decomposition is retarded, rather than because primary productivity is high. Waterlogging is one of the major causes of peat formation, although any factor that reduces the respiratory activity of aerobic microorganisms may promote peat accumulation. Such factors include low pH, mineral and nutrient deficiency and low temperatures.

Peatland is a generic term used to refer to all peat-accumulating wetlands from bogs to fens. According to Maltby *et al.* (1992), the term peatland is not synonymous with MIRE because it includes areas that may no longer carry peat-forming communities but where peat soils occur; for example, agricultural land. Peatlands cover around 3% of global land mass or $(385–410) \times 10^6$ ha. Most peatlands form in high latitudes, although about 20% of the total peatland area is recorded in warmer climates. On a global scale, around 28×10^6 ha of peatland have been exploited for agriculture and forestry (~7% of the total peatland area). The UK Joint Nature Conservation Committee estimates that, prior to disturbance, peatlands covered around 1,640,000 ha in the UK. Around 14% of this total formed in lowland areas. The former peatland area in Ireland is estimated to have covered 1,180,000 ha (Maltby *et al.*, 1992).

The main peat bog types are as follows.

* *Blanket mires*: rain-fed, terrain-covering peatlands, generally 1–3 m deep with a global extent of over 1×10^7 ha. Approximately 13% of the total global blanket mire area is found in the UK. They generally develop in cool climates with small seasonal temperature fluctuations and over 1000 mm of rainfall and over 160 rain days each year.
* *Raised mires*: rain-fed, potentially deep peatlands usually forming in complexes and having a domed microtopography. Primarily recorded in lowland areas across much of northern Europe, including the UK. Also recorded in the former USSR, North America and parts of the southern hemisphere.
* *String mires*: flat or concave peatlands usually showing a string-like pattern of hummocks. Primarily found in the cold northern forest (boreal) zone of Scandinavia, but also recorded in the western former USSR and North America. A few examples exist in northern Britain.
* *Tundra mires*: describe peatlands with a shallow peat layer, ~50 cm thick, dominated by sedges and grasses. They cover around $(11–16) \times 10^6$ ha in Alaska, Canada and the former USSR. This mire type forms in areas where the ground is continuously frozen (permafrost zone).
* *Palsa mires*: describe a peat microtopography with characteristic high peat mound with a permanently frozen core with wet depressions between the mounds. They develop where the ground surface is only frozen for part of the year and are common in the former USSR, Canada and parts of Scandinavia.
* *Peat swamps*: forested peatlands, including both rain- and groundwater-fed types. They are commonly recorded in tropical regions with high rainfall. This peat type covers around 35×10^6 ha, primarily in southeast Asia.

A wider definition of peatlands may include wet heath, marshes and meadows, but these are more strictly WETLANDS because they contain significant proportions of mineral material and are not necessarily accumulating peat. ALH

Reading and Reference
Hughes, J.M.R. and Heathwaite, A.L. (1995) *Hydrology and hydrochemistry of British wetlands*. Chichester: John Wiley & Sons, Ltd. · Maltby, E., Immirzi, C.P. and McLaren, D.P. (1992) *Do not disturb! Peatbogs and the greenhouse effect*. London: Friends of the Earth. · Mitsch, W.J. and Gosselink, J.G. (2011) *Wetlands*, 4th edition. Chichester: John Wiley & Sons, Ltd.

A rock-cut pediment in central Iran.
Photograph by David Thomas.

pedalfer   A well-drained soil that has had most of its soluble minerals leached from it. They occur in more humid situations than PEDOCALS.

pediment   A term applied by G. K. Gilbert (1880) to alluvial fans flanking the mountains near Lake Bonneville in Utah. Since then it has changed its meaning and is now defined (Adams, 1975) as follows: a smooth planoconcave upward erosion surface, typically sloping down from the foot of a highland area and graded to either a local or more general base level. It is an element of a piedmont belt, which may include depositional elements such as fans and playas. The pediment, as defined, excludes such depositional components, although an alluvial cover is frequently present. It is broadly synonymous with the French term *glacis*. Coalescing pediments create pediplains.

Most pediments have low-angle surfaces (generally less than 10°) and the junction between them and the hill mass behind is often marked by an abrupt change of angle. For this to happen, a sharp contrast is needed between the nature of sound rock and the weathering debris produced. In particular, it is necessary for the products of weathering to be relatively fine grained and to have a limited size range, so that they can be transported across the low-angle pediment slope, thereby permitting development or maintenance of an abrupt break of slope.   ASG

Reading and References

Adams, G. (ed.) (1975) *Planation surfaces*. Stroudsburg, PA: Dowden, Hutchinson & Ross. · Gilbert, G.K. (1882) Contributions to the history of Lake Bonneville. In *Second annual report of the United States Geological Survey to the Secretary of the Interior 1880–'81*. Washington, DC: US Government Printing Office; pp. 167–200. · Huggett, R.J. (2003) *Fundamentals of geomorphology*. Abingdon: Routledge.

pedocal   A poorly drained soil that has no soluble salts leached from it. Free calcium occurs in the profile, in the form of concretions, veins, nodules or layers.

pedogenesis   The process of soil formation, leading to the development of soil horizons as a consequence of such factors as additions, removals, mixing, transformations and translocations. Additions come from the WEATHERING of parent material, organic accumulation, and solutes and particles (e.g. dust) derived from the atmosphere. The main removals occur through wind and water erosion and leaching of solutes. Mixing is achieved by soil animals (including worm and termites), microbes and plant roots (BIOTURBATION), freezing and thawing of water, salt heave, and shrinking and swelling of the soil in response to moisture changes. Transformations of soil components can be achieved by chemical and biological processes, with, for example, some minerals precipitating and others dissolving. Translocation of material is associated with the movement of suspended and dissolved substances up and down through the soil profile. The nature and importance of these different processes will depend on the various 'factors of soil formation' (Jenny, 1941), which include climate, parent material, relief, organisms and time.   ASG

Reference

Jenny, H. (1941) *Factors of soil formation: a system of quantitative pedology*. New York: McGraw Hill.

pedology From Greek (πέδον, *pedon*, 'soil'; and λόγος, *logos*, 'study'), meaning the study of soils. It deals with pedogenesis, soil morphology and soil classification (Duchaufour, 1982). The study of fossil or ancient soils is called palaeopedology (Retallack, 2001).   ASG

References

Duchaufour P. (1982) *Pedology*. London: Alllen & Unwin. · Retallack, G.J. (2001) *Soils of the past: an introduction to paleopedology*. Oxford: Blackwell.

pedon A term used in soil science, and defined as the smallest volume that can be called soil. Its lateral area ranges from 1 to $10\,m^2$ and is large enough to permit the study of the nature of any horizon present, for a horizon may be variable in thickness and even discontinuous.

pedosphere The outermost layer of the Earth that is composed of soil and is subject to soil formation processes (PEDOGENESIS). It lies below the vegetative cover of the biosphere and above the hydrosphere and the lithosphere. It lies within the Critical Zone, a broader interfacial zone that includes also vegetation, groundwater aquifers, the regolith (saprolite) and the upper part of the lithosphere.   ASG

peds Soil aggregates; their strength, size and shape give the soil its structure. (See also SOIL STRUCTURE.)

pelagic Refers to that component of an aquatic ecosystem that excludes its margins and substrate. The pelagic environment corresponds to that of the main part of the body of water (Barnes, 1980).

The pelagic environment varies in depth. In deep water (e.g. of oceans), light is able to penetrate only a limited distance below the water surface. The lighted or euphotic zone is the site of abundant organic productivity. Here, phytoplanktonic primary producers are consumed by zooplankton (see PLANKTON) and small fish. Beneath the euphotic zone are the mesopelagic and bathypelagic zones, mainly populated by predatory nekton. These zones benefit from

downward-moving biological detritus and are visited by migratory organisms. In the bathypelagic zone, organisms are relatively infrequent. Numerous species emit their own light in order to obtain their prey (Isaacs, 1977).   RLJ

Reading and References

Barnes, R.S.K. (1980) The unity and diversity of aquatic systems. In R. S. K. Barnes and K. H. Mann (eds), *Fundamentals of aquatic ecosystems*. Oxford: Blackwell; pp. 5–23. · Isaacs, J.D. (1977) The nature of oceanic life. In H. W. Menard (ed.), *Ocean science: readings from Scientific American*. San Francisco, CA: W.H. Freeman; pp. 189–201. · Moss, B. 1980: *Ecology of fresh waters*. Oxford: Blackwell.

Penck and Brückner model Developed during the first decade of the twentieth century, A. Penck and E. Brückner (1909), working primarily in Bavaria, provided a framework for understanding the PLEISTOCENE history of the Alps. Four main glacial phases were initially identified (Günz, Mindel, Riss and Würm) together with various interglacials, including the GREAT INTERGLACIAL. A more complex Pleistocene history has now been determined on the basis of deep sea core studies, but the Penck and Brückner model has been immensely influential and extended uncritically worldwide.   ASG

Reference

Penck, A. and Brückner, E. (1909) *Die Alpen im Eiszeitalter*. Leipzig: C.H. Tauchnitz.

peneplain A term adopted by W. M. Davis, to describe 'an almost featureless plain, showing little sympathy with structure' (Davis, 1899), and considered as 'the penultimate form developed in a cycle of erosion' (Davis, 1902). It therefore invokes a cycle of uplift and lowering by DENUDATION of mountain ranges to form plains of low relief, gentle slope and extensive weathering cover close to *base level*.

The apparent absence of erosional plains, cut across structure, from many regional landscapes led to early criticism that peneplains were abstract forms nowhere observed intact and not seen close to present-day base levels. The central problem for Davis's theory was not his ideas about slope evolution, as often implied by later critics, but the nature and frequency of crustal movements and the time needed to effect peneplanation. Penck (1924) constructed an alternative model, relating planation to uplift rates, but this was widely neglected or misunderstood by anglophone readers, and interest waned in the topic after the 1950s. Modern plate tectonic theory and its links to mantle convection and gravity tectonics, together with a better understanding of the timescale involved in planation (Ahnert, 1970), suggest that extensive peneplains must date from the Mesozoic or before. Many authors use the term 'peneplain' in a generic rather than a genetic sense, to describe landscapes of low relief, without determining formational process or history. Such landscapes on the old Gondwana continents date from before the modern era (Mesozoic–Cainozoic) of continental movement and Alpine tectonics. MFT

Reading and References
Adams, G. (ed.) (1975) *Planation surfaces*. Benchmark Papers in Geology 22. Stroudsberg, PA: Dowden, Hutchinson & Ross. · Ahnert, F. (1970) Functional relationships between denudation relief and uplift in large, mid-latitude drainage basins. *American Journal of Science*, **268**, 243–263. · Davis, W.M. (1899) The peneplain. *American Geologist*, **23**, 207–239. · Davis, W.M. (1902) Base-level, grade and peneplain. *Journal of Geology*, **10**, 77–111. · Huggett, R.J. (2003) *Fundamentals of geomorphology*. Abingdon: Routledge. · Penck, W. (1924) *Die Morphologische Analyse: ein Kapitel der physikalischen Geologie*. Geographische Abhandlungen, vol. 2. Stuttgart: Engelhorn.

penetrometer A device used to measure soil strength. Hand-held penetrometers consist of a steel rod housed within a metal casing with either a line engraved 6 mm from the tip of the rod or a standard-sized metal cone attached to the end. The rod or cone is pressed into the soil until the engraved line or the base of the cone is reached. The rod compresses a calibrated spring and a pointer attached to the rod allows the soil strength in kilogram-force per square centimetre (1 kgf cm^{-2} is equal to 98 kPa) to be measured. Laboratory penetrometers are similar in principle but rely on gravity, with a standard cone positioned at the surface of a soil sample allowed to penetrate into the soil for 5 s. The depth of penetration is measured and converted to a value in kilogram-force per square centimetre. DJN

peninsula A headland or promontory surrounded by water but connected to the mainland by a neck or isthmus.

Penman formula See EVAPORATION.

perched block A boulder or block of rock that is balanced on another rock or outcrop having been deposited there by ice.

Reading
Patterson, E.A. (1984) A mathematic model for perched block formation. *Journal of Glaciology*, **30**, 296–301.

perched groundwater An isolated body of unconfined GROUNDWATER suspended by a discontinuous relatively impervious layer above the main saturated zone and separated from it by unsaturated rock. Groundwater that is perched also has a perched WATER TABLE. PWW

percolation The process of essentially vertical water movement downwards through soil or rock in the unsaturated (or vadose) zone. Percolation water is the water that has passed through the soil or rock by this process. It may be measured by a PERCOLATION GAUGE. PWW

percolation gauge An instrument for measuring the quantity of water that passes vertically through soil or rock by the process of PERCOLATION. Such gauges may be established beneath lysimeters to catch the excess water after losses due to evapotranspiration, and sometimes they are installed in caves to measure the response at percolation input points to rainfall at the surface. PWW

percoline A path or seepage line along which moisture flow becomes concentrated, is particularly well developed where soils are relatively deep, and usually presents a dendritic pattern tributary to surface stream channels. After water from precipitation infiltrates, it may flow throughout the soil profile as matrix flow; but once it becomes more concentrated – but before a definite stream channel is produced – there will be a percoline, which may be indicated by a broad linear depression on the surface, and this may be interrelated with SUFFOSION and the feature at the head of seasonally occupied channels such as DAMBOS in tropical landscapes. KJG

perennial stream A stream that flows all year. A dynamic drainage network also includes INTERMITTENT STREAMS and EPHEMERAL STREAMS, but there should always be flow in a perennial stream channel. For much of the time this flow may be in the form of BASE FLOW or DELAYED FLOW, except when QUICKFLOW occurs after rainstorms. KJG

pericline A dome produced by folding. An ANTICLINE that pitches at both ends.

perigee The point in its orbit at which the Moon is closest to the Earth.

periglacial A term first used by Walery von Lozinski in 1909 to describe frost weathering conditions in the Carpathian Mountains of central Europe. The concept of a 'periglacial zone' subsequently developed referring to the climatic and geomorphic conditions of areas peripheral to Pleistocene ice sheets and glaciers. Theoretically, it was a tundra region extending as far south as the treeline. Modern usage refers to a wide range of cold nonglacial conditions regardless of their proximity to glaciers, either in time or space (Table 2). Periglacial environments exist not only in high latitudes and tundra regions, but also below the treeline and in high-altitude (alpine) regions of temperate latitudes.

Approximately 20% of the Earth's land surface currently experiences periglacial conditions in the form of either intense frost action or the presence of permafrost, or both. There are all gradations between environments in which frost processes dominate, and where a whole or major part of the landscape is the result of such processes, and those in which frost action processes are subservient to others. Complicating factors are that certain lithologies are more prone to frost action than others, and no perfect correlation exists between areas of intense frost action and areas underlain by permafrost.

Unique periglacial processes are the formation of PERMAFROST, the development of thermal contraction cracks, the thawing of permafrost (THERMOKARST) and the formation of wedge and injection ice. Other processes, not necessarily restricted to periglacial regions, are important on account of their high magnitude or frequency. These include ICE SEGREGATION, seasonal frost

Table 2 Classification of periglacial climates

Polar lowlands	Mean temperature of coldest month less than −3 °C. Zone is characterized by ice caps, bare rock surfaces and tundra vegetation
Subpolar lowlands	Mean temperature of coldest month less than −3 °C and of warmest month >10 °C. Taiga type of vegetation. The 10 °C isotherm for warmest month roughly coincides with treeline in the northern hemisphere
Mid-latitude lowlands	Mean temperature of coldest month is less than −3 °C but mean temperature >10 °C for at least 4 months per year
Highlands	Climate influenced by altitude as well as latitude Considerable variability over short distance depending on aspect. Diurnal temperature ranges tend to be large

Source: Based on the classification presented by Washburn (1979: 7–8).

action, frost (i.e. cryogenic) weathering and rapid mass movements.

The most distinctive periglacial landforms are those associated with permafrost. The most widespread are tundra polygons, formed by thermal contraction cracking. They divide the ground surface into polygonal nets 20–30 m in dimension. Ice-cored hills, or pingos, are a less widespread but an equally classic periglacial landform; they form when water moves to the freezing plane under hydraulic or hydrostatic pressure. Other aggradational landforms, such as palsas and peat plateaux, are associated with ice segregation. Ground ice slumps, thaw lakes and irregular hummocky topography with enclosed depressions (thermokarst) result from the melt of ice-rich permafrost.

Many periglacial phenomena form by frost wedging and the cryogenic weathering of exposed bedrock. Frost wedging is associated with the freezing and expansion of water that penetrates joints and bedding planes. The details of cryogenic weathering are poorly understood. Coarse angular rock debris (block-fields), upthrust bedrock blocks, talus (scree) slopes and certain types of patterned ground are usually attributed to frost action. Angular bedrock masses (tors) may stand out above the debris-covered surfaces, reflecting more resistant bedrock. Flat erosional surfaces, termed cryoplanation terraces, are sometimes associated with tors but, equally, can occur quite independently.

The overall flattening of landscape and smoothing of slopes, thought typical of many periglacial regions, is generally attributed to mass wasting. Agents of transport include frost creep and solifluction. NIVATION, a combination of frost wedging, solifluction and sheetwash operating beneath and downslope of snowbanks, is often regarded as important. In areas dominated by extreme nival regimes and underlain by unconsolidated sediments, fluvial activity can be a significant landscape modifier.   HMF

Reading and Reference
French, H.M. (2013) *The periglacial environment*. Chichester: John Wiley & Sons, Ltd. · Von Lozinski, W. (1909) Uber die mechanische Verwitterung der Sandsteine im gemässigten Klima. *Bulletin International de l'Academie des Sciences et des Lettres de Cracovie, Classe des Sciences Mathématiques et Naturelles*, **1**, 1–25. · Washburn, A.L. (1979) *Geocryology*. London: Edward Arnold.

perihelion   The point in its orbit about the Sun that a planet or comet is closest to the Sun.

permafrost   The thermal condition in soil and rock where temperatures below 0 °C persist over at least two consecutive winters and the intervening summer. Permafrost is defined purely as a thermal condition: moisture, in the form of water and/or ice may or may not be present. The term was first introduced by S. W. Muller in 1947, as a shortened form of permanently frozen ground. Permafrost is not necessarily synonymous with 'frozen ground', however, since earth materials may be below 0 °C in temperature but essentially unfrozen on account of depressed freezing points due to mineralized groundwaters or other causes. One solution is to differentiate between 'cryotic' (i.e. below 0 °C) and 'non-cryotic' (i.e. above 0 °C) ground and to subdivide the former into 'unfrozen', 'partially frozen' and 'frozen' depending upon the amount of unfrozen water present. Equally, permafrost is not 'permanently frozen ground', since changes in climate and terrain may cause it to degrade.

The upper boundary of permafrost is known as the permafrost table, and the near-surface layer, which is subject to seasonal thaw, is called the active layer (see figure). The depth at which annual temperature fluctuations are minimized is termed the depth of zero annual amplitude; this usually varies between 10 and 20 m depending upon climate and terrain factors such as amplitude of annual surface temperature variation, snow cover and effective thermal diffusivity of the soil and rock. In polar regions, permafrost temperatures may be as low as −15 °C at the depth of zero annual amplitude. Permafrost underlies approximately 20–25% of the Earth's land surface. It is widespread in Siberia, northern Canada, Alaska and China. Zones of continuous, discontinuous and sporadic (isolated) permafrost are generally recognized, together with alpine, intermontane and subsea (offshore) permafrost. Thicknesses range from a few metres at the southern limits to approximately 500 m in parts of northern Canada and Siberia, where it may be relic. HMF

Reading and References

Cooke, R.U. and Doornkamp, J.C. (1974) *Geomorphology in environmental management*. Oxford: Clarendon Press. · French, H.M. (2013) *The periglacial environment*. Chichester: John Wiley & Sons, Ltd. · Muller, S.W. (1947) *Permafrost or permanently frozen ground and related engineering problems*. Ann Arbor, MI: J.W. Edwards. · Walker, H.J. (1983) *Mega-geomorphology*. Oxford: Clarendon Press.

permeability

permeability A measure of the capacity of a rock or soil to transmit fluids. It is often termed INTRINSIC PERMEABILITY or specific permeability to distinguish it from HYDRAULIC CONDUCTIVITY, which is sometimes (but now less frequently) called the coefficient of permeability.

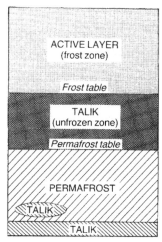

Terminology of some features associated with permafrost. *Source*: Cooke and Doornkamp (1974: figure 9.1). Reproduced with permission of Oxford University Press.

Permeability (or PERVIOUSNESS) depends upon the physical characteristics of the rock, whereas hydraulic conductivity also takes account of the physical characteristics of the fluid. PWW

Reading

Fitts, C.R. (2013) *Groundwater science*, 2nd edition. Waltham, MA: Academic Press.

permeameter An instrument for measuring the saturated hydraulic conductivity K of soils or other sediments. Two types of apparatus are in common use. One is a constant-head permeameter, the other a falling-head permeameter. In the constant-head case:

$$K = \frac{QL}{Ah}$$

and in the falling-head case:

$$K = \frac{aL}{aT} \ln \frac{h_0}{h_1}$$

Contant-head permeameters can be operated under negative pressure heads, by allowing air entry into the permeameter only via a bubble tube where a nominated head of water can be set. In this way, the part of the water entry into a soil that occurs via macropores can be eliminated, and so its value estimated. PPW/DLD

Reading

Clothier B.E. and White, I. (1981) Measurement of sorptivity and soil water diffusivity in the field. *Soil Science Society of America Journal*, 45, 241–245. · Fitts, C.R. (2013) *Groundwater Science*, 2nd edition. Waltham, MA: Academic Press. · Freeze, R.A. and Cherry, J.A. (1979)

Groundwater. Englewood Cliffs, NJ: Prentice-Hall. · Simunek. J, and van Genuchten M.T. (1996) Estimating unsaturated soil hydraulic properties from tension disc infiltrometer data by numerical inversion. *Water Resources Research*, **32**, 2683–2696.

persistence

The continued existence of an ecosystem without significant fluctuations in overall species composition or in the relative number of individuals in different species. Persistence represents one concept of STABILITY and has also been referred to as constancy or 'no-oscillation' stability. This aspect of stability does not take into account the ability of a system to adjust to disturbance.

The linkage of the term *persistence* to the notion of constancy of ecosystem species populations is unfortunate. Although major fluctuations in population size are thought to increase the risk of extinction during episodes of low population, there is considerable evidence that oscillations of population numbers may enhance the survival of many species by producing episodes of saturation and scarcity with respect to predation.

This term also has a meteorological usage, referring to the continued existence of a weather or climate anomaly. 'Persistence forecasting' is a forecast based on the assumption that a certain anomaly pattern will continue. It is used as a way to assess the performance of other methods of weather forecasting.   ARH/SEN

Reading
Dunbar, M.J. (1973) Stability and fragility in Arctic ecosystems. *Arctic*, **26**, 179–185. · Robinson, C.L.K. and Yakimishyn, J. (2013) The persistence and stability of fish assemblages within eelgrass meadows (*Zostera marina*) on the Pacific coast of Canada. *Canadian Journal of Fisheries and Aquatic Sciences*, **70**, 775–784.

perturbation

Variation, usually of small amplitude, about some well-defined, usually uninteresting, basic state. It is the conceptual equivalent of gently shaking an unknown package before embarking on more irreversible probing. Perturbation analysis allows all possible states similar to the initial one to be followed, at least for a short time. In contrast, numerical integration (see NUMERICAL MODELLING) allows a few possible states to be followed for a longer time. The term is also used to denote various types of weather disturbance, such as the waves of the trades and equatorial easterlies.   JSAG

perviousness

A property of soils and rocks indicating its capacity for transmitting fluids. It is equivalent to PERMEABILITY. Impervious rocks behave hydrologically like an AQUIFUGE or aquiclude.   PWW

pesticides

Poisonous chemicals used by humans to regulate or eliminate plant and animal pests. The term biocide is sometimes used in the same context, but, as Ware (1983) points out, is without a precise scientific meaning and is often used in a rather emotive manner to describe the lethal effects of chemicals on living organisms.

Pesticides are mainly synthetic organic chemicals that possess varying levels of toxicity. They may be selective or nonselective and are aimed at 'target' organisms that are economically or socially undesirable (Cremlyn, 1978).

Among elementary forms of life, viruses and bacteria are responsible for numerous plant and animal diseases. Of lower plants, algae that bloom in bodies of water and fungi parasitic on living tissue are pests. Fungi are also the cause of many diseases of flora and fauna. The 30,000 or so higher plants that are pests because they shade crops and/or diminish their supply of moisture and nutrients are termed *weeds*. Animal pests include some Nematoda (roundworms), Arthropoda (insects), Arachnida (spiders, mites, ticks), Mollusca (slugs and snails), birds and rodents.

There are three main categories of pesticide – herbicides, insecticides and fungicides – in the production of which four major groups of chemicals are employed. Herbicides may operate on either a systemic (penetrative) or nonsystemic (contact) basis. Phenoxyaliphatic acids (e.g. 2,4-D and 2,4,5-T) are important herbicides, the collapse, wilting and death of target weeds under their application being due to amended phosphate and nucleic acid metabolism and cell division.

Among insecticides, the best known, DDT, is an organochlorine (or chlorinated hydrocarbon) that causes death by upsetting sodium and potassium equilibrium in the nervous system, thereby altering its normal functions. Organophosphates (e.g. parathion) have largely supplanted organochlorines as insecticides. Additionally, these are very toxic to vertebrates as they also inhibit normal functioning of the nervous system.

Fungicides need to kill the parasite but not its host. Their usual function is to halt the germination of fungal spores. Inorganic compounds of sulphur, copper and mercury are long-established fungicides, but there is an increasing number of organic fungicides such as dithocarbamates. After metabolism these influence amino acid activity in disease cells and cause their elimination.

During the first millennium BC, sulphur was used to fumigate Greek houses. Around 150 years ago, sulphur had been joined by arsenic

399

and phosphorus as pesticides. The impetus for much pesticide use came during the Second World War. Some pesticides (e.g. DDT) were then used in the control of human disease carriers such as mosquitoes. At this time, research into nerve gas led to the discovery of the insecticidal property of organophosphates. Subsequent use of pesticides was often enthusiastic and without regard to their possible effects upon nontarget organisms and their environment (see BIOLOGICAL MAGNIFICATION).

Pests can also become resistant to pesticides. This occurs because a pesticide disrupts a single genetically controlled process in the metabolism of the pest (Ware, 1983). This resistance is often a sudden occurrence and takes place either by the natural selection of hardy individuals, which form the basis of subsequent populations, or by mutation to form resistant genotypes.　RLJ

Reading and References
Cremlyn, R. (1978) *Pesticides*. Chichester: John Wiley & Sons, Ltd. · Rathore, H.S. and Nollet, L.M.L. (2012) *Pesticides: evaluation of environmental pollution*. Boca Raton, FL: CRC Press. · Ware, G. (1983) *Pesticides: theory and application*, 2nd edition. New York: W.H. Freeman.

pF　Logarithm to base 10 of the negative head of water (in centimetres) indicating the strength of soil moisture SUCTION at a site. Soil moisture suction is often termed the matric or capillary potential, and it results from the attraction of solid surfaces (the soil matrix) for water and also of the water molecules for each other, so that water is held in the soil against gravity by adsorptive and CAPILLARY FORCES. These forces lower the potential energy of the soil water, and the degree of this reduction in potential energy in comparison with free water at the same elevation and under the same air pressure may be indicated by a negative head of water. The negative head is measured by a TENSIOMETER, and because the negative head may increase rapidly with quite small changes in soil moisture content, it is usually expressed in relation to a $\log_{10}$ base as a pF value. The graph of pF plotted against soil moisture content (by volume) for a soil sample that has been progressively drained of water is called the soil moisture characteristic curve or pF curve.　AMG

Reading
Schofield, R.K. (1935) The pF of water in the soil. In *Transactions of the third international congress on soil science. Volume II. Plenary session papers and presidential address*. London: Thomas Murby; pp. 37–48.

p-form　Smoothed, apparently plastically sculptured forms caused by the action of glaciers. They comprise *grooves* that may be slightly sinuous with soft flowing outlines and sometimes display an overhanging lip, crescentic-shaped depressions or SICHELWANNEN 1–10 m in length and 5–6 m wide, with horns pointing down glacier, and POTHOLES and a variety of shallower forms, ranging in size from depressions a few centimetres across to giant potholes 15–20 m deep and 16 m in diameter. Good descriptions are given by Dahl (1965).

Although Gjessing (1965) argued in favour of erosion by saturated till, it may be that the grooved features can be explained by normal glacial abrasion and the remaining forms by high-velocity subglacial meltwater action.　DES

References
Dahl, R. (1965) Plastically sculptured detail forms on rock surfaces in northern Nordland, Norway. *Geografisker Annaler, Series A: Physical Geography*, 47, 83–140. · Gjessing, J. (1965) One 'plastic scouring' and 'subglacial erosion'. *Norsk Geografisk Tiddskrift*, 20, 1–37.

pH　The measure of the acidity or alkalinity of a substance measured as the number of hydrogen ions present in one litre of the substance and reported as a figure on a scale the centre of which is 7, representing neutrality. Acid substances have a pH of less than 7 and alkaline substances have a higher pH.

phacolith　A lens-shaped igneous intrusion usually situated beneath an anticlinal fold or in the base of a syncline.

phenology　The scientific study of the timing of recurring natural phenomena in the life cycle of plants and animals in nature. While all natural phenomena may be included (e.g. the timing of animal migrations, flowering and fruiting of plants, harvest, ripening and seed-time), phenological observations are often restricted to the time at which certain plants come into leaf and flower each year, and to the date of the first and last appearance of birds and animals. Phenological observations have been used as a source of proxy climate change, in order to describe temperature patterns during pre-instrumental times, based on long-term observations of the growth and maturity of cultivated plants; for example, wine harvest records (Pfister, 1992).　AP

Reading and Reference
Pfister, C. (1992) Monthly temperature and precipitation in central Europe 1525–1979: quantifying documentary evidence on weather and its effects. In R. S. Bradley and P. D. Jones (eds), *Climate since AD 1500*. London: Routledge; pp. 118–143. · Schwartz, M. (ed.) (2013) *Phenology: an integrative environmental science*, 2nd edition. Dordrecht: Springer.

phoresy The transport by one animal of another of a different species to a new feeding site.

phosphate rock An indurated sedimentary deposit that is rich in apatite, a calcium phosphate. Much of it results from the interaction of guano (seabird excrement) and the calcium carbonate of reef sands and elevated limestones.   ASG

Reading
Stoddart, D. and Scoffin, T. (1983) Phosphate rock on coral reef islands. In A. S. Goudie and K. Pye (eds), *Chemical sediments and geomorphology.* London: Academic Press; chapter 12.

photic zone The surface zone of a lake, sea or ocean above the maximum depth to which sunlight penetrates.

photochemical smog See SMOG.

photogrammetry The science and technology of obtaining reliable measurements by means of photography. The measurement of aerial photographs to provide details of area, distance and height are skills readily practised by physical geographers. The task of the photogrammetrist is usually considered to be the measurement of aerial photographs both with sufficient accuracy in two dimensions to enable features to be plotted with respect to a national or international grid coordinate system, and with sufficient accuracy in three dimensions to enable them to be located in relation to their height above sea level.   PJC

Reading
Egels, Y. and Kasser, M. (2002) *Digital photogrammetry.* London: Taylor and Francis.

photosynthesis A chemical process that takes place in the cellular structures of green plants, blue–green algae, phytoplankton and certain other organisms and which transforms received solar energy into chemically stored foodstuffs through the conversion of carbon dioxide and water into carbohydrates with the simultaneous release of oxygen, as indicated in the following simplified formula:

$$6CO_2 + 6H_2O \rightarrow C_6H_{12}O_6 + 6O_2 + 673\,\text{kcal}$$

The carbohydrate in this case is glucose. The radiant energy reacts initially with the green pigment chlorophyll, with the result that one electron of this complex molecule is raised above its normal energy level for some 10^{-7} to 10^{-8} s, so triggering other chemical changes (Hutchinson, 1970). The 673 kcal represents a fairly large store of energy, which is then used for plant growth and respiration. Photosynthesis is solely a daytime phenomenon. It is thus the basis of the flow of energy through the TROPHIC LEVELS of ECOSYSTEMS and, indeed, the fundamental process that enables life itself to occur on this planet, with the exception of a few species (mainly bacteria) that obtain their energy from chemicals.   DW

Reading and Reference
Hutchinson, G.E. (1970) The biosphere. *Scientific American,* **223**, 44–53. · Raghavendra, A.S. (1998) *Photosynthesis: a comprehensive treatise.* Cambridge: Cambridge University Press.

photosynthetic pathway At its basic level, PHOTOSYNTHESIS is the biochemical bonding of carbon dioxide (CO_2) with water to make carbohydrates (sugar) using the Sun's energy. However, there are three pathways by which plants conduct this, depending on their metabolisms: C3, C4, and CAM photosynthetic pathways. C3 photosynthesis is the most common form among plant species and the only mechanism that was originally thought to exist. C4 and CAM photosyntheses are adaptations to arid conditions because they result in better water use efficiency.

C3 photosynthesis is so called because the CO_2 is first incorporated into a three-carbon compound. Its adaptive value is that it is more efficient than the other forms under cool and moist conditions and under normal light because it requires fewer enzymes and no specialized anatomy. Most plants have the C3 photosynthetic pathway. C4 photosynthesis is so called because the CO_2 is first incorporated into a four-carbon compound. Its adaptive value lies in the fact that it occurs faster under high light intensity and at higher temperatures because the CO_2 is delivered in such a way as to reduce photorespiration. It is associated with greater water use efficiency because of the faster rate of conversion so that leaf stomata do not have to stay open for as long, thus reducing water loss through transpiration. Common C4 plants include the tropical grasses and salt marsh plants. CAM stands for Crassulacean acid metabolism and is named after the plant family in which it was first described (Crassulaceae) and because the CO_2 is stored in the form of an acid before use in photosynthesis. In this case the leaf stomata are open at night (when evaporation rates are usually lower) and may remain closed during the day. The CO_2 absorbed at night is converted to malic acid and then released for photosynthesis the following day. This is an especially useful adaptation in hot and arid environments.

CAM plants include many succulents, and also some orchids and bromeliads.   SEN

Reference
Morton, O. (2009) *Eating the Sun: the everyday miracle of how plants power the planet*. London: Harper Perennial.

phreatic   GROUNDWATER may be divided into zones of aeration (interstices partially occupied by water and partially by air) and saturation (all interstices filled with water under hydrostatic pressure). Usually, a single zone of aeration overlies a single zone of saturation and extends up to the ground surface. Overlying impermeable strata usually bound the upper saturated zone, which extends down to underlying impermeable strata such as clay beds or bedrock. Where overlying impermeable strata are absent, the upper surface of the zone of saturation is the water table or phreatic surface, and is defined as the surface of atmospheric pressure. It would be revealed by the level at which water stands in a well penetrating the aquifer. Phreatic is derived from the Greek *phrear*, *-atos*, meaning a well.   ALH

phreatic divide   An underground watershed. The WATER TABLE or upper surface of the permanently saturated zone usually has an undulating topography that depends upon variations in HYDRAULIC CONDUCTIVITY in the bedrock and on the location of zones of RECHARGE and discharge. Water flows down the hydraulic gradient (see DARCY'S LAW) away from high points on the water table, which are therefore phreatic divides; that is, the watersheds of GROUNDWATER basins. But the pattern of groundwater divides need not mirror the surface pattern of topographic watersheds. Unlike the watersheds of surface catchment areas, which are permanent, phreatic divides can shift as the water table rises, lowers or changes its configuration because of localized recharge. (See also GROUNDWATER.)   PWW

phreatophytes   Plants that have developed root systems with the capability to penetrate great depths in order to draw water directly from groundwater reservoirs. The name is derived from Greek meaning 'well plant'. In arid and semi-arid regions they make up almost all the plants in riparian habitats, where they affect the geomorphological processes of the nearby stream by causing increased hydraulic roughness and concomitant sedimentation. Phreatophytes have important economic impacts on the hydrological cycle because they transpire moisture from the groundwater table that might otherwise be used for pumped irrigation water (Horton and Campbell, 1974).   WLG

Reference
Horton, J.S. and Campbell, C.J. (1974) *Management of phreatophyte and riparian vegetation for maximum multiple use values*. US Department of Agriculture Forest Service research paper RM-117. Ft Collins, CO: Rocky Mountain Forest and Range Experiment Station, Forest Service, US Department of Agriculture.

phylogenesis   The emergence of new taxa through the splitting of evolutionary lineages; the word is derived from the Greek *phylon* (tribe) and *genesis* (origin). Phylogeny is an approach to the classification of organisms based on the evolutionary history of a species or group of related species and describes, in effect, its 'family history'. The phylogeny is usually depicted in the form of a dendrogram, known as a phylogenetic tree, which illustrates the evolutionary relationship. Phylogenesis describes the process giving rise to the characteristic branching pattern of the evolutionary tree. The taxonomic group in question is said to be monophyletic if a single ancestor species gave rise to all the species placed within that taxon.   MEM

physical geography   Mary Somerville's *Physical geography* (first published in 1848) was one of the first and most influential of textbooks in physical geography and gave a clear definition of the field (Somerville, 1858):

Physical geography is a description of the earth, the sea, and the air, with their inhabitants animal and vegetable, of the distribution of these organized beings and the causes of that distribution . . . man himself is viewed but as a fellow-inhabitant of the globe with other created things, yet influencing them to a certain extent by his actions and influenced in return. The effects of his intellectual superiority on the inferior animals, and even on his own condition, by the subjection of some of the most powerful agents in nature to his will, together with the other causes which have had the greatest influence on his physical and moral state, are among the most important subjects of this science.

Somerville's view of physical geography had certain similarities with that of Arnold Guyot (1850), who thought that physical geography should be more than 'mere description'. 'It should not only describe, it should compare, it should interpret, it should rise to the *how* and the *wherefore* of the phenomena which it describes'. Guyot regarded physical geography as 'the science of the general phenomena of the present life of the globe, in reference to their connection and their mutual dependence'. As part of his concern with 'connection' and 'mutual dependence' he included a consideration of racial differences and performance in relation to environmental controls, a theme that was to be developed by the North American school of environmental determinists over the next six to seven decades.

Thus, both Guyot and Somerville saw a human dimension in physical geography and believed that it was more than an incoherent catalogue. This was a view shared by Thomas Huxley, who used the term 'physiography' in preference to 'physical geography'. Similarly, in France, Emmanuel de Martonne (1909), in addition to covering what he regarded as the four main components of physical geography – climatology, hydrography, geomorphology and biogeography – also included a consideration of environmental influences.

In the twentieth century, physical geography was dominated for much of the time by geomorphology, and as Stoddart (1987) remarked about the British situation during the inter-war years:

In research, if not teaching, 'physical geography' meant geomorphology: for while some attention was given to meteorology, climatology and to some extent pedology and biogeography, it was on the level of elementary service courses for students rather than as a contribution to new knowledge.

In contrast to the works of Guyot, Somerville and de Martonne, physical geography tomes tended increasingly to ignore human and environmental influences. So, for example, Pierre Birot (1966) saw physical geography as the study of 'the visible surface of natural landscapes as they would appear to the naked eye of an observer travelling over the globe before the interaction of mankind'.

Physical geography possibly reached its nadir within geography in the late 1960s and early 1970s, when spatial modellers, particularly of urban systems, saw little room for it in the discipline, and even some physical geographers doubted the role of the subdiscipline in a world where regional differences were seen to be declining and where many people were thought to be becoming progressively divorced from the reality of their immediate physical surroundings (Chorley, 1971).

In recent decades physical geography has become more concerned with the integration of its various elements (Goudie, 1994) and has re-established its concern with human issues (Gregory, 1985). Physical geography texts now tend to include the word 'environment' in their titles. The human impact (Goudie, 2000) has

become a major concern, and many of the great issues facing the world today have proved susceptible to geographical treatment, including acid deposition (Battarbee et al., 1988), forest decline (Innes, 1992), desertification (Thomas and Middleton, 1994), salinity problems (Goudie and Viles, 1997) and climate change (Williams et al., 1998).

Another major theme of modern physical geography is natural environmental change. It is remarkable, but partly uncoincidental, that at the same time that scientists have been concerned with anthropogenic impacts on the environment they have also become increasingly aware of the frequency, magnitude and consequences of natural environmental changes at a whole range of timescales from relatively short-lived events like El Niño Southern Oscillation phenomena, to events at the decade and century scales (e.g. Grove (1988) on the Little Ice Age), to the major fluctuations of the HOLOCENE (Roberts, 1998) and the Younger Dryas (Anderson, 1997), to the cyclic events of the PLEISTOCENE and to the longer term causes of the Cainozoic climate decline. Much of the reason for this concern arises from the development in the last four decades of new technologies for dating and environmental reconstruction, including the coring of ocean floors, lakes and ice sheets (Lowe and Walker, 1997).

There has also been an increasing concern with the application of physical geography to societal needs (Jones, 1980; Cooke and Doornkamp, 1992). Likewise, environmental management has become a major field in many branches of physical geography (O'Riordan, 1995), including management of water resources (Beaumont, 1988), water pollution (Burt et al., 1993) and coasts (Viles and Spencer, 1995). In recent years, physical geographers have made many contributions to the study of hazards (e.g. Smith, 1992; Jones, 1993) and disasters (Alexander, 1993), and in some cases this has also brought them to consider societal issues (e.g. see Chester (1993)) at the same time as they consider geomorphological, hydrological or climatic events.

Perhaps a major reason for a tendency towards integration of the elements of physical geography has been the resurgence of biogeography, which for too long, with climatology, was one of the less vibrant parts of the discipline. This has been seen in broad-ranging texts that attempt to give an integrated view of landscape types as diverse as ocean islands (Nunn, 1994), caves (Gillieson, 1996) and rain forests (Milington et al., 1995),

403

and also of whole continents (e.g. Adams *et al.*, 1996). Exciting new developments are taking place in our understanding of environments like savannas through an increasing concern with forces such as fire (natural and human), herbivores, soil nutrient status and soil hydrology, and a long history of human land-use practices (e.g. see Fairhead and Leach (1997)).

Building upon some of these tendencies, Slaymaker and Spencer (1998: 7) have sought to define physical geography so that it is redirected from an emphasis on 'the pot-pourri of information about the Earth and its atmosphere to a coherent integrating theme of global environmental change'. They believe that, to achieve that, physical geography should be redefined as that branch of geography concerned with:

1 identifying, describing and analysing the distribution of biogeochemical elements in the environment;
2 interpreting environmental systems at all scales, both spatial and temporal, at the interface between atmosphere, biosphere, hydrosphere, lithosphere and society; and
3 determining the resilience of such systems in response to perturbations, including human activities.

They also argue that a commitment to the understanding of human–environmental linkages is crucial to the sustainability of our planet and that this should be the mandate for physical geography in the twenty-first century. ASG

Reading and References

Adams, W., Goudie, A.S. and Orme, A.R. (1996) *The physical geography of Africa*. Oxford: Oxford University Press. · Alexander, D. (1993) *Natural disasters*. London: UCL Press. · Anderson, D. (1997) Younger Dryas research and its implications for understanding abrupt climatic change. *Progress in Physical Geography*, **21**, 230–249. · Anderson, M.G. and Brooks, S.M. (eds) (1996) *Advances in hillslope processes*. Chichester: John Wiley & Sons, Ltd. · Battarbee, R., Anderson, N.J., Appleby, P.G., *et al.* (1988) *Lake acidification in the United Kingdom 1800–1986*. London: Ensis. · Beaumont, P. (1988) *Environmental management and development in drylands*. London: Routledge. · Birot, P. (1966) *General physical geography*. London: Harrap. · Burt, T.P., Heathwaite, A.L. and Trudgill S.T. (eds) (1993) *Nitrates: processes, patterns and control*. Chichester: John Wiley & Sons, Ltd. · Chester, D. (1993) *Volcanoes and society*. London: Arnold. · Chorley, R.J. (1971) The role and relations of physical geography. *Progress in Geography*, **3**, 87–109. · Cooke, R.U. and Doornkamp, J.C. (1992) *Geomorphology in environmental management*, 2nd edition. Oxford: Oxford University Press. · De Martonne, E. (1909) *Géographie physique*. Paris: Colin. · Fairhead, J. and Leach, M. (1997) *Misreading the African landscape*. Cambridge: Cambridge University Press. · Gillieson, D. (1996) *Caves*. Oxford: Blackwell. · Goudie, A.S. (1994) The nature of physical geography: a view from the drylands. *Geography*, **79**, 194–209. · Goudie, A.S. (2000) *The human impact on the natural environment*, 5th edition. Oxford: Blackwell. · Goudie, A.S. and Viles, H. (1997) *Salt weathering hazards*. Chichester: John Wiley & Sons, Ltd. · Gregory, K.J. (1985) *The nature of physical geography*. London: Arnold. · Grove, J.M. (1988) *The Little Ice Age*. London: Routledge. · Guyot, A. (1850) *The Earth and man: lectures on comparative physical geography in its relation to the history of mankind*. New York: Scribners. · Huggett, R.J. (2009) *Physical geography: the key concepts*. London: Routledge. · Innes, J. (1992) Forest decline. *Progress in Physical Geography*, **16**, 1–64. · Jones, D.K.C. (1980) British applied geomorphology: an appraisal. In H. Hagedorn, M. Thomas and D. Busche (eds), *Perspectives in geomorphology*. Zeitschrift für Geomorphologie, Supplementbände, vol. **36**. Berlin: Borntraeger; pp. 48–73. · Jones, D.K.C. (ed.) (1993) Environmental hazards: the challenge of change. *Geography*, **78**, 161–198. · Lowe, J. and Walker, M. (1997) *Reconstructing Quaternary environments*, 2nd edition. Harlow: Longman. · Milington, A.C., Thompson, R.D. and Reading, A.J. (1995) *Humid tropical environments*. Oxford: Blackwell. · Nunn, P.D. (1994) *Oceanic islands*. Oxford: Blackwell. · O'Riordan, T. (ed.) (1995) *Environmental science for environmental management*. Harlow: Longman. · Roberts, N. (1998) *The Holocene*, 2nd edition. Oxford: Blackwell. · Slaymaker, O. and Spencer, T. (1998) *Physical geography and global environmental change*. Harlow: Longman. · Smith, K. (1992) *Environmental hazards: assessing risk and reducing disaster*. London: Routledge. · Somerville, M. (1858) *Physical geography*, 4th edition. London: Murray. · Stoddart, D.R. (1987) Geographers and geomorphology in Britain between the wars. In R. W. Steel (ed.), *British geography 1918–1945*. London: Institute of British Geographers; pp. 156–176. · Thomas, D.S.G. and Middleton, N.J. (1994) *Desertification: exploding the myth*. Chichester: John Wiley & Sons, Ltd. · Viles, H.A. and Spencer, T. (1995) *Coastal problems*. London: Arnold. · Williams, M.A.J., Dunkerley, D., de Deckker, P., *et al.* (1998) *Quaternary environments*, 2nd edition. London: Arnold.

physical meteorology That part of meteorology or the atmospheric sciences that deals with the physical properties of the atmosphere and the processes occurring therein. Usually included are atmospheric chemistry, electricity, radiation, thermodynamics, optics and acoustics, cloud and precipitation physics, AEROSOL physics, and physical climatology. Because all of these fields of study interact in one way or another with atmospheric motions, it is somewhat artificial to distinguish between physical meteorology and DYNAMICAL METEOROLOGY. This is especially true in the present day when so much research activity in the atmospheric sciences deals with the numerical modelling of physical processes and their interaction with the atmospheric circulation.

Of the various sub-branches of physical meteorology, atmospheric thermodynamics is perhaps the most basic and the most closely related to dynamical meteorology. Usually included under this heading are the gas laws (see EQUATION OF STATE), the HYDROSTATIC EQUATION, the first and second laws of thermodynamics, latent heats and water vapour in the air, ADIABATIC processes, static stability (see VERTICAL STABILITY/INSTABILITY) and entropy. Cloud and precipitation physics (see CLOUD MICROPHYSICS) includes not only studies of the formation and growth of cloud droplets, raindrops and ice crystals, but also studies involving the processes whereby CLOUDS may be modified to increase or decrease PRECIPITATION, to suppress HAIL, LIGHTNING, and HURRICANE winds, and to dissipate FOG.

The sub-branches of physical meteorology are not autonomous. For example, a study of the effect of AEROSOLS on clouds would involve aerosol physics, radiative transfer, cloud physics and possibly atmospheric chemistry. Modern research in physical meteorology may also use REMOTE SENSING, either from satellites or from the ground using radiometers, radar and lidar.   WDS

Reading
Wallace, J.M. and Hobbs, P.V. (2006) *Atmospheric science*, 2nd edition. New York: Academic Press.

physiography A word that has obscure origins, although in common currency in eighteenth-century Scandinavia, and in regular usage in the English-speaking world in the nineteenth century (Stoddart, 1975). It was defined by Dana (1863):

Physiography, which begins where geology ends – that is, with the adult or finished earth – and treats (1) of the earth's final surface arrangements (as to its features, climates, magnetism, life, etc.), and (2) its systems of physical movements and changes (as atmospheric and oceanic currents, and other secular variations in heat, moisture, magnetism, etc.).

One of the most notable exponents of physiography was T. H. Huxley (1877), who published the highly successful text *Physiography*. In the USA, W. M. Davis preferred the term to GEOMORPHOLOGY, but he used it without the catholicity of meaning that it had for Huxley. ASG

References
Dana, J.D. (1863) *Manual of geology: treating on the principles of the science*. Philadelphia: Bliss. · Huxley, T.H. (1877) *Physiography: an introduction to the study of nature*. London: Macmillan. · Stoddart, D.R. (1975) 'That Victorian science': Huxley's *Physiography* and its impact on

geography. *Transactions of the Institute of British Geographers*, **66**, 17–40.

phytogeography The scientific study of the distribution of plants on the Earth. Although traceable as a theme to the 'father of botany', Theophrastus (*c.* 370–287 BC), the first comprehensive studies in plant geography were the *Essai sur la géographie des plantes* of Humboldt and Bonpland (1805, German edition 1807) and the *Géographie botanique raisonnée* of Alphonse de Candolle (1855). Used narrowly, the term tends to be confined to the study of geographical or spatial distribution of plant species, genera and families over the surface of the Earth; more generally, it is often used, particularly by geographers, to include many aspects of plant ecology and plant biology as well.   PAS

Reading
Moore, D.M. (ed.) (1982) *Green planet: the story of plant life on Earth*. Cambridge: Cambridge University Press. · Stott, P.A. (1981) *Historical plant geography*. London: Allen & Unwin.

phytogeomorphology A concept reflecting landform/landscape–vegetation relationships that are visibly dominant on the landscape (Howard and Mitchell, 1985: 5).   ASG

Reading and Reference
Howard, J.A. and Mitchell, C.W. (1985) *Phytogeomorphology*. New York: Wiley-Interscience. · Kruckeberg, A.R. (2002) *Geology and plant life: the effects of landforms and rock types on plant life*. Seattle, WA: University of Washington Press.

phytokarst Features produced by the weathering and erosive action of plants and animals on limestone rocks. It is also called phytokarren or biokarst. The identification of erosive (boring and digestive) action or chemical weathering (chelation) on rocks is very problematic. While phytokarst can form spectacular features (Bull and Laverty, 1982), it usually produces random spongework forms (Folk *et al.*, 1973). The term BIOKARST is preferred to phytokarst (Viles, 1984). PAB

References
Bull, P.A. and Laverty, M. (1982) Observations on phytokarst. *Zeitschrift für Geomorphologie*, **26**, 437–457. · Folk, R.L., Roberts, H.H. and Moore, C.H. (1973) Black phytokarst from Hell, Cayman Islands. *Bulletin of the Geological Society of America*, **84**, 2351–2360. · Viles, H. (1984) A review of biokarst. *Progress in Physical Geography*, **8**, 523–543.

phytosociology In its widest sense, the study of plants as social or gregarious organisms, and thus the study of plant communities; more specifically, the floristic description, classification and naming of community types and the study of their distribution and chief ecological characteristics. The most fully developed approach to phytosociology is that of the Zürich–Montpellier school developed by Braun-Blanquet around 1913, now much refined and widely applied throughout the world. The basic unit of classification in this approach is the association (see ASSOCIATION, PLANT), which is a recurrent grouping of plant species recorded by means of sampling units termed *relevés*. In recent years, the application of quantitative methods and of computers has revolutionized the study of phytosociology.   PAS

Reading
Chapman, S.B. (ed.) (1976) *Methods in plant ecology.* Oxford: Blackwell. · Harrison, C.M. (1971) Recent approaches to the description and analysis of vegetation. *Transactions of the Institute of British Geographers*, 52, 113–127. · Randall, R.E. (1978) *Theories and techniques in vegetation analysis.* Oxford: Oxford University Press. · Shimwell, D.W. (1971) *The description and classification of vegetation.* London: Sidgwick & Jackson.

piedmont   An area of relatively gentle slopes and low relief flanking an upland area. The term is commonly applied to assemblages of planar alluvial surfaces flanking an area of mountains or rocky desert uplands. Desert piedmont surfaces may be underlain by sandy and gravelly ALLUVIUM tens of metres in thickness and of Quaternary and pre-Quaternary age, or may carry only an alluvial veneer 1–2 m thick. Alluvial veneers may overlie rock-cut surfaces. Commonly, piedmont surfaces are dissected by drainage systems, and the incisions made by these reveal that the piedmont surfaces often reflect a long history of AGGRADATION, pedogenesis and incision, the patterns of which may be quite complex and variable in different areas of the piedmont (Bourne and Twidale, 1998).

The development of desert piedmont surfaces by multiple episodes of aggradation and incision has been hypothesized to relate to many potential causes of changes in the run-off regime and associated erosional processes active in the uplands and upon the piedmont surfaces themselves. Triggering processes for such changes include: (1) CLIMATIC CHANGE (and associated vegetation change); (2) regional warping or tilting of the terrain; (3) deposition of dust following AEOLIAN transport – dust deposition may be involved in the development of silty surfaces that alter the infiltration properties of the surface of deposition, and so the nature of run-off generated from it; (4) pedogenesis, including the slow accumulation of carbonate materials in the subsurface materials of the piedmont, so building up carbonate hardpans, which again may alter the ability of the sediments to take in and transmit water.

Remnant surfaces of different ages within a desert piedmont often display differences in surficial rock weathering, subsoil carbonate accumulation, soil texture, and extent of stone pavement and desert varnish development as a result of differences in the age of the surfaces.   DLD

Reference
Bourne, J.A. and Twidale, C.R. (1998) Pediments and alluvial fans: genesis and relationships in the western piedmont of the Flinders Ranges, South Australia. *Australian Journal of Earth Sciences*, 45, 123–135.

piedmont glacier   A glacier that spreads out into a piedmont lobe as it debouches onto a lowland.

piezometer   An instrument for measuring pressure head at a point within a saturated porous medium. A piezometer consists of a tube of greater than capillary cross-section, placed in the soil or rock so that water may only enter the tube at a fixed level (usually at the base of the tube through a porous pot). Water enters the tube and the water level rises until the head of water in the tube balances the water pressure at the entry point. The depth of water in the piezometer is known as the pressure or piezometric head.   AMG

Reading
Curtis, L.F. and Trudgill, S. (1974) *The measurement of soil moisture.* British Geomorphological Research Group technical bulletin 13. Norwich: Geo Abstracts.

piezometric surface   An imaginary surface to which water levels rise in wells tapping confined aquifers. Confined aquifers (rock formations that contain and can yield significant quantities of water) occur where the aquifer contains GROUNDWATER that is confined at a pressure greater than air pressure by overlying, relatively impermeable rock formations. A well that penetrates a confined aquifer will experience water levels higher than the junction between the aquifer and the overlying rock formations, and the piezometric surface is the surface that passes through these well water levels. If the piezometric surface rises above the ground surface, a well located at that point will produce flowing water at the surface and is described as an artesian well.   AMG

Reading
Fitts, C.R. (2013) *Groundwater science*, 2nd edition. Waltham, MA: Academic Press.

pingo   An ice-cored hill that is typically conical in shape and can only grow and persist in permafrost. The term is of Inuit origin but has been widely adopted elsewhere.

Pingos form through the freezing of water that moves under a pressure gradient to the site of the pingo. If water moves from a distant elevated source, the pingo is hydraulic (i.e. open system) in nature. If water moves under pressure arising from local permafrost aggradation and associated pore water expulsion, the pingo is hydrostatic in nature (i.e. closed system). The greatest concentration, about 1450, and some of the largest in the world, occur in the Mackenzie Delta region of Canada, where they commonly form in drained lake basins.   HMF

Reading
Burr, D.M., Tanaka, K.L. and Yoshikawa, K. (2009) Pingos on Earth and Mars. *Planetary and Space Science*, 57, 541–555.

piosphere (or piosphere effect)   This was used by Andrew (1988) to describe the spatially variable impact on the environment of animals using a point (usually well or borehole) water source. The term is applicable to situations in DRYLANDS and SAVANNAS, where the development of ranching has occurred in environments with no surface water to support livestock. By tapping groundwater, and given that animals need to drink and therefore have to focus their movements on the water source, a radial pattern of pressure on the environment arises, which is greatest close to the water source and diminishes with distance away from it. The word is derived from *pios*, Greek for 'drink'. (See also HERBIVORE USE INTENSITY (HUI).)   DSGT

Reading and Reference
Andrew, M.H. (1988) Grazing impact in relation to livestock watering points. *Trends in Ecology and Evolution*, **3**, 336–339. · Perkins, J.S. and Thomas, D.S.G. (1993) Environmental responses and sensitivity to permanent cattle ranching, semi-arid western central Botswana. In D. S. G. Thomas and R. J. Allison (eds), *Landscape sensitivity*. Chichester: John Wiley & Sons, Ltd; pp. 273–286.

pipes   Subsurface channels up to several metres in diameter caused by deflocculation of clay particles in fine-grained, highly permeable soils. Pipes are commonly found in arid or semi-arid regions, less commonly elsewhere. They are usually formed where the soils contain significant amounts of swelling clays such as montmorillonite, illite or bentonite, and where there is a low water table with a steep hydraulic gradient in the near-surface environment. Locally, pipes are usually found in steep slopes and on gully and arroyo sides (Heede, 1971). The pipes carry sediment as well as water, and if erosion continues for a long enough period the conduit may enlarge so much that the roof collapses, forming a gully. Pipes are economically important because they are a sign of deteriorating soil conditions and represent accelerated erosion.   WLG

Reading and Reference
Heede, B. (1971) *Characteristics and processes of soil piping in gullies*. US Department of Agriculture Forest Service research paper RM-68. Ft Collins, CO: Rocky Mountain Forest and Range Experiment Station, Forest Service, US Department of Agriculture. · Jones, J.A.A. (1981) *The nature of soil piping: a review of research*. British Geomorphological Research Group monograph 3. Norwich: Geo Books.

pipkrakes   See NEEDLE ICE.

pisoliths   Spherical rock particles of around 5–6 mm in diameter that are formed by the gradual accretion of material around a nucleus. Laterites and calcretes often display pisolithic textures. ASG

pitometer (or Pitot tube)   A means of estimating flow velocity by measuring the pressure of the fluid as it hits an immersed rounded body. It was invented by Henri Pitot in 1732.   AMG

pixel   The basic spatial unit of a digital image or raster data set. Pixel is a contraction of picture element and is the standard unit for storing and displaying many digital data sets, such as satellite sensor images and digital elevation models. Such images are composed of a two-dimensional grid, which typically defines thousands to many millions of pixels. Each pixel will have a single digital number assigned to it that represents the variable of interest for the area it covers, such as reflectance in the case of a satellite sensor image or height above sea level for a digital elevation model. For example, the digital number in a digital elevation model may indicate a height above sea level of 100 m for the terrain represented by a pixel. This is clearly an approximation and does not mean that the whole area represented by the pixel is exactly at this height, but that this is a fair approximation. Individual pixels are not normally visible to the human eye unless the image has been considerably enlarged.

Pixels can be described in areal units (e.g. square metres), although it is far more common for the term pixel size to be taken as its X or Y dimensions. For example, the pixel size of an image in the visible wavebands acquired by the Landsat THEMATIC MAPPER (TM) is commonly stated as being 30 m, meaning each pixel is $30\,m \times 30\,m$ in dimensions with a ground area of $900\,m^2$. The pixel size is often used as a rough guide to the spatial resolution of the data.

The pixel is simply a convenient data structure and is not usually the framework within which the actual measurements were made or representative of the character of the area represented and its context. For example, a pixel in a satellite sensor image is an artificial spatial unit defined mainly by sensor properties without any regard to natural geographical units that may exist in the landscape being represented. The quality of this representation is dependent on the size of the pixels and the nature of the variable being represented.   GF

Reading
Cracknell, A.P. (1998) Synergy in remote sensing – what's in a pixel? *International Journal of Remote Sensing*, **19**, 2025–2047. · Fisher, P. (1997) The pixel: a snare and delusion. *International Journal of Remote Sensing*, **18**, 679–685.

planation surface   See EROSION SURFACE.

plane bed   Applied to the deformable bed of a fluid flow on which there are no organized BEDFORMS and where the relief is that of individual grains. Such quasi-flat beds can occur when sediment is hardly moving under close-to-threshold conditions of fluid shear (lower stage), and under high shear and high transport rate conditions (upper stage). The latter may be a transitional stage that develops when the Froude number approaches unity and dune forms are destroyed before upper regime bedforms develop.   JL

Reading
Allen, J.R.L. (1970) *Physical processes of sedimentation*. London: Allen & Unwin.

planimeter   An instrument for the measurement of areas on maps and plans, which is less time consuming and more accurate than counting squares or estimation but which is less efficient than digitizing methods associated with a mainframe or microcomputer.   KJG

plankton   Small freshwater and marine organisms, a substantial number of which are microscopic. While some plankton possess limited mobility, many are inactive. The movement of plankton mainly depends upon the motion of the water in which they are suspended.

There are three major planktonic categories: phytoplankton, zooplankton and bacterioplankton. Phytoplankton (algae) account for the bulk of primary production in aquatic ecosystems, and their BIOLOGICAL PRODUCTIVITY is conventionally high (Barnes, 1980). Zooplankton, which include mature and/or larval representatives of numerous important animal groups (e.g. Protozoa, Crustacea and Mollusca), may be herbivores, carnivores or omnivores, either filtering or seizing living planktonic or detrital organic matter (Parsons *et al.*, 1977; Parsons, 1980). Bacterioplankton (for instance, *Bacillus* and *Nitrosomonas*) are mainly decomposers. Some are able to perform photosynthesis and chemosynthesis, thereby contributing to primary production (Fogg, 1980).

The distribution and productivity of plankton vary in both space and time. This is due to a combination of environmental factors, among which nutrient availability and climate are of considerable importance.   RLJ

Reading and References
Barnes, R.S.K. (1980) The unity and diversity of aquatic systems. In R. S. K. Barnes and K. H. Mann (eds), *Fundamentals of aquatic ecosystems*. Oxford: Blackwell; pp. 5–23. · Fogg, G.E. (1980) Phytoplanktonic primary production. In R. S. K. Barnes and K. H. Mann (eds), *Fundamentals of aquatic ecosystems*. Oxford: Blackwell; pp. 24–45. · Parsons, T.R. (1980) Zooplanktonic production. In R. S. K. Barnes and K. H. Mann (eds), *Fundamentals of aquatic ecosystems*. Oxford: Blackwell; pp. 46–66. · Parsons, T.R., Takahashi, M. and Hargrave, B. (1977) *Biological oceanographic processes*. 2nd edition. Oxford: Pergamon Press. · Suthers, I.M. and Rissik, D. (2009) *Plankton: a guide to their ecology and monitoring for water quality*. Collingwood: CSIRO Publishing.

plastic limit   The water content of an unconsolidated material when it is at the point of transition from a plastic solid to a rigid mass.

plasticity   The behaviour under stress of weak materials such as moist clays and weak rocks. Such weak materials do not deform under very low magnitudes of stress, but above a critical magnitude, called a yield stress, they deform at a continuous rate if the level of stress is constant; materials exhibiting such behaviours are said to be plastic substances.   MJS

Reading
Selby, M.J. (1993) *Hillslope materials and processes*, 2nd edition. Oxford: Oxford University Press; chapter 4.

plate tectonics A theory of global TECTONICS that holds that the LITHOSPHERE forming the Earth's surface is divided into eight major and several minor internally rigid plates that are in motion with respect to each other and the underlying ASTHENOSPHERE. CONTINENTAL DRIFT is a consequence of plate motion, and earthquakes, volcanoes and mountain building are concentrated in the vicinity of, although are not entirely confined to, plate boundaries.

There are three main types of plate boundary. Divergent boundaries, which are mostly located along the extensive ridge system of the ocean basins, represent the sites of sea-floor spreading where new lithosphere is created and two plates move away from each other. Convergent plate boundaries occur when two plates move towards each other. This leads to the subduction (reabsorption into the sublithospheric mantle) of one of the plates as it plunges down under the leading edge of the other. Sites of subduction are marked by deep ocean trenches and are associated with intense seismicity and volcanic activity. Plate subduction involving only oceanic lithosphere leads to ISLAND ARC formation, but where subduction occurs along the margin of a

continent a mountain belt develops (such as the Andes). If two converging plates are capped by continental crust the two continental masses will eventually collide and subduction will be halted. A complex and extensive zone of crustal deformation results, as exemplified by the Himalayas and the Tibetan Plateau, which have been created as a result of the collision of India and Eurasia. The third major category of plate interaction occurs along a transform boundary where two plates slip horizontally past each other, such as along the San Andreas Fault System in California. This type of boundary is characterized by numerous earthquakes but low levels of volcanic activity.

Although convection currents in the mantle are clearly involved in plate motion, it does not seem that they are the main driving force. The most important mechanism is probably the pull exerted on the rest of a plate by those parts being actively subducted. Although the plate tectonics model has revolutionized our understanding of the oceans, research over the past decade has emphasized that it provides only a generalized understanding of the morphology and structure of the continents.   MAS

Plate tectonics. Map of the lithospheric plates. The various types of plate boundary are shown and the estimated current rates (mm a^{-1}) and directions of plate movements are indicated.

Source: Summerfield (1991). Reproduced with permission of John Wiley & Sons.

Reading

Frisch, W., Meschede, M. and Blakey, R. (2011) *Plate tectonics, continental drift and mountain building.* Berlin: Springer-Verlag. · Kearey, P., Klepeis, K.A. and Vine, F.J. (2009) *Global tectonics.* Chichester: John Wiley & Sons, Ltd. · Molnar, P. (1988) Continental tectonics in the aftermath of plate tectonics. *Nature*, **335**, 131–137. · Oreskes, N. (ed.) (2003) *Plate tectonics: an insider's history of the modern theory of the Earth.* Cambridge, MA: Westview Press. · Summerfield, M.A. (1991) *Global geomorphology.* London/New York: Longman/John Wiley & Sons, Inc.

plateau   An elevated area of relatively smooth terrain, frequently separated from adjacent areas by steep slopes. Plateaux are often composed of horizontally bedded rocks, and vary in size from subcontinental features such as the Deccan Plateau of India to small mesas, as in the American southwest.   PSH

plateau basalt   An extensive flow or flows of basalt rock that, owing to erosion of the surrounding less-resistant rocks, forms an upstanding plateau (e.g. the Deccan of India).

playa   A closed depression in a DRYLANDS area that is periodically inundated by surface water. The term is also used to refer solely to the salt flat that may occupy such a depression. The term is derived from the Spanish for beach and is thus incorrectly used by English-speaking geomorphologists. Terms such as pan, chott and kavir are used for the same features in some parts of the world.

Playas are highly variable in form and in terms of the sediments that they accumulate, but several general characteristics can be determined (Shaw and Thomas, 1997):

1 occupy regional or local topographic lows;
2 lack surface outflows;
3 ephemerally, not permanently, occupied by surface waters;
4 usually have very flat surfaces;
5 have a hydrological budget in which evaporation greatly exceeds inputs;
6 are usually vegetation free, or if vegetation occurs they have distinct vegetation assemblages.   DSGT

Reference

Shaw, P.A. and Bryant, R.G. (2011) Pans, playas and salt lakes. In D. S. G. Thomas (ed.), *Arid zone geomorphology: process, form and change in drylands.* Chichester: John Wiley & Sons, Ltd; pp. 373–402.

Playfair's law   J. Playfair (1802) suggested that every river will flow in a valley proportional to the size of the river and that their levels will be accordant where rivers join. This law of accordant tributary junctions came to be known as Playfair's law.   KJG

Reading and Reference

Kennedy, B. (1984) On Playfair's law of accordant tributary junctions. *Earth Surface Processes and Landforms*, **9**, 153–173. · Playfair, J. (1802) *Illustrations of the Huttonian theory of the Earth.* Edinburgh: William Creech.

Pleistocene   The Pleistocene is the first epoch of the Quaternary. It is preceded by the Pliocene and succeeded by the Holocene. The Pleistocene was composed of alternations of great cold (glacials, stadials) with stages of relatively greater warmth (interglacials, interstadials), during which worldwide sea levels fluctuated in response to the formation and melting of ice sheets. In glaciated regions, these eustatic changes were accompanied by isostatic depression under the weight of ice cover during glaciations and recovery during the interglacial phases.

The classic interpretation of the history of the Pleistocene, especially in the northern hemisphere, has been based on the study of the extent and character of these alternations of glacial and interglacial deposits on land. However, there is a marked degree of controversy over the number of glaciations, stadials, interglacials and interstadials. This is due to the problem of definition of these events, and also a lack of agreement with regard to correlations of events between different areas.

Until recent times, Pleistocene events were recorded in a chronology based on the location of evidence in geological or archaeological strata. Evidence was dated by correlation with known successions based on the typology, stratigraphy or prehistoric cultures. Increase in the use of new dating techniques and in deep sea core evidence have transformed Pleistocene studies. The traditional view from terrestrial studies indicated that four, five or possibly six glacials existed during the Pleistocene. Indications from ocean cores are that there have been no less than 17 glacial cycles in the past 1.6 million years.

However, stratigraphic terminology is still to a great extent understandardized, and often no clear distinction is made as to whether the classification system is lithostratigraphic (based on rock or sediment classification), biostratigraphic (based on the occurrence of fossil fauna and/or flora), chronostratigraphic or a combination of these. A major point of contention in attempts to construct a framework for dating subdivisions of the Pleistocene is the location of the Pleistocene–

Pliocene boundary and, as a corollary, the duration of the Pleistocene itself. Some authors suggest a short timescale (600,000 years); some favour a medium timescale –for example, on faunal grounds, the Pleistocene–Pliocene boundary has been placed at about 1.6 million years and thus conveniently coincides with a major geomagnetic reversal (the top of the Olduvai event). Others accept the long timescale (up to 3 million years) on the basis of some major climatic deterioration; namely, the marked appearance of mid-latitude, as opposed to polar, glaciers. The variations appear in the main to be due to differing forms of evidence, their interpretation and geographical location, especially marine versus terrestrial evidence.   AP

Reading
Lowe, J.J. and Walker, M.J.C. (2015) *Reconstructing Quaternary environments*, 3rd edition. Abingdon: Routledge.

Plinian eruption   An explosive volcanic eruption that is frequently so violent that the volcanic cone is destroyed.

plots, erosion/run-off   Plots are small areas of the landscape isolated from their surroundings by low walls that are embedded in the soil. The goal of this isolation is to be able to study the hydrologic and erosional response of the enclosed area and to apply the understanding gained from this to the larger landscape beyond. Plots receive no run-off or eroded sediment from areas upslope, so that they cannot truly encapsulate all of what would ordinarily take place over the area that they cover. On the other hand, many INTERRILL PROCESSES of the kind studied through the use of plots take place to a degree in isolation from upslope contributing areas, since the landscape is broken up into smaller subcatchments by microrelief. A rill, for example, may be a response to run-off from a local contributing area and not strongly linked to areas more than a few metres distant. Therefore, the behaviour of part of this contributing area can appropriately be studied by plot methods.

Small plots (having sizes of a few square metres) have the advantage that their hydrological response involves little delay between the generation of run-off on the plot and its arrival at a measuring or sampling structure located on the downslope edge of the plot. Likewise, there is limited scope for sediment eroded on the upper parts of the plot to be redeposited on the lower margins. Progressively larger or longer plots (up to hundreds of square metres) increasingly involve the possibility of significant delays in run-off response or of sediment deposition before the plot border is reached. This means that derived data may need to be processed using a flow routing procedure. It also means that the plot ceases to exhibit the very simple behaviour that can be revealed on small plots.

Plot methods are widely employed in agricultural research, where they are appropriate in the investigation of the effects on soil erosion of factors such as varying styles of tillage, mulching of the soil or fertilizer application (Andreu *et al.*, 1998; Ghidey and Alberts, 1998). They are also used in fundamental research, in work designed to understand the effects of changing gradient, plant canopy cover, soil type and other factors. The advantage that small plots possess in this kind of work is that a plot can exhibit quite uniform characteristics of soil and gradient, whereas a larger area (such as a small experimental drainage basin) would certainly not do so. This simplifies and facilitates the interpretation of derived data; and because of this, very large amounts of data derived from experimental plots have been used in the development and calibration of tools such as the USLE (universal soil loss equation). Plots are frequently applied to the assessment of hydrologic parameters such as infiltration rate. In this application, they have the advantage of being larger and potentially more representative than small areas tested by other techniques, such as cylinder infiltrometry. Moreover, plots can be operated under simulated rain, so that soil crusting and sealing phenomena can be more realistically maintained than in infiltration trails using shallow ponds.   DLD

Reading and References
Andreu, V. Rubio, J.L. and Cerni, R. (1998) Effects of Mediterranean shrub cover on water erosion (Valencia, Spain). *Journal of Soil and Water Conservation*, **53**, 112–120. · Ghidey, F. and Alberts, E.E. (1998) Runoff and soil losses as affected by corn and soybean tillage systems. *Journal of Soil and Water Conservation*, **53**, 64–70. · Lal, R. (1994) *Soil erosion research methods*, 2nd edition. Delray Beach, FL: St Lucie Press.

plucking   A process of glacial erosion describing the removal of discrete blocks of bedrock. It is commonly contrasted with the other main form of glacial erosion, ABRASION, which describes the process of rock wear. Plucking results from failure of the rock along joint planes and reflects two processes. The first is wedging by the pressure of overriding rock particles. The second is the freezing of blocks to overriding glacier ice in response to temperature fluctuations at the ice–rock interface as a result of pressure variations.   DES

Reading
Addison, K. (1981) The contribution of discontinuous rock-mass failure to glacier erosion. *Annals of Glaciology*, 2, 3–10. · Röthlisberger, H. and Iken, A. (1981) Plucking as an effect of water pressure variations at the glacier bed. *Annals of Glaciology*, 2, 57–62.

plume   See MANTLE PLUME.

pluton   A mass of rock that has solidified underground from intrusions of magma. Plutons have variable shapes, sizes and relationships with the country rock (the invaded rock) surrounding them. Batholiths, dykes, laccoliths, lopoliths, sills and stocks are the main forms.

plutonic   Refers to rock material that has formed at depth (e.g. igneous rocks such as granite) where cooling and crystallization have occurred slowly.

plutonism   A term used to describe the ideas of James Hutton in the late eighteenth century that some rocks had once been molten. In 1785 Hutton discovered that in Glen Tilt, Perthshire, granite veins were breaking and displacing local rocks; he postulated that the granites were formed by the solidification of molten material intruded into the crust from the Earth's hot interior. He invoked a similar igneous origin for basalt. In contrast to NEPTUNISM, he regarded many igneous rocks as younger than the surrounding strata.   ASG

pluvial   Time of greater moisture availability, caused by increased precipitation and/or reduced evapotranspiration levels. A term once widely used in Quaternary research but less used today because it is recognized that past environmental and climatic changes are spatially and temporally complex.

pluviometric coefficient   The ratio between the mean rainfall total of a particular month and the hypothetical amount equivalent to each month's rainfall were the total to be equally distributed throughout the year.

pneumatolysis   See HYDROTHERMAL ALTERATION.

podzol   In the traditional 'Great soil groups' system of taxonomy, a soil that has a distinctive and strongly leached upper horizon underlain by a genetically linked horizon strongly enriched in the materials carried downwards (see HORIZON, SOIL). The term is derived from Russian *pod* meaning 'under' and *zola*, 'ash', and refers to the distinctively pale, leached A horizon found in podzols. The translocated materials that accumulate in the B horizon may include organic matter, carbonates, iron and aluminium. Since they require intense mobilization of materials, podzols are characteristic of soils where acid leachates can dissolve and carry substances downward. The leachates are organic acids derived from a surface litter layer, and they are most active in permeable, sandy or loamy parent materials. Podzols are included in the Spodosol Order of the US system of soil taxonomy (from the Greek *spodos*, 'wood ash'), where the distinctive *spodic* horizon is one where amorphous organic matter and aluminium, with or without iron, have accumulated. (See LEACHING.)   DLD

Reading
USDA Soil Survey Staff. (1988) *Soil taxonomy: a basic system of soil classification for making and interpreting soil surveys*. Florida: Krieger Publishing Co.

point bar deposits   Sediments laid down on the inside of a meander bend or 'point' largely by LATERAL ACCRETION. Individual attached bars commonly form low arcuate ridges or scrolls. Units of accretion are added as the meander loop develops, eventually making up a complex of ridges separated by depressions or swales. The layout of the individual scroll bars may reveal the growth pattern of the point bar as a whole. In some environments, however, this pattern is either irregular or not even apparent at all on the inside of meander bends. The extent of point bar development depends on the amount of blanketing OVERBANK DEPOSIT and on the variable nature of the point bar accretion in conditions with contrasting sediments and river regimes.   JL

Reading
Reading, H.G. (ed.) (1986) *Sedimentary environments and facies*, 2nd edition. Oxford: Blackwell.

polar front   The front separating air of polar origin from that originating within the subtropics. In winter it can often be traced as a band of cloud over thousands of kilometres between 40° and 50° latitude, especially over the oceans. Extratropical cyclones may be initiated along the strong thermal gradient of the front. Bjerknes (1921) based his theory of frontal evolution upon the presence of this front. In summer the front is more variable in its location and the temperature gradient is weaker.   PS

Reference
Bjerknes, V. (1921) *On the dynamics of the circular vortex with applications to the atmosphere and atmospheric vortex and wave motions.* Geofysiske Publikationer, vol. 2, no.4. Kristiana: Kommission hos Cammermeyers bokhandel.

polar wander The progressive apparent shifting of the location of the Earth's magnetic poles. Polar positions at past epochs are estimated from the dip angle of magnetic minerals in igneous rocks of known age, with low dip angles indicating cooling at low latitudes and high dip angles at high latitudes. The polar wander that results when the position of the pole at a series of past times is plotted is largely an *apparent* wander, rather than an actual motion of the magnetic pole. The explanation is rather that the rocks recording former locations of the pole have themselves been moved by plate tectonic displacements. The magnetic poles do move by small amounts, but since the geomagnetic field has its origin in processes linked to the rotation of the planet, the magnetic and geographic poles do not diverge as widely as apparent polar wander curves suggest.   DLD

Reading
McElhinny, M.W. (1973) *Palaeomagnetism and plate tectonics.* Cambridge: Cambridge University Press. · Tarling, D.H. (1971) *Principles and applications of palaeomagnetism.* London: Chapman & Hall.

polder A low-lying area of land that has been reclaimed from the sea or a lake by artificial means and is kept free of water by pumping.

polje An extensive depression feature in karst, closed on all sides, mostly with an even floor, a steep border in places and a clear angle between the polje bottom and the slope. It has underground drainage and can be dry all year round, have ephemeral streams within it or be inundated continually. The very diverse nature of poljes prevents a definition based on genesis; they are truly polygenetic features.   PAB

Reading
Gams, I. (1977) Towards the terminology of the polje. In T. D. Ford (ed.), *Proceedings of the Seventh International Congress of Speleology, Sheffield.* Bridgwater: The British Cave Research Association; pp. 201–202.

pollen analysis The scientific analysis of microscopic pollen grains and spores preserved in sediments. Pollen analysis, or palynology, is utilized principally in the reconstruction of former vegetation types and, by extension, palaeoenvironments. The technique was first used in Sweden in 1916 by Lennart von Post, who recognized that relatively large numbers of pollen grains were preserved in Quaternary lake and mire sediments as microfossils. Subsequent statistical analysis of the inventory of various pollen types recorded with the aid of a microscope then facilitated a reconstruction of the relative proportions of the various plant taxa that produced the pollen types over the period of deposition. Today, pollen analysis is widely applied to Quaternary sediments in particular, but is utilized in geologically older sediments, as well as in applications that deal with, for example, the pollen content of the contemporary atmosphere in allergy studies and even in forensic science. Pollen analysis has proved especially useful in reconstructing the increasing level of human impact on vegetation communities and in elucidating the changes in plant distribution consequent upon climate change over the late Quaternary.

Pollen analysis is based on the following principles:

1 All seed-plants (ANGIOSPERMS and gymnosperms) produce pollen as the male gametophyte.
2 Many plant species produce large quantities of pollen and only small proportions are used in reproduction, so that much is released into the environment where it may accumulate in sediments.
3 Pollen grains, because of the chemistry of sporopollenin (the substance from which the pollen grain wall, or exine, is constructed), are extremely resistant to decay (especially, but not exclusively, in anaerobic conditions) and may be found in deposits in which other microfossils have been destroyed.
4 Because of their small size (5–200 μm), pollen grains are widely and relatively evenly dispersed.
5 The pollen grain is an indicator of the plant that produced it because its size, structure, morphology and sculpturing are taxon specific.
6 Quantities of pollen become preserved in sediments accumulating over time so that analysis of grains from particular layers of sediment yields information about the nature of the vegetation types in the surrounding area at the time of deposition.
7 The quantities of pollen accumulating in sediments is such that a statistical approach to data analysis is possible; pollen frequency data may be compiled into pollen diagrams and subject to various multivariate statistical analyses to reveal additional information about trends over time.

These principles have led to the wide application of pollen analysis in the reconstruction of palaeoenvironments, and the technique is arguably the most important in Quaternary ecology (see QUATERNARY and ECOLOGY). There are, nevertheless, several methodological problems that need to be considered. Among these is the fact that pollen is produced in varying quantities by different taxa; for example, wind-pollinated (anemophilous) plants produce pollen in greater quantities and disperse it over much greater distances than do those pollinated by insects or birds (zoophilous). Moreover, some pollen grains are more resilient than others, so that pollen diagrams require careful interpretation in the light of information, if possible, about contemporary pollen productivity and preservation. Although pollen is taxonomically specific, it is not always feasible, especially with fossil grains, to determine the precise taxonomic group beyond genus level, hence reducing the degree of resolution possible in the vegetation reconstruction.

Ultimately, the pollen diagram (see figure) is the fundamental interpretive tool in pollen analysis. Pollen data for the various stratigraphic layers are arranged in chronological sequence and plotted either as percentages or, in the case of so-called absolute pollen data, as pollen concentrations. Changing pollen frequencies over time are identified as distinctive pollen assemblage zones, and these are then used as the basis of the reconstruction of the vegetation history of the site in question. Key to the analysis is the development of a precise chronology; in the case of late Quaternary sediments (the most common application of pollen analysis), RADIOCARBON DATING is frequently applied to resolve the chronology.

Although initially applied only to localities subject to cool temperate climates favouring high moisture contents and the development of anaerobic conditions within the sediments, pollen analysis is now successfully employed in a wide range of other types of environment. For example, pollen grains are preserved in lake, pan and fluvial sediments in arid environments, in marine sediments at all latitudes, in cave sediments and in COPROLITES. Even in the tropics, where prolific plant species diversity makes accurate pollen taxonomy difficult and where warm, humid conditions were thought to promote pollen decay, the technique has been applied to great effect.   MEM

Reading

Behling, H. and Hooghiemstra, H. (1998) Late Quaternary palaeoecology and palaeoclimatology from pollen records of the savannas of the Llanos Orientales, Colombia. *Palaeogeography, Palaeoclimatology, Palaeoecology*, **139**, 251–267. · Faegri, K. and Iversen, J. (1989) *Textbook of pollen analysis*. Chichester: John Wiley & Sons, Ltd. · Lowe, J.J. and Walker, M.J.C. (2015) *Reconstructing Quaternary environments*, 3rd edition. Abingdon: Routledge. · Moore, P.D., Webb, J.A. and Collinson, M.E. (1991) *Pollen analysis*, 2nd edition. Oxford: Blackwell.

pollution A condition that ensues when environmental attributes become inimical to the normal existence of living organisms. A contaminant is a substance foreign to an environment and capable of pollution within it. A contaminant has a source from which it is dispersed, usually by means of an atmospheric or aquatic pathway. During this process it may be rendered harmless by transformation or dilution. If this does not occur, the contaminant becomes a pollutant which has a target (Holdgate, 1979; Newson, 1992). As Mellanby (1972) states, while there are numerous instances of natural pollution (volcanic emission that becomes toxic and inhibits the development of vegetation in the vicinity, for example), that resulting from human activity is the more significant.

Mature organisms are better able to cope with harmful effects than are young ones. However, as Bailey *et al.* (1978) point out, a substance may only need to reach a concentration of 1 ppm to become a pollutant, the presence of which could ultimately lead to the death of an organism. As pollutants are Earth materials, they comprise part of a finite quantity. Thus, their components may be changed from one state or position to another but not obliterated (Jørgensen and Johnsen, 1981). When change is possible, dilution in air or water (of pesticides, heavy metals and toxic gases, for example) or degradation on land (of garbage and sewage, for instance) are normally involved. Some pollutants, though (certain nuclear wastes and lethal chemicals, for example), are so hazardous and/or of low degradability that they must be sealed and interred rather than released to the environment.

Pollution can occur at a variety of scales and in numerous circumstances. Atmospheric pollutants, for example, may give rise to serious local conditions, and can also be circulated widely. Smog is a localized atmospheric condition formed by the combination of pollutants (such as carbon monoxide and sulphur compounds) and fog. More widespread effects on the atmosphere can be brought about by carbon dioxide, together with sulphur and nitrogen oxides (see ACID PRECIPITATION and GREENHOUSE EFFECT). In

Laguna Angel
Summary pollen diagram

Pollen analysis. Pollen diagram for Laguna Angel, Colombia. This is a summary pollen diagram (i.e. simplified to show changes in ecological groups rather than individual taxa) and displays many of the features typical of this kind of representation of pollen data. For example, the x-axis shows the frequency of the various pollen taxa as percentages, the y-axis represents depth and age; stratigraphy is shown in stylized form on the left of the diagram, statistical manipulation producing the various pollen assemblage zones is shown on the right.

Source: Behling and Hooghiemstra (1998). Reproduced with permission of Elsevier.

415

a similar vein, an increase in particulate concentration (by dusts, for example) lessens atmospheric transparency and affects the reflectivity of solar radiation.

Aquatic pollution can result from the addition of harmful substances such as acids or hydrocarbons. However, the gradual build-up of essential elements in fresh water subsequent to their application as terrestrial agricultural fertilizers (EUTROPHICATION) may also pollute. In the terrestrial environment, the major pollutant by volume is urban–industrial refuse, which, if treated, is either stored, or reduced – usually by BIODEGRADATION and burning.

Pollution is a significant and developing environmental problem. None the less, specific instances involving potentially harmful substances and circumstances often lead to disagreement. As Barbour (1983) notes, fact is frequently obscured by conjecture, while there are sometimes political and socioeconomic undertones to particular cases. Some such factors may be exemplified with reference to noise pollution, much of which (up to the 120 dB limit when human pain is felt) is a rather subjective experience.

Pollution control strategies vary. Control at source has been preferred in the USA and on mainland Europe, while in the UK it has been customary to manage the pathways of pollutants (Newson, 1992). RLJ

Reading and References

Bailey, R.A., Clark, H.M., Ferris, J.P., *et al.* (1978) *Chemistry of the environment.* New York: Academic Press. · Barbour, A.K. (1983) The control of industrial pollution. In R. M. Harrison (ed.), *Pollution: causes, effects and control.* London: Royal Society of Chemistry; pp. 1–18. · Holdgate, M.W. (1979) *A perspective of environmental pollution.* Cambridge: Cambridge University Press. · Hill, M.K. (2010) *Understanding environmental pollution.* Cambridge: Cambridge University Press. · Jørgensen, S.E. and Johnsen, I. (1981) *Principles of environmental science and technology.* Amsterdam: Elsevier. · Jacobson, M.Z. (2012) *Air pollution and global warming: history, science and solutions,* 2nd edition. Cambridge: Cambridge University Press. · Macrory, R. (1990) The legal control of pollution. In R. M. Harrison (ed.), *Pollution: causes, effects and control,* 2nd edition. Cambridge: Royal Society of Chemistry; pp. 277–296. · Mellanby, K. (1972) *The biology of pollution.* London: Edward Arnold. · Newson, M. (1992) The geography of pollution. In M. Newson (ed.), *Managing the human impact on the natural environment: patterns and processes.* London: Belhaven Press; pp. 14–36.

polyclimax A theory of vegetation that allows the coexistence of a number of stable plant communities in an area. According to polyclimax theory, all the SERES (community sequences) in an area do not converge to identify in a MONOCLIMAX, but SUCCESSION produces a partial convergence to a mosaic of different stable communities in different HABITATS. In polyclimax theory, all climax types are of equal rank rather than subordinate to the climatic climax as is required by monoclimax theory. (See also CLIMAX VEGETATION.) JAM

polygonal karst A limestone landscape entirely pitted with depressions that are smooth rimmed and soil covered, producing a crude polygonal network when viewed from the air. The term was introduced by Williams (1971) when reporting the features from New Guinea. PAB

Reference
Williams, P.W. (1971) Illustrating morphometric analysis of karst with examples from New Guinea. *Zeitschrift für Geomorphologie,* **15,** 40–61.

polynya A pool of open water within pack ice or an ice floe.

polypedon A collection of small columns that run through the soil zone.

polytopy Loosely used, a term to describe the occurrence of any organism in two or more completely separate geographical areas; more specifically, a term referring to the process by which an organism may evolve two or more times, quite independently, in differing geographical localities. If the polytopic populations have developed at different times they are also termed polychronic in origin. The term is primarily employed by plant geographers and, with our present understanding of genetics and evolution, such an explanation for a disjunct distribution would only be accepted in very exceptional circumstances. A classic example of polytopy is afforded by the separate, but closely related, sand dune ecotypes of *Hieracium umbellatum* found along the coasts of Sweden. (See also DISJUNCT DISTRIBUTION.) PAS

Reading
Stott, P. (1981) *Historical plant geography.* London: Allen & Unwin.

pool and riffle The pool-and-riffle sequence is a large-scale bedform characteristic of streams with gravelly, heterogeneous bed material. Pools are closed hollows scoured in the bed and commonly floored by relatively fine gravel and sand, while riffles are topographical highs representing accumulations of coarser pebbles and cobbles. These features are created by the pattern of scour and deposition at bankfull discharge, when bed velocity is higher in the pool than in the adjacent riffle, and the coarse sediment then in motion is removed from the pools and deposited on the riffles. At low discharges the flow adjusts to the bed topography, and the water surface slope is flat over the deep, sluggish pools and steep over riffles, where flow is shallow and rapid. Fine sediment mobile at this flow stage is removed from the riffles and deposited in pools. The pool–riffle sequence repeats with a mean wavelength of five to seven times the mean channel width, suggesting initial control by a large-scale turbulent eddy scaled to the channel size. Once formed, riffles and pools are fairly stable morphological features, although individual sediment grains move through the sequence from riffle to riffle. In regular meander patterns, riffles tend to occur at inflection points and pools at bends, and the meander wavelength is twice the pool–riffle wavelength. This suggests that the pool–riffle feature is a fundamental bedform common to 'straight' and meandering rivers.   KSR

population dynamics The study of changes in population size, relating not just to humans as studied in human geography but also to other organisms investigated in biogeography and ecology. Population dynamics involves a consideration of those factors that might give rise to population growth, and those that might lead to its decline. Growth may be achieved by an increase in rates of birth relative to death and/or by immigration; and decline normally results from an excess of deaths over births and/or emigration. The structure of a population may also influence its dynamics; for example, a population with a relatively high percentage of females of reproductive age is clearly in a favourable position for growth, whereas one without such a representation may not be.

The earliest theoretical studies of population dynamics date back to Malthus (1798), and the first mathematical representation of population growth, characterized as it is by a sigmoid curve, to P. F. Verhulst in Paris in 1838. A population will establish itself slowly in a new environment and then, once it is adjusted ecologically and competitively to it, will grow rapidly within its determined NICHE. This explosive phase of growth has been noted widely among both plants and animals, and is associated with r-selection (see r- AND K-SELECTION). Subsequently, as it nears the biotic potential for that niche and area, environmental resistance will flatten the growth curve until it reaches an equilibrium, which is equivalent to the CARRYING CAPACITY of the area for that

species. The process may be expressed by

$$\frac{\Delta N}{\Delta t} = \gamma N \frac{K - N}{K}$$

in which N refers to the numbers of a given population, t to a given period of time, γ to the rate of increase (as determined by the difference between specific birth and death rates for the population), and K is a constant that relates to the upper limit of population growth (in other words, the carrying capacity). Close to the top of the growth curve, most species will be associated with K-selection.

Once at equilibrium level, populations may adopt several strategies in order to maintain themselves. In laboratories most species will keep as close to equilibrium as possible, but under field conditions many of the larger organisms will display slight cyclical oscillations around it. These are density dependent (see DENSITY DEPENDENCE), in the sense that, when population densities become low enough for a real and permanent decline to become a possibility, in-built compensatory mechanisms (usually in the form of increased birth rates) set in to restore the balance; similar, though reverse, responses occur when densities become too high. Should environmental circumstances change and organisms fail to adapt sufficiently, populations may begin to decline in numbers in a density-independent manner, seriously enough for EXTINCTION to threaten, though if the change is only temporary (e.g. an exceptionally cold winter, which may affect birds which eat freshwater organisms, or a single chemical or heat pollution event in water), such populations may still recover.   DW

Reference
Malthus, T.R. (1798) *An essay of the principle of population as it affects the future improvement of society*. London: J. Johnson. (Various modern editions.).

pore ice A type of ground ice occurring in the pores of soils and rocks. It is sometimes referred to as cement ice. On melting, pore ice does not yield water in excess of the pore volume, in contrast to SEGREGATED ICE. In terms of the total global ground-ice volume, pore ice probably constitutes the most important ground-ice type, primarily because of its ubiquitous distribution.   HMF

Reading
Mackay, J.R. (1972) The world of underground ice. *Annals of the Association of American Geographers*, **62**, 1–22.

pore water pressure The pressure exerted by water in the pores of a soil or other sediment.

417

Pressure is positive when below the WATER TABLE and negative when above it. Negative pore water pressure in soil is referred to as soil moisture tension (or suction). Pore water pressure can be measured by a tensiometer connected to a mercury manometer, vacuum gauge or pressure transducer (Burt, 1978).   PWW

Reference
Burt, T.P. (1978) *An automatic fluid-scanning switch tensiometer system*. British Geomorphological Research Group technical bulletin 21. Norwich: Geo Abstracts.

porosity A property of a rock or soil concerned with the extent to which it contains voids or INTERSTICES. It is usually defined as a ratio of the aggregate volume of voids to the total volume of the rock or soil, and is expressed as a percentage. A distinction is sometimes made between primary porosity, arising from intergranular interstices at the time of deposition of the rock, and secondary porosity, arising from later jointing or corrosion. All interstices, whether primary or secondary, are included in the estimation of the aggregate volume of voids for the purpose of measuring porosity.   PWW

Reading
Davis, S.N. (1969) Porosity and permeability of natural materials. In R. J. M. De Wiest (ed.), *Flow through porous media*. New York: Academic Press; pp. 54–89.

positive feedback See SYSTEMS.

postglacial See HOLOCENE.

potamology The scientific study of rivers.

potassium argon (K/Ar) dating An isotopic dating technique that utilizes unaltered potassium-rich minerals of volcanic origin in basalts, obsidians and the like. It is particularly useful for materials more than 50,000 years old.   ASG

Reading
Miller, G.H. (1990) Miscellaneous dating methods. In A. S. Goudie (ed.), *Geomorphological techniques*, 2nd edition. London: Unwin Hyman; pp. 405–407.

potential energy The energy change when a system is reduced to some standard state, usually applied to gravitational potential energy, with mean sea level defining the reference state. The concept has very wide applicability in physics and, more particularly, in the physics of the natural environment. It has been used very frequently within meteorology for a greater understanding of the energetics and dynamics of atmospheric

motion. For example, using

$$v^2 = 2gh$$

where v is air velocity, g is acceleration due to gravity and h is the height above a specified datum, a parcel of air in the upper TROPOSPHERE has potential energy to reach a speed of $450\,\mathrm{m\,s^{-1}}$. Natural parcels do not acquire such speeds because they are unable to fall freely, needing to push other air out of the way. This notion eventually defines the available potential energy and gives

$$v^2 \simeq 2gh\frac{\delta T}{T}$$

where δT is the temperature difference, as they pass, of two parcels being exchanged between two levels in the atmosphere. Hence, a more realistic magnitude of air speed is $30\,\mathrm{m\,s^{-1}}$.   JSAG

potential evaporation The rate of water loss from a surface when water supply to the surface is sufficient to meet the evaporative demand. Since evaporation from a water surface will always be at the potential rate, potential evaporation is often called potential evapotranspiration and is the rate of water loss from a surface, other than a water surface, through evaporation and transpiration processes and when these processes are not limited by a water deficiency. In order to ensure comparability of estimates from different areas, Penman (1956) defined potential water loss from a vegetated surface (which he called potential transpiration) as evaporation from an extended surface of 'fresh green crop of about the same colour as green, completely shading the ground, of fairly uniform height and never short of water'.   AMG

Reading and Reference
Allen, R.G. (1986) A Penman for all seasons. *American Society of Civil Engineers, Journal of the Irrigation and Drainage Division*, **112**, 348–368. · Morton, F.I. (1983) Operational estimates of areal evapotranspiration and their significance to the science and practice of hydrology. *Journal of Hydrology*, **66**, 1–76. · Penman, H.L. (1956) Estimating evaporation. *Transactions of the American Geophysical Union*, **37**, 43–46. · Ward, R.C. and Robinson, M. (1989) *Principles of hydrology*, 3rd edition. London: McGraw-Hill.

potential evapotranspiration See EVAPOTRANSPIRATION.

potential sand transport See DRIFT POTENTIAL.

potential temperature (θ) The temperature an air parcel would possess if it were moved from its level to a level with a pressure of 1000 mb dry adiabatically (a rate of $9.8\,\mathrm{K\,km^{-1}}$). If ascending or descending air parcels are subject only to these ADIABATIC changes their potential temperature will remain constant. In fact, the motion of air within the atmosphere is often close to adiabatic and the θ value of a given sample of air is conserved and acts as a kind of 'label'. Thus, when an unsaturated parcel of air ascends or descends it will do so dry adiabatically and will move up or down an imaginary surface of constant potential temperature. If it ascends it will cool at $9.8\,\mathrm{K\,km^{-1}}$ and will warm at this rate in descent – so long as no CONDENSATION occurs.

Because it is a more conservative property than dry-bulb temperature it is often used to highlight frontal changes in vertical cross-section of the atmosphere. Lines of constant θ appear on THERMODYNAMIC DIAGRAMS.   RR

pothole   A deep, circular hole in the rocky bed of a river that has formed by abrasion by pebbles caught in eddies. Any vertical shaft in limestone.

prairie   North American term for mid-latitude grasslands, equivalent to the Eurasian STEPPES.

Pre-Boreal   See BLYTT–SERNANDER MODEL.

precession of the equinoxes   An orbital mechanism caused by the gravitational forces of the Sun, Moon, and other planets on the Earth's equatorial bulge. Precession has two components. The first component (axial precession) causes the Earth's axis of rotation to 'wobble' like a spinning top. As a result, the axis of rotation describes a circle in space. For example, the axis of rotation is currently oriented towards the North Star. In ~13,000 years' time the axis of rotation will point in the direction of the star Vega. A complete cycle takes ~26,000 years. The second component (precession of the ellipse) changes the elliptical orbit of the Earth around the Sun about one focus. A complete cycle takes ~22,000 years. The combined effect of the two precessional forces changes the calendar dates for aphelion, perihelion, the equinoxes and the solstices. For example, aphelion today occurs near 4 July and perihelion is near 3 January. The positions of aphelion and perihelion will be reversed in ~11,000 years. The climatic result of precession can cause a change in the distribution of seasonal insolation between the northern and southern hemispheres. The current orientation of precession enables the southern hemisphere to receive more radiation at the top of the atmosphere than the northern hemisphere.

Also, the southern hemisphere currently receives as much as 10% more insolation than it will in 11,000 years when the opposite conditions will exist.   JAS

precipitation   The deposition of water from the atmosphere in solid or liquid form. It covers a wide range of particle sizes and shapes, such as RAIN, SNOW, HAIL and dew (see DEW, DEWPOINT). In most parts of the world rain is the only significant contribution to annual precipitation totals and the terms are frequently used synonymously. In polar regions and at high altitudes snow will be the dominant type of precipitation. The processes whereby water vapour is converted into precipitation are explained in CLOUD MICROPHYSICS.   PS

Reading
Strangeways, I. (2006) *Precipitation: theory, measurement and distribution.* Cambridge: Cambridge University Press.

predation   The killing of one free-living animal by another for food. Technically, this may involve the total removal of a species, or several species, from an environment by a predator, though in mature and/or complex communities, and in natural circumstances, it is unlikely that this would ever happen, for the predator would then have eliminated a potential food resource; moreover, most such communities possess a large number of prey species for each predator, so that the demands on any one are never too heavy. Also, most predators exploit a range of different animals in their diet.

However, there is little doubt that continued predation modifies the patterns of COMPETITION in an area and often, in consequence, the local distribution of species. Through reducing their population densities predators tend to lower the competition pressure from prey species in similar NICHES; this may result in two competitors surviving where only one would without predation.
DW/MEM

Reading
Paine, R.T. (1966) Food web complexity and species diversity. *American Naturalist*, **100**, 65–75.

predator–prey relationships   The population and energy balances between predators and prey. An intimate relationship exists between these two groups of organisms, for while the former can easily reduce the population of the latter, they are themselves vulnerable to decline and possible EXTINCTION through starvation should prey become too few. Accordingly, balanced predator–prey interactions depend in large measure on the effective control of the population size of, and by, both sets of participants. They are also important in the EVOLUTION of new species forms by natural selection, selection favouring the efficient predator and the elusive prey: the development of a wide variety of cryptic and mimetic coloration in many species in the latter case.

Animals that may be listed as predators include both 'true predators' and insect parasitoids (often incorrectly termed insect parasites). The latter are extremely numerous and account for about 10% of the approximately 1 million known insect species. Most belong to the Diptera (flies) and Hymenoptera (ants, bees, wasps) families; for example, there are huge numbers of different species of parasitoid wasps, ranging in size from free-living forms to microscopic egg parasitoids. Most, too, are host specific; that is, they seek out one host species alone. Unlike true predators, only the females of insect parasitoids look for hosts, and then usually only to lay eggs in or on them. The larvae that subsequently emerge feed from the host either internally or externally, but an effective energy balance between them is maintained until the larvae approach maturity, at which point the host's vital organs are eaten, and the host is killed. In this way, insect parasitoids also differ from PARASITES, which tend to ensure that the host's life is secured. Relationships between the populations of both host and insect parasitoid are fairly simple, in the sense that host mortality depends solely on the ability of females of the prey species, at one particular stage alone in their life cycle, to search out a host; and the reproductive rate of insect parasitoids may be seen to relate clearly to the number of hosts that are colonized. Although some insect parasitoids control the size of their host populations quite closely, this is not always the case.

In contrast, most true predators have more diffuse interactions with their prey, and this is especially the case for vertebrate predators, many of which are not prey specific, having a wide range of preferred foods. Males and females, and often their young as well, must all search for prey throughout the year, unless they hibernate for part of it; and this is undertaken with different degrees of efficiency. Moreover, the reproductive rates of some true predators are more finely determined by the demographic characteristics of their own populations rather than by the numbers of prey, though this is not an invariable situation.   DW/MEM

Reading
Barbosa, P. and Castellanos, I. (2005) *Ecology of predator–prey interactions*. Oxford: Oxford University Press.

pressure, air   The force per unit horizontal area exerted at any given level in the atmosphere by the weight of the air above that level. At sea level the average air pressure is $14.7\,lb\,in^{-2}$, 760 mmHg, 29.92 inHg, or 1013.25 mbar (or hectopascals). The air pressure decreases most rapidly with height near sea level, where the air is most dense. It decreases by about 50% for every 5 km of ascent.   WDS

pressure melting point   The temperature at which a liquid becomes solid at a particular pressure. The concept is fundamental to glacial geomorphology since glacier ice may exist at the pressure melting point. Since the temperature at which water freezes diminishes under additional pressure, by a rate of $1\,°C$ for every 1400 kPa (140 bar), the melting point at depth will be below $0\,°C$. For example, water was discovered beneath 2164 m of ice at Byrd Station in West Antarctica at a pressure melting point of $-1.6\,°C$.   DES

pressure release   The process whereby large sheets of rock become detached from a rock mass owing to the continuing relaxation of the pressure within the mass that built up before it was exhumed. For example, the erosion of overlying sedimentary strata from above a granite intrusion may cause pressure release to open joints in the granite when it is exposed.

prevailing wind   This term is really an abbreviation for 'prevailing wind direction' and means the wind direction most frequently observed during a given period. The periods most frequently used are days, months, seasons and years.   BWA

primarrumpf   An upwarped dome that though still undergoing uplift is being eroded at an equal rate.

probable maximum precipitation   The rainfall depth, for a particular size of catchment, that approaches the upper limit that the present climate can produce.

process–response system   See SYSTEMS.

productivity   A general term referring to the rate of BIOLOGICAL PRODUCTIVITY and accounting for both primary and secondary productivity in an ecosystem. Productivity refers to the accumulation of organic matter resulting from energy assimilation and transfer processes, including its initial entry into the system via PHOTOSYNTHESIS and its subsequent dissemination through the various TROPHIC LEVELS. There are several identifiable components to productivity, such as NET PRIMARY PRODUCTIVITY (NPP) and gross primary productivity among the lower trophic levels, and secondary productivity at higher levels. MEM

profile See SOIL PROFILE.

proglacial See ENGLACIAL.

proglacial lake A lake impounded in a depression in front of a glacier. During glacial periods such lakes were well developed in front of the southern margins of the Laurentide and Eurasian ice sheets. The proglacial Lake Agassiz in the area northwest of Lake Superior had an area larger than that of the Great Lakes today and was over 1000 km long. Similar lakes were impounded in the lower valleys of the northward-flowing Asian rivers such as the Ob. DES

progradation The extension of a shoreline into the sea through sedimentation.

protalus rampart A narrow ridge, usually a metre or so high and tens of metres long in front of a mountain rock face, composed of rock fragments. It may look like a small moraine, with which they have sometimes been confused, but whereas the former are the result of glacier action a protalus rampart is generally considered to be formed by rock debris sliding over a snow patch. They have also been called 'winter talus ridges' and 'nival ridges', but protalus rampart (Bryan, 1934) is now the accepted name. WBW

Reading and Reference
Ballantyne, C.K. and Kirkbride, M.P. (1986) Characteristics and significance of some Lateglacial protalus ramparts in upland Britain. *Earth Surface Processes and Landforms*, **11**, 659–671. · Bryan, K. (1934) Geomorphic processes at high altitudes. *Geographical Review*, **24**, 655–656.

proximal trough A depression around a steep rock body caused by the increased velocity that can occur when a moving element (e.g. ice, water or wind) flows round an obstruction. ASG

Reading
Lassila, M. (1986) Proximal troughs and ice movements in Gotland, southern Sweden. *Zeitschrift für Geomorphologie*, **30**, 129–140.

proxy A substitute; a term often used in palaeoclimate and palaeoenvironmental studies to refer to the source of information used to reconstruct past conditions. The past conditions cannot be observed or measured directly, so alternate sources, from organic and inorganic sources, are used; for example, pollen records, tree ring data, isotope records. See also GEOPROXY. DSGT

pseudokarst Landforms produced in non-carbonate rocks that are morphologically similar to those normally associated with karst rocks. These non-calcareous rocks can produce features by solutional processes similar to those types of reaction found in limestone (karren features) or by processes entirely different from these (lava caves). PAB

Reading
Warwick, G.T. (1976) Geomorphology and caves. In T. D. Ford and C. H. D. Cullingford (eds), *The science of speleology*. London: Academic Press; pp. 61–126.

pumice A highly porous fine-grained volcanic rock produced when numerous gas bubbles are trapped within the lava when it solidifies.

puna A cold desert, especially one at high altitude as in the Andes.

punctuated aggradational cycles A stratigraphic model that states that most stratigraphic accumulation occurs episodically as thin (1–5 m thick) shallowing-upward cycles separated by sharply defined non-depositional surfaces. They are created by geologically instantaneous basin-wide relative base-level rises (punctuation events), with deposition occurring during the intervening periods of base-level stability. Glacial eustatic changes driven by orbital perturbations (see MILANKOVITCH HYPOTHESIS) may be a preferred mechanism. ASG

Reading
Goodwin, P.W. and Anderson, E.J. (1985) Punctuated aggradational cycles: a general hypothesis of episodic stratigraphic accumulation. *Journal of Geology*, **93**, 515–533.

push moraine See MORAINE.

pyroclastic Refers to fragmental rock products (e.g. ash, volcanic bombs, ignimbrite) ejected by volcanic explosions.

Q

quartz An extremely abundant crystalline silica mineral that is very resistant to breakdown. It is often the most common constituent of SAND and SILT deposits.

quasi-biennial oscillation (QBO) A quasi-cyclic variation in atmospheric circulation on timescales of roughly 2.3–2.4 years. The best known QBO is a shift in the stratospheric winds between easterlies and westerlies. These phase shifts propagate downwards from the top of the stratosphere to the tropopause, where they dissipate. Two other 'quasi-biennial' oscillations have been documented. The tropospheric QBO (TBO) is a quasi-regular shift in tropical atmospheric circulation and sea-surface temperature and is a prominent feature of the Asian–Australian monsoon. Quasi-periodicities on the order of 2 years are also evident in southern-hemisphere zonal winds, in rainfall over much of Africa, and other atmospheric parameters.   SJN

quasi-equilibrium The 'apparent' balance between opposing forces and resistances. In geomorphology it is applied often to the 'concept' of an apparent balance between the rate of supply, temporary storage and removal of material from a dynamic depositional landform such as a scree slope, alluvial fan, beach or dune. The geometry of the landform or depositional store is dependent on the nature of the balance, the length of time over which it is maintained and the overall volumes involved. In many cases the volume of material passing through the system may be much greater than that in the store at any one time.

The idea may be applied to a surface of transportation such as a hillslope or pediment where there is an apparently close relationship between the geometry of the debris mantle and the form of the surface over which it moves. This implies that there is a negative feedback between the balance of input–storage–output processes of the mantle and the adjustment of the bedrock surface. Thus, a decrease in storage thickness may increase weathering rates on the bedrock to restore the mantle to its previous condition.

In this way the balance of equilibrium conditions of the depositional layer may partially control overall landform development.

A major weakness of the concept is that the relevant timescales over which budgetary or geometrical relationships are achieved or maintained are unknown. No evaluations have been made of the relaxation times involved and it is not possible, at present, to understand fully the effects of environmental change or the status of landscape relics.   DB

Quaternary The Cainozoic era of geological time is divided into the Tertiary period, about 65–2 million years (Ma) BP and the Quaternary period, from about 2 Ma BP up to and including the present day. The major subdivisions of geological time have by tradition been named in accordance with fossil evidence of the stage reached in biological evolution; hence Cainozoic = 'recent life', and the epochs within it (Palaeocene, Eocene, Oligocene, Miocene, Pliocene, Pleistocene and Holocene) are derived from terms expressing the proportions of modern marine shells occurring in fossil form (e.g. 'Meion' less, 'Pleion' more, 'Pleiston' most and 'Holos' complete or whole). The Quaternary period, consisting of the PLEISTOCENE and HOLOCENE epochs, is today delimited on different criteria, and, depending upon the criterion in favour from time to time, the date of its commencement varies somewhat (see below). The Quaternary is the time during which modern human beings evolved from primate ancestors, becoming makers of primitive stone tools, developing agriculture and, finally, in the last half of the Holocene, modern mining, wheeled transport, agriculture and industry. It is also one of the intervals of geological time when extensive GLACIATION affected parts of the Earth. The environmental stresses of this period may have been part of the driving force underlying human cultural development.

Following slow environmental deterioration in the Tertiary, glaciers began to form in high northern latitudes towards the end of the Tertiary and the beginning of the Quaternary.

The Dictionary of Physical Geography, Fourth Edition. Edited by David S.G. Thomas.
© 2016 John Wiley & Sons, Ltd. Published 2016 by John Wiley & Sons, Ltd.

Reasons for this glacial onset remain unclear, but tectonic uplift of areas like the Himalayas, Sierra Nevada and Rocky Mountains may have provided areas suitable for the steady accumulation of snow and ice, whose reflectivity may have lessened the amount of solar heating that was absorbed. In bottom sediments of the North Atlantic, increasing amounts of siliceous rock debris are recorded at this time, reflecting transportation into the open ocean on floating ice calved from glaciers reaching down to sea level. Elsewhere, the commencement of major Quaternary environmental change was indicated in other ways. In China, rapid LOESS accumulation began, reflecting a strengthening of winds from the desert interior. On the basis of dating transitions of this kind, the start of the Quaternary has been placed at 2.58 Ma BP (the date of the Gauss–Matuyama magnetic reversal) (Suc *et al.*, 1997).

During the Quaternary, great environmental changes affected most of the Earth. With a somewhat changing periodicity that appears to be controlled astronomically (see MILANKOVITCH HYPOTHESIS), glaciers repeatedly extended over large areas of Europe, Canada, and North America, with smaller areas in South America and elsewhere, and again retreated. Sea ice also became periodically more extensive. Sufficient water was stored in the continental ice sheets, which reached thicknesses of more than 3 km, to lower sea level by 130 m or so, creating large areas of new land on the continental shelves as well as opening land bridges across former seaways that were traversed by animals, plants and people. The weight of ice depressed the crust and, together with the loss of ocean volume, created major perturbations in the relative levels of land and sea. The last ice melted by about 10 thousand years (ka) BP, and the return of warm conditions marks the start of the Holocene epoch of the Quaternary. In areas heavily loaded down at the time of the last peak in ice extent (at about 18 ka BP in radiocarbon years, termed the 'Last Glacial Maximum'), such as Scandinavia and central Canada, the now unloaded crust is still in the process of rising to its former level.

On land, climatic changes, involving swings in rainfall, temperature, windiness, continentality, and so on, led to parallel adjustments in flora, fauna and landscape processes, like erosion and soil formation. Global average temperatures periodically fell by perhaps 5 °C, but cooling was much more marked, in excess of 15 °C cooler than the present day, in the interiors of the continents. During the cool periods, termed GLACIALS, treelines were lowered and vegetation cover changed structurally and floristically. Over large areas, forest retreated and was replaced by much more open communities of grasses. The repeated advance and retreat of forests, in successive glacial episodes, modified the carbon balance and so the abundance of carbon dioxide, a major greenhouse gas, in the atmosphere. Similar changes were triggered by altered biological productivity in the oceans, and by the increased solubility of the gas itself in colder sea water. Carbon dioxide concentrations fell by 50–100 ppmv, which compares with the preindustrial level of about 280 ppmv.

Changes to the hydrological cycle were also marked. Cold oceans release less moisture to the atmosphere, so that, as conditions descended toward glacial cold, rainfall declined in many areas, though the pattern of response is quite variable. In some locations, cold also appears to have curtailed evaporative losses from lakes, so that lake levels fluctuated dramatically, some closed basins periodically overflowing and later drying out completely. Some Quaternary lakes of the 'high lake' phases, like the large Willandra Lakes system of inland southeast Australia, which are now dry, periodically provided sustenance for early people, and were among the sites where early human cultural development took place.

In the last 750,000 years, the kinds of environmental changes noted here came and went with a periodicity of about 100,000 years. Palaeoenvironmental indicators of various kinds (marine microfossils, terrestrial pollen, isotopes trapped in deeply buried layers of ice in Greenland, Antarctica and other remaining ice caps) demonstrate that, through most of this period, conditions have been colder than they are presently. The Earth is now in a Quaternary warm phase (INTERGLACIAL); the great ice sheets have long since retreated, and sea level is at about its highest level with respect to the land. However, these interglacial periods are typically brief, lasting perhaps 10,000 years. Given that the interglacial that we now enjoy has already persisted for this long, we must anticipate renewed cooling in the coming millennia. Indeed, predictions based on what is known of the orbital influence on terrestrial climates suggest that the next full glaciation will be reached in about 60,000 years. Human activity in the meantime may of course upset these ancient rhythms.   DLD

Reading and Reference
Anderson, D., Goudie, A. and Parker, A. (2013) *Global Environments through the Quaternary: exploring environmental change*. Oxford: Oxford University Press. · Lowe, J.J. and Walker, M.J.C. (2015) *Reconstructing Quaternary environments*, 3rd edition. Abingdon: Pearson. · Suc, J.-P., Bertini, A., Leroy, S.A.G. and Suballyova, D. (1997)

Towards lowering of the Pliocene/Pleistocene boundary to the Gauss–Matuyama reversal. *Quaternary International*, **40**, 37–42. · Williams, M.A.J., Dunkerley, D.L., De Deckker, P., *et al.* (1998) *Quaternary environments*, 2nd edition. London: Edward Arnold.

quick clay   Water-saturated clay that has insufficient cohesion to prevent heavy objects from sinking into its surface.

quickflow   The part of the stream hydrograph that lies above an arbitrary cut-off line drawn on the hydrograph, representing the most rapidly responding hydrological processes and parts of the catchment. The division between quickflow and DELAYED FLOW is usually made by a line that rises from the start of the hydrograph rise at a gradient of $0.551 \, \mathrm{s}^{-1} \, \mathrm{km}^{-2} \, \mathrm{h}^{-1}$ until the line meets the falling limb of the hydrograph.   MJK

Reading
Hewlett, J.D. (1961) *Soil moisture as a source of base flow from steep mountain watersheds*. Southeastern Forest Experimental Station paper 132. Asheville, NC: US Department of Agriculture – Forest Service.

quicksand   Water-saturated sand that is semi-liquid and cannot bear the weight of heavy objects.

R

r- and K-selection The influence of the physical environment on life history traits as expressed through two different kinds of pattern of population increase over time. Different kinds of habitat tend to favour species with different life history responses, especially with regard to the rate of reproduction. For example, in relatively stable habitats, where a given range and abundance of resources is consistently available for exploitation by resident species, populations of many of these species over time will tend to remain at or near the CARRYING CAPACITY of that habitat. By contrast, instability or uncertainty of conditions within a habitat will tend to favour those opportunistic species able to rapidly exploit the resources, which may be available only on a temporary basis. MacArthur and Wilson (1967) labelled the opportunists as r-selected, given their tendency to increase numbers rapidly from low initial populations, whereas the species found at population numbers approaching the carrying capacity were said to be K-selected.

K-selected populations, then, occupy equable habitats and, as a result, the populations of residents tend to be both relatively high and reasonably constant. Intense competition between fertile adult members of the population determines the survival rate, while the young must compete equally intensively and only proportionally few reach maturity and are themselves able to reproduce. The characteristics of individuals belonging to the K-selected populations are as follows: larger body size, deferred and extended reproductive maturity, a smaller allocation of time and energy to reproduction, larger offspring with a higher degree of parental care. The r-selected populations occupy unpredictable or temporary habitats and, as a result, the population may undergo periods of rapid growth during favourable periods followed by periods of equally rapid population decline when conditions become unfavourable – in essence, a kind of 'boom and bust' behaviour. The characteristics of individuals belonging to r-selected populations are as follows: smaller body size, earlier age at and shorter period of maturity, a high proportion of time and energy allocated to reproduction, smaller and greater numbers of offspring.

As with other generalizations, this dichotomy is oversimplistic, but a useful means of appreciating that there are different ways for species to respond to habitat constraints. Actually, the r- and K-selected life histories described are really special cases at opposing ends of what is probably a continuum. Certainly, it is a useful way to identify differences among taxa. Trees in the K-selected habitat of, say, a mature oak woodland in northwest Europe, do indeed exhibit longevity, delayed and extended maturity, large size and produce large seeds. On the other hand, the plants found commonly in arid and semi-arid regions tend to follow the trends exhibited by r-selected individuals; many are EPHEMERAL PLANTS with very short lifespans that exhibit mass flowering and seed-set to take advantage of the transient nature of the soil moisture. In other temporary or disturbed habitats (e.g. road or railway verges), there is a predominance of r-selected ruderal plants typically thought of as weeds.

It should be noted, though, that attempts to classify organisms by their life-history strategies can be misleading if interpreted too rigidly. Species make a trade-off in the allocation of resources to maintenance, growth and reproduction and may display life-cycle characteristics containing elements of both r- and K-categories. For example, in an important paper, Solbrig and Simpson (1974) demonstrated that a dandelion population in a highly disturbed grass lawn behaved as r-strategist with a high reproductive allocation, whereas the same species in a nearby undisturbed site set fewer seeds and allocated more resources to growth and maintenance. Different populations of the same species can therefore occupy different positions on the r–K continuum. In some cases, populations of species occupying variable habitats display dynamic life-history responses and are able to shift between relative r- and K-positions over time.   MEM

The Dictionary of Physical Geography, Fourth Edition. Edited by David S.G. Thomas.
© 2016 John Wiley & Sons, Ltd. Published 2016 by John Wiley & Sons, Ltd.

References
MacArthur, R.H. and Wilson, E.O. (1967) *The theory of island biogeography*. Princeton, NJ: Princeton University Press. · Solbrig, O.T. and Simpson, B.B. (1974) Components of regulation of a population of dandelions in Michigan. *Journal of Ecology*, **62**, 473–486.

radar Remote sensing devices that record electromagnetic radiation from the microwave part of the electromagnetic spectrum (wavelengths approximately from 1 mm to 1 m) are often active sensors that transmit microwave energy and then measure the backscattered signal. These active sensors are based on radio detection and ranging (radar) technology, which exploits the measured time taken for each echo to return to the sensor from a target in the scene of interest to determine its distance (range) from the sensor. Simultaneously, the intensity and polarization of backscattered radiation may also be detected. In addition to this, the relative motion between the target and receiver along the line of sight can be inferred from the frequency shift of the reflected waves using the Doppler effect. The prevalent use of radar sensing over passive microwave sensing (i.e. sensors that only record microwaves emitted from the scene of interest) is due to the low energy of microwaves, which results in a poor signal-to-noise ratio of the recorded radiation (unless sampling is at very coarse spatial resolutions – 5–60 km depending upon frequency). At certain wavelengths (particularly lower frequencies – less than 40 GHz) the microwave signal is unaffected by atmospheric attenuation caused by aerosols, clouds and ice, and since radar instruments are themselves the source of radiation illuminating a scene they can operate both day and night.

Radar systems typically consist of (i) a pulse generator, (ii) a transmitter, (iii) a duplexer, (iv) an antenna, (v) a receiver and (vi) a data recorder. If the antenna on the radar is used for both transmission and reception of microwaves this is known as a monostatic system. Systems having two antennae (one for transmission and one for reception or one or both doing both tasks) are referred to as bistatic systems. The antenna acts like a lens and is the region of transition between an electromagnetic wave propagating in free space and a guided wave propagating in a transmission line, or vice versa. Another important feature that distinguishes between systems is whether the transmitter and receiver of the energy pulses retain information on the phase of the signal in addition to its amplitude. If phase information is retained, an active microwave sensor is in coherence; if not, it is incoherent. There are three generic types of space-borne radar instruments: altimeters, scatterometers and imaging radars. Nadir- (downward-)looking radars, or altimeters, are used to derive elevation profiles under the orbit track. These profiles are constructed by accurately measuring the time delay for a radar echo (pulse) to propagate to the surface and return back to the sensor. This has applications in mapping surface topography and ocean and ice monitoring. Scatterometers measure the radar cross-section of a target (surface reflectivity), which is a function of how the target interacts with microwave radiation, and are typically used to measure wind speed and direction over water or detect rainfall. Imaging radars, such as synthetic aperture radar (SAR), are used to acquire high spatial resolution (few meters to a few tens of meters) images, measuring range, intensity and sometimes polarization of backscattered radiation over large areas. SAR instruments emit radar pulses as the platform moves and, by correcting the pulses for the transmission and reception times, an aperture can be synthesized that has a much greater size than the physical size of the antenna, which results in a finer spatial resolution.

Early examples of imaging radar systems include the Shuttle Imaging Radar experiments SIR-A (1981), SIR-B (1984) and SIR-C (1994), the European Remote Sensing Satellites ERS-1 (1991) and ERS-2 (1995) and the Canadian Space Agency's Radarsat-1 (1995), which has now been replaced with Radarsat-2 (2007), a fully polarimetric imaging radar. Other recently launched polarimetric imaging radar systems include the *Advanced Land Observation Satellite-Phased Array type L-band Synthetic Aperture Radar* ALOS-PALSAR (2006) by the Japanese Space Agency and TerraSAR-X (2007) launched by a public–private partnership of DLR and Astrium. Scatterometer missions include NASA's *Quick Scatterometer* QuikSCAT (1999) and ESA's *Advanced Scatterometer* ASCAT. Current satellite-borne radar altimeters include ESA's Cryosat mission (2010) and the joint CNES/NASA Jason-1 (2001) and Jason-2 (2008) instruments. In 2014, a C-band SAR was launched on the first of ESA's Sentinels (Sentinel-1) and should be joined by a sister Sentinel in 2016 providing unprecedented radar remote sensing of the planet. DSB

Reading
Kellndorfer, J.M. and McDonald, K.C. (2009) Active and passive microwave systems. In T. A. Warner, M. D. Nellis and G. M. Foody (eds), *The SAGE handbook of remote sensing*. London: SAGE; pp. 179–198. · Pitz, W. and Miller, D. (2010) The TerraSAR-X satellite. *IEEE*

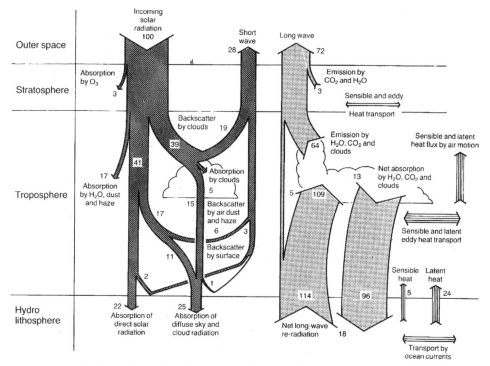

Schematic diagram showing the interactions that radiation undergoes in the atmosphere.
Source: Lockwood (1979).

Transactions on Geoscience and Remote Sensing, **48**, 615–622. · Skou, N. and Le Vine, D. (2006) *Microwave radiometer systems: design and analysis*. Norwood, MA: Artech House. · Woodhouse, I.H. (2006) *Introduction to microwave remote sensing*. Boca Raton, FL: Taylor and Francis.

radiation   Any object not at a temperature of absolute zero (−273 °C) transmits energy to its surroundings by radiation; that is, by energy in the form of electromagnetic waves travelling with the speed of light and requiring no intervening medium. This radiation is characterized by its wavelength, of which there is a wide spectrum extending from the very short X-rays through the ultraviolet and visible to infrared, microwaves and radio waves.

A valuable theoretical concept in radiation studies is that of the black body, which is one that absorbs all the radiation falling on it and which emits, at any temperature, the maximum amount of radiant energy. The term arises from the relation between darkness of colour and the proportion of visible light absorbed, since a body that appears white scatters most of the visible light falling on it. For a perfect all-wave black body, the intensity of radiation emitted and the wavelength distribution depend only on the absolute temperature, and in this case several simple laws apply. The Stefan–Boltzmann law states that the amount of energy F emitted in unit time from a unit area of a black body is proportional to the fourth power of its absolute temperature T:

$$F = \sigma T^4$$

where σ is Stefan's constant (5.6697×10^{-12} W cm^{-2} K^{-4}). The higher the temperature of an object the more radiation it will emit.

A black body does not radiate the same amount of energy at all wavelengths for any temperature. At a given temperature, the energy radiation reaches a maximum at some particular wavelength and then decreases for longer or shorter wavelengths. The Wien displacement law states that this wavelength of maximum energy λ_{max} is inversely proportional to the absolute temperature:

$$\lambda_{max} = \frac{\alpha}{T}$$

where α is a constant (0.2897 cm K if λ is in centimetres). As the temperature of an object increases, the wavelength of maximum energy decreases, passing from the infrared for objects at room temperature to the visible wavelengths for extremely hot objects.

If the Sun is assumed to be a black body, an estimate of its effective radiating temperature may be obtained from the Stefan–Boltzmann law, which suggests an effective surface temperature of 5750 K. For the Sun, the wavelength of maximum emission is near 0.5 μm, which is in the visible portion of the electromagnetic spectrum, and almost 99% of the Sun's radiation is contained in so-called short wavelengths from 0.15 to 4.0 μm. Observations show that 9% of this shortwave radiation is in the ultraviolet (<0.4 μm), 45% in the visible (0.4–0.6 μm) and 46% per cent in the infrared (>0.74 μm).

The surface of the Earth, when heated by the absorption of solar radiation, becomes a source of longwave radiation. The average temperature of the Earth's surface is about 285 K (12 °C), and therefore most of the radiation is emitted in the infrared spectral range from 4 to 50 μm, with a peak near 10 μm, as indicated by the Wien displacement law. This radiation may be referred to as longwave, infrared, terrestrial or thermal radiation. JGL

Reading and Reference

Budyko, M.I. (1974) *Climate and life*. New York: Academic Press. · Houghton, J.T. (2002) *The physics of atmospheres*, 3rd edition. Cambridge: Cambridge University Press. · Lockwood, T.G. (1979) *Causes of climate*. London: Edward Arnold. · Sellers, W.D. (1965) *Physical climatology*. Chicago, IL: University of Chicago Press.

radiative forcing

A change or perturbation that is imposed upon the climate system and modifies the radiative balance. The causes of such a change include changes in solar radiation input, cloud cover and character, ice, greenhouse gases, volcanic activity, and so on. Quantification of the effects of radiative forcing is one of the main goals of many climatic models. ASG

radiocarbon dating

The most widely used technique for dating carbon-bearing materials in the age range 0–40,000 years, and therefore used in studies of late QUATERNARY palaeoenvironmental changes. It is based on measuring the relative abundance of the radioactive carbon ISOTOPE (^{14}C) in comparison with a stable carbon isotope (^{12}C). ^{14}C is continuously formed at low, approximately equilibrium, levels in the atmosphere from the interaction of cosmic-ray neutrons with a stable isotope of nitrogen (^{14}N), where it is oxidized into carbon dioxide. Variations in the atmospheric concentration of ^{14}C are related to solar and other modulations of the cosmic-ray flux. Interactions in the atmosphere–Earth systems result in the fixation of carbon into a variety of biogenic and inorganic forms from where the radioactive decay of ^{14}C takes place. The radioactive decay of ^{14}C occurs, via emission of a beta (β) particle, as an exponentially declining trend with a HALF-LIFE of 5730 years.

The actual measurement procedure comprises either the counting of radioactive (β-particle) decay events (commonly referred to as conventional radiocarbon dating) or direct counting of abundances of stable (^{12}C) and unstable carbon atoms by accelerator mass spectrometric (AMS) methods. A major disadvantage of conventional radiocarbon dating is that only about 1% of the ^{14}C atoms in a sample will emit a β-particle in about 80 years – only a very small portion of the total sample carbon content is therefore measured. As all the carbon atoms present in a sample are counted in the AMS method, the required sample size is less by several orders of magnitude. Not only does the AMS method's small sample size requirements (minimum size ~100 μg) provide access to a wider range of sample types than conventional methods (e.g. individual seed grains, included organic debris from rock varnish layers), but it also allows determinations to be made on separate chemical components from within samples – some of which may be more reliable for dating than others. Ultimately, both methods are limited by the half-life of ^{14}C.

A variety of materials are datable by the radiocarbon methods, many of which are in fact relatively low in carbon content. While typically scarce, charcoal, wood and macrofossil fragments (including insects, chronomids and other aquatic organisms, bones, molluscs and snails) constitute the sample of preference for dating. Additional datable animal remains include faecal pellets, ivory, horn, hair and egg shells.

Application of radiocarbon methods to soil and sediment components has been an area of considerable investigation and debate. This relates to the open system nature of soils and the corresponding high likelihood of introduction of old or young carbon-bearing compounds and solutions via percolating groundwaters or modern root penetration. Application of radiocarbon methods to soil carbonates, and discrete CALCRETE and TUFA deposits have generally resulted in erroneous radiocarbon ages. This has been attributed to the incorporation of old carbonate in percolating groundwaters.

While the basic assumptions of the radiocarbon method are sound, a number of complicating factors must be considered in order to guarantee the generation of accurate absolute dates. Isotopic fractionation resulting from metabolic processes causes different kinds of samples to exhibit slightly different initial ^{14}C activities to that of the atmospheric carbon dioxide reservoir from where it was derived. As the fractionation of the stable carbon isotopes ($^{13}C/^{12}C$) is proportional to that of the ratio $^{14}C/^{12}C$, it is possible to use the ratio of the more abundant and more easily measured stable isotopes to estimate, and correct for, the degree of fractionation within any given dating.

A further complication to radiocarbon dating lies in the need for calibration of the dates produced to determine absolute ages. This is necessary because the assumed constant ^{14}C production rate (and therefore also the atmospheric ratio of $^{14}C/^{12}C$) is only approximately true due to subtle changes in the cosmic-ray flux. Calibration is achieved by comparison with absolute chronologies generated from tree rings, VARVES and from URANIUM SERIES DATING of coral–algal reefs (see CORAL REEFS AND CORAL–ALGAL REEFS).

In addition to correction for fractionation and calibration, it is necessary to consider nonsystematic factors that may influence radiocarbon evaluations for a given suite of samples. These include contamination from intrusion of younger material (e.g. roots, humic acids, bacteria) or the introduction of older materials; the 'hardwater effect', whereby ecological and hydrological systems fed by groundwater containing dissolved fossil carbonates produce organic materials (e.g. sapropels, shells) that are effectively depleted in ^{14}C; and the young age range of the method. Burning of ^{14}C-free fossil fuels and the generation of ^{14}C by atmospheric testing of nuclear devices has resulted in competing effects that limit the generation of radiocarbon dates over the last few centuries.   SS

Reading

Walker, M. (2005) *Quaternary dating methods*. Chichester: John Wiley & Sons, Ltd.

radiocarbon years The measure of time used in ages derived from RADIOCARBON DATING. Radiocarbon dates are usually expressed in the form x years plus/minus y before present (BP), where y represents a standard deviation and the present is defined as AD 1950. Initially, Libby, the pioneer of radiocarbon dating, had assumed that levels of atmospheric carbon had remained stable over the time range used and that a radiocarbon year would be constant. However, there have been significant variations in ^{14}C production, arising from flux in incoming cosmic radiation and from long-term changes in the carbon cycle through the late Quaternary, leading to ^{14}C ages that are significantly younger than true calendar ages. It has been possible to correct the radiocarbon dates using a calibration curve, based, for the Holocene, on radiocarbon assays of wood of known age, usually from dendrochronology studies, and back to ~50,000 years cal BP (Reimer *et al.*, 2009) using a comparison with uranium series ages from pristine corals and modelling (see URANIUM SERIES DATING). The IntCal09 and Marine09 calibration schemes were ratified at the 20th International Radiocarbon Conference in 2009. As the limitations of radiocarbon dating have become recognized, it has become common practice to quote a corrected (calendar) age alongside the radiocarbon date. Calibration can be achieved using various programs, including OxCal and CALIB. Once calibrated, ages should be expressed as cal BP (before 1950), cal BC or cal AD (these corresponding to real historical years).   PSH/DSGT

Reference

Reimer, P.J., Baillie, M.G.L., Bard, E., *et al.* (2009) Intcal09 and Marine 09 radiocarbon age calibration curves, 0–50,000 years cal BP. *Radiocarbon*, **51**, 1111–1150.

radioisotope The isotopes of some elements that undergo radioactive decay and alter either to the same or a different element. The decay process is a spontaneous process that involves the emission of alpha, beta or gamma radiation. The rate of decay of a radioisotope is proportional to the number of atoms remaining. Mathematically, this is referred to as the decay constant λ; this is specific to each radioisotope and is related to the HALF-LIFE $T_{1/2}$ by

$$T_{1/2} = \frac{\ln 2}{\lambda}$$

SS

Reading

Faure, H. (1986) *Principles of isotope geology*, 2nd edition. Chichester: John Wiley & Sons, Ltd.

radiolaria Single-celled amoebic protozoans that secrete a delicate siliceous framework to support soft body parts. They are usually circular in shape and are between 100 and 2000 μm in diameter. They are present throughout the world's oceans and are able to live in surface waters down to depths of more than 4 km. Radiolaria can live in and are tolerant of a range of salinity and temperature conditions. They are particularly useful in biostratigraphic studies where calcareous microfossil components (e.g. foraminifera, ostracods) have been removed due to post-burial chemical processes.   SLO

radiosonde An instrument for measuring the pressure, temperature and humidity of the air at heights of greater than 10 m above the ground. The instrument is attached to a balloon that rises through the TROPOSPHERE, usually bursting near the TROPOPAUSE. The meteorological data are transmitted to a ground station as the balloon ascends. Upper air winds are established by tracking the whole package by radar.   BWA

radon gas A colourless, odourless gas (radon-222) about eight times denser than air. It is derived largely from uranium, which is present in rocks such as granite. It poses an environmental problem when it accumulates in houses, for it may lead to cancer formation. Some areas of Britain are regarded as being at high risk (especially portions of Devon and Cornwall), but the threat can be reduced by appropriate building techniques and ventilation procedures.   ASG

Reading
Clarke, R. and O'Riordan, M. (1990) Rumours of radon. *Science and Public Affairs*, 5, 23–36.

rain Precipitation in the form of liquid water drops. Drop sizes vary up to a maximum of about 0.5 cm in diameter. The smallest diameter of a raindrop is sometimes defined rather arbitrarily as being 0.02 cm. Drops in the range of 0.02–0.05 cm are classed as drizzle and are fairly common on windward locations in temperate latitudes.   PS

rain day In Britain, a climatological period of 24 h from 09.00 UT within which at least 0.2 mm of precipitation is recorded. The average number of rain days ranges from about 160 in southeastern England to over 250 in the highlands of Scotland and the Outer Hebrides, where more than 300 rain days may be recorded in wet years. In some countries a different base is taken as the definition of a rain day, e.g. 0.01 inches in the USA.   PS

rain factor A measure of the relationship between temperature and precipitation designed to provide an indication of the climatic aridity of a region. The formula used is:

$$\text{rainfall factor} = \frac{\text{mean annual precipitation(mm)}}{\text{mean annual temperature}(^{\circ}\text{C})}$$

It is clearly inappropriate for polar deserts, where mean annual temperatures are below freezing. Elsewhere it gives an impression of the dryness of an area, although the seasonal distribution of rainfall will affect what plants can be grown. For example, Hobart in Tasmania and Kaduna in northern Nigeria have a similar rain factor, but the former site has rain each month and the latter only for 6 months.   PS

rain gauge An instrument used to measure rainfall amounts. Gauge characteristics differ in detail between countries, but basically most gauges consist of a funnel and storage or measurement system shielded from the free air to prevent evaporation. In the UK the standard gauge made of copper has a collection funnel of 12.7 cm diameter placed on a cylinder 30.5 cm above the ground surface. The water is stored in a glass bottle held within the copper cylinder. Any overflow from the bottle is collected in another container to prevent loss during heavy storms. The rainfall is normally read daily (at 0900 GMT) by emptying the contents of the collecting vessel into a cylindrical flask graduated to allow a direct reading of the rainfall total. Some gauges are read only weekly or even monthly in remote areas.

Rain gauges do not catch all the rain falling upon the surface because they present an obstruction to the airflow. In strong winds the actual catch may be depleted by up to 50% of the true catch or even more during snowfall. The height of the rain gauge funnel above the ground influences the catch. A taller cylinder will generate more turbulence and reduce the catch compared with an identical gauge close to the surface. Other errors may be caused by splashing, evaporation or even observer errors. The total obtained by a rain gauge is therefore only an approximation to the true fall. Where a large proportion of the annual precipitation falls as snow, as in Canada and parts of northern Europe, separate snow gauges are used as standard rain gauges are unsuitable. Because of the different types and sizes of gauge used, comparison of rainfall totals across international frontiers is difficult.

Some gauges make a continuous record of precipitation and its time of occurrence, allowing the calculation of rainfall intensity or rainfall rate. This is achieved by collecting the water in a tipping bucket, on a weighing system or by a float that measures water level, or by optical drop counting and sizing methods. Each type offers some advantages and disadvantages.   PS/DLD

Reading
Lanza, L.G. and Vuerich, E. (2009) The WMO field intercomparison of rain intensity gauges. *Atmospheric Research*, **94**, 534–543. · Sumner, G. (1988) *Precipitation: process and analysis*. Chichester: John Wiley & Sons, Ltd.

rain shadow An area experiencing relatively low rainfall because of its position on the leeward side of a hill or mountain range. Uplift and precipitation over the hills decreases the water content of the air, and this, coupled with the descent of air down the leeward slope, reduces the capacity of the air to produce rain. A good example in the UK is the Moray Firth area to the lee of the Scottish Highlands, where rainfall totals are relatively low.
PS

rain splash The process in which the impact of raindrops ejects water droplets and soil particles from a soil surface. Ejected splash droplets may be composed largely of sediment-free water or they may contain sediment particles derived from the soil surface. Rain splash, which has also been termed *airsplash* to emphasize that the particles follow a trajectory above the soil surface, is especially effective for silt and fine sand-sized particles. Particles carried in splash droplets may be carried more than 0.5 m in a single splash event. The splash process is driven by the energy possessed by raindrops, which commonly have masses of 5–30 mg and fall at 5–10 m s^{-1}. Even in a small storm, there may be millions of raindrop impacts on every square metre of exposed soil. The actual energy expended at the surface varies with rainfall intensity, because the distribution of drop sizes in rain shifts towards a larger median diameter at higher intensities, but is typically about 20 J m^{-2} per millimetre of rain.

Detailed investigations of the splash process have shown that many of the droplets thrown out from the point of impact of a raindrop are derived from the splash corona, which is a sheet of water forced upwards as the arriving droplet passes into any water film covering the surface. The ejected corona is rapidly drawn into a series of droplets by surface tension forces, and these continue on an outward trajectory. Droplets striking a dry surface, or one carrying only a very thin water film, cannot throw up a splash corona, and so they are less capable of rain splash. Splash is consequently accentuated when the surface is covered by a thin water film, but becomes relatively ineffective when the water layer exceeds about three drop diameters in thickness. At this point, the soil surface is sheltered from the impact-related forces by the water lying above it.

Net transport by rain splash is possible on a sloping surface, or when wind drives the rain. Splash processes are also of great importance in dislodging particles that may then be eroded by flowing water. DLD

Reading
Proffitt, A.P.B., Rose, C.W. and Hairsine, P.B. (1991) Rainfall detachment and deposition: experiments with low slopes and significant water depths. *Soil Science Society of America Journal*, **55**, 325–332. · Selby, M.J. (1993) *Hillslope materials and processes*, 2nd edition. Oxford: Oxford University Press. · Sharma, P.P., Gupta, S.C. and Rawls, W.J. (1991) Soil detachment by single raindrops of varying kinetic energy. *Soil Science Society of America Journal*, **55**, 301–307.

rainbeat crust see SOIL CRUST.

rainbow An optical effect consisting of an arc of the spectral colours. It is formed by the passage of sunlight through raindrops. As the beam of light enters the raindrop it is first refracted, then internally reflected from the far side before being refracted again as it leaves the drop. Some of the light may be reflected twice to produce a double rainbow effect. White light from the sun is composed of the colours of the spectrum which are separated in the refraction process because of their slightly different wavelengths. The degree of coloration of the rainbow depends upon the size of the drop and the intensity of the sunlight. PS

raindrop impact erosion Erosion that is attributable to, or accentuated by, physical forces that arise from the impact of raindrops at the soil surface. Raindrops are held quite strongly in an approximately spherical form by the surface tension of water, and consequently are not readily disrupted when they strike water ponded on or flowing over the surface. Rather, raindrops embed themselves into surface water, forcing a crater to be evacuated at the site of collision. Drops are progressively deformed as they strike the floor of the crater, and the water is thrown outward, merging with water ejected as the crater develops. Modelling studies (e.g. Harlow and Shannon, 1967) have shown that high-speed flows are created in this way, whose outward velocity may exceed that of the arriving raindrop by up to five times. These high-speed flows, which may last for only 0.1–0.2 s, are thought to result in shear forces at the soil surface that are capable of rupturing soil aggregates. In this way, the physical forces following droplet impact can disaggregate soil materials so that the resulting particles are more readily moved by RAIN SPLASH or surface run-off. Once the outward flow ceases, the flow direction reverses and the crater collapses. The inward flow of water towards the impact point once again creates shearing forces at the surface that may further contribute to particle breakdown. Repeated raindrop impacts continually move disaggregated particles and lift them away from the surface, to which they may rapidly resettle. However, these brief periods provide opportunities for lifted grains to be carried along in incremental steps in surface run-off, and this provides a second mechanism through which raindrop impact can increase the rate of erosion.   DLD

Reading and Reference

Al-Durrah, M.M. and Bradford, J.M. (1982) The mechanism of raindrop splash on soil surfaces. *Soil Science Society of America Journal*, **46**, 1086–1090. · Bradford, J.M. and Nearing, C. (1996) Splash and detachment by waterdrops. In M. Agassi (ed.), *Soil erosion, conservation, and rehabilitation*. New York: Marcel Dekker; pp. 61–76. · Harlow, F.H. and Shannon, J.P. (1967) The splash of a liquid drop. *Journal of Applied Physics*, **38**, 3855–3866.

rainfall   The total water equivalent of all forms of precipitation or condensation from the atmosphere received and measured in a RAIN GAUGE. Values are normally expressed as daily, monthly or annual totals. On a world scale, mean annual rainfall amounts are very variable, from almost zero in the driest deserts to above 10,000 mm in a few parts of India and Hawaii.   PS

rainfall intensity   The rate at which water arrives at the ground surface at any instant during rainfall. Rainfall intensity frequently exhibits rapid temporal variations, so that rates are reported as averages, or rainfall rates, over intervals from a few minutes to hourly or daily values. Owing to the inability of most rain gauges to capture intensity data, the phenomenon is not as well studied as rainfall rate.

Rainfall intensity appears to exhibit a complex dependence on temperature, and global average rainfall intensity is considered to increase with global temperature increases at about $23\% \, \text{K}^{-1}$ (Liu *et al.*, 2009). Modelling suggests that the nature of rainfall events will alter, such that low-intensity rainfall events may become less frequent, and intense events more frequent, towards 2100 (Chou *et al.*, 2012).   PS/DLD

References

Chou, C., Chen, C.A., Tan, P.H. and Chen, K.T. (2012) Mechanisms for global warming impacts on precipitation frequency and intensity. *Journal of Climate*, **25**, 3291–3306. · Liu, S.C., Fu, C., Shiu, C.-J., *et al.* (2009) Temperature dependence of global precipitation extremes. *Geophysical Research Letters*, **36**, L17702, doi: 10.1029/2009GL040218.

rainfall run-off   That part of the hydrological cycle that connects rainfall and channel flow. When rainfall exceeds the infiltration capacity of the surface materials water begins to collect in small surface depressions. Eventually these surface depressions fill and the water begins to move downslope. The entire overland process is termed Hortonian flow, after R. E. Horton who first outlined the process. (See also OVERLAND FLOW and RATIONAL FORMULA.)   WLG

rainfall simulator   A research tool designed to generate water droplets whose sizes correspond to those of natural rain, deliver them to the soil surface at speeds comparable to those reached by raindrops that have fallen from the cloud base, and operate at a range of simulated rainfall intensities. Rainfall simulators have been widely adopted in studies of erosion and flow mechanics because of the many advantages they confer, such as freeing the investigator from the inefficiency of having to wait for natural storms to pass over a study site and the ability to control rain intensity and duration. In many EXPERIMENTAL DESIGNS, another advantage is the ability to replicate as many identical rain events as are required while other factors, like the soil type exposed to rain, are changed. Rainfall simulators of many designs have been used, some for laboratory use and others for field operation. Among the means used to produce drops, the two principal techniques are *drop-formers*, often banks of syringes from whose tips drips are released, and *spray nozzles*, which are specially designed to yield known drop sizes when run at nominated water pressures (Bowyer-Bower and Burt, 1989). It is difficult to generate high intensities using drop-formers, and hard to produce low intensities with spray nozzles. Indeed, no simulator is capable of generating rain with drop size, intensity, and fall speed or drop kinetic energy all correctly matching natural rain. Nevertheless, the advantages offered by rainfall simulation see it in continued use as a research tool.   DLD

Reading and Reference

Bowyer-Bower, T.A.S. and Burt, T.P. (1989) Rainfall simulators for investigating soil response to rainfall. *Soil Technology*, **2**, 1–16. · Iserloh, T., Ries, J.B., Cerdà, A., *et al.* (2012) Comparative measurements with seven rainfall simulators on uniform bare fallow land. *Zeitschrift für Geomorphologie, Supplementary Issues*, **57** (1), 11–26.

raised beach   An emerged shoreline represented by stranded beach deposits, marine shell beds and wave-cut platforms backed by former sea cliffs. During the first decades of the twentieth century it was believed that many raised beaches were the result of eustatic changes brought about by changes in the volume of water stored in ice caps during glaciations and deglaciations, though it was also appreciated that in areas where the Earth's crust had been weighed down by the presence of an ice mass that ISOSTASY would have been an important process. In reality, though both these processes are important, there is a wide range of factors that can cause sea-level changes. Raised beaches, which can occur at heights of as much as tens or hundreds of metres above current sea levels, may be warped, so that care needs to be exercised in correlating on the basis of height alone.   ASG

Reading
Guilcher, A. (1969) Pleistocene and Holocene sea-level changes. *Earth Science Reviews*, 5, 69–97. · Rose, J. (1990) Raised shorelines. In A. S. Goudie (ed.), *Geomorphological techniques*, 2nd edition. London: Unwin Hyman.

raised channel
A channel whose form is preserved in the environment in negative relief long after it was active. Maizels (1987) described raised Plio-Pleistocene channels in Oman, where cementation of the original channel sediments had rendered them more resistant to erosion than the surrounding landscape. This had subsequently been lowered by erosion, leaving the raised channels as positive features in the landscape.   DSGT

Reference
Maizels, J.K. (1987) Plio-Pleistocene raised channel systems in the western Sharqiya (Wahiba), Oman. In L. Frostick and I. Reid (eds), *Desert sediments, ancient and modern*. Geological Society of London, Special Publications, vol. 35. London: The Geological Society; pp. 31–50.

raised mire
An acid peatland dominated by *Sphagnum* mosses and supplied by precipitation solely from atmospheric sources (rain, snow, fog, etc.). Raised mires form characteristic shallow domes of peat where the topography is typically convex, with a gently sloping rand away from its centre towards the surrounding moat-like drainage channel or lagg (Swedish terms describing margin of raised bog, typically with a stream and/or minerotrophic-poor en or fen woodland) surrounding the bog. This mire type is primarily a lowland system, and mainly occurs in broad, flat (ish) valleys or basins.

Raised mires have a wide distribution in Britain but predominate in the cooler, wetter north and west. Remnants of raised mires are recorded in Lincolnshire, the Cambridgeshire Fens, Somerset Levels and Amberley Wild Brooks in Sussex, but reclamation and ancient peat winning has removed most traces of ombrotrophic peat from these regions. In lowland Britain, raised mires are recorded in basins, floodplains and at the heads of estuaries; for example, Thorne Moors National Nature Reserve, which forms part of the Humberhead Peatlands. Upland examples of raised mires exist at Tarn Moss, Malham, North Yorkshire, and Dun Moss in Perthshire.   ALH

Ramsar Convention
Formally, the Convention on Wetlands of International Importance, especially as Waterfowl Habitat. This is an intergovernmental treaty that provides the framework for national action and international cooperation for the conservation and wise use of wetlands and their resources. It is named after the city of Ramsar in Iran, where the Convention was signed in 1971 by 21 countries; there are now 168 contracting parties, The Ramsar definition of wetlands is broad, including 'areas of marine water the depth of which at low tide does not exceed six metres' as well as fish ponds, rice paddies and salt pans. The Ramsar List now includes over 2000 sites (known as *Ramsar Sites*) covering over 200,000,000 ha.   TS

Reading
The Ramsar Convention Secretariat (2014) *Ramsar*. http://www.ramsar.org/ (accesed 16 July 2015).

randkluft
The gap between the back wall of a cirque and the glacial ice that fills the cirque.

random-walk networks
Drainage networks can be simulated by a random-walk stochastic model (see STOCHASTIC MODELS) in which the network evolves on a regular grid, some cells being randomly selected to contain a stream source. The direction of exit from the current cell is then chosen by reference to random number tables. The model may be purely random with equal probabilities ($P = 0.25$) of a move in each cardinal direction, or biased with, for example, $P = 0.25$ for left and right moves, $P = 0.5$ 'downhill' and $P = 0.0$ 'uphill' to simulate a regional slope. Spatial variation of bias can be allowed to simulate different topographical influences. Constraints to disallow triple junctions, reversals of direction and closed loops are necessary. These simulated networks commonly obey the laws of drainage network composition.   KSR

ranker
The name given to a soil that has undergone limited development and has started to have some organic accumulation. Such soils occur on

young geomorphological surfaces (e.g. recently deposited alluvium or dune sand).

rapids A steep section in a river channel where the velocity of flow increases and there is extreme turbulence.

raster One of the two main types of GEOGRAPHIC INFORMATION SYSTEM (GIS) for storing and analysing spatial data. A grid of square cells (PIXELS) is laid across the area of interest and values of one geographical variable (e.g. land use, temperature, population count) recorded for each pixel. One set of pixels forms a layer, with multiple variables being stored in further layers. A raster GIS is particularly good at handling information about phenomena that vary continuously across space (see VECTOR) and is widely used in the environmental sciences.   SMW

rated section A method of obtaining a continuous record of discharge for a river section by continuously recording water stage or level and establishing a relationship between stage and discharge to rate the section. The discharge measurements required to establish this relationship are commonly made using the VELOCITY–AREA METHOD. It is important to obtain a stable rating relationship, and a section with a bedrock control or artificially stabilized bed and banks is frequently employed. The records obtained are generally less accurate than those provided by WEIRS and FLUMES,

particularly for low flows. (See also DISCHARGE.)
DEW

rating curve A term frequently used to describe a relationship between discharge and water stage or between suspended sediment and solute transport and water discharge, which can be used to estimate values of the former variable from measurements of the latter. In the case of suspended sediment and solute transport, values of either concentration (milligrams per litre) or material discharge (kilograms per second) may be plotted against water discharge, and logarithmic axes are commonly employed. The characteristics of these plots, including slope, intercept and degree of scatter, have frequently been used to characterize the sediment and solute response of a drainage basin.   DEW

Reading
Gregory, K.J. and Walling, D.E. (1976) *Drainage basin form and process*. London: Edward Arnold; pp. 215–225.

rational formula A simple and long-used formula to estimate run-off Q in terms of a coefficient C depending upon the character of the surface of the drainage area, a measure of precipitation I, and an index of the basin area A in the form

$$Q = CIA$$

KJG

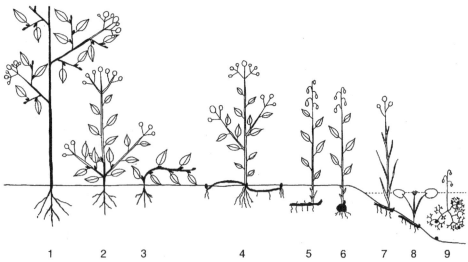

Raunkiaer's life forms. (1) Phanerophytes; (2–3) Chamaephytes; (4) Hemicryptophytes; (5–9) Cryptophytes. The parts of the plant that die during the unfavourable season are unshaded; the persistent portions and perennating buds are black.

Raunkiaer's life forms The life form of a plant is its gross structure. The best-known classification of life forms is that of C. Raunkiaer (1934), who arranged the life forms of plant species into a series primarily based on the position of the perennating buds. He suggested that this reflected an adaptation to climate, arguing that plants' environments were formerly more uniformly hot and moist than at present, that the most primitive life form is the one dominating tropical environments, and that the more highly evolved forms are found particularly in areas with colder climates. The main categories he distinguished are as follows:

- *Phanerophytes*, with the perennating buds on aerial shoots – the most primitive form, in Raunkiaer's view. Evergreen and deciduous phanerophytes are distinguished, and each of these categories can be subdivided:
  - *nanophanerophytes*, 2 m in height (shrubs);
  - *microphanerophytes*, 2–8 m;
  - *mesophanerophytes*, 8–30 m;
  - *megaphanerophytes*, over 30 m (the larger forest trees).
- *Chamaephytes*, with the perennating buds close to ground level. This category includes forms in which aerial shoots die away as the unfavourable season (winter or the dry period) approaches, plants in which the vegetative shoot grows along the ground and cushion plants.
- *Hemicryptophytes*, where the perennating buds are at ground level, almost all the aboveground material dying with the advance of unfavourable conditions. The group includes plants with stolons and rosette plants.
- *Cryptophytes*, where the perennating buds are below the ground surface or submerged in water. PHA

Reference

Raunkiaer, C. (1934) *The life forms of plants and statistical plant geography*. Translated by H. Gilbert-Carter, A. Fausboll and A. G. Tansley. Oxford: Clarendon Press.

reach A length of channel, as applied to a coastal inlet, the arm of a lake and river channels. The term may be used in more specialized senses, as for a relatively straight section of a navigation waterway, or alternatively for a short length of channel for which discharge or other hydraulic conditions are approximately uniform. JL

reaction time The time that elapses between the application of a change to, or a constraint upon, an Earth system and the beginning of adjustment of the system. Reaction time is therefore the period of time after the modification and before RELAXATION TIME begins. KJG

Reading

Graf, W.L. (1977) The rate law in fluvial geomorphology. *American Journal of Science*, 277, 178–191.

recession limb of hydrograph, recession curve The recession limb of the stream discharge hydrograph is the portion after the peak discharge and represents the time after precipitation has ceased when discharge gradually falls and is unaffected by rainfall (for diagram see HYDROGRAPHS). By analysing several hydrographs for the same gauging station it is possible to derive an average recession curve that may be expressed by a mathematical function, such as the exponential form:

$$Q_t = Q_o e^{-at}$$

where Q_t is discharge at time t, Q_o is the initial discharge, a is a constant, t is the time interval and e is the base of natural logarithms. KJG

recharge The process by which precipitation is absorbed through the soil or regolith, transmitted and ultimately added to the saturated or phreatic zone within the underlying soil or bedrock, thereby replenishing GROUNDWATER reserves. The term can also be used to describe mechanisms by which soil moisture is replenished after it has been depleted due to percolation loss to underlying horizons or by evapotranspiration. Recharge zones are those areas in a landscape that are particularly important for the replenishment of groundwater. The location of such zones varies according to climate, soil permeability and thickness, permeability of underlying bedrock and the distribution of geomorphological surface depressions such as rivers and lakes. In arid areas the main recharge areas are often the beds of large ephemeral rivers, whilst recharge may be more diffuse in humid temperate environments. DJN

recovery time A term sometimes used in the explanation of SYSTEMS to describe the length of time it takes a system to recover its original state after being disturbed by an external force or process. *Elasticity* is sometimes used as an alternative term. DSGT

recumbent fold A fold that is so overturned that the strata lie horizontally.

recurrence interval The recurrence interval for an event such as a riverine flood, tsunami, or

435

earthquake, is defined as the average period of years between occurrences of an event of a particular nominated magnitude. The recurrence interval can be defined from sufficiently long historical records, such as streamflow records, as

$$RI = \frac{n+1}{m}$$

where n is the number of years of record and m is the recorded number of events of the kind being tallied from the record (e.g. earthquakes of magnitude 6). Frequently, the limited available historical records need to be supplemented by proxy data, such as earthquake magnitudes estimated from damage to ancient buildings or from the morphologic evidence preserved in the landscape. The estimation of recurrence intervals proceeds on several key assumptions, including independence of events in time and unvarying average rate of occurrence. The latter may, however, be violated as a result of climate change, urban encroachment on catchment areas or the effects of fire. DLD

Reading
Molnar, P. (1979) Earthquake recurrence intervals and plate tectonics. *Bulletin of the Seismological Society of America*, **69**, 115–133. · Nott, J. (2012) Storm tide recurrence intervals – a statistical approach using beach ridge plains in Northern Australia. *Geographical Research*, **50**, 368–376.

red beds Sediments, soils or sedimentary rocks that possess red coloration because of the presence of finely divided ferric oxides, chiefly haematite. Most represent iron-stained or cemented clastic sediments. In the continental literature, equivalent terms for red beds are *couche rouge*, *rotschicht* and *capa roja* (Pye, 1983). A distinction may be drawn between *in-situ red beds* (formed in place, without transportation, by processes of direct chemical precipitation or by weathering, soil formation and diagenesis) and *detrital red beds* (formed by erosion, transportation and redeposition of existing red soils or sediments). Only sediments and soils having hues redder than 5YR on the Munsell soil colour chart should be described as red beds. Recent work suggests that, if a suitable source of iron and oxidizing conditions exist, reddening can occur very rapidly after deposition of a sediment. ASG

Reading and Reference
Pye, K. (1983) Red beds. In A. S. Goudie and K. Pye (eds), *Chemical sediments and geomorphology*. London: Academic Press. · Turner, P. (1980) *Continental red beds*. Amsterdam: Elsevier.

redox cycle When soils, whether mineral or organic, are inundated with water, anaerobic conditions usually result. As water fills the soil pore spaces, the rate at which oxygen can diffuse through the soil is around 10,000 times slower than diffusion through a porous medium such as a drained soil. Such low rates of diffusion lead to anaerobic or reduced conditions.

The redox cycle describes the response of a soil to inundation. Oxidation occurs during uptake of oxygen and/or when hydrogen is removed (e.g. $H_2S \rightarrow S^{2-} \rightarrow 2H^+$) or when a chemical gives up an electron (e.g. $Fe^{2+} \rightarrow Fe^{3+} + e^-$). Reduction is the opposite and is the process of giving up oxygen and gaining hydrogen, or gaining an electron.

The redox potential is shorthand for the *oxygen-reduction potential*. It may be measured on a hydrogen scale (similar to pH measurement) and is referred to as E_H. It represents the concentration of oxidants {ox} and reductants {red} in a redox reaction and is defined by the Nernst equation:

$$E_H = E^\circ + 2.3 \frac{RT}{nF} \log \frac{\{ox\}}{\{red\}}$$

where E° (mV) is potential or reference, R is the gas constant ($1.987\ cal\ K^{-1}\ mol^{-1}$), T (K) is temperature, n is number of moles of electrons transferred and F is the Faraday constant ($23,061\ cal\ mol^{-1}\ V^{-1}$). ALH

Reading
Mitsch, W.J. and Gosselink, J.G. (1993) *Wetlands*. New York: Van Nostrand Reinhold.

redox potential Reduction is the gain of an electron in a chemical reaction; the opposite is known as oxidation. Oxidation is most common in weathering processes. Some elements exist in several oxidation states; for example, the ferrous state, Fe(II), is oxidized to the ferric state, Fe(III). The stability of any state depends upon the ease with which oxidation or reduction can occur (i.e. electron transfer). This is often dependent on pH. Redox potential is a measure of this ease of transfer and is symbolized by E_H. Redox potential may be measured using a redox probe, which measures electric potential in millivolts relative to a hydrogen electrode or a calomel reference electrode. A plot of redox potential (measured in volts) versus pH is a useful tool in determining the stability fields of products of reactions, especially for geochemical environments.

The presence of free dissolved oxygen in solution gives a fairly stable redox potential around +400 to +700 mV. The redox potential of wetland soils is generally in the range +400 (most oxidized) to −400 mV (reduced). WBW/ALH

Reading

Yatsu, E. (1988) *The nature of weathering*. Tokyo: Sozosha.

reduced complexity models Modelling at the mesoscale (10^0–10^2 km and 10^1–10^3 years) with 'solutions embedded in physical theory, albeit simplified to appropriate levels of complexity in order to reduce computational and parametric overheads and thus allow for time-efficient simulation and uncertainty analysis' (Brasington and Richards, 2007). Examples include topography-based hillslope hydrological models (TOPMODEL), raster storage-cell models for floodplain inundation (LISFLOOD) and cellular automaton models of channel dynamics and landscape evolution (CHILD, CEASAR).   TS

Reference

Brasington, J. and Richards, K.S. (2007) Reduced-complexity, physically-based geomorphological modelling for catchment and river management. *Geomorphology*, 90, 171–177.

reef A rocky construction found at or near sea level, formed mainly from biogenically produced carbonates. The term bioherm is also used. Several forms of marine organisms are capable of precipitating calcium and other carbonates in skeletal or nonskeletal forms, and it is the accumulation of this material that gives rise to the characteristic form of reefs. Coral reefs, whose form is usually dominated by coral skeletons, are the most common form of reef, but other organisms may also form or contribute to the development of reefs; for example, algae, vermetids and serpulids. Reefs vary greatly in size from individual clusters of organisms (called patch reefs or microatolls) to large-scale rock masses. Large coral reefs exhibit a range of facies controlled by the positions of the corals relative to exposure. Four major forms of large-scale coral reefs have been identified: fringing reefs, barrier reefs, atolls and table reefs (Stoddart, 1969).   HAV

Reference and reading

Hopley, D. (ed.) *Encyclopedia of modern coral reefs: structure, form and process*. Dordrecht: Springer. · Stoddart, D.R. (1969) Ecology and morphology of recent coral reefs. *Biology Reviews*, 44, 433–498.

reflective beach A BEACH that reflects a substantial fraction of the wave energy impinging upon it. The beach profile is steep relative to the steepness of the WAVES that it reflects. The steep profile is usually the result of accretion during swell wave conditions. The reflective condition represents one end of the spectrum of morphodynamic beach states described by Wright *et al.*

(1979). A dissipative beach, where almost all wave energy is dissipated in the nearshore, is at the other end of the continuum, and intermediate beaches separate the extremes. Reflective beach profiles are morphologically simple. They are convex upward in shape, and are without nearshore bars (see BAR, NEARSHORE). They may be more complex in plan, as beach cusps (see CUSP, BEACH) are commonly present. Reflection is a necessary condition for the generation of EDGE WAVES on beaches.   DJS

Reading and Reference

Wright, L.D., Chappell, J., Thom, B.G., *et al.* (1979) Morphodynamics of reflective and dissipative beach and inshore systems, southeastern Australia. *Marine Geology*, 32, 104–140. · Wright, L.D. and Short, A.D. (1984) Morphodynamic variability of surf zones and beaches: a synthesis. *Marine Geology*, 56, 93–118.

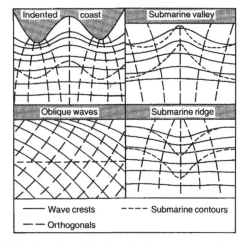

Examples of wave refraction showing wave crests and orthogonals relative to submarine contours.
Source: King (1980) Reproduced with permission of John Wiley & Sons.

refraction, wave Wave refraction is the bending of the wave front as water depth changes when it is less than half the wave length. The wave velocity decelerates as water depth decreases. The wave front, therefore, bends to become more nearly parallel to the bottom contours. Orthogonals, which are lines normal to the wave crests and between which energy is constant in deep water, converge on headlands and diverge in bays. Wave energy is concentrated on the headlands and dissipated in bays. Wave refraction diagrams can be drawn manually or by computer where the offshore relief, wave approach direction and wave period are known. These show zones of wave energy concentration and dissipation by constructing the orthogonals or wave rays.   CAMK

Reading and Reference
Authur, R.S., Munk, W.H. and Isaacs, J.D. (1952) The direct construction of wave rays. *American Geophysical Union Transactions*, **33**, 955–965. · Harrison, W. and Wilson, W.W. (1964) *Development of a method for numerical calculation of wave refraction. Technical memoir 6.* Washington, DC: Coastal Engineering Research Centre. · King, C.A.M. (1980) *Physical geography*. Oxford: Blackwell.

refugia Localities or habitats in which formerly widespread organisms survive in small restricted populations that are either disjunct or endemic in their distribution. Such localities usually possess some distinctive microclimatic, geomorphological, ecological or historic characteristic that accounts for the survival and persistence of the relict organism at that site. A classic example of such a refugium is the Upper Teesdale area of Yorkshire and Co. Durham in northeast England, which harbours survivals from the open-habitat flora that existed just after the last Ice Age (e.g. the arctic-alpine species, *Betula nana*).   PAS

Reading
Holmquist, C. (1962) The relict concept. *Oikos*, **13**, 262–292. · Pennington, W.A. (1974) *The history of British vegetation*, 2nd edition. London: English Universities Press.

reg A stony desert or any desert region where the surface consists of sheets of gravel.

regelation The refreezing of meltwater. This commonly occurs at a glacier surface to form superimposed ice and at the base of a sliding glacier to form regelation ice. In the latter case, ice at the base of a warm-based glacier melts on the upstream side of an obstacle where pressure is locally high and refreezes on the downstream side where pressure is lower.   DES

Reading
Paterson, W.S.B. (1981) *The physics of glaciers*. Oxford: Pergamon.

regeneration complex A term coined by Scandinavian botanists in the early twentieth century to describe the small-scale mosaic pattern of vegetation on raised-bog plateaux surfaces that was thought to be the mechanism whereby such OMBROTROPHIC bogs grew or regenerated themselves in an autogenic cyclic fashion. The mosaic of species – *Sphagna* (bog mosses) *Calluna* (heather), *Eriophorum* (cotton-grass), and so on – also reflects the microtopography of the mire surface, a pattern of hummocks and hollows,

which may contain open water. Each species, because of its tolerance to waterlogging and acidity and so on, has a microhabitat range within this topography. Of all the plant associations of an untouched raised bog it is this regeneration complex that is the most vigorous in growth and peat accumulation. The principle of the regeneration complex cycle was that, as a hummock became higher and drier than the surrounding hollows, its growth would slow down and it would become moribund. The hollows would then grow up through peat accumulation to overtop and drown the former hummock that would then in turn become a hollow or pool. This hummock–hollow cycle would be repeated all over the growing bog surface and was seen as the normal growth process over centuries and millennia. Specific testing of the theory, from peat sections of the past two millennia, was carried out by Barber (1981), who rejected the theory in favour of close climatic control of the hydrology, and therefore of the floral composition and peat stratigraphy, of raised bogs.   KEB

Reading and Reference
Barber, K.E. (1981) *Peat stratigraphy and climatic change: a palaeoecological test of the theory of cyclic peat bog regeneration.* Rotterdam: Balkema. · Godwin, H. (1981) *The archives of the peat bogs.* Cambridge: Cambridge University Press.

regime The average annual variations of climatic or hydrological variables. Seasonal fluctuations of river discharge, represented by the mean monthly flows, are river regimes that vary regionally in relation to precipitation, temperature, evapotranspiration and drainage basin characteristics (Beckinsale, 1969). Using the analogy of average climatic statistics, equilibrium channel morphology is also referred to as the regime state, which is constant over a period of years, is adjusted to prevailing hydrological and sedimentological influences, and is predicted using regime theory (Blench, 1969). The term is also applied to flow regimes, which are defined by the Froude and Reynolds numbers and associated bedform types.   KSR

References
Beckinsale, R.P. (1969) River regimes. In R. J. Chorley (ed.), *Water, earth and man*. London: Methuen. · Blench, T. (1969) *Mobile bed fluviology*. Edmonton: University of Alberta Press.

regime theory An approach to the design of river channel shape based upon theoretical and empirical assessment of the best shape for transporting a given DISCHARGE and sediment supply. It has its origins in engineering in early work on the design of canals through mobile bed sediments, notably for overseas irrigation. One of the most basic supporting concepts behind regime theory is the idea of HYDRAULIC GEOMETRY. This argued that width, depth and velocity in a channel all change in response to increases in discharge through the channel, something reflected in both: (1) trends in cross-section morphology from headwaters to the sea, as river discharge increases; and (2) changes in cross-section morphology at a fixed cross-section in response to discharge fluctuation through time.

Regime theory requires an assumption of EQUILIBRIUM, with the river being adjusted to channel discharge and the sediment supplied to it. The main support for this followed from magnitude and frequency effects, in which Wolman and Miller (1960) argued that intermediate events in rivers systems (e.g. the bankfull discharge) do most work. This leads directly into the concept of DOMINANT DISCHARGE, which describes the average discharge of a river to which its average form is most directly related. Thus, regime theory can be based upon identification of associations between the dominant channel-forming discharge combined with relationships identified in a hydraulic geometry analysis to design rivers whose shape is capable of transporting the discharge supplied to them.

Recent progress has recognized the problems of regime theory for a number of reasons. First, most river channel adjustments are essentially multivariate. In addition to changes in width, depth and velocity, it is also possible to change bed slope and boundary resistance. As a result, most river channel adjustments involve mutually interacting processes, where many variables may change simultaneously. As a result, rivers may display many different responses, and hence morphologies, according to the interaction of feedbacks during the adjustment process (e.g. Miller, 1984).

Second, many river channels will only show equilibrium characteristics over certain timescales. As discharge and sediment supply change, so the morphology required to achieve transport will change. Cross-section morphology rarely adjusts at a sufficient rate to a given change in discharge or sediment transport for the morphology to be always capable of transporting the water and sediment supplied to it; it is for this reason that rivers flood! Thus, even if it is possible to identify a stable morphology capable of transporting a particular discharge, this does not mean to say that the morphology will be stable at all discharges, particularly extreme ones.

Third, fluvial geomorphologists now recognize that river morphology does not just adjust to the discharge and sediment supplied to it, but can actively influence the processes operating within it. In a benchmark paper, Schumm and Lichty (1965) argued that variables changed their status according to the scale of interest. For instance, discharge is a product of catchment characteristics (e.g. basin relief, geology, hydrology) and is therefore a dependent variable. However, for smaller scale processes, it is an independent variable. For instance, sediment transport may be explained by discharge variation. At these smallest scales, Schumm and Lichty note that, rather than river channel morphology being a product of variables like discharge, morphology could be an active influence upon the interactions between flow and sediment transport processes. This 'bottom-up' view of river channel change emphasizes the dynamics of that change in response to imposed external factors (e.g. sediment supply, discharge variation) and is one of the reasons for more intensive investigation of particular case-study rivers. It implies that morphology influences process, as well as process causing changes in morphology.   SNL

Reading and References

Blench, T. (1969) *Mobile-bed fluviology*. Edmonton: University of Alberta Press. · Hey, R.D. (1997) Stable river morphology. In C. R. Thorne, R. D. Hey and M. D. Newson (eds), *Applied fluvial geomorphology for river engineering and management*. Chichester: John Wiley & Sons, Ltd; pp. 223–236. · Miller, T.K. (1984) A system model of stream-channel shape and size. *Bulletin of the Geological Society of America*, **95**, 237–241. · Schumm, S.A. and Lichty, R.W. (1965) Time, space and causality in geomorphology. *American Journal of Science*, **263**, 110–119. · Wolman, M.G. and Miller, J.P. (1960) Magnitude and frequency of forces in geomorphic processes. *Journal of Geology*, **68**, 54–74.

regional climate model (RCM – or a limited area model)

A discretized numerical rendition of the Earth's climate system for a limited region of the planet. The RCMs are based on the same set of primitive equations (including Newton's laws of motion, the laws of thermodynamics and/or mass continuity) as the general circulation (global) models. Because these RCMs are integrated over a limited spatial domain, they can afford to run at much higher spatial resolutions than the global models for given computational resources. At times, given the disposition of the available computational resources and or the size of the spatial domain, the RCM can be discretized to very high spatial resolutions that are cloud resolving, at which point the mass continuity equation is abandoned for the fully compressible Navier–Stokes set of equations that can resolve acoustic modes.

The RCMs are, however, based on an ill-posed lateral boundary condition (Staniforth, 1997; Warner *et al.*, 1997), wherein the global model solution is prescribed at the lateral (east, west, north and south) boundaries of the limited spatial domain. At the lateral boundaries, the RCMs have to reconcile with disparate spatial resolutions of the coarser global model solution and its own finer spatial resolution. This is often accomplished empirically in the so-called buffer or sponge zone of several grid points wide, wherein the regional model solution is gradually nudged from the interior of the sponge zone towards the global model solution at the exterior points of the sponge zone. This empirical adjustment in many instances is supplemented with spectral nudging of the large scale in the regional model solution to the corresponding global model solution, which is applied throughout the regional climate domain (von Storch *et al.*, 2000; Kanamaru and Kanamitsu 2007; Misra 2007). SJN

References
Kanamaru, H. and Kanamitsu, M. (2007) Scale selective bias correction in a downscaling of global analysis using a regional model. *Monthly Weather Review*, **135**, 334–350. · Misra, V. (2007) Addressing the issue of systematic errors in a regional climate model. *Journal of Climate*, **20**, 801–818. · Staniforth, A. (1997) Regional modeling: a theoretical discussion. *Meteorology and Atmospheric Physics*, **63**, 15–30. · Von Storch, H., Langenberg, H. and Feser, F. (2000) A spectral nudging technique for dynamic downscaling purposes. *Monthly Weather Review*, **128**, 3664–3673. · Warner, T.T.,. Peterson, R.A and Treadon, R.E. (1997) A tutorial on lateral boundary conditions as a basic and potentially serious limitation to regional numerical weather prediction. *Bulletin of the American Meteorological Society*, **78**, 2599–2617.

regolith A term coined by Merrill (1897: 299–300) to describe the 'superficial and unconsolidated portion of the earth's crust . . . the entire mantle of unconsolidated material, whatever its nature or origin'. Some later workers have tended to use it as a synonym for weathering products, but such a narrow usage is incorrect (Gale, 1992). ASG

References
Gale, S.J. (1992) Regolith: the mantle of unconsolidated material at the Earth's surface. *Quaternary Research*, **37**, 261–262. · Merrill, G.P. (1897) *A treatise on rocks, rock weathering and soils*. New York: Macmillan.

regosol Any weakly developed soil.

regur Black, clayey soils that develop in tropical regions of high rainfall and temperatures, notably the northwest Deccan of India.

rejuvenation The renewal of former activity, as of a fault that is reactivated by new tectonic movement or especially a stream whose erosional activity is redeveloped by uplift, a base-level fall or possibly a change in stream sediment load or discharge. The term may be applied to whole landscapes whose reduced relief may be redeveloped as a result of such changes, giving valley-in-valley forms.

The term especially derives from the interpretation of landforms developing in a cyclical fashion from youth to old age, with rejuvenation representing the redevelopment of 'young' surface forms in landscapes otherwise well advanced in a cycle of development. Some of the landforms that were so interpreted more than half a century ago might now be explained in alternative ways, with emphasis placed on equilibrium adjustments between forms, environments and processes. JL

relative age The age of an event or feature expressed not in terms of time units, such as years, but in relation to other phenomena (see ABSOLUTE AGE).

relative dating A very simple form of GEOCHRONOLOGY whereby a sediment, landform or other feature is ascribed an age relative to other sediments, landforms or features. Relative dating may have two forms: the *relative position* of a feature and the *relative change* of a feature. Relative position may be exemplified by the position of glacial end moraines in a formerly glaciated part of a valley, whereby the innermost (closest to source area) moraine is believed to be younger than another end moraine further away from the source area. If the period of time that lapsed between the two features is not known, and if the numerical age of one of the features is not known, then they are only differentiated by the position in the landscape. An example of relative change is when one soil is deemed to be older than another on the basis of mineral weathering or organic content. In this case, it is assumed that change is a function of time and that the sediments in which the two soils developed were initially the same. DSGT

relative humidity See HUMIDITY.

relaxation time   A term used in the study of SYSTEMS to describe the length of time it takes a system to establish a new quasi-stable state after an external disturbance has taken place.   DSGT

relic/relict   These words are variously used by different authors to describe landforms or features that are no longer experiencing the principal process that was responsible for their origin. The *Oxford English Dictionary* describes *relic* as a dead body and *relict* as a geological or other object that has survived in a primitive form. The terms are widely used in palaeoenvironmental studies.   DSGT

remanié   A glacier that is not directly connected to a snow field but receives ice from avalanches.

remote sensing   The means by which a scene of interest can be studied using a device separated from it by some distance. Within the context of physical geography, the remote sensing definition can be refined to focus on the measurement and recording of the electromagnetic radiation (EMR) emanating from the Earth's environment (surface and atmosphere) by sensors mounted on a platform at a vantage-point above the Earth's surface. The data acquired by remote sensing offers a continuous record of our environment that is consistent and can be more accurate than that derived through the interpolation of limited field observations. Remote sensing through the medium of EMR (commonly, but inaccurately, known as "light") comprises four components, all of which are linked through the transfer of that radiation. These components are source of EMR, Earth surface targets, atmospheric effects and the sensor. The first component of remote sensing, the source of EMR, can be natural, originating from the Sun or the Earth by emission, or artificial. The second component, Earth surface targets, determines the EMR received by the remote sensor. The essential physical, chemical and biological properties of the targets determine the amount of radiation from it and thus measured by a sensor at a particular wavelength. EMR may also interact with the atmosphere residing between the Earth's surface and the sensor, contributing to or modifying the radiation reaching the sensor, and this constitutes the third component of remote sensing. The final component, the sensor, is that which measures remotely the radiation from the Earth's environment. Commonly, there are three remote sensing approaches used, with each of these featuring the four components of remote sensing in different ways. The first of these exploits EMR from the Sun (source) that has been reflected from the Earth's surface and received by the sensor (also known as OPTICAL REMOTE SENSING), whereas the second of these uses EMR that has been emitted from the Earth's surface (as the source) and received by a sensor. Both of these are known as passive remote sensing. The third approach uses artificial EMR from a sensor (source) that is backscattered from the Earth's surface and received by a sensor and is therefore an active approach to remote sensing (common systems are lidar and radar).

EMR, on which remote sensing works, can be thought of as time-varying electric and magnetic fields that are orthogonal to each other, which travel through space and media in the form of a wave at the speed of light $(2.998 \times 10^8 \, \mathrm{m\,s^{-1}})$. These waves can be short or long in length with an associated high or low energy respectively. These properties are described by the electromagnetic spectrum, which is commonly divided into regions for convenience and by tradition within the field of remote sensing. The main regions being visible (short wavelengths of 0.4–0.75 μm), near infrared (NIR: 0.7–1.50 μm), middle/shortwave infrared (MIR: 1.5–5.0 μm), thermal infrared (TIR: 5.0–15 μm) and microwave (long wavelengths of 15.0–100 μm). The length, and thus the energy, of the electromagnetic waves principally determines how the radiation interacts with targets in the environment, thus allowing us to distinguish between the properties of our environment that vary not only by geography but also by time. All matter having a temperature above absolute zero (−273 °C, 0 K) emits EMR. In remote sensing, the Sun is the most obvious source of natural EMR (solar radiation). Generally speaking, solar-emitted EMR is used for remote sensing at visible to MIR wavelengths and Earth-emitted EMR is used for remote sensing at wavelengths beyond the mid-MIR region of the electromagnetic spectrum. Remote sensing exploits this natural phenomenon and, once the sensor is deployed, represents the cheapest approach per unit area in cost for measuring and monitoring the environment and its properties.

Once emitted from its source there are a number of ways in which radiation can interact with matter, depending on its wavelength and the properties of the matter. Examples include the high absorption of visible radiation by chlorophyll in leaves that varies as a function of the species of vegetation present and/or its biomass/leaf area, the high absorption of NIR radiation by organic-rich soil, the high backscattered microwave radiation from snow-covered sea ice, and the higher reflectivity of water at visible wavelengths due to the presence of suspended sediments. Since we understand how EMR interacts with different targets, using remote sensing we can also indirectly infer related properties and extrapolate over time and place to keep an 'eye' on our environment. Owing to improvements in technology, we can design remote sensing devices (sensors) that provide ever more accurate and precise information about our environment, allowing better informed decision-making. Owing to the commitment of countries around the globe to remote sensing programmes, we can look forward to more of these data. Remotely sensed records for the study of the Earth's environment are currently available from a plethora of sensors that are constantly evolving with respect to their size, sensing capabilities, platform type and costs. Sensors differ in their characteristics. They measure radiation at a range of pixel sizes (i.e. the spatial resolution), wavebands (i.e. the spectral resolution) and temporal frequencies (i.e. the temporal resolution). Much remote sensing uses sensors that are on board satellite platforms (satellite remote sensing) that provide a global perspective on the environment (e.g. the Landsat satellite). However, other platforms can be used, usually in order to acquire a very detailed perspective on a particular locality (including airplanes, helicopters and unmanned aerial vehicles). The deployment of different remote sensing devices on these various platforms ensures that our environment can be measured at the appropriate scale for the phenomenon under investigation. In order to facilitate the use of the data from these sensors, it is increasingly the case that remote sensing agencies (such as the European Space Agency, NASA) are producing higher order geospatial data sets derived from the imagery produced by remote sensing that serve to increase the integration of data for physical geographers to use. As such, remote sensing is now able to routinely produce a wide range of geospatial products about the environment, and it is an increasingly important approach to studying the Earth's environment.   DSB

Reading
Boyd, D.S. (2009) Remote sensing in physical geography: a twenty-first century perspective. *Progress in Physical Geography*, 33, 451–456. · Campbell, J.B. (2009) *Introduction to remote sensing*, 4th edition. New York: The Guildford Press. · Jones, H.G. and Vaughan, R.A. (2010) *Remote sensing of vegetation: principles, techniques and applications*. Oxford: Oxford University Press. · Lillesand, T., Kiefer, R.J. and Chipman, J. (2009) *Remote sensing and image interpretation*, 6th edition. New York: John Wiley & Sons, Inc. · Tatem, A.J., Goetz, S.J. and Hay, S.I. (2008) Fifty years of Earth-observation satellites. *American Scientist*, **96**, 390–398. · Warner, T.A., Nellis, M.D. and Foody, G.M. (eds) (2009) *The SAGE handbook of remote sensing*. London: SAGE.

rendzina Dark, organic-rich soil horizons developed upon unconsolidated calcareous materials in areas of chalk and limestone bedrock.

renewable resources See FLOW RESOURCES.

representative and experimental basins Small drainage basins where the representative basins are chosen to be representative of hydrological regions and should experience minimum change during monitoring and where the experimental basins are subjected to deliberate modification so that the impact of the modification on drainage basin dynamics may be identified. The establishment of a large number of representative and experimental basins was one of the major achievements of the International Hydrological Decade (see HYDROLOGY).

Studies in representative basins involve monitoring hydrological processes to improve understanding of these processes and their interrelationships. If plentiful data can be built up from catchments that are representative of a particular hydrological region, these data should help to improve estimation of the magnitude of hydrological processes, particularly discharge characteristics, in ungauged catchments in the same region. Benchmark catchments (Toebes and Ouryvaev, 1970) have been identified as a special type of representative basin that are in their natural state and, therefore, permit the observation of hydrological processes without the effects of humans. The Vigil Network also forms a network of representative basins that was originally conceived within the USA (Leopold, 1962; Slaymaker and Chorley, 1964). In the case of the Vigil Network, the basins are not protected from artificial change, but the aim is to monitor a large number of hydrological and geomorphological variables using simple and inexpensive techniques so that landscape and process change may be identified over long periods of time.

Work in experimental basins involves not only observation of hydrological processes, but also deliberate modification of the basin so that the impact of cultural changes on hydrological processes may be measured. Experimental basins are often subjected to a calibration period for several years before modification. This allows the magnitude and variability of processes before and after the change to be observed and compared. If more than one basin is available, where all the basins are virtually identical in size, physical character and vegetation cover, either a paired or a multiple watershed experiment can be carried out. In this type of experiment a long calibration period is desirable but not essential. One or more control basins can be monitored without receiving modification and the remaining basin or basins can be subjected to one or more experimental treatments, preferably using replicate basins for each treatment. In all the basins the same processes are observed, so that differences between the control and the treated basins may be identified.   AMG

Reading and References

IAHS (1965) *Bassins representatifs et experimentaux/Representative and experimental areas, Symposium of/de Budapest, 1965,* vols I and II. International Association of Scientific Hydrology publication no. 66. Wallingford: IAHS Press. · IAHS (1980) *The influence of man on the hydrological regime with special reference to representative and experimental basins. Proceedings of a symposium held at Helsinki, June 1980.* International Association of Hydrological Sciences publication no. 130. Wallingford: IAHS Press. · Leopold, L.B. (1962) The Vigil Network. *International Association of Scientific Hydrology. Bulletin,* 7 (2), 5–9. · Slaymaker, H.O. and Chorley, R.J. (1964) The Vigil Network system. *Journal of Hydrology,* 2, 19–24. · Swank, W.T. and Crossley, D.A. (1987) *Forest hydrology and ecology at Coweeta.* Ecological Studies, vol. 66. Berlin: Springer-Verlag. · Toebes, C. and Ouryvaev, V. (eds) (1970) *Representative and experimental basins: an international guide for research and practice.* UNESCO studies and reports in hydrology 4. Paris: UNESCO.

reptation   Transitional between the creep and saltation of sand grains in aeolian transport, reptation (Anderson, 1987) occurs when the high-velocity impact of a saltating grain sets several other grains moving through a low hopping process.   DSGT

Reference

Anderson, R.S. (1987) Eolian sediment transport as a stochastic process: the effects of a fluctuating wind on particle trajectories. *Journal of Geology,* **95**, 497–512.

resequent stream   A stream following the original direction of drainage, but developed at a later stage. This might apply to streams on the back slope of a cuesta of *resistant* rock that did not outcrop on the originally exposed land surface that guided initial stream development. The term is not now commonly used.   JL

reservoir rocks   Also see GROUNDWATER. Rocks, usually sedimentary, that act as a store of a liquid or gas that has resource potential. This may be water, but also oil and other hydrocarbons.

The release of this resource might require the use of FRACKING techniques.

reservoir, storage effects of   Reservoirs may be constructed to regulate river flows for supply purposes downstream (regulating reservoirs), to provide water supply directly from direct supply reservoirs or to control floods downstream, in which case the reservoir is maintained at less than capacity so that flood discharges may accumulate in the reservoir (flood control reservoirs). Immediately downstream of the dam, scour of the channel bed and banks may occur because the water released is comparatively free of sediment, and further downstream there can be major changes of CHANNEL CAPACITY or of channel planform as a response to the changes in FLOOD FREQUENCY and to the altered sediment transport. There will also be changes in the river ecology as a response to changes in flow and to the alterations of aquatic habitats.   KJG

Reading

Eschner, T., Hadley, R.F. and Crowley, K.D. (1983) *Hydrologic and morphologic changes in channels of the Platte River Basin in Colorado, Wyoming, and Nebraska: a historical perspective.* United States Geological Survey Professional Paper 1277-A. Washington, DC: US Government Printing Office. · Petts, G.E. (1979) Complex response of river channel morphology to reservoir construction. *Progress in Physical Geography,* **3**, 329–362. · Williams, G.P. and Wolman, M.G. (1984) *Downstream effects of dams on alluvial rivers.* United States Geological Survey Professional Paper 1286. Washington, DC: US Government Printing Office.

residence time   A concept employed in studies of chemical weathering and solute generation in drainage basins to describe the length of time between the input of water as precipitation and its output as run-off. The residence time is seen as a major control on the magnitude of solute uptake by water moving through the drainage basin system and, therefore, on solute concentrations in streamflow, since a significant period of time may be required for the water to reach chemical equilibrium with the soil and rock.   DEW

residual strength (or residual shear strength)   The strength of a soil (usually a clay soil) that is below the peak SHEAR STRENGTH. Large shear strains (deformations) produce a reorientation of the clay mineral particles in the FABRIC and allow a shear plane to develop. The shear strength along this plane is less than the peak strength. The residual strength can be determined in a SHEAR BOX (or a modified version of this called a ring shear) apparatus for soil testing.   WBW

resilience First introduced into the ecological literature by Holling (1973), it is a measure of the number of times a system can recover after a stress or disturbance. It measures the ability of an ecosystem to persist in the face of such perturbations as extreme weather events, fires, pollution and invasive species. ASG

Reading and Reference
Holling, C.S. (1973) Resilience and stability of ecological systems. *Annual Review of Ecology and Systematics*, 4, 1–23. · Gunderson, L.H., Allen, C.R. and Holling, C.S. (eds) (2009) *Foundations of ecological resilience*. Washington, DC: Island Press.

resistance *a.* Any property that opposes acceleration. Acceleration or movement of rock, soil, snow or ice is opposed by compressive strength, tensile strength and especially SHEAR STRENGTH. MASS STRENGTH is more important than INTACT STRENGTH. Components of strength include plane friction, interlocking and COHESION. More broadly, resistance is more than strength: it includes surface ROUGHNESS, vegetation and permeability. (See also FORCE and FACTOR OF SAFETY.) *b.* Term sometimes used in the study of SYSTEMS to explain the ability of a system to withstand disturbance by an external force or change. IE

Reading
Carson, M.A. and Kirkby, M.J. (1972) *Hillslope form and process*. Cambridge: Cambridge University Press; chapter 4.

respiration Chemical process in cells that is, for land plants, the reverse of PHOTOSYNTHESIS, whereby oxidation of nutrients releases energy for use in metabolism. Products of the process, carbon dioxide and water, are transferred back into the atmosphere. The reactions may be represented, in simplified form, by

$$2(C_3H_6O_3) + 6O_2 \rightarrow 6CO_2 + 6H_2O + \text{heat}$$

The released heat forms part of the longwave energy component of the Earth's atmosphere, and is unavailable thereafter for use in biological systems. The amount of the plant's available energy store that is used in respiration varies widely, but it is often in excess of 50% of that used in growth, and may be very much more than this (see also NET PRIMARY PRODUCTIVITY (NPP)). Unlike photosynthesis, respiration is a day-long, not just a day-time, phenomenon. MEM

resultant wind The vectorial average of all wind directions and speeds for a given level at a given place for a specified period, such as 1 month. It is obtained by resolving each wind observation into components from north and east, summing over the given period, obtaining averages and reconverting the average components into a single vector. BWA

resurgence An emergence point of underground water. The term is often used to describe karst SPRINGS, the headwaters of which are initially on the surface but are lost underground by disappearing down a stream-sink (or SWALLOW HOLE). A distinction is sometimes made between this kind of spring and that with no known surface headwaters, which is called an exsurgence. PWW

Reading
Bögli, A. (1980) *Karst hydrology and physical speleology*. Berlin: Springer-Verlag. · Smith, D.I., Atkinson, T.C. and Drew, D.P. (1976) The hydrology of limestone terrains. In T. D. Ford and C. H. D. Cullingford (eds), *The science of speleology*. London: Academic Press; pp. 179–212.

retention curve Or soil moisture retention curve, obtained by plotting soil moisture content against soil moisture suction. The curve will be of a different form according to whether the soil is drying or wetting. This is due to the phenomenon of capillary hysteresis. PWW

Reading
Childs, E.C. (1969) *An introduction to the physical basis of soil water phenomena*. New York: John Wiley & Sons, Inc. · Ward, R.C. and Robinson, M. (1990) *Principles of hydrology*, 3rd edition. Maidenhead: McGraw-Hill.

retention forces Forces responsible for holding water in the pores of a rock or soil against the force of gravity. The forces involved are capillarity (surface tension), adsorption and osmosis. PWW

Reading
Ward, R.C. and Robinson, M. (1990) *Principles of hydrology*, 3rd edition. Maidenhead: McGraw-Hill.

retrogradation The destruction of a beach profile by large breakers resulting in retreat of the shoreline.

return period The average time between events such as the flooding of a particular level. This information may also be expressed as the level that has a particular return period of flooding; for example, a hundred years. The inverse of the return period is the statistical probability of an event occurring in any individual year. DTP

reversed fault A fault whose fault plane is inclined towards the upthrown side.

reversed polarity The condition occurring when the Earth's magnetic poles change their polarities, the positive becoming the negative and vice versa. Generally applied when the positive magnetic pole lies in the southern hemisphere.

reversing dune A dune that tends to grow upwards but migrate only a limited distance because seasonal shifts in direction of the dominant wind cause it to move alternately in nearly opposite directions. See DUNE.

Reynolds number (Re) A dimensionless ratio defining the state of fluid motion as laminar or turbulent according to the relative magnitude of inertial and viscous forces. In general:

$$\text{Re} = \frac{\rho_f v L}{\mu} = \frac{vL}{\nu}$$

where ρ_f is fluid density, v is velocity, L is a characteristic length, μ is the dynamic viscosity and ν is the kinematic viscosity ($\nu = \mu/\rho_f$), which is a measure of molecular interference between adjacent fluid layers. The Reynolds number for channel flow is defined by using the hydraulic radius R or the mean depth as the characteristic length, and the mean flow velocity. Flow is turbulent if $\text{Re} > 750$; laminar flow ($\text{Re} < 500$) is rare in rivers and only occurs in shallow overland flow depths. Grain Reynolds numbers can be defined using grain diameter as the length and grain fall velocity in still water as the velocity. If the grain Reynolds number is <0.1, flow around the falling grain is streamlined and laminar, and the fall velocity varies as the square of the grain diameter (Stokes' law). This is the case for silt and clay particle sizes; grain Reynolds numbers are larger for sand particles, and turbulence and flow separation occur around the falling grain as inertial forces dominate. For such grains the fall velocity varies as the square root of grain diameter (Richards, 1982: 76–79). KSR

Reference
Richards, K.S. (1982) *Rivers: form and process in alluvial channels*. London: Methuen.

rheidity The capacity of some solid materials to flow under certain conditions.

rheology The study of the deformation and flow of matter. In particular, rheology is concerned with whether a substance behaves as an elastic solid or effectively as a 'viscous' solid or in a manner intermediate between these two. The type of deformation experienced by a particular material is influenced by the duration of the applied stress. This is expressed by the property of rheidity, which depends on the relationship between the resistance to viscous flow (viscosity) and the resistance to elastic deformation (rigidity) of a substance and is expressed in units of time. For ice a deforming stress must be applied for only a few weeks before it begins to flow, but for subcrustal material the time required is several thousand years. MAS

Reading
Hager, B.H. and O'Connell, R.J. (1980) Rheology, plate motions and mantle convection. In P. A. Davies and S. K. Runcorn (eds), *Mechanisms of continental drift and plate tectonics*. London: Academic Press. · Mörner, N.-A. (ed.) (1980) *Earth rheology, isostasy and eustasy*. Chichester: John Wiley & Sons, Ltd. · Ramberg, H. (1981) *Gravity, deformation and the Earth's crust*, 2nd edition. London: Academic Press.

rheotrophic See NUTRIENT STATUS.

rhexistasy See BIOSTASY.

rhizome The underground stem of some plants. The rootstock.

rhizosphere The portion of the soil zone that immediately surrounds the root systems of plants.

rhodoliths Free-living massive and branching spheroidal growths of calcareous red algae that occur in two different settings: (a) shallow lagoonal, reef-flat and back-reef environments, commonly in tidal channels and sea-grass beds, and (b) moderately deep water on fore-reef terraces or shelf edges. ASG

Reading
Scoffin, T.P., Stoddart, D.R., Tudhope, A.W. and Woodroffe, C. (1985) Rhodoliths and coralliths of Muri Lagoon, Rarotonga, Cook Islands. *Coral Reefs*, 4, 71–80.

rhourd A pyramid-shaped dune akin to a STAR DUNE.

ria An inlet of the sea formed by the flooding of river valleys, either by the rising of the sea during the Flandrian transgression (see TRANSGRESSION) or as a consequence of sinking of the land. They contrast with fjords, which are drowned glacial valleys. They are a feature of Pembrokeshire, the southwest peninsula of England, Brittany and Galicia (Spain).   ASG

ribbed moraine   See ROGEN MORAINE.

Richardson number   In the presence of a density gradient in a liquid, whether or not the liquid is stratified in a stable manner, stability can only be judged with respect to the level of turbulence, and hence the shear within the liquid. Thus, the Richardson number is a parameter that is used to classify a liquid in terms of its propensity to be stable and is expressed as the ratio of the stabilizing forces that exist in a fluid due to its density to the destabilizing forces that arise from turbulent shear.   SNL

Richter denudation slope   A straight rock-slope unit with an angle of inclination that is at the maximum angle for stability (usually 32–36°) of its thin talus cover. Such slopes are relatively common in the Transantarctic Mountains, and have been recognized in the Cape Mountains of southern Africa and the European Alps. They form as rock fragments fall from the cliff at the crest of the Richter slope. If the newly fallen material just covers a little of the base of the cliff, the next fall will be over the new talus, and hence the base of the cliff will then be higher, so the cliff will extend upwards by a series of minute steps at the angle of the talus. Eventually, the free face will be eliminated and a smooth rock slope of uniform inclination will underlie a talus sheet. The talus may subsequently be removed.   MJS

Reading
Selby, M.J. (1993) *Hillslope materials and processes*, 2nd edition. Oxford: Oxford University Press.

Richter scale   The Richter scale is a measure of EARTHQUAKE magnitude introduced in 1935 by Charles Richter, based on the amplitude of the largest wave form as recorded on a seismogram. To accommodate the wide range of values of wave form associated with earthquake amplitude the scale is logarithmic. Each unit on the Richter scale represents a 10-fold increase in wave amplitude and around a 30-fold increase in the energy released. The largest recorded earthquakes have a magnitude of 8.6, and earthquakes with a magnitude of more than 5.5 typically cause structural damage. The measured values are corrected for depth and distance from the epicentre. The

magnitude of an earthquake is a representation of the energy released, and this is reflected not only by the wave form amplitude but also by the duration of ground shaking. Richter magnitude is represented by the expression

$$M = \log\left(\frac{X}{T}\right) + Y$$

where M is Richter magnitude, X is waveform amplitude, T is frequency (measurement of duration) and Y is the difference in arrival time between P and S waves (a measure of the distance of the reading from epicentre). Though still used in the media and so on, the Richter scale has been superseded in science by the MOMENT MAGNITUDE SCALE.   AD

ridge and runnel topography   A series (typically $n = 2$ to 6) of alternating shore-parallel intertidal bars and troughs on beaches of overall low gradient in fine to medium sands. Such topography can be well developed in fetch-limited, macrotidal coastal settings. Low to moderate wave energy appears to favour ridge development and maintenance, as high wave energy events flatten the ridges. Traditionally, it has been argued that the ridges are formed by swash processes

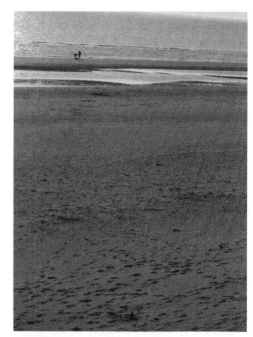

A ridge and runnel sequence on a low gradient sandy shoreline.

Photograph by Tom Spencer.

Rift valley. Schematic representation of contrasting rift structures: (A) classic symmetric graben structure with a downthrown block bounded by normal faults. (B) asymmetric, half-graben structure. In both cases the number of faults in real rifts and the complexity of their structure are much greater than indicated here. (C) Highly schematic illustration of alternating half-graben along a rift valley.
Source: Summerfield (1991: figures 4.9 and 4.10). Reproduced with permission of John Wiley & Sons.

acting at stationary tide levels. However, more recent studies have found that the highest ridges generally occur just above mid-tide level where tidal nonstationarity is greatest, suggesting that both swash zone and surf zone processes are important in ridge generation.   TS

Reference
Masselink, G. and Anthony, E.J. (2001) Location and height of intertidal bars on macrotidal ridge and runnel beaches. *Earth Surface Processes and Landforms* **26**, 759–774.

riedel shears   A special form of shear fracture that forms roughly perpendicular to the main direction of shear strain or movement. Although they are mainly found in brittle materials, they can occur in muds below the plastic limit.   WBW

riegel   A step in the rock floor of a glacial valley.

riffle   See POOL AND RIFFLE.

rift valley   A valley or linear trough formed by subsidence or downthrusting in areas of continental crust in plate interiors where tensional stresses predominate in the lithosphere. The classic interpretation of the morphology is that of a graben, with the rift floor being seen as a downthrown block bounded by normal faults, which create steep bounding escarpments. Seismic data suggest, however, that in many cases this is an oversimplification. The structure of many rifts appears to be asymmetric, with much of the downthrow occurring along a major boundary listric fault on one side only. These major faults tend to be discontinuous and may alternate along the rift, separated by transfer faults. Such rifts have a half-graben structure. A major example is the East African Rift Valley, an extensive and complex feature of the African landscape, extending from Ethiopia in the north to Mozambique in the south, with a westward extension into Botswana, and representing the progressive fracture of the African tectonic plate into two components.   ASG

Reading and Reference
Chorowicz, J. (2005) The East African rift system. *Journal of African Earth Sciences*, **43**, 379–410. · Summerfield, M.A. (1991) *Global geomorphology*. Harlow: Longman.

rifting The tearing apart of the LITHOSPHERE along a roughly linear zone. This is due to tectonic stresses from either thermal doming or plate motions. If doming from upwelling ASTHENO-SPHERE or MANTLE PLUMES comes first, the rifting is *active*, volcanism is abundant and uplift may extend for hundreds of kilometres. If stresses come from moving lithospheric plates or drag at the base of the lithosphere, the rifting is *passive*: uplift is confined to the shoulders of the rift zone and volcanism comes later as the crust and lithosphere are thinned. Production of a RIFT VALLEY by faulting leads to abundant clastic sedimentation as the steep margins are eroded.

In continental lithosphere, passive rifts may form at high angles to collision OROGENS; for example, the Rhine GRABEN in relation to the Alps, and even the Baikal rift and Shanxi graben system (northeast Asia) in relation to the distant Himalayas. Down-faulting in Baikal is 5 or 6 km. Other passive rifts relate to transform faults, as in southern California and western Turkey, and the pull-apart basin of the Dead Sea. Rifts such as the Red Sea may be active and develop into sea-floor spreading ridges; that is, new divergent plate boundaries (see PLATE TECTONICS). In the Red Sea, continental crust is stretched, thinned and injected by basalt DYKES. Points of high heat flow about 50 km apart become the initial sites of oceanic crust (mid-ocean ridge basalt) formation. Magnetic anomalies develop as this process propagates along the rift axis and, after a few million years of rifting, the segments join to form a continuous sea-floor rift valley at about 1500 m depth (Bonatti, 1985).   IE

Reference
Bonatti, E. (1985) Punctiform initiation of sea-floor spreading in the Red Sea during transition from a continental to an oceanic rift. *Nature*, **316**, 33–37.

rill Rills are small ephemeral channels that often form in sub-parallel sets on sloping agricultural land in response to intense run-off events. They are also common on steep and unprotected surfaces like road and other earthen embankments. The term 'rill' is also applied to small but longer lived channels that carry run-off from hillslopes and convey it to larger stream channels (see INTERRILL FLOW). Rills are primarily found on hillslopes, though thalweg rills do occur (e.g. Slattery *et al.*, 1994). On hillslopes, larger and deeper rills may capture the flow from adjoining ones, so that there is some integration of the rill network in the downslope direction. Rill spacing tends to diminish on steeper hillslope gradients. Rill dimensions lie in the centimetre to decimetre range for both

width and depth, but rill length may reach hundreds of metres. Being a form of channelized flow, rills carry relatively deep and fast-flowing water and are able to scour their beds efficiently and convey pebble-sized materials as bedload. Rills formed on tilled soils are ordinarily obliterated by ploughing, but can evolve again rapidly during subsequent run-off. Laboratory studies have shown that rill cutting can begin in shallow, laminar and subcritical flow, when a threshold of STREAM POWER is crossed that may be as low as $0.05\,W\,m^{-2}$. The retreat of rill headcuts and the onset of active sediment transport along a rill appear to require power levels about 10 times higher than this. DLD

Reading and Reference
Slattery, M.C., Burt, T.P. and Boardman, J. (1994) Rill erosion along the thalweg of a hillslope hollow: a case study from the Cotswold Hills, central England. *Earth Surface Processes and Landforms*, **19**, 377–385. · Vinci, A., Brigante, R., Todisco, F., *et al.* (2015) Measuring rill erosion by laser scanning. *Catena*, **124**, 97–108.

rime ice Ice formed by the accumulation of supercooled water droplets when they strike a cold object and freeze on impact. Rime ice builds up most rapidly in cool humid conditions on surfaces exposed to the wind. On mountain peaks it can accumulate as large cauliflower-shaped excrescences.   DES

Reading
Koerner, R.M. (1961) Glaciological observations in Trinity Peninsula, Graham Land, Antarctica. *Journal of Glaciology*, **3** (30), 1063–1074.

ring complex 'A petrologically variable but structurally distinctive group of hypabyssal or subvolcanic igneous intrusions that include ring dykes, partial ring dykes and cone sheets. Outcrop patterns are arcuate, annular, polygonal and elliptical with varying diameters ranging from less than 1 to 30 km or greater. The majority of ring complexes represent the eroded roots of volcanoes and their calderas' (Bowden, 1985: 17).   ASG

Reference
Bowden, P. (1985) The geochemistry and mineralization of alkaline ring complexes in Africa (a review). *Journal of African Earth Sciences*, **3**, 17–39.

ring-dyke A funnel-shaped or cylindrical intrusion of igneous rock usually surrounded by an older intrusive mass. The dyke appears as a ring of rocks when viewed from the air.

rip current A narrow, fast current flowing seaward through the breaker zone. The term, first introduced by F. P. Shepard in 1936, replaces undertow. Rip currents are fed by longshore currents in the surf zone. Their velocity ranges from about $1\,\mathrm{m\,s^{-1}}$ to over $5\,\mathrm{m\,s^{-1}}$ in severe storms. Channels 1–3 m deep can be scoured in the breaker zone by rip currents. Their spacing depends on variation of wave height alongshore, which may arise through wave refraction or may be related to the width of the surf zone. They may be associated with cuspate shoreline features and edge waves. CAMK

Reading
Bowen, A.J. (1969) Rip currents. *Journal of Geophysical Research* 74, 5467–5478. · Gruszczyński, M., Rudowski, S., Semil J., *et al.* (1983) Rip currents as a geological tool. *Sedimentology*, 40, 217–236.

riparian   A feature occurring in the region of a river bank. Thus, it is conventional to refer to riparian vegetation, riparian BUFFER STRIPS, conservation of the riparian zone, and so on. The condition of the riparian zone is of great importance to the stability of the river banks and to the quality of the stream habitat, and provides a focus for management in catchments where extensive land use has changed the hydrologic and erosional behaviour of the broader landscape. In such cases, the riparian zone forms a link between the catchment and the stream. A number of specific values are associated with the riparian zone. These include the maintenance of water quality, especially in surface run-off and groundwater that pass over and through the soils of the riparian zone. Grass and forest can provide very efficient removal of excess nutrients such as nitrogen, perhaps sourced from fertilizers, that would otherwise enter the stream system. Riparian environments may also provide valuable remnant ecosystems, a corridor for the movement of animals and an environment of considerable recreational and scenic amenity value. The condition of in-stream habitats may also be related to events in the riparian zone, which may release important organic materials that nourish the aquatic ecosystem, and also moderate water temperatures where shading of the stream occurs. Riparian vegetation may assist in the stabilization of stream banks and also contribute large fragments of organic debris (e.g. branches) that, singly or in debris jams, may temporarily obstruct flow in the stream and contribute to the diversity of habitats there.   DLD

Reading
Malanson, G.P. (1993) *Riparian landscapes*. Cambridge: Cambridge University Press.

ripple   See CURRENT RIPPLES.

rising limb   The portion of the stream hydrograph for a storm event between the beginning of the rise in discharge and the peak of the hydrograph. (For diagram, see HYDROGRAPH.)

river basin planning (integrated basin planning)   The United Nations first advocated integrated river basin development in 1958, and its report was reissued in 1970. The process of managing water resources within the drainage basin in a manner that optimizes water use throughout the basin and minimizes deleterious effects for water, river channels and land use. Although the idea of an integrated approach to river basin development is long established, experience of piecemeal development that has induced subsequent feedback effects in other parts of the drainage basin has fostered the movement towards a more integrated view. This movement has also been encouraged both by the need to design multiple-purpose water resource projects that will, for example, supply water and power and reduce floods, and by the use of the drainage basin as the most appropriate unit for analysis in relation to planning.   KJG

Reading and Reference
Saha, S.K. and Barrow, C.J. (1981) *River basin planning: theory and practice*. Chichester: John Wiley & Sons, Ltd. · Molle, F. (2006) *Planning and managing water resources at the river-basin level: emergence and evaluation of a concept*. IWMI Comprehensive Assessment Research Report 16. Colombo: International Water Management Institute.

river classification   Rivers can be classified according to various purposes, such as geomorphic analysis, hydrological description and regionalization (Haines *et al.*, 1998), categorization for river health and ecosystem purposes (Holmes *et al.*, 2008), river management, and others. Early morphologic groupings included simple descriptions such as braided, meandering or straight. Rosgen (1994) proposed a classification that considered form of the longitudinal profile, degree of entrenchment, nature of the channel materials and other factors. Some classifications work at the reach level and assess the presence of a consistent set of geomorphic attributes, and hence are more reflective of landscape processes than simply of morphology or morphometry (Brierley *et al.*, 2002).   DLD

References
Brierley, G., Fryirs, K., Outhet, D. and Massey, C. (2002) Application of the River Styles framework as a basis for river management in New South Wales, Australia. *Applied Geography*, 22, 91–122. · Haines, A.T., Finlayson, B.L.,

Table 1 River metamorphosis: potential channel adjustments

Potential adjustments of	Fluvial landform		
	River channel cross-section	River channel pattern	Drainage network
Size	INCREASE OR DECREASE OF RIVER CHANNEL CAPACITY Erosion of bed and banks can produce a larger channel that maintains the same shape. Sedimentation can produce a smaller channel that maintains the same shape.	INCREASE OR DECREASE OF SIZE OF PATTERN Increase or decrease of meander wavelength while preserving the same planform shape.	INCREASE OR DECREASE OF NETWORK EXTENT AND DENSITY Extension of channels or shrinkage of perennial, intermittent and ephemeral streams.
Shape	ADJUSTMENT OF SHAPE Width/depth ratio may be increased or decreased.	ALTERATION OF SHAPE OF PATTERN A change from regular to irregular meanders.	DRAINAGE PATTERN CHANGE IN SHAPE Inclusion of new stream channels after deforestation.
Composition	CHANGE IN CHANNEL SEDIMENTS Alteration of grain size of sediments in bed and banks possibly accompanied by development of berms and bars.	PLANFORM METAMORPHOSIS Change from single to multi-thread channel or converse.	NETWORK COMPOSITION CHANGE The replacement of channels with no definite stream channel (dambos in West Africa) by a clearly defined channel.

General changes are indicated in upper case and examples are given in lower case.

McMahon, T.A. (1988) A global classification of river regimes. *Applied Geography*, **8**, 255–272. · Holmes, N.T.H., Boon, P.J. and Rowell, T.A. (2008) A revised classification system for British rivers based on their aquatic plant communities. *Aquatic Conservation: Marine and Freshwater Ecosystems*, **8**, 555–578. · Rosgen, D. (1994) A classification of natural rivers. *Catena*, **22**, 169–199.

river discharge See DISCHARGE.

river metamorphosis Refers to the change of river channel morphology that can occur when changes of discharge and sediment exceed a threshold condition (see THRESHOLD, GEOMORPHOLOGICAL). Channel changes that occur can be from a multithread to a single-thread channel, and S. A. Schumm (1969) introduced the term *river metamorphosis* to signify the range of changes that may arise. Metamorphosis may involve changes of size, shape or composition of aspects of river channel morphology, as shown in Table 1. It is important to be able to predict the degrees of freedom that a river system possesses: width, depth, slope, velocity and plan shape. To predict the character of river metamorphosis it is necessary to suggest what changes will occur, where they will obtain, and when the changes will begin and end.   KJG

Reading and Reference
Gregory, K.J. (1981) River channels. In K. J. Gregory and D. E. Walling (eds), *Man and environmental processes*.

London: Butterworth; pp. 123–143. · Schumm, S. A. (1969) River metamorphosis. *Proceedings of the American Society of Civil Engineers, Hydraulic Division*, **95**, 251–273. · Schumm, S. A. (1977) *The fluvial system*. Chichester: John Wiley & Sons, Ltd.

river regime See REGIME.

river restoration Restablishing structure and function of a river or stream to a predisturbance condition. Defining the reference condition or the state that restoration seeks to attain can be challenging in regions where examples of undisturbed systems are uncommon. Moreover, determining the timescale of a restoration goal can be challenging, especially when there are multiple potential options (i.e. pre-urbanization, pre-agricultural or pre-colonial) and if the hydrologic and sediment delivery processes and amounts have changed. Attempting to reestablish a predisturbance morphology without re-establishing the processes that formed and maintained that morphology is not a recipe for self-sustaining restoration. As many river systems have had extensive historical modifications to their discharge and sediment regimes, river restoration has been distinguished from both river rehabilitation (improving river conditions to a goal that does not necessarily re-establish predisturbance conditions) and from river reclamation (actions designed to simply improve the biological capacity or geomorphological function of the system). DRM

Reading
Montgomery, D.R. (1997) River management: What's best on the banks?, *Nature*, **388**, 328–329. · Montgomery, D.R. and MacDonald, L.H. (2002) Diagnostic approach to stream channel assessment and monitoring. *Journal of the American Water Resources Association*, **38**, 1–16. · Thorne, C.R., Hey, R.D. and Newson, M.D. (1997) *Applied fluvial geomorphology for river engineering and management*. Chichester: John Wiley & Sons, Ltd. · Riley, A.L. (1998) *Restoring streams in cities: a guide for planners, policymakers, and citizens*. Washington, DC: Island Press. · Waal, L.C., Large, A.R.G. and Wade, P.M. (1998) *Rehabilitation of rivers: principles and implementation*. Chichester, John Wiley & Sons, Ltd.

river terrace A nearly flat area formed at river level but now above the river, separated at least by an eroded slope. Such a landform implies that production of a *valley floor* or alluvial plain, by a phase of deposition or lateral erosion by the river, was followed by a phase of *downcutting* (incision). Terraces may be *rock floored* (though a veneer of river deposits may remain) or *alluvial* (accumulation terraces). The simplest alluvial terraces consist of deposits from one phase, but often there are patches of older deposits: just as a terrace may be cut into rock, it may be cut into any previous deposit, fluvial or otherwise. Thus, the age of the terrace materials must not be confused with the age of the landform.

Initially, a river terrace should slope down-valley rather than toward the river, but later modification by compaction, erosion or deposition (slopewash or aeolian) commonly produces a riverward slope. There are other types of terrace or bench that might, after further erosion, be confused with river terraces. Colluvial terraces are formed of slope deposits and have greater slopes towards the river; they are common in British uplands, in soliflucted till. PEDIMENTS are erosional forms at the foot of highlands; they are usually attributed to slope retreat, but lateral planation by rivers has also been invoked. Pediment slopes are usually steeper than those of rock-floored terraces. Structural benches relate to differential erosion of weaker rocks and are floored by more resistant rocks. In flat-lying rocks (e.g. most of the Grand Canyon) they may resemble river terraces, but their upslope margins are more gradual (not former river bluffs) and if traced down-valley they maintain their relation to a single bed of rock. Agricultural terraces are artificial, which may not be obvious long after their abandonment when even the remains of terrace-front walls may become obscure. KAME TERRACES are a special type of river terrace whose formation was influenced by the presence of a glacier.

Traced downstream, a terrace provides the former gradient (as modified by tectonism) and its surface deposits show the competence of the former river, and perhaps the channel type. Later lateral erosion may completely destroy a terrace, but in a valley the alluvial terrace remnants may be *rock defended* as the river encounters the sloping surface of underlying bedrock. Terraces on either side of a river will be *paired* at the same relative height if downcutting is rapid, but *unpaired* (at varied heights) if it is steady, especially by a meandering stream. A series of river terraces provides valuable information on landscape development.

The alternation of alluviation or lateral erosion with downcutting may be produced in various ways, which must be worked out for each region. Regional uplift is the most obvious possibility, producing, for example, the mainly rock-floored terraces along the middle Rhine (Ahnert, 1998), though the existence of a staircase of 12 terraces probably relates to climatic variation rather than interruptions of uplift. Changes in BASE LEVEL commonly cause downcutting or alluviation. A fall in sea level (see EUSTASY) has been assumed to cause river downcutting, working headward from the newly exposed course; the assumption that the new course has a steeper gradient than the old lower course may not apply to smaller rivers where continental shelves are broad (see CONTINENTAL SHELF). River diversion or capture may produce a single terrace set.

River aggradation or downcutting is controlled by the balance between the river's capacity to erode and transport, and the supply of material (mainly bedload, since suspended silts and clays are unlikely to settle out in flowing water – see BEDLOAD, BEDLOAD EQUATION). Both capacity and supply are changing continuously, giving ample opportunity for phases of aggradation and downcutting to alternate; in fact, this may occur in a single flood, with metres of scour on the rising stage followed by deposition as discharge wanes. Average discharge is less important than the frequency of major FLOODS. Delivery of material from slopes and tributaries may be on a different timescale to their onward transport by the main stream, which may leave such material as terraces after a major flood. Minor floods may cause aggradation, but major ones take material farther down-valley, leaving terraces upstream. Major phases of aggradation may come from glaciation, periglaciation, climatic changes or (anthropogenic) soil erosion.

Idealized transverse sequence through the terraces of the lower Thames, England.
Source: Bridgland *et al.* (1995). Reproduced with permission of Dave Brigland.

Downcutting is likely to follow the reduction of such high rates of sediment supply. When rivers are blocked by landslides, they initially aggrade but soon cut down as the slide material is eroded (see figure).

There have been debates (e.g. concerning the Thames, England) on whether terraces were built up in cold phases of the Quaternary because of sediment supply or in warm phases because of high sea levels. The work of Bridgland (1994) shows that the Thames terraces are more complex. Although each terrace suite seems to relate to a major warm/cold cycle, it contains both fossiliferous fine-grained interglacial deposits and gravels that Bridgland relates to the beginning and end of cold phases. The common assumption that each terrace level relates to one change in base level or climate is challenged also by the concepts of *episodic* erosion and sediment supply, and of *complex response* in the relation of sediment transport to supply. Schumm *et al.* (1987: 123–126, 211) have shown by experimentation that deposition and downcutting may overshoot sustainable levels, producing a damped oscillation; thus, two or more terraces may relate to a single change. They give examples from Colorado and California of flights of four or more terraces formed in a century, over a 10 m range of altitude. The thresholds involved may be intrinsic to development of the river system, and some terraces may be unrelated to external changes.   IE

Reading and References

Ahnert, F. (1998) *Introduction to geomorphology*. London: Arnold; chapter 15, pp. 183–190. · Bridgland, D.R.

(1994) *Quaternary of the Thames*. Geological Conservation Review Series, 7. London: Chapman & Hall. · Bridgland, D.R., Allen, P. and Haggart, B.A. (eds) (1995) *The Quaternary of the lower reaches of the Thames: field guide*. Durham: Quaternary Research Association. · Richards, K. (1982) *Rivers: form and process in alluvial channels*. London: Methuen; pp. 272–275. · Schumm, S.A., Mosley, M.P. and Weaver, W.E. (1987) *Experimental fluvial geomorphology*. New York: John Wiley & Sons, Inc.

roche moutonnée An asymmetric rock bump with one side ice-moulded and the other side steepened and often cliffed, generally recognized as the hallmark of glacial erosion. The term was introduced by H. B. de Saussure in 1787 in recognition of the similarity of the rocks to the rippled appearance of wavy wigs styled *moutonnées* at the time. The morphology of roches moutonnées seems to reflect the contrast between ABRASION on the smoothed up-ice side and PLUCKING on the lee side.   DES

Reading

Sugden, D.E., Glasser, N. and Clapperton, C.M. (1992) Evolution of large roches moutonnées. *Geografiska Annaler, Series A: Physical Geography*, **74**, 253–264.

rock coatings Some 14 different classes of rock coatings have been identified (see Table 2), occurring on rock surfaces in all environments and altering their surface appearances. These are distinct from DURICRUSTS, which form within the soil profile, but which may have similar chemical compositions. Rock coatings are usually nanometres think only.   RID

Table 2 Nomenclature of rock coatings, in alphabetical order

Term	Summary description	Related terms
Carbonate skin	Coating composed primarily of carbonate, usually calcium carbonate, but could be combined with magnesium or other cations	Caliche, calcrete, patina, travertine, carbonate skin, dolocrete, dolomite
Case-hardening agents	Addition of cementing agent to rock matrix material; the agent may be manganese, sulphate, carbonate, silica, iron, oxalate, organisms, or anthropogenic	Sometimes called a particular type of rock coating
Dust film	Light powder of clay- and silt-sized particles attached to rough surfaces and in rock fractures	Gesetz der Wüstenbildung, clay skins, clay films, soiling
Heavy metal skins	Coatings of iron, manganese, copper, zinc, nickel, mercury, lead and other heavy metals on rocks in natural and human-altered settings	Described by chemical composition of the film
Iron film	Composed primarily of iron oxides or oxyhydroxides; unlike orange rock varnish because it does not have clay as a major constituent	Ground patina, ferric oxide coating, red staining, ferric hydroxides, iron staining, iron-rich rock varnish, red–brown coating
Lithobiontic coatings	Organic remains form the rock coating; for example, lichens, moss, fungi, cyanobacteria, algae	Organic mat, biofilms
Nitrate crust	Potassium and calcium nitrate coatings on rocks, often in caves and rock shelters in limestone areas	Saltpetre, nitre, icing
Oxalate crust	Mostly calcium oxalate and silica with variable concentrations of magnesium, aluminium, potassium, phosphorus, sulphur, barium and manganese. Often found forming near or with lichens. Usually dark in colour, but can be as light as ivory	Oxalate patina, lichen-produced crusts, patina, scialbatura
Phosphate skin	Various phosphate minerals (e.g. iron phosphates or apatite) that are mixed with clays and sometimes manganese	Organophosphate film, epilithic biofilm
Pigment	Human-manufactured material placed on rock surfaces by people	Pictograph, paint, sometimes described by the nature of the material
Rock varnish	Clay minerals. Manganese and iron oxides, and minor and trace elements; colour ranges from orange to black in colour produced by variable concentrations of different manganese and iron oxides	Desert varnish, desert lacquer, patina, manteau protecteur, Wüstenlack, Schutzrinden, cataract films
Salt crust	The precipitation of sodium chloride on rock surfaces	Halite crust, efflorescence, salcrete
Silica glaze	Usually clear white to orange shiny lustre, but can be darker in appearance, composed primarily of amorphous silica and aluminium, but often with iron	Desert glaze, turtle-skin patina, siliceous crusts, silica–alumina coating, silica skins
Sulphate crust	Composed of the superposition of sulphates (e.g. barite, gypsum) on rocks; not gypsum crusts that are sedimentary deposits	Gypsum crusts, sulphate skin

Reading

Dorn, R.I. (1998) *Rock coatings*. Amsterdam: Elsevier.

rock drumlin   A rock hill streamlined by the passage of overrunning glacier ice.

rock flour   The fine debris created by ABRASION beneath a glacier. The material, generally less than 0.2 mm in diameter, may be flushed away to give the characteristic brown and blue–green colour of glacial meltwater streams and lakes. (See GLACIER MILK.)

DES

Reading

Haldorsen, S. (1981) Grain-size distribution of subglacial till and its relation to glacial crushing and abrasion. *Boreas*, 87, 1003–1015.

rock glacier A term that has been used to describe a feature, comparatively common in many alpine areas, which looks like a glacier (with apparent flow structures, etc.) but is composed of rock debris. Some workers use the term to imply that such a feature has a specific mode of origin. There are two main theories: that the rock debris has ice mixed in the spaces between the rock (the interstitial ice model) or that the debris is a thick covering on a thin, probably decaying true glacier (the glacier ice model). There is much disagreement as to which model is correct. One major feature is that rock glaciers usually exhibit slow movement, often less than a metre per year; that is, at least an order of magnitude less than most true ice glaciers. WBW

Reading
Giardino, J.R., Shroder, J.F. and Vitek, J.D. (1987) *Rock glaciers*. Boston, MA: Allen & Unwin. · Martin, H.E. and Whalley, W.B. (1987) Rock glaciers a review: part I. *Progress in Physical Geography*, **11**, 260–286. · Whalley, W.B. and Martin, H.E. (1992) Rock glaciers: part II. Models and mechanisms. *Progress in Physical Geography*, **16**, 127–186.

rock mass strength An important concept in geomorphology, developed by Selby (1980) and others in an attempt to gain a quantitative measure of the resistance of a rock mass to erosion (Table 3). It involves giving a rank of importance to a range of different rock parameters and then summing them to come up with a total rating of strength. ASG

Reference
Selby, M.J. (1980) A rock mass strength classification for geomorphic purposes: with tests from Antarctica and New Zealand. *Zeitschrift für Geomorphologie*, **24**, 31–51.

rock quality indices Measures used to relate the numerical intensity of fractures in a rock to the quality of the unweathered rock. One measure is the relationship between the compressional wave velocity measured in situ in massive rock and on a core of the intact rock (see INTACT STRENGTH). The fewer the joints, the nearer the ratio is to unity. Rock quality designation is the relationship between intact cored rock length to the total length drilled, and the 'fracture index' is the frequency of fractures occurring within a rock unit. WBW

Reading
Bell, F.G. (1992) *Engineering properties of soils and rocks*, 3rd edition. Oxford: Butterworth Heinemann.

rock step See RIEGEL.

rock varnish A type of a rock coating that is dark in colour and is characterized by clay minerals (~40–60%) cemented to rock surfaces by oxides and hydroxides of manganese (birnessite) and iron (goethite and haematite) that typically comprise 20–40% (see ROCK COATINGS). The oxides are nanometres in thickness and they are a by-product of the weathering of bacterial casts. Rock varnish forms only where and when the nanometre-scale remnants of bacteria manoeuvre in between the broken and decayed fragments of clay minerals weathered at the nanometre scale. Most varnishes seen at the surface today tend to start in the subsurface in fissures. It forms in all terrestrial environments, but the term 'desert varnish' is common because varnishes are most geochemically stable in the subaerial environment in deserts, including Antarctica (Giovanna and Baroni, 2007). Aeolian DUST may contribute to varnish development (Dorn et al., 2013). Typical varnish thicknesses are less than 100 μm. Usually dull in lustre, its occasional sheen comes from a smooth surface micromorphology in combination with manganese enrichment at the very surface of the varnish. The constituents in varnish accrete on the host rock. The most volumetrically significant post-depositional modification is the leaching of cations from rock varnish. Varnishes are also eroded by lithobionts, but in places where varnishes are intact, black (manganese-rich) and orange (iron-rich) microlaminae may indicate past changes in climate. Rock varnishes sometimes act as case-hardening agents for the underlying rock. Many varnishes also encapsulate organic matter; unfortunately, radiocarbon dating of these organics is not useful because the entombed organics do not form an open system of older and younger organics. RID/DSGT

References
Dorn, R.I., Krinsley, D.H., Langworthy, K.A., et al. (2013) The influence of mineral dust on rock varnish formation. *Aeolian Research*, **10**, 61–76. · Giovanna, G. and Baroni, C. (2007) High-resolution analysis of silica and sulphate-rich rock varnishes from Victoria Land (Antarctica). *European Journal of Mineralogy*, **19**, 381–389.

roddon A sinuous, silty ridge that snakes about above the general level of the peat fenland of East Anglia (England). Roddons represent ancient river systems that may initially have flowed between levées above the general level of the surrounding land or which have subsequently become relatively elevated as a consequence of peat wastage. They are favoured sites for settlement. ASG

Table 3 Classification of rock mass strength

Variable	Weighting (%)	Very strong	Strong	Moderate	Weak	Very weak
Intact rock strength (Schmidt hammer rebound value)	20	100–60 $r=20$	60–50 $r=18$	50–40 $r=14$	40–35 $r=10$	35–10 $r=5$
Weathering	10	Unweathered $r=10$	Slightly weathered $r=9$	Moderately weathered $r=7$	Highly weathered $r=5$	Completely weathered $r=3$
Joint spacing	30	>3 m $r=30$	3–1 m $r=28$	1–0.3 m $r=21$	300–50 mm $r=15$	<50 mm $r=8$
Joint orientations	20	Very favourable Steep dips into slope, cross joints interlock $r=20$	Favourable Moderate dips into slope $r=18$	Fair Horizontal dips or nearly vertical dips (hard rocks only) $r=14$	Unfavourable Moderate dips out of slope $r=9$	Very unfavourable Steep dips out of slope $r=5$
Joint width	7	<0.1 mm $r=7$	0.1–1 mm $r=6$	1–5 mm $r=5$	5–20 mm $r=4$	>20 mm $r=2$
Joint continuity and infill	7	None, continuous $r=7$	Few, continuous $r=6$	Continuous, no infill $r=5$	Continuous, thin infill $r=4$	Continuous, thick infill $r=1$
Groundwater outflow	6	None $r=6$	Trace $r=5$	Slight <40 ml s^{-1} m^{-2} $r=4$	Moderate 40–200 ml s^{-1} m^{-2} $r=3$	Great >200 ml s^{-1} m^{-2} $r=1$
Total rating		100–91	90–71	70–51	50–26	<26

Source: Modified from Selby (1980: table 6, pp. 44–45).

Reading
Fowler, G. (1932) Old river beds in the Fenlands. *Geographical Journal*, **79**, 210–212. · Godwin, H. (1938) The origin of roddons. *Geographical Journal*, **91**, 241–250.

rogen moraine Landform assemblage of numerous, parallel, closely spaced ridges consisting of glacial drift, usually TILL. The ridges are formed transverse to ice flow in a subglacial position and are usually found in the central portions of former ice sheets. Individual ridges are typically 10–30 m high, 300–1200 m long and 150–300 m wide, have a straight to arcuate planform, concave in the down-ice direction. Their name derives from Lake Rogen in Sweden. *Ribbed moraine* is the North American and, perhaps, preferred term. Their formation is uncertain but is often closely linked to that of DRUMLINS. (See SUBGLACIAL BED-FORMS.) CDC

Reading
Hättestrand, C. and Kleman, J. (1999) Ribbed moraine formation. *Quaternary Science Reviews*, **18**, 43–61.

rollability A property related to the angle of a slope down which a given sedimentary particle will roll. The concept was introduced by Winkelmolen to account for the ease with which particles (usually of sand size) can be rolled in unidirectional fluid flow. It is indexed by the time it takes for grains to travel down the inside of a revolving cylinder inclined at a slight angle. The time taken is a measure of the rollability potential. Usually, a few grams of the sediment is used in a test. The measure is said to correlate with the influence of grain form on the particle settling velocity. Rollability can be considered as amalgamating several aspects of PARTICLE FORM. WBW

Reading
Winkelmolen, A.M. (1971) Rollability, a functional shape property of sand grains. *Journal of Sedimentary Petrology* **41**, 703–714.

Rossby waves Wave motions in the atmosphere of planetary scale. They take the form of vast meanders of airflow around a hemisphere and are most clearly identifiable at upper levels. They owe their existence to the variation with the latitude ϕ of the Coriolis parameter ($2\Omega \sin \phi$, where Ω is the angular velocity of rotation of the Earth).

In non-divergent, large-scale motion, absolute VORTICITY ζ_a can be considered constant, given by

$$\frac{d}{dt}(\zeta_a) = \frac{d}{dt}(\zeta_r + 2\Omega \sin \phi) = 0$$

where ζ_r is the relative vorticity. In the northern hemisphere, if a uniform westerly current with zero initial relative vorticity is displaced poleward, then, as latitude increases, the relative vorticity must become negative (i.e. anticyclonic) so that the air turns southward. After moving towards the equator of its original latitude its relative vorticity becomes positive (cyclonic) so that it turns northward. The current thus oscillates about its original latitude giving a series of waves called long waves or Rossby waves (named after Carl-Gustav Rossby, a Swedish-American meteorologist) usually numbering between two and six around a latitude circle.

Rossby wave theory predicts that, in a basic westerly current and for a particular number of troughs and ridges around a latitude circle, there is a critical flow speed for which the waves are steady and stationary. If the flow is less than this critical speed the waves drift westward, whereas if it is greater they drift eastward.

Rossby waves are forced by three principal mechanisms: by orographic forcing resulting from a basic westerly flow impinging on a mountain range (especially the Rockies or the Andes); by thermal forcing due to differential heating of the oceans and continents; and by interaction with smaller scale disturbances, such as extratropical cyclones.

Rossby waves have also been identified in the oceans. KJW

Reading
Gill, A.E. (1982) *Atmosphere–ocean dynamics*. London: Academic Press. · Houghton, J. (2002) *The physics of atmospheres*, 3rd edition. Cambridge: Cambridge University Press.

rotational failure The name given to failure, usually in clays, although also in weak rocks, where the shape of the slip surface (akin to a shear plane), which forms the boundary between the stable ground and the mass that has moved, is curved. At the top it frequently forms a step, where tension cracks develop in the early stages of failure. At the base there is usually a bulge. If the mass is rather mobile, as is frequently the case with QUICK CLAY, then there may be spreading at the foot, perhaps to give a secondary earthflow. In some cases the production of one rotational failure may give rise to a second failure. WBW

Reading
Selby, M.J. (1993) *Hillslope materials and processes*, 2nd edition. Oxford: Oxford University Press.

rotor streaming A condition of unsteady flow over a mountain in which lee eddies are generated, then blow away. It probably involves a mechanism similar to that influencing the vortex streets noted by and named after von Karman. JSAG

Reading
Atkinson, B.W. (1981) *Meso scale atmospheric circulations.* London: Academic Press.

roughness As a fluid moves over a surface (e.g. water over a river bed or wind over the Earth's surface) it is retarded by friction at its base. The amount of friction at the interface between surface and fluid is dependent upon the roughness of the surface, and the resultant retardation develops the fluid velocity profile in the BOUNDARY LAYER. Very close to the surface the drag imposed by the roughness results in a fluid velocity of zero, and the depth of this zero-velocity layer is called the ROUGHNESS LENGTH. Surface roughness is a very important parameter in sediment transport modelling in both wind and water, because, by initiating a roughness length, it partly controls the gradient of the velocity profile, and hence the shear velocity of the fluid.

The friction resulting from surface roughness can be affected by a number of parameters across a wide range of scales. In both wind (Bagnold, 1941) and fluvial environments (Bathurst, 1993) the roughness of the surface has been shown to increase with the mean or median particle size of the surface material. This is commonly termed *grain roughness*. The size, shape and arrangement of vegetation may also significantly contribute to the roughness evident in both wind and fluvial environments. At a larger scale, bedforms (such as ripples or dunes) may provide a *form roughness*. Similarly, *free-surface resistance* may stem from surface waves and hydraulic jumps in rivers (Knighton, 1998). At the largest scale, bends and islands in rivers may contribute to the total fluvial roughness, and in meteorological studies roughness may be imparted by forests, cities or mountain fronts.

In fluvial studies, roughness is commonly measured by friction coefficients such as Manning's n (see MANNING EQUATION), Chézy C or Darcy–Weisbach f, whilst in aeolian studies the aerodynamic roughness length z_0 is measured. A particular problem with the assessment of roughness is the great spatial and temporal variability that the property exhibits. For example, the form roughness provided by bedforms may alter substantially with increasing or decreasing flow velocities as the bedforms themselves grow or diminish. The relationships between surface roughness, roughness length, flow velocities and sediment transport are still not completely understood. GFSW

Reading and References
Bagnold, R.A. (1941) *The physics of blown sand and desert dunes.* London: Methuen. · Bathurst, J.C. (1993) Flow resistance through the channel network. In K. Beven and M. J. Kirkby (eds), *Channel network hydrology.* Chichester: John Wiley & Sons, Ltd; pp. 69–98. · Knighton, D. (1998) *Fluvial forms and processes: a new perspective.* London: Arnold. · Wiggs, G.F.S. (2011) Sediment mobilisation by the wind. In D. S. G. Thomas (ed.), *Arid zone geomorphology: process, form and change in drylands,* 3rd edition. London: John Wiley & Sons, Ltd; pp. 455–486.

roughness length Fluid velocity over a surface (e.g. water over a river bed or wind over the earth's surface) is retarded by the surface ROUGHNESS, and very close to that surface the velocity of the fluid becomes zero. The depth of this zero-velocity layer (termed z_0) is called the roughness length and is controlled by the surface roughness that develops the fluid velocity profile. The roughness length is a particularly important parameter in sediment transport modelling in both fluvial and aeolian studies as it partly controls the gradient of the velocity profile, and hence the shear velocity of the fluid. The relationships between fluid velocity, roughness length and shear velocity are described by the LAW OF THE WALL.

More commonly measured in aeolian studies (in fluvial studies relative resistance coefficients such as Manning's n are usually identified) the aerodynamic roughness length has been shown to vary widely both temporally and spatially. Typical values for z_0 may be 0.0007 m for stationary sand surfaces or 0.2 m and above for vegetated or semi-vegetated desert surfaces. Bagnold (1941) identified a relationship between z_0 and surface grain diameter d, where $z_0 = d/30$. More recent experiments have shown that z_0 is also a function of roughness element (e.g. large stones) spacing, where z_0 may reach a maximum value of $d/8$ for widely spaced elements before returning to $d/30$ as element spacing is increased further (Greeley and Iversen, 1985). Roughness length can also be determined dynamically from measured velocity profiles and using the law of the wall to calculate the height intercept at zero velocity. Such an approach works well for flat and simple terrain, but there is still no agreed method for determining z_0 on surfaces that are complex in character and may involve a changing mixture of sediment sizes and/or topography (Blumberg and Greeley, 1993). GFSW

Reading and References

Bagnold, R.A. (1941) *The physics of blown sand and desert dunes*. London: Methuen. · Blumberg, D.G. and Greeley, R. (1993) Field studies of aerodynamic roughness length. *Journal of Arid Environments*, 25, 39–48. · Greeley, R. and Iversen, J.D. (1985) *Wind as a geological process*. Cambridge: Cambridge University Press. · Wiggs, G.F.S. (2011) Sediment mobilisation by the wind. In D. S. G. Thomas (ed.), *Arid zone geomorphology: process, form and change in drylands*, 3rd edition. London: John Wiley & Sons, Ltd; pp. 455–486.

roundness Tending towards rounded edges, describing the degree of abrasion of a clastic fragment as shown by the sharpness of its edges and corners, independent of shape. Spherical particles are perfectly rounded, but well-rounded objects (such as an egg), need not be spherical (Waddell, 1933).   ASG

Reference

Waddell, H. (1933) Sphericity and roundness of quartz particles. *Journal of Geology*, 41, 310–331.

routing, flood See FLOOD ROUTING.

r-selection See *r*- AND *K*-SELECTION.

ruderal vegetation Plants that grow in waste land, on or among rubbish and debris. The term was used originally in the sense of stone waste but has been extended to include roadsides, verges and old fields. The plant succession is usually rapid, with profusely seeded and rapidly growing annuals being replaced by hardy perennials, particularly grasses. Naturalized plants are often found in the early seral stages because of the relative absence of competition, but in time they tend to be eliminated in the struggle with native plants.   PAF

run-off Water that drains from an area of the land is referred to as run-off. The process may be considered at various scales, spanning continental run-off, the run-off from a river basin or other catchment area, or the run-off from a small area such as a roof, a field or an experimental run-off plot (see PLOTS, EROSION/RUN-OFF). In all cases, the run-off forms a link in the global hydrologic cycle, in which about 40,000 km^3 of water drain from the land areas of the Earth annually.

Run-off follows both above-ground and subsurface pathways, and may then be referred to explicitly as surface run-off, subsurface run-off or groundwater run-off. The partitioning of water delivered as precipitation among these various possible flow pathways is influenced partly by the nature of the precipitation itself, and also by a range of land surface and soil characteristics (Brammer and McDonnell, 1996; Newman *et al.*, 1998). The intensity of rainfall exerts an important control on the flow pathway that run-off follows. Intense rain may exceed the ability of the soil to take in water, and in this case surface run-off is promoted. This may take the form of infiltration-excess or Hortonian surface run-off (see HORTON OVERLAND FLOW MODEL), a form that is particularly common in desert areas, where soils often display low infiltration rates, but is also characteristic of the extensive impervious surfaces of urban areas. However, surface run-off may also arise in rain of lower intensity and can be promoted where soils are partly saturated, such as hollows or low-lying areas flanking streams, or where soils are thin so that their ability to hold water is low. Surface run-off generated in such locations is referred to as saturation overland flow. Soils that are very permeable, such as the humus-rich soils of some forested regions, may have the capacity to take in almost all arriving rain water, so that surface run-off may be rare or absent, and the run-off dominantly follows the subsurface pathways. Groundwater run-off in particular is an enormously important pathway, and this slow flow provides the continuous flow that sustains many perennial streams even through long periods of dry weather.

Storm characteristics are also important in run-off production. Small events can lose a significant fraction of the water they deliver to canopy interception, for example, and so their potential to generate run-off is diminished. Intense storm events, in contrast, can trigger widespread soil saturation and so generate large amounts of run-off.

The efficiency with which an area produces run-off can be expressed in terms of the run-off ratio that it exhibits. This is the ratio of the volume of run-off released to the volume of precipitation delivered over the area. Impervious urban areas can exhibit run-off ratios as high as 0.95. In contrast, a forested area with permeable soils may have a run-off ratio of only 0.1–0.2. The balance of the water is commonly lost to evaporation and transpired by vegetation.   DLD

Reading and References

Brammer, D.D. and McDonnell, J.J. (1996) An evolving perceptual model of hillslope flow at the Maimai catchment. In M. G. Anderson and S. M. Brooks (eds), *Advances in hillslope processes*, vol. 1. Chichester: John Wiley & Sons, Ltd; pp. 35–60. · Dingman, S.L. (1994) *Physical hydrology*. Engelwood Cliffs, NJ: Prentice-Hall. · Newman, B.D., Campbell, A.R. and Wilcox, B.P. (1998) Lateral subsurface flow pathways in a semiarid ponderosa pine hillslope. *Water Resources Research*, 34, 3485–3496.

run-off plot An area of the landscape generally enclosed by leakproof barriers so that all water and sediment leaving it is known to have come from within the borders of the plot. Run-off plots are used for such purposes as the comparison of the effects of land management treatments on run-off and erosion, the investigation of the erosional behaviour of differing soil types, and monitoring the effects of various experimental treatments such as vegetation removal by clipping or by fire. Dimensions range from $0.5\,m^2$ to hundreds of square metres. Run-off water and sediment shed from a plot are either measured continuously by appropriate apparatus, or directed into buried storage drums whose contents are later examined and subsampled. Plots may be exposed to natural or simulated rain. DLD

Reading
Mohamoud, Y.M., Ewing, L.K. and Boast, C.W. (1990) Small plot hydrology. I. Rainfall infiltration and depression storage determination. *Transactions of the American Society of Agricultural Engineers*, **33**, 1121–1131.

ruware A low, dome-shaped exposure of bedrock projecting from a cover of alluvium or weathered bedrock. It is either an incipient or a relict INSELBERG.

S

sabkha (also sabkhah) *a.* a closed depression, often with a saline surface (akin to a PAN or PLAYA), in an arid environment.
b. A saline flat in arid areas that is above the mean high tide level but subject to periodic inundation. Both terms are commonly used in Arabic countries.   DSGT

Sahel   The Sahel (from Arabic ساحل, *sāḥil*, shore, border or coast) is a semi-arid tropical savanna portion of Africa and forms the transitional zone between the Sahara Desert to the north and the more humid savanna belt or Sudan zone to the south. It stretches from the Atlantic Ocean on the west to the Red Sea coast in the east, a distance of ~3862 km. It is a belt that varies in width from several hundred to a thousand kilometres, The Sahelian climate is tropical. Summers are hot, with maximum mean temperatures varying from 33 to 36 °C. Monthly mean minimum temperatures fall in the range 18-21 °C. Rainfall in the southern portions of the zone is around 600 mm, but this declines to around 200 mm in the north. The quantity of rainfall is affected by the INTER-TROPICAL CONVERGENCE ZONE (ITCZ). When this penetrates far enough to the north, a long rainy season follows. If it fails to move sufficiently north, rainfall is more limited. During the winter months (the end of November to the middle of March), the dusty Harmattan wind blows from the north. Since the mid-1960s, the Sahel has suffered from a persistent drought, leading to the desiccation of Lake Chad, an increase in dust storms, a decrease in flow along the Niger and Senegal, and severe problems for the local inhabitants. One of the prime drivers of this drought was a change in sea surface temperatures in the North Atlantic Ocean (Hagos and Cook, 2008). It is also possible that the drought was reinforced by increasing levels of dust in the atmosphere causing a diminution of rainfall (Hui *et al.*, 2008) and by land cover changes that altered surface albedo, temperatures and convectional activity (Lauwaet *et al.*, 2009). In addition, on the north side of the Sahara, in central Tunisia, there is another area known as Sahel.   ASG

References
Hagos, S. and Cook, K. (2008) Ocean warming and late twentieth-century Sahel drought and recovery. *Journal of Climate*, 21, 3797-3815 · Hui, W., Cook, B., Ravi, S., *et al.* (2008) Dust-rainfall feedback in the West African Sahel. *Water Resources Research*, 44, 1–6. · Lauwaet, D., van Lipzig, N.P.M. and de Ridder, K. (2009) The effect of vegetation changes on precipitation and mesoscale convective systems in the Sahel. *Climate Dynamics*, 33, 521–534.

salars   Basins of inland drainage in a desert which are only occasionally inundated with saline water. They are also known as PLAYAS, *sabkhas* or *chotts*.

salcrete   A light-coloured surface crust of halite-cemented beach sand caused by the concentration by evaporation of swash or spray blown onshore by breaking waves (Yasso, 1966).   ASG

Reference
Yasso, W.E. (1966) Heavy mineral concentration and sastrugi-like deflation furrows in a beach salcrete at Rock-away Point, NY. *Journal of Sedimentary Petrology*, 36, 836–838.

salinization   The process of accumulation of soluble salts in upper soil horizons, thereby compromising plant growth. Though there are many plants that can grow in salt-rich soils, many plants grown as crops cannot do so effectively. In DRY-LANDS especially, attempts to increase crop production by irrigation can, in a matter of a few years, result in salinization, either because water application exceeds the amount that plants can use, and drainage is poor, or/and because high EVAPOTRANS-PIRATION rates cause salts to be precipitated in the soil. Salinization is in some areas (e.g. the San Joaquin basin of California and parts of Pakistan and India) a major form of DESERTIFICATION (Thomas and Middleton, 1994). Though remedial measures may be attempted, including efforts to flush excess salts out of the soil, salinization may simply be a facet of attempting to increase crop yields in environments generally unsuited to agriculture. It is not a new problem, and may have occurred in conjunction with early

Mesopotamian agriculture (Jacobsen and Adams, 1958). DSGT

References
Jacobsen, T. and Adams, R.M. (1958) Salt and silt in ancient Mesopotamian agriculture. *Science*, **128**, 1251–1258. · Thomas, D.S.G. and Middleton, N.J. (1994) *Desertification: exploding the myth*. Chichester: John Wiley & Sons, Ltd.

salt marsh Salt marshes are vegetated mud flats in the high intertidal zone found commonly on many low-lying coasts in a wide range of temperate environments. On tropical coastlines salt marshes tend to be replaced by mangrove swamps (see MANGROVE, MANGROVE FOREST), although sometimes they are found together. Salt marshes, which support a range of halophytic vegetation, grade seawards into mud- or sand-flats. Salt-marsh plants themselves play an important role in trapping sediment and in building up the marsh surface. In turn, the development of the marsh encourages a succession of plants from early colonizers such as *Salicornia* spp. and *Spartina* spp. to plants that are less tolerant of frequent inundation by seawater. Salt marshes vary greatly throughout the world in both ecology and geomorphology, but are often characterized by intricate creek systems and salt pan development. HAV

Reading
Adam, P. (1990) *Saltmarsh ecology*. Cambridge: Cambridge University Press. · Allen, J.R.L.and Pye, K. (eds) (1992) *Saltmarshes: morphodynamics, conservation and engineering significance*. Cambridge: Cambridge University Press.

salt tectonics Deformation of the Earth's crust by the flow of salts from deep-seated evaporite deposits to form salt domes, salt pillows and associated structures. The intrusion into and disturbance of existing strata is also known as diapirism. Salt diapirs are economically important as oil reservoirs and as sources of sulphur. PSH

salt weathering The breakdown of rock by HALOCLASTY; it is caused primarily by physical changes produced by salt crystallization, salt hydration or the thermal expansion of salts. Among the most effective salts are sodium sulphate, sodium carbonate and magnesium sulphate. Salt weathering has been recognized as an important process in desert, coastal, polar and urban areas, and is a serious hazard to concrete structures in saline regions. ASG

Reading
Cooke, R.U. (1981) Salt weathering in deserts. *Proceedings of the Geologists' Association of London*, **92**, 1–16. · Viles,

H.A. (2011) Weathering. In D. S. G. Thomas (ed.), *Arid zone geomorphology: process form and change in drylands*. Chichester: John Wiley & Sons, Ltd; pp 85–99.

saltation Has two main meanings in physical geography. The first refers to the hopping motion of sand grains transported by a fluid (water or air). The grains are ejected from the bed in a near-vertical trajectory by lift forces, accelerate in the flow direction when affected by fluid drag, then fall to the bed again on a path inclined at 6–12°, which is the result of gravitational and drag forces. Sand is only two to three times more dense than water, so the inertia of the rising grain only carries it to a height of about three grain diameters. The viscosity of water allows the grain to settle gently back to the bed. Saltation in air, however, involves trajectories up to 2-3 m high, especially after bouncing impacts on rock or pebble surfaces. The return impact also splashes other grains into the air, allowing a mechanical chain reaction of accelerated transport once motion has been initiated by fluid forces.

The second usage occurs in biogeography and refers to a theory that postulates the origin of major new taxonomic groups by the occurrence of single massive mutations. KSR

salt-dome A rounded hill produced by the upward doming of rock strata as a result of the diapiric movement of a halite bed or other evaporite deposit. ASG

Reading
Goudie, A.S. (1989) Salt tectonics and geomorphology. *Progress in Physical Geography*, **13**, 597–605.

salt-flat A near horizontal stretch of salt crust representing the bed of a former salt lake.

saltwater intrusion See GHYBEN-HERZBERG PRINCIPLE.

sand Both a type of SOIL TEXTURE and a PARTICLE SIZE class, ranging from 63 to 2000 μm. Since sand is a particle size that is readily transported by both water and wind, it is commonly found in fluvial and aeolian deposits. DSGT

sand banks These form significant depositional features in many coastal regions and on continental shelves, and are often fed and capped by sand waves. Two main types have been identified (Dyer, 1986). Linear sand banks or ridges occur in shallow tidal seas where sand is present and current velocities exceed about $0.5\,\mathrm{m\,s^{-1}}$. They can be up to 80 km long, and typically

1-3 km wide and tens of metres high. Headland (or banner) banks develop in association with promontories, again where current strengths exceed about $0.5\,\mathrm{m\,s^{-1}}$. They are only a few kilometres in length and have an elongated pear-shaped form, the broad end being directed towards the top of the headland.   ASG

Reference

Dyer, K.R. (1986) *Coastal and estuarine sediment dynamics.* Chichester: John Wiley & Sons, Ltd.

sand lens A discontinuous layer of sand in a sedimentary sequence representing the remnant of a former channel infill or overbank deposit comprising dunes, ripples, plane-bedded sands or sand sheets, or channel margin sediments.   JM

sand ramp An accumulation of sediment, usually upwind, but sometimes downwind, of a topographical obstacle and along a sand transport pathway that comprises interdigitated aeolian and slope or fluvial deposits. Palaeosols may also be present. Sand ramps may exceed 100 m in thickness and contain within their sediments a record of environmental changes, since sediments accumulate under fluctuations in conditions that favour aeolian transport and those favouring slope or fluvial processes. However their development may also be controlled by sediment availability; that is, from a fluvial system downwind of the feature. Bateman *et al.* (2012) suggest that their formation may be rapid, but other work suggest slow development can also occur. In some circumstances, distinguishing aeolian from colluvial accumulation may be challenging.   DSGT

Reading and Reference

Bateman, M.D., Bryant, R.G., Foster, I.D.L., *et al.* (2012) On the formation of sand ramps: a case study from the Mojave Desert. *Geomorphology*, **161-162**, 93–109. · Telfer, M.W., Thomas, Z.A. and Breman, E. (2012) Sand ramps in the Golden Gate Highlands National Park, South Africa: evidence of periglacial aeolian activity during the last glacial. *Palaeogeography, Palaeoclimatology, Palaeoecology*, **313-314**, 59–69.

sand ridge A type of low LINEAR DUNE that may have a pronounced vegetation cover on its lower slopes and only limited slip face development.

sand rivers See SANDBED CHANNELS.

sand rose A circular histogram that depicts the amount of sand that can potentially be moved by winds from various compass directions. A sand rose can be produced from surface wind velocity data collected at meteorological stations or from data collected in, for example, a desert dune field using a portable weather station. A sand rose differs from a WIND ROSE in that only wind speeds above the threshold velocity for sand transport ($\sim$5 m s^{-1} for sand with a mean particle diameter of 2 mm) are used in the calculation of the potential sand transport (usually expressed as vector units) from each direction for which wind data are recorded. A method for their construction is described by Fryberger (1979). Above the threshold velocity, wind data are usually grouped into several velocity classes, the median value of each class being multiplied by a weighting factor representing the rate of sand transport for each class and the percentage time the wind blew from that direction and in that velocity class. The arms of a sand rose are proportional to the potential sand drift from each direction, while total potential drift and the overall resultant vector, and its strength, can also be computed. Sand roses can be calculated for daily, weekly, monthly or annual data and are a useful tool for evaluating the aeolian environments in which sand drift is a problem or in which different dune types occur (e.g. Bullard *et al.*, 1996). Since sand roses can be calculated for any environmental setting for which suitable data are available, and since their construction averages out many key variables that control sediment entrainment by the wind, ignoring surface factors such as soil moisture and vegetation cover, they only represent the generalized *potential* for sand transport, rather than actual rates of sand drift.   DSGT

References

Bullard, J.E., Thomas, D.S.G., Livingstone, I. and Wiggs, G.F. (1996) Wind energy variations in the southwestern Kalahari desert and implications for linear dunefield activity. *Earth Surface Processes and Landforms*, **21**, 263–78. · Fryberger, S. (1979) Dune forms and wind regime. In E. D. McKee (ed.), *A study of global sand seas*. United States Geological Survey Professional Paper 1052. Washington, DC: US Government Printing Office; pp. 137–169.

sand sea The largest unit of aeolian deposition; an extensive (10^2-10^6 km^2) area of aeolian sand that may comprise sand dunes and/or sand sheets. In excess of 99% of dryland aeolian sand deposits are found in sand seas (also called ergs). Various definitions exist in the literature; for example, Fryberger and Ahlbrandt (1979) cite a minimum size of 125 km^2, with at least a 20% aeolian sand cover. Sand seas may be currently active with regard to aeolian processes or relict or fixed in terms of present-day climates. For a sand sea to develop it is necessary for there to be sufficient sand supply and winds strong enough to

Sand seas in Africa, Asia and Australasia, today and during the Last Glacial Maximum, when their extent may have been significantly greater due to enhanced aridity, windiness, and continentality.
Source: Thomas (2011). Reproduced with permission of John Wiley & Sons.

transport sand to a location where net accumulation occurs. Wilson (1971) noted that, at the regional scale, sand sea development is favoured in locations where sand flow is convergent and winds are decelerating (or where the actual net sand transport is decreasing). In the Sahara, for example, many sand seas occur in intermontane basins. Fryberger (1979) classified sand seas into low-, intermediate- and high-energy environments, with many high-energy sand seas occurring in the trade wind belt and low-energy sand seas near the centre of semi-permanent high- and low-pressure cells.   DSGT

Reading and References
Fryberger, S. (1979) Dune forms and wind regime. In E. D. McKee (ed.), *A study of global sand seas*. United States Geological Survey Professional Paper 1052. Washington, DC: US Government Printing Office; pp. 137–169. · Fryberger, S. and Ahlbrandt, T.S. (1979) Mechanisms for the formation of eolian sand seas. *Zeitschrift für Geomorphologie, NF*, **23**, 440–460. · Thomas, D.S.G. (2011) Aeolian landscapes and bedforms. In D. S. G. Thomas (ed.), *Arid zone geomorphology: process, form and change in drylands*, 3rd edition. Chichester: John Wiley & Sons, Ltd; pp. 427–453. · Wilson, I.G. (1971) Desert sand flow basins and a model for the development of ergs. *Geographical Journal*, **137**, 180–197.

sand sheet An accumulation of aeolian sand that is without significant morphological

expression. While sand dunes are generally absent, sand sheets may possess aeolian ripples or even low ZIBAR dunes. Sand sheets are, in fact, very common in dryland environments and range in size from small features of a few square kilometres to the vast $100,000\,km^2$ Selima sand sheet in the eastern Sahara. Five major controls on sand sheet development were recognized by Kocurek and Nielson (1986): vegetation, particularly grasses, which encourages low-angle laminations to develop; coarse sand, which does not readily form dunes other than zibar; a water table close to the surface (see STOKES' SURFACE) that may act as a base level to wind scour or an encouragement to 'sticky surface' deposition; periodic or seasonal flooding, which inhibits dune formation; and surface crusts such as SALCRETE, which inhibit deflation.   DSGT

Reference
Kocurek, G. and Nielson, J. (1986) Conditions favourable for the formation of warm-climate eolian sand sheets. *Sedimentology*, **33**, 795–816.

sand trap A device to measure the rate of sand transport (or flux) by the wind by efficiently removing airborne sediment over a known period of time whilst simultaneously allowing the free passage of airflow.

463

Simple but effective rotating BSNE (Big Spring Number Eight) horizontal flux sand traps. Heights of individual traps often relate to those of accompanying anemometers, which record wind velocities.
Photograph by David Thomas.

Horizontal traps consist of a partitioned rectangular box dug into the sand flush with the surface. Sand in SALTATION is trapped in the box, which is then retrieved and weighed. Typically, the rate is calculated in kilograms per width per minute. Horizontal traps do not disturb the upstream airflow, but sand can jump over the box if care is not taken in selecting its length.

Many vertical sand traps are based on a simple design (commonly referred to as the Aarhus type) whereby particles are collected through a vertical slot about 0.01 m wide, 0.5 m high with lengths up to 0.3 m. Commonly, the traps are partitioned horizontally so that vertical flux can also be established. To limit back pressure within the trap, air is bled from the back through a thin gauze. A popular version is described by Leatherman (1978). A wedge-shape design (widening downwind), minimizes back pressure through the Bernoulli effect by 'sucking' air through the trap (Nickling and McKenna Neuman, 1997). Traps can also be mounted on rotating masts so that they continuously face into the oncoming sand stream. Sophisticated designs

are now in use that include electronic load cells at their base giving a continuous high-frequency recording of the mass of sand in the trap. Trap efficiencies (the proportion of moving sand collected) vary from as low as 20% to 70%, depending upon design and field operation.   GFSW

References
Leatherman, S.P. (1978) A new aeolian sand trap design. *Sedimentology*, **25**, 303–306. · Nickling, W.G. and McKenna Neuman, C.K. (1997) Wind tunnel evaluation of a wedge-shaped aeolian sediment trap. *Geomorphology*, **18**, 333–345.

sand volcano
A small mount of sand with a smaller conical depression at the apex. Dimensions range from less than 2.5 cm in diameter up to 5 cm, and heights rarely exceed 3 cm. They are a surface expression of sediment compaction and dewatering. As the sediment settles, interstitial water may be expelled as small springs, at the mouth of which sand and mud particles may be deposited in a cone.   ASG

Reading
Picard, M.D. and High, L.R. (1973) *Sedimentary structures of ephemeral streams*. Amsterdam: Elsevier; pp. 139–141.

sand wedge
A relict ICE WEDGE, where the cast of the ice wedge has been infilled by windblown or washed-in sediment, usually sand. They can be found in environments that have experienced periglaciation and permafrost; for example, in parts of southern England where they are inherited from ice wedges formed during the last glacial cycle.   DSGT

sandbed channels
These are common in semi-arid environments. They have bed material in the 0.0625–2 mm sand size-range that is mobile even during low sub-bankfull flows, and is transported by SALTATION, or in suspension at high flow stages. During transport the sand travels in bedforms (ripples, dunes, antidunes) that change systematically with the flow regime (see FLOW REGIMES), strongly influencing the bed ROUGHNESS, and which may be preserved as sedimentary structures in the sand river deposits. Sand-bed rivers have characteristically high rates of bedload transport, are migratory on their floodplains and are often morphologically unstable with multiple, changing, braided channels and sand bars.   KSR

sandstone
An indurated sedimentary rock composed of cemented particles of sand, with a range of grain sizes between 0.0625 and 2 mm. They can be subdivided into various types on the basis of grain size and mineral composition. Types

with a limited matrix content (<15%) are called *arenites*, whereas those with a greater matrix content are termed *wackes*. Within the arenites, distinction must be made between *quartz* (<5% feldspar or rock fragments), *lithic* (>25% rock fragments, excluding feldspar), *arkose* (>25% feldspar) and *volcanic* (>50% volcanic fragments) subtypes. Sandstones cover very approximately the same area of the continents as do granites and carbonates, but have been much less the scene for the development of any particular geomorphological approach than the other two rock types. None the less, sandstone landscapes do have some distinctive geomorphological features that have been the subject of a review by Young and Young (1992). ASG

Reference

Young, R. and Young, A. (1992) *Sandstone landforms*. Berlin: Springer-Verlag.

sandstorm An atmospheric phenomenon occurring when strong winds entrain particles of dust and sand and transport them in the atmosphere.

sandur (pl. sandar; Icelandic) An extensive plain of glaciofluvial sands and gravels deposited in front of an ice margin by a system of braided or anastomosing meltwater streams that migrate across the sandur surface. The whole sandur is rarely flooded except during jökulhlaup events. A valley sandur is confined between valley walls. A pitted sandur forms on an ice margin that melts out to produce kettle holes. For morphological and sedimentary characteristics, see OUTWASH. JM

Reading

Bluck, B.J. (1974) Structure and directional properties of some valley sandur deposits in southern Iceland. *Sedimentology*, **21**, 533–554. · Hjulström, F. (1952) The geomorphology of the alluvial outwash plains (sandurs) of Iceland, and the mechanics of braided rivers. In *International Geographical Union, seventeenth congress proceedings*. Washington, DC; pp. 337–342. · Krigström, A. (1962) Geomorphological studies of sandur plains and their braided rivers in Iceland. *Geografiska Annaler, Series A: Physical Geography*, **44**, 328–345. · Price, R.J. (1969) Moraines, sandar, kames and eskers near Breidamerkurjökull, Iceland. *Transactions of the Institute of British Geographers*, **46**, 17–43.

sapping The undermining of the base of a cliff, rock or sediment face, usually associated with the subsequent failure of the face. It can occur in a number of ways, including undercutting by wave action, lateral stream erosion and erosion by emerging (exfiltrating) groundwater flow.

Erosion by emerging GROUNDWATER is often referred to as 'spring sapping' but is actually made up of two sets of processes: tunnel scour and seepage erosion. The relative operation of these processes depends upon whether groundwater emergence is focused upon one point or is along a more diffuse seepage zone, the latter most commonly occurring where permeable soil, sediment or rock overlies less permeable material. Tunnel scour operates when stress is applied to the walls of a pre-existing macropore by near-surface groundwater flow, commonly within a partially or fully consolidated material, leading to its widening and eventual collapse. Seepage erosion occurs when sufficient drag force is generated as water seeps through and exfiltrates from a porous material to entrain particles, cause failure or liquefy the material. It may also take place by the operation of biological, chemical and resultant physical weathering processes in the zone of groundwater emergence (e.g. algal growth or the precipitation of salts in pore spaces) that weaken the material and lead to mass wasting. Spring sapping and seepage erosion have been held responsible for the development of the steep heads of some DRY VALLEYS, scarp faces, ALCOVES and the amphitheatre-shaped heads of certain canyons (Howard *et al.*, 1988). DJN

Reading and Reference

Higgins, C.G. and Coates, D.R. (eds) (1990): *Groundwater geomorphology: the role of subsurface water in Earth-surface processes and landforms*. Special Paper 252. Boulder, CO: Geological Society of America. · Howard, A.D., Kochel, R.C. and Holt, H.E. (1988) *Sapping features of the Colorado Plateau: a comparative planetary geology field guide*. NASA Publication SP-491. Washington, DC: NASA. · Nash, D.J. (2011) Groundwater as a geomorphological agent in drylands. In D. S. G. Thomas (ed.), *Arid zone geomorphology: process, form and change in drylands*, 3rd edition. Chichester: John Wiley & Sons, Ltd; pp. 403–424.

saprolite Weathered or partially weathered bedrock that is in situ.

sapropel Amorphous organic compounds that collect in various water basins (lakes, lagoons, shallow marine basins and estuaries) are termed sapropels. The sapropel is formed by predominantly anaerobically decomposing remains of phyto- and zooplankton. These are richer in fatty and protein substances than is peat. The decomposition and putrefaction of the organic content leads to the formation of various hydrocarbons, which are believed to be the basis of the origin of petroleum and natural gas compounds that form after compression under accumulated sediment. The progressive accumulation of sapropel is governed largely by rapid multiplication of the organisms responsible for it.   AP

Reading
Pettijohn, F.J. (1984) *Sedimentary rock*, 3rd edition. Delhi: CBS Publishers. · Rossignol-Strick, M. (1985) Mediterranean Quaternary sapropels, an immediate response of the African monsoon to variation of insolation. *Palaeogeography, Palaeoclimatology, Palaeoecology*, **49**, 237–263.

sarsen A block of silica-cemented sandstone, breccia or conglomerate found in many parts of southern England, notably on the margins of the London and Hampshire basins. Sometimes called 'grey wethers' or 'pudding stone', sarsens are thought to be the result of weathering processes in a surface or near-surface environment under conditions of Tertiary warmth. Humans have often employed such blocks to make monuments, such as Avebury stone circle or Windsor Castle. Sarsens are in effect a fossil silcrete duricrust, and comparable deposits in the Paris Basin are called *meulières*.   ASG

Reading
Summerfield, M.A. and Goudie, A.S. (1980) The sarsens of southern England: their palaeoenvironmental interpretation with reference to other silcretes. In D. K. C. Jones (ed.), *The shaping of southern England*. Institute of British Geographers, Special Publication no. 11. London: Academic Press; pp. 71–100.

sastrugi Furrows and ridges in the surface of ice and snow accumulations through the action of wind.

satellite meteorology Meteorology that depends largely, or completely, on data generated by METEOROLOGICAL SATELLITES. As such it is a relatively young branch of the parent science, but one that grew very rapidly after the launching of the first specialized meteorological satellite in 1960. With meteorological satellite platforms and sensor systems still undergoing active development as part of the present-day rise of environmental remote sensing, satellite meteorology continues to increase in both its scope and sophistication. Each meterological satellite or satellite system is supported by ground facilities that receive satellite data for dissemination to the user community. There, initial preprocessing (e.g. removal of extraneous data contents) and processing (e.g. geographical rectification) are carried out, and ranges of satellite products are prepared to meet the needs of the primary user community. These include:

1 Manual and basic products, including unenhanced and enhanced visible and infrared imagery; facsimile maps (e.g. nephanalyses, and snow and ice charts); and alphanumeric outputs (e.g. satellite weather bulletins, moisture analyses and plume winds).
2 Man-machine combined products, including cloud motion vectors and precipitation fields.
3 Computer-derived image products, including period minimum brightness composites, and cloud field analyses (cloud types, cloud top heights, etc.).
4 Computer-derived digital products; for example, vertical temperature and moisture profiles, and sea surface temperature data.
5 Archival products (e.g. magnetic tapes of raw and/or processed data, and photographic images) for use in atmospheric research.

These products form a very important part of the complete data pool for use in weather forecasting. Some data (e.g. cloud images) may be used *qualitatively* in the analysis of synoptic situations and forecasting for synoptic, subsynoptic or mesoscale regions, either on the ground or even aloft. Others are used *quantitatively*; for example, satellite-derived vertical profiles and satellite winds; such data may be added to the available conventional (in-situ) observations to provide improved data arrays for numerical (computer-based) forecasting procedures. The satellite inputs are particularly important for tropical and polar regions, and for some of the more remote continental areas in middle latitudes; that is, those regions traditionally least well observed by surface weather stations. Satellite-derived quantitative data are also important in vertical profiling of the atmosphere in these regions because of the sparseness of their upper air weather observatories - but satellite soundings are of value almost everywhere because they penetrate the atmosphere downwards, whereas radiosondes and similar conventional devices inspect it from the bottom upwards, usually to the tropopause, and rarely far beyond it.

Equally important and varied uses are made of satellite data in *meteorological research*. On the global scale, evaluations of Earth/atmosphere radiation and related energy budgets have benefited greatly from satellite radiation data. Since satellite sensor systems orbit above the top of the Earth's atmosphere, a number of budget components that were previously amenable only to estimation can now be measured. These include the Earth/atmosphere albedo, longwave radiative fluxes towards space and the net radiation budget for the globe, or any selected area, and their attendant columns of the atmosphere. Studies of spatial and temporal changes of these and other quantities are yielding vital insights into the behaviour of the atmosphere, involving both long-distance interactions ('teleconnections') and vertical interrelationships among its different layers. Satellite imagery has revolutionized our knowledge and understanding of many regions and types of weather systems, ranging in the first case from tropical latitudes to polar regions, and in the second from synoptic down to mesoscale features. Virtually no region or type of weather system has not been elucidated in some way. The First GARP Global Experiment (FGGE) in 1979 depended heavily on satellite data, and is likely to be the forerunner of other large and complex meteorological research projects in the future.

As the runs of data from meteorological satellites have lengthened, so increasing attention has been paid to *climatological analysis* of these types of data sets. Particular advantages have accrued in studies of the climatologies of cloud distributions, synoptic weather systems (e.g. the intertropical cloud band and associated features, hurricanes, jet streams and mid-latitude depressions) and the structure and behaviour of the upper atmosphere. As the data sets lengthen further, so they may be expected to support an increasing number and range of studies of climatic change. It is also likely that microwave data will feature increasingly in both meteorological and climatological satellite applications.   ECB

Reading
Barrett, E.C. and Martin, D.W. (1981) *The use of satellites in rainfall monitoring*. London: Academic Press. · Cracknell, A.P. (1981) *Remote sensing in meteorology, oceanography and hydrology*. Chichester/New York: Ellis Horwood/Halsted Press.

saturated wedge and zone   Consists of a layer of saturated soil at the base of a hillslope, thickening downslope and in some areas reaching the soil surface to define an area of surface saturation.

Saturation develops in response to THROUGH-FLOW collecting from the hillside above. Its total volume generally increases downslope, whereas its rate of flow may decrease if the slope profile is concave or if water is backed up from the stream. The magnitude of this effect may be seen for conditions of steady net rainfall at intensity I, falling in a collecting area of a per unit contour width. The total outflow along 1 m of contour is then $Q = Ia$. This flow may also be expressed in terms of the depth h of saturated flow in the soil and the slope (or strictly the total potential) gradient S, in the form $Q = Sf(h)$, where f is a function that increases with h and with soil permeability. Equating these expressions for the flow, the saturated wedge has thickness h given by

$$f(h) = \frac{Ia}{S}$$

The effect of slope profile concavity (decreasing S) and of a large collecting area or converging flow (large a) may readily be seen. It is also evident that saturated zones/wedges only persist seasonally in areas where net rainfall (rainfall minus evapotranspiration) remains positive seasonally. Saturated wedges rise to the soil surface where h is greater than the soil water storage, giving rise to a DYNAMIC SOURCE AREA of surface saturation.   MJK

Reading
Kirkby, M.J. (1978) *Hillslope hydrology*, Chichester: John Wiley & Sons, Ltd; especially chapters 7 and 9.

saturation coefficient (or degree of saturation S)   The water content of a rock after free saturation, expressed as a percentage of the maximum water content. The free saturation is the water that can be absorbed into the pore spaces of a sample when it is just immersed in water, while the maximum water content is the amount of water absorbed when it is forced in under vacuum. The saturation coefficient is important in the testing of the mechanical breakdown of rocks.   WBW

saturation deficit   The depth of water required to bring saturation up to the soil surface and initiate overland flow. Since THROUGHFLOW is not necessarily connected to the groundwater table at depth, a zero saturation deficit need not imply complete saturation of the soil profile. The saturation deficit is important in establishing areas of current overland flow production and gives directly the amount of storm rainfall needed to initiate overland flow. The deficit is also important as one of the controls on evaporation, particularly from unvegetated soil surfaces.   MJK

saturation overland flow   See OVERLAND FLOW.

savanna (savannah)   A grassland region of the tropics and subtropics. The word is probably of American Indian origin, signifying treeless grasslands. Savannas today are taken to mean plant formations dominated by grasses and grass-like species (graminoids) with herbaceous non-grass species (forbs), often possessing a light to dense scattering of trees. Woody savannas prevail where grasses are at least co-dominant, grading to savanna woodland where trees assume dominance. Seasonal climates impose water stresses during the dry period, which inhibit tree growth and encourage drought-resistant plants. Burning, both natural and human induced, also favours a herbaceous formation with xeromorphic characteristics.   PAF

Reading
Bourlière, F. (ed.) (1983) *Tropical savannas*. Amsterdam: Elsevier. · Huntley, B.J. and Walker, B.H. (1982) *Ecology of tropical savanna*. Berlin: Springer-Verlag. · Mistry, J. (2000) *World savannas: ecology and human use*. London: Pearson.

scabland   An area where, as a result of erosion, there is no soil cover and the land surface consists of rock and rock fragments.

scald   Localized destruction of vegetation, in particular ground cover, caused by increased soil salinity. Scalding is a common, if ephemeral, feature of semi-arid regions.   PSH

scar   A cliff or very steep slope or a rocky outcrop.

Schmidt hammer   Originally, this device was invented to test the curing of concrete and as a measure of its 'strength'. It has been used in geomorphology to provide a measure of weathering of rock and stones as well as durability of building materials and as a surrogate measure of the age of deposits. It consists of a barrel that holds a steel rod that is made to impact upon the target area by a calibrated coil spring. The rod's rebound gives a value on a scale that can be read off. This is the R value (a measure of the coefficient of restitution between rod and target). For a calibrated rod and spring with (usually normal) direction of impact, the R value can be used as a measure of the compressive strength of the near-surface material. Crystal type, grain size and distribution, as well as small-scale irregularities, provide a scatter in the results, thus requiring careful sampling and statistical analysis. It is best used for comparative measures at a site. It has been used, where rock type is 'constant', to show a progression of weathering over time; moraine chronologies have been developed by this method.   WBW

Reading
Goudie, A.S. (ed.) (2005) *Geomorphological technique*, 2nd edition. London: Taylor and Francis; pp. 174–175, 218–219.

sciophyte   A plant that lives in a well-shaded environment.

sclerophyllous   Refers to species of evergreen woody vegetation with toughened, leathery and sometimes waxy leaves. The characteristic is an adaptation to drought, as the hard surfaces with embedded leaf stomata, act to reduce moisture loss through transpiration. Sclerophylls are especially common in the Mediterranean-type climate regions of the world (e.g. olive, cork oak).   MEM

scoria   A volcanic rock or slag consisting of angular rock fragments and numerous voids that were originally filled with volcanic gases.

scour and fill   The processes of cutting and subsequent filling of fluvial channels.

scree   An accumulation of primarily angular clasts that lies at an angle of around 36° beneath an exposed free face or cliff. The prime cause of deposition is rock fall, but other processes, such as debris flows, may contribute to their development. The largest clasts occur at the base of the scree.   ASG

Reading
Statham, I. (1973) Scree slope development under conditions of surface particle movement. *Transactions of the Institute of British Geographers*, 59, 41–53.

S-curve   A method of extending a unit hydrograph. It represents the surface run-off hydrograph caused by an effective rainfall of intensity I/T mm h^{-1} applied indefinitely, where T is the duration of the effective rainfall (see UNIT HYDROGRAPH).

sea cliffs   Steep slopes that border ocean coasts. They are ubiquitous and occur along approximately 80% of the ocean coasts of the Earth (Emery and Kuhn, 1982). They can be classified into three main types according to their stage of development: *active cliffs*, where bedrock is exposed by continuous retreat under the influence of both marine and subaerial forces; *inactive*, where their bases are mantled by talus, and where there is some vegetation cover; and *former*, where the influences of marine erosion have disappeared, so that subaerial erosion rounds the crests and provides material for stream deposition beyond their bases. Among the important factors in their development are the nature of the landward topography, the structure and lithology of the materials in which they have been eroded, the nature of marine erosion and the power of subaerial processes (rockfalls, mass movements, frost weathering, etc.). Many cliff profiles display complex forms brought about by long, continued changes in sea level, climate and the balance between subaerial and marine processes (Steers, 1962). ASG

References
Emery, K.O. and Kuhn, G.G. (1982) Sea cliffs: processes, profiles, classification. *Bulletin of the Geological Society of America*, **93**, 644–653. · Steers, J.A. (1962) Coastal cliffs: report of a symposium. *Geographical Journal* **128**, 303–320.

sea ice Ice that forms on the sea surface when water temperatures fall to around −1.9 °C. Such conditions apply to considerable areas of the Arctic and Antarctic. In polar latitudes the equilibrium thickness of sea ice is around 3 m, and any surface melting is equalled by bottom freezing. When the sea ice is attached to the land it is known as *fast* ice, and when floating free under the influence of currents and wind it is called *pack* ice. The latter consists of ICE FLOES centred on relatively resistant ice, leads of open water that may freeze over rapidly in winter, and irregular fractured and crumpled ice-forming features known as *pressure ridges* or *keels* (see illustration). DES

Reading
Lewis, E.L. and Weeks, W.F. (1971) Sea ice: some polar contrasts. In G. Deacon (ed.), *Symposium on Antarctic ice and water masses, Tokyo, September 1970*. Cambridge: Scientific Committee for Antarctic Research; pp. 23–31. · Nansen, F. (1897) *Farthest north*, 2 vols. London: Constable. · Sugden, D.E. (1982) *Arctic and Antarctic: a modern geographical synthesis*. Oxford/Totowa, NJ: Blackwell/Barnes & Noble. · Zwally, H.J. and Gloersen, P. (1977) Passive microwave images of the polar regions and research applications. *Polar Record*, **18** (116), 431–450.

sea/land breeze The sea and land breezes form part of a diurnally varying, vertical circulation of air that is ultimately induced by temperature differences between land and sea. Given a morning of near calm conditions and clear skies, a land area heats up more rapidly than does the adjacent water body. Consequently, as a result of turbulent transfer and convection of heat, the air over the land becomes warmer than that at the same height over the sea. In turn, this means that the land air is less dense than the sea air and thus, by the HYDROSTATIC EQUATION, the vertical gradient of pressure is less over land than over sea. Assuming that pressure was initially uniform over land and sea, the difference in vertical gradients must result in a higher pressure over the land at a height H (say) than at the same height over the sea. Consequently, at height H, air moves from land to sea.

Sea ice photographed on the Imperial Trans-Antarctic expedition, 1914–1916.

Schematic diagram of a sea breeze and a land breeze.
Source: Nicholson (2011). Reproduced with permission of Cambridge University Press.

In turn, this disturbs the hydrostatic equilibrium of the air columns over both land and sea and also generates a horizontal pressure gradient from sea to land near the surface. The air motion resulting from this gradient force is the sea breeze. At the landward and seaward extremities of this breeze the air rises and sinks respectively, creating a complete vertical cell. Often, at the landward extreme of the sea breeze, cloud and occasionally precipitation form along with what is termed the sea-breeze front. The sea breeze frequently reaches as far as 50 km inland and 1000 km in the vertical, reaching speeds of $4\text{-}7\,\mathrm{m\,s^{-1}}$ or more. The land breeze results from a reversal of

these processes when the land is cooler than the sea. It is generally a nocturnal phenomenon and weaker and smaller in scale than the sea breeze. Like the sea breeze, it can result in cloudiness and precipitation in the region of rising motion (i.e. over the sea).   BWA/SEN

Reading and Reference
Crosman E.T. and Horel, J.D. (2010) Sea and lake breezes: a review of numerical studies. *Boundary-Layer Meteorology* **137**, 1–29. · Nicholson, S.E. (2011) *Dryland climatology*. Cambridge: Cambridge University Press. · Simpson, J.E. (1994) *Sea breeze and local winds*. Cambridge: Cambridge University Press.

sea level   The elevation of the ocean surface relative to a vertical datum. The term is often used synonymously with mean sea level, although the latter is derived from a series of observations of the former. Relative sea level changes at a location can involve changes in either the absolute elevation of the ocean or the land, or both. Instantaneous measurements of sea level exhibit cyclical and secular changes, and the averaging duration will vary depending upon a particular purpose. Monthly averages may be used to remove short-term cycles, such as those associated with waves and tides, and to allow seasonal comparisons of sea level. Averages comprising a year (or more) of data

Global mean sea level, 1860–2009. Blue: estimated from coastal data. Red: Church and White (2007) estimates. Black: satellite altimeter data from 1993 onwards.
Source: Church and White (2011). Reproduced with permission of Springer.

are appropriate for revealing seasonal cycles and anomalies, and for the compilation of data series for the analysis of long-term sea-level change.

Several geophysical processes affect sea level. Waves and tides cause regular, short-term fluctuations. There are also local fluctuations associated with atmospheric low-pressure systems (the inverted barometer effect), storm surges and flood discharges from rivers. These transient phenomena can cause local sea-level rises exceeding 5 m and may have catastrophic human impacts (e.g. Pugh, 1987). There are also seasonal changes, on the order of 0.1 m, caused by variations in the location and intensity of semi-permanent high- and low-pressure systems and changes in ocean circulation. At return intervals of several years, atmospheric phenomena such as El Niño are capable of changing regional sea levels by as much as 0.5 m. In order to establish a stable estimate of mean sea level for purposes of tracking long-term global changes, averaging periods must be long enough to minimize the impacts of such local/regional effects.

The study of long-term sea-level changes involves several academic disciplines, including climatology, engineering, geography, geology, geophysics and oceanography. The focus of such research may extend back through geological time for several billion years, and into the future for several centuries. In the former case, the motivation is to understand the evolution of the Earth as a planet. In the latter case it is to understand Earth as the home of humanity. Sea-level studies for Earth history are based largely on estimations of continental freeboard and are concerned, necessarily, with eustatic, isostatic and epeirogenic changes. Eustatic changes are caused by changes in the volume of the ocean basins, or in the volume of the water in the oceans. Changes in ocean basin volume occur over timescales of 10^4 to 10^8 years, and are associated with sea-level changes of the order of 1-100 m (Geophysics Study Committee et al., 1990). Epeirogenic changes, vertical movements of the crust, occur at similar scales. Isostatic changes result from crustal deformation caused by the redistribution of mass over the Earth's surface and occur at timescales of 10^2-10^8 years. The shorter time frame is associated with redistribution of ice and water, while the latter are caused by sedimentation.

Changes in ocean volume can occur over substantially shorter time frames usually caused by shifts in the volume of water (including ice) stored on the land, and by steric or thermohaline changes. There are also longer term changes in the total mass of water on the Earth. According to Meier (1990), there are about 30×10^6 km^3 of water currently stored as ice, and it is estimated that the volume was two to three times larger during the Last Glacial Maximum, about 20,000 years ago (e.g. Pirazzoli, 1996). The change in volume was caused by melting of the Pleistocene ice sheets and was accompanied by sea-level rise of more than 100 m.

Changes in ocean volume can arise from changes in water storage in the continental groundwater systems (Baur et al., 2013, Fasullo et al., 2013).

The density of seawater varies with temperature and salinity. Large-scale salinity changes will occur slowly, but ocean temperature responds relatively quickly to global atmospheric temperature changes. If the upper 500 m of the ocean were warmed uniformly by 1 °C, thermal expansion of the water would raise sea level by about 0.06 m in high latitudes and about 0.15 m in the tropics (Wigley and Raper, 1993). Current rates of sea-level rise, estimated to be approaching 2 mm per year (Gregory et al., 2013), are attributed largely to thermal expansion of seawater, and to a lesser degree to continued melting of alpine glaciers. Melting of Pleistocene ice also caused substantial isostatic adjustment, with land elevation exceeding 800 m in some regions (Pirazzoli, 1996).

Local changes in sea level may also be caused by subsidence, faulting and folding. Many deltaic coastlines, (e.g. in Louisiana, Bangladesh, Venice or Bangkok) are experiencing rapid sea-level rise caused by compression of sediments and crustal downwarping. Vertical movement along faults may cause almost instantaneous, metre-scale rise or fall of local sea level. Folding also causes vertical movements that may translate to a localized change in sea level.

Rates of sea-level rise are projected to increase, perhaps doubling in the next century, caused by greenhouse effect warming and the resulting warming of the oceans and additional ice melting. The potential inundation of populous, low-lying coastlines has become an issue of international concern (Cazenave and Le Cozannet, 2014).

DJS/DLD

References

Baur, O., Kuhn, M. and Featherstone, W.E. (2013) Continental mass change from GRACE over 2002-2011 and its impact on sea level. *Journal of Geodesy*, **87**, 117–125. · Cazenave, A. and Le Cozannet, G. (2014) Sea level rise and its coastal impacts. *Earth's Future*, **2** (2), 15–34, doi: 10.1002/2013EF000188. · Fasullo, J.T., Boening, C., Landerer, F.W. and Nerem, R.S. (2013) Australia's unique influence on global sea level in 2010-2011. *Geophysical Research Letters*, **40**,

4368–4373. · Geophysics Study Committee, Commission on Physical Sciences and Resources and National Research Council (1990) Overview and recommendations. In Geophysics Study Committee, Commission on Physical Sciences and Resources and National Research Council (eds), *Sea-level change*. Washington, DC: National Academy Press; pp. 3–34. · Gregory, J.M., White, N.J., Church, J.A., *et al.* (2013) Twentieth-century global-mean sea level rise: is the whole greater than the sum of the parts? *Journal of Climate*, **26**, 4476–4499. · Meier, M.F. (1990) Role of land ice in present and future sea-level change. In Geophysics Study Committee, Commission on Physical Sciences and Resources and National Research Council (eds), *Sea-level change*. Washington, DC: National Academy Press; pp. 171–184. · Pirazzoli, P.A. (1996) *Sea-level changes: the last 20,000 years*. Chichester: John Wiley & Sons, Ltd. · Pugh, D.T. (1987) *Tides, surges and mean sea level*. Chichester: John Wiley & Sons, Ltd. · Wigley, T.M.L. and Raper, S.C.B. (1993) Future changes in global mean temperature and sea level. In R. A. Warrick, E. M. Barrowand T. M. L. Wigley (eds), *Climate and sea level change: observations, projections and implications*. Cambridge: Cambridge University Press; pp. 111–133.

sea mount   A mountain or other area of high relief on the sea floor that does not reach the surface. Flat-topped sea mounts are called GUYOTS.

sea surface temperature (SST)   The sea surface temperature is the subsurface bulk temperature in the top few metres of the ocean, measured by ships, buoys and drifters. From ships, measurements of water samples in buckets were mostly switched in the 1940s to samples from engine intake water. Satellite measurements of *skin temperature* (uppermost layer; a fraction of a millimetre thick) in the infrared or the top centimetre or so in the microwave are also used, but must be adjusted to be compatible with the bulk temperature.   TS

secondary depression   A region of low pressure that forms within the circulation of a pre-existing depression. It sometimes occurs at the 'triple point' where the occlusion and warm and cold fronts meet, or along the individual fronts as warm- or cold-front waves. The development of the secondary is often preceded by an increase in the spacing of the isobars locally; once formed, the secondary can even absorb the parent low.

Secondary depressions also form occasionally in the unstable airstreams as polar lows.   RR

secondary flows   Those currents in moving fluids that have a velocity component transverse to the local axis of the primary, or main, flow direction. In river channels they are associated with the longitudinal vortex known as helical flow.

Secondary flows in straight channels are probably caused by nonuniform distribution of boundary shear stress and by turbulence generated at the base of the channel banks. Two circulatory cells occur side by side, with flow alternately *converging* at the surface and then downwelling, and *diverging* at the surface where upwelling occurs in mid-channel, at locations spaced apart at five to seven times the channel width. Downwelling intensifies the local bed shear stress exerted by the flow and encourages scour, while upwelling results in a low shear stress and deposition. This secondary circulation flow pattern thus relates to the POOL AND RIFFLE sequence. In curved channels, secondary flows result from skewing of the main flow towards one bank, and the consequent transverse HYDRAULIC GRADIENT then drives a transverse current that crosses the channel at the bed after plunging near the outer bank and assisting bed scour and undermining of the bank. Secondary flows represent an important control of the spatial distribution of erosion and deposition within a river channel, and of channel pattern change.   KSR

sediment, fluvial   Has been defined as particles derived from rock or biological material that are, or have been, transported by water. It has provided the focus for numerous studies by physical geographers, although a dichotomy exists between studies of contemporary *transport* and studies of the *deposits* associated with contemporary and past fluvial activity. The movement of fluvial sediment also has important practical implications for the problems of channel management, river intake and irrigation canal maintenance, reservoir and harbour siltation, debris dumping and floodplain accretion that stem directly from its deposition. Furthermore, there is an increasing awareness of the important role of fine sediment in the transport of contaminants such as heavy metals and pesticides by rivers.

Information on the properties of fluvial sediment is of significance in all these contexts, and particle size must rank as the most important since it exerts a major control on entrainment, transport and deposition processes. The size of particulate material transported by rivers ranges between fine clay and colloidal particles of less than $0.5\,\mu m$ in diameter to large boulders moved during flood events. Documentary evidence points to boulders weighing up to $7.5\,t$ being transported by the Lynmouth flood, which occurred in North Devon, UK, in 1953. A useful distinction may be made between material moving as bedload (see BEDLOAD, BEDLOAD EQUATION) and that transported as SUSPENDED LOAD. The latter generally involves particles $<0.2\,mm$ in

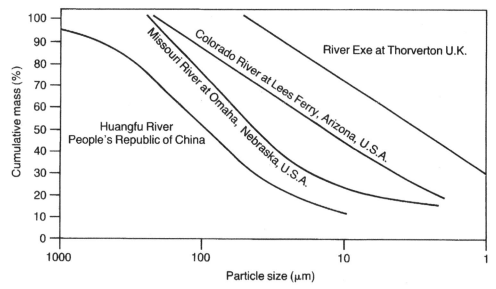

Examples of particle size distribution in fluvial sediment.

diameter, and the diagram provides several examples of typical particle size distributions of suspended sediment from rivers in different areas of the world. Although such particle size analyses commonly relate to the size of the discrete sediment particles, it is important to recognize that many of these will be transported within larger aggregates. In the case of suspended sediment, the particle size characteristics are largely governed by the nature of the source material. With the larger particles comprising the bedload, the boundary shear stress or force available to initiate movement is the dominant control, although the range of sizes available may also be important. The degree of sorting will vary according to the precise character of the fluvial environment and the transport distances involved. Measurements of particle shape and roundness have also been employed to demonstrate downstream changes in bed material character.

Fluvial sediment transported as bedload consists almost entirely of inorganic material, and this will closely resemble the parent rock in terms of mineral composition. The finer sediment transported in suspension may, however, incorporate a considerable proportion of organic material. Its mineralogy may differ considerably from that of the parent rock due to the chemical weathering processes involved in the disintegration of the rock and the selectivity of the detachment and transport processes. This selectivity will result in the suspended sediment showing considerable enrichment in clay-sized particles and inorganic matter, when compared with the source material. Organic matter contents are typically of the order of 10%, although values as high as 40% have been encountered in some rivers. Enrichment in fine material has important implications for sediment-associated transport of contaminants, because clay-sized particles exhibit considerably greater specific surface areas (typical values are $200\,\mathrm{m^2\,g^{-1}}$ for clay, $40\,\mathrm{m^2\,g^{-1}}$ for silt and $0.5\,\mathrm{mm^2\,g^{-1}}$ for sand) and cation-exchange capacities.

Information on the average chemical composition of inorganic suspended sediment transported by world rivers and its relation to that of surficial rocks, abstracted from Martin and Meybeck (1979) and Meybeck (1981), is listed in Table 1. These data indicate that, in general, suspended sediment is enriched in aluminium, iron and titanium with respect to parent rock, while sodium, calcium and magnesium are strongly depleted. This tendency is more marked for tropical rivers than for rivers in temperate and Arctic regions because of the greater efficacy of chemical weathering in tropical areas.

Several workers have attempted to estimate the total annual sediment load transported from the land surface of the globe to the oceans, although in the absence of data on bedload transport these values relate to suspended sediment. The

473

Table 1 Average chemical composition of inorganic suspended sediment and surficial rocks

	Concentration (mg g^{-1})							
	Al	Ca	Fe	K	Mg	Na	Si	Ti
Tropical rivers	114	7.5	62	18	9.6	5.1	264	7.3
Temperate and Arctic rivers	72	36	45	23	12	8.6	293	4.9
World rivers	90	25	52	21	11	7.1	281	5.8
Surficial rocks	70	45	36	24	16	14	275	3.8

estimate of Milliman and Meade (1983) pointed to a mean annual transport to the ocean of $13.5 \times 10^9 \, t a^{-1}$. This is equivalent to a sediment yield of approximately $135 \, t \, km^{-2} a^{-1}$ from the land surface of the globe and is 3.6 times greater than the total dissolved load transport to the oceans of $3.72 \times 10^9 \, t a^{-1}$ suggested by Meybeck (1979).

Existing estimates of the relative importance of bedload and suspended load in the total transport of fluvial sediment to the oceans are extremely tentative, but they place the bedload contribution at about 10% of the suspended load. This proportion varies markedly for individual rivers. In Arctic streams on Baffin Island, measurements suggest that coarse material or bedload constitutes 80–95% of the total transport of fluvial sediment, whereas data from the Volga river in the former USSR indicate that 98–99.7% of the sediment transported is composed of fine material in suspension.

Studies of fluvial sediment deposits have been largely concerned with the coarse fraction that moves relatively slowly through a river system and which may come to rest temporarily or permanently in depositional sinks or stores. These deposits are commonly classified on a genetic basis and they include, for example, point, longitudinal and marginal bars, lag deposits, splays and alluvial fans. Whereas most channel deposits are composed of relatively coarse material (bedload), finer material may be deposited by rivers on floodplains and in lakes or reservoirs, and where fine sediment is transported downslope by unconcentrated surface run-off this may frequently be deposited before reaching the stream channel.

The dichotomy between studies of fluvial sediment transport and of the associated deposits can usefully be bridged by considering the *sediment budget* of a drainage basin. With this concept, the transport of sediment out of the basin is seen as the result of the various processes involved in mobilizing sediment within the basin and of the deposition and storage of the sediment

within the basin. Only a small proportion of the sediment mobilized may be transported out of the basin, and some material may be deposited in temporary storage to be remobilized on a subsequent occasion; for example, during a higher magnitude event. The deposits are therefore treated as an integral part of the conveyance system.

DEW

Reading and References

Abrahams, A.D. and Marston, R.A. (eds) (1993) Drainage basin sediment budgets. *Physical Geography*, **14**, 221–320. · American Society of Civil Engineers (1975) *Sedimentation engineering.* New York: American Society of Civil Engineers. · Lehre, A.K. (1982) Sediment budget of a small coast range drainage basin in north-central California. In F. J. Swanson, R. J. Janda, T. Dunneand D. N. Swanson (eds), *Sediment budgets and routing in forested drainage basins.* US Forest Service general technical report PNW-141. US Department of Agriculture Forest Service, Pacific Northwest Forest and Range Experiment Station; pp. 67–77. · Martin, J.M. and Meybeck, M. (1979) Elemental mass-balance of material carried by major world rivers. *Marine Chemistry*, 7, 173–206. · Meybeck, M. (1979) Concentration des eaux fluviales en éléments majeurs et apports en solution aux océans. *Revue de Géologie Dynamique et de Géographie Physique*, **21**, 215–246. · Meybeck, M. (1981) Pathways of major elements from land to ocean through rivers. In J. M. Martin, D. Burtonand D. Eisma (eds), *River inputs to ocean systems.* UNEP/UNESCO report. Paris: UNEP/UNESCO. · Milliman, J.D. and Meade, R.H. (1983) World-wide delivery of river sediment to the oceans. *Journal of Geology*, **91**, 1–21. · Richards, K. (1982) *Rivers: form and process in alluvial channels.* London: Methuen. · Statham, I. (1977) *Earth surface sediment transport.* Oxford: Clarendon Press. · Trimble, S.W. (1981) Changes in sediment storage in the Coon Creek Basin, Driftless Area, Wisconsin, 1853 to 1975. *Science*, **214**, pp. 181–183. · UNESCO. (1982) *Sedimentation problems in river basins.* Paris; UNESCO.

sediment budget A sediment budget is the difference between sediment input to a given area and sediment output from that area, over a given amount of time. Thus, it can apply to any geomorphological process that involves the transport of sediment, whether in the atmosphere, on land or in water. It follows directly from the CONTINUITY

EQUATION, in recognizing that mass cannot be created or destroyed: if the budget is negative, there must have been erosion in the system; if the budget is positive, there must have been deposition in the system. One definition of an EQUILIBRIUM system is one where the sediment budget is zero, implying that the system is just capable of transporting all the sediment applied to it. However, it is exceptionally rare for inputs to be equal to outputs over all timescales, and this is where the sediment budget can be useful. As a nonzero sediment budget implies that the area delimited by the budget must be eroding or depositing, identification of those timescales at which the sediment budget is either negative or positive can indicate the timescales associated with landform change within that system.

This diagnostic property is of immense use to the geomorphologist and may be central to management: Pethick (1996) shows how cliff recession along the Holderness coast in the medium term is regulated by the difference between sediment inputs from cliff recession and sediment outputs driven by longshore drift. Determination of the sediment budget can be either process based or morphologically based. Process-based approaches are suitable where sediment transport rates are readily measured. For instance, by measuring the concentration of SUSPENDED LOAD and DISCHARGE at two points in a river with no tributary inputs, it is possible to estimate a budget for suspended load. However, the transport of coarser sediment as bedload (see BEDLOAD, BEDLOAD EQUATION) in rivers and processes such as landsliding are much more difficult to measure directly. However, it may be possible to estimate a sediment budget by repeat monitoring of landform morphology. This is sometimes referred to as an inverse treatment of the problem. An example is provided by Lane (1997), who repeatedly surveyed the surface of a gravel-bed river in the Swiss Alps to determine sediment budgets and then used this information to estimate the bedload transport rate upstream required to produce these budgets.   SNL

References

Pethick, J.S. (1996) Coastal slope development: temporal and spatial periodicity in the Holderness cliff recession. In M. G. Anderson and S. M. Brooks (eds), *Advances in hillslope processes*. Chichester: John Wiley & Sons, Ltd. · Lane, S.N. (1997) The reconstruction of bed material yield and supply histories in gravel-bed streams. *Catena*, 30, 183–196.

sediment delivery ratio (SDR) The rate of sediment yield at a specified point in a channel network, expressed as a fraction of the rate of erosion in the contributing catchment. They tend to be highest in catchments where channel and valley side slopes are steep, and where relief and drainage density are high. They are lower where the sediment sources are far from channels, or are separated from them by sediment-trapping zones (i.e. where connectivity is low). The SDR also tends to decrease as drainage area increases because, as one moves downstream, channel and valley gradients become gentler and floodplains and footslopes typically become wider and hence provide greater opportunities for sediment storage.   ASG

Reading

Fryirs, K. (2013) Connectivity in catchment sediment cascades: a fresh look at the sediment delivery problem. *Earth Surface Processes and Landforms*, **38**, 30–46.

sediment fabric This refers to the composition and organization of a sedimentary deposit. Identification of sediment fabrics follows from the idea that different generating processes produce markedly different sediment fabrics. For instance, glacial, fluvial and aeolian deposits are each typically associated with different fabrics, and thus identification of fabric in the field is central to reconstructing the environment in which the sediment was deposited.   SNL

sediment routing The process through which fluvial sediment (including both suspended and bedload) is transported downstream following a specific path or route. This route may be the course of the natural channel, an artificial channel or canal, or a restored channel. Sediment routing may involve a consideration of storage and transit times within the fluvial system (Blöthe and Korup, 2013).   ASG

Reference

Blöthe, J. H. and Korup, O. (2013) Millennial lag times in the Himalayan sediment routing system. *Earth and Planetary Science Letters*, **382**, 38–46.

sediment yield The total mass of particulate material reaching the outlet of a drainage basin. Values of sediment yield are commonly evaluated on an annual basis (tonnes per year) and may also be expressed as specific sediment yields or yields per unit area (tonnes per square kilometre per year). The total sediment yield comprises material transported both as SUSPENDED LOAD and bedload (see BEDLOAD, BEDLOAD EQUATION) and separate measurements of the two components will generally be necessary. However, where the sediment load of a river is deposited in a lake or reservoir, it may be possible to estimate the total yield directly

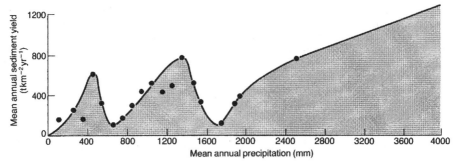

Sediment yield. The relationship between mean annual sediment yield and mean annual precipitation. *Source*: Walling and Kleo (1979). Reproduced with permission of IAHS.

by monitoring the volume of deposited sediment. The sediment yield from a drainage basin will commonly represent only a small proportion of the gross erosion within the basin. Much of the eroded material will be deposited before reaching the outlet of the basin, and the ratio of sediment yield to gross erosion is termed the sediment delivery ratio.

The magnitude of the sediment yield from a drainage basin will reflect control by several factors, including climate, topography, lithology, vegetation cover and land use. Maximum values in excess of $20,000 \, t \, km^{-2} \, a^{-1}$ have been recorded in the severely eroded loess areas of the Middle Yellow River basin in China and in the high rainfall areas of South Island, New Zealand. DEW

Reading and Reference
Laronne, J.B.and Mosley, M.P. (eds) (1982) *Erosion and sediment yield.* Benchmark papers in geology 63. Stroudsburg, PA: Hutchinson Ross. · Walling, D.E. and Webb, B.W. (1983) Patterns of sediment yield. In K. J. Gregory (ed.), *Background to palaeohydrology.* Chichester: John Wiley & Sons, Ltd. · Walling, D.E. and Kleo, A.H.A. (1979) Sediment yields in areas of low precipitation: a global view. In *Canberra symposium, 1979, hydrology of areas of low precipitation.* IAHS Publication 128. Wallingford: IAHS Press; pp. 479–493.

sedimentary rock Rock composed of the fragments and particles of older rocks that have been eroded and the debris deposited by wind or water, often as distinct strata. Some sedimentary rocks may be of organic origin.

seeding of clouds See WEATHER MODIFICATION.

segregated ice Ice formed by the migration of pore water to the freezing plane where it forms into discrete lenses, layers or seams ranging in thickness from hairline to greater than 10 m.

Segregated ice commonly occurs in alternating layers of ice and soil. Ice structure tends to be parallel to the freezing surface (i.e. usually dominantly horizontal) with air bubbles tending to be elongated and aligned normal to the horizontal layering. When dealing with massive icy bodies, as in the Mackenzie Delta, or within pingos, it is sometimes difficult to differentiate segregated ice from injection ice. HMF

Reading
Mackay, J.R. (1972) The world of underground ice. *Annals of the Association of American Geographers,* **62**, 1–22.

seiche The oscillation of a body of water at its natural period. Coastal measurements of sea level often show seiches with amplitudes of a few centimetres and periods of a few minutes due to oscillations of the local harbour, estuary or bay superimposed on the normal tidal changes. DTP

seif Arabic term, meaning *sword*, for a LINEAR DUNE.

seismicity The intensity and frequency of earthquakes in an area. Earthquake intensity is measured on the logarithmic MOMENT MAGNITUDE SCALE. Most seismic activity is located along plate boundaries (see PLATE TECTONICS) and is particularly concentrated in the circum-Pacific belt, which accounts for about 80% of global seismicity. MAS

Reading
Bolt, B.A. (1978) *Earthquakes: a primer.* San Francisco, CA: W.H. Freeman. · Gubbins, D. (1990) *Seismology and plate tectonics.* Cambridge: Cambridge University Press. · Wyss, M. (ed.) (1979) *Earthquake prediction and seismicity patterns.* Contributions to Current Research in Geophysics 8. Basel: Springer. (reprinted from *Pure and Applied Geophysics,* 117 (6)).

self-mulching A process whereby swelling and shrinking in soils, resulting either from alternate wetting and drying or from freezing and thawing, gives rise to a surface layer, composed of well-aggregated granules or fine blocks, that does not crust.

self-organized criticality (SOC) An approach to the study of complex systems. The concept of SOC arose in the late 1980s as a means to explain the behaviour of simple cellular-automata models. The latter were able to produce key characteristic features observed in natural complexity, including scale-invariant phenomena such as fractal geometry and power laws, in a manner that was linked to critical-point phenomena. The complexity observed emerges spontaneously from simple local interactions and, without any external agent imposing order, is, therefore, self-organizing. SOC has been widely used as a means to attempt to explain a range of complex phenomena, including earthquakes, forest fires, river networks and the spread of diseases.

A range of SOC models have been used, but two broad categories may be identified: stochastic and external models. In the former, stochastic dynamics operate within a deterministic environment, while in the latter the environment is random and the dynamics deterministic. Although SOC has been widely used to study a range of complex natural phenomena and has value in research, it has been criticized, notably as being a gross oversimplification that does not provide a general explanation for complex phenomena.   GF

Reading
Frigg, R. (2003) Self-organised criticality - what it is and what it isn't. *Studies in History and Philosophy of Science*, 34, 613–632. · Malamud, B.D. and Turcotte, D.L. (1999) Self-organised criticality applied to natural hazards. *Natural Hazards*, **20**, 93–116. · Turcotte, D.L. (1999) Self-organized criticality. *Reports on Progress in Physics*, **62**, 1377–1429.

semi-arid These areas are the most extensive DRYLANDS, covering almost 18% of the world's land surface, including large parts of the western interior USA and Canada, interior Asia, southern Africa, parts of the Sahel, Australia and, in Europe, Spain. They comprise areas where P/PET values fall between 0.2 and 0.5 (see DRYLANDS for methodology). Semi-arid areas have distinctively seasonal rainfall regimes and are DROUGHT prone, but because mean annual rainfall amounts can be up to 800 mm in summer rainfall areas and 500 mm in winter rainfall zones, they can and do support sizeable human populations. This, in turn, makes their populations and livelihoods drought prone and the environments susceptible to DESERTIFICATION.   DSGT

semi-desert An imprecise and loosely used term, equated with SEMI-ARID.

sensible heat A measure of the heat content of a substance; physically, it is termed enthalpy. Climatologically, it is most frequently used as a term in the HEAT BUDGET equation where the available NET RADIATION is utilized in terms of a sensible heat flux and LATENT HEAT fluxes. Sensible heat transfer includes both conduction and convection, with conduction providing the main transfer into the ground and convection (or turbulent motions) providing the main transfer into the atmosphere. Sensible heat fluxes are particularly important over hot, dry surfaces, such as deserts, where there is a steep temperature gradient between the surface and the air above.   PS

Reading
Wallace, J.M. and Hobbs, P.V. (2006) *Atmospheric science - an introductory survey*. San Diego, CA: Academic Press.

sensible temperature Used in the context of thermal comfort, it is the indoor temperature that would produce the same sense of comfort or discomfort to a lightly clothed person as the actual outdoor weather environment. It is dependent upon wind speed, humidity and the radiation balance, as well as the air temperature.

Sensible heat is the same as enthalpy, which represents the total heat or total energy content of a substance per unit mass. In the atmosphere, a change of sensible heat of a mass of gas is the heat gained or lost by the gas in an exchange at constant pressure. The transport of sensible heat horizontally in the atmosphere is of fundamental importance to the general circulation of the atmosphere, as heat is transferred from the tropics towards the poles.   JET

sensitivity Used in various ways in physical geography, often referring to the susceptibility of a system (e.g. geomorphological, ecological) to disturbance, or the ability of a system to respond to external disturbance. The sensitivity of landscapes to environmental change is, for example, considered in Thomas and Allison (1993). There are also very specific meanings. For example, in a clay sediment or soil, sensitivity is the ratio of undisturbed, undrained strength to the remoulded, undrained strength when tested at the same moisture content and (dry) density. A reduction in strength on remoulding is usually seen for most normally consolidated soils (see CONSOLIDATION), but it is rarely greater than 10% and for overconsolidated soils it is usually near zero. A catastrophic decrease in clay strength can be caused in the field by an earthquake, for example, and has led to some very large slope failures. LOESS may also exhibit a certain amount of sensitivity. DSGT/WBW

Reading
Maerz, N.H. and Smalley, I.J. (1985) *The nature and properties of very sensitive clays: a descriptive bibliography #12.* Ontario: University of Waterloo. · Thomas, D.S.G. and Allison, R.J. (1993) *Landscape sensitivity.* Chichester: John Wiley & Sons, Ltd.

serac Pinnacles and cuboid masses of ice associated with rapid glacier flow, as for example found on ice falls and SURGING GLACIERS.

seral community A development or successional plant or animal community. In a community sequence or SERE, all the communities not at the stable terminal (climax) stage are seral. Seral communities tend to be dominated by opportunistic species adapted for rapid dispersal and growth, often small and capable of completing their life cycle comparatively rapidly (r-selection). They may also be short-lived communities, of low diversity, small biomass, high net production, with short food chains and poorly developed homeostatic mechanisms. According to E. P. Odum (1969), many of their features are also characteristic of agricultural plant communities that are similarly unstable ecologically. (See also SUCCESSION.) JAM

Reading and Reference
Drury, W.H. and Nisbet, I.C.T. (1973) Succession. *Journal of the Arnold Arboretum,* **54**, 331–368. · Odum, E.P. (1969) The strategy of ecosystem development. *Science,* **164**, 262–270.

serclimax A plant community forming a relatively stable stage in a SERE. It is a long-persisting SERAL COMMUNITY that gives at least an impression of permanence, as particular species or environmental factors may stabilize a community well before the climax stage. (See also ALLELOPATHY, CLIMAX VEGETATION and SUCCESSION). JAM

Reading
Connell, J.H. and Slatyer, R.O. (1977) Mechanisms of succession in natural communities and their role in community stability and organization. *American Naturalist,* **111**, 1119–1144. · Moravec, J. (1969) Succession of plant communities and soil development. *Folia Geobotanica et Phytotaxonomica,* **4**, 133–164.

sere A sequence of communities, usually plant communities, at a particular site. It is a series of stages that follows from the process of SUCCESSION. Each sere is made up of a seral communities and may eventually terminate in a stable community (see SERAL COMMUNITY). Different seres in a landscape have for long been viewed as at least partially convergent; thus, the later stages in a dry environment (xerosere) are supposed to become increasingly similar to the later stages in a wet environment (hydrosere). Many factors, both natural and anthropogenic, may arrest, disturb or deflect such a simple pattern of development. JAM

Reading
Daubenmire, R. (1968) *Plant communities: a textbook of plant synecology.* New York: Harper & Row. · Matthews, J.A. (1979) A study of the variability of some successional and climax plant assemblage-types using multiple discriminant analysis. *Journal of Ecology,* **67**, 255–271.

serir A REG. A desert with a surface mantled by sheets of pebbles.

seston A term describing all of the mineral particles and living and nonliving organic matter that is suspended in water bodies, including rivers, lakes and oceans.

shakehole A roughly circular depression in the landscape in which water drains into an underground limestone cave system. The term is used synonymously with DOLINE and swallow-hole, but should be restricted in usage to a depression formed by the collapse of underlying limestone strata (Warwick, 1976). In practice, shakeholes, and so on are produced by a combination of solution and collapse processes. The term derives from Derbyshire, England, but is now used extensively for any circular depression in limestone regions. PAB

Reference
Warwick, G.T. (1976) Geomorphology and caves. In C. H. D. Cullingford and T. D. Ford (eds), *The science of speleology.* London: Academic Press; pp. 61–126.

shale A compacted sedimentary rock composed of fine-grained particles usually clay sized. Shales are characteristically fissile in the plane of their bedding.

shape In general, the shape of an object is its external appearance. It is often rather loosely defined; for example, fan shaped, cumulus cloud, prismatic peds. A more precise definition is usually required, and the ratios of three orthogonal axes that run through the object are conveniently used. How these axes are defined and used depends upon the complexity and size of the object as well as the purpose. A spherical ball has the three orthogonal axis lengths equal, although an object with such three axes equal in length may be much more complex in detail. A golfball is one such example. A Zingg diagram is often used to show relative shapes of sedimentary particles. Here, for each particle, the ratio of intermediate *b* axial length to long axis *a* length is plotted as ordinate against the ratio of short *c* axis to intermediate *b*. This procedure is hard to achieve with very small particles. Similar plots of ratio lengths can be used to define landforms for analysis of MORPHOMETRY.

As the external morphology of an object is surprisingly difficult to define, more than one parameter may have to be used; for example, angularity, roundness and surface texture. In one scheme 'shape' is only one component of a particle's 'form'. For very complex forms, techniques such as FOURIER ANALYSIS or FRACTAL dimension may give a better indication of shape rather than axial length ratios. WBW

Reading
Orford, J.D. and Whalley, W.B. (1991) Quantitative grain form analysis. In J. P. M. Syvitski (ed.), *Principles, methods, and applications of particle size analysis*. Cambridge: Cambridge University Press; pp. 88–108.

shear box
An apparatus used to measure the shear strength of a soil. In its simplest form it consists of a split box into which the soil to be tested is placed. A normal stress is applied to the top of the sample via weight and, while one half of the box is moved parallel to the base to provide the shear stress, the other half resists the movement, the magnitude of the resistance being measured by a 'proving ring'. In this form the test is strictly called the direct shear test as it differs from simple shear which is an angular rather than a linear displacement. At least three tests are performed with different values of normal stress; a plot of the resisting stress corresponding to the normal stress gives values of COHESION and FRICTION that can be used in the MOHR-COULOMB EQUATION. The device can also be used to measure the RESIDUAL STRENGTH of the soil. Different sizes of box are required - the larger the size of grain under test the larger the size of the box; thus, 30 cm square boxes may be required for testing gravelly soils. Although mainly a laboratory test, some versions are for in-situ field determinations. The TRIAXIAL APPARATUS does a similar job but has advantages over the direct shear box. WBW

shear strength
A measure of the ability of a material to resist shear stress. This is an important parameter in determining the engineering and geomorphic properties of materials. The shear strength of a soil is controlled by components of the MOHR-COULOMB EQUATION. Soils have a maximum strength value (peak strength) usually determined by a triaxial test or SHEAR BOX test. Various types of test conditions may be used to determine the shear strength. These relate to consolidation and drainage conditions. Particularly important are the 'undrained test, where no PORE WATER PRESSURE dissipation is allowed during shearing, and the 'drained test', where such drainage is allowed. Generally speaking, the undrained test, which is done with a high strain rate, gives a lower shear strength value. WBW

Reading
Atkinson, J.H. and Bransby, P.L. (1978) *The mechanics of soils*. London: McGraw-Hill. · Whalley, W.B. (1976) *Properties of materials in geomorphological explanation*. Oxford: Oxford University Press.

shear stresses
Two perpendicular loads or stresses (force per unit area) applied parallel (tangential) to the surface of a body. They are themselves perpendicular to the NORMAL STRESS. The diagram illustrates their orientation with respect to a cube of material. Shear stresses produce angular deformation (shear strain) in the body. WBW

Reading
Statham, I. (1977) *Earth surface sediment transport*. Oxford: Clarendon Press.

sheet erosion
The stripping of soil material from a relatively broad area such as a field or hillside to a uniform depth and without the incision of discrete channels as large as rills or gullies. In reality, all hillslope run-off involves some degree of concentration of flow into small

proto-channels or preferred flow lines, but if the soil material is resistant to scour or the shearing forces created by the water flow remain below the level required to entrain soil particles on a large scale, then the surface may be lowered without significant incision taking place. The primary eroding force in sheet erosion is probably raindrop splash (SEE RAINDROP IMPACT EROSION), and for this to be the case the erosion process must be acting on a surface of low gradient. Steep gradients are associated with faster flow in proto-channels, and hence with a greater likelihood that sheet erosion will give way to rill erosion and other forms of channelized run-off. Once flow becomes channelized, flow depth is increased, the surface roughness that retards the flow becomes less effective and flow speeds increase. Increasing flow speeds are in turn associated with increasing shear stresses at the soil surface, and a greater likelihood that incision will be initiated. It has been suggested (Moss and Green, 1983) that, on low gradients, the impact of raindrops may result in the creation of widespread local mixing of the flow that tends to restrict the development of flow concentrations, and which may thus help to account for the occurrence of sheet erosion. (See also RILL and GULLY.) DLD

Reading and Reference
Moss, A.J. and Green, P. (1983) Movement of solids in air and water by raindrop impact. Effects of drop-size and water-depth variations. *Australian Journal of Soil Research*, **21**, 257–269. · Moss, A.J., Walker, P.H. and Hutka, J. (1979) Raindrop-stimulated transportation in shallow water flows: an experimental study. *Sedimentary Geology*, **22**, 165–184.

sheet flow   Surface run-off that is not concentrated into rills or other small channels, but which traverses the surface at shallow depth. Sheet flow is one of the forms of flow that define the interrill phase of surface run-off. Such flow can only occur on relatively smooth surfaces having little MICROTOPOGRAPHY. Shallow flows are relatively slow, owing to frictional drag, and thus they generate limited shearing forces at the soil surface that restrict soil entrainment. The flow is also susceptible to the disturbing influence of raindrops, which can generate marked agitation of the moving water, so further slowing its motion and limiting its potential to scour the surface. Nevertheless, INTERRILL FLOW or sheet flow provides the means by which eroded particles are delivered from the unchannelized parts of the landscape into rills. The primary dislodgment of particles is caused by RAIN SPLASH processes, and the drop impacts, by agitating the flow and resuspending particles that settle out, also assist in the transportation of the eroded particles. These mechanisms are jointly referred to as raindrop-induced flow transportation or rain-flow transportation (Moss, 1988), and they account for much of the erosion that takes place in sheet flow.   DLD

Reading and Reference
Gascuel-Odoux, C., Cros-Cayot, S. and Durand, P. (1996) Spatial variations of sheet flow and sediment transport on an agricultural field. *Earth Surface Processes and Landforms*, **21**, 843–851. · Moss, A.J. (1988) Effects of flow-velocity variation on rain-driven transportation and the role of rain impact in the movement of solids. *Australian Journal of Soil Research*, **26**, 443–450.

sheeting   The formation of joints in a massive rock such that the outer layers of the rock separate in shells or spalls and exfoliate (i.e. peel away from the parent rock mass). The shells are large, with minimum areas of several square metres and thicknesses of tens of centimetres. They commonly form, and may be responsible for, the shape of rounded outcrops and boulders. Successive exfoliation of concentric rock sheets results in the maintenance of rounded forms of some tors and bornhardts.

Sheeting commonly occurs in massive rocks that have once been deeply buried and have subsequently been brought to the ground surface by removal of overburden or overlying ice. The uncovered rock undergoes release of confining stresses and expands outwards towards the ground surface and bounding joints. Extension fractures form perpendicular to the minimum stress direction at the time of failure; this occurs because the original maximum stress direction is controlled by the overburden. As the overburden is eroded, the maximum stress converts to being the minimum stress and sheeting develops nearly parallel to the unloading surface. Well-known examples occur in the granites of Yosemite Valley, California, and curved forms in sandstones of the Colorado Plateau are described by Bradley (1963).   MJS

Reading and Reference
Bradley, W.C. (1963) Large-scale exfoliation in massive sandstones of the Colorado Plateau. *Bulletin of the Geological Society of America*, **74**, 519–527.

shelf   The continental shelf. The sea floor lying between the coast and the steeper slope down to the deep ocean.

shell middens   See MIDDEN.

shell pavements   Accumulation of shell valves that occur in all places where mixtures of sand and shell are subjected to selective erosional winnowing, leaving a superficial lag deposit of the coarse shell material. Pavements of this type are common in coastal situations where, for example, fine sediment can be moved by tidal currents or by aeolian processes above high-water mark. ASG

Reading
Carter, R.W.G. (1976) Formation, maintenance, and geomorphological significance of an aeolian shell pavement. *Journal of Sedimentary Petrology*, **46**, 418–429.

shield A continental area of exposed Precambrian rocks within a CRATON and bordered by a platform area covered by post-Precambrian sedimentary strata. Shields are highly stable low-lying areas that have experienced little deformation or volcanic activity since the Precambrian. Examples include the Canadian Shield of North America and the Baltic Shield in northern Europe. MAS

Reading
Spencer, E.W. (1977) *Introduction to the structure of the Earth*, 2nd edition. *New York*: McGraw-Hill. · Windley, B.F. (1984) *The evolving continents*, 2nd edition. *Chichester*: John Wiley & Sons, Ltd.

shield volcano Shield volcanoes are igneous rock constructs, with low gentle slopes rarely exceeding 6°. They are typically composed of predominantly basaltic lava flows. They range in size from diameters of less than 10 km, to Hawaiian edifices that have diameters on the sea floor in excess of 300 km. Viewed from the side, shield volcanoes have the convex profile of a shield lying flat on the ground. Shield volcanoes form from the effusive eruption of fluid (low viscosity) LAVAs that flow away from the summit or are erupted from flank fissures that ensure dispersal of the flows over a wide area. Most shield volcanoes have summit *calderas* that are thought to form by collapse, when magma is withdrawn from high-level reservoirs to feed flank eruptions. The largest shield volcanoes on Earth occur in the Hawaiian islands. The big island of the Hawaiian archipelago is composed of five shield volcanoes, of which Mauna Loa (4169 m asl) and Mauna Kea (4206 m asl) are the largest. Mauna Loa, reaches a height of 10 km above the sea floor and is a very recent feature, having been built during the Quaternary. It is one of the largest topographic features on the surface of the Earth. The massive Hawaiian shield volcanoes cause sagging of the Pacific LITHO-SPHERE. Shield volcanoes are important topographic features on Mars and Venus as well as on Earth. On Mars, the shield volcano Olympus Mons is the largest known volcano in the Solar System, with a height of 23 km and a summit caldera complex 80 km in diameter. AD

Shields parameter A dimensionless shear stress, relating the fluid drag on a particle to the particle's immersed weight; sometimes called Shields beta:

$$\beta = \frac{\tau_0}{(\rho_s - \rho)gD}$$

where τ_0 is bed shear stress, ρ_s and ρ are sediment and fluid density respectively, g is acceleration due to gravity and D is grain diameter. The most common application of the Shields parameter is in the estimation of the threshold shear stress for the initiation of sediment movement. This is done using a Shields diagram and Shields curve to define the threshold conditions for different grain sizes. DJS

Reading
Middleton, G.V. and Southard, J.B. (1984) *Mechanics of sediment movement*, 2nd edition. Tulsa, OK: Society of Economic Paleontologists and Mineralogists.

shoal Area of shallow water in a lake or sea. A sand bank that lies just beneath the surface of a lake or sea.

shore The area of land immediately adjacent to a body of water.

shore platforms Intertidal rock surfaces of low slope angle. The term is preferred to 'wave-cut platform' because processes other than mechanical wave action can play a role in their formation, not least weathering and bio-erosion. The form of the platforms depends on the nature of the main processes operative upon them, the nature of the rocks and their structures, on tidal characteristics, on their age and on their history. ASG

Reading
Trenhaile, A.S. (1980) Shore platforms: a neglected coastal feature. *Progress in Physical Geography*, **4**, 1–23.

shrink–swell soils Soils that exhibit marked changes in volume when drying or wetting. The shrink-swell behaviour results from the presence of clay minerals, such as montmorillonite, that have sheet structures. When these clays become wet, water molecules diffuse into the spaces between the lamellae, so causing swelling. The reverse occurs as the soils become drier. Vertisols are among the soil groups where shrink-swell behaviour is very marked, and which can consequently exhibit large desiccation cracks, soil collapse processes, gilgai and other phenomena.

According to standard shrinkage tests, linear shrinkage in clay-rich soils can exceed 30%, and shrinkage may be larger in the subsoil than close to the soil surface, because of greater clay content at depth. Considerable damage to roads, fences, buildings and other infrastructure can result from the instability of foundations in shrink–swell soils.   DLD

Reading

Fityus, S.G., Cameron, D.A. and Walsh, P.F. (2005) The shrink swell test. *Geotechnical Testing Journal,* **28**, 1–10. · Stewart, R.D., Abou Najm, M.R., Rupp, D.E., *et al.* (2014) Hillslope run-off thresholds with shrink-swell soils. *Hydrological Processes,* **29**, 557–571, doi: 10.1002/hyp.10165.

sial A term introduced by Suess to describe that part of the Earth's crust with a granitic-type composition dominated by minerals rich in silicon and aluminium. It is contrasted with the term SIMA. Sial forms at least the upper part of the crust of the continents and has a mean density of about $2700\,\mathrm{kg\,m^{-3}}$ and a silica content of between 65 and 75%.   MAS

sichelwannen Bow-shaped furrows, with arms generally pointing in the direction of flow of the glacier that created them. They are of the order of 1-2 m long, relatively shallow and tend to occur in large localized assemblages on glaciated surfaces. (See also P-FORM.)   ASG

Reading

Allen, J.R.L. (1984) *Sedimentary structures.* Amsterdam: Elsevier; pp. 264–266.

sideways-looking airborne radar

(SLAR) A microwave remote sensor used to derive images of the Earth's surface.

The SLAR senses the terrain to the side of an aircraft's track. It does this by pulsing out long, up to radio, wavelengths of ELECTROMAGNETIC RADIATION and then recording, first, the strength of the pulse return to the aircraft, to detect objects, and, second, the time it takes for the pulse to return, to give the range of objects from the aircraft. The name radar is the acronym of these functions of radio detection and ranging. As these pulses are emitted at right angles to the aircraft track, the movement of the aircraft enables pulse lines to be built up to form an image.

Like a MULTISPECTRAL SCANNER or THERMAL INFRARED LINESCANNER, the SLAR possesses collectors, detectors and recorders and, in addition, a transmitter and antenna.

The transmitter produces pulses of microwave energy that are timed by a synchronizer and standardized to a known power by a modulator. For a fraction of a second the transmit/receive switch is switched to transmit, as the transmitter releases a microwave pulse from the antenna. The transmit/receive switch then returns to its original position and the antenna continues to receive pulses that have been back-scattered from the Earth's surface. These pulses are converted to a form suitable for amplification and further processing by a mixer and local oscillator before being passed to a receiver. The receiver amplifies the signal before passing it to the detector that produces an electronic signal suitable for recording on to photographic film or analogue or digital tape. To improve the spatial resolution of this sensor, a larger antenna can be synthesized electronically. This is termed a synthetic aperture radar (SLAR), and is the primary SLAR for environmental research. The SAR is carried by aircraft and several unmanned Earth resources satellites, most notably ERS-1 and JERS-1.

Four characteristics of SLAR imagery determine its fields of application: its relatively high cost, its rapid rate of data acquisition (as it is unhindered by cloud or nightfall) and its sensitivity to both surface roughness and surface moisture content. SLAR was used initially for geological exploration as the likely financial returns were high, the areas to be covered were large and surface roughness and moisture content often varied between the areas of interest.

Today, SLAR imagery is used in geomorphology, the mapping of soil moisture and vegetation, the estimation of forest biophysical properties and the location of oil pollution and sea ice.   PJC

Reading

Trevett, J.W. (1986) *Imaging radar for resources surveys.* London: Chapman & Hall. · Ulaby, F.T., Moore, R.K. and Fung, A.K. (1981) *Microwave remote sensing, active and passive. Volume 1: Fundamentals and radiometry.* Reading, MA: Addison-Wesley.

sieve analysis Procedure used to assess the particle size distribution of a granular material. It is particularly suited to sand-sized materials. A typical sieve analysis involves a nested column of sieves. A representative weighed sample is poured into the top sieve, which has the largest woven wire mesh screen openings. Each lower sieve in the column has smaller openings than the one above, with a pan at the bottom of the nest. The entire nest is then agitated, and the material whose diameter is smaller than the mesh opening passes through the sieves. After a set shaking period, the amount of material retained in each sieve, and in the pan, is then weighed. The weight of the sample of each sieve is then divided by the total weight to give a percentage retained on each sieve. In most analyses using sieves, grain-size data is given in phi (ϕ) intervals rather than in micrometres, millimetres or inches. Phi diameter is computed by taking the negative log of the diameter in millimetres. Statistical computations and graphic presentations are much simpler when phi diameters are used. There are four standard statistical measurements for sieved samples: a measure of central tendency (including median, mode and mean); a measure of the degree of scatter or sorting; kurtosis, the degree of peakedness of the distribution; and skewness, the presence or absence of a coarse or fine tail to the distribution. Various formulae have been defined for these parameters, of which the most widely used are the Folk and Ward (1957) statistics. Sieve analysis results are often plotted as cumulative probability curves. This has the advantages that the plot is independent of sieve interval used, the grain size parameters are easy to extract from the plot (and a normal distribution plots as a straight line), and the steeper the slope of the line the better sorted the sample. Sieve analysis assumes that all particles will be spherical and will pass through the square openings when the particle diameter is less than the size of the square opening in the screen. For elongated and flat particles a sieve analysis will not yield reliable mass-based results, as the particle size reported will assume that the particles are spherical, where in fact an elongated particle might pass through the screen end-on, but would be prevented from doing so if it presented itself side-on. Sieve analysis is also problematic when sieving chambered particles (such as gastropod shells) as the size : weight relations are violated in such instance. This technique is now being replaced by laser diffraction measures of sediment size, which has the advantages of speed (particularly when used in conjunction with semi-automated sampling lines) and the use of very small sample sizes. For larger materials, sediment size analysis through digital image processing is likely to replace sieve analysis in the near future.   TS

References

Folk, R.L. and Ward, W.C. (1957) Brazos River bar, a study in the significance of grain size parameters. *Journal of Sedimentary Petrology*, 27, 3–26 · Carver, R.E. (ed.) (1971) *Procedures in sedimentary petrology*. New York: Wiley-Interscience.

sieve deposits

On alluvial fans, these are characterized by a very steep lobe front, coarse materials throughout the deposit, but especially at the lobe front, and a very high infiltration capacity for the deposit. Transportation of the debris ceases when the flow infiltrates. 'Because water passes through rather than over such deposits, they act as strainers or sieves by permitting water to pass while holding back the coarse material in transport' (Hooke, 1967: 454). Sieve deposits lack the fine matrix characteristic of debris-flow deposits. Although named in deserts, sieve deposits have also been noted in alpine settings.

The existence of sieve deposits poses problems for the current paradigm of rheology in debris-flow generation. This is because the transportation of large clasts by just water does not fit current models that require the role of fine materials as a transporting agency. The existence of sieve deposits lacking fines has forced some investigators to argue that fines have been washed out of sieve deposits over time. But this argument ignores the overwhelming evidence from studies of STONE PAVEMENT formation that desert sieve deposits are effective dust traps.
RID

Reading and Reference

Hooke, R.LeB. (1967) Processes on arid-region alluvial fans. *Journal of Geology*, 75, 438–60. · Krainer, K. (1988) Sieve deposition on a small modern alluvial fan in the Lechtal Alps (Tyrol, Austria). *Zeitschrift für Geomorphologie NF*, 32, 289–298.

silcrete

A highly siliceous indurated material formed at, or near, the Earth's surface through the silicification of bedrock, weathering products or other deposits by low-temperature physico-chemical processes. Silcrete of Cainozoic age is particularly well developed in areas of inland Australia, in southern Africa and in northwest Europe. It may attain a thickness in excess of 5 m, and through its resistance to weathering and erosion it plays an important role in armouring erosion surfaces. It forms in areas of minimal local relief under both semi-arid and humid climatic regimes.   MAS

Reading

Langford-Smith, T. (ed.) (1978) *Silcrete in Australia*. Armidale, NSW: Department of Geography, University of New England. · Summerfield, M.A. (1983) Silcrete. In A. S. Goudie and K. Pye (eds), *Chemical sediments and geomorphology*. London: Academic Press; pp. 59–91.

sill

A tabular sheet of igneous rock injected along the bedding planes of sedimentary or volcanic formations.

silt

Fine sediment contained in a soil. More specifically, silt is a SOIL TEXTURE and a PARTICLE SIZE between CLAY and SAND. The precise size range regarded as silt varies according to different classification schemes. The lower size limit in all schemes is 2 µm, but the upper limit ranges from 20 µm in the Atterberg system, 50 µm in the USDA system and 63 µm in the British Standards, Soil Service of England and Wales, and MIT schemes.

Silt is readily transported in suspension in wind and water, and is a common component in fluvial and lacustrine deposits, and in DUST and LOESS.   DSGT

siltation The accumulation of fine sediment (strictly speaking of silt) in a body of water. Applied technically to the settling out of fine particles in water and more generally to the filling or choking of lakes, reservoirs or water courses.   JL

sima A term introduced by Suess to describe that part of the Earth's crust with a basaltic-type composition dominated by minerals rich in silicon and magnesium. It is contrasted with the term SIAL. It forms the crust of the ocean basins and the lower portion of the crust of the continents. Sima has a higher mean density than sial (2800-3400 kg m^{-3}) and a lower silica content (<55%).   MAS

simulation Hypothesis testing in physical geography often makes use of the vicarious experiment procedure of simulation in which genuine phenomena are represented by scale, analogue or mathematical models. Scale models are small-scale hardware replications such as flumes, wave tanks or kaolin glaciers. Analogue models involve equivalent systems; for example, an electrical potential model of groundwater is feasible because of the mathematical equivalence of the laws governing current flow in a circuit and flow in a porous medium. Computer-based mathematical simulation models (Thornes and Brunsden, 1977: 157-171), which may be deterministic or probabilistic, are often used to test alternative hypotheses by comparing model outputs under different assumptions with actual behaviour. Links between system components are represented by mathematical relations, whose constants are adjusted by PARAMETER-IZATION procedures until the best fit with reality is achieved. (See also RANDOM-WALK NETWORKS.) KSR

Reference
Thornes, J.B. and Brunsden, D.B. (1977) *Geomorphology and time*. London: Methuen.

singing sands When in motion, certain dune sands generate clearly audible sounds that have been variously reported as roaring, booming, squeaking, musical and singing (Curzon, 1923). A unique combination of granulometric properties appears to be responsible, including a high

sorting value, uniform grain size and a high degree of roundness of grains (Van Rooyen and Verster, 1983).   ASG

Reading and References
Curzon, G.N. (1923) The singing sands. In *Tales of travel*. London: Hodder & Stoughton; chapter 11. · Haff, P.K. (1986) Booming dunes. *American Scientist*, 74, 376–381. · Van Rooyen, T.H. and Verster, E. (1983) Granulometric properties of the roaring sands in the south-eastern Kalahari. *Journal of Arid Environments*, 6, 215–222.

sinkhole A roughly circular depression in the landscape into which water drains and collects. Specifically, it is a depression in limestone terrain, often connecting with an underground cave system through which the water drains. It is used synonymously with SHAKEHOLE and DOLINE and was originally an American term.   PAB

Reading
Beck, B.F. and Wilson, W.L. (1987) *Karst hydrology: engineering and environmental applications*. Rotterdam: Balkema.

sinter A precipitate of silica or calcium carbonate associated with geysers and hot springs.

sinuosity The degree of wandering or winding, applied especially to river channels. It may be defined as the ratio of actual channel distance between identified points compared with the straight or down-valley distance.   JL

siphon A vertical or inverted U-shaped portion of a subterranean stream channel in which the water is in hydrostatic equilibrium.

skerry A rocky islet shaped primarily by glacial erosion, common along many high-latitude coastlines. They show little reworking by marine processes, although they may be submerged at high tides. Skerries often occur as chains or fields of islands, and are widespread along the STRANDFLAT coasts of Norway, Iceland and Greenland and across the mouths of fjords. Skerry coasts also occur in arctic Canada and Finland.   DJS

Reading
Bird, E.C. and Schwartz, M.L. (eds) (1985) *The world's coastline*. New York: Van Nostrand Reinhold.

skewness A statistical measure that is widely used for PARTICLE SIZE analysis. It measures the degree of asymmetry of a statistical distribution as well as whether the distribution has an asymmetrical tail to the left or right.   ASG

Reading
Folk, R.L. (1974) *Petrology of sedimentary rocks.* Austin, TX: Hemphill.

slab failure   A term usually used of strong rocks (although it can occur in muds and weak rocks) where the TRANSLATIONAL SLIDE is along discontinuities (cracks, joints, etc.) that dip outwards from the face. Failure is largely controlled by the FRICTION between the blocks, so that the angle of the discontinuities needs to be high enough to allow most of the frictional strength to be exceeded but less than the cliff slope angle. The final 'trigger' that causes failure may be caused by ice wedging, where the blocks moved are small, or by (cleft) water pressure that changes the EFFECTIVE STRESS conditions. Quantitative assessment of cliff stability can be made with various ROCK QUALITY INDICES. (See also TOPPLING FAILURE.)   WBW

Reading
Attwell, P.B. and Farmer, I.W. (1976) *Principles of engineering geology.* London/New York: Chapman & Hall/John Wiley & Sons, Inc. · Selby, M.J. (1993) *Hillslope materials and processes,* 2nd edition. Oxford: Oxford University Press.

slack   A depression or hollow in an area of sand dunes or mud banks.

slackwater deposit   A deposit of riverine sediment laid down during a flood at a location sheltered from the main current, and often later employed to estimate the size of the flood. Slackwater deposits may accumulate in the regions of a river bend where the main current follows one bank, leaving the other more sheltered, or near the junctions of a trunk stream and lesser tributaries whose flow is held back by the depth of flow in the main channel. These deposits are primarily used to infer the water depth or *stage* reached by major floods, especially floods occurring through the last few thousand years, in order to extend the effective period of observational data on flood magnitudes and their relation to catchment area (Kochel and Baker, 1988). Datable materials within or buried by the slackwater sediments are used to determine the age of the flood event that created them. In some areas, the analysis of slackwater deposits laid down by palaeofloods has indicated the former passage of flows considerably larger than have been monitored during modern times, but elsewhere the magnitudes of inferred palaeofloods fall within the range of modern events.   DLD

Reference
Kochel, R.C. and Baker, V.R. (1988) Paleoflood analysis using slackwater deposits. In V. R. Baker, R. C. Cochel and P. C. Patton (eds), *Flood geomorphology.* New York: John Wiley & Sons, Inc.; chapter 21, pp. 357–376.

slaking   The disintegration of a loosely consolidated material on the introduction of water or exposure to the atmosphere.

slickenside   A polished or scratched rock surface produced by the friction during faulting.

slide   A landslide. The landform produced by mass movement under the influence of gravity.

slip face   When sand accumulation has built up to a critical angle on the lee side of a SAND DUNE, it fails under the influence of gravity to form a slip (or avalanche) face. Individual grain fall, grain flow or avalanching dominate the downslope movement of sand on dry slip faces, while slumping can occur if sand is damp. Slip faces form at the angle of repose for sand.   DSGT

slip-off slope   The more gently sloping bank of a river on the inside of a meander.

slope   A word with two applications in physical geography:

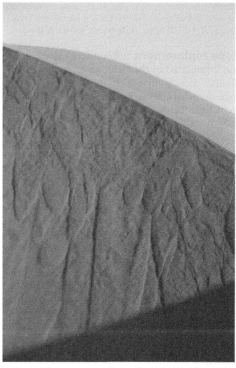

Avalanche flow on a steep linear dune slip face, Namibia. Photograph by David Thomas.

1 In a general sense, it is used to refer to the angle that any part of the Earth's surface makes with a horizontal datum. Synonyms for this usage include inclination, declivity and gradient.

2 In geomorphology, *slope* refers to any geometric element of the Earth's solid surface whether that element is above or below sea level. Slope elements thus form entire landscapes. In a more restricted use, the term is often applied to escarpments and valley sides, and thus excludes floodplains, terrace surfaces and other nearly horizontal elements. To avoid confusion, the more explicit word *hillslope* is in common use.

Hillslopes are regarded as three-dimensional forms produced by weathering and erosion with basal elements that may be either depositional or erosional in origin. The development of hillslopes is consequently the principal result of denudation, and the study of such features is a major part of geomorphology.   MJS

Reading
Carson, M.A. and Kirkby, M.J. (1972) *Hillslope form and process*. Cambridge: Cambridge University Press. · Finlayson, B. and Statham, I. (1980) *Hillslope analysis*. London: Butterworth. · Selby, M.J. (1993) *Hillslope materials and processes*, 2nd edition. Oxford: Oxford University Press. · Young, A. (1972) *Slopes*. Edinburgh: Oliver & Boyd.

slope replacement   A model of slope evolution formulated by the German geomorphologist Walther Penck, in which the maximum slope angle decreases through time as a result of replacement from below by gentler slopes, causing the majority of the slope profile to become occupied by a concavity.

slope wind   See ANABATIC FLOWS and KATABATIC FLOWS.

smog   A term originally used to describe a combination of smoke and FOG but now used for any visibly polluted air. Dr Harold Antoine Des Voeux first used the word in 1911 to describe a series of pollution episodes in Glasgow, Scotland, during the autumn of 1909. Photochemical smog forms when hydrocarbons, originating from vaporized gasoline and other petroleum products, combine with nitrogen oxide molecules, also emitted by combustion engines, and water vapour in the presence of ultraviolet sunlight.   WDS

Reading
Lewis, H.R. (1965) *With every breath you take*. New York: Crown.

snout, glacial   The terminus of a glacier.

snow   Solid precipitation composed of single ice crystals or aggregates known as snowflakes. Ice crystals are most frequent when temperatures are much below freezing and the moisture content of the air is small. As temperatures increase towards 0 °C, the ice crystals grow and cluster into flakes. Snow is difficult to measure accurately as it tends to block standard rain gauges or be blown out.   PS

snow line   The altitudinal limit on land separating areas in which fallen snow disappears in summer from areas in which snow remains throughout the year. The altitudinal distribution over the globe is the same as for the FIRN LINE, but unlike the latter it is not restricted to glaciers.   DES

snowball Earth   Times earlier than 630 million years ago when glaciers were present near sea level in tropical latitudes and sea ice was globally extensive. During the Cryogenian Period of the Neoproterozoic (towards the end of the Precambrian) some major glacial episodes occurred, including the Sturtian (between 750 and 700 million years ago) and the Marinoan or Varangian (between c. 660 and 635 million years ago). Proposed causes include reductions in greenhouse gases, lower total solar radiance, continental reorganization, shifts in Earth's orbit, volcanic emissions and various positive feedbacks, including a runaway ice-albedo feedback, whereby a covering of ice and snow would increase the Earth's albedo and lead to greater reflectance of incoming solar radiation.   ASG

Reading
Macdonald, F.A. Schmitz, M.D., Crowley, J.L., *et al.* (2010) Calibrating the Cryogenian. *Science*, **327**, 1241–1243.

snowblitz theory   A popular and extreme version of the view that glaciations may start rapidly as a result of positive feedback processes related to the increased albedo of a high-latitude continent covered with persistent snow. A few years of excess snow could modify atmospheric circulation patterns and enhance snow accumulation (Lamb and Woodroffe, 1970).

Developing the concept for Britain, Calder (1974: 118) wrote:

In the snowblitz the ice sheet comes out of the sky and grows, not sideways, but from the bottom upwards. Like airborne troops, invading snowflakes seize whole counties in a single winter. The fact that they have come to stay does not become apparent, though, until the following summer. Then the snow that piled up on the meadows fails to melt completely. Instead it lies through the summer and autumn, reflecting the sunshine. It chills the air and guarantees more snow next winter. Thereafter, as fast as the snow can fall, the ice sheet gradually grows thicker over a huge area.

Yet evidence of rapid ice sheet build-up as a result of sudden changes in atmospheric and oceanic circulation is emerging. (See also ICE AGE.)   DES

Reading and References
Calder, N. (1974) *The weather machine and the threat of ice.* London: BBC Publications. · Denton, G.H. and Hughes, T.J. (1983) Milankovitch theory of ice ages: hypothesis of ice-sheet linkage between regional insolation and global climate. *Quaternary Research*, **20**, 125–144. · Imbrie, J. and Imbrie, K.P. (1979) *Ice ages.* London: Macmillan. · Ives, J.D., Andrews, J.T. and Barry, R.G. (1975) Growth and decay of the Laurentide ice sheet and comparisons with Fenno-Scandinavia. *Die Naturwissenschalt*, **62**, 118–125. · Lamb, H.H. and Woodroffe, A. (1970) Atmospheric circulation during the last ice age. *Quarternary Research*, **1**, 29–58.

snowmelt   The part of run-off that is generated by melting of a snowpack on the ground surface. Portions of the snowpack that do not melt may remain on the surface long enough to be compressed by subsequent snowfalls into glacial ice. It may not melt over a period of many years and thus become a semi-permanent snow body, or its mass may be lost through sublimation and deflation.

The quality of the snowpack refers to the potential amount of run-off that may be generated by snowmelt and is measured as the weight of ice divided by the total weight of a unit volume of the snowpack. When the snowpack nears the time for melting, its quality is usually in the 0.90–1.00 range (US Army Corps of Engineers, 1960).   WLG

Reference
US Army Corps of Engineers (1960) *Runoff from snowmelt.* US Army Corps of Engineers engineering manual 1110-2-1406. Washington, DC: US Army Corps of Engineers.

soil   The material composed of mineral particles, gases, liquids and organic remains that overlies the bedrock and supports the growth of rooted plants. Soil is the principle element of the pedosphere and is vital in being the fundamental medium for the establishment, growth, survival and reproduction of flowering plants, as well as providing a habitat for many other lifeforms, such as bacteria, earthworms, and so on. Soil also acts as a means of water storage. Soils evolve over time through the interaction of Physical and chemical weathering with biological processes. A temporal sequence

may be identified from immature, shallow and simply structured soils to those with complex configurations of different horizons. Soils are classified (see SOIL CLASSIFICATION) on the basis of their characteristic SOIL PROFILES in which properties such as SOIL TEXTURE, SOIL STRUCTURE, colour and other physical and chemical variables interact to produce a characteristic sequence of soil horizons.   MEM

Reference
Hartemink, A.E., McBratney, A.B. and White, R.E. (eds) (2009) *Soil science.* London: Earthscan.

soil classification   Systems of soil classification are oriented toward treating soils as either an engineering material or a natural system. Geotechnical engineers generally classify soil in relation to properties relevant to foundation design, recognizing coarse-grained soils (sand and gravel), fine-grained soils (silts and clays) and organic soils (peat). The *Unified Soil Classification System* further subdivides these major classifications.

A system for the classification of the varieties of soil types recognized in the USA, but applied elsewhere, has been outlined in the USDA's *Soil Taxonomy* (Soil Survey Staff, 1999). The classification system recognizes 12 soil orders, together with suborders (55), great groups (238), subgroups (>1200), families (>7500) and series (>18,500). The names given to the soils in this system are generally based on Greek and Latin roots and are used to convey information on the properties and development of the soil.

This hierarchical system of classification is based upon the identification of a series of diagnostic features in a soil. These include horizons of various kinds (see HORIZON, SOIL), both surface (termed *epipedons*) and subsurface, whose identification is based upon colour, texture, chemical composition and other properties. Examples are the *mollic* epipedon (Latin, *mollis*, 'soft'), a surface layer rich in organic matter (the definition requires >0.6% organic carbon) and hence soft. Soils displaying this feature are widespread under native grasslands and are grouped into the *Mollisol* order.

Because of their importance to plant growth, to the degree of oxygenation of the soil and for the breakdown of organic materials, both the temperature and the wetness of soils are used as criteria for classification. In terms of wetness, soils may range from *aquic*, having features related to saturation and poor aeration, to *xeric*, with features diagnostic of long periods of dryness. In a similar way, the temperature criterion recognizes soils ranging from *pergelic*, where the

Table 2 Soil orders and their characteristics in the USDA classification

Soil order	Primary characteristics
Entisols	Very little profile development
Inceptisols	Embryonic profile differentiation
Mollisols	Mollic epipedon
Alfisols	Argillic or natric horizons, with moderate to high base saturation
Ultisols	Argillic horizon with low base saturation
Oxisols	Oxic horizon; highly weathered
Vertisols	Rich in fine, swelling clays
Aridisols	Dry soils, often with a pale epipedon
Spodosols	Spodic horizon (strongly eluviated)
Histosols	Peaty soils (>20% organic matter)
Gelisols	Permafrost near surface, cryoturbated
Andisols	Soils on volcanic cinders or ash

mean annual soil temperature is below freezing, to *hyperthermic* if the mean annual soil temperature is above 22 °C. The annual range of temperature is also taken into consideration.

The soil *orders*, the highest level of the taxonomy, are based upon dominant soil-forming processes, recognized primarily from diagnostic soil horizons. They are set out in Table 2. Note that all the order names end in -*sol*, from the Latin, *solum*, 'soil'.

In naming suborders and subsequent subdivisions, part of the order name is retained, thereby indicating immediately the order to which they belong. For example, subdivisions of the *Molli*sols retain the -*oll* component. These include the Aqu*olls* suborder, the Argiaqu*olls* great group and the typic Argiaqu*olls* subgroup.

In general, the suborders are groupings of soils belonging to one order that all show some common developmental process, and terms relating to the temperature and wetness regimes described above are thus used in naming suborders.

Great groups are identified within a suborder primarily by the recognition of diagnostic horizons. Great groups are further subdivided into subgroups, of which the primary representatives form the *typic* suborder.

The soil families are a finer subdivision of the subgroups and identify soils whose kindred properties result in similar management needs when cultivated. Particular instances of the soil families, often named to record the location where they were first mapped, form the soil *Series*.　　　DRM

Reading and Reference

Soil Survey Staff (1999) *Soil taxonomy: A basic system of soil classification for making and interpreting soil surveys*, 2nd edition. Agriculture Handbook No. 436. Washington, DC: United States Department of Agriculture, Natural Resources Conservation Service. · Brady, N.C. and Weil, R.R. (2007) *The nature and properties of soils*, 14th edition.

Upper Saddle River, NJ: Prentice-Hall. · Schaetzl, R. and Anderson, S. (2005) *Soils: genesis and geomorphology*. Cambridge: Cambridge University Press.

soil crust　A thin layer at the soil surface that may be due to algal growth ('algal crust', which may occur in dryland areas on sandy soils), raindrop impact (sometimes called a 'rainbeat crust') and/or the accumulation of fine particles that seal the surface (e.g. a 'clay crust'). All such crusts may be only a fraction of a millimetre in thickness. A crust due to raindrop effects can result from structural changes in the arrangement of soil particles. The upper surface is composed of aligned clay particles that are packed tightly together and only weakly permeable to water. Below this, there may be a layer of clay particles 1-3 mm in thickness formed by the washing-in of clay particles that partially obstruct the soil pores. Soil crusts may permit very little infiltration of rainwater, and therefore promote ponding or surface run-off. Crusts may be short lived or more durable, and may be disturbed by the impact of animals, which may result in increased erosion.　　　DLD

soil erosion　The natural process of removal of topsoil by water and wind. It is a process whose rates may be magnified by humans (accelerated erosion). On a global scale, the fastest rates occur in zones with highly seasonal precipitation, as in monsoonal, Mediterranean and semi-arid climates. There is a long history of the study of accelerated soil erosion (e.g. Marsh, 1864; Bennett, 1938), and in spite of the introduction of soil conservation measures such as those discussed by Hudson (1971), it continues to be a serious environmental problem - see Boardman and Poeson (2006). In the USA, soil erosion on agricultural land operates at a rate of about 30 t ha^{-1} a^{-1}. Water run-off delivers around 4 billion tonnes of soil to the rivers of the 48 contiguous states, and three-quarters of this comes from agricultural land. Another billion tonnes of soil are eroded by the wind, a process that created the Dust Bowl of the 1930s. In addition to the soil erosion caused by deforestation and agriculture, urbanization, fire, war and mining are often significant in accelerating erosion of the soil.　　　ASG

Reading and References

Bennett, H.H. (1938) *Soil conservation*. New York: McGraw-Hill. · Boardman J and Poesen J (eds) (2006) *Soil erosion in Europe*. Chichester: John Wiley & Sons, Ltd. · Carter, L.J. (1977) Soil erosion: the problem still persists despite the billions spent on it. *Science*, **196**, 409–411. · Hudson, N. (1971) *Soil conservation*. London: Batsford. · Marsh, G.P. (1864) *Man and nature*. New York: Scribner. · Morgan, R.P.C. (1986) *Soil erosion and conservation*. Harlow: Longman.

soil moisture deficit 'Soil moisture deficits are considered to have been set up when evapotranspiration exceeds precipitation and vegetation has to draw on reserves of moisture in the soil to satisfy transpiration requirements' (Grindley, 1967).

The evaluation of soil moisture deficit is essential to the estimation of irrigation need since it provides an estimate of the degree to which soil moisture content has dropped below field capacity. Field capacity is the soil moisture condition when excess water has drained out of a saturated or near-saturated soil and it is thought to be the soil moisture condition that will promote maximum plant growth, with transpiration occurring at the potential rate (i.e. transpiration is not limited by moisture availability). Any reduction of soil moisture content below field capacity will create a soil moisture deficit that may be removed by irrigation.

Soil moisture deficits can be estimated using field instruments or evaporation estimation equations combined with evaluation of the soil WATER BALANCE. LYSIMETERS and NEUTRON PROBES allow changes in soil moisture storage to be directly observed so that the deficit may be calculated, but other instruments, including evaporation pans and atmometers, as well as the majority of evaporation equations, provide estimates of open water evaporation (Eo) or potential evapotranspiration (PEt), which may be actual evapotranspiration (Et) and a soil water budget for the site. In calculating Et from PEt or Eo, it is necessary to take into account the degree to which a decrease in soil moisture content will reduce the Et rate below the PEt rate. This is a controversial topic that is reviewed by Baier (1968), but the method adopted by the UK Meteorological Office (Grindley, 1967) employs the concept of root constants and related drying curves described by Penman (1949). Grindley (1967) explains the way in which the difference between PEt and precipitation in consecutive time periods can be partitioned to calculate Et and soil moisture surplus or soil moisture deficit.   AMG

Reading and References
Baier, W. (1968) Relationship between soil moisture, actual and potential evapotranspiration. In *Soil moisture*. Proceedings of the Hydrology Symposium 6, University of Saskatchewan, 15–16 November 1967. Ottawa: Queen's Printer; pp. 155–191. · Calder, I.R., Harding, R.J. and Rosier, P.T.W. (1983) An objective assessment of soil moisture deficit models. *Journal of Hydrology*, **60**, 329–355. · Grindley, J. (1967) The estimation of soil moisture deficits. *Meteorological Magazine*, **96**, 97–108. · Penman, H.L. (1949) The dependence of transpiration on weather and soil conditions. *Journal of Soil Science*, **1**, 74–89.

soil profile   The full sequence through the soil zone from the surface down to the unaltered bedrock.

soil structure   The grouping of aggregates within a soil. Individual structures have a variety

Soil structure: some common forms.

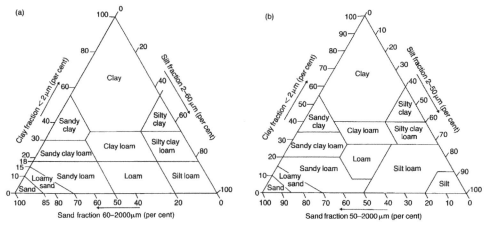

Soil texture. Triangular classification based on the limits laid down by (a) Soil Survey of England and Wales and (b) US Department of Agriculture.

of forms (see diagram) and may range in size from tiny granules to large blocks. Among the controls of soil structure are the presence of clay, humus and soluble silts. Soil aggregates are sometimes referred to as peds.   ASG

soil texture The character of the soil imparted by the proportions of sand, silt and clay within a sample. Two examples of soil textural classification schemes are shown in the diagram.   ASG

solar constant The rate at which solar radiation is received outside the Earth's atmosphere on a surface normal to the incident radiation, and at the Earth's mean distance from the sun. Its exact value is still a little uncertain but it is nearly $1380\,\mathrm{W\,m^{-2}}$. Despite its name it is probably slightly variable in time.   BWA

solfatara A small volcanic vent through which acid gases are emitted, usually in areas where violent volcanism has ceased.

solifluction A term first used by J. G. Andersson (1906) to describe the 'slow flowing from higher to lower ground of waste saturated with water' that he observed in the Falkland Islands. It has subsequently been applied elsewhere to the slow gravitational downslope movement of water-saturated, seasonally thawed materials. In contrast to gelifluction, solifluction does not require permafrost for its occurrence, but modern use of the term does imply the existence of cold climate conditions. It is a form of mass wasting (i.e. viscous flow), faster than soil creep, often in the order of $0.5\text{-}5\,\mathrm{cm\,a^{-1}}$. Features produced by solifluction include uniform sheets of locally derived materials, tongue-shaped

lobes and alternating stripes of coarse and fine sediment. When associated with the active layer (i.e. in permafrost regions) the term gelifluction should be used.   HMF

Reading and Reference
Andersson, J.G. (1906) Solifluction, a component of sub-aerial denudation. *Journal of Geology*, **14**, 91–112. · French, H.M. (2007) *The periglacial environment*, 3rd edition. Chichester: John Wiley & Sons, Ltd.

solonchak A group of soils that are the result of salinization and occur where there is an accumulation of soluble salts of sodium, calcium, magnesium and potassium in the upper horizon. The anions found are mostly sulphate and chloride. In contrast to SOLONETZ soils, which are highly alkaline, solonchaks, often called white alkali soils, are only slightly alkaline, their pH seldom rising much above pH 8.   ASG

solonetz An intrazonal group of soils that have surface horizons of varying degrees of friability underlain by dark, hard soil characterized by a columnar structure. The hard layer is usually highly alkaline, with the high pH resulting from the adsorbed sodium and the presence of sodium carbonate. The soil colloids, both inorganic and organic, become dispersed and tend to move slowly down the profile, while the frequently observed dark colour of the surface crust is due to dissolved organic matter. These soils, sometimes known as black alkali soils, occur in semi-arid and subhumid areas.   ASG

solstice The day of maximum or minimum declination of the sun. Either the longest or shortest day of the year.

Table 3 Average solute content of precipitation

	Concentration (mgYl^{-1})					
	Ca^{2+}	Mg^{2+}	Na	K	Cl$^-$	SO$_4{}^{2-}$
Coastal	0.29	0.45	3.45	0.17	6.0	1.45
Inland	0.43	0.19	0.37	0.15	0.75	1.73

Source: Meybeck (1983).

solum The soil zone above the weathered parent material, in effect the A and B horizons.

solutes All natural waters contain organic and inorganic material in solution or solutes. The oceans constitute about 97% of the hydrosphere, and its average chemical composition is therefore essentially that of seawater, with a total solute concentration of approximately 34,558 mg l^{-1}. The solute content of the remaining water associated with the terrestrial phase of the hydrological cycle exhibits considerable spatial and temporal variation and is of greater interest to the physical geographer. It has been widely studied as a means of investigating chemical weathering processes and rates of chemical denudation, evaluating nutrient cycling by vegetation communities, and elucidating the processes and pathways involved in the movement of water through the drainage basin system. The major solutes contained in these waters are Ca^{2+}, Mg^{2+}, Na$^+$, K$^+$, Cl$^-$, HCO$_3{}^-$, SO$_4{}^{2-}$, NO$_3{}^-$ and SiO$_2$.

Precipitation inputs to the land surface contain significant concentrations of solutes as a result of rain-out and wash-out of atmospheric aerosols. The magnitude and composition of this solute content will vary according to the relative importance of terrestrial and marine aerosols. Highest total solute concentrations are found in coastal areas, where marine aerosols dominate and where Na$^+$, Cl$^-$, Mg^{2+} and K$^+$ are the dominant ions. The overall solute content of precipitation declines inland and, at distances in excess of about 100 km from the coast, terrestrial aerosols predominate as solute sources. Here, the major ions are Ca^{2+} and SO$_4{}^{2-}$. Estimates of the average

solute content of precipitation over coastal and inland zones as shown in Table 3.

As it moves through the vegetation canopy and the soil and rock of a drainage basin the solute content of the water will increase and its chemical composition will frequently change. Solutes will be leached and washed from vegetation and solute levels within the soil will be influenced by concentration and precipitation mechanisms, interactions with the soil matrix, release of solutes through chemical weathering, and biotic uptake and release of nutrients. Further evolution of the solute content may occur within the groundwater body.

The solute content of streamflow will therefore reflect the characteristics of the upstream drainage basin, including its geology, topography and vegetation cover, and the pathways and RESIDENCE TIME associated with water movement through the basin. Concentrations will vary through time in response to hydrological conditions and will frequently exhibit a DILUTION EFFECT during storm run-off events.

At the global scale, climate and lithology exert a major influence on the solute content of river water. Solute concentrations commonly exhibit an inverse relationship with mean annual run-off and demonstrate marked contrasts between major rock types. Total solute concentrations in rivers draining basins underlain by sedimentary rocks are on average about five times greater than from basins underlain by crystalline rocks and about 2.5 times greater than from basins underlain by volcanic rocks. Typical ranges of concentrations associated with individual solute species in major world rivers are listed in Table 4.

Table 4 Range of concentrations of solutes in major world rivers

	Concentration (mg l^{-1})							
	Ca^{2+}	Mg^{2+}	K	Na	HCO$_3{}^-$	Cl$^-$	SO$_4{}^{2-}$	SiO$_2$
Minimum	2.0	1.0	0.5	1.0	10	1.0	1.5	2.0
Maximum	55	15	4	40	170	45	65	20

Source: Meybeck (1983).

Calcium is the dominant cation and HCO_3^- the dominant anion in nearly all major rivers. A greater degree of variation in solute concentrations is to be found when considering data from small streams.

Interest in the solute content of natural waters has necessitated the development of a wide range of analytical methods, which are documented in a number of laboratory manuals. Many of these methods are now semi-automated and provide a means of dealing with the large numbers of samples produced by automatic samplers or intensive manual sampling programmes. Measurements of specific conductance (see CONDUCTANCE, SPECIFIC) are widely employed as a simple means of estimating the total solute content of a sample, and this parameter may be continuously recorded using simple equipment. Progress has also been made in the development of apparatus for continuous monitoring of individual solute species using specific ion electrodes. DEW

Reading and Reference
Eamus, D., Hatton, T, Cook, P. and Colvin, C. (2006) *Ecohydrology: vegetation function, water and resource management*. Collingwood: CSIRO Publishing. · Meybeck, M. (1983) Atmospheric inputs and river transport of dissolved substances. In B. W. Webb (ed.), *Dissolved loads of rivers and surface water quantity/quality relationships*. IAHS Publication no. 141. Wallingford: IAHS Press; pp. 173–192.

solution, limestone The change of limestone from the solid state to the liquid state by combination with water. When water charged with carbon dioxide comes in contact with limestone (either as free carbon dioxide or as HCO_3^-) it dissolves the rock (CORROSION). When this occurs beneath the water table in the phreatic zone it dissolves bedding planes or joints to produce characteristically oval cave passages. When it occurs in normal stream situations it produces notches at the side of the stream on a line approximating the water level. Limestone solutional processes can also produce many different microfeatures (KARREN). PAB

Reading
Picknett, R.G., Bray, L.G. and Stenner, R.D. (1976) The chemistry of cave water. In T. D. Ford and C. H. D. Cullingford (eds), *The science of speleology*. London: Academic Press; pp. 213–266. · Trudgill, S.T. (ed.) (1986) *Solute processes*. Chichester: John Wiley & Sons, Ltd.

sorptivity S The mechanism responsible for water entry into soils during the very early stage of wetting-up, during which limitations relating to the saturated hydraulic conductivity of the bulk soil do not limit water entry. Sorptivity describes the water entry that occurs primarily owing to capillarity

effects during the approach to saturation. During the phase of soil wetting dominated by sorptivity, the rate of water entry is typically rapid, but declines as the square root of elapsed time toward a lower, final infiltrability, which is approximately equal to the saturated hydraulic conductivity K_{sat}. Saturated hydraulic conductivity is governed by the gravity-driven entry of water into the soil. S is expressed in $cm\ h^{-0.5}$, whilst K_{sat} is expressed in $cm\ h^{-1}$. DLD

Reading
Culligan, P.J., Ivanov, V. and Germaine, J.T. (2005) Sorptivity and liquid infiltration into dry soil. *Advances in Water Resources*, **28**, 1010–1020. · Philip, J.R. (1957) The theory of infiltration: 4. Sorptivity and algebraic infiltration equations. *Soil Science*, **84**, 257–264.

sorting A measure of the standard deviation of the sample of a particle size distribution. It relates to the way in which material is differentially removed by particular geomorphic agencies. For example, wind action tends to leave a well-sorted residual that is represented by a low sorting value; that is, a predominance in a narrow size range roughly around the mean for a log-normal distribution. It can be obtained graphically from phi (ϕ) percentiles by

$$So = \frac{\phi 90 + \phi 80 + \phi 70 - \phi 30 - \phi 20 - \phi 10}{5.3}$$

A method of calculation using moment measures is also available. WBW

Reading
Tucker, M.E. (2001) *Sedimentary petrology: an introduction*. Oxford: Blackwell.

source area The area of a catchment that is physically producing OVERLAND FLOW at any time. This area is changing dynamically during and after storms. Its estimation is central to PARTIAL AREA MODELS, and the term is generally used in the context of overland flow produced by subsurface saturation rather than because the INFILTRATION capacity has been exceeded. Before a storm, the previous rainfall establishes the pattern of the saturated zone within a catchment (see SATURATED WEDGE AND ZONE). Storm rainfall is added to this layer of saturated water and increases the source area during the storm. After the storm, THROUGHFLOW gradually diminishes the saturated wedge and the source area declines with it. The source area typically consists of river floodplains, together with a narrow strip along the base of concave hillsides and a larger area of converging flow in stream-head hollows. The total area involved varies from 1 to 3% of catchments under dry conditions up to 10–50% during and immediately after major storms, although there are wide differences between catchments.   MJK

Reading
Kirkby, M.J. (1978) *Hillslope hydrology*. Chichester: John Wiley & Sons, Ltd.

source bordering dune A sand DUNE that is close to the source of sediment from which material is deflated by aeolian processes; the term is usually used with reference to dunes bordering fluvial systems in drylands but can equally apply to a LUNETTE DUNE on the margin of a PAN or to COASTAL DUNES.   DSGT

Southern Oscillation

The fluctuation of atmospheric mass (pressure) between the eastern and western hemispheres of the tropical South Pacific Ocean. Atmospheric pressure varies in a see-saw fashion between the two areas, and when one experiences high pressure the other tends to have lower pressure. The oscillation itself occurs on an irregular, interannual basis and is strongly associated with sea surface temperatures.

Under 'normal' conditions, sea-level pressure tends to be relatively high in the tropical South Pacific Ocean and comparatively low in the western Pacific and eastern Indian Oceans. Similarly, sea surface temperatures are usually much higher in the western Pacific than in the east. Under these conditions the Walker circulation operates in its normal fashion: the net transport of air at the surface is from east to west, and rainfall is much higher in the western Pacific.

At irregular intervals the sea level pressure difference across the Pacific decreases dramatically or even reverses. This weakens the surface easterlies and causes them to retreat eastward, disrupting the Walker circulation. These developments are usually, but not always, accompanied by EL NIÑO conditions, wherein the warm pool of water (and thus rainfall) normally located in the western Pacific migrates eastward.

The state of the Southern Oscillation is expressed in the form of a numerical index. This Southern Oscillation index (SOI) is defined as the difference between the standardized sea level pressure at Papeete, Tahiti, and Darwin, Australia. Low index values are typically associated with EL NIÑO conditions. See also ENSO.   RSV

speciation See DARWINISM and EVOLUTION.

species–area curve A graph of the relationships between plant or animal numbers and the area of sample plots. In general, the number of species present will increase as area increases within any given community. Eventually, the number of new species found in successively larger plots will become progressively fewer and the species–area curve will flatten and become approximately horizontal. Species–area curves are nevertheless useful guides in the determination of a satisfactory quadrat size for sampling and for comparing the size of the fauna or flora of different islands or various sized landmasses.   PAS

Reading
Hopkins, B. (1957) The concept of minimal area. *Journal of Ecology*, 45, 441–449. · Krebs, C.J. (2014) *Ecology: the experimental analysis of distribution and abundance*, 6th edition. Abingdon: Pearson.

specific conductance See CONDUCTANCE, SPECIFIC.

specific retention The volume of water that a rock or soil retains against the influence of gravity if it is drained following saturation. The difference between POROSITY and specific retention is the SPECIFIC YIELD.   PWW

Reading
Ward, R.C. and Robinson, M. (1990) *Principles of hydrology*, 3rd edn. Maidenhead: McGraw-Hill.

specific yield The volume of water that a water-bearing rock or soil releases from storage under the influence of gravity. In an unconfined AQUIFER, it is expressed as the volume per unit surface area of aquifer per unit decline in the WATER TABLE.   PWW

Reading
Freeze, R.A. and Cherry, J.A. (1979) *Groundwater*. Englewood Cliffs, NJ: Prentice-Hall.

spectral analysis A mathematical tool for evaluating the time domain of a signal, such as a climatic time series. The amount of variance of the time series as a function of frequency is represented by what is termed a frequency spectrum or power spectrum. The power spectrum is often calculated as a Fourier transform. In each frequency domain, an amplitude and a phase are determined. As an example, the Southern Oscillation–EL NIÑO phenomenon tends to recur on average roughly every 5–6 years, so that its power spectrum has a strong peak at this frequency.   BJS

Reading
Chatfield, C. (1980) *The analysis of time-series: an introduction*. London: Chapman and Hall.

speleology The scientific study of caves, their formation and processes. It includes studies of speleogenesis (cave formation processes), cave survey, biology, geology and chemistry. The science used to be conducted by amateurs or semi-professionals, but since 1970, in response to the increasing technicalities of the subject, most of the pertinent research is carried out by full-time scientists and researchers.   PAB

Reading
Bögli, A. (1980) *Karst hydrology and physical speleology*. New York: Springer-Verlag. · Ford, T.D.and Cullingford, C.H.D. (eds) (1976) *The science of speleology*. London: Academic Press.

speleothem A general term for depositional features that include stalactites, stalagmites, columns, flowstone, helictites and curtains. Speleothems are commonly calcareous, crystalline deposits but can be made of a number of different materials; silica, gypsum, peat and ice have all been recorded. Calcareous speleothems are formed when rainwater seeps through organic-rich soils and absorbs carbon dioxide. On contact with limestone it dissolves some of the rock, but when it reaches a cave roof it comes in contact with air that is not charged with as much carbon dioxide. The air absorbs some of the carbon dioxide, the water becomes less aggressive (indeed, supersaturated) and deposits some calcium carbonate in the cave.   PAB

Reading
Warwick, G.T. (1962) Cave formations and deposits. In C. H. D. Cullingford (ed.), *British caving*. London: Routledge & Kegan Paul; pp. 83–119.

sphagnum PEAT mosses are characteristic peat formers in ombrotrophic mires, particularly raised mires. In undisturbed raised mires, the characteristic hummock–hollow microtopography is characterized by specific *Sphagnum* species such as *Sphagnum cuspidatum* in peat hollows and *Sphagnum magellanicum* in peat hummocks.

Sphagnum leaf cells are not of uniform size but consist of narrow chlorophyll cells for assimilation sandwiched between larger hyaline cells. *Sphagnum* cells take up water through their pores until they are full and release it very slowly; in effect, they function like a sponge: the anatomical structure of *Sphagnum* cells allows water absorption up to 15 to 30 times dry weight. *Sphagnum* grows as billions of plantlets side by side in a raised MIRE forming a mattress of moss growing upward by a few millimetres each year and decaying below the ground surface as a result of lack of light. Other species surviving in the raised mire environment must adapt to this upward growth habit.   ALH

sphenochasm and sphenopiezm The former is the triangular gap of oceanic crust separating two cratonic blocks with fault margins converging to a point and is interpreted as having originated by the rotation of one of the blocks with respect to the other (e.g. the Bay of Biscay). By contrast, the latter is a wedge of crust caused by the squeezing together of blocks (e.g. the Pyrenees).   ASG

sphericity The degree to which a particle tends toward the shape of a sphere.

spheroidal weathering Exfoliation, onion-weathering. The disintegration of a rock by the peeling of the surface layers that tends to round boulders and cobbles.

spits Spits are generally linear desposits of beach material attached at one end to land and free at the other. There are many different types, including single, recurved, looped, hooked, complex and double spits. They occur where LONG-SHORE DRIFT carries material beyond a change in orientation of the coast or at river mouths. They usually have a narrow proximal part and a broader distal end, where recurves are common. Spits may be called bay head, mid bay or bay mouth according to their position in an embayment. They occur mainly on indented coasts where abundant sediment can move alongshore freely. Nearly all spits have formed since sea level stabilized about 4000 years ago, and many are much younger.   CAMK

Reading
Schwartz, M.L. (ed.) (1972) *Spits and bars*. Stroudsburg, PA: Dowden, Hutchinson and Ross. · Zenkovich, V.P. (1967) *Processes of coastal development*. Edinburgh: Oliver & Boyd.

spring line See SPRINGS.

spring mounds These are formed by the evaporation of mineralized water that reaches the ground surface via artesian springs. The salts that are precipitated on evaporation build up a raised mound, which may be several metres high. Different forms exist, including conical, pinnacles and ridges, which may or may not have a central crater. They are well described from Lake Eyre, Australia, and central Tunisia.   DSGT

Reading
Roberts, C.R. and Mitchell, C.W. (1987) Spring mounds of central Tunisia. In L. Frostick and I. Reid (eds), *Desert*

sediments, ancient and modern. Geological Society of London special publication 35. London: The Geological Society; pp. 321–334.

springs Concentrated point discharges of groundwater outflow, which can occur in terrestrial, intertidal and submarine settings. Springs form one end of a spectrum of ways in which groundwater emergence takes place, with more diffuse seepage areas forming the other. Springs are sometimes called resurgences if they represent the point of emergence of a known surface stream and exsurgences if the headwaters are unknown.

Springs can be classified in a number of ways. For example, cold water springs can be distinguished from geothermal springs, and mineral, hardwater and saline springs can be identified on the basis of water quality. Spring discharge can also be used as a classificatory criterion. Discharge from the majority of springs is variable, with those that flow throughout the year termed perennial whilst those that flow for part of the year are classed as intermittent. Perennial springs can, in turn, be termed as ebbing and flowing if there are alternating high and low pulsating flows on a diurnal or seasonal basis, whilst ARTESIAN springs tend to have more constant flow.

Variations in the amount of flow are usually a direct result of variations in the amount of water in GROUNDWATER storage. This varies due to a combination of climatic conditions, which determine the seasonal or annual height of the WATER TABLE, and the characteristics of the AQUIFER. Spring flow from thick, highly porous aquifers tends to be relatively constant because the volume of seasonal storage change is small when compared with the aquifer's total storage volume. Flows from some thin superficial aquifers (e.g. scree or glacial gravels) may be highly variable, with discharge only occurring for short periods after rainfall. Springs may occur in groups or clusters, often in lines along the foot of hills and where lithologies of differing permeabilities are juxtaposed. In the case of those draining the same aquifer, the springs at lower altitude are likely to be larger and faster flowing (underflow springs), whilst those at higher elevation are more likely to be intermittent or overflow springs that act as outflow points for the aquifer when the water table is high.

Less commonly, springs may experience a reversal in flow direction and become an inflow point rather than a water outflow. Such springs are called estavelles and are found in KARST terrain. Under normal circumstances, the water table rises upstream of the spring, beneath the slope from which spring water emerges. However, sometimes a low-lying or enclosed basin (a POLJE in karst terrain) becomes so full of water that the water level is temporarily higher than the water table in the adjacent hill. If this is the case, the hydrological gradient is reversed, with subterranean fractures and joints allowing the waters to exit the depression via the estavelle.   DJN

Reading
Price, M. (1996) *Introducing groundwater.* 2nd edition. London: Chapman & Hall.

squall line A few cumulonimbus storms in a row, organized to produce strong along-line winds and heavy rain. It may be hundreds of kilometres long but only a few wide, so wind speeds increase very rapidly, followed quickly by heavy rain over a large transverse distance, causing widespread damage. Two broad categories of 'tropical' and 'mid-latitude' are not necessarily confined to those regions. Extensive cirrus sheets seen on satellite pictures help diagnosis in regions where data are sparse.   JSAG

Reading
Ludlam, F.H. (1980) *Clouds and storms.* Englewood Cliffs, NJ: Pennsylvania State University Press.

stability The ability of an ecosystem to maintain or return to its original condition following a natural or human-induced disturbance. This concept of stability has been widely used by scientists in recent years, but many other meanings have been attached to the term 'stability' (Orians, 1975). For example, the term has been used in reference to the constancy or PERSISTENCE of species populations or ecosystems. Two major aspects of ecosystem stability in relation to disturbance have received most attention. Even in this context, a confusing variety of terms has been used. The first property, which is often labelled *resistance*, is the ability of a system to remain unaffected by disturbances. This property is referred to as *inertia* by Orians (1975) and Westman (1978) and as *resilience* by Holling (1973). The second attribute is usually termed *resilience* and is the ability of the system to recover to its original state following a disturbance. The more general term *stability* has also been applied to this property by May (1973) and Holling (1973).

The concepts of resistance and resilience are of considerable interest not only to scientists engaged in basic research, but also to environmental planners and managers. Knowledge of the varying ability of ecosystems to resist change, or to recover quickly following disturbance, is of obvious value in planning development projects and assessing potential damage from pollutants.

Resistance and resilience can be measured in a variety of ways. The particular ecosystem characteristics analysed will depend on the nature of the disturbance and the goals of the research, and may range from a focus on individual species populations to overall system properties such as species diversity, primary production and nutrient losses in drainage water. Resistance to a disturbance can be measured by the magnitude of the system response and by the time delay before a response occurs. These parameters provide an assessment of the relative resistance of an ecosystem to different types of stress or alternatively of different ecosystems of the same stress. For example, a study by Vitousek *et al.* (1981) of nitrate losses from disturbed forest plots revealed a major peak in nitrate losses within 6 months of disturbance in Indiana maple and oak forests, whereas much smaller losses occurred in hemlock and Douglas-fir forests in Oregon. A pine forest in Indiana exhibited an extended delay in response, with substantial losses beginning almost 2 years later.

Westman (1978) examines a variety of measurement problems associated with resilience and suggests that four aspects of this component can be evaluated. Elasticity refers to the rapidity of the system's return to its original state; amplitude to the zone from which the system can recover; hysteresis, the extent to which the recovery pathway differs from the pattern of disruption that occurred in response to the disturbance; and malleability, the degree to which the new stable state established after disturbance differs from the original steady state. The amplitude aspect of resilience is of particular interest because it deals with a threshold beyond which the system cannot recover to its initial state. It may be possible to suggest that a system is reaching a threshold by studying the rate of change of various characteristics in relation to the range of intensity of a particular stress. Baker (1973) has studied the amplitude response of saltmarsh vegetation to oil pollution. His data suggest that recovery was good when the vegetation was exposed to not more than four oil spillages. Substantial damage and very slow recovery occurred after 8-12 successive oilings, suggesting the presence of a threshold of recovery.

Ecosystem resistance and resilience may not necessarily be closely linked. The initial degree to which a system is altered by disturbance may in some cases be a poor indicator of the ultimate ability of the system to cope with stress. In the Great Lakes, many species of fish, including herring, walleye and lake trout, withstood fishing pressure for many years without any obvious

signs of decline, but all these species experienced sudden collapses of populations to near-extinction levels without any advance warning (Holling, 1973).

Much research has been devoted to the study of relationships between stability and other ecosystem properties, particularly species diversity (Goodman, 1975; Pimm, 1984). The notion that more complex systems involving large numbers of interacting species should be more stable than simple systems with few species is intuitively attractive since there should be more alternative pathways for feedback and adjustment to disturbance in the complex system. MacArthur's (1955) hypothesis that stability was a function of the complexity of feeding linkages between organisms in an ecosystem was therefore rapidly accepted, and several lines of evidence were used to support the relationship. This evidence, which is reviewed by Elton (1958), involved data from laboratory experiments with one prey-one predator systems, which revealed the occurrence of large population fluctuations followed by rapid extinction. Emphasis was also placed on the vulnerability of simple agricultural systems to pest outbreaks and the contrast between prominent population oscillations in Arctic tundra and the apparent lack of such fluctuations in the complex and species-rich tropical rain forests. A more critical evaluation of the linkage between stability and diversity in recent years has underlined the weakness of this evidence. For example, instability in the laboratory predator-prey system does not provide a valid analogy to the real world since even very simple ecosystems contain many different species. Similarly, the instability of crop monocultures can probably be attributed to the absence of coevolution over long time periods of the species involved.

Empirical studies suggest that there is no simple link between diversity and stability (Goodman, 1975). The total species diversity of an ecosystem may not be an adequate measure of complexity, which can take a variety of forms in relation to the trophic level involved and the spatial organization of the system. Watt (1968) has proposed that stability at any herbivore or carnivore trophic level increases with the number of competitor species at that level, decreases with the number of competitor species that feed upon it and decreases with the proportion of the environment containing useful food. The question of the relationship between diversity and stability is further complicated by evidence that some ecosystems contain a single species high in the food web that can influence the system structure (Paine, 1969). The stability of such systems would be highly dependent on the effects of disturbance on the 'keystone' species rather than on the overall species diversity of the system. Extensive examination of the diversity-stability hypothesis has been undertaken using mathematical models (May, 1973). Defining stability as the ability of the system to return to equilibrium after stress, May found that complex model systems are less stable than simple ones. In view of the complexities revealed by recent research and the fact that existing evidence is contradictory, many ecologists now regard the equating of stability with diversity as a tentative hypothesis rather than an axiom. The influence of other ecosystem properties on stability has generally been neglected, although it is likely that the resistance and resilience of systems to human-induced stresses can be affected by the spatial organization of the ecosystem and a variety of linkages between biological and abiotic components (Hill, 1975).   ARH

References

Baker, J.M. (1973) Recovery of salt marsh vegetation from successive oil spillages. *Environmental Pollution*, 4, 223–230. · Elton, C.S. (1958) *The ecology of invasions by animals and plants*. London: Methuen; chapter 8, pp. 143–153. · Goodman, D. (1975) The theory of diversity-stability relationships in ecology. *Quarterly Review of Biology*, 50, 237–266. · Hill, A.R. (1975) Ecosystem stability in relation to stresses caused by human activities. *Canadian Geographer*, 19, 206–220. · Holling, C.S. (1973) Resilience and stability of ecological systems. *Annual Review of Ecology and Systematics* 4, 1–23. · MacArthur, R.H. (1955) Fluctuations of animal populations and a measure of community stability. *Ecology*, 36, 633–636. · May R.M. (1973) *Stability and complexity in model ecosystems*. Princeton, NJ: Princeton University Press. · Orians, G.H. (1975) Diversity, stability and maturity in natural ecosystems. In W. H. Van Dobben and R. H. Lowe-McConnell (eds), *Unifying concepts in ecology*. The Hague: W. Junk; pp. 139–150. · Paine, R.T. (1969) A note on tropic complexity and community stability. *American Naturalist* 103, 91–93. · Pimm S.L. (1984) The complexity and stability of ecosystems. *Nature*, 307, 321–326. · Vitousek, P., Reiners, W.A., Melillo, J.M *et al.* (1981) Nitrogen cycling and loss following forest perturbation: the components of response. In G. W. Barrett and R. Rosenberg (eds), *Stress effects on natural ecosystems*. Chichester: John Wiley & Sons, Ltd; pp. 114–127. · Watt, K.E.F. (1968) *Ecology and resource management*. New York: McGraw-Hill; chapter 3, pp. 39–50. · Westman, W.E. (1978) Measuring the inertia and resilience of ecosystems. *Bioscience*, 28, 705–710.

stability analysis   The procedure for examining the likelihood of failure of a soil or rock slope. The types of analysis and the way they are carried out depend largely upon the nature of the materials to be investigated, but they generally require a knowledge of the COHESION and FRICTION properties of the slope material as well as the jointing characteristics if the material is a rock slope. Slope geometry is also crucial, as is a knowledge of the water availability as pore water pressure (in soils) or cleft water pressure (in rocks). Many stability analyses are two-dimensional, but digital computing methods can now make three-dimensional analyses relatively easy. In soil mechanics such analyses are used to determine a factor of safety so that a safe design can be produced, but they can be used in a more geomorphological way to help determine the characteristics once a slope has failed.   WBW

Reading

Bell, F.G. (1992) *Engineering properties of soils and rocks*, 3rd edition. Oxford: Butterworth-Heinemann. · Lambe, T.W. and Whitman, R.V. (1981) *Soil mechanics*. New York: John Wiley & Sons, Inc.

stable equilibrium   A condition of a system in which very limited displacement in any direction is followed by a return to a persistent state or condition. Huggett (1980: figure 1.3) employs the mechanical analogy (from Spanner (1964)) of a ball resting in a deep cup: the ball may move from side to side or round and round if the cup is shaken, but will always ultimately return to the bottom of the cup as long as the stable equilibrium condition prevails.   BAK

References

Huggett, R. (1980) *Systems analysis in geography*. Oxford: Clarendon Press. · Spanner, D.C. (1964) *Introduction to thermodynamics*. London: Academic Press.

stable isotope geochemistry   Stable isotopes are atoms of chemical elements that have the same number of protons but a different number of neutrons. Most chemical elements have isotopes, many of which are classed as stable; in other words, they do not alter their state over time or - at least if they are radioactive - have very long half-lives. Since all isotopes of a given element contain the same number and arrangement of electrons, they frequently behave similarly to each other in terms of chemical activity. The relative abundance of such stable isotopes can be measured experimentally, yielding an isotope ratio that can be used as a research tool because, as a result of fractionation, the ratio may shift in predictable ways during exchange reactions or kinetic processes. Stable isotope geochemistry is concerned with these variations in isotopic compositions arising from fractionation. For example, plants that produce C4 hydrocarbon chains during photosynthesis (see PHOTOSYNTHETIC

PATHWAY) fractionate carbon differently from C3 plants, and this fractionation is retained up the food chain. Stable light isotope ratios are therefore different between C3 and C4 plants, and this difference in ratio is maintained among the herbivores and carnivores in the food chain. This gives rise to an application of stable carbon isotope studies in reconstructing palaeodiet and palaeoenvironments, since the relative abundance of C3 and C4 plants in the environment can be estimated over time if the organic material is preserved in sediments. There are now many applications of stable isotope geochemistry used in Earth sciences employing the elements hydrogen (H), lithium, boron, nitrogen (N), oxygen (O), silicon, sulphur (S) and chlorine, as well as carbon (C). Of these, by far the most widely applied are those of O, H, C, N and S. For example, fractionation of the two common stable isotopes of oxygen (^{16}O, ^{18}O) can be used to reconstruct global changes in temperature over extended periods due to the fact that the lighter of the two isotopes evaporates at a proportionally higher rate at higher temperatures (OXYGEN ISOTOPE) and that microscopic marine organisms may preserve these ratios in ocean sediments.   MEM

Reading
Hoefs, J. (2009) *Stable isotope geochemistry*, 6th edition. Berlin: Springer.

stack   A free-standing pinnacle of rock, usually in the sea, that represents an outlier of a coastal cliff.

stadial   A short cold period with smaller ice volumes than the full glacial stages of an ice age. The warmer intervals between them are called interstadials. See also HEINRICH EVENTS.

staff gauge   An instrument for determining water depth at a site on a river system. It is often located at a stream gauging site to provide a datum for setting and checking continuous stage recorders. The staff gauge consists of a plate or board with painted or engraved elevation divisions, which is firmly fixed in the water at a river cross-section so that water levels can be read by eye from the divisions on the staff gauge at all flow stages, usually to an accuracy of 0.5 cm. Single staff gauges are usually installed to stand vertically and close to one bank of the river, but in some locations more than one staff gauge will be installed to cover different ranges of stage, and occasionally the staff gauge will be inclined to lean against the bank so that flow disturbance is minimized and

so that the depth scale may be more easily read. (See also DISCHARGE and HYDROMETRY.)   AMG

stage   The term used to describe water depth or the elevation of the water surface at a location on a river system. Instruments that continuously monitor water surface elevation at a gauging station site are called stage recorders. The river stage fluctuates through time in response to precipitation events. In a stable and straight river section a relationship can be established between discharge and stage so that discharge can be estimated for any water stage, and this relationship between stage and discharge is known as the discharge rating curve for that site. (See also DISCHARGE.)   AMG

stagnant ice topography   The view that many glacial deposits in formerly glaciated areas are the result of stagnant ice downwasting in situ was championed in North America in the 1920s (Cook, 1924; Flint, 1929), in Scandinavia in the 1940s and 1950s (Mannerfelt, 1945; Hoppe, 1959) and in Scotland (Sissons, 1967). The idea is that the ice, by virtue of shallow surface gradients or by its isolation from the main ice mass, no longer flows actively but melts from the surface downwards, the last remnants being preserved in depressions. The landforms that are formed depend both on the shape of the underlying topography and on the debris characteristics of the glacier. In places, MORAINE landforms of disintegration are dominant and consist of irregular mounds and kettles (see KETTLE, KETTLE HOLE) built of varying quantities of basal and supraglacial TILL. The basal till may have reached the glacier surface as a result of COMPRESSING FLOW before being redistributed through the action of slumping and surface sediment flows as the ice melts away. Such conditions also favour considerable meltwater activity. One fine example of such topography with a relief of 30-45 m occurs on the Canadian prairies and covers areas of thousands of square kilometres (Prest, 1983).

In hillier parts of the world the role of valleys in influencing the location of the stagnant remnants of ice seems to favour another association of landforms. Once isolated from the ice sheet, such stagnant ice masses are plugged by glaciofluvial deposits associated with meltwater streams from both the ice surface and the surrounding hills, and their courses are marked by KAME TERRACES, KAMES and associated ESKERS and irregular mounds. Mixed with the glaciofluvial deposits may be irregular mounds of till. Such landscapes abound in upland Scandinavia and Scotland.   DES

Reading and References

Cook, J.H. (1924) The disappearance of the last glacial ice-sheet from eastern New York. *Bulletin of the New York State Museum*, **251**, 158–176. · Flint, R.F. (1929) The stagnation and dissipation of the last ice sheet. *Geographical Review* **19**, 256–289. · Hoppe, G. (1959) Glacial morphology and inland ice recession in north Sweden. *Geografiska Annaler*, **41**, 193–212. · Mannerfelt, C.M. (1945) Några glacialmorfologiska Formelement och deras vittnesbörd om inlandsisens avsmältningsmekanik i Svensk och Norsk fjällterräng. *Geografiska Annaler*, **27**, 3–239 (English summary and figure captions). · Moran, S.R., Clayton, L., Hooke, R.LeB., *et al.* (1981) Glacier-bed landforms of the prairie region of North America. *Journal of Glaciology*, **25** (93), 457–476. · Prest, V.K. (1983) *Canada's heritage of glacial features*. Geological Survey of Canada miscellaneous report 28. Ottawa: Minister of Supply and Services Canada. · Sissons, J.B. (1967) *The evolution of Scotland's scenery*. Edinburgh: Oliver & Boyd.

stalagmite, stalactite   See SPELEOTHEM.

star dune

A type of sand dune that has three or more radial arms extending in various directions from a central high point. Star dunes develop in multidirectional wind regimes, where seasonal changes in the overall sand-transporting direction cause sand to accumulate vertically, sometimes through the merging of other dune types. Star dunes are therefore sand-accumulating forms - which tend to occur at the depositional centres of sand seas or where airflow is modified by topographic barriers - and can attain heights up to 300-400 m. The morphology of the individual arms of star dunes can vary seasonally in response to sand transport shifts such that they can behave like transverse or linear dunes. Star dunes are sometimes found in chains, and if one transport direction is stronger than the others, slow migration may result. (See DUNE.)   DSGT

Reading

Lancaster, N. (1989) Star dunes. *Progress in Physical Geography*, **13**, 67–91.

steady flow

The flow of water in open channels is classified according to its temporal and spatial variability, and steady flow occurs when the water depth, the discharge and therefore the velocity are temporally constant. Since open channels have a free surface exposed to the atmosphere, the flow responds to rainfall and run-off inputs and is naturally temporally unsteady during storm hydrographs. The rate of change of depth and discharge is often sufficiently slow for a steady flow to be assumed during the time interval under consideration, and this forms the basis for the development and application of simplified FLOW EQUATIONS. Temporally steady flow may be spatially uniform or varied, and varied flow may be gradually varied or rapidly varied.   KSR

steady state

Dynamic systems with interacting feedbacks may lead to development of steady-state systems that retain characteristics or form over time even as material is cycled through them. For example, in mountain building the dynamic system of tectonics and erosion contains important feedback mechanisms such that orogenic systems tend towards a steady state. However, four distinct types of steady state may characterize orogenic systems: (1) flux steady state, (2) topographic steady state, (3) thermal steady state and (4) exhumational steady state; they respectively refer to the erosional flux, the topographic form, the subsurface temperature field and the spatial pattern of cooling ages. Numerical models and common sense suggest that a perfect topographic steady state is unlikely to be achieved at short length scales, but that, under consistent tectonic and climate forcing, topography will reach a steady mean form at the scale of an orogenic belt. Thermal steady state and exhumational steady state are related, in that the former is a precondition for the latter.   DRM

Reference

Willett, S.D. and Brandon, M.T. (2002) On steady states in mountain belts, *Geology* **30**, 175–178.

steam fog

Fog formed when cold air passes over warm water. Both heat and water vapour are added to the air, which quickly becomes saturated. Any additional water vapour evaporated from the warm water will rapidly condense, forming a swirling, steam-like fog. Steam fog is common over heated swimming pools in winter, above lakes in autumn and early winter mornings, over thermal ponds, such as those in Yellowstone National Park, all the year round, and above open water in polar regions, where it is called arctic sea smoke.   WDS

stem flow

The drainage of intercepted water down the stems of plants. Precipitation may be intercepted by vegetation and subsequently lost through evaporation (INTERCEPTION or interception loss), it may drip through the vegetation canopy (THROUGHFALL) or it may drain across the leaves and stems of plants or down the branches and trunks of trees as stem flow. Stem flow is a means by which precipitation may reach the ground surface, and it may form quite an important route for water, particularly if the structure of the vegetation encourages this form of drainage.   AMG

Reading
Courtney, F.M. (1981) Developments in forest hydrology. *Progress in Physical Geography*, 5, 217–241. · Sopper, W.E. and Lull, H.W. (eds) (1967) *International symposium on forest hydrology*. Oxford: Pergamon.

step-pool systems   These commonly occur in mountain streams with steep gradients and coarse bed materials. They are formed by high-magnitude, low-frequency flood events, and the staircase-like structure tends to be relatively stable for long periods of time during low flows. The steps and pools alternate to produce a characteristic, repetitive sequence, with the steps composed of an accumulation of cobbles and boulders that are transverse to the channel. Finer materials fill the pools. Logs may contribute to the formation of steps.   ASG

Reading
Chin, A. (1989) Step pools in stream channels. *Progress in Physical Geography* **13**, 391–407.

steppes   Mid-latitude grasslands with few trees. The Russian equivalent of the North American prairies and Argentinian pampas.

stick slip   The jerky motion by which glaciers slide over bedrock. Sudden slip phases of 1-3 cm displacement are interspersed with longer quiescent phases. Each slip phase is highly localized beneath a glacier and probably relates to the failure of a local bond between part of the glacier and the bed.   DES

Reading
Robin, G.deQ. (1976) Is the basal ice of a temperate glacier at the melting point? *Journal of Glaciology* **16**, 183–196.

stilling well   A large-diameter tube installed at the edge of a river channel, or in a river bank, in order to obtain accurate measurements of river level from a still water surface. The tube is connected to the river by intake pipes, in order to ensure that the water level in the well is identical to that in the channel. The diameter of these pipes should be sufficiently small to damp out turbulence or short-term oscillations, but must be large enough to permit instantaneous response to changes in river level. Provision must be made for flushing or clearing the intake pipes. A FLOAT RECORDER is often associated with a stilling well.   DEW

stillstand   A period of stability between two phases of tectonic activity in the Earth's history.

Also a period during which mean sea level is constant.

stochastic models   Any mathematical model that represents a stochastic process - a phenomenon whose temporal or spatial sequence is characterized by statistical properties - is a stochastic model. The incorporation in the model of a sequential time or space function of the probability of occurrence is what distinguishes a stochastic model from a purely probabilistic model (such as the probability distribution of floods of different magnitude). The selection of an appropriate model depends on the nature of the process being modelled and the type of data collected to summarize the process numerically (Thornes and Brunsden, 1977: 5-7, 70-87). A continuous process is observed continuously through time or space, even if the measured variable only takes a discrete set of values, as in the case of a binary variable denoting presence or absence. For convenience, a continuous process is often represented by data obtained at discrete time or space intervals (usually equal), either as discrete sampled data read at specific points or as discrete aggregate data that are summed or averaged over a period (e.g. daily rainfall). Another class of stochastic phenomena involves point processes, in which either the frequency of *events* in successive discrete time periods is counted or the distribution of time *intervals* between events is assessed.

Discrete approximations of continuous time or space series may be modelled using the general linear random model:

$$z_t - \phi_1 z_{t-1} - \cdots - \phi_p z_{t-p}$$
$$= e_t - \theta_1 e_{t-1} - \cdots - \theta_q e_{t-q}$$

where z are values of a variable measured at times $t, t-1$, and so on, e are random 'shocks' at these times, and the ϕ_1 and θ_1 are coefficients. In fitting this mixed autoregressive moving-average model, the objective is to minimize its complexity by reducing the 'order' of dependency defined by the lags p and q (Chatfield, 1975; Richards, 1979). The stochastic point process (Cox and Lewis, 1966) may be described by appropriate probability distributions such as the binomial or Poisson distributions, and modelled as a series using the theory of queues.   KSR

Reading and References
Chatfield, C. (1975) *The analysis of time series: theory and practice*. London: Chapman & Hall. · Cox, D.R. and Lewis, P.A.W. (1966) *The statistical analysis of series of events*. London: Methuen. · Pinsky, M. and Karlin, S. (2011) *An introduction to stochastic modelling*, 4th edition. Burlington, MA: Academic Press. · Richards, K.S. (1979)

Stochastic processes in one-dimensional series: an introduction. Concepts and Techniques in Modern Geography, no. 23. Norwich: Geo Abstracts. · Thornes, J.B. and Brunsden, D. (1977) *Geomorphology and time.* London: Methuen.

stochastic process

A statistical phenomenon in which the evolutionary sequence in time and/or space follows probabilistic laws. 'Stochastic', from the Greek word meaning 'guess', implies a chance process that contrasts with a deterministic phenomenon whose future values can be predicted with certainty if the existing values of the controlling variables are known. In a stochastic process, exact prediction is impossible because dependence is partly on past conditions and partly on random influences. Nevertheless, stochastic models can be used to represent the process mathematically, and to provide both efficient forecasts and the distribution of forecast errors. Natural phenomena may be inherently stochastic, but often the randomness apparent in their behaviour reflects the scientist's incomplete understanding and inaccurate measurement (Mann, 1970). In practice, many phenomena display the mixed deterministic-stochastic behaviour typified by climatic and hydrological processes (Yevjevich, 1972). For example, daily river flows vary seasonally with a fundamentally deterministic cycle related to annual variation of radiation receipt and evaporation loss. Superimposed on this is the random occurrence of sharp increases of flow caused by individual storm inputs, followed by the gradual decrease of discharge in the flood recession curve, caused by water retention in the drainage basin and consequent slow outflow. Thus, the stochastic component of the hydrological process involves both random 'shocks' and a system 'memory', which can be modelled by autoregressive STOCHASTIC MODELS. Note that the hydrologist treats rainfall as a random input, whereas the meteorologist seeks to explain rainfall deterministically.   KSR

References

Mann, C.J. (1970) Randomness in nature. *Bulletin of the Geological Society of America*, **81**, 95–104. · Yevjevich, V. (1972) *Stochastic processes in hydrology.* Fort Collins, CO: Water Resources Publications.

stock

A large, irregularly shaped intrusion of igneous rock.

stock resources

NATURAL RESOURCES that have finite availability within the Earth system, relative to the time they have taken to form and human lifespans and expectations. In their broadest definition they include 'all minerals and land'

(Rees, 1990: 14) in which three further subdivisions can be recognized. First, there are all mineral elements that are theoretically recoverable but in practice are not because of their geographical distribution or their dispersion amongst other material. Second, there are those resources that are consumed by use; that is, the process of using them as a resource destroys the mineral that is being used. This category includes all minerals that are used for fuel (e.g. coal and oil). Third are minerals that, whilst used as a resource, are theoretically recoverable, even though usage has changed their form. This group includes all metal minerals, which can, if disposal and economic factors are favourable, be recycled. Finally, through unsustainable usage, some FLOW RESOURCES can become stock resources.   DSGT

Reference

Rees, J. (1990) *Natural resources: allocation, economics and policy*, 2nd edition. London: Routledge.

Stokes' surface

A sedimentary unconformity, recognized by Stokes (1968), where wind scour deflates sediment down to the level of a near-surface water table. Applies principally to deflation from dry lake beds or in coastal settings.   DSGT

Reading and Reference

Stokes, W.L. (1968) Multiple parallel truncation bedding planes: a feature of wind-deposited sandstone formations. *Journal of Sedimentary Petrology*, **38**, 510–515. · Fryberger, S.G., Schenk, C.J. and Krystinik, L.F. (1988) Stokes surfaces and the effects of near surface groundwater table on aeolian deposition. *Sedimentology*, **35**, 21–41.

stone line

A horizon of gravel-sized rock fragments within a soil profile or accumulation of relatively fine-grained sediments.

stone pavement

An area, often planar and level or only gently sloping, across which stone materials resembling paving cover most of the surface. In DESERTS, stone pavements (also termed desert pavements) are composed of abundant stone fragments of pebble size in which the stones are often embedded within a loamy soil material. The stones, which may carry a desert varnish, are very closely packed so that little soil material is exposed at the surface. In many areas, pavements of this kind overlie deep silty soils that contain few or no stones. This seems to require that the stones have been concentrated at the surface following upward migration through the underlying material, probably in response to forces caused by wetting and drying within the soil. In some areas, frost action may be involved (Wainwright *et al.*,

1995). Deflation of surface fines by wind and erosion by water may also be involved.

Stone pavements also occur within glaciated limestone landscapes. Large areas of flat-lying bare rock, crossed by solutionally enlarged joints, reflect glacial stripping to the level of a bedding plane, and possibly the subsequent removal of overlying soil materials. Landforms of this kind are referred to as limestone pavements. DLD

Reading and Reference
Cooke, R.U. (1970) Stone pavements in deserts. *Annals of the Association of American Geographers*, **60**, 560–577. · Wainwright, J., Parsons, A.J. and Abrahams, A.D. (1995) A simulation study of the role of raindrop erosion in the formation of desert pavements. *Earth Surface Processes and Landforms*, **20**, 277–291.

storage Describes the stores of sediment, carbon and water included in global biogeochemical and hydrological cycles. The size and stability of these stores are critical for the understanding the linkages between atmospheric, oceanic and terrestrial processes and scenario modelling of the likely magnitude and rate of global climate change. At more regional and local levels, changes in sources and sinks can have important implications for landscape stability and for the removal and release of contaminants in the environment.

For water, we can think of surface storage, soil moisture storage and groundwater storage as locations where dynamic reservoirs of water may exist in a drainage basin. In hydrological modelling, complex distributions of stored water may be represented as one or more mathematically defined stores or reservoirs, the simplest of which is the linear store. In a linear store, water outflow Q is directly proportional to water storage S:

$$S = kQ$$

where k is the storage coefficient.

The storage equation is combined with a continuity equation that states that the difference between inflow I to the store and outflow from the store is accommodated by a change in the amount of water in storage:

$$\frac{\partial S}{\partial t} = I - Q$$

AMG/TS

Reading
Kirkby, M.J. (1975) Hydrograph modelling strategies. In R. Peel, M. Chisholm and P. Haggett (eds), *Progress in physical and human geography*. London: Heinemann. · Kirkby, M.J. (ed.) (1978) *Hillslope hydrology*. Chichester: John Wiley & Sons, Ltd.

storm beach A BEACH with a profile flattened by storm-wave-induced erosion. Wave steepness during storm conditions is relatively high, and this condition is considered erosive, or destructive. Beach material is eroded from the foreshore and moved offshore to form a nearshore bar. The storm beach has a low-gradient, concave-upward profile, and its landward margin may be marked with a storm berm. Storm beaches are usually in the dissipative morphodynamic regime, because most wave energy is dissipated through waves breaking in the SURF zone. DJS

Reading
Davis, R.A. Jr and Fitzgerald, D.M. (2004) *Beaches and coasts*. Malden, MA: Blackwell.

storm run-off See HYDROGRAPHS and RUN-OFF.

storm surges Changes in sea level generated by extreme weather events. They appear on sea-level records as distortions of the regular tidal patterns and are most severe in regions of extensive shallow water. When maximum surge levels coincide with maximum high-water levels on spring tides, very high total sea levels result. Low-lying coastal areas are then vulnerable to severe flooding. In tropical regions, severe surges are occasionally generated by cyclones, hurricanes or typhoons; the actual levels depend on the intensity of the meteorological disturbance, the speed and direction with which it tracks towards the coast and the simultaneous tidal levels. Areas at risk include the Indian and Bangladesh coasts of the Bay of Bengal, the southeast coast of the USA and the coast of Japan. Satellite and radar tracking of the weather patterns are used to give advanced warning of imminent flood danger. Extratropical surges, generated by meteorological disturbances at higher latitudes, usually extend over hundreds of kilometres, whereas the major effects of tropical surges are confined to within a few tens of kilometres of the point where the hurricane meets the coast. Flood-warning systems for extratropical surges must take account of the total response of a region to the weather patterns. DTP

Reading
Davis, R.A. Jr and Fitzgerald, D.M. (2004) *Beaches and coasts*. Malden, MA: Blackwell.

stoss The direction from which wind, water or ice moves. The windward side of a sand dune.

strain A measure of the deformation of a body when a load or STRESS is applied. It is usually expressed as a dimensionless value (ratio or percentage) as, for linear strain, it is the change in length divided by the original length. Similarly, areal and volumetric strains can be defined. In the SHEAR BOX and TRIAXIAL APPARATUS, the tests are normally done at a constant deformation rate and are called 'strain-controlled' tests.　　　　WBW

strain rate The rate at which a body deforms in response to stress. It is often represented by the symbol $\dot{\varepsilon}$. (See also STRAIN.)

strandflat The term was introduced to describe an undulating rocky lowland up to 65 km wide in western Norway (Reusch, 1894; Nansen, 1922). It is partly submerged, forming an irregular outer belt of skerries, and is backed by a steeply rising coast. Similar features may have been recognized in Iceland, Svalbard, Novaya Zemlya, East and West Greenland, Baffin Island and the Antarctic Peninsula. Nansen's view that the strandflat was cut by freeze–thaw processes adjacent to the shoreline has survived to the present day, but there is an alternative view; namely, that it is a marine or subaerial lowland, subsequently modified by glacial action.　　　　DES

Reading and References
Nansen, F. (1922) *The strandflat and isostasy*. Matematisk Naturvitenskapelig klasse nr II. Oslo: Dybwad Kristiana. · Reusch, H. (1894) The Norwegian coast-plain. *Journal of Geology*, 2, 347–349.

strandline Name sometimes given to an old abandoned shoreline of a lake.

stratigraphy The study of the order and arrangement of geological strata. Lithostratigraphy (rock stratigraphy) is concerned with the

Shell-rich ('coquina') bed in the strandline deposits at Williams Point, Lake Eyre, Australia.
Photograph by David Thomas.

organization of strata into units based on their lithological characteristics. Biostratigraphy is concerned with the organization of strata into units based on their fossil content. Chronostratigraphy (time stratigraphy) is concerned with the organization of strata into units based on their age relationships.　　　　ASG

Reading
Catuneanu, O. (2006) *Principles of sequence stratigraphy*. Amsterdam: Elsevier.

stratocumulus See CLOUDS.

stratosphere The layer from heights of about 10-30 km immediately above the TROPOSPHERE, in which temperature is nearly independent of height. Heat transfer is dominated by thermal RADIATION, which tends to eliminate temperature differences. Emden's theory of radiative equilibrium suggests a stratospheric temperature some $2^{-1/4}$ that of the troposphere, which, at 215 K (0.84 times a tropospheric temperature of 255 K), is about right.　　　　JSAG

Reading
Goody, R.M. and Walker, J.C.G. (1972) *Atmospheres*. Englewood Cliffs, NJ: Prentice-Hall.

stratovolcano A composite landform generally made up of intermediate composition lava flows, pyroclastic debris and debris flow deposits. Stratovolcanoes are steep due to the relatively high silica content, and consequent high viscosity, of the andesitic lava they typically erupt. After an explosive eruption, stratovolcanoes slowly rebuild through the gradual extrusion of viscous lava. Stratovolcanoes typically occur along subduction zones.　　　　DRM

stream capture The process by which one stream erodes more aggressively than a neighbouring stream during the course of drainage basin evolution so that it intersects the channel of that stream and captures its discharge. The removal of discharge from the stream that has been 'beheaded' by capture usually results in its source moving down the original valley, leaving an abandoned valley, or 'wind gap', with its floor marked by older fluvial deposits. The beheaded stream may also be left as a misfit stream as it is now too small for the valley.

The capturing stream may be eroding more aggressively for a variety of reasons, which may act singly or in combination. For example, it may have a higher discharge due to differences in precipitation and vegetation cover, as might occur on either side of a mountain range, or climate change may have increased local rainfall. It may have a steeper gradient, possibly due to a change in base level, which will increase its erosive potential, or it may be cutting through a less resistant lithology. Capture takes place as a result of progressive headward erosion by the capturing stream, with headwater channels cutting into the hillside so that valleys and channels extend towards and eventually beyond a drainage divide. Many rivers or streams that have experienced capture as part of their evolution exhibit an abrupt change in channel direction at the point of capture, frequently of the order of 90°, termed the 'elbow of capture'.

River capture has been important in the evolution of many of the world's river systems. For example, the easternmost tributary of the Indus was captured by the Ganges in geologically recent times, diverting drainage from a large area of the Himalayas from Pakistan to India. Understanding and reconstructing sequences of river capture may be of economic importance. For example, if an ancient river system contained mineral placer deposits but has been beheaded, it will be necessary to be able to trace the old river course to find areas of abandoned alluvial fill in headwater regions.   DJN

stream ordering   See ORDER, STREAM.

stream power   The rate of energy expenditure in flowing water. The energy possessed by flowing water is held by virtue of elevation above some BASE LEVEL towards which the water can flow, and the elevation of water above this in turn is derived from the solar-driven atmospheric processes that lift water vapour and deliver precipitation over the land. Along a stream, potential energy is progressively transformed into other forms, notably the kinetic energy of the flowing mass, together with energy dissipated as frictional heat, sound and in moving sediment particles.

Some workers have employed the total stream power per unit length of channel, $1/2$, as a useful measure of the ability of a stream to do landscape work (e.g. Graf, 1983). Total power is given by

$$\frac{1}{2} = \rho g Q s$$

where ρ is the density of water, g is the gravitational acceleration, Q is discharge and s is the energy slope, often approximated as the channel gradient.

As originally presented by R. A. Bagnold, however, stream power was expressed in terms of amount of energy expended per unit area of the bed, a measure that was sought for studies seeking to explain sediment transport. This was written

$$\omega = \rho g Q \frac{s}{w}$$

so that

$$\omega = \rho g d v s$$

in which d is flow depth and v is flow velocity (and $Q = wdv$). The value of ω ranges from $<1\,\mathrm{J\,m^{-2}\,s^{-1}}$ in interrill flow (Parsons et al., 1998) to $>12,000\,\mathrm{J\,m^{-2}\,s^{-1}}$ in riverine flood flows (Rajaguru et al., 1995), the latter sufficient to move boulders metres in diameter.

Another widely used way to express the stream power is in terms of the power per unit weight of water. This is termed unit stream power and is written

$$\omega^* = \rho g \frac{Qs}{wd} = \rho g v s$$

Various attempts have been made to employ stream power in sediment transport equations (Yang and Stall, 1976; Bagnold, 1977), where it forms an intuitively appealing alternative to the many other flow parameters that have been evaluated in attempts to understand sediment motion, which include stream discharge, velocity and bed shear stress. The stream power approach has also been applied to shallow overland flows (e.g. Moore and Burch, 1986).   DLD

References
Bagnold, R.A. (1977) Bed load transport by natural rivers. Water Resources Research, 13, 303–312. · Graf, W.L. (1983) Downstream changes in stream power in the Henry Mountains, Utah. Annals of the Association of American Geographers, 73, 373–387. · Moore, I.D. and Burch, G.J. (1986) Sediment transport capacity of sheet and rill flow: application of unit stream power theory. Water Resources Research, 22, 1350–1360. · Parsons, A.J., Stromberg, G.L. and Greener, M. (1998) Sediment-transport competence of rain-impacted interrill overland flow. Earth Surface Processes and Landforms, 23, 365–375. · Rajaguru, S.N., Gupta, A., Kale, V.S., et al. (1995) Channel form and processes of the flood-dominated Narmada River, India. Earth Surface Processes and Landforms, 20, 407–421. · Yang, C.T. and Stall, J.B. (1976) Applicability of unit stream power equation. Journal of the Hydraulics Division, American Society of Civil Engineers, 102 (HY5), 559–568.

streamline   A line whose tangent at any point in a fluid is parallel to the instantaneous velocity of the fluid at that point. A map of streamlines gives an instantaneous 'snapshot' of the flow. Alternatively, one may think of such a map as being one frame in a moving film of the flow. The streamline pattern changes with time. Only in a steady-state flow do the streamlines coincide with the trajectories of the fluid particles.   BWA

strength See INTACT STRENGTH and MASS STRENGTH.

strength equilibrium slopes These are formed on exposed bedrock that has an inclination adjusted to the MASS STRENGTH of the rock. This type of slope is controlled by processes of erosion and by geomorphic resistance operating at a scale of individual joint blocks. A distinction is made between (1) equilibrium slopes and (2) those that are formed with critical angles for stability dipping out of the slope - these slopes fail by large-scale landsliding along the critical joints and have forms controlled by this process.   MJS

Reading
Selby, M.J. (1982) Controls on the stability and inclinations of hillslopes formed on hard rock. *Earth Surface Processes and Landforms*, 7, 449–467.

stress Produced by a system of forces in equilibrium tending to produce STRAIN in a body.

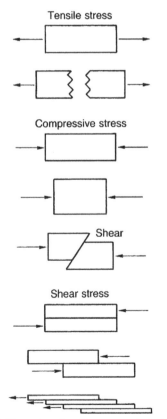

Schematic illustration of tensile, compressive and shear stresses.
Source: Summerfield (1991). Reproduced with permission of John Wiley & Sons.

Stress can be produced in tension or compression, by hydrostatic pressure or by shear stress. The units of stress are (force per unit area) newtons per square metre ($N m^{-2}$). In most geomorphological examples the forces will give values as kilonewtons per square metre or the equivalent kilopascals (kPa) where 1 Pa is equal to 1 $N m^{-2}$. Tensile stress is an extensional force that tends to stretch or pull material apart. Compressive stress is a force that tends to compress material and thereby change its shape. Shear stress is a force that deforms a mass of material by one part sliding over another along one or more failure plains.   WBW

Reading and Reference
Summerfield, M.A. (1991) *Global geomorphology*. London/New York: Longman/John Wiley & Sons, Inc. · Whalley, W.B. (1976) *Properties of materials and geomorphological explanation*. Oxford: Oxford University Press.

striated soil (Also called needle ice, striped ground and striated ground.) Consists of a miniature pattern characterized by a distinct alignment of the surface soil particles. The orientation of the stripes does not necessarily coincide with slope gradient. Wind direction and the alignment of the early Sun's rays may play a role. NEEDLE ICE is the predominant formative process.   ASG

striation Scratches etched onto a rock surface by the passage over it of another rock of equal or greater hardness. Striations are characteristic of erosion by glaciers but may also occur beneath snow patches (Jennings, 1978) and on coasts affected by sea ice (Laverdière *et al.*, 1981; Hansom, 1983). Glacial striations are generally up to a few millimetres in width and rarely more than a metre in length. Larger striations grade into grooves. Striations are best displayed on rock surfaces that face up-ice, mainly because pressure melting in these locations forces the rock tools against the bedrock.   DES

References
Hansom, J.D. (1983) Ice-formed intertidal boulder pavements in the Sub-Antarctic. *Journal of Sedimentary Petrology*, 53, 135–145. · Jennings, J.N. (1978) The geomorphic role of stone movement through snow creep, Mount Twynam, Snowy Mountains, Australia. *Geografiska Annaler: Series A, Physical Geography*, 60, 1–8. · Laverdière, C., Guimont, P. and Dionne, J.C. (1981) Marques d'abrasion glacielles en milieu littoral Hudsonien. Québec Subarctique. *Géographie physique et Quaternaire*, 35, 269–275.

Strickler equation Strickler (1923) analysed data from Swiss gravel-bed rivers lacking bed undulations to develop an equation permitting estimation of the ROUGHNESS coefficient n in the MANNING EQUATION from the measured bed material particle size. It can be shown theoretically that

$$n = 0.0132k_s^{1/6}$$

where k_s is a grain roughness height in millimetres. Strickler shows that if k_s is taken as the median grain diameter D_{50}, this relationship becomes

$$n = 0.0151D_{50}^{1/6}$$

This can be used to estimate the Manning coefficient if grain roughness is the primary source of flow resistance, in wide, flat-bed gravel-floored channels. However, it is often preferable to define a friction coefficient in terms of a depth : grain size ratio, since the resistance of a grain is partly dependent on the water depth covering it (Richards, 1982: 66).   KSR

References
Richards, K.S. (1982) *Rivers: form and process in alluvial channels*. London: Methuen. · Strickler, A. (1923) Beiträge zur Frage der Geschwindigkeitsformel und der Rauhigkeitszahlen für Strome, Kanäle und geschlossene Leitungen. Mitteilungen des Eidgenössischen Amtes für Wasserwittschaft 16. Bern: Amtes für Wasserwittschaft.

strike   The direction of a horizontal line drawn in the same plane as strata are bedded but at right angles to their dip.

string bog   An area of waterlogged land characterized by ridges of peat separated by water-filled troughs.

stromatolite (stromatolith)   A term first used by E. Kalkowsky (1908) to describe some sedimentary structures in the Bunter of North Germany. A currently favoured definition (Walter, 1976: 1) is that stromatolites are 'organosedimentary structures produced by sediment trapping, binding and/or precipitation as a result of the growth and metabolic activity of microorganisms, principally cyanophytes'. They can develop in marine, marsh and lacustrine environments and, though they form today where conditions permit, they reached the acme of their development in the Proterozoic (Hofmann, 1973). The largest known forms are mounds several hundreds of metres across and several tens of metres high. Gross morphologies vary in the extreme and range from stratiform crustose forms, through nodular and bulbous mounds and spherical oncoids, to long slender columns, erect to inclined, and with various styles of branching.   ASG

References
Hofmann, H.J. (1973) Stromatolites: characteristics and utility. *Earth Science Reviews*, **9**, 339–73. · Kalkowsky, E. (1908) Oolith und Stromatolith im norddeutschen Buntsandstein. *Zeitschrift der Deutschen Geologische Gesellschaft*, **60**, 68–125. · Walter, M.R. (ed.) (1976) Stromatolites. *Developments in Sedimentology*, **20**, 1–790.

Strouhal number   A dimensionless number relating the frequency of vortex shedding f from a flow obstruction with a diameter (or other characteristic length) l and the flow speed V: the Strouhal number $S = fl/V$. For a range of Reynolds numbers from about 103 to 105, S is about 0.2. The Strouhal number has applications for the study of flow over BEDFORMS, for example.   DJS

Reading
Schlichting, H. (1979) *Boundary-layer theor*, 7th edition. New York: McGraw-Hill.

sturzstrom   Very large rock avalanches (with volumes $>5 \, \mathrm{Mm}^3$) initially falling or avalanching from high cliffs may develop low coefficients of internal friction and, therefore, travel large horizontal distances (5–30 km) at velocities of 90–350 km h^{-1}. They are the most powerful forms of mass-wasting and may be major, but rare, causes of erosion in very high mountain ranges and on large volcanoes. These very large avalanches were named *sturzstroms* by Hsü (1975).   MJS

Reading and Reference
Hsü, K.J. (1975) Catastrophic debris streams (sturzstroms) generated by rockfalls. *Bulletin of the Geological Society of America*, **86**, 129–140. · Selby, M.J. (1993) *Hillslope materials and processes*, 2nd edition. Oxford: Oxford University Press; chapter 14.

subaerial   Occurring or existing at the land surface.

suballuvial bench   The lower portion of a rock pediment where it is overlain by alluvial sediments.

Sub-Atlantic   See BLYTT-SERNANDER MODEL.

Sub-Boreal   See BLYTT-SERNANDER MODEL.

subclimax   Any plant community related to and closely preceding the true climax community for an area. Usage is normally in the sense of a stable community resembling the climax but prevented from developing towards it by some disturbance or other arresting factor. If the arresting factor is removed, a subclimax is expected to proceed to the climax stage. The term is also used simply for a long-persisting SERAL COMMUNITY that appears to be climax. Many subclimaxes are the result of the activities of humans and domesticated animals, particularly burning and grazing. (See also CLIMAX VEGETATION, DISCLIMAX and MONOCLIMAX.) JAM

Reading
Eyre, S.R. (1966) *Vegetation and soils: a world picture.* London: Edward Arnold. · Oosting, H.J. (1956) *The study of plant communities.* San Francisco, CA: W.H. Freeman.

subduction zones These occur where oceanic lithosphere is consumed into the mantle at convergent plate margins. The inclined plane, where the subducting oceanic lithospheric plate slides past the overlying mantle, is seismically active and this inclined plane is called a Benioff zone. Subduction zones involve the destruction of oceanic plate material, and this is complementary to the formation of new oceanic plate at constructive plate margins (active ocean ridges); this allows the Earth to maintain a constant surface area.

There are two main types of subduction zone: (1) island arcs, where oceanic lithosphere is subducted beneath oceanic lithosphere (e.g. the Tonga island arc in the Pacific Ocean) and (2) active continental margins, where oceanic lithosphere is subducted beneath continental lithosphere (e.g. the Andes along the western coast of South America). Subduction zones are characterized by an oceanic trench, and the arcuate plan of the trench and distribution of volcanic islands in an island arc are a function of the geometry of pushing part of the surface of a sphere inwards into its interior. The subducted oceanic lithosphere remains coherent as a distinct relatively rigid slab for depths of up to 700 km (where the deepest earthquakes are recorded) before being consumed into the mantle. The subducted oceanic crust, which forms the uppermost part of the slab, undergoes progressive dehydration as it moves down, and this promotes melting in the overlying mantle wedge. These melts segregate as ascending magmas that give rise to arc volcanoes, and subduction zones are associated with typically explosive andesitic volcanism. Magmatism at active continental margins can also involve the emplacement of large granitic BATHOLITHS. AD

subglacial The environment beneath a glacier.

subglacial bedforms A generic term for the range of longitudinal and transverse landforms produced at the base of a GLACIER or ICE SHEET as a result of active ice flow across a sediment base. Whilst individual types of subglacial landforms have often been investigated in isolation, there has been an increasing realization that they all belong to a single family of related landforms, best thought of as a continuum of subglacial bedforms (Rose, 1987). This view uses the concept of a single set of processes that may lead to a wide range of landform shapes and characteristics, only some of which conform to our 'classical text book' landform types such as the DRUMLIN. As with aeolian and fluvial BEDFORMS, we would expect a combination of distinctive landforms and a range of intermediate morphologies.

Specific landform types that are part of the subglacial bedform system are FLUTES, drumlins and mega-scale glacial lineations, which are all longitudinally streamlined landforms at different scales from metres to tens of kilometres and are commonly referred to as glacial lineations. ROGEN MORAINES or RIBBED MORAINES are also subglacial bedforms but are sculptured ridges formed transverse to the ice flow direction. A strong characteristic of all subglacial bedforms is that they occur in discrete and fairly well defined fields or swarms comprising hundreds to thousands of individuals, and that different types, such as drumlins and rogen moraines, are often found in close association with each other.

An alternative school of thought considers that subglacial bedforms arise from large subglacial flood events, with water rather than ice being the shaping medium. (See DRUMLIN for further details.) CDC

Reading and Reference
Benn, D.I., and Evans, D.J.A. (2010) *Glaciers and glaciation,* 2nd edition. Abingdon: Hodder. · Rose, J. (1987) Drumlins as part of a glacier bedform continuum. In J. Menzies and J. Rose (eds), *Drumlin symposium.* Rotterdam: Balkema; pp. 103–116.

subhumid One of the five basic humidity provinces recognized in Thornthwaite's (1948) climate classification. The classes are defined by calculating a PRECIPITATION efficiency (P/E) index, which is the sum of 12 monthly values of the ratio of mean precipitation to mean EVAPORATION. Subhumid falls in the middle of the range so defined, with a P/E value of 32–63 and characterizes areas of grassland-type vegetation. The savanna regions of Africa that experience an extensive dry season every year fall in this category. RR

Reference
Thornthwaite, C.W. (1948) An approach to a rational classification of climate. *Geographical Review*, **38**, 55–94.

sublimation   The process of direct deposition of atmospheric water vapour on to an ice surface or evaporation from an ice surface. Cirrus clouds are sometimes formed as a result of a direct phase change from water vapour to ice crystals.   RR

sublittoral   The area of the seas between the intertidal zone and the edge of the continental shelf. Also the deeper parts of a lake in which plants cannot root.

submarine canyon   A canyon-like valley form cut into the CONTINENTAL SHELF, continental slope or continental rise. Its resemblance to a subaerial valley extends to the existence of tributary valleys and the presence of knickpoints along its profile. It can extend beyond the continental slope almost as far as the deep ocean floor. Most submarine canyons contain sediments apparently deposited by high-density flows initiated by submarine slides known as turbidity currents. These are probably capable of preventing the canyons from filling with sediment, but the initial formation of most submarine canyons is probably attributable to fluvial erosion prior to subsidence.   MAS

Reading
Barnes, N.E., Bouma, A.H. and Normark, W.R. (1985) *Submarine fans and related turbidity systems*. New York: Springer-Verlag. · Shanmugan, G. and Miola, R.J. (1988) Submarine fans: characteristics, models, classification and reservoir potential. *Earth-Science Reviews*, **24**, 383–428. · Shepard, F.P. and Dill, R.F. (1966) *Submarine canyons and other sea valleys*. Chicago, IL: Rand McNally.

subsequent stream   A stream that follows a course determined by the structure of the local bedrock.

subsere   A secondary successional sequence of plant communities. It is a series of community stages, the result of SUCCESSION on incompletely bared surfaces, or beginning with a community not truly climax in status (e.g. a SERAL COMMUNITY, SUBCLIMAX or DISCLIMAX). The essential characteristic of any subsere is its initiation from at least the vestige of a previous community at the site. This may involve residual species, seedlings, existing soil with its seed bank, or a complete community from which some controlling factor has been removed. In many instances, therefore, the subsere may be viewed as a manifestation of the recovery process in damaged ecosystems.   JAM

Reading
Cairns, J. (ed.) (1980) *The recovery process in damaged ecosystems*. Ann Arbor, MI: Ann Arbor Science. · Fontaine, R.G., Gomez-Pompa, A. and Ludlow, B. (1978) Secondary successions. In *Tropical forest ecosystems: a state of knowledge report prepared by UNESCO/UNEP/FAO*. Paris: UNESCO; pp. 216–232.

subsidence   Land-surface sinking resulting from such processes as the withdrawal of groundwater, geothermal fluids, oil and gas; the extraction of coal, salt, sulphur and other solids through mining; the hydrocompaction of sediments; oxidation and shrinkage of organic deposits (notably peats); the development of thermokarst in areas underlain by permafrost; and karstic collapse. ASG

Reading
Johnson, A.I. (ed.) (1991) *Land subsidence*. IAHS Publication No. 200. Wallingford: IAHS Press.

succession   The sequential change in both the form and composition of an ecological community over time. The concept was originally developed in North America by, among others, H. C. Cowles (1901), but F. E. Clements, whose monograph entitled *Plant succession: an analysis of the development of vegetation* was published in 1916, is regarded as most influential in the consolidation and widespread adoption of the concept. Clements observed changes in lakeshore dune vegetation with time and developed an explanation and corresponding comprehensive system of terminology to describe the process. Clements defined succession as a sequence ('sere') of plant communities ('seral stages') characterized by increasing complexity and culminating in the CLIMAX VEGETATION. The model recognized various 'subclimaxes' in which, because of prevailing local conditions such as soil chemistry, succession was delayed, although it would proceed eventually to the true climax. Human activity was seen to influence the course of succession and could result in a 'deflected sere' with a 'plagioclimax', while subsequent 'secondary succession' might ultimately restore the regional climax vegetation with time.

A key element of the theory is the emphasis on reaction; the establishment of later colonizers is facilitated by previous occupation of a site by species that modify their habitat to the extent that they bring about their own replacement. Attractively simple though the concept is, and it was certainly widely adopted, there are numerous inherent problems.

Most fundamental of these is the idea that succession is progressive, deterministic and, in following a regular sequence and ultimately producing a climax community, predictable. One development of the initial idea mooted that uniform climax communities do not occur across large areas and, alternatively, there is a mosaic of different forms known as the POLYCLI-MAX. This extension of the theory fails to explain the really dynamic nature of vegetation in time and space. Although vegetation change per se is undeniable, the ecological critique of Clements' concept has centred on the reality that vegetation communities are fundamentally dynamic entities. It is now clearly recognized that different species assemblages develop through time in any climatic region, forming a complex mosaic pattern depending on age since disturbance, on intrinsic and extrinsic ecological factors and on particular site characteristics.   MEM/TS

Reading and References
Burrows, C.J. (1990) *Processes of vegetation change.* London: Unwin Hyman. · Clements, F.E. (1916) *Plant succession: an analysis of the development of vegetation.* Washington, DC: Carnegie Institute Publications 242. · Cowles, H.C. (1901) The physiographic ecology of Chicago and vicinity. *Botanical Gazette*, **31**, 73–108, 145–182. · Luken, J.O. (1990) *Directing ecological succession.* London: Chapman and Hall.

suction   Describes the energy state of soil moisture under unsaturated conditions. Soil moisture suction is often called the matric potential or capillary potential of the soil moisture, and it is described as a suction because it is a negative pressure potential resulting from the capillary and adsorptive forces due to the soil matrix holding moisture in the soil at a pressure less than atmospheric pressure. The magnitude of soil moisture suction can be indicated by a negative head of water (or mercury) that can be determined using a tensiometer and which is often expressed as a pF value. (See also CAPILLARY FORCES.)   AMG

Suess effect   The relative change in the $^{14}C/C$ or $^{13}C/C$ ratio of any carbon pool or reservoir caused by the addition of fossil-fuel carbon dioxide (CO_2) to the atmosphere. Fossil fuels are devoid of ^{14}C because of the radioactive decay of ^{14}C to ^{14}N during long underground storage and are depleted in ^{13}C because of isotopic fractionation eons ago during photosynthesis by the plants that were the precursors of the fossil fuels. CO_2 produced by the combustion of fossil fuels is thus virtually free of ^{14}C and depleted in ^{13}C. The term 'Suess effect' originally referred to the dilution of the $^{14}C/C$ ratio in atmospheric CO_2 by the admixture of fossil-fuel-produced CO_2, but the definition has been extended to both the ^{14}C and

^{13}C ratios in any pool or reservoir of the carbon cycle resulting from human disturbances.   ASG

Reading
Keeling, C.D. (1979) The Suess effect: ^{13}carbon–^{14}carbon interrelations. *Environment International*, **2**, 229–300.

suffosion   An erosional process occurring in areas where limestone bedrock is overlain by unconsolidated superficial materials. The sediments slump down into widened joints and cavities in the bedrock surface, producing an irregular land surface. It has been likened to an egg-timer effect.

sulphation   The reaction between materials containing calcium carbonate and sulphur dioxide in humid atmospheres. The sulphur dioxide is oxidized to sulphur trioxide in a reaction that can be catalysed, for example, by vanadium oxide produced by internal combustion or by iron oxides on the material surface. Sulphur trioxide then reacts with the calcium carbonate to form gypsum and carbon dioxide.

$$SO_2(g) + \tfrac{1}{2}O_2(g) \rightarrow SO_3(g)$$
$$CaCO_3(s) + SO_3(g) + 2H_2O(g) \rightarrow$$
$$CaSO_4 \cdot 2H_2O(s) + CO_2(g)$$

This reaction particularly occurs on calcareous stonework in polluted urban environments that are protected from rainwash. A layer of gypsum can form on these surfaces that can incorporate combustion particles to form a *black crust*. Gypsum from these films may be washed into underlying porous stonework by periodic wetting and can contribute to stone disruption by *salt weathering* mechanisms.   BJS

summit plane   See EROSION SURFACE.

sunspots   Vortex-like temporary disturbances with large associated magnetic fields that afflict the Sun. They are relatively dark regions on the disc of the Sun, with an inner 'umbra' of effective radiation temperature about 4500 K and an outer 'penumbra' of somewhat higher temperature. Varying in size from a few kilometres to more than 150,000 km in diameter, sunspots are associated with solar flares and so-called magnetic storms, a term originally coined by Alexander von Humboldt. Sunspot frequency is quasi-periodic, with an average period of around 11 years. Various relationships between sunspot activity and climatic fluctuations have been proposed, and the Maunder Minimum, a period of significantly reduced sunspot activity between about AD

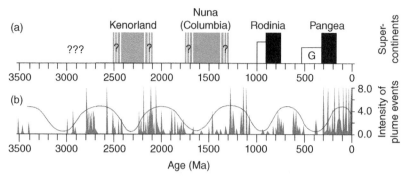

Distribution through time of supercontinents.
Source: Revised from Li and Zhong (1979). Reproduced with permission of Elsevier.

1645 to 1715 has been proposed to have contributed to the phase of global cooling known as the LITTLE ICE AGE.   MEM

supercontinent An assemblage of two or more continents. The present continents are North America, Eurasia, Africa, Australia, Antarctica and South America. Because of plate tectonics, this arrangement is temporary. In the late Palaeozoic, the four latter continents were joined together (with Arabia and India) as the supercontinent of Gondwana; North America and Eurasia were joined as Laurasia. At least from 260 Ma to 180 Ma (million years) ago, these two were joined as Pangaea (also Pangea), incorporating most of the continental crust on Earth. Momentous though Pangaea's rifting apart to form the present continents was, it is unlikely to have been a unique event, since it is now known that plate tectonic processes started in the first half of the 4600 Ma history of the Earth. Evidence exists for earlier supercontinents about 600 Ma ago (Gondwana), 1100 Ma (Rodinia) and before.

Supercontinents are likely to be metastable because of the amount of geothermal heat trapped in the underlying mantle: heat escapes more easily through (thin) oceanic crust than through (thick) continental crust. Build-up of heat leads to thermal uplift, active rifting and fragmentation, creating new ocean basins by seafloor spreading. The fragmented continents move apart, only to collide with each other on the other side of the Earth, where a new supercontinent is formed. A further instability is that as oceanic crust moves away from the mid-ocean ridges where it is generated it cools and becomes denser. No present oceanic crust is older than 200 Ma, and it may be that the asthenosphere cannot support the weight of crust any denser; this may start a new subduction zone, turning a

passive margin into an active margin and reversing the enlargement of the spreading ocean. The hypothetical supercontinent cycle seems to take about 500 Ma, and there may have been as many as seven cycles in the Earth's history. This has many implications for mountain building, glaciation, volcanism, sea level, ocean currents and climatic change.

Indeed, the assembly and break-up of supercontinents has been suggested to have contributed to burst of photosynthesis in the oceans, and hence to the rise in the level of oxygen in the Earth's atmosphere (Campbell and Allen, 2008). Emplacement of kimberlites and their contained diamonds has also been linked to the development of supercontinents (Torsvik *et al.*, 2010).   DLD

Reading and references
Campbell, I.H. and Allen, C.M. (2008) Formation of supercontinents linked to increases in atmospheric oxygen. *Nature Geoscience*, 1, 554–558. · Condie, K.C. (1997) *Plate tectonics and crustal evolution*, 4th edition. Oxford: Butterworth-Heinemann. · Dalziel, I.W.D. (1995) Earth before Pangea. *Scientific American*, 272, 58–73. · Li, Z.X. and Zhong, S. (1979) Supercontinent–superplume coupling, true polar wander and plume mobility: plate dominance in whole-mantle tectonics. *Physics of the Earth and Planetary Interiors*, 176, 143–156. · Murphy, J.B. and Nance, R.D. (1992) Mountain belts and the supercontinent cycle. *Scientific American*, 266, 84–91. · Torsvik, T.H., Burke, K., Steinberger, B., *et al.* (2010) Diamonds sampled by plumes from the core-mantle boundary. *Nature*, 466, 352–357.

supercooling Denotes the presence of a substance in the liquid phase at a temperature that is below its normal freezing point. It is a common occurrence in the atmosphere, where cloud droplets often occur in the supercooled state down to −20 °C and even down to −35 °C in rare cases.   RR

superimposed drainage   The pattern of a drainage network that developed on a landscape or bedrock that has since been removed by erosion, the network being preserved on the new land surface. With antecedence, it is one of the main reasons why drainage may appear to be unadjusted to present structures.

superimposed ice   Formed when water comes into contact with a cold glacier surface and freezes. Under such circumstances it is a process of glacier accumulation, and it is common in dry continental climates, such as northern Canada, where 90% of the glacier ice may have formed in this way. Here, summer temperatures are high enough to melt the winter snow, but it refreezes when it comes into contact with the glacier surface, which has been chilled by low winter temperatures.   DES

Reading
Koerner, R.M. (1970) The mass balance of the Devon Island ice cap, NWT, Canada, *Journal of Glaciology*, **9** (57), 325–336.

superposition, law of   The law that states that the upper strata in sedimentary sequences postdate those that they overlie.

supersaturation   A metastable state occurring when a solution contains more of a solute than is necessary to saturate it. The term is usually applied to a solution of limestone and is important in the process of SPELEOTHEM growth in caves.

supervised classification   A form of numerical analysis that groups together cases with similar properties into a set of predefined classes. This type of analysis is commonly used to generate a THEMATIC MAP from imagery acquired by REMOTE SENSING. The classification is led by the analyst, who specifies the set of possible classes to which cases may be allocated. The process can be considered to comprise three stages: training, allocation and testing. In the training stage, each class is characterized. For example, the analyst may identify areas of known class membership within a satellite sensor image and use these to derive descriptive statistics for each class. These training statistics are then used in the allocation stage to allocate each case (e.g. image PIXEL) to the class with which it has greatest similarity. In essence, the allocation stage converts the remotely sensed image that depicts the spectral response of the terrain into a THEMATIC MAP that depicts the spatial distribution of the classes of interest to the analyst.

In the final stage, testing, the ACCURACY of the derived classification is evaluated.

The analysis is often undertaken on a per-pixel basis, which makes an implicit assumption that each pixel represents a homogeneous region of a single class. This may be the case when pixels are small. In such situations, pixels may also be grouped together to form more natural objects (e.g. fields) and an object-based classification undertaken. As the pixel size and/or landscape complexity increases, the proportion of pixels representing heterogeneous regions increases. This can be a problem, as standard classification methods assume each pixel represents a pure area (i.e. a single land-cover class), but mixed pixels, which contain two or more classes, may be common. The composition of mixed pixels may sometimes be estimated using unmixing or soft classification techniques and mapped through a super-resolution analysis.   GF

Reading
Lillesand, T.M., Kiefer, R.W. and Chipman, J.W. (2008) *Remote sensing and image interpretation*, sixth edition. New York: John Wiley & Sons, Inc. · Tso, B. and Mather, P.M. (2009) *Classification methods for remotely sensed data*, second edition. Boca Raton, FL: CRC Press.

supraglacial   See ENGLACIAL.

surf   Surf is formed on coasts as waves initially break on the shore. It consists of complex, transitional forms of waves and develops in the surf zone on beaches, between the breaker and swash zones. The surf zone is also where longshore currents develop.

surface detention   The part of precipitation that remains in temporary storage during or immediately after a storm as it moves downslope by overland flow. The low flow speed of overland flow, especially through grass or organic litter, or over rough soil surfaces, means that significant volumes of water can be detained in transit and ultimately, perhaps, reach the slope foot sometime after the end of rainfall. The majority of surface detention water will form a part of the storm hydrograph, but some of the water may infiltrate the soil or may be evaporated after the storm ends. Detention storage is distinct from depression storage, in that the latter is composed of small stores of non-flowing water trapped in small ponds and surface depressions (See also DEPRESSION STORAGE and SURFACE STORAGE.)   AMG/DLD

Reading
Antoine, M., Chalon, C., Darboux, F., *et al.* (2012) Estimating changes in effective values of surface detention,

depression storage and friction factor at the interrill scale, using a cheap and fast method to mold the soil surface microtopography. *Catena*, **91**, 10–20.

surface run-off See RUN-OFF.

surface storage Water stored on the ground surface within a drainage basin. In some circumstances the volume of water stored in this way may be very large, so surface storage is of particular importance in drainage basins containing large reservoirs, natural lakes or swamps. Frozen surface water in the form of ice and snow also forms part of surface storage. In relation to the Horton (1933) model of run-off generation, surface storage is considered to be the sum of DEPRESSION STORAGE and SURFACE DETENTION (Chorley, 1978).   AMG

References
Chorley, R.J. (1978) The hillslope hydrological cycle. In M. J. Kirkby (ed.), *Hillslope hydrology*. Chichester: John Wiley & Sons, Ltd. · Horton, R.E. (1933) The role of infiltration in the hydrological cycle. *Transactions of the American Geophysical Union*, **14**, 446–460.

surface tension The force required per unit length to pull the surface of a liquid apart. It is the result of the change in orientation of molecular bonds as the interface with another substance is approached. A molecule beneath the surface of a liquid is attracted in all directions by surrounding molecules within its 'sphere of molecular attraction', and so the resultant force on it is zero. However, a molecule on the surface of the liquid has a resultant force towards the main body of the liquid because there are more molecules within the part of its 'sphere of molecular attraction' within the liquid than there are in the part that falls within the overlying vapour.   AMG

Reading
Baver, L.D., Gardner, W.H. and Gardener, W.R. (1991) *Soil physics*, 5th edition. New York: John Wiley & Sons, Inc.

surge See STORM SURGES.

surging glacier A glacier that flows at a velocity of an order of magnitude higher than normal. Whereas normal glaciers flow at $3–300\,\mathrm{m\,a^{-1}}$, surging glaciers may flow at velocities of $4–12\,\mathrm{km\,a^{-1}}$. Some glaciers flow permanently at surging velocities, as is the case of the Jakobshavn Isbrae in West Greenland, which, nourished by the Greenland Ice Sheet, flows at $7–12\,\mathrm{km\,a^{-1}}$. Others experience periodic surges whereby a wave of ice moves downglacier at velocities of $4–7\,\mathrm{km\,a^{-1}}$ and may represent a velocity of 10–100 times higher than the pre-surge velocity. In such cases, the jump from normal to surging flow takes place very suddenly. The wave of fast-moving ice may plough into formerly stagnant ice near the glacier margin or extend beyond it. The wave is associated with strong COMPRESSING FLOW in front and extending flow behind and, following its passage, has the effect of lowering the glacier long profile. After a surge, the glacier experiences a quiescent phase, often of several decades, while it builds up to its original profile before surging once more.

One possible explanation of surging behaviour is that there are two modes of basal sliding: one normal and one fast. Budd (1975) has related the discharge of glaciers to glacier velocity and suggested that large outlet glaciers from big ice sheets have sufficient ice supply to maintain the fast mode of sliding permanently. Normal glaciers only have enough ice to maintain normal flow. Periodically surging glaciers occupy an intermediate position and may have enough ice supply for them to be able to cross the threshold from normal to fast flow, but then they cannot provide enough ice to maintain the fast mode of flow and so revert to normal flow. The two modes of flow probably relate to water thicknesses beneath the glacier and the development of cavities (Hutter, 1982).   DES

Reading and References
Budd, W.F. (1975) A first simple model for periodically self-surging glaciers. *Journal of Glaciology*, **14**, 3–21. · Hutter, K. (1982) Glacier flow. *American Scientist*, **70**, 26–34. · Sharp, M. (1988) Surging glaciers: behaviour and mechanisms. *Progress in Physical Geography*, **12**, 349–370.

susceptible drylands Those DRYLANDS areas that are susceptible to reduced productivity through the impact of unsustainable land-use practices and other human actions. In the context of DESERTIFICATION, the term has been applied within the United Nations Environment Programme (UNEP) (see Middleton and Thomas (1997)) to dry–SUBHUMID, SEMI-ARID and arid environments, HYPER-ARID areas; the true deserts, are omitted, since they have extremely low natural productivity. Thus, with only a few exceptions, human actions are unlikely to make productivity even lower than natural levels (and, therefore, conditions even more 'desert like').   DSGT

Reference
Middleton, N.J. and Thomas, D.S.G. (1997) *World atlas of desertification*, 2nd edition. London: UNEP/Edward Arnold.

suspended load Sediment transported by a river in suspension. The material is carried within the body of flowing water, with its weight supported by the upward component of fluid turbulence. Particles are commonly less than 0.2 mm in diameter, and in many rivers the suspended load will be dominated by silt- and clay-sized particles (i.e. <0.062 mm diameter). This fine-grained material is frequently referred to as wash load and is supplied to the river by erosion of the catchment slopes. The coarser particles usually represent suspended bed material and are derived from the channel perimeter. Transport rates are supply controlled and the suspended load is therefore a noncapacity load. Measurements of suspended sediment concentration expressed in milligrams per litre or kilograms per cubic metre are required to calculate transport rates. (See also SEDIMENT YIELD.)   DEW

sustainability Often used in environmental contexts with respect to human use of natural resources (soil, biota, etc.) that does not alter or compromise the future existence (or, somewhat selfishly, human use of) that feature. Sustainability can be defined in a plethora of ways, and it is in economics and the wider social sciences. These can include monetary values, value in terms of wider sets of attributes (e.g. in terms of the landscape this can include aesthetic value), as well as a temporal component. Unfortunately, we tend to hear the term *unsustainability* more commonly that sustainability, with respect to forest use and the fate of species (e.g. tigers and elephants, as well as more mundane but vital creatures). Human impacts on the atmosphere through CO_2 emissions, linked to GLOBAL WARMING, are also closely linked to sustainability debates.

In 2001 the International Council for Science and other bodies embraced the scientific context of sustainability by proposing a new discipline, 'sustainability science' (Kates *et al.*, 2001; Clark and Dickson, 2003).   DSGT

References

Kates, R., Clark, W., Corell, R., *et al.* (2001) Sustainability science. *Science*, 292, 641–642. · Clark, W.C. and Dickson, N.M. (2003) Sustainability science: the emerging research program. *Proceedings of the National Academy of Science*, 100, 8059–8061.

sustained yield Some resources are, in theory, renewable. They can be infinitely recycled through the biosphere and through human societies, either because they are basically unchanged by their use (e.g. water) or because they are self-regenerating (e.g. plants and animals). Sustained yield is a management concept that aims to regulate the system providing the resources so as to maintain the yield at a desired level into the foreseeable future and perhaps longer. In the case of agroecosystems, for example, it means the maintenance of soil structure and nutrient status, the control of weeds, pests and diseases, and the selection of biota to respond to, for example, climatic change. In the case of whaling, it means avoiding the rate of cull that undermines the species' ability to reproduce themselves, an attempt that has so far been notably unsuccessful. Calculation of rates of sustained yield requires a thorough empirical knowledge of the ecosystem in question, and possibly sophisticated modelling as well. It is made much more complicated when, as in the case of fish populations, for example, the natural condition seems to be one of considerable cyclic fluctuation. The optimum yield for one year, then, may be markedly different from the next, and neither science nor politics may be able to cope with this amplitude of coming and going.   IGS

Reading

Dasmann, R.F., Milton, J.P. and Freeman, P.H. (1973) *Ecological principles for economic development.* Chichester: John Wiley & Sons, Ltd.

Sverdrup A unit of discharge used to describe ocean currents, named for oceanographer H. U. Sverdrup. It is abbreviated sv, and represents a flow of $1,000,000\,\mathrm{m^3\,s^{-1}}$. The discharge of the Gulf Stream, for example, is 18.5 sv.

swale A depression in regions of undulating glacial moraine, a trough between beach ridges produced by erosion or an area of low ground between dune ridges.

swallet A sinkhole. A hole in limestone bedrock that has been produced by solution and through which stream water disappears.

swallow hole (or swallet) A feature through which surface water goes underground in a limestone area. They have local names (e.g. AVEN) and a range of forms, from deep shafts (like Gaping Gill in Yorkshire) to less obvious zones in a stream bed where discharge is lost as a result of downward percolation.   ASG

swamp A type of wetland, dominated by woody plants, and including swamp forests such as mangroves. By comparison, the term *marsh* tends to be reserved for wetland dominated by graminoids, grasses or herbs. See also BOG and MIRE.

sympatry Originating in or occupying the same geographical area. The term sympatric is used to describe species or populations with overlapping geographical ranges that are, therefore, not spatially isolated. Where the ranges are adjacent, but not exactly overlapping, the term used is parapatric. Species isolated are said to be allopatric. When brought together artificially, allopatric species often hybridize quite freely. (See also ISOLATION, ECOLOGICAL.)   PAS

Reading
Stebbins, G.L. (1950) *Variation and evolution in plants.* New York: Columbia University Press. · Stebbins, G.L. (1977) *Processes of organic evolution,* 3rd edition. Englewood Cliffs, NJ: Prentice-Hall.

syncline A trough in folded strata.

synforms See ANTIFORMS.

synoptic climatology An aspect of climatology concerned with the description of local or regional climates in terms of the properties and motions of the atmosphere rather than with reference to arbitrary time intervals, such as months. There are two stages to a synoptic climatological study: the determination of categories of atmospheric circulation type (often referred to as WEATHER TYPE) and the assessment of mean, modal and other statistical parameters of the weather elements in relation to these categories.

Although the first investigations of synoptic patterns were made in the nineteenth century, longer standing popular weather lore had already associated cold, heat or precipitation with particular wind directions in many countries. Modern synoptic climatology developed during and after the Second World War in response to the needs of military operations to assess likely weather conditions (Barry and Perry, 1973). Since then, both subjective and objective means of classifying the totality of weather patterns have been widely used.

With the advent of high-speed computers and digitized grid-point data sets of sea-level pressure and geopotential height fields, there has been a transformation in procedures for preparing catalogues of synoptic types since the 1960s (Perry, 1983). Two approaches have been widely adopted in developing classifications: (1) a determination of pattern similarity based on correlation methods; (2) the use of one of a range of statistical techniques to extract components of the fields, perhaps combined with a clustering approach to obtain pattern types. New types of data – for example, satellite imagery (see SATELLITE METEOROLOGY) – are now being employed in the classification process and are helping to extend the synoptic approach to the tropics, where hitherto it has been little used.

Classification catalogues exist for several areas of the world for the available period of synoptic weather maps (Smithson, 1986). In addition to being used in the description and analysis of such persistence and recurrence models as singularities, weather spells and natural seasons, they have found applications in studies of air chemistry, climatic fluctuations and the reconstruction of past climatic states. Synoptic climatology can also serve as an important check on computer-derived numerical climate studies, and because the method relates local climate conditions to the atmospheric circulation it provides a realistic basis for much climatological teaching and project work.   AHP

Reading and References
Barry, R.G. and Carleton, A.M. (2001) *Synoptic and dynamic climatology.* London: Routledge. · Barry, R.G. and Perry, A.H. (1973) *Synoptic climatology: methods and applications.* London: Methuen. · Perry, A.H. (1983) Growth points in synoptic climatology. *Progress in Physical Geography,* 7, 90–96. · Smithson, P.A. (1986) Dynamic and synoptic climatology. *Progress in Physical Geography* 12, 119–129.

synoptic meteorology An aspect of meteorology concerned with the description of current weather and the forecasting of future weather using synoptic charts that provide a representation of the weather at a particular time over a large geographical area.

Brandes was the first to develop the idea of synoptic weather mapping by comparing meteorological observations made simultaneously over a wide area. It was not until the invention of the electric telegraph that the rapid preparation of a map of weather observation became possible; the first British daily weather map was sold at the Great Exhibition of 1851. The loss of numerous lives at sea encouraged the issue of regular gale warnings by Admiral Fitzroy, who was appointed to the first official meteorological post in the UK. From the 1860s onwards national weather services were established in many countries, and by the 1890s the first upper air soundings gave a better understanding of the vertical structure of weather systems (Bergeron, 1981). Just after the First World War the Norwegian meteorologists J. Bjerknes and H. Solberg produced a synoptic model of the mid-latitude frontal cyclone, using observations from a dense network of observing stations. A knowledge of such synoptic models, and a combination of experience, skill and judgement, was the mainstay of the weather forecaster up to the

1950s. The current weather situation was analysed on surface maps by drawing isobars and fronts by hand and distinguishing areas of significant weather, while on upper air charts the pressure contours and THICKNESS lines were indicated. Examination of the circulation patterns and extrapolation of the movement and development of the weather systems was also part of the forecasting process.

The development of weather prediction by numerical methods using high-speed electronic computers has transformed synoptic meteorology in the past 40 years, and the role of the meteorologist today is that of monitoring the output of computer-based forecasts, modifying its products in the light of experience of atmospheric behaviour.   AHP

Reading and Reference
Bergeron, T. (1981) Synoptic meteorology: an historical review. In G. H. Liljequist (ed.), *Weather and weather maps*. Stuttgart: Birkhauser Verlag. · Hardy, R., Wright, P., Gribbin, J. and Kingston, J. (1982) *The weather book*. London: Michael Joseph. · Lackmann, G. (2011) *Midlatitude synoptic meteorology: dynamics, analysis and forecasting*. Berlin: Springer.

systems Simply 'sets of interrelated parts' (Huggett, 1980: 1). They are defined as possessing three basic ingredients: 'elements, states, and relations between elements and states'. There can be both concrete and abstract systems; for example, the hot water system of a house or the set of moral values of a society; and the elements may therefore be real things or concepts, each of which is held to possess a variety of properties (or can be said to exist in a variety of states). The overall state of the whole system is then defined by the character of these properties at a given moment. Because the system is defined as a *set* of parts, it follows that there is some boundary that distinguishes that particular set from all other possible sets; and the boundedness of systems is both an important theoretical attribute and a source of immense practical difficulty.

While it is possible to view some systems as completely isolated from all external influences, which in physical examples implies that no movement of energy or mass can occur across their boundaries, most are defined as either *open* or *closed*. An open system exchanges both energy and mass with its surroundings, whereas a closed system is open only to the transfer of energy. These *inputs* of mass and/or energy are termed *forcing functions*, and they are generally of considerable importance in the concrete systems of interest to physical geographers. The *throughput* of energy and/or mass creates the linkages or relations between the system elements, which

may adjust in the process, either by *negative* feedback mechanisms (*homeostasis*) so that the system state remains unchanged, or by *positive* feedback, so that a net change in the system state results. The outcome of the transfers is termed the system *output*, which may be energy and/or mass and/or a new system state.

Finally, systems are regarded as hierarchical sets: the whole system at any one level is merely a component or element of some higher order set and its own elements being, in reality, smaller scale systems in their own right. To give an example: a drainage basin with a single stream channel may be studied as a first-order drainage system; yet, its slopes and stream channel may equally well be viewed as individual systems; and the whole basin readily becomes just one element of a larger drainage network.

From this description of systems it should be clear that the concept may be applied to an infinitely wide range of phenomena. This, indeed, was seen as its chief methodological merit by Von Bertalanffy (1950), who introduced the notion, with the explicit hope that a focus on such a general feature of the physical and mental environment would encourage the sciences, in particular, to adopt a unified methodology. From this came the idea of GENERAL SYSTEM THEORY. The concept of the open system was introduced into geomorphology by Strahler (1950a,b), and its merits were widely advocated by Strahler's pupil, Chorley (e.g. Chorley, 1960, 1967; Chorley and Kennedy, 1971; Bennett and Chorley, 1978). An extremely persuasive and influential use of the concept was made by yet another of Strahler's pupils, Schumm (1977), and several other substantial works of a more general nature have also appeared (e.g. Huggett, 1980; Sugden, 1982).

More influential, however, has been the application of the idea of systems in the field of ecology, where the concept of the ecosystem – as a formal statement – goes back to Tansley (1935), although its basics can readily be traced into the far earlier views of Darwin and Haeckel (Stoddart, 1967). Studies such as H. T. Odum's (1957) monumental analysis of the Silver Springs ecosystem remain classic examples of the possibilities and limitations of what has come to be known as systems analysis. It seems clear that it was in large measure the apparent success with which the systems concept was used as an analytical device by ecologists in the 1950s that spurred its adoption in mainstream physical geography. The growing realization of the problems posed by identification and isolation of ecosystems as objects of analysis has, similarly, been accompanied by a reduction in the emphasis placed upon their investigation in other fields.

The system concept seems to have made four different kinds of appearance in physical geography. First, it is now widely and loosely used as a source of jargon and terminology: the ideas of various forms of equilibrium are particularly persistent and poorly defined borrowings. Second, the idea has been adopted as a pedagogic framework into which the results of earlier studies, or the fruits of different concepts, can be slotted for ease of exposition: Chorley and Kennedy (1971) and Sugden (1982) are examples of the category. Third, the concept may be taken as the basis for a substantive investigation of the workings of some portion of reality, in line with ecological studies such as that of Odum (1957); these inquiries are most commonly directed towards drainage basin hydrology, ecology and climatology, and few of them have actually been undertaken by geographers (although human geographers have made more explicit use of systems in this particular way). An exception is the study of soils (e.g. Huggett, 1975). Finally, the idea of systems has been adopted as a useful framework in which to view questions relating to environmental management and planning; a fairly early, clear and comprehensively explained example is Hamilton et al. (1969), and the whole approach has been exhaustively discussed by Bennett and Chorley (1978).

The application of the systems concept to any substantive investigation of the actual or potential workings of portions of the physical environment encounters two principal difficulties. The first is the need to define system boundaries. While arbitrary lines can, of course, be drawn anywhere to define a system of any magnitude, the very nature of the concept is of an interrelated set of elements that is in some real sense functionally or at least morphologically distinguishable from all adjacent sets. In practice, even lakes, islands and river basins – the most clearly definable of the physical units of interest to geographers – do not actually possess sharp and precise boundaries. The resulting difficulties of definition are horribly but understandably akin to those related to the definition of regions; they are equally hard to resolve.

The second problem stems from the need to evaluate and interpret the results of system analyses – see Kennedy (1979). Langton (1972) points out that the only effective way of comparing different systems is with reference to their success in attaining some preferred or most efficient state of operation or morphology. It is clear that we can use this criterion very readily to deal with engineered, planned or 'control' systems, since it is possible to specify the preferred or 'best' outcome.

It is very far from clear that it is justifiable to regard natural systems in such a light. What is the 'goal' of an ecosystem? Or a drainage basin? Or the general circulation of the atmosphere? Could this be the reason why substantive studies of the workings of the natural systems of interest to the physical geographer are actually rather thin on the ground?

The concept of the system is, in essence, simple, and it can be widely applied. The actual applications in physical geography to date have been most numerous in pedagogic and applied areas, and little fresh insight has been gained there into the actual workings of the natural environment. The greatest impact seems to have been made in the wholesale introduction of the terminology of systems analysis.   BAK

References

Bennett, R.J. and Chorley, R.J. (1978) *Environmental systems: philosophy, analysis and control.* London: Methuen. · Chorley, R.J. (1960) *Geomorphology and general systems theory.* US Geological Survey Professional Paper 500-B. Washington, DC: United States Government Printing Office. · Chorley, R.J. (1967) Models in geomorphology. In R. J. Chorley and P. Haggett (eds), *Models in geography.* London: Methuen; pp. 59–96. · Chorley, R.J. and Kennedy, B.A. (1971) *Physical geography: a systems approach.* London: Prentice-Hall International. · Hamilton, H.R., Goldstone, S.E., Milliman, J.W., *et al.* (eds) (1969) *Systems simulation for regional analysis. An application to river-basin planning.* Cambridge, MA: MIT Press. · Huggett, R.J. (1975) Soil landscape systems: a model of soil genesis. *Geoderma,* **13,** 1–22. · Huggett, R.J. (1980) *Systems analysis in geography.* Oxford: Clarendon Press. · Kennedy, B.A. (1979) A naughty world. *Transactions of the Institute of British Geographers,* **4,** 550–558. · Langton, J. (1972) Potentialities and problems of a systems approach to the study of change in human geography. *Progress in Geography,* **4,** 125–179. · Odum, H.T. (1957) Trophic structure and productivity of Silver Springs, Florida. *Ecological Monographs,* **27,** 55–112. · Schumm, S.A. (1977) *The fluvial system.* Chichester: John Wiley & Sons, Ltd. · Stoddart, D.R. (1967) Organism and ecosystem as geographical models. In R. J. Chorley and P. Haggett (eds), *Models in geography.* London: Methuen; pp. 511–548. · Strahler, A.N. (1950a) Equilibrium theory of erosional slopes approached by frequency distribution analysis. Part I. *American Journal of Science,* **248,** 673–696. · Strahler, A.N. (1950b) Equilibrium theory of erosional slopes approached by frequency distribution analysis. Part II. *American Journal of Science,* **248,** 800–814. · Sugden, D.E. (1982) *Arctic and Antarctic: a modern geographical synthesis.* Oxford/Totowa, NJ: Blackwell/Barnes & Noble. · Tansley, A.G. (1935) The use and abuse of vegetational concepts and terms. *Ecology,* **16,** 284–307. · Von Bertalanffy, L. (1950) The theory of open systems in physics and biology. *Science,* **111,** 23–29.

syzygy   One of the two points at which the Moon or a planet is aligned with the Earth and the Sun.

T

tafoni The plural of tafone, which refers to a weathering hollow in a vertical rock face. The term tafoni is frequently used to refer to large (square metre) features, while alveoli (honeycomb features) is used for smaller (square centimetre) weathering hollows that may cover rock surfaces, usually in sandstone or granite, in profusion. Tafoni and alveoli are often seen in coastal environments and DRYLANDS, such that salt weathering has been viewed as a major formative process. However, microorganisms and diurnal thermal contrasts may also contribute to their development, which may exploit lines or zones of weakness in rocks.   DSGT

taiga The most northerly coniferous forest of cold temperate regions. It does not exist as a zone in the southern hemisphere, and generally refers to open woodland lying to the south of TUNDRA and to the north of the dense BOREAL FOREST. It has also been used more broadly to include the entire area covered by coniferous forest of high latitudes and high mountain slopes. More literally (from the Russian), it is characterized by open, rocky landscapes dominated by conifers but with scattered deciduous trees, such as birch and alder, locally dense along river valleys, with a fairly continuous carpet of lichens and heathy shrubs. The area is often poorly drained and peat filled.   PAF

Reading
Larsen, J.A. (1980) *The boreal ecosystem*. New York: Academic Press.

takyr A desert soil with a bare, parquet-like surface, broken up by a network of splits into numerous polygonal aggregates. Takyrs are typical landscape elements of the deserts of Central Asia, but comparable claypans are found in the deserts of Australia, Iran and North Africa. They have no higher vegetation, a crusted surface, occur in the lower parts of piedmont plains, and their formation requires a seasonal flooding of the surface by a thin layer of water, carrying suspended clay material and soluble salts (Kovda *et al.*, 1979).   ASG

Reference
Kovda, V.A., Samoilova, E.M., Charley, J.L. and Skujinš. J.J. (1979) Soil processes in arid lands. In D. W. Goodhall, R. A. Perry and K. M. W. Howes (eds), *Arid-land ecosystems: structure, functioning and management*, vol. I. Cambridge: Cambridge University Press; pp. 439–470.

talik A layer of unfrozen ground below the seasonally frozen surface layer and above or within PERMAFROST.

talsand ('valley sand') A widely used concept in north German Quaternary geology. Talsands are large-scale sandy infillings of ice marginal valleys that consist of fluvial beds with a capping of windborne sands.   ASG

Reference
Schwan, J. (1987) Sedimentologic characteristics of a fluvial to aeolian succession in Weichselian Talsand in the Emsland (FRG). *Sedimentary Geology*, 52, 273–298.

talus A SCREE slope; the term is sometimes also used simply to refer to scree material.

tank A man-made pond or lake. A natural depression in bare rock that is filled with water throughout the year.

tarn A small mountain lake (northern England). Compare Scottish 'lochan' and Welsh 'llyn'.

taxonomy The study, description and classification of variation in organisms and also in soils, including the causes and consequences of such variation. This is a modern wide definition that makes taxonomy synonymous with the now interchangeable term *systematics*. Traditionally, taxonomy was often restricted to the narrower activity of the classification and naming of organisms; in this sense, it was only a part of systematics.

The origins of taxonomy lie in human need to classify the discontinuities of variations to be seen in nature. Scientific taxonomy reached its great flowering in the eighteenth century with the still fundamental contributions of the Swedish

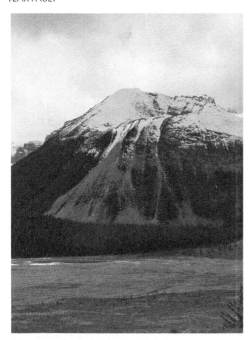

Talus cones, Rocky Mountains, Canada.
Photograph by David Thomas.

biologist Carolus Linnaeus (1707–1778). These included his *Systema Naturae* (1735 and many later editions), in which he classified all the then known animals, plants and minerals.

Today, two major approaches to biological classification are recognized. The first is *phenetic*, which expresses the relationships between organisms in terms of their similarities in characters, without taking into account how they came to possess these. The second is *phylogenetic*, or evolutionary, in which some aspects of their evolution are taken into account when making the classification, including their cladistic relationship, which refers to the pathways of ancestry (see CLADISTICS). These pathways are usually expressed in the form of a tree-diagram or cladogram. (See also VICARIANCE BIOGEOGRAPHY.)

For soils, a range of taxonomic schemes exist, but perhaps the most widely used for agriculture and plant growth purpose is the United Sates Department of Agriculture scheme, and for engineering purposes the United soil Classification System.

tear fault A fault with a vertical fault plane, the blocks either side of which move horizontally. A transcurrent or strike fault.

tectonics The study of the broad structures of the Earth's lithosphere and the processes of faulting, folding and warping that form them. Tectonics focuses on structures at the regional scale and above, the investigation of smaller scale structural features usually being described as structural geology. Tectonics is concerned not only with the determination of the three-dimensional form of geological structures, but also their history, origin and relationship to each other. Tectonic landforms are those, such as faultscarps, produced directly by tectonic mechanisms and those larger features, such as warped erosion surfaces, that owe their general, though not detailed, form to such processes. (See also PLATE TECTONICS.) MAS

Reading
Kearey, P., Kleipeis, K.A. and Vine, F.J. (2009) *Global tectonics*, 3rd edition. Oxford: John Wiley & Sons, Ltd-Blackwell. · Ollier, C. (1981) *Tectonics and landforms*. London: Longman. · Spencer, E.W. (1977) *Introduction to the structure of the Earth*, 2nd edition. New York: McGraw-Hill.

tektites Small glassy spherules (<1 mm to 0.2 m diameter) that show distinct effects of streamlining during their molten stage. They are found mainly in five 'strewnfields', designated Australasian, Ivory Coast, Czechoslovakian, North American or Libyan. Tektites within a given strewnfield are related mineralogically, and in terms of age and shape. The most common shape is teardrop, but spheres, discs and dumbbell shapes are also found. The source of tektites has been debated, but consensus is that the particles represent air-cooled, molten ejecta from meteorite or comet impacts with the Earth. Tektites found in Haiti and Mexico have been linked to the Chicxulub impact. DJS

Reading
King, E.A. (1977) The origin of tektites: a brief review. *American Scientist*, **65**, 212–218.

teleconnections Fundamentally, these are long-distance linkages between the weather patterns of widely separated regions of the Earth. Teleconnections are normally described by statistical relationships depicting the amount of shared variance between two regions or between two atmospheric phenomena. Descriptively, a teleconnection might associate changes in the surface temperatures or in the upper circulation of one region with, for example, the precipitation regime of another part of the Earth such that variations in the first create changes in the second. The identification and analysis of teleconnections is currently one of the most important areas of climate research.

Many of the most studied teleconnections in climate since the late 1970s have been those associated with the South Pacific phenomena known as ENSO. Interest in these teleconnections has been fuelled by the marked global changes linked to the powerful ENSO event of 1982–1983, the subsequent long-lasting event in the early 1990s and the high-amplitude event in 1997–1998. One extreme of ENSO, EL NIÑO, a massive atmospheric–oceanic event characterized by abnormally warm sea surface temperatures in the equatorial Pacific and corresponding changes in the upper air circulation of that region, promotes many corollary changes in weather across the planet. For example, the high-magnitude El Niño event of 1997–1998 has been linked to abnormally wet winter conditions in the southern USA, to massive floods in western South America, to extensive severe drought in Indonesia and a substantial reduction in the number of Atlantic hurricanes in 1997. All of these climate linkages with ENSO are characterized as teleconnections.

Recently, scientists have also identified teleconnections existing between the other extreme of ENSO, LA NIÑA, abnormally cool waters in the equatorial Pacific, and various regions of the Earth. In general, those teleconnections are opposite of those associated with El Niño – drier conditions in the southern USA and western South America, wetter conditions in Australia and Indonesia and higher numbers of hurricanes in the Atlantic Ocean.

Teleconnections research has focused on two main areas of investigation. The first is the identification of mechanisms that influence and initiate the core phenomena, such as ENSO, while the second is the identification of other weather phenomena that may teleconnect to the weather of distant parts of the world.

With regard to the second area of study, many other weather phenomena are being analysed for their teleconnections to other regions. For example, the so-called Pacific North America pattern is a circulation pattern defined as having a large upper level ridge over the western part of the continent and a trough over the eastern half of the continent. Such a pattern leads to a dichotomy of conditions between the two halves of the continent, with generally warmer and drier conditions in the west and cooler and wetter conditions in the east. Another cyclic regional climate phenomenon thought to have teleconnective linkages to the climate of other parts of the world is the North Atlantic Oscillation, a periodic circulation existing between Iceland and the Azores.

Long-term climate teleconnections have also been hypothesized. A regional weather circulation known as the 'Greenland Above' pattern has been suggested as a possible favourable circulation for initiating an ice age climate regime (Crowley, 1984). Development of a large ridge over Greenland forces trough development to the west. Such a circulation is favourable for sustained snow production over eastern Canada. Crowley hypothesizes that, if such a pattern were to remain in place for a long period of time, enough ice could accumulate over eastern Canada to force permanent climate change.

The concept of teleconnections is grounded in those associated with chaos theory as applied to climate change by Edward Lorenz. The general principle of climatic chaos theory is that internal instabilities in the climate system can cause complex behaviour elsewhere in the system. Such would appear to be the case with teleconnections. We are now discovering that many internal instabilities, specifically regional weather phenomena like ENSO, can indeed lead to disruptions of the entire climate system. This concept, carried to its most extreme theoretical form (the 'butterfly' principle), suggests that even the small-scale influence of butterfly wings could conceivably produce large-scale disturbances elsewhere on the planet. Such causative linkages between remote locations on the planet can be termed teleconnections and remain a central focus of current scientific investigation. RSC

Reference

Crowley, T.J. (1984) Atmospheric circulation patterns during glacial inception: a possible candidate. *Quaternary Research*, **21** (1), 105–110.

temperate ice   Ice that is at the pressure melting point. In temperate glacier ice, water is present throughout and generally amounts to between 0.1 and 2% of the total volume (Lliboutry, 1976). *Temperate* ice is contrasted with *cold* ice that is below the pressure melting point. It is common to classify whole glaciers as temperate or cold, but this is misleading since both types of ice are common in most glaciers. For example, a glacier that consists wholly of temperate ice in summer may have a cold surface layer of ice in winter, while it is also likely that cold patches exist at the bottom as a result of pressure changes around obstacles (Robin, 1976). Furthermore, many 'cold' glaciers have ice at the pressure melting point at depth.   DES

Reading and References

Lliboutry, L. (1976) Physical processes in temperate glaciers. *Journal of Glaciology*, **16**, 151–158. · Paterson,

W.S.B. (1981) *The physics of glaciers*. London: Pergamon.
· Robin, G.deQ. (1976) Is the basal ice of a temperate glacier at the pressure melting point? *Journal of Glaciology*, **16**, 183–196.

temperature

To the general public temperature is a confusing parameter. When used to measure how warm or how cold substances feel it is confusing to have different temperature scales, so that, for instance, the temperature at which pure ice melts can be stated as 0 °C or 32°F or 273.15 K. In science, the Kelvin scale is used because it avoids negative temperatures and is a more accurate reflection of the energy possessed by the molecules in a substance. Temperature is really a measure of the molecular kinetic energy of a substance; in other words, the average speed at which the molecules are moving in gas, or vibrating in a solid. The Celsius and Fahrenheit scales are used for convenience, in that they give easier numbers for people to handle for the range of temperatures normally encountered at the Earth's surface.   JET

temperature–humidity index

Originally called the discomfort index when it was introduced by E. C. Thom (1959) in the USA, the temperature–humidity index (THI) is a simplified form of effective temperature:

$$THI = 0.4(T_{DB} + T_{WB}) + 4.8$$

when the dry-bulb and WET-BULB TEMPERATURES are recorded in degrees Celsius. Wind speeds and solar radiation are not taken into consideration, and so the index must be used with caution. The THI is widely used in weather forecasts for the general public in the USA during summer. At a THI below 70 there is no discomfort. When the THI reaches 80 everyone feels uncomfortable.   DGT

Reference
Thom, E.C. (1959) The discomfort index. *Weatherwise*, **12**, 57–60.

temperature inversion

An increase of temperature with height; the inverse of the normal decrease of temperature with height that occurs in the TROPOSPHERE. When the temperature is getting warmer with height, natural buoyancy is limited and the air is described as being stable. Under these conditions, vertical dispersion of air pollution and visibility are both restricted. Inversions can form in a variety of different ways.

1 Most frequently, inversions form near the surface as a result of radiative cooling during clear nights; more energy is lost from the surface than is gained by counter-radiation from the atmosphere, so most cooling takes place at the surface. They are normally destroyed by surface heating the following morning.

2 When warm air passes over a cold surface an inversion will form. If the air is moist or the surface very cold, persistent fog may form. This often happens when warm air is drawn from an ocean over a snow-covered surface.

3 Inversions can form dynamically as a result of ADIABATIC subsidence of air associated with ANTICYCLONES. The temperature may decrease initially from the surface but later increases as the effects of the descending warming air are felt. The trade wind inversion of the subtropics is the best example of this type (see TRADE WINDS). As it may persist for days or weeks, dispersal of air pollution can be a problem, as in Los Angeles.

4 Radiative heating in the upper atmosphere can produce strong temperature inversions. The absorption of ultraviolet light by ozone in the stratosphere is a good example.   PS

Reading
Barry, R.G. and Chorley, R.J. (2003) *Atmosphere, weather and climate*, 8th edition. London: Routledge. · McGregor, G.R. and Nieuwolt, S. (1998) *Tropical climatology*. Chichester: John Wiley & Sons, Ltd.

tensiometer

An instrument for estimating the matric or capillary potential of soil moisture. A tensiometer consists of a porous pot at the required point in the soil and connected through watertight tubing to a manometer or pressure gauge. The whole instrument is filled with water before the cup is placed in a carefully augured hole, refilled as near to its former condition as possible. Water may pass through the walls of the pot in response to suction forces in the surrounding soil. As water passes into the soil a partial vacuum builds up inside the instrument, and this is monitored as a negative head of water by the vacuum gauge or manometer. Water will move into or out of the pot until the soil moisture suction is balanced by the strength of the partial vacuum or the weight of a negative head of water (or, more usually, mercury). Tensiometers are only suitable for use in comparatively moist soil because of problems in keeping the instrument airtight at high suctions.   AMG

Reading
Burt, T.P. (1978) *An automatic fluid-scanning switch tensiometer system*. British Geomorphological Research Group technical bulletin 21. Norwich: Geo Abstracts. · Curtis, L.F. and Trudgill, S. (1974) *The measurement of soil*

moisture. British Geomorphological Research Group technical bulletin 13. Norwich: Geo Abstracts.

tepee An overthrust sheet of limestone that appears as an inverted V in a two-dimensional exposure, so named because of its two-dimensional resemblance to the hide dwellings of early American Indians. Tepees are found in tidal areas, around salt lakes and in CALCRETE, developing as a result of deformation or desiccation and contraction processes related to fluctuations in water levels and in the nature of chemical precipitation. ASG

Reading
Kendall, C.G.StC. and Warren, J. (1987) A review of the origin and setting of tepees and their associated fabrics. *Sedimentology*, **34**, 1007–1027. · Warren, J.K. (1983) Tepees, modern (Southern Australia) and ancient (Permian – Texas and New Mexico) – a comparison. *Sedimentary Geology*, **34**, 1–19.

tephigram A meteorological thermodynamic chart used for plotting and analysing dry-bulb and dewpoint temperature from a RADIOSONDE ascent, usually from the surface to the lower STRATOSPHERE. The vertical axis is atmospheric pressure (see PRESSURE, AIR) decreasing upwards on a logarithmic scale and TEMPERATURES are plotted with reference to the horizontal axis, although the isotherms are skewed to run from bottom left to top right of the diagram. Dry adiabats, saturation adiabats and lines of constant humidity mixing ratio are also printed on the chart. The tephigram is basically an aid to forecasting; for example, fog formation or shower development. RR

tephra The solid material ejected from a volcano that includes dust, ash, cinders and volcanic bombs.

tephrochronology The study of volcanic ash layers (tephra) for dating. Each ash layer provides a time-equivalent marker than can be correlated from one site to another. Different ash falls may be recognizable and can be 'fingerprinted' on the basis of petrology and chemical composition. They may also be susceptible to dating by K/Ar and other methods. Large-magnitude eruptions can disperse tephra up to thousands of kilometres from the vent, producing a near-instantaneous marker horizon. Many eruption episodes, especially very explosive, tephra-generating phases, typically last for only minutes, hours, or days to perhaps weeks or months, and rarely extend beyond a year or two. The resultant tephras thus represent short-lived isochronous timelines in the context of geological time – at

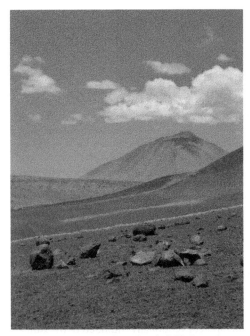

Volcanic bombs, Patagonian volcanic field, Argentina. Photograph by David Thomas.

the most, only months or so in duration, commonly considerably less (Lowe, 2011). ASG

Reference
Lowe, D. J. (2011) Tephrochronology and its application: a review. *Quaternary Geochronology*, **62**: 107–153.

terlough Depressions, with a sinkhole, that fill with water when the water table rises. The rise may be associated with tidal effects. They are a feature of parts of western Ireland.

terminal grades The fine fraction due to a glacial comminution that reflects the mineral components of the debris. Dreimanis and Vagners (1971) suggested that once a glacial TILL has been broken down to mineral-sized fragments it experiences relatively little further breakdown, thus meriting the use of the term terminal grades. This grain-size group comprises one of the bimodal grain-sized products of abrasion, the other being rock fragments. Under some circumstances, however, fine material fragments do break down further to finer material (Haldorsen, 1981). DES

References
Dreimanis, A. and Vagners, U.J. (1971) Bimodal distribution of rock and mineral fragments in basal tills. In R. P. Goldthwait (ed.), *Till: a symposium*. Columbus, OH: Ohio

State University Press; pp. 237–250. · Haldorsen, S. 1981: Grain-size distribution of subglacial till and its relation to glacial crushing and abrasion. *Boreas* **10**, 91–105.

terminal moraine   The moraine at the terminus of a glacier. (See also MORAINE.)

terminal velocity   An object falling through a fluid (wind or water) is retarded by viscous drag forces at its surface. Initially, the force of GRAVITY exceeds the drag force and the object accelerates downwards. As velocity increases, viscous drag forces also increase and acceleration continues until the gravity and drag forces become equal. At this point the object is said to have reached terminal velocity and no further acceleration will take place without an additional external force being applied.   GFSW

terminations   The boundaries in deep sea core sediments that separate pronounced oxygen isotopic maxima from exceptionally pronounced minima. They are in effect rapid deglaciations, and are conventionally numbered by Roman numerals in order of increasing age. The segments bounded by two terminations are called glacial cycles. Nine terminations have occurred in the past 0.7 million years.   ASG

Reading
Kukla, G.J. (1977) Pleistocene land–sea correlations. I. Europe. *Earth Science Reviews*, **13**, 307–374.

termites   Of which there are several thousand species, are insects of the Isoptera order, and about four-fifths of the known species belong to the Termitidae family (Harris, 1961). They vary in size according to their species, from the large African Macrotermes, with a length of around 20 mm and a wing span of 90 mm, down to the Middle Eastern Microcerotermes that are only around 6 mm long with a wing span of 12 mm. Major recent taxonomic and ecological surveys include that of Brian (1978), while Lee and Wood (1971) provide a detailed study of the effects of termites on soils, and Goudie (1988) reviews their geomorphological impact.

Termites, though 'fierce, sinister and often repulsive' (Maeterlinck, 1927), are remarkable for having been adapted to living in highly organized communities for as long as 150–200 million years, and much of their success is due to their development of elaborate architectural, behavioural, morphological and chemical strategies for colony defence. They occur in great numbers: $2.3 \times 10^6 \, \mathrm{ha^{-1}}$ in Senegal and $9.1 \times 10^6 \, \mathrm{ha^{-1}}$ in the Ivory Coast (UNESCO/UNEP/FAO, 1979).

Maeterlinck (1927) regarded them as 'the most tenacious, the most deeply rooted, the most formidable, of all the occupants and conquerors of this globe'. The vast majority of termite species are found in the tropics.   ASG

References
Brian, M.V. (ed.) (1978) *Production ecology of ants and termites.* Cambridge: Cambridge University Press. · Goudie, A.S. (1988) The geomorphological role of termites and earthworms in the tropics. In H. A. Viles (ed.), *Biogeomorphology.* Oxford: Blackwell; pp. 166–192. · Harris, W.V. (1961) *Termites: their recognition and control.* London: Longman. · Lee, K.E. and Wood, T.G. (1971) *Termites and soils.* London: Academic Press. · Maeterlinck, M. (1927) *The life of the white ant.* London: Allen & Unwin. · UNESCO/UNEP/FAO (1979) *Tropical grazing and land ecosystems.* Paris: UNESCO.

terra rossa   Red soils developed on the iron-oxide-rich residual material on limestone bedrock, particularly in warm temperate regions.

terracette   Miniature terrace or ridge extending across a slope, usually normal to the direction of maximum slope. Terracettes are rarely more than 0.5 m wide and deep. Their origin is still a matter for debate. Some may be animal tracks, but as others occur in areas where animals are very rare it would seem that some other mechanisms are involved. They are probably a consequence of *soil* mantle instability on steep slopes.   ASG

terrane   'A mappable structural entity that has a stratigraphic sequence and an igneous, metamorphic and structural history quite distinct from those of adjacent units. Each terrane is separated from its neighbours by a structural break that may take the form of a normal fault, a reverse fault, a wrench fault, or an overthrust' (Barber, 1985: 116).

During the 1980s the terrane concept became an important one for many Earth sciences (Howell, 1989). It was developed most notably in the context of the North American Cordillera system, where, it was maintained, substantial portions of the Cordillera were 'exotic' blocks and slithers that had 'docked' onto the North American Craton. Subsequently, the importance of terranes has been recognized with respect to areas as diverse as Highland Scotland and the archipelagos of Southeast Asia. Terranologists hold that major oceanic belts often consist of 'collages' of fault-bounded crustal and/or lithospheric fragments, of diverse origins and different sizes. However, as Sengör and Dewey point out, reactions to the concept range from enthusiastic applause to abusive rejection. They themselves express some cogent doubts, suggesting, for example (Sengòr and Dewey, 1991: 6) that 'terranology not only does not go beyond plate tectonics, it takes a backward step; by confusing primary and secondary collage components, it confuses also their genetic implications.' They continue (Sengòr and Dewey, 1991: 17):

the word terrane is a lump term for a number of older and more informative non-genetic (block and sliver) and genetic (fragment, nappe, strike-slip duplex, microcontinent, island arc, suture, etc.) terms. Because it is less informative, it is less useful than any of these and also because, historically, the term "terrane" has a number of different meanings it is best avoided. Terrane analysis is neither a new way of looking at orogenic belts, nor a particularly helpful one.

ASG

References

Barber, A. (1985) A new concept of mountain building. *Geology Today*, 1, 116–121. · Howell, D.G. (1989) *Tectonics of suspect terranes*. London: Chapman & Hall. · Sengòr, A.M.C. and Dewey, J.F. (1991) Terranology: vice or virtue? In J. F. Dewey, I. G. Gass, G. B. Curry *et al.* (eds), *Allochthonous terranes*. Cambridge: Cambridge University Press.

terrestrial magnetism   The natural magnetism of the Earth, also referred to as geomagnetism. The shape of the Earth's mainly dipole magnetic field suggests that it is related to circular electrical currents flowing approximately normal to the axis of rotation. These electrical currents may be induced by slow convective movements within the partially molten iron-rich core of the Earth, with large-scale eddies producing the regional variations in the main field. Secular changes in the field include the continuous movement of magnetic north, variations in the strength of the field and periodic reversals of polarity.   MAS

Reading

Parkinson, W.D. (1983) *Introduction to geomagnetism*. Edinburgh/Amsterdam: Scottish Academic Press/Elsevier. · Smith, P.J. (1981) The earth as a magnet. In D. G. Smith (ed.), *The Cambridge encyclopaedia of Earth sciences*. Cambridge/New York: Cambridge University Press/Crown Publishers; pp. 109–123.

Tertiary   This geological term remains in common use, although officially it is now redundant, replaced by the terms Palaeogene and Neogene. Refers to the first part or period of the CAINOZOIC era, comprising the Palaeocene, Eocene, Oligocene, Miocene and Pliocene epochs, which span 65.5 million to 2.6 million years ago. See GEOLOGICAL TIMESCALE.   DSGT

Tethys Sea   An enormous seaway initially formed in the Palaeozoic era and attaining its maximum development during the Mesozoic, which extended from what is now the Mediterranean eastwards as far as Southeast Asia. Beginning about 75 million years ago, extensive volcanism, followed by intense deformation and uplift of sediments accumulated in the Tethys Ocean throughout the Mesozoic, led to the formation of the present-day Alpine–Himalayan mountain belt. The opening and closure of the Tethys Ocean (see MESSINIAN SALINITY CRISIS) is currently interpreted in terms of CONTINENTAL DRIFT and the operation of PLATE TECTONICS.   MAS

Reading

Sonnenfeld, P. (ed.) (1981) *Tethys: the ancestral Mediterranean*. Stroudsburg, PA: Dowden, Hutchinson & Ross.

thalassostatic   A term used to describe river terraces that are produced by aggradation during periods of rising or high sea level and by incision at times of low sea level.

thalweg   The line of maximum depth along a river channel. It may also refer to the line of maximum depth along a river valley or in a lake.

thaw lake   A shallow, rounded lake occupying a depression resulting from the melting of ground ice (see also THERMOKARST). Thaw lakes are extremely common on the North American and Siberian Arctic lowlands and are ubiquitous wherever there is a flat lowland with silty alluvium and a high ice content. Most lakes are less than 300 m in diameter and less than 3–4 m deep. Following random exposure of ground ice, a water-filled depression soon develops into a roughly circular lake. Eventually, vegetation protects the banks from further thawing and in a matter of 2000–3000 years or so the lake is infilled and the cycle of development complete. Many thaw lakes are elongate in shape with a systematic orientation of the long axis at right angles to the prevailing wind (Carson and Hussey, 1963).   DES

References

Carson, C.E. and Hussey, K.M. (1963) The oriented lakes of Arctic Alaska: a reply. *Journal of Geology*, 71, 532–533.

thematic map   A representation of the spatial distribution of a specific phenomenon. In this type of map the focus is upon the particular theme of interest, with other features such as coastlines or urban areas only depicted to aid interpretation. Popular examples of thematic maps include those representing soil types, geology, precipitation and land cover. A variety of different types of thematic map have been produced, including choropleth, dot and isopleth maps. Thematic maps are

sometimes derived from images acquired by REMOTE SENSING, often through the use of a SUPERVISED CLASSIFICATION analysis.   GF

Thematic Mapper (TM)

A sensor that acquires multispectral satellite images of use in a wide range of geographical applications. The name derives from its suitability for the production of land-cover maps divided into themes such as arable, forestry and built-land (i.e. thematic maps). Thematic Mapper images have been continuously acquired from the American-owned Landsat satellites since 1982, originally run by NASA and now in commercial ownership. This image type is commonly referred to as Landsat TM. The TM sensor was carried aboard the Landsat satellites 4 and 5, which orbited the Earth at an altitude of 705 km providing complete Earth coverage between 81°N and 81°S. The minimum revisit period for any point on the ground is 16 days, although in practice the presence of cloud cover could reduce this periodicity considerably. An enhanced TM was launched aboard Landsat 7 in 1999, also orbiting at 705 km above the Earth, while Landsat 8, launched in 2013, increases capabilities further.

TM images cover a ground area of approximately 185 km × 185 km and with a PIXEL size of 30 m in most wavebands. The thermal waveband, band 6, is an exception with an original pixel size of 120 m. The amount of reflected and emitted radiation was initially measured in seven regions of the electromagnetic spectrum: three in the visible portion (blue, 0.45–0.52 μm; green, 0.52–0.60 μm; red, 0.63–0.69 μm), and one in each of the near infrared (0.76–0.90 μm), mid-infrared (1.55–1.75 μm) and thermal infrared (10.42–12.50 μm) and a further in mid-infrared (2.08–2.35 μm). These wavebands are numbered from 1 to 7 respectively. An eighth panchromatic waveband (0.52–0.9 μm) was added in Landsat 7, with a resolution of 15 m, while the resolution of the thermal waveband 6 was also reduced, to 60 m. These improvements collectively led the system to be renamed 'Enhanced Thematic Mapper plus' (ETM+).

The broad sampling of the electromagnetic spectrum makes TM/ETM+ a versatile tool and is perhaps the most widely used of satellite images in geographical research. Pixel values are recorded as 8-bit numbers, providing a data range from 0 to 255, which can be calibrated to radiance or reflectance units if required. Further improvements have occurred with the launch of Landsat 8 in 2013, where two new instruments provide coverage of the Earth's surface (see OPERATIONAL LAND IMAGER).   CDC

Reading

Lillesand, T.M., Kiefer, R.W. and Chipman, J. (2014) *Remote sensing and image interpretation*, 7th edition. New York: John Wiley & Sons, Inc.

thermal depression

A region of low surface pressure that is generated by strong solar heating on fine days over land in the summer. Thermal depressions can range from the large, seasonal thermal low of the South Asian monsoon to the short-lived lows that sometimes form over England on hot summer days. Their cyclonic inflow tends to be strongest near the surface and to weaken with height since they are shallow features, usually a few kilometres deep.

Fine summer weather in Britain occasionally breaks down in association with an extending thundery thermal depression over France.   RR

thermal efficiency

The term was used by C. W. Thornthwaite (1931) in his first classification of climate. The thermal efficiency index (TEI) indicates the plant growth potential of a location and is calculated by summing the 12 monthly values of $(T - 32)/4$, where T (°F) is the mean monthly temperature. The index ranges from zero on the polar limit of the tundra ('frost climate') to over 127 in the tropics. Six temperature provinces were defined. In this classification Thornthwaite made moisture, in the form of precipitation effectiveness (PEI), the primary factor for a T–E index of over 31 (taiga/cool temperate boundary). In his second classification of climate, Thornthwaite (1948) used potential evapotranspiration as a measure of thermal efficiency.   DGT

Reading and References

Thornthwaite, C.W. (1931) The climates of North America according to a new classification. *Geographical Review*, **21**, 633–655. · Thornthwaite, C.W. (1933) The climates of the Earth. *Geographical Review*, **23**, 433–440. · Thornthwaite, C.W. (1948) An approach toward a rational classification of climate. *Geographical Review*, **38**, 55–94.

thermal equator

Variously defined, but most commonly the line that circumscribes the Earth and connects all points of highest mean annual temperature for their longitude. Sometimes the seasonal or monthly variation of this line is considered. The annual mean position departs by as much as 20° latitude from the equator. In recent years the term thermal equator or heat equator has also been taken as synonymous with the INTERTROPICAL CONVERGENCE ZONE (ITCZ), the belt along which the thermally driven trade winds of the two hemispheres converge and are forced to rise.   WDS

thermal infrared linescanner An optical remote sensor used to derive thermal images of the Earth's surface. Early thermal infrared linescanners have two thermal detectors and record an image onto a photographic film, while more recent thermal infrared linescanners are often part of a MULTISPECTRAL SCANNER that records data digitally. With both these systems the physical geographer must choose the waveband to be used, the time of day when the image is to be recorded and the method of approximate calibration. The two thermal infrared wavebands used by thermal infrared linescanners are defined by the two transmitting atmospheric 'windows' located between the wavelengths of 3–5 µm and 8–14 µm. The choice of which of these wavebands to employ depends upon the application. The peak of radiant emission from the Earth's surface occurs in the 8–14 µm region, and this waveband has proved to be the most popular in physical geography.

Thermal infrared linescanners are used most frequently at night when there is no interference from reflected solar radiation. The usual flying time is just before dawn, when the effects of differential solar heating are minimized. Flights are occasionally made during daylight hours, either because AERIAL PHOTOGRAPHY is to be taken or because it is advantageous to have terrain details enhanced by differential solar heating and shadowing. Owing to the variation in emissivity within a scene and the presence of a thermally variable atmosphere between the sensor and the ground, it is not possible to calibrate absolutely the radiant temperature sensed by the detector. Various approximate calibration methods have been employed, of which the most accurate involves repeated flights at a range of altitudes. By using the temperature of thermally stable objects of known emissivity as standards, a graph is constructed of temperature versus height. This is extrapolated to the ground surface to give the temperature of enough points to enable calibration of the imagery.

Thermal infrared linescanner data can be presented in image form to be interpreted like an aerial photograph. The interpreter generally uses the images not to map an area but to search for thermal patterns that give a clue to some past, present or future environmental process, such as soil movement, frost hollows, water stress in crops, volcanism or thermal pollution of water. PJC

Reading
Cracknell, A.P. and Hayes, L.W.B. (1991) *Introduction to remote sensing*. London: Taylor & Francis. · Lillesand, T.M., Kiefer, R.W. and Chipman, J. (2014) *Remote sensing and image interpretation*, 7th edition. New York: John Wiley & Sons, Inc.

Thermal Infrared Sensor See OPERATIONAL LAND IMAGER.

thermal pollution The pollution of water by increasing its temperature. Many fauna are affected by temperature, so that this environmental impact has some significance. Among the main sources of thermal pollution of stream waters are condenser cooling water released from electricity generating stations, the urban 'heat island effect', reservoir construction, shade removal by deforestation, and changes in the width/depth ratios of channels. The effects of thermal pollution are especially severe at times of low flow. ASG

Reading
Langford, T.E.L. (1990) *Ecological effects of thermal discharges*. London: Elsevier. · Pluhowski, E.J. (1970) *Urbanization and its effects on the temperature of the streams on Long Island, New York*. United States Geological Survey Professional Paper 627-D. Washington, DC: United States Government Printing Office.

thermal wind v_t The vector difference between the geostrophic wind at two levels in the atmosphere. It is calculated by subtracting the lower level wind v_l from the upper level wind v_u and is therefore not a real wind, which would be observed in the intervening layer, but an expression of the shear of the horizontal wind within the layer.

It is termed 'thermal' because its strength and direction are dependent on the thermal pattern of the layer involved. Thus, v_t is aligned parallel to the layer's mean isotherms with cold air to its left and warm air to its right. Its magnitude or length is proportional to the strength of the layer's mean temperature gradient, so that thermal winds are strongest in regions where steep horizontal temperature gradients occur in depth; for example, in the polar front zone.

Thermal winds are also therefore parallel to THICKNESS contours and can in fact be calculated precisely with reference to thickness variations across a map. RR

thermistor A type of semiconductor resistor that has a high (usually negative) temperature coefficient of resistance. Thus, they are often used as temperature sensors or measuring devices. The resistance response to temperature is not linear but can be linearized either in hardware or software in the measuring instrument or subsequent data processing. Despite this disadvantage, it is often preferred to the linear response of the platinum resistance thermometer. WBW

thermoclasty   See INSOLATION WEATHERING.

thermocline   A layer of water within a lake or ocean through which the rate of decrease of temperature with depth is much greater than in adjacent layers. It is particularly well developed in tropical oceans, where a *permanent* thermocline, up to a few hundred metres in thickness, lies with its upper boundary some 25–150 m below the ocean surface. The temperature gradient may reach 10 °C per 100 m in the upper part of the layer. Towards mid-latitudes the permanent thermocline becomes thicker and less intense, with its upper surface as much as 600 m deep; it is entirely absent at latitudes greater than about 60°. The relatively stable stratification of the thermocline layer inhibits interchange between the warm waters of the surface mixed layer and the deeper cold waters forming the main body of the ocean.

A *seasonal* thermocline readily develops in both lakes and oceans whose surface temperatures undergo a significant annual variation. Lying close to the surface, its depth and intensity will depend on the amount of summer insolation and extent of turbulent mixing by the wind. JEA

Reading
Wells, N.C. (2012) *The Atmosphere and ocean*, 3rd edition. Chichester: John Wiley & Sons, Ltd. · Open University Oceanography Course Team (1989) *Seawater: its composition, properties and behaviour*. Oxford/Milton Keynes: Pergamon/Open University Press.

thermocouple   A temperature-measuring device. It employs the phenomenon that when a metal wire is interposed in a wire of a different metal then if one junction is heated relative to the other an electromotive force (emf) is produced (Seebeck effect). Generally, one junction is used as the sensor, the other being held at 0 °C. Various metal combinations produce different emfs and are used for different purposes. WBW

thermodynamic diagram   A chart on which are plotted observations of pressure, temperature and humidity from a given RADIOSONDE ascent. In its simplest form it is a diagram whose vertical axis is pressure and horizontal axis temperature, with the former being logarithmic with pressure decreasing upwards and the latter a linear scale with temperature increasing to the right.

Operational thermodynamic diagrams are more complex, however, and are designed so that area on the chart represents energy; the TEPHIGRAM is one example. This is useful when a forecaster has to assess the likelihood of fog clearance on the basis of the amount of energy required to evaporate it – and whether solar heating at various times of the year will be great enough to meet the requirement.

A variety of derived quantities are easily obtained once the temperatures (dry-bulb and dewpoint) have been plotted; for example, the POTENTIAL TEMPERATURE and both relative and absolute humidity.

Thermodynamic diagrams are also used in the graphical estimation of the nature and intensity of vertical stability in the atmosphere for the time and place represented (see VERTICAL STABILITY/INSTABILITY). RR

Reading
Atkinson, B.W. (1968) *The weather business*. London: Aldus Books.

thermodynamic equation   Expresses a fundamental relationship in meteorology in which the time rate of change of an air parcel's TEMPERATURE as it moves through the atmosphere is related to both ADIABATIC expansion or compression and to diabatic heating. It is a predictive equation.

In a typical middle-latitude disturbance (e.g. cyclone), air parcels at middle levels undergo adiabatic temperature changes of about 30 K day^{-1} while the diabatic changes (due to absorption of solar radiation, absorption and emission of infrared radiation, latent heat release, etc.) tend to compensate one another and have a net value of about 1 K day^{-1}. RR

thermograph   A meteorological instrument that is housed in a weather screen and provides a continuous record of air temperature. The response of a sensor (e.g. a bimetallic coil) to fluctuations of air temperature is magnified by a long pen to which it is connected mechanically. The pen traces a line on a daily or weekly strip chart that is fixed to a slowly revolving clockwork-driven drum. RR

thermohaline circulation (THC)   A slow-moving, but extensive, current or conveyor of water linking the oceans and the surface and deep currents globally. The formation of deep or bottom currents is driven by temperature and salinity differences (density differences), with cold, saline water in some locations (high-latitude North Atlantic and the margins of Antarctica) causing strong overturning. Within the North Atlantic, the Gulf Stream and North Atlantic Drift currents carry warm and relatively salty surface water from the Gulf of Mexico up to the seas between Greenland, Iceland and Norway. As these waters migrate northwards they cool and become dense enough to sink into the deep ocean.

The 'pull' exerted by this dense, sinking water may help to maintain the strength of the warm Gulf Stream, ensuring a current of warm tropical water moves into the North Atlantic that, along with prevailing westerly winds, causes mild air masses to cross the European continent. This keeps winters in the British Isles far milder than what would otherwise be expected for its latitude.

It has proved possible to reconstruct past variations in the THC from oceanic sediment cores. This has shown that the nature of the conveyor can change abruptly in response to the pulses of meltwater from decaying ice sheets and that the switching on and off of the THC helps to explain rapid climatic change events. Times of weakened THC correlate with reduced North Atlantic sea surface temperatures and the cold phases of Dansgaard–Oeschger cycles (see DANSGAARD–OESCHGER (D–O) EVENTS). Ocean circulation modelling studies suggest that a relatively small increase in freshwater flux to the Arctic Sea and critical regions of the high-latitude North Atlantic could cause deep water production to cease, thus providing the trigger for a sudden 'switching off' of deep water formation in the North Atlantic. The drainage of freshwater from the melting of the Laurentide ice sheet of North America, and in particular a drainage event from glacial Lake Agassiz during the Late Glacial, is thought to have triggered the YOUNGER DRYAS. The widespread 8200 cal year BP climate event is thought to be related to a later, and much larger, sudden drainage of Lake Agassiz that disrupted the THC. Such massive pulses of freshwater dilute the dense, salty 'Atlantic Conveyor', forming a temporary lid of fresh water that stops the sinking and pulling of water that drives the Gulf Stream. In consequence, the Gulf Stream weakens or switches off altogether, breaking the 'conveyer belt' and allowing a sea ice cap to form. The expansion of sea ice then exerts a further cooling effect through an albedo feedback and by lowering atmospheric water vapour content. Evidence from ocean sediments indicates that deep water formation in the North Atlantic was diminished during the sudden cold HEINRICH EVENTS and the Younger Dryas phase.   ASG

Reading

Alley, R.B. (2007) Wally was right: predictive ability of the North Atlantic "conveyor belt" hypothesis for abrupt climate change. *Annual Review of Earth and Planetary Sciences*, **35**, 241–263. · Anderson, D.E., Goudie, A.S. and Parker, A.G. (2013) *Global environments through the Quaternary*, 2nd edition. Oxford: Oxford University Press; pp. 317–322. · W.S. Broecker (2010) *The great ocean conveyor: discovering the trigger for abrupt climate change*. Princeton, NJ: Princeton University Press.

thermokarst   Topographical   depressions resulting from the thawing of ground ice (Washburn, 1979). There are many kinds of thermokarst, including collapsed PINGOS, ground-ice mudslumps, linear and polygonal troughs, THAW LAKES and ALASES. Some thermokarst features result from climatic change, but most relate to minor environmental changes that promote thawing; for example, the shift of a stream channel, natural and human-induced disturbance to tundra vegetation. Thermokarst is sometimes used in a wider sense also to include thermal erosion by flowing water at a temperature above 0 °C (French, 1974). As such, it would include thermo-erosional niches and overhangs and various features associated with slopewash.   DES

References

French, H.M. (2013) *The periglacial environment*. Chichester: John Wiley & Sons, Ltd. · Washburn, A.L. (1979) *Geocryology*. London: Edward Arnold.

thermoluminescence   The light emitted in addition to incandescence from an insulating crystal when a sample is heated. Thermoluminescent properties of a material accumulate progressively following exposure to continuous radiation. As a result, the thermoluminescence intensity at any given heating temperature is a product of the amount of radiation received. Specific minerals (e.g. quartz, sodium and potassium feldspar) are found to emit at characteristic excitation temperatures, and in spectra of specific wavelengths. The thermoluminescence signal is usually determined by heating the sample at a constant rate from room temperature up to 500 °C (producing a characteristic 'growth curve'). Independently determining the amount of radiation received means that it is possible to establish the age of the commencement of signal accumulation, providing that a time-zero event is identified (i.e. a time at which all previously accumulated radiation damage has been removed). (See also LUMINESCENCE DATING METHODS.)   SS

Reading

Aitken, M.J. (1990) *Science-based dating in archaeology*. London: Longman.

thermopile   A radiation-detecting device that uses a series of thin-wire THERMOCOUPLES to measure radiation incident on the hot junctions of the thermocouples. The cold junctions are shielded and kept at a uniform, measured temperature. This device differs from a bolometer, where a blackened platinum foil is heated by the radiation and the increase in resistance is measured.   WBW

thickness The difference in height above mean sea level of two pressure surfaces above a given point (see PRESSURE, AIR). It is proportional to the mean TEMPERATURE of the layer in question so that the larger values indicate relatively warm air and smaller values relatively cold air. Thickness values are obtained from RADIOSONDE observations and are plotted on a base map that commonly depicts the variations for the 1000–500 mb layer. Contours are drawn, say, every 60 m to produce a thickness analysis (which is also effectively an isotherm analysis of the layer involved) that is of importance in synoptic meteorology.   RR

Reading
Atkinson, B.W. (1968) *The weather business*. London: Aldus Books.

Thiessen polygon Defines the horizontal area that is nearer to one rain gauge than to any other in a rain-gauge network. The areas of the polygons are used to weight rain-gauge catches when calculating the average areal precipitation. Thiessen polygons are constructed from the perpendicular bisectors of the horizontal projections of straight lines joining adjacent rain gauges. These perpendicular bisectors are extended and joined to leave each rain-gauge site at the centre of a polygon, and the area of the polygon is known as its Thiessen weight. The method is widely applied because it is easy to use and makes allowance for the uneven distribution of rain gauges.   AMG

Reading
Damant, C., Austin, G.L., Bellon, A. and Broughton, R.S. (1983) Errors in the Thiessen technique for estimating areal rain amounts using weather radar data. *Journal of Hydrology*, **62**, 81–94. · Thiessen, A.H. (1911) Precipitation averages for large areas. *Monthly Weather Review*, **39**, 1082–1084.

tholoid A volcanic cone situated within a large volcanic crater or caldera.

threshold, geomorphological A condition that, if exceeded, leads to a dramatic change in a landform or landform stability. A threshold may be exceeded either by intrinsic change of the landform itself, or by a progressive change of an external variable (Schumm, 1979). A factor of safety (FS) equal to 1.0 in hillslope stability analyses is an example of a geomorphic threshold. When the FS (the ratio of resisting to driving stresses) exceeds 1 the slope is stable – slope failure occurs when FS ≤ 1 and the state of the system changes.

An *intrinsic* threshold implies that changes can take place within the system without a change in an external variable. An example is a SURGING GLACIER

that may exhibit periodic surges although the input of snow remains identical over many decades. In this case there is a build-up of snow and ice to a critical level that causes a sudden change in the process of basal sliding. The sudden transition to a fast mode of flow lowers the ice surface until the glacier reverts to a slow mode of flow and the cycle starts once more. In this particular case the threshold probably relates to different modes of sliding related to critical basal conditions (Budd, 1975). Other geomorphological examples of intrinsic thresholds involve river channel patterns and river terrace changes (Schumm, 1973, 1979).

An *extrinsic* threshold describes an abrupt change in landform characteristics triggered by a progressive change in an external variable. Well-known examples are the threshold velocities required to set in motion particles of a given size from the bed of a river. With a continuous change in velocity the response of river channel bedforms or of aeolian forms will suddenly change. Another example is the progressive degradation of vegetation that may allow the threshold of gully formation and thereby active soil erosion to be crossed suddenly.   DES/DRM

Reading and References
Budd, W.F. (1975) A first simple model for periodically self-surging glaciers. *Journal of Glaciology*, **14**, 3–21. · Chorley, R.J. and Kennedy, B.A. (1971) *Physical geography: a systems approach*. Englewood Cliffs, NJ: Prentice-Hall. · Hutter, K. (1982) Glacier flow. *American Scientist*, **70**, 26–34. · Schumm, S.A. (1973) Geomorphic thresholds and complex response of drainage systems. In M. Morisawa (ed.), *Fluvial geomorphology*. Binghampton, NY: State University of New York Press; pp. 299–310. · Schumm, S.A. (1979) Geomorphic thresholds: the concept and its applications. *Transactions of the Institute of British Geographers*, **4** (4), 485–515. · Coates, D.R. and Vitek, J.D. (1980) *Thresholds in geomorphology*. London: George Allen & Unwin. · Montgomery, D.R. and Dietrich, W.E. (1994) Landscape dissection and drainage area–slope thresholds. In M. J. Kirkby (ed.), *Process models and theoretical geomorphology*. Chichester: John Wiley & Sons, Ltd; pp. 221–246.

threshold slopes Hillslopes with inclinations controlled by the resistance of their soil cover to a dominant degradational process. Such slopes are recognized within areas of consistent rock and soil types and erosional processes by nearly uniform inclinations of characteristic hillslope units. It has been postulated that these units are at the maximum inclination, for temporary stability, permitted by the soil strength and pore water pressures within the soil cover.

Three characteristic maximum hillslope angles have been identified for areas prone to erosion by landsliding processes (Carson, 1976):

(1) a frictional threshold slope angle where the soil is a dry rock rubble and the angle of inclination equals the angle of plane sliding friction of the rubble; (2) a semifrictional threshold slope angle for cohesionless soils where the water table can rise to the surface and seepage is parallel to it; the slope angle then approximates to half that of the effective angle of plane sliding friction of the soil; and (3) an artesian condition in which water flows out of the soil and the slope angle is less than that of case (2).

The concept of threshold slopes applies primarily to straight slope segments between upper convexities and lower concavities. In areas of rapid uplift and incision of stream channels, slope angles may be steeper than threshold angles until landsliding reduces the slope angle to the threshold angle. Threshold slopes will eventually be eliminated by creep, wash and other processes. They are consequently temporary features of a landscape dominated by landsliding.   MJS

Reading and Reference
Carson, M.A. (1976) Mass-wasting, slope development and climate. In E. Derbyshire (ed.), *Geomorphology and climate*. Chichester: John Wiley & Sons, Ltd; pp. 101–130. · Carson, M.A. and Kirkby, M.J. (1972) *Hillslope form and process*. Cambridge: Cambridge University Press. · Selby, M.J. (1993) *Hillslope materials and processes*, 2nd edition. Oxford: Oxford University Press.

throughfall   The net rainfall below vegetation cover excluding STEM FLOW. Throughfall comprises both precipitation that falls straight through the vegetation canopy and precipitation that has been intercepted by the vegetation but then drips onto the ground.   AMG

Reading
Ford, E.D. and Deans, J.D. (1978) The effects of canopy structure on stemflow, throughfall and interception loss in a young Sitka spruce plantation. *Journal of Applied Ecology*, 15, 905–917. · Sopper, W.E. and Lull, H.W. (eds) (1967) *International symposium on forest hydrology*. Oxford: Pergamon.

throughflow (or subsurface flow)   Downslope flow within the soil. Where there are well-defined aquifers, downward percolation is commonly rapid enough to prevent appreciable throughflow. Where the bedrock is not highly permeable, lateral flow within the soil is, for many sites, the most effective form of downslope flow because the soil is more permeable and more porous than the bedrock. This is particularly so in open-structured soils like those beneath many mature woodlands. Infiltrated water percolates downwards until it meets an impeding horizon, where it is diverted laterally as saturated throughflow. Impedance may be due to general saturation below, or to the reduction in permeability with depth that is a normal feature of soils, with the greatest reduction near the base of the 'A' horizon.   MJK

Reading
Kirkby, M.J. (ed.) (1978) *Hillslope hydrology*. Chichester: John Wiley & Sons, Ltd.

throw of a fault   The vertical displacement of a fault.

thrust   A low-angle reverse fault. Also a fault occurring on the overturned limb of a fold.

thufur   A soil hummock found in periglacial environments.

Reading
Schunke, E. and Zoltai, S.C. (1988) Earth hummocks (thufur) In M. J. Clark (ed.), *Advances in periglacial geomorphology*. Chichester: John Wiley & Sons, Ltd; pp. 231–245.

thunderstorm   Storm accompanied by lightning, and therefore thunder; rarely heard more than about 1 min after the flash. Separation of electrical charge probably demands, at some stage, water droplets going down colliding with ice crystals going up and reinforcing an initial electrostatic field. More comprehensive theories are numerous, elaborate and uncertain. The name is often used for any severe convective storm of middle latitudes. Tropical storms are rarely accompanied by thunder.   JSAG

Reading
Lane, F.W. (1966) *The elements rage*. Newton Abbot: David & Charles. · Strangeways, I. (2006) *Precipitation: theory, measurement and distribution*. Cambridge: Cambridge University Press.

tidal currents   The periodic horizontal motions of the sea, generated by the gravitational attraction of the Moon and Sun. They are linked hydrodynamically with tidal changes of sea level and have similar spring to neap modulations. Places that have a large TIDAL RANGE also have large tidal currents. Large tidal currents may also occur where tidal ranges are small; for example, near amphidromes, or through narrow straits that separate two regions having different tidal regimes. Typically, tidal currents on the continental shelf have speeds of $1\,\mathrm{m\,s^{-1}}$.   DTP

tidal prism   The total amount of water that flows in or out of a coastal inlet with the rise and fall of the tide, excluding any freshwater discharges. For any given period it is the product of the mean and the high- and low-water surface areas of the bays behind the inlet entrance and the TIDAL RANGE in each segment. ASG

tidal range The vertical distance between tidal low water and high water. It varies between spring and neap tides over a period of 14 days. Ocean tidal ranges are usually less than a metre, but ranges increase as the tides spread onto the shallower continental shelves. Here, typical ranges are 2–5 m, but there are many local variations. In exceptional cases where large spring tides excite local hydrodynamic resonance, as in the Minas Basin of the Bay of Fundy, ranges in excess of 15 m can occur. Tidal amplitude is a half of the tidal range. DTP

tides The regular movements of the oceans and seas, generated by the gravitational attraction of the Moon and Sun. They are most easily observed as changes in coastal sea levels, but the associated horizontal currents are equally important for mariners. There are also tidal movements of the atmosphere and of the solid Earth that are not apparent to the casual observer.

On average, the gravitational attraction between the Earth and Moon balances the orbital centrifugal force. On the side of the Earth nearest to the moon the gravitational force is slightly greater than the centrifugal force, whereas on the opposite side it is weaker. This gives two tidal maxima each day (semi-diurnal tides) as the Earth rotates about its axis; however, the times of the maximum lunar tides are later by an average of 52 min each day because of the advance of the Moon on its orbit. Changes of declination cause daily (diurnal) tides. Longer period tides are generated by varying lunar and solar distances.

Solar tides have average amplitudes that are 46% of the lunar tides, but their maximum values at a site occur at the same times each solar day. Every 14 days, at new and full moon, when the lunar and solar tidal maxima coincide, the combined spring tidal range is large. Between, small neap tide ranges occur when the solar tides tend to cancel the lunar tides.

Tides calculated directly from gravitational theory, the equilibrium tides, are not observed in the ocean because of several additional effects: these include the land boundaries that prevent their westward movement, the deflection of tidal currents caused by the Earth's rotation, the tidal movements of the solid Earth, and the natural periods of oscillation of the oceans and shelf seas. If the natural period is near to a period in the tidal

forcing, large resonant tides are produced. Because the oceans have natural periods close to 12 h, the observed tides are predominantly semi-diurnal.

From the oceans the tides spread to the adjacent continental shelf regions. Here, they are modified by amplification, by local resonances and by reflections at land boundaries. A reflected tidal wave can interfere with an incoming wave to produce zero tidal range at a distance of a quarter wavelength from the reflecting coast; because of the Earth's rotation, the tides circulate around an amphidromic point of zero tidal amplitude. Tidal energy is eventually dissipated by the frictional drag of the sea bed in shallow water. Schemes to use tidal power have been developed over many centuries. Early examples of tidal mills are found in East Anglia and along the coast of New England. Large schemes have been proposed for the Bristol Channel and the Bay of Fundy. The first scheme to use modern technology is La Rance, near Saint Malo in France, which began operating in 1966. DTP

Reading
Cartwright, D.E. (1977) Oceanic tides. *Reports on Progress of Physics*, **40**, 665–708. · Pugh, D.T. (1987) *Tides, surges and mean sea level*. Chichester: John Wiley & Sons, Ltd.

till A Scottish word, popularly used to describe a coarse, bouldery soil, which was adopted to describe unstratified glacier deposits by A. Geikie (1863). The term is now understood to refer to the sedimentary material deposited directly by the action of a glacier. As such, it supplants the former, oversimplistic term boulder clay.

Till covers the land surface in many former land-bound sectors of mid-latitude ice sheets in the former USSR, northern and northwestern Europe, Canada and the northern USA. It has been the focus of much interest among geomorphologists and geologists both because of its importance in the understanding of glacier activity and because of its engineering implications. Till has spawned an enormous literature, including several significant symposium volumes (Goldthwait, 1971; Legget, 1976; Johansson, 1977).

Table 1 shows a classification of till based on the processes of debris release and the position of the debris deposition. Following Lawson (1981), the processes are subdivided into those that are primary and influence the nature of the sediment directly and those that are closely related secondary processes that modify the sediment.

Meltout till forms by the direct release of debris from a body of stagnant, debris-rich ice by melting of the interstitial ice. When it occurs

Till. Generalized relationships of ice and debris in a temperate glacier; vertical scale greatly exaggerated. Two cases of end moraine formation shown: (a) glacier piles up ridge as it slides over bedrock; (b) glacier rests on thick saturated till and produces 'squeeze-up' moraines.
Source: E. Derbyshire, K.J. Gregory and J.R. Hails 1979: *Geomorphological processes*. London: Butterworth. Figure 5.56. Reproduced with permission of Elsevier.

subglacially, the debris is let down onto the underlying bed without modification. Thus, it retains a fabric (see TILL FABRIC ANALYSIS) inherited from transport in the ice, with preferentially oriented pebbles. The deposit frequently consists of structureless, pebbly, sandy silt, discontinuous laminae, and bands of sediment that may be deposited over large clasts. These latter characteristics are the direct result of the melting of debris-bearing REGELATION ice. Meltout also occurs at the surface, but here secondary processes of flow and slumping may affect the sediment.

Table 1 Classification of till based on the processes of debris release and position in relation to the glacier

Primary	Secondary
Subglacial	
Meltout	Deformation
Sublimation	Settling through standing water
Lodgement	
Supraglacial	
Meltout	Sediment flow
Sublimation	Gravitational slumping
	Settling through standing water

Source: Lawson (1981).

In cold, arid environments, such as Victoria Land in Antarctica, the interstitial ice may be lost by sublimation rather than by melting (Shaw, 1979). This second primary process produces *sublimation* till, which also retains textural and structural characteristics inherited from glacial transport.

The third primary depositional process is subglacial and produces *lodgement* till. Particles lodge when the frictional resistance with the bed exceeds the drag of the moving ice. Since the tills are deposited beneath the weight of overlying ice, they tend to be characterized by a high degree of compaction, high SHEAR STRENGTH and low porosity. Foliation, slip planes, fine lamination and horizontal joints are also typical. The fabric of the till usually shows a preferred orientation with elongated pebbles parallel to the direction of ice flow. Occasionally, large boulders may have been moulded into minor ROCHE MOUTONNÉE shapes as a result of overriding during lodgement (Sharp, 1982).

The secondary processes are intimately associated with till deposition. *Deformation* till has been deformed by glacier movement after its primary deposition. The most common occurrence is beneath glaciers where lodgement occurs. Such a layer of deformation till is

531

massive, relatively poorly consolidated and, when saturated, easily collapses beneath the weight of a person! This is the 'mud' surrounding so many glaciers, and it is commonly up to 70 cm thick. Deformation till and lodgement till are two types of subglacial till that have in the past been called *ground moraine*.

Sediment flows and *gravitational slumps* are the result of surface processes that modify supraglacial debris. Sediment flows occur when fine-grained debris is exposed on the surface of a glacier (see COMPRESSING FLOW). They are frequently called *flow tills* (Hartshorn, 1958; Boulton, 1968). The till becomes saturated and flows down the local ice slope. Lawson (1982) studied such processes on the Matanuska Glacier in Alaska and recognized four main types of sediment flow, depending on the proportion of water they contained and the amount of sorting present. On Matanuska Glacier, sediment flows account for 95% of till deposition. Gravitational slumps occur when the surface debris contains insufficient fine material to flow. Instead, it is stable on ice slopes up to ~35°, but further steepening causes it to slump down the ice slope.

A further secondary process is settling through a water column. This is the situation common around the Antarctic where basal debris melts out from the bottom of an ICE SHELF or ICEBERG and falls to the sea floor. Anderson *et al.* (1980) distinguish such a glaciomarine deposit from normal basal tills on several criteria. The most diagnostic characteristics are the horizontal randomly oriented pebble fabrics of the dropstones and the distinctive marine micro-fauna.   DES

References
Anderson, J.B., Kurtz, D.D., Domack, E.W. and Balshaw, K.M. (1980) Glacial and glacial marine sediments of the Antarctic continental shelf. *Journal of Geology*, **88**, 399–414. · Johansson, H.G. (ed.) (1977) A symposium on the genesis of till. *Boreas*, **6** (2), 71–227. · Boulton, G.S. (1968) Flow tills and related deposits on some Vestspitzbergen glaciers. *Journal of Glaciology*, **7**, 391–412. · Geikie, A. (1863) On the phenomena of the glacial drift of Scotland. *Transactions of the Geological Society of Glasgow*, **1**, 1–190. · Goldthwait, R.P. (ed.) (1971) *Till, a symposium*. Columbus, OH: Ohio State University Press. · Hartshorn, J.H. (1958) Flowtill in southeastern Massachusetts. *Bulletin of the Geological Society of America*, **69**, 77–82. · Lawson, D.E. (1981) Distinguishing characteristics of diamictons at the margin of the Matanuska Glacier, Alaska. *Annals of Glaciology*, **2**, 78–84. · Lawson, D.E. (1982) Mobilization, movement and deposition of active subaerial sediment flows, Matanuska Glacier, Alaska. *Journal of Geology*, **90**, 279–300. · Legget, R.F. (ed.) (1976) *Glacial till: an interdisciplinary study*. Royal Society of Canada Special Publication no. 12. Ottawa: Royal Society of Canada. · Sharp, M.J. (1982)

Modification of clasts in lodgement tills by glacial erosion. *Journal of Glaciology*, **28**, 475–481. · Shaw, J. (1979) Tills deposited in arid polar environments. *Canadian journal of Earth Sciences*, **14**, 1239–1245.

till fabric analysis Measurement of the direction and dip of elongated stones in glacial till (see FABRIC). Elongated stones in lodgement TILL and subglacial meltout till tend to be deposited with their long axes parallel to the direction of ice flow.   DES

Reading
Andrews, J.T. (1971) *Techniques of till fabric analysis*. British Geomorphological Research Group Technical Bulletin 6. Norwich: Geo Abstracts. · Glen, J.W., Donner, J.J. and West, R.G. (1957) On the mechanism by which stones in till become orientated. *American Journal of Science*, **255**, 194–205.

tillite A consolidated sedimentary rock formed by LITHIFICATION of glacial till, especially pre-Pleistocene till.

time-transgressive Applies to boundaries of 'climatostratigraphic' units that are not time equivalent (i.e. are diachronous) over large distances. For instance, depending on location and other factors, geological and biological responses to palaeoclimatic changes exhibit time lags of differing lengths. For example, it takes time for glaciers and ice sheets to respond to warming and for plants to change their range. This is particularly true with respect to responses to the often abrupt climatic changes of the Quaternary. This means that proxy data used to reconstruct the timing of a past climatic event in one place may not correlate in time with similar evidence in another place (Walker, 2005: 198).   ASG

Reference
Walker, M. (2005) *Quaternary dating methods*. Chichester: John Wiley & Sons, Ltd.

timebound A data set covering a known period of time. The term is useful when, in climatic data for example, the mean values of a certain parameter are being compared. In a timebound data set, the means for different locations are derived for data covering the same period. In an un-timebound set, the means may be derived for total data sets covering a range of periods, such that the effect of a widespread climatic event (e.g. a 10-year drought) is included in the data for some locations but not all.   DSGT

tolerance The ability of organisms to withstand environmental conditions. Plants and animals within a particular environment have limits of tolerance beyond which they cannot exist. These tolerance limits reflect a range between minimum and maximum values for essential materials, such as heat, light, water and nutrients, that are necessary for growth and reproduction. The relative degree of tolerance is expressed by a series of terms that utilize the prefixes 'steno', meaning narrow, and 'eury', meaning wide. For example, stenothermal and eurythermal refer to narrow and wide temperature tolerance respectively.   ARH

Reading
Odum, E.P. (1971) *Fundamentals of ecology*, 3rd edition. Philadelphia, PA: W.B. Saunders; chapter 5, pp. 106–139.

tombolo   A bar or spit connecting an island to the mainland or to another island.

topographic dune   A DUNE that accumulates where a sand-carrying wind encounters a hill or other obstacle that causes the sand carrying capacity to be reduced, through flow separation, leading to the deposition of sediment and the accumulation of a dune. In some circumstances a valley may also cause the transport capacity of a wind to fall, leading to aeolian deposition. (See CLIMBING DUNE, ECHO DUNE, FALLING DUNE, LEE DUNE and SAND RAMP.)   DSGT

toposequence   A sequence or grouping of related soils that differ from each other on account of their topographical position. (See also CLINOSEQUENCE and CATENA.)

toppling failure   A type of slope failure (usually in rocks) characterized by overturning of columns of rock as they fall from a cliff. The mode of failure is common where bedding planes and joints are inclined to the valley side but dip downwards only to a maximum of around 35°. Beyond this value, sliding is more common, in which case SLAB FAILURE results. Triggering of falls may be due to water pressures in the joints or, for small blocks, ice wedging.   WBW

Reading
De Freitas, M.H. and Watters, R.J. (1973) Some examples of toppling failure. *Geotechnique*, **23**, 495–514.

topset beds   Horizontal sedimentary layers laid down on the surface of inclined beds, as in deltaic environments and aeolian sands.

tor   'An exposure of rock *in situ*, upstanding on all sides from the surrounding slopes . . . formed by the differential weathering of a rock bed and the removal of the debris by mass movement' (Pullan, 1959: 54). This definition is basically the same as that used by Caine (1967): 'residuals of bare bedrock usually crystalline in nature, isolated by freefaces on all sides, the result of differential weathering followed by mass wasting and stripping'. The definition employed by Linton (1955: 476) introduced some genetic implications that have not been accepted as desirable:

a residual mass of bedrock produced below the surface by a phase of profound rock rotting effected by groundwater and guided by joint systems, followed by a phase of mechanical stripping of the incoherent products of chemical action.

Some workers would not recognize a stage of prior deep chemical weathering as being a *sine qua non* for tor development, pointing to the possible role of physical disintegrative processes or the concurrent operation of weathering and stripping processes on rock of variable strength (e.g. Palmer and Radley, 1961).

Although Linton's definition has problems, his description of what tors are like is indeed graphic (Linton 1955: 470):

They rise as conspicuous and often fantastic features from the long swelling skylines of the moor, and dominate its lonely spaces to an extent that seems out of all proportion to their size. Approach one of them more closely and the shape that seemed large and sinister when silhouetted against the sunset sky is revealed as a bare rock mass, surmounted and surrounded by blocks and boulders; rarely will the whole thing be more than a score or so feet high.

Though he was talking about the granite tors of Dartmoor, southwest England, comparable forms occur on a wide range of rock types elsewhere in Britain.   ASG

References
Caine, N. (1967) The tors of Ben Lomond, Tasmania. *Zeitschrift für Geomorphologie NF*, **4**, 418–429. · Linton, D.L. (1955) The problem of tors. *Geographical Journal*, **121**, 470–487. · Palmer, J. and Radley, J. (1961) Gritstone tors of the English Pennines. *Zeitschrift für Geomorphologie NF*, **5**, 37–51. · Pullan, R.A. (1959) Notes on periglacial phenomena: tors. *Scottish Geographical Magazine*, **75**, 51–55.

toreva blocks   Large masses of relatively unfractured rock that have slipped down a cliff or mountain side, rotating backwards towards the cliff in doing so.

tornado   A violent rotating storm with winds of $100 \, \mathrm{m\,s^{-1}}$ circulating round a funnel cloud some 100 m in diameter that includes aerial debris such as doors, bushes and frogs. It is associated with violent cumulonimbus of right-hand parity, identifiable on radar, and is a menace to the midwestern USA. Houses with closed windows may explode, due to sudden imposition of low external pressure.   JSAG

Reading

Lane, F.W. (1966) *The elements rage*. Newton Abbot: David & Charles.

torrent   A swift, turbulent flow of water or lava.

tower karst   Residual limestone hills rising from a flat plain. They are distinguished from KEGELKARST in that the hills have near-vertical slopes and are separated from each other by an alluvial plain or swamp. The extremely steep sides of the hills may be caused by marginal solution or fluvial erosion at the edge of the swampy plains.   PAB

Reading

McDonald, R.C. (1976) Hillslope base depressions in tower karst topography of Belize. In M. Sweeting and K.-H. Pfeffer (eds), *Karstprocesses*. Zeitschrift für Geomorphologie, Supplementbände, vol. **26**. Berlin: Borntraeger; pp. 98–103.

tracers   A general term given to any substance that can be measured, tagged or retrieved in order to infer the operation of an environmental process. A variety of techniques have been developed to investigate the mechanisms, rates and pathways of the transfer of Earth surface materials. The essence of a good tracer is that it closely mimics the environmental behaviour of the property of interest but is considerably more efficient to measure or identify. Strategies involve either seeding the phenomenon of interest with a suitable tracer substance or using existing distinct physical or geochemical properties to interpret system dynamics. For example, river gravels can be seeded with identifiable CLASTS that can be traced downstream. At its simplest this might involve painting a number of clasts that are mapped in subsequent surveys. Higher rates of retrieval are likely if magnets are inserted within clasts and can be detected using a search loop. Further information about the conditions under which bedload movement occurs might be obtained if individual clasts can be radio-tagged (Schmidt and Ergenzinger, 1992). Process dynamics can also be inferred from pre-existing properties in environmental systems. For example, the radionuclide caesium-137, produced from nuclear weapons testing, is strongly adsorbed near the soil surface, such that its environmental mobility is largely controlled by movement of the soil material. Consequently, the measurement of caesium-137 can be used to infer rates of SOIL EROSION and deposition (Ritchie and McHenry, 1990). Other properties of sediments, such as mineralogy, PARTICLE SIZE, organic content, mineral magnetic characteristics and radionuclide inventories, are often strongly related to either geological or source conditions. The properties of sediments transported can therefore be used to 'fingerprint' the relative contribution from different parts of the catchment or different sources (e.g. surface or subsurface soils, contrasting land-use types).   DH

Reading and References

Foster, I.D.L. (ed.) (2000) *Tracers in geomorphology*. Chichester: John Wiley & Sons, Ltd. · Ritchie, J.C. and McHenry, J.R. (1990) Application of radioactive fallout cesium-137 for measuring soil erosion and sediment accumulation rates and patterns: a review. *Journal of Environmental Quality*, **19**, 215–233. · Schmidt, K.-H. and Ergenzinger, P. (1992) Bedload entrainment, travel lengths, rest periods – studied with passive (iron, magnetic) and active (radio) tracer techniques. *Earth Surface Processes and Landforms*, 17, 147–165.

tractive force   The drag force exerted when a fluid moves over a solid bed. In UNIFORM STEADY FLOW in open channels this force is the effective component of the gravity force acting on the water body in the direction of flow. For a reach of length L, cross-section area A and slope S, this is $\gamma_f ALS$, where γ_f is the unit weight of water. The average value of tractive force per unit of bed area (the mean bed shear stress τ_0) is

$$\tau_0 = \frac{\gamma_f ALS}{PL} = \gamma_f RS \approx \gamma_f dS$$

if the wetted perimeter is P. The simplification follows because the HYDRAULIC RADIUS $R = P/A$ and is approximated by depth d in wide channels. Channel perimeter sediments have a maximum permissible tractive force or shear stress, the critical or threshold shear stress τ_{0c}. If the flow exerts a stress in excess of this, entrainment of erosion (traction) occurs at a rate dependent on the excess stress (see DU BOYS EQUATION). It is theoretically possible to design channels to carry clear water with no sediment transport by ensuring that the perimeter sediments are everywhere at or below the threshold state; this is the tractive force theory of channel design (Richards, 1982: 281–286).   KSR

Reference

Richards, K.S. (1982) *Rivers: form and process in alluvial channels*. London: Methuen.

trade winds   Winds with an easterly component that blow from the subtropical high-pressure areas around 30° of latitude towards the equator.

Although only surface winds, they are a major component in the general circulation of the atmosphere as they are the most consistent wind system on Earth. Together, the northeast and southeast trade winds occupy most of the tropics.   BWA

transfer function   The transfer function S is the operator that defines the relationship between the time-series (or space-series) of inputs X_t to a SYSTEM and the time-series (or space-series) of outputs Y_t from the system:

$$Y_t = SX_t$$

Thus, S represents the effects of a given system upon the variable Y. Transfer functions may be developed through a BLACK BOX approach, in which two variables are related statistically, under the assumption that one variable Y responds to X. However, transfer functions may have a strong physical basis and may be derived from experimentation as parameters within equations derived from first physical principles.   SNL

Reading
Bennett, R.J. and Chorley, R.J. (1978) *Environmental systems: philosophy, analysis and control.* London: Methuen.

transgression   Sometimes used to describe a marine incursion resulting from deglaciation. The term is most widely used in the context of the 'Flandrian transgression' that occurred as the ice sheets of the last glacial melted in the Late Pleistocene and Holocene. World sea levels rose, possibly by as much as 170 m. The continental shelves were flooded, and river and glacial valleys were transformed into rias and fiords respectively. This transgression was more or less complete by 6000 years ago and may locally have reached a few metres above the present level. It caused the separation of Ireland from mainland Britain, and of Britain from the continent of Europe. In areas of very gentle slope, like the Arabian Gulf, the rapid lateral spread of the transgression (by as much as $100 \, \text{m} \, \text{a}^{-1}$ at its peak) may have contributed to the biblical story of the flood.   ASG

translational slide   Occurs where the failure of the soil or rock is along planes of weakness (such as bedding planes or joints) that are approximately parallel to the ground surface. This term is often used in a broad sense and can include a variety of types of mass movement, such as mudflows, debris flows, and so on. Solid rock movements, such as wedge failures, can also be translational.   WBW

transmission loss   Rivers that flow from a subhumid area into a more arid climate may suffer an attenuation of flow downstream, which is exacerbated by transmission losses into sediments. Knighton and Nanson (1994) found that transmission losses amounted to over 75% of discharge at certain flows along Cooper's Creek, Australia, while in the Kairouan area of Tunisia, Besbes (2006) reported transmission losses of 40–50% of flood volume. Transmission losses occur because of percolation into channel beds, and, at high flows into channel terraces palaeochannels and the like. They may also occur because of evapotranspiration where floodwaters become disconnected in low-lying areas, such as billabongs (McMahon *et al.*, 2008). Transmission losses increase with increasing channel wetted perimeter and also with the permeability of bed and bank materials, which in turn relates to whether they are sand or clay rich (Dunkerley, 2008).   ASG

References
Besbes, M. (2006) Aquifer recharge by floods in ephemeral streams. In I. Tchiguirinskaia, K. N. N. Thein and P. Hubert (eds), *Frontiers in flood research.* IAHS Publication 305. Wallingford: IAHS Press; pp. 43–72. · Dunkerley, D.L. (2008) Bank permeability in an Australian ephemeral dry-land stream: variation with stage resulting from mud deposition and sediment clogging. *Earth Surface Processes and Landforms,* **33**, 226–243. · Knighton, D. and Nanson, G.C. (1994) Flow transmission along an arid zone anastomosing river, Cooper Creek, Australia. *Hydrological Processes,* **8**, 137–154. · McMahon, T.A., Murphy, R.E., Peel, M.C., *et al.* (2008) Understanding the surface hydrology of the Lake Eyre Basin: part 2 – streamflow. *Journal of Arid Environments,* **72**, 1869–1886.

transpiration   Plant perspiration, or the loss of water vapour mainly from the cells of the leaves through pores (stomata) but also from the leaf cuticle and through lenticels of the stem. The cooling effect is secondary to the fundamental role of the transpiration, bringing a stream of water, and dissolved mineral nutrients, from the root hairs through the stem vessels (xylem) to the leaves, which is maintained by the vapour pressure gradient of the transpiring cell surfaces. The velocity of the transpiration stream varies from 1–$2 \, \text{m} \, \text{h}^{-1}$ in coniferous trees to $60 \, \text{m} \, \text{h}^{-1}$ in some herbs. The main control mechanism on transpiration is the opening and closing of the stomata, induced by osmotic pressure changes consequent upon water balance changes due to high temperatures or other factors.   KEB

Reading
Etherington, J.R. (1982) *Environmental and plant ecology,* 2nd edition. Chichester: John Wiley & Sons, Ltd.

transport limited   A condition where the rate of material transport is lower than regolith formation (supply unlimited). Weathering and soil formation rates are faster than rates of removal. Slope form is greatly controlled by mass movement processes, and slope wash. Slopes may tend to be convexo-concave. By contrast in WEATHERING-LIMITED SLOPES (supply limited), the rate of regolith formation is slower than transport. Debris removal by erosional processes, such as mass wasting, slope wash, fluvial activity, and so on, is faster than weathering (soil-forming) processes. Under this situation, slopes are steep and have little to no soil, they may evolve through parallel retreat, and structure and lithology control their shape. Recently, there has been an increasing interest in how weathering affects rates of erosion and vice versa (e.g. West *et al.*, 2005).   ASG

Reading and Reference

Carson, M.A. and Kirkby, M.J. (1972) *Hillslope form and process*. Cambridge: Cambridge University Press; pp. 104–105. · West, A.J., Galy, A. and Bickle, M. (2005) Tectonic and climatic controls on silicate weathering. *Earth and Planetary Science Letters*, **235**, 211–228.

transverse dune

A type of sand dune, found in desert and sometimes coastal environments, that forms perpendicular to the dominant sand-transporting wind direction, in environments where one general direction of sand transport occurs. Isolated BARCHAN forms or more continuous barchanoid or transverse ridges may occur, depending on the supply of sediment for dune building and the wind energy in the environment. Transverse dunes usually have distinct slip-faces and gentle stoss slopes, and migrate in a downwind direction, the rate of transport depending on wind energy and the volume of sand in the dune, with, generally, large dunes moving more slowly than small ones. Net migration rates of up to $63\,\mathrm{m\,a}^{-1}$ (for 3 m high dunes) have been measured, while in a dune field small barchan dunes may migrate on to, or coalesce with, larger, slower forms, even forming compound megabarchans (tens of metres high) in extreme cases. In cases where extreme (~180°) seasonal wind direction changes occur, the direction of migration may reverse, resulting in a reversing dune. See DUNE.   DSGT

Reading

Lancaster, N. (2011) Sand seas and aeolian bedforms. In D. S. G. Thomas (ed.), *Arid zone geomorphology: process, form and change in drylands*, 3rd edition. Chichester: John Wiley & Sons, Ltd; pp. 486–515.

transverse rib

A narrow ridge of well-imbricated pebbles and cobbles, which lies transverse to streamflow direction. Transverse ribs normally form a series of regularly spaced ridges, apparently associated with the development of antidune breaking waves.   JM

treeline

The upper altitudinal limit to which trees can grow. This depends on such factors as latitude, aspect, exposure and soil type. In arid areas there may be a lower treeline, the position of which is largely controlled by moisture availability, a commodity that tends to become scarcer at lower altitudes.

triaxial apparatus

A device used to measure the SHEAR STRENGTH of a soil according to the MOHR–COULOMB EQUATION criterion. Unlike the SHEAR BOX, failure is not along a predetermined line but takes place in a cylinder loaded axially across the ends by a compressive stress. In its simplest form (for cohesive soils only), the curved surfaces of the cylinder are at atmospheric pressure and failure produced by the axial compression gives an 'unconfined' strength. More usually, a surrounding pressure is applied to the sample, tested in a water-filled cell. This surrounding pressure is changed so that three tests give failure at different axial loads. A plot involving the applied load and cell pressure is a Mohr circle construction from which the strength of the soil can be derived. In clays, a shear plane can develop in the soil, but granular materials tend to bulge and a given STRAIN value is used to indicate failure (often 10%). WBW

Reading

Lambe, T.W. and Whitman, R.V. (1981) *Soil mechanics*. New York: John Wiley & Sons, Inc.

Trombe's curves

A graph portraying the relationship between the calcium content of saturated solutions at different temperatures and the pH. As pH falls from alkali to acid conditions, Trombe suggests that there is a curvilinear increase in the amount of calcium able to be held in solution. These curves are now superseded by more recent work.   PAB

Reading

Sweeting, M.M. 1972: *Karst landforms*. London: Macmillan.

trophic levels

Literally, 'nourishment' or feeding levels within a biological system, which represent stages in the transfer of energy through it: the concept links in with that of the food chain or food web (see FOOD CHAIN, FOOD WEB). Thus, in the grazing food chain, all producer organisms (green plants, blue–green and other algae, phytoplankton, etc.) are placed in the first trophic level, which contains the maximum store of energy that is available for use in any given system; the second

To next trophic level ($kcal/m^2/yr$)				Respiratory loss ($kcal/m^2/yr$)
Exported	Retained			
8	0	21	(top Carnivores)	13
46	21	383	(Carnivores)	316
1555	383	8428	(Herbivores)	6490
405	8428	20810	(Producers)	11977

is comprised of HERBIVORE consumers, the third of CARNIVORE consumers, the fourth of top carnivore consumers, and so on.

The amount of energy present in any trophic level results in a characteristic trophic structure that is determined by the constraints set by the first two laws of thermodynamics, the first of which (the law of the conservation of energy) may be expressed by

$$\Delta E = Q + W$$

where ΔE refers to changes in the internal energy of that level, Q represents the heat given off by it (mainly, the heat associated with RESPIRATION) and W is the work done within it (i.e. the energy retained in cells for growth). The second states that most energy will eventually degrade to heat. The application of the theoretical relationships to real-life situations may be seen by reference to the classic series of experiments conducted at Silver Springs, Florida, by H. T. Odum (1957), a summary of the results of which is given in the diagram. In this small aquatic system the amount of energy E fixed, mainly through photosynthesis, in the first trophic level amounted to $20{,}810\,kcal\,m^{-2}\,a^{-1}$; respiration Q was equivalent to the loss of $11{,}977\,kcal\,m^{-2}\,a^{-1}$, and a small quotient went into a subsidiary, DECOMPOSER food chain. Some $8428\,kcal\,m^{-2}\,a^{-1}$ were stored at this level (W), this being

its NET PRIMARY PRODUCTIVITY (NPP), or, in other words, the quantity available for passage to the second trophic level, at which point a new 'first law' balance comes into operation. The process is then repeated at subsequent trophic levels, causing a further diminution in the amount of available energy in each, until eventually all the energy within the system has been utilized.

In view of the above, the number of trophic levels that any biological system can support is clearly limited. The usual maximum on land is four or five. But in marine environments, where the ratio of plant to animal BIOMASS is much more balanced, and energy transfer efficiencies are greater, there may be up to seven. Unstable and unpredictable communities tend in general to have fewer trophic levels than do stable ones (May, 1975); and so also do polluted ones. (See also BIOLOGICAL PRODUCTIVITY and ECOLOGICAL ENERGETICS.) DW/MEM

References

May, R.M. (1975) Stability and complexity in model ecosystems. Princeton, NJ: Princeton University Press. ·
Odum, H.T. (1957) Trophic structure and productivity of Silver Springs, Florida. *Ecological Monographs*, **27**, 55–112.

tropical cyclones A collective term that refers to the intense cyclonic vortices that are observed principally across tropical oceans and exhibit maximum sustained surface winds of $33\,m\,s^{-1}$ (64 kn).

Tropical cyclones. Distribution, frequency and movement (hurricanes). Encircled numbers indicate approximate number in each area per year.
Source: Goudie (1984) Reproduced with permission of Andrew Goudie.

They are known as hurricanes in the Caribbean, typhoons in the northwest Pacific and cyclones in the Indian Ocean. They are formed from pre-existing disturbances most frequently in the 10–15° latitude band and are known to be favoured by the following conditions:

1 Strong low-level cyclonic relative VORTICITY.
2 A reasonably large CORIOLIS FORCE in order for an organized circulation to develop.
3 A small difference between the disturbance velocity and the vertical profile of the horizontal wind of the large-scale surrounding current (this small 'ventilation' aids the concentration of heating in a vertical column).
4 Sea-surface temperature warmer than 27 °C.
5 An unstable LAPSE RATE from the surface to middle levels.
6 High humidity at mid-tropospheric heights.

Tropical cyclones are characterized by cyclonic inflow (counterclockwise in the northern hemisphere) that is strongest in the lowest 2 km, with inward-spiralling cloud bands hundreds of kilometres long and ten kilometres or so wide. These bands converge towards the deep wall cloud that surrounds the eye of the system and is the zone of strongest WINDS (up to $100 \, m \, s^{-1}$ in extreme cases) and heaviest PRECIPITATION ($50 \, cm \, day^{-1}$ in vigorous cyclones). The eye is a circular region in which the air subsides, winds decrease and precipitation ceases and it is typically 10–15 km in diameter.

These systems are classically asymmetric in plan view because one flank (right in the northern and left in the southern hemisphere) is characterized by stronger flow as a result of the compounding effect of the large-scale steering current when 'added' to the disturbance's wind pattern. The elevated water levels of storm surges (e.g. in the Gulf of Mexico and Bay of Bengal) are associated with a cyclone's low central pressure (typically 920–950 mb), and damage is also caused by the high winds and heavy precipitation that commonly occur in tropical cyclones.

The strength of the inflow decreases to a minimum at middle levels (near 6 km), where the ascent is strongest, while at the top of the circulation (above 9 km) air spirals out anticyclonically (clockwise in the northern hemisphere) to balance roughly the mass flowing in at low levels. The high-level outflow is marked by an extensive shield of cirrus that is clearly visible from weather satellites. In general, tropical cyclones are around 650 km in diameter, and thus substantially smaller than middle-latitude cyclonic disturbances.

For the period 1958–1977 an annual average of 54 tropical cyclones were observed in the northern hemisphere and 24 in the southern hemisphere. Of this total, 33% occurred in the northwest Pacific, 17% in the northeast Pacific, 13% in the Australasian area, 11% in the northwest Atlantic, 10% in the South Indian Ocean and 8% in both the South Indian and Pacific Oceans. The time of maximum frequency coincides with the period of highest sea-surface temperatures: 72% occur from July to October in the northern hemisphere (21% in September) and 68% occur from January to March in the southern hemisphere (25% in January). RR

Reading and Reference
Goudie, A.S. (1984) *The nature of the environment*. Oxford: Blackwell. · Pielke, R.A. (1990) *The hurricane*. London: Routledge.

tropical forest Defined literally, an area lying between the lines of the tropics, with trees as the dominant life form. In practice, similar forests extend outside the tropics. There are three broad groups of tropical forest: evergreen, deciduous and mangrove. Evergreen forests are the most widespread, characterized by leaf-exchange mechanisms and little or no bud protection. They occur in lowlands below around 1000 m, in mountain and cloud formations and in water-saturated areas. Deciduous forests range from areas with predominantly evergreen subcanopy trees to dominantly deciduous trees throughout the vertical structure of the forest. Mangrove forests have a range of adapations to saline conditions and waterlogging. PAF

Reading
Golley, F.B., Leith, H. and Werger, M.J.A. (eds) (1982) *Tropical rain forest ecosystems*. Amsterdam: Elsevier. · Longman, K.A. and Jenik, N. (1987) *Tropical forest and its environment*, 2nd edition. London: Longman. · Mather, A.S. (1990) *Global forest resources*. London: Belhaven Press. · Whitmore, T.C. (1975) *Tropical rainforests of the Far East*. Oxford: Clarendon Press.

tropical rain forest Forest formations of the permanently moist, perhumid, tropics and subtropics. The term was originally coined (as *tropische Regenwald*) by the German botanist A. W. F. Schimper in 1898 (English translation 1903). Schimper in fact identified four major forest types within the tropics, these being thorn and savanna forest at the drier end of the climatic continuum and monsoon and rain forest at its wetter end. Rain forests, then, are typically associated with areas that lack seasonality in terms of either temperature or rainfall. The tropical rain forest climate is consistently warm (mean temperature of the coolest month exceeding 18 °C) and moist (with mean annual precipitation usually at least 2000 mm and no month with less than 100 mm). Tropical rain forests occur in all three tropical land areas: the Amazon and Orinoco basins of South America, the Zaire basin of West and Central Africa, and the eastern tropics of Indo-Malesia. Strictly speaking, they are confined to areas between the tropics of Cancer and Capricorn, although there are localities immediately beyond the tropics that, owing to particular combinations of deeper soil and groundwater, may support closed forest of equivalent diversity and structure.

Tropical rain forest is the most luxuriant of all plant communities; trees are tall, perhaps in excess of 45 m and even up to 80 m (*Araucaria hunsteinii* of the New Guinean rain forest is one of the world's tallest trees at 89 m), and form a closed, evergreen canopy. Conventionally, the forest is identified as having three tree layers or strata, a group of emergents lying over a main stratum at around 20 m down to a layer dominated by smaller, shade-tolerant trees below 15 m. In terms of physiognomy, the large trees are dominated by those with clear, cylindrical boles, often with plank or stilt buttresses. Although there is great diversity in leaf shape and size, they are most frequently mesophyll in structure; pinnate forms are prominent, and so-called drip-tips are also common features. Cauliflory (i.e. flowers and fruits borne directly on stems or boles) is an intriguing feature of a number of species. Large woody climbers, either free-hanging or bole-climbing, are especially abundant in these forests.

Perhaps more than any other biogeographical characteristic, it is the species diversity of tropical rain forests that has attracted most attention. It is estimated that these regions contain more than 100,000 flowering plant species, more than 40% of the global flora. Locally, extraordinary richness is evident; for example, there may be as many as 180 different tree species per hectare and as many as 330 species if the other forms of plant life are included. Animal diversity reaches extraordinary proportions in the tropical rain forest, although precise numbers of species are impossible to calculate because most are insects found in the tree canopy and remain undiscovered. Terry Erwin's work in the rain forest suggests insect diversity at an unfathomable scale, with as many as 163 different species of beetle within the canopy of a single type of tree (Erwin, 1983). If this is representative of insect diversity throughout the rain forests, then most estimates of global BIODIVERSITY would have to be increased by an order of magnitude or more. The factors underlying these levels of diversity have been the object of considerable debate. Earlier interpretations were based on the presumed ancient origin and environmental stability of these forests, although it is now realized that environmental fluctuations over time have impacted rain forests. Perhaps, then, the nature of such changes, together with the relatively sedentary nature of many of the constituent forest species, has favoured ALLOPATRIC speciation and resulted in elevated species numbers.

The issue as to how such luxuriant and diverse forests are sustained is also of interest and importance. Much has been made of the fact that high levels of BIOLOGICAL PRODUCTIVITY appear to be based on relatively infertile soils, certainly in the case of forests developed on geologically ancient land surfaces or highly leached soils. This has led to suggestions that nutrient cycling in the rain forest must be highly efficient and that the majority of nutrients are maintained within the BIOMASS itself and recirculated rapidly and efficiently by decomposition. More recently, however, it has been realized that there is great variability in the distribution of mineral nutrients in these ecosystems, and a substantial proportion of the nutrients may even be supplied from external sources.

Human impact on the tropical rain forests has been very problematic and has, especially since the late 1970s, taken the form of wholesale clearance in many areas, often for subsistence or plantation agriculture. In forests associated with impoverished soils, this means not only the loss of BIODIVERSITY, but also the removal of the physically protective vegetation cover and reduction in total nutrient content. As a consequence, such agriculture is often unsustainable for more than a few seasons and the land may subsequently be abandoned and exposed to erosion. Rates of clearance are subject to much speculation, but it is clear that the effects of development in many remote areas of, for example, Rondoñia State in Amazonia are threatening biological diversity of arguably the world's most important ecological entity.   MEM

Reading and References
Corlett, R.T. and Primack, R.B. (2011) *Tropical rain forests: an ecological and biogeographical comparison*. Chichester: John Wiley & Sons, Ltd. · Erwin, T. (1983) Beetles and other insects of tropical forest canopies at Manaus, Brazil, sampled by insecticidal fogging. In S. L. Sutton, T. C. Whitmore and A. C. Chadwick (eds), *Tropical rain forest ecology and management*. Oxford:

Blackwell; pp. 59–75. · Park, C.C. (1992) *Tropical rainforests*. London: Routledge. · Proctor, J. (ed.) (1989) *Mineral nutrients in tropical forest and savanna ecosystems*. Oxford: Blackwell. · Whitmore, T.C. (1991) *An introduction to tropical rain forests*. Oxford: Clarendon Press.

tropopause The boundary between the TRO-POSPHERE and the STRATOSPHERE, usually revealed by a fairly sharp change in the LAPSE RATE of temperature. The change is in the direction of increased atmospheric stability from regions below to regions above the tropopause. The height of the tropopause is about 20 km in the tropics and 10 km in polar regions, the decline with increasing latitude being step-like, with jet streams occupying the steep rises. The tropopause is frequently difficult to locate, largely due to its comprising several 'leaves', giving rise to the idea of a multiple tropopause, rather than a single continuous surface.   BWA

troposphere The portion of the atmosphere lying between the Earth's surface and the TROPO-PAUSE. Owing to the varying height of the tropopause, tropospheric depths vary, on average, from 10 to 20 km, the low values being in polar regions and the high in tropical regions. Within the troposphere, the mean values of temperature, water vapour content and pressure decrease with height. Horizontal WIND speeds increase with height, and VERTICAL MOTION is substantial. As a result, the troposphere is the part of the atmosphere that contains all the WEATHER we experience from day to day.   BWA

trottoir (from the French word for a pavement or sidewalk) A narrow organic reef constructed by such organisms as *Lithophyllum tortuosum*, *Vermetidae* and *Serpulidae*. Trottoirs are common in the Mediterranean and in low latitudes and develop in the intertidal zone.   ASG

Reading
Tzur, Y., and Safriel, U.N. (1979) Vermetid platforms as indicators of coastal movements. *Israel Journal of Earth Sciences*, 27, 124–127.

trough In meteorology, this term virtually always relates to pressure (see AIR PRESSURE). Thus, a pressure trough is an elongated area of relatively low pressure; the opposite is a ridge, of relatively high pressure. The words trough and ridge clearly derive from the valleys and ridges familiar to us in the solid earth. Troughs occur on scales ranging from mesoscale to continental, the most familiar being those appearing on the synoptic weather map on television. The very largest troughs, found within the planetary waves, or

ROSSBY WAVES, are critical to the formation of EXTRATROPICAL CYCLONES.   BWA

truncated spur Steepened bluff on the side of a glacial trough in between tributary valleys. It arises through the widening and straightening of a pre-existing sinuous river valley by glacial action. Such features have long been regarded as characteristic of glacial erosion.   DES

tsunami Sea surface waves generated by submarine earthquakes (most transoceanic tsunamis are generated by massive subduction-zone earthquakes), volcanic eruptions, landslides (and perhaps large bolide impacts) that are generally imperceptible in deep water but may be very destructive when striking the shoreline. The Indian Ocean tsunami of 26 December 2004 was the most catastrophic such event in recent history, killing more than 230,000 people in the near field and a further 70,000 in the Indian Ocean far field. This death toll was far in excess of the estimated 40,000–50, 000 deaths associated with the Lisbon earthquake and tsunami of 1 November 1755 and the 36,500 deaths associated with the tsunami generated by the cataclysmic explosion of Krakatau on 26–27 August 1883.

Following generation, tsunami dynamics relate to three sets of processes: (1) propagation, either from open ocean to more restricted coastal waters, entirely within shallow nearshore waters or within lakes; (2) inundation of onshore areas; and (3) traction, the tsunami-generated backwash current from the shoreline into deeper waters. Processes (2) and (3) are often accompanied by the transport of sediments and large amounts of both natural and artificial debris. Fast-moving tsunamis have a small amplitude and a very long wavelength; in the open ocean, the 2004 Asian tsunami had a wave height of 1 m, a wavelength of 430 km, a wave period of 37 s and a wave velocity of 200 m s^{-1}. As the tsunami approaches the coast and the waters become shallow, wave shoaling compresses the wave. Its speed decreases below 80 km h^{-1}, wavelength diminishes to <20 km and amplitude grows enormously. Except for the very largest tsunamis, the approaching wave does not break, but rather appears like a fast-moving tidal bore. Open bays and coastlines adjacent to very deep water may shape the tsunami further into a step-like wave with a steep-breaking front. If the first part of the tsunami to arrive at the coast is the wave trough, then the arrival phase may show 'drawback', an exposure of the sea floor that can extend for hundreds of metres out to sea before a rapid rise in water level as the crest of the tsunami wave arrives. In the 2004 Asian tsunami, shorelines close to the earthquake epicentre on the island of Sumatra were characterized by run-up elevations of 15–30 m. Wave scour and subsidence set back the shoreline at Banda Aceh by up to 1.5 km; eroded sand was deposited in overwash-type deposits over 70 cm thick in places. In southeast Sri Lanka, run-up levels reached 11 m, whilst on the east coast of India and in the Maldives run-up was of the order of 3–4 m. The tsunami waves associated with the 11 March 2011 Tohoku earthquake reached heights of over 40 m at Miyako, Iwate Prefecture, and in the Sendai area, and travelled up to 10 km inland, slightly exceeding the run-up elevations calculated for the 1896 Sanriku earthquake in the same region.

There has been considerable debate as to whether or not coastal ecosystems, such as mangrove (see MANGROVE, MANGROVE FOREST), provide effective natural protection against tsunami waves. Whether or not a coral reef enhances a tsunami during inundation, has little or no effect, or reduces the impact of the tsunami depends on the reef morphology, coastal planform and bathymetry and the nature of the tsunami waves themselves. Narrow (<100 m) fringing reefs cause more wave shoaling than energy dissipation, resulting in a change in an increase in wave height and the tsunami inundating farther inland. Wide fringing reefs cause greater dissipation of energy through bottom friction, resulting in lower wave heights by the time the wave passes the shoreline and reduced inundation distances compared with if there were no reef. A balance between wave shoaling and wave dissipation such that the reef has little or no impact on the tsunami appears to occur at a reef width of about a few hundred metres.

Tsunami wave velocities are such that they can transport fine clay to gravel size materials and, because the periods of tsunami waves are very long, any coastal zone subject to tsunami inundation will experience a series of tsunami waves over a period of hours; each wave can deposit new sediments and rework those from previous wave arrivals. Sediment erosion, transport and deposition is associated with both onshore run-up and offshore backwash. The resulting sedimentation pattern is therefore likely to be complex. Although there has been debate about the separation of storm deposits from tsunami deposits, sand sheets that can be related to known earthquakes or landslide tsunami triggers have been described from, amongst other locations, the coasts of the northwest Pacific Ocean, the eastern boundary to the North Sea and southwest Portugal. The evidence from Holocene sediment sequences along subduction-zone coasts is that tsunamis that leave a geological trace appear to strike every few centuries to millennia. In northeast Japan, the geological record suggests the impact of a mega-tsunami comparable to the 2011 event is likely to recur once every 1000 years.

Around the Pacific Ocean, which is particularly vulnerable to tsunamis, a network of tide gauges is coordinated to give advanced warning of their arrival; a similar system is being implemented in the Indian Ocean in the wake of the 2004 tsunami. TS

Reading
Bryant, E. (2007) *Tsunami: the underrated hazard*, 2nd edition. Chichester: Springer/Praxis. · Chatenoux, B. and Peduzzi, P. (2007) Impacts of the 2004 Indian Ocean tsunami: analysing the potential protecting role of environmental features. *Natural Hazards*, 40, 289–304. · Dawson, A.G. and Stewart, I. (2007) Tsunami deposits in the geological record. *Sedimentary Geology*, 200, 166–183. · Gelfenbaum, G., Apotsos, A., Stevens, A.W. and Jaffe, B.E. (2010) Effect of fringing reefs on tsunami inundation; American Samoa. *Earth-Science Review*, 107, 12–22. · Merrifield, M.A., Firing, Y.L., Aarup, T., *et al.* (2005) Tide gauge observations of the Indian Ocean tsunami, December 26, 2004. *Geophysical Research Letters*, 32, L09603, doi: 10.1029/2005GL022610. · Spencer, T. (2007) Coral reefs and the tsunami of 26 December 2004: generating processes and ocean-wide patterns of impact. *Atoll Research Bulletin*, 544, 1–36. · Tsuji, Y., Satake, K., Ishibe, T., *et al.* (2014) Tsunami heights along the Pacific Coast of northern Honshu recorded from the 2011 Tohoku earthquake and previous great earthquakes. *Pure and Applied Geophysics*, 171, 3183–3215.

tufa A freshwater carbonate deposit that, according to Pentecost (1981: 365), is 'a soft, porous, calcareous rock formed in springs, waterfalls and lakes in limestone regions'. The term is often used interchangeably with *travertine*, although some authors consider one as being a special case of the other. Pentecost (1981: 365), for example, states that 'travertine is identical in composition to calcareous tufa but is a hard non-porous variety used for building'. *Sinter* or *calc-sinter* is another commonly used term but is usually restricted to inorganically precipitated deposits. Tufas can form significant landforms (terraces, barrages, dams, etc.) and may contain much palaeoenvironmental information. Organic processes (e.g. precipitation by blue–green algae) probably play a major role in their development. ASG

Reading and Reference
Pentecost, A. (1981) The tufa deposits of the Malham District, North Yorkshire. *Field Studies*, 5, 365–387. · Viles, H., and Pentecost, A. (2007) Tufa and travertine.

In D. J. Nash and S. McLaren (eds), *Geochemical Sediments and Landscapes*. Chichester: John Wiley & Sons, Ltd; 173–199.

tuff Consolidated clastic material ejected from volcanoes with a predominance of fragments less than 2 mm in diameter.

tundra Vast, level, treeless and marshy regions, usually with permanently frozen subsoil. Originally derived from northern Eurasia, the term has expanded to include all Arctic and Antarctic areas polewards of the TAIGA, and also to similar alpine environments above the timberline on mountains. Drier tundra sites are characterized mainly by herbaceous perennials, with occasional trees and scattered heath plants, grasses, lichens and mosses, while cotton grass, hygrophytic sedges and willows are typical of wet sites. Cryptophytic communities develop in snow and ice.　　PAF

Reading
Bliss, L.C., Heal, D.W. and Moore, J.T. (eds), (1981) *Tundra ecosystems: a comparative analysis*. Cambridge: Cambridge University Press. · Ives, J.D. and Barry, R.T. (eds) (1979) *Arctic and alpine environments*. London: Methuen.

tunnel valleys Formed by subglacial stream action. They tend to have flat floors, steep sides and irregular long profiles, and are a feature of northern Germany and Denmark.

tunnelling, tunnel gully erosion A form of erosion, initiated by subsurface water movement, that often causes surface collapse, leading to open gullying. Water movement through soil cracks eluviates material, thereby leading to the development of tunnels that continue to erode as gullies, following tunnel collapse. It is thus related to PIPES.　　ASG

Reading
Lynn, I.H. and Eyles, G.O. (1984) Distribution and severity of tunnel gully erosion in New Zealand. *New Zealand Journal of Science*, 27, 175–186.

turbidite A sedimentary sequence, fining upwards from coarse sands to clays, deposited by a turbidity current. These submarine flows of sediment are usually triggered on continental slopes by earthquakes, with deposition covering extensive areas of the continental rise and abyssal plains.　　PSH

turbidity current A density current. A sinking mass of sediment-laden air or water. Their erosive activity is thought to contribute to the formation of some submarine canyons on the continental shelves. Sediments deposited by turbidity currents are known as turbidites.

turbulence In wind or water, in contrast to LAMINAR FLOW that occurs in parallel layers, turbulence consists of a series of apparently random, quasi-periodic eddies of differing size and velocity that mix the flow over a wide range of scales. Deemed by Bradshaw (1971) the most common, most important and most complicated kind of fluid motion, it is significant in all natural fluid flows at scales ranging from the ENTRAINMENT of individual DUST particles in the atmosphere to large-scale turbulent currents in the oceans.

Turbulence may be viewed as a cascading energy transfer (Clifford and French, 1993) whereby energy from the mean flow is extracted by large eddies and dissipated into small (microscale) eddies. The range in eddy sizes allows the local flow direction at a point within TURBULENT FLOW to be different from that of the mean flow direction.　　GFSW

References
Bradshaw, P. (1971) *An introduction to turbulence and its measurement*. Oxford: Pergamon. · Clifford, N.J. and French, J.R. (1993) Monitoring and modelling turbulent flow: historical and contemporary perspectives. In N. J. Clifford, J. R. French and J. Hardisty (eds), *Turbulence: perspectives on flow in sediment transport*. Chichester: John Wiley & Sons, Ltd; pp. 1–34.

turbulent flow A fluid flow (air or water) characterized by a mean forward direction but consisting of a series of eddies of various sizes moving in a random manner. The exchange of momentum throughout a turbulent BOUNDARY LAYER is achieved through the mixing motion of gusts and turbulent eddies. Such momentum exchange is far more efficient than the molecular exchange seen in LAMINAR FLOW and this is represented by a steeper velocity gradient, and hence higher SHEAR STRESS toward the surface. Turbulent flow is characterized by a logarithmic increase in velocity away from the surface as a result of bed ROUGHNESS producing a drag at the base of the boundary layer.

Turbulent flow arises when the flow inertia swamps the effects of fluid viscosity and the REYNOLDS NUMBER exceeds 2000. Natural wind flow is almost always turbulent because air has a low viscosity and boundary layer depths are high. Streamflow may not be turbulent in thin or very slow flows.

The measurement of turbulent flow requires time-averaged measurements of velocity that smooth out the instantaneous peaks and troughs of the eddies, which are themselves a measure of the turbulence intensity of the flow. A fluid flow with a high turbulence intensity is more efficient at eroding and transporting sediment as maximum peaks in velocity will be much greater than the mean velocity.　　　　　　　　　　GFSW

Reading
Clifford, N.J., French, J.R. and Hardisty, J. (1993) *Turbulence: perspectives on flow in sediment transport*. Chichester:

John Wiley & Sons, Ltd. · Wiggs, G.F.S. (2011) Sediment mobilisation by the wind. In D. S. G. Thomas (ed.), *Arid zone geomorphology: process, form and change in drylands*, 3rd, edition. London: John Wiley & Sons, Ltd; pp. 455–486.

typhoon　See TROPICAL CYCLONES.

U

ubac The side of a hill or valley that is most shaded from the sun.

unconformity A discontinuity between sedimentary strata that testifies to a temporary interruption in the process of accumulation.

underfit stream A stream that is much smaller than expected from the size of its valley. An underfit stream could have occurred as a result of river capture when the beheaded CONSEQUENT STREAM would be smaller than expected. However, Dury (1977) has shown that underfit streams are a widespread occurrence, that they reflect the impact of climatic change, and that the wavelength of VALLEY MEANDERS may be 3 to 10 times greater than the wavelengths of the underfit stream meanders. *A manifestly underfit stream* is an underfit stream that meanders within a more amply meandering valley, and an *osage type* of stream has a much smaller pool–riffle spacing than would be expected from the size of the valley meanders. KJG

Reference
Dury, G.H. (1977) Underfit streams: retrospect, perspect and prospect. In K. J. Gregory (ed.), *River channel changes*. Chichester: John Wiley & Sons, Ltd; pp. 281–293.

underplating A potentially highly important tectonic process caused when magma generated over a mantle plume is accreted to the base of the crust. According to the underplating model, the addition of volcanic rock in this way thickens the crust and the resulting isostatic adjustment leads to the formation of a broad *hot-spot* swell up to 2000 km across and with an increase in surface elevation of up to 2000 m. This has substantial implications for river network evolution. ASG

Reading
Summerfield, M.A. (1991) *Global geomorphology*. London/New York: Longman/John Wiley & Sons, Inc.

uniclinal Pertaining to a formation of rock strata that dip uniformly in one direction.

uniclinal shifting The process whereby a stream or river flowing in an asymmetric valley in an area of gently dipping rocks migrates down the dip slope of the valley, cutting back the steeper scarp slope.

uniform steady flow Exists when the water depth is equal at every section in a channel reach. Unsteady uniform flow would require the water surface to remain parallel to the channel bed as discharge changes, which is practically impossible. Accordingly, spatially uniform flow is temporally steady. Discharge and flow depth, width, cross-section area and velocity are all constant from section to section, and the ENERGY GRADE LINE, water surface and bed profile are all parallel. This is rare in natural channels with variable width and POOL AND RIFFLE bedforms.

The CHÉZY EQUATION and MANNING EQUATION define the mean velocity of uniform flow as a function of depth, slope and ROUGHNESS, and are often applied to short natural river reaches where uniform flow can be assumed. If local velocities at every point in the cross-section are constant along a reach, the entire velocity distribution is uniform, the turbulent BOUNDARY LAYER is fully developed and the logarithmic vertical velocity profile occurs. KSR

Uniformitarianism A practical tenet held by all modern sciences concerning the way in which we should choose between competing explanations of phenomena. It rests on the principle that the choice should be the simplest explanation that is consistent both with the evidence and with the known or inferred operation of scientific laws. Uniformitarianism is therefore applicable both to historical inference (or 'postdiction') and to prediction of the future outcome of the operations of natural processes (Goodman, 1967). It is, in consequence, as Shea (1982: 458) has forcibly emphasized, a concept 'with no substantive content – that is, it asserts nothing whatever about nature. Uniformitarianism must be viewed as telling us how to behave as scientists and not as telling nature how it must behave.'

The Dictionary of Physical Geography, Fourth Edition. Edited by David S.G. Thomas.
© 2016 John Wiley & Sons, Ltd. Published 2016 by John Wiley & Sons, Ltd.

In physical geography, Uniformitarianism is usually linked with James Hutton's (1788) demonstration that the simplest explanation for the nature of the Earth's surface topography and rock strata was not the invocation of divine intervention at a single moment of creation and then again by the biblical flood, but rather the assumption that processes of erosion, lithification and uplift comparable to those whose operations could be observed or inferred in the modern world, acting over immensely long periods of time, were responsible. Shea (1982) points out that Hutton was not the first to adopt this viewpoint, but as he certainly was the first to provide an extensive working out of its implications it is reasonable to regard him as the founder of modern Earth science. Hutton's views conflicted sharply with those of other natural philosophers, notably Werner and Kirwan, who came to be termed Catastrophists (see Chorley *et al.* (1964)). These latter produced interpretations, often incredibly complex, of rocks and relief that derived from an implicit belief that the biblical chronology was sacrosanct and that God had intervened directly to control the mechanisms of Earth sculpture.

One of the major sources of confusion that has come to enshroud discussions of Uniformitariansim in Earth science in general, and in geomorphology in particular, has derived from a change in the interpretation of 'Catastrophism'. Increasingly, the term has been taken to imply a belief that large, sudden and (to human eyes) 'catastrophic' events have more significance in Earth history than the slow and virtually continuous operation of 'normal' processes (e.g. see the extraordinary influential paper by Wolman and Miller (1960)). This change in meaning has left Uniformitarianism apparently standing for a view in which the *simplest* explanation is equated with that which requires the slowest and/or most *uniform* rate of operation of processes; this fallacy – termed 'gradualism' (see Hooykaas (1963)) – is a complete misinterpretation of the Uniformitarian tenet. Moreover, any careful reading of Hutton, or Playfair (1802), or any edition of Lyell's (1830–1833) *Principles of geology*, make it abundantly clear that all those early Uniformitarians ascribe a very important role to what would now be termed 'high-magnitude, low-frequency' events. This tendency becomes particularly marked in the later editions of Lyell's *Principles of geology* (e.g. Lyell, 1853), as his congenital reluctance to believe that 'normal' fluvial processes are *actually* responsible for substantial Earth sculpture leads him to an increasing emphasis on the role of sudden, large and

locally 'catastrophic' occurrences. It cannot be emphasized too forcibly that Uniformitarianism does not, as a principle, require any presupposition about the rates of operation of processes, other than those limits apparently fixed by the laws of physics and chemistry.

Nor does the concept involve – as another fallacy proposes – the belief that only processes that can actually be observed in operation may be properly invoked as explanations. In consequence, it equally does not assume that the nature and rates of operation of processes have remained unchanged over time. It is, for example, both apparent and entirely consistent with the Uniformitarian tenet that the nature and rates of processes on the Earth must have been very different from today either before the evolution of land plants or at the height of one of the Pleistocene glacial advances.

Probably the most instructive example of the application and misapplication (or misconstruction) of the Uniformitarian principle, and of the conflict that can be generated, is the case of J. H. Bretz (1923) and his interpretation of the channelled scabland of the northwest USA as the product of an almost unimaginably large flood. Baker (1981) has provided a fascinating (and sobering) collection and discussion of the major papers in the channelled scabland debate, which should be required reading for all geomorphologists.

The essence of this controversy concerns the most probable origin of the huge complex of deep channels cut through loess and basalt in the Columbia Plateau region of eastern Washington State (including the site of the Grand Coulee Dam). Using the evidence from painstaking field studies, Bretz in a series of papers from 1923 onwards argued that the simplest interpretation of the data called for a great flood or 'debacle', which cut channels over a 40,000 km^2 area. The source of the water for this 'Spokane flood' was, Bretz found, in the site of glacial Lake Missoula; it was suggested that the ice damming the lake had been suddenly breached, releasing some 500 cubic miles of water into the scabland tract. Bretz's explanation was entirely consonant with Uniformitariansim. He observed scabland features that, while huge, were clearly products of running water, and all that was required to explain them, therefore, was a way of providing a very large flow of water in a very short period. Glacial lake dam bursts are well-documented occurrences.

The Spokane flood theory was attacked – and very viciously attacked – on the grounds both that the proposed explanation smacked far too much of the Diluvial Catastrophists' interpretations of *all* valleys in terms of the mighty waters of Noah's flood and, in addition, because no flood as large as the one hypothesized had ever been observed. Both lines of argument depend upon fallacious interpretations of the Uniformitarian principle. Their proponents, in attempting to provide 'permissible' explanations for the channelled scablands without Bretz's flood, tied themselves in increasingly complicated knots.

Ironically, it was ultimately rather small-scale and geomorphologically undramatic evidence to which the Uniformitarian principle was correctly applied, which led to the vindication of Bretz's earlier hypothesis. The vital evidence consisted of the discovery of giant current ripple marks both within the Grand Coulee area and on the floor of the former Lake Missoula. The hydraulic and hydrodynamic relationships between depth and velocity of water movement and the dimensions of bedforms such as current ripples are, in fact, so well established that the simplest, and therefore Uniformitarian, explanation for the giant examples was a water body with all the characteristics of Bretz's Spokane flood.

While the details of Bretz's explanation of the channelled scabland have been modified by later studies (there were, it seems, several different dam bursts and floods), in essence his 1923 views have been accepted. Indeed, his 'outrageous' mechanism has since been used to explain the apparently similar 'channelled scablands' seen on the planet Mars (see Baker (1981)), as it is a very sound application of the Uniformitarian tenet to assume that direct analogy may be the simplest explanation of apparently directly analogous forms.

Nevertheless, it must be emphasized again that Uniformitarianism is a guiding tenet of science and *not* a rule of nature. As theories about the operation of nature change, so it is possible – and, indeed, inevitable – that one 'Uniformitarian' explanation will come to replace another.   BAK

References

Baker, V.R. (ed.) (1981) *Catastrophic flooding*. Stroudsburg, PA: Dowden, Hutchinson & Ross. · Bretz, J.H. (1923) The channelled scablands of the Columbia Plateau. *Journal of Geology*, **31**, 617–649. · Chorley, R.J., Dunn, A.J. and Beckinsale, R.P. (1964) *The history of the study of landforms*, vol. I. London: Methuen. · Goodman, N. (1967) Uniformity and simplicity. In C. C. Albritton, Jr (ed.), *Uniformity and simplicity: a symposium on the principle of the uniformity of nature*. Geological Society of America, Special Paper 89. Boulder, CO: The Geological Society of America; pp. 93–100. · Hooykaas, R. (1963) *Natural law and divine miracle: the principle of uniformity in geology, biology and theology*. Leiden: E.J. Brill. · Hutton, J.

(1788) Theory of the Earth; or an investigation of the laws observable in the composition, dissolution and restoration of land upon the globe. *Transactions of the Royal Society of Edinburgh*, I (part II), 209–304. · Lyell, C. (1830–1833) *Principles of geology*, 3 vols. London: John Murray. · Lyell, C. (1853) *Principles of geology*, 9th edition. London: John Murray. · Playfair, J. (1802) *Illustrations of the Huttonian theory of the Earth*. London: Cadell & Davies. · Shea, J. (1982) Twelve fallacies of Uniformitarianism. *Geology*, **10**, 455–460. · Wolman, M.G. and Miller, J.P. (1960) Magnitude and frequency of forces in geomorphic processes. *Journal of Geology*, **68**, 54–74.

unit hydrograph   A characteristic or generalized hydrograph for a particular drainage basin. A unit hydrograph of duration t is defined as the hydrograph of direct run-off resulting from a unit depth of effective rainfall generated uniformly in space and time over the basin in unit time. The unit depth was originally 1 inch but is now usually 1 cm, and t is chosen arbitrarily according to the size of the basin and to the response time to major events and can be 1, 6 or 16 h, for example. The technique was developed by L. K. Sherman (1932), and it has been used to predict hydrographs for the engineering design of reservoirs, flood detention structures and urban stormwater drainage. Since many streams are still ungauged, the discharge records from all stations in an area can be analysed and synthetic unit hydrographs developed from the relations between unit hydrograph parameters and drainage basin characteristics. The drainage basin characteristics of the basin above an ungauged site can then be used to obtain the synthetic unit hydrograph for that site. To compare drainage basins of different areas, dimensionless unit hydrographs can be constructed with the discharge ordinate expressed as the ratio to the peak discharge and the time ordinate expressed as the ratio to the lag time. The instantaneous unit hydrograph is a mathematical abstraction produced when the duration of the effective precipitation becomes infinitesimally small, and this is used in the investigation of rainfall–run-off dynamics. Unit hydrograph theory depends upon a number of assumptions, including the HORTON OVERLAND FLOW MODEL and, with the advent of the PARTIAL AREA MODEL of run-off formation, it has been necessary to revise the use and analysis of the unit hydrograph.   KJG

Reading and Reference

Dunne, T. and Leopold, L.B. (1978) *Water in environmental planning*. San Francisco, CA: W.H. Freeman; pp. 329–50. · Shaw, E.M. (1983) *Hydrology in practice*. Wokingham: Van Nostrand Reinhold; pp. 326–344. · Sherman, L.K. (1932) Stream flow from rainfall by the unit graph method. *Engineering News Record*, **108**, 501–505.

unit response graph The theoretical quick-flow hydrograph produced by 1 inch of effective rainfall and derived from the actual quickflow hydrography by assuming a linear extension such as is carried out in the derivation of simple unit hydrographs (Walling, 1971). The derivation of a unit response graph is similar to the derivation of a unit hydrograph, but flow separation is based upon the method proposed by Hibbert and Cunningham (1967) (see HYDROGRAPHS) and each unit response graph will vary in shape in relation to the contributing area generating storm run-off.

Classic unit hydrograph theory assumes that the whole catchment contributes to storm run-off, and so the form of the hydrograph will reflect the characteristics of the whole catchment and any variation in hydrograph form will result entirely from variations in the time distribution of effective rainfall. Hydrograph separation can be achieved using any consistent technique, although Linsley *et al.* (1982) stress the importance of using a separation technique that ensures that the time base of direct or storm run-off remains relatively constant from storm to storm. An hour unit hydrograph will be storm run-off response to a unit of effective rainfall (usually 1 cm or 1 inch) falling with even intensity over the entire catchment during a period of *n* hours. Because of the assumed linear relationship between effective rainfall and storm run-off, it is possible to derive standard hydrographs for different rainfall intensities and for different rainfall durations by applying simple transformations to a unit hydrograph or to a derivative of a unit hydrograph such as an S-curve or an instantaneous unit hydrograph. The unit hydrograph concept is explained in detail by Linsley *et al.* (1982).

In the case of the unit response graph, the magnitude of the ordinates of the quickflow hydrograph are adjusted so that there is a volume of run-off equivalent to a unit of effective rainfall over the catchment, but unit response graphs to storms of the same duration would not be expected to have the same form because of the influence of the size and shape of the contributing area of the speed with which water can drain from the catchment.   AMG

References

Hibbert, A.R. and Cunningham, G.B. (1967) Streamflow data processing opportunities and application. In W. E. Sopper and H. W. Lull (eds), *International symposium on forest hydrology*. Oxford: Pergamon; pp. 725–736. · Linsley, R.K., Kohler, M.A. and Paulhus, J.L.H. (982) *Hydrology for engineers*, 3rd edition. New York: McGraw-Hill. · Walling, D.E. (1971) Streamflow from instrumented catchments in south-east Devon. In K. J. Gregory and W. I. D. Ravenhill (eds), *Exeter essays in geography*. Exeter: University of Exeter; pp. 55–81.

unloading The stripping of rock or ice from a landscape and the resulting effects the release of pressure has on the exhumed land surface.

unstable channels A river or tidal channel that is shifting through erosion and deposition. Some writers restrict the term to those that are shifting rapidly, changing their pattern or adjusting to changed conditions, because many channels quite normally shift their courses without being in a state of imbalance or disequilibrium with environmental controls. By contrast, engineering design commonly aims to achieve stable channels or canals that remain fixed in position and which will not require costly maintenance works. (See also CHANNELIZATION.)   JL

unstable equilibrium If we had two spheres, one larger than the other, and we placed the smaller one upon the larger one in such a way that, upon letting it go, it remained where we put it, then, in the absence of any disturbing force, the two spheres would be in equilibrium. But we are all aware that it would be extraordinarily difficult to achieve the above result and that, even should we succeed, the merest hint of a breath would disturb the equilibrium, sending the smaller sphere increasingly further away from its original position in equilibrium. Thus, we had initially a state of unstable equilibrium. In many natural systems, and particularly in fluids, this type of equilibrium may exist. For example, very warm parcels of air may remain near the ground until some small disturbance triggers their release.   BWA

unsteady flow Occurs in an open channel (e.g. a river or canal) when the depth and discharge of water at different points along a reach change through time because of the passage of a flood wave or surge along the channel. Simplified FLOW EQUATIONS cannot be applied to such translatory wave processes. Analysis of changes of flow conditions at a section, or of the shape of the flood wave as it travels downstream, therefore require the application of wave theory or FLOOD ROUTING methods.   KSR

upper level westerlies The prevailing westerly winds in the mid-latitudes (30–60°) of both hemispheres that occur above the boundary layer. With the boundary layer, friction significantly influences the atmospheric circulation. Above the boundary layer, beginning anywhere from 850 to 500 mb, westerly wind is produced by a balance (geostrophic balance) between the Coriolis force and the pressure gradient force. The westerly wind increases with height until a level is reached at which there is no north–south temperature gradient. The polar front and subtropical jet streams are imbedded in the upper level westerlies. Frequently, stationary or transient waves (called Rossby waves or longwaves) develop upon the westerly flow. The upper level westerlies are critical in both the development and steering of mid-latitude storm systems.   SEN

upwelling

The vertical movement of deeper water towards the sea surface. It occurs where a divergence of surface currents must be compensated for by vertical flow; for example, wind-driven offshore currents may be balanced by coastal upwelling. This deeper water is often rich in nutrients, which allow a high productivity of phytoplankton near the surface. As a result, many of the world's most important fisheries are in areas of upwelling. These include the seas off northwest Africa, Oregon and Peru. Every 5 years or so the Peruvian upwelling is inhibited when the tropical Pacific Ocean responds to a relaxation of the trade winds. The phenomenon, known as the EL NIÑO effect, has serious consequences for the Peruvian fishing industry.   DTP

uranium series dating

Uranium series dating methods are numerical dating techniques that exploit the properties of the radioisotopic decay of uranium (see RADIOISOTOPE). They are applicable to a variety of materials over the time range from 10^4 to 10^6 years and represent a valuable component of GEOCHRONOLOGY investigations. They may be used to date a range of inorganic (e.g. secondary carbonate and salt) and biogenic (e.g. enamel, bone, shell) deposits. The principle of the method is based on the observation that uranium in surface sediments is typically bound in silicate and oxide minerals. Over time (a few million years) it will approach secular equilibrium with its decay chain protégé products. Weathering of uranium-bearing minerals produces water-soluble complexes that may become separated from their less soluble protégé products. When this mobile uranium precipitates (inorganically or biogenically) as a trace constituent of surficial minerals it develops a new succession of protégé isotopes. There are two uranium decay series, each commencing with a different parent uranium ISOTOPE: ^{238}U and ^{235}U. Each series undergoes a series of transitions to a stable isotope of lead. Methods employing ^{230}Th (the protégé isotope of ^{234}U from within the ^{238}U decay series) are most frequently used owing to its convenient HALF-LIFE

($T_{1/2}$) of 75.4 ka, and corresponding age range of approximately 350 ka. By measuring the activity ratio of protégé isotopes to parents (e.g. $^{230}Th/^{234}U$) and knowing the half-lives of the daughter products it is possible to determine the time that has elapsed since precipitation. The method assumes that a sample has behaved as a closed system since the time of formation/deposition (i.e. there has been neither loss nor gain of any isotopes except by radioactive decay) and that the activity ratios of the isotopes being used to determine the age were either zero or some determinable level at the time of formation.

The conventional means of activity measurement is by alpha spectroscopy (the particles emitted during the decay of key isotopes being measured). More recently, ratios of uranium isotopes have been directly counted by thermal ionization mass spectrometric techniques. This improves precision of analyses, increases the age range of the method, reduces minimum sample size requirements and makes possible the measurement of entirely new types of precipitated terrestrial deposits.   SS

Reading
Taylor, R.E. and Aitken, M.J. (1997) *Chronometric dating in archaeology: advances in archaeological and museum science*, vol. 2. New York: Plenum Press.

urban ecology

A branch of ecology dealing with the environment and its organisms specifically in the context of urbanized areas. Although once thought of as ecological deserts, it is now recognized that heavily built-up areas support a variety of plant and animal species, some well adapted to the environmental constraints. The impacts of human activities on the urban environment, together with the identification of the conservation potential of open (or 'green') spaces in urban areas, provide the focus of urban ecology. In recent years, urban planners have adopted a still broader definition in recognizing the potential of ecological principles in promoting the sustainable management of urban ecosystems. This involves, for example, revising land-use practices, restoring degraded habitats and promoting recycling. MEM

Reading
Breuste, J., Feldmann, H. and Ohlmann, O. (eds) (1998) *Urban ecology*. Berlin: Springer-Verlag.

urban hydrology

The study of the hydrological cycle and of the water balance within urban areas. Extensive impervious areas mean that surface storage is reduced, infiltration is not possible and evapotranspiration is much less than in rural areas. Impervious areas increase the amount of surface run-off, and this is accentuated by the stormwater drainage system that collects water from roads, roofs and other impervious surfaces. Modern stormwater drainage systems are installed separately from foul water drainage systems, but in the past a single system was often employed in urban areas. Stream discharge from urban areas tends to have higher peak flows and lower base flows than discharge from rural areas, and the FLOOD FREQUENCY along a river draining urban areas will be significantly changed from the time when the urban area did not exist. Increased flooding has often been observed within and downstream from urban areas as urbanization has occurred, and the larger and more frequent floods may have led to increased river channel erosion. Problems of increased frequency and extent of flooding have often been mitigated by engineering works. The area of Moscow, Russia, was shown by Lvovich and Chernogaeva (1977) to have a decrease of evapotranspiration of 62%, a decrease of groundwater run-off of 50% and an increase in total run-off of 155%. In urban hydrology, it is not simply a modification of the rural hydrological cycle; there can also be a series of other components supplying or reducing water. Urban areas also generate a characteristic water quality, with water temperatures often higher than those of rural areas, with higher solute concentrations reflecting additional sources, including pollutants, and suspended sediment concentrations particularly high during building activity but much lower when the urban area is established and the sources of suspended sediments are no longer exposed. (See also HYDROLOGICAL CYCLE.) KJG

Reading and Reference

Hollis, G.E. (1979) *Man's impact on the hydrological cycle in the United Kingdom*. Norwich: Geo Books. · Lazaro, T.R. (1990) *Urban hydrology*. Lancaster, PA: Technomic. · Lvovich, M.I. and Chernogaeva, G.M. (1977) The water balance of Moscow. In *Amsterdam symposium, 1977, effects of urbanization and industrialization on the hydrological regime and on water quality*. International Association of Hydrological Sciences publication 123. Pp. 48–51..

urban meteorology

The study of atmospheric phenomena attributable to the development of human settlements. It encompasses work on the process involved (physical, chemical and biological), the resulting climate effects and the application of this knowledge to the planning and operation of urban areas. It is one of the clearest examples of the human role in climate modification.

Urban development disrupts the climatic properties of the surface and the atmosphere. These, in turn, alter the exchanges and budgets of heat, mass and momentum that underlie the climate of any site. Every land clearance, drainage, paving and building project leads to the creation of a new microclimate in its vicinity, and the collection of these diverse, human-affected microclimates is what constitutes the urban climate in the air layer below roof level, henceforward called the urban canopy layer (UCL; Oke 1987). These very localized effects tend to be merged by turbulence above roof level, where they form the urban boundary layer (UBL), which appears like a giant urban plume over and downwind of the city.

A city exerts both roughness and thermal influences on winds. When synoptic winds are strong, the greater roughness produces greater turbulence (by 10–20%), increased frictional drag, slower winds (by about 25%), cyclonic turning and a general tendency towards uplift. In the downwind rural area the near-surface flow recovers its original characteristics, but urban effects are detectable in the elevated UBL plume for tens of kilometres. The drag may even retard the passage of weather fronts. In windy conditions, flow in the UCL is extremely variable. While some areas are sheltered, others may be experiencing strong across-street vortices, gustiness or jets (especially near tall buildings). When synoptic winds are light or absent, thermal effects associated with the heat island (see below) become evident. The city may generate its own thermal circulation, analogous to SEA/LAND BREEZE, with 'country' breezes converging on the city centre, rising and diverging aloft to form a counter flow. Urban thermal effects can also lead to acceleration near the surface both as a result of the heat pressure field and because thermal turbulence helps transport momentum downwards.

Considering the major changes in the physical environment wrought by urban development, the changes in the energy (heat) budget are surprisingly small. For example, despite all the radiant fluxes being altered (by pollution or changed surface properties), the net radiation in cities is usually within 5% of that of their rural surroundings. It is true that the city's energy budget is supplemented by heat released by combustion, but though this heat source may be important to climate in some locations, in most places it is minor (Kalma and Newcombe, 1976). Usually, more important is the fact that the city channels more energy into sensible rather than LATENT HEAT. This is because of the removal of many sources of water for EVAPOTRANSPIRATION. As a result, more heat is used to warm the air and ground (including buildings, etc.). The relative warmth of the city is called its 'heat island' (Landsberg, 1981; Oke, 1982). On an annual basis, the canopy layer of a large city (10 million inhabitants) is typically 1–3 °C warmer than its surrounding countryside. This may seem small, but the magnitude of the heat island varies diurnally (largest near midnight, the smallest in the afternoon) and in response to weather (largest with calm and no cloud). The difference is also related to city size (measured by its population, or better by the geometry of its central street canyons; Oke, 1982). On the most favourable nights in a large city, differences of 10 °C and more have been recorded. Spatial variation of temperature within the UCL bears a strong relation to land use and building density, and there is a sharp gradient at the urban–rural boundary. The city's warmth extends down into the underlying ground and upward into the UBL above. At night the heat island maintains a weak mixed layer above the city (tens to hundreds of metres deep), when rural areas are stable.

The exchange of moisture between the surface and the air is altered by changes in the availability of water and energy, and in the perturbed airflow. Normally, values of atmospheric moisture in the daytime UCL are lower than in the country (on account of less evapotranspiration and greater mixing), but the reverse holds at night (because of decreased dewfall and the release of water vapour from combustion). The effects seen in the UCL are also evident in the UBL plume. An exception is provided by high-latitude cities in winter, where evaporation from frozen surfaces is very small so that humidity is largely governed by vapour from combustion, with the result that the city is more humid by both day and night. At temperatures below −30 °C ice fog is a common, and unpleasant, fact of urban life. Above freezing, urban effects on fogs are complex: extra warmth may decrease their frequency, but extra condensation nuclei may increase their density and severity. AEROSOL is also responsible for a general increase in daytime haze in the subcloud layer of the UBL and a deterioration of visibility (Braham, 1977).

Urban modification of PRECIPITATION is a subject that has received considerable research study, especially through Project METROMEX in St Louis (Changnon, 1981). There seems to be a consensus view that cities enhance precipitation in their downwind areas. These effects seem to be most marked in relation to summer convective rainfall, especially heavy rain, and severe weather (thunder- and hailstorms) rather than frontal precipitation. Annual increases of up to 10% are commonly reported, but the exact role of urban versus nonurban influences is often hard to determine. There is also difficulty in isolating the most important causes. It is possible that the microphysics of urban clouds is altered (e.g. cloud droplet sizes and numbers) and/or that cloud dynamics are changed by the UBL (e.g. strength of uplift, height of mixed layer) leading to more favourable precipitation conditions. The latter changes seem most important in St Louis, but much more work is needed.

The field, which began in the early nineteenth century with descriptive studies, is now engaged in the study of meteorological processes and attempts to devise models that link processes and effects. Its most significant deficiencies are in having little knowledge of tropical urban climates and its failure to develop applied science aspects (Page, 1970). TRO

References
Braham, R.R. Jr (1977) Overview of urban climate. In G. M. Heisler and L. P. Herrington (cochairmen), *Proceedings of the conference on metropolitan physical environment: use of vegetation, space and structures to improve amenities for people*. USDA Forest Service general technical report NE-25. Upper Darby, PA: US Department of Agriculture, Forest Service, Northeastern Forest Experiment Station; pp. 3–17. · Changnon, S.A. Jr (ed.) (1981) *METROMEX; a review and summary*. Meteorological Monographs, vol. 18, no. 40. Boston, MA: American Meteorological Society. · Kalma, J.D. and Newcombe, K.J. (1976) Energy use in two large cities: a comparison of Hong Kong and Sydney, Australia. *Environmental Studies*, **9**, 53–64. · Landsberg, H.E. (1981) *The urban climate*. New York: Academic Press. · Oke, T.R. (1982) The energetic basis of the urban heat island. *Quarterly Journal of the Royal Meteorological Society*, **108**, 1–24. · Oke, T.R. (1987) *Boundary layer climates*, 2nd edition. London: Methuen. · Page, J.K. (1970) *The fundamental problems of building climatology considered from the point of view of decision-making by the architect and urban designer*. WMO technical note 109. Geneva: World Meteorological Organization.

urstromtäler An anastomosing pattern of meltwater channels in northern Germany. Individual channels may be hundreds of kilometres long and often more than 100 m deep with irregular long profiles. Some channel patterns are buried by later glacial deposits (Ehlers, 1981). Although there is still discussion about their origin, it seems that they were cut primarily by subglacial meltwater erosion. DES

Reference
Ehlers, J. (1981) Some aspects of glacial erosion and deposition in northern Germany. *Annals of Ecology*, **2**, 143–146.

uvala A depression or large hollow in limestone areas produced when several sinkholes coalesce.

V

vadose That zone of the groundwater system in which the available pore spaces are not fully saturated; that is, located between the ground surface and the WATER TABLE. The zone below the water table is correspondingly termed the *phreatic* zone. In the vadose zone, water-filled pores in the soil may be tension-saturated; that is, the water is not under positive hydrostatic pressure, but is tightly held to the walls of the pore spaces. Below the water table, all pressures are positive. DLD

Reading
Dingman, S.L. (1994) *Physical hydrology*. Upper Saddle River, NJ: Prentice-Hall.

valley bulges (valley-bottom bulges)
These consist of strata that have bulged up in the base of a valley as a result of erosive processes. They are widespread in the sedimentary rock terrains of the English Midlands, where limestones, sandstones and clays occur in close juxtaposition. The mechanism of formation invoked for those of the Stroud area in Gloucestershire (Ackermann and Cave, 1977) is that, during the Pleistocene, severe erosion and valley incision occurred at a time when permafrost conditions pertained. At the end of the cold period the rocks thawed out and the susceptible clays, silts and sands, highly charged with water, became plastic and, under the weight of the more competent limestones above, were extruded through the weakest points of the recently developed valley floors. Cambering of strata would occur on the valley sides. ASG

Reference
Ackermann, K.J. and Cave, R. (1977) Superficial deposits and structures, including landslip, in the Stroud district, Gloucestershire. *Proceedings of the Geologists' Association of London*, **78**, 567–86.

valley meanders Meanders that are usually cut in bedrock and that usually have a greater wavelength than that of the contemporary river pattern. Dury (1977) has shown that valley meanders were produced during periods of higher runoff and higher peak discharges before stream shrinkage that led to contemporary underfit streams. Some writers have suggested that the valley meanders may not indicate stream shrinkage but are rather related to rare high-magnitude events, to the contrast between bedrock and fluvial deposits, and to the pattern of stream migration (see Dury (1977) and papers cited therein). KJG

Reading and Reference
Dury, G.H. (1976) Discharge prediction, present and former from channel dimensions. *Journal of Hydrology*, **30**, 219–245. · Dury, G.H. (1977) Underfit streams: retrospect, perspect and prospect. In K. J. Gregory ed., *River channel changes*. Chichester: John Wiley & Sons, Ltd; pp. 281–293.

valley wind The upvalley flow that develops during the day, especially in north–south-orientated valleys during anticyclonic weather in summer. The flow is induced by strong heating of the valley air, making it much warmer than the air at the same elevation over the adjacent plain. Valley winds are usually as reliable as the mountain, or downvalley, winds that develop at night. WDS

valloni The drowned river valleys of a dalmatian-type coastline.

Van't Hoff's rule The rule that states that when a system is in thermodynamic equilibrium a lowering of temperature will promote an exothermic reaction and a raising of the temperature an endothermic one.

vapour pressure The pressure exerted by the molecules of a given vapour. In meteorology, the vapour in question is usually water vapour and the pressure is the partial pressure; that is, the contribution by water vapour to the total pressure of the atmosphere. It may be calculated by using a humidity slide rule or tables in conjunction with values of dry-bulb and WET-BULB TEMPERATURE and is expressed in millibars. Water vapour's concentration is highly variable in the atmosphere, so vapour pressure changes substantially in time and space, with the highest values

(15–20 mb) being found in the humid tropics and the lowest (1–2 mb) across wintertime high-latitude continents.   RR

varves Traditionally defined as being sedimentary beds or lamina deposited in a body of still water within the course of 1 year. The term has normally been applied to thin layers, usually deposited by meltwater streams in a body of water in front of a glacier. A glacial varve normally includes a lower 'summer' layer consisting of relatively coarse-grained sand or silt, produced by rapid ice melt in the warmer months, which grades upwards into a thinner 'winter' layer composed of finer material deposited from suspension in quiet water while the streams feeding the lake are frozen. However, it is becoming increasingly clear that varves may be deposited in a wide range of environments, both lacustrine and marine, and an alternative term, *rhythmite*, is now widely used.   ASG

Reading
O'Sullivan, P.E. (1983) Annually laminated sediments and the study of Quaternary environmental changes – a review. *Quaternary Science Review*, **1**, 245–313. · Schlüchter, Ch. (1979) *Moraines and varves: origins, genesis, classification*. Rotterdam: Balkema.

vasques Wide (up to several decimetres), shallow pools with flat bottoms that form a network consisting of a tiered, terrace-like series of steps on limestone coastal platforms, especially in aeolianite. The pools are separated from each other by winding, narrow, lobed ridges, 10–200 mm in height, and running continuously for tens of metres. They develop between high and low tide levels, especially in intertropical and Mediterranean climatic regions.   ASG

Reading
Battistini, R. (1983) La morphogénèse des plateformes de corrosion littorale dans les grès calcaires (plateforme supériere et plateforme à vasques) et le problème des vasques, d'après des observations faites à Madagascar. *Revue de Géomorphologie Dynamique*, **30**, 81–94.

vauclusian spring See SPRINGS.

vector Beyond its general use as a word for a line with a fixed length and direction but no fixed location, vector is now used to describe one of the two main types of GEOGRAPHIC INFORMATION SYSTEM (GIS) for storing and analysing spatial data. All the objects of interest are represented as either points, lines or areas, each of which can have attributes recorded. In the case of streams, for example, each section of stream would be stored as a line, which might have information on pH and

water temperature recorded. A vector GIS is particularly good at handling information about discrete objects (see RASTER) and is used in areas such as local government and the utilities.   SMW

veering See WIND.

velocity–area method A widely used method of measuring the discharge of a river. A series of verticals is used to subdivide the cross-section into a number of segments. The discharge of each segment is then determined as the product of average *velocity* and cross-sectional *area*, and the total discharge is calculated as the sum of the values for the individual segments. Verticals are spaced at intervals of no greater than 1/15th of the width. Measurements of mean velocity in the vertical are obtained by using a rotating current meter, and these are assumed to be representative of the average velocity in the adjacent segment. (See also DISCHARGE.)   DEW

Reading
British Standards Institution (2007) *BS EN ISO 748: 2007 Hydrometry. Measurement of liquid flow in open channels using current-meters or floats*. London: British Standards Institution.

velocity profile and measurement Any fluid (e.g. water or air) moving over or adjacent to a fixed boundary will illustrate a vertical variation in velocity, increasing from zero or near zero at the boundary, to higher values in the main body of the flow. Understanding how velocity varies for a given flow over a boundary is crucial to estimating erosion and deposition of sediment. Most natural flows have a high REYNOLDS NUMBER (>2000) and can therefore be classified as turbulent. The zone of influence of the boundary upon the flow is called the BOUNDARY LAYER. If the flow is of limited depth, and the boundary layer extends throughout this depth, the flow is said to be fully developed. Close to the bed, there is a thin zone where LAMINAR FLOW occurs, called the laminar sublayer. Above this, there is a transitional buffer layer and then the fully turbulent zone. Field measurements have suggested that in this zone the velocity increases regularly with elevation above the bed (see BOUNDARY LAYER). This is supported by basic theory (Prandtl, 1952) and is commonly known as the law of the wall:

$$\nu = \frac{1}{\kappa}\sqrt{\frac{\tau_o}{\rho_w}}\ln\frac{y}{y_o}$$

where ν is the velocity at elevation y, κ is the Von Karman constant (which is often assumed to be 0.4), τ_o is the shear stress exerted by the flow upon the bed, ρ_w is the density of water and y_o is the roughness height. Applying this equation to a measured velocity profile allows estimation of bed SHEAR STRESS, although there is debate over how far up into the flow this equation holds. If it is assumed that it holds throughout the flow, integrating the equation shows that the depth-averaged velocity is approximately $0.38h$ above the bed, where h is water depth. This can be used in the VELOCITY–AREA METHOD for DISCHARGE estimation in rivers where a CURRENT METER is available that can measure point velocities.   SNL

Reading and Reference
Carson, M.A. (1971) *The mechanics of erosion*. London: Pion. · Herschy, R.W. (ed.) (1978) *Hydrometry, principles and practices*. Chichester: John Wiley & Sons, Ltd. · Prandtl, L. (1952) *Essentials of fluid dynamics: with applications to hydraulics, aeronautics, meteorology and other subjects*. London: Blackie.

ventifact   A stone that has been shaped by the wind, especially in arid and polar areas. Abrasion is achieved by sand, dust or snow, and the stones become shaped (often into three-sided DREIKANTER), and have surface textures that may be polished, pitted or fluted. They have some utility for estimating past and present wind directions. ASG

Reading
Whitney, M.I. and Dietrich, R.V. (1973) Ventifact sculpture by windblown dust. *Bulletin of the Geological Society of America*, **84**, 2561–2581.

vertical motion (in the atmosphere)
The vertical component of air velocity. Persistent net horizontal flow out of a region, called horizontal DIVERGENCE, would result in the depletion of air in the region, so that its density would decrease; but observed density changes are small, so that vertical motion must result. Such persistent outflow leads to downward motion above a flat surface (where the vertical motion is necessarily zero). In the upper troposphere, persistent horizontal divergence usually leads to upward motion because the great stability of the stratosphere above is unfavourable for vertical motions. Persistent horizontal convergence leads to vertical motions of the opposite sign.

Vertical motion in the troposphere is normally upwards in CYCLONES and downwards in ANTICYCLONES. Magnitudes of vertical motions in such systems are typically only a few centimetres per second, but, because they persist, large vertical displacements of air are involved. The vertical motion in frontal zones is greater than the average values within depressions, being typically a few tens of centimetres per second.

Vertical motions within cumulus clouds are of the order of a metre per second, but in cumulonimbus cloud they may be as much as several tens of metres per second.   KJW

Reading
Battan, L.J. (1974) *Fundamentals of meteorology*. Englewood Cliffs, NJ: Prentice-Hall.

vertical stability/instability   In the atmosphere, terms that characterize the response of a parcel of air to its vertical displacement. A return of the parcel to its original position indicates vertical stability, while continued vertical movement of the parcel indicates vertical instability. When initially possessing a temperature that is equal to that of the surrounding environment, a parcel of rising air that cools at a faster (slower) rate than does the surrounding environment will be colder (warmer) than the environment, and stable (unstable). The rate at which an air parcel cools is directly related to the amount of water vapour that the parcel contains.   AWE

vertisol   One of the 11 orders of the US system of soil taxonomy. Vertisols are soils rich in clays (>30%), especially 2 : 1 lattice clays belonging to the smectite group, which shrink and swell upon drying and wetting. Consequently, vertisols are soils that frequently display deep cracking, especially in environments where seasonal moisture conditions are variable. Materials from the upper parts of the soil profile may mix downward into the soil by falling into these cracks. Distinctive polished surfaces (SLICKENSIDES) may be produced in vertisols by the rubbing of adjacent blocks of material.   DLD

Reading
Brady, N.C. and Weil, R.R. (1996) *The nature and properties of soils*, 11th edition. Upper Saddle River, NJ: Prentice-Hall.

vesicular   Possessing numerous large pores and internal voids.

vicariance biogeography   Vicariance is a process by which the geographical range of an individual taxon, or a whole biota, is split into discontinuous parts by the formation of a physical or biotic barrier to gene flow or dispersal. Vicariance biogeography is therefore the explanation of disjunct distributions of taxa that have been caused by processes that resulted in spatial isolation, in particular continental drift. It is a 'school of thought' that traces its ancestry through the writings of de Candolle, Croizat and Hennig, focusing on allopatric differentiation (see SYMPATRY) and ENDEMISM, linking the study of cladistics and taxonomy with that of biogeography.

This classic concept of vicariance involves allopatric species that have descended from a common ancestral population and which attained some degree of spatial isolation. This traditional form of vicariance is described as horizontal or geographical vicariance. The differences in the isolated pairs of groups will have arisen through the lack of gene exchange, chance differences in genetic composition and, possibly, genetic drift.

Three other forms of vicarious distribution have also been recognized. The first of these is altitudinal vicariance, in which the related taxa form lowland/highland pairs. The second is habitat or ecological vicariance, where the species pairs occupy different environmental niches. This is well exemplified by the sea arrow-grass (*Triglochin maritima*) and the marsh arrow-grass (*Triglochin palustris*), which inhabit respectively saltwater and freshwater marshes. Paradoxically, the term has also been used to describe taxa unrelated from the evolutionary standpoint, but which appear to be adapted to the same ecological niche in different locations. Finally, phenological vicariants have been recognized in which the separation is seasonal, with the related taxa flowering or breeding at different times in the year. Such taxa may actually be sympatric.

The proponents of the new school of vicariance biogeography seek a far more comprehensive philosophy and methodology and see themselves in direct and crusading opposition to what they term 'dispersalist' biogeography, in which, they argue, the traditional explanations for allopatric species depend on the long-range dispersal of taxa from centres of origin. They claim that this 'misguided' approach is based on the views and writings of Charles Darwin, whose *On the origin of species* (1859) and ideas have been dismissed as 'piffle' by one of the main protagonists of the new school, Leon Croizat. Unfortunately, the whole debate was marred by totally unnecessary and often gratuitous *ad hominem* abuse, so much so that it has frequently proved very difficult to grasp the thrust of the arguments involved. The contrived opposition of vicariance versus dispersal also seems to many biogeographers somewhat forced, as these are by no means the only processes involved in the development of plant and animal distribution patterns.

Two particular developments underpinned the growth of the new school, which is most fully documented in the pages of the journal *Systematic Zoology*. The first is traced through the phylogenetic systematics, now normally referred to as cladistics, of the entomologist Willi Hennig

(1966). Cladistics is simply a method of classification, giving rise to graphs (cladograms) of relative affinity. They make no *a priori* assumptions about the nature of the relationships involved. On replacing the taxa represented by cladograms with the localities they inhabit, cladograms of affinities of areas or biological area-cladograms result. These are crucial to the new approach, for, as Nelson (in Nelson and Rosen (1981)) stresses, vicariance biogeography begins by asking the question: 'Is there a cladogram of areas of endemism?'

Second, the new biogeography implicitly accepts that the motor of vicariance is continental movement, brought about through seafloor spreading and plate tectonics, and it is this geological or geophysical model of the Earth that has given the concept of vicariance biogeography such a boost since the geologists' crucial synthesis of plate tectonics in the 1960s. Nevertheless, most vicariance biogeographers insist that the sequence of distributional events must be derived from the taxa themselves *before* the geological evidence is introduced, and that the biological evidence must stand whether or not it matches the 'fashionable' geology.

Thus, the new biogeography attempts above all to link recent developments in systematics with biogeography. It is first and foremost concerned with the coincidence of pattern, a point clearly emphasized by the essential character of Croizat's original methods in what are seen as founder publications, particularly *Panbiogeography* (Croizat, 1958) and *Space, time, form: the biological synthesis* (Croizat, 1964). His idea was to map distributions of whole varieties of taxa and then to establish concordances of pattern in order to identify ancestral biotas. This then necessitated explanations for the fragmentation of previously more extensive distributions. The phenomenon of fragmentation was vicariance. If a given type of distribution – 'an individual track' – recurs in group after group of organisms, the region delineated by the coincident distributions – 'the generalized track' – becomes statistically, and thus geographically, significant.

Many biogeographers objected to this rejection of independent sources of information and regard the methods of vicariance biogeography as inductive, mechanical, unrelated to process, lacking in any true sense of geography beyond that of simple geographical coordinates, and overencumbered with jargon. Undoubtedly, the case was not helped by the stridency of some of its proponents. It is probably fair to say that there is no 'one size fits all' when it comes to explaining biotic disjunct distributions. Recently, even exceptionally long-distance dispersal has been shown to underlie some disjunctions with extremely large gaps. For example, le Roux *et al.* (2014) argue that a marine bird brought a seed of the *koa* tree from Hawaii to Réunion in its stomach or stuck to its feet in a one-off event some 1.4 million years ago that has resulted in a disjunction over some 18,000 km. The remarkable finding is but one of several that indicates the potency of rare, random events in determining biogeographic pattern. Other examples include the proposed drifting of New World (flat-nosed) monkeys on a raft from Africa to South America within the last 50 million years ago, long after the two continents drifted apart and the transfer of sundew carnivorous plants (*Drosera* species) from western Australia to Venezuela, probably by birds (Marris, 2014). However, it is more often the combination of both vicariance and jump dispersal that explains biogeographic patterns; see, for example, an explanation of the composition of the fauna of Sulawesi, the largest island in the Indonesian archipelago (Stelbrink *et al.*, 2012). Molecular methods are now being used to determine the relative importance of each (see Bartish *et al.* (2011)). Vicariance remains an important feature of biogeographic pattern, and, along with dispersal, needs to be considered as a mechanism underlying biotic distributions. PAS/MEM

Reading and References

Bartish, I.V., Anonelli, A., Richardson, J.E. and Swenson, Y. (2011) Vicariance or long-distance dispersal: historical biogeography of the pantropical subfamily Chrysophylloideae (Saptoaceae). *Journal of Biogeography*, **38**: 177–190. · Croizat, L. (1958) *Panbiogeography, or an introductory synthesis of zoogeography, phytogeography, and geology*. Caracas: privately published. · Croizat, L. (1964) *Space, time, form: the biological synthesis*. Caracas: privately published. · Hennig, W. (1966) *Phylogenetic systematics*. Urbana, IL: University of Illinois Press. · Le Roux, J.J., Strasbourg, D., Rouget, M., *et al.* (2014) Relatedness defies biogeography: the tale of two island endemics (*Acacia heterophylla* and *A koa*). *New Phytologist*, **204** (1), 230–242. · Marris, E. (2014) Tree hitched a ride to island. *Nature*, **510**, 320–321. · Nelson, G. (1978) From Candolle to Croizat: comments on the history of biogeography. *Journal of the History of Biology*, **11**, 269–305. · Nelson, G. and Platnick, N.I. (1981) *Systematics and biogeography: cladistics and vicariance*. New York: Columbia University Press. · Nelson, G. and Rosen, D.E. (eds) (1981) *Vicariance biogeography: a critique*. New York: Columbia University Press. · Stelbrink, B., Albrecht, C., Hall, R. and von Rintelen, T. (2012) The biogeography of Sulawesi revisited: is there evidence for a vicariant origin of taxa on Wallace's 'anomalous island'? *Evolution*, **66** (7), 2252–2271.

vigil network See REPRESENTATIVE AND EXPERIMENTAL BASINS.

virgation The formation of trails of ice crystals falling from a cloud.

viscosity The property of fluids by virtue of which they resist flow. Newton's law of viscous (nonturbulent) flow is given by

$$F = \eta A \frac{dv}{dx}$$

where F is the tangential force between two parallel layers of liquid of area A, a distance dx apart, moving with relative velocity dv, and η is the coefficient of viscosity, dynamic viscosity or just viscosity. It is measured in newton seconds per square metre. It should be carefully distinguished from the kinematic viscosity ν, which is η divided by the fluid density ρ; the units here are in square metres per second. WBW

Reading

Whalley, W.B. (1976) *Properties of materials and geomorphological explanation*. Oxford: Oxford University Press.

vlei A term, Afrikaans in origin, used in southern Africa, especially South Africa, to describe valley-bottom areas such as floodplains, valley grasslands and other low-lying wetlands. Vlei is used to describe permanently, seasonally or sporadically flooded land, most frequently in direct association with river valleys, although the presence of a river channel per se is not a precondition, in which case use of the term is equivalent to that of DAMBO more commonly used further north in Africa. MEM

void ratio The ratio of the volume of interstitial voids in a portion of sedimentary rock to the volume of that portion.

volcano An opening or vent through which magma, molten rock, ash or volatiles erupt on to the Earth's surface, or the land form produced by the erupted material. Volcanoes tend to be conical in shape but can have a variety of forms, depending on the nature of the erupted material (particularly its viscosity), the character of recent eruptive activity and the extent of post-erupted modification by erosion (see Table 1 and illustration). Most volcanoes are concentrated at convergent and divergent plate boundaries (see PLATE TECTONICS) but others, located in the interior of plates, are associated with HOT SPOTS. MAS

Table 1 Types of volcanic eruptions

Type	Characteristics
Icelandic	Fissure eruptions, releasing free-flowing (fluidal) basaltic magma; quiet, gas-poor; great volumes of lava issued, flowing as sheets over large areas to build up plateaux (Colombia).
Hawaiian	Fissure, caldera and pit crater eruptions; mobile lavas, with some gas; quiet to moderately active eruptions; occasional rapid emission of gas-charged lava produces fire fountains; only minor amounts of ash; builds up lava domes.
Strombolian	Stratocones (summit craters); moderate, rhythmic to nearly continuous explosions, resulting from spasmodic gas escape; clots of lava ejected, producing bombs and scoria; periodic more intense activity with outpourings of lava; light-coloured clouds (mostly steam) reach upward only to moderate heights.
Vulcanian	Stratocones (central vents); associated lavas more viscous; lavas crust over in vent between eruptions, allowing gas build-up below surface; eruptions increase in violence over longer periods of quiet until lava crust is broken up, clearing vent, ejecting bombs, pumice and ash; lava flows from top of flank after main explosive eruption; dark ash-laden clouds, convoluted, cauliflower-shaped, rise to moderate heights more or less vertically, depositing tephra along flanks of volcano. (Note: ultravulcanian eruption has similar characteristics but results when other types, e.g. Hawaiian, become phreatic and produce large steam clouds, carrying fragmental matter.)
Vesuvian	More paroxysmal than Strombolian or Vulcanian types; extremely violent expulsion of gas-charged magma from stratocone vent; eruption occurs after long interval of quiescence or mild activity; vent tends to be emptied to considerable depth; lava ejects in explosive spray (glow above vent), with repeated clouds (cauliflower) that reach great heights and deposit tephra.
Plinian	More violent form of Vesuvian eruption; last major phase is uprush of gas that carries cloud rapidly upward in vertical column for miles; narrow at base but expands outward at upper elevations; cloud generally low in tephra.
Peléean	Results from high-viscosity lavas; delayed explosiveness; conduit of stratovolcano usually blocked by dome or plug; gas (some lava) escapes from lateral (flank) openings or by destruction or uplift of plug; gas, ash and blocks move downslope in one or more blasts as nuées ardentes or glowing avalanches, producing directed deposits.
Katmaian	Variant of a Peléean eruption characterized by massive outpourings of fluidized ash flows; accompanied by widespread explosive tephra; ignimbrites are common end products, also hot springs and fumaroles.

Source: Short (1986: table 3-1, p. 187).

Reading and References
Francis, P.W. (1976) *Volcanoes.* Harmondsworth: Penguin. · Goudie, A.S. (2001) *The nature of the environment,* 4th edition. Oxford: Blackwell. · Ollier, C.D. (1988) *Volcanoes.* Oxford: Blackwell. · Marti, J. and Ernst, G.G.J. (eds) (2005) *Volcanoes and the environment.* Cambridge: Cambridge University Press. · Short, N.M. (1986) Volcanic landforms. In N. M. Short and R. W. Blair (eds), *Geomorphology from space: a global overview of regional landforms.* NASA SP 486. Washington, DC: NASA; pp. 185–253. · Williams H. and McBirney, A.R. (1979) *Volcanology.* San Francisco, CA: Freeman, Cooper.

Von Karman constant A universal constant that relates the mixing length in turbulent flow to the distance from a solid boundary. The mixing length is a measure of the size of turbulent eddies, or in other words of the average distance travelled by small volumes of water following random paths that are superimposed on the main forward flow direction of a stream.

The relation between the mixing length l and distance from the boundary y is

$$l = \kappa y$$

where κ is the von Karman constant. This is taken to have a value of 0.4 in clear water, but its value is thought to be lowered in the presence of high concentrations of suspended sediment particles because these impede the free movement of the turbulent eddies.　　　　DLD

Reading
Richards, K. (1982) *Rivers.* London: Methuen. · Wang, S. (1981) Variation of Karman constant in sediment-laden flow. *Proceedings of the American Society of Civil Engineers, Journal of the Hydraulics Division,* 107, 407–417.

vorticity A microscopic measure of rotation in a fluid. It is a vector quantity defined as the curl of the velocity and has dimensions of $(time)^{-1}$. In

Some major volcanic landforms: (a) types of eruption; (b) a lava plateau – the Deccan of India; (c) acid lava (viscous) extrusion forms; (d) the stages in the formation of a caldera by collapse: (i) initial volcano; (ii) explosion; (iii) collapse. *Source*: Goudie (2001).

Cartesian coordinates the x, y and z components of vorticity ζ are given by

$$\zeta_x = \frac{\partial w}{\partial y} - \frac{\partial v}{\partial z}; \quad \zeta_y = \frac{\partial u}{\partial z} - \frac{\partial w}{\partial x}; \quad \zeta_z = \frac{\partial v}{\partial x} - \frac{\partial u}{\partial y}$$

where u, v and w are the x, y and z components of velocity. The absolute vorticity is given by the curl of the absolute velocity, and relative vorticity is given by the curl of the velocity relative to the Earth. In the northern hemisphere, relative vorticity in a cyclonic sense is positive and in an anticyclonic sense is negative. Fluid in solid rotation with angular velocity ω has vorticity 2ω.

In large-scale motions in the atmosphere the vertical components of absolute and relative vorticity are of chief importance, and these terms are often used without the explicit modifier 'vertical component of'. The difference between the vertical component of absolute and relative vorticity is given by the vertical component of vorticity of the Earth due to its rotation, being $2\Omega \sin \phi$, where Ω is the angular velocity of the Earth and ϕ is latitude.

In adiabatic, frictionless motion a quantity called potential vorticity is conserved, this being expressed by

$$(\text{absolute vorticity}) \times \frac{\partial \theta}{\partial p} = \text{constant}$$

where θ is potential temperature and p is pressure. KJW

Reading
Atkinson, B.W. (ed.) (1981) *Dynamic meteorology: an introductory selection*. London: Methuen. · Gill, A.E. (1982) *Atmosphere–ocean dynamics*. London: Academic Press. · Houghton, J. (1986) *The physics of atmospheres*. Cambridge: Cambridge University Press.

557

vugh   A void or cavity within a rock that can be lined with mineral precipitates.

vulcanism   The movement of magma or molten rock and associated volatiles onto or towards the Earth's surface. Extrusions of material onto the Earth's surface can occur through a vent known as a VOLCANO or through linear openings in the crust, called fissures. Intrusion of magma or molten material into the upper part of the crust can give rise to large masses of igneous rock, such as BATHOLITHS, which may cause domal uplift of the overlying strata. The term *vulcanism* is also used as a synonym of volcanism, and in this sense refers only to extrusion of material onto the Earth's surface.   MAS

Reading
Williams, H. and McBirney, A.R. (1979) *Volcanology*. San Francisco, CA: Freeman, Cooper.

W

---

wadi An Arabic word, sometimes also spelt *oued*, generally used as a term for ephemeral river channels in desert areas. Wadis may flow only occasionally, and then sometimes discontinuously, along their courses.

Wadi Bhi, a major desert channel emanating in the Omani mountains and extending into Ras Al-Khaimah, UAE. Despite its great width and in places depth, this is an extremely ephemeral feature in terms of channel flow, which may owe some of its characteristics to tectonic uplift and past wetter climates. Photograph by David Thomas.

Walker circulation Large atmospheric circulation cells oriented east–west along the equator. The most strongly developed Walker cell is in the Pacific. Under 'normal' conditions, it involves rising air (low pressure) over the warm maritime continent of the western tropical Pacific, sinking air (high pressure) over the cold ocean currents of the eastern Pacific, westerly winds aloft between these centres of vertical motion, and easterly winds at the surface. The surface easterlies of the Walker circulation result in the upwelling of cold ocean water along the South American coast, as well as higher sea surface temperatures and higher mean sea levels in the western Pacific. Rainfall is also higher in the western part of the basin.

The term 'Walker circulation' usually refers to the east–west wind regime over the tropical Pacific Ocean. However, similar zonal wind systems exist in both the tropical Atlantic and Indian Oceans during some seasons and some longitudes. Technically, the term 'Walker cell' applies to these circulation patterns as well. These zonal cells vary in magnitude and structure over time, and some evidence indicates that the Pacific cell has intensified in recent decades.

The El Niño phenomenon is a good example of changes in the Walker circulation. It weakens during EL NIÑO events but is enhanced during LA NIÑA. In an El Niño, the pressure gradient between the eastern and western Pacific

Schematic cross section of the Walker circulation during December to February.
Source: Webster (1983). Reproduced with permission of Elsevier.

The Dictionary of Physical Geography, Fourth Edition. Edited by David S.G. Thomas.
© 2016 John Wiley & Sons, Ltd. Published 2016 by John Wiley & Sons, Ltd.

decreases and the easterly trade winds relax. The weakened easterlies allow the warm surface waters of the western Pacific to migrate eastward. As the warm water migrates, the ascending branch of the Walker cell (and thus the zone of maximum precipitation) moves into the normally arid central and eastern Pacific. In theory, the weakening of the Walker circulation can both cause and be caused by changes in sea surface temperatures.   SEN

Reading and Reference

Hastenrath, S. (2007) Equatorial zonal circulations: historical perspectives. *Dynamics of Atmospheres and Oceans*, **43**, 16–24. · Webster, P.J. (1983) Large-scale structure of the tropical atmosphere. In B. J. Hoskins and R. P. Pearce (eds), *Large-scale dynamical processes in the atmosphere*. London: Academic Press; pp. 235–275.

Wallace's line   A zoogeographical boundary, originally put forward in 1858 by Alfred Russel Wallace, that runs through the middle of the Malay archipelago and that, Wallace argued, marked the meeting of the Asian and Australian faunas. The original Wallace line ran between the islands of Bali and Lombok and then between Borneo and Celebes (Sulawesi). In 1910 Wallace moved his line to the east, so that it lay between Celebes and the Moluccas. Many alternative lines have been suggested, and in 1928 the whole island area with its complex distributions was designated by Dickerson a separate zoogeographical region called Wallacea.

Wallace's realms   A division of the world into six zoogeographical regions, defined by their distinctive faunas and proposed by Alfred Russel Wallace in his classic of zoogeography *The geographical distribution of animals* (1876); also called the Sclater–Wallace system of zoogeographical regions. The names of the six regions were adapted from the continents or were of a classical form, and are normally given as: the Pal(a)earctic, the Nearctic, the Neotropical, the Ethiopian, the Oriental and the Australian (see FAUNAL REALMS for a full description and illustration). Wallace based his system on the earlier work of Philip Lutley Sclater (1858), who had likewise recognized six regions based on his study of the distribution of birds. Wallace chose mammals to verify Sclater's six divisions and, in doing so, he drew on his wealth of personal exploration, which ranged from his visits to the Amazon and Rio Negro (1848–1852) to southeast Asia (1854–1862). Though frequently modified and expanded, the Sclater–Wallace regions remain the basis of our understanding of the great 'realms of life'. (See also FLORISTIC REALMS and WALLACE'S LINE.)   PAS

Reading and Reference

George, W. (1962) *Animal geography*. London: Heinemann. · George, W. (1964) *Biologist philosopher: a study of the life and writings of Alfred Russel Wallace*. London: Abelard-Schuman. · Sclater, P.L. (1858) On the general distribution of the members of the class *Aves. Journal of the Linnean Society of London*, **2**, 130–145.

Walther's law   A key to the interpretation of vertical stratigraphic sequences and environmental reconstructions. Only laterally contiguous FACIES (sedimentary subenvironments) can form vertically contiguous facies. Walther's law assumes that facies migrate laterally with local changes in depth, and thereby transpose horizontal associations into vertical associations. This relationship is useful for reconstructing the lateral associations from vertical sequences obtained from cores or exposures.   DJS

Reading

Middleton, G.V. (1973) Johannes Walther's law of the correlation of facies. *Geological Society of America Bulletin*, **84**, 979–988.

waning slopes   Depositional hillslope units formed at the base of a talus (scree) slope as weathering and rain wash fine-grained particles from the talus and deposit them as a concave unit that may become progressively flatter (Wood, 1942). L. C. King (1957) adopted the basic definitions of slope units, proposed by Wood, and the waning slope is identified as a rock-cut pediment with a veneer of sediment produced by surface wash.   MJS

References

King, L.C. (1957) The uniformitarian nature of hillslopes. *Transactions of the Edinburgh Geological Society*, **17**, 81–102. · Wood, A. (1942) The development of hillside slopes. *Proceedings of the Geologists' Association*, **53**, 128–140.

warm front   A frontal zone in the atmosphere where, from its direction of movement, cool air is being replaced by rising warm air. As the front approaches, cirrus clouds are gradually replaced by cirrostratus, altostratus and finally nimbostratus clouds. Precipitation usually occurs within a wide belt up to about 400 km ahead of the surface front. As the surface front passes, temperatures and dewpoint increase, winds veer and pressure stops falling. Most fronts exhibit considerable differences from this standard model outlined, partly depending upon the rate of uplift within the warm air. Well-developed warm fronts are relatively rare in the southern hemisphere extratropical cyclone belt, where sources of warm air for the warm sector are limited. Kinematically and dynamically there is no fundamental difference between cold and warm fronts, despite the views of the Norwegian School of Meteorology that made a sharp distinction between their roles in the precipitation process. PS

warm ice   See TEMPERATE ICE.

warm occlusion   See OCCLUSION.

warm sector   The area of warm air lying between the warm and cold fronts of an EXTRATROPICAL CYCLONE. It eventually disappears from the surface as the cyclone evolves to become occluded. Temperatures in the warm sector are noticeably higher than in the preceding and following air streams. Cloud and precipitation are very variable. With a strong ridge of high pressure there may be clear skies, but more frequently the skies are overcast. Heavy rain is likely over upland areas if the warm sector is potentially or conditionally unstable. PS

warping   The bending and deformation of extensive areas of the Earth's crust without the formation of folds or faults.

washboard moraine   See DE GEER MORAINE.

water balance   The water balance or water budget of an area over a period of time represents the way in which precipitation during the time period is partitioned between the processes of evapotranspiration and run-off, taking account of changes in water storage. The water balance summarizes the changes in the components of the HYDROLOGICAL CYCLE during a particular time period and may be expressed for a drainage basin as

$$P = Q + \text{Et} \pm \Delta\text{SS} \pm \Delta\text{SMS} \pm \Delta\text{AZS} \pm \Delta\text{GS} \pm \text{DT}$$

where all the variables are expressed as a depth of water over the catchment area for the time period studied. P is precipitation, Q is run-off, Et is evapotranspiration losses, ΔSS is change in surface storage, ΔSMS is change in soil moisture storage, ΔAZS is change in aeration zone storage, ΔGS is change in groundwater storage and DT is deep transfer of water across the watershed.

The water balance can be evaluated by direct field measurements, or a climatic water balance can be calculated by making simplifying assumptions about the role and operation of different stores. A water balance can be calculated for any size of area from small LYSIMETERS up to whole continents, and it can be used to isolate the hydrological effects of human activities, but water balance evaluation is most widely applied at the drainage basin scale. A water balance can also be calculated for any time period, but it is simplest to evaluate for a time period where storage is approximately the same at the start and end of the period. In the UK the *water year* that runs from 1 October to 30 September is often used for water balance calculations because this starts and ends at the period of minimum storage in most years. If the storage can be assumed to be the same at the start and end of the required period, the water balance equation simplifies so that precipitation is only partitioned between evapotranspiration and run-off.

Estimation of the water balance of individual basins is often used to check or to estimate the magnitude of a water balance component that is difficult to measure. Evapotranspiration loss has been frequently studied in this way and formed the basis of the Institute of Hydrology's original Plynlimon catchment study, where losses from an afforested catchment and a grassland catchment were estimated using a water balance approach and were compared with potential evapotranspiration rates calculated from observations from automatic weather stations. Water balance studies are also useful in identifying the impact of catchment modifications on hydrological processes. This is well illustrated by Clarke and Newson's (1978) analysis of the impact of the 1975–1976 drought in the Institute of Hydrology's research catchments and the implications for the water balance of drainage basins under different vegetation cover.

Calculation of the water balance from climatic data was first attempted by Thornthwaite (1948) in his classification of world climates. The evaluation of a climatic water balance requires observations or estimates of precipitation and potential evapotranspiration (see POTENTIAL EVAPORATION) and a means of correcting potential evapotranspiration to an actual evapotranspiration rate by taking account of the degree to which evapotranspiration is limited by availability of soil moisture. Baier (1968) reviewed various proposals for the relationship between the ratio of actual to potential evapotranspiration and percentage soil moisture availability. The UK Meteorological Office (Grindley, 1967) assumes a mix of riparian, short-rooted and long-rooted vegetation. The riparian vegetation is assumed continually to transpire at the potential rate and the short-rooted and long-rooted vegetation are assumed to transpire at the potential rate up to soil moisture deficits of 75 mm and 200 mm respectively, and then at a gradually

Table 1 Water balance components in millimetres for world continents

Continent	Precipitation	Evapotranspiration	Run-off
Europe	657	375	282
Asia	696	420	276
Africa	696	582	114
Australia (with islands)	803	534	269
Australia (without islands)	447	420	27
North America	645	403	242
South America	1564	946	618
Antarctica	169	28	141

Source: Baumgartner and Reichel (1975).

reducing rate. Thus, the water budget that partitions precipitation between evapotranspiration, soil moisture storage and rainfall excess can be calculated from an area and for consecutive time periods, most frequently producing a monthly water budget.

The evaluation of the water balance provides an essential stage in the estimation of the water resources of an area, and so mapping of water balance components, notably precipitation, evapotranspiration and run-off, has received a great deal of attention. Gurnell (1981) reviews attempts at evapotranspiration mapping and many of these maps fall within a context of water balance mapping, ranging from maps of the water balance of administrative areas within countries, to national maps and continental maps (Doornkamp *et al.*, 1980), to maps for the whole world (Baumgartner and Reichel 1975; UNESCO 1978a,b). Table 1 summarizes the water balance of the continents according to Baumgartner and Reichel (1975). Mapping the water balance for such large areas requires the combined use of measurements of precipitation, run-off and climatic variables and allows the estimation of evapotranspiration, the evaluation of climatic water balances and the use of regional relationships between altitude and each of the water balance variables. In this way, isoline maps of the water balance components over large areas may be produced.

Water balance information is a useful tool in water resources planning because it can represent both long-term average and extreme conditions at a site. Extreme values of soil moisture deficit and water surplus may give an indication of the drought- and flood-producing potential of an area, and temporal trends in the water balance may reflect climatic change or the influence of catchment modification. The water balance equation and the concept of the hydrological cycle are the foundations of hydrological studies.   AMG

References

Baier, W. (1968) Relationship between soil moisture, actual and potential evapotranspiration. In Proceedings of the hydrology symposium: no. 6: soil moisture: held at University of Saskatchewan, on 15 & 16 Nov. 1967. Ottawa: Queen's Printer. · Baumgartner, A. and Reichel, E. (1975) *The world water balance: mean annual global, continental and maritime precipitation, evaporation and run-off.* New York: Elsevier. · Clark, R.T. and Newson, M.D. (1978) Some detailed water balance studies of research catchments. *Proceedings of the Royal Society of London, Series A: Mathematical and Physical Sciences,* **363**, 21–42. · Doornkamp, J.C., Gregory, K.J. and Burn, A.S. (eds) (1980) *Atlas of drought in Britain 1975–1976.* London: Institute of British Geographers. · Grindley, J. (1967) The estimation of soil moisture deficits. *Meteorological Magazine,* **96**, 97–108. · Gurnell, A.M. (1981) Mapping potential evapotranspiration: the smooth interpolation of isolines with a low density station network. *Applied Geography,* **1**, 167–183. · Thornthwaite, C.W. (1948) An approach toward a rational classification of climate. *Geographical Review,* **38**, 55–94. · UNESCO (1978a) *World water balance and water resources of the Earth.* Paris: UNESCO. · UNESCO (1978b) *Atlas of the world water balance.* Paris: UNESCO.

water mass A body of water having approximately uniform characteristics typically acquired in a source region in contact with the atmosphere. Oceanic water masses are usually identified from their values of temperature and salinity that are conservative properties, although nonconservative properties – for example, the concentrations of dissolved oxygen or of nutrients such as phosphate and silica – can be useful in tracing their movements. Upper water masses are found above the THERMOCLINE, where their formation is controlled by the pattern of surface currents. For example, the subtropical GYRES enclose central waters of different types all of which have relatively high temperatures and salinities. The source regions of deeper water masses are predominantly at high latitudes where the thermocline is weak or absent. Most of the Atlantic Ocean basin is occupied by North Atlantic

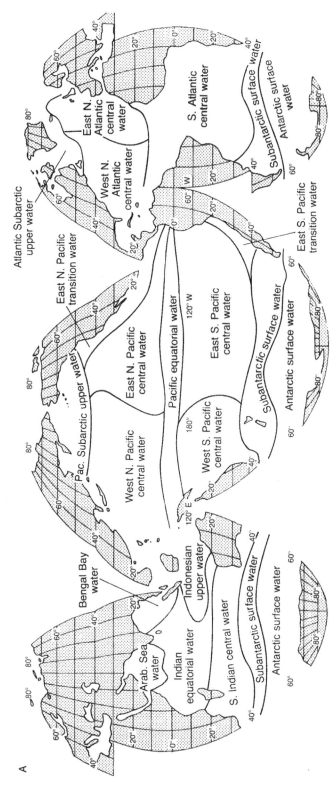

Water mass 1. The global distribution of upper water masses.

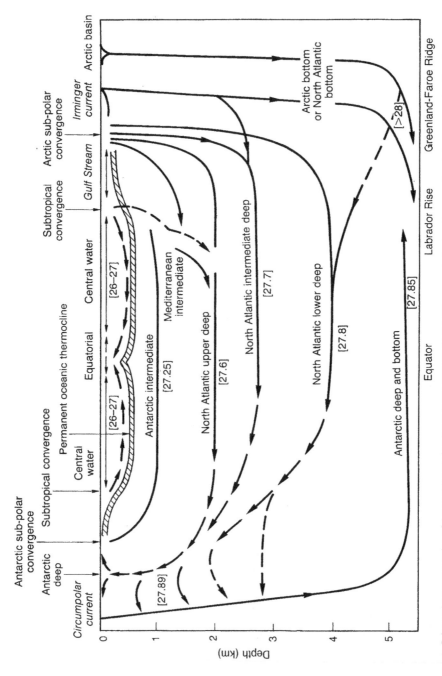

Water mass 2. Schematic representation of the meridional circulation in the Atlantic Ocean. The numerical values in square brackets are mean density values σ_t.
Source: Tolmazin (1985).

Table 2 Major water masses of the world oceans and their T–S characteristics

Atlantic Ocean			Indian Ocean			Pacific Ocean		
Name	T (°C)	S (%)	Name	T (°C)	S (%)	Name	T (°C)	S (%)
Central waters								
North Atlantic	20.0	36.5	Bay of Bengal	25.0	33.8	Western North Pacific	20.0	34.8
South Atlantic	18.0	35.9	Equatorial	25.0	35.3	Eastern North Pacific	20.0	35.2
			South Indian	16.0	35.6	Equatorial Western		
						South Pacific	20.0	35.7
Intermediate waters								
Atlantic subarctic	2.0	34.9	—			Pacific subarctic	5–9	33.5–33.8
Mediterranean intermediate	11.9	36.5	Red Sea intermediate	23.0	40.0	North Pacific intermediate	4–10	34.0–34.5
Antarctic intermediate	2.2	33.8	Timor Sea intermediate	12.0	34.6	South Pacific intermediate	9–12	33.9
			Antarctic intermediate	5.2	34.7	Antarctic intermediate	5.0	34.1
Deep and bottom waters								
North Atlantic deep and bottom	2.5	34.9	Antarctic deep and bottom	0.6	34.7	Antarctic deep and bottom	1.3	34.7
Antarctic deep	4.0	35.0						
Antarctic bottom	−0.4	34.66						

Source: Tolmazin (1985: table 7.1). After Mamayev (1975) and Sverdrup *et al.* (1942).

Deep Water; south of about 30°N Antarctic Bottom Water is found close to the sea bed and Antarctic Intermediate Water at depths in the range 500–1500 m. These deep water masses of the Atlantic are characterized by values of temperature T and salinity S that vary within very narrow limits; such water masses may be designated *water types*. Other water masses, including the central waters mentioned above, have a characteristic range of values of T and S (see Table 2). Investigations of the stratification of the oceans and the identification, movement and mixing of water masses is greatly facilitated by the plotting of data values on a graph of T as ordinate against S as abscissa (a T–S diagram) on which lines of constant density (isopycnals) are convex upwards. (See Water mass 1 and 2 diagrams.)　　　JEA

Reading and References
Harvey, J.G. (1976) *Atmosphere and ocean*. London: Artemis Press. · Mamayev, O.I. (1975) *Temperature–salinity analysis of world ocean waters*. Amsterdam: Elsevier. · Open University Oceanography Course Team (1989) *Seawater: its composition, properties and behaviour*. Oxford/Milton Keynes: Pergamon/Open University Press. · Sverdrup, H.U., Johnson, M.W. and Fleming, R.H. (1942) *The oceans*. Englewood Cliffs, NJ: Prentice-Hall. · Tolmazin, D., (1985) *Elements of dynamic oceanography*. Boston, MA: Allen & Unwin.

water table　The surface defined by the level of free-standing water in fissures and pores at the top of the saturated zone. It is an equilibrium surface at which fluid pressure in the voids is equal to atmospheric pressure. The equivalent term in continental European literature is the PIEZOMETRIC SURFACE. A potential source of confusion arises because the latter is sometimes used in English literature to describe the water-level elevations in wells tapping a confined artesian AQUIFER. The term *potentiometric surface* is preferred for this (Freeze and Cherry, 1979).　　　PWW

Reading and Reference
Fitts, C.R. (2013) *Groundwater science*, 2nd edition. Waltham, MA: Academic Press. · Freeze, R.A. and Cherry, J.A. (1979) *Groundwater*. Englewood Cliffs, NJ: Prentice-Hall. · Jones, J.A.A. (1997) *Global hydrology: processes, resources and environmental management*. Harlow: Longman.

water transfer　The movement of water away from natural water courses, to satisfy human needs. This is a common feature in DRYLANDS, to provide water for agricultural irrigation. It is also carried out to meet the demands of urban areas. The term *water transfer* is increasingly used, however, to refer to the movement of water from one drainage basin or catchment, where there is a perceived surplus related to human needs, to another where there is a deficit. To achieve such transfers it is necessary not only to divert and store water, but to allow its passage across, or through, the divides or watersheds that separate catchments. Water transfers are a feature of the drier western USA, especially in order to satisfy the demands of heavily populated areas in, for example, California.   DSGT

waterfall   A stream that falls from a height. Waterfalls are often the sites of greatest concentration of energy dissipation along the course of a stream and have generally been regarded as forming where a soft rock is eroded from beneath a harder rock (the caprock model). However, in reality, waterfalls have more diverse forms than this simple model would suggest. It is likely that it is applicable only in areas with gently dipping strata. In addition to such structural control, waterfalls depend for their development on such factors as glacial overdeepening, tectonic changes and base-level change.   ASG

Reading
Young, R.W. (1985) Waterfalls: form and process. In H. Bremer (ed.), *Fluvial geomorphology*. Zeitschrift für Geomorphologie, Supplementbände, vol. 55 Berlin: Borntraeger; pp. 81–95.

watershed   The boundary that delimits a drainage basin as the basic hydrological unit. On large-scale topographical maps it is often drawn as a line according to the contour information, but this surface watershed may not correspond with the subsurface boundary of the basin delimited according to the WATER TABLE as the PHREATIC DIVIDE. Sometimes, especially in North America, watershed is used synonymously with catchment or drainage basin area.   KJG

waterspout   A vortex disturbance that forms in the atmosphere over a body of water when the atmosphere is unstable and the air stagnant. Waterspouts are favoured by conditions of high temperature, high humidity and still air, and can cause damage to, and even the foundering of, large boats.   ASG

Reading
Gordon, A.H. (1951) Waterspouts. *Weather*, 6, 364–371.

wave-cut platform   See SHORE PLATFORMS.

wavelet transform   A mathematical tool for representing the variance of a time series as a function of frequency. In contrast to spectral analysis, it is a time–frequency transformation, so that temporal changes in the contribution of individual frequencies to the total variance can be discerned.   SEN

waves   Regular oscillations in the water surface of large water bodies. They are produced by the pushing effect of wind on the water surface. Energy is transferred from the wind to the water and eventually dissipated when the waves hit a shoreline. Waves are of great geomorphological importance because of the role they play in coastal sedimentation and erosion processes. In deep water, wave form can be predicted by a variety of equations, and movement of individual water particles takes the form of circular orbits within the wave. On approaching shallow water, wave forms become more complex and there is a general forward trend to the movement of water particles.   HAV

Reading
Carter, R.W.G. (1988) *Coastal environments*. London: Academic Press.

waxing slopes   Hillslope units forming crests of convex profile at the intersection of the cliff and the hilltop or plateau surface. It is assumed by Wood (1942) that these slope units will increase in significance in the landscape as cliffs (free faces) and talus (scree) slopes are eliminated, and that they will lengthen to become major parts of the total hillslope as a result of weathering and soil creep.   MJS

Reference
Wood, A. (1942) The development of hillside slopes. *Proceedings of the Geologists' Association*, 53, 128–140.

weather   The overall state of the atmosphere on a timescale of minutes to months, with particular emphasis on those atmospheric phenomena that affect human activity. Thus, sunshine, temperature, rainfall, wind and cloud contribute to weather, whereas air density does not. In contrast to weather, CLIMATE is concerned with the long-term behaviour of the atmosphere.   BWA

weather forecasting   The science of predicting the future state of the atmosphere from very short periods of less than 1 h up to 7 or even 10 days ahead. In recent years the science has been revolutionized by the application of modern technology (e.g. by satellites and radar) and by the development of advanced mathematical models that can be run on the most powerful of modern electronic computers.

In many of the more advanced forecasting centres operations are carried out as an automatic routine with many thousands of observations flowing in continuously from around the globe. These come from manned surface and upper air stations, automatic data buoys in remote areas, flight level reports from commercial aircraft, and polar orbiting and geostationary satellites. The data are accepted automatically into the computing system and are first assimilated from their real geographically scattered pattern into a regular global network of so-called gridpoints at typically 15 levels from the surface to the STRATOSPHERE. These points number around 36,000 at a given level or approximately 500,000 in total. Given that seven variables are assimilated at each point (including TEMPERATURE, WIND and HUMIDITY), it is apparent that about 3.5×10^7 numbers are stored to define the state of the atmosphere at the given observation time.

At every point a set of predictive equations is solved in order to calculate the values of temperature, wind, humidity, and so on for a sequence of 'time-steps' to produce a forecast for, say, 12, 24, 48 and 72 h ahead. The longer term or medium range (up to 7 days) forecasts are based on observations taken at the main synoptic hours of 00 and 12 GMT.

Modern weather forecasting by this means provides a vast range of products for appraisal by forecasters, who are still required to interpret and to communicate the information to the user. Predicted fields include the traditional mean sea-level pressure charts and upper air charts along with more modern fields of 'large-scale' and 'convergence-scale' precipitation totals on a regular grid and forecast TEPHIGRAMS. Very short -term prediction of precipitation (nowcasting) in the period 0 to 6 h ahead is underway in Britain, based on the linear extrapolation of radar network observations of precipitation patterns.

Weather forecasting services are provided for many users; for example, offshore oil developments (sea and swell, deck-level winds), civil aviation (wind, temperature and clear-air turbulence along the flight), ship routeing (optimum course for a given journey) and agriculture (e.g. crop spraying operations and weather-related disease outbreak).   RR

Reading
Aktinson, B.W. (1968) *The weather business.* London: Aldus Books. · Browning, K.A. (ed.) (1982) *Nowcasting.* London: Academic Press. · Gadd, A.J. (1981) Numerical modelling of the atmosphere. In B. W. Atkinson (ed.), *Dynamical meteorology; an introductory selection.* London: Methuen; pp. 194–204. · Wickham, P.G. (1970) *The practice of weather forecasting.* London: HMSO.

weather modification
This refers to both intentional and inadvertent alteration of natural atmospheric processes by human intervention. Most commonly, intentional weather modification research involves cloud seeding to enhance precipitation, suppress hail formation, lessen tropical storm intensity, disperse supercooled fog and increase understanding of cloud physics and dynamics. Additionally, increasing evidence is linking anthropogenic activities to unintentional weather modification. For example, urban development has replaced natural surfaces with high thermal mass materials, such as asphalt and concrete, causing minimum temperatures to increase in the heat island effect. Industrial and automobile pollutant gases act as condensation nuclei, increasing rainfall downwind of urban and industrial centres. The weekly build-up of pollutant gases has also been linked to changes in the temporal patterns of precipitation along the Atlantic coast of the USA. Excessive grazing, DEFORESTATION and DESERTIFICATION decrease moisture-retaining surface vegetation, leading to higher temperatures and drier conditions. A recent study has linked anti-radar chaff releases by the military to the suppression of lightning in intense convective storms in the southwest USA.   NJS

weather type
The categorization of the principal modes or patterns of atmospheric circulation, on a local, regional or hemisphere scale, produces a series of classes known as weather types. Each type tends to be associated with a particular type of weather, and weather typing represents the first stage in a SYNOPTIC CLIMATOLOGY study.

weathering
One of the most important of geomorphological and pedological processes, occurring when the rocks and sediments in the top metres of the Earth's crust are exposed to physical, chemical and biological conditions much different from those prevailing at the time the rocks were formed. Two main types are recognized (see Table 3). Physical or mechanical weathering (e.g. by frost or salt crystallization) involves the breakdown or disintegration of rock without any substantial degree of chemical change taking place in the minerals that make up the rock mass, while chemical weathering involves the decomposition or decay of such minerals. In most parts of the world both types of weathering may operate together, though in differing proportions, and one may accelerate the other. For example, the mechanical disintegration of a rock will greatly increase the surface area that is then exposed to chemical attack. Biological weathering can be both chemical and physical in nature. The

Table 3 Classification of weathering processes

Processes of disintegration (physical or mechanical weathering)
Crystallization processes
  salt weathering (by crystallization, hydration and thermal expansion)
  frost weathering
Temperature change processes
  insolation or thermal fatigue weathering (heating and cooling)
  fire
Expansion of dirt in cracks (dirt cracking)
Wetting and drying, especially of shales (HYGRIC WEATHERING)
Pressure release by erosion of overburden
Organic processes (e.g. root wedging)
Processes of decomposition (chemical weathering)
Hydration and hydrolysis
Oxidation and reduction
Solution and carbonation
Chelation
Biological chemical changes

rate at which weathering occurs will depend both on climatic conditions (e.g. the frequency of frost, temperature, the amount of water available for chemical reactions) and rock character (e.g. porosity, the density of jointing, and the susceptibility of the minerals). Consequences of weathering include DURICRUST formation, the formation of DEEP WEATHERING profiles (regolith and saprolite), the development of KARST, the creation of TAFONI and weathering pits (gnammas) and rinds, the generation of talus, the supply of much of the solute load of rivers, and the provision of material for removal by erosive processes. It is gaining increasing attention because of the role of chemical weathering in affecting the CARBON CYCLE and global climates. Weathering also attacks engineering structures and leads to the decay of buildings, including many of great cultural heritage value.

ASG

Reading
Goudie A.S. and Viles, H.A. (1997) *Salt weathering hazards.* Chichester: John Wiley & Sons, Ltd. · Goudie, A.S and Viles, H.A. (2012) Weathering and the global carbon cycle: geomorphological perspectives. *Earth-Science Reviews*, **113**, 59–71. · Pope, G.A. (ed.) (2013) *Weathering and soils geomorphology. Treatise on geomorphology*, volume 4. Amsterdam: Elsevier. · Siedel, H., Siegesmund, S. and Sterflinger, K. (2011) Characterisation of stone deterioration on buildings. In S. Siegesmund and R. Snethlage (eds), *Stone in architecture.* Berlin: Springer; pp. 347–410. · Yatsu, E. (1988) *The nature of weathering.* Tokyo: Sozosha.

weathering front The limit of the zone of weathering of bedrock beneath the land surface.

weathering index A quantitative or semi-quantitative indicator of the degree of weathering. Indices of both chemical and physical weathering have been proposed. Chemical indices of the alteration of rocks, W_I, may be expressed as the ratio of unweathered to weathered minerals in a volume of material, or as the ratio of the chemically more mobile to the chemically less mobile species, in the general form

$$W_I = \frac{\text{Proportion of chemical reactants}}{\text{Proportion of residual products}}$$

or more specifically in a form such as

$$W_I = \frac{\text{feldspar} + \text{mica} + \text{calcite}}{\text{clay minerals} + \text{quartz}}$$

Physical indices of weathering have been expressed in a number of ways, such as: capacity of unweathered material to take up water compared with the capacity of weathered material; the softening effect of water on material; the degree of swelling and slaking that results from water absorption; changes in strength or hardness of minerals; and velocity of ultrasound (seismic) waves through materials.

MJS

Reading
Selby, M.J. (1993) *Hillslope materials and processes*, 2nd edition. Oxford: Oxford University Press; chapter 8.

weathering potential index This is a measure of the ease with which rocks decay chemically. It is the ratio of the mole percentage of alkali (K) and alkaline earth metals (Ca, Na, Mg) less combined water to the mole percentage total metals (including Fe and Ti) less water. It provides a means of assessing the Goldich weathering stability series in numerical form. However, the computed value of quartz (zero) is not in agreement with its generally stable nature.

WBW

Reference
Wahlstrom, E.E. (1948) Pre-Fountain and recent weathering on Flagstaff Mountain near Boulder, Colorado. *Geological Society of America Bulletin*, **59**, 1173–1189.

weathering profile The complete cross-section of the chemically altered zone above fresh bedrock; essentially developed in situ by weathering processes. Combined with the near-surface transported layer, also termed REGOLITH. It commonly extends below the conventional *soil profile* with its A, B, C horizons contained within 1–5 m of the ground surface, and may reach depths of 100 m or more. At the base of the profile is found the *weathering front*, where fundamental physico-chemical changes cause the rock to become fissile, and disaggregate, transforming up-profile to a

Table 4 Scale of weathering grades of rock mass

Term	Description	Grade
Fresh	No visible sign of rock material weathering; perhaps a slight discoloration on major discontinuity surfaces.	I
Slightly weathered	Discoloration indicates weathering of rock material and discontinuity surfaces. All the rock material may be discoloured by weathering.	II
Moderately weathered	Less than half of the rock material is decomposed or disintegrated to a soil. Fresh or discoloured rock is present either as a continuous framework or as corestones.	III
Highly weathered	More than half of the rock material is decomposed or disintegrated to a soil. Fresh or discoloured rock is present either as a discontinuous framework or as corestones.	IV
Completely weathered	All rock material is decomposed and or disintegrated to soil. The original mass structure is still largely intact.	V
Residual soil	All rock material is converted to soil. The mass structure and material fabric are destroyed. There is a large change in volume, but the soil has not been significantly transported.	VI

mixture of sand and clay usually called *saprolite*. In susceptible rock affected by intense weathering, as below well-drained sites within the forested tropics, almost complete transformation to saprolite can occur within a narrow zone on the order of 10 cm, and the weathering front can appear as a *basal weathering surface*. In other situations a zone of gradual transition from saprolite to bedrock may occur over a few metres. Such transitional materials with a low clay content (<2–7%) are called *saprock* or *grus* (German).

Chemical alteration generally increases towards the surface and the content of fresh rock declines. Metal cation mobility during weathering allows DURICRUSTS to form under stable conditions. These crusts are dominated by one or more metallic oxides; aluminium (alucrete or bauxite), iron (ferricrete or laterite), silica (silcrete) and calcium (calcrete or caliche) are the most common. A sixfold division of the weathering profile is often used (especially in engineering) (see Table 4 and figure). MFT

Reading
Fookes, P.G. (1997) *Tropical residual soils*. Geological Society Professional Handbooks. London: The Geological Society. · Ollier, C.D. (1984) *Weathering*, 2nd edition. London: Longman. · Gerrard, A.J. (1988) *Rocks and landforms*. London: Unwin Hyman. · Thomas, M.F. (1994) *Geomorphology in the tropics*. Chichester: John Wiley & Sons, Ltd.

weathering rinds Oxidation phenomena that stain the parent rock red–yellow when exposed to air or near-surface groundwater for some time. They may extend for more than a millimetre into the rock, whereas DESERT VARNISH is generally much thinner. ASG

weathering-limited slopes Hillslopes on which the potential rate at which weathered soil and debris can be removed by erosional processes exceeds the rate at which the material can be produced by weathering. The rate of ground loss is consequently controlled by the rate of weathering, and the form of the hillslope unit is controlled by the relative resistance to weathering of the rock masses on which the slope is formed. MJS

Reading
Young, A. (1972) *Slopes*. Edinburgh: Oliver & Boyd.

wedge failure The removal of blocks of rock (or, more unusually, clay) where two or three fractures, often joints, intersect at a high angle in a cliff, and dip down towards the valley. The size of blocks from this kind of fall is rarely greater than a few $10\,m^3$. (See also SLAB FAILURE and TOPPLING FAILURE.) WBW

weir A structure built across a river or stream channel in order to measure the flow. It may be built of concrete, metal or wood and possesses two major features. First, it ponds back the flow to create a pool with flow velocities. Second, it incorporates a crest or notch over or through which the water flows freely from the upstream pool. A unique and stable relationship exists between the depth or head of water above the weir crest and water discharge. Values of discharge may be obtained from measurements of water stage by using published formulae or calibration curves established using other methods of flow measurement. (See also DISCHARGE.) DEW

		All rock material converted to soil; mass structure and material fabric destroyed. Significant change in volume.
VI	Residual soil	
V	Completely weathered	All rock material decomposed and/or distintegrated to soil. Original mass structure still largely intact.
IV	Highly Weathered	More than 50% or rock material decomposed and/or disintegrated to soil. Fresh/discoloured rock present as discontinuous framework or corestones.
III	Moderately weathered	Less than 50% of rock material decomposed and/or disintegrated to soil. Fresh/discoloured rock present as continuous framework or corestones.
II	Slightly weathered	Discolouration indicates weathering of rock material and discontinuity surfaces. All rock material may be discoloured by weathering and may be weaker than in its fresh condition.
IB	Faintly weathered	Discolouration on major discontinuity surfaces
IA	Fresh	No visible sign of rock material weathering

A. Idealized weathering profiles without corestones (left) and with corestones (right)   B. Example of a complex profile with corestones

Rock decomposed to soil
Weathered / disintegrated rock
Rock discoloured by weathering
Fresh rock

Reading
Ackers, P., White, W.R., Perkins, J.A. and Harrison, A.J.M. (1978) *Weirs and flumes for flow measurement*. Chichester: John Wiley & Sons, Ltd.

wells   Vertical dug or bored shafts penetrating to the saturated zone for the purpose of exploiting GROUNDWATER. In some old literature and on old maps, wells may refer to SPRINGS used as water supplies.   PWW

westerlies   Belts of winds, generally southwesterly in the northern hemisphere and northwesterly in the southern hemisphere, with average position between about latitudes 35° and 60°. These belts move poleward in the winter and equatorward in summer and are part of the GENERAL ATMOSPHERIC CIRCULATION, being the lowlevel branch of the FERREL CELL.

wet-bulb temperature   The temperature at which a sample of air will become saturated by evaporating pure water into it at constant pressure. It is measured by a thermometer whose bulb is covered by muslin that is kept constantly wet with distilled water. The lower the humidity is, the stronger the evaporation from the muslin and the stronger the cooling of the thermometer bulb. Wet-bulb temperature is an indicator of absolute humidity, and its difference from the dry-bulb is a measure of relative humidity.   RR

wetland
A collective term for ecosystems of high biodiversity whose formation has been dominated by water, and whose processes and characteristics are largely controlled by water. A wetland is a place that has been wet enough for a long time to develop specially adapted vegetation and other organisms. They include areas of marsh, fen, peatland or water, whether natural or artificial, permanent or

A wetland in the Upper Zambezi Valley, Zambia. Photograph by David Thomas.

temporary, with water that is static or flowing, brackish or salt, including marine water the depth of which at low tide does not exceed 6 m. Wetlands play a number of roles in the environment, principally water purification, flood control and shoreline stability. The UN Millennium Ecosystem Assessment determined that environmental degradation is more prominent within wetland systems than any other ecosystem on Earth; they are the focus of considerable international conservation efforts to preserve and restore them.   TS

Reading
Keddy, P.A. (2010) *Wetland ecology: principles and conservation*, 2nd edition. Cambridge: Cambridge University Press. · Mitsch, W.J. and Gosselink, J.G. (2007) *Wetlands*, 4th edition. New York: Van Nostrand Reinhold.

wetted perimeter   The perimeter of a river channel that is covered by water at a specific stage of flow. Such a wetted perimeter p is used together with the cross-sectional area a of the water at the same stage to give the HYDRAULIC RADIUS as a/p. KJG

wetting front   The subsurface limit to which a soil, particularly in a desert region, becomes saturated by infiltrating rainwater.

whirlwind   Loosely used term to describe rotating winds of scales up to that of a tornado. They are usually manifestations of intense convection over very small areas. The winds are frequently strong enough to raise surface materials (causing dust devils) and can become extremely hazardous when associated with large fires, such as may result from heavy bombing, or as occur in semi-arid bush areas.   JSAG

width/depth ratio   A simple measure of the shape of a river channel cross-section usually obtained as w/d, where w is the top width of the cross-section to be characterized and d is derived as an average value by dividing the cross-sectional area a by w in the form

$$d = \frac{a}{w}$$

The ratio is conventionally evaluated for the bank-full stage, and Riley (1976) suggested two alternative measures: namely bed width and the exponent x in the equation

$$w_1 = cd^x$$

where w_1 is width at stage d_t.   KJG

Reference
Riley, S.J. (1976) Alternative measures of river channel shape and their significance. *Journal of Hydrology (New Zealand)*, **15**, 9–16.

wilderness   A wild, uncultivated area. Ideally, wilderness areas should never have been subject to human activity that has resulted in manipulation, either deliberate or unconscious, of the ecology of the area (Simmons, 1981: 62).   ASG

Reference
Simmons, I. (1981) *The ecology of natural resources*, 2nd edition. Oxford: Blackwell.

Wilson cycle   One of the hypothesized long-term tectonic cycles related to the mechanisms of plate tectonics, and named in honour of J. Tuzo Wilson. Modern ocean basins are no more than 200–300 million years old, and many will close, as a result of plate movement, within a similar period. Plate tectonic processes, though, are thought to have been active over 2 billion years, allowing sufficient time for many cycles of opening and closing of ocean basins. When oceans close and continents are sutured together, the larger mass of continental crust insulates the mantle beneath, which warms slowly until convection is triggered. Continental fragmentation then begins anew, with rifting, sea-floor spreading and ultimately the next closing of the oceans occurring in a sequential or cyclic manner. This sequence of events, spanning 500 million years or so, is the Wilson cycle.   DLD

Reading
Kearey, P., Klepeis, K.A. and Vine, F.J. (2013) *Global tectonics*, 3rd edition. Oxford: Blackwell.

wind   Air in motion relative to the surface of the Earth. It is important to note that this definition covers more than the layman's view of simple horizontal airflow. First, the motion may also have a substantial vertical component, and indeed it is this component that is a primary cause of all the WEATHER we experience on this planet. Second, it is vital to appreciate that the motion is *relative* to the surface of the Earth. Thus, in 'calm' conditions, when air is stationary relative to the surface, it is still moving through space at the same velocity as the surface of the Earth. This characteristic has important repercussions on the direction of airflows and upon the momentum budget of the atmosphere.

Wind is specified in terms of speed and direction. Speeds are reported in knots or metres per second, the reading being taken over a few minutes to reduce the effects of very short duration gusts, and direction (which is the direction from which the air flows) is reported to the nearest 10° of the compass. If wind direction varies with time, a clockwise shift is called veering and an anticlockwise one backing. At the surface of the Earth winds are measured by ANEMOMETERS, with the actual sensor preferably being at a height of 10 m. At higher levels winds are measured by pilot balloons and RADIOSONDES. In both cases the measurements are made of only the horizontal component of the wind. Hence, we have no routine direct measurements of vertical winds, largely because they are notoriously difficult to measure at all, let alone accurately. This is so because of the smallness of vertical wind speeds: the speed of vertical winds is usually of the order of centimetres per second, whereas that of horizontal winds is of the order of metres per second. This lack of information on vertical motion is most frustratin, both for the theoretician and for the practical meteorologist, such as a forecaster. It is precisely the phenomenon that we wish to measure that, up to now, has proved unmeasurable.

Although observations of winds have been, and continue to be, vital to our knowledge of the atmosphere, it is true to say that deeper understanding has emerged from theoretical analyses of airflow. These analyses all rest on Newton's second law, which states that if a force is applied to a mass, it will accelerate that mass in the direction of the force. This law can readily be applied to a particle of air in the atmosphere. The value in doing this lies in the acceleration part of the law. Acceleration is the change of velocity with time.

Hence, if we can derive an acceleration of air at a given time from the application of Newton's law, we can then derive an air velocity at a future time. This has the double advantage not only of giving information about the velocity of the airflow, which has been shown to be fundamental to weather and climate, but also of giving its value at a future time; that is, a forecast. The acceleration of the air largely results from the application of three forces: the pressure gradient force, the CORIOLIS FORCE and the FRICTION force. Hence, the equation of motion describing the flow in a horizontal direction, say x, is

$$\underbrace{\frac{\partial u}{\partial t}}_{\text{Acceleration}} = \underbrace{\frac{1}{\rho}\frac{\partial p}{\partial x}}_{\substack{\text{Pressure} \\ \text{gradient force}}} + \underbrace{2\Omega v \sin\phi}_{\substack{\text{Coriolis} \\ \text{force}}} + \underbrace{\frac{\partial}{\partial z}\left(\frac{K\partial u}{\partial z}\right)}_{\text{Frictional force}}$$

where u is the air velocity in the x direction, t is time, ρ is air density, p is air pressure, Ω is the angular velocity of the Earth's rotation, v is the air velocity along a direction perpendicular to x, ϕ is latitude, z is the vertical coordinate and K is the EDDY DIFFUSIVITY of momentum.

In large-scale meteorology it is customary to make the x-direction east–west, the y-direction

north–south and the z direction vertical. The equation is thus complemented by similar ones in the y and z directions, but for present purposes it suffices.

The fact that four terms appear in the equation does not mean that all four are important to all airflows at all times. This was appreciated by Jeffreys in the 1920s, and by scrutiny of each term he was able to classify winds dynamically. In addition, Jeffreys showed that certain types of wind were associated with circulations of certain sizes. This classic paper deserves to be more widely known among non-meteorologists. Jeffreys argues that the pressure gradient force is an important term, being the force responsible for most air motion relative to the Earth's surface. If, of the remaining three terms, the Coriolis force is much larger than both the acceleration and frictional terms, then, in the equation, the pressure gradient term is balanced by the Coriolis term and the resultant wind is known as the geostrophic wind, a term suggested by Sir N. Shaw. Such a wind blows parallel to the isobars, with low pressure to the left and high pressure to the right in the northern hemisphere and the converse in the southern hemisphere. If, secondly, the Coriolis and frictional terms are small in comparison with the acceleration term, then the latter balances the pressure gradient term to give Eulerian winds. If, thirdly, the frictional terms exceed the Coriolis and accelerational terms, then friction must balance the pressure gradient. The wind blows along the pressure gradient and the friction, assumed to act opposite to the flow, is sufficient to prevent the velocity of a particular mass of air from increasing steadily throughout its journey. Jeffreys termed such a wind antitriptic.

Crude as Jeffreys' classification was, it provided a firm, rational foundation for an understanding of air motion per se and the organization of air flow into recognizably different configurations. Jeffreys himself noted that there would be many cases in which the number of terms in the equations of motion comparable to the pressure gradient term is two or three. Such a case is the wind in the lowest kilometre or so of the atmosphere, where the pressure gradient, Coriolis and frictional forces are in balance. In this boundary layer the wind speed is zero at the ground, but increases in value (and direction veers) with height until the geostrophic condition is reached at the top of the friction layer. A further case occurs when the isobars are curved, as opposed to straight. This means that a centripetal force enters the equation and the resultant (friction-free) wind is known as the

GRADIENT WIND, a flow resulting from a balance of pressure gradient, Coriolis and centripetal forces. Clearly, one can further complicate each of these basic types by allowing the relative magnitude of any term to change.

Observations have confirmed the basic validity of Jeffreys' results. The general atmospheric circulation comprises entities ranging in characteristic horizontal dimension from thousands of kilometres (the ROSSBY WAVES) to a few kilometres or even a few hundreds of metres (locally induced flows, such as cold air drainage). For the Coriolis term to have a major influence upon airflows, their characteristic horizontal size must be greater than 70 km (at the poles) and 400 km (10° from the equator). Thus, cyclones, anticyclones and Rossby waves are manifestations of geostrophic and gradient flows. Systems smaller than the critical size (mesoscale and smaller) are largely unaffected by the Coriolis force (the sea breeze is the notable exception). They fall into Jeffreys' 'antitriptic' category if friction is an important force; sea breezes and mountain breezes exemplify this type. If, however, friction may be ignored but the accelerational terms may not, then these small systems fall into Jeffreys' Eulerian category. Severe local storms, such as thunderstorms and tornadoes, exemplify this type of airflow. BWA

Reading
Atkinson, B.W. (1981) *Dynamical meteorology – an introductory selection*. London: Methuen; especially chapter 1. · Atkinson, B.W. (1986) *Meso-scale atmospheric circulations*. New York: Academic Press. · Atkinson, B.W. (1982) Atmospheric processes. In J. M. Gray and R. Lee (eds), *Fresh perspectives in geography*. Special publication 3. London: Department of Geography, Queen Mary College, University of London. · Barry, R.G. (1992) *Mountain weather and climate*, 2nd edition. London: Routledge. · Barry, R.G. and Chorley R.J. (2003) *Atmosphere, weather and climate*, 8th edition. London: Routledge. · Defant, F. (1951) Local winds. In T. F. Malone (ed.), *Compendium of meteorology*. Boston, MA: American Meteorological Society; pp. 655–672. · Jeffreys, H. (1922) On the dynamics of winds. *Quarterly Journal of the Royal Meteorological Society*, **48**, 29–46. · Meteorological Office (1978) *A course in elementary meteorology*, 2nd edition. London: HMSO. · Yoshino, M.M. (ed.) (1976) *Local wind bora*. Tokyo: University of Tokyo Press.

wind chill An index of the degree of atmospheric cooling experienced by a person. The amount of heat loss or gain can be measured either by the actual heat exchange in watts per square metre of skin exposed or as an equivalent temperature; that is, the temperature in still air that would correspond to the cooling (or heating) generated by the particular combination of temperature and wind speed. Other indices have been devised to take account of clothing type and thickness. PS

Reading
Dixon, J.C. and Prior, M.J. (1987) Wind chill indices – a review. *Meteorological Magazine*, **116**, 1–17.

wind rose Illustrates graphically the climatic characteristics of *wind* direction, and also frequently wind speed, at a particular location. Radii are drawn outwards from a small circle, proportional to the frequency of wind from each direction. The percentage frequency of calms is usually indicated inside the circle. Wind speed can be depicted by varying the thickness of the radiating lines. There are many variations of this basic method. A wind rose can be adapted to show the relationship between wind direction and other meteorological variables, such as precipitation or the incidence of thunderstorms. DGT

Reading
Monkhouse, F.J. and Wilkinson, H.R. (1971) *Maps and diagrams*. London: Methuen; pp. 240–243.

wind shadow The region downwind of an obstacle, shielded from the wind. Hedges are less of an obstacle than walls and hence reduce the strength of lee eddies while still providing shelter. The effect persists a distance downwind about 30 times the height of the obstacle in neutrally stratified conditions, and much further for strong inversions. It may lead to complete stagnation of air in city basins, as in Los Angeles, Rome and London. JSAG

Reading
Monteith, J.L. (1973) *Principles of environmental physics*. London: Arnold.

wind shear A change of wind speed and/or direction with horizontal or vertical distance. Its magnitude is important in the development of dynamic instabilities in the atmosphere, such as baroclinic and barotropic instability.

The term is also applied to a situation more properly referred to as a downburst. This occurs when a cold and dense air parcel (such as from a thunderstorm) drops to the surface, spreading intense outflow in all directions. This creates wind shear near the surface, sometimes so intense than an airplane cannot safely fly through. The plane encounters an extreme head wind upon entering the shear zone, forcing the plane upward, and an extreme tail wind upon leaving the shear zone, forcing the plane downward. SEN

World Heritage Site Site considered officially to be of great significance to global cultural

573

or natural heritage and designated as such under the Convention Concerning the Protection of the World Cultural and Natural Heritage (more commonly known as 'The World Heritage Convention'). The convention was adopted by the General Conference of UNESCO in 1972, although it came into force only in 1975. To date, more than 150 countries are signatories to the Convention, making it one of the most universal international legal instruments for the protection of the world's cultural and natural heritage. The Convention operates through a World Heritage Committee that establishes, keeps up to date, and publishes a World Heritage List of cultural and natural properties, known as World Heritage Sites. There are three international non-governmental or intergovernmental organizations named in the Convention to advise the Committee in its deliberations. These bodies are the IUCN, the World Conservation Union (for natural sites), and the International Council on Monuments and Sites (ICOMOS) together with the International Centre for the Study of the Preservation and Restoration of Cultural Property (ICCROM) (for cultural sites). Nominations for such sites are submitted by the member states and must be considered to be of universal cultural or natural value in order to be accorded this status. The Convention is legally binding on signatory countries, who must help identify, protect, conserve, and transmit to future generations World Heritage properties. When a site is nominated, experts conduct a careful investigation into its merits. The World Heritage Fund helps give technical cooperation, and gives emergency assistance in the case of properties severely damaged by specific disasters or threatened with imminent destruction.

For the purposes of the Convention, the following are considered as 'cultural heritage':

- monuments – architectural works, works of monumental sculpture and painting, elements or structures of an archaeological nature, inscriptions, cave dwellings and combinations of features that are of outstanding universal value from the point of view of history, art or science;
- groups of buildings – groups of separate or connected buildings that, because of their architecture, their homogeneity or their place in the landscape, are of outstanding universal value from the point of view of history, art or science;
- sites – works of humans or the combined works of nature and humans, and areas including archaeological sites, that are of

outstanding universal value from the historical, aesthetic, ethnological or anthropological point of view.

For the purposes of the Convention, the following are considered as 'natural heritage':

- natural features consisting of physical and biological formations or groups of such formations that are of outstanding universal value from the aesthetic or scientific point of view;
- geological and physiographical formations and precisely delineated areas that constitute the habitat of threatened species of animals and plants of outstanding universal value from the point of view of science or conservation;
- natural sites or precisely delineated natural areas of outstanding universal value from the point of view of science, conservation or natural beauty.

Articles 4 and 5 of the Convention deal with the responsibilities of the signatory states. Each must recognize that the duty of ensuring the identification, protection, conservation, presentation and transmission to future generations of the cultural and natural heritage situated on its territory belongs primarily to that state and that it is charged with doing all it can to this end. To ensure that effective and active measures are taken for the protection, conservation and presentation of the cultural and natural heritage situated on its territory, each signatory must endeavour, as far as possible:

1 to adopt a general policy that aims to give the cultural and natural heritage a function in the life of the community and to integrate the protection of that heritage into comprehensive planning programmes;
2 to set up within its territories, where such services do not exist, one or more services for the protection, conservation and presentation of the cultural and natural heritage with appropriate human resources;
3 to develop scientific and technical studies and research and to work out operating methods to counteract the dangers that threaten cultural or natural heritage;
4 to take the appropriate legal, scientific, technical, administrative and financial measures necessary for the identification, protection, conservation, presentation and rehabilitation of this heritage; and

5 to foster the establishment or development of national or regional centres for training in the protection, conservation and presentation of the cultural and natural heritage and to encourage scientific research in this field.

There are now more than 500 World Heritage Sites, including Australia's Great Barrier Reef, India's Taj Mahal, the Grand Canyon of the United States, South America's Iguaçu Falls and the Giant's Causeway in Britain. There are approximately three times as many cultural sites as those identified for their natural significance. The Convention has its headquarters in Paris under the auspices of UNESCO.   MEM

Reference

UNESCO. (1975) *Convention concerning the protection of the world cultural and natural heritage.* Paris: UNESCO.

X

X-ray diffraction Also referred to as X-ray crystallography, this is a nondestructive analytical technique used to examine the crystalline structure, chemical composition and physical properties of minerals. The science owes its origin to William Lawrence Bragg, who became the youngest ever Nobel Laureate in Physics when he realized that X-rays could be used to detect the arrangement of individual atoms inside solid crystals. The technique is widely applied in the analysis of clay minerals in particular (see CLAY).   MEM

xerophyte A plant adapted to living in conditions of aridity such that it is able to survive and reproduce in the long-term absence or scarcity of water. A range of specific metabolic and morphological adaptations are associated with xerophytes and include drought avoidance and drought-tolerance strategies. Adaptations include the presence of deep roots (so-called tap-roots, e.g. *Welwitschia mirabilis* of the Namib Desert), the ability to store water (e.g. numerous cactus species and succulents), the ability to imbibe water from condensed atmospheric moisture (as in some highly specialized Chilean desert plants) and the ability, as in EPHEMERAL PLANTS, to complete the life cycle in the short period of moisture availability following rain.   MEM

Y

yardangs Streamlined wind-erosion forms with their long axes parallel to the wind. They are often described as resembling an inverted ship's hull, although in many cases they are flat-topped. Length-to-width ratios are commonly 3:1 or greater. Height varies from a few metres to 200 m and length from several metres to several kilometres. The windward face of the yardang is typically blunt-ended, steep and high, whereas the leeward end declines in elevation and tapers to a point. However, yardangs may take on many forms. Yardangs form parallel to one another and are separated by either U-shaped or flat-bottomed troughs.

Yardangs form in a broad range of geological materials, including sandstones, limestones, clay-stones, granites, gneisses, schists and lacustrine sediments. Variations in form result from differences in material. Yardangs develop by abrasion, deflation or a combination of these processes, and are further modified by fluvial erosion and mass movement. The ultimate form varies according to lithology, structure and regional climatic history.

Yardangs are uncommon, as they require conditions of great aridity, unidirectional or seasonally reversing winds and, in some cases, a favourable material and some assistance from weathering to form. In Africa, yardangs occur in the Arabian Peninsula, Egypt, Libya, Chad, Niger and the Namib Desert. Asia has several yardang fields, located in the Taklimakan Desert in the Tarim Basin, the Qaidam depression of Central Asia and the Lut Desert, Iran. Yardangs are found along the coastal desert of Peru in South America and as minor groups in North America. Large yardang fields also occur on Mars.   JL

Reading
Goudie, A.S. (2007) Mega yardangs: a global analysis. *Geography Compass*, 1, 65–81.

yazoo A tributary stream that runs parallel to the main river for some distance.

Younger Dryas Between around 12,800 and 11,500 BP there appears to have been a cold phase of climatic deterioration characterized by major ice advance in northwest Europe. This phase has been called the Younger Dryas (termed the Loch Lomond stadial in the British Isles) named after the Alpine plant Mountain Avens (*Dryas octopetula*), the distribution of which was more southerly in northwest Europe. It has traditionally been referred to as a northwest European climatic anomaly; however, indications of this episode of cooler climatic conditions are found eastwards into Russia, Spain and Portugal, North Africa and North America. This has been interpreted as having resulted from precipitation variations due to latitudinal changes in the positions of the polar atmospheric and oceanic fronts in the North Atlantic. Outside the glaciated areas in the Younger Dryas time, periglaciation, with discontinuous permafrost, was both extensive and highly effective.   AP

Reading
Dawson, A.G. (1992) *Ice age Earth: late Quaternary geology and climate*. London: Routledge. · Muscheler, R., Kromer, B., Björck, S., *et al.* (2008) Tree rings and ice cores reveal ^{14}C calibration uncertainties during the Younger Dryas. *Nature Geoscience*, 1, 263–267.

Z

zeuge A tabular mass of rock perched on a pinnacle of softer rock as a result of erosion, usually by wind, of the underlying materials. Although common in textbooks describing desert landscapes, zeuges are in reality rather rare.

zibar A low dome-shaped sand dune that is usually formed of coarse sand and has no slip face.

zonal circulation Any flow along latitude circles. The term is used in a more specific sense to indicate a high-index circulation in mid-latitudes with a strong westerly wind between 35° and 55° and little meridional air mass exchange.

At the surface, pressure systems have a dominantly east–west orientation and in middle latitudes unsettled weather with alternating troughs of low pressure and ridges of high pressure move in quick succession, giving changeable conditions in any specified area. AHP

zonal soil A soil that occurs over a wide area and owes its characteristics to climatic rather than local topographical or geological factors.

zonality The division of the world into zones. At a global scale the world has been divided up into major climatic zones (equatorial, dry tropical, desert, Mediterranean, temperate, boreal, tundra and polar, etc.), major vegetation zones (rainforest, savanna, desert, tundra, etc.) and major soil zones (latosols, chernozems, podzols, etc.). Climatic geomorphogists have attempted to divide the world into morphogenetic zones. Zones can also be related to altitude (e.g. sub-montane, montane, subalpine, alpine, nival, etc.). Zones also occur at more local scales, as for example on a slope or on a beach. ASG

Reading
Huggett, R. (2010) *Physical geography, the key concepts.* London: Routledge.

zonation In biogeography and ecology, one of the most commonly occurring patterns of plant and animal distribution. On a world scale, characteristic groups of organisms occupy different latitudinal zones of the terrestrial and aquatic surface, which are primarily related to climate (see BIOME). Each zone forms an idealized band around the Earth, although this is frequently strongly modified by other factors. There are also approximately concentric zones around MOUNTAINS, consisting of plant and animal communities typical of the changing environment with elevation. PAF/MEM

zoogeography The branch of BIOGEOGRAPHY that deals with the distribution of animals, linking the subject matter of the discipline of zoology with the viewpoint of geography. The first important organizer of zoogeographical knowledge was the German biologist and philosopher Ernst Haeckel, who introduced the term *ecology* into the subject (1866). Early synthesists of historical zoogeography were P. L. Sclater (1858), T. H. Huxley (1868) and J. A. Allen (1871), but the modern 'Father of zoogeography' is usually regarded as Alfred Russel Wallace, who published his now classic work, *The geographical distribution of animals*, in 1876 (see WALLACE'S LINE and WALLACE'S REALMS).

With the evolutionary thinking of Darwin, Wallace, Huxley, Lamarck and many others and with the publication of *On the origin of species by means of natural selection* (1859), zoogeography grew in stature as a scientific discipline. The earliest writers in this new mould (e.g. Wallace) still believed in the permanency of the continents, a view soon to be challenged by Wegener's theory of continental drift, developed in the early part of the twentieth century. The essential vindication of this theory in the 1960s with the models of plate tectonics and sea-floor spreading has, in contrast, provided a completely new insight into the causes of animal distributions and has led to a lively and often bitter debate on the relative merits of dispersal and vicariance, the latter term meaning the separate development of faunas from common ancestors on the separating continents. This development has been accompanied by the rise of a new school of zoogeography, called VICARIANCE BIOGEOGRAPHY

The Dictionary of Physical Geography, Fourth Edition. Edited by David S.G. Thomas.
© 2016 John Wiley & Sons, Ltd. Published 2016 by John Wiley & Sons, Ltd.

which, in turn, is linked with the development of phylogenetic systematics and CLADISTICS in systematics (see TAXONOMY). PAS/MEM

Reading
Nelson, G. and Rosen, D.E. (1981) *Vicariance biogeography: a critique.* New York: Columbia University Press. · Udvardy, M.D.F. (1969) *Dynamic zoogeography.* New York: Van Nostrand Reinhold. · See also the journal *Systematic Zoology.*

zoogeomorphology The study of the geomorphic effects of animals, ranging in size from small invertebrates (e.g. worms, termites, ants) to terrestrial and aquatic vertebrates, including fish, amphibians, reptiles, birds and mammals (both wild and domesticated). Animals dig, burrow, turn over stones to seek food, create nests, build mounds, wallow and, in the case of beavers, build dams. In addition, their trampling and overgrazing may change soil infiltration capacities and vegetation characteristics, which may influence surface run-off and erosion. ASG

Reading
Butler, D.R. (1995) *Zoogeomorphology – animals as geomorphic agents.* Cambridge: Cambridge University Press.

Index

Page numbers in **bold** refer to definitions; those in *italics* refer to illustrations or tables

A
abiotic components **1**
ablation **1**, 107, 134
abrasion **1**, 411, 451
absolute age **1**
absolute drought 156
abyss **2**
abyssal plain *2, 2*
accommodation space **2–3**
accordant junctions, laws of **3**
accordant summits **3**
accretion **3**
  lateral 314, 411
accumulated temperature **3**
accuracy **3–4**
acid
  deposition **4**
  neutralizing capacity **6**
  precipitation **4**, 5–6, 144, 267, 415
  rocks **6**
acidity **4**, *5*
  profile **6**
active layer **6**, 123
active margin **7**
actualism *see* uniformitarianism
adaptation **7**
adaptive capacity **7**
Adelie Land Bottom Water 25
adhesion ripple **7**
adhesion warts **7**
adiabatic processes **7**, 102, 296, 303,
      404, 519
adsorption **7–8**
advection *8*, **8–9**, 113, 263
advective processes **9**
aeolation **9**
aeolian dunes 161–162
aeolian processes **9**
aeolianite *9*, **9**, 77, 285
aerial photography **10**
aerobic processes **10–11**, 23
aerobiology **11**
aerodynamic ripples **11**
aerosols **11**, 39, 404
aestivation **11**
affluent **11**
afforestation **11**
African easterly waves (AEWs) 167

aftershocks **11**
age
  absolute **1**
  relative **439**
ageostrophic motion **11–12**
agglomerate **12**
aggradation **12**, 405
aggregation ratio **12**
  punctuated aggradational cycles **420**
aggressivity **12**
agonic line **12**
agricultural drought **12**, 156
agroclimatology **12–13**
agroforestry **13**
agrometeorology 12, **13**, 345
aiguille **13**
air mass **13–14**
  classification 13–14, *14*
  compressional 108
  convergence 114
  Ferrel's law **209**
air parcel **14**
air pollution **14**
air pressure **419**
air wave theory **14–15**
airborne laser scanning (ALS) 317
airsplash 430
aklé **15**
alas **15**
albedo **15**, *16*, 343
alcoholic fermentation 23
alcove **15**
alcrete **16**
alfisol **16**
algae **16**
  blue–green 125
  diatoms **144**
  green 319
  rhodoliths **444**
algal bloom **16**, 125, 196
algal mat **16**
algal ridge/rim **16**, *17*
alien species **17**
    see also introduction, ecological; invasive species
allelopathy **17**
Allen's rule *17*
Allerød interstadial **17**
allochthonous sediments **18**

The Dictionary of Physical Geography, Fourth Edition. Edited by David S.G. Thomas.
© 2016 John Wiley & Sons, Ltd. Published 2016 by John Wiley & Sons, Ltd.

allogenic
  stream **18**, 42
  succession **18**, 42
allometric growth **18**
allometry 18
allopatric species **18–19**
allophane 19
alluvial channel **19**
alluvium **19–20**, 105, 405
  alluvial fans *see* fans
  alluvial fill 19
alp **20**
alpha diversity **20**
alpine orogeny **20**
alpine zone **20**
altiplanation **21**
  terrace 246
altithermal **21**, 278
aluminium 49
alveolar structures **21**
ambient temperature **21**
amensalism **21**
amino acid racemization **21–22**
amphidromic point **22**
ana-front **23**
anabatic flows **22**, 357
anabranching channel **22–23**, *23*
anaclinal features **23**
anaerobic conditions **23**
analemna **23**
analogue approach **23**
anamolistic cycle **23**
anaseism **23**
andosols **23**
andromy **24**
anemograph **24**
anemometer **24**, 570
angiosperms **24**
angle
  of dilation **24**
  of initial yield **24**, 25, 259
  of internal shearing resistance **24**, 25
  of plane sliding friction **25**
  of repose **25**, 246, 259
  of residual shear *25*, 25
angular momentum **25**, 351
angular unconformity **25**
animals
  areography **35**
  autecology **41**
  biomass **62**
  carnivores **80–81**
  competition 107
  dispersal **150**
  dominant organism 152–153
  endemism 178
  exotic **201**
  faunal realms **209**
  herbivores 218, **260–261**
  invasive species 295
  island biogeography **297–299**
  kingdom **306**
  migration 347

nanism **361**
phenology **400**
phoresy **400**
predation **418**, 419
soil formation 394
species–area curve **492**
zoogeography **577–578**
zoogeomorphology **578**
annual series **25**
annular drainage **25**
Antarctic Bottom Water (AABW) **25–26**, *26*, 368
Antarctic Circumpolar Current (ACC) **26–27**
Antarctic cold reversal (ACR) **27**
Antarctic Ice Sheet **27–28**, *28*
Antarctic Intermediate Water (AAIW) 368
Antarctic Surface Water 26
antecedent drainage **29**
antecedent moisture **29**, 271
antecedent precipitation index (API) **29**
anteconsequent stream **29**
Anthropocene **29**, 236, 243
anthropochore **29**
anthropogeomorphology *30*, **30**, 124, 267
anthrosols **30–31**
antibiosis **31**
anticentre **31**
anticline **31**, 396
anticyclone **31**, 519, 552
  blocking 157
antidune **31**
antiforms **31**
antipleion **31**
antipodal bulge **31**
antipodes **31**
antitrade winds **31**
aphanitic rock texture **31**
aphelion **31**
aphotic zone **31**
aphytic zone **32**
apogee **32**
aposematic coloration **32**
applied geomorphology **32**
applied meteorology **32–33**, *33*
aquaculture **33**
aquatic macrophyte **34**, 325
aquiclude **34**, 253
aquifer **34**, 155
  confined 406
aquifuge **34**
aquitard **34**
arboreal **34**
arches, natural **34**
archipelago **34**
arctic **34**
Arctic Circle **34**
arctic haze **35**
arctic smoke 225
arctic-alpine flora **34**
areic **35**
arena **35**
arenaceous rocks **35**
arenites 464
areography **35**

arête 35
argillaceous rocks 35
aridisols 35
aridity 160
arkose 35
armoured mud balls 35
armouring 35–36, 78, 284
arroyo 36
artesian water 36, 155, 253
  springs 494
aspect *36*, **36**
association, plant **36–37**, *37*
  see also community
asthenosphere **38**, 320, 408
astrobleme 38
asymmetric fold 38
asymmetric valley 38
Atlantic coastlines 38
Atlantic Meridional Overturning Circulation
      (AMOC) 27, 338
Atlantic Multidecadal Oscillation (AMO)
      338
atmometer 38
atmosphere **38–39**
  composition 38, *39*, **39**
  energetics **39**
  layers **39–40**
  pollutants 39
  temperature structure *38*, *38*
atmospheric instability **39**, 339
atmospheric waves 40
atoll **40**, 116
Atterberg limits **40–41**
aufeis *see* icing
auge 41
aulacogens 41
aureole, metamorphic 41
aurora borealis 41
autecology 41
autochthonous sediment 18, **41**
autocorrelation **41–42**, 237
autogenic stream 42
autogenic succession 42
autotrophic organisms 42, 64
autovariation 42
avalanche **42–43**
  debris avalanche 43
  rock avalanche 43
    sturzstrom 505
  snow avalanche 43
avalanche tarns 43
aven **43**, 512
avulsion 43
azimuth 43
azoic period 43
azotobacter 43

B
backshore 44
backswamp 44, 213
backwall 44
backwash 44, 322
backwearing 44

bacteria
  cyanobacteria **125**
  subsurface lithoautotrophic 42
badlands **44–45**
bajada **45**, 207
bank erosion 19, **45**
bank storage 45
bankfull discharge **45–46**, 86, 213
banner cloud 46
bar, nearshore 46
barchan **46**, 535
barchanoid ridge 46
baroclinicity 46
barometer 46
barrier island 47, 103, 381
barrier reef 47, 116
barrier spit 47
basal ice 47
basal sapping 35, **48**, 91
basal sliding **48**, 107, 282
basalt **48**
  plateau 409
base exchange 48
base flow **48**, 160
  base-flow recession curve 139
base level **48–49**, 126, 395, 450
base saturation 49
basement complex 49
basic rocks 49
basin
  drainage 154
  experimental **441–442**
  exudation 205
  planning 448
  representative **441**
  salar 459
basin-and range 49
batholith **49**, 557
bathymetry 49
bauxite 49
beach **49**, *50*
  cell circulation *85*, **85**
  cusp **125**
  dissipative **150**, 436
  equilibrium 192
  nourishment **49–50**, 85
  raised **431–432**
  reflective **436**
  retrogradation **443**
  ridge 50
    chenier **89–90**
  rock **50**, 77, 285
  sandy, sea level rise effect 72–73
  storm 46, **501**
beaded drainage 50
bearing capacity **50–51**
Beaufort scale *51*, **51**
bed
  deformable **133–134**
  forest **220**
  plane 407
  red **435**
  roughness **51**, 541

bedding
  cross **122**
  current 19, **124**
  false **207**
  graded **246**
bedding plane **52**
bedform 52, 124
  cluster **103**
  pool and riffle **416**, 471
  reconstitution **52**
  subglacial 216, **506**
bedload **52**
  tracers **53**
  transport 161
bedload pit trap **53**
bedrock **53**
  channel 19
Beer's law **53**
behavioural adaptations 7
benchmark **53**
Benioff zone **53**
benthic zone **53**
benthos 53
berg wind **53**
berghlaup **53–54**
Bergmann's rule 54
bergschrund **54**
berm **54**
Bernoulli's theorem/effect **54**
beta diversity 20
bifurcation ratio **54**
bioaccumulation 54–55, 61
biochar **55**
biochemical cycle **125**
biochemical oxygen demand (BOD) **55**
bioclastic rock **55**
bioclimatology **55**
biodegradation **55–56**
biodiversity **56**
  hot spot 265–266
  rain forest 190
bioerosion **56–57**
biofilm 57, *58*
  biostabilization 64
biofuel **57**
biogeochemical cycle **57–58**, *58*
biogeocoenosis **59**
biogeography 35, **59–60**
  island 110, **297–298**
  vicariance 93, **552–554**, 577–578
biogeomorphology **60**
bioherm **60**
biokarst **60**, 405
bioleaching **60**
biological control **60**
biological magnification **60–61**, 399
biological productivity **61**, 170, 364, **420**,
    535–536
  plankton 407
  rain forest 190
bioluminescence **61**
biomantle **61–62**
biomass *62,* **62**

biome **62**, *63*
biometeorology **63**
biosphere **63–64**
  energy flow 64
  nutrient recycling 64
biostabilization **64**
biostasy **64**
biota **65**
biotope **65**
bioturbation 61, **65**, 394
bipolar seesaw 27
birth rate, density-dependent 81, *81*
black body 426–427
black box **65**
blanket bog **65**, 392–393
blind valley **65**, 336
blizzard **65**
block faulting **65**
block fields **65**
block streams **65**
blocking **65–66**
blocking anticyclone 157
blow-hole **66**
blowouts **66**
bluehole **66**
Blytt–Sernander model *66,* **66**, 68
bog bursts **67**
bogaz **67**
bogs **66–67**
  blanket **65**, 392–393
  floating **211**
  peat 392–393
  raised 41, **437**
  string **505**
  see also mires
Bølling interstadial **67**
bolson **67**
bombardment **67**
Bond cycles **67**
bora **67**, 304
bore **67**
boreal
  climate **67**
  forest **67–68**
  period **68**
bornhardt **68**
boss **68**
botryoidal **68**
bottom-sets **68**
Bouguer anomaly **68**
boulder clay *see* till
boulder train **68**
boulder-controlled slopes **68**
boundary conditions **68–69**
boundary layer **69–70**, *70,* 147, 215, 223, 269, 270,
    456, 541, 551
  atmospheric 174
  dune 162
bounding surface **70**
bourne **70**
Bowen ratio **71**
Bowen's reaction series **71**
brackish water **71**

braided river   19, **71**, 87, 284, 314
  cellular automata model   86
brash   **71**
Braun-Blanquet scale   **71–72**, 152
breccia   **72**
brodel   **72**
brousse tigrée   **72**
Brune curve   **72**
Brunhes–Matuyama   **72**
brunizem   **72**
Bruun rule   **72–73**
buffer   **73**
buffer strip   **73**, 137
buffering capacity   **73–74**
bulk density   **74**
bush encroachment   **74**
bushveld   **74**
butte   **74**
Buys Ballot's law   **74**
  geostrophic wind   237
  gradient wind   247
bypass flow   **74–75**, 334

C
caatinga   **76**
caballing   **76**
caesium-137 (^{137}Cs) analysis   **76**
Cainozoic   *76*, **76**
calcicole   **76**
calcifuge   **76**
calcite compensation depth   **76–77**
calcite saturation index   **77**
calcrete   **77**, 520
caldera   **77**, 480
caliche   **77**
calms   **77**
calving   **77**
cambering   **77–78**, 254
canopy   **78**
cap-rock   **79**
capacity load   *78*, **78**
capillary forces   **78**, 399
capture   **79**
  stream   **502–503**
carapace   **79**
carbon cycle   **79**
carbon dating *see* radiocarbon dating
carbon dioxide (CO_2)
  ice   280
  removal techniques   234
carbon sequestration   **79**
carbonate budget, coral reefs   **79–80**, *80*
carbonate compensation depth   **80**
carbonation   **80**
carnivores   **80–81**, 218, 536
carrying capacity   **81**, 170, 416, 424
carse   **81**
case hardening   **81**, 285
cataclasis   **81**
cataclinal   **81**
catastrophe   **81–82**
catastrophism   **81–82**
catchment   **82**

contributing area   **113**
control   **82**
  experimental   **202**
  source area   **165**
catena   **82–83**, 128
cation exchange   48, **83**
cation-exchange capacity (CEC)   **83**
  soil   49, 83
cation-ratio dating   **83**
causality   **83–84**
causse   **84**
  see also karst
cave   **84**
  speleology   **493**
cavern *see* cave
cavitation   **84**
celerity   **84–85**
cell circulation   *85*, **85**
cellular automata   **86**
Cenozoic   *76*, **76**
cerrado   **86**
chamaephytes   *433*, **434**
channel
  alluvial   **19**
  bedrock   **19**
  bourne   **70**
  chute   **91**
  classification   *22*, **87–88**
  cross-profile   **123**
  distributary   **151**
  lateral migration   **314**
  loss of capacity   **212**
  marginal   **330**
  monumented sections   **353**
  overflow   **380–381**
  palaeochannel   **384**
  raised   **432**
  reach   **434**
  regime theory   **437–438**
  reservoir effects   **442**
  resistance   **88**
  rills   **447**
  sandbed   215, **463**
  sinuosity   **483**
  storage   **88**
  unstable   **546**
  urstromtäler   **549**
  wadi   **558**
  wetted perimeter   **570**
  see also river; stream
channel capacity   45, **86**
  reservoir effects   **442**
channelization   **88**
chapada   **88**
chaparral   **88**
char   **88–89**
character species   **37**
charcoal   **89**
chattermarks   **89**
cheiorographic coast   **89**
chelation   **89**
cheluviation   **89**
chemoautotrophs   **42**

chemosphere **89**
chenier **89–90**
chernozem **90**
chert **90**
Chézy equation 51, **90**, 215, 329, 543
chine **90**
chinook **90**, 217
chlorofluorocarbons (CFCs) **90**, 259, 383
chott **91**
chronosequence **91**
chute **91**
circadian rhythm **91**
circulation index **91**
cirque **91**, 259
cirriform clouds, albedo 15
cirrus clouds 102, *102*
cladistics *92*, **92–93**
clastic **94**
clasts **94**, 379–380, 533
clay 35, **94**
    armoured mud balls 35
    boulder *see* till
    cation exchange 83
    cohesion **104–105**
    cutan **125**
    dune **94**, 324, 387
    fabric 206
    gumbo 254
    kaolin 303
    parna 390
    pellet **94**, 324
    quick 423
clay-with-flints **94**
claypan **94**
clay–humus complex **94**
clear-water erosion **94**
cleavage **94**
cliffs **467–468**
CLIMAP study **94**
climate **95**
    boreal 67
    classification **97–98**
        Köppen 307, *308*, 335
    continental 111
    epiclimate 190
    local 338
    Mediterranean 335
    megathermal 336
    mesoclimate **338–339**
    mesothermal 340
    models 96
        ensemble climate modelling 179–180
        regional climate model (RCM) **438–439**
    scales *343*
    statistics 95
climate change **95–97**, 271, 361
    abrupt **1**, 129
    Intergovernmental Panel on Climate Change
        (IPCC) **296**
    Kyoto Protocol **308–309**
    see also global warming; greenhouse effect
climate sensitivity parameter **97**
climatic cycle **125–126**

climatic geomorphology **98–99**, 191
climatic hinge **99**
climato-genetic geomorphology **99**
climatology **99**
    agroclimatology **12–13**
    bioclimatology 55
    synoptic **513**, 566
climax vegetation 37, **99**, 352, 507
    disturbance 149
climbing dune **99–100**, 207
climogram **100**
cline **100**
clinometer **100**
clinosequence **100**
clint **100**
clisere **100**
clitter **100**
closed system **514**
cloud computing **100**
cloud dynamics **100–101**, 102
cloud forest **101**
cloud microphysics **101**, 102, 404
cloud streets **101–102**
clouds *102*, **102**, 274
    albedo 15
    banner 46
    cirriform 15
    cloud cover measurement 376
    cumuliform 15
    formation 101
    fractus **222**
    glowing 371
    modification 404
    nephanalysis 364
    noctilucent **367–368**
cluse **102–103**
cluster bedform **103**
CMIP *see* Coupled Model Intercomparison Project
coast
    cheiorographic **89**
    concordant **108**
    dalmation **128**
    pacific-type **384**
    strandflat 483
coastal dunes **103**, 492
    foredune **219–220**
    see also dune
coastal flood 212
coastlines
    Atlantic **38**
    self-similar 221
coccoliths **104**
cockpit karst **104**
coefficient of variation (CV) **104**
coevolution **104**, 199
cohesion **104–105**, 478
    Mohr–Coulomb equation 351, 478
col **105**
cold front **105**
cold pole **105**
Coleoptera **105**
colloid substance **105**
colluvium **105**

colmation **105**
colonization **105**
colour display 146
combe/coombe **106**
comfort zone **106**
commensalism **106**
comminution **106**
community **106**
  mature 99
  plant 71–72
    disclimax **149**
    plagioclimax **149**
    subclimax **505–506**
    see also climax vegetation
  seral **477**
  see also succession
compaction **106**
compensation flows **106**
competence **106–107**
competition **107**, 171
  predation relationship 418
competitive exclusion 203
complex response **107**
compressing flow **107**, 497, 511
compressional air masses **108**
computational fluid dynamics **108**
conchoidal fracture **108**
concordance **108**
concordant coast **108**
concretion **108**
condensation **108**, 251
  fog formation 217
conductance, specific **108**, 491
conduction, heat budget 259
cone of depression **108–109**, 155
cone-karst 305
conglomerate **109**
coniology 307
connate water **109**
connectivity **109**
consequent stream **110**, 543
conservation **110–111**, 203
consociation **111**
consolidation 106, **111**
constant slope **111**
contaminant 413
  see also pollution
continental
  climate **111**
  drift **111**, 265, 408
  freeboard **112**
  islands **112**
  shelf **112**
    ocean sediments 375
  slope **112**
continuity equation 86, **112–113**, 190, 305, 371
contour *113*, **113**
contrails **113**
contributing area **113**
control structures 148
  flumes 148
  weirs 148
convection 102, **113–114**, 339

heat budget 259
mesoscale convective systems **342**
  mesoscale cellular convection (MCC)
    339–340, *340*, 342
convective precipitation *114*, **114**
convergence **114**
convergent evolution 199
coordination number **114**
coprolite **114**
coquina **115**
coral reef **115–117**, 436
  bleaching **115**
  bluehole **66**
  carbonate budget **79–80**, *80*
  coral disease **115**
  coral–algal reefs *116*, **116–117**
  Daly level **128**
corallith **117**
corange lines 22
corestone **117**
Coriolis force 22, **117**, 173, 228–229
  geostrophic wind 237
  gradient wind 247
  gyre 254
Coriolis parameter **118**, 174
cornice **118**
corniche **118**
corrasion **118**
correlogram 41–42
corrie **91**
corrosion **118**
  limestone 491
  mixing **350**
cosmogenic isotope **118**
cotidal lines 22
coulée **118**
couloir **118**
coupled general circulation models (GCMs)
  118–119
Coupled Model Intercomparison Project
  (CMIP) **103**
coversand **119**
crab-holes **119**
crag and tail landform **119–120**
  depositional 120, *120*
  erosional 120, *120*
crater **120**
  volcanic 77, 120
craton **120**, 350, 480
creationism 92, **120–121**, 130
creep 94, **121**, 282, 333
  frost **225**
crevasse **121**
critical load **121**
critical rationalism 132
critical velocity **121–122**, 180
critical zone *122*, **122**
cross-bedding **122**
cross-lamination **123**, 379
cross-profile **123**
cross-valley moraine 134
crumb structure **123**
crust **123**, 487

cryergic features and processes   123
cryoplanation   21, **123**
   terrace   246
cryosphere   **123**
cryostatic pressures   **123**
cryoturbation   **123**
cryovegetation   **124**
cryptophytes   *433*, **434**
cryptovolcano   **124**
cuesta   **124**
cuirass   **124**
cultural geomorphology   **124**
cultural services   171
cumec   **124**
cumulative soil profiles   **124**
cumuliform clouds, albedo   15
cumulonimbus *see* clouds
cumulonimbus convection   113
cumulus clouds   102, *102*
cup-anemometer   24
cupola   **124**
current
   bedding   19, **124**
   density   137
   rip   46, **447**
   ripples   **124**, 125
   tidal   **528**
   turbidity   217, **541**
current meter   **124**
cusp, beach   **125**
cut-off   **125**
   neck   **363**
cutan   **125**, 177
cutter   **125**
cwm   91
cyanobacteria   **125**
cycle
   anamolistic   23
   biochemical   **125**
   biogeochemical   **57–58**, *58*
   Bond   67
   carbon   79
   climatic   **125–126**
   erosion   **126**, 193
   freeze–thaw   **222–223**
   hydrological   148, 252, 271, **272–273**, *273*
   K-cycle   **303**
   nitrogen   **366–367**
   punctuated aggradational cycles   **420**
   redox   **435**
   Wilson   **570**
cyclic time   247
cyclone   39, **126–127**, 339, 552
   extratropical   105, **203–205**, *204*, 223,
      539, 560
   tropical   *536*, **536–537**
cyclone waves   **40**
cyclostrophic forces   **127**
cymatogeny   **127**

D
dalmatian coast   **128**
Daly level   **128**

dam   **128–129**
   clast, transverse   103
   collapse   212
   ice   **281–282**
   litter   **320**
dambo   **128**
Dansgaard–Oeschger (D–O) events   67, **129**, 250,
      260, 526
Darcy's law   *129*, **129**, 192, 268, 270, 334
Darcy–Wiesbach friction coefficient   51
Darwinism   **130**
data compression, digital image processing
      146
data logger   **130–131**
dating
   absolute   1
   cation-ratio   **83**
   fission track   **211**
   luminescence   175, **322–323**
   obsidian hydration (OHD)   **373**
   potassium argon (K/Ar)   **417**
   radiocarbon   118, 257, **427–428**
   relative   **439**
   tephrochronology   **520**
   thermoluminescence   **322–323**
   uranium series   **547**
daya   **131**
DDT bioaccumulation   61
De Geer moraine   **134**
débâcle   **131**
debris
   avalanche   **43**
   flow   **131**, 333
deciduous forest   **131**
decomposer organisms   **131–132**, 218
deductive science   **132**, 285
Deep Sea Drilling Project (DSDP)   **374–375**
deep weathering   **132**, 567
deflation   **132**, 253
deforestation   **132–133**, 566
   consequences of   133
deformable beds   **133–134**
deformation   **134**
deformation till   **530–531**
deglaciation   **134**
degradation   **134**
degree day   **134**
delayed flow   **134**
delta   *135*, **135–136**, *136*
   fan   **207**
demoiselle   **137**
denitrification   **137**, 366
density current   **137**
density dependence   **137**
denudation   **137–138**, 395
   chronology   **138**, 193
   dambo   128
   parallel retreat   **389**
   rates   **139**, 151
   Richter denudation slope   **445**
   see also erosion
deoxygenation   **139**
depletion curve   48, *139*, **139–140**, 271

deposition
  dry   4, 5, **159**
  occult   **373**
  wet   4, 5, 373
depression   126–127
  geosyncline   **237**
  pan   94, *387*, **387**
  playa   94, 253, 387, **409**
  sabkha   **459**
  secondary   **471**
  shakehole   305, **477**
  sinkhole   104, **483**
  slack   **484**
  swale   **512**
  terlough   **520**
  thermal   **523**
  thermokarst   **526**
  uvala   **549**
depression storage   **140**
depth hoar   280
depth–duration curve   **140**
desalinization   **140**
desert   **140–141**, *141*
  aeolation processes   9
  erg   **192**
  hamada   **258**
  hyper-arid   **277**
  puna   **420**
  reg   **437**
  semi-desert   **476**
  serir   **477**
  stone pavement   **500**
desertification   134, **141–142**, 186, 297, 362,
      459, 566
  savanna   **392**
  susceptible drylands   **511**
desiccation   *143*, **143**
desiccation event   157
design discharge   **143**, 213
desquamation   **143**
desulphurization   **143**
dew   **143**
dewpoint   **143**
diachronous sediment   **143**
diaclinal rivers   **143**
diagenesis   **144**
diamictitie   **144**
diamicton   **144**
diapir   **144**
diastrophism   **144**
diatoms   **144**
diatreme   **144**
die-back   132, **144**
differential species   37
diffluence   **144–145**
diffusion equation   **145**
diffusive processes   **145**
digital elevation model (DEM)   **145**
digital image processing   **145–146**
digital terrain model (DTM)   **146**, 235
dikaka   **146**
dilation (dilatation)   **146–147**
  angle of   24

dilution effect   147, 150, **490**
dilution gauging   147, **148**
diluvialism   **147**
dimensionless number   **147**
dimethylsulphide (DMS)   **147**
dip   **147**
dipslope   **147**
dipwell   **147–148**
dirt cone   **148**
discharge   **148–149**, 474
  bankfull   **45–46**, 86, 213
  design   **143**, 213
  dominant   45, 86, **152**, 438
  flash flood   **211**
  flashiness   **211**
  measurement   **148–149**
    rated section   **433**
  peak   **392**
  rating curve   **433**
  record   149
  regime theory   **437–438**
disclimax   **149**
disconformity   **149**
discordance   **149**
disjunct distribution   **149–150**
dispersal   **150**
dissection   **150**
dissipative beach   **150**, 436
dissolved load   **150**, 151
dissolved oxygen   **150–151**
dissolved solids   *151*, **151**
distributary   **151**
distribution
  disjunct   **149–150**
  extreme value   205
diurnal tides   **151**
divergence   **151–152**, 552
diversity
  alpha   **20**
  beta   **20**
  gamma   **20**
  geodiversity   **234**
doab   **152**
doldrums   **152**, 191
doline   104, **152**, 305, 477
dolocrete   **152**
dome
  ice   **282**
  salt   **460**
domin scale   **152**
dominant discharge   45, 86, **152**, 438
dominant organism   **152–153**
dominant wind   **153**
donga   *153*, **153**
dormant volcano   **154**
double mass analysis   **153**
downbursts   **153**
downscaling   **153–154**
downwelling   **154**
draa   **154**
drainage   **154**
  annular   **25**
  antecedent   **29**

drainage (*Continued*)
  beaded   50
  density   154–155
  depletion curve   *139*, **139–140**
  endoreic systems   178
  field   210
  superimposed   510
drainage basin   154
drainage network   154, *155*, **155**, 221
  structure   155
draw down   108, **155**
dreikanter   **156**
drift
  continental   **111**, 265
  genetic   200, **233**
  longshore   322
  potential, sand   **156**
dripstone   **156**
drop size   **156**
drought   **156**, 476
  absolute   156
  agricultural   **12**, 156
  meteorological   156
  partial   156
drought index   **157**
drumlin   52, **157–159**, 506
  rock   451
dry deposition   4, 5, **159**
dry valley   **159–160**
dry weather flow   **160**
dry-bulb temperature   106, 519, 550
drylands   140, 142, *160*, **160–161**
  degradation   142
  hyper-arid   277
  pan   94, *387*, **387**
  piosphere   406
  playa   94, 253, 387, **409**
  semi-arid areas   476
  susceptible   511
  see also desert
du Boys equation   **161**, 533
dune   52, **161–162**
  aeolian   161–162
  barchan   46
  blowouts   **66**
  clay   **94**, 324, 387
  climbing   **99–100**, 207
  coastal   **103**, 492
    foredune   **219–220**
  complex   162
  compound   162
  draa   154
  echo   **168**
  falling   207
  lee   **316–317**
  linear   162, **319**, 475
  longitudinal   322
  lunette   94, **323–324**, 387, 492
  mega   336
  nebkha   **362–363**
  parabolic   220, **388**
  reversing   444
  rhourd   444
  seif   319
  simple   162
  singing sands   **483**
  slip face   319, *484*, **484**
  source bordering   **492**
  star   162, **498**
  topographic   99, 162, 207, **532**
  transverse   46, 162, 324, **535**
  zibar   462, **577**
dune network   **162**
durability   **163**
duricrust   77, 123, **163**, 285, 451, 567
duripan   **163**
dust   **163–164**, 258, 307
  devil   **164–165**
  haze   164
  storm   163–164, **165**
    haboob   **257**
  veil index   **165**
dyke   **165**
dynamic allometry   18
dynamic equilibrium   **165**, 246
  river   289
dynamic source area   **165**, 390
dynamical downscaling   153
dynamical meteorology   **166**, 404
dzud   **166**

E
Earth
  albedo   15, *16*
  expanding   **201**
  observation   **167**, 346
  snowball   **485**
earth pillar   **167**
Earth system science (ESS)   **167**
earthquake   **167**
  aftershocks   **11**
  epicentre   **190**
  Mercalli scale   *337*, **337**
  moment magnitude scale   **351**
  Richter scale   **445**
  seismicity   **475**
East Antarctic Ice Sheet (EAIS)   27
East Pacific Rise (EPR)   346
easterly wave   **167**
eccentricity   **168**
echo dune   **168**
ecological energetics   **168–169**
ecological footprint   **169**
ecological replacement principle   203
ecology   **169**
ecosystem   59, **169–171**, 172
  abiotic components   1
  goods and services   *170*, 171, 348, *349*
  nutrient addition   196
  stability   **494–496**
ecotone   **171**
ecotope   59, 65, **171**
ecotoxicology   **171**
ecotype   **171–172**
ecozone   **172**
edaphic factors   **172**, 256

eddies 172
  diffusivity (viscosity) **172–173**, 174, 571
edge waves 125, **173**, 436
effective
  precipitation **173**
  stress **173**
  temperature 106
effluent **173**
  stream **173**
egre **173**
Ekman layer **173–174**
Ekman spiral **174**
El Niño events **175–176**, *176*, 518, 558
El Niño Southern Oscillation (ENSO) **180**, 325, 518
elastic rebound theory **174**
elastic waves **40**
elasticity 434
electromagnetic distance measurement **174**
electromagnetic gauging, discharge
      measurement 149
electromagnetic radiation (EMR) **174**
  remote sensing 440–441
electron spin resonance (ESR) **174–175**
ellipsoid **175**
eluviation **177**, 264
eluvium **177**
Emerson test **177**
emissions scenario **177**
endangered species **178**
endemism **178**, 233
  hot spot 265–266
endogenetic processes **178**, 201
endoreic drainage systems **178**
energy 179
  geothermal **237**
  kinetic **306**
  potential 179, **417**
energy balance models (EBMs) 96
energy flow **179**
  biosphere 64
  trophic levels 169, 170
energy grade line **179**
engineering geomorphology **179**
englacial conditions **179**
enhanced basal creep 48
ensemble climate modelling **179–180**
ENSO *see* El Niño Southern Oscillation
entrainment 121, **180–181**
  dust 163–164, 165
entrenched meander *181*, **181**
entropy **181**
envelope curve **182**
environmental
  assessment **182**
  economics **182**
  impact **183–184**, *184*
    assessment (EIA) 184, *185*
    statement (EIS) **184–186**
  issue **186**
  magnetism **186–187**
  management **188–189**
epeiric sea **189**
epeirogeny 127, **189**

ephemeral
  plants *189*, **189**, 424, 575
  stream **190**
    floodout 213
epicentre **190**
epiclimate **190**
epilimnion **190**, 278
epimerization 21
epipedon **190**, 486
epiphyte **190**
epoch **190**
equation of state **190**, 404
equations of motion **190**, 371
equator, thermal **523**
equatorial rain forest **190**
equatorial trough **190–191**
equifinality **191**
equilibrium **191**, 371, 438, 474
  beach **192**
  dynamic **165**, 246, 289
  stable 81, 316, **496**
  unstable **546**
equilibrium line **191**, 380
  glaciers **191–192**
equilibrium shoreline **192**
equinox precession **418**
equipotentials **192**, 271
era **192**
erg **192**
ergodic hypothesis **192**
erodibility **192–193**
erosion 118, **193**
  badlands 44–45
  bank 19, **45**
  base level **48–49**
  beach, sea level rise response 72–73
  caesium-137 (^{137}Cs) analysis **76**
  cavitation 84
  critical velocity **121–122**
  cycle of **126**, 193
  evorsion **200**
  glacial 411
  gully 254
  headward **259**
  micro-erosion meter **344**
  pin **193**
  raindrop impact **430–431**, 479
  sapping **464**
  sheet **478–479**
  soil 110–111, 186, 362, **487**
  surface **193**
  tunnel gully **541**
  water 44
    clear-water **94**
  see also denudation
erosivity **193**
erratic (glacial) **193–194**
eruption **194**
  fission **211**
  Hawaiian **258**
  Plinian **410**
escarpment **194**
  great **249**

esker **194**
estuary *194*, **195**
　macrotidal　195
　mesotidal　195
　microtidal　195
eulittoral zone **195**
euphotic zone **195**
eustasy　112, **195–196**, 450
　geoidal　196
　glacial　195–196
eutrophic **196**, 372
　lakes　196, 311
eutrophication　*196*, **196–197**, 311, 367, 415
evaporation　**197–198**, 274, 314
　assessment　324
　atmosphere-controlled　197
　potential **417**
　soil-controlled　197
evaporite **199**
evapotranspiration　197–198, **199**, 459
　combination methods　197
　potential　197
　precipitation relationship　160–161
evolution　130, **199–200**
　coevolution　199
　convergent　199
evorsion **200**
exaration **200**
exchangeable sodium percentage (ESP) **200–201**
exfoliation **201**
exhaustion effects **200**
exhumation **201**
exogenetic processes **201**
exosphere **39**
exotic **201**
expanding Earth **201**
expansive soils **201–202**
experimental
　basins **441–442**
　catchment **202**
　design **202–203**
extending flow **282**
external forcing **203**
extinct species **178**
extinction **203**
extracellular polymeric substances (EPSs) **57**
extratropical cyclone　105, **203–205**, *204*, 223, 539, 560
extreme value theory **205**
extremophiles **7**
extrusion flow of glaciers **205**
extrusion, volcanic **205**
exudation basic **205**
eye **205**
eyot **205**

F
fabric **206**
　sediment　206, 474
facet **206**
facies **206**, 284
factor of safety **206**
fadama **128**

fairy circles　*207*, **207**
falling dune **207**
fallout **207**
false-bedding **207**
fan delta **207**
fanglomerate **207**
fans **207–208**
　megafan **336**
　sieve deposits **482**
fatigue failure **208**
fault　*208*, **208**
　normal **368**
　oblique-slip　208, *208*
　reverse　208, *208*, **444**
　strike-slip　208, *208*
　　megashear **336**
　tear **517**
　throw of **528**
　thrust　208, **528**
faulting
　block **65**
　elastic rebound theory **174**
　gravity **248**
fauna *see* animals
faunal realms **209**, 559
feather edge **209**
feedback **514**
feldspars **209**
felsenmeer *see* block fields; block streams
fen **209**
fermentation **23**
ferrallitization **209**
Ferrel cell **209**, 569
Ferrel's law **209**
ferricrete **209**
fetch length **209**
fiard **209**
field capacity **210**
field drainage **210**
Finger Lakes **210**
fire **210–211**
firn **211**, 485
fish farming **33**
fission track dating **211**
fissure eruption **211**
fjord **210**
flash flood **211**
flashiness **211**
flatiron **211**
float recorder **211**
floating bogs **211**
flocculation　94, **212**
flood **212**
　celerity **84–85**
　coastal **212**
　dam collapse **212**
　flash **211**
　frequency **212–213**, 548
　　analysis **392**
　　reservoir effects **442**
　precipitation **212**
　protection **212**
　routing **214**

Muskingum method   **359**
snowmelt   212
floodout   **213**
floodplain   **213**
 backswamp   **44**, 213
 formation   213
 genetic   213
 hydraulic   213
flora *see* plants
floristic realms   *214*, **214**
flow
 anabatic   **22**, 357
 base   **48**, 160
 boundary layer   146
 bypass   **74–75**, 334
 celerity   **84–85**
 channel resistance   **88**
 compensation   **106**
 competence   **106–107**
 compressing   **107**, 497, 511
 debris   **131**, 333
 delayed   **134**
 diffluence   **144–145**
 diffusion equation   145
 dry weather   **160**
 energy
  biosphere   64
  loss of   179
 extending   **282**
 gradually varied   179, 215, **247–248**
 helical   **260**, 319
 hillslope flow processes   **261–262**, 286, 381
 hydrometry   **274**
 hyperconcentrated   78, **277**
 ice   **282**
 interflow   **290**
 interrill   **291**
 katabatic   **303–305**
 laminar   **311**, 541, 551
 lateral   **314**
 lava   **322**
 low flow analysis   **322**
 matrix   **334**
 meridional   229
 minimum acceptable   **349**
 overbank   **380**
 overland   237, 261, 287, **380**
  Horton model   **265**, 286, 290, 389, 457, 545
  partial area model   **390**, 545
 pathways   109, *109*
 preferential   75
 rate of   217
 regimes   **215**, 463
 Reynolds number   215, 311, **444**
 secondary   **471**
 sheet   **479**
 steady   **498**
 stem   226, 290, **498–499**, 528
 throughflow   261, 286, 290, 390, **528**
 tills   **531**
 transient   145
 turbulent   **541–542**
 uniform steady   145, 179, 215, 223, 533, **543**
 unsteady   **546**
 visualization   **216**
 zonal   229
flow duration curve (FDC)   **214–215**
flow equations   90, **215**, 270, 498
 Manning equation   90, 215, **329**
flow resources   **215–216**, 362
 critical zone   216
 noncritical zone   215
fluid dynamics, computational   **108**
 see also flow
fluid mechanics   **216**, 270
fluid potential   **270**
flume   148, *216*, **216**
flute   **216**, 506
fluting   **216–217**
fluvial   **217**
 sediment   **471–473**, *472, 473*
 see also river
fluviokarst   **217**
flux   **217**
flyggberg   **217**
flysch   **217**
fog   **217**
 ice   **282**
 steam   **498**
föhn   **217**
fold   **217–218**, *218*
 asymmetric   **38**
 recumbent   **434**
foliation   **218**
food chain/web   64, **218**, *219*, 535–536
foraminifera   **218–219**
Forbes bands   **375**
force   **219**
 capillary   **78**
 Coriolis   22, **117**, 173
 cyclostrophic   **127**
 geostrophic   117
 retention   **443**
 tractive   **533**
forebulge   **219**
foredune   **219–220**
forest   133
 afforestation   **11**
 biomass   62
 boreal   **67–68**
 cloud   **101**
 deciduous   **131**
 deforestation   **132–133**
 die-back   **144**
 ecological importance of   133
 gallery   **228**
 hydrology   **220**
 krummholz   **307–308**
 mangrove   **328–329**
 monsoon   **353**
 rain   **537–538**
  equatorial   190
 taiga   **516**
 tropical   190, 353, **537**
forest beds   **220**
form line   **220**

form ratio   220
form roughness   456
founder effect   233
Fourier analysis   220
fracking   221
fractal   221–222
   dimension   222
   microtopography   345
fractional Brownian motion   221
fracture   222
   conchoidal   **108**
fractus   222
fragipan   222
frazil ice   222
free face   222
free-surface resistance   456
freeze–thaw cycle   222–223, 225
freezing front   223
freezing index   223
frequency-dependent magnetic susceptibility   327
friction   223, 478
   Mohr–Coulomb equation   351, 478
   see also roughness
front   223–224, *224*
   cold   **105**
   freezing   223
   ice   **282**
   kata-front   **305**
   polar   **412**
   temperature inversion   295–296
   warm   **559–560**
   weathering   567
   wetting   570
frontogenesis   224
frost
   ground   251
   hoar   262
   see also permafrost
frost action   225
frost creep   225
frost heave   225, 345
frost smoke   225
frost weathering   225–226
Froude number   215, **226**, 270, 284
fulgurite   226
fulje   226
fumarole   226
fungi   226, 306
   mycorrhizal   **359–360**
funnelling ratio   226–227
fynbos   227

G
gabbro   228
Gaia hypothesis   64, 228
gallery forest   228
gamma diversity   20
gaping gores   201
garrigue   228
gauging stations   228
GCM *see* general circulation modelling
geest   **228**
gelifluction   225, **228**

genecology   **228**
general atmospheric circulation   **228–231**, *230*, 325, 569
general circulation modelling (GCM)   96, **231–232**
   coupled GCM   **118–119**
general system theory   **232–233**, 514
genetic drift   200, **233**
genetic floodplain   213
geo   **233**
geochronology   **233**, 322, 439, 547
   see also dating
geoconservation   **233**
geocryology   **233**
geode   **233**
geodesy   **234**
geodiversity   **234**
geoengineering   **234**
geographic information system (GIS)   146, **234–235**
   raster GIS   234, **433**
   vector GIS   234, **551**
geographical information science (GISc)   **234**
geohazard   **235**, 361
geoheritage   **235**
geoid   186, **235**
geological timescale   *236*, **236**
geomorphological threshold   **448**, 527
geomorphology   138, **236**
   applied   **32**
   climatic   **98–99**, 191
   climato-genetic   **99**
   cultural   **124**
geomorphometry   355–356
geoparks   **236**
geophyte   **237**
geoproxy   **237**, 385
geostatistics   **237**
geostrophic force   117
geostrophic wind   40, 174, **237**
geosyncline   **237**
geothermal energy   **237**
Gerlach trough   **237**
geyser   **237**, *238*
Ghyben–Herzberg principle   **237–239**, *238*
gibber   **239**
Gibbsian distribution   **239**
gilgai   202, **239**
gipfelflur   **239**
glacial
   erosion   411
      plucking   **411**
   horn   **264**
   landscape   **239**
   overdeepening   **380**
   period   422
   reservoir   240
glaciation   **240**
   Quaternary   421–422
glaciation level   **240**
glacier   **240**
   ablation   1, 107, 134
   basal ice   **47**
   basal sliding   **48**, 107
   calving   77

crevasse 121
deglaciation 134
equilibrium line **191–192**
flow
    compressing 107
    extrusion 205
grounding line **251–252**
ice cap 281
mass balance (GMB) **240–241**
melting 1
meltwater 336
milk 241
outlet 379
piedmont 405
remanié 440
rock 452
snout 485
stick slip motion **499**
surging 282, 355, **511**
table 241
glacieret 241
glacierization 241
glaciofluvial activity 241, 303, 379
    deposits *242*
glaciosostasy **241–242**
glaciotectonism 242
glacis 242, 393
glei (gley) **242**
Glenn's law **242–243**, 282
glint line **243**
global dimming **243**
global environmental change 243, 266
Global Geoparks Network 236
Global Positioning System (GPS) **243–244**, 317
    differential (DGPS) 244
global warming 243, **244–245**, *245*, 249, 267, 362
    see also climate change; greenhouse effect
GLOSS (Global Sea-Level Observing System) **246**
glowing cloud **371**
gnammas 21, **246**
gneiss **246**
goletz terrace **246**
Gondwana 509
gorge **246**
gouffres **246**
graben **246**, 446
grade **246**
graded bedding **246**
graded slopes **246–247**
graded time **247**
gradient wind 247, 572
gradually varied flow 179, 215, **247–248**
grain roughness 456
granite **248**
granulometry **248**
grassland **248**
    bush encroachment 74
    prairie **418**
    steppes **499**
    see also savanna
gravel **248**
    lag **310**
gravimetric method **248**

gravity **248**
    drops 156
    faulting **248**
    wave 40, 173, **248**
great escarpment **249**
great interglacial **249**, 394
greenhouse effect 96, 244, **249**, 415
    Kyoto Protocol **308–309**
    see also climate change; global warming
Greenland Ice Sheet 240, **249–250**, *250*
greywacke **251**
grike 100, **251**
gross primary productivity (GPP) 61, 420
ground
    frost **251**
    ice **251**, 280
    moraine 531
ground-penetrating radar (GPR) **251**
grounding line **251–252**
groundwater 34, 108, *252*, **252–253**, 401
    confined 406
    Darcy's law **129**
    denitrification 137
    equipotentials **192**
    meteoric **341**
    perched 253, **395**
    phreatic surface **401**
    recharge **434**
    vadose zone 252, **550**
    watersheds 401
growan **253**
growth
    allometric **18**
    isometric 18
groyne **253**
grumusol **253**
grus **254**
guano **254**
Gulf Stream 525–526
gull **254**
gully **254**
gully erosion **254**
Gumbel extreme value theory 213, **254**
gumbo **254**
gustiness factor **254**
guyot **254**, 471
gymnosperms 24
gypcrete **254**
gypsum **254**
gyre **254–255**
gyttja **255**

H
habitat **256**
haboob **257**
hadal zone **257**
Hadley Cell 230, **257**
    anticyclone 31
Hadley-type circulation 229
haff **257**
hagg **257**
hail **257**
haldenhang **257**

half-life   *257*, **257–258**
halocarbons   **258**
haloclasty   **258**, 460
halons   **258**
haloturbation   **258**
hamada   **258**
hamra   **258**
hanging valley   **258**
hardness   **258**
hardpan   **258**
harmattan   **258**
Hawaiian eruption   **258**
haystack hill   351
haze   **258–259**
   arctic   35
headcut   **259**
headwall   **259**
headward erosion   **259**
headwater   **259**
heat
   budget   **259–260**, 476
   latent   343, 476
   sensible   343, **476**
heathlands   **260**
Heinrich events   **260**
helical flow   **260**, 319
helictite   **260**
hemera   **260**
hemicryptophytes   *433*, 434
herbivore use intensity (HUI)   **261**
herbivores   218, **260–261**, 536
heterosphere   **261**
heterotrophs   **261**
hiatus   **261**
hierarchical rhythmic topography   **261**
high field magnetic susceptibility   327
high-energy window   **261**
hill
   bornhardt   **68**
   butte   74
   drumlin   52, **157–159**
   flyggberg   217
   haystack   351
   hum   **266**
   inselberg   68, 146, *288*, **288**
   monadnock   352
hillslope   485
   flow processes   **261–262**, 286, 381
   mass movement types   **332–334**, *335*
   threshold slopes   **527–528**
   waxing   **565**
   see also slope
hindcasting   **262**
Hjulström curve   107, *262*, **262**
hoar frost   **262**
hodograph   **262–263**, *263*
hogsback   **263**
Holocene   **263**, 421
   Blytt–Sernander model   *66*, **66**
   hypsithermal   **278**
holokarst   **263**
homoclines   **263**
hoodoo   44, **263**, *264*

horizon
   marker   **332**
   soil   **264**, 486
horn, glacial   **264**
horse latitudes   **264**
horst   **264–265**
Horton overland flow model   **265**, 286, 290, 389, 457, 545
Horton's laws   **265**, 356
hot spot   **265–266**, 330
hot spring   **266**
hot-wire anemometers   24
hum   **266**
human appropriation of net primary production (HANPP)   266
human impact   **266–267**
humate   **267**
humic acid   **267–268**
humidity   **268**
   subhumid climate   **506**
   temperature–humidity index   **519**
humus   **268**
   clay–humus complex   **94**
hurricane   537
hydration   **268**
hydraulic(s)   216, **270–271**
   conductivity   129, 145, **268**, 293, 397, 401
   diffusivity   **268**
   floodplain   213
   force   **268–269**
   geometry   *269*, **269**, 438
   gradient   **269–270**
   head   129, 145, 155, 268, *270*, **270**
   jump   **270**, 271
   potential   192, **270**
   radius   90, **270**
hydrochlorofluorocarbons (HCFCs)   258
hydrodynamic levelling   **271**
hydrofacturing   **271**
hydrofluorocarbons (HFCs)   **271**
hydrogeological map   **271**
hydrograph   139, 149, **271–272**, *272*, 392, **545**
hydroisostasy   **272**
hydrolaccolith   **272**
hydrological cycle   148, 252, 271, **272–273**, *273*, 274, 276, 560
hydrology   272, **273–274**
   palaeohydrology   **385–386**
   urban   **547–548**
hydrolysis   **274**
hydrometeorology   **274**
hydrometer   **277**
hydrometry   **274**, *275*
hydromorphy   **274–275**
hydrophobic soils   **275**
hydrophyte   **275**
hydrosphere   **275–276**
hydrostatic equation   **276**, 404, 468
hydrostatic pressure   **276**
hydrothermal alteration   **276**
hydroxyapatite   175
hyetograph   **276**
hygric weathering   **276–277**

hygrograph 277
hypabyssal rock 277
hyper-arid areas 277
hyperconcentrated flow 78, 277
hyperspectral imagery 277
hyphae 226
hypogene 278
hypolimnion 278
hypothesis 132, 278
  ergodic 192
  Gaia 228
  Milankovitch 280, *347*, **347–348**, 420, 422
  multiple working hypotheses 358
  testing 202
hypsithermal 278, 363
hypsographic curve 278
hypsometry 278
hysteresis 278–279

I
ice 280
  basal 47
  blink 281
  cap 240, 281, 372
  contact slope 281
  core *see* ice core analysis
  dam 281–282
  dome 282
  fall 107, 282
  fast 468
  field 282
  floe 282, 468
  flow 282
  fog 282
  frazil 222
  front 282
  ground 251, 280
  island 284
  jam 282
  needle 225, 363
  ogives 375
  pack 468
  pore 417
  rime 280, 447
  rind 282
  sea 280, *468*, **468**
  segregated 475
  segregation 225, **283**, 396
  sheet *see* ice sheet
  shelf 240, 283
  stagnant 497
  stream 283, 379
  superimposed 280, **510**
  temperate 518
  wedge 283, 463
ice age 280–281
  Little (LIA) 320–321
ice core analysis 27, 281
  acidity profile 6
  altithermal 21
  Antarctic Cold Reversal 27
  D–O events 67, 129
ice sheet 240, 242, **283**, 372

Antarctic **27–28**, *28*
  East (EAIS) 27
  West (WAIS) 27–28
  Greenland 240, **249–250**, *250*
  Laurentide 242, **315**
icebergs 283, **284**
  calving 77
icehouse 284
icing 284
igneous rock 284
  basic 49
illuviation 177, 284
image
  classification 146
  enhancement 146
  restoration and correction 146
imbrication 284
impermeable structure 284
impervious structure 284
in and out channel 284–285
inconsequent stream 285
indeterminate species 178
Indian Ocean zonal mode (IOZM) 285
inductive science 285
induration 285
infiltration 285–286
  capacity 261, 491
inflorescence 287
influent 287
infragravity wave 173, 287
infrared thermometer/thermometry 287
inheritance 287–288
initial yield, angle of 24, 25, 259
inlier 288
inselberg 68, 146, *288*, **288**
insequent stream 288
insolation 288–289
  weathering *288*, 289
insufficiency known status 178
intact strength 289, 443, 453
integrated basin planning 448
integrity, of rivers 289
interannual variability 289
interbasin transfers 289
interception 289–290, 498
  capacity 290
interdecadal Pacific oscillation (IPO) 384
interflow 290
interfluve 290
interglacial 239, 290, 422
  great 249, 394
Intergovernmental Panel on Climate Change
  (IPCC) 296
intermittent spring 290
intermittent stream 290–291, 396
internal shearing resistance, angle of 24, 25
interpluvial 291
interrill flow 291, 447, 479
interrill processes 291, 410
interstadial 129, 291
  Allerød 17
  Bølling 67
  Late Glacial 27

interstices **291**, 417
intertropical convergence zone (ITCZ)   99, 230, 257,
    292, **292**, *293*, 338, 523
intrazonal soil **292**
intrenched meander **292**
intrinsic permeability 268, **293**, 397
introduction, ecological **294**
    see also alien species; invasive species
intrusion **294**
invasive species 17, **294–295**
    see also alien species; introduction, ecological
inversion of temperature **295–296**, 343, **519**
inverted relief **296**
involution **296**
ion concentrations **296**
ionosphere 39, **296**
iridium layer **296**
irrigation **296–297**
island arc **297**, 408
island biogeography 110, **297–298**
islands **299**
    barrier 47, 103, 381
    continental **112**
    ice 284
    oceanic 112
isochrones **299**
isocline **299**
isolation, ecological **299**
isometric growth 18
isopleths *299*, **299**
isostasy 112, 242, **299–300**, 364, 432
isotope **300**

J
jet stream **301–302**, 325
joint probability estimates **302**
jökulhlaup 212, 241, 281, **302**
juvenile water **302**

K
K-cycle **303**
K-selection *see* r- and *K*-selection
kame **303**, 450, 497
    terrace **303**, 497
kamenitza **303**
kaolin **303**
karren 100, **303**, 345
karst 45, 253, **303**, *304*
    biokarst **60**, 405
    cockpit **104**
    doline **152**
    phytokarst 405
    polje **412**
    polygonal **415**
    springs 443
    tower **533**
kata-front **305**
katabatic flows **303–305**
kavir **305**
kegelkarst **305**, 533
Kelvin wave 22, **305**
kettle hole **305**, 497
keystone species **305**

khamsin **305**
kinematic wave **305**, 371
kinematics **306**
kinetic energy 179, **306**
kingdom, animal and plant **306**
klippe **306**
knickpoint **306**
knock-and-lochan topography **306–307**
kolk **307**
koniology **307**
kopje *307*, **307**
Köppen climatic classification **307**, *308*, 335
krotovina **307**
krummholz **307–308**
kumatology **308**
kunkur 77
kurtosis **308**
Kyoto Protocol **308–309**

L
La Niña **311–312**
laccolith 272, **310**
lactic acid formation 23
lacustrine **310**
lag gravel **310**
lag time **310**
lagoon **310**
    haff 257
lahar **310**
lake **310–311**
    chott **91**
    eutrophic **196**, 311
    ice-dammed, drainage of 212
    mere **337**
    oxbow 125, **381–382**
    paternoster **392**
    proglacial **420**
    tarn **516**
    thaw **522**
Lambert Glacier 27
laminar flow **311**, 541, 551
land capability **312**
land systems **312**
land-bridge **312**
landform memory capacity **337**
Landsat program **312**, 376, 406
    Thematic Mapper 312, 407, **523**
landscape ecology **312**
landslide 259, *313*, **313**
Laplace equation 145
lapse rate **313**
laser anemometers 24
Late Glacial **314**
Late Glacial Interstadial 27
latent heat **314**, 343, 476
lateral
    accretion **314**, 411
    flow **314**
    migration **314**
laterite **314**
latosol **314–315**
Laurasia **315**
Laurentide Ice Sheet 242, **315**

lava *315*, **315**
  caves   84
  coulée   **118**
  louderback   **322**
law of the wall   **315**, 456
Le Chatelier principle   **316**
leachate   **316**
leaching   **316**
  requirement   **316**
leaf area index (LAI)   **316**
lee
  depression   **316**
  dune   **316–317**
  eddy   **317**
  waves   40, **317**, 339
Leopold matrix   184
lessivage   **317**
levée   213, **317**
  breaching   212
lichenometry   **317**
lichens   226
lidar (light detection and ranging)   **317–318**
life form   **318**
  Raunkiaer's classification   189, 318
lightning   **318**
limestone
  caves   84
  clint   **100**
  fluviokarst   **217**
  grike   **251**
  holokarst   **263**
  karren   **303**
  kegelkarst   **305**
  makatea   *328*, **328**
  solution   **491**
  solutional microforms   *304*
  tepee   **520**
  see also karst
limited area model   438–439
limiting angles   **318–319**
limiting factors   **319**
limnology   **319**
  see also lake
limon   **319**
lineament   **319**
linear dune   162, **319**, 475
linear wave theory   14–15
liquid limit   **320**
liquid limit (LL)   40
liquidity index (LI)   40–41
lithification   **320**
lithology   **320**
lithosol   **320**
lithosphere   **320**, 330, 379, 386, 408
  rifting   446
litter   **320**
  dam   **320**
Little Climatic Optimum   **320**, 335
Little Ice Age (LIA)   **320–321**
littoral zone   **321**
load
  capacity   *78*, **78**
  critical   212

non-capacity   **78**
  river, dissolved   **150**
  stream   78, **321**
  structures   **321**, 379–380
  suspended   471, 474, **512**
  see also bedload
local climate   338
local winds   **321**
lodgement till   530
loess   164, **321–322**
logan stone   **322**
long profile, river   **322**
longitudinal dune   **322**
longshore drift   **322**, 493
lopolith   **322**
louderback   **322**
low flow analysis   **322**
luminescence dating   175, **322–323**
  loess   322
lunette dune   94, **323–324**, 387, 492
lynchet   **324**
lysimeter   **324**, 488, 560

M
maar   **325**
macroecology   **325**
macrofossils   **325**
macrometeorology   **325**
macrophyte   **325**
  aquatic   **34**, 325
macropore   **326**
macrotidal estuary   195
Madden–Julian Oscillation (MJO)   **326**
maelstrom   **326**
magma   **326**
magnetic
  anomaly   **326**
  declination   **326**
  pole positions   412
  storm   **326**
  susceptibility   **326–327**
magnetic field
  environmental magnetism   **186–187**
  reversal   72
magnetism, terrestrial   **522**
makatea   *328*, **328**
mallee   *328*, **328**
mammilated surface   **328**
mangrove forest   **328–329**
manned Earth resources satellites   **329**
Manning equation   90, 215, **329**, 456, 505, 543
mantle   320, **330**
  plume   265, 330, **330**
maquis   **330**
margalitic soils   **330**
margin
  active   7
  passive   330
marginal channel   **330**
marine isotope stage (MIS)   **330–331**
marine pollution   **331**
marker horizon   **332**
Markov process   **332**

marl **332**
marsh 512
mass
  balance **332**
  movement types **332–334**, *333*
  specific susceptibility 327
  strength **334**, 443, **453**, *454*, 504
matorral **334**
matrix flow **334**
mature community 99
maturity **334**
Maunder minimum **334**
mean sea level 112, **334–335**, *469*
meander
  entrenched *181*, **181**
  intrenched **292**
  valley 543, **550**
meandering **335**
Medieval Warm Period 320, **335**
Mediterranean climate **335**
mega dune **336**
megafan **336**
megashear **336**
megathermal climate **336**
mekgacha **336**
meltout till 529–530
meltwater **336**
  deposits 194
  glacial 336
  regelation **437**
  stream 194, 330
memory capacity **337**
Mercalli scale *337*, **337**
mere **337**
meridional circulation 209, 229, **338**, *563*
meridional overturning circulation (MOC) **338**
  Atlantic (AMOC) 27, 338
mesa **338**
mesoclimate **338–339**
mesometeorology 338, **339**
mesophyte **339**
mesoscale **342**
  convective systems **342**
    cellular convection (MCC) **339–340**, *340*
  precipitation areas (MPAs) 224
mesosphere 39, **340**
mesothermal climate **340**
mesotidal estuary 195
mesotrophic 372
Messinian salinity crisis **340**, 522
meta-analysis **341**
metamorphism **341**
metamorphosis, river 384, **448**, *449*
metasomatism **341**
meteoric water **341**
meteorite crater 120
meteorological drought 156
meteorological satellite **341**, 364
meteorology **343**
  agrometeorology **13**
  applied **32–33**, *33*
  biometeorology **63**
  dynamical **166**, 404

mountain **357**
  physical **404**
  satellite **465–466**, 513
  synoptic **513–514**
  urban **548–549**
micro-erosion meter **344**
microclimate **343–344**
microcracks **344**
microfossils **344**
micrometeorology **344–345**
microphytes **345**
microsomia **361**
microtidal estuary 195
microtopography **345**
microwave remote sensing **346**
Mid-Atlantic Ridge (MAR) **346**
mid-ocean ridge **346**
midden **346**
migration **347**
Milankovitch hypothesis 280, *347*, **347–348**, 420, 422
Millennium Ecosystem Assessment (MA) **348**
mima mounds **348–349**, 392
mineral fracture, conchoidal **108**
minimum acceptable flow **349**
miombo woodland **350**
mirage **349**
mires **349–350**, 392
  blanket **392–393**
  ombrotrophic 41, 66, 372
  palsa 393
  raised 393, **432**
  spring 393
  tundra 393
  see also bogs
mist **350**
Mistral 304–305, **350**
mixing corrosion **350**
mixing models **350**
mobile belt **350**
moder **351**
mogotes 305, **351**
Mohorovicic discontinuity **351**, 375
Mohr–Coulomb equation 24, 173, *351*, **351**, 478, 535
moisture, antecedent **29**, 271
  soil *see* soil water
molard **351**
molasse **351**
mollisol 486, 487
moment magnitude scale **351**, 475
momentum
  angular **25**, 351
  budget **351–352**
monadnock **352**
Monera 306
monoclimax **352**
monocline **352**
monsoon 325, **352–353**
  forest **353**
monumented sections **353**
mor **354**
moraine *353*, *354*, **354–355**, 497

De Geer   134
fluted   216
ground   531
ice cored   **281**
ribbed   453–455, 506
rogen   354, **453–455**, 506
terminal   **521**
morphodynamics   **355**
morphogenetic regions   **355**
morphological mapping   **355**
morphometry   **355–356**, 478
morphotectonics   **356**
mortality rate, density-dependent   81, *81*
mosaic vegetation   **356**
mottled zone   **357**
moulin   336, **357**
mountain   **358**
   sea mount   **471**
   summit, concordant   **108**
mountain meteorology   **357**
mountain/valley wind   338, 339, **357–358**
mud lumps   **358**
mud volcano   **358**
mudflats   **358**
mull   **358**
multifractality   221
multiple working hypotheses   **358**
multispectral scanner   **358–359**, 524
muskeg   **359**
Muskingum method   214, **359**
mutation   **359**
mutualism   **359**
mycorrhizal fungi   226, **359–360**

N
nanism   **361**
nanoclimate   **190**
nappe   **361**
natural disaster   **361**
natural hazard   **361**
natural remanent magnetization (NRM)   187
natural resources   **361–362**, 500
natural selection   130, 199–200
   r- and *K*-selection   105, 416, 424
natural vegetation   **362**
Navier–Stokes equation   108
nearshore bar   **46**
nebkha   **362–363**
neck   **363**
   cut-off   **363**
needle ice   225, **363**
nehrung   **363**
neo-Darwinism   **363**
neocatastrophism   82, **363**
neoglacial   320, **363**
neotectonics   20, **363–364**
nephanalysis   **364**
nephroscope   **364**
neptunism   **364**
neritic   **364**
ness   **364**
net primary productivity (NPP)   61, 62, **364**, 420,
   536

net radiation   **365**, 476
network   **365**
   random-walk   **432**
   stream   377
neutron probe   **365–366**, 488
nevé   **366**
niche   107, 256, **366**, 418
nick point   **366**
nimbus clouds   102, *102*
nitrate   **366**
nitrification   **366**
nitrogen   137
   cycle   **366–367**
nivation   **367**, 397
nivometric coefficient   **367**
noctilucent clouds   **367–368**
non-capacity load   **78**
nonconformity   **368**
nonlinear system   **368**
normal fault   **368**
normal stress   **368**, 478
normalized difference vegetation index (NDVI)   **368**
North Atlantic Deep Water (NADW)   26, **368–369**,
   369
North Atlantic Oscillation (NAO)   241, **369–370**
Northern Lights   **41**
notch   **370**
nubbins   **370**
nuclear waste   **370–371**
nuée ardente   **371**
numerical modelling   **371–372**
nunatak   35, **372**
nutrient   **372**
   recycling   **64**
   status   **372**

O
oasis   **373**
oblique-slip fault   208, *208*
obliquity   **373**
obsequent stream   **373**
obsidian hydration dating (OHD)   **373**
occlusion   **373**
occult deposition   **373**
ocean   *374*, **374**
   acidification   **374**
   sediments   **375**
Ocean Drilling Programme (ODP)   **374–375**
oceanic crust   **375**
oceanic islands   **112**
oceanography   **375**
ogives   **375**
oil-shale   **376**
okta   **376**
oligotrophic   372
ombrotrophic bogs/mires   41, 66, 372, **376**, 437
onion weathering   **376**
ontogeny   **376**
ooids   **376**
oolite   **376**
ooze   **376**
open system   **514**
Operational Land Imager   **376**

opisometer 376
optical remote sensing 376–377, 440
optically stimulated luminescence (OSL) dating 323
organic weathering 378
orocline 378
orogens 378, 446
orogeny 127, 189, 196, 350, **378–379**, *379*
  alpine 20
orographic precipitation 379
osmosis 379
ostracods 379
outlet glacier 379
outlier 379
outwash 379–380
  terrace 380
overbank
  deposit 19, **380**
  flow **380**
overdeepening, glacial **380**
overflow channel 380–381
overland flow 261, 287, **381**
  Horton model 265, 286, 290, 389, 457, 545
  partial area model **390**, 545
  sampling 237
overthrust **381**
overtopping **381**
overwashing **381**
oxbow lake 125, **381–382**
oxidation **382**, 435
oxygen
  biochemical oxygen demand (BOD) **55**
  dissolved **150–151**
  isotope analysis 281, 300, 331, 375, 382, **382**, 497
ozone **382–383**
  destruction 90, 271, 382–383
  hole **383**
ozonosphere 39

P
p-form **399**
Pacific decadal oscillation (PDO) **384**
pacific-type coast **384**
pack ice 468
padang **384**
palaeobotany **384**
palaeochannel **384**, 385
palaeoclimatology 95, **384–385**
palaeoecology **385**
palaeoenvironment **385**
palaeoenvironmental indicators **385**
Palaeogene **385**
palaeogeography **385**
palaeohydrology **385–386**
palaeomagnetism 111, 385, **386**
  Brunhes–Matuyama 72
palaeosol *386*, **386**
palaeovelocity **386**
pali ridge **387**
Palmer Drought Severity Index (PDSI) 156, 157
palsa **387**
palsa mires 393
paludal sediments **387**

paludification **387**
palynology *see* pollen analysis
pan 94, *387*, **387**
panarchy **387**, *388*
panfan **387**
Pangaea 201, *388*, 509
panplain **388**
pantanal **388**
parabolic dune 220, **388**
paraglacial conditions **388–389**
parallel retreat **389**
parameterization **389**
páramo **389**
parasite **389–390**
parna **390**
partial area models 286, **390**, 491, 545
partial drought 156
partial duration series **390**
particle form **390**
particle size **390–391**, *391*
  skewness **483**
passive margin 330, **391–392**
pastoralism **392**
paternoster lake **392**
patterned ground **392**
peak discharge **392**
peat 325, **392–393**, 493
  swamps 393
  see also bogs; mires
pedalfer **393**
pediment *393*, **393**, 450
pedocal **394**
pedogenesis **394**
pedology **394**
pedon **394**
pedosphere **394**
peds **394**
pelagic **394**
Penck and Brückner model 249, **394**
peneplain 126, **394–395**
penetrometer **395**
peninsula **395**
Penman equation 198
perched block **395**
perched groundwater 253, **395**
perched water table **395**
percolation **395**
  gauge **395**
percoline **395–396**
perennial stream **396**
pericline **396**
perigee **396**
periglacial *396*, **396–397**
perihelion **397**
permafrost 283, 396, *397*, **397**
permeability 114, 268, **397**
  intrinsic 268, **293**, 397
  rock 108
  specific **293**
permeameter 268, **397**
persistence **398**, 494
pertubation **398**
perviousness 397, **398**

pesticides **398–399**
  bioaccumulation **61**
pF **399**
ph 4, *5*, **399**
phacolith **399**
phanerophytes *433*, 434
phenology **400**
phoresy **400**
phosphate rock **400**
photic zone **400**
photoautotrophs 42, 64
photogrammetry **400**
photosynthesis 61, 168, 218, 372, **400**
  net primary productivity 61, 364, 420
  photosynthetic pathway **400–401**, 496–497
phreanophytes **401**
phreatic divide 154, **401**, 565
phreatic zone 252, **401**, 550
phylogenesis **401**
phylogenetic systematics *92*, **92–93**
physical geography **401–403**
physical meteorology **404**
physiography **404**
physiological adaptations 7
phytogeography **404**
phytogeomorphology **404**
phytokarst **405**
phytosociology 37, **405**
piedmont **405**
piedmont glacier **405**
piezometer 147, **405**
piezometric surface 252, **406**, 564
pingo 272, **406**
piosphere **406**
pipes **406**
pipkrakes *see* needle ice
pisoliths **406**
pitometer **406**
pixel **406–407**
plagioclimax 149
plain, abyssal *2*, **2**
planation surface 193
plane bed **407**
plane sliding friction, angle of **25**
planetary albedo 15
planimeter **407**
plankton 394, **407**
plant cover **119**
  Braun-Blanquet scale **71–72**
  domin scale **152**
  see also vegetation
plants
  aquaculture 33
  aquatic 34
  arctic-alpine flora 34
  areography **35**
  association **36–37**, *37*
  autecology **41**
  biomass **62**
  biosphere evolution 64
  calcicole **76**
  communities **71–72**
    disclimax **149**

ecotone **171**
  plagioclimax 149
  serclimax **477**
  see also climax vegetation
competition 107, 171
cover *see* plant cover
die-back **144**
dispersal **150**
dominant organism 152–153
endemism **178**
ephemeral *189*, **189**, 424, 575
epiphytes **190**
exotic **201**
floristic realms *214*, **214**
geophyte **237**
invasive species 294–295
island biogeography **297–299**
kingdom **306**
macrophytes 34, **325**
mesophytes **339**
nanism **361**
phenology **400**
phreanophytes **401**
phytogeography **404**
phytosociology 37, **405**
pollen analysis **412–413**, *414*
Raunkiaer's life forms *433*, **434**
sciophyte **467**
species–area curve **492**
sphagnum **493**
xerophyte **575**
see also plant cover; vegetation
plastic limit (PL) 40, **408**
plasticity **408**
plasticity index (PI) 40
plate tectonics 111, 320, 330, 375, 388, *408*,
    408–409
  active margin 7
  asthenosphere **38**
  convergence 114
plateau **409**
  basalt **409**
playa 94, 253, 387, **409**
Playfair's law **409**
Pleistocene 394, **409–410**, 421
Plinian eruption **410**
plots **410**
  run-off **457–458**
plucking **411**
plume, mantle 265, 330, **330**
pluton **411**
plutonic rock **411**
plutonism **411**
pluvial period **411**
pluviometric coefficient **411**
podzol **411**
point bar deposits 314, **411**
Polar Front 26, **412**
polar wander **412**
polder **412**
polje **412**
pollen analysis 385, **412–413**, *414*
  coprolites 114

pollution 362, **413–415**
  acid deposition 5–6
  adverse effects 14
  air **14**
  atmospheric 39
  bioaccumulation **54–55**, 61
  critical load **121**
  dry deposition 4, 5, **159**
  marine **331**
  smog **485**
  thermal **524**
  wet deposition 4, 5
  see also eutrophication
polyclimax **415**
polygonal karst **415**
polynyas 25, **415**
polypedon **415**
polytopy **415–416**
pool-and-riffle sequence **416**, 471
population dynamics **107**, **416**
population regulation 81, *81*
pore ice **417**
pore water pressure 43, **417**
porosity 24, 114, **417**
potamology **417**
potassium argon (K/Ar) dating **417**
potential
  energy 179, **417**
  evaporation **417**
  temperature **418**, 525
potentiometric surface 253
pothole 43, 399, **418**
prairie **418**
pre-fractals 221
precession of the equinoxes **418**
precipitation 274, **418**
  acid 4, **5–6**, 144, 267, 415
  convective *114*, **114**
  cyclone **537**
  effective **173**
  evapotranspiration relationship 160–161
  interception **289–290**
  orographic **379**
  probable maximum precipitation 274, **419–420**
  storm 212
  see also hail; rain; rainfall; snow
predation **418**
predator–prey relationships **419**
preferential flow 75
pressure
  air **419**
  anemometers 24, 570
  cryostatic **123**
  hydrostatic **276**
  melting point **419**
  pore water **417**
  release 146, **419**
  vapour 268, **550–551**
prevailing wind **419**
primarrumpf **419**
probable maximum precipitation 274, **419–420**
productivity **420**
  biological **61**, 170, 364, 420

gross primary productivity (GPP) 61, **420**
net primary productivity (NPP) 61, 62, **364**, 420, 536
proglacial lake **420**
progradation **420**
protalus rampart **420**
Protista 306
provisioning services 171
proximal trough **420**
proxy **420**
pseudokarst 84, **420**
pumice **420**
puna **420**
punctuated aggradational cycles **420**
pyroclastic **420**

Q
quartz **421**, 464
quasi-biennial oscillation (QBO) **421**
quasi-equilibrium **421**
Quaternary 385, **421–422**
quick clay **423**
quickflow 113, 134, **423**
quicksand **423**

R
r- and *K*-selection 105, 416, **424**
racemization 21
  amino acid **21–22**
radar **425**
  sideways-looking airborne (SLAR) **481**
radiation *426*, **426–427**
  electromagnetic **174**
  heat budget 259
  net **365**, 476
  see also solar radiation
radiative forcing 97, **427**
radiative–convective models 96
radiocarbon dating 118, 257, **427–428**
  calibration 428
  pollen analysis 413
radiocarbon years 428
radioisotopes 257, 300, **428**
radiolaria **428–429**
radiosonde **429**, 520, 525
radon gas **429**
rain **429**
  day **429**
  factor **429**
  gauge **429**, 431
  shadow **430**
  splash **430**
  see also precipitation; rainfall
rain forest, tropical **537–538**
rainbow **430**
raindrop impact erosion **430–431**, 479
rainfall 114, **431**
  convective 114
  depth–duration curve **140**
  drop size **156**
  intensity **431**
  run-off **431**
  simulator **431**

stratiform   114
   see also precipitation
raised beach   431–432
raised bog/mire   393, **432**, 437
raised channel   **432**
ramp, sand   **461**
Ramsar Convention   **432**
randkluft   **432**
random-walk networks   **432**
ranker   **432–433**
rapids   **433**
rare species   178
raster GIS   234, **433**
rated section   **433**
rating curve   **433**
rational formula   **433**
Raunkiaer's life forms   189, 318, *433*, **434**
reach   **434**
reaction time   **434**
recession curve   139, **434**
recharge   **434**
recovery time   **434**
recumbent fold   **434**
recurrence interval   **434–435**
red beds   **435**
redox cycle   **435**
redox potential   **435**
reduced complexity models   **436**
reduction   435
reef   **436**
   barrier   **47**
   see also coral reef
reflective beach   **436**
refraction   *436*, **436–437**
refugia   **437**
reg   **437**
regelation   48, **437**
regeneration complex   **437**
regime   **437**
   flow   **215**, 463
   theory   246, 271, **437–438**
regional climate model (RCM)   **438–439**
regolith   122, **439**, 567
regosol   **439**
regulating services   171
regur   **439**
rejuvenation   **439**
relative age   **439**
relative dating   **439**
relaxation time   **440**
relic/relict landforms   385, **440**
remanié   **440**
remote sensing   145, 174, 235, 404, **440–441**, 510
   Earth observation   **167**, 346
   Landsat program   **312**
   lidar   317
   microwave   **346**
   optical   440
   vegetation   368
rendzina   **441**
renewable resources   215
repose, angle of   25, 246, 259
representative basins   **441**

reptation   94, **442**
resequent stream   **442**
reservoir   128
   Brune curve   **72**
   compensation flow   **106**
   glacial   240
   rocks   **442**
   storage effects   **442**
residence time   **442**
residual shear, angle of   *25*, **25**
residual strength   **442–443**, 478
resilience   **443**, 494–495
resistance   219, **443**, 494–495
   free-surface   456
resistance to force ratio   206
resources
   flow   **215–216**, 362
   natural   **361–362**, 500
   renewable   215
   stock   216, 362, **500**
respiration   168, **443**
resultant wind   **443**
resurgence   **443**
retention curve   **443**
retention forces   **443**
retrogradation   **443**
return period   **443–444**
reverse fault   208, *208*, **444**
reversed polarity   326, **444**
reversing dune   **444**
Reynolds number   215, 311, **444**, 541, 551
rheidity   **444**
rheology   **444**
rheotrophic   372
rhexistasy   64
rhizome   **444**
rhizosphere   **444**
rhodoliths   **444**
rhourd   **444**
rhythmite   551
ria   **444–445**
ribbed moraine   453–455, 506
ribs, transverse   103, 379, **535**
Richardson number   **445**
Richter denudation slope   **445**
Richter scale   **445**
ridge
   barchanoid   **46**
   beach   **50**
      chenier   **89–90**
   cuesta   **124**
   levée   213, **317**
   pali   **387**
   protalus rampart   **420**
   ridge and runnel topography   **445**
   sand   319, **461**
riedel shears   **445**
riegel   **445**
rift valley   445–446, *446*
rifting   **446–447**
rills   291, 410, **447**
rime ice   280, **447**
ring complex   **447**

ring-dyke 447
rip current 46, 447
riparian 447–448
ripples 52
  adhesion 7
  aerodynamic 11
  current 124, 125
rising limb 448
river
  bank erosion 19, 45
  bankfull discharge 45–46
  basin planning 448
  bedload 52
  braided 19, 71, 87, 284, 314
    cellular automata model 86
  capture 79, 503
  cataclinal 81
  channelization 88
  classification 448
  diaclinal 143
  discharge measurement 433
  dissolved load 150
  dissolved solids 151, 151
  glaciofluvial 241
  hydrometry 274, 275
  hyperconcentrated flow 78
  integrity 289
  knickpoint 306
  long profile 322
  meandering 335
  metamorphosis 384, 448, 449
  restoration 449
  sinuosity 483
  terrace 213, 449–451, 451
  transmission loss 534
  see also channel; dam; flow
roche moutonnée 451, 530
rock
  acid 6
  aphanitic textures 31
  arenaceous 35
  argillaceous 35
  avalanche 43
  basic 49
  beach 50, 77, 285
  bedrock 53
  bioclastic 55
  breccia 72
  cap-rock 79
  case hardening 81, 285
  cataclasis 81
  chattermarks 89
  coatings 451, 452
    see also rock varnish
  conglomerate 109
  coquina 115
  corrasion 118
  corrosion 118
  desquamation 143
  erosion measurement 344
  flour 451
  fracture, conchoidal 108
  gabbro 228

  glacier 452
  granite 248
  growan 253
  hypabyssal 277
  igneous 284
  intact strength 289, 443, 453
  mass strength 334, 443, 453, 454
  microcracks 344
  oolite 376
  permeability 108
  phosphate 400
  plutonic 411
  porosity 24, 114, 417
  pressure release 146, 419
  pumice 420
  quality indices 453
  reservoir 442
  sedimentary 475
  stability analysis 496
rock drumlin 451
rock varnish 239, 453
  cation-ratio dating 83
roddon 453
rogen moraine 354, 453–455, 506
rollability 455
Ross Sea Bottom Water 25
Rossby waves 40, 126, 325, 455, 539, 572
Rossby-type circulation 229
rotational failure 455
rotor streaming 455
roughness 443, 455–456, 463
  coefficient 51, 329, 330, 505
  form 456
  grain 456
  length 315, 456
roundness 456–457
ruderal vegetation 457
run-off 457
  connectivity 109
  plots 457–458
  rainfall 431
  rational formula 433
  storm 113

S
S-curve 467
sabkha 459
safety, factor of 206
Sahel 459
salars 459
salcrete 459
salinization 297, 459–460
salt
  marsh 460
  tectonics 460
  weathering 21, 460
salt-dome 460
salt-flat 460
saltation 94, 121, 180, 284, 460, 463
sand 460
  banks 460–461
  coastal dunes 103
  coversand 119

drift potential **156**
lens 379, *461*
quicksand **423**
ramp **461**
reptation **442**
ridge 319, **461**
rose 156, 162, **461**
seas 162, **461–462**, *462*
sheet **462**
singing **483**
trap **462–463**, *463*
valley **516**
wedge **463**
see also beach
sand volcano **463**
sandbed channels 215, **463**
sandstone **463–464**
arkose **35**
sandstorm **464**
sandur **464**
sapping **464**
basal 35, **48**, 91
spring 253, **464**
saprolite 132, **464**
sapropel **464–465**
sarsen **465**
sastrugi **465**
satellite
meteorology **465–466**, 513
nephanalyses 364
saturated wedge and zone **466**, 491
saturation
coefficient **466**
deficit 261, 286, 390, **466**
savanna 74, 248, 392, **466–467**
cerrado **86**
desertification 392
piosphere **406**
Sahel **459**
scabland **467**
scald **467**
scar **467**
Schmidt hammer 258, **467**
sciophyte **467**
sclerophyllous vegetation **467**
scoria **467**
scour and fill **467**
scree **467**, 516
scrubland, garrigue **228**
sea cliffs **467–468**
sea, epeiric **189**
sea ice 280, *468*, **468**
sea level **469–470**
change 195–196
sandy beach response 72–73
mean 112, **334–335**, *469*
monitoring 246
sea mount **471**
sea surface temperature (SST) **471**
sea/land breezes 338, 339, **468–469**, *469*
seaweeds **34**
secondary depression **471**
secondary ferrimagnetic minerals (SFMs) 187

secondary flows **471**
sediment
aggradation **12**
allochthonous **18**
autochthonous 18, **41**
biostabilization **64**
bioturbation **65**
bounding surface **70**
budget **473–474**
bulk density **74**
coccoliths **104**
crust **123**
delivery ratio (SDR) **474**
diachronous **143**
entrainment **180–181**
fabric 206, **474**
fluvial **471–473**, *472*, *473*
geest **228**
greywacke **251**
lateral accretion 314, 411
load 78, 321
capacity load *78*, **78**
non-capacity load **78**
suspended **512**
marl **332**
ocean **375**
ooze **376**
paludal **387**
particle form **390**
particle size **390–391**, *391*
determination 248
routing **474**
silt **482–483**
slackwater deposit **484**
transport, diffusive processes 145
trap, Gerlach trough 237
varves **551**
yield **474–475**, *475*
see also alluvium; clay; till
sedimentary rock **475**
arenaceous **35**
beach **50**
sedimentation
caesium-137 (^{137}Cs) analysis **76**
siltation **483**
segregated ice **475**
seiche **475**
seif dune 319
seismicity **475**
selection
natural 130, 199–200
r- and *K*-selection 105, 416, 424
self-affine surfaces 221
self-mulching **476**
self-organized criticality (SOC) **476**
self-similar coastlines 221
semi-arid areas **476**
semi-desert **476**
sensible heat 343, **476**
sensible temperature **476**
sensitivity **476–477**
serac **477**
seral community **477**

serclimax 477
sere 415, 477
serir 477
seston 477
shakehole 305, 477
shale 477
shape 477–478
shear box 351, 478–479
shear strength 24, 114, 478
  measurement 535
  Mohr–Coulomb equation 351, 535
  residual 442–443
shear stress 180, *478*, 478
  Glenn's law 242–243
sheet erosion 478–479
sheet flow 479
sheeting 479
shelf 479
  continental 112, 479
  ice 240, 283
shell pavements 479–480
shield 120, 480
shield volcano 480
Shield's parameter 480
shoal 480
shore 480
  platforms 480
shoreline, equilibrium 192
shrink-swell soils 480–481
shrinkage limit (SL) 40
sial 481
sichelwannen 399, 481
sideways-looking airborne radar (SLAR) 481
sieve analysis 481–482
sieve deposits 482
silcrete 482
sill 482
silt 379, 482–483
siltation 483
sima 483
simulation 483
singing sands 483
sinkhole 104, 483, 512, 520
sinter 483
sinuosity 483
siphon 483
sixth-power law 107
skerry 483
skewness 483
slab failure 484, 532
slack 484
slackwater deposit 484
slaking 484
slickensides 202, 484, 552
slide 484
  translational 484, 534
SLIME (subsurface lithoautotrophic microbial
      ecosystems) 42
slip face 162, 319, *484*, 484
slip-off slope 484
slope 484–485
  boulder-controlled 68
  constant 111
  continental 112
  dipslope 147
  graded 246–247
  ice contact 281
  limiting angle 318–319
  replacement 485
  Richter denudation slope 445
  rollability 455
  slip-off 484
  stability analysis 496
  strength equilibrium slopes 504
  threshold slopes 527–528
  waning 111, 559
  waxing 565
  weathering-limited 535, 568
  see also hillslope
slope–area method, discharge measurement
      148–149
smog 485
snow 485
  blizzard 65
  firn 211
snow avalanche 43
snow line 485
snowball Earth 485
snowblitz theory 485–486
snowmelt 212, 486
soil water
  antecedent moisture 29
  bypass flow 74–75
  field capacity 210
  field drainage 210
  gravimetric determination 248
  irrigation 296–297
  pore water pressure 417
  retention curve 443
  soil moisture deficit 488
  soil moisture suction 399
  throughflow 261, 286, 290, 390, **466**, **528**
  water balance 488
soil(s) 486
  adsorption 7–8
  aggregate stability 177
  aggregation ratio 12
  andosols 23
  anthrosols 30–31
  aridisols 35
  bearing capacity 50–51
  bioturbation 65
  brodel 72
  brunizem 72
  bulk density 74
  catena 82–83
  cation-exchange 83
    capacity 49, 83
  chernozem 90
  chronosequence 91
  classification 486–487, *487*
  clinosequence 100
  cohesion 104–105
    Atterberg limits 40
  compaction 106
  consolidation 111

crumb structure   123
crust   **123**, 487
denitrification   137
edephic factors   172
eluviation   177
erosion   110–111, 186, 362, **487**
expansive   **201–202**
glei (gley)   242
gumbo   254
hamra   258
horizon   **264**, 486
hydrophobic   275
infiltration   285–286
intrazonal   292
latosol   **314–315**
leaching   316
macropores   326
margalitic   330
moisture *see* soil water
osmosis   379
palaeosol   *386*, 386
pedogenesis   394
pedology   394
pedon   394
pedosphere   394
peds   394
podzol   411
porosity   24, 114, 417
profile   488
   cumulative   124
ranker   **432–433**
regosol   439
regur   439
salinization   297, **459–460**
shrink-swell   **480–481**
silt   379, **482–483**
solonchak   489
solonetz   489
strength   351
   measurement   395
   residual   **442–443**
striated   504
structure   *488*, **488–489**
takyr   516
terra rossa   521
texture   *489*, 489
till   284, 354, 453, **529–531**, *530*
vertisol   552
water *see* soil water
zonal   577
see also clay
solar constant   **489**
solar radiation   427
   insolation   **288–289**
      weathering   *288*, 289
   management techniques   234
solfatara   **489**
solifluction   345, **489**
solonchak   **489**
solonetz   **489**
solstice   **489**
solum   **490**
solutes   *490*, **490–491**

sonic anemometers   24
sorptivity   **491**
sorting   **491**
source area   262, **491–492**
source bordering dune   **492**
Southern Antarctic Circumpolar Current Front
      26–27
Southern Oscillation   **492**
   see also El Niño Southern Oscillation
Southern Oscillation index (SOI)   **492**
species–area curve   **492**
specific
   conductance   **108**, 491
   retention   **492**
   yield   **492**
spectral analysis   **492**
speleology   **493**
speleothem   77, 260, **493**
spericity   **493**
sphagnum   **493**
sphenochasm   **493**
sphenopiezm   **493**
spheroidal weathering   **493**
spit   **493**
   barrier   47
spring   **494**
   hot   **266**
   intermittent   **290**
   karst   443
   sapping   253, 464
spring mires   393
spring mounds   **493**
squall line   **494**
stability   170, **494–496**
   analysis   496
   vertical   552
stable equilibrium   81, 316, **496**
stable isotope geochemistry   **496–497**
stack   **497**
stadial   **497**
staff gauge   **497**
stage   **497**
stagnant ice topography   **497**
stalactite   **493**
stalagmite   493
star dune   162, **498**
static allometry   18
statistical downscaling   153–154
steady flow   **498**
   uniform   145, 179, 215, 223, 533, **543**
steady state   **498**
steady time   247
steam fog   **498**
stem flow   226, 290, **498–499**, 528
step-pool systems   103, **499**
steppes   **499**
stick slip   **499**
stilling well   **499**
stillstand   **499**
stochastic models   **499**
stochastic process   **500**
stock   **500**
   resources   216, 362, **500**

Stoke's surface   462, **500**
stone line   **500**
stone pavement   **500–501**
storage   **501**
  bank   45
  channel   88
  depletion curve   *139*, **139–140**
  depression   **140**
  equation   112
  surface   273, **511**
storm   339
  dust   163–164, **165**
    haboob   257
  magnetic   326
  run-off *see* run-off
  sandstorm   **464**
  squall line   **494**
  thunderstorm   **528**
storm beach   46, **501**
storm surge   212, **501**
stoss   **501**
strain   **501–502**
  rate   **502**
strandflat   483, **502**
strandline   *502*, **502**
stratiform rainfall   114
stratigraphy   **502**
stratosphere   39, **502**
stratovolcano   **502**
stratus clouds   102, *102*
stream
  allogenic   **18**, 42
  anteconsequent   **29**
  autogenic   **42**
  bourne   70
  cataclinal   81
  consequent   **110**, 543
  distributary   **151**
  effluent   **173**
  ephemeral   **190**, 213
  ice   **283**, 379
  inconsequent   **285**
  insequent   **288**
  intermittent   **290–291**, 396
  law of stream numbers   265
  meltwater   194, 330
  network   377
  obsequent   **373**
  order   *377*, **377–378**
  perennial   **396**
  pool-and-riffle bedform   **416**
  power   447
  resequent   **442**
  sediment load   78, **321**
  subsequent   **507**
  underfit   **543**
  see also channel; flow
stream capture   384, **502–503**
stream power   87, **503**
streamline   **503**
streamlines   172
strength equilibrium slopes   **504**
stress   219, *504*, **504**

effective   **173**
normal   **368**, 478
see also shear stress
striated soil   **504**
striation   **504**
Strickler equation   **504–505**
strike   **505**
strike-slip fault   208, *208*
  megashear   **336**
string bog   **505**
stromatolite   77, 125, **505**
Strouhal number   **505**
structural adaptations   7
structure from motion (SfM) technique   10
sturzstrom   **505**
subaerial   **505**
suballuvial bench   **505**
Subantarctic Front   26
subclimax   **505–506**
subduction zones   201, 408, **506**
subglacial bedforms   216, **506**
subglacial environment   **506**
subhumid climate   **506**
sublimation   **507**
  till   530
sublittoral zone   **507**
submarine canyon   **507**
subsequent stream   **507**
subsere   **507**
subsidence   **507**
succession   42, 352, 415, 477, **507–508**
  allogenic   **18**, 42
  autogenic   **42**
suction   **508**
Suess effect   **508**
suffosion   **508**
sulphation   **508**
summit, accordant   3
sunspots   **508–509**
supercontinent   388, *509*, **509**
supercooling   **509**
superimposed drainage   **510**
superimposed ice   280, **510**
superposition, law of   **510**
supersaturation   **510**
supervised classification   **510**
surf   **510**
surface
  detention   **510**
  storage   273, **511**
  tension   78, **511**
surging glaciers   282, 355, **511**
susceptible drylands   **511**
suspended load   471, 474, **512**
sustainability   **512**
sustained yield   **512**
Sverdrup   **512**
swale   **512**
swallet   **512**
swallow hole   **512**
swamp   **512**
  backswamp   44
  peat   393

sympatry   299, **513**
syncline   **513**
synforms   31
synoptic climatology   **513**, 566
synoptic meteorology   **513–514**
system albedo   15
systematics   516
systems   **514–515**
   general system theory   **232–233**, 514
syzygy   **515**

T
tafoni   21, **516**
taiga   67, **516**
takyr   **516**
talik   **516**
talsand   **516**
talus   **516**, *517*
tank   **516**
tarn   **516**
   avalanche   **43**
taxonomy   92, **516–517**
tear fault   **517**
tectonic geomorphology   356
tectonics   **517**
   see also plate tectonics
tektites   **517**
teleconnections   157, 180, **517–518**
temperate ice   **518**
temperature   **519**
   accumulated   3, **3**
   ambient   **21**
   atmosphere temperature structure   38, *38*
   dry-bulb   106, 519, 550
   effective   106
   inversion   **295–296**, 343, **519**
   potential   **418**, 525
   sea surface temperature (SST)   **471**
   sensible   **476**
   wet-bulb   106, 519, 550, **569**
temperature–humidity index   **519**
tensiometer   78, 147, 399, **519**
tepee   **520**
tephigram   **520**, 525
tephra   **520**
tephrochronology   **520**
terlough   **520**
terminal grades   **520**
terminal moraine   **521**
terminal velocity   **521**
terminations   **521**
termites   **521**
terra rossa   **521**
terrace
   altiplanation   246
   cryoplanation   246
   goletz   **246**
   kame   **303**, 450, 497
   outwash   **380**
   river   213, **449–451**, *451*
   thalassostatic   **522**
terracette   345, **521**
terrane   **521–522**

terrestrial magnetism   **522**
Tertiary   **522**
Tethys Sea   340, **522**
thalassostatic   **522**
thalweg   **522**
thaw lake   **522**
thematic map   510, **522–523**
Thematic Mapper   312, 407, **523**
thermal
   depression   **523**
   efficiency   **523**
   equator   **523**
   pollution   **524**
   susceptibility measurement   327
   wind   263, **524**
thermal infrared linescanner   287, **524**
thermistor   **524**
thermocline   **525**, 561
thermocouple   **525**
thermodynamic diagram   39, **525**
thermodynamic equation   **525**
thermograph   **525**
thermohaline circulation (THC)   369, **525–526**
thermokarst   **526**
thermoluminescence   **526**
   dating   322–323
thermopile   **526**
thermosphere   39
thickness   **527**
Thiessen polygon   **527**
tholoid   **527**
threatened species   178
threshold
   geomorphological   **448**, **527**
   slopes   **527–528**
throughfall   498, **528**
throughflow   261, 286, 290, 390, **466**, **528**
throw of a fault   **528**
thrust fault   208, **528**
thufur   **528**
thunderstorm   **528**
tidal
   currents   **528**
   flats   358
   prism   **528–529**
   range   **529**
tides   **529**
   diurnal   **151**
   rotation   22
till   284, 354, 453, **529–531**, *530*
   classification   **530**
   deformation   530–531
   flow   531
   lodgement   530
   meltout   529–530
   sublimation   530
till fabric analysis   206, 530, **531**
tillite   **531**
time
   cyclic   247
   graded   **247**
   lag   **310**
   steady   247

time-transgressive boundaries **531**
timebound data set **531**
tolerance **531–532**
tombolo **532**
topographic dune 99, 162, 207, **532**
toposequence **532**
toppling failure **532**
topset beds **532**
tor **532**
toreva blocks **532**
tornado **532–533**
torrent **533**
total dissolved solids (TDS) 108, 151
tower karst **533**
tracers **533**
tractive force 269, **533**
trade winds 229–230, 519, **533–534**
transfer function **534**
transgression **534**
translational slide 484, **534**
transmission loss **534**
transpiration 256, 274, **534**
transport limited condition **534–535**
transverse
    clast dams 103
    dune 46, 162, 324, **535**
    rib 103, 379, **535**
travertine 77
treeline **535**
triaxial apparatus test 351, 478, **535**
Trombe's curves **535**
trophic levels 64, 168, **535–536**, *536*
    energy flow 169, 170, 420
tropical cyclone *536*, **536–537**
tropical forest 190, 353, 537
    rain forest **537–538**
tropopause **539**
troposphere 39, 429, **539**
tropospheric quasi-biennial oscillation (TBO) 421
trottoir **539**
trough **539**
    equatorial **190–191**
    Gerlach **237**
    proximal **520**
trucated spur **539**
tsunami 212, **539–540**
tufa 77, **540**
tuff **541**
tundra **541**
    mires 393
    páramo **389**
tunnel valleys **541**
tunnelling **541**
turbidite **541**
turbidity current 217, **541**
turbulence 172, 315, **541**
    kolk **307**
turbulent flow **541–542**

U
ubac **543**
ultrasonic gauging, discharge measurement 149
unconformity **543**

angular **25**
underfit stream **543**
underplating **543**
uniclinal **543**
    shifting **543**
uniform steady flow 145, 179, 215, 223, 533, **543**
uniformitarianism 82, 363, **543–545**
unit hydrograph 149, 271–272, **545**, 546
unit response graph 272, **546**
unloading **546**
unstable channels **546**
unstable equilibrium **546**
unsteady flow **546**
upper level westerlies **546–547**
upwelling **547**
uranium series dating **547**
urban
    ecology **547**
    hydrology **547–548**
    meteorology **548–549**
urstromtäler **549**
uvala **549**

V
vadose zone 252, **550**
valley
    asymmetric **38**
    blind **65**, 336
    bulges 78, **550**
    cluse **102–103**
    combe/coombe **106**
    cross-profile **123**
    dry **159–160**
    gorge **246**
    graben **246**
    hanging **258**
    meanders 543, **550**
    mekgacha **336**
    rift **445–446**, *446*
    sand **516**
    tunnel **541**
    wind 22, 338, 339, 357, **550**
valloni **550**
Van't Hoff's rule **550**
vapour pressure 268, **550–551**
variogram analysis 221
varves **551**
vasques **551**
vector **551**
vegetation
    banding 72
    buffer strip **73**
    cerrado **86**
    chaparral **88**
    climax 37, **99**, 352, 507
    consociation **111**
    cover 71–72
    cryovegetation **124**
    heathland **260**
    mallee *328*, **328**
    maquis **330**
    matorral **334**
    monoclimax **352**

mosaic 356
natural 362
normalized difference vegetation index
(NDVI) 368
phytogeomorphology 404
polyclimax 415
regeneration complex 437
ruderal 457
sclerophyllous 467
succession 415
see also forest; grassland; plant cover; plants;
succession; woodland
velocity profile and measurement 551–552
velocity–area method, discharge measurement 148,
551
ventifact 552
vertical motion 552
vertical stability/instability 552
vertisol 552
vesicular structure 552
vicariance biogeography 93, 552–554, 577–578
virgation 554
viscosity 554
vlei 128, 554
void ratio 24, 554
volcanic bombs *520*
volcanic eruption 194
dust veil index 165
extrusion 205
fissure 211
Hawaiian 258
Plinian 410
types of *555, 556*
volcano 554, *556*, 557
crater 77, 120
maar 325
dormant 153
mud 358
sand 463
shield 480
solfatara 489
stratovolcano 502
see also volcanic eruption
volume susceptibility 327
volumetric gauging 148
Von Karman constant 555
vorticity 555–556
vugh 557
vugs 291
vulcanism 557
vulnerable species 178

W
wackes 464
wadi 558
Walker circulation 230–231, *558*, 558–559
Wallace's line 559
Wallace's realms 559
Walther's law 559
waning slopes 111, 559
warm front 559–560
warm sector 560
warping 560

warts, adhesion 7
wash-board moraine 134
water
abstraction 2
artesian 36
bank storage 45
brackish 71
caballing 76
condensation 108
connate 109
erosion 44
clear-water 94
critical velocity 121–122
groundwater 108
juvenile 302
meteoric 341
specific retention 492
specific yield 492
transfer 564–565
waves 15
see also drainage; soil water; storage
water balance 273–274, 324, **560–561**, *561*
water mass **561–564**, *562, 564*
water table 252, 271, 401, **564**
cone of depression **108–109**
draw down 108, **155**
perched 395
position measurement 147–148
water vapour 39
waterfall **565**
watershed 154, **565**
groundwater basins 401
waterspout **565**
wave theory 14–15
wavelet transform **565**
waves **565**
atmospheric **40**
backwash **44**
bore **67**
cyclone **40**
easterly **167**
edge 125, **173**, 436
elastic **40**
gravity **40**, 173, **248–249**
infragravity **173**, 287
Kelvin **305**
kinematic **305**, 371
lee 40, **317**, 339
refraction *436*, **436–437**
Rossby **40**, 126, 325, **455**, 539, 572
tsunami 212, **539–540**
waxing slopes **565**
weather **565**
forecasting **565–566**
modification **566**
type 513, **566**
weathering **566–567**
classification of processes *567*
deep **132**, 567
front **567**
frost **225–226**
hygric **276–277**
index **567**

weathering (*Continued*)
  insolation   *288*, **289**
  onion   **376**
  organic   **378**
  pedogenesis   394
  potential index   **567**
  profile   **567–568**, *569*
  rinds   **568**
  salt   21, **460**
  scale of   *568*
  sphenoidal   **493**
weathering-limited slopes   535, **568**
Weddell Sea Bottom Water (WSBW)
   25
wedge failure   **568**
weirs   148, **568**
wells   **569**
West Antarctic Ice Sheet (WAIS)   27–28
westerlies   **569**
  upper level   **546–547**
wet deposition   4, 5, 373
wet-bulb temperature   106, 519, 550, **569**
wetlands   393, *569*, **569–570**
  fen   **209**
  Ramsar Convention   **432**
  swamp   **512**
  see also bogs; mires
wetted perimeter   **570**
wetting front   **570**
whirlwind   **570**
  dust devil   164–165
width/depth ratio   **570**
wilderness   **570**
Wilson cycle   **570**
wind   **570–572**
  anabatic   22
  Beaufort scale   *51*, **51**
  berg   **53**
  chinook   **90**, 217
  dominant   **153**
  erosion, critical velocity   **121–122**
  fetch length   **209**
  föhn   **217**
  geostrophic   40, 174, **237**
  gradient   **247**, 572
  gustiness factor   **254**
  harmattan   **258**

hodograph   **262–263**, *263*
jet stream   **301–302**
katabatic flows   **303–305**
khamsin   **305**
local   **321**
mesoscale   338
mountain/valley   338, 339, **357–358**
prevailing   **419**
resultant   **443**
thermal   263, **524**
trade   229–230, 519, **533–534**
valley   22, 357, **550**
westerlies   **569**
  upper level   **546–547**
wind chill   **572**
wind rose   **572**
wind shadow   **572**
wind shear   **572**
wind speed measurement   24
woodland
  caatinga   **76**
  canopy   **78**
  krummholz   **307–308**
  miombo   **350**
World Heritage Site   **572–574**

X
X-ray diffraction   **575**
xerophyte   **575**

Y
yardangs   **576**
yazoo   **576**
Younger Dryas   129, **576**

Z
zeuge   **577**
zibar dune   462, **577**
Zingg diagram   478
zonal
  circulation   **577**
  flow   229
  soil   **577**
zonality   **577**
zonation   **577**
zoogeography   **577–578**
zoogeomorphology   **578**

Printed and bound by CPI Group (UK) Ltd, Croydon, CR0 4YY

27/10/2024

14580310-0004